THE
ALBUQUERQUE
MEETING

Sunset Ranchos ▼ Robert Daughters

VOLUME 1

THE
ALBUQUERQUE
MEETING

▼▼▼▼▼▼▼▼▼▼▼▼▼▼▼

August 2 - 6, 1994
The University of New Mexico

▼

Edited by Sally Seidel

Proceedings of the 8th Meeting
Division of Particles and Fields
of the American Physical Society

World Scientific
Singapore • New Jersey • London • Hong Kong

Published by

World Scientific Publishing Co. Pte. Ltd.

P O Box 128, Farrer Road, Singapore 9128

USA office: Suite 1B, 1060 Main Street, River Edge, NJ 07661

UK office: 57 Shelton Street, Covent Garden, London WC2H 9HE

Library of Congress Cataloging-in-Publication Data

American Physical Society, Division of Particles and Fields. Meeting
   (8th : 1994 : University of New Mexico)
      The Albuquerque meeting : August 2–6, 1994, the University of New
   Mexico : Proceedings of the 8th Meeting, Division of Particles and
   Fields of the American Physical Society / edited by Sally Seidel.
         p.   cm.
      ISBN 9810220723 (set), -- ISBN 9810229895 (vol. 1), -- ISBN 9810229909 (vol. 2)
      1. Particles (Nuclear physics) -- Congresses.   I. Seidel, Sally,
   1958-   .   II. Title.
   QC793.A47   1994
   539.7'2--dc20   95-18376
                                                                           CIP

Painting, Sunset Ranchos
    ROBERT DAUGHTERS, Tucson, Arizona

Logo and Poster Design
    M. T. HYATT & COMPANY, Albuquerque, New Mexico

Book Design (Cover, Jacket, Title Pages, and Photo Layout)
    TERRI WRIGHT, Santa Barbara, California

Photography
    J. E. TOWNLEY, RAINFOREST STUDIO, Albuquerque, New Mexico

Contributed Photographs
    MARK FRAUTSCHI, Albuquerque, New Mexico

Editorial Development and Manuscript Organization
    SANDBAR PRODUCTIONS, Albuquerque, New Mexico

Printed in the United States of America.

CONTENTS

Volume 1

*Oral Presentation Only

PARALLEL SESSIONS

ELECTROWEAK
Convener: JAMES G. BRANSON, California (San Diego)

CHARM PHYSICS
Convener: JANIS A. McKENNA, British Columbia

CHARM PRODUCTION AT FIXED TARGET AND HADRON COLLIDERS

CHARMONIUM - J/Ψ'S FOR CHARM AND BEAUTY TAGGING

CHARM PRODUCTION AND DECAY AT e⁺e⁻ COLLIDERS

SEMILEPTONIC DECAYS OF CHARMED PARTICLES AND CHARM SUMMARY

TAU PHYSICS
Convener: TOMASZ SKWARNICKI, Southern Methodist

TAU LEPTON HADRONIC BRANCHING RATIOS

LIFETIME, LEPTONIC DECAYS, AND TAU NEUTRINO

TOP PHYSICS
Convener: STEPHEN J. PARKE, Fermilab

B PHYSICS
Convener: MARINA ARTUSO, Syracuse

SEMILEPTONIC DECAYS AND LIFETIMES

THE HADRONIC MATRIX ELEMENT IN B DECAYS: THEORETICAL APPROACHES
AND EXPERIMENTAL STUDIES

CP VIOLATION, RARE DECAYS, AND B PHYSICS
Conveners: MARINA ARTUSO, Syracuse
GERMAN E. VALENCIA, Iowa State
GEORGE G. GOLLIN, Illinois (Urbana-Champaign)

B DECAYS CONFRONT THE STANDARD MODEL

CP VIOLATION AND RARE DECAYS

Conveners: GERMAN E. VALENCIA, Iowa State
             GEORGE G. GOLLIN, Illinois (Urbana-Champaign)

CONTENTS

Volume 2

PARALLEL SESSIONS

BEYOND THE STANDARD MODEL
Conveners: JOANNE L. HEWETT, SLAC
ANDREW WHITE, Texas (Arlington)

SUSY

STRONG ELECTROWEAK SYMMETRY BREAKING

HEAVY IONS

COSMOLOGY AND DARK MATTER
Conveners: KIM GRIEST, California (San Diego)
                JOEL R. PRIMACK, California (Santa Cruz)

PLANCK SCALE PHYSICS AND NEW THEORETICAL DEVELOPMENTS
Conveners: BRIAN GREENE, Cornell
JORGE PULLIN, Pennsylvania State

ACCELERATORS
Convener: JOHN MARRINER, Fermilab

NON-ACCELERATOR, ASTROPARTICLE, AND NEUTRINO MASS PHYSICS
Convener: CARL W. AKERLOF, Michigan (Ann Arbor)

QCD

MEASUREMENT OF α_s AND TESTS OF THE STRUCTURE OF QCD

RAPIDITY GAPS

THEORY

STRUCTURE FUNCTIONS

DIRECT PHOTONS AND TWO-GAMMA REACTIONS

INSTRUMENTATION
Conveners: ANNA PEISERT, CERN
RICHARD WIGMANS, Texas Tech

PREFACE

The Eighth Meeting of the American Physical Society Division of Particles and Fields was held on the campus of the University of New Mexico on August 2-6, 1994. The conference was attended by over 600 physicists from 16 countries. The meeting was organized such that eight sessions ran in parallel every morning and four plenary sessions were held every afternoon but the last. A total of 398 talks were presented in the parallel sessions, including summary talks by the parallel session conveners which nicely complemented the plenary talks.

Special conference events included the opening reception, which was held at the Maxwell Museum of Anthropology; a public lecture by Murray Gell-Mann entitled "Simplicity and Complexity," followed by a reception in the Fine Arts Center; and a trip to the Santa Fe Opera. The conference dinner was held at the New Mexico Museum of Natural History.

I would like to thank the members of the Local Organizing Committee for their help, most especially John Matthews and David Wolfe, who offered assistance with many aspects of the planning, and Terry Goldman and Uriel Nauenberg, who reviewed and organized the abstracts. I thank the members of the International Advisory Committee for their suggestions concerning the physics program. I appreciate the work that Michael Barnett contributed in arranging the HEP Outreach and Education plenary session. Anatoly Klypin also provided special assistance with the Cosmology and Dark Matter parallel session organization.

The combined talents and professional skills of several individuals are reflected throughout the proceedings. Artist Robert Daughters, Tucson, Arizona, generously granted permission for reproduction of his painting, *Sunset Ranchos*, for the conference poster and proceedings. M. T. Hyatt & Company, Albuquerque, produced the conference poster and logo. Terri Wright Design, Santa Barbara, California, incorporated the established conference image into the design and production of book covers, title pages, and photographic illustrations for the two volumes. Conference photographer J. E. Townley, Rainforest Studio, Albuquerque, produced hundreds of photographs from which a pictorial record of conference leaders and main events was selected. Mark Frautschi, Albuquerque, voluntarily contributed additional photographs of conference activities and participants. Conference managers Barbara Harral Goldston and Sandra Broome Weeke of Sandbar Productions supervised production of the final manuscript.

Timothy Thomas, Lawrence B. Crowell, Paul Kingsberry, and Ling Yu contributed their time during the meeting to ensure that services to conference delegates ran smoothly. The

Office and Technical Staff of the Department of Physics and Astronomy of the University of New Mexico provided consistently strong support throughout the planning period as well as during the meeting itself.

Sally Seidel
Albuquerque, New Mexico

The primary sponsors of DPF'94 were the University of New Mexico, the U.S. Department of Energy, The National Science Foundation, and Los Alamos National Laboratory.

PLENARY SESSIONS

Physics at the Fermilab Collider

MELVYN J. SHOCHET, *Chicago, Enrico Fermi Institute*

PHYSICS AT THE FERMILAB COLLIDER

MELVYN J. SHOCHET

Enrico Fermi Institute and Department of Physics
University of Chicago, 5640 S. Ellis Ave., Chicago,
Illinois 60637, USA

ABSTRACT

The CDF and D0 experiments at the Fermilab Tevatron Collider have produced many results from the search for the top quark, the study of both the electroweak and strong interactions, the production and decay of b quarks, and the search for new high mass objects. A sample of recently obtained results are presented.

1. Introduction

The Fermilab Tevatron Collider provides $\bar{p}p$ collisions at a center of mass energy of 1.8 TeV. During the 1992-93 data run, the initial store luminosity was typically 5×10^{30} cm^{-2}-sec^{-1}, five times the design luminosity of the Collider. The integrated luminosity recorded on tape was 19.3 (13.4) pb^{-1} for the CDF (D0) detectors.

Many results have been produced from these data. For example, 28 papers were submitted to the recent Rochester Conference by the CDF Collaboration. The time limitation prevents me from reporting on all the new CDF and D0 results. Instead, I present selected topics from the search for the top quark, tests of the Standard Model of electroweak interactions, studies of the strong interaction, the production and decay of states containing a b quark, and the search for new massive objects. Much of the material presented here is preliminary.

The CDF and D0 collaborations each have over 400 collaborators from the Americas, Europe, and Asia. The D0 detector, in its first data taking run, featured uranium liquid-argon calorimeters and a muon system with large rapidity coverage. CDF installed for this run a silicon vertex detector, which, when combined with precision tracking in the magnetic field, provided significant new physics capability.

2. Top Quark

If the Minimal Standard Model is correct, consistency of all electroweak data (Z decay, M$_{\text{W}}$, ν scattering, etc.) with the model requires[1]

$$M_{\text{top}}^{\text{SM}} = 177 \pm 11^{+18}_{-19} \text{ GeV/c}^2$$

An object this massive can only be produced at the Tevatron Collider. The dominant production mechanism is shown in Fig. 1. For an integrated luminosity of 20 pb^{-1},

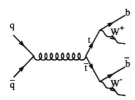
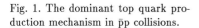

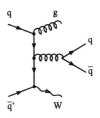

Fig. 1. The dominant top quark production mechanism in $\bar{p}p$ collisions.

Fig. 2. The dominant background in
the single lepton top search.

approximately 200 $t\bar{t}$ pairs should be produced. Decay branching ratios and detection efficiency reduces this to a small number of expected events.

Since both the top quark and W boson are very short lived, the experimental signature is determined by the decay of the two W bosons in each event. In order not to be overwhelmed by background, both CDF and D0 searched in two modes. The dilepton search seeks events in which both W's decay into $e\nu$ or $\mu\nu$ ($t\bar{t} \to ll\nu\nu b\bar{b}$). The product branching ratio is 5%. The single lepton search seeks events in which one W decays into $e\nu$ or $\mu\nu$ and the other decays into a light quark pair ($t\bar{t} \to l\nu q\bar{q}b\bar{b}$). Here the product branching ratio is 30%.

2.1. CDF Dilepton Search

The CDF top quark analysis was recently published.[2] In the dilepton mode, they search for ee, $e\mu$, and $\mu\mu$ pairs of opposite electric charge. The minimum lepton P_T is 20 GeV/c, at least one of the leptons must be isolated, and ee and $\mu\mu$ pairs are removed if they are in the Z mass region - (75, 105) GeV/c^2. The missing E_T ($\not{E}_T$) must be at least 25 GeV, with the azimuthal angle between the missing E_T vector and the nearest jet or lepton required to be greater than 20° if $\not{E}_T < 50$ GeV. Finally, there must be at least 2 jets of $E_T > 10$ GeV and pseudorapidity $|\eta| < 2.4$.

The dominant backgrounds are WW production, $Z \to \tau\tau$, fake leptons, $b\bar{b}$ production, and lepton pairs coming from Drell-Yan production of γ^*/Z. The total background is $0.56^{+0.25}_{-0.13}$, compared to an expected signal of 1–2 events for a heavy top quark (140-180 GeV/c^2). CDF observes 2 events in this channel, as shown in Fig. 3.

2.2. CDF Single Lepton Search

The single lepton search begins with a selection of $W \to l\nu$ candidates: events containing an electron or muon with $P_T > 20$ GeV/c and $\not{E}_T > 20$ GeV. To suppress background without greatly reducing the top sensitivity, 3 or more jets with $E_T > 15$ GeV and $|\eta| < 2.0$ are required. There are 52 events in this sample. The dominant background is the production of a W recoiling against multiple jets (Fig. 2).

With these selection criteria, the expected signal to background ratio for a heavy top quark is approximately $\frac{1}{2} - \frac{1}{4}$. CDF further reduces the background by requiring that at least one of the b jets in the final state be identified. Two techniques are employed:

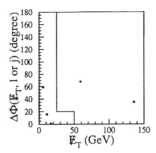

Fig. 3. The missing E_T vs. its angle relative to the nearest jet or lepton. The two events to the right of the cut line are top candidates.

finding the secondary vertex from the b decay, and finding an additional lepton from the semileptonic decay of the b or its daughter c quark.

For secondary vertex tagging, CDF uses jet data to determine the tagging efficiency and the background rate in high P_T jets. Figure 4 compares the proper decay distance ($c\tau$) distribution for b-tags in a b enriched sample of moderate P_T inclusive electron data with the prediction from a b Monte Carlo calculation. A similar distribution for data from a generic jet sample (Fig. 5) shows the contributions from false tags as well as the heavy quark content in such a sample. The efficiency for tagging at least one b jet in a $t\bar{t}$ event is $22 \pm 6\%$.

The dominant background comes from tagging errors plus the process shown in Fig. 2 in which the $q\bar{q}$ is a $b\bar{b}$ pair. The generic jet sample is used to measure the tag rate from these processes as a function of jet E_T and track multiplicity. This function is applied to each jet in the 52 event $W + \geq 3$ jet sample to obtain the expected number of background tags. The total background is 2.30 ± 0.29 tags compared to 1.5–5 expected from heavy top. There are 6 tags observed in the data.

The second b tagging method searches for electrons or muons with $P_T > 2$ GeV/c, in order to be sensitive to leptons from both b decay and the decay of the daughter c quark. The efficiency for finding electrons is determined from a data sample of conversion electrons, while the muon efficiency is measured in $J/\psi \to \mu\mu$ data. The efficiency for tagging at least one b jet in a $t\bar{t}$ event is $16 \pm 2.5\%$. The dominant background is measured with generic jet events. The total background is 3.1 ± 0.3 tags, with 1–3.5 expected from a top quark signal. Seven tags are observed, 3 of which are in events that also have an identified secondary vertex. Figure 6 shows the total number of observed tags in W events as a function of the number of jets in the event. There is an excess over background in the top signal region (≥ 3 jets).

2.3. CDF Top Search Summary

In total, CDF observes 15 counts (dilepton events, secondary vertex tags, soft lepton tags) in 12 events with $5.96^{+0.49}_{-0.44}$ counts expected from background. The probability that the background would fluctuate up to 15 counts is 0.26%, which is the 2.8

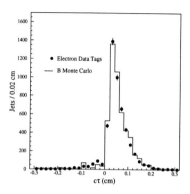

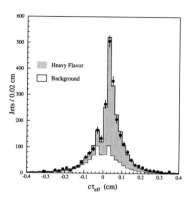

Fig. 4. The secondary vertex proper decay distance distribution from a b enriched sample of inclusive electrons compared to a Monte Carlo prediction.

Fig. 5. The secondary vertex proper decay distance distribution from a sample of generic jet events. The contributions from tagging errors and heavy flavor jets are shown.

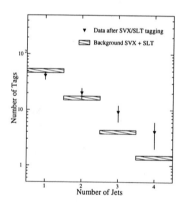

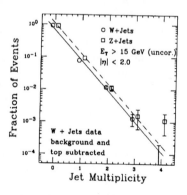

Fig. 6. The total number of observed b-tags (secondary vertex, SVX, and soft lepton, SLT) as a function of the number of jets in W events.

Fig. 7. The jet multiplicity distributions for Z events and for W events after subtracting the expected top quark and background contributions. The straight lines are to guide the eye.

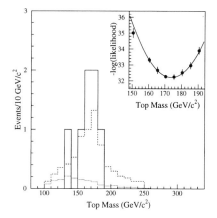

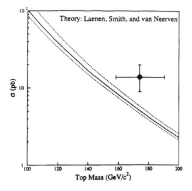

Fig. 8. The top quark mass from single lepton plus 4 jet events (solid), the expected background (dotted), and the expected shape for 175 GeV/c² top plus background (dashed). The insert shows the likelihood as a function of top mass.

Fig. 9. The CDF measurement of the top quark mass and production cross section compared to a partial NNLO calculation.

σ point on a Gaussian distribution.

Single lepton events ($l\nu q\bar{q}b\bar{b}$) containing 4 jets can be reconstructed (2-C fit) to obtain the top quark mass. Of the 10 single lepton events with a b-tag, 7 have a fourth jet with $E_T > 8$ GeV. Figure 8 shows the mass from these 7 events along with the expected distribution from background and 175 GeV/c² top plus background. The best mass, extracted from a likelihood fit (see insert in Fig. 8), is

$$M_{top} = 174 \pm 10^{+13}_{-12} \text{ GeV/c}^2.$$

The background subtracted number of events in each of the searches is used to calculate the $t\bar{t}$ production cross section, $\sigma = 13.9^{+6.1}_{-4.8}$ pb. Figure 9 compares the CDF mass and cross section with a partial next-to-next-to-leading-order calculation.[3]

There are additional features of the data that support the top hypothesis and others that do not. On the negative side, the estimated contributions from the top quark and other backgrounds account for all the observed W + ≥ 4 jet events, leaving little room for the dominant background (Fig. 2). Also in the control sample of Z + ≥ 3 jet events, there are two events with b-tags, compared to an expected 0.64 event.

Both of these effects can be seen in Fig. 7. It is clear that additional statistics will help a great deal.

On the positive side, one of the dilepton events has a jet containing both a secondary vertex and soft lepton tag. This is very unlikely to come from background sources that don't contain b-jets. In single lepton events, the kinematic distributions of the jets support the top hypothesis, and the mass fit prefers top plus background to background alone by a factor of 50 in relative likelihood.

2.4. D0 Dilepton Search

Early this year, the D0 collaboration published a lower top mass limit of 131 GeV/c^2 at the 95% CL.[4] At the end of April, they presented a status report on their analysis. Recently the analysis was updated, with slightly modified analysis cuts and recalculated backgrounds.

In the dilepton mode, D0 searches for $e\mu$, ee, and $\mu\mu$ pairs. For the $e\mu$ case, the electron (muon) transverse momentum must be greater than 15 (12) GeV/c. The missing E$_T$ in the calorimeter must be greater than 20 GeV, and after correcting for the muon momentum it must be greater than 10 GeV. In the ee case, both electrons must have P$_T$> 20 GeV/c, and the missing E$_T$ must be greater than 25 GeV. For muon pairs, the muon P$_T$ must be greater than 15 GeV/c and the azimuthal angle between the muons must be less than 140° if $\not{E}_T$< 40 GeV. Finally, all dilepton events must have at least 2 jets with E$_T$> 15 GeV.

The total estimated background in the dilepton mode is 0.76± $\sim$ 0.15 events.* For a heavy top quark, 0.5–1.5 events are expected. D0 observes 1 dilepton event.

2.5. D0 Single Lepton Search

For the single lepton mode, D0 separately considers events with and without a b-tag. Their b-tagging algorithm is applied in electron plus jet events and requires an additional muon of P$_T$> 4 GeV/c.

In the single lepton search without b-tagging, D0 requires either an electron with P$_T$> 20 GeV/c or a muon with P$_T$> 15 GeV/c. $\not{E}_T$ must be greater than 25 GeV for electron events, while for muon events both the calorimeter $\not{E}_T$ and the muon corrected $\not{E}_T$ must be greater than 20 GeV. To suppress background, three event topology requirements are made. There must be at least 4 jets of E$_T$> 15 GeV and $|\eta|$ < 2, the aplanarity of the event must be greater than 0.05, and the event transverse energy variable, H$_T$, must be above 140 GeV. In electron events, H$_T$ is the scalar sum of the transverse energies of the jets and the W. For muon events, only the jets are included.

The total background is 1.8± $\sim$ 1 events, compared to 1.5–4 expected from a heavy top quark. D0 observes 4 events in this mode.

The D0 b-tag search includes events with an electron of P$_T$> 20 GeV/c, missing E$_T$ in the calorimeter greater than 20 GeV, 3 jets with E$_T$> 20 GeV and $|\eta|$ < 2, and a muon with P$_T$> 4 GeV/c. The contribution of b decay in a low P$_T$ inclusive muon data sample can be seen in Fig. 10, which shows the P$_T$ of the muon relative to the nearest jet.

---

*The quoted uncertainties in the dilepton and single lepton without b-tag modes are my estimates, since I have combined a few of their search modes. In the D0 top search summary, however, the total background and its uncertainty are as quoted by D0.

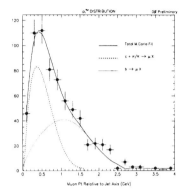

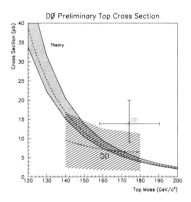

Fig. 10. The transverse momentum of the muon relative to the nearest jet axis. The D0 data are compared to a fit (solid) to the sum of two shapes: b-decay (dotted), and charm/π/K decay (dashed).

Fig. 11. The D0 top quark production cross section compared with theory and the CDF result. The D0 cross section varies slightly with top mass because the efficiency of the event selection criteria depends on the mass.

The total background in this mode is 0.55 ± 0.15 events. Between 0.5 and 1.5 events are expected from a heavy top quark. D0 observes 2 events.

2.6.   D0 Top Search Summary

The total background for all the D0 modes is 3.2 ± 1.1 events, while 2.5-6.5 events are expected for a top quark of mass between 140 and 180 GeV/c^2. D0 observes 7 events. As seen in Fig. 11, the D0 result is consistent with either the no-top hypothesis or the CDF result.

2.7.   Top Prospects

In the current Tevatron Collider run, each experiment should accumulate approximately four times the data they acquired in the 1992-93 run. That should not only allow for confirmation of the CDF result, but also provide a top mass measurement with an uncertainty of 8-10 GeV/c^2.

The Fermilab Main Injector is now under construction. By the end of the decade, there should be 500–1000 top events between the two experiments. With this quantity of data, the mass uncertainty could drop to 5 GeV/c^2. A sensitive search for non-standard decay modes of the top quark will be under way, and the systematic study of top quark couplings will have begun.

3.   W Mass

Precision measurements of Z^0 decay provide a prediction of the W mass to approximately ± 100 MeV/c^2, assuming that the Minimal Standard Model is correct.[1]

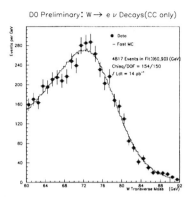

Fig. 12. The D0 W→ $e\nu$ transverse mass plot. The histogram is the best fit Monte Carlo line shape.

Since the W production cross section is approximately 20 nb at the Tevatron, there are thousands of leptonic W decay events (W→ $l\nu$) in each experiment even after applying the tight fiducial cuts needed for a precision mass measurement. The $l\nu$ invariant mass cannot be calculated because of the undetected neutrino. Instead the two dimensional analog, transverse mass, is used, with the neutrino transverse momentum taken as balancing the visible transverse momentum in the event.

$$M_T^W = \sqrt{2P_T^l P_T^\nu (1 - \cos\phi_{l,\nu})}$$
$$\vec{P}_T^\nu = -\vec{P}_T^l - \vec{P}_T^{\text{hadrons}}$$

The D0 measurement uses 5830 W→ $e\nu$ events. CDF has 6421 W→ $e\nu$ and 4090 W→ $\mu\nu$ events. Figure 12 shows the transverse mass distribution from D0, while Fig. 13 and Fig. 14 show the CDF distributions.

The critical issues for the measurement, as seen in the equations above, are the lepton energy scale and resolution, and the response of the detector to the hadron system recoiling against the W. Much of the needed information comes from $Z^0 \rightarrow l^+l^-$ events.

Both groups obtain the lepton energy resolution from test beam studies and the measured $Z^0 \rightarrow l^+l^-$ line shape. D0 obtains its absolute electron energy scale from the reconstructed Z mass (Fig. 15). Although their absolute scale is low by 4%, the W mass is close enough to the Z mass that they can rescale without incurring a very large systematic uncertainty. The CDF energy scale is determined in two steps. First the calorimeter electron scale is tied to the magnetic spectrometer using the ratio of the calorimeter energy to the tracking chamber momentum (Fig. 16). Second, the absolute scale of the spectrometer is checked with the J/ψ, Υ, and Z^0 masses (Fig. 17 and Fig. 18).

Modeling the response of the calorimeters to the recoil hadron system is handled

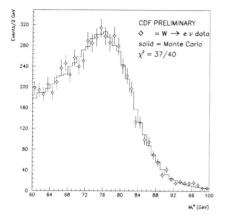

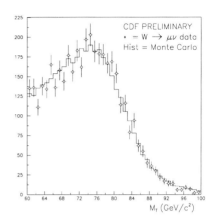

Fig. 13. The CDF W→ $e\nu$ transverse mass plot. The histogram is the best fit Monte Carlo line shape.

Fig. 14. The CDF W→ $\mu\nu$ transverse mass plot. The histogram is the best fit Monte Carlo line shape.

differently in the two experiments. In the D0 Monte Carlo, the W recoil system is treated as a single jet, with the D0 jet energy resolution applied and the scale determined from Z^0 events (Fig. 19). CDF notes that the detector response to the hadrons recoiling against a vector boson of $\vec{P}_T^V$ is the same whether the boson is a W or a Z^0. Thus for each Monte Carlo W, the rest of the event is taken from a Z^0 **data** event of the same P_T (Fig. 20).

Both CDF and D0 obtain the W mass by fitting Monte Carlo templates to the W→ $l\nu$ transverse mass distributions (Fig. 12, Fig. 13, Fig. 14). The systematic uncertainties are given in Fig. 21 and Fig. 22. All of them can be studied with the data. Even the structure function uncertainty (from the u to d quark ratio) can be directly measured with the CDF W asymmetry data (Fig. 23).

The masses obtained by the two collaborations are:

$$\text{D0}: \quad M_W = \quad 79.86 \pm 0.16 \pm 0.16 \pm 0.26(\text{scale}) \text{ GeV/c}^2$$
$$79.86 \pm 0.345 \text{ GeV/c}^2$$

$$\text{CDF}: \quad 80.47 \pm 0.15 \pm 0.25 \text{ GeV/c}^2 \ (W \rightarrow e\nu)$$
$$80.29 \pm 0.20 \pm 0.24 \text{ GeV/c}^2 \ (W \rightarrow \mu\nu)$$
$$80.38 \pm 0.23 \text{ GeV/c}^2 \ (\text{CDF combined})$$

The new world average, combining these results with earlier results from CDF[5] and

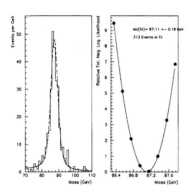

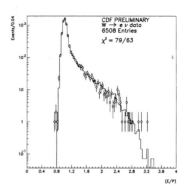

Fig. 15. The D0 Z→ ee mass peak with the fit line shape and the negative log-likelihood plot used to determine the mass.

Fig. 16. The ratio of calorimeter energy to tracking chamber momentum in CDF W→ eν electrons compared to the prediction of a radiative W Monte Carlo. The high side tail is due to electron bremsstrahlung.

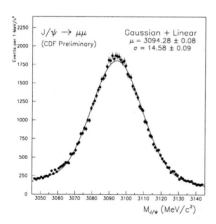

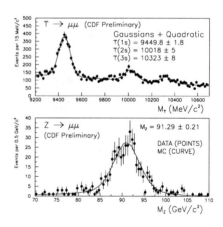

Fig. 17. The CDF $J/\psi \rightarrow \mu\mu$ mass peak.

Fig. 18. The Υ resonances and the Z peak from CDF.

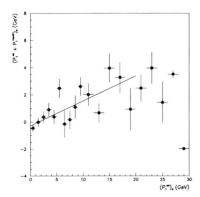

Fig. 19. The sum of the $\vec{P}_T$ of the recoil hadrons and the two electrons in D0 Z→ ee events as a function of the P_T of the electron pair. Plotted is the component of these vectors along the bisector of the two electron directions.

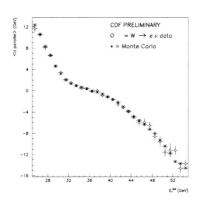

Fig. 20. The component of the net recoil hadron transverse momentum along the electron direction in CDF W→ eν events as a function of the electron E_T.

Source of Uncertainty	MeV/c^2
Trigger Efficiency	20
Resolution and neutrino E_T scale	149
Energy underlying electron	50
$U_{//}$ efficiency	10
Hadronic to EM scale	80
QCD background	30
Theoretical model uncertainty	86
W width	20
Fitting Error	30
Total	200

Resolution and neutrino E_T scale	
Electron energy resolution	70
Neutrino E_T scale, resolution (W underlying event)	130
Jet energy resolution	20

Theoretical model uncertainty	
Structure function	70
P_{tw} spectrum	50

Fig. 21. The D0 W mass systematic uncertainties.

	$W \to e\nu$	$W \to \mu\nu$	Correlated	Uncorrelated	
				$W \to e\nu$	$W \to \mu\nu$
Statistical	150	200		150	200
Momentum Scale	130	60	60	120	
Systematics	210	220			
Momentum Resolution	140	120		140	120
P_T^W	90	110	80	40	70
$u_\parallel$	70	90		70	90
Backgrounds	50	50		50	50
Fitting	20	20		20	20
Structure Functions	100	100	100		
Total	290	300	140	260	270

Fig. 22. The CDF W mass uncertainties.

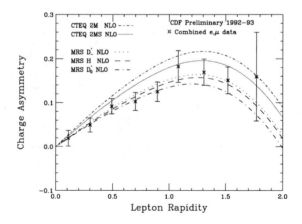

Fig. 23. The CDF lepton charge asymmetry in W decay compared to several modern structure functions.

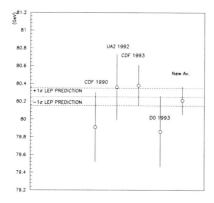

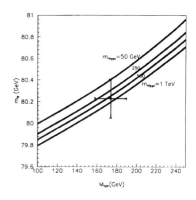

Fig. 25. The W and top quark mass measurements compared to the predictions of the Minimal Standard Model for different Higgs boson masses.

Fig. 24. The experimental measurements of the W mass.

UA2,[6] is (Fig. 24)

$$80.23 \pm 0.18 \text{ GeV}/c^2 \text{ (CDF, D0, UA2)}$$

We now have the first experimental point on the M_W vs. M_{top} plot (Fig. 25). Since the systematics of the M_W measurement are statistics limited (Z^0 events, W asymmetry data), the uncertainty should continue to be reduced with increasing integrated luminosity. It seems quite possible that ± 50 MeV/c^2 could be reached by the end of the decade. When combined with a top quark mass measurement of ± 5 GeV/c^2, we would have a data point on Fig. 25 that would severely test the Minimal Standard Model. If the model is correct, then the data could discriminate between a light and a heavy Higgs boson.

4.   Vector Boson Pair Production

In the Standard Model, the triboson couplings are specified. In some non-standard models, the vector bosons are composite, resulting in anomalous triboson couplings which modify the electric and magnetic multipole moments of the bosons. These can be determined by looking at the production rate and kinematic properties of boson pairs ($W\gamma$, $Z\gamma$, WW, WZ, ZZ) produced in $\bar{p}p$ collisions.

As an example, I consider $W\gamma$ production. If CP conservation is assumed, there are two anomalous couplings, usually written as $\Delta\kappa$ and λ. The magnetic dipole and electric quadrupole moments of the W are related to these couplings:

$$\mu_W = \frac{e}{2M_W}(2 + \Delta\kappa + \lambda)$$

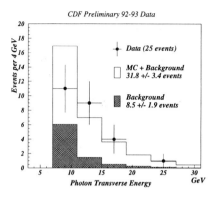

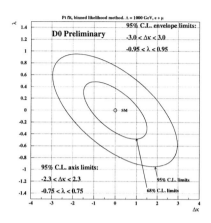

Fig. 26. The photon P_T in CDF W events compared to the calculated background and the expected signal from Standard Model couplings.

Fig. 27. The D0 limit on the anomalous couplings $\Delta\kappa$ and λ from $W\gamma$ data.

$$Q_W^e = \frac{-e}{M_W^2}(1 + \Delta\kappa - \lambda)$$

CDF and D0 have searched for high P_T photons in W events. In D0 the minimum E_T is 10 GeV, while in CDF it is 7 GeV. In both experiments, the photon is required to be at least a distance 0.7 (in $\eta - \phi$ space) away from the charged lepton in order to suppress the contribution from e/μ bremsstrahlung. CDF (D0) finds 25 (19) events, which after background subtraction becomes 16.5 (13.8). The photon P_T distribution is shown in Fig. 26. The shape of the distribution is used to extract the anomalous couplings. Figure 27 shows the D0 contours for $\Delta\kappa$ and λ. The CDF limits are very similar. Limits for CP violating anomalous $WW\gamma$ couplings as well as CP conserving and violating $ZZ\gamma$ anomalous couplings have also been extracted by the two experiments.

5. Inclusive Photon Production

Inclusive photon production in high energy $\bar{p}p$ collisions is a simple process at leading order ($qg \to \gamma q$) with a well defined and measured final state parton. The cross section can be used to determine the gluon distribution function in the proton. In addition, events with a final state photon and charm quark (through its daughter e or μ) provide a measure of the charm content of the proton.

The major experimental background comes from jets that fragment largely into a π^0 or η, which decay into photon pairs. CDF uses two methods to separate single

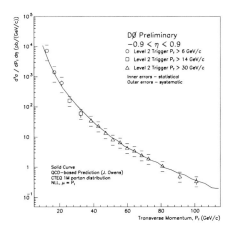

Fig. 28. D0 inclusive photon cross section as a function of the γ P_T compared with a theoretical prediction.

photons from the multiple photon background. In the profile method, they use the transverse shower shape approximately six radiation lengths into the shower to differentiate a single photon narrow shower from the broader shower induced by multiple photons. In the conversion method, they use the fraction of γ candidates that convert in the one radiation length solenoid coil. That fraction depends on whether one or multiple photons passed through the coil. For both methods, they measure the efficiency and separation power using reconstructed wide angle $\eta \rightarrow \gamma\gamma$ and $\rho^{\pm} \rightarrow \pi^0\pi^{\pm}$ events. The D0 group counts the number of candidates that convert in their tracking chamber.

The D0 data (Fig. 28) agree well with the predicted cross section. The CDF results have significantly smaller uncertainties. Although qualitatively they agree with next-to-leading-order QCD[7] (Fig. 29), when the comparison is made on a linear scale (Fig. 30) the agreement is not so good. At high P_T, there is no problem; however at low P_T the slope of the data is at variance with the theory. There are several possible explanations for the disagreement. The contribution from gluon-charm scattering may not be correct. The bremsstrahlung diagrams, where a photon is radiated from an initial state or final state parton, first appear in next-to-leading-order. There thus may be important NNLO contributions. Another possibility is that the assumed gluon distribution is not correct. A light gluino (1–5 GeV/c^2) could affect the cross section. Finally, the calculated K_T smearing (the effective initial state parton transverse momentum) may be too small. At the quantitative level, the last possibility may well have the largest effect. K_T smearing with a sigma that increases from 1 GeV/c at ISR energy to about 3 GeV/c at Tevatron energies improves the agreement between theory and the ISR, $S\bar{p}pS$, and Tevatron data.

6.   b Production Cross Section

Studying the b production cross section is interesting because, at Tevatron energies, higher order QCD diagrams (ex, $gg \rightarrow gg \rightarrow gb\bar{b}$) can dominate over the leading order diagrams (ex, $gg \rightarrow b\bar{b}$). Also, as we shall see, it has led us to reexamine the dominant mechanisms for producing charmonium states.

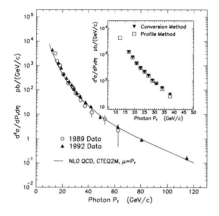

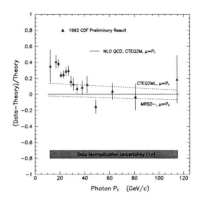

Fig. 29. CDF inclusive photon cross section compared to a QCD prediction.

Fig. 30. The difference between the CDF inclusive photon cross section and theory as a fraction of the theoretical prediction.

D0 has measured single muon and dimuon production rates. Figure 31 shows their muon production cross section as a function of muon P_T. It is compared with the sum of the dominant sources of muons: b decay, c decay, π/K decay, and W decay. They extract the b production cross section using the ISAJET prediction of the fraction of muons coming from b decay. This prediction is checked in the data by looking at the P_T of the muon relative to the nearest jet (Fig. 10).

CDF measures the b production cross section by studying inclusive e, μ, J/ψ, and ψ', the semi-exclusive modes eD, μD, and μD^*, as well as the exclusive final states $J/\psi K$ and $J/\psi K^*$. Figure 32 shows the right sign and wrong sign $\mu D\pi$ combinations. The ψ and ψ' mass peaks are shown in Fig. 33 and Fig. 34.

There has been a significant change since results were presented by CDF in 1990, namely the b production cross section deduced from the inclusive J/ψ and ψ' rates have gone down by a factor of 2–3, even though the inclusive J/ψ and ψ' cross sections have not changed. In 1990, theoretical guidance was used to go from the charmonium cross section to the b cross section. It was believed that there are two important sources of J/ψ: B decay and χ_c decay. Because the χ mass is below that of the ψ', it was expected that the only significant source of ψ' would be B decay.

The installation of the silicon vertex detector for the 1992-93 run made it possible to separate prompt production from B decay. Figure 35 shows the proper decay length distribution for the ψ' sample. After background subtraction, it is found that only $22.8\pm3.5\%$ of the ψ' comes from B decay, compared to the expected $\sim 100\%$. Figure 36

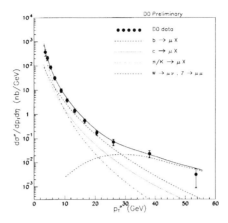

Fig. 31. D0 inclusive muon cross section as a function of the muon P_T. Also shown are the expected contributions from b decay, c decay, π/K decay, and W decay.

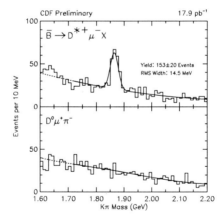

Fig. 32. The $K\pi$ mass distribution in right sign and wrong sign $\mu K\pi\pi$ events. The D peak is seen for the charge combination allowed in B decay.

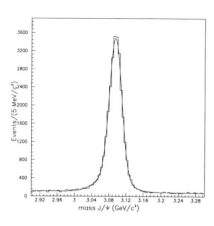

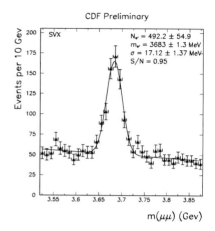

Fig. 33. The $\mu^+\mu^-$ mass distribution in the J/ψ region.

Fig. 34. The $\mu^+\mu^-$ mass distribution in the ψ' region.

shows the direct and B decay contributions to the ψ' cross section. The B component is in the range predicted by theory. The direct component, however, is much larger than predicted.[8]

A number of theorists are working on this problem.[8–10] Higher order fragmentation of a gluon or c quark into charmonium seems to be important. Figure 37 and Fig. 38 compare the CDF direct charmonium data with the leading order calculation as well as with the addition of fragmentation.[8] In the case of the J/ψ, the new calculation is close to the data. For the ψ', however, the calculated cross section is still much too small. There is clearly much to be done by the theorists to understand these discrepancies. The experimenters also have some issues to resolve, for example the difference between the CDF and D0 b cross sections at $P_T < 10$ GeV/c (Fig. 39). These issues will be easier to study now that sufficiently high statistics data are available to produce true differential cross sections (Fig. 40).

7. B Lifetimes

For many of the important B physics goals of the future, a large production cross section is not sufficient. It has to be shown that precision measurements can be made with the B mesons. To show that this can be done in a high luminosity hadron collider environment, I want to briefly present recent work measuring the B meson lifetimes.

CDF has published measurements of the average B lifetime[11] and the lifetimes for the charged and neutral B mesons using fully reconstructed decays.[12] Figure 41 shows the proper decay length distribution used to obtain the average B lifetime.

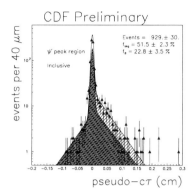

Fig. 35. The lifetime distribution from ψ''s. The dark cross-hatched area is the background obtained from the mass sidebands.

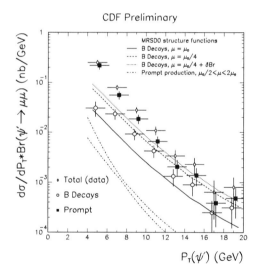

Fig. 36. The differential cross section for producing ψ' separated into the prompt and B decay contributions and compared with theoretical predictions.

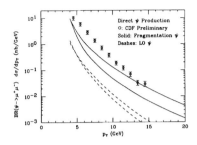

Fig. 37. The comparison made by Braaten *et al.* of CDF data with the leading order and fragmentation contributions to direct J/ψ production.

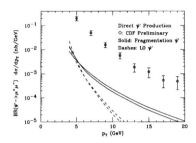

Fig. 38. The comparison made by Braaten *et al.* of CDF data with the leading order and fragmentation contributions to direct ψ' production.

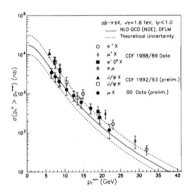

Fig. 39. The D0 and CDF b cross section as a function of P_T. Each point represents the integrated cross section above that P_T.

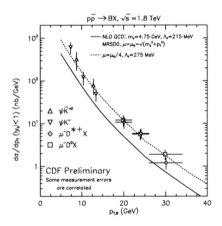

Fig. 40. The differential b production cross section measured by CDF.

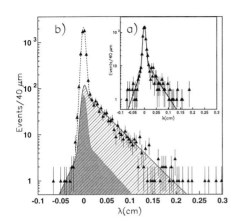

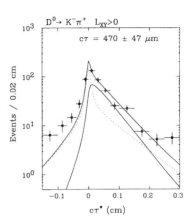

Fig. 42. The B lifetime from CDF eD^0 events. The inner solid line is the signal contribution, while the outer solid line is the signal plus the background as determined from the mass sidebands (dotted).

Fig. 41. The CDF B proper lifetime from inclusive J/ψ's.

Recently CDF has included the semi-exclusive eD and eD* final states to help measure the individual B meson lifetimes (Fig. 42). The values obtained for the B^- are:

$e^- D^0 X \qquad \tau_- = 1.52 \pm 0.21^{+0.09}_{-0.10}$ psec
$J/\psi\, K^- \qquad \tau_- = 1.61 \pm 0.16 \pm 0.05$ psec

For the $\overline{B}^0$, the results are:

$e^- D^{*+} X \qquad \tau_0 = 1.63 \pm 0.16 \pm 0.09$ psec
$J/\psi\, K^{*0} \qquad \tau_0 = 1.57 \pm 0.18 \pm 0.08$ psec

The resulting ratio of the charged to neutral B lifetime is

$\frac{\tau(B^-)}{\tau(\overline{B}^0)} = 0.98 \pm 0.13$

Figure 43 shows this measurement compared to those at LEP.

CDF has also measured the lifetime of the B_s meson using both exclusive and semi-exclusive (Fig. 44) final states. The results below are compared with other measurements in Fig. 45.

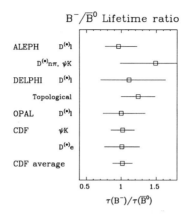

Fig. 43. The charged to neutral B lifetime ratio.

$e/\mu^- D_s^+$ $\tau_s = 1.42^{+0.27}_{-0.23} \pm 0.11$ psec
$J/\psi\,\phi$ $\tau_s = 1.74^{+0.90}_{-0.60} \pm 0.07$ psec

8.   Searching for New Heavy Objects

The Fermilab Tevatron Collider is at the "energy frontier" and should remain there for the next decade. It is thus important to search for new massive objects. Because of a lack of time, I will only present the current limits, all at the 95% CL.

Both CDF and D0 have new limits for heavy vector bosons. Assuming standard couplings, CDF (D0) find that the mass of a W′ must be greater than 652 (620) GeV/c², while a Z′ must have a mass greater than 505 (480) GeV/c².

D0 has a new limit on first generation leptoquarks.[13] If its spin is 0 and its branching ratio into eq is 1, the mass must be greater than 130 GeV/c² (Fig. 46). The mass limit for a spin 1 leptoquark is a factor of 1.5–2 higher. CDF has searched for second generation leptoquarks and finds that if BR(LQ→ μq)=1.0, the mass must be greater than 133 GeV/c² (Fig. 47).

D0 has a new limit on gluinos. They select the Minimal Supersymmetric Standard Model parameters $\tan\beta = 2$, $\mu = -250$ GeV, and $m_{H+} = 500$ GeV. They find that the gluino mass must be > 146 GeV/c² @ 90% CL for very large squark mass, and it must be > 205 GeV/c² @ 90% CL for equal squark and gluino masses.

Finally, CDF has a set of limits for particles that decay into jet pairs. They exclude axigluons in the range 200<M<920 GeV/c², a color octet technirho[14] in the range 260<M<470 GeV/c² if the branching ratio of the technirho into dijets is 1,

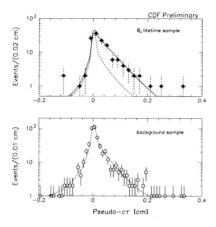

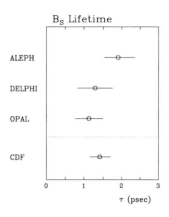

Fig. 44. The B_s lifetime from CDF lD_s events.

Fig. 45. B_s lifetime measurements.

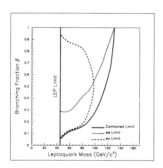

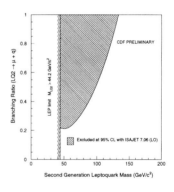

Fig. 46. The D0 1st generation lepto-quark mass limit as a function of its branching ratio into eq

Fig. 47. The CDF 2nd generation lep-toquark mass limit as a function of its branching ratio into μq

and excited quarks with mass below 620 GeV/c^2 if they have standard couplings. The excited quark search looks at three modes: q* →qW and q* →qγ in addition to the dijet mode (q* →qg).

9. Conclusions

The most exciting result of the past year is the evidence for the top quark. At a mass of 174 GeV/c^2, the top quark Yukawa coupling is 1. This makes the top quark a potentially powerful laboratory for learning about mass generation.

There were many other analyses in the past year that addressed important issues in electroweak interactions, QCD, b physics, and the search for new massive objects. Hopefully, much more information will be forthcoming in all of these areas with the data from the current Tevatron Collider run, which should increase the existing data sample by a factor of 4.

References

1. B. Pietrzyk, *Proceedings of the XXIXth Rencontres de Moriond* (1994).
2. F. Abe *et al.*, *Phys. Rev. Lett.* **73** (1994) 225. The details of this analysis are presented in F. Abe *et al.*, *Phys. Rev.* **D50** (Sept. 1, 1994).
3. E. Laenen *et al.*, *Phys. Lett.* **B321** (1994) 254.
4. S. Abachi *et al.*, *Phys. Rev. Lett.* **72** (1994) 2138.
5. F. Abe *et al.*, *Phys. Rev. Lett.* **65** (1990) 2243.
6. J. Alitti *et al.*, *Phys. Lett.* **B276** (1992) 354. The W mass is calculated using 91.187 ± 0.007 GeV/c^2 for the Z mass.
7. J. Ohnemus *et al.*, *Phys. Rev.* **D42** (1990) 61.
8. E. Braaten *et al.*, FERMILAB-PUB-94/135-T (1994).
9. M. Cacciari and M. Greco, FNT/T-94/13 (1994).
10. D.P. Roy and K. Sridhar, CERN-TH.7329/94 (1994).
11. F. Abe *et al.*, *Phys. Rev. Lett.* **71** (1993) 3421.
12. F. Abe *et al.*, *Phys. Rev. Lett.* **72** (1994) 3456.
13. S. Abachi *et al.*, *Phys. Rev. Lett.* **72** (1994) 965.
14. K. Lane and M. Ramana, *Phys. Rev.* **D44** (1991) 2678; E. Eichten and K. Lane, *Phys. Lett.* **B327** (1994) 129.

Rare Decays/CP–The Perspective of a Kaon Physicist

LAURENCE LITTENBERG, *Brookhaven National Laboratory*

RARE DECAYS/CP - THE PERSPECTIVE OF A KAON PHYSICIST

LAURENCE LITTENBERG

Physics Department
Brookhaven National Laboratory
Upton, NY 11973

ABSTRACT

The status and future prospects of searches for and studies of forbidden and highly suppressed K decays are reviewed as is the search for CPT violation in K decays.

1. Introduction

The study of K decays has played an enormous role in the history of particle physics. No one would dispute this, but even the most glorious history offers little protection against the question "what have you done for me lately?" Perhaps more germane is the question "what can K's still do for me in a world where the decays of heavier quarks are becoming more and more accessible?" In the interest of addressing this, I review the current activity in this area and discuss the future possibilities.

2. Forbidden Decays

The appeal of forbidden K decays is the lure of high scales. This is quite a rich subject, but I will confine myself to two areas of current activity in which kaon decays can access scales beyond anything possible in studies of heavy quarks for at least the next few years.

2.1. Lepton flavor violation

Although the violation of separate lepton flavor (LFV) is not allowed in the Standard Model, there is no known fundamental reason why such processes should not exist[1] and in fact most proposed extensions[2] of the Standard Model predict LFV at some level. The search for the LFV in the decays of muons and kaons has a long history, and has been pursued to prodigious sensitivities. Fig. 1 illustrates $K_L \to \mu e$ via the exchange of a hypothetical horizontal gauge boson, compared to $K^+ \to \mu\nu$ via W exchange.

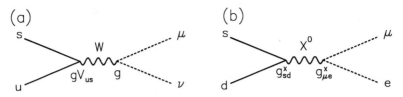

Fig. 1. (a) $K^+ \to \mu^+\nu$ decay, (b) $K_L \to \mu e$ via the exchange of a boson, X.

Assuming a V-A interaction, one can readily show:

$$B(K_L \to \mu e) = \left(\frac{g_{sd}^X g_{\mu e}^X}{g^2 sin\theta_C}\right)^2 \left(\frac{M_W}{M_X}\right)^4 \frac{\tau_{K_L}}{\tau_{K^+}} B(K^+ \to \mu^+\nu)$$

$$= 3.3 \times 10^{-11} \left(\frac{91\ TeV}{M_X}\right)^4 \left(\frac{g_{sd}^X g_{\mu e}^X}{g^2}\right)^2 \tag{2.1}$$

Since the latest limit[3] on $B(K_L \to \mu e)$ is in fact 3.3×10^{-11}, assuming that $g_{sd}^X \approx g_{\mu e}^X \approx g$, one is already probing scales in the 100 TeV ballpark for an LFV interaction. Sensitivity to $B(K_L \to \mu e)$ is expected to reach $\sim 10^{-12}$/event in the next couple of years. For comparison, a similar exercise on $B^0 \to \mu e$ yields:

$$B(B^0 \to \mu e) = 10^{-14} \left(\frac{91\ TeV}{M_X}\right)^4 \left(\frac{g_{bd}^X g_{\mu e}^X}{g^2}\right)^2 \tag{2.2}$$

This comparison can also be expressed as[4]:

$$\frac{B(B^0 \to \mu e)}{B(K_L \to \mu e)} \approx \frac{\tau_B}{\tau_{K_L}} \frac{m_B}{m_K} \left(\frac{g_{bd}^X}{g_{sd}^X}\right)^2 \tag{2.3}$$

$$\approx 3 \times 10^{-4} \left(\frac{g_{bd}^X}{g_{sd}^X}\right)^2 \tag{2.4}$$

Thus unless g_{bd}^X is significantly larger than g_{sd}^X, to compete with $B(K_L \to \mu e) < 3.3 \times 10^{-11}$, one would need to reach $B(B^0 \to \mu e) < 10^{-14}$. A similar comparison of three-body decays, where the helicity suppression that penalizes heavy parent mass is absent, yields a result more favorable to B's[4]:

$$\frac{B(B \to K\mu e)}{B(K \to \pi\mu e)} \approx \frac{1}{10} \frac{\tau_B}{\tau_{K_L}} \left(\frac{m_B}{m_K}\right)^5 \left(\frac{g_{bd}^X}{g_{sd}^X}\right)^2 \tag{2.5}$$

$$\approx \left(\frac{g_{bd}^X}{g_{sd}^X}\right)^2 \tag{2.6}$$

The factor $\frac{1}{10}$ reflects the B's preference for decaying to heavy hadronic systems. Thus for the three-body decays, equal branching ratios have about equal reach for LFV.

What is the comparative experimental situation? Fig. 2 shows the most recent data on $B^- \to K^- e\mu$ from CLEO[5] and on $K^+ \to \pi^+ e\mu$ from AGS-777.[6] In some ways these plots look remarkably similar. Each is a distribution of the final state effective mass versus a second variable that also distinguishes signal from background. In each case background events threaten or actually encroach upon the signal region. The difference is that the K result represents a sensitivity roughly five orders of magnitude beyond that of the B result (soon to be almost seven orders of magnitude, if AGS-865 meets its goal). This is primarily due to the relative availability of Ks and Bs, but it should also be realized that even if one had the requisite 50,000 fold increase in B-statistics, the signal box would contain 10^5 background events. This underlines the advantages of a dedicated detector, in which particle identification factors of 10^7 or even more are available.

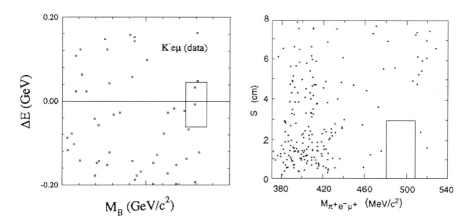

Fig. 2. Search for lepton flavor violation in Bs and Ks. The left hand plot shows the result of CLEO search for $B^- \to K^- e\mu$. The ordinate is the difference between the energy of the beam and that of the final state particles and the abscissa is the effective mass of the B candidates. The right hand plot shows the result of AGS 777 search for $K^+ \to \pi^+ e\mu$. The ordinate is the rms distance of closest approach of the three tracks to a common vertex and the abscissa is the three body effective mass.

Of course K's can't touch things like $B \to \tau\mu$. The situation is also quite different for SM-forbidden processes mediated by scalars, and for mechanisms involving t-quarks loops, where there's a B/K enhancement $\sim |\frac{V_{ts}^* V_{tb}}{V_{td}^* V_{ts}}|^2 \sim \frac{1}{A^2 \lambda^6} \sim 10^5$.

Table 1 gives a reasonably up-to-date summary of searches or lepton flavor violation. One can observe that the sensitivity to LFV attained in muon decays is quite comparable to that reached in K decays, and that this field is still very active.

Table 1. Status of searches for lepton flavor violation

Mode	Current u.l.	Experiment	ref.	Date	(Near-)future aim
$K^+ \to \pi^+ e^- \mu^+$	2.1×10^{-10}	AGS777	6	1990	3×10^{-12}(AGS865)
$K_L \to \mu e$	3.3×10^{-11}	AGS791	3	1993	2×10^{-12}(AGS871)
$\pi^0 \to \mu e$	8.6×10^{-9}	FNAL799	7	1994	
$B \to K \mu e$	1.2×10^{-5}	CLEO II	5	1994	
$B \to K^* \mu e$	2.7×10^{-5}	CLEO II	5	1994	
$B \to \mu e$	5.9×10^{-6}	CLEO II	8	1994	
$B \to \tau e$	5.3×10^{-4}	CLEO II	8	1994	
$B \to \tau \mu$	8.3×10^{-4}	CLEO II	8	1994	
$\mu^+ \to e^+ \gamma$	4.9×10^{-11}	Xtal Box	9	1988	7×10^{-13} (MEGA)
$\mu^+ \to e^+ e^- e^+$	1.0×10^{-12}	SINDRUM	10	1988	
$\mu^- Ti \to e^- Ti$	4.3×10^{-12}	SINDRUM II	11	1993	$few \times 10^{-14}$
$\mu^+ e^- \to \mu^- e^+$	3.3×10^{-7}	PSI R-89-06	12	1994	3×10^{-9}
$Z^0 \to \mu e$	3.9×10^{-6}	L3	13	1994	
$Z^0 \to e\tau$	8.6×10^{-6}	L3	13	1994	
$Z^0 \to \mu\tau$	1.1×10^{-5}	L3	13	1994	
$\tau \to \mu\gamma$	4.2×10^{-6}	CLEO II	14	1993	
$\tau \to \mu\mu\mu$	4.3×10^{-6}	CLEO II	15	1994	
$\tau \to \mu\mu e$	3.5×10^{-6}	CLEO II	15	1994	
$\tau \to \mu ee$	3.4×10^{-6}	CLEO II	15	1994	
$\tau \to eee$	3.3×10^{-6}	CLEO II	15	1994	

2.2. CPT violation

The observation of CPT violation would be much more of an upset to current dogma than LFV, since it would imply that we were not dealing with a local quantum field theory. In the flurry of theoretical activity prompted by the discovery of CP-violation, a number of tests for CPT violation in the K were proposed.[17] Experiment soon indicated that CPT violation in K decay was at most small in comparison to CP-violation, and interest in the subject waned. Recently there has been renewed activity, both theoretical and experimental. Stimulated by Hawking's proposal that the generalization of quantum mechanics which includes gravity allows the evolution of pure into mixed states,[16] and Page's demonstration that this leads to CPT-violation,[18] Ellis *et al.* suggested experimental tests of the idea in the $K^0 - \bar{K}^0$ system.[19] More recently, this was developed further by Ellis, Mavromatos, and Nanopoulos who showed that CPT is generally violated in a non-quantum-mechanical manner in the effective low-energy theory derived from string theory. They suggested that this CPT-violation might even account for the observed CP-violation in the $K^0 - \bar{K}^0$ system.[20] This is quite a different conclusion than is reached concerning "normal", quantum mechanically allowed, CPT-violation in the traditional phenomenological analysis.[21] Finally

Huet and Peskin analyzed the dynamics of the K system allowing both varieties of CPT-violation.[22] They conclude that the observed CP-violation in this system is "dominantly quantum mechanical in nature and of CPT-conserving origin." They propose a number of tests for CPT and quantum mechanics violation at ϕ factories. Another possible vulnerability of CPT was emphasized by Kobayashi and Sanda,[23] who pointed out that the proof of the CPT theorem may not apply to QCD because quarks and gluons are not asymptotic states.

There is new experimental data which bears on the possibility of CPT violation both in the neutral kaon mass matrix and in the decay amplitudes. The theoretical work discussed above suggests the former will be more important than the latter. "Traditional" CPT-violation in the state mixing can be parameterized as follows:

$$|K_S> \quad \sim \quad (1+\epsilon_S)|K^0> +(1-\epsilon_S)|\bar{K}^0>$$

$$|K_L> \quad \sim \quad (1+\epsilon_L)|K^0> -(1-\epsilon_L)|\bar{K}^0>$$

$$\text{where } \epsilon_S \quad \equiv \quad \epsilon_M + \Delta$$

$$\epsilon_L \quad \equiv \quad \epsilon_M - \Delta \tag{2.7}$$

In the absence of CPT-violation, $\epsilon_M = \epsilon_S = \epsilon_L$ and all are equivalent to the CP-violation parameter normally denoted simply as ϵ. The quantity Δ parameterizes the CPT-violation. CPLEAR measures the real parts of ϵ_M and Δ by measuring the time development of asymmetries in $Ke3$ decays from initially pure K^0 and $\bar{K}^0$ states. These states are strangeness tagged in $\bar{p}p \rightarrow K^-\pi^+K^0$ or $K^+\pi^-\bar{K}^0$. In particular

$$Re(\epsilon_M) = \frac{1}{4} \frac{R(\bar{K}^0_{t=0} \rightarrow e^+\pi^-\nu) - R(K^0_{t=0} \rightarrow e^-\pi^+\bar{\nu})}{R(\bar{K}^0_{t=0} \rightarrow e^+\pi^-\nu) + R(K^0_{t=0} \rightarrow e^-\pi^+\bar{\nu})} \tag{2.8}$$

is a direct measurement of T-violation.

Fig. 3 shows the time distribution of the asymmetry A_T $(= 4Re(\epsilon_M)$ if the $\Delta S = \Delta Q$ Rule is obeyed) from CPLEAR.[24] It is consistent with a flat time dependence and yields $A_T = (4.5 \pm 2.1_{stat} \pm 2.0_{sys} \pm 1.5_{MC}) \times 10^{-3}$ or $Re(\epsilon_M) = (1.1 \pm 0.5_{stat} \pm 0.5_{sys} \pm 0.4_{MC}) \times 10^{-3}$. The value of $Re(\epsilon_M)$ can be compared with the world average value of $Re(\epsilon_L)$ obtained from the $K_L\ell3$ charge asymmetry,[25] $(1.63 \pm 0.06) \times 10^{-3}$, which ought to be the same in the absence of CPT-violation. There's no significant disagreement, but this comparison is clearly not yet a very demanding test of CPT. However, the ϵ_M result is important as a direct demonstration of T-violation. It is also possible to form an asymmetry that directly measures Δ (again, assuming $\Delta S = \Delta Q$):

$$A_{CPT} = \frac{R(\bar{K}^0_{t=0} \rightarrow e^-\pi^+\bar{\nu}) - R(K^0_{t=0} \rightarrow e^+\pi^-\nu)}{R(\bar{K}^0_{t=0} \rightarrow e^-\pi^+\bar{\nu}) + R(K^0_{t=0} \rightarrow e^+\pi^-\nu)} \tag{2.9}$$

After a few τ_S, $A_{CPT} = 4Re(\Delta)$. The CPLEAR result is $Re(\Delta) = (-0.1 \pm 0.5_{stat} \pm 0.6_{sys}) \times 10^{-3}$. This is statistically consistent with 0, but also less than 3σ from $Re(\epsilon)$.

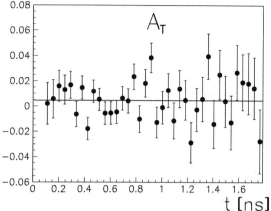

Fig. 3. The asymmetry A_T from CPLEAR.

There is also a new result in the search for CPT-violation from FNAL-773.[26] They measured the relative phase of η_{+-} and η_{00}. The presence of a finite value for Δ does not affect this phase difference, since to a good approximation

$$\eta_{+-} = \epsilon_M + \epsilon' - \Delta$$

$$\eta_{00} = \epsilon_M - 2\epsilon' - \Delta \qquad (2.10)$$

Since $\epsilon' << \epsilon$ and is nearly aligned with it, a measurable value of $\phi_{00} - \phi_{+-}$ implies CPT-violation in the decay amplitudes. The experiment measured the time dependence of 2π decays from a K_L beam after it traversed a regenerator. This is given by:

$$\Gamma \sim |\rho|^2 e^{-\Gamma_S \tau} + |\eta|^2 e^{-\Gamma_L \tau} + 2|\rho\eta| e^{-\bar{\Gamma}\tau} cos(\Delta m \tau + \phi_\rho - \phi_\eta) \qquad (2.11)$$

where η refers to the $\pi^+\pi^-$ or $\pi^0\pi^0$ K_L/K_S amplitude ratio, ρ is the regeneration amplitude, $\bar{\Gamma}$ is the average of the K_S and K_L decay rates, and Δm is the $K_L - K_S$ mass difference. The experiment was designed to minimize systematics via a two-beam technique. One beam passed through a thin regenerator, and the subsequent 2π decays were measured in the small τ region where the sensitivity of the interference term to ϕ_η is maximum. The second beam passed through a thick regenerator placed about 11 m upstream of the thin regenerator. Decays from this beam emanated from a region where the argument of the interference term of Eq. 2.11 is near $0°$ so that the 2π rate is very insensitive to the phase. Normalization uncertainties, rate effects, imperfections in acceptance modeling, and other systematics tend to cancel when rates in the two beams are compared. The preliminary result, $\Delta\phi = (0.67 \pm 0.85_{stat} \pm 1.1_{sys})°$, is consistent with other recent measurements,[27,28] all of which are consistent with CPT-conservation. Making a weighted average of the new result with the PDG fit[25] yields

$(-0.43 \pm 0.81)°$. Thus, in contrast to the situation of a few years ago, there is no evidence for CPT-violation in this quantity.

There are also tests of CPT which can reflect violations in either the mass matrix or in decay amplitudes. It can be shown from the Bell-Steinberger unitarity relation[29] and CPT-conservation, that ϕ_ϵ should be very nearly equal to $\phi_{SW} \equiv tan^{-1}(\frac{2\Delta m}{\Gamma_S - \Gamma_L})$. Since in the absence of CPT-violation $\eta_{+-} \approx \epsilon_M + \epsilon'$, and $\phi_{\epsilon'} \approx \phi_\epsilon$ (and anyway $|\epsilon'| \ll |\epsilon|$), then $\phi_{+-} \approx \phi_{SW}$. Here, until recently, one had an apparent 2.4σ disagreement[30]: $\phi_{+-} = 46.6° \pm 1.2°$ versus $\phi_{SW} = 43.73° \pm 0.15°$. There are now new measurements of ϕ_{+-} from FNAL-773[26] and CPLEAR.[24] The value of Δm influences this test in a critical way. Not only does it directly enter the determination of ϕ_{SW}, but since ϕ_{+-} occurs in oscillation formulae in the combination $\Delta m \tau - \phi_{+-}$, the errors on Δm and ϕ_{+-} tend to be highly correlated.

Fitting Eq. 2.11 with $\eta = \eta_{+-}$, FNAL-773 obtains:

$$\phi_{+-} = (43.35 \pm 0.35_{stat} \pm 0.79_{sys})° \quad - \quad 0.62° \frac{\tau_S - 0.8922 \times 10^{-10}s}{0.0020 \times 10^{-10}s}$$
$$+ \quad 0.38° \frac{\Delta m - 0.5286 \times 10^{10}s^{-1}}{0.0024 \times 10^{10}s^{-1}} \quad (2.12)$$

The value of Δm implied in Eq. 2.12 is that obtained in a fit to Eq. 2.11 constraining $\phi_{+-} \equiv tan^{-1}(2\Delta m\tau_S)$ and is equal to that found by FNAL-731.[27] If they fit simultaneously to ϕ_{+-}, Δm, and τ_S, they obtain:

$$\phi_{+-} - \phi_{SW} = (-0.84 \pm 1.42_{stat} \pm 1.22_{sys})°$$
$$\Delta m = (0.5286 \pm 0.0041_{stat} \pm 0.0029_{sys}) \times 10^{10}s^{-1}$$
$$\tau_S = (0.8942 \pm 0.0026_{stat} \pm 0.0018_{sys}) \times 10^{-10}s \quad (2.13)$$

so that there is no significant CPT-violation.

CPLEAR measures the asymmetry:

$$A_{+-}(t) = \frac{R(\bar{K}^0_{t=0} \to \pi^+\pi^-)(t) - R(K^0_{t=0} \to \pi^+\pi^-)(t)}{R(\bar{K}^0_{t=0} \to \pi^+\pi^-)(t) + R(K^0_{t=0} \to \pi^+\pi^-)(t)}$$
$$= 2Re(\epsilon_L) - \frac{2|\eta_{+-}|e^{\Delta\Gamma t/2}}{1 + |\eta_{+-}|^2 e^{\Delta\Gamma t}} cos(\Delta mt - \phi_{+-}) \quad (2.14)$$

The presence of the positive exponential in Eq. 2.14 leads to the impressive asymmetry shown in Fig. 4. Their result is $\phi_{+-} = (44.7 \pm 0.9_{stat} \pm 1.1_{sys} \pm 0.7_{\Delta m})°$. Aside from the Δm dependence, this result has rather different systematics than those of FNAL-773. Once again, there's no significant deviation from the prediction of CPT.

It has been known for some time[31] that when the value of Δm is corrected to $0.5286 \times 10^{10}s^{-1}$, all previous determinations of ϕ_{+-} are within one σ of $\phi_{SW} = 43.35°$

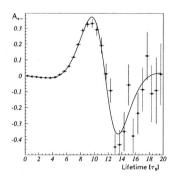

Fig. 4. The asymmetry A_{+-} from CPLEAR. The line is a fit to the data.

(the value of ϕ_{SW} corresponding to the $\Delta m = 0.5286 \times 10^{10} s^{-1}$ and $\tau_S = 0.8928 \times 10^{-10} s$). Thus a long-standing discrepancy has been cleared up.

One can use the current data to obtain to improve our knowledge of the CPT-violating quantity $\left| \frac{m_{\bar{K}^0} - m_{K^0}}{m_K} \right|$. If CPT-violation stems from Planck-scale physics, one might expect this to be about $\frac{m_K}{m_{Planck}} \sim 4 \times 10^{-20}$. One can obtain from perturbation theory, the quantity $\Delta_\perp \approx \frac{m_{\bar{K}^0} - m_{K^0}}{2\Delta m} sin(\phi_{SW})$, where $\Delta_\perp$ is the component of Δ transverse to ϕ_{SW}. From Eq. 2.10 and the phenomenology, $\Delta_\perp \approx |\eta_{+-}|(\phi_\epsilon - \frac{2}{3}\phi_{+-} - \frac{1}{3}\phi_{00})$. Equating these formulae for $\Delta_\perp$, one obtains:

$$\frac{m_{\bar{K}^0} - m_{K^0}}{m_K} \approx \frac{\Delta m}{m_K} \frac{2|\eta_{+-}|}{sin(\phi_{SW})}(\phi_\epsilon - \phi_{+-} - \frac{\Delta\phi}{3}) \qquad (2.15)$$

This was evaluated in 1990 by Ref. 28, who found $\left| \frac{m_{\bar{K}^0} - m_{K^0}}{m_K} \right| < 5 \times 10^{-18}$ (@ 95%c.l.). With a **very** rough and ready treatment of the errors, adding in the new results for Δm, ϕ_{+-}, and, $\Delta\phi$, I get $\sim 1.6 \times 10^{-18}$ for the current limit. There's still a factor ~ 50 to go before the Planck scale can be probed, but where else are we as close as this?

The discussion of checks of CPT has so far assumed that quantum mechanics is still valid. As mentioned above, Huet and Peskin[22] have re-examined the consequences of relaxing this assumption for the $K^0 - \bar{K}^0$ system. They make a number of well-motivated simplifying assumptions which facilitate the confrontation of this approach with the data: probability must be conserved, mixed states should not evolve into pure states, the new interaction conserves strangeness, the CPT-violation is confined to the kaon propagation, and the $\Delta S = \Delta Q$ Rule is obeyed. The resulting time evolution equation for the $K^0 - \bar{K}^0$ density matrix gains three new (real) parameters, α, β, and γ, which satisfy $\alpha, \beta > 0$ and $\alpha\gamma > \beta^2$. An example of the extra complication introduced into the formalism is given by the time distribution of $\pi^+\pi^-$ decays from an initially pure K^0 beam:

$$R_{+-}(\tau) \ \sim \ e^{-\Gamma_S\tau} + Ae^{-\Gamma_L\tau} + B\cos(\Delta m\tau + \phi_{+-})e^{-\bar{\Gamma}\tau} \qquad (2.16)$$

If CPT is conserved, the coefficients are simply $A = |\eta_{+-}|^2$ and $B = 2|\eta_{+-}|$. If non-quantum mechanical CPT-violation only is allowed, the coefficients become:

$$
\begin{aligned}
A &= |\epsilon - \beta/d|^2 + \frac{\gamma}{\Delta\Gamma} + \frac{4\beta}{\Delta\Gamma}Im[(\epsilon - \beta/d)\frac{d}{d^*}] \\
B &= 2|\epsilon - \beta/d|
\end{aligned}
\qquad (2.17)
$$

where $d \equiv \Delta m + i\Gamma/2$. In addition, the exponential decay constant of the interference term is changed slightly, from $\bar{\Gamma}$ to $\bar{\Gamma} + \alpha - \gamma$

Since the experimental data aren't normally analyzed in this formalism, Huet and Peskin were able to use a very limited subset of the potentially available results. They managed to extract constraints on β and γ from 1992 PDG averages[30] for ϕ_{+-} and the $K_L\ell3$ charge asymmetry, twenty-year old results from CERN-Heidelberg[32] on $R_{+-}(\tau)$ at large τ, and the initial CPLEAR result[33] on η_{+-}. They obtained[34]:

$$\beta = (0.12 \pm 0.44) \times 10^{-18} GeV$$

$$\gamma = (-1.1 \pm 5.6) \times 10^{-21} GeV \qquad (2.18)$$

If they allow for possible CPT-violation in decay amplitudes, they can still get useful constraints, and they find that in the absence of unnatural cancelations among CPT-violating parameters, the magnitude of CPT-violation is $\leq$ 10% of that of CP-violation. Some of Huet and Peskin's input has recently been superseded. In particular, there's a new result from CPLEAR,[24] $|\eta_{+-}| = (2.163 \pm 0.045_{stat} \pm 0.064_{sys} \pm 0.010_{\Delta m}) \times 10^{-3}$. In addition, there have been the above-mentioned new results for ϕ_{+-} and $\phi_{00} - \phi_{+-}$. Using the same cavalier treatment of correlated errors as above, I recalculate the numbers in Eq. 2.18 as $\beta = (0.20 \pm 0.32) \times 10^{-18}$ GeV and $\gamma = (3.3 \pm 3.5) \times 10^{-21}$ GeV.

There is likely to be continued progress in probing CPT-violation The FNAL-773 results should improve significantly when their analysis is complete. CPLEAR promises to eventually do two or three times better in precision for most of their measurements. The upcoming round of CP-violation experiments should allow further advances. Eventually $DA\Phi NE$ should also play a large rôle.

3.   CP-violation

It's important to keep in mind how little we actually know about CP-violation if we refrain from begging the question by assuming that the Standard Model is right. All confirmed observations are consistent with the superweak model,[35] which is to say

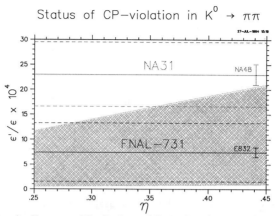

Fig. 5. CP-violation in $K_L \to \pi\pi$. The horizontal lines give the experimental central values and errors. The error bars show the projected precision of the successor experiments now in preparation. The shaded region indicates the range of SM predication as a function of the CKM parameter η.

that they can be explained by state-mixing alone. There is no shortage of more modern theoretical approaches[36] that have consequences very similar to superweak, at least for the K and B systems. Moreover, even if the superweak model is disproved and the KM mechanism[37] turns out to successfully describe the CP-violation we observe, it probably can't account for the baryon asymmetry of the universe.[38]

3.1. $K_L \to 2\pi$

The $K_L \to 2\pi$ system is the best bet for producing the first confirmed observation of direct CP-violation. The marginally conflicting results of FNAL-731[39] and NA31[40] have been with us for some time, and each group is presently constructing a successor experiment. Fig. 5 summarizes the present experimental and theoretical[41] situation. Unfortunately, since the SM allowed range of ϵ'/ϵ extends so close to 0, it is possible that the upcoming experiments will not be definitive even if they achieve their goals for precision $(1 - 2 \times 10^{-4})$.

3.2. $K_S \to 3\pi$

Two experiments, CPLEAR and FNAL-621 have recently reported results on $K_S \to \pi^+\pi^-\pi^0$. The 3π state is to K_S more or less what the 2π state is to K_L, so why has there been so little work on on the former relative to the latter? The reasons are not hard to find. In the case of K_S one must work close to the production target, usually a poor environment. In the case of $K_L \to 2\pi$, one can easily remove K_S "background" by moving a few τ_S downstream, whereas even in the first τ_S, $K_S \to 3\pi$ is swamped by $K_L \to 3\pi$. Moreover, only the $3\pi^0$ state is pure CP $-$. The spin-0 $\pi^+\pi^-\pi^0$ system is actually quite complicated and has a CP+ component which, although suppressed by angular momentum (it has $\ell_{\pi^+\pi^-} = 1, 3, ..$) and the $\Delta I = \frac{1}{2}$ rule (it has $I = 2$), still dominates the CP-violating ($\ell = 0$, $I = 1$) component. From state-mixing alone, one expects that the CP-violating branching ratio, $B(K_S \to \pi^+\pi^-\pi^0)_\epsilon = |\epsilon|^2 \frac{\Gamma_L}{\Gamma_S} B(K_L \to \pi^+\pi^-\pi^0) \approx 10^{-9}$, while theoretical estimates[42] of the CP-conserving branching ratio are in the $few \times 10^{-7}$ range. In practice, one measures the interference between the K_S and K_L, so that one can remove the CP-conserving component by integrating over the Dalitz plot, as long the acceptance is understood well enough. Then, by separately analyzing the proper time distribution of the halves of the Dalitz plot with $E_{\pi^+} - E_{\pi^-} > 0$ and < 0, one can extract the CP-conserving branching ratio. Figs.6 and 7 show the results of this procedure for the two experiments. In each case evidence for a positive signal can be seen at early proper time. The branching ratios calculated from the observed interference amplitudes are: $B(K_S \to \pi^+\pi^-\pi^0)_{CPcons.} = (3.9^{+5.4\ +0.9}_{-1.8\ -0.7}) \times 10^{-7}$ (FNAL 621[43]) and $B(K_S \to \pi^+\pi^-\pi^0)_{CPcons.} = (8.2^{+6.0\ +7.3}_{-4.4\ -4.9}) \times 10^{-7}$ (CPLEAR[24]) where in each case the first error is statistical and the second systematic. Thus, it is probably safe to say that $K_S \to \pi^+\pi^-\pi^0$ has been seen for the first time.

However, the main object of these exercises is the hunt for the CP-violating component of this decay. The Standard Model predicts[45] direct CP-violation at a level equal or greater to that in $K \to 2\pi$, with $\epsilon'_{\pi^+\pi^-\pi^0}$ ranging up to at least $10\epsilon'$. Prior to the present round of experiments, the limit on CP-violation in this channel was extremely poor, i.e. $|\eta_{+-0}|^2 < 0.12$ @ 90%c.l. ($\eta_{+-0} \equiv A(K_S \to \pi^+\pi^-\pi^0)_{CPviol.}/A(K_L \to \pi^+\pi^-\pi^0)$). The present experiments have improved this situation substantially, as can be seen in Fig.8, where both the old and new results are displayed. Each experiment promises further improvement. However, as indicated in Table 2, there is quite a long way to go before reaching even ϵ, much less ϵ'.

Table 2. CP-violation in $K_S \to \pi^+\pi^-\pi^0$

Experiment	$Re(\eta_{+-0})$	$Im(\eta_{+-0})$	ref.
FNAL 621	0.019 ± 0.027	0.019 ± 0.061	46
"	$\equiv Re(\epsilon)$	$-0.015 \pm 0.017_{stat.} \pm 0.025_{syst.}$	46
CPLEAR	$0.005 \pm 0.022_{stat.} \pm 0.007_{syst.}$	$0.016 \pm 0.024_{stat.} \pm 0.018_{syst.}$	24
ϵ	0.0016	0.0017	25

4. One-loop decays

Most of the current interest in rare K decays is directed toward the study of one-loop effects. Diagrams such as those shown in Fig. 9 are expected to dominate, or at least contribute significantly to decays of the form $K^{0,\pm} \to (\pi^{0,\pm})\ell\bar{\ell}$. The SM calculation of these diagrams is very clean, with QCD corrections small or moderate, and with hadronic matrix elements reliably obtainable from the rates of common K decay modes. The presence of t-quarks in the loops makes the processes sensitive to parameters like m_t and V_{td}. The decays $K_L \to \pi^0\ell\bar{\ell}$ are CP-violating to first order, and are potentially very useful probes of this phenomenon. One should also keep in mind that over the years, numerous non-SM contributions to $K^{0,\pm} \to (\pi^{0,\pm})\ell\bar{\ell}$ have been predicted. Since for most of these modes experiments have not yet reached the Standard Model level, a potential remains for the discovery of new physics.

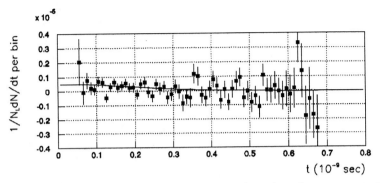

Fig. 6. FNAL-621: difference of time distributions of $K \to \pi^+\pi^-\pi^0$ with $E^{cm}_{\pi^+} > E^{cm}_{\pi^-}$ and those with $E^{cm}_{\pi^-} > E^{cm}_{\pi^+}$

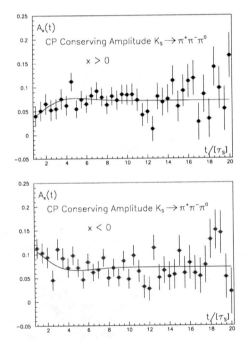

Fig. 7. CPLEAR: Decay rate asymmetry for $K \to \pi^+\pi^-\pi^0$ with $E^{cm}_{\pi^+} > E^{cm}_{\pi^-}$ and for $E^{cm}_{\pi^-} > E^{cm}_{\pi^+}$

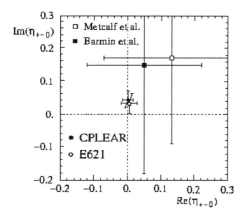

Fig. 8. Results of searches for CP-violation in $K_S \to \pi^+\pi^-\pi^0$

The exploitation of these decays motivates the study of a number of other modes. In some cases these constitute experimental backgrounds to the one-loop processes (e.g. $K_L \to \ell^+\ell^-\gamma\gamma$ as background to $K_L \to \pi^0\ell^+\ell^-$). Others can be used to evaluate the strength of possible long-distance competition to the one-loop effects (e.g. $K_L \to \pi^0\gamma\gamma$, which scales the CP-conserving contribution to $K_L \to \pi^0\ell^+\ell^-$).

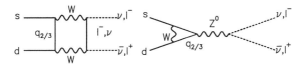

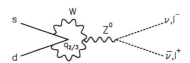

Fig. 9. One-loop contributions to $K^{0,\pm} \to (\pi^{0,\pm})\ell\bar{\ell}$

4.1. $K^+ \to \pi^+\nu\bar{\nu}$

The prototypical one-loop processes is $K^+ \to \pi^+\nu\bar{\nu}$. Here long-distance 'background' is negligible[44] so that the short-distance contributions (Standard or non-Standard) are completely unobscured. The QCD corrections are relatively small and well determined. Recent calculations by Buchalla and Buras[47] beyond leading logarithms (using a two-loop renormalization group analysis for the charm contribution and $O(\alpha_s)$ for top) indicate theoretical uncertainties no greater than 7% for typical

SM parameters. The branching ratio is particularly sensitive to the value of $|V_{td}|$, and offers the theoretically cleanest way of determining this parameter.

The branching ratio is given by:

$$B(K^+ \rightarrow \pi^+ \nu\bar{\nu}) \;=\; \frac{\alpha^2 B(K^+ \rightarrow \pi^0 e^+ \nu)}{V_{us}^2 2\pi^2 sin^4\theta_W} \sum_{\ell} |V_{cs}^* V_{cd} X_{NL}^\ell + V_{ts}^* V_{td} X(x_t)|^2 \quad (4.1)$$

where $x_t \equiv (m_t/m_W)^2$. If CDF is right about m_t, $X(x_t) \approx 1.6$. Note the occurrence of the product $V_{ts}^* V_{td}$. If one assumes a unitary CKM matrix, Eq. 4.1 can be rewritten as

$$B(K^+ \rightarrow \pi^+ \nu\bar{\nu}) \;=\; 1.2 \times 10^{-10} A^4 [\eta^2 + \frac{2}{3}(\rho_o^e - \rho)^2 + \frac{1}{3}(\rho_o^\tau - \rho)^2] \quad (4.2)$$

where $\rho_o^\ell \equiv 1 + \frac{X_{NL}^\ell}{A^2\lambda^4 X(x_t)}$ and $X_{NL}^\ell \approx 10^{-3}$.

In leading order in Wolfenstein parameters, $B(K^+ \rightarrow \pi^+ \nu\bar{\nu})$ determines a circle in the ρ, η plane with center $(\rho_o, 0)$; $\rho_o \equiv \frac{2}{3}\rho_o^e + \frac{1}{3}\rho_o^\tau \approx 1.4$ and radius $= \frac{1}{A^2}\sqrt{\frac{B(K^+\rightarrow\pi^+\nu\bar{\nu})}{1.2\times10^{-10}}}$

Detecting this decay mode at the 10^{-10} level is rather challenging. Since two of the final state particles are undetected, there is no kinematic constraint (other than $p_{\pi^+}^* \leq 227\ MeV/c$). The one detectable particle, the π^+, is a very common decay product of the K^+. Since the major backgrounds are the copious two-body decay modes $K^+ \rightarrow \pi^+\pi^0$ (21.2%) and $K^+ \rightarrow \mu^+\nu$ (63.5%), excellent kinematic resolution is essential. In addition, to eliminate $K^+ \rightarrow \pi^+\pi^0$, one needs a γ veto capable of suppressing π^0s by a factor of $\sim 10^6$. To eliminate $K^+ \rightarrow \mu^+\nu$ one needs particle identification capability to reject μ^+ by $> 10^8$. AGS-787 has been on the track of this decay mode for several years now. This experiment, currently a collaboration of BNL, INS/Tokyo, KEK, Osaka, Princeton, and TRIUMF, searches for $K^+ \rightarrow \pi^+\nu\bar{\nu}$ and related decays in a solenoidal spectrometer[48] situated in a stopping K^+ beam. New preliminary results are now available. Fig. 10 shows a plot of π^+ kinetic energy versus range for $K^+ \rightarrow \pi^+\nu\bar{\nu}$ candidates from three years' running. No candidates survive, allowing a 90% c.l. upper limit of 3.6×10^{-9} to be set. Combining this with the limit[49] 1.8×10^{-8} previously obtained by E787 for a different kinematic region ($p_{\pi^+} < 205$ MeV/c), yields $B(K^+ \rightarrow \pi^+\nu\bar{\nu}) < 3.0\times10^{-9}$ (@ 90% c.l.). One can also extract a limit on the process $K^+ \rightarrow \pi^+ + X^0$, where X^0 is a single massless unseen neutral. The new limit on this intrinsically non-SM process is $B(K^+ \rightarrow \pi^+ X^0) < 6.1 \times 10^{-10}$ (@ 90% c.l.). These limits represent a factor 50 improvement in the sensitivity to $K^+ \rightarrow \pi^+ +$ 'nothing' with respect to previous experiments. However, there is still an additional factor 10 in sensitivity needed to reach the prediction of the Standard Model for $K^+ \rightarrow \pi^+\nu\bar{\nu}$. Therefore an upgraded detector and beam have been built to continue this search. The first run in the new configuration has just been completed.

At this conference we have seen two results from B-decays that purport to bound $|\frac{V_{td}}{V_{ts}}|$. One is a limit of $B(B \rightarrow \rho(\omega)\gamma)/B(B \rightarrow K^*\gamma) < 0.34$ by CLEO.[50]

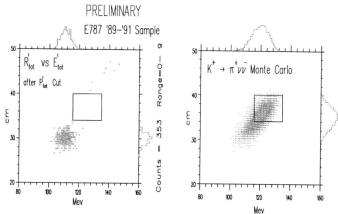

Fig. 10. Results of '89-91 search for $K^+ \to \pi^+ \nu\bar{\nu}$(left). Ordinate is π^+ range and abscissa is π^+ energy. A cut has also been placed on the π^+ momentum. Signal region is outlined. Peak at lower left is $K^+ \to \pi^+\pi^0$. Events at upper right are from tail of $K^+ \to \mu^+\nu$ and $K^+ \to \mu^+\nu\gamma$. The right hand plot shows a Monte Carlo simulation of the signal. The falloff of events with range < 30cm is due to the trigger.

On the assumption that these decays are mediated purely by short-distance penguin diagrams like those of Fig. 11, this results in a limit of $\left|\frac{V_{td}}{V_{ts}}\right| < \frac{0.58}{\sqrt{\xi}}$, where ξ is an SU(3)-breaking parameter in the range $0.58 - 0.81$. Taking the most pessimistic value of ξ gives $\left|\frac{V_{td}}{V_{ts}}\right| < 0.76$. However, as pointed out by Atwood, Blok, and Soni,[51] the interpretation of this measurement may be complicated by possible long distance effects in $B \to V\gamma$. It is believed however that the long distance contributions, although not negligible, are somewhat smaller than the short distance contributions, and that further experimental and theoretical work will eventually make this a viable method of measuring $\left|\frac{V_{td}}{V_{ts}}\right|$.

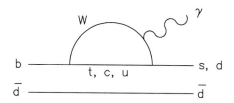

Fig. 11. Penguin diagram for $B \to V\gamma$.

A second limit comes from recent work on $B_s - \bar{B}_s$ mixing at LEP.[52] $B_d - \bar{B}_d$ mixing is proportional to $|V_{td}|^2$, but the presence of a difficult-to-calculate hadronic matrix element makes extracting $|V_{td}|$ problematic (until recently one also had to worry about a dependence on m_t). A popular response to this is to try to measure

$B_s - \bar{B}_s$ mixing so that one can reduce the theoretical problem to that of determining the ratio of B_d and B_s matrix elements. This ratio is alleged to be known to about 10%, and one has[53] $\frac{\Delta m_d}{\Delta m_s} = \frac{1}{(1.19\pm0.10)} \times |\frac{V_{td}}{V_{ts}}|^2$. The numerator here has been known pretty well for a while through time-averaged measurements, and recently there have been beautiful results from LEP exploiting the time dependence. These have now developed to the point where meaningful limits on Δm_s can be extracted. The ALEPH result discussed by Sharma at this conference is $\Delta m_s > 6ps^{-1}$. Using $\tau_{B_S} = (1.56 \pm 0.14)ps$, they obtain $|\frac{V_{td}}{V_{ts}}| < 0.33$. A more conservative way to express this is: $|\frac{V_{td}}{V_{ts}}| <$ $0.36\frac{f_{B_s}\sqrt{B_{B_s}}}{f_{B_d}\sqrt{B_{B_d}}}$. Progress beyond this at LEP is expected to be slow.

By making perhaps invidious assumptions about $|V_{ts}|$, one can beat the $K^+ \rightarrow \pi^+\nu\bar{\nu}$ result into a comparable form: $|\frac{V_{td}}{V_{ts}}| < 1.38$. This is not yet good as the B measurements, but it is very clean theoretically and is limited only by the experiment. Rapid progress is expected if AGS running time is adequate. It's also good to keep in mind that the B and K experiments are not quite measuring the same thing, and that it's possible that non-SM effects will show up as discrepancies between the two.

4.2. $K_L \rightarrow \pi^0\nu\bar{\nu}$

Perhaps the most enticing process in the K system is $K_L \rightarrow \pi^0\nu\bar{\nu}$, because in the Standard Model it is virtually 100% direct CP-violating.[54] There are no significant long distance contributions, no significant QCD corrections, and virtually no indirect CP-violating "background". The branching ratio is given by:

$$B(K_L \rightarrow \pi^0\nu\bar{\nu}) = \frac{\tau_{K_L}}{\tau_{K^+}} \frac{\alpha^2 B(K^+ \rightarrow \pi^0 e^+\nu)}{V_{us}^2 2\pi^2 sin^4\theta_W} \sum_l |Im V_{ts}^* V_{td} X(x_t)|^2 \qquad (4.3)$$

$$= 5 \times 10^{-10} A^4\eta^2 \qquad (4.4)$$

where Eq. 4.4 assumes CKM unitarity and the CDF value for m_t.

By the principle of the conservation of agony, the experimental difficulty is inversely proportional to the theoretical appeal of this process. In general one cannot measure the initial state energy, there is no visible vertex, two of the final state particles cannot be detected so that there are no kinematic constraints, and many backgrounds produce apparently unaccompanied π^0's – yet one needs to get to the 10^{-12} level. The best one can say is that probably there soon will be enough K_L available. The latest limit is from FNAL 799 who looked for the Dalitz converted version, $K_L \rightarrow \pi^0\nu\bar{\nu} \rightarrow e^+e^-\gamma\nu\bar{\nu}$, thereby obtaining a vertex at a cost of a factor 80 in statistics. Their results are shown in Fig. 12 which is a plot of $ee\gamma$ p_T vs mass. The signal box is centered on the π^0 mass and on a p_T greater than that possible for π^0 from most K_L decays. The background just below the signal box comes from $\Lambda \rightarrow n\pi^0$. $Ke3$ events populate the region above and to the right of the signal region. It's clear that future progress is contingent on detector improvements as well as on increased K flux. With no events in the signal box, a 90% c.l. limit of $B(K_L \rightarrow \pi^0\nu\bar{\nu}) < 5.8 \times 10^{-5}$ is obtained.[55] Thus there are only 7 orders of magnitude to go. So far the only proposal for dedicated experiment has been at KEK.[56] To my mind this process presents a great opportunity to do an important experiment that doesn't require a cast of thousands.

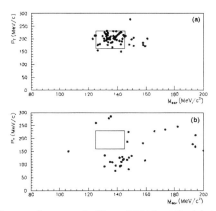

Fig. 12. Results of $K_L \rightarrow \pi^0 \nu \bar{\nu}$ search (from FNAL-799). (a) is Monte Carlo, (b) are candidates

4.3. $K_L \rightarrow \pi^0 \ell^+ \ell^-$

If $K_L \rightarrow \pi^0 \nu \bar{\nu}$ is so hard to detect, why not just replace the elusive neutrinos with charged leptons? For the direct CP-violating component, the theory is complicated by the fact that the final state leptons can now couple to photons. However techniques for dealing with this have been developed over the past few years[57,58] and recently Buras and collaborators have calculated the rate, taking QCD effects into account consistently in the next-to-leading order.[59] Thus, if $B(K_L \rightarrow \pi^0 e^+ e^-)_{dir}$ could be measured in isolation, it would indeed be similarly useful to $K_L \rightarrow \pi^0 \nu \bar{\nu}$. However this is not the end of the story. Assuming CKM unitarity and the CDF value for m_t:

$$B(K_L \rightarrow \pi^0 e^+ e^-)_{dir} \approx 7 \times 10^{-11} A^4 \eta^2 \tag{4.5}$$

Thus it is several times smaller than $B(K_L \rightarrow \pi^0 \nu \bar{\nu})$, making it difficult to pick out from certain competing contributions to $K_L \rightarrow \pi^0 e^+ e^-$. These once again arise because the final state leptons can couple to photons, as in Fig. 13 (a).

To begin with, there is an indirect CP-violating term, $\epsilon A(K_S \rightarrow \pi^0 e^+ e^-)$. Unfortunately the approach used to calculate the direct contribution breaks down here because the real parts of the Wilson coefficients get large contributions from regions below m_c, where perturbative QCD can't be trusted. This contribution could be completely determined by a measurement of the CP-conserving decay $K_S \rightarrow \pi^0 e^+ e^-$. However, the current upper limit on this branching ratio,[60] 1.1×10^{-6}, falls at least three orders of magnitude short of the required sensitivity. One can get a rough idea of the expected size of this component from the well-determined[61] value of

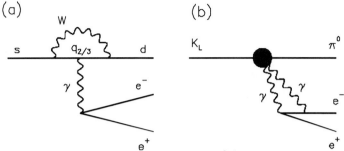

Fig. 13. Photon-mediated contributions to $K_L \rightarrow \pi^0 e^+ e^-$ (a) CP-violating, (b) CP-conserving.

$\Gamma(K^+ \rightarrow \pi^+ e^+ e^-)$. This gives $B(K_L \rightarrow \pi^0 e^+ e^-)_\epsilon \approx 6 \times 10^{-12}$, a number very close to that implied by Eq. 4.5. The indirect contribution is far larger relative to the direct piece than in the case of $K_L \rightarrow \pi^0 \nu \bar{\nu}$, but still far smaller than in $K_L \rightarrow 2\pi$. However at the moment the precise size of this contribution remains very uncertain. More sophisticated attempts[62,63] to relate $\Gamma(K_S \rightarrow \pi^0 e^+ e^-)$ to $\Gamma(K^+ \rightarrow \pi^+ e^+ e^-)$ via chiral perturbation theory (χPT) have run into a number of problems.[64,65] To further complicate matters, there is also a CP-conserving two photon contribution of roughly the same order that arises from the intermediate state $K_L \rightarrow \pi^0 \gamma \gamma$ (see Fig. 13 (b)). In principle the size of this contribution can be predicted from data on the latter process. Previous to the observation of $K_L \rightarrow \pi^0 \gamma \gamma$, the theoretical situation was quite muddled (see e.g. Ref. 66 for a summary). The observations by NA31[67] and FNAL-731[68] have been a mixed success for χPT. The Dalitz Plot distribution, which differs markedly from phase space, is pretty well predicted by χPT.[63] However the observed branching ratio, $\sim 1.7 \times 10^{-6}$, is about three times higher than that originally predicted. Refinements to the theory[69] have considerably reduced this discrepancy, but this is not as convincing as a correct prediction would have been. Moreover, even if the refined theory is accepted at face value, the implications for $K_L \rightarrow \pi^0 e^+ e^-$ are not unambiguous. Ref. 69 predicts $B(K_L \rightarrow \pi^0 e^+ e^-)_{CPcons} = 1.8 \times 10^{-12}$, while Ref. 65 including dispersive effects obtains $B(K_L \rightarrow \pi^0 e^+ e^-)_{CPcons} = 4.9 \times 10^{-12}$.

As if the above were not sufficiently challenging, there is also a true background which is very difficult to eliminate. This is $K_L \rightarrow e^+ e^- \gamma \gamma$, which is basically a radiative correction to $K_L \rightarrow e^+ e^- \gamma$. The impact of this background, which was discovered[70] in the course of the first dedicated search for $K_L \rightarrow \pi^0 e^+ e^-$, was studied in detail by Greenlee.[71] In spite of the give-away tendency of the radiated photon to line up with its parent electron, and the low odds of its pairing with the other photon to make a spurious π^0, this background appears very hard to beat at the 10^{-11} level. Of course efforts are continuing to verify this. At this conference FNAL-799 has reported new results on this process.[72] On the basis of 58 events they find $B(K_L \rightarrow e^+ e^- \gamma \gamma; E_\gamma^* > 5MeV) = (6.5 \pm 1.2_{stat} \pm 0.6_{sys}) \times 10^{-7}$, which is quite consistent with the theoretical expectation of 5.8×10^{-7}.

Recent data on $K_L \rightarrow \pi^0 \ell^+ \ell^-$ has also been reported by FNAL-799. Fig. 14

shows 2-d plots of final state p_T^2 versus invariant mass of candidates for $K_L \to \pi^0 e^+ e^-$ and $K_L \to \pi^0 \mu^+ \mu^-$. Since no events are observed in either signal box, 90% c.l. upper limits of $B(K_L \to \pi^0 e^+ e^-) < 4.3 \times 10^{-9}$ [73] and $B(K_L \to \pi^0 \mu^+ \mu^-) < 5.1 \times 10^{-9}$ [74] are obtained. The former limit is only a small advance over previous work, but the latter represents nearly a factor 250 improvement in sensitivity. The $K_L \to \pi^0 \mu^+ \mu^-$ signal region seems remarkably clean and one can anticipate very significant further progress.[75] The same cannot be said for $K_L \to \pi^0 e^+ e^-$, but upgrades to FNAL-799 now in progress should help significantly. The experimenters aim at eventual single event sensitivities of $10^{-10} - 10^{-11}$ for both processes. This will not be sufficient to see S.M. CP-violation, but it should greatly clarify future prospects for these modes in the study of CP-violation.

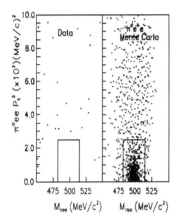 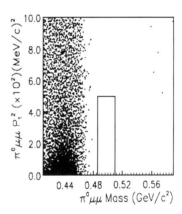

Fig. 14. Results of $K_L \to \pi^0 \ell^+ \ell^-$ search (from FNAL-799).

4.4. $K_L \to \mu^+ \mu^-$

The theoretical connection between the short distance contribution to this branching ratio and CKM parameters has been reviewed by Buras and Harlander,[76] and more recently by Buchalla and Buras.[47] Although the short-distance contribution tends to be obscured by long-distance effects about an order of magnitude larger, a measurement of sufficient precision to overcome this difficulty seems to be possible. The CKM parameter ρ could then be extracted from such a measurement. Both this and the closely related decay $K_L \to e^+ e^-$ are also potentially sensitive to new physics beyond the Standard Model.

The short distance contribution to this process comes from the box and electroweak penguin diagrams of Fig. 9. The hadronic matrix element $< 0|\bar{s}\gamma_\mu(1+\gamma_5)d|K_L >$ is given by that occurring in $K^+ \to \mu^+ \nu$. In leading logarithmic approximation, the top-quark contribution can be computed to be[77]:

$$\frac{B(K_L \to \mu^+\mu^-)_{SD}}{B(K^+ \to \mu^+\nu)} = \frac{\tau_{K_L}}{\tau_{K^+}} \left(\frac{\alpha}{\pi \sin^2 \theta_W}\right)^2 A^4 \lambda^8 (1-\rho)^2 \left[Y(x_t)\right]^2 \quad (4.6)$$

where:

$$Y(x_t) = \frac{x_t}{8}\left(\frac{x_t - 4}{x_t - 1} + \frac{3x_t}{(x_t - 1)^2}\ln x_t\right) \quad (4.7)$$

The contribution of the charmed quark, with perturbative QCD corrections can be found in Ref. 58. These authors provide us with the approximate expression for the complete result:

$$\begin{aligned} B(K_L \to \mu^+\mu^-)_{SD} &= 1.7 \times 10^{-10} x_t^{1.56} A^4 (\rho_0 - \rho)^2 \\ &= 1.9 \times 10^{-9} A^4 (\rho_0 - \rho)^2 \text{ (for } m_t = 174) \quad (4.8) \end{aligned}$$

where deviations of ρ_0 from 1 measure the charmed quark contribution with QCD corrections. For typical values of all the parameters involved, $\rho_0 \sim 1.27$. Then, for example, if $A = 0.85$ and $\rho = 0.27$, $B(K_L \to \mu^+\mu^-)_{SD} = 10^{-9}$. (Note that for the case of $K_L \to e^+e^-$, this BR is reduced by a factor $\sim (m_e/m_\mu)^2$.) Buchalla and Buras[47] discuss the intrinsic theoretical uncertainty in the relationship between $B(K_L \to \mu^+\mu^-)_{SD}$ and the fundamental parameters once the calculation has been taken to the next to leading logarithmic order. They conclude that for typical values of the parameters, the uncertainty on the branching ratio is $\sim 8.5\%$. This suggests a target for the precision of possible future measurements.

Unlike most of the one-loop processes discussed in this paper, high statistics data are available for $K_L \to \mu^+\mu^-$. Fig. 15 shows the result of a single year's run of a recent BNL experiment,[78] in which hundreds of signal events are evident. There

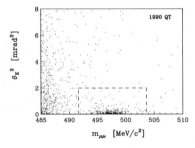

Fig. 15. Square of the collinearity angle vs. $m_{\mu\mu}$ for $K_L \to \mu^+\mu^-$ candidates from 1990 data set of AGS-791.

are roughly 1000 events in the world sample of $K_L \to \mu^+\mu^-$, and about ten times this number are expected within a few years from AGS-871. In the absence of long-distance effects, this would be more than adequate according to the criterion of Ref 47.

However, the long-distance contribution to this decay is far from negligible, and it determines the extent to which $K_L \to \mu^+\mu^-$ can be exploited to determine ρ. The long distance contribution is dominated by the two photon intermediate state as in Fig. 16. Therefore one can compute the absorptive part of the amplitude by using the

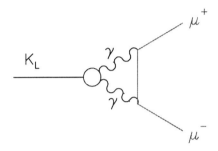

Fig. 16. Long distance contribution to $K_L \to \mu^+\mu^-$

experimental rate for $K_L \to \gamma\gamma$. The result[79] is:

$$B(K_L \to \ell^+\ell^-)_{abs} = \frac{1}{2}\alpha^2 r_\ell^2 \frac{1}{\beta}\left|\ln\frac{1+\beta}{1-\beta}\right|^2 B(K_L \to \gamma\gamma)$$
$$= (1.2022 \times 10^{-5})B(K_L \to \gamma\gamma) \qquad (4.9)$$

where $\beta \equiv \sqrt{1-4r_\ell^2}$ and $r_\ell^2 \equiv m_\ell^2/m_K^2$. This can be compared to the latest measurements:

$$B(K_L \to \mu^+\mu^-) = \begin{cases} (6.81 \pm 0.32) \times 10^{-9} & \text{absorptive} \\ (7.9 \pm 0.7) \times 10^{-9} & \text{KEK-137}[80] \\ (6.86 \pm 0.37) \times 10^{-9} & \text{AGS-791}[78] \end{cases}$$

$$B(K_L \to e^+e^-) = \begin{cases} (3.0 \pm 0.1) \times 10^{-12} & \text{absorptive} \\ < 4.1 \times 10^{-11} & \text{AGS-791}[81] \end{cases} \qquad (4.10)$$

where the absorptive contribution is derived by inserting $B(K_L \to \gamma\gamma) = (5.70 \pm 0.27) \times 10^{-4}$ into Eq. 4.9.[25] Now if this were the only long-distance effect, the prospects for a measurement of ρ would be easy to evaluate. However, there is also a possible long distance dispersive contribution. This adds in amplitude with the weak contribution, and unfortunately even the relative sign of the two contributions is controversial.

Setting this problem aside for the moment, we can get an idea of the potential of a precision measurement of $B(K_L \to \mu^+\mu^-)$ under the most favorable possible assumptions. Table 3 compares the precision on $B(K_L \to \mu^+\mu^-)_{dsp}$ available a decade ago, that available at the present (the two latest experiments are averaged), and that which *could* be available after the result of AGS-871, if a very precise measurement of $B(K_L \rightarrow \gamma\gamma)$ were also made. For the purposes of the Table, the 1997 value of $B(K_L \rightarrow \gamma\gamma)$ was assumed to be the same as the present PDG value. Also, the dispersive BR is assumed to be 10^{-9}. The precision of the 1997 $B(K_L \rightarrow \mu^+\mu^-)$ measurement is assumed to be a little worse than that given by the anticipated statistics alone. Also the assumption is made that the precision on the BR measurement of the relatively copious $K_L \rightarrow \gamma\gamma$ mode can be improved to better than 1%.[82] Applying Eq. 4.8 to the 1997 value of Table 3 gives a precision on ρ of about $\pm.06$ under the assumption that A is known. This precision is a function of the value of the dispersive BR, *e.g.* for $B(K_L \rightarrow \mu^+\mu^-)_{dsp} = 0.23 \times 10^{-9}$ it rises to ±0.12, while for $B(K_L \rightarrow \mu^+\mu^-)_{dsp} = 2 \times 10^{-9}$ it falls to ±0.04. This level of precision on ρ would be quite valuable in the anticipated time frame. Thus in the scenario that the long-

Table 3. Precision with which dispersive part of $B(K_L \rightarrow \mu^+\mu^-)$ can be extracted

Date	$B(K_L \rightarrow \mu\mu)$	$B(K_L \rightarrow \gamma\gamma)$	$B(K_L \rightarrow \mu\mu)_{abs}$	$B(K_L \rightarrow \mu\mu)_{dsp}$
1984	$(9.1 \pm 1.9)10^{-9}$	$(4.9 \pm .4)10^{-4}$	$(5.9 \pm .5)10^{-9}$	$(3.2 \pm 2.0)10^{-9}$
1994	$(7.09 \pm .33)10^{-9}$	$(5.70 \pm .27)10^{-4}$	$(6.85 \pm .32)10^{-9}$	$(.23 \pm .46)10^{-9}$
1997	$(7.85 \pm .10)10^{-9}$	$(5.70 \pm .05)10^{-4}$	$(6.85 \pm .06)10^{-9}$	$(1.00 \pm .12)10^{-9}$

distance dispersive contribution is negligible, if $B(K_L \rightarrow \mu^+\mu^-)_{dsp}$ is $\sim 10^{-9}$, the ongoing experiments could come close to meeting the criterion of Ref 47.

Unfortunately there is little reason to believe that the long-distance dispersive contribution is negligible. The difficulties in calculating it are outlined in Ref. 64. There have been a number of model calculations,[83-85] which disagree with one another. However these approaches also tend to make predictions about other decays, such as $K_L \rightarrow \ell^+\ell^-\gamma$, $K_L^{*} \rightarrow 4$ leptons, $K^+ \rightarrow \pi^+\ell^+\ell^-$, etc. which are under active study and which can possibly serve to distinguish among them. In any case there is hope that further experimental and theoretical work can refine the prediction of the long-distance dispersive contribution. To get an idea of the experimental progress that might be necessary, let us focus on one model calculation, that of Ref. 84. This model contains a parameter, α_K which characterizes the relative strength of K^*-mediated diagrams. This parameter (whose absolute value was predicted to be $0.2-0.3$ several years before the measurements) was measured to be $-0.28\pm0.083^{+0.054}_{-0.034}$ [86] and -0.28 ± 0.13 [87] in two studies of $K_L \rightarrow e^+e^-\gamma$. At this conference, FNAL-799 reports a new measurement[88] of $K_L \rightarrow \mu^+\mu^-\gamma$, based on 200 events. Their result, $B(K_L \rightarrow \mu^+\mu^-\gamma) = (3.55 \pm 0.25_{stat} \pm 0.23_{sys}) \times 10^{-7}$, implies $\alpha_K = -0.15^{+0.14}_{-0.12}$, consistent with the $K_L \rightarrow ee\gamma$ results. In the model of Ref 84 with $\alpha_K = -0.28$, the long distance contribution to the real part is extremely small, but this is somewhat of a numerical accident $(B(K_L \rightarrow \mu\mu)^{l.d.}_{dsp} = 0.76 \times 10^{-9}(1 + \alpha_K/0.27)^2)$. If one attributes an error of ±0.09 to α_K, the corresponding error in the dispersive amplitude is $\pm1.0\times10^{-5}$ (about 1/3 the size of the short distance amplitude if $B(K_L \rightarrow \mu\mu)_{sd} = 10^{-9}$). It corresponds to about ±0.32 in ρ. To reduce it to ±0.06, would require α_K to be determined to ±0.017. Scaling from AGS-845,[86] this would require about 30,000 $K_L \rightarrow e^+e^-\gamma$ events, which is not out of the question for the upcoming round of CP-violation experiments. Much larger samples of $K_L \rightarrow \mu^+\mu^-\gamma$ should also be available.

There has also been significant progress on other decays which bear on this issue. FNAL-799 recently published results[89] on $K_L \to e^+e^-e^+e^-$. Based on 28 events, they measured $B(K_L \to e^+e^-e^+e^-) = (3.96 \pm 0.78_{stat} \pm 0.32_{sys}) \times 10^{-8}$ (vs. a theoretical expectation of $(3.55 \pm 0.17) \times 10^{-8}$). Whereas this sample was sufficient to verify that the final state is mainly $CP = -1$, is it too small to distinguish among theoretical approaches to off-mass shell behavior of the photons in $K_L \to \gamma\gamma$. However it is a harbinger of much larger samples that will be available in the near future. If a well-motivated theoretical approach can give a good account of high statistics samples of this process plus $K_L \to \ell^+\ell^-\gamma$, $K_L \to e^+e^-\mu^+\mu^-$, $K_L \to \ell^+\ell^-\gamma\gamma$, etc., we will have the basis we need for extracting very valuable information on ρ from $K_L \to \mu^+\mu^-$.

4.5. $K^+ \to \pi^+\ell^+\ell^-$

Although it was originally believed that $K^+ \to \pi^+\ell^+\ell^-$ would be short-distance dominated,[90] it is now understood to be very difficult to distinguish one-loop effects in this decay, because of long distance photon exchange contributions that are known to be large but which are very difficult to calculate precisely.[91] AGS-777/851 collected more than 1000 events of $K^+ \to \pi^+e^+e^-$; results based on ~ 500 have been published.[61] These results are consistent with the predictions of χPT,[62] but are not yet precise enough to constitute a real challenge. This situation is likely to change with the advent of AGS-865 which anticipates collecting tens of thousands of both $K^+ \to \pi^+e^+e^-$ and $K^+ \to \pi^+\mu^+\mu^-$. The latter decay is particularly interesting because measurements of the $P-$violating polarization asymmetry of the μ^+ do potentially allow access to the short-distance contribution.[92] Such measurements are potentially sensitive to ρ. Thus there is an alternative to $K_L \to \mu^+\mu^-$ for determining ρ with kaon decays. Recently Buchalla and Buras[94] have generalized the previous analyses beyond the leading logarithmic approximation. For $-0.25 \leq \rho \leq 0.25$, $V_{cb} = 0.040 \pm 0.004$, and $m_t = (170 \pm 20)$ GeV, they find $3.0 \times 10^{-3} \leq |\Delta_{LR}| \leq 9.6 \times 10^{-3}$ ($\Delta_{LR} \equiv \frac{\Gamma_R - \Gamma_L}{\Gamma_R + \Gamma_L}$, where Γ_R and Γ_L are the rates to produce right- and left-handed μ^+). To attain this precision in a process with a branching ratio predicted to be a few $\times 10^{-8}$ is very challenging indeed. If, in addition, one could measure the polarization of the μ^- as well as that of the μ^+, it is also possible in principle to measure η in this process.[93] However a measurement sufficient to do this is far beyond current experimental capabilities. In any case we must walk before we can run and the first step in exploiting $K^+ \to \pi^+\mu^+\mu^-$ is to observe it. This step has recently been taken by AGS-787. Fig. 17 shows a plot of p_T^2 vs three-body effective mass for events selected as $K^+ \to \pi^+\mu^+\mu^-$ candidates on the basis of particle ID, photon vetoing, and other cuts. A clear accumulation of candidates is visible in the signal region near $p_T^2 = 0$ and $m_{\pi\mu\mu} = m_K$.

4.6.   One-loop decays and the Standard Model parameters

It's implicit in the discussion above that one-loop K decays can be very useful in constraining the unitarity triangle. With the addition of a single measurement from the B system (e.g. a measurement of $|V_{cb}|$ or of $|V_{ub}/V_{cb}|$), they can actually over-determine the triangle. This is graphically illustrated in Fig. 18. The information

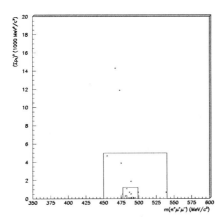

Fig. 17. $K^+ \to \pi^+\mu^+\mu^-$ candidates from AGS-787. Smaller box is signal region. Region between smaller and larger boxes is used to estimate background.

from K decays can be used in a number of different ways; several have been suggested in the literature. For example, Buchalla and Buras[95] estimate that $sin(2\beta)$ could be determined to ±0.11, given 10% accuracy on $B(K^+ \to \pi^+\nu\bar{\nu})$ and $B(K_L \to \pi^0\nu\bar{\nu})$.

5. B's and K's

Clearly the information available from B's and K's is complementary and it can be combined in various ways, as stressed by Buras and his collaborators.[95–97] One perhaps surprising example is the proposal to extract a precise value for $|V_{cb}|$ from $B(K_L \to \pi^0\nu\bar{\nu})$ plus measurements of CP-asymmetries in B decays. Since, as evident in Eq. 4.4, $B(K_L \to \pi^0\nu\bar{\nu})$ is proportional to $A^4\eta^2$, once η is well-determined in B-decays, one can choose to assume its value is the same in the K decay and get a precision on A that's $\sim 4\times$ better than that on $B(K_L \to \pi^0\nu\bar{\nu})$. One can also take a different point of view and use the comparison of K and B-determinations of quantities that ought to be the same to check the consistency of the Standard Model. Basically one can check consistency or assume consistency and get precision.

6. Conclusions

Although searches for forbidden K decays are accessing formidably high scales, unless a dramatic discovery is made in the current round of experiments, there may not be sufficient motivation to launch another major initiative in this area in the near future. This is because there is currently no particular theoretical "hard target", according to Eq. 2.1 one progresses in reach only as $\propto (BR)^{-\frac{1}{4}}$, and the experiments

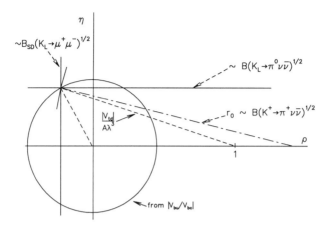

Fig. 18. Rare K decays and the unitarity triangle.

are getting increasingly difficult. On the other hand, the motivation to pursue GIM-suppressed one-loop decays is very strong even in the era of B-factories, and should be pushed until and unless it founders on backgrounds or on the supply of kaons. The opportunity presented by an alternative, theoretically very clean, method of determining the unitarity triangle that is complementary to studies in the B-system is just too good to miss. Experiments to accomplish this almost inevitably also embody opportunities to pursue the search for forbidden decays, so that progress in that area is likely to continue even in the absence of new dedicated experiments.

The $K_L \to 2\pi$ experiments are still the best prospect for the first confirmed observation of CP-violation that is inconsistent with the superweak model. Odds are they will succeed in settling this issue. However, even if they do not, they may serve a very valuable purpose in developing beams and techniques for probing CPT-violation to a new level of sensitivity. CPT in the K system has been thus far tested to a level of roughly 10% of the observed (ϵ) CP-violation. If the sensitivity relative to ϵ could be improved to that already attained for ϵ', one would be at least numerically approaching the Planck scale, surely an intriguing benchmark. This is certainly more difficult than getting to $\epsilon'/\epsilon \leq 10^{-4}$, but it may not be out of the question. In this regard it's fortunate that DAΦNE is on the way, and it's too bad that there's no successor in the works for CPLEAR.

Finally, we should always remember that in pursuing rare decays and precision measurements to new levels of sensitivity we may find something completely unexpected that challenges our conception of particle physics. The K system has been remarkable for this kind of surprise in the past, and another one could be as near as the next data set.

7. Acknowledgments

I would like to thank J. Frank, G. Gollin, P. Huet, K. Jon-And, S. Kettell, T. Numao, M. Peskin, V. Sharma, A. Soni, G. Thomson, Y. Wah, and C. Witzig for discussions, corrections, access to data and other assistance with this paper. This work was supported by the U.S. Department of Energy under Contract No. DE-AC02-76CH00016.

References

1. J. Ritchie and S. Wojcicki, *Rev. of Mod. Phys.* **65** 1149 (1993).
2. W. Buchmüller and D. Wyler, *Nucl. Phys.* **B268** 621 (1986); R. Cahn and H. Harari, *Nucl. Phys.* **B176** 135 (1980); J. Pati and H. Stemnitzer, *Phys. Lett.* **172B** 441 (1986): O. Shanker, *Nucl. Phys.* **B206** 253 (1982); A. Acker and S. Pakvasa, *Mod. Phys. Lett.* **A7** 1219 (1992); W. Marciano and A. Sanda, *Phys. Rev. Lett.* **38** 1512 (1977); W. Marciano, *Phys. Rev.* **D45** R721 (1992); A. Barroso, G. Branco and M. Bento, *Phys. Lett.* **134B** 123 (1984); P. Langacker, S. Uma Sankar and K. Schilcher, *Phys. Rev.* **D38** 2841 (1988)
3. K. Arisaka *et. al.*, *Phys. Rev. Lett.* **70** 1049 (1993).
4. W.J. Marciano, *Vancouver Rare Decay*, 1 (1988).
5. R. Balest *et al.*, **CLEO CONF 94-4** Submitted to the 1994 Glasgow ICHEP.
6. A. M Lee *et al.*, *Phys. Rev. Lett.* **64** 165 (1990).
7. P. Krolak *et al.*, *Phys. Lett.* **B320** 407 (1994).
8. R. Ammar *et al.*, *Phys. Rev.* **D49** 5701 (1994).
9. R. Bolton *et al.*, *Phys. Rev.* **D38** 2121 (1988).
10. U. Bellgardt *et al.*, *Nucl. Phys.* **B299** 1 (1988).
11. C. Dohmen *et al.*, *Phys. Lett.* **B317**, 631 (1993).
12. R. Abela *et al.*, PSI preprint (1994).
13. S. Shotkin, this conference.
14. A. Bean *et al.*, *Phys. Rev. Lett.* **70** 138 (1993).
15. J. Bartelt *et al.*, *Phys. Rev. Lett.* **73** 1890 (1994).
16. S. Hawking, *Phys. Rev.* **D14**, 2460 (1975); *Commun. Math. Phys.* **87**, 395 (1982).
17. T.D.Lee and C.S. Wu, *Ann. Rev. Nucl. Sci.* **16**, 511 (1966) and references therein.
18. D.N. Page, *Gen. Rel. Grav.* **14**, (1982).
19. S. Ellis, J.S. Hagelin, D.V. Nanopoulos and M. Srednicki, *Nucl. Phys.* **B241**, 381 (1984).
20. S. Ellis, N.E. Mavromatos, and D.V. Nanopoulos, *Phys. Lett.* **B293**, 142 (1992); CERN-TH.6755/92 (1992).
21. V.V. Barmin *et al.*, *Nucl. Phys.* **B247** 293 (1984).
22. P. Huet and M. Peskin, SLAC-PUB-6454, March 1994.
23. M. Kobayashi and A.I. Sanda, *Phys. Rev. Lett.* **69** 3139 (1992); M. Hayakawa and A.I. Sanda, *Phys. Rev* **D48**, 1150 (1993).
24. K. Jon-And, *Recent Results of CP, T, and CPT Tests with the CPLEAR Experiment*, this conference (1994); T. Ruf, *Measurements of CP and T-violation Parameters in the Neutral Kaon System at CPLEAR*, Glasgow Conference (1994).
25. Particle Data Group 1994, *Phys. Rev.* **D50**, 1173 (1994).
26. G.D. Gollin and W.P. Hogan, P-94-05-033, submitted to the Glasgow Conference (1994)

27. L.K. Gibbons *et al.*, *Phys. Rev. Lett.* **70** 1199 (1993).
28. R. Carosi, *et al.*, *Phys. Lett.* **B237**, 303 (1990).
29. J.S. Bell and J. Steinberger, *Proc. Intl. Conf. on Elem. Part., (Oxford 1965)*, 195.
30. Particle Data Group 1992, *Phys. Rev.* **D45**, S1 (1992).
31. B. Schwingenheuer, *CPT Tests in the Neutral Kaon System by FNAL E773*, Conference on the Intersections of Particle and Nuclear Physics, St. Petersburg, Florida, June 1994.
32. C. Geweniger, *et al.*, *Phys. Lett.* **B48**, 483 (1974); *ibid.* 487 (1974); S. Gjesdal, *et al.*, *Phys. Lett.* **B52**, 113 (1974).
33. R. Adler, *et al.*, *Phys. Lett.* **B286**, 180 (1992).
34. Due to an arithmetic error, the preprint version of Ref. 22 has $\gamma = (-1.1 \pm 3.6) \times 10^{-21}$ GeV.
35. L. Wolfenstein *et al.*, *Phys. Rev. Lett.* **13** 562 (1964).
36. L. Hall and S. Weinberg, *Phys. Rev.* **D48**, 979 (1993); L. Lavoura, CMU-HEP92-17 (1992); S.M. Barr and E.M. Friere, *Phys. Rev.* **D41**, 2129 (1990); S.M. Barr *Phys. Rev.* **D34**, 1567 (1986); S.M. Barr and D. Seckel, *Nucl. Phys.* **B233**, 116 (1984).
37. M. Kobayashi and T. Maskawa, *Prog. Theor. Phys.* **49**, 652 (1973).
38. A.G. Cohen, D.B. Kaplan, and A.E. Nelson, *Ann. Rev. Nucl. Part. Phys.* **43**, 27 (1993).
39. L.K. Gibbons *et al.*, *Phys. Rev. Lett.* **70** 1203 (1993).
40. G.D. Barr *et al.*, *Phys. Lett.* **B317** 233 (1993).
41. A.J. Buras and M.E. Lautenbacher, *Phys. Lett.*, **B318** 212 (1993);A.J. Buras, *et al.*, *Nucl. Phys.* **B408**, 209 (1993); M. Ciuchini *et al.*, *Phys. Lett.* **B301**, 263 (1993).
42. J. Kambor *et al.*, *Phys. Lett.* **B261**, 496 (1991), S. Fajfer and J.-M. Gerard, *Z. Phys.* **C42**, 425 (1989); H.-Y. Cheng, *Phys. Lett.* B238, 399 (1990).
43. G.B. Thomson *et al.*, RU-94-14 June 1994.
44. J. Hagelin and L. Littenberg, *Prog. Part. Nucl. Phys.* **23** 1 (1989);D. Rein and L.M. Sehgal, Phys. Rev. **D39** 3325 (1989); M. Lu and M. B. Wise, Phys. Lett. **B324** 461 (1994).
45. L.-F. Li and L. Wolfenstein, *Phys. Rev.* **D21**, 178 (1980); J.F. Donoghue *et al.*, *Phys. Rev.* **D36**, 798 (1987); A.A. Belkov *et al.*, *Nucl. Phys.* **B359**, 322 (1991); H.-Y. Cheng *et al.*, *Mod. Phys. Lett.* **A4**, 869 (1989); H.-Y. Cheng, *Phys. Rev.* **D43**, 1579 (1991); H.-Y. Cheng, *Phys. Lett.* **B238**, 399 (1990); Erratum **B248**, 474 (1990);S. Faijfer and J.M. Gérard, *Z. Phys.* **C42**, 425 (1989).
46. Y. Zou, *et. al*, *Phys. Lett.* **B329**, 519 (1994).
47. G. Buchalla and A.J. Buras, *Nucl. Phys.* **B412** 106 (1994).
48. M. S. Atiya et al., *N.I.M* **A321**, 129 (1992)
49. M. S. Atiya, *et. al.*, *Phys. Rev.* **D48** 48 (1993).
50. M. Athanas, *et. al.*, CLEO CONF 94-2.
51. D. Atwood, B.Block, and A. Soni, SLAC-PUB-6635, BNL-60709, TECHNION-PH-94-11.
52. V. Sharma, this conference.
53. A. Ali, *Proc. of XXVI int. Conf. on High Energy Physics, Dallas (1992)* **Vol. I**, 484.

54. L. Littenberg, *Phys. Rev.* **D39** 3322 (1989).
55. M.B. Weaver, *et. al*, *Phys. Rev. Lett.* **72**, 3758 (1994).
56. T. Inagaki, *et. al*, KEK Proposal 324, 1994.
57. F. Gilman and M. Wise, *Phys. Rev.* **D21** 3150 (1980); C. Dib, I. Dunietz, and F.J. Gilman, *Phys. Rev.* **D39** 2639 (1989); C. Dib, I. Dunietz, and F.J. Gilman *Phys. Lett.* **218B** 487 (1989); J.M. Flynn, *Nucl. Phys.* **B13** 474 (1990); J.M. Flynn and L. Randall, *Nucl. Phys.* **B326** 31 (1989), E.-*ibid.* **B334** 580 (1990).
58. G. Buchalla, A. Buras and M. Harlander, *Nucl. Phys.* **B349** 1 (1991).
59. A. Buras, *et. al*, *Nucl. Phys.* **B423** 349 (1994).
60. G.D. Barr, *et. al*, *Phys. Lett.* **B304** 381 (1993).
61. C. Alliegro *et. al*, *Phys. Rev. Lett.* **68** 278 (1992).
62. G. Ecker, A. Pich and E. de Rafael, *Nucl. Phys.* **B291** 692 (1987).
63. G. Ecker, A. Pich and E. de Rafael, *Nucl. Phys.* **B303** 665 (1988).
64. L.Littenberg and G. Valencia, *Ann. Rev. Nucl. Part. Sci.* **43** 729 (1993).
65. J.F. Donoghue and F. Gabbiani, UMHEP-410, Aug 94.
66. B. Winstein and L. Wolfenstein, *Rev. of Mod. Phys.* **65** 1113 (1993).
67. G. D. Barr *et al.*, *Phys. Lett.* **242B** 523 (1990);G. D. Barr, *et, al.*, *Phys. Lett.* **284B** 440 (1992).
68. V. Papadimitriou *et. al.*, *Phys. Rev.* **D44** R573 (1991).
69. A.G. Cohen, G. Ecker, and A. Pich, *Phys. Lett.* **304B**, 347 (1993).
70. W.M. Morse *et al.*, *Phys. Rev.* **D65**, 36 (1992).
71. H.B. Greenlee, *Phys. Rev.* **D42** 3724 (1992).
72. T. Nakaya, *et al.*, *Phys. Rev. Lett.* **73**, 2169 (1994).
73. D.A. Harris *et. al.*, *Phys. Rev. Lett.* **71** 3914 (1993).
74. D.A. Harris *et. al.*, *Phys. Rev. Lett.* **71** 3918 (1993).
75. Note that $B(K_L \to \pi^0 \mu^+ \mu^-)_{CP-viol}$ is expected to be about five times smaller than $B(K_L \to \pi^0 e^+ e^-)_{CP-viol}$. Also, relative to $B(K_L \to \pi^0 \mu^+ \mu^-)_{CP-viol}$, the CP-conserving 2 γ component is expected to be quite a bit larger and the $K_L \to \mu^+ \mu^- \gamma \gamma$ background about a factor two smaller than in the electron case.
76. A. Buras and M. Harlander, *Review Volume on Heavy Flavors*, ed. A. Buras and M. Lindner, World Scientific, Singapore (1992).
77. T. Inami and C. S. Lim, *Prog. Theo. Phys.* **65** 297 (1981); E.**65** 1772 (1981).
78. A. P. Heinson *et al.*, Princeton/hep/94-15, Aug 94.
79. L.M Sehgal, *Phys. Rev.* **183** 1511 (1969).
80. T. Akagi, *et al.*, *Phys. Rev. Lett.* **67** 2618 (1991).
81. K. Arisaka *et. al.*, *Phys. Rev. Lett.* **71** 3910 (1993).
82. Note that one can actually do somewhat better than is indicated in the table, because the normalization uncertainties on $B(K_L \to \mu^+ \mu^-)$ and $B(K_L \to \gamma\gamma)$ are correlated (see Ref. 78).
83. C. Quigg and J. D. Jackson, UCRL-18487 unpublished (1968).
84. L. Bergström, E. Massó, P. Singer, *Phys. Lett.* **131B** 229 (1983); L. Bergström *et al.*, *Phys. Lett.* **134B** 373 (1984).
85. P. Ko, *Phys. Rev.* **D45** 174 (1992).
86. K.E. Ohl *et al.*, *Phys. Rev. Lett.* **65**, 1407 (1990).
87. G. D. Barr *et al.*, *Phys. Lett.* **240B** 283 (1990)
88. M. B. Spencer, *A measurement of the branching ratio $K_L \to \mu^+ \mu^- \gamma$*, this conference.
89. P. Gu, *et al.*, *Phys. Rev. Lett.* **72**, 3000 (1994).

90. M.K. Gaillard and B.W. Lee, *Phys. Rev.* **D10**, 897 (1974).
91. A.I. Vainshtein, *et al.*, *Yad. Phys.* **24**, 820 (1976) [*Sov. J. Nucl. Phys.* **24**, 427 (1976)].
92. M. Savage and M. Wise, *Phys. Lett.* **250B** 151 (1990): M. Lu, M. Wise and M. Savage, *Phys. Rev.* **D46** 5026 (1992);G. Bélanger, C. Q. Geng and P. Turcotte, *Nucl. Phys.* **B390**, 253 (1993).
93. P. Agrawal *et. al.*, *Phys. Rev. Lett.* **65** 537 (1991); *Phys. Rev.* **D45** 2383 (1992).
94. G. Buchalla and A.J. Buras, *Phys. Lett.* **B336**, 263 (1994).
95. G. Buchalla and A.J. Buras, *Phys. Lett.* **B333**, 221 (1994).
96. A.J. Buras, M.E. Lautenbacher, and G. Ostermaier, *Waiting for the Top Quark Mass*, $K^+ \rightarrow \pi^+ \nu \bar{\nu}$, $B_s^0 - \bar{B}_s^0$ *Mixing and CP Asymmetries in B-Decays*, MPl-Ph/94-14, TUM-T31-57/94.
97. A.J. Buras, *Phys. Lett.* **B333**, 476 (1994).

Cosmology at the Turn of the Millennium

ANGELA OLINTO, *Chicago, Enrico Fermi Institute*

COSMOLOGY AT THE TURN OF THE MILLENNIUM[*]

ANGELA OLINTO[†]

Department of Astronomy and Astrophysics
and Enrico Fermi Institute,
University of Chicago
Chicago, IL 60637

ABSTRACT

I review the current status of cosmology and the expectations for the field in the near future as the millennium comes to a close.

1. Introduction

During the 20th Century, cosmology has moved from the realm of philosophy to that of science and as observers and experimentalist start to outnumber theorists, cosmology is becoming a mature data driven science. By the end of this century there should be enough data to rule out all the theories. The great success of the second half of the century is mainly due to the strong base provided by the standard cosmological model, the Big Bang.

Supported by the observational pillars of the Hubble expansion, the cosmic background radiation, and primordial nucleosynthesis, the Big Bang model provides a reliable account of the Universe from $\sim 10^{-2}$ sec to ~ 15 Gyr, temperatures of 10 MeV to 2.726 K. Building on the robustness of the Big Bang, cosmologists have expanded the frontiers to much earlier times and higher energies where models of high energy particles can be related to the latest development in the Universe, the formation of galaxies. In the following, I give an overview of how the connections between the early universe and present observations contribute to the understanding of both particle physics and astrophysics.

2. The Big Bang Pillars

The Big Bang is a description of the evolution of the universe as a whole. It assumes that the Universe is homogeneous and isotropic on large scales and that it has been expanding from a hot dense past according to a Friedmann-Roberston-Walker cosmology. The Hubble expansion, the homogeneity and isotropy of the cosmic background radiation (CBR), and the synthesis of light elements probe the global nature of the Universe and are consistent the Big Bang's predictions. Today, the interest is on the deviations from homogeneity and the formation of structure building upon the Big Bang framework. Before discussing the deviations, I review the basics.

---

[*]Talk presented at DPF-94 in Albuquerque, NM.
[†]Research supported in part by the DOE and NASA

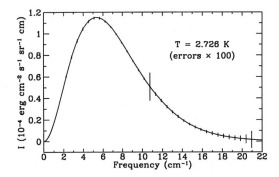

Fig. 1. The spectrum of the CMBR from [6].

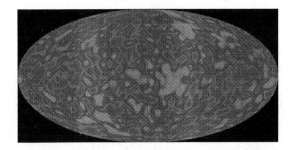

Fig. 2. The map of fluctuations in the CMBR as seen by COBE.

2.1. The CBR

The discovery of the microwave background radiation marks the beginning of the establishment of the Big Bang as the standard cosmological model. Today, we have the remarkably precise measurements of the spectrum of the cosmic background radiation by the FIRAS instrument in the COBE satellite that can be perfectly fit by a Planck spectrum as seen in Fig. 1 and give a temperature for the universe of $T_0 = 2.726 \pm 0.01$ K.[1] The COBE/DMR detector confirmed the isotropy of the background radiation to the level of $\Delta T/T \sim 10-5$ after subtracting the dipole due to our motion.[2] The isotropy, also seen in very large galaxy catalogues,[3] confirms the hypothesis that the universe is globally homogeneous and isotropic and can be described by a Frieman-Roberton-Walker metric.

DMR has also shown that in smaller scales the universe is not homogeneous and isotropic.[2] The discovery of anisotropies at the level of $\frac{\Delta T}{T} \sim 10-6$ is consistent with the observe structure in the universe being formed by gravitational collapse of small perturbations through the Jeans instability. As it has been much publicized, the anisotropy measurements give more information than a consistency check since they can probe directly the primordial spectrum. We return to this in section §3.4.

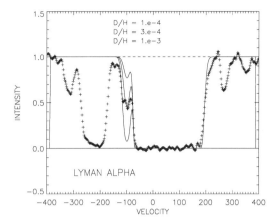

Fig. 3. Spectrum of Q0014+813 with deuterium absorption fits

2.2. *Primordial Nucleosynthesis*

The final pillar in the foundation of the Big Bang model comes from the synthesis of light elements during the first few minutes of the universe. The abundance of ^{4}He, D, ^{3}He, and ^{7}Li produced in the early universe can be predicted solely as a function of the baryon to photon ratio at the time of nucleosynthesis. The observed light element abundances show a remarkable agreement with the primordial predictions if the baryon to photon ratio is such that the fraction of the critical density in baryons $\Omega_B \equiv \frac{\rho_B}{\rho_c}$ lies on the range $0.010 \lesssim \Omega_B h^2 \lesssim 0.015$,[4] where $h \equiv H_0/100$ km/sec/Mpc. The latest development in the observations of primordial elements is the possibility of measuring deuterium at high redshifts through quasar absorption lines.[5]

Several constraints on particle physics models can be obtain by Big Bang nucleosynthesis (BBN). The ^{4}He abundance is very sensitive to the number of degrees of freedom during BBN, which can be used to constrain the number of neutrinos or any other hypothetical particle present during BBN. In addition, recent studies of inhomogeneous nucleosynthesis have shown that the universe was very close to homogeneous during BBN. In turn, limits can be set on the degree of inhomogeneity generated by the dynamics of the QCD and EW phase transitions.[6]

As the millennium comes to its closing, the Big Bang model provides a solid base from which we can progress. Some questions remain unanswered within the Big Bang framework. How do galaxies form? What is the universe made of? And, what is the origin of the initial perturbation? Answering these questions will keep us busy beyond the end of the decade.

3. The Machine

The questions left open by the Big Bang are bound to have complex answers. We believe that galaxies form from some initial perturbations that grow and become non-linear. There are many steps to the process; initial perturbations provided by theories at the GUT scale and beyond evolve in a universe whose main matter component and rate of evolution are still uncertain, and finally galaxies and larger scale structures form. The non-linear evolution of structure in small scales adds another degree of sophistication to inverting the problem back to the initial perturbations as hydrodynamic effects become important. Fortunately, a lot can be studied in the linear regime as studies of galaxy clustering on very large scales and CBR anisotropy are now being developed.

Below, I discuss each step on the machine that takes initial perturbations, evolves them and gives us galaxies at the end of the process. Each step on this machine is now being systematically studied and soon we may run the machine backwards.

3.1. Initial Perturbations

The main contenders for theories of initial density perturbations are inflation and defect models. Both involve extensions of the particle physics standard model to the GUT scale. Testing their predictions through astronomical observations may probe physics at the GUT scale and beyond.

In inflation models the universe goes through a de Sitter phase in which vacuum energy dominates the energy density. The exponential expansion during the de Sitter phase has profound consequences for the "global" character of the universe and also gives rise to initial density perturbations. An inflationary epoch naturally makes the universe homogeneous, isotropic, flat ($\Omega = 1$), and free of monopoles. In addition, it naturally predicts a scale invariant spectrum for density perturbations which arise from quantum fluctuations of the inflaton field.[7]

The particle physics motivation for inflation is still somewhat incomplete, to be kind. There are many models: old,[8] new,[9] chaotic,[10] extended, double, designer, and, my favorite, natural inflation,[11] to name a few. But no particular model at present seems to be a necessity for particle physics theories at the GUT scale. Although GUT scale physics is far from being settled, in general there is no need for a scalar field with extremely small self coupling such as the inflaton field.

In contrast, topological and non-topological defects are natural products of particle physics models when spontaneous symmetry breaking is involved. In unified gauge theories, the underlying gauge symmetry is generally larger than $SU(3)_C \otimes U(1)_{EM}$, and is restored at high energies. As the universe cools from temperatures above the symmetry breaking scale, the unbroken phase undergoes spontaneous symmetry breaking and defects may form through the Kibble mechanism, which is based on the misalignment of fields on scales larger than the causal horizon. For instance, if the high temperature symmetry group G breaks to group H and the homotopy group of the quotient, $\Pi_0(G/H) \neq I$, domain walls may form, while cosmic strings result if $\Pi_1(G/H) \neq I$, and monopoles if $\Pi_2(G/H) \neq I$. Defects can also arise in the breaking of global symmetries. If a global $O(N)$ model breaks to a $O(N-1)$, global walls form for $N = 1$, for $N = 2$ strings, for $N = 3$ monopoles, for $N = 4$ textures, and for $N > 4$ Kibble gradients.

Although defect models have a natural origin in particle physics models, they loose out to inflation when addressing the problem of cosmological initial conditions; the observation that the universe is homogeneous, isotropic, and flat, remains unexplained. Therefore, in the realm of aesthetic values, the two proposals have complementary strengths and weaknesses, and today, a preference is left to the reader.

When compared on the grounds of predictive power, inflation takes the lead by being more prone to generic predictions that can be more easily tested. Inflation generically predicts ($\Omega = 1$) and a scale free Harrison-Zel'dovich spectrum of random phase (Gaussian) density perturbations. Defect models abstain from making statements about Ω and generically predict a Harrison-Zel'dovich power spectrum on large scales and non-Gaussian fluctuations still to be better quantified. Testing the spectrum and Gaussianity of initial perturbations through the observations of anisotropies of the microwave background and the clustering of galaxies will ultimately be able to test these predictions.

If we define the density perturbations at some point in space and time by, $\delta\rho(\mathbf{x}, t) \equiv \rho(\mathbf{x}, t) - \bar{\rho}(t)$, where $\bar{\rho}(t) = \Omega\rho_C = \Omega\, 3H_0^2/8\pi G$ is the mean density of the universe at time t, we can describe the spectrum of density perturbations by expanding $\delta\rho/\bar{\rho}$ in Fourier modes:

$$\frac{\delta\rho}{\bar{\rho}} \propto \int d^3k \, e^{-i\mathbf{k}\cdot\mathbf{x}} \, \delta_{\mathbf{k}} \ . \tag{1}$$

The power spectrum is defined as $P(k, t) = |\delta_k|^2$, and the variance $\Delta(k, t) = |\delta_k| k^{3/2}$, where the dependence solely on the magnitude of the wavenumber is a consequence of the isotropy of the universe. It is also convenient to define the power spectrum today, $P_0(k) = P(k, t_0)$, as the product of an initial power spectrum, $P_i(k) = P(k, t_i)$, times a transfer function, $T(k)$, such that

$$P_0(k) = T(k)P_i(k) \ . \tag{2}$$

P_i gives the density perturbations from inflation or the symmetry breaking phase transition early in the radiation era, while $T(k)$ evolves this initial spectrum given the values of the cosmological parameters and the nature of the dark matter up to today.

The Harrison-Zel'dovich spectrum was proposed to avoid the divergence of gravitational perturbations on the Hubble scale at any time (past or future) of the universe. It corresponds to the requirement that density perturbations have the same amplitude on the Hubble radius scale, $H^{-1}(t)$, for all times, i.e., the density perturbations are constant on the scale, $\lambda(t)$, that is crossing the Hubble radius at a given time t,

$$\left(\frac{\delta\rho}{\rho}\right)_{\lambda=H^{-1}} = \text{constant} \ . \tag{3}$$

This requirement can be translated to a fixed time for all scales and gives $P_i(k) \propto k$ for the initial spectrum of density perturbations.

The slow roll phase of inflationary models implies:

$$\left(\frac{\delta\rho}{\rho}\right)_{\lambda=H^{-1}} \simeq \frac{H^2}{\dot{\phi}} \simeq \text{constant} \ , \tag{4}$$

where ϕ is the inflaton field, while for defects

$$\left(\frac{\delta\rho}{\rho}\right)_{\lambda=H^{-1}} \simeq \frac{(\nabla\Phi)^2}{T^4} \simeq \left(\frac{f}{m_{Pl}}\right)^2 \sim 10^{-6} \ , \tag{5}$$

where Φ is a Higgs field with symmetry breaking scale f. Both models generically predict a Harrison-Zel'dovich spectrum on large scales, with small deviations from it for particular models. The H-Z spectrum contains remarkable power on large scales when compared to models that generate density perturbations within the Hubble radius where $P \propto k^4$ or steeper.

As shown above, the spectral dependence on large scales is not powerful enough to decide between the models, but the amplitude of the power spectrum as well as the evolved power spectrum and the statistical properties of the density field can be more easily differentiated. Inflation predicts Gaussian density perturbations and, given a choice of dark matter, makes a well defined prediction for $T(k)$, while defects have non-Gaussian initial fluctuations and $T(k)$ is not easily obtained since defects evolve with the universe. With the prospects of mapping the CBR anisotropies in small scales, we will be able to test not only the generic predictions but detailed properties of the fluctuations.

3.2. Dark Matter

The dark matter problem may still be a challenge into the next millennium. There are many dark matter problems, which increase in magnitude as the cosmological scales probed increase. From rotational curves of galaxies, through cluster mass estimates to large scale peculiar velocity flows, the need for dark matter is ubiquitous. Two questions remain to be answered: how much dark matter and what is it. The first question is been addressed by a variety of new techniques as well as larger data sets of traditionsl estimators (see recent review by Dekel.[12]) Estimates from peculiar velocity flows give $\Omega \gtrsim 0.3$ while estimates from redshift distortions indicate $\Omega \sim 1$. Weak lensing of background galaxies by clusters as well as distortions in redshift space are likely to become powerful estimators of the mass distribution.

When compared to the density in luminous matter $\Omega_{lum} \sim 0.007$, the range of allowed Ω_B implies that a large fraction of baryons are dark. A search for dark baryons in the form of low mass stars or planets has been undertaken by three microlensing experiments. By monitoring stars in the Magellanic Clouds as well as the Bulge of our Galaxy the MACHO, EROS, and OGLE collaborations have found a number of microlensing events. The rate of events towards the Bulge is much higher than that towards the LMC, which indicates a disk rather than a halo population. The exact mass in machos is still unclear, but the technique has proven to be a very effective probe of the mass in low mass stars. It is still unclear how much mass is in machos, but it does not seem likely that Ω_{machos} may be large enough to account for the missing baryons. Where are the rest of the baryons?

With observations requiring $\Omega \gtrsim 0.3$ and inflation (as well as aesthetics) predicting $\Omega = 1$, while BBN limits $\Omega_B \lesssim 0.015 h^{-2}$, there is clearly a need for non-baryonic dark matter. Candidates for non-baryonic dark matter are generally based on physics beyond the standard model, some examples are massive neutrinos, axions, and the lightest supersymmetric particle. Dark matter candidates are systematically being search for in different ways: direct searches for cosmic dark matter particles, indirect searches for decay or annihilation products of the dark matter, and experiments that probe theories which predict dark matter particles.

If one of the neutrinos had a mass $\sim 25eV$, it could be the dark matter.

Neutrino masses maybe found in neutrino oscillation and tritium beta decay experiments. Low-temperature low-background detectors are being built to search directly for a class of candidate dark matter particles which are weakly interacting and massive, named WIMPS. WIMP cross-sections are expected to be around $\sim 10^{-38}$ cm^2 and present experiments are still far from reaching the expected cosmological rate. Microwave cavity detectors are being used to search for axions, and soon should be reaching the cosmological rate. Underground experiments work as indirect detectors by searching for high energy neutrinos from wimp anti-wimp annihilations in the Sun. With all this effort, by the end of the millennium we may have learned about more than one dark matter component.

For cosmological models the dynamical behavior of the dark matter is more important then their exact origin. In general, candidates can be divided into two categories: cold and hot dark matter, according to whether they are relativistic or non-relativistic.

3.3. Cosmological Parameters

Although cosmological parameters do not involve physics beyond the standard model, they have consistently eluded definitive measurements. These parameters determine the evolution of the universe in which initial fluctuations are to be evolved and compared to observations. A better determination of these parameters will lift some degeneracies in the predictions of early universe cosmology for the observed structure in the universe. Inflation also predicts a flat universe, which in principle is verifiable.

The Friedmann-Robertson-Walker spacetime metric can be written as:

$$ds^2 = dt^2 - R(t)^2 \left(\frac{dr^2}{1 - kr^2} + r^2(d\theta^2 + sin^2\theta d\phi^2) \right) , \qquad (6)$$

where $R(t)$ is the global scale factor which describes the overall expansion or contraction of the universe, r, θ, ϕ, are the fixed coordinates carried by fundamental observers, t is the proper time , and $k = 0, -1, +1$ is the sign of the spatial curvature ($k = 0$ corresponds to flat, $k = -1$ open, and $k = 1$ closed universes). The dynamics of $R(t)$ is determined by the matter content of the universe through Einstein's equations,

$$H^2 \equiv \left(\frac{\dot{R}}{R} \right)^2 = \frac{8\pi G\rho}{3} - \frac{k}{R^2} + \frac{\Lambda}{3} \qquad (7)$$

and

$$\frac{\ddot{R}}{R} = -\frac{4\pi G}{3}(\rho + 3p) + \frac{\Lambda}{3} \qquad (8)$$

The principal observable cosmological parameters are the age of the Universe, t_0, the Hubble parameter, $H_0 = (\dot{R}/R)_0$, the present mass density relative to the critical, $\Omega_0 = \rho_0/\rho_{crit} = 8\pi G\rho_0/3H_0^2$, the deceleration parameter, $q_0 = -\ddot{R}R/\dot{R}^2|_{t_0}$, and the contribution of the cosmological constant to the present expansion rate $\Omega_\Lambda = \Lambda/3H_0^2$

From the Friedmann's equations for the matter dominated era, we can relate

$$\Omega_0 + \Omega_\Lambda - \frac{k}{R_0^2 h_0^2} = 1 \qquad (9)$$

and

$$q_0 = \frac{\Omega_0}{2} - \Omega_\Lambda .$$ (10)

The age of the universe is related to the other parameters through $t_0 = H_0^{-1} f(\Omega_0, \Omega_\Lambda)$, were f is a fuction of order unity. Thus, the Hubble parameter $H_0^{-1} = 9.8 \times 10^9 h^{-1}$ yr sets the timescale for the age of the universe.

Traditionally, three methods have been used to infer the age of the Universe, t_0. Nuclear cosmochronology is based on radioactive dating of $r-$process elements, that is, heavy elements formed by rapid neutron capture, most probably in supernovae. The element ratios Re/Os and Ur/Th generally indicate $t_0 = 10-20$ Gyr,[13] with a large uncertainty due to the unknown element formation history (e.g., the star formation rate over time). A second method involves the cooling of white dwarfs: when low-mass stars exhaust their nuclear fuel, they become degenerate white dwarfs, gradually cooling and becoming fainter. The number of white dwarfs as a function of luminosity drops dramatically for $L_{wd} < 3 \times 10^{-5} L_\odot$, suggesting that there has not been sufficient time for them to fade below this value. Coupled with models of white dwarf cooling, this implies that the age of the galactic disk is about $t_0 \simeq 10 \pm 2$ Gyr.[14]

The most extensively studied technique for constraining t_0 is the determination of the ages of the oldest globular clusters in the galaxy. When stars finish burning hydrogen, they turn off the main sequence, characteristically reddening and brightening. By observing the color-magnitude (color vs. apparent brightness) diagram for a cluster, one can determine the apparent brightness of stars in the cluster that are now leaving the main sequence. Knowing the distance to the cluster then gives the absolute luminosity of stars at the turn-off. On the other hand, stellar evolution theory relates stellar luminosity to the time a star spends on the main sequence. The turn-off luminosity in the oldest globular clusters in the galaxy is slightly below that of the sun, yielding the age estimate $t_{gc} = (13 - 15) \pm 3$ Gyr.[15] This argument is a gross oversimplification of the complex process of stellar modelling and cluster isochrone fitting, but it gives a heuristic understanding of the resulting age estimate. The largest source of error is apparently the uncertainty in the distances to the globular clusters. It is hoped that observations with the corrected Hubble Space Telescope mirror, successfully repaired in late 1993, could reduce the uncertainty in t_{gc} to as little 10%. There may also be residual systematic model uncertainties associated with the initial hydrogen (or helium) fraction, opacities, metallicity, etc.

The Hubble parameter relates the observed recession velocity v_r or redshift z of a galaxy to its distance d: for $v_r \ll c$, the recession velocity is $v_r = cz = H_0 d + v_p$, where v_p is the peculiar radial velocity of the galaxy with respect to the Hubble flow, usually assumed to arise from gravitational clustering. Galaxy redshifts can be measured quite accurately, so all the difficulty in determining H_0 resides in finding reliable distance indicators for extragalactic objects at distances large enough that the Hubble term dominates over the peculiar motion. Observed peculiar velocities are typically of order 300 km/sec, so that distance measurements well beyond 40 Mpc or more are required for reasonable accuracy.

A wide variety of techniques has been used to establish an extragalactic distance scale,[16] and this is reflected in the spread of results for H_0, roughly $40 - 100$ km/sec/Mpc. Distance estimates made using methods such as the Tully-Fisher relation between 21-cm rotation speed and infrared luminosity for spiral galaxies, calibrated by observations of Cepheid variable stars in several nearby galaxies, have yielded high values for the expansion rate, roughly $H_0 = 80 \pm 10$ km/sec/Mpc. Two newer methods, planetary nebula luminosity functions[17] and galaxy surface brightness fluctuations[18] yield values for H_0 in this range as well, and are being further developed. On the other hand, methods using Type Ia supernovae as standard candles have yielded low values, $H_0 \simeq 50 \pm 10$ km/sec/Mpc. However, a recent extension of this method using light curve shapes finds[19] $h \simeq 0.65$.

There are also a variety of methods being employed to measure the distances of extragalactic objects directly, bypassing the extragalactic distance ladder built up from Cepheids. Using the expanding photosphere method, Schmidt, Kirshner, and Eastman[20] have determined the distances to a number of type II supernovae at large distances, and find good agreement with the Tully-Fisher distances for these galaxies. Other 'direct' methods which hold future promise include measurement of the Sunyaev-Zel'dovich effect, due to the Compton upscattering of CMBR photons by hot gas in rich clusters,[21,22] and the differential time delay between images in gravitationally lensed quasars.[23]

Recently, observations at the CFHT telescope[24] and with the Hubble Space Telescope (HST)[25] have found respectively 3 and 20 Cepheids in two galaxies in the Virgo cluster. The inferred distance to Virgo (assuming these galaxies are close to the center of the cluster) are in accord with the high values of the Hubble constant, $h \simeq 0.8 \pm 0.17$. In the future, HST observations of Cepheids in other nearby galaxies (as well as other Virgo cluster galaxies) which are hosts to SNe Ia or which can be used as Tully-Fisher calibrators should reduce this uncertainty by a factor of order two.

As we discuss in the previous section, Ω_0 is at least $\gtrsim 0.3$ and most of the dark matter is non-baryonic. Inflation usually gives $k = 0$ and, therefore, $\Omega_0 + \Omega_\Lambda = 1$. If inflation is correct, either $\Omega_\Lambda = 0$ and $H_0 \lesssim 65$ km/sec/Mpc for $t_0 \lesssim 10$ Gyr, or $\Omega_\Lambda \neq 0$. A cosmological constant is Higher values of H_0 combined with values of $t_0 \gtrsim 10$ Gyr, limit

If an extended period of inflation took place in the early universe, then the spatial geometry should now be observationally indistinguishable from $k = 0$. If the cosmological constant vanishes, $\Omega_0 = 1$, and $t_0 = 2/3H_0 = 6.5h^{-1} \times 10^9$ yr. This is uncomfortably low compared to globular cluster ages unless $h \lesssim 0.65$. However, a non-vanishing λ_0 is certainly allowed at some level by observations and has sporadically come into vogue, most recently to alleviate both this age problem and the large-scale power problem for inflationary cosmology. For example, in a flat model with $\Omega_0 = 0.25$, $\lambda_0 = 0.75$, this gives $t_0 \simeq 1/H_0 = 9.75h^{-1}$ Gyr, consistent with the lower bound $t_{gc} > 10$ Gyr for the entire observed range of H_0, and yielding a healthy, $t_0 > 13$ Gyr-old universe for $h < 0.75$. This lower value of Ω_0 is consistent with the dynamical mass estimates from clusters, but somewhat below the recent estimates from large-scale flows. Since there is currently little theoretical guidance as to why $|\lambda_0|$ is as small as it is, and no firm proof that it should vanish, it is probably best to keep an open mind, although the fact that we would be living just at the epoch when Ω_0 is comparable to λ_0 might seem to beg for explanation. The third possibility is that theoretical prejudice is wrong, and that we live in an open, low-density universe with negligible Λ, in which case the globular cluster age range is also compatible with somewhat larger values of H_0. The challenge in this case is to explain the large-scale flows and to form large-scale structure without violating CBR anisotropy constraints. In this case, it also would be an odd coincidence that we live just at the epoch when the curvature term is becoming appreciable compared to the matter term. By the end of the millennium, new techniques, new large telescopes, and dedicated telescopes should be able to beat down the systematics and decrease the range of uncertainties in the cosmological parameters to a level where one could lift the degeneracy between different initial perturbations models and different evolution of the Universe.

3.4. Cosmic Background Radiation - Anisotropies

The key to initial perturbations is the study of the CBR anisotropies. As hydrogen combined at redshift $z = 1100$, the opacity of the universe dropped precipituously and photons decoupled from the matter. This photon background redshifted as the universe expanded while retaining their blackbody spectrum (Fig.1). Inhomogeneities in the universe at that time left an imprint in the CBR which were first detected by COBE on large angular scales and now have been reported by a number of balloon and ground-based observations on smaller angular scales at the level of $\frac{\delta T}{T} \simeq 10^{-5} - 10^{-6}$. As the spectrum and statistical properties of the CBR anisotropies get determined, a powerful probe of the universe at redshift 1100 materializes.

The anisotropies in the CBR can be described as temperature fluctuations on the celestial sphere, and can be expanded as $Y_{l,m}$ modes:

$$\frac{\Delta T}{T}(\theta, \phi) = \Sigma_{l,m} a_{lm} Y_{lm}(\theta, \phi) \ . \tag{11}$$

In analogy to the three dimensional power spectrum, we can define $C_l \equiv \langle |a_{lm}|^2 \rangle$ to be the power for each l-mode.

Different models for the initial density perturbations give us different predictions for the C_l's. The anisotropies in the CBR due to the gravitational potential fluctuations at the surface where the photons last scatter in called the Sachs-Wolfe effect. This effect is the dominant component of CBR anisotropies on large angular scales (small l) after the dipole due to our motion is subtracted. The Sachs-Wolfe effect for inflationary models with the Harrison-Zel'dovich spectrum gives $C_l \propto (l(l+1))^{-1}$ at large angular scales or small l, as seen in Fig. 4. The COBE satellite anisotropy measurements probe the large angular scales with an angular resolution of ~ 7 degrees which corresponds to $l \lesssim 13$. Within the range $2 \lesssim l \lesssim 13$, COBE results confirmed the flat behaviour of $l(l + 1)C_l$ which is predicted by the Harrison-Zel'dovich spectrum. The amplitude of the fluctuations is in also good agreement with the expected perturbations in the gravitational potential at $z = 1100$.

The higher l-modes contain more discerning power on the theoretical predictions. On smaller angular scales, the detailed dynamics at the surface of last scattering starts to play an important role. Inflationary models predict a so-called Doppler peak around $l \sim 200$ and a few lower amplitude peaks for even higher l before the power is damped away. The position and amplitude of these peaks should give information on the initial power spectrum as well as the matter content and cosmological parameters. As the small angular scale experiments improve their sensitivity, we may learn how the spectrum of these anisotropies behave and ultimately what theory created it. In the case of inflation, it maybe

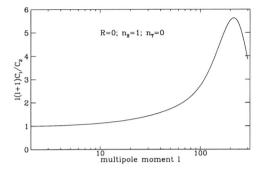

Fig. 4. The spectrum of temperature fluctuations from inflation

even possible to determine the form of the inflaton potential by studying their contribution to the CBR spectrum from both scalar and gravitational fluctuations.

There are other sources of anisotries which get imprinted as the photons travel from the last scattering surface to us. For instance, Sachs and Wolfe also studied the effect of time varying gravitational potentials from the last scattering surface to the present. This source of anisotropies is important for defect models where the generation of gravitational fluctuations continues after decoupling. Another source of anisotropy is the change in gravitational potential due to the formation of clusters of galaxies. In additional, the scattering of CBR photons off hot electrons in cluster gas generate temperature fluctuations and this is called the Sunyaev-Zel'dovich effect.

Measurements of the C_l's are not the only information that can be used to constrain models. Since one of the important distinctions between defect models and inflationary ones is in the detailed statistical properties of the fluctuations it is crucial to have small angular resolution maps to study the higher order modes of the distribution of fluctuations. A future satellite with small angular resolution should be able to map the C_l's to a few percent all the way to $l = 10^4$. In the mean time, balloon experiments probe scales between $5 \lesssim l \lesssim 200$; while ground based experiments cover the range $15 \lesssim l \lesssim 2000$.

3.5. Observations of Large Scale Structure

Finally, I would like to mention the traditional form of studying large scale fluctuations which has been experiencing a tremendous growth. The technique is that of using galaxy as tracers of the mass distribution. Galaxy clustering seen in the CfA,[26] QDOT,[27] IRAS,[28] APM,[29] EDSGC[30] catalogs has been now measured in large surveys reaching several hundred megaparsec scales. The main assumption here is that light traces mass which may not be necessaraly the case. Higher order correlations in these more powerful galaxy catalogs are now being used to study non-gaussianity and to test if there is a bias in galaxy distribution relative to the mass. The dynamical range tested by galaxy clustering is just entering the scales where CBR measurements overlap and the broad brush picture of structure in the Universe is very consistent as can be seen in Fig. 5. By the

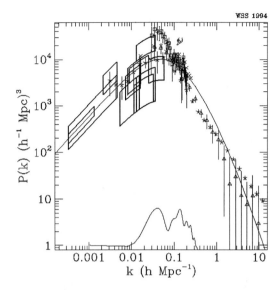

Fig. 5. The power spectrum from CBR and galaxy surveys.[31]

end of the millennium, the CBR plus the galaxy spectra should give us an unified picture of the structure in the universe.

As different models for structure formation are contrasted against the data, we find that when normalized to the COBE scale, inflation with cold dark matter does not have the right shape on intermediate scales (too much power) while hot dark matter has no power on small scales (see Fig. 6). A recently popular alternative is to employ a mixture of hot and cold dark matter, with $\Omega_{hot} \sim 1 - \Omega_{cold} \simeq 0.3$, which can be achieved with a 7 eV neutrino. As these models are studied in the non-linear regime with numerical simulations, questions of how small scale non-linear objects form become important. At this point, it is not clear if the mixed dark matter alternative has enough power to produce collapsing objects at high redshift as are observed for example in damped Lyman alpha systems which are observed in high redshift quasar spectra.

The Sloan Digital Sky Survey should be able to address this unified picture effectively. The plan is to take redshifts of one million galaxies, 150000 quasars, and photometric survey of π sterad area of the sky which should contain about 10 million galaxies up to the magnitude limit in B $\sim$ 25.5. Together with surveys like Las Canpanas and AAT2dF galaxy correlation data should reach over the Hump.

4.   Tying it all together:

I hope I succeeded in giving a small sample of the exciting ongoing and future projects in cosmology. The perspective of learning how the largest structures in the universe formed and relating that to the smallest scales are quite realistic at this point. The theoretical predictions in the linear regime ($\delta\rho/\rho << 1$) for inflationary models are well understood and anxiously await scrutiny by data. The linear regime is probed on the largest scales by galaxy clustering and in

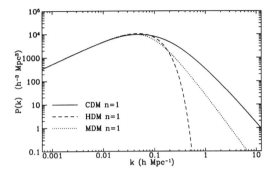

Fig. 6. Power spectra: CDM - solid, HDM - dashed, MDM - dotted.

the full range of the CBR anisotropy measurements. The smaller scale clustering of galaxies is in the non-linear regime where a lot of interesting physics beyond the initial perturbation spectrum can be studied. The nature of the non-linear regime requires large simulations of structure formations which are now in their infancy. In particular, hydrodynamic effects are starting to be implemented and soon we may actually be able to make a Milky Way in a numerical laboratory.

To summarize, I would like to make a "spoiled theorist's" wish list. By the end of the millennium we want: all the dark matter; all the "naughts" with the right values: $H_0 = 65 km/sec/Mpc$, $\Omega_0 = 1$, $t_0 = 2H_0/3$, $q_0 = 1/2$ and $\Omega_\Lambda = 0$; CBR anisotropy measurements of C_l's to few to $l = 10^4$, maps to test the statistical nature of the fluctuations, the galaxy power spectrum and higher order correlations, or, even better, the mass power spectrum; and, finally, simulations that include all the necessary physics for galaxy formation. The odds are not in favor of fulfilling all my wishes but efforts to reach these seemingly unrealistic goals are taking place now. Obviously, there will be something left to do in the next millennium.

Acknowledgements

I would like to thank the organizers of DPF'94 and J. Frieman and E. Kolb for giving me some of the figures. This work was supported by DOE and NASA at the University of Chicago.

References

1. J. C. Mather et al, *Ap. J. Lett.* **354**, l37 (1990).
2. G. Smoot et al, *Ap. J. Lett.* **396** (1992) L1.
3. P. J. E. Peebles, *Principles of Physical Cosmology* (Princeton University Press, Princeton, 1993).
4. T. Walker et al., *Ap. J.* **376** (1991) 51.
5. A. Songalia et al., *Nature* bf 368, 599 (1994).
6. R. Malaney and G. Mathews, *Phys. Rep.* **229**, 145 (1993).
7. E. W. Kolb and M. S. Turner, *The Early Universe* (Addison Wesley, New York, 1989).
8. A. Guth, *Phys. Rev. D* **23** (1981) 347.
9. A. Albrecht and P. Steinhardt, *Phys. Rev. Lett.* **48** (1982) 1220, and A. Linde, *Phys. Lett. B.* **108** (1982) 389.

10. A. Starobinskii, *Phys. Lett. B* **117**, 175 (1982)
11. K. Freese, J. Frieman, and A. Olinto, *Phys. Rev. Lett.* **65** (1990) 3233.
12. A. Dekel, *Ann. Rev. Astron. Astrophys.*, in press (1994).
13. D. Schramm, in *Astrophysical Ages and Dating Methods*, eds. E. Vangioni-Flam etal, (Editions Frontieres, Gif-sur-Yvette, 1990).
14. D. Winget etal, *Ap. J. Lett.* **315** (1987) L77.
15. A. Renzini, in *Proc. 16th Texas Symposium on Relativistic Astrophysics and 3rd Symposium on Particles, Strings, and Cosmology*, eds. C. Akerlof and M. Srednicki, (New York Academy of Sciences, New York, 1992).
16. G. Jacoby etal, *Pub. Astr. Soc. Pac.* **104** (1992) 559.
17. G. Jacoby etal, *Ap. J.* **356** (1990) 322.
18. J. Tonry, *Ap. J. Lett.* **373** (1991) L1.
19. A. G. Riess, W. H. Press, and R. P. Kirshner, CfA preprint.
20. B. P. Schmidt, R. P. Kirshner, and R. G. Eastman, *Ap. J.* **395** (1992) 366.
21. M. Birkinshaw, in *Physical Cosmology*, Proc. of the 2nd Rencontres de Blois, ed. A. Blanchard etal, (Editions Frontieres, Gif-sur-Yvette, 1991).
22. T. Wilbanks, P. Ade, M. Fischer, W. Holzapfel, and A. Lange, *Ap. J. Letts*, in press.
23. R. Blandford and R. Narayan, *Ann. Rev. Astron. Astrophys.* **30** (1992) 311.
24. M. J. Pierce, etal., *Nature* **371** (1994) 385.
25. W. Freedman, etal., *Nature*, in press.
26. M. Geller and J. Huchra, *Science* **246** (1989) 897.
27. G. Efstathiou etal, *MNRAS* **247** (1990) 10P.
28. K. B. Fisher etal, *Ap. J.* **402** (1993) 42.
29. S. J. Maddox etal, *MNRAS* **246** (1990) 433.
30. C. A. Collins, R. C. Nichol, and S. L. Lumsden, *MNRAS* **254** (1992) 295.
31. M. White, D. Scott, and J. Silk, Ann. Rev. Astron. and Astrophys. **32** (1994) 319.

Panel: High Energy Physics Outreach and Education

Moderator: GEOFFREY WEST,
Los Alamos National Laboratory

Outreach and Education
on High Energy Physics

R. MICHAEL BARNETT, *Lawrence Berkeley Laboratory*

OUTREACH AND EDUCATION ON HIGH ENERGY PHYSICS

R. MICHAEL BARNETT*

Physics Division, Lawrence Berkeley Laboratory,
Berkeley, CA 94720

ABSTRACT

We review ongoing efforts and discuss possible future directions in informing the public and educating students about Particle Physics.

1. Introduction

In the post-Cold War, post-SSC era, many of us in High Energy Physics feel we need to redouble our efforts to communicate what we do and why we do it. While we believe that physicists should devote more time to outreach and education, each of us feels that time is a scarcer resource than money. Nonetheless very substantial efforts are already underway, and we just need to build upon these existing activities. Ongoing efforts range from those done by individuals, to those done by the labs and other organized groups. Inevitably there are questions about whether such efforts are or should be done for selfish or altruistic motives. I don't find such discussions useful. Many physicists have spent many years in education and outreach; the public and our field have both benefitted.

The Division of Particles and Fields of the APS, as part of its current study of future directions for our field, has created a working group on "Structural Issues." This working group formed a subgroup to examine ongoing outreach and education efforts and seek proposals for new directions. It called the "Public Outreach and Education Team" (or POET). This group has had six activities so far:

1. An evening meeting at Fermilab on February 14, 1994 at which general issues were discussed.

2. A survey of ongoing ideas and of proposals was conducted via e-mail. The results of the survey and the Fermilab meeting were distributed via e-mail.

3. An electronic bulletin board was set for discussion of these issues (but it has not been used much).

4. At the Structural Issues Working Group meeting at LBL on July 9, 1994 a report was given and feedback obtained.

*This work was supported by the Director, Office of Energy Research, Office of High Energy and Nuclear Physics, Division of High Energy Physics of the U. S. Department of Energy under Contract DE-AC03-76SF00098, and by the U.S. National Science Foundation under Agreement No. PHY-9320551.

5. At the DPF'94 meeting in Albuquerque on August 2, 1994, a plenary session on Public Outreach and Education was held at which four reports on these issues were given.

6. At the DPF'94 meeting in Albuquerque on August 3, 1994, the POET group met with conference attendees and had an intensive discussion of proposals for future directions.

The DPF'94 meeting was the first general meeting at which a plenary session was devoted to outreach and education in High Energy Physics. Despite the late hour of the session (4:30-6:00 PM) half of the attendees remained for the entire session. The moderator of the session was Geoffrey West (LANL). The speakers included: Malcolm Browne (Pulitzer-Prize-winning writer for the New York Times) on "High Energy vs. Low Education: A National Challenge," Julia Thompson (Pittsburgh) on "Outreach to Women and Minorities," Ernest Malamud (SciTech and Fermilab) on "Using Science Centers to Expose the public to the Microworld." This report is a summary of my talk in this session.

In general, we have heard many comments about the need to convey the excitement of physics, and not just the big discoveries but the controversies too. There seems to be a consensus that we need to find means to recognize and reward outreach and education activities by physicists. The holding of a plenary session at a major conference was a good first step.

I am reporting here not on my own activities or ideas but on those received through the survey and various meetings. In the following sections, I will briefly review selected activities and proposals concerning general lay audiences, students, the government, the news media, etc. The topics of targeted groups such as women and minorities and of science museums were covered by other speakers. I will discuss the use of the World Wide Web, the creation of a catalog, and activities by collaborations.

Regrettably, the lack of space and color make it impossible to reproduce the many color transparencies, figures, and photographs that were sent to me. Clearly many of us pursue physics education/outreach for physics as a whole, but I will limit the scope of this report to HEP-related activities.

2.   Reaching Lay Audiences

Many people have told us about their talks and classes for lay audiences. In a later section I will discuss resources that are available to people interested in presentations to any non-technical audience. There are a variety of books available to the public about Particle Physics topics. There have been suggestions that someone should produce a large-size "coffee table" photo book showing some of the detectors, accelerators, events, and even people of High Energy Physics.

Two recent books brought to our attention were: Cindy Schwarz's *A Tour of the Subatomic Zoo*, which was written for the interested layperson/ undergraduate/ high school teacher or student. It assumes no prior physics background and can serve as an introduction to the basics. It was published by AIP. Lawrence Krauss' *Fear of Physics*, tries to reach out to a broad popular audience, in order to explain what physicists are interested in and why.

The Florida State University Physics Department has been producing mall exhibits for a number of years and report that they are quite popular. CERN has set up its own science museum, MICROCOSM, and has now built a separate building for it. They say its purpose is "to let the public see the research work carried out by the physicists in their quest to understand the laws of Nature." The number of visitors has remained consistently very high.

3.   News Media

In connection with the announcement of the initial top quark candidates, Fermilab carried out an excellent program to inform the news media about the physics and the experiments in a manner that allowed the media to report the news accurately. They put together a substantial package of information well-suited for the target audience. They had excellent results in getting good coverage in many media outlets throughout the country.

We do not always have big news to report. However, many people have reported success at getting media coverage for aspects of their experiments. Some sample comments: "We had a press conference when we did the last touches on the experimental hardware... It works, but takes a lot of work and courage... We were on the evening news. One key point was good contact to the public relations office. One has to be extremely careful about the scope of their press release... I think a press conference after the publication of key result or a press conference when an experiment takes first data is the best approach."

One comment was "It is often easy to interest journalists in (well-defined) stories, but it needs a significant effort to establish the 'networking' links to them." Some thought should be given to how we can do this.

A common problem felt by many physicists "is the NEGATIVE peer pressure to go public." The culture of our field has equated talking to the press about one's research to "publishing in the New York Times." Clearly one should continue to publish in the standard journals, but in the world we now live in, we are obligated to communicate our results, our conclusions, and the benefits of our work to a broad audience. Physicists should be *encouraged* to describe what we are doing and why. We are excited by the theories and experiments of our field, and we should not be ashamed to share that excitement. These lessons have not been lost on the astrophysicists; their stories appear weekly in the press (even the less-glamorous stories).

A number of people at Fermilab have proposed a national meeting of science writers and physicists to discuss the reporting of science. Clearly many of us feel that both the quantity and quality of reporting about particle physics are not adequate. It is a difficult subject about which many writers may feel insecure. Such a meeting might not only help break down some these barriers, but would help foster contacts between writers and physicists.

4.   Radio, TV, and Cable TV

In general it is difficult to present science on television because of the cost. However, Bernice Durand (Wisconsin) teaches modern physics for nonscientists very successfully on Madison area cable TV where watchers know her as the "physics lady."

PBS has recently begun a new television series called the *Magic School Bus*. Several people has asked whether we might be able to interest the producers in an episode on Particle Physics. It is a fully animated children's educational series. It features a teacher named Ms. Frizzle (played by Lily Tomlin), who takes her students on a magically powered bus for scientific field trips into the human body, around the solar system, or back to the time of dinosaurs. "Children's interest in science starts to erode in the elementary grades," project organizers say. *The Magic School Bus* project is designed to keep children's curiosity alive.

People have noted that other sciences seem to be featured in 60-second science profiles on the radio and have asked why not HEP.

5.   Government

It is generally agreed that our field could do a better job of informing Washington officials about what we do and why we do it, about the benefits of our research, and about the excitement of particle physics. Other areas of physics participate in APS' congressional visits programs much more than we do. My own experience is that many Members of Congress and their aides have never seen an HEP physicist and are happy for the opportunity. The recent Drell Panel report had a significant impact, in part because of significant followup in Washington by members of the panel and others.

It has been suggested that the DPF should sponsor occasional Congressional Fellows similar to those from the APS and AAAS. The cost is about $50,000 each in salary, moving expenses, etc. Unfortunately I doubt that the DPF can afford this. One should not underestimate the impact of Congressional Fellows. I have been told by aides in other offices that these Fellows are regarded as "gurus" on science issues. Unfortunately they have usually not been from our field, and in fact, they have even campaigned *against* our interests. It has been suggested that we should simply push to end APS's program which we pay for and which some believe may have done more harm than good with respect to HEP. These Fellows in no sense represent our field, nor is it clear to some of our respondents that we get the best qualified people to accept such positions.

A former Congressional aide has suggested that we would benefit more by sponsoring quarterly receptions for Members and aides (from the House and Senate) at which leading figures in HEP would discuss HEP physics issues and developments. He estimated that these evening receptions would typically attract 15-20 people (assuming food was provided), and felt that such numbers were well worthwhile. This is already done by other fields including chemistry and biological sciences.

6.   Science Community

One of the lessons of the SSC debacle is that we could benefit from better relations with the rest of the science community. A British correspondent reported that they have made great strides in improving their relations with other communities and that it is greatly benefitted them.

A proposal has been made to hold a meeting in Washington on the benefits of basic research for America, cosponsoring it with biologists, chemists, medical researchers, geologists, astronomers, etc. Leading researchers from each field would speak about the importance of basic research. Reporters would be invited to attend. Later participants in the meeting could visit the Capitol to relate this message to whatever committees or individuals are interested. The purpose of the meeting would be general and not to promote any particular projects. It would serve the dual purpose of reaching out to these other fields and explaining to the public the value of basic research.

7.   Documenting the value of basic research in HEP

A number of people have urged a new effort to document the impact of basic research in areas ranging from education to technology transfer to medical benefits to economic impact. One suggestion is to trace the history of particular technologies. We have not received any specific proposals on how to coordinate this.

8.   College students

A recurrent theme from many respondents is that there are enormous numbers of young people taking introductory physics courses in our own universities and that we are wasting a tremendous opportunity by not turning them on to physics and basic research as much as possible. These people will be the congressional aides, opinion leaders, etc. in a few years.

Others have proposed that we should spend more time giving talks at neighboring colleges.

9.   Teachers and school children

Many physicists are currently active in bringing particle physics to high school students. This can be done through presentations, workshops, open houses, the creation of materials, etc. The national laboratories all have such programs which I will discuss later. One very active national group is the Contemporary Physics Education Project (CPEP) which consists of teachers,educators, and physicists (among the physicists are Cahn, Goldhaber, Quinn, Riordan, Schwarz, and myself). This group has created the wall chart on *Fundamental Particles and Interactions* (in three sizes) and distributed more than 100,000 copies of it. It also has very popular color software for high school/college students in both Mac and PC versions. It mailed a packet of classroom activities about particle physics to every high school physics teacher in the US. They are completing a book on the subject of particle physics, detectors, accelerators, and astrophysics. CPEP conducts many workshops for teachers on how to use CPEP materials to teach particle physics. CPEP has been featured in *Science, Physics Teacher*, and even on the BBC World Service.

The American Chemical Society together with AIP periodically publishes booklets for students with cartoons, etc. The April 1993 issue was on particle accelerators.

The book published by Cindy Schwarz with AIP (described earlier) is intended for high school students.

A popular suggestion has been the idea of creating a catalog of resources, materials, workshops, etc. on particle physics. This would be made available (for free) not only to teachers but to physicists to aid and stimulate them in joining education/outreach efforts. The catalog would be available both in printed form and on the World-Wide Web. Some people propose mailing it to all high school teachers, but others feel that would not be useful.

A number of people are currently making presentations and giving workshops at teachers meetings such as the American Association of Physics Teachers (AAPT) and the National Science Teachers Association (NSTA). These organizations have national, regional, and state meetings. Those involved in these presentations find them well received and advocate that more people do it.

Another proposal is that we set up a national referral service (via telephone and e-mail) that would direct high school and college teachers with HEP questions to physicists who are willing to answer questions. The idea would be to refer the teachers to physicists in or near their own state. They might call a number such as 1-800-PARTICLE (extra digits are ignored). This service may also provide a list of speakers.

Finally, physicists can and are working with local school districts and state agencies. In addition, there are university, college, high school alliance programs (organized via the APS).

10.   Resources available to physicists

Many of the national laboratories such as Fermilab, SLAC, Brookhaven, and CEBAF have substantial education departments that sponsor workshops and programs for both students and teachers, and material development. They are anxious for the involvement of additional physicists.

Fermilab opened the Leon M. Lederman Science Education Center in September 1992. They have their own building with many exhibits. They have 45 precollege programs serving over 40,000 teachers and 8,000 teachers per year. In addition they have many college programs. They sponsor workshops for Latin American countries and create Spanish versions of instructional materials. Physicists are involved in Fermilab programs as research mentors, seminar speakers, role models, question & answer sessions with school kids, consultants on science content, hands-on-science in the classroom, museum volunteers, and SBIR proposals. CEBAF programs emphasize "Teach science by doing science."

Existing materials include transparencies, slides, comics, software, etc. These will be included in the catalog discussed above in the section on teachers and school children. The public relations staff at laboratories and universities often have resources available for physicists.

We should continue to report on outreach/education at DPF meetings to inform physicists about resources and ongoing activities. Many have suggested that we should work through the DPF and other organizations. We can also communicate about these activities via Internet bulletin boards and newsgroups.

11.   Using the Information Superhighway

More and more public schools are gaining access to the Internet. One suggestion is that the labs should set up files from which events pictures, detector designs, accelerator pictures, etc. can be obtained by anonymous ftp. These should be appropriately annotated.

The World-Wide Web (WWW) presents tremendous opportunities as use is growing by 300 percent a year. Major news media are searching the Web for stories, among them the New York Times. Even the sheriff of Tulsa, Oklahoma has listed Tulsa's most wanted criminals on the Web

An example of the impact of WWW can be seen in the interest generated by LBL's "Whole Frog" link-up. Users can examine many three-dimensional images of the frog with or without skin, from any angle. Different organs can also be seen separately. In half a year 160,000 users from 56 countries have connected to it (http://george.lbl.gov/ITG.hm.pg.docs/dissect/info.html).

CERN organized a major WWW Workshop on Teaching and Learning with the Web in May 1994. They had speakers and participants from throughout Europe but few from the US.

NASA has placed on WWW tremendous numbers of images from the Hubble Space telescope and elsewhere including pictures of supernova, comets, galaxies, planets, etc. These are annotated and sometimes very useful for education. There are also a variety of animations. A prime focus of NASA pages is always on hot and current topics. They have coordinated the efforts of their many different labs and facilities.

Fermilab has made great strides in making a major presence on the Web with some excellent educational pages and a coherent, organized approach. They cover the physics, the detectors, the accelerators, the benefits, and more. I suggest you look at it.

Clearly HEP (like NASA) should have a coordinated approach to the Web with a single homepage for the public that points to the labs and other relevant sites. This effort may require a meeting of the interested groups. This page should contain short items summarizing the current excitement and controversies in particle physics and point to lab and university homepages for more information.

Physicists may also need to make some effort to aid schools and libraries getting onto WWW. Many are already on the Web (even some elementary schools classes have their own pages), but most are not.

Other suggested approaches are to create multimedia CD-ROM programs about particle physics or even Nintendo-type games.

The AIP has an e-mail news service on physics education. It summarizes information on resources, national initiatives, outreach programs, grants, publications, etc. To subscribe to AIP's PEN, send an e-mail message to listserv@aip.org . Leave the "Subject" line blank. In the body of the message, enter the following command: <add pen>.

12.   Outreach by experimental collaborations

One suggestion is that experimental collaborations should be responsible for creating WWW and ordinary printed materials about their experiment. These should describe the physics motivations of their experiment and explain how the experiment might accomplish these goals. There are people who believe that any experimental collaboration that cannot explain these basic concepts to the public should not be funded.

Several people have suggested that experimental collaborations can do much more. A very interesting proposal is one under which traditionally non-research colleges (and possibly high schools) could become "affiliates" of experimental collaborations. Arrangements would be made whereby they would "participate" in research activities. Their work might involve a small scale hardware study (table top) or a simulation study. They would need computer time or the loan of some small hardware system for a few months.

An incentive for these schools would be very important: some degree of recognition of being part of the experiment. The institution names might be listed on scientific papers under the banner "educational institutions."

One possibility with CDF or D0 data for a college senior lab experiment would be to do some data analysis and event reconstruction for particles such as Z, W, and top.

Once such educational material is developed, it could be distributed to other colleges. Later it might be distributed to high-level high schools as a test.

Astrophysicists have already developed such a program, and it has been very successful. It is called "Hands on Universe." The organizers feel it gives high school teachers and their students the opportunity to become collaborators on real scientific research. The program provides them with access to professional grade telescopes, analytical tools and the training to use them. It is currently delivered to high schools across the United States. Students can request telescope time to obtain images of the moon, planets, galaxies, or supernovae.

The program recently made national news (ABC Nightly News, Associated Press, etc.) when two 17-year-old juniors at a Pennsylvania high school while searching for a galaxy photographed a supernova (1994I). While they did not, of course, recognize this, their photograph was the earliest one taken and therefore quite valuable. Both the publicity for science and the impact on young people were also valuable.

13.   Conclusions

There is no doubt that there are some exciting things happening in high energy physics outreach and education, carried out by educators and by physicists. However, the reality is that extremely few physicists spend any time at all on these efforts. They heartily endorse these programs, but find that they lack either the time or the inclination to join in.

This plenary session was an attempt to change attitudes, and we thank the conference organizers for their precedent-making initiative. It is important to show by our actions that we value public awareness. We should make communication a priority and reward it. We need a mechanism to make this happen, and motivation for people to do it.

As the conference's summary speaker (Howard Georgi) said, we need to think of speaking with the news media as a means of informing the public about the impact of public money spent on high energy physics, and we need to stop calling it "publishing in the New York Times."

Conference participants who attended our POET meeting seemed especially interested in the following proposals:

1) Create a catalog of HEP resources (materials, workshops, etc.) for teachers and for physicists. It would be printed and on the World-Wide Web.

2) Together with basic researchers from other fields, organize a meeting in Washington on the impact and importance of basic research.

3) Organize a unified approach to presenting Particle Physics on the World-Wide Web, presenting the highlights and controversies of our field.

4) Begin a program of educational affiliates of experimental collaborations who would perform specially designed analysis or experiments.

5) Find means to better inform Washington staff and officials about HEP (quarterly receptions at the Capitol, congressional fellows, etc.).

6) Organize a national science writers meeting with physicists.

7) Encourage more HEP participation in science museum programs and find means to present our subject in museum-type settings.

For these and other efforts to succeed, the DPF needs to give them some priority and provide vital organizational support. Moral support is welcome, but if we wish for outreach and education activities to progress, meaningful action by the DPF would be more beneficial.

Outreach to Women and Minorities

JULIA THOMPSON, *Pittsburgh*

OUTREACH TO WOMEN AND MINORITIES

JULIA THOMPSON*

1. Introduction

Each of us is influenced by "role models", from whom we learn – how to parent, how to play basketball, how to sew, how to interact with different cultures, how to choose scientific topics for investigation, how to design and carry out experiments. And each of us in turn is a role model for others. While it is powerfully important for some role models to be of the same gender and ethnicity, it is an important responsibility for each of us to encourage others in attaining the highest competence and welcome others different from ourselves into our field. Increasing the variety of participants in our field helps to keep it vigorous. Including a larger variety in the broad technical training our field requires helps break down the compartmentalization in our society which has led to a loss of hope for full participation by many young people of minority ethnicity. Those already in the "system" have a chance, and an obligation, to change it, and open it up. We are fortunate to be paid for our work of chasing quarks and Higgses. Improving the broad technical training of young people in our society may be as important a spinoff as particle beam cancer treatment facilities, faster computers, detectors for medical physics applications, and practical superconductivity.

2. Context of Talk

In this note I will use "minority" to refer to groups under- represented in physics. In this sense, women (about 50 % of the population, but only 11% of the physics PhD's) are included, as well as African Americans (about 12% of the population, less than 1% of the physics PhD's), Hispanics (about 8% of the population, less than 1% of the physics PhD's), and groups such as Native Americans, Pacific Islanders, and even some white males with sketchy early academic training. In detail the problems of these groups are different, and I will not attempt to deal with these differences here, although for illustrative purposes I will occasionally draw on examples from the two groups with which I have worked most closely – women of all races, and African Americans of various backgrounds. I invite you to substitute examples from other groups.

But here I would like to focus on common aspects of these groups– a sense of intrusion, coming from their lack of representation among current practitioners, and not infrequently different cultural and personal characteristics and habits which may mask the recognition of their ability by those from different cultural backgrounds.

3. Role Models – How Much Like Us?

One of my students draws the distinction between role models and mentors, using mentor to include a role model not necessarily of the same gender and ethnicity, but someone who helps you to thread the paths of learning and supports you in going further alone. All of us can be mentors or role models in this sense – visiting elementary or secondary schools, talking with our neighbors or those in our churches, in scout troops, or other community organizations, paying particular attention to minority students in our classes, talking about the excitement of our field.

*Department of Physics and Astronomy, University of Pittsburgh, Pittsburgh, Pa. 15260, USA

In my own career there were very few female role models. But my students, among whom women are much more highly represented, have made it clear to me that female role models (and, by extension, role models of matching gender and ethnicity generally) play an important role in many cases in attracting or repelling students from our field and from technical careers generally. One example of the "role model effect" comes from one of the Scandinavian countries, Iceland I believe. A woman had been prime minister for several years. Finally she was voted out, and succeeded by a man. The correspondent telling the story recounted the reaction of his young son to the first appearance of the new prime minister on evening news: "Father, come here! something is wrong. There is a man on television saying he is the prime minister! But only women can be prime minister!" Maybe it is easier to see the effect when the usual connections are reversed.

Some examples of special problems which are easier to face when one can talk with role models of the same ethnicity and gender come quickly to mind:

1. child care

2. pregnancy, PMS (pre-menstrual stress), other special physical conditions.

3. sexist/racist comments, and deciding if offensive behavior is really bias or racist or just thoughtless

4. if offensive behavior is really bias or racist, how to deal with it. can it be defused or avoided? is it best to ignore it or fight it head on?

5. two-career constraints (stronger still for women, though the progress of the last 20 years has brought this to be a significant issue also for men)

6. sexual harrassment; emotionally/ socially/ sexually ambiguous situations

7. racial or other harrassment

8. how to maintain both one's professional and personal connections, and both one's professional and personal life and obligations.

4.   How To Achieve More Variety? I: Flexibility

Given that there are not so many candidates, for a given institution there is a problem in how to recognize future stars early. In the present climate this may be a problem for the institution rather than the candidate, because of affirmative action pressures. However, minority candidates often have a less direct career path, and rigid rules about age or time past PhD, or time in a given position (such as Assistant Professor) may disproportionately impact minority candidates.

The minority candidates who advance to the candidate stage (to be considered for possible permanent positions) have all survived a rather tough process and have evolved their own individual ways of dealing with the obstacles, real and sometimes also imagined, which they have passed through and overcome. This can leave them with personal characteristics which may not seem to "fit in" or be promising for the "personal chemistry" or "good collaborator" which is necessary for success in our highly collaborative field. They may be too quiet and withdrawn, too abrasive and aggressive, too militant, too passive, not quick enough with back of the envelope calculations because they prefer to check their calculations first, preferring rap to classical music,.....

Because often the final decision is made from a few comparable candidates, such peripheral characteristics shape our thinking and selection processes. It is true that the new candidate must be able to work well with those already present. But this may take some flexibility on the part of those already present, both in the collaborative process and in the selection process.

For minorities, family responsibilities may be more pressing, because the support of the family is more central to the candidate's confidence and success, because the responsibility for family support may fall disproportionately on the minority candidate, and because the family problems may be more serious and central (family members without health insurance, being evicted, in jail...) for minority candidates. To combat these problems, and also the two-career family problem, options like part-time work, flex-time work, telecommuting, and shared positions have been discussed and bear serious consideration, despite obvious problems with implementation. These options may be particularly important to convince young women of child-bearing age that a technical career is compatible with having their own family. Lest we downgrade this desire, we should think carefully about the implications of forcing our technically competent and gifted women to choose between a technical career and raising their own families. From some aspects it is a loss of both genetic and social resources.

A final problem in bringing minorities into our field in greater numbers is the challenge of recognizing and controlling behavior which is harrassing, bigoted, or sexually insistent or ambiguous. This point may seem obvious but is not so simple to achieve as to say.

FOR ALL THE ABOVE REASONS, MY OWN PERSONAL CONCLUSION IS THAT AFFIRMATIVE ACTION IS STILL A USEFUL ANTIDOTE TO THE ABOVE PROBLEMS. By affirmative action I do not mean filling quotas regardless of ability, but rather that for candidates plausibly equal within errors bars (1-2 σ, eg), one should give the minority candidate the edge. It is very hard to be sure otherwise that one has properly accounted for these other factors. And the combination of the determination which has brought a candidate thus far, and the role model advantage for the field and society, may reasonably tip the balance.

5.   How to Achieve More Variety? II: More Candidates

There are several facets to this endeavor:

1. Reactions and influence of parents and friends. Family and peer pressure can be important supports or important deterrents. Family support is often more important for minorities who may be going against stereotypes to enter technical fields.

2. High school guidance counselors. Some counselors strongly dissuade women and minorities, in some cases even block them from pursuing the necessary technical courses.

3. Teachers and faculty. Give equal recognition and be equally demanding with your good minority students as with your good mainstream students.

4. Special programs, usually involving some research experience. These range from early childhood, through middle school to high school, college undergraduate, and on to special support at the graduate student level. Sponsors include the government through NSF and Department of Education; all the DOE-sponsored

national labs: ANL, BNL, FNAL, LANL, ORNL, SLAC; many businesses; and many universities, either individually, or in partnership with state and federal support. The minority participation in these programs is a range, from mostly mainstream students with a few minorities, to all or mostly minority students with a few disadvantaged mainstream students.

I will describe, as an example, the program (sponsored jointly by the NSF and the University of Pittsburgh) which I direct: Research Experiences for Undergraduates in Physics at the University of Pittsburgh - Focus on Minorities.

We aim for 80% African American and 30% female participation, with a conscious intent to include about 15% high quality mainstream students. It was developed after a trip by me to about 9 historically black colleges and universities, to determine whether yet another program would be useful.

A large drop in science-bound students occurs at the end of the freshman year in college. Partly for that reason, our program is a three year program, targeted at beginning students just after their freshman year, to help encourage students to stay on in this important period. Students are picked in conjunction with home school faculty for motivation and determination, and contact is maintained during the following year, through the home school advisors, in some cases with substantial input into students' courses or supplementary reading at their home school.

The program is a mixture of free-form academic work and participation at as realistic a level as possible in a physics-related research group. We have included high energy, intermediate energy, MRI, condensed matter, astrophysics, and in some cases physical chemistry (infrared studies of protein folding), electrical engineering (optical wave guides), and computer science (ray-tracing and graphics). The academic work is a mixture of basic math and physics and more advanced work targeted at topics necessary for the student's research work. This year a "GRE" workshop and a machine shop workshop were included, in response to student interest.

The two T.A.'s follow the research experiences of the students; provide academic support; help with extra curricular activities and preparation of presentations and papers; and generally help the students succeed by identifying and overcoming obstacles, including communications problems between the student and the research group.

Weekly conferences help the director, students and T.A.'s follow the student's problems and progress. Group meetings help build morale, introduce other research and discuss larger questions like ethics. In addition to the academic and research work, there are mentoring activities, e.g., presentation of physics experiments to an on-site YMCA program. Alumni/ae visits help the students see the fulfillment of later work and the importance of persistence in followthrough in later work.

The minority inversion gives both mainstream and minority students practice in interacting with those of different cultures, in what we hope will be a useful intermediate step towards full participation in a researchenvironment still dominantly white male. This variety also helps students to confront and go beyond stereotypes of their own and other ethnic and gender groups.

6.   Conclusions

In closing, I would like to repeat my main thesis: role models are important in shaping the way we act. We are all role models and all react to role models. We who are now "insiders" should act in accordance with that responsibility.

Using Science Centers to Expose the General Public to the Microworld

ERNEST MALAMUD, *Fermilab / The Science and Technology Interactive Center*

USING SCIENCE CENTERS TO EXPOSE THE GENERAL PUBLIC TO THE MICROWORLD

ERNEST MALAMUD

Fermilab and
The Science and Technology Interactive Center, Aurora, Illinois

ABSTRACT

Despite the remarkable progress in the past decades in understanding our Universe, we particle physicists have failed to communicate the wonder, excitement, and beauty of these discoveries to the general public. I am sure all agree there is a need, if our support from public funds is to continue at anywhere approximating the present level, for us collectively to educate and inform the general public of what we are doing and why. Informal science education and especially science and technology centers can play an important role in efforts to raise public awareness of particle physics in particular and of basic research in general. Science Centers are a natural avenue for particle physicists to use to communicate with and gain support from the general public.

1. POET Goals

Lets begin by restating some goals of POET:

- To encourage HEP physicists to support and participate in public outreach and education,

and by so doing to help

- rebuild the support that basic science has had not only from people in Washington but from the American public in general.

The rationale for these efforts can be stated as follows:

- We have an obligation to explain to them what we are doing and why.

- The general public, including your congressmen and your next door neighbor, finds science in general, and particle physics and astrophysics in particular tremendously exciting if they are exposed to them.

In this talk I will suggest a way, <u>not exclusive of other suggestions</u>, for members of DPF to aid in achieving these goals.

2. What Is A Science Center?

Most of you are familiar with science centers [1,2,3,4] but let me summarize briefly to set the stage for what follows.

"Informal science education," happens through channels that include television, newspapers, advocacy organizations, museums and science centers. Each of these channels reaches a different audience and presents a different kind of information. Informal science education plays a significant role in shaping public awareness of science and technology. Science centers have been recognized as important players in this arena.

In the past several decades, people all over the world have been starting science centers. There are now about 300 of these new-style museums, two-thirds of them in North America. Science centers have visitations of over 60 million people annually nationwide. Their emphasis is on presenting subject matter in a non-threatening, friendly, exploratory, <u>fun</u> manner.

Schools alone can't ensure that an increasing number of youngsters will be prepared to learn science, or even be motivated to try. Students' attitudes and choices are influenced by family, friends, media, and the possibilities revealed by their own experience. They need many varied and comprehensible encounters with science in order to want more of it. People today growing up in urban settings and absorbing information passively from screens, do not labor with ropes, levers, fire and liquids as farmers and artisans did in the last century, and so do not have well-formed intuitions about how the natural world behaves.

Science centers provide access to science, they improve our comfort level with technical ideas, artifacts and numbers and they enable us to obtain first-hand experience with phenomena. Operating in three dimensions, unlike print or electronic media, science centers provide opportunities for students, parents and teachers to play around first-hand with materials and experimental apparatus, and so to build intuitions about the natural world. Science centers provide experiential learning. As parents and children handle the displays, they act as both experimenters and experimental subjects. They watch demonstrations and ask questions about the dramatic and intriguing things they see. They enjoy the time spent in a safe, upbeat environment where their curiosity is both stimulated and rewarded.

Science centers are playing an increasingly important role in the nation's efforts to stimulate and motivate our youth to take an interest in science and mathematics, to learn high order thinking skills, so they can lead meaningful lives in the next century. As organizations science centers are not as concerned about the elite, the talented, the gifted, but more with the general population. We are populist organizations. We are deeply disturbed by the growing underclass in America.

Although there are many similarities between science centers, you find that each have their own particular character and philosophy. Some approach theme parks in their atmosphere; others are less glitzy. But they all share a similar mission. Each new science center has to define its own aims, methods, audience, and content within the context of local possibilities.

What is accomplished? In none of these interactions have adults or children mastered a body of theory, or rarely have they used mathematics to make a prediction. They have not followed an informed sequence of experiment -- What they have done is to share in some of the marvels of the natural world, made accessible to them. One of the problems in science centers is that we can't predict when learning will happen. A child may see something in an exhibit that enables him, months later, to grasp a concept the first time it is taught at school.

3. The Microworld-How to Go There and Whom to Bring?

The essence of the science center is interactive exploration of scientific phenomena. A typical science center offers exhibits on mechanics, electricity, optics, perception, health, and transportation that can appeal to children as well as their parents. On weekdays during the school year, hundreds of elementary school children visit the center with their teachers and chaperones. In addition, as individuals they participate in workshops, camp-ins and more structured programs using the rich resource of the center's exhibits.

> Visitors experiment with the laws of reflection.
> They appreciate the speed of sound by saying "woof" into a 100 meter long
>   PVC pipe and listening to the echoes.
> Children work cooperatively to construct a Roman arch.
> And they are fascinated to watch the build-up of a normal distribution
>   curve.

Most of contemporary science is microscopic--particle physics, DNA, and quantum mechanics--and difficult to present in the exhibit medium. How do we show something that is too small to see? How do we motivate visitors to care about something they can't experience directly? Does the apparatus hinder visitors' entry into the microworld? Can "quantum weirdness" spur interest, or only frustrate?

4. Past and ongoing efforts

4.1. A word about SciTech

SciTech is patterned on the philosophy of San Francisco's Exploratorium where I worked for a few months in 1982. SciTech has grown from an all volunteer effort in 1989 and a budget that year of $20,000, to a organization currently employing 20 FTE plus an additional half a dozen scientists, engineers and other professionals paid by other organizations, or "high-tech grass-roots" volunteers. The 1994 operating budget is about $600,000 We are currently serving 100,000 people per year and still growing.

4.2. Building Blocks of the Universe

The Building Blocks of the Universe Project[5] (Fig. 1) officially began with a grant from the National Science Foundation, in June 1990. The project was a collaboration between SciTech and COSI, Ohio's Center Of Science & Industry in Columbus. The partnership drew on the years of experience of the COSI staff. COSI is one of the leaders in interactive science exhibit development. It also took advantage of the wealth of talent amassed around our newly formed SciTech, drawn from nearby scientific institutions and industries, including Fermilab, Argonne, AT&T Bell Labs, and Amoco Research Center.

Who needs exhibits on the basic mechanisms of the Universe? Our experiences are naturally based on the macroscopic - that which we can see, feel and manipulate. These experiences generate but also limit our concepts, and this applies as much to the formalized concepts of a scientist as the intuitive concepts of others. In general, we are unfamiliar with the workings of the microscopic world. However, it is precisely this world which defines our own world, gives it its rules and exceptions, creates its possibilities and probabilities, and holds the secrets of the universe in which our existence is immersed. To break through to the microscopic world we - scientist and non-scientist alike - need new concepts, freed from the limits of our immediate experience. The Building Blocks of the Universe exhibition seeks to bring everyone at least to the gateway of a world which otherwise exists only in a guide book written in mathematics.

Present day physics, which seeks to explore the Big Bang, the top quark, the Higgs boson, and other building blocks seemed too complex to explain and describe in terms that an eleven year old could understand, not to mention the

Fig. 1. Entrance to the 120 m^2 Building Blocks of the Universe Exhibition at SciTech in Aurora, Illinois. *Photo Credit: Fermilab Visual Media Services.*

non-physicist adult. However, through all the calculus, perturbation theory, and QCD, we must make every effort to find simple analogs and solutions.

We attempt with these interactive exhibits to convey to a wide audience some of the fascination of the world of the very small. In both centers, SciTech and COSI, existing exhibits already offer a rich variety of experiences of nature at the human scale. On entering this new group of exhibits, the visitor is led from the human scale to progressively smaller scales of size, discovering a sequence of new ways that matter is constituted.

We had to weigh the value and necessity of classical analogs of quantum behavior against their inevitable shortcomings. The challenge is to engender a sense of wonder and excitement, but not of mystery. The visitor must feel that the exhibits represent gateways that open up, not present barriers to surmount; that the new worlds discovered do not undermine common experience, but rather form an exciting extension of that experience.

Exhibit components incorporate the "hands-on" style which has proved so attractive and effective in science centers worldwide. The exhibits are linked and enhanced by an effective museum environment, using graphics, accurate non-technical text, and artistic displays to create an atmosphere in which visitors can learn about phenomena beyond the range of direct perception. Placing such exhibits in a general science environment is important. E.g. at SciTech 12 visitor experiences dealing with particle and nuclear physics are embedded in an environment of over 200 exhibits dealing with physics, astronomy, mathematics, and technology. Some of the elements of Building Blocks are described below.

4.3. An Investigation Into The Very Small

The notion of "small" generally conjures up pictures of paper clips, a strand of hair, a sliver of cake, or a grain of salt. To those of us who are fortunate to have used a microscope, those images might be extended down to mitochondria and proteins.

To prepare the visitor for "an investigation into the very small" we start with an activity suggested by Professor Ken Wilson. The participant is offered scissors and a piece of paper, and challenged to divide it in half as many times as possible. Most people think that dividing something in half is trivial, and can be done numerous times if desired. This part of the exhibit gives an appreciation of how fast factors of two multiply, as people realize that they can barely cut the paper ten times. A magnifying glass or a simple microscope might help, but then what? The participant, still able to hold the smallest piece of paper easily in his hands, surely realizes that there is still something smaller. Pictures of the paper down to the next few factors, are displayed, as are pictures of other common objects not normally viewed at that magnification. The participant is then led through a thought experiment, imagining what might be smaller. Is there some smallest unit that we can still call paper? Given the right tools, can that smallest unit be divided into yet smaller non-paper things? Can those be divided further? How far can we go? Is there some fundamental limit? How many cuts to reach the size of an atom?

The goal of this exhibit is to set the stage for the entire exhibition, and to slowly draw the visitor into the world of sub nuclear realities. It starts out with the things that he knows, and can touch, see and manipulate. It then provides an introduction, a gateway of sorts, into the microscopic universe, as the layers are peeled back, one after another.

Cutting the paper is followed by a projection microscope. If we have prepared the slide carefully the visitor can see Brownian motion. It would be more effective if we also had additional projection microscopes showing a variety of microscopic living and non-living objects, as at the New York Hall of Science.

4.4. The Atom

The Hydrogen atom is the simplest atom that is possible. It consists of a single proton in the central nucleus, and a single electron satellite, dancing around the center. Unlike the macroscopic world, the bound electron can only have one of a set number of possible energy levels. Inside an atom, there is no energy continuum. The electron is not allowed to have any energy it wants, take any position it wants. Instead, it is limited to be in one of a few possible energy levels.

Three concentric Plexiglas spheres, each covered with an array of red LED's, are used in our model, 10^9 the size of a real Hydrogen atom. Each sphere represents a single allowable energy level, and each light represents a possible position for an electron. Having but one electron, only one light can ever be illuminated at a time. But, since the electron is never stationary, this point of light flickers on and off randomly around a particular sphere. The flashing is done quickly so that the observer can never zero in on the electron position, and sees only a cloud of light where the electron might be found. Only the electron position is changing, not its energy.

While in one of the lower energy levels, the participant is asked to excite the atom into one of its higher energy configurations. There are three allowable excitations: inner to middle, middle to outer, and inner to outer. Each transition is made possible by the participant imparting energy of a characteristic frequency. The visitor is asked to whistle, or for those that can't whistle to use a toy Xylophone. If the note matches one of the three allowable frequencies, then the electron jumps to the indicated higher level. After some time, the electron is allowed to relax back down to a lower energy level. Again, there are three possible transitions, each with a characteristic frequency. A relaxation event is indicated by the automatic firing of one of three colored strobes, positioned inside the nucleus. The blast of light represents the single photon that would be released by the atom for such an event.

There are four major goals for this exhibit. First, it conveys the concept of quantized energy levels. Second, it shows that the bound electron is never stationary, and that its position at a particular energy is uncertain. Third, it demonstrates that it is the frequency which determines the nature of a transition. Finally, it shows that energy is radiated from the atom only when an electron drops into a lower level.

Nearby we have an exhibit with real atoms called "atomic fingerprints" The visitor views various gases in discharge tubes with a hand held diffraction grating.

4.5. Rubber Ball Nuclei

Hard rubber balls, 2 inches in diameter represent the individual nucleons. Those that are dyed red are protons. Those that are dyed blue are neutrons. The nucleons are held together by a piece of clear plastic tubing, extending from the center of one ball to the next. A plastic green bead rolls back and forth within each tube. It represents a meson, the messenger particle that is exchanged between the two objects, delivering the attraction of the strong nuclear force, keeping them bound together. It is important to note that the bonds are not rigid and fixed. Instead, they are rubbery and flexible, allowing angles and distances to change slightly, providing a more accurate representation of the atomic nucleus. This exhibit can be easily picked up, twisted and bounced. It appeals to the younger visitors.

4.6. Half Life Cluster

There are several stations which provide models for radioactive decay. In one of them the participant is handed a container of six-sided dice. One of the sides on each die is painted a distinctive color. In the first turn, the dice are spilled onto the playing surface. All dice with the colored side up are removed from the group, and stacked one upon the other on the board. The second turn is the same, with the remaining dice being spilled, and the few with the colored sides being stacked, next to those already present. This cycle continues until all dice have been stacked. The result is a graph, made by the individual stacks, that represents the "decay" of the dice. With each toss, 1/6 of all dice decay.

4.7. Valley Of The Isotopes (Fig. 2)

A map of the isotopes is usually three dimensional. Proton number is along the x-axis, with neutron number along the y-axis. The third dimension represents the binding energy of the isotope. Standard wall charts, unable to use 3 spatial dimensions, choose their third dimension to be mapped as a color, or a shade on the grid. However, on the museum floor, we have no such restriction, and we are free to build in x, y and z, as we please. Therefore, we have chosen to translate the binding energy into a depth along the z-axis. As binding energy increases, the depth in z increases. This is similar to digging a hole. The deeper the hole is, the higher the side walls are, the greater is the amount of energy needed to get out.

Fig. 2. The Valley of the Isotopes Light Sculpture and Interactive Exhibit. *Photo Credit: Fermilab Visual Media Services.*

As the table is mapped, a central depression is formed, with high walls on either side. This is called the Valley of the Isotopes. The entries within the valley have high binding energies and are therefore very stable. The entries that are on the high ridges, have smaller binding energies, making them much less stable and prone to decay. For instance, Iron-56 is the most stable of all isotopes, and is found at lowest point in the valley, while Sodium-22 is precipitously balanced on the ridge, a prime candidate for radioactive breakdown.

There are thousands of known isotopes, and to have a functional and easily maintainable exhibit for all of these would require a Herculean effort. Instead, we have chosen to reproduce a section of the valley. Our initial effort includes all isotopes through to Vanadium. The x-y grid is made with lights, mounted on circuit boards, and controlled by a computer. A light pipe is mounted over each light. The height of the pipe, and the color of the light represent the binding energy of the isotope. The result is a beautiful and dynamic light sculpture.

A voice in the computer invites the visitor to explore fusion processes, flashing the appropriate lights as a guide through the narrative, programmed to explain stellar evolution and how our sun burns Hydrogen to make Helium and energy. It details Carbon dating, and the behavior of some medical isotopes. It works better as a teaching tool than as an unstaffed exhibit.

4.8. Quark Machines And Particle Scoreboard (Fig. 3)

All matter of which we are commonly aware, including all elements in the periodic table, all nuclei, and all isotopes in the valley, can be made of three things: protons, neutrons, and electrons. We further know that protons and neutrons are composed of up and down quarks. The goal of this exhibit is to show that there are actually a total of six quarks, and that there are many other particles that can be created through their various combinations.

The basic component of these exhibits resembles a three-wheel slot machine. A color computer screen is used to simulate the wheels and the spinning motion and an arm mimics the look and feel of a true slot machine. Instead of lemons, cherries, and Lucky 7's popping up, quarks will appear, and baryons will be formed. In the case of the antiquark machine, antiquarks will appear, and antibaryons will be formed. The machines can be played, side by side, by two competitive would-be physicists, and the resulting baryons can be compared.

The probability of a particular quark appearing on a wheel is based on its existence in the real world. That means, it is most likely that an up or down quark will appear, and that a proton or neutron be formed. After that, there is a slim chance that either a strange or a charm quark will appear. Finally, there is a tiny remote chance that a bottom or top quark appears.

Fig. 3. The Quark Machine, The Antiquark Machine, and the Particle Scoreboard.
Photo Credit: Fermilab Visual Media Services.

5. Next steps

The goal of the effort is to assist in the quest to promote science literacy, and to produce a set of interactive experiential science exhibits, detailing the world of high energy physics. It focuses on the peculiar realities of atoms, subnuclear particles, quantum mechanics, and the fundamental principles that cooperate to tie our understanding of the cosmos, the quarks, and the coffee table all together into a single construct.

5.1. $E=mc^2$

We are continuing this program to bring 20th century physics into science centers with a current project funded by the U.S. Department of Energy called $E=mc^2$.

Everyone has heard of Einstein's famous formula "$E=mc^2$", but few people ever expect to know what it means. This exhibition will provide a wide audience with the opportunity to experience its meaning and understand it in simple terms. The subject of the exhibition is the equivalence and inter convertibility of energy and matter. The formula will come alive for museum visitors by providing them the experience of manipulating two real subatomic physics processes: positron-electron annihilation into two photons and photon conversion into electron-positron pairs.

Fig. 4.   An early prototype of Matter into Energy, part of the $E=mc^2$ Exhibition.
   Photo Credit: Fermilab Visual Media Services.

Detectors used in basic physics research will be employed to "visualize" these transformation processes. Accompanying graphics panels, lecture and demonstration scripts, and computer animation will enhance the exhibits with explanations, historical notes about the development about our concepts of energy, and the relation of the exhibits to modern energy issues -- nuclear power, radiation, and medical applications.

In the matter-into-energy exhibit, (Figs. 4 and 5) positrons and electrons (the "matter") annihilate into photon pairs (the "energy"). We use a weak encapsulated Na22 positron emitter and detect the photons with two scintillation photo tube counters, both mounted on movable arms so the visitor can search for the antimatter annihilation process. A chirp signals a new event. The visitor can insert lead collimators to block the particles and stop the chirps.

Fig. 5. The finished exhibit element, Matter into Energy for the $E=mc^2$ Exhibition. Photo taken at Argonne's May 21, 1994 Open House. More than 18,000 visitors attended.
Photo Credit: Argonne National Laboratory.

In Energy into Matter, photons from natural thorium having an energy of 2.6 MeV will enter a diffusion cloud chamber. The visitor can move a foil into the photon beam and so create electrons and positrons via pair production. These charged particles make visible tracks in the chamber. In this process the photon disappears and its energy reappears as a particle and an antiparticle. A small rare earth magnet will show that the two particles that make the tracks are of opposite electric charge.

The visitor will embark on a "treasure hunt" using a freeze frame video setup coupled to the cloud chamber to find, and be rewarded, for the rather infrequent pairs. Graphic aids will help the visitor recognize the various phenomena.

5.2. The Quantum World

On the drawing board is a new proposal, the Quantum World being developed by a partnership of current and retired Argonne physicists and some of the top high school physics teachers in our area.

6. Is There A Role For Scientists In A Science Center?

Despite the remarkable progress in the past decades in understanding our Universe, we particle physicists have failed to communicate the wonder, excitement, and beauty of these discoveries to the general public. I am sure all agree that there is a need, if our support from public funds is to continue at anywhere approximating the present level, for us collectively to educate and inform the general public of what we are doing and why.

Informal science education and especially science and technology centers can play an important role in efforts to raise public awareness of particle physics in particular and basic research in general. Science Centers are a natural milieu where particle physicists can communicate with and gain support from the general public.

Physicists can make a difference by volunteering at their local Science Center. Science centers do great 19th century physics, but they need help with the 20th century. And can probably help with 19th century physics too! This education effort is not just reserved for the retired or almost retired!

The most promising area of growth for science centers is in their relationship to teachers and schools. Because centers are outside the school bureaucracy, because they are entrepreneurial, and because they have accumulated resources for science teaching, they are well-positioned to deliver innovative programming. There is much recent emphasis on teacher training. And a great deal of effort in science centers is going into diversifying both the audience and the staff.   This evolution of the role of science centers provides many opportunities for DPF members to get involved and help.

6.1. The Chasm between the Science Center and Research Communities

On the one hand there is the high-tech, ivory tower, high powered research establishment. On the other side is the science center community.

From the research community side we see oversimplification, lack of rigor, an environment sometimes approaching that of Disneyland.

The science center community sees Ph.D.'s, busy, important people rushing off to meetings, wheeling and dealing for grants, machine time or whatever.

The people who run science centers are by and large very bright, creative, dedicated, hard-working, humanistic, with a mission to accomplish.

Suppose your local science center has just completed an exhibit on radar tracking of tornadoes and needs help with a simple, clear explanation of the Doppler effect. But they are hesitant to call up their local physics department and ask for help.

6.2. How Do We Bridge This Chasm?

I urge DPF members to visit your local science center, wander around with a notebook, spot errors or obscurity in labels. E.g. I have discovered that total internal reflection is explained incorrectly in many large science centers. Or suppose you see a set of wonderful exhibits on angular momentum, but then remember that last year in your freshman class you did this neat demonstration that might be turned into a robust, hands-on exhibit.

Meet with the science center director and offer to help. Commit to a half-day a week for a few months. Be helpful. Write signs. Build prototype exhibits in your University shop or in your basement. Become a part of the science center team as a regular volunteer. At first, the staff might not understand how you can help or that they need you. But they will soon see that you can help them present more accurate science to their visitors. You may even find they will ask you to become a member of their regular staff. (Attention graduate students: in this shrinking job market, science centers are a growth industry.) Gain their confidence. You will have a lot of fun and you will get a lot of very positive feedback. Then start discussing bringing modern physics into the center. Offer suggestions on a hands-on atom where the visitor can excite it to different levels, or some models of isotopes and signs that connect to real-life concerns.

Another area hungry for scientifically trained people are the teacher training programs at science centers. As a volunteer, or as a contract teacher, or as a member of the science center staff, you can participate in developing education programs and teaching classes or workshops. Not only can it be an avocational or vocational opportunity, but it can be a way to make a significant contribution to solving the deepening crisis of science illiteracy in our country.

I am more than happy to serve as a "marriage broker" and help form such partnerships. Please contact me at
    MALAMUD@FNAL.GOV or MALAMUD@SCITECH.COM

Acknowledgments

I appreciate the advice of Olivia Diaz, President of the Board of Directors of SciTech, in the preparation of this paper. I am grateful for the support of this work by Dr. John Peoples, Director of Fermilab, and Dr. Alan Schriesheim, Director of Argonne National Laboratory.

References

1.   Frank Oppenheimer, "A Rationale for a Science Museum," reprinted from Curator, November 1968.

2.   George W. Tressel, "A Museum is to Touch," 1984 Encyclopedia Britannica Yearbook of Science and the Future.

3.   Sheila Grinell, "Science Centers Come of Age," in *Issues in Science and Technology* Spring 1988.

4.   Sheila Grinell, "A New Place for Learning Science," published by the Association of Science-Technology Centers (1992.)

5.   Miriam E. Bleadon et al, "Building Blocks of the Universe," presented at the Fourth IISSC, March 4-6, 1992 and published in the conference proceedings (Plenum, 1992.)

High Energy vs. Low Education:
A National Challenge

MALCOLM W. BROWNE, *The New York Times*

HIGH ENERGY VS. LOW EDUCATION

MALCOLM W. BROWNE
Senior Science Writer, The New York Times

Despite such milestones as the probable discovery of the top quark, physics in general and particle physics in particular are in trouble. The collapse of Congressional support for the Superconducting Supercollider is but a symptom of the malaise, which goes deeper than financial funding alone. Many physicists these days express a sense of isolation from a badly educated American public--a public that includes members of physicists' own families, as well as the politicians who control the funding of large science projects. Advancement opportunities, pay scales, tenure, health benefits and most of the other material measures of well-being are notably poor for physicists: a fact that affects not only their standard of living but the social status of their profession in the eyes of the very young people physics hopes to attract.

Paradoxically, Americans are interested in physics, even if their understanding of many of the most important ideas is naive or non-existent. Because words like "renormalizable" require a lot of explanation, many print publications and broadcast outlets contend that science is beyond the interest and understanding of most people. The Los Angeles Times abandoned a short-lived science section, science magazines that briefly flourished during the 1980s have mostly closed, and the coverage of science by commercial television is virtually non-existent. The New York Times, however, has found that science news can be both popular and remunerative. Polls have shown that our Tuesday inside section called "Science Times" is rated the best read and most widely circulated special section The Times publishes on weekdays--even ahead of sports. Moreover, our science section attracts ample advertising to sustain it, mostly from computer dealers. Admittedly, The New York Times attracts readers who are better educated than the norm, but even allowing for that, news directors and editors elsewhere tend to underestimate their readers and viewers.

Unfortunately, the education of most young people at the grade school and college undergraduate levels does not prepare them to understand the physical sciences; only a small minority are introduced to calculus or the basic algebraic tools essential even to catching the flavor of important scientific ideas. This is a problem that plagues science writers as well as scientists: we, too, are often at a loss to explain things when we are barred from using equations. It has been estimated, moreover, that between 20- and 40-million Americans are functionally illiterate, and therefore beyond the reach of even the most skillful science writing.

Out-reach initiatives by physicists are doubtless steps in the right direction. The spectacular successes of science in our own time have hooked countless thousands of young people on science, and some have embarked on the hard road to careers. But there remains a gigantic gap between the dinosaurs, comet impacts and other gee-whiz marvels portrayed by the printing and broadcasting industries as science, and the long, cold plunge into mathematics and complex problem-solving demanded by real science.

In Russia, high-energy experimental physics has gone into hibernation, and there is danger that a decade from now, the field may become moribund in the United States as well.

I think that physicists who want to keep the torch burning would do well to heed a letter to Physics Today by James B. Cole of the Naval Research Laboratory, who suggested that for Ph.D. candidates in physics, "a few drama and psychology courses might help. Introverted geniuses had better hire a good agent." The main people who need to be sold on particle physics, of course, are the Appropriations Committee members who make the vital (and all-too-often woefully badly informed) decisions.

Particle physics explores the ultimate roots of universal existence, and its discoveries are vital to astrophysical research into such questions as the origin of a lumpy universe, the ultimate fate of the universe, and other matters that once were monopolized by theologians. Particle physicists are truly successors to the priesthoods of all religions that have sought to explain the universe.

I believe that ordinary people are willing to accept the bonafides of physicists as latter-day priests, even though the education of ordinary people is inadequate to understand what physics is really about. This may smack of elitism, but it's a fact that every physicist, as Newton put it, sees farther than plain folk, because he or she stands on the shoulders of giants.

But the whole trick lies in convincing potential believers that physicists really are priests of a sort, however spartan their salaries may be--that the knowledge and horizons to which physicists are privy are treasures of the human intellect that must not be allowed to slip away.

It may be that no amount of public relations will help much, and that science is entering a fallow period, as it did during the dark ages in Europe when the seminal ideas of mathematics and physics were kept alive mainly by the Arabs. Physicists may face a long period of privation before we again see a scientific flowering like that of the 20th century. But to keep the flame alive should be the goal of any man or woman who values the chance given our species to understand our ultimate roots.

e$^+$ e$^-$ Physics

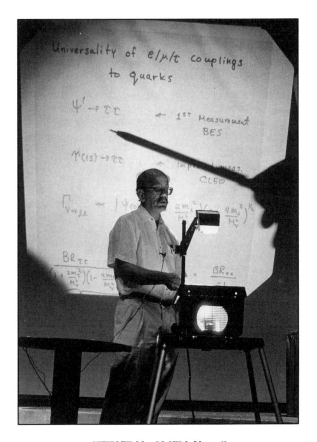

STEPHEN L. OLSEN, *Hawaii*

e⁺e⁻ PHYSICS

STEPHEN L. OLSEN

Department of Physics and Astronomy, University of Hawaii
2505 Correa Road, Honolulu, HI 96822, U.S.A.

1. Introduction

At this meeting there are over a hundred parallel session talks on various aspects of experimental e^+e^- physics—about a quarter of all of the contributions. In contrast, twenty years ago this summer, a few months before the famous November Revolution, of the 1110 papers submitted to the London meeting there were a grand total of three were on experimental e^+e^- physics. The explosion of experimental efforts associated with e^+e^- colliders is a result of the especially powerful capabilities that the e^+e^- technique has for probing many different areas of particle physics. In this review, I will try to illustrate this with results associated with those elementary particles of the Standard Model where the e^+e^- reaction has provided some unique insight. These include the τ-lepton, the c-, b-, and t-quarks, the Z-boson, and the Higgs scalar.

2. The τ–lepton.

I start by discussing recent measurements involving the τ-lepton. If there were no e^+e^- experiments, we might be aware of the τ-lepton's existence, but we would have virtually no knowledge of any of its properties.

Experimentally, the key question concerning the τ is the validity of the notion of *lepton universality*, i.e., are the τ interactions the same as those of the muon and the electron, as postulated by the Standard Model, or are there measureable differences between them? There are recent results that test universality in the couplings of the τ both to leptons and to quarks.

2.1. τ couplings to leptons.

The partial width for the decay $\tau^- \to \nu_\tau \ell^- \overline{\nu}_\ell$ is given by the familiar expression

$$\Gamma(\tau \to \ell \nu \bar{\nu}) = \frac{G^2_{\tau \to \ell \nu \bar{\nu}} m^5_\tau}{192\pi^3} \times f(m^2_\ell/m^2_\tau) r_{ew}, \tag{1}$$

where $f(m^2_\ell/m^2_\tau)$, a kinematic factor, and r_{ew}, the radiative correction, are well understood and nearly equal to one. Experimentally, the partial width is determined from $\Gamma(\tau \to \ell \nu \bar{\nu}) = Br(\tau \to \ell \nu \bar{\nu})/t_\tau$, the ratio of the τ's leptonic branching ratio and lifetime. From measurements of m_τ, $Br(\tau \to \ell \nu \bar{\nu})$, and t_τ, the decay constant $G_{\tau \to \ell \nu \bar{\nu}}$

can be determined. Universality implies that this should be the same as the Fermi constant $\mathcal{G}_F = G_{\mu \to e\nu\bar{\nu}}$, which is determined from analogous measurements for the muon.

In the past there has been some discrepancy between $G_{\tau \to \ell\nu\bar{\nu}}$ and $G_{\mu \to e\nu\bar{\nu}}$, the former being 2.4σ lower than the latter. This discrepancy was partially eliminated by a precision measurement of m_τ reported by the BES group in 1992, which was about two (of the previous) standard deviations lower than the world average at that time.[1] That measurement, based on 14 e vs μ events collected near the $\tau^+\tau^-$ threshold, gave $m_\tau = 1776.9^{+0.4}_{-0.5} \pm 0.2 \; MeV$.

At this meeting, the BES group has reported the results based on 64 events from a variety of different 1 vs 1 final states that were observed in the same threshold scan. The BES group's new result is consistent with their previous measurement but with substantially reduced statistical errors:[2]

$$m_\tau = 1776.96^{+0.18}_{-0.19} \; {}^{+0.20}_{-0.16} \; MeV. \tag{2}$$

The BES group's excitation curve for τ-pair production is shown in Fig. 1.

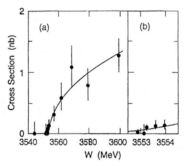

Figure 1: The cross section for $e^+e^- \to \tau^+\tau^-$ near threshold (from BES).

This result, coupled with the averages of new precision measurements from LEP,[3] SLD,[4] and CLEO[5] of the τ lifetime, $t_\tau = 291.6 \pm 1.7 fs$, and the leptonic branching ratio, $Br(\tau \to e\nu\bar{\nu}) = 17.66 \pm 0.08\%$, yields

$$\frac{G^2_{\tau \to \ell\nu\bar{\nu}}}{G^2_{\mu \to e\nu\bar{\nu}}} = 0.989 \pm 0.009, \tag{3}$$

quite close to universality expectations.

2.2. τ couplings to quarks.

The universality of the couplings between τ's and quarks can be tested in the leptonic decays of heavy quarkonium states. This summer, the BES group has reported the first measurement of the branching ratio for $\psi' \to \tau^+\tau^-$ and the CLEO group has reported a new result for the decay $\Upsilon(1S) \to \tau^+\tau^-$.

The partial widths for the decay of a vector $Q\overline{Q}$ state V to a pair of leptons is expected to be of the form

$$\Gamma(V\to\ell^+\ell^-) \propto |\psi(0)|^2(1 + \frac{2m_\ell^2}{M_V^2})(1 - \frac{4m_\ell^2}{M_V^2}), \tag{4}$$

where $\psi(0)$ is the value of the $Q\overline{Q}$ wave function at the origin. For the cases of $\ell = e$ and $\ell = \mu$, the kinematic factor involving m_ℓ/M_V is indistinguishable from unity and the implications of universality are:

$$\frac{Br(\psi'\to\tau^+\tau^-)}{0.39} = Br(\psi'\to\mu^+\mu^-) = Br(\psi'\to e^+e^-) \tag{5}$$

$$\frac{Br(\Upsilon(1S)\to\tau^+\tau^-)}{0.99} = Br(\Upsilon(1S)\to\mu^+\mu^-) = Br(\Upsilon(1S)\to e^+e^-). \tag{6}$$

The BES group's measurement of $Br(\psi'\to\tau^+\tau^-)$ is based on approximately 9K events with e vs μ, e vs π, e vs ρ, and μ vs ρ final states.[6] The result,

$$Br(\psi'\to\tau^+\tau^-) = (0.37 \pm 0.07 \pm 0.07)\%, \tag{7}$$

is in good agreement with expectations of lepton universality, namely $0.39\times <B_{\mu\mu,ee}> = 0.39 \times (0.84 \pm 0.10)\% = (0.33 \pm .04)\%$.

The new CLEO measurement of $Br(\Upsilon(1S)\to\tau^+\tau^-)$ is based on about 5K events with a 1 vs 1 topology, where one of the tracks is positively identified as an electron and the other is positively identified as not being an electron.[7] From these events, they derive the branching ratio

$$Br(\Upsilon(1S)\to\tau^+\tau^-) = (2.59 \pm 0.12 \,{}^{+0.13}_{-0.16})\%, \tag{8}$$

which agrees well with $0.99\times <B_{\mu\mu,ee}> = 0.99 \times (2.48 \pm 0.06)\% = (2.46 \pm 0.06)\%$. Previous results for this branching fraction were about 1.5σ higher than expectations.

3. The c–quark.

A central question in the area of heavy quark decays is the value of the decay constants for the psuedoscalar mesons. In non-relativistic terminology, these constants correspond to $q\overline{Q}$ wave function at the origin. There is poor agreement between different theoretical ways of calculating these decay constants. For example, in the case of the D_d and D_s mesons, calculated values of the f_D and f_{D_s} decay constants range from 90 to 350 MeV.[8] The B-meson psuedoscalar decay constants appear quadratically in expressions that relate measured quantities, such as $B\overline{B}$ mixing, to KM matrix elements and are, therefore, of considerable importance.

Experimentally, these constants can be determined from measurements of the purely leptonic decay branching ratios. The partial width for the decay of a psuedoscalar meson M to an $\ell\nu$ lepton pair is related to M's decay constant f_M by the expression

$$\Gamma(M \to \ell\nu) = \frac{G_F^2 |V_{ij}|^2}{8\pi} f_M^2 m_M m_\ell^2 (1 - \frac{m_\ell^2}{m_M^2})^2, \tag{9}$$

where V_{ij} is the KM matrix element between the quarks constituting M. For the case of the Cabbibo-favored D_s mesons, this translates into the decay branching ratios

$$Br(D_s \to \tau\nu) = 2.7\%(\frac{f_{D_s}}{200 \ MeV})^2 \tag{10}$$

and

$$Br(D_s \to \mu\nu) = 0.30\%(\frac{f_{D_s}}{200 \ MeV})^2. \tag{11}$$

Leptonic decays for B mesons are expected to have very small branching ratios, for example

$$Br(B_u \to \tau\nu) = 1.2 \times 10^{-4}(\frac{f_{B_u}}{200 \ MeV})^2 |\frac{V_{ub}}{0.005}|^2, \tag{12}$$

and it is unlikely that these will be measureable in the near future. The hope is that measurements of the more accessible charm decays will discriminate among the different theoretical approaches and result in more reliable estimates for the B-meson systems.

3.1. Measurements of f_{D_s}.

The first measurement of f_{D_s} was reported by the WA75 group using D_s mesons produced in nuclear emulsion targets.[9] They measure the p_T distribution for 1-prong muon decays of short lived particles and find six events with $p_{T\mu} > 0.9 GeV/c$, the position of the Jacobian peak for two body $\mu\nu$ decays. From this event yield they determine $f_{D_s} = 232 \pm 45 \pm 20 \pm 48 \ MeV$, where the last error is an estimate of the uncertainty introduced by the normalization, which is rather difficult in this experiments.

The CLEO group used D_s mesons produced in the reaction $e^+e^- \to D_s^* + X$, followed by the decay chain $D_s^* \to \gamma D_s$; $D_s \to \mu\nu$.[10] Using the missing momentum in the hadron jet that contains the D_s candidate as an estimate of the neutrino momentum, the CLEO group sees a D_s signal in a background-subtracted $M(\gamma\mu\nu) - M(\mu\nu)$ mass difference plot shown in Fig. 2. (The background is measured using electrons instead of muons: the electron decay is suppressed relative to that for muons by a factor of $(m_e/m_\mu)^2$.) From the number of signal events, the CLEO group determines the value $f_{D_s} = 344 \pm 37 \pm 52 \pm 42 \ MeV$, where the last error is due to the normalizing branching fraction uncertainty.

The BES group has recently reported preliminary results for a measurement of f_{D_s} using a sample of tagged D_s mesons produced in the reaction $e^+e^- \to D_s^+ D_s^-$ at $\sqrt{s} = 4.03 \ GeV$, just above $D_s\overline{D_s}$ threshold but below the threshold for $D_s^*\overline{D_s}$ production. In this measurement, they select events where one of the D_s meson decays is fully reconstructed, thus tagging the identity of the other particle in the

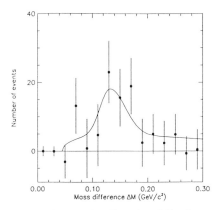

Figure 2: The background subtracted $M(\gamma\mu\nu) - M(\mu\nu)$ mass difference distribution (from CLEO).

event. The branching ratios for the decays of the tagged sample can then be determined with no ambiguities introduced by normalization or branching fraction uncertainties.

In a 22.3 pb^{-1} data sample, the BES group finds 95.1 ± 13.5 events where a $D_s \to \phi\pi$, $K^{*0}K$, or $K_S^0 K$, is reconstructed.[11] In this tagged sample are three candidate leptonic decay events; two are consistent with $D_s \to \tau\nu$ and one with $D_s \to \mu\nu$. This event rate, together with an assumption of lepton universality is used to determine the branching ratio $Br(D_s \to \mu\nu) = 1.4^{+1.2}_{-0.7}\%$, which corresponds to

$$f_{D_s} = 434^{+153+35}_{-133-33} MeV. \tag{13}$$

The small systematic errors reflect the insensitivity of this technique to the normalization difficulties inherent in the other measurement techniques, an advantage that comes at a cost of reduced statistics. Although the BES group plans to include more branching ratios in its tagging analysis, a precision measurement of these fundamental parameters will likely be a major task of a future tau-charm factory.

4. The inclusive $b \to s\gamma$ decay rate.

This summer the CLEO group reported the first measurement of the branching fraction for the inclusive decay $b \to s\gamma$.[12] This inclusive nature of the measurement is important for comparisons with theory; the measurement of such a rare inclusive process involving a γ-ray is probably only possible at an e^+e^- machine.

The interest in this decay stems from the fact that the lowest order Standard Model process for it is the penguin diagram shown in Fig. 3. Here, the particles in the internal loop are off mass shell so any particle combinations that conserve the appropriate quantum numbers at the vertices can contribute. Thus, for example, if there are charged Higgs scalars, as suggested by a number of models for spontaneous symmetry breaking, diagrams with a H^- replacing the W^- in Fig. 3 also contribute. In supersymmetric models there are a number of particle combinations that can occur in the loop, all of which could cause the actual rate to differ from that of the

Standard Model diagram. In addition, anomalous $\gamma - W$ couplings could also show up as a discrepancy in the rate for this process.

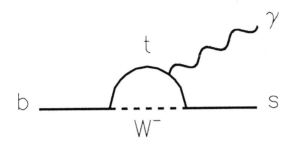

Figure 3: The lowest-order Standard Model graph for $b{\rightarrow}s\gamma$.

The signal for $b{\rightarrow}s\gamma$ is an excess of γ's in the energy range between 2.2 and 2.7 GeV from decays of B-mesons at the $\Upsilon(4S)$. The main conventional sources of such γ's are decay products of hadronic states in continuum $q\bar{q}$ events and the initial-state-radiation process $e^+e^-{\rightarrow}\gamma q\bar{q}$. The CLEO group used different techniques to supress these background processes including neural-net-generated cuts and a quasi-reconstruction procedure. The level of background remaining after the suppression procedure is estimated from data accumulated off the $\Upsilon(4S)$ resonance. The observed γ spectrum after continuum suppression is shown as the solid histogram in Fig. 4a. The "data points" in this figure are the remaining continuum background and Monte Carlo expectations for conventional B-meson decays. The clear excess of events in the signal region, evident in the background-subtracted histogram in Fig. 4b, is used to determine the inclusive branching ratio

$$Br(b{\rightarrow}s\gamma) = (2.32 \pm 0.51 \pm 0.29 \pm 0.32) \times 10^{-4}, \tag{14}$$

where the first systematic error is an additive error reflecting the uncertainty in the yield and the second is a multiplicative error reflecting model-dependent uncertainties. This result is in good agreement with the Standard Model calculation of $(2.75 \pm 0.80) \times 10^{-4}$ (for $m_t = 175$ GeV).

This result has been used to set limits on charged Higgs masses in a two-Higgs-doublet model, as a function of $\tan\beta$, the ratio of the vacuum expectation values v_1/v_2. These limits, shown in Fig. 5, exclude charged Higgs masses of less than ~ 270 GeV for all values of $\tan\beta$.

5. The Z gauge boson.

An enormous amount of experimental effort, both at LEP and at SLAC, is being applied to the study of the Z boson. In these experiments, Z bosons are produced in the s-channel annihilation process $e^+e^-{\rightarrow}Z{\rightarrow}f\bar{f}$, where $f\bar{f}$ denotes

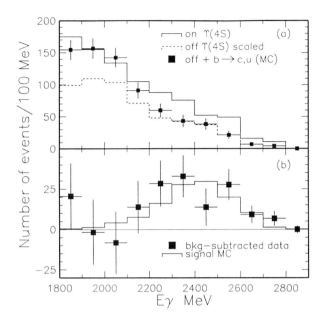

Figure 4: (a) The inclusive γ spectrum; (b) the background subtracted spectrum.

fermion-antifermion final states. The lowest-order Feynman diagram for this process is indicated in Fig. 6.

In the Standard Model, the fermion-Z couplings involve vector and axial-vector terms, and the corresponding coupling constants are all specified to within a single parameter $\sin^2 \theta_W$ as

$$g_V^f = \frac{T_3}{2} - Q_f \sin^2 \theta_W; \qquad g_A^f = \frac{T_3}{2}, \tag{15}$$

where T_3 is the third component of the fermion's weak ispin. ($T_3 = +\frac{1}{2}$ for the neutrinos and u-type quarks and $-\frac{1}{2}$ for the charged leptons and d-type quarks.) Experimentally, $\sin^2 \theta_W$ is nearly equal to $1/4$ ($\sin^2 \theta_W \simeq 0.23$), thus the vector coupling for the charged leptons is quite small ($g_V^\ell \sim -0.02$); the couplings to the quarks are larger ($g_V^d \sim -0.17$; $g_V^u \sim 0.10$).

5.1. The mass and total width of the Z

A fundamental parameter of the Standard Model is M_Z, which is related (at tree level) to $\sin^2 \theta_W$ as

$$M_Z^2 = \frac{\pi \alpha(M_Z^2)}{\sqrt{2} G_F \sin^2 \theta_W \cos^2 \theta_W}, \tag{16}$$

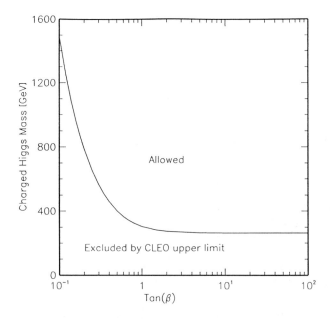

Figure 5: Charged Higgs mass limit vs $\tan\beta$ from $b\to s\gamma$.

where $\alpha(M_Z^2)$ is the fine structure constant evaluated at $Q^2 = M_Z^2$. The effects of vacuum polarization cause $\alpha(M_Z^2)$ to differ from its familiar zero momentum transfer value of $1/137.04$ by about 7%, namely $\alpha(M_Z^2) = 1/127.8$

The LEP energy scale has been calibrated to a precision level of ± 3 parts in 10^5 using resonant depolarization of transversely polarized beams.[13] A change in energy at this level corresponds to a change in the 27 km circumference of an orbiting particle of only $\sim 500\mu m$! Since this is approximately one quarter of the tidal changes in the LEP circumference caused by the gravitational attraction of the Moon, the LEP energy monitoring has been validated in a spectacular way by the observation of its variation as a function of the Moon's attitude (see Fig. 7).

Each of the four LEP groups have determined M_Z and the total width Γ_Z

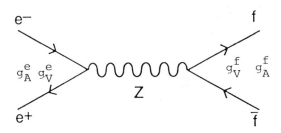

Figure 6: The lowest-order diagram for $e^+e^-\to Z\to f\bar{f}$.

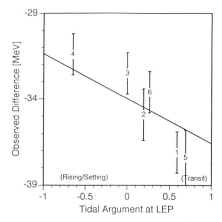

Figure 7: The LEP energy scale *vs* the tidal argument of the Moon.

from combined fits to the hadronic and lepton-pair cross sections made for different energy points around the peak of the Z resonance.[14] The radiative-corrected cross sections at each energy are fit to a modified Breit-Wigner expression

$$\sigma(E) = \sigma_0 \frac{E^2 \Gamma_Z^2}{(E^2 - M_Z^2)^2 + E^4 \Gamma_Z^2/M_Z^2}, \tag{17}$$

where σ_0 is the cross section at $E = M_Z$. The combined results for M_Z and Γ_Z from the four experiments are

$$M_Z = 91.1899 \pm 0.0044 GeV \qquad \Gamma_Z = 2.4971 \pm 0.0038, \tag{18}$$

where the correlations between the different measurements have been taken into account.

5.2. Forward-backward lepton asymmetries

The non-zero values of both g_V^e and g_A^e in the production vertex of the Z boson in Fig. 6 causes a difference between the left-handed ($g_L = g_V + g_A$) and right-handed ($g_R = g_V - g_A$) couplings, resulting in a net longitudinal polarization of the Z bosons. The interference of the vector and axial-vector couplings at the decay vertex results in a parity-violating forward-backward asymmetry in the decay of the polarized Z's. The differential cross section for $e^+e^- \to f\bar{f}$ (for unpolarized e^+ and e^- beams) at the Z peak is

$$\frac{d\sigma}{d\cos\theta} \propto [1 + \cos^2\theta + 2A_e A_f \cos\theta], \tag{19}$$

where the parameters A_e and A_f ($\equiv 2g_V^f g_A^f/[(g_V^f)^2 + (g_A^f)^2]$) denote the polarization of the Z and the analyzing power of the $Z \to f\bar{f}$ decay, respectively. This product of coefficients is measureable as a forward-backward asymmetry for the fermion f,

$$A_{FB}^f = \frac{\sigma_f(\cos\theta > 0) - \sigma_f(\cos\theta < 0)}{\sigma_f(\cos\theta > 0) + \sigma_f(\cos\theta < 0)} = \frac{3}{4} A_e A_f. \tag{20}$$

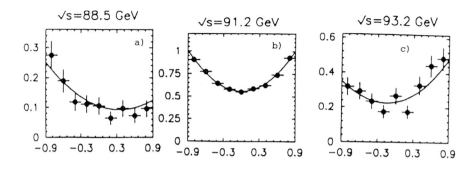

Figure 8: ALEPH measurements of $e^+e^- \to \mu^+\mu^-$ below (a), at (b), and above (c) the Z peak.

In the Standard Model, $A_{f=\ell}$ is the same for all charged leptons and $A_{FB}^\ell = 3/4 A_\ell^2$.

Since, as noted above, g_V^ℓ is close to zero, A_ℓ is small and the forward-backward asymmetry for leptons, which is quadratic in A_ℓ, is a very small number. This is illustrated in Fig. 8, which shows ALEPH measurements of the $e^+e^- \to \mu^+\mu^-$ differential cross section below, on, and above the Z peak.[15] Forward-backward asymmetries are readily apparent in the off-peak distributions, reflecting the interference of the Z-boson's axial-vector coupling with the vector EM coupling of the photon. At the peak, where only the Z couplings interfere, the asymmetry is barely apparent. The experimental groups determine the peak asymmetry from fits to the energy dependence; the L3 group's measurements and fit[16] are shown in Fig. 9. The combined results from the four LEP experiments using the $e, \mu,$ and τ channels are[14]

$$< A_{FB}^\ell > = 0.0170 \pm 0.0016; \implies A_\ell = 0.151 \pm 0.006. \qquad (21)$$

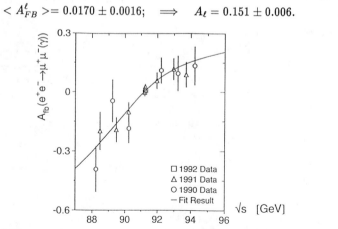

Figure 9: L3 results for the energy variation of A_{FB}^μ.

5.3. Forward-backward b-quark asymmetry

Since the b-quark vector coupling g_V^b is much larger than that for the leptons, the b-type quark forward-backward asymmetry is considerably bigger than that for leptons. Moreover, the value of A_b is rather insensitive to $\sin^2\theta_W$—ten times less sensitive than A_ℓ—so, in practice, b-type quark asymmetry measurements at the Z peak are sensitive to A_ℓ.

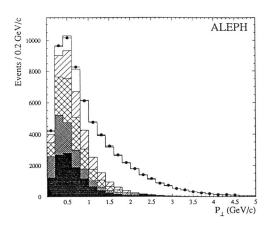

Figure 10: The $p_{T\mu}$ distribution for muons as measured by ALEPH.

The reaction $e^+e^- \to b\bar{b}$ can be distinguished by leptons originating from the semileptonic decays and/or by the displacement of the decay vertex from the beam-beam collision point. The former technique is illustrated in Fig. 10, where the ALEPH group's measurement of the transverse momentum distribution of muons (relative to the thrust axis) is shown.[17] Muons from b-quark semileptonic decays dominate the distribution for $p_{T\mu} > 1.25$ GeV. The asymmetry in the ALEPH group's on-peak $b\bar{b}$ differential cross section, shown in Fig. 11a, is clearly visible. The energy dependence of the asymmetry is shown in Fig. 11b.

The combined LEP results are[18] A_{FB}^b ($= 3/4 A_e A_b$) $= 0.096 \pm 0.004 \pm 0.002$. Using the nominal value of $A_b = 0.935$, we extract an independent measurement of $A_e = 0.137 \pm 0.007$.

5.3. τ polarization measurements

Even in the case of unpolarized incident beams, the final-state fermions in $e^+e^- \to Z \to f\bar{f}$ are polarized, with a $\cos\theta$ dependence of the polarization given by

$$\mathcal{P}_f(\cos\theta) = \frac{A_f + A_e \frac{2\cos\theta}{1+\cos^2\theta}}{1 + A_f A_e \frac{2\cos\theta}{1+\cos^2\theta}}. \tag{22}$$

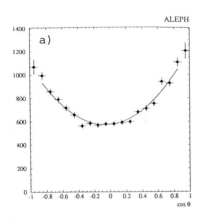
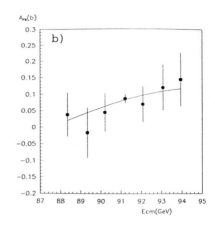

Figure 11: (a) The angular distribution for $e^+e^- \to b\bar{b}$. (b) The energy dependence of A_{FB}^b. Both distributions are from ALEPH.

For the case where $f = \tau$, the subsequent decay of the τ provides a convenient and well understood analyzer for the τ polarization. Moreover, the different dependence of the angular dependence on A_τ and A_e permits nearly independent determinations of A_τ and A_e. The angular dependence of the polarization is apparent in the ALEPH group's measurements[19] of $\mathcal{P}_\tau(\cos\theta)$ shown in Fig. 12.

Averaging Eq. 22 over all $\cos\theta$ gives $< \mathcal{P}_\tau > = -\mathcal{A}_\tau$. The forward-backward polarization asymmetry, defined as

$$A_{FB}^{pol} = \frac{< \mathcal{P}_\tau(\cos\theta > 0) > - < \mathcal{P}_\tau(\cos\theta < 0) >}{< \mathcal{P}_\tau(\cos\theta > 0) > + < \mathcal{P}_\tau(\cos\theta < 0) >} = -\frac{3}{4}A_e, \tag{23}$$

provides a measurement of A_e. The combined LEP results from τ polarization measurements are[20]

$$A_\tau = 0.150 \pm 0.010 \qquad A_e = 0.120 \pm 0.012. \tag{24}$$

5.3. Measurement of A_{LR} using polarized electrons in the SLC

The SLC has the unique capabilty for provided longitudinally polarized electrons to the SLD intersection point. With the application of a strained GaAs crystal photocathode electron source, the polarization during the past year has averaged 63%. The polarization-dependent asymmetry A_{LR}, the normalized difference in the cross section for left-handed and right-handed electrons, for *any* on-peak, s-channel process is equal to A_e:

$$A_{LR} = \frac{\sigma_L - \sigma_R}{\sigma_L + \sigma_R} = A_e = \frac{1}{\mathcal{P}_e} \frac{N_L^{ev} - N_R^{ev}}{N_L^{ev} + N_R^{ev}}, \tag{25}$$

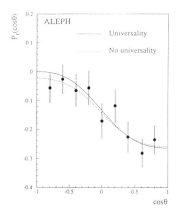

Figure 12: The angular dependence of the ALEPH group's τ polarization measurements.

where $N_{L,R}^{ev}$ are the number of hadronic, μ-, and τ-pair events observed with left- and right-handed helicity electron beams, respectively. Since the normalized cross section difference depends only on the polarization direction and is independent of the fiducial volume or efficiency of the detector, the A_{LR} measurement is essentially a counting experiment and relatively free of systematic errors. Of course, the precision of the result depends directly on the precision with which the polarization is known. The SLD group measures the polarization with a 1.3% relative precision using Compton scattering from a laser beam in a polarimeter located downstream of the intersection region.[21]

The resulting asymmetry is $A_{LR}(= A_e) = 0.1637 \pm 0.0075$. This is 2.1$\sigma$ higher than the combined LEP results of $A_\ell = 0.144 \pm 0.006$.

6. $\sin^2 \theta_W$ and the t-quark mass.

Until now, the discussion of the Z-boson has been in the context of the tree-level diagram in Fig. 6. However, experimental measurements include the effects of all higher order processes. Higher-order processes involving photons can be evaluated with considerable precision and the data presented by the experimental groups are corrected for these effects. Corrections can also be computed for higher-order processes involving gluons, although here the poorly known quark-gluon coupling strength α_s introduces some uncertainties. Higher-order electroweak processes include virtual t-quarks and Higgs bosons; the unknown masses of these particles introduce uncertainties in the calculations of these diagrams.

The effect of these uncertainties is a modification of the tree-level Standard Model relation given in Eq. 16 to the form

$$M_Z^2 = \frac{\pi \alpha(M_Z^2)}{\sqrt{2} G_F \sin^2 \theta_W \cos^2 \theta_W} \times \frac{1}{1 - \Delta r}, \tag{26}$$

where Δr is an electro-weak correction factor that depends on the t-quark mass, m_t, the Higgs mass, m_H, and differs according to the experimental definition of $\sin^2 \theta_W$.

6.1. $\sin^2 \theta_W^{eff}$ *and* $\sin^2 \theta_W^S$.

In the Z boson measurements described above, where g_V^ℓ and g_A^ℓ are readily measureable, it is convenient to use the *effective* Weinberg angle defined as

$$\sin^2 \theta_W^{eff} = \frac{1}{4}\left(1 - \frac{g_V^\ell}{g_A^\ell}\right). \tag{27}$$

On the other hand, if one is using measurements of the W-mass to determine the Weinberg angle, it is more convenient to use the *Sirlin* definition

$$\sin^2 \theta_W^S = \left(1 - \frac{M_W^2}{M_Z^2}\right). \tag{28}$$

The m_t dependence for $\sin^2 \theta_W$ for the two definitions is shown in Fig. 13.[22] From the figure it can be seen that the two definitions of $\sin^2 \theta_W$ differ, especially in the $m_t \sim 175$ GeV region favored by the CDF group's recent direct observation of the t-quark.[23]

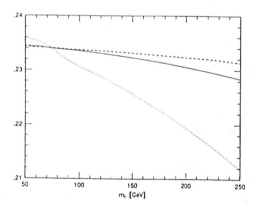

Figure 13: The dependence of $\sin^2 \theta_W$ on the m_t for cases where it is inferred from $\Gamma_{b\bar{b}}$ (dashed), A_ℓ (solid), and M_W/M_Z (dotted).

Figure 14 shows a summary of the $\sin^2 \theta_W^{eff}$ measurements from LEP and SLD.[24] There is good agreement among the many different determinations used by the LEP groups; the average value from LEP is $\sin^2 \theta_W^{eff}(LEP) = 0.2322 \pm 0.0005$. A comparison this value with the solid curve in Fig. 13 indicates a t-quark mass in the vicinity of $m_t \simeq 160~GeV$. In contrast, the SLD group's A_{LR} measurement translates to $\sin^2 \theta_W^{eff}(SLD) = 0.2294 \pm 0.0010$, and prefers a t-quark mass around $m_t \simeq 230~GeV$. The W mass measurements from CDF,[25] D0,[26] and UA2,[27] when combined, give a value of $M_W = 80.22 \pm 0.16~GeV$, corresponding to $\sin^2 \theta_W^S = 0.2261 \pm 0.0031$. This value crosses the dotted line in Fig. 13 around a t-quark mass value of $m_t \simeq 150~GeV$, consistent with the LEP measurements and the CDF group's direct observations.

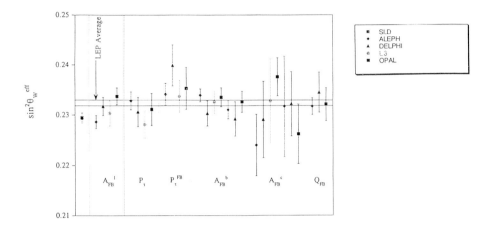

Figure 14: Values of $\sin^2 \theta_W^{eff}$ derived from various LEP measurements and the SLD measurement of A_{LR}.

6.2. Combined fit to the data.

The LEP Electroweak Working Group[34] has made a global fit of average LEP measurements of the Z lineshape parameters, partial widths, and asymmetries, together with the M_W measurements, the neutral current νN cross section, and the A_{LR} measurement from SLD. Fits were made to Standard Model constraints with the Higgs mass fixed and m_t and $\alpha_s(M_Z^2)$ left as free parameters. The results from the fit are

$$m_t = 178^{+11}_{-11}{}^{+18}_{-19}~GeV \quad \text{and} \quad \alpha_s(M_Z^2) = 0.125 \pm 0.005 \pm 0.002, \qquad (29)$$

with a $\chi^2/(d.o.f.) = 15/12$. Here the central values and the first error refer to $M_H = 300~GeV$ and the second errors reflect the range of changes that occur when M_H is varied from 60 to 1000 GeV. The main contributors to the χ^2 of this fit are A_{LR}, which is 2.6σ high, and $\Gamma_{b\bar{b}}$, which is 2.2σ high. (If the A_{LR} measurement is left out of the fit, the value of the t-quark mass drops to $m_t = 171~GeV$ and the $\chi^2/(d.o.f.)$ improves to 7.7/11.) The χ^2 curves for $M_H = 60$, 300, and 1000 GeV

are shown in Fig. 15, where it can be seen that the fit likes smaller values of the Higgs mass.

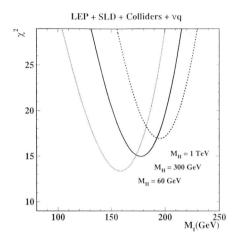

Figure 15: Results for m_t from the fit of Z, W, νN, and A_{LR} measurements.

The results of the fit are in excellent agreement with the mass determined from the direct observation of the t-quark,[23] $m_t = 174 \pm 10^{+13}_{-12}$ GeV. This agreement between the direct measurement of the t-quark mass and its value inferred from higher-order contributions to the properties of the Z and W gauge bosons is a dramatic confirmation of the Standard Model.

7. Search for the Higgs particle

The elucidation of the spontaneous symmetry-breaking mechanism responsible for mass generation is a fundamental goal of particle physics today. In the Minimal Standard Model (MSM), all particle masses are generated by one doublet of Higgs fields that produces a single neutral Higgs scalar particle. If its mass is below about 65 GeV it could be produced at LEP via the bremsstrahlung process $e^+e^- \to Z \to Z^* H^0$. Given the mass of the MSM Higgs, its production cross section and decay branching ratios are specified; for $M_H < 2M_W$, the decays $H \to b\bar{b}$ and $\tau^+\tau^-$ are expected to dominate. LEP-200 could access higher masses via the process $e^+e^- \to Z^* \to ZH^0$, where the final-state Z is on-mass-shell. MSM Higgs particles with masses as high as $\sim \sqrt{s} - M_Z - 10$ GeV could be produced with measureable cross sections.

7.1. Limits on the MSM Higgs from LEP

The LEP groups have searched for $Z^* H$ final states using leptonic decays to tag the Z^*. Here the decay $Z^* \to \nu \bar{\nu}$, which is important because of its large branching ratio, is seen as an event with a massive hadronic system (from the $H \to b\bar{b}$ decay) with a large energy and transverse-momentum imbalance. The backgrounds in this channel come primarily from ordinary Z decays where there are either energetic final-state neutrinos where particles are lost in cracks of the detector. The LEP groups also use the $Z^* \to e^+ e^-$ and $\mu^+ \mu^-$ channels, here the main source of background are two-photon processes.

Figure 16 shows the results of the DELPHI group's search;[28] the curves indicate the expected numbers of events for the different Z^* decay channels as a function of M_H and the horizontal line indicates the experimental upper limit that obtains from the non-observation of any event candidates in a sample of 730K hadronic Z decays. The resulting 95% confidence level mass limit is $M_H > 58.3 \ GeV$. The other LEP groups derive similar limits; a reasonable estimate of the combined (95% CL) limit from all four groups is $M_H > 64.5 \ GeV$.

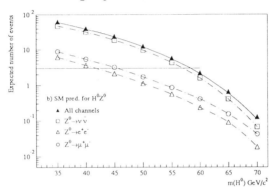

Figure 16: DELPHI results on the MSM Higgs search.

7.2. Future prospects for Higgs searches

Although the LEP experiments expect to more than triple their data samples, the steeply falling cross section will result in only a $\sim 5 \ GeV$ increase in the accessible M_H range. Significant further improvements will await LEP-200 where the expected 190 GeV cm energy will extend the accessible mass range to about $M_H \sim 90 \ GeV$.

Experiments at the Large Hadron Collider (LHC) will be well suited to find the MSM Higgs if its mass is greater than twice M_Z in which case the dominant decay $H \to ZZ$ will have a clean experimental signature. If, however, the Higgs mass is below $2M_Z$, the dominant decay modes will be $H \to b\bar{b}$ and $\tau\tau$ which will be difficult to distinguish in hadron collisions. LHC experiments plan to cover the lower mass range using the $H \to \gamma\gamma$ and $H \to ZZ^*$ decay channels, both of which have low rates and large backgrounds.

This lower mass region would be better covered by a high energy $e^+ e^-$ collider. Considerable effort has gone into the development of techniques to build a high luminosity $(\sim 10^{33} cm^{-2} s^{-1})$ linear e^+e^- collider with a cm energy of 500 GeV and above.[29] In 500 GeV e^+e^- collisions, two processes contribute significantly to Higgs production: the bremsstrahlung process $e^+e^- \to HZ$ discussed above, and the WW fusion process $e^+e^- \to \nu\bar{\nu}H$. Both processes provide clean experimental signatures for Higgs production and in a cannonical year (i.e. 10 fb^{-1}) at 500 GeV an experiment at a linear e^+e^- collider would comfortably cover the mass region $M_H < 200$ GeV that is difficult for the LHC.[30]

7.2. Non-MSM Higgs

There is a general suspicion that the spontaneous symmetry-breaking mechanism responsible for mass generation is more complex than the MSM Higgs discussed above. A commonly considered variant is a model with two Higgs doublet fields that produces five Higgs scalar particles: h^0, H^0, A^0, and $H^\pm$. A characteristic of these models is that one of the Higgs doublet fields couples only to the $T_3 = -\frac{1}{2}$ fermions while the other couples only to those with $T_3 = \frac{1}{2}$. The ratio of vacuum expectation values for the neutral components of the two doublets, $v2/v1$, is a parameter in the model and usually denoted by $\tan\beta$.

The two-Higgs-doublet scenario occurs in the Minimal Super Symmetric Model, MSSM. Here, all five Higgs masses and couplings depend only on two parameters, usually taken as m_A and $\tan\beta$. This model has the interesting feature of an upper limit on the mass of the h^0 of $m_{h^0} < 146$ GeV. Moreover, in specific versions of these models this limit is found to be rarely saturated; it is typically found to be in the range $70 < m_{h^0} < 110$ GeV, tantalizingly close to the capabilities of LEP-200.[31]

Figure 17 illustrates the regions in the $\tan\beta - m_A$ parameter space ruled out by LEP and accessible by LEP-200. Also indicated in the figure are those regions of the parameter space that could be accessed by LHC experiments.[30] The LHC experiments could rule out large values on m_A, but only by means of a $h^0 \to \gamma\gamma$ search with ideal capabilities and a $100 fb^{-1}$ data sample. They could also access some large $\tan\beta$ regions by looking for $A, H \to \tau\tau \to e\mu$, another very difficult experiment.[32] Even with full capabilities, the combined LEP-200/LHC reach still leaves a large region of the parameter space uncovered.

In contrast, a 500 GeV linear e^+e^- collider would easily cover all of the parameter space and either confirm or rule out decisively the MSSM hypothesis. Moreover, if some of the MSSM Higgs particles are seen, the cleanliness of the signals inherent in the e^+e^- production process will permit careful study of the properties of the Higgs particles.[30]

8. Summary and Future Prospects

In addition to the continued operation of BES, CESR, LEP, SLC TRISTAN, and VEPP4, there are a number of new e^+e^- facilities that I am aware of that are under construction, being planned, or being considered. In order of increasing energy:

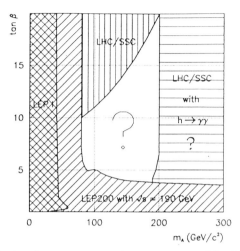

Figure 17: Regions in $\tan\beta$ *vs* m_A parameter space covered by LEP, LEP-200, and the LHC.

ϕ-factories: There are ϕ-factories under construction at Frascati and Novosibirsk.

τ-charm factories: IHEP-Beijing and Argonne National Laboratory are preparing proposals for *tau*-charm factories.

B-factories: KEK and SLAC are both constructing asymmetric B-factories. CESR will be upgraded to a symmetric B-factory.

Z-factories: Although there are no specific proposals, a post-LEP Z-factory is an obvious and frequently mentioned possibility.[33]

LEP-200: LEP-200 is scheduled for operation at cm energies sufficient for producing W^+W^- pairs around 1996.

New Linear Collider: As I tried to point out above, a high luminosity e^+e^- collider in the 500 GeV cm energy region would nicely complement the LHC and have real discovery potential.

With this impressive assortment of new, proposed, and dreamt-of facilities, physics studies using e^+e^- collisions will continue to study a broad range of important topics for the foreseeable future.

9. Acknowledgements

I wish to express my gratitude to Sally Seidel and rest of the conference organizers for an interesting and congenial meeting in a very pleasant setting, and for inviting me to give this talk, which forced me to learn about the large amount of interesting work going on in this field. I especially thank Tom Rizzo for the enthusiastic help he gave me in preparing this talk, suggesting topics to be discussed, and for educating me on the intricasies of Z-boson physics. In addition I received a considerable amount of detailed instruction from Young Kee Kim and my Hawaii colleagues Sandip Pakvasa and Xerxes Tata.

1. J.Z. Bai et al. (BES), *Phys. Rev. Lett.* **69** (1992) 3021.

2. E. Soderstrom (BES), *these proceedings.*

3. A. Lusiani (ALEPH), *ibid.;* P. Weber (OPAL) *ibid.*

4. S. Hedges (SLD), *ibid.*

5. R. Kass (CLEO) *ibid.*

6. Li Jin (BES) *Proceedings of the 27th International Conference on High Energy Physics, Glasgow 1994.*

7. D. Cinabro et al. (CLEO), *contributed paper to the 27th International Conference on High Energy Physics, Glasgow 1994.*

8. Jonathon L. Rosner, *Proceedings of TASI-90, Boulder Colorado (1990); World Scientific.*

9. S. Aoki et al. (WA75), *Prog. Theor. Phys.* **89** (1993) 131.

10. D. Acosta et al. (CLEO), CLNS93/1238, CLEO93-14 (1993) *Submitted to Phys. Rev. D.*

11. M. Kelsey (BES), *these proceedings.*

12. B. Barish et al. (CLEO), *contributed paper to the 27th International Conference on High Energy Physics, Glasgow 1994.*

13. L. Arnaudon et al. (LEP Energy Group), *Phys. Lett.* **B307** (1993) 187.

14. P. Clarke et al. (LEP Lineshape and FB Asymmetry Working Group), LEPLINE/94-01, *contributed paper to the 27th International Conference on High Energy Physics, Glasgow 1994.*

15. D. Buskulic et al. (ALEPH), CERN-PPE/93-52 (1993) *Submitted to Z. Phys. C.*

16. M. Acciarri et al. (L3), CERN-PPE/94-45 (1994) *Submitted to Z. Phys. C.*

17. D. Buskulic et al. (ALEPH), CERN-PPE/94-17 (1994) *Submitted to Z. Phys. C.*

18. D. Abbaneo et al. (LEP Heavy Flavor Working Group), LEPHF/94-03, *contributed paper to the 27th International Conference on High Energy Physics, Glasgow 1994.*

19. D. Buskulic et al. (ALEPH), CERN-PPE/93-39 (1993) *Submitted to Z. Phys. C.*

20. J. Harton et al. (LEP tau-polarization Working Group), LEPTAU/94-02, *contributed paper to the 27th International Conference on High Energy Physics, Glasgow 1994.*

21. K. Abe et al. (SLD), *Phys. Rev. Lett.* **73** (1994) 25.

22. Wolfgang Hollik *Proceedings of the 16th International Symposium on Lepton and Photon Interactions,* Ithaca, NY 1993; P. Drell and D. Rubin editors, AIP

press, p.352.

23. F. Abe et al. (CDF), *Phys. Rev. Lett.* **73** (1994) 225.

24. A. Lath, *private communication.*

25. R.M. Keup (CDF), *these proceedings.*

26. A.I. Mincer (D0), *ibid.*

27. J. Alitti et al. (UA2), *Phys. Lett.* **B276** (1992) 354.

28. R. Møller (DELPHI), *these proceedings.*

29. R. Sieman, *ibid.*

30. Patrick Janot*Proceedings of the 2nd International Workshop on Physics and Experiments with Linear e^+e^- Colliders,* Waikoloa, HI 1993; F.A. Harris, S.L. Olsen, S. Pakvasa and X. Tata editors, World Scientific, p.192.

31. G.L. Lane, *ibid., p.218.*

32. A. Rubbia, *ibid., p.610.*

33. T. Omori, *ibid., p.791.*

34. A. Blondel et al. (LEP Electroweak Working Group), LEPEWWG/94-02, *contributed paper to the 27th International Conference on High Energy Physics, Glasgow 1994.*

Current Directions in String Theory and Abstract Supersymmetry

JEFFREY A. HARVEY, *Chicago, Enrico Fermi Institute*

CURRENT DIRECTIONS IN STRING THEORY AND ABSTRACT SUPERSYMMETRY

JEFFREY A. HARVEY

Enrico Fermi Institute and Department of Physics, University of Chicago
5640 S. Ellis Ave., Chicago, IL 60637, USA

ABSTRACT

I discuss some current thoughts on how low-energy measurements and consistency constrain theories at high energies with emphasis on string theory. I also discuss some recent work on the dynamics of supersymmetric gauge theories.

1. Introduction

In her talk at this meeting Angela Olinto used both broad and fine brush strokes to paint her view of cosmology in the next millennium. In this talk I intend to use only the broadest of brushes, perhaps spray painting would be a better analogy, to discuss our ability to probe Planck scale physics using only low-energy experiments and self-consistency. I will also discuss in very general terms some exciting recent results on the dynamics of supersymmetric gauge theories.

There is a great deal of doom and gloom in particle physics at the moment as a result of the cancellation of the SSC and the meager job market for young particle physicists. There is also great frustration at the success of the Standard Model and our seeming inability to probe what lies beyond it. This has led to statements in the popular press implying that particle physics is approaching a dead end as a field of scientific inquiry. While there is certainly cause for concern, I think such pronouncements are short-sighted and seriously underestimate the ingenuity of both theorists and experimentalists over the long term. To counter this I would like to discuss some reasons for optimism and some new developments which show that our bag of tricks is not yet exhausted.

One valid concern about current particle theory is the vast chasm which has opened up between experiment and some of the theoretical frontiers, especially string theory. There is a tendency in many quarters to regard string theory as a failure, not because of any internal problems, but because it only addresses physics at the Planck scale and hence is inherently untestable in the foreseeable future. While direct tests are certainly hard to imagine, I would like to discuss a number of indirect ways in which physics at the Planck scale and particularly string theory are already playing a significant role in our view of low-energy physics.

Before beginning let me mention a remark which I have heard with varying attributions to the effect that the problem is not that we take our theories too seriously, but rather that we do not take them seriously enough. Examples abound, ranging from black holes to quarks to gauge theories and electroweak theory. In each case the theory was initially viewed as either a mathematical curiosity or abstraction or as a toy model.

In each case the theory was more real than physicists could bring themselves to believe. As a mature field, particle physics is more tightly constrained than newer fields and often a small amount of data is sufficient to develop a very rigid theory. Electroweak theory is a particularly good example of this. I would like to encourage you to take string theory as seriously as I think it deserves to be taken, not as a mathematical abstraction or a toy model of quantum gravity, but rather as a real theory of the world that may shed light on real phenomena.

2.   Large-scale small-scale connections in particle physics

2.1.   The Standard Model

We are currently both blessed and cursed by the incredible success of the Standard Model as reviewed at this meeting by Jon Rosner.[1] In spite of its success, it is certainly true that it is not a final theory of everything, but rather only a low-energy effective theory of something, namely the light particles which include the three generations of quarks and leptons, the gauge bosons of $G = SU(3) \times SU(2) \times U(1)$, and presumably one or more Higgs bosons. There are many reasons to believe that the Standard Model is only a stepping stone on the way to a more complete theory. First of all, it does not include gravity. Second, it almost certainly does not exist formally as a quantum field theory due to the lack of asymptotic freedom of some of the coupling constants. Finally, it has too many loose ends, requiring the specification of some 20 odd parameters. One of the main goals of particle physics at the end of the 20th century is to figure out what lies beyond the Standard Model.

Since we do not believe the Standard Model is a final, self-contained theory, we should view it only as a low-energy effective theory that describes physics in an approximate (but nonetheless remarkable accurate) way at low-energies. According to the philosophy of effective field theory[2] if the Standard Model is effective below a scale Λ then in the effective Lagrangian we should write down all interactions involving the light degrees of freedom which are compatible with the symmetries of the problem. Furthermore, all dimensionfull couplings should be given by appropriate powers of Λ up to factors which are (roughly) of order one. Very schematically this gives

$$\mathcal{L}_{eff} = (\Lambda^4 + \Lambda^2 \phi^2) + ((D\phi)^2 + \bar{\psi} \not{D} \psi + F_{\mu\nu} F^{\mu\nu} + \phi \bar{\psi}\psi + \phi^4) + (\frac{\bar{\psi}\psi\bar{\psi}\psi}{\Lambda^2} + \cdots) \quad (1)$$

in a notation where ϕ stands for generic scalar field, ψ for generic fermion fields, and $F_{\mu\nu}$ for the gauge fields. The first set of terms have positive powers of Λ, they are the most relevant terms in understanding the low-energy theory. They consist of the cosmological constant term and a mass term for any scalar particles. Note that a mass term for fermions is forbidden in the Standard Model by gauge invariance. The second set of terms are independent of Λ (or have at most log dependence on Λ) and consist of all the interactions of the usual Standard Model except for the Higgs mass term. The final set of terms is really an infinite expansion which includes all terms with negative powers of Λ. These terms are non-renormalizable in the usual sense, and are irrelevant in very low-energy processes, but are nonetheless there in a generic low-energy effective field theory.

Now one beautiful thing about the standard model is that *all* the renormalizable terms which are consistent with the gauge symmetries are present. It is true that some of the couplings are small for poorly understood reasons (e.g. the electron and up and down quark Yukawa couplings), but we are not forced to set any couplings to zero in order to agree with experiment.

The natural question to ask is, what is the value of Λ? Historically the first answer was that $\Lambda = \Lambda_{GUT} \sim 10^{15} GeV$ based on unification of the gauge couplings as well as on the absence of proton decay, flavor changing neutral currents, and other non-standard model processes. The point being that such effects are generally contained in the non-renormalizable interactions but are sufficiently small if suppressed by powers of Λ_{GUT}. This answer leads to two great embarrassments which involve the two relevant terms in $\mathcal{L}_{eff}$. The first is the cosmological constant problem. The value of the cosmological constant is *much* less than Λ_{GUT}^4 for reasons that are completely mysterious. The relevance of this term is clear. If the cosmological constant were of order Λ_{GUT}^4 the universe would have come and gone in a Planck time without any physicists to worry about the problem. The second embarrassment is usually called the hierarchy problem, namely that the Higgs mass is not Λ_{GUT}^2 but must be more of order M_W^2. There is of yet no good answer to the first problem, but at least we can appeal to an ignorance of quantum gravity. The second problem is a problem of particle physics and most be faced up to.

If we are not allowed unnatural fine tuning then the only possibility is that Λ is not Λ_{GUT} but rather is not far from the weak scale, say generically $\Lambda \sim 1TeV$. There are currently two main speculations as to the new physics at this scale. The first is that there are new strong interactions, that is technicolor, top quark bound states or whatever; the second is that this is the scale of new supersymmetric particles. Whatever the new physics, we know that it must have special features. In particular, we know that many of the non-renormalizable interactions in $\mathcal{L}_{eff}$ cannot be present if they are suppressed only by powers of $\Lambda \sim 1TeV$. Of course this is not a surprise. It is well known that technicolor and supersymmetric models are strongly constrained by the demands that proton decay, CP violation and flavor changing neutral currents be compatible with present experimental limits. But it is worth emphasizing that this is a positive feature. It tells us that the new dynamics at the TeV scale is not generic but must have special structures or symmetries which forbid these effects.

2.2. The MSSM

Currently the most popular proposal for going beyond the standard model is based on supersymmetry. There is now a standard minimal model incorporating supersymmetry, the MSSM.[3] We should first understand how supersymmetry solves the naturalness problem. In the early days of supersymmetry one sometimes heard the following idea. Supersymmetry if unbroken requires the equality of fermion and boson masses. Since fermions are required to be massless in the Standard Model before weak symmetry breaking, bosons would also have to be massless before electroweak breaking. A Higgs mass of order the weak scale would then be generated provided that supersymmetry was effectively broken near the weak scale.

This idea does not work in the most naive sense. The reason is as follows. In $N = 1$ supersymmetry chiral fermions are paired with complex scalars. Thus one standard Higgs doublet is paired with a chiral fermion doublet. Now adding such a multiplet to the Standard Model is not possible without upsetting the delicate cancellation of anomalies between chiral fermions. So what is done is to add two multiplets corresponding to two Higgs doublets H_1, H_2 with opposite hypercharge. Their fermion partners are then non-chiral and there is no problem with anomalies. But then there is a supersymmetric and gauge invariant mass term allowed for the multiplets which in a supersymmetric notation takes the form

$$W = \mu H_1 H_2. \tag{2}$$

Since this term is supersymmetric and gauge invariant there is no reason for it not to be of order the Planck scale or whatever scale comes beyond the MSSM and thus the hierarchy or naturalness problem is not really solved. In the context of the MSSM this is usually called the " μ problem ".

It is tempting to try to forbid this term by some symmetry, but this also leads to problems, in particular with $\mu = 0$ the MSSM has a standard Peccei-Quinn axion which is ruled out experimentally. So we need μ to be non-zero and of order the weak scale. This suggests that the value of μ should be connected with supersymmetry breaking, even though it is by itself supersymmetric.

There are in fact solutions to the μ problem in supergravity and string theory where this is precisely what happens. μ is zero initially for reasons that have to do with the precise high-energy structure of the theory and a non-zero value is induced only through supersymmetry breaking effects.[6] Thus at least in one class of models the consistency of the MSSM requires the existence of special features at the Planck scale.

It was argued earlier that the theory beyond the standard model would have to have special features in order to ensure that baryon number violation and flavor changing neutral currents (FCNC) are sufficiently small. In the MSSM one can say quite specifically what is required. The presence of scalar partners of quarks (squarks) means that there are renormalizable couplings which violate baryon number. These must be very small to be phenomenologically acceptable. The simplest solution is to use a discrete symmetry to force them to be zero. This is conventionally done by defining a Z_2 discrete symmetry called R-parity which has eigenvalue $(-1)^{3(B-L)+2S}$ when acting on a state of baryon number B, lepton number L and spin S. Equivalently, it is $+1$ on all standard model particles and -1 on all their superpartners. R parity also forbids all renormalizable couplings which violate lepton number. R parity also implies that the lightest supersymmetric partner (LSP) is absolutely stable since there is nothing it can decay into while preserving R-parity. Thus R-parity has very important phenomenological consequences. There are other discrete symmetries which do not forbid all these couplings but which nonetheless are phenomenologically acceptable. Perhaps the most attractive is a Z_3 matter parity discussed by Ibanez and Ross.[4] The central position played by such discrete symmetries makes it important to ask whether we really expect exact discrete symmetries to exist.

Global discrete symmetries may be approximate symmetries of low-energy effective Lagrangians, but it is difficult to see why they should not be violated at sufficiently high energies, say by gravitational effects. However discrete symmetries can be exact if they are gauge symmetries. The distinction between gauge and global symmetries applies whether the symmetry is continuous or discrete. Global symmetries are true symmetries of the configuration space of a theory while gauge symmetries are only a redundancy in our description of the theory or equivalently of its configuration space. This is why one does not find Nambu-Goldstone bosons in "spontaneously broken" gauge theories. Gauge symmetries cannot be spontaneously broken since they are just a redundancy of our description, equivalently there are no flat directions in the configuration space due to gauge symmetries which could give rise to massless modes.

With this in mind it should be clear that discrete symmetries can also be classified in the same way. One simple way such symmetries can arise is by "breaking" of a $U(1)$ gauge symmetry down to a Z_N subgroup. The low-energy theory will have a Z_N discrete symmetry which is a gauge symmetry since it is a subgroup of the original $U(1)$ gauge symmetry. Such discrete gauge symmetries can arise by other means and in fact they are very common in compactifications of string theory. There are also consistency conditions for discrete gauge symmetries coming from a discrete version of anomaly cancellation,[5] these conditions are satisfied both for the usual Z_2 R-parity and for the Z_3 matter parity. Thus we again find that a phenomenological requirement on the MSSM leads us to conclude that it should be imbedded in a theory at high energies which can contain discrete gauge symmetries.

Probably the outstanding problem in the MSSM is understanding the mechanism responsible for supersymmetry breaking. Ultimately the hierarchy problem must be solved by eplaining why the effective scale of supersymmetry breaking is near the weak scale. Also, without an understanding of supersymmetry breaking we can only parametrize the breaking by soft terms in the low-energy Lagrangian. Although only four such terms are often used based on certain "minimal" assumptions, these assumptions are totally unjustified from a theoretical point of view and there are really 60 odd parameters needed to specify the most general soft breaking terms. As a result the model has little predictive power. At present we don't know whether supersymmetry is broken by the dynamics of some hidden strongly coupled gauge sector, by non-perturbative effects in the visible sector, or by some intrinsically stringy mechanism. Up until recently the most popular models have been hidden sector models where supersymmetry is actually broken at a rather large scale, but only communicated to the visible world by gravitational effects, leading to an effective scale of supersymmetry breaking around the TeV scale.

Thus in the MSSM there are many hints as to the structure of whatever theory lies beyond the MSSM, perhaps at the Planck scale. It must solve the μ problem, allow for discrete symmetries, preferably gauged, and perhaps provide a mechanism for dynamical supersymmetry breaking. Such a mechanism is further constrained by the allowed values of the soft supersymmetry breaking parameters.

2.3.   String Theory

In the previous two sections I have tried to argue that the low-energy structure we have observed and hope to observe in the future carries important clues about the behavior of physics at very short distances, e.g. the Planck scale. Currently the most successful model of Planck scale physics is superstring theory. It is not my intention to review string theory or even current progress in string theory in this talk. Instead I would just like to mention a rather simple but important way in which string theory itself provides a connection between physics on vastly different scales.

This comes about by asking the question " How big is a string?". Consider for simplicity a closed bosonic string which traces out a closed loop in space parametrized by a coordinate σ with $0 < \sigma \leq 2\pi$. We can write a normal mode expansion for the coordinate of the string as

$$X(\sigma) = x_{cm} + \sum_n \left(\frac{x_n}{n} e^{in\sigma} + \frac{\tilde{x}}{n} e^{-in\sigma} \right) \tag{3}$$

with x_{cm} the center of mass of the string and x_n, $\tilde{x}_n$ being the Fourier coefficients. When time dependence is added the x_n correspond to excitations running around the string counterclockwise and the $\tilde{x}_n$ to excitations running around the string clockwise.

One measure of the size of the string is the average deviation of any point on the string from the location of the center of mass, that is

$$R^2 = \langle (X(\sigma) - x_{cm})^2 \rangle \tag{4}$$

We want to calculate this average quantum mechanically in which case the x_n and $\tilde{x}_n$ become harmonic oscillator creation and annihilation operators for $n < 0$ or $n > 0$ respectively. Clearly for a very excited string R will be very large, but what is rather strange is that due to zero-point fluctuations R is large even in the string ground state. In the string ground state one easily finds that the nth oscillator contributes a factor of $1/n$ and

$$R^2 \sim \sum_n \frac{1}{n} \tag{5}$$

To make sense of this divergent sum we have to understand what it means physically to measure the size of a string. Any real measuring device will have some finite time resolution τ_{res}. On the other hand the divergence comes from modes with large n and these modes have frequencies which grow as n. Since a real measuring device cannot measure frequencies greater than $1/\tau_{res}$ it makes sense to cut off the sum at a mode number corresponding to the time resolution of the measuring device. Doing this somewhat more carefully than described here and reinstating dimensions gives[7]

$$R^2 \sim l_s^2 \log(P_{tot}/\tau_{res}) \tag{6}$$

where l_s is the string scale which is of order the Planck scale (it is less than the Planck scale by a power of the dimensionless string coupling constant) and P_{tot} is the total transverse momentum of the string.

This equation is quite remarkable when one thinks about it. It says that the size of a string depends on how fast you can measure it, and that the faster you can measure it, the bigger it appears. We are used to the idea, which follows from the uncertainty principle, that we need high energies (short times) to probe short distance scales. But as we approach the Planck scale things change, at least in string theory. We need high energies (short times) to probe large distance scales, or at least the large size behavior of strings.

Considerations like this lead to a "string uncertainty principle"[9] which relates the uncertainty in the size of a string Δx to its energy E: (in units with $\hbar = c = 1$)

$$\Delta x \geq \frac{1}{E} + \frac{E}{M_{Pl}^2}. \tag{7}$$

This equation incorporates both the usual quantum mechanical uncertainty and the less well understood uncertainty due to the growth of string states at high energies.

This behavior is responsible for many qualitative features of string theory. For example it is generally believed that string theory has no ultraviolet divergences and this is certainly backed up by many explicit calculations. The above behavior explains why. At very high energies strings become very large and floppy and there is no large concentration of energy at a point which could lead to bad high-energy behavior. It also sounds temptingly like what one wants in order to solve the cosmological constant problem. There one needs some peculiar connection between large scale physics (as governed by the smallness of the cosmological constant which explains the largeness of the present universe) and small scale physics (which should govern the leading contributions to the cosmological constant.) However I do not know of any serious attempt to solve this problem in string theory which utilizes this feature of string theory.

2.4. Black Holes

Over the last few years there has been a resurgence of interest in the quantum mechanics of black holes, inspired in part by the hope that string theory or tools derived from string theory may help to solve some of the outstanding puzzles. The basic outstanding questions are

1. What is the correct treatment of the black hole singularity? Is the breakdown of classical general relativity a sign that quantum gravity is important, or that new short distance classical effects are present (e.g. classical string theory)?

2. Is the Hawking evaporation of black holes consistent with the formalism of quantum mechanics? In particular, can pure states evolve to mixed states due to black hole intermediate states?

3. Is there a microscopic description in terms of state counting of the Bekenstein-Hawking black hole entropy $S = A/4$ with A the area of the event horizon?

In spite of much activity, there is as yet no consensus on the answers to these questions. If black hole entropy is associated with new states on the horizon of a black hole then these states must either show up in a more careful treatment of quantum gravity in the presence of black holes or they must be associated to new states in a more complete theory of quantum gravity such as string theory. The first point of view has been explored recently in the context of toy models in $2 + 1$ dimensions by Carlip, Teitelboim and collaborators.[8] The second point of view has been explored by Susskind and collaborators[10] and fits in nicely with the behavior of string theory discussed earlier and with the general idea of looking for connections between short and long distance physics.

Following Susskind and collaborators consider the difference between a point particle and a fundamental string falling into a black hole as seen by an outside observer. In either case, the outside observer sees any radiation or signal coming from the infalling object redshift exponentially as it approaches the horizon. Thus a measurement made at a fixed frequency probes the structure of the infalling object on increasingly shorter time scales. According to the previous description this should mean that the outside observer in effect "sees" a string increase in size as it falls into a black hole. This suggests that as it reaches the horizon the string is effectively smeared over the whole horizon of the black hole. It seems plausible that such smeared string states could account for the entropy of the black hole, but there has so far been no explicit calculation which shows this to be the case. Also, the word "see" should be taken with a grain of salt. Because of the exponential redshift in the the radiation emitted by an object falling into a black hole, it becomes harder and harder to actually see the physics of the string spreading. Presumably it eventually becomes mixed up with the Hawking radiation in some complicated way.

The question of loss of coherence or information in black hole formation and evaporation should in some sense be tied to the microscopic description of entropy. If it is possible to account for the states and their thermal behavior in an honest way, then we would expect that there would be no information loss once the detailed interaction with the horizon states is take into account. On the other hand, the whole notion of new states on the horizon is rather peculiar and contrary to many of the developments in general relativity over the last twenty years. In particular, to an infalling observer there is absolutely nothing special about the horizon and no reason to expect any new states, nor anyplace they would be expected to be. Thus there would have to be some very strange new kind of complementarity between the description of physics by different observers.

There have also been many attempt to resolve these problems in toy models based on $1+1$ dimensional versions of gravity, but again there seems to be no consensus. An up to date overview of this problem can now be found on the Web.[11]

2.5. Inflation

I will be very brief here since Inflation has already been discussed at this conference in the lecture by Angela Olinto.[12] There are several reasons why inflation provides a particularly compelling example of relations between small scale and large scale physics. First of all, there is so far no microscopically compelling model of inflation, that is a natural particle physics model which gives rise to an inflaton field and potential that leads to the proper amount of inflation and the density perturbations of the required amplitude. Such a model would provide a direct link between physics near the GUT scale and the size of the current universe. More spectacularly, in inflation models the density perturbations which eventually grow into the stuff we see about us are supposed to have their origin in the quantum mechanical fluctuations of the inflaton field, what more dramatic connection between microscopic and macroscopic physics could we hope for? Finally, our current understanding of inflation rests on the idea that there was a past epoch of the universe where the vacuum energy density, a.k.a. the cosmological constant was non-zero. While this seems inescapable in many models given that the present cosmological constant is zero, one cannot escape an uneasy feeling that our whole picture of inflation may change dramatically if we ever understand the cosmological constant problem.

3.   Dynamics of Supersymmetric Gauge Theories

3.1.   Motivation

During the last 15 years two theories have been developed in some detail that involve supersymmetry in an essential way. The first is superstring theory. Superstring theory allows us for the first time to address the physics of the Planck scale in a well defined way. It is also well understood by now that there are many solutions to superstring theory that leave us with an effective four-dimensional supersymmetric theory which resembles the standard model in the sense that it can contain some number of chiral fermion generations in representations of a gauge group which is usually somewhat larger than the gauge group of the standard model. One the other hand there is so far no understanding of which (if any) of these vacua is picked out dynamically as the true vacuum (they are equally good in perturbation theory). There may also be many inequivalent vacua left after all dynamical effects are included in which case one needs some further principle (e.g. quantum cosmology) in order to make contact with our particular world. Another aspect of this problem is that most superstring vacua contain many massless scalar fields called moduli which have no potential. Different vacuum expectation values for these moduli correspond to different choices of superstring vacua. The "moduli problem" in superstring theory is the problem of how a suitable potential is chosen by the dynamics for these fields.

The second theory where supersymmetry has played a crucial role is of course the supersymmetric extension of the standard model (MSSM) discussed earlier. There the main roadblock to detailed predictions is our lack of understanding of supersymmetry breaking.

One of the main themes in particle theory over the last few years has been the attempt to tie together the structure needed in the MSSM with the sort of structures that arise in superstring theory. This not only provides further constraints on the MSSM but also may shed some light on both the moduli problem and the problem of supersymmetry breaking.

Over the last year there has been dramatic progress, building on work in the early 1980's, in understanding the dynamics of supersymmetric gauge theories in four dimensions. Although this work has not yet answered either of the above questions, it has introduced new techniques, and also promises to shed light on some old problems in non-supersymmetric gauge theories. As a result I would like to give a brief discussion of these new developments following a recent paper by Seiberg and Witten.[13]

3.2. $N = 1$ Supersymmetry

Before discussing the results of Seiberg and Witten it will be useful to very briefly review a few facts about theories with $N = 1$ supersymmetry. These theories are the basis for supersymmetric extensions of the standard model (the MSSM) because only theories with $N = 1$ supersymmetry are compatible with having chiral fermions. Lagrangians with $N = 1$ supersymmetry are most easily constructed in terms of superfields in superspace. The coordinates of superspace consists of the usual spacetime coordinates x^μ and fermionic coordinates θ which can be thought of as the two complex components of a Weyl fermion. The basic matter multiplet, consisting of a Weyl fermion and a complex boson can be packaged into a complex scalar function on superspace, $\Phi(x,\theta)$ called a chiral superfield. The general Lagrangian then has two types of terms, written as

$$\mathcal{L} = \left(\int d^2\theta d^2\bar\theta K(\Phi, \Phi^*)\right) + \left(\int d^2\theta W(\Phi) + h.c.\right) \tag{8}$$

and referred to as D terms and F terms respectively. For a renormalizable Lagrangian the D terms contain the kinetic energy terms while the F terms contain potential terms and Yukawa interactions. The important point to note is that the F terms are determined purely by functions of Φ while the D terms involve functions of both Φ and Φ^*. This is peculiar to supersymmetry and has no analog in ordinary non-supersymmetric field theories. It has been known some time that F terms are not renormalized in perturbation theory. The proofs of this rely on the detailed structure of superspace perturbation theory. More recently it has been realized that there is a very simple and powerful argument for this non-renormalization which can also be extended to prove results about non-perturbative effects.

The argument is best illustrated with a simple example.[14] For a single scalar superfield the most general renormalizable choice of W is

$$W(\Phi) = \frac{m}{2}\Phi^2 + \frac{\lambda}{3}\Phi^3 \tag{9}$$

When $m = \lambda = 0$ this theory would have two $U(1)$ symmetries, consisting of changing the phase of the scalar and fermion components of Φ separately. The idea is to view the parameters m and λ as expectation values of chiral superfields and these two symmetries as being spontaneously broken. Thus we replace

$$m \rightarrow \langle\Phi_m\rangle \tag{10}$$
$$\lambda \rightarrow \langle\Phi_\lambda\rangle \tag{11}$$

and find that the theory is invariant under the two $U(1)$ symmetries

$$S: \quad \Phi_m \rightarrow e^{-2i\alpha}\Phi_m \tag{12}$$
$$\Phi_g \rightarrow e^{-3i\alpha}\Phi_g \tag{13}$$
$$\Phi \rightarrow e^{i\alpha}\Phi \tag{14}$$

and

$$R: \quad \Phi_m(\theta, x) \rightarrow \Phi_m(\theta, x) \tag{15}$$
$$\Phi_g(\theta, x) \rightarrow e^{-i\beta}\Phi_g(e^{i\beta}\theta, x) \tag{16}$$
$$\Phi(\theta, x) \rightarrow e^{i\beta}\Phi(e^{-i\beta}\theta, x) \tag{17}$$

Now since we can view the symmetry as being spontaneously broken, we know that it must be respected by the one-particle irreducible effective action. But the only terms allowed in the 1PI effective action by these two symmetries are the original terms! Thus there can be no renormalization of the potential terms. It should be mentioned that there are some subtleties with this argument when there are massless particles.[15] Similar arguments can be used to determine various non-perturbative effects in these theories, but only through the contribution to F terms. The argument does not work for D terms since they depend on both Φ and Φ^*. If there were an additional symmetry relating D and F terms then one would have very strong constraints on the dynamics of the theory. This is precisely what happens in theories with $N = 2$ supersymmetry.

3.3. $N = 2$ Supersymmetry

The simplest Yang-Mills theory with $N = 2$ supersymmetry contains a single supermultiplet of fields consisting of what we might call a gauge boson and gaugino plus a Higgs boson and Higgsino. These fields are all in the adjoint representation of the gauge group. Taking the gauge group to be $SU(2)$ they are all isotriplet fields. The Higgs ϕ is a complex field. $N = 2$ supersymmetry dictates that the potential term for ϕ is

$$V(\phi) = \text{Tr}[\phi, \phi^*]^2. \tag{18}$$

Clearly $V = 0$ for any configuration of ϕ which lies in a single direction in group space, say $\phi = a\tau_3/2$ with a an arbitrary complex number. A gauge invariant way of describing this is to say that there is a zero of V for every complex value of $u = \text{Tr}\phi^2 = a^2/2$.

The fact that classically there is a continuous set of vacua specified by u is a feature which is common to many supersymmetric theories. In general it is a disaster for phenomenological applications of supersymmetry. This is both because it reflects the presence of massless scalars in the low-energy theory and because it makes it unclear which vacuum is the right one for doing physics. This latter problem is particularly onerous in low-energy supersymmetric string theory. The beauty of the recent ideas is to take this bad feature of supersymmetric theories and to make use of it to study the dynamical structure of these theories.

To see how this works first consider this theory at some non-zero value of u. We then have $SU(2)$ broken down to a $U(1)$ which I will call electromagnetism. The spectrum of the theory consists of a massless photon supermultiplet, massive $W^{\pm}$ supermultiplets, and also massive monopole supermultiplets. At very low-energies the theory just looks like a supersymmetric form of electromagnetism with some electric coupling e_{eff}. I first want to explain how e_{eff} is related to the value of u. This theory is asymptotically free at high energies, that is $\beta_e < 0$ at scales above the masses of all particles. As we drop below the masses of charged particles they decouple from loops and the coupling constant stops running. Clearly the asymptotic value of the coupling constant at low energies depends on the scale of masses in the theory. These masses are determined by the value of u. So, specifying a value of u is equivalent to specifying the coupling e_{eff} in the low-energy theory.

As long as we only consider the massive electrically charge particles and the photon then the low-energy physics is completely specified by e_{eff}. When we include the monopoles as well it is necessary to also specify the effective value of the θ angle, θ_{eff}. This parameter is the electromagnetic analog of the θ angle in QCD. In QED it also leads to CP violation by giving magnetic monopoles of magnetic charge n_M an electric charge given by

$$Q_{el} = \frac{n_M}{2\pi}\theta_{eff} \tag{19}$$

The upshot of this is that the complex parameter u in fact determines two real low-energy parameters (e_{eff}, θ_{eff}) or equivalently, one complex low-energy parameter.

Now we would like to know what happens to the theory and the spectrum of states as we move around between classical vacua labelled by u, or equivalently move in the space of couplings (e_{eff}, θ_{eff}) that govern the low-energy field theory. First of all it can be shown fairly easily that at weak coupling $\theta_{eff} \propto \text{Im} u$. Thus at weak coupling the effect of shifting the imaginary part of u is simply to shift the electric charge of magnetic monopole states. Clearly it would be much more interesting to determine what happens when g_{eff} changes since this could tell us about the behavior of the theory at strong coupling.

Unfortunately the analysis get quite a bit more complicated and I cannot do it justice in this talk. Let me just mention the following points.

1. The one-loop beta function predicts that the coupling e^2 should blow up in the infrared and then become negative. Clearly this is physically unacceptable. Of course there is no reason to believe the one-loop result at strong coupling.

2. The possible behavior of the theory as a function of u is highly constrained by the $N = 2$ supersymmetry. In particular the u plane must be a special kind of complex manifold, a Kahler manifold.

3. By utilizing the supersymmetry and analytic structure one can find a simple guess for the behavior of the theory as a function of u which passes many non-trivial consistency tests.

The resulting structure found by Seiberg and Witten has many remarkable features. First of all, there is a kind of "dual" behavior at strong coupling where as one moves in u the electrically charged states pick up magnetic charge, just as the magnetic states can pick up charge at weak coupling. Second, one finds singular points where monopole states are becoming massless. It is possible to perturb the theory by breaking $N = 2$ supersymmetry to $N = 1$ supersymmetry in such a way that the monopoles mass squared becomes negative, indicating an instability to monopole condensation. It has long been thought that confinement may be described by a "dual " superconductor in which magnetic charges condense. In these models one has for the first time a concrete realization of these ideas.

There seem to be two major areas where these results may have a broad impact. The first is in the construction of new models for dynamical supersymmetry breaking. As I mentioned earlier, the lack of a convincing mechanism for supersymmetry breaking is one of the major obstacles to obtaining predictions from the MSSM. The second area is the dynamics of strongly coupled gauge theories, both supersymmetric and non-supersymmetric. The models discussed by Seiberg and Witten are in a sense toy models, but they contain most of the features of non-trivial gauge theories including running couplings, non-trivial scattering, confinement, and chiral symmetry breaking. In addition the way these effects are realized is rather dramatic involving an infinite resummation of non-perturbative instanton effects. It seems possible that some of these features may extend beyond these specific models.

4. General Remarks

I would like to end this talk with a few general remarks about the prospects for progress in string theory in the upcoming years.

The string revolution is now ten years old. The discovery of anomaly cancellation by Green and Schwarz was discussed at the November 1984 DPF meeting. Roughly a year later the basics of superstring models of unification were established through Calabi-Yau compactifications of the heterotic string. Since then there has been a great deal of effort and progress in formal areas, but little that an experimentalist would find applicable to current or future experiments. As a result a certain amount of pessimism and scepticism has arisen regarding the future of string theory.

It is worthwhile recalling the situation before string theory. As is the case now, in 1984 there were no startling new experimental results and the leading idea for unification, Kaluza-Klein theory, had run into insurmountable obstacles. However there were a number of novel theoretical ideas and structures floating around which did not fit into any coherent whole. As examples I might mention connections between the renormalization group and geometry, the beautiful structures of two-dimensional current algebras or Kac-Moody algebras, the development of realistic supersymmetric extensions of the standard model with hidden sector supersymmetry breaking, as well as a number of other purely theoretical constructions. Many of these are now regarded as either part of the structure underlying string theory or as possible low-energy consequences of string theory.

If we think of the situation today there are again many new ideas floating around, many are classified in a general way as "string theory" but in fact many have very little to do with string theory in any direct way. For example there are matrix models of two-dimensional string theory, new ideas concerning the structure of black holes with toy models in $1 + 1$ and $2 + 1$ dimensions, topological field theory with its connection to many deep mathematical structures, new ideas about the structure and dynamics of supersymmetric gauge theory, a deeper understanding of special two-dimensional systems, etc. Of course this is not to say that another revolution comparable to string theory is around the corner, but rather that there has been continuing theoretical progress and that there are new connections and structures which will undoubtedly fit into a more coherent picture in the future.

I have tried to take an optimistic, but I hope not unreasonable point of view in this talk. My point of view is that in spite of the dearth of new experimental results, there are new interesting theoretical ideas and tools and that these plus consistency and constraints from the Standard Model provide us with non-trivial information about the structure of string theory or whatever theory governs physics at the Planck scale. I am certainly not suggesting that we can find the theory of everything without additional experimental input. I am suggesting that as our theories become more tightly constrained each additional piece of experimental information carries much more weight. I think this holds out hope that we will eventually understand Planck scale physics without building a Planck scale accelerator.

5. Acknowledgements

I would like to thank the organizers for the invitation to speak and the other speakers for their lively and interesting presentations.

References

1. J. Rosner, plenary talk, this conference.
2. For a review see J. Polchinski, "Effective Field Theory and the Fermi Surface" in *Recent Directions in Particle Theory* (Proceedings of the 1992 Theoretical Advanced Study Institute in elementary Particle Physics), edited by J. Harvey and J. Polchinski (World Scientific, Singapore, 1993).
3. H. Haber, "Introductory Low-Energy Supersymmetry" in *Recent Directions in Particle Theory* (Proceedings of the 1992 Theoretical Advanced Study Institute in elementary Particle Physics), edited by J. Harvey and J. Polchinski (World Scientific, Singapore, 1993).
4. L. E. Ibanez and G. G. Ross *Nucl. Phys.* **B368** (1992) 3
5. L. E. Ibanez and G. G. Ross *Phys. Lett.* B **260** (1991) 291; T. Banks and M. Dine *Phys. Rev.* D **45** (1992) 1424
6. J.A. Casas and C. Munoz *Phys. Lett.* B **306** (1993) 288; V. Kaplunovsky and J. Louis *Phys. Lett.* B **306** (1993) 269; G.F. Giudice and A. Masiero *Phys. Lett.* B **206** (1988) 480
7. L. Susskind *Phys. Rev. Lett.* **71** (1993) 2367
8. S.Carlip, "The statistical mechanics of the (2+1)-dimensional black hole," gr-qc/9409052 and references therein.
9. D. J. Gross and P. F. Mende *Phys. Lett.* B **197** (1987) 129; *Nucl. Phys.* **B303** (1988) 407
10. L. Susskind *Phys. Rev.* D **50** (1994) 2700; L. Susskind, L. Thorlacius and J. Uglum *Phys. Rev.* D **48** (1993) 3743; A. Mezhlumian, A. Peet, and L. Thorlacius, *Phys. Rev.* D **50** (1994) 2725
11. S. Giddings, "Quantum Aspects of Gravity" page in the Virtual Review http://www.het.brown.edu/physics/review/index.html.
12. A. Olinto, plenary talk, this conference.
13. N. Seiberg and E. Witten *Nucl. Phys.* **B426** (1994) 19
14. N. Seiberg, hep-th/9408013.
15. M. Dine and Y. Shirman *Phys. Rev.* D **50** (1994) 5389

High-Rate, Position-Sensitive Radiation Detectors: Recent Developments and Applications in Particle Physics, Medicine and Biology

FABIO SAULI, *CERN*

HIGH-RATE, POSITION-SENSITIVE RADIATION DETECTORS: RECENT DEVELOPMENTS AND APPLICATION IN PARTICLE PHYSICS, MEDICINE AND BIOLOGY

FABIO SAULI

CERN, CH-1211 Geneva, Switzerland

ABSTRACT

After some examples of applications of classic multi-wire proportional chambers, this paper describes the basic principles of operation of two advanced position-sensitive electronic radiation detectors: the micro-strip gas chamber and the silicon micro-strip and pixel detector. Some representative examples of use of the devices as trackers for particle physics, as well as for imaging radiation in other fields are reviewed, and promising developments and perspectives of use are discussed.

1. Introduction

The multi-wire proportional chamber (MWPC), developed in the late sixties by G. Charpak[1], has thereafter been refined and diversified in a large variety of devices exploiting various gas properties and improving on performances. A comprehensive description of the detector can be found in dedicated review articles and books[2-5]. In its basic configuration, with anode and cathode wire planes segmented and equipped with recording electronics, the MWPC allows to determine with sub-millimetre accuracy the co-ordinates of an avalanche generated by the multiplication processes in the gas in the region of an ionizing encounter (fig. 1).

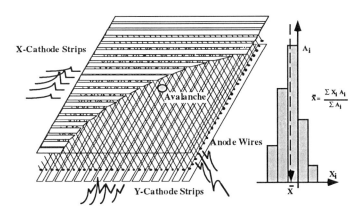

Fig. 1: MWPC with two-dimensional readout of induced charge on cathodes.

To provide a context for the following discussion, fig. 2 shows an electron-positron collision event at LEP, recorded with the DELPHI detector[6]; the main tracking information is obtained from a time projection chamber (TPC), a large-volume gaseous tracking device (over 2 m in diameter and about 3 m long) developed originally by D. Nygren[7,8]. Ionizing trails are sampled many times in short segments, each segment being accurately localized in space. As can be seen, complex events (tens of tracks) are well identified; the energy loss information (or dE/dx) can also be recorded and exploited for attempting the identification of particles.

The main limitation in the otherwise very powerful TPC, preventing its use for experiments at the new generation of high-luminosity colliders comes from the long electron drift time in the sensitive volume (tens of μs).

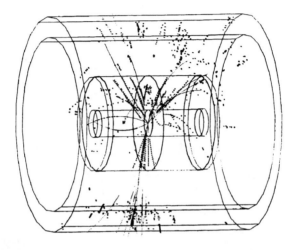

Fig. 2: An event recorded by the DELPHI detector at CERN. The main tracking device is a TPC, the central cylinder outlined in the figure.

Owing to their good efficiency and localization accuracy, the use of gaseous detectors has spread also in fields other than particle physics, and namely medicine and biology. Substantial improvements over standard diagnostics methods can be obtained with the new devices used as X-ray imagers; in many cases, comparable contrast levels can obtained at a much reduced doses with obvious advantage for the patients. While most of these applications are still in the early experimental phase, some (as for example a digital radiography system, see later) have already been installed and used in a real hospital environment. For a recent review, see Sauli[9].

To achieve good detection efficiency for hard X-rays, one has to use rather thick conversion layers and possibly high pressures. Xenon mixtures are preferred to reduce the physical thickness of the detector: fig. 3 shows the computed efficiency as a function of energy and gas thickness for pure xenon.

Various methods have been devised to eliminate the dispersive effects of thick conversion layer for non-parallel X-ray beams (the parallax error). In the digital radiography chamber[10], shown schematically in fig. 4, a thin, collimated x-ray fan beam is sent through the body under examination, and the transmitted radiation is detected in a special MWPC. Generator and detector move together vertically to realize a body scan. In order to allow fast imaging, the simplest method of electronics detection (counting on fast scalers) is used to permit high acquisition rates.

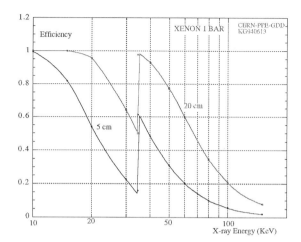

Fig. 3: Detection efficiency of xenon as a function of X-ray energy.

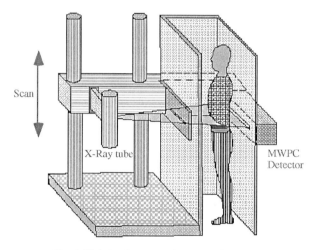

Fig. 4: Digital radiography with a MWPC detector.

The particular conception of the detector (fig. 5) with non-parallel anode wires aiming at the focal point of the generator allows to use a thick absorption volume without a parallax error; the uniformity of gain along the anodes is guaranteed by an increasing spacing between anode wires and cathode planes.

Digital radiography systems as described above are presently evaluated in hospital environment, both in Russia and in the West; recently, a detailed comparison of performances as a function of dose to the patient, made by an independent group clearly demonstrates the advantages of the electronic system over classic photographic plates, particularly for low-dose exposures[11].

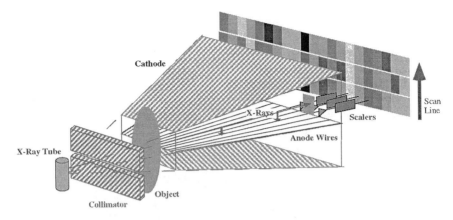

Fig. 5: Schematics of the MWPC detector for the digital radiography.

The picture in fig. 6 shows an example of digital radiography of a woman's pelvic region (hysterosalpingography), obtained at an entrance skin dose of 0.030 mGy; a contrast medium was injected to enhance the features under study (a partially obstructed fallopian tube). Even at low doses, the counting rate capability of the device has to be large in order to decrease the exposure time and to reduce artefacts due to the movements of the body or of internal organs during the scan; using fast scalers, the counting rate in the present design is limited to 600 kHz per wire.

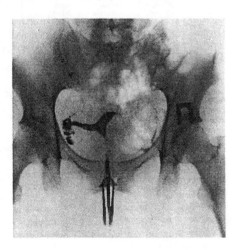

Fig. 6: Digital radiography of the pelvis.

Both particle physics experimentation and medical applications however require today rate capabilities and resolutions exceeding those achievable with standard MWPCs. Multiwire chambers have an intrinsic limitation in resolution determined by the minimum practical wire spacing (around one mm); an even more fundamental limit to high-rate operation is an outcome of the production of (slow) positive ions in the avalanche process. With drift times of several hundred μs or more, ions tend to accumulate in the anode-to cathode gap modifying the electric field and affecting the detector efficiency. Fig. 7 shows an example of gain reduction as a function of particle rate in a MWPC[12].

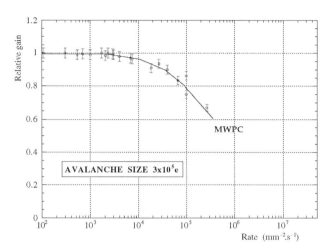

Fig. 7: Space-charge gain decrease as a function of rate in a MWPC.

2. Micro-strip gas chambers

The micro–strip gas chamber (MSGC, fig. 8) was introduced in 1988 by A.Oed[13]. With anode and cathode strips etched on an insulating support using high–accuracy photo–lithography, MSGCs allow the realization of gaseous detectors with distances between anodes an order of magnitude smaller than in conventional multiwire proportional chambers, with the consequent improvements in rate capability, localization accuracy and multi-particle resolution[14-16]. Typically, anodes and cathodes are 10 and 80 μm wide, respectively, at a pitch of 100 μm centre to centre. Much as in a MWPC, a negative signal generated by an avalanche on the anode is accompanied by positive induced signals on the neighbouring cathode strips; readout of the induced signals improves the localization accuracy of the detector. When using thin supports, the induced signals can be detected also on strips etched on the back of the plate, thus allowing two-dimensional localization of ionizing events. Proportional gains between 10^3 and 10^4 can be reached, with a very good energy resolution (in fact, better than most MWPCs); this has suggested the use of the device in applications where both position and energy of the detected radiation (X-ray) are desired. The detector is intrinsically capable of withstanding counting rates much larger than a MWPC, since most of the ions produced at the anode are collected in a very short time (1 μs) by the adjacent cathodes. A great effort is at present being undertaken to fully understand the operating characteristics of the new device and to find cheap and reliable manufacturing methods; fig. 9 shows an assembly scheme making use of thin glass frames glued with vacuum-grade epoxy to the MSGC plate and a drift electrode[17]. MSGC prototype having an active area of 120 by 250 mm^2 have been manufactured and successfully tested[18].

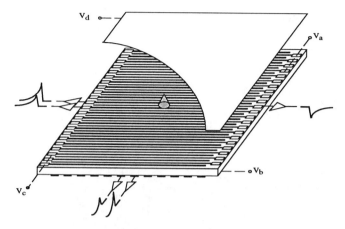

Fig. 8: The micro-strip gas chamber.

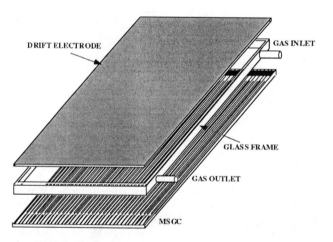

Fig. 9: Schematic assembly of a light, all-glass MSGC module.

In order to operate the detector efficiently at high radiation fluxes, the use of too well insulating supports should be avoided. Indeed, a fraction of the charge produced in the gaseous amplification process sticks to the surface between strips, modifying the electric field with a consequent fast, rate–dependent gain drop. This "charging up" effect can be overcome using as support electron–conducting glass[19] with bulk resistivity in the range 10^9 to 10^{12} Ω cm; the same result can be obtained coating the plates with a thin semi–conducting layer which reduces the surface resistivity to 10^{14}–10^{15} Ω/square. It has been experimentally demonstrated[19,20] that MSGCs made on resistive supports can withstand, without gain modification, radiation rates up to and above 10^6 mm^{-2} s^{-1} at an avalanche size of around 10^5 electrons. In fig. 10, the measured rate capability of MSGCs manufactured on a range of supports is summarized (compiled from the quoted references); comparing with the similar measurement realized with a MWPC one can see that MSGCs manufactured on low resistivity supports can tolerate radiation fluxes about two orders of magnitude larger. It has also been demonstrated that, at least in very clean gas conditions, MSGCs can withstand long-term irradiation, up to and above 10 MRad of charged particles, without deterioration of performances[15].

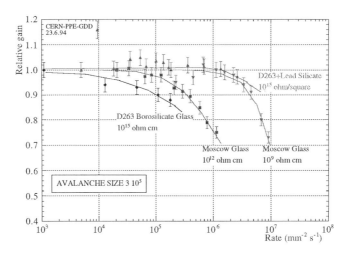

Fig 10: Rate capability of MSGCs manufactured on supports with different resistivity.

In view of their superior rate capability and multi-particle resolution, MSGCs have been adopted in the design of the detectors to be operated at CERN's large hadron high-luminosity collider (LHC); fig. 11 shows as an example the present conception of the vertex detector for the Compact Muon Solenoid (CMS) experiment[21]. It consists in arrays of MSGC plates, mounted on modular supporting wheels in the forward directions and on cylindrical barrels for the large angle region. About 20,000 independent plates of sizes from 10 by 10 to 25 by 12 cm^2 constitute the gas tracking detector, surrounding a smaller but more accurate silicon micro-strip detector array.

Applications of MSGCs are foreseen in many fields, taking advantage of their excellent energy resolution and rate capability. In particular, the group that has developed the digital radiography system is considering replacing the MWPC with a dedicated MSGC; operation at high pressures, necessary for efficient detection of hard X-rays, has been demonstrated[22]. The plates can be manufactured with the fan-out geometry necessary to avoid the parallax error, adjusting the strips' width to achieve uniform gain when varying the anode-to cathode distance, in the so-called keystone geometry (originally developed for use in trackers at high luminosity colliders[23]). Another application, fully exploiting the extreme rate capability and proportional response of MSGCs, could be for beam diagnostics in the TERA project, a dedicated light ions accelerator for cancer therapy[24]. In this case, one wants to build a "totally active" detector to study the micro-dosimetry of energy loss within a phantom, at very high radiation levels; an attractive possibility is to build a sandwich of thin, plastic MSGCs with tissue-equivalent absorbers (the so-called "magic cube") with on-line electronic measurement of the tracks' distribution and energy loss.

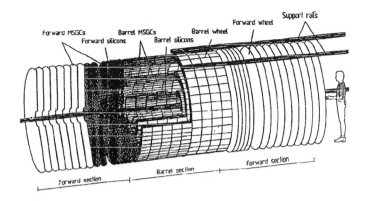

Fig. 11: CMS Vertex detector.

3. Silicon micro-strip detectors

Semi-conductors have been used to detect heavily ionizing particles and nuclear radiation form the early fifties. However, detection of minimum ionizing particles was not feasible before substantial progress in micro-electronics technology and the development of low-noise highly integrated amplifiers. Fast particles release about 3×10^4 electron-hole pairs in a typical silicon diode and, taking into account Landau fluctuations, efficient detection requires the use of a amplifiers with a noise of few thousand electrons or less. From the early eighties several groups started the development of silicon micro-strip detectors and of the associated electronics taking advantage of large scale integration technologies[25,26]; the first application of the new detectors was for tagging the secondary vertex of short-lived charmed particles. Recent reviews of the developments of solid state detectors can be found in the articles by A. Peisert[27] and G. Hall[28]; a classic paper by V. Radeka[29] analyses in detail performances of detectors and of associated electronics.

Fig. 12 shows schematically a silicon micro-strip detector. A high-resistivity wafer, 300 µm thick, is processed to provide the multiple diode structure shown. The n-type silicon is implanted along thin strips with boron ions on the rectifying side, to create p-type junctions; the electrical contacts are realized by evaporation or sputtering of aluminium strips on the junction side and of a uniform conductor on the back (ohmic) side as shown. Distances between strips in the range 25 to 50 µm are commonly used. On application of a reverse voltage, the wafer can be fully depleted and becomes sensitive to ionizing radiation; on the ohmic side, an n+ implant is used to prevent massive electron injection when reaching the full depletion voltage. Electrons and holes created by ionization at the passage of charged particle are collected by the electrodes, and the resulting charge signal are amplified and detected.

The physical width of the ionized trail, due both to long-range delta electrons and to the diffusion of migrating careers, results in more than one strip providing a signal even for perpendicular tracks; a measurement of charge profile and calculation of the centre-of gravity then yields a localization accuracy better than the strip's pitch.

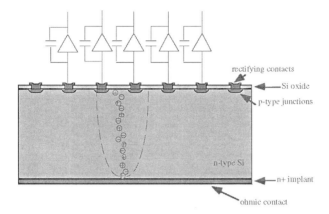

Fig. 12: Single-sided silicon micro-strip detector.

An example of charge distribution on strips (at 25 μm pitch) and of localization accuracy for minimum ionizing tracks is given in fig. 13[30]. Often not each strip is readout, but only every second or third; the capacitive coupling between adjacent strips is sufficient to share the charge between strips proportionally to the average position of the ionization.

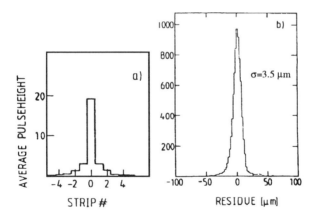

Fig. 13: Single-track charge distribution on strips at 25 μm pitch (a) and localization accuracy (b).

The first generation of silicon trackers was realized using single-sided detectors, the second co-ordinate being either measured with other devices (gaseous detectors), or by mounting a second layer of silicon detectors with strips at angles with the first. Recent progress in manufacturing technology has allowed to realize double-side readout detectors, providing both co-ordinates from the same wafer. A straightforward implementation of readout strips on the ohmic side however cannot be used, because of problems with the accumulation of positive ions in the oxide layer attracting negative charge at the interface oxide-silicon and creating a low-resistivity bridge between readout strips, as shown in fig. 14a. One solution is the insertion of intermediate acceptor-doped strips (p$^+$ implants), or p-stops[31], interrupting the conduction channel between readout strips on the ohmic side (Fig. 14b). Another method that can be used to prevent the short-circuit between readout strips on the ohmic side is the creation of a MOS gate structure, or field plate, obtained extending the metal strips over the oxide[32].

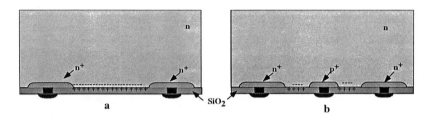

Fig. 14: Readout of the ohmic side: charge-induced short circuit (a) and solution with a p-stop implant (b).

A practical problem arises when realizing double-side readout detectors: one of the sides is necessarily at high voltage (100 V or so), thus requiring decoupling capacitors to connect to the (grounded) amplifiers. The capacitor can be integrated in the structure interleaving a thin layer of silicon oxide between the diode implant and the metallic strip used for the readout (fig. 15a); on one end of the strips, a complex structure integrating a polysilicon resistor is used to provide the bias voltage to the diodes[30] (fig. 15b).

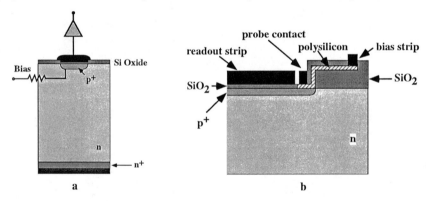

Fig. 15: Integrated readout capacitor (a) and polysilicon bias resistor (b).

An AC-coupling of the diodes to the readout electronics is also essential to avoid the effect on the amplifiers operation of the leakage current, varying for different detectors and strongly affected by exposure to radiation.

A second problem with double-side readout is purely mechanical: in the cylindrical geometry of vertex detectors it is very inconvenient to mount the readout electronics along the axial direction, and signals have to be extracted from the ends. Various schemes have been developed to this purpose, with inclined strips or connection to flat kapton cables; the most elegant solution (but also technologically the most difficult to implement) is a direct integration on the wafer itself of a set of conductors perpendicular to the readout strips, the so-called double metal layout (fig. 16). This is the solution adopted for the DELPHI upgrade vertex detector[27,33] at LEP, and is being considered for the CDF upgrade at Fermilab[34].

N-SIDE

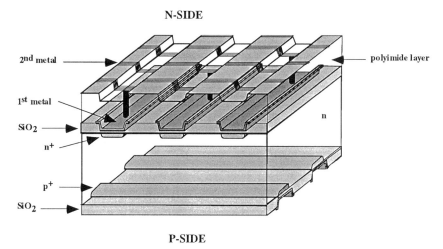

P-SIDE

Fig. 16: Two-sides, double-metal silicon micro-strip detector.

Presently, due to high costs and technological limitations in manufacturing, the size of silicon micro-strip detectors does not exceed about 8 by 8 cm^2 (four-inch wafers). A vertex detector for experiments around colliders, that needs to be up to one meter long, is then built assembling hundreds of silicon microstrip plates on layers in a cylindrical barrel geometry. In most cases, in order to reduce the number of readout channels two or more plates are bonded together in ladders; needless to say, very stringent requirements on the mechanical accuracy and stability of the support have to be satisfied.

Silicon micro-strip detectors have been assembled and operated on all the major set-ups on colliding beam experiments, and a new generation of detectors is in preparation or in advanced design to cope with higher luminosity; fig 17 shows, as an example, the design of the silicon vertex upgrade for the CDF detector at Fermilab[33]. Individual plates are mounted in four concentric layers and three independent barrels, for a total of 260,000 readout channels; the active length of the detector is close to one meter. Fig. 18 shows a simulated event at the Large Hadron Collider (LHC), the future machine at CERN, as seen by the CMS central detector[35]. Clearly exceptional performances in terms of rate capability and multi-particle resolution are imperative for such detectors.

Another fundamental but separate issue concerning silicon detectors is their resistance to high radiation flux[36]; recent progress in radiation hard devices guarantees continuous operation with tolerable degradation up to several MRad of exposure to charged particles and 10^{14}-10^{15} neutrons cm^{-2}; this is barely good enough for medium term operation at high luminosity colliders, and replacement of the complete detector after a few years of operation has been envisaged as integral part of the detector design. The effort in improving the radiation hardness of the devices continues in other directions, for example with the development of gallium arsenide and diamond detectors.

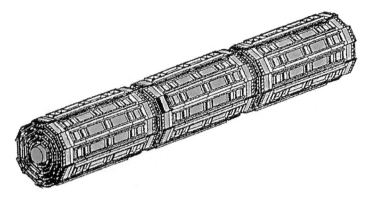

Fig. 17: Silicon vertex detector for the CDF upgrade.

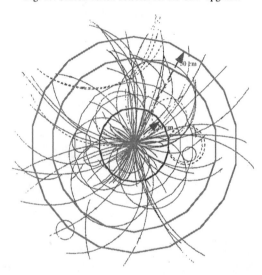

Fig. 18: A simulated event in the CMS detector at LHC.

Silicon detectors have been used for applications in medical diagnostics. For the detection of X-rays however one encounters a problem with rather low values of efficiency (fig. 19). Even assuming a depleted thickness of one mm, rather difficult to realize, at energies around 50 keV (required by applications in full-body radiography) the efficiency of detection is only a few percent; on the lower end of the energy scale, the use of a silicon detector is limited by the intrinsic noise of the system (5 and 10 keV rms).

A summary of applications of solid state detectors in medicine and biology can be found for example in ref. 28; I will discuss here only one recent development.

In detecting a well collimated, high intensity beam an elegant solution to the efficiency problem is to enter the detector from the edge, each diode strip acting as individual, long "pixel" with good efficiency, an expedient devised by a group working at the ELETTRA dedicated synchrotron radiation accelerator[37] (see fig 20).

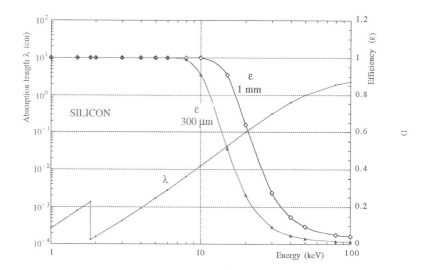

Fig. 19: Efficiency of silicon detectors as a function of x-ray energy.

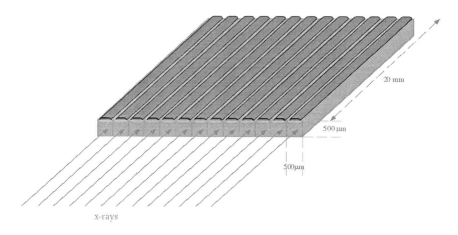

Fig. 20: Silicon "line pixel" detector for x-rays.

The silicon strip plate acts as a line detector: each channel is connected to an amplifier-discriminator and to a scaler, and a scan through the object allows to obtain a two-dimensional image. Fig. 21 shows an example of digital radiography of a standard mammography phantom (a 75 µm thick aluminium disk, 6 mm in diameter, embedded in 16 mm thick plastic). The scanning step was 250 µm, and the skin dose 130 µGy; the actual counting rate corresponds to about 7×10^5 counts per pixel.

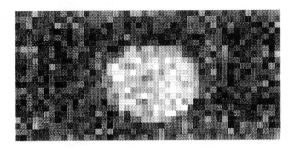

Fig. 21: Scanning radiography of a standard mammography phantom.

Clearly, with the advent of cheap, large area solid state bi-dimensional detectors intrinsically simpler to operate than gaseous detectors, one can foresee an increase of their use in medicine and biology; for high-rate applications however remains the problem of developing dedicated electronic circuits capable to handle very large throughputs, in fields where (unlikely what happens in particle physics) all or most of the events have to be recorded. A digital radiography system based on single photon counting with silicon micro-strip detectors has been recently described, making intensive use of transputers and parallel processing for analysing and recording the data[38].

4. Silicon pixel detectors

Despite their fine granularity, at the highest luminosity of the future generation of colliders and for heavy ions interactions even silicon micro-strips are easily saturated in the innermost layers of the detectors. This has motivated the development of solid state pixel devices, where one can reach the ultimate multi-track resolution corresponding to the pixel size, a fraction of mm^2. The first generation of pixel detectors for particle physics was developed using charge coupled devices (CCD), improved versions of commercially available components of solid state cameras[39]. Extremely powerful in terms of resolution but limited by their low readout speed, these devices are used in the vertex detector of the SLD experiment at the Stanford Linear Collider (having low repetition rates).

Full exploitation of the multi-track resolution power of a solid state pixel device at high luminosity demands the development of dedicated readout electronics, coping with the extremely large data throughput; moreover, since detector and readout electronics are manufactured on separate wafers, a suitable interconnection technology has to be implemented, reliable and cheap enough to allow readout of millions of channels. Pioneering work in this direction has been done by various groups using a bump-bonding method to connect large arrays of silicon pixels to the readout elctronics[40-42]. Fig 22 shows the basic concept of the devices. A detector-grade (high resistivity) silicon wafer is etched to create a matrix of diodes, individually acting as detectors, and having a rectangular metallic contact on one side. On a separate wafer, a custom-made large scale integrated chip has an array of active elements (amplifiers, discriminators) with individual size corresponding to the detector's pixels. On each cell of the electronics wafer a small pad holds a metal "bump" made of soft soldering alloy; after careful alignment, the two wafers are contact-bonded together heating above the melting temperature of the alloy. Readout and power lines (not shown in the figure) allow to operate the chip and to transfer selected events to the data acquisition system.

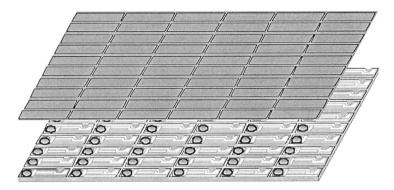

Fig. 22: Silicon pixel detector with the readout electronics before bump bonding.

The RD19 collaboration at CERN has recently successfully tested in a beam a silicon micro-pattern detector realized with several arrays having about 40,000 pixels each, fully instrumented[43]; the pixel size of 75x500 μm^2 provides a localization accuracy in one direction of about 20 μm for minimum ionizing particles. Applications in other fields are already envisaged; optical coupling to a stack of scintillating fibers may allow to overcome the problem of low intrinsic conversion efficiency for X-rays of the thin silicon detector itself[44].

5. Wat's next?

The striking developments of fast position-sensitive detectors in particle physics are matched by an increasing interest in the use of such sophisticated devices in other fields, and namely in medical diagnostics and biology research. The relatively slow diffusion of gaseous detectors based on the multiwire proportional chamber in applied fields outside high energy physics might be explained by the relative complexity of operation (that involves the use of flammable gases and elaborated electronics) and by the fragility of the detectors, prone to fail in case of electrical breakdown. Solid state detectors, intrinsically simpler to operate and more reliable, suffer from a high price tag even for small sensitive areas, and are limited by their low efficiency for X-ray energies above a few tens of keV. The newly introduced the micro-strip gas chamber solves in part the reliability problem of gaseous devices (no thin wires), and competes very favorably in cost with solid state devices; coupled to a thick conversion volume of high-Z gas it allows to obtain high detection efficiencies for energetic X-rays. Its tolerance to very high radiation fluxes has been demonstrated, at least in laboratory conditions. For very high rate operation, however, the basic problem remains the data acquisition and storage: contrary to the general case met in particle physics, where the interesting events rate is a small fraction of the total flux, in most medical applications all events have to be recorded. Only a pixel device, with individual readout channels for each detection element and a sophisticated data acquisition/compression system can allow substantial progress.

To overcome the low efficiency limitation of the silicon pixel detector for X-ray detection, I propose to couple a gaseous detector, and namely a gas micro-strip chamber, to a pixel readout circuit such as the one described above; small pads etched on the back side of the MSGC plate can be bump-bonded to the readout circuit using the same technology as for the silicon detector. As a matter of fact, a physical bonding might even not be necessary: as described in section 2 and shown in fig. 8, an electrode on the back plane senses an induced signal through the dielectric substrate simply as a consequence of the avalanche growth and ions movement in the gas. One could imagine the readout pad to be an integral part of the readout circuit (a large bump bonding pad....without the bump), therefore not requiring mechanical or electrical connections between the active detector and the readout electronics. The advantage for replacement of faulty elements is obvious.

Fig. 24 shows an artist's view of this innovative device, that I propose to name the bump-less induction pad (BLIP) detector.

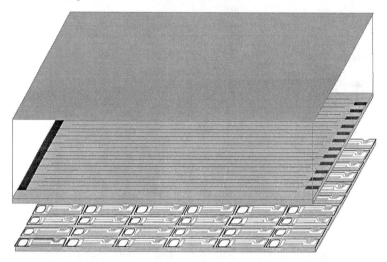

Fig. 24: A new digital imaging detector realized coupling a thin-plate micro-strip gas chamber with a pixel readout electronic chip.

REFERENCES

1. G. Charpak et al, Nucl. Instrum. Methods 62 (1968) 235.
2. G. Charpak, Ann. Rev. Nucl. Sci. 20 (1970) 195.
3. F. Sauli, Principles of operation of Multiwire proportional and drift chambers, CERN 77-09 (1977).
4. G. Charpak and F. Sauli, Ann. Rev. Nucl. Sci. 34 (1984) 285.
5. W. Blum and G. Rolandi, Particle detection with drift chambers (Springer-Verlag 1993).
6. C. Brand et al, Nucl. Instr. Methods A283 (1989) 567.
7. D. Nygren, PEP 197 (1975).
8. T. Lohse and W. Witzeling, The Time Projection Chamber, *in* Instrumentation in High Energy Physics, ed. by F. Sauli (World Scientific, 1992).
9. F. Sauli, Nucl. Instr. Methods A323 (1992) 1.
10. A.E. Babichev et al, Nucl. Instr. Methods A310 (1991) 449.
11. A. Martínez-Dávalos et al, Phys. Med. Biol. 38 (1993) 1419.
12. A. Walenta, Phys. Scripta 23 (1981) 353.
13. A. Oed, Nucl. Instrum. Methods A263 (1988) 351.
14. M. Geijsberts et al, Nucl. Instrum. Methods A313 (1992) 377.

15. R. Bouclier et al, Nucl. Instrum. Methods A332 (1993) 100.
16. RD–28 Status Report, CERN/DRDC/93–34 (13 August, 1993).
17. R. Bouclier et al, CERN-PPE/94-63 (1994).
18. Yu.N. Pestov and L. I. Shekhtman, Nucl. Instrum. Methods A338 (1994) 368.
19. R. Bouclier et al, CERN–PPE/93–192 (1993).
20. M. Salomon et al, TRI-PP-94-24 (1994).
21. CMS: The Compact Muon Solenoid, CERN/LHCC 92-3 (1992).
22. G.D. Minakov et al, Nucl. Instrum. Methods A326 (1993) 566.
23. SDC Technical Desigh Report, SDC-92-201 (1992)
24. The TERA project and the Centre for Oncological Hadrontherapy, ed. by U. Amaldi and M. Silari (INFN-LNF, 1994).
25. J. Kemmer, Nucl. Instrum. Methods 169 (1980) 499.
26. B. Hyams et al, Nucl. Instrum. Methods 205 (1983) 99.
27. A. Peisert, Silicon microstrip detectors, *in* Instrumentation in High Energy Physics, ed. by F. Sauli (World Scientific, 1992).
28. G. Hall,Semiconductor particle tracking detectors, IC/HEP/93/12 (1993). To be published in Report on Progress in Physics.
29. V. Radeka, Ann. Rev. Nucl. Part. Sci. 38 (1988) 217.
30. M. Caccia et al, Nucl. Instrum. Methods 260 (1987) 124.
31. P. Holl et al, Nucl. Instrum. Methods A257 (1987) 587.
32. J. Kemmer and G. Lutz, Nucl. Instrum. Methods A273 (1988) 588.
33. S. Masciocchi et at, IEEE Trans. Nucl. Sci. NS-40 (1993) 328.
34. See for example the contributions by S. Seidel and J. Skarha in these proceedings.
35. CMS, The Compact Muon Solenoid, CERN/LHCC 92-3.
36. H.J. Ziock et al, IEEE Trans. Nucl. Sci. NS-40 (1993) 344.
37. F. Arfelli et al, Silicon X-ray detector for synchrotron radiation digital radiography, INFN/TC-94/09 (1994).
38. W. Bencivelli et al, Silicon microstrip detectors for X-ray imaging, pres. at the 6th Pisa meeting on advanced detectors (Elba, May 1994).
39. C. Damerell et al, Nucl. Instrum. Methods A253 (1987) 478.
40. S. Shapiro et al, Nucl. Instrum. Methods A275 (1989) 580.
41. F. Krummenacher et al, Nucl. Instrum. Methods A288 (1990) 176.
42. M. Campbell et al, Nucl. Instrum. Methods A342 (1994) 52.
43. E.Heijne et al, First operation of a 72 k element hybrid silicon micropattern pixel detector array, CERN ECP EH1 (1994). Subm. Nucl. Instrum. Methods.
44. C. Da Già et al, Imaging of visible photons using hybrid silicon pixel detectors. CERN/ECP 93-18 (1993). Subm. Nucl. Instrum. Methods.

Physics with Relativistic Heavy Ions: QGP and Other Delicacies

GLENN R. YOUNG, *Oak Ridge National Laboratory*

PHYSICS WITH RELATIVISTIC HEAVY IONS:
QGP AND OTHER DELICACIES*

GLENN R. YOUNG

Physics Division, Oak Ridge National Laboratory,
Oak Ridge, Tennessee 37831-6372, U.S.A.

ABSTRACT

Conditions favorable to formation and observation of a deconfined state of quarks and gluons (often called the quark-gluon plasma) are thought to exist following the collision of very heavy nuclei at center-of-mass energies exceeding several tens of GeV/nucleon. The Relativistic Heavy Ion Collider under construction at BNL since 1991 is designed to provide such collisions at energies up to $\sqrt{s}/A = 200$ GeV. Two large dedicated experiments are being built to operate there; these two experiments take rather different approaches to the problem of classifying such collisions and probing for signals of QGP formation. Two smaller experiments are proposed to focus on specific aspects of these collisions. Recent developments in the understanding of the initial state formed in such collisions include, particularly, the possible rapid equilibration of the gluon density, leading in an equilibrium picture to such high temperatures that sizable thermal excitation of charm becomes probable. Recent theoretical conjectures have focussed on the possible formation of a disordered chiral condensate following chiral symmetry restoration in heavy-nucleus collisions, which might be a consequence of nonequilibrium deexcitation of a dense partonic state.

1. Introduction

This talk addresses the general considerations for formation of a deconfined state of quarks and gluons in collisions of very heavy nuclei (colloquially known as "heavy-ion collisions") at energies of several tens of GeV/nucleon in the center of mass. Remarks on the present status of the Relativistic Heavy Ion Collider (RHIC) under construction at BNL since 1991 and the two large experiments already under construction to run at RHIC are given next. This is followed by a discussion of likely conditions formed in the initial stages of a collision at RHIC, plus a sketch of the recent theoretical conjectures concerning nonequilibrium decay of such a system, particularly the chance of forming and observing a chirally disordered phase following chiral symmetry restoration.

It has long been conjectured[1,2] that, were it heated and/or compressed sufficiently, strongly-interacting matter would undergo a transition from a system of quarks and gluons confined into bound mesons and baryons to a system of deconfined quarks and gluons free to move over some region of space much larger than a typical hadron.

*Research sponsored by the U.S. Department of Energy under contract DE-AC05-84OR21400 with Martin Marietta Energy Systems, Inc.

A phase diagram resulting from viewing this transition as involving heating and/or compressing nuclear matter is depicted in Fig. 1. More recent theoretical study suggests viewing a collision of heavy nuclei at relativistic energies as being a large set of partonic collisions resulting in many nearby (but probably causally disconnected) regions of large parton density, whose subsequent evolution necessarily involves expansion and "cooling" and may involve crossing a phase boundary. Whether there is an actual phase transition is the subject of intense debate; if there is one, the present view from lattice QCD studies as well as general considerations of the behavior of QCD in the two-flavor limit, is that it is likely to be of second order.[2-4]

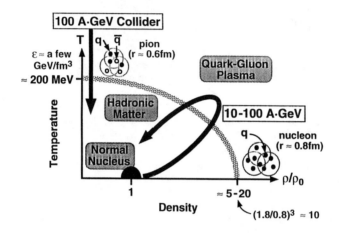

Fig. 1. Phase diagram of nuclear matter and nuclear collisions (figure from Ref.[7]).

Estimates of the energy densities of a pion gas vs a deconfined system of quarks and gluons (popularly dubbed a quark-gluon plasma, or QGP) made in a Stefan–Boltzmann picture can be compared to see what increase in energy density must be achieved before formation of a QGP is even possible. The estimate illustrated in Fig. 2, which only sums degrees of freedom for a two-flavor case, suggests some 2–3 GeV/fm^3 must be added to the normal rest-energy density of an atomic nucleus of some 0.15 GeV/fm^3. That such an energy density is a likely occurrence in a relativistic heavy-ion collision can be estimated using a simple formula suggested by Bjorken,[5] specifically $\epsilon = \frac{1}{\pi R^2 \tau_0} \times \frac{dE_T}{dy}$. For Au nuclei, $R = 6.6$ fm and dE_T/dy is estimated (using versions of, e.g., ISAJET or the Lund Model Monte Carlo codes which have been adapted to the case of collisions of heavy nuclei) to be of the order of 700 GeV for 100 GeV/nucleon + 100 GeV/nucleon collisions in a collider. The formation time τ_0 is estimated (e.g., from $e^+e^- \to$ hadrons results) to be of the order of 1 fm/c, although it is rather uncertain. This leads to energy density estimates of 3–6 GeV/fm^3, well in excess of the expected 2 GeV/fm^3 energy density expected to be required to achieve deconfinement.

	Pion Gas	Quark-Gluon Plasma	
	pion	quark	gluon
Spin/Polarization	1	2	2
Isospin	3	2	1
Color	1	3	8
Total Degrees of Freedom (f)	3	12	16

$$\text{Boson: } \varepsilon = f_B \times \frac{\pi^2 T^4}{30} \qquad q\bar{q}: \varepsilon_q + \varepsilon_{\bar{q}} = (f_q + f_{\bar{q}}) \times \frac{7}{8} \frac{\pi^2 T^4}{30}$$

$$\varepsilon_{\text{pion gas}} = 3 \times \frac{\pi^2 T^4}{30} \qquad \varepsilon_{QGP} = 37 \times \frac{\pi^2 T^4}{30} \approx 2\text{-}3 \text{ GeV/fm}^3 \text{ at } T_C$$

Fig. 2. Degrees of freedom for pion and quark-gluon gasses; u and d quarks alone (figure from Ref.[7]).

2. RHIC Facility at BNL

The premier facility for studying such collisions will be the Relativistic Heavy Ion Collider now under construction (since 1991) at Brookhaven National Laboratory. It will use the existing AGS/Booster/Tandem Van de Graaff/proton linac accelerator complex as injector and makes use of the tunnel, helium refrigerator, and experimental halls constructed for the ISABELLE/CBA project in the early 1980s. Beams of Au ions can be accelerated to energies of 100 GeV/nucleon. Lighter ions such as ^{12}C and ^{28}Si can be accelerated to somewhat higher energies of 125 GeV/nucleon due to their larger charge-to-mass ratios, Z/A, and protons can be accelerated to 250 GeV. The machine involves two rings of superconducting magnets, so collisions between any pair of nuclear species can be arranged, in principle. A site plan for RHIC and the associated accelerators is shown in Fig. 3. The total project cost is of the order of $550M, including the experimental program. The first industrially-produced accelerator dipole magnet was delivered from Northrop–Grumman to BNL and passed its quench test with 40% margin on May 9 of this year. A dozen dipoles have been delivered since and have passed quench tests; all have passed field-quality checks also. Quadrupole, sextupole, corrector, and insertion quadrupole construction are all progressing well. Most civil construction is complete for the project, including preparing the two out of six experimental areas not completed as part of ISABELLE/CBA. Installation of dipoles into the ring has commenced and will continue over the next two years.

The expected luminosity for RHIC is rather low, and must seem particularly so to those who have been engaged in designing SSC and LHC experiments. The design luminosity for Au–Au at top energy is 2×10^{26}/cm^2/s, and for $p - p$ it is 10^{31}/cm^2/s. The proton luminosity in fact does not make full use of available AGS proton intensity and could be increased by an order of magnitude. The Au–Au luminosity could also be increased by a similar amount with a straightforward program of upgrades planned by

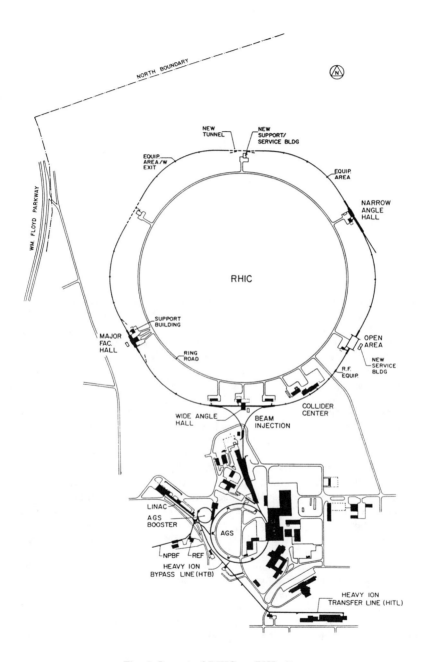

Fig. 3. Layout of RHIC on BNL site.

the RHIC staff. It is worth noting, however, that production rates for massive objects (e.g., J/ψ and Υ mesons, Drell–Yan pairs) are well-known to scale with mass A of the colliding nuclei as $\sim A^{1.8-2.0}$ in nuclear collisions, which compensates a seemingly low luminosity by more than four orders of magnitude when count rates are computed.

There are two major experiments now approved and under construction for RHIC. These are named STAR and PHENIX and are described further below. Various smaller experiments are proposed, with two (BRAHMS and PHOBOS) now moving through the initial technical and budgetary review process.

There are also other areas of study that can or will be pursued at RHIC but which will not be discussed at any length here. These include:

a. Study of two-photon collisions at RHIC (and also at LHC) when heavy ions are collided. This offers a chance to study pseudoscalar mesons with little background yet good cross section; it is even possible to talk of W, Z, and Higgs production (for a Higgs of mass less than twice the W mass) at such colliders. The nuclei do not come within a strong-interaction radius; rather, the strong Z^4 dependence of the photon flux and thus of the two-photon luminosity lets one select only distant collisions of the ions, meaning no particles are formed from strong-interaction collisions of the entrance channel nucleons.[6]

b. Polarized proton operation of RHIC. A "magic" momentum exists for the injection lines of RHIC, such that polarized protons accelerated in the AGS can be transferred to RHIC without losing their polarization. A series of Siberian snakes and spin rotators then lets a program of spin physics be carried out for $p-p$ collisions up to $\sqrt{s} = 500$ GeV, and for $p+A$ collisions up to $\sqrt{s} = 250$ GeV. Measurements would include polarized Drell–Yan to study quark helicity and transversity, heavy flavor (J/ψ, ψ', Υ) and direct photon production to study gluon helicity, and $W^{\pm}$ and Z^0 production from polarized beams.

c. Operation of RHIC as a facility for $B\overline{B}$ physics. This has been addressed in previous RHIC summer studies, and the case made for this physics at a $p-p$ machine with luminosity up to $10^{32}/\mathrm{cm}^2/\mathrm{s}$ or more at $\sqrt{s} = 500$ GeV.

3.   Experiments at RHIC

The first experiment approved for RHIC is known as STAR (Solenoidal Tracker at RHIC). It consists of a large Time Projection Chamber (TPC), surrounded by a time-of-flight barrel and an electromagnetic calorimeter. The TPC will have a silicon-drift microvertex detector at its center and external TPCs at high rapidity. An isometric view of STAR is shown in Fig. 4.

STAR emphasizes full track reconstruction and particle identification for all particles emitted in the central two units of rapidity at RHIC. Since rapidity densities of charged particles can exceed $dN/dy = 1000$, the design and event analysis for STAR are quite challenging tasks. Prototype TPC tests have shown that required chamber performance can be achieved, and event/track simulation Monte Carlos have shown that track reconstruction in the dense environment can be accomplished. The track density is not unlike that found in a jet which has fragmented into hadrons, but occurs

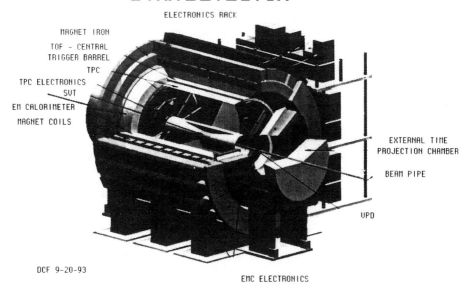

Fig. 4. STAR detector.

over nearly the entire detector volume. STAR can perform particle identification for π, K, and p over the central two units of pseudorapidity and expects to obtain momentum spectra for $\pi^\pm$, $K^\pm$, K_s^0, p, $\overline{p}$, d, $\overline{d}$, ϕ, $\Lambda/\overline{\Lambda}$, Ξ^+, Ξ^-, and Ω^-.

The identification of multistrange hyperons and antihyperons is of particular interest and is a prime motivation for including the silicon-drift microvertex detector in STAR in order to spot displaced decay vertices. An illustration of the underlying physics motivation is shown in Fig. 5 (see Ref.[7]). The basic physics idea is that for a system at a temperature of 200 MeV (i.e., the QCD critical temperature), the thermal excitation of strange quarks is greatly enhanced. This coupled with strange quarks' lack of Pauli-blocking (unlike the case for the valence u and d quarks) makes creation of $s\overline{s}$ pairs quite likely, largely via $gg \rightarrow s\overline{s}$ reactions. It is easy for the $\overline{s}$ quarks to form mesons and the s quarks to form baryons, as shown in the figure, due to the large number densities of u, d quarks and the low densities for the corresponding antiquarks. Formation of multistrange antihyperons in particular is sensitive to the temperature reached due to the need to form not only several antiquarks, but for the mean temperature to favor formation of antistrange quarks.

Given its large coverage, STAR can perform fluctuation analyses of single events, which is a means of searching, e.g., for the analog in a QGP of the critical opalescence expected for a second-order phase transition. This large coverage also makes it possible

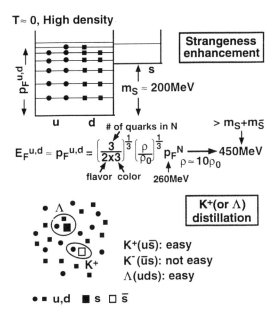

Fig. 5. Strangeness enhancement and K^+ (or Λ) distillation (figure from Ref.[7]).

to extract the spectral shape parameters for identified particles; these can be related to the temperature reached during a given collision. This large coverage also means that small-relative-momentum correlations can be studied for single events, making it possible to study single event shapes, which might provide evidence for rapid or even violent cooling of the reaction volume, as might be expected following a first-order transition. STAR can also measure jet, photon, photon-jet, and jet-jet cross sections for all types of collisions at RHIC. These distributions are conjectured to provide information on the propagation of partons in a deconfined QCD phase.[8,9]

The PHENIX experiment, which is also under construction, emphasizes detection of lepton pairs, photons, and identified hadrons over a wide range in p_T and mass. PHENIX has emphasized running at full possible RHIC luminosity, coupled with selective triggering, and high event- and data-rates given that it is pursuing infrequently-produced objects. One philosophy guiding the design of the experiment has been to be as inclusive as possible regarding potential signatures of QGP formation (since there are no "guaranteed" signatures for QGP formation agreed upon by the interested community) and to look for correlated changes in several observables as a function of some global parameter such as charged-particle rapidity density or $dE_T/d\eta$. The list of possible signatures which PHENIX will be able to observe includes:

Debye-screening of strong interactions among partons due to large density of partons, resulting in a "suppression" of the formation probability of certain massive vector mesons [such as the J/ψ, ψ', $\Upsilon(2s)$, $\Upsilon(3s)$] at low p_T

Chiral symmetry restoration, resulting in changes in branching ratios and widths, particularly for the ϕ meson, and possibly a dramatic narrowing of very wide mesons such as the σ

Thermal radiation from a hot gas of partons, resulting in copious emission of lepton pairs and direct photons with transverse mass in the range 1–4 GeV/c^2

Signals of phase transition phenomena, either for a first order transition with a large latent heat and resulting marked change in observables such as mean p_T, or for a second order transition with resultant fluctuation phenomena, such as non-Gaussian distribution of the neutral to total pion number density

Production of strange and charmed quarks in great numbers, observed via copious production of kaons, ϕ mesons, and charmed mesons

Jet quenching, occurring due to increased energy loss for a jet trying to penetrate a QGP, and observed via the leading high-p_T products from jet fragmentation to hadrons

"Nonstandard" space-time evolution of the system produced, resulting in very long emission times for hadrons as could happen if a long proper time were needed for a QGP to cool to a hadronic state, observed via HBT correlations of identified pions and kaons produced at small ΔQ

An isometric view of PHENIX is shown in Fig. 6. In contrast to STAR, PHENIX is better thought of as a collection of spectrometers with very fine granularity and including an array of specialized detectors to handle electron, muon, and photon detection. Those for electron detection are the most extensive, because the momentum range covered for electrons (150 MeV/c – 4 GeV/c and above) does not lend itself to the use of only one or two electron detection methods, instead requiring several with overlapping ranges of applicability as a function of electron momentum.

The luminosity of RHIC is not nearly as high as that of present high-energy colliders, leading to some concern about counting rates for the infrequently created objects PHENIX is designed to observe. The luminosity is compensated by the scaling of production rates as $\sim A^{1.8-2.0}$, as noted above. The resulting estimated counting statistics for various vector mesons, Drell–Yan pairs, and for charmed-meson $D\overline{D}$ pairs, are given in Table 1, from which it can be seen that statistics should not be a limiting factor. It should be kept in mind that p_T coverage out to $p_T/M > 2$ is usually required to distinguish QGP from hadron-gas models of the collision. PHENIX was thus designed to ensure counting rates of several tens of thousand per vector meson species of interest in order to assure good counting statistics for $p_T/M > 2$.

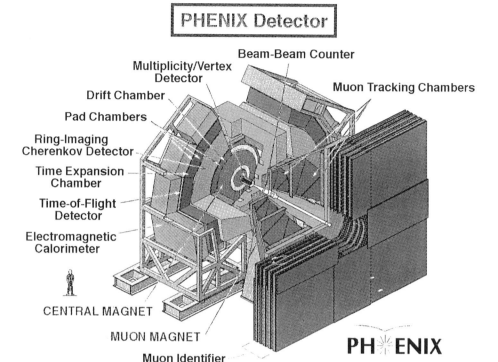

Fig. 6. Isometric view of the PHENIX detector, showing the various spectrometer elements.

4.   Recent Theoretical Results Concerning a "Hot Glue" Scenario

There has been a significant shift of theoretical viewpoint over the past three years concerning the likely energy densities (resp. temperatures) reached during the initial stages of a relativistic collision of heavy nuclei. (For example, compare the summary talk from the QM'91 conference[1] and the theoretical talks in the QM'93 conference.[2]) It has been realized that the initial stages of these collisions, at center-of-mass energies which RHIC and the LHC will reach, will be dominated by semihard partonic processes. This has two main sources: the cross section for $gg \rightarrow gg$ is large, and the measured structure functions indicate that most of the incoming partons of interest are gluons. This means that processes with momentum transfers $Q \sim 1 - 3$ GeV/c are not at all "rare" or "isolated" in a sea of soft collisions, but are instead so frequent that a cascade calculation is needed to describe the situation properly, due to the large probability that a parton will undergo multiple collisions with other partons. Such calculations are now available (see, for example, Ref.[10]) from several groups. One significant outcome is that the estimates for $T_{initial}$ achieved must be revised upwards by more than a factor of two, which means energy densities reached increase more than an order of magnitude (recall that a Stefan–Boltzmann gas has $\epsilon \propto T^4$).

Table 1. Typical yields per RHIC-year for various species into the PHENIX muon acceptance.

	Au + Au (Central*)	p + Au (Min. bias)
$\rho \to \mu^+\mu^-$	309K	920K
$\phi \to \mu^+\mu^-$	139K	500K
$J/\psi \to \mu^+\mu^-$	390K	1060K
$\psi' \to \mu^+\mu^-$	5.7K	15K
$\Upsilon \to \mu^+\mu^-$	1.2K	1.4K
$\Upsilon' \to \mu^+\mu^-$	0.12K	0.14K
Drell-Yan $\to \mu^+\mu^-$ at 3 GeV	15K	26K
$D\overline{D} \to \mu^+\mu^-$ at 3 GeV	26K	73K
$D\overline{D} \to e - \mu$	140K	370K

*Au + Au central collisions correspond to 1% most central collisions for 2×10^{27} luminosity, or to top 10% for 2×10^{26} ("Blue Book") luminosity.

The Drell-Yan and $D\overline{D}$ yields are quoted per GeV/c². Yields into the electron arms are 3–10 times smaller due to acceptance and solid angle.

In the picture which has held sway earlier, namely the Bjorken scaling scenario,[5] one expected that QGP formation would be complete by $\tau_0 \sim 1$ fm/c. This leads to estimates of $T_{initial} \sim 240$ MeV (for RHIC) and ~ 290 MeV (for LHC). The present view (which has its roots in suggestions by Shuryak some time ago[11,12]) leads to a description of the proper time evolution of the collision in brief as follows:

$0 < \tau < 0.3$ fm/c
- entropy production
- parton momentum equilibration via $gg \to gg$ and $gq \to gq$

$0.3 < \tau < 3$ fm/c
- gluon chemical equilibration via $gg \to Ng$ (where $N = 3, 4, 5, \ldots$) (described by the usual gluon splitting of QCD)

$3 < \tau < \tau_{crit} (\sim 5 - 7$ fm/c)
- expansion of a gluon-dominated plasma, which then evolves ("cools") first to a state of mixed QGP and hadrons and then finally into a purely hadronic state

Such considerations result in estimates of $T_{initial} \sim$ 400–500 MeV (for RHIC) and 600–900 MeV (for LHC).

Predictions for other observables change significantly enough to require some rethinking of experimental strategy in searching for evidence of deconfinement. Three obvious areas to consider are:

- The predicted cross sections for thermal direct photon and lepton pair emission in the range of $M_{transverse} = 2 - 4$ GeV/c^2 increase substantially, since they depend on T^3 or T^4.

- Over 50% of the E_T is expected to be carried by minijets (already a subject of considerable study, see Refs.[13–16]), as opposed to hadrons formed late in a cascade.

- Open charm production is greatly enhanced, since the mean temperature is now within a factor of 2 or 3 of the charmed quark mass. One expects "direct" open charm production at the level of $dN/dy \sim 0.5$/event (for Au–Au central collisions at RHIC). For thermal production from the initial phase, one estimate[17] reaches $dN/dy \sim 0.1$, 1.0, or 10.0/event for $T_{initial}$ of 300, 400, or 550 MeV. Wang and Müller[18] have pointed out that charm production now becomes a useful experimental observable to diagnose preequilibrium times in the collision, a regime which was felt earlier to be quite difficult to probe.

5. Modification of Hadrons at Finite Temperature

It has long been speculated that observable properties of hadrons should change if they are immersed in a dense medium of strongly-interacting particles. Speculation related to QGP formation has been particularly intense concerning the properties of vector mesons; in this case a decay channel to lepton pairs remains open and presumably unaffected by an environment which might modify properties of hadronic decays of vector mesons.

A simple view is to take the constituent quark model quite literally and assume that a baryon mass is three times that of a constituent quark mass and that a meson mass is twice the constituent mass. Then one naively expects that if the constituent mass tends towards zero, so will the hadrons' masses. (That is, this would be another consequence of chiral symmetry restoration.) A prescription for the temperature dependent behavior of a hadron's mass could then be

$$m_{hadrons}(T) \sim f_\pi(T) \sim |\langle \psi\psi \rangle (T)|^{(1/3)} \sim (1 - T^2/T_{crit}^2)^{(1/6)} \ ,$$

which then tends towards zero as T approaches the critical temperature T_{crit}.

A better view[19] then takes into account the following:

- Not all mesons are alike.

- Instanton effects (e.g., the η is heavy and experiences a repulsive interaction, while the π is light and experiences an attractive one) are suppressed at high T. This leads to the question if masses for different species first tend to come together to a common value before all tend to decrease towards zero around $T_{critical}$. That is, does the η mass decrease in the same manner as do those of the (ρ, ω, ϕ)?

– Effects due to resonance interaction (e.g., the ρ for $\pi\pi$ and the K^* for $K\pi$).

– Mesons (and nucleons?) act as quasiparticles in a strongly interacting environment and thus are not absorbed strongly $\rightarrow$ a collective potential exists, and this potential is momentum dependent. Thus pionic matter is not an ideal gas, but rather more closely resembles a liquid with attractive interactions. (At low wave number k, the interactions of a pion vanish as $k \rightarrow 0$.)

These modifications are compensated away from $T_{critical}$, but are possibly significant near $T_{critical}$.[19]

A favored area for speculation along these lines regarding observables is the decay properties of the ϕ meson. Its lifetime is $\tau = 45$ fm/c, meaning that some fraction should decay inside any "fireball" produced (although this same lifetime means some ϕ's will almost invariably decay "outside," but this apparent shortcoming might become an experimental advantage in deciding whether any observable effect exists, and what fraction of either the collisions observed or of the proper lifetime of the system formed, exhibits any medium modification effects at all.) The ϕ mass is only 30 MeV above the KK mass, yet $\phi \rightarrow KK$ is its dominant decay mode. Thus its decay should be quite sensitive to any relative shift at all of the dispersion curves of the ϕ and K in hot hadronic matter, as noted by R. Pisarski and E. Shuryak. Additionally, the ϕ has isospin 0. Since π scattering at low energies is proportional to isospin, one expects no strong resonances decaying into $\phi + \pi$. Thus one expects that the ϕ is modified in-medium less than is the K.

Experimental consequences of this could be:

1. If the K mass increases just a little (15 MeV is only 3% of the K mass), then the main ϕ decay channel is blocked. (The K^+K^- channel takes 49.1% of the ϕ decays and the $K_L^0 K_S^0$ channel takes 34.3%, with $\rho\pi$ being next at 12.9%.)

2. If the K mass decreases, then Γ_ϕ can increase by a large amount.

3. Since the $\phi \rightarrow e^+e^-$ and $\mu^+\mu^-$ channels should not be affected, they can be used as a control.

An open question is whether the mass peak in $\phi \rightarrow l^+l^-$ also shifts. A schematic picture of two differing views is shown in Fig. 7. Similar arguments have been presented for the ρ meson.[20] The ρ and A_1 are chiral partners, so their masses might be expected to become equal if chiral symmetry is restored. An open question pointed out by these authors is whether the phase transition dynamics do, indeed, lead to critical points in the (m, T) plane which actually reach, as opposed to just approach, one another for the QCD deconfinement and the chiral symmetry restoring phase transitions.

6.   Disordered Chiral Condensates

Much of what has been discussed above and in the body of literature concerning this subject has dealt with equilibrium behavior of the system. It is reasonable to ask what might the consequences be for relativistic heavy nucleus collisions should the system formed evolve in a highly nonequilibrium manner. One such area which has

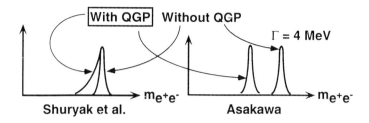

Fig. 7. Two views on possible in-medium behavior of the ϕ meson (figure from Ref.[7]).

received intensive theoretical attention in the past several years is that of disordered chiral condensates,[21-26] which is discussed briefly below.

The general idea of these investigations is to study the chiral order parameter $\langle \bar{q}_L q_R \rangle$ in a region far from equilibrium following a QCD phase transition. Lattice QCD calculations lead us to expect that chiral symmetry is restored under similar (or the same) conditions as when deconfinement occurs. Near $T_{critical}$, the chiral order parameter $\langle \bar{q}_L q_R \rangle$ and the QCD pion-like parameter $\langle \bar{q}_a \gamma_5 q_b \rangle$ become equivalent. (As $T \to 0$, domains form with nonzero value of this latter parameter, given nonzero pion content.) One wants then to probe the correlation length of the chiral order parameter. (The condensed matter analog is the size of domains in a spin system. Similarly, the study of QCD Debye-screening via heavy vector meson suppression is a probe of the color correlation length; in this case the condensed matter analog is the Mott metal-to-insulator transition.) If chiral symmetry is restored, all directions in $\vec{\pi}$ (the direction in isospin) are equally likely. One then wants to know if one (or a few) large domains form with any favored direction in $\vec{\pi}$. (If many small domains form, even with a favored direction in $\vec{\pi}$ for each, the gaussian limit for the aggregate of all domains is approached, and an experiment will only find the familiar equal numbers of each isospin type.) One can further ask the question whether long-wavelength modes are excited in the system, whether there are limits to the wavelength of the modes (m_π^{-1} or larger?), and whether there are expected experimental consequences. This all has its roots in the long-term speculation whether regions of "misaligned" vacuum occur. Defining $\phi^\alpha \equiv (\sigma, \vec{\pi})$ (which has the expectation value $(\nu, 0)$ in the ground state), if ϕ^α becomes partly aligned along one of the $\vec{\pi}$ directions in such a misaligned region, then due to explicit chiral symmetry breaking (i.e., $m_\pi \neq 0$) in such a region, ϕ^α will oscillate about the σ direction. Such a region can be expected to decay by coherent pion emission, leading to clusters of pions, bunched in rapidity, with possibly highly non-Gaussian charge distributions. Defining R to be the ratio of neutral pions to all pions, then, for example, Rajagopal and Wilczek[27] find a probability distribution for such clusters' value of R

$$P(R) = 0.5 R^{-1/2} ,$$

which is not at all Gaussian, and leads to the unusual expectation that there is a 10% chance for the π^0 fraction to be less than 1%!

A particular example has been put forth by Rajagopal and Wilczek,[27] who model the central region of a heavy nucleus collision, assume that the baryon density is locally negligible, and assume the chiral order parameter $\langle \bar{q}_L q_R \rangle = 0$, i.e., is disordered after the expected transition to a chirally symmetric state. They first ask if this is likely for an equilibrium transition. For QCD with two massless flavors and a second order phase transition (as now seems likely from the lattice QCD work of, for example, Christ et al.[4]), formation of such a state initially seems possible because second order transitions have long-wavelength critical fluctuations. However, since m_u and m_d are nonzero, this is not likely in reality. Near $T_{critical}$, the correlation length in the π channel is less than $T_{critical}^{-1}$, meaning misaligned regions are small, leading to small or no chance to radiate the copious numbers of pions needed to observe this phenomenon in an experiment.

Their model thus assumes nonequilibrium evolution; the system is postulated to undergo a "quench." This means the $(\sigma, \vec{\pi})$ field is suddenly removed from contact with a high-temperature heat bath, and its subsequent evolution modelled using the $T = 0$ equations of motion. (For a real heavy nucleus collision, life is somewhere in between the extremes of an equilibrium phase transition, where T decreases arbitrarily slowly, and a quench, where thermal fluctuations stop instantaneously. The verdict will be out on this at least until data from RHIC are available and subject to analysis.)

Rajagopal and Wilczek performed numerical simulations in the linear sigma model, using $T = 1.2 T_{critical}$ as a starting point. They found large power concentrated in the long wavelength modes of their model. At $\tau_{initial}$, the power spectrum of these modes is white, and at late τ, the energy is equipartitioned among the modes. But, for intermediate times, $\tau \sim$ a few times m_π^{-1}, the low wave number modes can reach 100 (or more) times the amplitude of the modes in the initial white power spectrum. This leads to large fluctuations in the value of R. A similar analysis examining possible fluctuations in electric charge alone [i.e., between $\pi^1 = 1/\sqrt{2}(\pi^+ + \pi^-)$ and $\pi^2 = 1/\sqrt{2}(\pi^+ - \pi^-)$] concluded that long wavelength modes for charge were not amplified. This means an experimentalist needs to measure both charged and neutral pions in the same kinematical region. Rajagopal and Wilczek also found that m_π^{-1} did not form an obvious limit to the wavelength. The model results were limited only by the size of the lattice on which they were run, suggesting in an experiment that the modes could grow to wavelengths on the order of the transverse size of the fireball formed. Longitudinal modes are damped by the rapid longitudinal expansion of the fireball.

As a closing experimental note on this subject, recall that two quite odd and rare types of cosmic ray events are known that might find their explanation in this idea:

– Centauro events, exhibiting for example 50 π^+, 50 π^{-1}, and 1-2 π^0

– anti-Centauro events, exhibiting 32 photons and only 1-2 $\pi^\pm$

7.   Concluding Remarks

There has been an active experimental program in this area at the BNL AGS and CERN SPS since 1986. Tantalizing hints of the expected behavior characteristic of a QGP have been found in several observables,[1,2] but there is also a healthy industry engaged in trying to understand whether all these observations can be understood in the framework of conventional hadronic physics. The energy densities and proper lifetimes reached at RHIC should be much larger (and the running time much longer) than those reached at the AGS and SPS. The expectation is that this leads to a favorable experimental situation for resolving the conjectures about existence and properties of the QCD confinement transition and chiral symmetry restoration transition.

References

1. Proceedings of the Quark Matter '91 Conference, *Nucl. Phys.* **A544** (1992) and references therein.
2. Proceedings of the Quark Matter '93 Conference, *Nucl. Phys.* **A566** (1994) and references therein.
3. F. Wilczek, *Nucl. Phys.* **A566** (1994) 123c.
4. N. H. Christ, *Nucl. Phys.* **A544** (1992) 81c and private communication.
5. J. D. Bjorken, *Phys. Rev. D* **27** (1983) 140.
6. *Workshop on Can RHIC be Used to Test QED?*, edited by M. Fatyga and M. J. Tannenbaum, April 20-21, 1990, BNL, Upton, New York, Report BNL-52247.
7. Figure courtesy of Shoji Nagamiya (private communication).
8. Xin-Nian Wang, Miklos Gyulassy, and Michael Plümer, Report LBL-35980.
9. *Proceedings of the Workshop on Pre-Equilibrium Parton Dynamics*, edited by Xin-Nian Wang, August 23–September 3, 1993, Berkeley, Calif., Report LBL-34831, CONF-9308181, and references therein.
10. K. Geiger and B. Müller, *Nucl. Phys.* **B369** (1992) 600.
11. E. V. Shuryak, *Sov. J. Nucl. Phys.* **28** (1978) 408.
12. E. V. Shuryak, *Phys. Rep.* **61** (1980) 72.
13. K. Kajantie, P. V. Landshoff, J. Lindfors, *Phys. Rev. Lett.* **59** (1987) 2517.
14. K. J. Eskola, K. Kajantie, J. Lindfors, *Nucl. Phys.* **B323** (1989) 37.
15. X.-N. Wang and M. Gyulassy, *Phys. Rev. D* **44** (1991) 3501.
16. X.-N. Wang and M. Gyulassy, *Phys. Rev. Lett.* **68** (1992) 1480.
17. E. V. Shuryak, *Nucl. Phys.* **A566** (1994) 559c.
18. X.-N. Wang and B. Müller, *Nucl. Phys.* **A566** (1994) 555c.
19. E. V. Shuryak, *Nucl. Phys.* **A544** (1992) 65c.
20. R. Pisarski, private communication (1994); R. Pisarski, S. Gavin, and R. Goksch, BNL preprint (1994).
21. A. Anselm and M. Ryskin, *Phys. Lett. B* **226** (1991) 482.
22. J.-P. Blaizot and A. Krzywicki, *Phys. Rev. D* **46** (1992) 246.
23. J. D. Bjorken, *Int. J. Mod. Phys.* **A7** (1992) 4189; *Acta Phys. Pol.* **B23** (1992) 561.
24. K. L. Kowalski and C. C. Taylor, preprint hepph/9211282 (1992).
25. J. D. Bjorken, K. L. Kowalski, C. C. Taylor, preprint SLAC-PUB-6109 (1993).
26. K. Rajagopal, *Nucl. Phys.* **A566** (1994) 567c.
27. K. Rajagopal and F. Wilczek, *Nucl. Phys.* **B399** (1993) 395.

Proton Structure, Diffraction and Jets in ep Scattering at HERA

JAMES WHITMORE, *Pennsylvania State*

PROTON STRUCTURE, DIFFRACTION AND JETS IN ep SCATTERING AT HERA

J. WHITMORE

Department of Physics, Pennsylvania State University, University Park, PA 16802, USA

and

DESY, Notkestrasse 85, 22603 Hamburg, Germany

ABSTRACT

The two HERA experiments, H1 and ZEUS, had their second running period during 1993, each collecting a data sample corresponding to an integrated luminosity of 0.5 pb^{-1}, a twentyfold increase in statistics over the previous running period. This increase together with an improved understanding of the detectors has brought a wide range of physics questions within the reach of the experiments. In this report we give a brief overview of some of the studies recently performed by these two collaborations.

1. Introduction

From May through October 1993 the HERA electron-proton collider at DESY provided the two collaborations H1 and ZEUS with their second data-taking period. During this time HERA delivered over 1 pb^{-1} of luminosity, twenty times more than during the first runs in 1992. This increase in statistics permits a wide range of investigations in a kinematic range inaccessible to previous fixed target experiments. Some highlights of the HERA experimental program are described in this report:

- the measurements of the proton structure function F_2 in the range $8.5 < Q^2 < 10^4$ GeV^2 and $x > 3 \cdot 10^{-4}$;

- the measurements of the gluon distribution in the proton;

- a measurement of the charged current cross section in a totally new kinematic regime of high Q^2 and a comparison with the neutral current cross section;

- the measurement of the total γp cross section at center of mass energies around 200 GeV and the separation of elastic, diffractive and non-diffractive components;

- the observation of vector meson production in photoproduction ($Q^2 \approx 0$ GeV^2);

- the observation of point-like (*direct*) and non-point-like (*resolved*) hard interactions of the photon with the proton; and

- the observation of hard scattering in large-rapidity-gap events in both photoproduction and DIS and the study of pomeron factorization in the latter.

2.   Experimental setup

2.1.   HERA machine conditions

The two experiments, H1 and ZEUS, are located at the electron-proton collider HERA at DESY. During 1993 HERA operated with bunches of electrons of energy $E_e = 26.7$ GeV colliding with bunches of protons of energy $E_p = 820$ GeV, with a time between bunch crossings of 96 ns. HERA is designed to run with 210 bunches in each of the electron and proton rings. For the 1993 data-taking, 84 paired bunches were filled for each beam and in addition 10 electron and 6 proton bunches were left unpaired for beam-gas background studies. The electron and proton beam currents were typically 10 mA. The design currents are 58 mA for electrons and 163 mA for protons, resulting in a design luminosity of $1.5 \cdot 10^{31}$ cm^{-2}s^{-1}.

2.2.   Kinematics of Deep inelastic scattering

The kinematic variables used to describe deep inelastic scattering (DIS) events

$$e\,(k)\,+\,p\,(P) \rightarrow e\,(k')\,+\,anything \tag{1}$$

are as follows. In lowest order QCD, the space-like photon with four-momentum q and virtuality $Q^2 = -q^2 = -(k - k')^2$ scatters off a parton in the proton carrying a fraction $x = Q^2/(2P \cdot q)$ of the proton's momentum P and produces a single jet in the final state in addition to the proton remnant. This picture is modified in higher order where multiple gluons may be radiated off the incoming or outgoing partons. In addition, use will be made of the variable which describes the energy transfer to the hadronic final state: $y = \frac{q \cdot P}{k \cdot P}$; and W, the center of mass (CM) energy of the $\gamma^* p$ system, with $W^2 = (q + P)^2 = \frac{Q^2(1-x)}{x} + m_p^2$, where m_p is the proton mass.

The large CM energy available at HERA allows measurements to be made for x values as small as 10^{-4} at Q^2 values of the order of 10 GeV^2, at least two orders of magnitude below the lowest values measured in fixed target experiments[1-3] at this Q^2. In addition, HERA covers a new kinematic regime with Q^2 ranging up to 10^5 GeV^2. The kinematic regions spanned by the various experiments are compared in Fig. 1.

2.3.   The H1 and ZEUS detectors

Detailed descriptions of the H1[4] and ZEUS[5,6] detectors exist so only a brief overview of the two multipurpose magnetic detectors is given here. In H1, a tracking system of jet and multiwire proportional chambers surrounds the interaction point (IP). This system and the liquid argon calorimeter are arranged inside a superconducting coil with a diameter of 6 m providing a magnetic field of 1.2 T. In ZEUS, charged particles are tracked by two detectors: the vertex detector surrounding the beampipe followed by a central drift chamber. A homogeneous axial magnetic field of 1.42 T is provided by a superconducting coil which immediately surrounds the tracking system.

In both H1 and ZEUS the tracking systems are enclosed by electromagnetic and hadronic calorimeters. H1 uses liquid argon with lead absorbers in the electromagnetic sections and steel absorbers in the hadronic sections. This system extends over the pseudorapidity range $3.4 > \eta > -1.4$, where the pseudorapidity η is defined as $-\ln(\tan\frac{\theta}{2})$

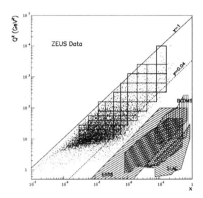

Figure 1: The Q^2 vs. x plane showing the ZEUS data and the region measured by BCDMS, NMC, SLAC and E665.

and the polar angle θ is taken with respect to the proton beam direction. The backward region of the detector, $-1.35 > \eta > -3.6$, is closed by a lead-scintillator calorimeter which is used predominantly to trigger on and to measure the scattered electrons in DIS processes with Q^2 ranging from 5 to 100 GeV^2. In ZEUS, outside the solenoid is a high-resolution uranium-scintillator calorimeter which is divided into three parts: forward, covering the pseudorapidity region $4.3 \geq \eta \geq 1.1$; barrel, covering the central region $1.1 \geq \eta \geq -0.75$; and rear, covering the backward region $-0.75 \geq \eta \geq -3.8$. The resulting solid angle coverage is 99.7% of 4π.

Both experiments include devices to measure the luminosity and to tag electrons scattered at very small angles. These devices consist of two electromagnetic calorimeters located upstream of the IP with respect to the proton direction in HERA.

2.4. Reconstruction of kinematic variables

Because the HERA detectors are almost hermetic, the kinematic variables can be determined from the scattered electron, from the hadronic system or from a combination of electron and hadronic system energies and angles. For example, the double angle (DA) method employs only the angles of the scattered electron (θ'_e) and the hadronic system (γ_H). In the naïve quark-parton model γ_H is the scattering angle of the struck quark and is determined from the hadronic energy flow as measured in the detector. The variable y, calculated from the electron variables, is given by the expression $y_e = 1 - E'_e(1 - \cos\theta'_e)/(2E_e)$ where E'_e and θ'_e denote the energy and angle of the scattered electron. Alternatively, y can be determined from the hadronic system, using the Jacquet-Blondel method: $y_{JB} = \sum_i(E_i - p_{zi})/(2E_e)$ where $p_{zi} = E_i \cdot \cos\theta_i$, E_i is the energy of an individual calorimeter cell and θ_i is the polar angle determined using the z-coordinate of the reconstructed IP. The sum runs over calorimeter cells associated with the hadronic system. The ability to measure x and Q^2 in different ways offers a powerful systematic check on the resulting F_2 values.

3.   Neutral Current scattering

The first measurements of the proton structure function F_2 in ep neutral current (NC) DIS at HERA were reported previously by the H1 and ZEUS collaborations.[7,8] The most striking feature of the data was the strong rise of F_2 as x decreases. Since then there has been much discussion[9] about the significance of this observation both for the standard evolution of structure functions and for the behavior of the gluon density distribution in the proton at small x.

3.1.   The proton structure function F_2

In DIS, the differential cross section for inclusive ep scattering mediated by virtual photon and Z^0 exchanges is given in terms of the structure functions $\mathcal{F}_i$ by:

$$\frac{d^2\sigma^{NC}}{dx\,dQ^2} = \frac{2\pi\alpha^2}{xQ^4}\left[Y_+\mathcal{F}_2(x,Q^2) - y^2\mathcal{F}_L(x,Q^2) + Y_-x\mathcal{F}_3(x,Q^2)\right](1 + \delta_r(x,Q^2)) \qquad (2)$$

where $Y_\pm = 1 \pm (1-y)^2$, $\mathcal{F}_L$ is the longitudinal structure function, $\mathcal{F}_3$ is the parity violating term arising from Z^0 exchange and δ_r is the radiative correction. F_2 and F_L will denote the familiar ep structure functions containing only contributions from the exchange of a virtual photon. At the highest mean Q^2 of 4200 GeV^2 in the selected ZEUS bins the contribution from Z^0 exchange to the cross section is equivalent to a 20% correction to F_2. For the next lowest Q^2 bin the corrections are less than 6%. A correction for F_L must also be made, particularly in the small x (high y) bins where it can amount to a 12-15% correction. The radiative corrections are of the order of $3-5\%$ for $x < 10^{-2}$ and low Q^2. Since these corrections are small, Eq. 2 can be rewritten as

$$\frac{d^2\sigma^{NC}}{dx\,dQ^2} = \frac{2\pi\alpha^2 Y_+}{xQ^4}F_2(x,Q^2)(1 - \delta_{F_L} + \delta_Z)(1 + \delta_r) \qquad (3)$$

where the corrections δ_i are functions of x and Q^2 but, to a good approximation, independent of F_2. Monte Carlo event simulation is used to correct for acceptance and resolution effects. The detector simulation incorporates the best knowledge of the apparatus, test beam results and trigger.

3.2.   Results for F_2

Figure 2 shows the 1993 H1[10] and ZEUS[11] F_2 measurements as a function of x for fixed Q^2. There is excellent agreement between the two sets and the strong rise of F_2 with decreasing x observed in the 1992 HERA data is confirmed. This behavior is in striking contrast to the almost constant behavior seen at larger values of x. The steep rise persists from the lowest Q^2 values up to a Q^2 of 1000 GeV^2. Figure 2 also shows some representative curves from recent parton density function (PDF) param-eterizations. MRSD$'_-$ (solid line) and MRSD$'_0$ (dashed line)[12] have a starting scale for the GLAP evolution of $Q_0^2 = 4$ GeV^2. D$'_-$ has a singular parametrization for the sea quark and gluon distributions, $xf \sim x^{-n}$ as $x \to 0$, where $n = 0.5$. D$'_0$ has a constant behavior $xf \sim$ const as $x \to 0$ and clearly does not describe the HERA data. At the

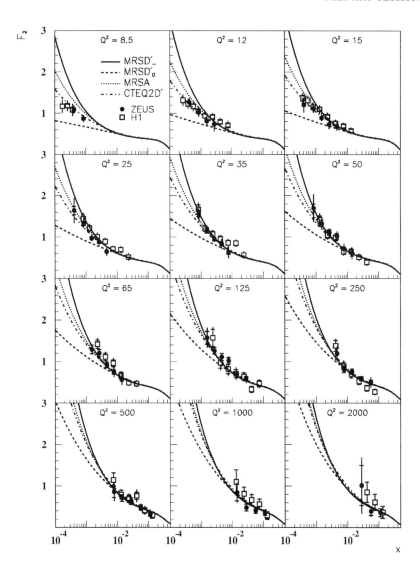

Figure 2: The values of F_2 of the proton as a function of x at different values of Q^2 in GeV^2. Note that for the data labelled $Q^2 = 125, 250, 500, 1000,$ and 2000 GeV^2, the H1 values are $120, 200, 400, 800$ and 1600 GeV^2, respectively.

lowest values of Q^2 these two PDFs span the data, and at Q^2 of 35 GeV^2 and above the data agree with the D$'_-$ curves. The dash-dotted curve shows CTEQ2D$'$[13] and the dotted curve MRSA.[12] Both of these PDFs have singular gluon distributions but with $n \sim 0.3$, $Q_0^2 = 4$ GeV^2 and are typical of recent PDF evaluations that have included the 1992 HERA data in determining their parameters. Both give good representations of the HERA data for $Q^2 \gtrsim 35$ GeV^2. In the lowest Q^2 bins the MRSA curves lie slightly above the data, whereas the CTEQ2D$'$ curves follow the data. The GRV(HO)[14] PDF (not shown) lie in between MRSA and MRSD$'_-$. They are obtained by evolving valence-like gluon and sea quark distributions at $Q_0^2 = 0.3$ GeV^2. For all these PDF evaluations four quark flavors were used and all quarks were assumed massless.

Figure 3 shows the HERA results for F_2 as a function of Q^2 at different values of x and a comparison with the NMC[2] and E665[3] fixed target experiments. The logarithmic scaling violations become clearly visible at low values of x and are in accord with the expected QCD scale breaking. The scaling violations are a direct consequence of QCD and it is now possible to extract the gluon density from them.

3.3.   Gluon density distribution

Although about half of the proton's momentum is carried by gluons, the determination of the momentum density of gluons in the proton has proven to be a difficult task. The gluon density $g(x, Q^2)$ is defined so that $g(x, Q^2)dx$ gives the number of gluons in the proton between x and $x + dx$. The gluon momentum density is then given by $xg(x, Q^2)$. The principal difficulty in measuring the gluon density in DIS is that the gluons do not contribute in zeroth order QCD (Quark-Parton Model) where the structure functions scale with Q^2. In the next order (leading order, LO) scaling violations occur by gluon bremsstrahlung from quarks and quark pair creation from gluons. At small x ($< 10^{-2}$) the latter process dominates the scaling violations. This property was exploited[15] to extract the gluon density from the slope of F_2 using the approximations of Prytz[16] in LO. The slope in F_2 is given by:

$$\frac{dF_2}{d\ln Q^2} = \frac{\alpha_s(Q^2)}{2\pi} \sum_{q,q'} e_q^2 \int_x^1 \frac{dz}{z} [P_{qq'}(\frac{x}{z})q_{q'}(z, Q^2) + P_{qg}(\frac{x}{z})g(z, Q^2)], \qquad (4)$$

where e_q is the charge of quark q and $q_q(x, Q^2)$ is its density in the proton; $\alpha_s(Q^2)$ is the strong coupling constant and P_{qq}, P_{qg} are the quark and gluon splitting functions. The sum runs over all active quark and antiquark flavours.

In Prytz LO the quark contribution is neglected and the LO result for P_{qg} is used. The gluon momentum density $zg(z, Q^2)$ is expanded in a Taylor series around $z = x/2$. Prytz has recently included the next-to-leading order (NLO) corrections.[16] The quark contribution was again neglected and the result for four flavors in the $\overline{MS}$ scheme is:

$$xg(x, Q^2) \approx \frac{dF_2(x/2, Q^2)/d\ln Q^2}{(40/27 + 7.96\alpha_s/4\pi)(\alpha_s/4\pi)} - \frac{(20/9)(\alpha_s/4\pi)N(x/2, Q^2)}{40/27 + 7.96\alpha_s/4\pi}. \qquad (5)$$

The correction function $N(x, Q^2)$ depends on the gluon density and has been estimated[16] using the MRSD$'_0$ and MRSD$'_-$ PDFs.[12] Note that the gluon density at x is given by the slope of F_2 (and by N) at $x/2$.

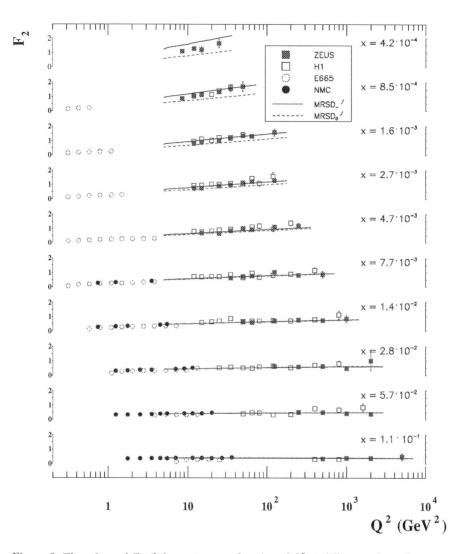

Figure 3: The values of F_2 of the proton as a function of Q^2 at different values of x.

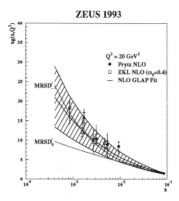

Figure 4: The gluon momentum density as a function of x at $Q^2 = 20$ GeV^2 as determined from the ZEUS data with Prytz and EKL in NLO. The shaded band shows the result of the global fit to the ZEUS data.

The method of Ellis et al. (EKL)[17] is based on the solution of the GLAP equations in moment space. This calculation includes all NLO quark and gluon contributions as well as those from the gluon up to next-to-next-to-leading order (NNL). The parton distributions are assumed to behave as x^{-w_0}, with $w_0 = 0.4$, and the non-singlet contributions are neglected. The expression for the gluon momentum density is

$$xg(x, Q^2) = \frac{18/5}{P^{FG}(w_0)} \left[\frac{dF_2(x, Q^2)}{d \ln Q^2} - P^{FF}(w_0) F_2(x, Q^2) \right]. \qquad (6)$$

In contrast to Prytz, the quark contribution is contained in the second term of Eq. 6. The evolution kernels P^{FF} and P^{GF} are expanded up to second (NLO) or third (NNL) order in α_s. Note that in the EKL method the gluon density x is given by the structure function F_2 and its logarithmic slope at the same value of x.

A QCD fit to the ZEUS data on F_2 has also been performed.[18] To constrain the fit at large x, NMC data[2] on F_2^p and F_2^d measurements were included for $Q^2 > 4$ GeV^2. The PDFs were parameterised as functions of x at a fixed value of $Q_0^2 = 7$ GeV^2. Optimal values for the parameters were obtained from a χ^2 minimization procedure and the normalisation of the gluon was constrained by the momentum sum rule.

Figure 4 shows the resulting gluon distribution[18] at a value of $Q^2 = 20$ GeV^2 from the NLO fit compared to the values obtained with Prytz and EKL in NLO. The results of the three methods are consistent with each other. The band in Fig. 4 indicates the uncertainty of the gluon density obtained from the fit. As shown in Fig. 4, the gluon distribution from MRSD$'_-$ is consistent with the central value of the global fit whereas MRSD$'_0$ is at the lower limit of the total error band. The parameterizations GRV(HO),[14] CTEQ2M[13] and MRSA[12] (not shown) are consistent with the central value of the fit. The gluon density is seen to increase as x decreases.

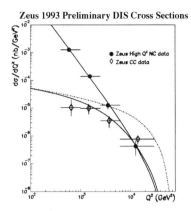

Figure 5: The cross sections $d\sigma/dQ^2$ for neutral and charged current ep interactions. The solid circles are the NC data and the open diamonds are the CC data from ZEUS.

3.4.   Charged current cross section

At HERA the charged current (CC) reaction $ep \to \nu_e + hadrons$ can be studied in a kinematic regime where the W propagator plays an important role. In particular, $< Q^2 >$ becomes comparable to M_W^2. The CC cross section may be written in the form:

$$\frac{d^2\sigma^{CC}}{dx\,dQ^2} = \frac{G_F^2}{2\pi x}\frac{1}{\left(1+\frac{Q^2}{M_W^2}\right)^2}\left[y^2 x F_1(x,Q^2) + (1-y)F_2(x,Q^2) + (y-\frac{y^2}{2})x F_3(x,Q^2)\right] \quad (7)$$

H1 made an early measurement[20] of the charged current cross section at HERA:

$$\mathbf{H1}:\ \sigma_{CC}(p_t > 25\ \text{GeV})\ =\ 55 \pm 15(stat.) \pm 6(sys.)\text{pb},$$

in good agreement with that predicted for $M_W = 80$ GeV and far from that predicted (about 140 pb) with an infinite mass propagator.

Figure 5 compares the Q^2 behavior of the NC and CC cross sections as measured[21] by ZEUS. For the first time in a single experiment one sees that at high Q^2 ($\sim 10^4$ GeV^2) the charged weak cross section is comparable to the electromagnetic cross section. The solid curves are the predictions of the Born-level calculations from a standard DIS Monte Carlo and are in good agreement with the data. The dotted line shows the effect of an infinite mass for M_W in Eq. 7 and is in clear disagreement with the data.

4.   Photoproduction

The high center of mass energy available at HERA permits the study of the physics of quasi-real photons at energies up to $W_{\gamma p} = 250$ GeV. In the rest frame of the proton this corresponds to incident photon energies of up to 20 TeV.

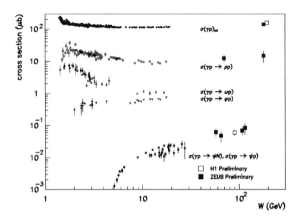

Figure 6: The total photoproduction cross section as a function of W compared to the cross sections for the elastic production of the vector mesons: ρ, ω, ϕ and J/ψ.

4.1. Total cross section

The total photoproduction cross sections, as measured by H1[22] and ZEUS,[6] are shown in Fig. 6 and exhibit only a moderate rise with energy. ZEUS has also determined[6] the fractions of the total cross section at $< W_{\gamma p} > = 180$ GeV due to the non-diffractive, inelastic diffractive and elastic reactions:

$$\mathbf{ZEUS} : (Non - diffractive) = 64.0 \pm 0.9 \pm 3.6\% \text{ of the total,}$$
$$\mathbf{ZEUS} : (Inelast. \, Diffract.) = 23.3 \pm 1.5 \pm 4.7\% \text{ of the total,}$$
$$\mathbf{ZEUS} : (Elastic) = 12.7 \pm 2.7 \pm 4.3\% \text{ of the total.}$$

These fractions are in good agreement with the predictions of Schuler and Sjöstrand[23] (not shown).

4.2. Vector meson production

Photoproduced vector mesons have been extensively studied in fixed target experiments up to energies of about $W_{\gamma p} = 20$ GeV. HERA extends the energy range for these studies by about a factor of five. Figure 6 compares the total cross section with vector meson cross sections for both low energy and HERA data. New ZEUS measurements[24] exist for $\gamma p \to \rho p$ and $\gamma p \to$ J/ψp. H1 has also determined[25] the J/ψ cross section. As shown in Fig. 6, the J/ψ cross section appears to be still increasing over the HERA energy range.

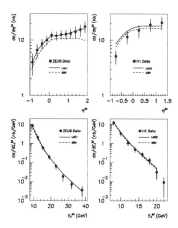

Figure 7: Inclusive jet cross sections as measured by ZEUS and H1.

5. Hard scattering in photoproduction

At HERA energies the interactions of quasi-real photons (with virtuality $Q^2 \approx 0$) with protons can produce jets with high transverse energy (E_T^{jet}). The presence of a 'hard' energy scale means that perturbative QCD should be applicable for comparison with data. In LO, $\mathcal{O}(\alpha_S)$, two types of processes are responsible for jet production: the photon may interact as a point-like particle with a parton in the proton or it may first fluctuate into an hadronic state. In the first case, known as the *direct* contribution, the hadronic final state consists of two jets and the proton remnant. The LO processes which contribute to this are Boson-Gluon-Fusion ($\gamma g \to q\bar{q}$) and the QCD Compton reaction ($\gamma q \to gq$). In the second case, known as the *resolved* contribution, the photon acts as a source of partons which can scatter off partons in the proton. In addition to the hard jets and the proton remnant one may then also observe a photon remnant. The two contributions may be partially distinguished by measuring x_γ, the fraction of the γ momentum flowing into the hard interaction:

$$x_\gamma = \frac{E_T^{jet1} e^{-\eta^{jet1}} + E_T^{jet2} e^{-\eta^{jet2}}}{2E_\gamma}. \tag{8}$$

5.1. Inclusive jet production

Figure 7 shows the inclusive jet cross sections measured by ZEUS[26] and H1.[27] Previous papers have shown the dominance of the resolved component[28,29] as well as the existence of the direct contribution.[28] These data demonstrate a consistency with LO QCD calculations at high scales ($E_T^2 \gtrsim 300$ GeV^2) and a sensitivity to the photon structure functions (see the curves from GRV[30] and LAC1,2[31]).

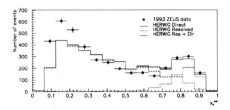

Figure 8: The (uncorrected) distribution of x_γ measured by the ZEUS collaboration.

5.2. Di-jet production and the gluon in the photon

Figure 8 shows the distribution of x_γ for two-jet events in photoproduction with $E_T^{jet} > 6$ GeV. The peak near $x_\gamma = 0.8$, stemming from the direct contribution, can be clearly distinguished from the resolved contribution which rises rapidly at small x_γ.

From the measured distribution of x_γ one can extract an effective gluon density in the photon since the gluonic component should become important at values of $x_\gamma \ll 1$. However, as x_γ decreases the jets go more into the foward direction and get entangled with the proton remnant complicating the measurement. A first estimate[27] by H1 is shown in Fig. 9, where the effective gluon density has been extracted by measuring down to $x_\gamma \approx 10^{-2}$ and unfolding back to the partonic level. The systematic errors are large but there are first indications that (as in the proton) the gluon density in the photon may be increasing at low values of x_γ. Also shown are three parametrizations of the gluon density from GRV[30] and LAC1,3.[31] However, a distinction between different parametrizations is not yet possible. Furthermore, additional study is needed to determine the sensitivity to the systematic effects introduced by different Monte Carlo programs for the direct and quark contributions and their fragmentation effects.

At HERA, events with two jets ($E_T^{jet} \gtrsim 6$ GeV, $\eta^{jet} \lesssim 2$) are sensitive down to $x_\gamma \sim 10^{-1}$. For jets at equal rapidities ($\eta_1^{jet} = \eta_2^{jet} = \bar{\eta}$), the differential cross sections in $\bar{\eta}$ have been measured at high and low x_γ by the ZEUS collaboration.[32] The cross section for $x_\gamma > 0.75$ is dominated by direct photon interactions and so the shape is insensitive to the parton distribution in the photon. This cross section is sensitive to the gluon distribution in the proton and is shown in Fig. 10a compared to LO QCD calculations using various PDFs for the proton. Effects such as non-zero k_T of incoming partons could explain why the measured cross section lies below most curves at low $\bar{\eta}$.[33] However, other higher order or hadronization corrections have not yet been fully considered.

The cross section for $x_\gamma < 0.75$ is dominated by resolved photon interactions and is sensitive to the gluon distribution in the photon. The x_p values probed here are in the region where the proton PDFs are well constrained by other measurements. In Fig. 10b, MRSA is used, whereas in Fig. 10c GRV(LO) is used. The sensitivity to different photon PDFs is large. Again, higher order QCD calculations are necessary before strong conclusions can be drawn.

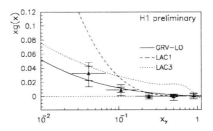

Figure 9: Effective gluon parton distribution in the photon, as determined by H1.

6. Large-rapidity-gap events

It has been shown above that at HERA energies the contributions of inelastic diffractive processes account for about 23% of the total photoproduction cross section. Such diffractive processes are believed to proceed via the t-channel exchange of a color-singlet object with vacuum quantum numbers carrying energy-momentum, called the pomeron. However, the true nature of the pomeron is still far from clear. Ingelman and Schlein[34] proposed that the pomeron may have a partonic substructure which could be probed by a hard scattering process. The UA8 experiment at CERN later observed events containing two high-P_t jets in $\bar{p}p$ interactions which are tagged with leading protons (or antiprotons).[35] These events could be explained in terms of a partonic structure with a hard parton distribution in the pomeron, i.e. a large fraction of the pomeron's momentum participating in the hard scattering.

In general, the hadronic final state in both DIS and photoproduction is character-ized by a large energy deposition in the forward (i.e. proton) direction and significant hadronic activity in the rapidity range between the direction of the quark jet and the proton remnant. This energy presumably stems from gluons radiated in the color field between the struck quark and the outgoing partons from the proton remnant. However, in the absence of color flow, it is expected that at sufficiently high energies diffractive processes should produce a number of events with a large rapidity gap (LRG) in the forward direction. Indeed, the HERA experiments have observed a number of such events.[36-38]

6.1. Monte Carlo event samples

The final states from DIS and photoproduction were modelled using different classes of generators, the first to describe standard processes and the second to model pomeron exchange. The term 'pomeron exchange' is used here as a generic name to describe the process which is responsible for creating diffractive LRG events.

To model the hadronic final states with a LRG, Monte Carlo events were gen-erated by POMPYT.[39] In this code the proton emits a pomeron whose partonic con-stituents subsequently take part in a hard scattering process with the photon or its constituents. The cross section for diffractive hard scattering is expressed in terms of the pomeron 'flux' factor; the parton densities for the pomeron and the photon; and the differential cross section $d\hat{\sigma}/d\hat{t}$ for the appropriate two-body hard scattering subpro-

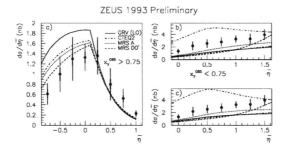

Figure 10: Di-jet cross sections, $d\sigma/d\bar{\eta}$, as determined by the ZEUS collaboration for $|\Delta\eta| <$ 0.5, $E_T^{jet} > 6$ GeV, $0.2 < y < 0.8$, and $\theta_{e'} > 176.5°$. Corrected ZEUS data (solid circles) compared to LO QCD calculations. (a): $x_\gamma^{OBS} > 0.75$. For the proton, the PDFs used are GRV (solid line), CTEQ2 (dash-dotted line), MRSA (dotted line) and MRSD0' (dashed line). The GS2 set was used for the photon in all cases. (b) and (c): $x_\gamma^{OBS} < 0.75$. The PDFs used for the photon are LAC3 (high, dash-dotted line), GS2 (dotted line), GRV (solid line), LAC1 (dashed line) and DG (low, dash-dotted line). The proton PDF used is MRSA in (b) and GRV(LO) in (c).

cess. For photoproduction, the photon contribution contains both direct and resolved processes. The PDF of the pomeron is parameterised according to the hard distribution[34,39] $\beta f(\beta) = constant \cdot \beta(1 - \beta)$ where β denotes the fraction of the pomeron's momentum involved in the scattering.

Another sample was generated following the Nikolaev-Zakharov (NZ) model[40] which assumes that the exchanged virtual photon fluctuates into a $q\bar{q}$ pair which interacts with a colorless two-gluon system emitted by the incident proton. The resulting effective β distribution is somewhat softer than that chosen for POMPYT.

6.2.   Large-rapidity-gap events in DIS

The distribution of η_{max} is shown in Fig. 11 for DIS events, where η_{max} is the maximum pseudorapidity (i.e. closest to the proton direction) of all condensates in an event. A condensate is defined as an isolated set of adjacent calorimeter cells with summed energy above 400 MeV. The data presented are not corrected for detector acceptance and smearing.[37] The bulk of the events cluster around $\eta_{max} \sim 4$. In addition, events are observed which have $\eta_{max} < 1.5$, called LRG events. The standard DIS model, shown as a solid histogram in Fig. 11, gives a reasonable account for $\eta_{max} > 2$, but does not account for the excess of events at lower values. The hard POMPYT model is shown as a dotted histogram. A qualitative description of the data can be obtained with about 10% of all events coming from either POMPYT or the NZ model (not shown). The background of DIS events in the large rapidity gap sample is about 7%.

The invariant mass of the hadronic system detected in the calorimeter, M_X, can be determined from the calorimeter data.[37] The LRG events are distinct from the

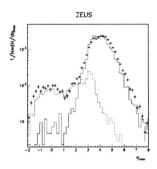

Figure 11: The distribution of η_{max} for DIS events from ZEUS.

standard events; they are characterized by small values, typically $M_X \leq 20$ GeV. For both the NZ and POMPYT models the acceptance due to the η_{max} cut is about 40% and high values of M_X are preferentially suppressed.

6.3. Jet studies in the ep system

To see if the process leading to LRG events contains a hard component, a jet search was performed in the (ep) laboratory system using a cone-based jet algorithm with a cone radius $R = (\Delta\phi^2 + \Delta\eta^2)^{\frac{1}{2}} = 1$. The transverse (with respect to the beam axis) energy weighted mean pseudorapidity (η_{jet}) and azimuth (ϕ_{jet}) were evaluated and jets were accepted if the transverse energy (E_T^{jet}) was larger than 4 GeV and $\eta_{jet} \leq 2$, corresponding to polar angles larger than 15°. Of the 1973 LRG events,[37] 294 events were of the 1-jet type and 6 events of the 2-jet category.

The total hadronic transverse energy distribution is shown in Fig. 12a for all LRG events and for those with ≥ 1 jet (hashed) and 2 jets (cross-hashed). For $E_T \geq 10$ GeV practically all events are of the 1-jet type. In Fig. 12b, the distribution of the azimuthal angular difference $\Delta\phi_{e-jet}$ between the scattered electron and the jet is displayed, showing that the electron and jet are preferentially back-to-back in azimuth. Fig. 12c shows the transverse jet energy. Both the hard POMPYT and NZ models give reasonable accounts of the data.

6.4. Jet studies in the $\gamma^* p$ system

Whether or not the observed jets are a mere kinematic artifact of the necessity to balance the transverse momentum of the scattered electron is investigated in the $\gamma^* p$ system. For this jet study the minimum transverse jet energy required, with respect to the γ^* axis, was 2 GeV. In Fig. 12d the E_T^* distribution exhibits transverse energies as large as 15 GeV. The events with ≥ 1, ≥ 2 and 3 jets are shown as hashed, cross-hashed and solid histograms, respectively. Jet production is the dominant mechanism for E_T^* values larger than 7 GeV. In Fig. 12e the difference in azimuth for the 2-jet events ($\Delta\phi_{jet-jet}$) is displayed, showing that the jets are preferentially back-to-back. In

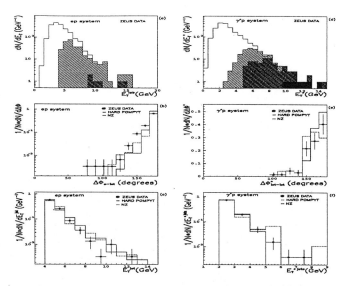

Figure 12: (a) The distribution of E_T; (b) the difference in azimuthal angle between the scattered electron and the jet; (c) the jet transverse energy in the laboratory; (d) the distribution of E_T^* in the $\gamma^* p$ system; (e) the difference in azimuthal angle between the two jets in the $\gamma^* p$ system; (f) the distribution of the jet energy transverse to the γ^* direction.

Fig. 12f the jet transverse energy distribution shows that jets with transverse energies, E_T^{*jet}, as large as 7 GeV are observed. Hence, there is strong evidence for hard scattering processes in both the ep and $\gamma^* p$ systems, producing LRG events ($\eta_{max} < 1.5$) in *ep* scattering.[37]

6.5. Large-rapidity-gap events in hard photoproduction

For this study[41] ZEUS selected events with $E_T > 5$ GeV (ie. hard photoproduction) to study the interaction of the pomeron. The distribution of η_{max} (not shown) is similar to that in DIS and yields 6678 LRG events with $\eta_{max} \leq 1.5$, corresponding to 1.6 % of the hard photoproduction sample. The excess of events with $\eta_{max} \leq 1.5$ is not accounted for by standard Monte Carlo simulations for hard photoproduction processes. It is, however, in good agreement[41] with the expectations of the POMPYT model. According to POMPYT, for events with $E_T > 5$ GeV the acceptance after the $\eta_{max} \leq 1.5$ cut is about 10%.

The invariant mass of the hadronic system detected in the ZEUS central detector can be determined from the calorimeter: $M_X^2 = E_H^2 - p_H^2$, where E_H and p_H are the energy and momentum of the hadronic system. The M_X distribution for LRG events is steeply falling for M_X values above 12 GeV (not shown). According to POMPYT, for events with $E_T > 5$ GeV, the η_{max} cut suppresses events at large values of M_X.

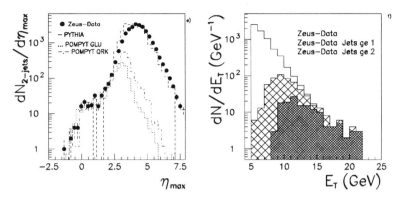

Figure 13: (a) The distribution of η_{max} for the photoproduction sample with $E_T > 5$ GeV and two or more jets along with Monte Carlo comparisons from PYTHIA and POMPYT. (b) The distribution in the total transverse energy E_T for the LRG photoproduction sample.

6.6. Jet structure in hard photoproduction

Evidence for multijet structure in hard photoproduction has been presented elsewhere.[28,29] A search for jets in LRG events with $E_T > 5$ GeV again uses the cone-based algorithm. Figure 13a displays the η_{max} distribution for all two-jet events. A clear excess with $\eta_{max} \leq 1.5$ is observed, corresponding to 0.63%, and is not accounted for by the standard hard photoproduction processes as modelled by PYTHIA which predicts fewer than 0.1% of the two-jet events with $\eta_{max} \leq 1.5$. Therefore, the two-jet LRG events are not just the tail of 'standard' hard photoproduction with two jets.[28] Both of the POMPYT samples (in which the pomeron is dominantly quark or gluon) give a good representation of the shape of the η_{max} distribution.

The E_T distribution is shown in Fig. 13b for all LRG events (open histogram), for events with at least one jet (cross hatched) and for those events with two or more jets (shaded). For $E_T \geq 8$ GeV, the minimum E_T for which two-jet production is possible, 9.9% of the events are of the two-jet type. For $E_T \geq 12$ GeV the majority of the events are of the two-jet type. For two-jet events the distribution of the transverse jet energies reaches up to values of 10 GeV.

6.7. Conclusions for large-rapidity-gap events

Studying the inclusive distributions in x and Q^2 of the DIS events, one finds that they are very similar to the standard DIS events.[37,38] In particular, the rate of LRG events compared to the total DIS rate is approximately constant with $W_{\gamma^* p}$ and, at fixed x, it does not depend on Q^2. Furthermore, clear jets are observed[37] in these events with E_T^{jet} ranging up to 7 GeV. This indicates that in these events the photon couples to a point-like object inside the proton in much the same way as the photon couples to the partons in 'ordinary' DIS events.

ZEUS has also observed photoproduction events with a large rapidity gap and large transverse energy. Their production rate as a function of $W_{\gamma p}$ is consistent with a diffractive process. Hard scattering, with jets having transverse energies up to 10 GeV, has been observed in these LRG events. The two-jet events in this sample show little energy outside the jets suggesting the absence of color flow between the outgoing proton and the struck partons.

Standard DIS and photoproduction Monte Carlo models fail to predict an adequate rate of LRG events with values of $\eta_{max} < 1.5$. However, models for diffractive scattering (e.g. POMPYT and NZ) are able to describe the properties of these events. A natural interpretation of these events is the interaction of the photon with the partons of a colorless object inside the proton.

7. F_2^{DIFF} and factorization in DIS

H1 has determined[42] the contribution to F_2 from diffractive events with the fraction of the proton's momentum carried by the pomeron $x_{\mathcal{P}om} < 10^{-2}$ for Q^2 between 8.5 and 60 GeV^2. The cross section may be written as:

$$\frac{d^2\sigma(ep \to epX)}{dx_{\mathcal{P}om}dtdx\,dQ^2} = \frac{4\pi\alpha^2}{xQ^4}\left(1 - y + \frac{y^2}{2}\right)F_2^D(x, Q^2; x_{\mathcal{P}om}, t). \tag{9}$$

F_2^D (not shown) rises as x decreases and is about 10% of the inclusive F_2. These measurements can be used to study the possiblity of pomeron factorization. If factorization holds then one can write

$$F_2^D(x, Q^2; x_{\mathcal{P}om}, t) = f(x_{\mathcal{P}om}, t)F_2^{\mathcal{P}om}(z, Q^2),$$

where $z = x/x_{\mathcal{P}om} = Q^2/(Q^2 + M_X^2)$. This implies that $\int F_2^D dt$ should have the same $x_{\mathcal{P}om}$ dependence for any z and Q^2. Figure 14 shows $\int F_2^D dt$ vs. $x_{\mathcal{P}om}$ at fixed $Q^2 = 15$ GeV^2 for four different values of z. A good fit is obtained to the form

$$\int F_2^D dt = F_2^D(z, Q^2; x_{\mathcal{P}om}) = F_2^{\mathcal{P}om}(z, Q^2) \cdot x_{\mathcal{P}om}^\alpha, \tag{10}$$

with $\alpha = -1.3 \pm 0.1$, indicating consistency with the idea of pomeron factorization.

8. Summary and Outlook

The HERA experiments have each collected ~ 0.5 pb^{-1} of data during 1993. This has brought forth a wide range of new results in a new kinematic region.

The 1993 HERA data confirm, with improved precision, the strong rise in F_2 as x decreases that was observed in 1992. The data have been compared with recent PDFs and are best described by those with singular gluon and sea-quark distributions at small x. Logarithmic scaling violations are observed in a new regime of low x and high Q^2 and the gluon density increases substantially between $x = 10^{-1}$ and 10^{-3} at $Q^2 = 20$ GeV^2.

The photoproduction total cross section and the heavy quark production continues to rise slowly up to energies as high as 200 GeV. Hard photoproduction yields information on the structure of both the photon as well as the proton so that we have a more complete description of the structure of the photon. Diffraction is not only a soft

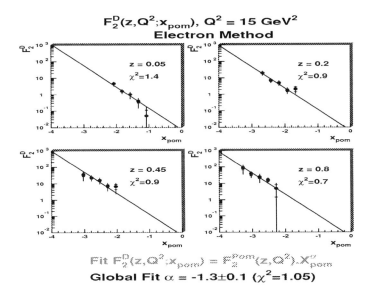

Figure 14: Pomeron factorization study from H1.

phenomenon but has been observed at both low and high Q^2 and may be accompanied by jets.

These are just the highlights among many interesting results. Yet they open up new questions: Does F_2 continue to rise at yet lower Q^2 and lower x? At what values will it start to saturate and by what mechanism? Does the photoproduction cross section start to rise more steeply at some yet higher scale as some theories predict? How can the hard diffractive processes seen at HERA be described and brought into line with the soft diffraction seen at much lower energies? What is the pomeron?

All of these questions and many more will be investigated at HERA as the machine reaches its design goals for the luminosity, produces polarized beams and runs at different energies. The HERA program is indeed a rich one and one can hope for many more interesting results in the near future.

9.   Acknowledgements

I would like to acknowledge the work of both the H1 and ZEUS collaborations and the many discussions with members of both experiments. I greatly appreciate the generous support from the DESY Directorate for my stays at DESY.

References

1. BCDMS Collab., A.C. Benvenuti et al., *Phys. Lett.* **B237** (1990) 592.
2. NMC Collab., P. Amaudruz et al., *Phys. Lett.* **B295** (1992) 159.
3. E665 Collab., A. Kotwal, Invited talk at the Workshop on *'The Heart of the Matter'*, Blois, France, June 1994, to be published in the proceedings.

4. H1 Collab., I. Abt et al., 'The H1 Detector at HERA', DESY 93-103(1993); B. Andrieu et al., *Nucl. Inst. and Meth.* **A336** (1993) 460; and *ibid* **A336** (1993) 499.
5. The ZEUS Detector, Status Report 1993, DESY (1993).
6. ZEUS Collab., M. Derrick et al, *Z. Phys.* **C63** (1994) 391.
7. H1 Collab., I. Abt et al, *Nucl. Phys.* **B407** (1993) 515.
8. ZEUS Collab., M. Derrick et al., *Phys. Lett.* **B316** (1993) 412.
9. E. M. Levin, A. D. Martin, Invited talks at the *'Workshop on DIS and Related Topics'*, Eilat, Israel, February 1994, to be published in the proceedings.
10. H1 Collab. V. Brisson, Talk at the 1994 IHEP Conference, Glasgow.
11. ZEUS Collab., M. Derrick et al., DESY 94-143, to be publ., *Z. Phys.* **C** (1994).
12. A. D. Martin et al., *Phys. Lett.* **B306** (1993) 145; *Phys. Rev.* **D47** (1993) 867; Univ. of Durham preprint DTP/94/34 (and RAL 94-055) June 1994.
13. CTEQ Collab., J. Botts et al., *Phys. Lett.* **B304** (1993) 159.
14. M. Glück, E. Reya and A. Vogt, *Phys. Lett.* **B306** (1993) 391.
15. ZEUS Collab., J.F. Martin, Rapporteur talk, *1993 Lepton - Photon Conf., Cornell*; G. Wolf, DESY 94-022; H1 Collab., I. Abt et al., *Phys. Lett.* **B321** (1994) 161.
16. K. Prytz, *Phys. Lett.* **B311** (1993) 286; K. Prytz, RAL-94-036 (1994).
17. R.K. Ellis, Z. Kunszt and E.M. Levin, Fermilab-PUB-93/350-T, ETH-TH/93/41.
18. ZEUS Collab., M. Lancaster, Talk at the 1994 IHEP Conference, Glasgow.
19. H1 Collab., I. Abt et. al., *Phys. Lett.* **B324** (1994) 241.
20. ZEUS Collab., G. Wolf, Invited talk at the *'Workshop on DIS and Related Topics'*, Eilat, Israel, February 1994, to be published in the proceedings.
21. H1 Collab., A. De Roeck, Proc. *Europhysics Conf. on HEP*, Marseille (1993).
22. G. A. Schuler and T. Sjöstrand, *Phys. Lett.* **B300** (1993) 169.
23. ZEUS Collab., M. Derrick et al., Papers 0672 and 0688, submitted to the 1994 IHEP Conference, Glasgow.
24. H1 Collab., T. Ahmed et al., DESY 94-153, to be publ. (1994).
25. ZEUS Collab., M. Derrick et al., Paper 0683, submitted to the 1994 IHEP Conference, Glasgow.
26. H1 Collab., H. Hufnagel, Talk at the 1994 IHEP Conference, Glasgow.
27. ZEUS Collab., M. Derrick et al., *Phys. Lett.* **B322** (1994) 287.
28. H1 Collab., I. Abt et al., *Phys. Lett.* **B314** (1993) 436.
29. M. Glück, E. Reya and A. Vogt, *Phys. Rev.* **D47** (1993) 867.
30. H. Abramowicz, K. Charchula and A. Levy, *Phys. Lett.* **B269** (1991) 458.
31. ZEUS Collab., M. Derrick et al., Paper 0682, submitted to the 1994 IHEP Conference, Glasgow.
32. J. R. Forshaw and R. G. Roberts, *Phys. Lett.* **B319** (1994) 539.
33. G. Ingelman and P. Schlein, *Phys. Lett.* **152B** (1985) 256.
34. UA8 Collab., A. Brandt et al., *Phys. Lett.* **B297** (1992) 417.
35. ZEUS Collab., M. Derrick et al., *Phys. Lett.* **B315** (1993) 481.
36. ZEUS Collab., M. Derrick et al., *Phys. Lett.* **B332** (1994) 228.
37. H1 Collab., T. Ahmed et al., DESY 94-133, to be publ. (1994).
38. P. Bruni and G. Ingelman, Proc. *Europhysics Conf. on HEP*, Marseille (1993), p. 595.
39. N.N. Nikolaev and B.G. Zakharov, *Z. Phys.* **C53** (1992) 331.
40. ZEUS Collab., M. Derrick et al., Paper 0686, submitted to the 1994 IHEP Conference, Glasgow.
41. H1 Collab., J. Feltesse, Invited Talk at the 1994 IHEP Conference, Glasgow.

Ten Milestones in Lattice Gauge Theory

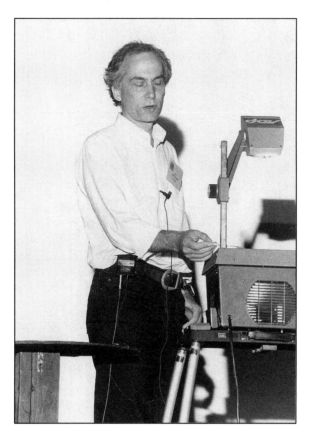

PAUL B. MACKENZIE, *Fermilab*

TEN MILESTONES IN LATTICE GAUGE THEORY

PAUL B. MACKENZIE*

*Theoretical Physics Group, Fermi National Accelerator Laboratory,
P.O. Box 500, Batavia, IL 60510, USA*

ABSTRACT

I discuss several recent applications of lattice QCD to standard model phenomenology, along with some of the most important previous advances in lattice gauge theory which made them possible.

1. Introduction

The last few years have seen significant advances in our ability to apply lattice gauge theory to standard model phenomenology. Lattice calculations of such quantities as the kaon mixing parameter B_K, the strong coupling constant α_s, and the mass of the b quark m_b are now believed to be as accurate or more accurate than the best conventional determinations.

The organizers have asked me to present my view of the current status of the field and its future prospects, rather than to present a normal review of the most important progress of the last twelve months. I cannot hope to present the entire history of lattice gauge theory in a single lecture. I will instead concentrate on ten significant topics. My focus will be several of the current crop of phenomenological lattice QCD calculations for which the understanding of the calculational uncertainties is most complete. These are B_K, α_s, and m_b. I will precede discussing them with a discussion of some of the pieces of work which have been most instrumental in making them possible.

2. Ten Milestones

2.1. Foundation of Lattice Gauge Theory

In classical mechanics, the principle of least action determines a unique path for an electron. In real life, the electron travels over all possible paths (sometimes in a nearly classical way, dominated by paths near the classical path). Feynman showed how to derive the Schrödinger wave function by integrating over all paths, weighting each path by the exponential of i times its classical action. Likewise, in classical electrodynamics, Maxwell's equations may be obtained by minimizing the classical action of the gauge field: $A = 1/2 \int (E^2 - B^2) \equiv 1/4 \int F_{\mu\nu}^2$. In real life, all classical field configurations contribute to physical processes, and these quantum fluctuations are very important at short distances. Quantum field theory may be obtained by integrating over all possible field configurations, each

---

*mackenzie@fnal.gov

weighted by the exponential of its classical action. This prescription has an infinite number of degrees of freedom (one for the field strength at every point in space-time), so a limit must be defined.

Wilson formed such a limit by defining variables for the quark and gluon fields on a discrete space-time lattice.[1] This gives a path integral with a large but finite number of degrees of freedom. The physical predictions of the theory are defined to be those in which the lattice spacing a goes to zero. If the theory is renormalizable, this limit should exist. Wilson took for the gauge variables of the theory link variables,

$$U_{x\mu} \equiv \exp(ig_0 A_{x\mu}), \tag{1}$$

SU(3) matrices which can be viewed as the path ordered products of the vector potentials $A_{x\mu}$ from one lattice site to another, rather than the vector potentials themselves which are normally used in perturbative QCD. (g_0^2 is the bare coupling constant of the lattice theory.) This enabled him to maintain an exact gauge invariance in the lattice theory. In the limit $a \to 0$, the action for the gauge theory can be taken to be the product of the link variables over the smallest squares on the lattice (the "plaquettes"). It is easy to show that this definition is equivalent to the continuum definition plus terms which vanish like a^2:

$$A = \frac{1}{g_0^2} \sum (1 - TrUUUU) \tag{2}$$

$$\equiv \frac{1}{g_0^2} \sum U_{1x1} \tag{3}$$

$$\to 1/4 \int F_{\mu\nu}^2 + \mathcal{O}(a^2). \tag{4}$$

Wilson proposed a criterion for quark confinement in his original paper. The theory is formulated with a "Wick" rotation to Euclidean time. The time dependence of quantum states is thus changed: $\exp(iE\ t) \to \exp(-E\ t_{\text{Euc}})$, and the exponential weight in the path integral becomes real: $\exp(iA) \to \exp(-A)$). Quark propagators contain a phase factor $\exp(iA \cdot dl)$, the exponential of the classical action for a charged particle traveling in a gauge field. On the lattice, this is simply the product of link matrices $U_{x\mu}$ along the path. Amplitudes for quark–antiquark pairs to be created at some point, separate a distance r and propagate some length t through Euclidean time, and then annihilate contain the factor of the product of the link matrices along the closed path traveled by the pair, averaged over gauge fields weighted by the exponential of the action:

$$W = \int_{[U]} TrUUUUUU \ldots \exp(-A). \tag{5}$$

The energy in the time dependence of the two–quark state contains a term $V(r)$, the static potential for quark separated a distance r. If the quarks are far apart and we have linear confinement, $V(r) \to \sigma r$, the loop average contains the factor

$$\exp(-E\ t) \to \exp(-V(r)\ t) \tag{6}$$

$$\to \exp(-\sigma r\ t) \tag{7}$$

$$= \exp(-\text{Area } \sigma). \tag{8}$$

Thus, the criterion for quark confinement is the presence of an area law for large Wilson loops.

2.2. The String Tension, and the Introduction of Monte Carlo Methods for Path Integrals

In his first paper, Wilson considered virtually every possible way of evaluating the path integrals of lattice gauge theory except the one that is in the widest use today, the Monte Carlo method. This was introduced in 1979 by Wilson and by Creutz and collaborators.[2,3] The idea is to replace the integral over all four-dimensional gauge configurations C with a finite set of configurations $\{C_1, C_2, C_3, \ldots\}$ chosen in such a way that the probability $P(C_i)$ of reaching any particular configuration is proportional to its desired weight in the path integral $\exp(-A)$. A simple way to do this is the Metropolis method. Make a random change in one of the links. If $\exp(-A(C_{i+1})) \geq \exp(-A(C_i))$, accept the new configuration. Else, accept it with probability $\exp(-A(C_{i+1}))/\exp(-A(C_i))$. You can then show that as the number of configurations approaches infinity, the probabability of reaching any given configuration becomes proportional to the desired probability $\exp(-A)$.

Creutz used Monte Carlo methods to find evidence for confinement. He evaluated ratios of Wilson loops of different sizes to look for evidence of the area law. The lattice spacing is varied implicitly in lattice calculations by varying the bare coupling constant

$$g_0^2 = \frac{1}{\beta_0 \ln(a\Lambda_0)^{-2} + \ldots}. \tag{9}$$

By checking that the string tension obtained in units of the lattice spacing a, scales in accordance with Eqn. 9 as the coupling constant is varied, one checks for the simultaneous presence of asymptotic freedom and confinement in QCD.

This was the first calculation of a (theorist's) physical quantity with methods that were unambiguously systematically improvable. Many improvements in methods have been made since the initial calculations, and today the string tension is one of the more accurately known quantities in lattice QCD without light quarks.

2.3. Hadron Spectrum with Monte Carlo Methods

The lines of U matrices pointing in the time direction in Wilson loops may be thought of as propagators for infinitely heavy quarks. To develop methods for light hadron propagators, those lines must be replaced by the lattice propagators of light quarks. Methods for such calculations were proposed around 1980 by Weingarten and by Parisi and their various collaborators.[4,5] Light quark propagators are inverses of the lattice Dirac operator, discrete versions of $\not{D} - m$. Discretized differential operators are sparse matrices. They can be inverted with sparse matrix methods such as the conjugate gradient or the minimum residual methods.

Another element in the early hadron calculations was the omission of the effects of sea quarks, the "quenched" or "valence" approximation. (I will return to this in Sec. 2.6.)

To calculate hadron properties in this approach, then, on must

- make a set a gauge configurations with Monte Carlo methods,
- calculate quark propagators on each configuration with sparse matrix methods,
- make hadron propagators on each configuration by combining quark propagators, and
- fit the averaged hadron propagators for large time separations to the expected asymptotic time dependence, $C(t) \to A \exp(-m\, t)$.

The m's are the predicted hadron masses. Decay constants for mesons may be obtained from the A's.

The earliest hadron mass calculations were done on rather small lattices, around 6^3. The lattice spacings used were around 0.2 fermi, so that the physical volumes were only about $(1.2 \text{ fm})^3$, not as large as the diameter of a proton. In quantitative accuracy, they were probably not improvements on, say, the bag model. (They could be thought of as a "periodic box" model of hadrons, as opposed to a bag model.) Their importance lies in their being the first calculations of (experimentalists') physical quantities with methods that were unambiguously systematically improvable (unlike bag model and quark model calculations). The methods used in the phenomenological calculations that I will discuss at the end of this talk are more or less straightforward refinements of the methods introduced here.

2.4. Temperature of the Quark–Gluon Deconfining Transition

The last of the pioneering calculations forming the foundation for subsequent developments that I will discuss is the calculation of the temperature of the quark–gluon deconfining phase transition, T_c, which were performed in the mid 1980's.[6] This is a bulk property of matter and is easier to calculate accurately than the hadron spectrum. It was therefore one of the first calculations which achieved a reasonably serious understanding of the calculational uncertainties.

Calculations of QCD at finite temperature make use of the fact that a system periodic in Euclidean time with period T_E has the same physics as a system at temperature $1/T_E$. One could imagine investigating the vanishing of confinement by looking for the vanishing of the string tension at high temperature using methods like those of Sec. 2.2. A simpler method is available, however. The Euclidean time dependence of a static quark propagator (the expectation value $< UUUU \ldots >$ of a product of U matrices in the time direction) has the form $\exp(-ET)$. These expectation values are called Polyakov lines. For confined quarks, the energy of a single, isolated quark becomes infinite, so that we have

$$< UUUU \ldots > \sim \exp(-ET) \to 0. \tag{10}$$

When quarks are not confined, a single quark can be isolated without raising its energy to infinity, so that we have

$$< UUUU \ldots > \sim \exp(-ET) \to \text{finite}. \tag{11}$$

The deconfinement temperature is determined by looking for the temperature (that is, the box size in the periodic Euclidean time direction) at which this expectation value vanishes.

These were the first numerical calculations in a new field, QCD thermodynamics, with applications in cosmology, astrophysics, and heavy ion collisions. From the standpoint of the standard model phenomenology toward which this review is heading, they were important in that they contained the most systematic study up to then of the calculational uncertainties: finite volume errors, finite lattice spacing errors, and statistical errors.

2.5. Symanzik "Improvement" of the Lattice Action

Much of the story of the path from the pioneering calculations to today's calculations lies in the development of improved calculational methods and of improved understanding of sources of calculational error. I will choose, somewhat arbitrarily, a couple of the most important to discuss. Other important refinements that could easily be added to the list include the use of improved operators which connect more strongly to the hadron state of interest, and calculations of the expected approach to the large volume limit.

Improvement of the lattice action refers to the reduction of finite lattice spacing errors through the use of more complicated actions. In numerical analysis of classical field theories, it is well known how to use next-nearest neighbor information to reduce dependence on the discretization scale. For example, the errors produced by the trapezoidal rule for numerical integration, which approximates functions with interpolating lines, vanish as the "lattice spacing" cubed. The errors produced by Simpson's rule, which approximates functions with interpolating parabolas, vanish as the fifth power of "lattice spacing". An analogous result in classical (tree level) lattice gauge theory is a gauge field action which approximates the continuum action with errors vanishing as a^4 by incorporating $1x2$ rectangles in the action.[7] This means replacing Eqn. 2 with

$$A = \frac{1}{g_0^2}\left(5/3\sum U_{1x1} - 1/12\sum U_{1x2}\right) \qquad (12)$$

$$\to 1/4\int F_{\mu\nu}^2 + \mathcal{O}(a^4). \qquad (13)$$

The coefficients 5/3 and -1/12 above have quantum corrections in quantum field theory. By inclusion of more and more terms in the action, it is in principle possible to make the predictions of a cut-off quantum field theory (such as lattice gauge theory with a finite lattice spacing) approximate the predictions of the continuum theory (the $a \to 0$ limit) with arbitrary precision. Wilson calls this the "statistical continuum limit". It is studied with the renormalization group, and can be contrasted with the classical continuum limit which is studied with calculus (a more mature and better understood subject). A well-defined program for evaluating the required coefficients in perturbation theory has been proposed by Symanzik.[8,7,9] Nonperturbative effects must be taken into account in calculating the coefficients if it is desired to improve the action to a sufficiently high power of the lattice spacing. It is not yet known what this power is. (See the end of Sec. 2.8.)

The improvement program is important because it provides increased understanding of sources of lattice spacing error and uncertainty. It is also important for the completely pragmatic reason that it saves computer time. Suppose one wants to reduce discretization errors completely by brute force, taking $a \to 0$. Consider for specificity Wilson fermions, whose improvement has received the most practical attention. Discretization errors fall linearly in the lattice spacing, as $a\Lambda_{QCD}$. The CPU time required to compute quark propagators rises as $1/a^5$, four powers of a for the increased number of sites on a four-dimensional lattice, plus an extra power due to the slowing down of matrix inversion algorithms as the lattice spacing is reduced ("critical slowing down"). CPU time thus rises as $1/\text{error}^5$.

$\mathcal{O}(a)$ improvement of Wilson fermions requires the addition of a single additional operator to the action, $\bar{\psi}\Sigma_{\mu\nu}F_{\mu\nu}\psi$.[10] This roughly doubles the CPU time required per site. It reduces the discretization errors from $\mathcal{O}(a\Lambda_{QCD})$ to $\mathcal{O}(\alpha_s a\Lambda_{QCD})$. If $\Lambda \sim 300$ MeV and a typical lattice spacing is $a^{-1} \sim 2$ GeV, typical errors might be a factor of five or ten. Reducing by a factor of $\alpha_s \sim 0.2$ by brute force would require an increase in computer time of $5^5 \sim 3000$!

The current state of the art in phenomenological calculations is $\mathcal{O}(a)$ improvement. It has been shown to be crucial in some quantities, and may be important in many cases. The current frontier is $\mathcal{O}(a^2)$ improvement. Many operators must be considered in this case. Computing power per site is multiplied by an order of magnitude instead of a factor of two (although fewer sites should be needed, since larger lattice spacings can be used).

2.6.   Hybrid and Hybrid Monte Carlo Algorithms for Light Quark Loops

The effects of light quark loops have proven to among the most difficult to control, in part because there is no practical way of including their effects gradually. A loop expansion for the quark effective action may be defined, but no practical method has yet been developed for calculating the individual terms in the expansion any more efficiently than the the the effects of the full series.

The effects of fermions are more difficult to deal with in computer calculations than those of bosons because of the different ways they are represented in path integrals. Bosons are represented by complex numbers, which are straightforward numerically. Fermions are represented by *anticommuting* complex numbers, which are well-defined mathematically, but have no straightforward numerical representation. Their integrals are quite well-defined, however. They lead to an effective action for the fermions which is a nonlocal function of the gauge fields. To update a link variable with a Monte Carlo algorithm in the pure gauge theory, one needs to look up the neighboring links and perform a few thousand numerical operations. To update a link in the most naive way when fermions are included, one needs to include the effects of quark paths throughout the entire lattice. This requires a few thousand operations per site times the number of sites in the lattice times a few hundred or a few thousand sweeps through the lattice to calculate the quark propagator.

More efficient algorithms for including light quark loops may be obtained by replacing the integral over fermion fields with the fermion action $\overline{\psi} M \psi$ with the equivalent integral over boson fields with boson action $\phi^* M^{-1} \phi$. (The matrix M is the discretized Dirac operator, $\not{D} - m$.) The bose variables can now be represented on the computer, but the problem is still highly nonlocal because the bose action is the inverse Dirac operator rather than the Dirac operator itself. The number of matrix inverse calculations can be reduced from one per link update to one per sweep through the lattice by the use of small step-size updating algorithms.

Giant progress was achieved between 1981 and 1986 along these lines. A complete set of references may be found in Ref. ([11]). The most efficient are the "hybrid" algorithm,[12,13] a combination of the microcanonical and Langevin algorithms, and the "hybrid Monte Carlo" algorithm,[14] a version of the hybrid algorithm with a correction step which eliminates the errors arising from the small but finite step size. These algorithms are orders of magnitude faster than the early algorithms for light quark loops. However, they are still much slower than algorithms for pure gauge theory, by factors of hundreds or thousands, depending on the parameters in the calculation. Progress following this approach ground to a halt in 1986, though, and little more has been achieved in the eight years since. Recently, Lüscher has proposed a new approach to the problem, in which the single Boson field with a nonlocal action is replaced by a large number of Boson fields with local actions.[15] If we are lucky, this may set off new wave of progress in this area.

2.7. f_π Unquenched

This calculation is not a "milestone" in the sense of being in some way a finished product, as many of the other topics are. It is included because it is the cleanest calculation in the light hadrons so far which includes the least controlled systematic effect in lattice QCD phenomenology: the effects of light quark loops. Unquenched calculations of the light hadron spectrum in general are at present still rather messy. Volume dependence is observed in the proton and rho masses even on relatively large volumes. The extrapolation to the physical light quark mass is problematical. The particles for which ones expects these problems to be controlled most easily are the pions and kaons. They are expected to be the smallest of the light hadrons, and their properties have relatively small and well-understood extrapolations to the light quark mass limit, in contrast to the baryons and vector mesons. They also have by far the the best statistical errors, since the signal drops as $\exp(-mt)$, and they have the smallest masses.

These expectations have been born out by work so far on pions in unquenched calculations. The decay constant of the pion, f_π, is a dimensionful quantity which may be one of the best quantities among the light hadrons to set the energy scale of lattice calculations. Recent results[16] for f_π are shown as a function of quark mass in Fig. 1. Careful checks have been performed showing the independence of the results on the type of operator used to represent the decay current. The

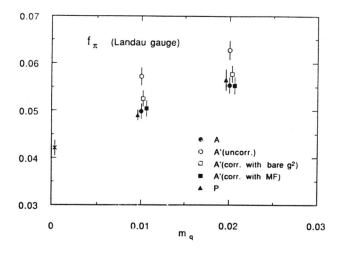

Fig. 1. The decay constant of the pion calculated in full QCD, with light quark loops.

mass of the pion has been shown to be independent of the volume (at volumes at which the corrections to the ρ and proton masses are still five to ten per cent). The quark masses used are roughly the strange quark mass m_s and $m_s/2$. The lighter of the two meson masses is therefore approximately the kaon mass. If the results at the two quark masses are extrapolated linearly to the chiral limit at $m_q = 0$, the 20% negative extrapolation is what is expected on the basis of chiral perturbation theory and the experimental values of f_K and f_π. Further tests which remain to be done are checks of the volume dependence of f_π itself, better checks on dependence on the lattice spacing, and checks of the linearity of the extrapolation to the chiral limit.

The potential importance of this calculation is that unquenched calculations of the properties of pions and kaons (as well such comparably simple mesons as B's, ψ's and Υ's) should become reliable at a stage when calculations of other light hadrons are still messy. If the further checks on calculational uncertainties in f_π do not turn up problems, it would mean that we have reached that stage already, with existing computers and algorithms.

The happiest prognosis for unquenched calculations would be for the new avenue in unquenched algorithms mentioned at the end of the preceding section to provide another big boost in algorithm speed. Even if that does not occur, it is possible that existing computers and algorithms have reached the stage at which realistic calculations are possible for sufficiently simple systems.

2.8.   The Strong Coupling Constant

The remaining three sections are calculations required for extracting for extracting standard model parameters from experimental data. They are among the easiest of such calculations. In part, this is because they are in one way or another special cases, in that for one reason or another we have some idea in advance about which sources of corrections and errors are likely to be large and which are likely to be small.

Lattice calculations for extracting standard model parameters from experimental data consist generically of two pieces. A long distance calculation determines what parameters of the bare lattice Lagrangian correctly reproduce hadron physics. These are done with Monte Carlo simulations. A short distance calculation is used to relate the parameters of the bare lattice Lagrangian to the parameters of the dimensionally regularized standard model Lagrangian. One way of doing this is relate the parameters of the lattice Lagrangian to those of another regularization by demanding that quark–gluon Green's functions match in the two regularizations. In other words, the lattice Lagrangian which gives the same short–distance physics as a given Lagrangian in dimensional regularization is determined, and then that lattice Lagrangian is used to calculate hadron physics using Monte Carlo simulations. A second way of obtaining standard model parameters from lattice calculations is to use nonperturbative definitions of the parameters which can be obtained directly from short distance Monte Carlo calculations, without reference to the bare lattice parameters or perturbation theory, as discussed below. Since the extant to which nonperturbative effects persist at short distances is not rigorously known at present, such nonperturbative determinations will be required everywhere eventually.

A determination of the strong coupling constant requires a lattice calculation of the coupling constant at short distances and a determination of the lattice energy scale (the inverse lattice spacing, a^{-1}) in GeV. The lattice spacing may be determined by calculating a hadron mass or mass splitting in lattice units, and then comparing the result with experiment. We are free to use the simplest hadrons for this purpose. These are the quarkonia, the mesons in the ψ and Υ systems. Likewise, more is known both perturbatively and nonperturbatively about the strong coupling constant on the lattice than about any other term of the standard model Lagrangian. These facts make α_s among the easiest parameters to determine well with lattice calculations.

Heavy quark-antiquark bound states are among the easiest of lattice calculations for a variety of reasons. They are small, so that the finite volumes of lattice calculations are less of a problem. Lattice Green's functions of heavy quarks can be calculated much faster than those of light quarks. The most important reason, however, is that the quarks in these mesons are slow moving, so that nonrelativistic reasoning and methods can be used to monitor the accuracy of the approximations (finite a, finite V, quenched, ...), to make corrections, and to guide physics expectations.

The fact that so much can be understood about these systems using such relatively simple methods as nonrelativistic potential models made them seem initially less challenging and therefore less interesting to lattice gauge theorists. It is this fact, though, that makes them the most important systems for some purposes, including understanding the accuracy of lattice approximations and extracting certain types of standard model information, such as the heavy quark

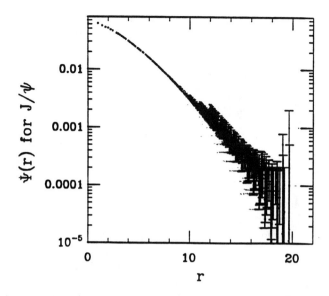

Fig. 2. The wave function of the J/ψ meson calculated in the Coulomb gauge on a 24^4 lattice.

masses and the strong coupling constant.

Fig. 2 shows the wave function of the J/ψ meson calculated in the Coulomb gauge on a 24^4 lattice. For nonrelativistic systems, such $Q\overline{Q}$ are reasonably good approximations to the full quantum state.

Such wave functions may be put to practical use in estimating the size of correction terms in lattice calculations before calculating them from first principles. The NRQCD collaboration in particular has made use of them to obtain very strong control over discretization errors. Their calculations were based on a nonrelativistic action for the fermions.[20] Coefficients of correction operators for this action were calculated in perturbation theory. The corrections were then estimated in advance of Monte Carlo simulations by evaluating the corrections perturbatively within the wave functions calculated with the unimproved theory. These estimates were then checked and improved by including the required correction operators in the Monte Carlo simulations. This type of analysis allows quarkonium systems to provide unusually accurate determinations of mass splittings, and thereby of the lattice spacing.

Once the lattice energy scale has been set, the second element in the determination of the strong coupling constant is the determination of the coupling constant at a given lattice energy or distance scale. One way of doing this is to relate the bare lattice coupling to the coupling of another regulator (say the $\overline{MS}$ coupling) by calculating perturbative Green's functions in the two regulators and demanding that they be equal.[17] The most straightforward perturbative determi-

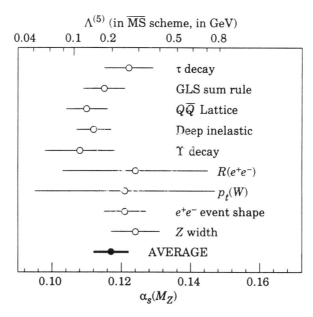

Fig. 3. The Particle Data Group's compilation of determinations of the strong coupling constant by various methods.

nation following this approach is not very accurate for reasons which are now well understood.[18] Just as in dimensional regularization, the most natural coupling in terms of the regulator (the MS coupling) is not the most useful coupling for expressing perturbative results (the $\overline{MS}$ coupling makes perturbation series more convergent), perturbation series expressed in terms of the bare lattice coupling have uniformly large second order coefficients (and when truncated, agree poorly with Monte Carlo calculations of short distance quantities). The reason for the large difference between the bare lattice coupling and renormalized couplings is also understood: it arises from large tadpole terms arising from higher dimension operators present in Wilson's lattice action but not present in continuum QCD actions.

The upshot of these considerations is that it is better to directly extract the strong coupling directly from nonperturbative short distance calculations than to rely on perturbation theory. A simple and direct way of doing this might be to calculate the static quark potential nonperturbatively at short distance and to fit the result to an asymptotically free Coulomb potential. This turns out to be numerically very difficult. A related but more tractable method is to extract the coupling constant from nonperturbative calculations of small Wilson loops.

Recent results[21,23] following this approach are shown along with other determinations of the strong coupling constant are shown in the Particle Data Group figure[19] reproduced in Fig. 2.8. The errors in the lattice determinations are dominated by uncertainties associated with the quenched approximation, followed by perturbation theory uncertainties. As far as is now known, uncertainties associated with other source are all under one per cent.

In the last few months, the first unquenched calculations, including the effects of light quark loops from first principles, have appeared.[23] They yield the result $\alpha_s(M_Z) = 0.115 \pm 0.002$, at the upper end of the error bar shown in Fig. 2.8, with an error bar about a factor of three smaller. The remaining known 2% uncertainties are those of perturbation theory. They will be removed eventually by third order calculations relating lattice couplings to continuum couplings which are now in progress.[22]

Once this has occurred there will be no known sources of uncertainty in lattice determinations of the strong coupling constant larger than a per cent. At this point the game will shift to a further examination of aspects of the calculations which can be better understood, whose numerical uncertainties have been assumed small. These include the reliability of methods used for including the effects of the light quark loops, and the perturbative calculability of correction terms in the lattice Lagrangian to $O(a^2)$. The second of these is suggested by the condensate picture of short distance nonperturbative effects, but is not known from first principles. This assumption will be tested by the methods being developed in Ref. ([22]) which allow a cleaner separation of perturbative short distance effects (logarithmic in the lattice spacing) from nonperturbative short distance effects (thought to be power law in the lattice spacing) than has been possible with methods used so far.

This was the first determination of a parameter of the standard model with lattice methods in which a numerical estimate was made of all large sources of calculational uncertainty. In the earliest of these calculations, some corrections and error estimates (for example, the effects of light quark loops) were made with phenomenological input, analogously to the way hadronization effects are estimated with hadronization Monte Carlos in collider determinations of α_s. As discussed above, these are now in the process of being systematically replaced with first principles calculations.

2.9. The b Quark Mass

Heavy quark masses can also be determined from quarkonium systems, and so share the advantages which make α_s easier to compute than many other quantities. In one way they are actually easier than the α_s determination. Although quark masses "run" for momentum transfers larger than the quark mass, perturbation theory suggests no running for momenta below the quark mass. Therefore, the large correction known to be present in quenched calculations for the running of α_s between the lattice spacing scale and the scale of ψ or Υ physics is not expected for the quark masses.

The best b quark mass determination[24] has been obtained from the Υ spectrum in two different ways: from the energy gap between the Υ system and the vacuum, and from the variation in the energy of the Υ as its momentum is varied.

The results are consistent. The mass has been determined with and without light quark loops, with little observed difference, as expected.

The current result is

$$m_b^{\overline{MS}}(m_b) = 4.0(1) \text{ GeV}. \tag{14}$$

The uncertainty is dominated by perturbation theory. The mass of the c quark can in principle be calculated by identical methods, but perturbative uncertainties are larger for the smaller quark mass.

When these calculations first appeared, they were the first determination of a standard model parameter which had no large corrections or uncertainties which were unestimated or treated phenomenologically. The results for the pole mass do not differ much from conventional wisdom. The result in Eqn. 14 is lower than some older values mostly as a result of the fact that the third order term in the relation between the pole mass and the $\overline{MS}$ mass is now known, and not because of surprises in the lattice calculations.

2.10. B_K

It is likely that quite a few of the most important weak matrix element calculations will be done in the near future to the level of accuracy already obtained for α_s and m_b. These include heavy meson decay constants, exclusive semileptonic decays, and B parameters. Progress is somewhat slower for these because one has less idea in advance about what to expect for the various sources of uncertainty.

The weak matrix element which is best understood at present is the B_K parameter of $K\overline{K}$ mixing. There are several technical reasons why. Kaons are pseudoscalars which have the best signal to noise ratio in numerical calculations (since the signal one is after goes like $\exp(-Mt)$). They are also the smallest hadrons, so that finite volume errors are the smallest. Moderately heavy mesons like the kaons are easier to work with numerically than very light pions because quark propagator algorithms begin to fail in the zero mass limit. No extrapolation to the physical quark mass is required for the kaons. (Instead, one has a much more benign extrapolation in $m_s - m_d$). Finally, the B parameter in a ratio of two very similar amplitudes, so that some corrections (such as perturbative corrections) are relatively small.

One source of uncertainty turns out not to be small and is in fact larger for B_K than for most other quantities: dependence on the lattice spacing. Fig. 4 shows the results obtained for several lattice spacings. The significant dependence on the lattice spacing can arise from logarithmic perturbative corrections, and from $O(a)$ and $O(a^2)$ effects arising from discretization errors and perhaps nonperturbative effects. The perturbative corrections are expected and calculated to be small. The most accurate numerical calculations[25] have been done with "staggered" fermions, which preserve an exact chiral symmetry at the expense of containing some extra unphysical flavors (as opposed to Wilson fermions which have the right number of flavors, but whose chiral symmetry must be recovered by fine

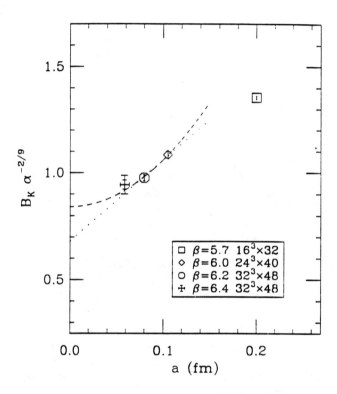

Fig. 4. B_K, the $K\overline{K}$ mixing paramter.

tuning). Staggered fermions have discretization errors in their propagators which start as $O(a^2)$. Discretization errors in multiquark operators may go as either $O(a)$ or $O(a^2)$. A year ago, it was finally determined that the errors in the operators occurring in B_K go like $O(a^2)$, allowing an extrapolation to the $a = 0$ limit.[26] The current result including uncertainties is

$$
\begin{aligned}
B_K(NDR, 2\text{GeV}) = 0.616 \quad &\pm \quad 0.020 \; (stat) \\
&\pm \quad 0.014 \; (g^2) \\
&\pm \quad 0.009 \; (scale) \\
&\pm \quad 0.004 \; (operator) \\
&\pm \quad 0.002 \; (correction) \\
= 0.616 \quad &\pm \quad 0.020 \pm 0.017.
\end{aligned}
$$

in the naive dimensional reduction definition of B_K, or

$$
\begin{aligned}
\widehat{B}_K &\equiv B_K(NDR, 2\text{GeV}) \, \alpha_s(2\text{GeV})^{-6/25} \\
&= 0.825 \pm 0.027 \pm .023.
\end{aligned}
$$

for the renormalization group invariant B_K parameter.

This was the first important standard model quantity calculated with no recourse to phenomenological arguments in calculating corrections which achieved a large improvement on the uncertainties in the best conventional determinations.

3. Outlook

The three examples discussed have achieved relatively small and understood error estimates earlier than other calculations because they are in one way or another special cases. In the generic case, one uses brute force error analysis, in which the lattice spacing, volume, and light quark mass are pushed to zero, infinity, and M_{phys}, respectively, and the results are fit to asymptotic forms. In the above cases, various sorts of arguments lead to advance expectations for the sizes of the various errors.

The brute force error analysis of many other interesting quantities, including the remaining quark masses, heavy meson decay constants, and semileptonic decay amplitudes, will probably not be qualitatively more difficult. However, the smaller advance knowledge about the errors will require a longer period of disentangling the various sources of uncertainty. Generally speaking, processes with a single meson on the lattice at a time (preferably a pseudoscalar) have the best chance of being solidly calculated over the next few years.

Acknowledgements

Fermilab is operated by Universities Research Association Inc. under contract with the U.S. Department of Energy.

References

1. K.G. Wilson, *Phys. Rev.* **D10** (1974) 2445.
2. M. Creutz, L. Jacobs, and C. Rebbi, *Phys. Rev. Lett.* **42** (1979) 1390; M. Creutz, *Phys. Rev. Lett.* **43** (1979) 553.
3. K. Wilson, in *Recent Developments in Gauge Theories*, edited by G. 't Hooft et al. (Plenum, New York, 1980).
4. H. Hamber and G. Parisi, *Phys. Rev. Lett.* **47** (1981) 1792; E. Marinari, G. Parisi, and C. Rebbi, *Phys. Rev. Lett.* **47** (1981) 1795.
5. D.H. Weingarten, *Phys. Lett.* **B109** (1982) 57.
6. L. D. McLerran and B. Svetitsky, *Phys. Lett.* **98B** (1981) 195; J. Kuti, J. Polonyi, and K. Szlachanyi, *Phys. Lett.* **98B** (1981) 199; S. A. Gottlieb et al., *Phys. Rev. Lett.* **55** (1985) 1958; N. H. Christ and A. E. Terrano *Phys. Rev. Lett.* **56** (1986) 111.
7. P. Weisz, *Nucl. Phys.* **B212** (1983) 1.
8. K. Symanzik, *Nucl. Phys.* **B226** (1983) 187, **B226** (1983) 205.
9. M. Lüscher and P. Weisz, *Phys. Lett.* **158B** (1985) 250.
10. B. Sheikholeslami and R. Wohlert, *Nucl. Phys.* **B259** (1985) 572.
11. For a complete set of references on this topic, see D. Weingarten, *Nucl. Phys. B (Proc. Suppl.)* 9 (1989) 447.
12. S. Duane, *Nucl. Phys.* **B257** (1985) 652.
13. S. Gottlieb. et al., *Phys. Rev.* **D35** (1987) 2531.
14. S. Duane, A. D. Kennedy, B. J. Pendleton, and D. Roweth, *Phys. Lett.* **B195** (1987) 216.
15. M. Lüscher, *Nucl.Phys.* **B418** (1994) 637.
16. M. Fukugita, N. Ishizuka, H. Mino, M. Okawa, and A. Ukawa, *Phys. Lett.* **B301** (1993) 224.
17. A. Hasenfratz and P. Hasenfratz, *Phys. Lett.* **93B** (1980) 165.
18. G. P. Lepage and P. B. Mackenzie, *Phys. Rev.* **D48** (1993) 2250.
19. *Phys. Rev.* **D 50** (1994) 1297.
20. G.P. Lepage and B.A. Thacker, *Nucl. Phys. B (Proc. Suppl.)* 4 (1988) 199; B.A. Thacker and G.P. Lepage, *Phys. Rev.* D43 (1991) 196.
21. A.X. El-Khadra, G. Hockney, A.S. Kronfeld, and P.B. Mackenzie, *Phys. Rev. Lett.* **69** (1992) 729.
22. M. Lüscher, R. Sommer, P. Weisz, U. Wolff, *Nucl. Phys.* **B413** (1994) 481, and to appear.
23. C. T. H. Davies et al. (the NRQCD Collaboration), A Precise Determination of α_s from Lattice QCD, OHSTPY-HEP-T-94-013 (Aug 1994) 11p, [HEP-PH 9408328].
24. C. T. H. Davies et al. (the NRQCD Collaboration), A New Determination of M_b Using Lattice QCD, OHSTPY-HEP-T-94-004 (Apr 1994) 8p, [HEP-LAT 9404012].
25. G.W. Kilcup, S.R. Sharpe, R. Gupta, and A. Patel, *Phys. Rev. Lett.* **64** (1990) 25, N. Ishizuka et al, *Phys. Rev. Lett.* **71** (1993) 24.
26. S. R. Sharpe, *Nuc. Phys.* **B (Proc. Supl.)** **34** (1994) 403; G.W. Kilcup, S.R. Sharpe, and R. Gupta, to appear.

Status and Future Prospects for United States Accelerators and Accelerator Physics

ROBERT H. SIEMANN, *SLAC*

STATUS AND FUTURE PROSPECTS FOR UNITED STATES ACCELERATORS AND ACCELERATOR PHYSICS

R. H. SIEMANN*

Stanford Linear Accelerator Center, Stanford University, Stanford, CA 94309

ABSTRACT

The recent performance and future prospects of accelerators in the United States are reviewed. The next decade promises significant improvements and major new facilities. There is uncertainty beyond that because of the SSC cancellation and the new, enhanced importance of international accelerator projects.

1. Introduction

This paper is a review of the status and future prospects of accelerators in the United States. These accelerators cover a wide range: electrons and protons, circular and linear, colliders and fixed target. Each type has its own peculiarities, but much of the underlying physics and language is common. It is worthwhile beginning by summarizing this.

Particles in an accelerator are focused in the directions transverse to their motion by quadrupoles, and they are bunched into short bunches along the direction of motion by the RF system. When n_1 particles per bunch in one beam collide with n_2 particles per bunch in the other beam with a collision frequency f_c, the luminosity is

$$L = \frac{1}{2\pi} \frac{n_1 n_2 f_c}{\sqrt{\sigma_{x1}^2 + \sigma_{x2}^2}\sqrt{\sigma_{y1}^2 + \sigma_{y2}^2}}.$$

The parameters σ_{x1}, σ_{x2}, σ_{y1} and σ_{y2} are the rms horizontal and vertical beam sizes at the interaction point, and the denominator is the effective area of overlap of the two beams. Usually $\sigma_{x1} = \sigma_{x2} = \sigma_x$ and $\sigma_{y1} = \sigma_{y2} = \sigma_y$ giving

$$L = \frac{1}{4\pi} \frac{n_1 n_2 f_c}{\sigma_x \sigma_y}.$$

The horizontal and vertical beam envelopes vary along the accelerator. They are described by amplitude functions, usually called β-functions, that follow from solutions of two Hill's equations with driving terms based on the specific magnet configuration. The rms beam size at a location s along the accelerator is given by

* Work supported by the Department of Energy, contract DE-AC03-76SF00515.

$$\sigma = \sqrt{\beta(s)\varepsilon_n / \gamma}$$

where $\beta(s)$ is the β-function at s, and γ is the beam energy in units of mc^2. The invariant emittance, ε_n, is a constant inversely proportional to the phase space density of the beam. The β-functions are minimized at the interaction point to maximize the luminosity, but there are limits to this. A tightly focused beam has a large angular divergence, and the bunch length, σ_L, and interaction point β-functions, β_x^* and β_y^*, must satisfy

$$\sigma_L \leq \min\left(\beta_x^*, \beta_y^*\right).$$

The bunch length itself is determined by a combination of the RF system, the magnet configuration, and the longitudinal phase space density of the beam.

Liouville's theorem states that phase space density is constant in Hamiltonian systems. Electron storage rings and $\bar{p}$ sources are not Hamiltonian because of synchrotron radiation and stochastic cooling, respectively. The phase space density and emittance of electron rings are determined by the properties of synchrotron radiation, and a wide range of emittances are possible depending on the magnet configuration. Anti-proton sources use feedback to reduce emittance. The bandwidth and power of the cooling electronics strongly influence intensity and emittance.

Proton storage rings are Hamiltonian, but nonlinearities and instabilities can lead to significant phase space distortions. The rms phase space density which is the density averaged over local distortions is the one of practical interest. This density can remain constant at best, and it is determined by the particle source and any effects that distort phase space during acceleration.

Linear colliders combine aspects of electron storage rings and proton accelerators. The particle sources are damping rings, and synchrotron radiation determines the emittance there. Once the beam is extracted from the damping rings and accelerated careful attention must be paid to minimizing instabilities and nonlinearities to preserve the rms invariant emittance.

Different factors that can limit the luminosity. These include the availability of particles, single beam instabilities, and beam-beam effects. The availability of particles applies most strongly to $p\bar{p}$ colliders where $\bar{p}$ production is the dominant performance limitation.

Single beam instabilities are caused by beam generated electromagnetic fields, called wakefields, acting back on the beam that caused them. Wakefields can impose hard limits on the number of particles with the beams being lost if those limits are exceeded, or they can impose soft limits through intensity dependent emittance or background increases. At low energies wakefields are electrostatic fields commonly called space-charge fields while at high energies they are caused by variations in vacuum chamber geometry and resonant modes in RF cavities or other structures. Short range wakefields act on a single bunch and affect the number of particles/bunch. They are difficult to control with feedback because frequencies comparable to the inverse of the bunch length are involved. In addition, the study of single bunch instabilities is at the forefront of accelerator theory and experiment, and recent experience at the SLC has taught us that we cannot be confident with calculations.

The bunches can interact with each other through long range wakefields thereby putting a limit on the total current, $I_{tot} = nef_c$. Feedback can be used to control multibunch instabilities, and, their effects are easier to estimate reliably.

The primary interaction between beams is through their electromagnetic fields at the interaction point. There is a maximum field strength, and when it is exceeded backgrounds increase and/or luminosity decreases. These effects are parametrized in storage rings by the beam-beam strength which is often called the beam-beam tune shift,

$$\xi = \frac{r}{2\pi} \frac{n\beta_y^*}{\gamma\sigma_y(\sigma_x + \sigma_y)};$$

$r = r_e = 2.82 \times 10^{-15}$ m for e^+e^- colliders and $r = r_p = 1.54 \times 10^{-18}$ m for proton colliders. Roughly speaking, e^+e^- storage rings have a limit per interaction region of $\xi < 0.03$ to 0.05, and proton storage rings have a limit summed over all interaction regions of $\Sigma\xi < 0.02$. The luminosity can be rewritten in terms of ξ when the beam-beam effect is a limit

$$L = \xi \frac{I_{tot}}{e} \frac{\gamma}{r\beta_y^*}.$$

(If the beams have unequal currents, etc., the parameters of one of the beams should be used in this equation.)

The fields at the interaction point of a linear collider can be much stronger than those of a storage ring because the beams only collide once. These fields lead to focusing during the collision (disruption), photon radiation (beamstrahlung), e^+e^- pair production, and low invariant mass hadronic events. The consequences of the beam-beam interaction in linear colliders is discussed in the section on future linear colliders, but, in contrast to experimentally established limits on ξ, there is almost no experience with the beam-beam interaction in linear colliders, and these consequences are based on theory and conjecture.

2. The AGS

The AGS (Alternating Gradient Synchrotron) at Brookhaven National Laboratory is a venerable accelerator with a long list of accomplishments in accelerator physics and in the particle physics experiments performed there.   It is the only US accelerator that does not have colliding beams as the mainstay of its operation. The high energy physics research program has become dominantly the search for rare or forbidden Kaon decays, and this has required a large increase in the AGS intensity.

The central element of that intensity increase was the addition of the AGS Booster synchrotron that was completed in 1991. The Booster overcame an intensity limit caused by space-charge effects by raising the injection energy from 200 MeV to 1.5 GeV. In addition to the Booster, improvements were needed to the AGS itself to handle higher current. The performance shown in Table 1 for 1994 is the result of improvements in the RF system, implementation of a pulsed quadrupole system that speeds up the passage through the transition energy, and the damping of longitudinal and transverse instabilities. This is an ongoing effort and further improvements are expected from additional work on the RF system and the feedback systems that control instabilities and from corrections of the AGS magnetic field at low energy. When this work is completed the design intensity of

Date	Proton Intensity per Pulse ($\times 10^{13}$)	Repetition Period (sec, w/o Slow Spill)
1990 (Before Booster)	1.8	1.7
1994 Peak	4.0	2.8
8 hr Shift Avg.	3.6	
1995 (Projection)	6.0	2.0

Table 1. AGS Performance.[1]

6×10^{13} protons/pulse for the Booster/AGS improvement program would have been reached. Further improvements are foreseen from increasing injection, acceleration, and transfer efficiencies and from increasing the Booster intensity.

In addition to these intensity improvements, acceleration of polarized protons is being studied in the AGS with the goals of maintaining the 80% injection polarization to high energy. Recent experiments have shown that the underlying spin dynamics are well understood, and a polarized beam at the normal AGS extraction energy is expected in 1995.

3. CESR

The CESR interaction region geometry has been modified to allow beams to collide at an angle of ±2 mrad. This makes it possible to use bunch trains, closely spaced bunches, where unwanted collisions at the interaction region are avoided by the crossing angle, and, in an extension of the "pretzel" technique developed at CESR, collisions between different bunch trains in the arcs are avoided by the use of electrostatic separation. Until recently only single bunches rather than bunch trains were being used, and people were gaining experience with this new mode of operation. Routine operation with multiple bunches per bunch train will require improvements in the interaction region apertures and superconducting RF to store higher currents. Recent CESR performance along with short term and longer term goals are presented in Table 2.

Parameter	1993	October, 1994
Peak Luminosity (10^{32}cm^{-2}s^{-1})	2.9	2.4
Best Daily Luminosity (pb^{-1})	15.2	12.8
Best Monthly Luminosity (pb^{-1})	284	
Integrated Luminosity (pb^{-1})	1362	
Current per Beam (Bunches $\times$ mA)	7×16 mA	9×11 mA
Beam-Beam Parameter (ξ)	0.04	0.04
Crossing Angle (mrad)	0	±2
Goals, Luminosity & Number of Bunches		6×10^{32}cm^{-2}s^{-1}, 27 10×10^{32}cm^{-2}s^{-1}, 45

Table 2. CESR Performance[2]

Parameter	High Energy Ring	Low Energy Ring
Luminosity (10^{33} cm^{-2}s^{-1})	3.0	
Beam Energy (GeV)	9.00	3.11
Number of Bunches	1658	
Single Bunch Current (mA)	0.60	1.29
Total Current (A)	0.98	2.14
Beam-Beam Parameter (ξ)	0.03	0.03
β_y^* (cm)	2.0	1.5
σ_x, σ_y (μm)	155, 6.2	
Bunch Length (cm)	1.0	1.0
Collision Geometry	Head-on with Magnetic Separation	
First Colliding Beams	Summer, 1998	

Table 3. PEP-II Parameters[3]

4. PEP-II

PEP-II will be a high luminosity e^+e^- collider operating at a center-of-mass energy, E_{CM} = 10.58 GeV, the mass of the Υ(4S). The beam energies will be unequal to give the center-of-mass a boost that will allow measurement of B-meson decay times with vertex detectors. Parameters are given in Table 3.

The total currents must be large since L $\propto$ I$_{tot}$, and storing these currents is the major issue for the vacuum and RF systems. The synchrotron radiation powers are 5.29 MW and 2.66 MW in the High Energy and Low Energy Rings, respectively. The vacuum chambers must absorb this power while maintaining a good vacuum and providing shielding to prevent radiation damage to equipment in the accelerator enclosure. The copper vacuum chamber technology pioneered at DESY is being used in the High Energy Ring, and the Low Energy Ring employs localized photon beam stoppers similar to those in third generation synchrotron light sources. Cleanliness during manufacturing and vacuum pump design are receiving special attention because dust particles can be trapped by a high current electron beam. This was discovered at HERA where the recent increase in luminosity came from switching from electrons to positrons to avoid this.

The RF system must make up the radiated power efficiently without causing instabilities. Wakefields are minimized by using a high accelerating gradient to reduce the number of cavities and by damping unwanted, high frequency cavity modes to decrease long range wakefields. In addition to control multibunch instabilities, PEP-II will have a feedback system that measures and corrects the positions of bunches that are only 4.2 nsec apart using Digital Signal Processing techniques. Another feedback system will control instabilities caused by the fundamental, accelerating mode.

The primary collision is head-on; the bunches are separated almost immediately after the interaction point using permanent magnet dipoles and taking advantage of the energy difference. Despite this the beams are still close enough to experience each other's fields at the parasitic crossing point. This complicates the beam-beam interaction and prevents further increasing the number of bunches.

Most of the global decisions affecting the PEP-II design, such as those described above, have been made, and construction is well underway. Collisions are expected in the summer of 1998, and studies of CP violation should start soon afterward when the BaBar detector is ready.

5. The Tevatron

In the Tevatron the horizontal and vertical β-functions at the interaction point are equal, $\beta_x^* = \beta_y^* = \beta^*$, the horizontal and vertical emittances of each beam are equal, $\varepsilon_{nx} = \varepsilon_{ny}$, but the p and $\bar{p}$ beams can have different emittances, $\varepsilon_{np} \neq \varepsilon_{n\bar{p}}$. The luminosity is limited by the number of $\bar{p}$'s and the $\bar{p}$ beam-beam tune shift. Writing the luminosity in terms of these $\bar{p}$ parameters

$$L = \xi_{\bar{p}} \frac{I_{tot,\bar{p}}}{e} \frac{2\gamma}{r_p \beta^* (1 + \varepsilon_{n\bar{p}} / \varepsilon_{np})}.$$

where $\xi_{\bar{p}}$ depends on the phase space density of the proton beam

$$\xi_{\bar{p}} = \frac{r_p}{4\pi} \frac{n_p}{\varepsilon_{np}}.$$

It has a maximum value $\xi_{\bar{p}} < 0.02 / N_{ip}$ where N_{ip} is the number of collision points. While these equations shouldn't be taken literally because there are subtle beam-beam effects for unequal emittance beams, they show the interplay that dominates the Tevatron luminosity. The number of protons should be increased until the beam-beam limit is reached; the number of collision points should be minimized, and, once that is done, the luminosity depends directly on the $\bar{p}$ current.

Table 4 summarizes the Tevatron performance for Run Ia in 1992/1993, Run Ib, the present run, and that expected in the era of the Main Injector. It illustrates the Tevatron improvement program that is centered on these points. Prior to Run Ia the beams collided at twelve points around the ring while they produced useful luminosity only at the CDF detector (D0 was not installed yet). The beam-beam strength parameter summed over all collision points reached the beam-beam limit and limited the luminosity. The improvement for Run Ia was operation with electrostatically separated orbits that avoided unwanted collisions and allowed the number of protons/bunch to be increased without exceeding the beam-beam limit.

The linac kinetic energy was increased from 200 MeV to 400 MeV for Run Ib to reduce the space-charge intensity limit at injection into the Booster. The Booster intensity and $\bar{p}$ production rate have increased by 1.7 and 1.5, respectively, from this improvement.

The Main Injector will significantly increase the number of protons that can be accelerated and targeted for $\bar{p}$ production. The number of $\bar{p}$'s will increase so much that the beam-beam limit would be reached for protons, and to avoid that, the number of bunches has to be increased also from 6 to 36. The Main Injector is scheduled to be commissioned in the summer of 1998, and collider operation is expected in the fall of that year.

Further Tevatron energy and luminosity upgrades are being actively considered. Parameters are still evolving, and the DiTevatron in Table 5 is a snapshot taken last summer that shows the key features of Tevatron upgrades: energy increases will require

Parameter	Run Ia 1992/1993	Run Ib 1994/1995	Main Injector
Peak Lum. (10^{30} cm^{-2} s^{-1})	5.4	16	123
Integrated Lum./Week (pb^{-1})	1.1	3	25
Beam Energy (GeV)	900	900	1000
Dipole Magnetic Field (T)	4.0	4.0	4.4
Total # of protons (10^{12})	0.8	1.2	13.7
Total # of $\bar{p}$'s (10^{12})	0.19	0.3	1.30
Total Beam-Beam Strength Parameter, $\Sigma \xi_{\bar{p}}$	0.011	0.013	0.020
$\bar{p}$ Stacking Rate (10^{10}/hr)	4	6	15
Bunches	6	6	36
Interactions/Crossing (45 mb)	1.1	3.3	3.2
Bunch Spacing (nsec)	3493	3493	395

Table 4. Tevatron Parameters.[4] The first two columns are achieved performance, and the third is a projection in the Main Injector era.

replacing the present Tevatron magnets with new ones that take advantage of developments made for the SSC and LHC, and luminosity upgrades depend on increasing the $\bar{p}$ accumulation rate and on a large number bunches to keep below the beam-beam limit and to reduce the number of interactions per crossing.

Doing that requires a new place to store $\bar{p}$'s between fills and improving the $\bar{p}$ flux from the production target. The present Accumulator can not hold the total number of $\bar{p}$'s needed, and a new accumulator would have to be constructed. A promising idea is building a fixed energy, permanent magnet ring in the Main Injector tunnel. In addition to providing storage between fills this ring offers the possibility of recovering $\bar{p}$'s from the previous store. The present Accumulator would still be used to cool the beam before injecting into the new accumulator.

There are two ways to improve the $\bar{p}$ flux from the target: increase the number of incident protons or increase the $\bar{p}$ acceptance downstream. Both approaches are being considered and both are difficult. The issue is associated with manipulations of longitudinal emittance which is proportional to the product of bunch length and momentum spread. The proton beam is tightly bunched just before hitting the $\bar{p}$ production target, and the acceptance of beamlines downstream of the target determine the momentum spread. This momentum spread is much larger than could be accommodated by the Accumulator, so the beam is "debunched", the momentum spread is decreased at the price of increasing the bunch length. The Debuncher Ring does this in the Fermilab $\bar{p}$ source.

The equivalent process must be performed in an improved source. If the entire Main Injector beam were targeted, a new, large debuncher would have to be constructed in the Main Ring tunnel because the present Debuncher has only one-seventh the circumference of the Main Injector. If only one-sixth of the Main Injector beam were targeted as is planned now, the debunching could be done with a ~ 1 GeV high gradient linac that would have a large momentum acceptance. An alternative to either of these would be replacing

Parameter	Main Injector	DiTevatron
Peak Lum. (10^{30} cm^{-2} s^{-1})	123	2000
Integrated Lum./Week (pb^{-1})	25	400
Beam Energy (GeV)	1000	2000
Dipole Magnetic Field (T)	4.4	8.8
Total # of protons (10^{12})	13.7	25.7
Total # of $\bar{p}$'s (10^{12})	1.30	9.8
Total Beam-Beam Strength Parameter, $\Sigma\xi_{\bar{p}}$	0.020	0.019
$\bar{p}$ Stacking Rate (10^{10}/hr)	15	100
Bunches	36	108
Interactions/Crossing (45 mb)	3.2	17
Bunch Spacing (nsec)	395	132

Table 5. Performance with the Main Injector is compared with DiTevatron parameters from the Hadron Collider Workshop, University of Indiana, July, 1994.

the Booster with a new, high intensity injector for the Main Injector and thereby increase the intensity of the proton beam itself. Each approach has advantages, difficulties and drawbacks, and implications for simultaneous operation of collider and fixed target programs. Whichever route is taken, a substantial accelerator, in addition in the new accumulator ring, would have to be constructed.

There are also crucial accelerator physics issues related to operating the Tevatron as a collider with a large number of bunches. Fortunately, many of these can be studied experimentally, and once that is done they should introduce relatively little uncertainty into Tevatron upgrade plans.

Of course, any Tevatron upgrade beyond the Main Injector must be justified by an outstanding particle physics program in the era of the LHC, and that is being actively discussed among high energy physicists.

6. The SLC

The SLC has been delivering luminosity to the SLD detector for several years, and it is planned to continue until PEP-II physics starts. The goal is greater than 5×10^5 polarized Z's. Two major projects were completed recently to improve on the 1993 performance (Table 6). A single bunch instability in the damping rings that limited the beam current was removed with the installation of new vacuum chambers designed specifically to reduce wakefields. (The characteristics of the instability changed in unexpected ways; this is an interesting accelerator physics problem with implications for future colliders.) The final focus was improved by removing the dominant optical aberrations and adding improved diagnostics. The result has been a reduction of the beam height from 0.8 μm in 1993 to 0.5 μm today. Commissioning these two improvements together took some time, but that is over and the SLC is well on the way to producing 10^5 Z's in the 1994/95 run.

Parameter	1993	1994
Luminosity (10^{29}cm^{-2}sec^{-1})	3	7
Polarization	62%	80%
Inter. Point Current (10^{10}/pulse)	3.0	3.5
RMS Spot Size (μm^2)	2.6×0.8	2.6×0.5
Integrated Lum. (pb^{-1}, Z's)	1.7, 50,000	Goal is 10^5 Z's

Table 6. Typical SLC Parameters.

7. Future Linear Colliders

Future linear collider development has focused on an $E_{CM} = 0.5$ TeV collider with the potential for being expanded to 1 TeV or more. The luminosity can be written in terms of the power of a single beam, $P_B = \gamma mc^2 nf_c$, the vertical spot size (it is assumed that $\sigma_x \gg \sigma_y$), and a factor related to detector backgrounds. Those backgrounds come from the strong electromagnetic fields at the interaction point. The number of beamstrahlung photons per incident particle serves as a measure and is given by

$$n_\gamma \approx \frac{2\alpha r_e n}{\sigma_x}.$$

The luminosity is

$$L = \frac{1}{8\pi\alpha r_e mc^2} \frac{n_\gamma}{\gamma} \frac{P_B}{\sigma_y}.$$

Roughly speaking, there are two different approaches to high energy linear collider design. In one the beam power and vertical spot size are large while in the other they are both small. Alignment and stability tolerances can be relaxed when the spot is large, and that is the attraction of that approach. The designers of colliders employing small spots agree with this but consider their tolerances reasonable.

Selected parameters from LC-93 are given in Table 7. The colliders are:

TESLA which is based on superconducting RF. All the others would use room temperature RF.

SBLC which uses 3 GHz RF where there is extensive operating experience. TESLA and SBLC are large beam power, large spot designs while the others rely on a nanometer vertical beam size for good luminosity.

NLC which uses higher frequency, 11.4 GHz, RF in configuration similar to conventional linacs.

JLC-I which has three RF frequency options. Multiple bunches are accelerated in each RF pulse as they are in TESLA, SBLC, and NLC.

VLEPP which employs a single high intensity bunch rather than multiple bunches.

CLIC which is a "two-beam" accelerator with klystrons replaced by an RF power source based on a high-current, low-energy beam traveling parallel to the high energy beam.

The two basic, interrelated issues are putting a high energy linear collider on solid technical footings and deciding between these different options. Work is going on worldwide with close coordination and frequent workshops, and in many cases that work is

Parameter	TESLA	SBLC	JLC-I (S)	JLC-I (C)	JLC-I (X)	NLC	VLEPP	CLIC
L $(10^{33}\text{cm}^{-2}\text{s}^{-1})$	7	4	4	7	6	8	15	2 - 9
RF Freq (GHz)	1.3	3.0	2.8	5.7	11.4	11.4	14	30
Loaded Gradient $(\text{MV/m})^c$	25	17	19	33	31	38	96	78 - 73
Rep Rate (Hz)	10	50	50	100	150	180	300	1700
Bunches per RF pulse	800	125	55	72	90	90	1	1 - 4
σ_{x0}/σ_{y0} (nm)	1000/64	670/28	300/3	260/3	260/3	300/3	2000/4	90/8
P_B (MW)	16.5	7.3	1.4	2.9	3.4	4.2	2.4	.4 - 1.6
n_γ	2.7	2.0	1.6	1.4	1.0	0.9	5.0	4.7
AC Power $(\text{MW})^a$	137	114	106	193	86	141	91	175
$2P_B/P_{AC}$	0.24	0.13	0.03	0.04	0.09	0.06	0.05	0.02

a) Linac power only (damping ring, detector, utility power, etc. not included)

Table 7. Selected Linear Collider Parameters for $E_{CM} = 0.5$ TeV.[5]

being done by collaborations similar to those that are common in experimental high energy physics.

One such collaboration, the Final Focus Test Beam (FFTB) Collaboration, has demonstrated optics with a demagnification comparable to that required for a next generation linear collider. Their goal was to focus a 47 GeV beam from the SLC to a 60 nm high spot. The beam was commissioned this spring, and during the last three hours of the first extended run they achieved a 70 nm spot that was stable and reproducible over several hours.

Other important recent developments include: a test of an accelerating structure designed specifically to reduce long range wakefields; accelerating gradients exceeding 50 - 100 MV/m in room temperature structures; gradients of 25 MV/m in superconducting cavities; and prototype klystrons reaching the performance needed for some of the colliders in Table 7. In addition, there are prototype facilities planned and under construction at several laboratories. These include linac prototypes at KEK, SLAC, DESY and Protvino, and a damping ring prototype well underway at KEK.

Energy reach and energy expandability have come to the fore recently. Ten years ago it was hoped that the next linear collider would have $E_{CM} \sim 1$ - 2 TeV. However, as work progressed it was realized that this would be too large a step from the SLC, and designs concentrated on $E_{CM} = 0.5$ TeV. While there is a strong physics program at that energy, the attractiveness of a linear collider increases significantly if the energy could be increased as a second stage.

Table 8 gives some preliminary parameters for $E_{CM} = 1$ TeV. The energy of the room temperature accelerators, SBLC and NLC, is increased by doubling the gradient which requires four times the RF power. The TESLA energy increase would come by increasing the length since the gradient is near that which can be obtained with superconducting RF. At 1 TeV everyone must rely on small spots, and the colliders based on large beam power must have the alignment, beam position monitor precision, vibration isolation, etc. to meet tight tolerances even if they are unneeded at 0.5 TeV.

Parameter	TESLA[6,7]		SBLC[6,7]		NLC[8]	
E_{CM}	0.5	1.0	0.5	1.0	0.5	1.0
L	7	10	4	6	8	20
Load Grad.	25	25	17	34	38	74
Linac Length	20	40	29.4	29.4	14	14
Rep Rate	10	5	50	50	180	120
Bunches/pulse	800	4180	125	50	90	75
σ_{x0}/σ_{y0}	1000/64	325/8	670/28	742/6.3	300/3	425/2
P_B	16.5	15.3	7.3	5.8	4.2	9.4
P_{AC}	137	153	114	230	141	144
$2P_B/P_{AC}$	0.24	0.20	0.13	0.05	0.06	0.13

Table 8. Comparison of Parameters for $E_{CM} = 0.5, 1.0$ TeV. Units are same as Table 7.

The combination of progress on individual components and the anticipated success of prototypes has lead to optimism that a technically sound proposal for a future linear collider could be completed in the next several years. The energy range of that collider should be an important part of the considerations.

8. Accelerator Physics

Continuing progress in accelerator science and in particle physics are inextricably linked. From AGS beam dynamics to the Tevatron $\bar{p}$ source to klystrons for a future linear collider, all of the accelerators in the US are at the forefront of accelerator physics and technology. The foundation for much of this work is the design, operation, and improvement of present and previous generations of accelerators.

A year ago high energy physics in the United States suffered a tremendous loss with the cancellation of the SSC, and we are still struggling to recover from that loss. International facilities appear to be part of that recovery. Roughly ten years from now the US will no longer be at the energy frontier, and the foundation of past successes will begin to erode as people and facilities age and become outdated. Future opportunities will erode along with the foundation, but this must not happen if there is to be any realism in thoughts about multi-TeV linear colliders, hadron colliders beyond the LHC, plasma accelerators, or $\mu\mu$ colliders. It is our challenge to avoid this by making accelerator science as international and collaborative as experiments have become.

References

1. W. T. Weng and T. Roser, private communication.
2. D. H. Rice, private communication.
3. J. T. Seeman, private communication.
4. D. Finley, G. P. Jackson, S. Werkema, private communication.
5. G. Loew, Proc Fifth Int Workshop on Next Generation Linear Colliders, p. 341 (1993).
6. R. Brinkmann, "Low Frequency Linear Colliders", 1994 European Particle Accel Conf.
7. B. Wiik, Proc Fifth Int Workshop on Next Generation Linear Colliders, p. 463 (1993).
8. R. Ruth, private communication.

Particle Astrophysics
with High Energy Neutrinos

FRANCIS HALZEN, *Wisconsin*

PARTICLE ASTROPHYSICS WITH HIGH ENERGY NEUTRINOS

F. HALZEN

Department of Physics, University of Wisconsin, Madison, WI 53706

ABSTRACT

Doing astronomy with photons of energies in excess of a GeV has turned out to be extremely challenging. Efforts are underway to develop instruments that may push astronomy to wavelengths smaller than 10^{-14} cm by mapping the sky in high energy neutrinos instead. Neutrino astronomy, born with the identification of thermonuclear fusion in the sun and the particle processes controlling the fate of a nearby supernova, will reach outside the galaxy and make measurements relevant to cosmology. The field is immersed in technology in the domains of particle physics to which many of its research goals are intellectually connected. To mind come the search for neutrino mass, cold dark matter (supersymmetric particles?) and the monopoles of the Standard Model. While a variety of collaborations are pioneering complementary methods by building telescopes with effective area in excess of 0.01 km^2, we show here that the natural scale of a high energy neutrino telescope is 1 km^2. With several thousand optical modules and a price tag unlikely to exceed 100 million dollars, the scope of a kilometer-scale instrument is similar to that of experiments presently being commissioned such as the SNO neutrino observatory in Canada and the Superkamiokande experiment in Japan.

Introduction

The photon sky has been probed with a variety of instruments sensitive to wavelengths of light as large as 10^4 cm for radio-waves to 10^{-14} cm for the GeV-photons detected with space-based instruments. Astronomical instruments have now collected data spanning 60 octaves in photon frequency, an amazing expansion of the power of our eyes which scan the sky over less than a single octave just above 10^{-7} cm. Doing gamma ray astronomy at TeV energies and beyond has, however, turned out to be a considerable challenge. Not only are the fluxes expected to be small, as one can demonstrate by extrapolating measured photon fluxes of MeV and GeV sources, they are dwarfed by a flux of cosmic ray particles which is larger by typically two orders of magnitude. The discoveries of the Crab supernova remnant and the active galaxy Markarian 421 at TeV-energy have proven that the problems are not insurmountable, more about that later. They are, however, sufficiently daunting to have encouraged a vigorous effort to probe the high energy sky by detecting neutrinos instead.

It is important to realize that high energy photons, unlike weakly interacting neutrinos, do not carry information on any cosmic sites shielded from our view by more than a few hundred grams of intervening matter. The TeV-neutrino could reveal objects with no counterpart in any wavelength of light. High energy neutrinos are the decay products of pions and are therefore a signature of the most energetic cosmic processes. Their energy exceeds those of neutrinos artificially produced at existing or planned accelerators. As was the case with radiotelescopes, for instance, unexpected discoveries could be made. Forays into new wavelength regimes have historically led to the discovery of unanticipated phenomena. As is the case with new accelerators, observing the predictable will be slightly disappointing. Neutrino telescopes have the advantage of looking at a large fraction of the sky all the time, an important astronomical advantage over gamma ray detectors such as Cherenkov telescopes which can at best scan a few degrees of the sky 10% of the time. The natural background for observing cosmic neutrinos in the TeV range and above consists of atmospheric neutrinos which are the decay products of pions produced in cosmic ray-induced atmospheric cascades. We will see that the expected sources actually dominate the neutrino sky in stark contrast to the situation, previously mentioned, where the cosmic ray flux exceeds those from high energy gamma sources by orders of magnitude.

1. Guaranteed CosmicNeutrino Beams

In heaven, as on Earth, high energy neutrinos are produced in beam dumps which consist of a high energy proton (or heavy nucleus) accelerator and a target in which gamma rays and neutrinos are generated in roughly equal numbers in the decays of pions produced in nuclear cascades in the beam dump. For every π^0 producing two gamma rays, there is a charged π^+ and π^- decaying into $\mu^+\nu_\mu$. As a rule of thumb the dump produces one neutrino for each interacting proton. If the kinematics is such that muons decay in the dump, more neutrinos will be produced. It should be stressed immediately that in efficient cosmic beam dumps with an abundant amount of target material, high energy photons may be absorbed before escaping the source. Therefore, the most spectacular neutrino sources may have no counterpart in high energy gamma rays.

By their very existence, high-energy cosmic rays do guarantee the existence of definite sources of high energy cosmic neutrinos.[1] They represent a beam of known luminosity with particles accelerated to energies in excess of 10^{20} eV. Cosmic rays produce pions in interactions with i) the interstellar gas in our galaxy, ii) the cosmic photon background, iii) the sun and, finally, iv) the Earth's atmosphere which represents a well-understood beam dump. These interactions are the source of fluxes of diffuse photons and neutrinos. The atmospheric neutrino beam is well understood and can be used to study neutrino oscillations over oscillation lengths varying between 10 and 10^4 km.[1]

A rough estimate of the diffuse fluxes of gamma rays and neutrinos from the galactic disk can be obtained by colliding the observed cosmic ray flux with interstellar gas with a nominal density of 1 proton per cm^3. The target material is concentrated in the disk of the galaxy and so will be the secondary photon flux. The gamma ray flux has been identified by space-borne gamma ray detectors. It is clear that a roughly equal diffuse neutrino flux is produced by the decay of charged pion secondaries in the same collisions. Conservatively, assuming a detector threshold of 1 TeV one predicts three neutrino-induced muons per year in a 10^6 m^2 detector from a solid angle of 0.07 sr around the direction of Orion. There are several concentrations of gas with similar or smaller density in the galaxy. The corresponding number of neutrino events from within 10 degrees of the galactic disc is 50 events per year for a 10^6 m^2 detector at the South Pole which views 1.1 steradian of the outer Galaxy with an average density of 0.013 grams/cm^2.

A guaranteed source of extremely energetic diffuse neutrinos is the interaction of ultra high energy, extra-galactic, cosmic rays on the microwave background. The major source of energy loss is photoproduction of the Δ resonance by the cosmic proton beam on a target of background photons with a density of $\sim$400 photons/cm^3 and an average energy of

$$\epsilon = 2.7 \times k_B \times 2.735° \simeq 7 \times 10^{-4} \text{ eV} . \tag{1}$$

For cosmic ray energies exceeding

$$E_p \approx \frac{m_\Delta^2 - m_p^2}{2(1 - \cos\theta)\epsilon} \approx \frac{5 \times 10^{20}}{(1 - \cos\theta)} \text{ eV} , \tag{2}$$

where θ is the angle between the proton and photon directions, the photopion cross-section grows very rapidly to reach a maximum of 540 μb at the Δ^+ resonance ($s = 1.52$ GeV^2). The Δ^+ decays to $p\pi^0$ with probability of 2/3, and to $n\pi^+$ with probability 1/3. The charged pions are the source of very high energy muon-neutrino fluxes as a result of decay kinematics where each neutrino takes approximately 1/4 of the parent pion energy. In addition neutrons decay producing a flux of lower energy $\bar{\nu}_e$.

The magnitude and intensity of the cosmological neutrino fluxes is determined by the maximum injection energy of the ultra-high-energy cosmic rays and by the distribution of their sources. If the sources are relatively near at distances of order tens of Mpc, and the maximum injection energy is not much greater than the highest observed cosmic ray energy (few $\times 10^{20}$ eV), the generated neutrino fluxes are small. If, however, the highest energy cosmic rays are generated by many sources at large redshift, then a large fraction of their injection energy would be presently contained in γ-ray and neutrino fluxes. The reason is that the energy density of the microwave radiation as well as the photoproduction cross-section scale as $(1 + z)^4$. The effect would be even stronger if the source luminosity were increasing with z, i.e. cosmic ray sources were more active at large redshifts – 'bright phase' models. Early speculations on bright phase models led to the suggestion of kilometer-scale neutrino detectors over a decade ago.[2]

The other guaranteed extraterrestrial source of high energy neutrinos is the Sun. The production process is exactly the same as for atmospheric neutrinos on Earth: cosmic ray interactions in the solar atmosphere. Neutrino production is enhanced because the atmosphere of the Sun is much more tenuous. The scaleheight of the chromosphere is $\sim$115 km, compared with 6.3 km for our upper atmosphere. A detailed calculation of the neutrino production by cosmic rays in the solar atmosphere shows a neutrino spectrum larger than the angle averaged atmospheric flux by a factor of $\sim$2 at 10 GeV and a factor of $\sim$3 at 1000 GeV. The decisive factor for the observability of this neutrino source is the small solid angle (6.8×10^{-5} sr) of the Sun. Although the rate of the neutrino induced upward going muons is higher than the atmospheric emission from the same solid angle by a factor of $\sim$5, the rate of muons of energy above 10 GeV in a 10^6 m^2 detector is only 50 per year. Taking into account the diffusion of the cosmic rays in the solar wind, which decreases the value of the flux for energies below one TeV, cuts this event rate by a factor of 3. Folded with a realistic angular resolution of 1 degree, observation of such an event rate requires, as for the previous examples, a 1 km^2 detector.

2.   Active Galactic Nuclei: Almost Guaranteed?

Although observations of PeV (10^{15} eV) and EeV (10^{18} eV) gamma-rays are controversial, cosmic rays of such energies do exist and their origin is at present a mystery. The cosmic-ray spectrum can be understood, up to perhaps 1000 TeV, in terms of shockwave acceleration in galactic supernova remnants. Although the spectrum suddenly steepens at 1000 TeV, a break usually referred to as the "knee", cosmic rays with much higher energies are observed and cannot be accounted for by this mechanism. This failure can be understood by simple dimensional analysis because the EMF in the supernova shock is of the form

$$E = ZeBRc, \tag{3}$$

where B and R are the magnetic field and the radius of the shock. For a proton Eq. (3) yields a maximum energy

$$E_{\text{max}} = \left[10^5\,\text{TeV}\right]\left[\frac{B}{3\times10^{-6}\,\text{G}}\right]\left[\frac{R}{50\,\text{pc}}\right] \tag{4}$$

and therefore E is less than 10^5 TeV for the typical values of B, R shown. The actual upper limit is much smaller than the value obtained by dimensional analysis because of inefficiencies in the acceleration process.

Cosmic rays with energy in excess of 10^{20} eV have been observed. Assuming that they are a galactic phenomenon, the measured spectrum implies that 10^{34} particles are accelerated to 1000 TeV energy every second. We do not know where or how. We do not know whether the particles are protons or iron or something else. If the cosmic accelerators indeed exploit the 3μGauss field of our galaxy, they must be much larger than supernova remnants in order to reach 10^{21} eV energies. Equation (3) requires that their size be of order 30 kpc. Such an accelerator exceeds the dimensions of our galaxy. Although imaginative arguments exist to avoid this impasse, an attractive alternative is to look for large size accelerators outside the galaxy. Nearby active galactic nuclei (quasars, blazars...) distant by order 100 Mpc are the obvious candidates. With magnetic fields of tens of μGauss over distances of kpc near the central black hole or in the jets, acceleration to 10^{21} eV is possible; see Eq. (3).

One can visualize the accelerator in a very economical way in the Blanford-Zralek mechanism. Imagine that the horizon of the central black hole acts as a rotating conductor immersed in an external magnetic field. By simple dimensional analysis this creates a voltage drop

$$\frac{\Delta V}{10^{20}\text{volts}} = \frac{a}{M_{\text{BH}}}\frac{B}{10^4\text{G}}\frac{M_{\text{BH}}}{10^9 M_\odot}, \tag{5}$$

corresponding to a luminosity

$$\frac{\mathcal{L}}{10^{45}\,\text{erg s}^{-1}} = \left(\frac{a}{M_{\text{BH}}}\right)^2 \left(\frac{B}{10^4\,\text{G}}\right)^2 \left(\frac{M_{\text{BH}}}{10^9 M_\odot}\right)^2. \tag{6}$$

Here a is the angular momentum per unit mass of a black hole of mass M_{BH}.

All this was pretty much a theorist's pipe dream until recently the Whipple collaboration reported the observation of TeV (10^{12} eV) photons from the giant elliptical galaxy Markarian 421.[3] With a signal in excess of 6 standard deviations, this was the first convincing observation of TeV gamma rays from outside our Galaxy. That a distant source such as Markarian 421 can be observed at all implies that its luminosity exceeds that of galactic cosmic accelerators such as the Crab, the only source observed by the same instrument with comparable statistical significance, by close to 10 orders of magnitude. More distant by a factor 10^5, the instrument's solid angle for Markarian 421 is reduced by 10^{-10} compared to the Crab. Nevertheless the photon count at TeV energy is roughly the same for the two sources. The Whipple observation implies a Markarian 421 photon luminosity in excess of 10^{43} ergs per second. It is interesting that these sources have their highest luminosity above TeV energy, beyond the wavelengths of conventional astronomy. During May 1994 observations of the flux of Markarian 421 was observed to increase by a factor 10 in one day, strongly suggesting the catastrophic operation of a high energy hadronic accelerator.

Why Markarian 421? Whipple obviously zoomed in on the Compton Observatory catalogue of active galaxies (AGNs) known to emit GeV photons. Markarian, at a distance of barely over 100 Mpc, is the closest blazar on the list. As yet TeV gamma rays have not been detected from any other AGNs. Although Markarian 421 is the closest of these AGNs, it is one of the weakest; the reason that it is detected whereas other, more distant, but more powerful, AGNs are not, must be that the TeV gamma rays suffer absorption in intergalactic space through the interaction with background infrared photons. TeV gamma rays are indeed efficiently absorbed on infra-red starlight and this most likely provides the explanation why astronomers have a hard time observing much more powerful quasars such as 3C279 at a redshift of 0.54. Production of e^+e^- pairs by TeV gamma rays interacting with IR background photons is the origin of the absorption. The absorption is, however, minimal for Mrk 421 with $z = 0.03$, a distance close enough to see through the IR fog. This implies that all of the AGNs may have significant very high energy components but that only Markarian 421 is close enough to be detectable with currently available gamma-ray telescopes.

This observation was not totally unanticipated. Many theorists[1] have identified blazars such as Markarian 421 as powerful cosmic accelerators producing beams of very high energy photons and neutrinos. Acceleration of particles is by shocks in the jets (or, possibly, also by shocks in the accretion flow onto the supermassive black hole which powers the galaxy) which are a characteristic feature of these radio-loud active galaxies. Many arguments have been given for the acceleration of protons as well as electrons. Inevitably beams of gamma rays and neutrinos from the decay of pions appear along the jets. The pions are photoproduced by accelerated protons on the target of optical and UV photons in the galaxy which reaches densities of 10^{14} per cm^3.

The latter are the product of synchrotron radiation by electrons accelerated along with the protons.

Powerful AGNs at distances of order 100 Mpc and with proton luminosities of 10^{45} erg/s or higher are obvious candidates for the cosmic accelerators of the highest energy cosmic rays. Their luminosity often peaks at the highest energies and their proton flux, propagated to Earth, can quantitatively reproduce the cosmic ray spectrum above spectrum 10^{18} eV.[4] Some have argued that all cosmic rays above the "knee" in the spectrum at 10^{15} eV may be of AGN origin. The neutrino flux from such accelerators can be calculated by energy conservation:

$$\mathcal{L}_p N \epsilon_{\text{eff}} = 4\pi d^2 \int dE[E \, dN_\nu/dE] \,, \tag{7}$$

where N_ν is the neutrino flux at Earth, d the average distance to the sources, N the number of sources and ϵ_{eff} the efficiency for protons to produce pions (or neutrinos, assuming the production of 1 neutrino per interacting proton) in the AGN beamdump. This yields

$$E\frac{dN_\nu}{dE} = \frac{N \epsilon_{\text{eff}}}{4\pi} \frac{7.5 \times 10^{-10}}{E \, (\text{TeV})} \, \text{cm}^{-2} \, \text{s}^{-1} \, \text{sr}^{-1} \tag{8}$$

for $\mathcal{L}_p = 10^{45}$ erg/s and $d = 100$ Mpc. We here assumed an E^{-2} energy spectrum extending to 10^{20} eV energy. With ϵ_{eff} of order 10^{-1} to 10^{-3} and the number of relatively nearby sources N in the range 10 to 1000, it is a reasonable estimate that $N\epsilon_{\text{eff}} = 1$. The total energy in excess of 1 EeV (10^{18}-10^{20} eV) is 5×10^{-9} erg/cm^2/s. This number nicely matches the energy density of the extra-galactic cosmic rays in the same interval of energy as it should assuming, again, that 1 neutrino is produced for every proton in the AGN dump. The flux of Eq. (8) is at the low end of the range of fluxes predicted by Biermann et al. and by Protheroe et al. and Stecker et al. in models where acceleration is in shocks in the jet[4] and accretion disc,[5,6] respectively.

The above discussion suggests a very simple estimate of the AGN neutrino flux that finesses all guesses regarding the properties of individual sources:

$$4\pi \int_{10^{17} \text{ eV}} dE[E \, dN_\nu/dE] \simeq \mathcal{L}_{\text{CR}}$$
$$\simeq 7.2 \times 10^{-9} \, \text{erg} \, \text{cm}^{-2} \, \text{s}^{-1} \,, \tag{9}$$

which simply states that AGNs generate 1 neutrino for each proton. $\mathcal{L}_{\text{CR}}$ is obtained by integrating the highest energy $E^{-2.71}$ component of the cosmic ray flux above 10^{17} eV. Assuming an E^{-2} neutrino spectrum we recover the result of Eq. (8). Is is now clear that our flux is a lower limit as protons should be absorbed in ambient matter in the source or in the interstellar medium.

3. Intermezzo: The Case for a Kilometer-Scale Detector

Observing AGNs has become a pivotal goal in the development of high energy neutrino telescopes. Neutrinos are observed via the muons they produce in the detector volume. At high energy it is possible to enhance the effective volume of detectors by looking for neutrino-induced muons generated in charged-current interactions of ν_μ in the water or ice outside the instrumented detector volume. The effective detector volume is then the product of the detector area and the muon range in rock R_μ. TeV muons have a typical range of one kilometer, which leads to a significant increase in effective detector volume. The average muon energy loss rate is

$$\left\langle \frac{\mathrm{d}E}{\mathrm{d}X} \right\rangle = -\alpha(E) - \beta(E) \times E , \tag{10}$$

where X is the thickness of material in g/cm^2. The first term represents ionization losses, which are approximately independent of energy, with $\alpha \sim 2$ MeV g^{-1}cm^2. The second term includes the catastrophic processes of bremsstrahlung, pair production and nuclear interactions, for which fluctuations play an essential role. Here $\beta \sim 4 \times 10^{-6}$ g^{-1}cm^2. The critical energy above which the radiative processes dominate is

$$E_{\mathrm{cr}} = \alpha/\beta \approx 500 \text{ GeV}. \tag{11}$$

To treat muon propagation properly when $E_\mu > E_{\mathrm{cr}}$ requires a Monte Carlo calculation of the probability P_{surv} that a muon of energy E_μ survives with energy $> E_\mu^{\mathrm{min}}$ after propagating a distance X. The probability that a neutrino of energy E_ν on a trajectory through a detector produces a muon above threshold at the detector is

$$P_\nu(E_\nu, E_\mu^{\mathrm{min}}) = N_A \int_0^{E_\nu} \mathrm{d}E_\mu \frac{\mathrm{d}\sigma_\nu}{\mathrm{d}E_\mu}(E_\mu, E_\nu) \\ \times R_{\mathrm{eff}}(E_\mu, E_\mu^{\mathrm{min}}) , \tag{12}$$

where

$$R_{\mathrm{eff}} = \int_0^\infty \mathrm{d}X \, P_{\mathrm{surv}}(E_\mu, E_\mu^{\mathrm{min}}, X) . \tag{13}$$

The flux of ν_μ-induced muons at the detector is given by a convolution of the neutrino spectrum ϕ_ν with the muon production probability (12) as

$$\phi_\mu(E_\mu^{\mathrm{min}}, \theta) = \int_{E_\mu^{\mathrm{min}}} \left\{ \mathrm{d}E_\nu \, P_\nu(E_\nu, E_\mu^{\mathrm{min}}) \\ \times \exp[-\sigma_t(E_\nu) X(\theta) N_A] \phi_\nu(E_\nu, \theta) \right\} . \tag{14}$$

The exponential factor here accounts for absorption of neutrinos along the chord of the Earth, $X(\theta)$. Absorption becomes important for $\sigma(E_\nu) \gtrsim 10^{-33}$ cm^2 or $E_\nu \gtrsim 10^7$ GeV.

The event rate in a detector is obtained by multiplying Eq. (14) by its effective area. From Eqs. (8),(14) we obtain order 300 upcoming muon events per year in a 10^6 m^2 detector. It is not a comfortably large rate however as the flux is indeed distributed over a large number of sources. There is, however, no competing background. Hopefully one will be able to scrutinize a few nearby sources with good statistics. We should recall at this point that our back-of-the-envelope estimate yields a flux at the lower end of the range of fluxes predicted by detailed modeling. Optimistic predictions exceed our estimate by over an order of magnitude and are, possibly, within reach of the telescopes now being commissioned.

The neutrino sky at GeV-energy and above is summarized in Fig. 1. Shown is the flux from the galactic plane as well as a range of estimates (from generous to conservative) for the diffuse fluxes of neutrinos from active galaxies and from the interaction of extra-galactic cosmic rays with cosmic photons. At PeV energies and above all sources dominate the background of atmospheric neutrinos. In order to deduce the effective area of an instrument required to study the fluxes in the figure, the detection efficiency must be included using Eq. (14). At the highest energies this efficiency approaches unity and 1 event per km^2 per year corresponds to the naive estimate of 10^{-18} neutrinos per cm^2 second. At TeV–PeV energy the one event level per year corresponds to a flux of 10^{-14}–10^{-15} per cm^2 second. As before, we conclude that the diffuse flux from AGN yields order 10^3 events in a kilometer-size detector per year in the TeV-energy range.

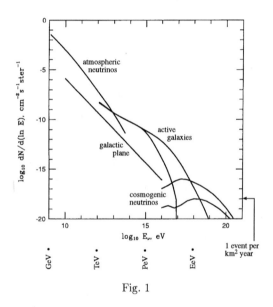

Fig. 1

It should be emphasized that high energy neutrino detectors are multi-purpose instruments. Their science-reach touches astronomy, astrophysics and particle physics. Further motivations for the construction of a km^3 deep underground detector include[1]:

1. The search for the t'Hooft-Polyakov monopoles predicted by the Standard Model.

2. The study of neutrino oscillations by monitoring the atmospheric neutrino beam. One can exploit the unique capability of relatively shallow neutrino telescopes, i.e. detectors positioned at a depth of roughly 1 km, to detect neutrinos and muons of similar energy. In a ν_μ oscillation experiment one can therefore tag

the π progenitor of the neutrino by detecting the muon produced in the same decay. This eliminates the model dependence of the measurement inevitably associated with the calculation of the primary cosmic ray flux. Surface neutrino telescopes probe the parameter space $\Delta m^2 \gtrsim 10^{-3}$ eV^2 and $\sin^2 2\theta \gtrsim 10^{-3}$ using this technique. Recently underground experiments have given tantalizing hints for neutrino oscillations in this mass range.

3. The search for neutrinos from the annihilation of dark matter particles in our galaxy.

4. The capability to observe the thermal neutrino emission from supernovae[7] (even though the nominal threshold of the detectors exceeds the neutrino energy by several orders of magnitude!). The detector will be able to monitor our galaxy over decades in a most economical fashion.

5. Further study of the science pioneered by space-based gamma ray detectors such as the study of gamma ray bursts and the high energy emission from quasars.

It is intriguing that each of these goals point individually at the necessity to commission a kilometer-size detector. In order to illustrate this I will discuss the search for the particles which constitute the cold dark matter.

4. Indirect Search for Cold Dark Matter

It is believed that most of our Universe is made of cold dark matter particles. Big bang cosmology implies that these particles have interactions of order the weak scale, i.e. they are WIMPs.[8] We know everything about these particles except whether they really exist. We know that their mass is of order of the weak boson mass, we know that they interact weakly. We also know their density and average velocity in our galaxy as they must constitute the dominant component of the density of our galactic halo as measured by rotation curves. WIMPs will annihilate into neutrinos with rates whose estimate is straightforward. Massive WIMPs will annihilate into high energy neutrinos. Their detection by high energy neutrino telescopes is greatly facilitated by the fact that the sun, conveniently, represents a dense and nearby source of cold dark matter particles.

Galactic WIMPs, scattering off protons in the sun, lose energy. They may fall below escape velocity and be gravitationally trapped. Trapped dark matter particles eventually come to equilibrium temperature, and therefore to rest at the center of the sun. While the WIMP density builds up, their annihilation rate into lighter particles increases until equilibrium is achieved where the annihilation rate equals half of the capture rate. The sun has thus become a reservoir of WIMPs which annihilate into any open fermion, gauge boson or Higgs channels. The leptonic decays from annihilation channels such as $b\bar{b}$ heavy quark and W^+W^- pairs turn the sun into a source of high energy neutrinos. Their energies are in the GeV to TeV range, rather than in the familiar KeV to MeV range from its thermonuclear burning. These neutrinos can be detected in deep underground experiments.

We illustrate the power of neutrino telescopes as dark matter detectors using as an example the search for a 500 GeV WIMP with a mass outside the reach of present accelerator and future LHC experiments. A quantitative estimate of the rate of high energy muons of WIMP origin triggering a detector can be made in 5 easy steps.

Step 1: The halo neutralino flux ϕ_χ.
It is given by their number density and average velocity. The cold dark matter density implied by the observed galactic rotation curves is $\rho_\chi = 0.4\,\text{GeV/cm}^3$. The galactic halo is believed to be an isothermal sphere of WIMPs with average velocity $v_\chi = 300\,\text{km/sec}$. The number density is then

$$n_\chi = 8 \times 10^{-4} \left[\frac{500\ \text{GeV}}{m_\chi} \right] \text{cm}^{-3} \tag{15}$$

and therefore

$$\phi_\chi = n_\chi v_\chi = 2 \times 10^4 \left[\frac{500\ \text{GeV}}{m_\chi} \right] \text{cm}^{-2}\,\text{s}^{-1} . \tag{16}$$

Step 2: Cross section σ_{sun} for the capture of neutralinos by the sun.
The probability that a WIMP is captured is proportional to the number of target hydrogen nuclei in the sun (i.e. the solar mass divided by the nucleon mass) and the WIMP-nucleon scattering cross section. From dimensional analysis $\sigma(\chi N) \sim (G_F m_N^2)^2 / m_Z^2$ which we can envisage as the exchange of a neutral weak boson between the WIMP and a quark in the nucleon. The main point is that the WIMP is known to be weakly interacting. We obtain for the solar capture cross section

$$\Sigma_{\text{sun}} = n\sigma = \frac{M_{\text{sun}}}{m_N}\sigma(\chi N)$$
$$= [1.2 \times 10^{57}]\,[10^{-41}\ \text{cm}^2] . \tag{17}$$

Step 3: Capture rate N_{cap} of neutralinos by the sun.
N_{cap} is determined by the WIMP flux (16) and the sun's capture cross section (17) obtained in the first 2 steps:

$$N_{\text{cap}} = \phi_\chi \Sigma_{\text{sun}} = 3 \times 10^{20}\,\text{s}^{-1} . \tag{18}$$

Step 4: Number of solar neutrinos of dark matter origin.
The sun comes to a steady state where capture and annihilation of WIMPs are in equilibrium. For a 500 GeV WIMP the dominant annihilation rate is into weak bosons; each produces muon-neutrinos with a branching ratio which is roughly 10%:

$$\chi\bar\chi \to WW \to \mu\nu_\mu . \tag{19}$$

Therefore, as we get 2 W's for each capture, the number of neutrinos generated in the sun is

$$N_\nu = \frac{1}{5}N_{\text{cap}} \tag{20}$$

and the corresponding neutrino flux at Earth is given by

$$\phi_\nu = \frac{N_\nu}{4\pi d^2} = 2 \times 10^{-8}\,\text{cm}^{-2}\text{s}^{-1} , \tag{21}$$

where the distance d is 1 astronomical unit.

Step 5: Event rate in a high energy neutrino telescope.

For (19) the W-energy is approximately m_χ and the neutrino energy half that by two-body kinematics. The energy of the detected muon is given by

$$E_\mu \simeq \frac{1}{2} E_\nu \simeq \frac{1}{4} m_\chi . \tag{22}$$

Here we used the fact that, in this energy range, roughly half of the neutrino energy is transferred to the muon. Simple estimates of the neutrino interaction cross section and the muon range can be obtained as follows

$$\sigma_{\nu \to \mu} = 10^{-38} \, \text{cm}^2 \, \frac{E_\nu}{\text{GeV}} = 2.5 \times 10^{-36} \, \text{cm}^2 \tag{23}$$

and

$$R_\mu = 5 \, \text{m} \, \frac{E_\mu}{\text{GeV}} = 625 \, \text{m} , \tag{24}$$

which is the distance covered by a muon given that it loses 2 MeV for each gram of matter traversed. We have now collected all the information to compute the number of events in a detector of area $10^4 \, \text{m}^2$, typical for those presently under construction.

For the neutrino flux given by (21) we obtain

$$\# \, \text{events/year} = 10^6 \times \phi_\nu \times \rho_{H_2O} \times \sigma_{\nu \to \mu} \times R_\mu \simeq 1000 \tag{25}$$

for a 1 km^2 water Cherenkov detector, where R_μ is the muon range and $\phi_\nu \times \rho_{H_2O} \times \sigma_{\nu \to \mu}$ is the analog of Eq. (14).

The above exercise is just meant to illustrate that high energy neutrino telescopes compete with present and future accelerator experiments in the search for dark matter and supersymmetry; see below. The above exercise can be repeated as a function of WIMP mass. The result is shown in Fig. 2 (the two branches as well as the structure in the curves are related to details of supersymmetry. These are, for all practical purposes, irrelevant). Especially for heavier WIMPs the technique is very powerful because underground high energy neutrino detectors have been optimized to be sensitive in the energy region where the neutrino interaction cross section and the range of the muon are large. Also, for high energy neutrinos the muon and neutrino are aligned along a direction pointing back to the sun with good angular resolution. A kilometer-size detector probes WIMP masses up to the TeV-range beyond which they are excluded by cosmological considerations. The technique fails for low masses only for those mass values already excluded by unsuccessful accelerator searches. Competitive direct searches for dark matter will have to deliver detectors reaching better than 0.05 events/kg/day sensitivity.

Particle physics provides us with rather compelling candidates for WIMPs. The Standard Model is not a model: its radiative corrections are not under control. A most elegant and economical way to revamp it into a consistent and calculable framework is to make the model supersymmetric. If supersymmetry is indeed Nature's extension of the Standard Model it must produce new phenomena at or below the TeV scale. A very attractive feature of supersymmetry is that it provides cosmology with a natural dark matter candidate in form of a stable lightest supersymmetric particle.[8] This is, in fact, the only candidate because supersymmetry completes the Standard Model all the way to the GUT scale where its forces apparently unify. Because supersymmetry logically completes the Standard Model with no other new physics threshold up to the GUT-scale, it must supply the dark matter. So, if supersymmetry, dark matter and accelerator detectors are on a level playing field. The interpretation of above arguments in the framework of supersymmetry are explicitly stated in Ref. 9.

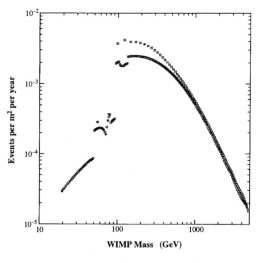

Fig. 2

5.   DUMAND et al.: Complementary Technologies

We have presented arguments for doing neutrino astronomy on the scale of 1 kilometer. In order to achieve large effective area it is, unfortunately, necessary to abandon the low MeV thresholds of detectors such as IMB and Kamiokande. One focuses on high energies where: i) neutrino cross sections are large and the muon range is increased; see Equation (14), ii) the angle between the muon and parent neutrino is less than 1 degree and, iii) the atmospheric neutrino background is small. The accelerator physicist's method for building a neutrino detector uses absorber, chambers with a few x, y wires and associated electronics with a price of roughly 10^4 US dollars per m². Such a 1 km² detector would cost 10 billion dollars. Realistically, we are compelled to develop methods which are more cost-effective by a factor 100 in order to be able to commission neutrino telescopes with area of order 1 km². Obviously, the proven technique developed by IMB, Kamiokande and others cannot be extrapolated to kilometer scale. All present telescopes do however exploit the well-proven Cherenkov technique.

In a Cherenkov detector the direction of the neutrino is inferred from the muon track which is measured by mapping the associated Cherenkov cone travelling through the detector. The arrival times and amplitudes of the Cherenkov photons, recorded by a grid of optical detectors, are used to reconstruct the direction of the radiating muon. The challenge is well-defined: record the muon direction with sufficient precision (i.e., sufficient to reject the much more numerous down-going cosmic ray muons from the up-coming muons of neutrino origin) with a minimum number of optical modules (OM). Critical parameters are detector depth which determines the level of the cosmic ray muon background and the noise rates in the optical modules which will sprinkle a muon trigger with false signals. Sources of such noise include radioactive decays such as potassium decay in water, bioluminescence and, inevitably, the dark current of the photomultiplier tube. The experimental advantages and challenges are different for each experiment and, in this sense, they nicely complement one another as engineering projects for a large detector. Each has its own "gimmick" to achieve neutrino detection with a minimum number of OMs:

1. AMANDA uses sterile ice, free of radioactivity;
2. Baikal triggers on pairs of OMs;
3. DUMAND and NESTOR shield their arrays by over 4 km of ocean water.

Detectors under construction will have a nominal effective area of 10^4 m^2. The OMs are deployed like beads on strings separated by 20–50 meters. There are typically 20 OMs per string separated by roughly 10 meters. Baikal is presently operating 36 optical modules, 18 pointing up and 18 down, and the South Pole AMANDA experiment started operating 4 strings with 20 optical modules each in January 94. The first generation telescopes will consist of roughly 200 OMs. Briefly,

1. AMANDA is operating in deep clear ice with an absorption length in excess of 60 m similar to that of the clearest water used in the Kamiokande and IMB experiments. The ice provides a convenient mechanical support for the detector. The immediate advantage is that all electronics can be positioned at the surface. Only the optical modules are deployed into the deep ice. Polar ice is a sterile medium with a concentration of radioactive elements reduced by more than 10^{-4} compared to sea or lake water. The low background results in an improved sensitivity which allows for the detection of high energy muons with very simple trigger schemes which are implemented by off-the-shelf electronics. Being positioned under only 1 km of ice it is operating in a cosmic ray muon background which is over 100 times larger than deep-ocean detectors such as DUMAND. The challenge is to reject the down-going muon background relative to the up-coming neutrino-induced muons by a factor larger than 10^6. The group claims to have met this challenge with an up/down rejection which is similar to that of the deep detectors.

Although residual bubbles are found at depth as large as 1 km, their density decreases rapidly with depth. Ice at the South Pole should be bubble-free below 1100–1300 m as it is in other polar regions. The effect of bubbles on timing of photons has been measured by the laser calibration system deployed along with the OMs. After taking the scattering of the light on bubbles into account reconstruction of muons has been demonstrated by a successful measurement of the characteristic fluxes of cosmic ray muons.

The polar environment turned out to be surprisingly friendly but only allows for restricted access and one-shot deployment of photomultiplier strings. The technology has, however, been satisfactorily demonstrated with the deployment of the first 4 strings. It is clear that the hot water drilling technique can be used to deploy OM's larger than the 8 inch photomultiplier tubes now used to any depth in the 3 km deep ice cover. AMANDA will deploy 6 more strings in 1995 at a depth of 1500 meters.

2. BAIKAL shares the shallow depth of AMANDA and large background counting rate of tens of kHz from bioluminescence and radioactive decays with DUMAND. It suppresses its background by pairing OMs in the trigger. Half its optical modules are pointing up in order to achieve a uniform acceptance over upper and lower hemispheres. The depth of the lake is 1.4 km, so the experiment cannot expand downwards and will have to grow horizontally.

The Baikal group has been operating an array of 18(36) Quasar photomultiplier (a Russian-made 15 inch tube) units deployed in April 1993(94). They have reached a record up/down rejection ratio of 10^{-4} and, according to Monte Carlo, will reach the 10^{-6} goal to detect neutrinos as soon as the full complement of 200 OMs is deployed. They expect to deploy 97 additional OMs in 1995.

3. DUMAND will be positioned under 4.5 km of ocean water, below most biological activity and well shielded from cosmic ray muon backgrounds. A handicap of using ocean water is the background light resulting from radioactive decays, mostly K^{40}, plus some bioluminescence, yielding a noise rate of 60 kHz in a single OM. Deep ocean water is, on the other hand very clear, with an attenuation length of order 40 m in the blue. The deep ocean is a difficult location for access and service. Detection equipment must be built to high reliability standards, and the data must be transmitted to the shore station for processing. It has required years to develop the necessary technology and learn to work in an environment foreign to high-energy physics experimentation, but hopefully this will be accomplished satisfactorily.

The DUMAND group has successfully analysed data on cosmic ray muons from the deployment of a test string. They have already installed the 25 km power and signal cables from detector to shore as well as the junction box for deploying the strings. The group will proceed with the deployment of 3 strings in 1995.

4. NESTOR is similar to DUMAND, being placed in the deep ocean (the Mediterranean), except for two critical differences. Half of its optical modules point up, half down like Baikal. The angular response of the detector is being tuned to be much more isotropic than either AMANDA or DUMAND, which will give it advantages in, for instance, the study of neutrino oscillations. Secondly, NESTOR will have a higher density of photocathode (in some substantial volume) than the other detectors, and will be able to make local coincidences on lower energy events, even perhaps down to the supernova energy range (tens of MeV).

5. Other detectors have been proposed for near surface lakes or ponds (e.g. GRANDE, LENA, NET, PAN and the Blue Lake Project), but at this time none are in construction.[10] These detectors all would have the great advantage of accessibility and ability for dual use as extensive air shower detectors, but suffer from the 10^{10}–10^{11} down-to-up ratio of muons, and face great civil engineering costs (for water systems and light-tight containers). Even if any of these are built it would seem that the costs may be too large to contemplate a full kilometer-scale detector.

6.   Sketch of a Kilometer-Size Detector

In summary, there are four major experiments proceeding with construction, each of which has different strengths and faces different challenges. For the construction of a 1 km scale detector one can imagine any of the above detectors being the basic building block for the ultimate 1 km^3 telescope. The redesigned AMANDA detector (with spacings optimized to the absorption length of 60 m), for example, consists of 5 strings on a circle of 60 meter radius around a string at the center (referred to as a 1 + 5 configuration). Each string contains 13 OMs separated by 15 m. Its effective volume for TeV-neutrinos is just below 10^7 m^3. Imagine AMANDA "supermodules" which are obtained by extending the basic string length (and module count per string) by a factor close to 4. Supermodules would then consist of 1 + 5 strings with 51 OMs separated by 20 meters on each string, for a total length of 1 km. A 1 km scale detector then may consist of a 1 + 7 + 7 configuration of supermodules, with the 7 supermodules distributed on a circle of radius 250 m and 7 more on a circle of 500 m. The full detector then contains 4590 phototubes, which is less than the 9000 used in the SNO detector. Such a detector (see Fig. 3) can be operated in a dual mode:

1. it obviously consists of roughly 4×15 the presently designed AMANDA array, leading to an "effective" volume of $\sim 6 \times 10^8$ m^3. Importantly, the characteristics of the detector, including threshold in the GeV-energy range, are the same as those of the AMANDA array module.
2. the 1 + 7 + 7 supermodule configuration, looked at as a whole, instruments a 1 km^3 cylinder with diameter and height of 1000 m with optical modules. High-energy muons will be superbly reconstructed as they can produce triggers in 2 or more of the supermodules spaced by large distance. Reaching more than one supermodule (range of 250 m) requires muon energies in excess of 50 GeV. We note that this is the energy for which a neutrino telescope has optimal sensitivity to a typical E^{-2} source (background falls with threshold energy, and until about 1 TeV little signal is lost).

Alternate methods to reach the 1 km scale have been discussed by Learned and Roberts.[11]

What are the construction costs for such a detector? AMANDA's strings (with 10 OMs) cost \$150,000 including deployment. By naive scaling the final cost of the postulated 1 + 7 + 7 array of supermodules is of order \$50 million, still below that of Superkamiokande (with 11,200 $\times$ 20 inch photomultiplier tubes in a 40 m diameter by 40 m high stainless steel tank in a deep mine). It is clear that the naive estimate makes several approximations over- and underestimating the actual cost.

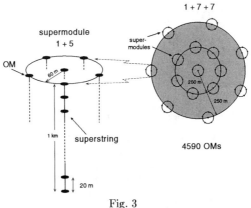

Fig. 3

Acknowledgements

I would like to thank Tom Gaisser, Todor Stanev, John Jacobsen, Karl Mannheim and Ricardo Vazquez for discussions. This work was supported in part by the University of Wisconsin Research Committee with funds granted by the Wisconsin Alumni Research Foundation, and in part by the U.S. Department of Energy under Contract No. DE-AC02-76ER00881.

References

1. T. K. Gaisser, F. Halzen and T. Stanev, *Physics Reports*, in press.
2. V.S. Berezinsky *et al.*, *Proc. of the Astrophysics of Cosmic Rays*; Elsevier, New York (1991).
3. M. Punch *et al.*, *Nature* **358**, 477–478 (1992).
4. K. Mannheim and P.L. Biermann, *Astron. Astrophys.* **22**, 211 (1989); K. Mannheim and P.L. Biermann, *Astron. Astrophys.* **253**, L21 (1992); K. Mannheim, *Astron. Astrophys.* **269**, 67 (1993).
5. A.P. Szabo and R.J. Protheroe, in *Proc. High Energy Neutrino Astrophysics Workshop* (Univ. of Hawaii, March 1992, eds. V.J. Stenger, J.G. Learned, S. Pakvasa and X. Tata, World Scientific, Singapore).
6. F.W. Stecker, C. Done, M.H. Salamon and P. Sommers, *Phys. Rev. Lett.* **66**, 2697 (1991) and **69**, 2738(E) (1992).
7. F. Halzen, J. E. Jacobsen and E. Zas, *Phys. Rev.* **D49**, 1758 (1994).
8. J. R. Primack, B. Sadoulet, and D. Seckel, *Ann. Rev. Nucl. Part. Sci.* **B38**, 751 (1988).
9. F. Halzen, M. Kamionkowski, and T. Stelzer, *Phys. Rev.* **D45**, 4439 (1992).
10. J.G. Learned, *Proc. of the 13th European Cosmic Ray Symposium*, Geneva 1992, CERN (1992).
11. J. G. Learned and A. Roberts, *Proceedings of the 23rd International Cosmic Ray Conference*, Calgary, Canada (1993); F. Halzen and J.G. Learned, *Proc. of the Fifth International Symposium on Neutrino Telescopes*, Venice (1993), ed. by M. Baldo-Ceolin.

Review of Heavy Flavor Physics*

VIVEK SHARMA, *Wisconsin*

*Oral contribution only

Physics Beyond the Standard Model: Prospects and Perspectives

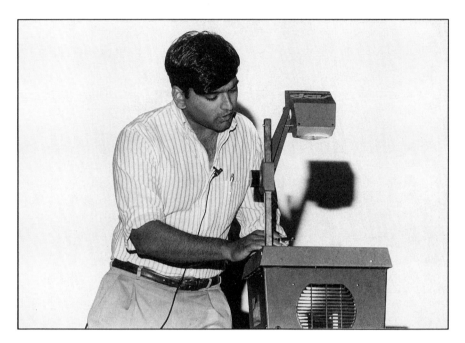

R. SEKHAR CHIVUKULA, *Boston*

PHYSICS BEYOND THE STANDARD MODEL:
PROSPECTS AND PERSPECTIVES

R. SEKHAR CHIVUKULA *

Department of Physics, Boston University
Boston, MA 02215, USA

ABSTRACT

In this talk I discuss the effects of physics beyond the standard model on the process $Z \to b\bar{b}$. I argue that, because the top-quark is heavy, this process is susceptible to large corrections from new physics.

1.   Introduction

In terms of the painting metaphor which has been used in several of the other theory talks at this conference, I am going to use a very narrow brush to paint a more detailed picture of a small part of physics beyond the standard model. In particular, instead of trying to describe all possible constraints on all proposed models of new physics, I will concentrate on a single process, $Z \to b\bar{b}$, and consider the contributions to this process from different types of physics. I will close with some perspectives and an advertisement for some other talks at this conference on related topics. For a more

---

*Talk presented at DPF '94, Albuquerque, New Mexico, Aug. 2-6, 1994.

conventional survey, I refer the reader to the contributions of Jon Rosner[1] and Jeff Harvey[2] in this conference, as well as my talk at the Lepton-Photon conference last year.[3]

Let me begin by describing why I chose the process $Z \to b\bar{b}$. First and foremost, it was because of the extraordinarily precise measurement reported by the LEP collaborations at this meeting by Richard Batley[4]:

$$R_b = 0.2192 \pm 0.0018, \tag{1}$$

where

$$R_b = \frac{\Gamma(Z \to bb)}{\Gamma(Z \to hadrons)}. \tag{2}$$

Given the reported CDF results on evidence for the top[5]

$$m_t = 174 \pm 10^{+13}_{-12} \tag{3}$$

we find a standard model prediction for R_b[6]:

$$R_B^{SM} = 0.2157 \pm 0.0004 \tag{4}$$

for a top in the mass range from 163 to 185 GeV. To be sure, no one would suggest that the apparent discrepancy between the calculated value and the reported LEP value is grounds to dismiss the standard model, especially given that there are of the order of 25 precisely measured electroweak quantities and only a few disagree by more than one sigma.

Nonetheless, as I will show in the rest of this talk, the $Z \to b\bar{b}$ branching ratio is particularly susceptible to contributions from new physics. For this reason, this discrepancy though not decisive, is certainly intriguing*. In addition, unlike flavor-changing neutral-currents, this process does not require GIM violation and, since it refers to an inclusive rate, does not suffer from uncertainties due to hadronic matrix elements or fragmentation.

2. Standard Model

First, let us consider the process in the standard model. At tree-level, we have the diagram shown in Fig. 1. The couplings of the b quark are

$$g_L = -\frac{1}{2} + \frac{1}{3}\sin^2\theta_W \tag{5}$$

and

---

*For a model-independent analysis of R_b, see ref. 7.

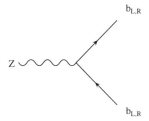

Fig. 1.

$$g_R = \frac{1}{3}\sin^2\theta_W. \tag{6}$$

Using these (and the corresponding expressions for the other quarks) we find

$$R_b^0 \simeq 0.2197 \tag{7}$$

at tree-level (with R_b defined as before).

As with other precisely-measured quantities at LEP, we also need to consider the leading (one-loop) radiative corrections to this quantity. The advantages of the *ratio* R_b now become clear: both the flavor-independent ("oblique") corrections[8] and the leading QCD corrections (which together are generally the most important radiative corrections) largely cancel in this ratio.[9] Therefore, the leading corrections to R_b are the *non-universal* corrections to the $Z \to b\bar{b}$ vertex. In t'Hooft–Feynman gauge, these vertex corrections, along with the corresponding wave-function renormalization diagrams, are shown in Fig. 2.[10,11] The results of this computation, shown as a fractional change in the partial width of the Z to b quarks, is shown in Fig. 3.[11] As we see, this correction varies from a little less than 1.5% to a little less than 2.5% as the top mass varies from 150 to 200 GeV.

In the limit $m_b \to 0$, there is no change in the right-handed b quark coupling, and the result of the calculation may be written

$$\delta g_L^b = A\frac{m_t^2}{16\pi^2 v^2} + B\frac{g^2}{16\pi^2}\log(\frac{m_t}{M_W})^2 + \cdots^\dagger \tag{8}$$

where A and B are computable constants.

---

†Note that while such an expansion in (inverse) powers of the top quark mass is useful for the purposes of illustration, one must go to quite high order in order to obtain an accurate result.[11]

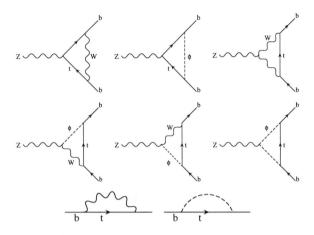

Fig. 2. The leading standard-model corrections to R_b, in t'Hooft-Feynman gauge.

Some features of the standard model calculation are of particular note. First, the result does not go to zero as $m_t \to \infty$. That is, the contribution does not decouple in m_t. The reason for this is that the couplings of the unphysical Goldstone bosons (and more generally of the longitudinal gauge bosons) to the t_R and b_L, Fig. 4, are proportional to m_t. Hence the Appelquist-Carazzone decoupling theorem[12] does not apply.

The fact that this coupling is proportional to m_t is not restricted to the standard model. As emphasized by Peccei and Zhang,[13] this result follows from the electroweak generalization of the Goldberger-Treiman (GT) relation.[14] In QCD, the GT relation reads

$$g_{\pi NN} = \frac{g_A m_N}{f_\pi} \tag{9}$$

where $g_{\pi NN}$ is the pion-nucleon coupling, m_N the mass of the nucleon, f_π the pion decay constant, and g_A is the renormalization of the axial-vector couplings (approximately 1.25 in QCD). In the electroweak theory, this relation reads

$$g_{W_L t_R b_L} = \sqrt{2} \frac{g_A m_t}{v}, \tag{10}$$

where $v \approx 246$ GeV, and $g_A = 1$ at tree-level in the standard model. Non-decoupling contributions appear in *all* theories.

Finally, we note that in order to have $\delta g_L \neq 0$ we must have $SU(2) \times U(1)$ breaking. In an unbroken gauge theory, the gauge currents are not renormalized: this,

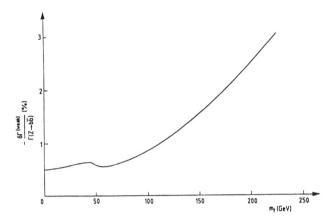

Fig. 3. Fractional change in $\Gamma(Z \to b\bar{b}$ (in %) as a function of m_t (in GeV). From ref. 11.

after all, is the reason why the $\bar{p}$ and e charges are the same – independent of the effects of QCD. In the absence of electroweak symmetry breaking, the current to which the Z couples *cannot* be renormalized and R_b would not change. This is why (in the limit $m_b \to 0$) there are no strong vertex corrections like the ones depicted in Fig. 5

3. Two Scalar Doublet Models

The simplest extension to the standard model is one in which the electroweak symmetry breaking sector involves two fundamental scalar doublets, φ_1 and φ_2, instead of one ‡. The new scalar degrees of freedom result in the appearance of an extra pair of charged scalar particles, $H^\pm$, as well as a pseudo-scalar and an additional scalar particle. The expectation values of the two scalars may be written

$$\langle \varphi_1 \rangle = \begin{pmatrix} v_1 \\ 0 \end{pmatrix} \tag{11}$$

and

$$\langle \varphi_2 \rangle = \begin{pmatrix} v_2 \\ 0 \end{pmatrix} \tag{12}$$

In order for the W and Z masses to be correct, we require that

$$v_1^2 + v_2^2 = v^2 \tag{13}$$

---

‡For a general review of these models, see ref. 15.

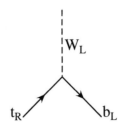

Fig. 4.

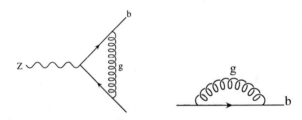

Fig. 5. Potential strong corrections to R_b which vanish in the limit $m_b \to 0$.

where v is the usual weak scale. Given the relation above, it is natural to define an angle β such that

$$v_1 = v \cos \beta \quad v_2 = v \sin \beta. \tag{14}$$

Then, the relationship between the charged scalar fields in the mass eigenstate fields is

$$\pi^{\pm} = \cos \beta \; \varphi_1^{\pm} + \sin \beta \; \varphi_2^{\pm} \tag{15}$$

for the "eaten" Goldstone boson and

$$H^{\pm} = -\sin \beta \; \varphi_1^{\pm} + \cos \beta \; \varphi_2^{\pm} \tag{16}$$

for the extra physical charged scalars.

Conventionally, it is expected that only one of the original scalar doublets (which we take to be ϕ_1) couples to the t_R so as to avoid flavor-changing neutral-currents [§].

[§]Though this may not be strictly necessary.[16]

This results in the couplings

$$\frac{m_t}{v_2}\bar{t}_R\varphi_1^+b_L \;\to\; \frac{m_t}{v\sin\beta}\bar{t}_R[\pi^+\sin\beta + H^+\cos\beta]b_L \tag{17}$$

to the mass eigenstate fields. Examining this expression, we see that the Goldstone boson field π^+ couples to $\bar{t}_Rb_L$ with the same strength as the standard model, while the coupling of the H^+ differs from this by a factor of $\cot\beta$. Since the coupling of the Goldstone boson field is the same as in the standard model, the calculations of the previous section still apply. This is a general result: unlike many weak radiative corrections, in the limit $m_b \to 0$ the standard model correction to the $Zb\bar{b}$ vertex does not involve the Higgs boson, only the longitudinal gauge bosons. And therefore, to the extent that $g_A \approx 1$, these contributions arise in *all* theories.

There are, however, additional contributions coming from the exchange of the extra charged scalars. These corrections are shown in Fig. 6.

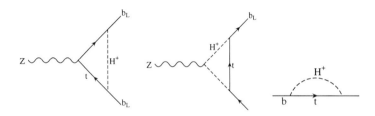

Fig. 6. Corrections to R_b in the two-scalar doublet models.

Note that these diagrams are a subset of the diagrams shown in Fig. 2, with the replacement $\pi^+ \to H^+$, resulting in the couplings changing by a factor of $\cot^2\beta$ and the replacement of M_W by M_{H^+}. For $\tan\beta \approx 1$ and $M_{H^+} \approx M_W$ therefore, we expect an effect of the same order of magnitude as the standard model.[17] This effect is shown in Fig. 7.[18] ¶

Note that, as in the standard model, this effect tends to *reduce* the width of the Z to $b\bar{b}$. This tendency holds in all two-scalar doublet models except in the limit where $\tan\beta$ is *very* large: there the Yukawa coupling of the b quark can be comparable to that of the t quark. Processes involving intermediate b quarks and neutral scalars become important, and can result in an *increase* of R_b.[19]

Two features of this calculation are of particular note. First, because the Yukawa coupling of the charged scalar is proportional to m_t,

---

¶ These values of $\tan\beta$ and M_{H^+} are chosen for the purposes of illustration only. Recent results from CLEO on $b \to s\gamma$ require that the charged scalar mass be greater than about 230 GeV.[40]

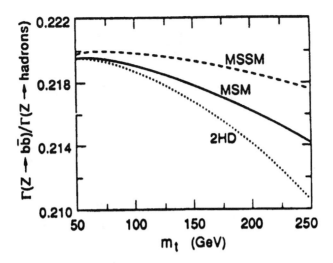

Fig. 7. R_b as a function of m_t (in GeV) in the standard model (MSM), two-scalar doublet model (2HD), and the minimal supersymmetric standard model (MSSM), assuming $\tan\beta = 1$, $m_{\tilde{t}} = M_{H^+} = 100$ GeV, $\mu = 30$ GeV and $M = 50$ GeV. From ref. 18.

$$\lambda_t \sim \frac{m_t}{v\tan\beta},\tag{18}$$

the effect on R_b does not decouple in m_t. Second, the effect on R_b *does* vanish in the limit that $m_{H^+} \to \infty$. This is because there is an $SU(2) \times U(1)$ preserving mass term for the two scalar doublets,

$$-\mu^2(\varphi_1^\dagger\varphi_2 + h.c.)\tag{19}$$

which can be introduced in the Lagrangian. In the limit that $\mu^2 \to \infty$, the theory reduces precisely to the standard model. For this reason, the extra contributions can be made *arbitrarily small*, independent of the t and W masses.

4. Supersymmetry

To judge by the volume of submissions on hep-ph, the most popular extensions of the standard model involve low-energy supersymmetry[||]. In the minimal version of

---

[||]For a review, see ref. 20.

this scenario, one introduces superpartners (a fermionic partner for every boson and visa versa) for all of the ordinary standard model particles

$$
\begin{aligned}
q_L &\rightarrow \tilde{q}_L \\
u_R &\rightarrow \tilde{u}_R \\
d_R &\rightarrow \tilde{d}_R \\
l_L &\rightarrow \tilde{l}_L \\
e_R &\rightarrow \tilde{e}_R \\
g &\rightarrow \tilde{g} \\
W^{\pm} &\rightarrow \tilde{W}^{\pm} \\
Z &\rightarrow \tilde{Z} \\
\gamma &\rightarrow \tilde{\gamma}
\end{aligned}
\tag{20}
$$

In addition supersymmetry requires that the theory involve (at least) two weak-doublet chiral superfields to perform the role of the standard model Higgs doublet.

$$
\begin{aligned}
H_1 &\rightarrow \tilde{H}_1 \\
H_2 &\rightarrow \tilde{H}_2
\end{aligned}
\tag{21}
$$

The primary attraction of supersymmetric theories is that corrections to the Higgs mass are no longer quadratically dependent on the cutoff, as we see (for example) in Fig. 8.

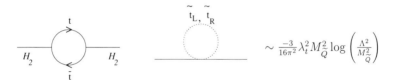

$$
\sim \frac{-3}{16\pi^2} \lambda_t^2 M_{\tilde{Q}}^2 \log\left(\frac{\Lambda^2}{M_{\tilde{Q}}^2}\right)
$$

Fig. 8.

Quadratic divergences are absent because the mass of the Higgs boson is related by supersymmetry to the mass of its fermionic partner, and the mass of this fermionic partner can be protected by a chiral symmetry. A light Higgs can be (technically) natural in SUSY.

Of course, SUSY cannot be exact. None of the extra particles required by supersymmetry have been observed. If SUSY is broken softly, the symmetry breaking does not reintroduce the quadratic divergences of an ordinary fundamental scalar theory. In a theory with soft SUSY breaking, the radiative corrections to the Higgs masses end up being proportional to the masses of the SUSY partners. Since we want the Higgs to "naturally" have a mass of order 1 TeV, SUSY is relevant to the hierarchy problem if the masses of the superpartners are of order 1 TeV (or less).

In SUSY theories, in addition to the contributions discussed in the last two sections, we have contributions coming from intermediate states involving the superpartners. The relevant vertices are shown in Fig. 9 and the new contributions in Fig. 10. Notice that the first set of vertices in Fig. 9 are proportional to m_t/v while the second are proportional to $m_t/v \tan \beta$. For a particular choice of superpartner and Higgs masses, the results of this computation are plotted in Fig. 7. As shown, for those relatively light superpartner masses (of order M_W) the result is of the same order of magnitude as the correction in the standard model, but has the *opposite* sign: the effects of radiative corrections involving superpartners tend to *increase* R_b.

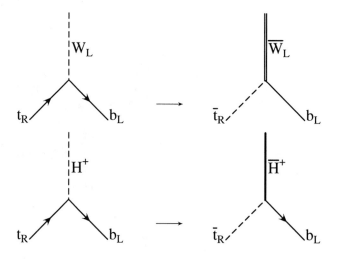

Fig. 9. SUSY interactions which will contribute to R_b.

In terms of the analysis presented before, these couplings are non-decoupling in m_t, but decoupling in the superpartner (top squark & chargino) masses. In the limit where the superpartner masses are large, but the charged-scalar masses are small, the total effect on R_b can approach that of the two-scalar model presented in the last section. The overall contribution, therefore, could be anywhere between the two-scalar and MSSM contributions shown on Fig. 7.[18]

If we take the central value of R_b reported at LEP and assume that the discrepancy with the standard model value is due to SUSY, the superpartners must be

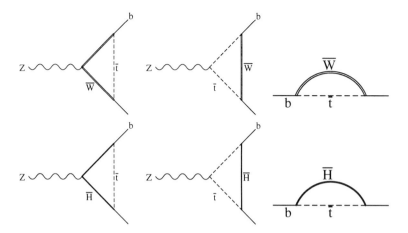

Fig. 10. Corrections to R_b in supersymmetric models.

quite light. This has recently been analyzed in detail by Wells, Kane, and Kolda.[21] For $\tan \beta < 30$, the bounds on the chargino and top squark masses are shown in Fig. 11. They conclude that (1) if the reduction in R_b is due to SUSY, superpartners must be discovered either at LEP II or the Tevatron and (2) the mass spectrum required cannot be accommodated in the popular "constrained" minimal grand-unified supersymmetry scenarios.

Finally, we should note that there are other new contributions to R_b in SUSY, including even some strong corrections involving the gluino, as shown in Fig. 12. These have recently been calculated by Bhattacharya and Raychaudhuri[22]; however they are very small: the contributions are entirely decoupling (they are not proportional to m_t) and vanish in the limit that there is no $\tilde{b}_L \leftrightarrow \tilde{b}_R$ mixing, which is the only $SU(2) \times U(1)$ breaking contribution to this process.

5. Technicolor

We move now to a completely different sort of theory, one with dynamical electroweak symmetry breaking. In these theories, the electroweak symmetry is broken due to the vacuum expectation value of a fermion bilinear instead of that of a fundamental scalar particle

$$\langle \varphi \rangle \rightarrow \langle \bar{\psi}_L \psi_R \rangle. \tag{22}$$

In the simplest theory[23] one introduces doublet of new massless fermions

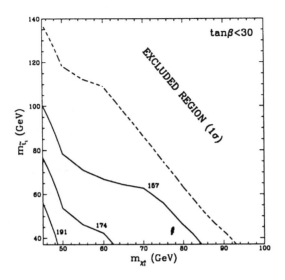

Fig. 11. One-sigma limits on chargino and top-squark masses coming from the measured value of R_b for various (191, 174, & 157 GeV) top-quark masses. The dashed line represents the upper-bound for a top-quark mass of 174 GeV and $R_b \geq 0.2172$. From ref. 21.

$$T_L = \begin{pmatrix} U \\ D \end{pmatrix}_L \qquad U_R, D_R \tag{23}$$

which are N's of an (asymptotically-free) technicolor gauge group $SU(N)_{TC}$. In the absence of electroweak interactions, the Lagrangian for this theory may be written

$$\begin{aligned} \mathcal{L} &= \bar{U}_L i\not{D} U_L + \bar{U}_R i\not{D} U_R + && (24) \\ &\quad \bar{D}_L i\not{D} D_L + \bar{D}_R i\not{D} D_R && (25) \end{aligned}$$

and thus has an $SU(2)_L \times SU(2)_R$ chiral symmetry. In analogy with QCD, we expect that when technicolor becomes strong,

$$\langle \bar{U}_L U_R \rangle = \langle \bar{D}_L D_R \rangle \neq 0, \tag{26}$$

which breaks the global chiral symmetry group down to $SU(2)_{L+R}$, the vector subgroup (analogous to isospin in QCD).

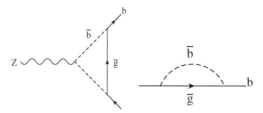

Fig. 12. Potential strong SUSY corrections to R_b.

If we weakly gauge $SU(2) \times U(1)$, with the left-handed technifermions forming a weak doublet and identify hypercharge with a symmetry generated by a linear combination of the T_3 in $SU(2)_R$ and technifermion number, then chiral symmetry breaking will result in the electroweak gauge group's breaking down to electromagnetism. The Higgs mechanism then produces the appropriate masses for the W and Z bosons if the F-constant of the technicolor theory (the analog of f_π in QCD) is approximately 246 GeV. (The residual $SU(2)_{L+R}$ symmetry insures that the weak interaction ρ-parameter equals one at tree-level.[24])

While this mechanism works wonderfully for breaking the electroweak symmetry and giving rise to masses for the W and Z bosons, it does not account for the non-zero masses of the ordinary fermions. In order to do so, one generally introduces additional gauge interactions, conventionally called "extended technicolor" (ETC) interactions,[25] which couple the chiral symmetries of the technifermions to those of the ordinary fermions (see Fig. 13).

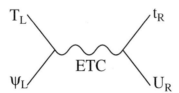

Fig. 13. ETC gauge-boson responsible for t-quark mass.

At low energies, below the mass of the ETC gauge boson, these interactions may be approximated by local four-fermion interactions, and include a coupling of the following form

$$\frac{g^2_{ETC}}{M^2_{ETC}}(\overline{T}_L U_R)(\bar{t}_R \psi_L). \tag{27}$$

After technicolor chiral symmetry breaking, this interaction leads to a mass for a quark (in this case the top-quark) of order

$$m_t \cong \frac{g^2_{ETC}}{M^2_{ETC}}\langle \bar{U}_L U_R\rangle. \tag{28}$$

It is the introduction of these extended technicolor interactions that is the source of many of the problems of theories with dynamical electroweak symmetry breaking. For example, in an ordinary QCD-like technicolor theory it is difficult to arrange for the strange-quark mass without introducing unacceptably large flavor-changing neutral-currents. There are various ways around this and other difficulties of ETC theories, for a review see.[26]

The ETC interactions produce corrections to the $Zb\bar{b}$ branching ratio. The ETC gauge boson pictured in Fig. 13 also mediates the interaction shown in Fig. 14.

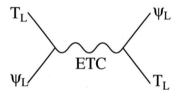

Fig. 14. ETC interactions which gives rise to correction to R_b.

At energies below the ETC gauge-boson mass, this interaction includes a coupling that can be approximated as

$$\xi^2 \frac{g^2_{ETC}}{M^2_{ETC}}\left(\overline{T}_L \gamma^\mu \frac{\vec{\tau}}{2} T_L\right)\left(\bar{\psi}_L \gamma_\mu \frac{\vec{\tau}}{2}\psi_L\right) \tag{29}$$

where ξ is a model-dependent Clebsch-Gordon coefficient equal to one in the simplest models. At energies below the technicolor chiral symmetry breaking scale, this gives rise to the interaction shown in Fig. 15 and results in a change in g^b_L.[27] Assuming for the moment that technicolor is QCD-like, we can estimate the size of this effect and find

$$\frac{\delta\Gamma}{\Gamma} = -6.5\,\%\,\xi^2\left(\frac{m_t}{175\text{GeV}}\right) \tag{30}$$

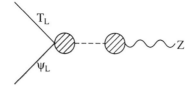

Fig. 15. ETC correction to R_b.

From this we find $\delta R_b^{ETC} = -0.011\xi^2$, which results in a total $R_b \approx 0.205$ (for $\xi = 1$). This is approximately eight-σ from the reported value.

Of course, ordinary technicolor theories were already in trouble for the flavor-changing neutral-current problems that were mentioned previously. Unfortunately, even in the more popular walking technicolor theories (so-called because the technicolor coupling runs slowly at energies above the technicolor scale, and capable of accommodating reasonable s and c quark masses without unreasonably large flavor-changing neutral-currents) the effect is perhaps a factor of two smaller[28] and is still hard to reconcile with experiment.

As with other new physics, TC/ETC theories give contributions to R_b which do not decouple with m_t. Furthermore, unlike the theories discussed previously, we *cannot* take M_{ETC}/g_{ETC} to be arbitrarily high**: its value is set by the mass of the top quark. In this sense, the contributions from TC/ETC theories are *completely* non-decoupling – their scale is set by the masses of the gauge bosons or quarks. It is this fact which makes the construction of phenomenologically acceptable TC/ETC theories so difficult.

In the discussion above we have implicitly assumed that the gauge bosons of the ETC theory do not carry electroweak quantum numbers. Recently, we have begun[31] to investigate the properties of theories containing ETC bosons which carry weak charge. In this case, it is possible for the correction to be the same order of magnitude, but *positive*. Such a correction may be too large in the opposite direction (it would be off by four-σ) – however such theories also include extra Z-bosons with flavor-dependent couplings. As we argue in the next section, these extra effects may possibly bring TC/ETC theories into a phenomenologically acceptable range.

6.   Extra Gauge Bosons

The last class of physics beyond the standard model which I will discuss concerns theories with extra weak gauge bosons. For simplicity, let us consider theories with an extra $U(1)$ gauge symmetry, resulting in an extra gauge boson X which will mix with the ordinary Z. Following the notation of Holdom,[32] the terms in the Lagrangian responsible for mixing include

---

**Unless we include additional low-energy scalar degrees of freedom.[29,30]

$$\frac{1}{2}M_Z^2 Z^\mu Z_\mu + \frac{1}{2}M_X^2 X^\mu X_\mu + x M_Z^2 X^\mu Z_\mu, \tag{31}$$

where we have chosen a basis so that the field Z_μ has the conventional gauge-couplings to the ordinary fermions and where we have neglected additional kinetic-energy mixing terms (which are small in weakly-coupled theories). In the limit $M_Z^2 \ll M_X^2$, this mixing results in a change in the coupling of the light mass-eigenstate (which we identify as the Z)

$$\delta g_{L,R}^f \approx -x\, \frac{M_Z^2}{M_X^2} g_{X_{L,R}}^f. \tag{32}$$

The mixing, therefore, results in a change in the width of the Z to various fermions (including, in particular, the b). In addition it also results in potentially dangerous changes in the relationship between $\sin^2\theta_W$ to α, G_F, and M_Z. For this reason, care must be taken in extracting limits on extra gauge bosons from precisely measured electroweak quantities.[33]

In ordinary extra gauge-boson models, of a type "inspired" by superstrings or $SO(10)$ GUT models, the X is usually assumed to couple to up- and down-quarks in a flavor universal fashion. In the limit $M_X \to \infty$, the theory reduces to the standard model. In this case, constraints on δR_b give constraints on M_X^2 and g_X.

In ETC/TC inspired models, however, the X can be related to the gauge boson responsible for generating the top-quark mass. For example, the X may be a "diagonal" generator associated with ETC breaking at the scale of top-quark mass generation. Therefore, in such theories it is natural to assume that such a gauge boson couples more strongly to the b_L, t_L, and t_R (and perhaps, though more dangerously, also to b_R, $\tau_{L,R}$, and ν_τ). In such theories it is not possible to take $M_X \to \infty$, since the mass of the X is related to the size of m_t – in this sense, the contributions are again completely non-decoupling.

The effects of such an extra family-dependent gauge boson are model dependent. In theories where the ETC gauge-boson responsible for generating the top-quark mass carries electroweak quantum numbers, the extra gauge bosons result in a *decrease* to R_b – perhaps by an amount sufficient to reconcile the ETC theory with experimental results. In a four-generation ETC model introduced by Holdom,[34] the theory does *not* give rise to an ETC contribution of the type discussed in the previous section but an extra weak-singlet X boson can *increase* R_b. These theories need to be studied in greater detail, perhaps by the time of DPF '96 we will understand them more fully and be able to determine whether they can be consistent with the experimentally measured value of R_b.

7. Perspectives

Although we have discussed only one process in detail, there is a general point that applies to the effects of physics beyond the standard model to any precisely measured electroweak quantity. Namely, that there are two types of theories of physics beyond the standard model.

First, there are *decoupling theories* that reduce to the standard model in the limit where some parameter with the dimensions of mass is taken to infinity. Generally, theories of this type investigated in the literature are *weakly coupled*. Examples include:

- Two-Higgs Doublet Models (decouple when $m_{H^+} \to \infty$)
- Supersymmetric Theories (decouple when $m_{\widetilde{H}}, m_{\widetilde{q}} \to \infty$)
- Extra Gauge-Boson Theories, if it is possible for $M_X \to \infty$

Theories of this sort have both good and bad points. On the one hand, because they reduce to the standard model in the decoupling limit, these theories *cannot* be ruled out on the basis of precision electroweak measurements (at least to the extent that these measurements are consistent with the standard model). On the other hand, it is disappointing that the answers to many of the interesting questions (SUSY breaking, Higgs masses, the origin of flavor, etc.) may be hidden at *very* high (M_{GUT} or M_{Pl}) energy scales. In his talk, Jeff Harvey[2] put the best face on this issue by arguing that there may be enough clues in low-energy parameters (e.g. the superpartner spectrum) to infer the properties of the high-energy physics. However, it is also possible that there will not be enough clues at low-energies to shed light on the high-energy physics.

Second, there are *non-decoupling theories*, whose scales are fixed by the masses of the observed particles. Theories of this type generally discussed in the literature are *strongly* coupled, and this makes them somewhat difficult to analyze. Examples include

- Technicolor/ETC Theories (the technicolor scale, Λ_{TC} – the analog of Λ_{QCD} in the ordinary strong interactions, is fixed at the weak scale).
- Extra Gauge-Boson Theories, in which M_X is fixed by the top-quark or some other mass.

If one succeeds in constructing a theory of this sort, one has made an enormous amount of progress – such a theory would explain a lot with physics at accessible energies (of order a TeV). However, there are no fully realistic models of this sort. Generically, these theories predict *large* low-energy effects of a sort that are excluded experimentally.

With luck, in time we will have experimental evidence to decide which type of theory is operative in the real world. However, given how little we actually know about the dynamics of electroweak and flavor symmetry breaking, we should be ready for either possibility.

8.   Prospects

In this talk I have concentrated on corrections to the coupling of the Z to b-quarks. This coupling is particularly susceptible to corrections because the left-handed b, being in the same weak doublet as the left-handed t, couples to the physics responsible for generating the large t-quark mass. However, one would like to probe the couplings of the t-quark *directly*.

One possibility is that the physics of electroweak symmetry breaking and flavor *could* lead to an enhancement of the cross section $\sigma(pp \to t\bar{t} + X)$ at the Tevatron. Two proposals of this sort have been put forward recently. In the first, due to Hill and Parke,[35] there is an additional color-octet or singlet gauge-boson, perhaps from "top-color" interactions,[36] which is produced in $q\bar{q}$ annihilation. In the second, due to Eichten and Lane,[37] a color-octet pseudo-Goldstone boson (a colored analog of the η in QCD, expected in some technicolor models) is produced in gluon fusion. In both cases, the new particle is associated with electroweak symmetry breaking and therefore couples most strongly and decays preferentially to the top quark.

These scenarios are particularly interesting because the rate of top quark production, $\sigma = 13.9^{+6.1}_{-4.8}$ pb (assuming that the excess of leptons plus jet events observed at the Tevatron is due to a 174 GeV top quark), is somewhat higher than the theoretically predicted value, $\sigma^{t\bar{t}}_{QCD} = 5.10^{+0.73}_{-0.43}$ pb.[5] With only a handful of events, we cannot be sure that the top-quark production cross section is, in fact, higher than expected from QCD. However, these models (1) demonstrate the often neglected possibility that the electroweak symmetry breaking sector couples to QCD and, more important, (2) will be tested in the near future as more data is collected at the Tevatron. These issues are discussed further by Steve Parke[38] in these proceedings.

A second possibility is that one may directly probe the couplings of the top-quark to the W and Z gauge bosons. A preliminary analysis by Barklow and Schmidt shows that it may be possible at an e^+e^- collider, with $\sqrt{s} = 500$ GeV and an integrated luminosity of 50 fb^{-1}, to measure these couplings to an accuracy of 5%-10%. Details of this work may be found in the contribution by Schmidt[39] in these proceedings.

9.   Summary

In conclusion let me reiterate that, because the top quark is heavy, it couples more strongly to the symmetry breaking sector. In general, this may be viewed as due to the Goldberger-Treiman relation, eqn. 9. Therefore, the top quark may provide a window on *both* electroweak and flavor symmetry breaking. Furthermore, because of $SU(2)_W$ symmetry, the physics responsible for generating the large t-quark mass also couples to the left-handed component of the b, resulting in contributions to the $Zb\bar{b}$ branching ratio which are generically *non-decoupling* in m_t (and therefore enhanced).

Will the measured value of R_b remain above the standard model value? Only time will tell.

10.   Acknowledgements

I thank Mike Dugan, Kenneth Lane, Elizabeth Simmons, and John Terning for discussions and for comments on the manuscript and gratefully acknowledge the support of an Alfred P. Sloan Foundation Fellowship, an NSF Presidential Young Investigator Award, and a DOE Outstanding Junior Investigator Award. This work was supported in part under NSF contract PHY-9218167 and DOE contract DE-FG02-91ER40676.

References

1. J. Rosner, talk presented at DPF '94, Albuquerque, New Mexico, Aug. 2-6, 1994

2. J. Harvey, talk presented at DPF '94, Albuquerque, New Mexico, Aug. 2-6, 1994

3. "Beyond the Standard Model", R. Sekhar Chivukula, talk presented at the *XVI International Symposium on Lepton-Photon Interactions*, Cornell University, Ithaca NY, Aug. 10-15, 1993, P. Drell & D. Rubin, eds., American Institute of Physics, New York, 1994.

4. R. Batley, talk presented at DPF '94, Albuquerque, New Mexico, Aug. 2-6, 1994

5. CDF Collaboration (F. Abe, et al.), Phys. Rev. Lett. **73** (1994) 225.

6. P. Langacker, private communication and PDG'94.

7. T. Takeuchi, A. K. Grant, and J. Rosner, talk presented by Takeuchi at DPF '94, Albuquerque, New Mexico, Aug. 2-6, 1994.

8. B.W. Lynn, M.E. Peskin and R.G. Stuart, SLAC-PUB-3725 (1985); in *Physics at LEP*, CERN Yellow Book 86-02, Vol. I, p.90.

9. See, for example J.M. Benlloch, E. Cortina, A.M. Llopis, J. Salt, C. De la Vaissiere, Z. Phys. **C59** (1993) 471.

10. A.A. Akhundov, D.Yu. Bardin and T. Riemann, Nucl. Phys. **B276** (1986) 1;

    W. Beenakker and W. Hollik, Z. Phys. C40 (1988) 141.

11. J. Bernabeu, A. Pich and A. Santamaria, Phys. Lett. **B200** (1988) 569

12. T. Appelquist and J. Carazzone, Phys. Rev. **D11** (1975) 2856.

13. R.D. Peccei and X. Zhang Nucl. Phys. **B337** (1990) 269.

14. M. Goldberger and S. B. Teiman, Phys. Rev. **110** (1958) 1478

15. *The Higgs Hunter's Guide*, John F. Gunion, Howard E. Haber, Gordon L. Kane, Sally Dawson, Addison-Wesley, New York (1990).

16. L. Hall and S. Weinberg, Phys. Rev. **D48** (1993) 979.

17. W. Hollik, Mod. Phys. Lett. **A5** (1990) 1909.

18. M. Boulware and D. Finnell, Phys.Rev. **D44** (1991) 2054.

19. A. K. Grant, Enrico Fermi Institute preprint EFI 94-24, June 1994.

20. H. E. Haber, Santa Cruz preprint SCIPP-92-33, Apr 1993. Presented at Theoretical Advanced Study Institute (TASI 92): *From Strings to Particles*, Boulder, CO, 3-28 Jun 1992.

21. J. D. Wells, Chris Kolda, G.L. Kane, UM-TH-94-23, Jul. 1994. **hep-ph 9408228**.

22. G. Bhattacharyya and A. Raychaudhuri, Phys. Rev. **D47** (1993) 2014.

23. S. Weinberg, Phys. Rev. **D19**, (1979) 1277.

    L. Susskind, Phys. Rev. **D20** (1979) 2619.

24. M. Weinstein, Phys. Rev. **8** (1973) 2511.

25. E. Eichten and K. Lane, Phys. Lett. **B90** (1980) 125.

    S. Dimopoulos and L. Susskind, Nucl. Phys. **B155** (1979) 237.

26. K. Lane, BUHEP-94-2, to appear in *1993 TASI Lectures* (World Scientific, Singapore).

    S. King, SHEP 93/94-2, **hep-ph 9406401**.

    M. Einhorn in *Perspectives on Higgs Physics*, G. Kane ed. (World Scientific, Singapore 1993) 429.

    Report of the "Strongly Coupled Electroweak Symmetry Breaking: Implications of Models" subgroup of the "Electroweak Symmetry Breaking and Beyond the Standard Model" working group of the DPF Long Range Planning Study. R. S. Chivukula, R. Rosenfeld, E. H. Simmons, and J. Terning, convenors.

27. R. S. Chivukula, S. B. Selipsky, and E. H. Simmons, Phys. Rev. Lett. **69** (1992) 575; N. Kitazawa Phys. Lett. **B313** (1993) 395 .

28. R. S. Chivukula, E. Gates, E. H. Simmons, and J. Terning, Phys. Lett. **B311** (1993) 157.

29. T. Appelquist, M. B. Einhorn, T. Takeuchi, and L. C. R. Wijewardhana, Phys. Lett. **B220** (1989) 223; V. A. Miransky and K. Yamawaki, Mod. Phys. Lett. **A4** (1989) 129; K. Matumoto, Prog. Theor. Phys. Lett. **81** (1989) 277; V. A. Miransky, M. Tanabashi, and K. Yamawaki, Phys. Lett. **B221** (1989) 177; V. A. Miransky, M. Tanabashi, and K. Yamawaki, Mod. Phys. Lett. **A4** (1989) 1043.

30. R.S. Chivukula, A.G. Cohen and K. Lane, Nucl. Phys. **B343** (1990) 554; T. Appelquist, J. Terning and L.C.R. Wijewardhana, Phys. Rev. **D44** (1991) 871.

31. R. Sekhar Chivukula, E. H. Simmons, and J. Terning, Phys. Lett. **B331** (1994) 383.

32. B. Holdom, Phys. Lett. **B259** (1991) 329.

33. G. Altarelli, *et. al.*, Phys. Lett. **B318** (1993) 139.

34. B. Holdom, University of Toronto preprint UTPT-94-18.

35. C. T. Hill and S. J. Parke, Phys. Rev. **D49** (1994) 4454.

36. C. T. Hill, Phys. Lett. **B266** (1991) 419.

    S. P. Martin, Phys. Rev. **D46** (1992) 2197 and Phys. Rev. **D45** (1992) 4283.

    M. Lindner and D. Ross, Nucl Phys. **B370** (1992) 30.

37. E. Eichten and K. Lane, Phys. Lett. **B327** (1994) 129.

    See also T. Appelquist and G. Triantaphyllou, Phys. Rev. Lett. **69** (1992) 2750.

38. S. Parke, talk presented at DPF '94, Albuquerque, New Mexico, Aug. 2-6, 1994.

39. C. Schmidt, talk presented at DPF '94, Albuquerque, New Mexico, Aug. 2-6, 1994.

    See also D. O. Carlson, E. Malkavrai and C.–P. Yuan, Michigan State preprint MSUHEP–94/05, May 1994, and **hep-ph 9405277** and **9405322**.

40. "First Measurements of the Inclusive Rate for the Radiative Penguin Decay $b \to s\gamma$", CLEO Collaboration preprint CLEO–CONF–94–1, July 1994.

Current Issues in Perturbative QCD

IAN HINCHLIFFE, *Lawrence Berkeley Laboratory*

CURRENT ISSUES IN PERTURBATIVE QCD

IAN HINCHLIFFE*

Lawrence Berkeley Laboratory
Berkeley CA 94720
USA

ABSTRACT

This review talk discusses some issues of active research in perturbative QCD. Among the topics discussed are, heavy flavor and prompt photon production in hadron-hadron collisions, "small x" phenomena and the current status of α_s.

1.   The current value of α_s

I will present a brief update of the value of $\alpha_s(M_Z)$. For a review of the results prior to this meeting see the article in QCD in the 1994 edition of the Review of Particle Properties.[1] The methodology adopted there will be followed here. Results from experiments using similar methods that have common systematic errors are first combined. These results are then extrapolated up to the Z mass using the renormalization group. An average of these values is then made to give the final result which is quoted as a value for $\alpha_s(M_Z)$. The new results will now be discussed.

1.1.   Lattice Gauge Theory.

Lattice gauge theory calculations can be used to calculated the energy levels of a $Q\overline{Q}$ system and then extract α_s. The FNAL group[2] uses the splitting between the 1S and 1P in the charmonium system $(m_{h_c} - (3m_\psi + m_{\eta_c})/4 = 456.6 \pm 0.4$ MeV). to determine α_s. The result quoted is $\alpha_s(M_Z) = 0.108 \pm 0.006$. The splitting is almost independent

*This work was supported by the Director, Office of Energy Research, Office of High Energy and Nuclear Physics, Division of High Energy Physics of the U.S. Department of Energy under Contract DE-AC03-76SF00098.

of the charm quark mass and is therefore dependent only on α_s. The calculation does not rely on perturbation theory or on non-relativistic approximation. The main errors are systematic associated with the finite lattice spacing (a), the matching to the perturbatively defined α_s, and quenched approximation used in the calculation. The extrapolation to zero lattice spacing produces a shift in Λ of order 5% and is therefore quite small. The quenched approximation is more serious. No light quarks are allowed to propagate and hence the extracted value of Λ corresponds to the case of zero flavors. $\alpha_s(M)$ is evolved down from the scale (~ 2.3 GeV) of the lattice used to the scale of momentum transfers appropriate to the charmonium system (~ 700 MeV). The resulting coupling is then evolved back up with the correct number of quark flavors. Perturbative running of $\alpha_s(M)$ has to be used at small M.

A recent calculation[3] using using the strength of the force between two heavy quarks computed in the quenched approximation obtains a value of α_s that is consistent with this result.

Calculations based on the Υ spectrum using non relativistic lattice theory give $\alpha_s(M_Z) = 0.115 \pm 0.003$.[4] This result includes relativistic corrections up to order $m_b(v/c)^4$. This recent result does not rely on the quenched approximation. Calculations are performed with two massless flavors. Combining this with the result in quenched approximation enables the result to be extracted for the physical case of three (u, d and s) light quarks. It is gratifying that this result is within the error quoted on the quenched calculations.[5] Averaging the lattice results then yields $\alpha_s(M_Z) = 0.113 \pm 0.003$

1.2.   Jet counting

A recent result from CLEO[6] measuring jet multiplicities at in e^+e^- annihilation at $\sqrt{s} = 10$ GeV, i.e. below the $b\bar{b}$ threshold gives a result of $\alpha_s(M_Z) = 0.113 \pm 0.006$. As with all measurements of this type, the dominant errors are systematic and arise from ambiguities in the scale at which α_s is evaluated and from the algorithms used to define a jet. This result is consistent with that from higher energies, in particular those from LEP[7] and SLC.[8]

Data at $\sqrt{s} = 29$, 58 and 91 GeV have been fit with the same set of Monte-Carlo (fragmentation) parameters. A consistent fit is obtained providing direct evidence for the running of $\alpha_s(Q)$.[9]

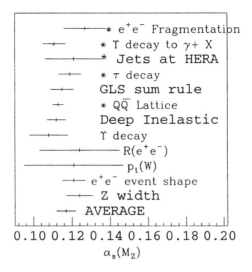

Fig. 1. The values of $\alpha_s(M_Z)$ determined by various methods. The symbol * denotes a result that has been updated from that in Ref.[1]

The H1 collaboration working at HERA[10] has determined α_s from a fit to the $2+1$ jet rate.[11] At lowest order in QCD the final state in deep inelastic scattering contains $1+1$ jets, one from the proton beam fragment and one from the quark that is struck by the electron. At next order, the struck quark can radiate another gluons giving rise to the $2+1$ jet final state. The determination involves data over a large Q^2 range and hence there is some correlation between the value of α_s and the structure functions that enter the computation of the event rate. The value quoted is $\alpha_s(M_Z) = 0.121 \pm 0.015$.

1.3. Upsilon decay.

The Cleo group[12] has determined α_s from a measurement of the ratio of Upsilon decay rates $\frac{\Upsilon \to \gamma + hadrons}{\Upsilon \to hadrons}$. In lowest order QCD this is given by $\frac{\Upsilon \to \gamma gg}{\Upsilon \to ggg}$. They quote $\alpha_s(M_Z) = 0.111 \pm 0.006$. There is non-perturbative contribution to this final state from the fragmentation of a gluon jet into a photon; this will introduce additional systematic errors into the result.

1.4. Scaling violations in Fragmentation functions.

The probability for a quark produced at scale Q (for example in $e^+ e^-$ annihilation at $\sqrt{s} = Q$) and energy E to decay into a hadron of energy zE is parameterized by a fragmentation function $d(z, Q)$. Just as in the case of the structure functions, the Q dependence of this fragmentation function is given by perturbative QCD and depends only on α_s. The QCD evolution of this fragmentation function also involves the fragmentation function of a gluon ($g(z,Q)$). Hence in order to determine α_s both $d(z,Q_0)$ and $g(z,Q_0)$ must be determined at some reference point Q_0. The ALEPH collaboration[13] uses three jet events from the decay of a Z. Two of the jets are tagged to be from b–quarks using the vertex detector and the finite b–quark lifetime. The third is then known to be due to a gluon. This method also determines the fragmentation functions for charm and bottom quarks which do not have the same form at Q_0 as the light quarks. It is worth recalling that whereas higher twist corrections in deep inelastic scattering are of order $1/Q^2$, here they can be order $1/Q$. These are parameterized in the ALEPH fit by replacing z by $z + c(z)/Q$. ALEPH quotes a value of $\alpha_s(M_Z) = 0.127 \pm 0.011$. The DELPHI collaboration, using a different method quotes $\alpha_s(M_Z) = 0.118 \pm 0.005$.[14] This result does not use the independent measurements of heavy quark and gluon fragmentation functions but rather fits to a Monte-Carlo. Its error could be underestimated.

1.5. Hadronic Width of the Tau Lepton

The hadronic width (or branching ratio) of the tau lepton can be used to determine α_s.[15] In the decay $\tau \to \nu_\tau + hadrons$, the decay rate, $R(M)$, can be measured as a function of the invariant mass M of the hadronic system. The inclusive hadronic width is then obtained by integrating over M, viz. $\Gamma = \int dM\, R(M)$. There are non-perturbative (higher twist) contributions that can be calculated using QCD sum rules.[16] Alternatively the data can be used to determine these quantities, which have different M dependence, from the data by using $\gamma(n) = \int dM\, M^n R(M)$. The values obtained in this are consistent with the estimates from the sum rules.[1718] There is an new result from CLEO[17] that gives $\alpha_s(M_Z) = 0.114 \pm 0.003$, a value somewhat below the old world average. A new result from ALEPH[19] of $\alpha_s(M_Z) = 0.124 \pm 0.003$ is larger than the old world average. The difference between these results is due to different values of the branching ratios R_e and R_μ measured for $\tau \to e\bar{\nu}\nu$ and $\tau \to \mu\bar{\nu}\nu$. The hadronic width is then inferred from these via $R_h = 1 - R_e - R_\mu$ as this results in a smaller error than that gotten by using Γ directly.

1.6.   Average value of $\alpha_s(M_Z)$

After taking into account this new data, the average of $\alpha_s(M_Z) =$ 0.117 quoted in RPP 1994 is left unchanged. If we assume that the systematic errors associated with the different methods are uncorrelated, then we obtain an error of ± 0.002. In view of the fact that most of the dominant errors are theoretical, involving such things as estimates of non-perturbative corrections and the choice of scale μ where $\alpha_s(\mu)$ is evaluated for the process in question, it is more reasonable to quote $\alpha_s(M_Z) = 0.117 \pm 0.005$ as the "world average".

2.   Heavy Quark Production in Hadron Collisions

New results are available from the CDF[20] and D0[21] collaborations on the production rate of bottom quarks in $p\bar{p}$ collisions. Several methods are used. The least subject to ambiguities involves the use of fully reconstructed decays of a B meson. In this case in order to get from the observed B meson rate to that of $b-$quarks, the fragmentation function of a $b-$quark needs to be known. This is well constrained from data at LEP so this method of measuring the $b-$quark production rate should be quite reliable. However there are rather few fully reconstructed events and hence this method is limited and does not permit measurement over a large range of transverse momenta.

The next method involves the use of inclusive Ψ production. This method can only be used if a vertex system is available to disentangle the Ψ's that come from b-decay from those produced directly. Here the systematic errors are a larger since one needs a model of the $b-$quark fragmentation and of the subsequent $b-$meson and $b-$baryon decays to ΨX.

Finally, there is the method with the largest statistical sample and hence the greatest range in transverse momentum. Here one searches for jets which have muons or electrons associated with them. The leptons are required to have some transverse momentum of order 1 GeV or greater with respect to the jet direction (p_a). Most of these leptons then arise from bottom decay: there is a small contribution from charm decay, the relative fraction being a function of p_a. A model of the lepton spectrum

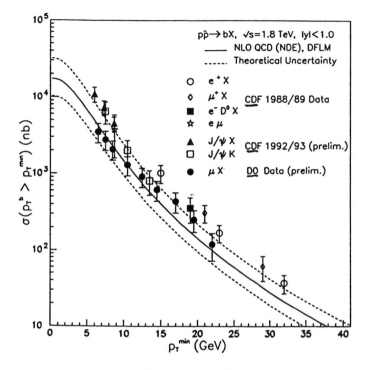

Fig. 2. The inclusive cross section for the production of b-quarks in $p\bar{p}$ collisions a $\sqrt{s} = 1.8$ TeV. The produced quark is required to have transverse momentum greate than p_T^{min} and the rate is shown as a function of p_T^{min}. See text for discussion.

from charm and bottom quarks is needed before the b-quark cross section can be extracted.

The measured rates from the different methods are shown in figure 2 which shows the cross-section for the production of a b-quark of transverse momentum greater that p_T^{min}. It can be seen from this figure that the rates measured by the $D0$ collaboration which uses only the last method are systematically lower than those of the CDF collaboration which uses all of the methods for b-quark transverse momenta of less than 15 GeV. Above this value the experiments are consistent with each other. Note that in the region of disagreement CDF is able to use what should be the most reliable method. The figure also shows the theoretical expectation for this rate.[22] It is rather uncertain since the QCD predictions are not stable with respect to the choice of the scale μ at which the parton distributions and $\alpha_s(\mu)$ is evaluated in the expression for the production rate. By lowering this to $\mu = \sqrt{m_b^2 + p_t^2/4}$ consistency with the CDF data can be achieved. The D0 data can be accommodated by using larger and *a priori* more reasonable value of μ. The cross-section now reported by CDF is lower than the values that were obtained from observation of Ψ production and the assumption, now known to be wrong, that almost all ψ's at large transverse momenta arise from b-quark decay.

The top cross-section quoted by CDF[23] is somewhat larger than that predicted by QCD for a mass of 170 GeV,[24] the value given by the CDF fit. Since D0 has not yet confirmed this rate,[25] it is premature to claim that there is a problem with QCD or that physics beyond the standard model has been discovered.

3. Production of Ψ and Υ in $p\overline{p}$ collisions.

At low transverse momentum the production of ψ's in $p\overline{p}$ collisions is expected to proceed dominantly via the production of χ states followed by their decay i.e. $g + g \rightarrow \chi \rightarrow \psi + X$.[26] The analogous process at large transverse momentum is $gg \rightarrow g\chi$. This process generates a cross section that falls off at large transverse momentum much faster than, say, the jet rate. This observation led to an assumption that ψ's produced at large transverse momentum came almost exclusively from b-quark decay. A measurement of the rate for ψ production could then be used to infer the b-quark rate. This assumption is now known to be false. By detecting whether or not the ψ's come from the primary event vertex, CDF is now able to test this assumption. The fraction of ψ's produced directly is almost independent of transverse momentum and the rate of direct Ψ production at large p_t is larger than had been expected.

The dominant production mechanism of ψ's at large transverse momentum is now believed to be the fragmentation of light quark and gluon jets into χ's that then decay to ψ.[27] CDF now has data on ψ, ψ, ψ', Υ, Υ' and Υ''.[30] While the rate for ψ production is in agreement with expectations, given the inherent theoretical uncertainties, the rate for ψ' is approximately a factor of 20 above the theoretical expectation.[28] The calculation does not include the possibility of ψ' production from the decay of 2P states. These states are above the $D\overline{D}$ threshold, however a branching ratio of a few percent to ψ' could be enough to explain the deficit. A recent paper investigates this possibility quantitatively[29]

The predicted rates for Υ production should have less uncertainties due to the larger value of the Υ mass. Preliminary data from CDF indicate that the agreement with theoretical expectations is poor.[30]

4.   Prompt Photon Production.

The production of photons in $p\bar{p}$ proceeds, at lowest order in QCD, via the parton process $qg \rightarrow \gamma q$. The process provides a direct probe of the gluon distribution and can be measured more reliably than the jet cross-section whose value depends on a jet definition and upon measurements of both hadronic and electromagnetic energy. The produced photon, provided that is is produced at large transverse momentum, is well isolated from other produced particles. At higher orders in α_s, the situation changes. Processes such as $qq \rightarrow qq\gamma$ start to contribute. This process is largest when the photon is collinear with one of the outgoing quarks. Since experiments cannot easily measure photons within jets, they search for isolated photons defined by having less than some amount ϵ of other energy in a cone of radius ΔR in rapidity-azimuth space around the photon direction. The rate then depends on ϵ and ΔR; the selection criteria discriminate against the bremsstrahlung component. The fragmentation of a quark into a photon is a non-perturbative phenomena which must be modeled by a fragmentation function into which the collinear singularity is absorbed.

A theoretical prediction of the prompt photon rate then depends on, the gluon and quark distributions, the fragmentation functions, and the scales μ and Q at which these functions and $\alpha_s(Q)$ are evaluated (these scales need not be the same). The dependence on these scales is an indication of the uncertainties in the theoretical predictions; if the process were calculated to all orders in perturbation theory, the dependence on μ and Q would, at least in principle, disappear. μ and Q should be of order p_t, beyond that theory provides no guidance. There is a longstanding problem in that at small values of $x_\perp = 2p_t/\sqrt{s}$, the data tend to be larger than the theoretical predictions. At values of $x_\perp$ that are probed at the Fermilab collider a smaller value of μ results in a larger predicted cross-section.

There have been several new developments in this field. Measurements of structure functions at small x at HERA[37] have indicated that the

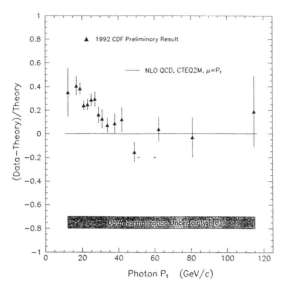

Fig. 3. A comparison of the CDF data on prompt photon production. Data refers to the quantity $\frac{d\sigma}{dp_t d\eta}$ at $\eta = 0$. Theory refers to the calculation of Ref[32] using the CTEQ2M structure functions[36] and $\mu = p_t$.

gluon distribution is larger at small values of x than used to be assumed. This change increases the predicted rate at small $x_\perp$. There has been a reassessment of the importance of fragmentation[31] and finally new data are available.

Figure 3 shows that data from the CDF collaboration.[34] The data fall very rapidly with increasing p_t so, to facilitate comparison with theory, the data are shown relative to the calculation of ref.[32] The tendency for the data to have a steeper dependence on p_t than the theory can be seen in this figure. A reduction in the scale to $\mu = p_t/2$ brings the data into better agreement with the theory. Figure 4 shows the same data compared to the calculation of ref.[31] Here the scale $\mu = p_t/2$ has been used along with the GRV structure functions.[33] The predictions of $MRSD-'$ structure functions[39] are almost identical. This theoretical result has a slightly larger fragmentation component. The tendance for the the data to have a steeper p_t slope than the theory is still apparent in this plot, although the agreement is quite good. A reduction in the scale μ to the, possibly unreasonable, value of $p_t/3$ improves the agreement further.

Preliminary results presented at this meeting from $D0$[35] lie somewhat below those of CDF and are therefore in better agreement with theoretical

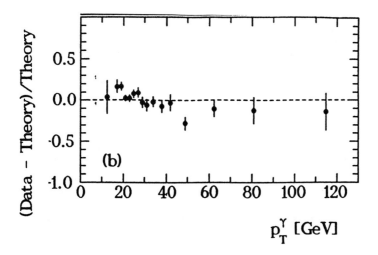

Fig. 4. As Figure 3 except that the theory refers to the calculation of Ref[31] at the GRV structure functions[38] and $\mu = p_t/2$.

estimates. Figure 5 shows a comparison with theory. Note that that theory is the same as that used in figure 3. The tendency for the data to have a steeper p_t slope than the theory is not evident in the D0 results although the systematic errors are such that there is no significant disagreement with CDF.

There is a preliminary measurement of the di-photon rate from CDF.[41] This measurement is important since, at the LHC, one of the decay mechanisms proposed to search for the Higgs boson is its decay to two photons.[42] The very large background is expected to occur from $q\bar{q} \rightarrow \gamma\gamma$ and $gg \rightarrow \gamma\gamma$. The rate observed by CDF is consistent with the expectation from calculations using these mechanisms.[43] We can now have more confidence in the ability of LHC to see the Higgs signal.

5. Small-x and related phenomena

QCD perturbation theory is an expansion powers of $\alpha_s(Q)$. Two conditions must be satisfied to have a reliable prediction. First, the scale Q must be large and second the perturbation series must contain no large coefficients. The value of some measured dimensionless quantity P and be expressed as a power series.

$$P = A\alpha_s^n(Q)(1 + b\alpha_s(Q) + \cdots)$$

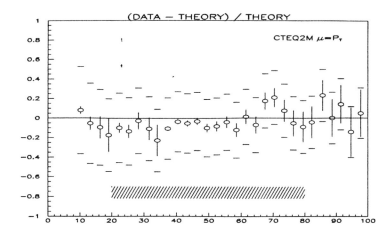

Fig. 5. As Figure 3 except that the experiment refers to the results of the D0 collaboration[40]

If $b\alpha_s(Q) \sim 1$, then the perturbation series useless. A physical prediction can be recovered if the large terms can be isolated order by order in perturbation theory and the terms summed up. Once these pieces are absorbed into A, the resulting series may be well behaved and a prediction possible.

The simplest example of a resummation of this type is that of the Altarelli-Parisi (DGLAP) equation which sums terms of the type $(\alpha_s \ln(Q^2))^n$ that arise in Deep Inelastic Scattering.[44] At very small values of Bjorken-x, b can contain terms of the type $\log(1/x)$. These terms can be resummed using the BFKL equation.[45] The result of this resummation is a structure function that rises very rapidly at small-x. While this behaviour is seen at HERA,[37] it cannot be used to distinguish between evolution expected from BFKL and DGLAP.[46]

The behaviour of the structure functions at very small x is connected with attempts to calculate the total cross-section in perturbative QCD. The same resummation that leads to BFKL is also responsible for the appearance in perturbative QCD of the pomeron.[47] This connection has recently been clarified.[48] I will now discuss some phenomena related to the pomeron.

5.1. Jets with Large Rapidity Separation

Events are selected in $p\bar{p}$ collisions having a pair of jets with transverse momenta p_1 and p_2 and rapidities η_1 and η_2 with azimuthal angle ϕ

between then. At lowest order in perturbative QCD, $p_1 = p_2$, $\cos(\pi - \phi) = 1$ and the rate is given by

$$\frac{d\sigma}{dp_1 d\eta_1 d\eta_2} = \frac{1}{16\pi s} g(x_1, Q^2) g(x_2, Q^2) \frac{d\sigma(gg \to gg)}{dt}$$

Here I have assumed that only gluons contribute. If $\eta_1 = -\eta_2 = y$, then $x_1 = x_2 = \frac{2p_1}{\sqrt{s}} \cosh y / 2$. If y is very large, the center of mass energy of the parton system ($= x_1 x_2 s$) becomes large and the partonic cross section can be approximated by

$$\frac{d\sigma}{dt} = \frac{9\pi\alpha_s^2}{2p_1^2 p_2^2}$$

If we now integrate over p_1 and p_2 greater than some scale M

$$\frac{d\sigma}{d\eta_1 \eta_2} \sim \frac{\alpha_s^2}{M^2} x_1 g(x_1) x_2 g(x_2)$$

At order α_s^3 in perturbation theory, several phenomena occur. A third jet is emitted and the correlation in azimuth and equality of p_1 and p_2 is lost. More important is that this order α_s^3 process modifies the result for $\frac{d\sigma}{dp_1 d\eta_1 d\eta_2}$ by a factor of $(1 + 3\alpha_s |\eta_1 - \eta_2| / \pi)$ (for large values of $\eta_1 - \eta_2$). If $\eta_1 - \eta_2$ is large enough this factor can be so large that perturbation theory is not reliable. In this case the leading terms at all orders in perturbation theory can be resummed to give a factor of $exp(3\alpha_s |\eta_1 - eta_2| / \pi)$.[49] This growth is not observable at the Tevatron since it is more than compensated by the drop off caused by the falling structure functions. (Note that x_1 and x_2 increase as y increases.). It may be observable at LHC.[50] However the other effects should be observable. The rapidity region between the two jets is filled with many mini-jets since there is no penalty of α_s to pay for each emission. The correlation in ϕ between the two trigger jets should show a rapid fall off as y is increased. The D0 collaboration[51] has searched for this effect by selecting events with two jets one of which has $p_t > 20$GeV and the other has $p_t > 50$ GeV. The ϕ correlation is then plotted as a function of the rapidity separation. The data show a decorrelation. However it is a much slower fall off than predicted and is consistent with that expected from a fixed order α_s^3 calculation or from showering Monte-Carlos such as HERWIG.[52] It is possible that the rather asymmetric trigger could be masking the effect in this case.

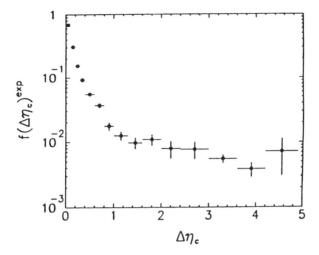

Fig. 6. The fraction of events with no particles in the rapidity interval between the two produced jets in a $p\bar{p}$ collision as a function of the rapidity interval. Data from the D0 collaboration.[54]

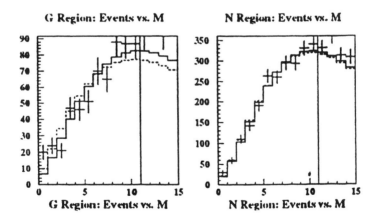

Fig. 7. The event rate plotted against particle multiplicity. Two jets are selected, separated in rapidity by 2.8 units of rapidity. The G (N) region is defined as the interval between the jets (between each jet and the end of the physical region closest to it). There is clear evidence for an excess of events in the zero multiplicity bin in the G region over that expected from a KNO fit (solid curve). No such excess is visible in the N region. See ref[53] for more details.

5.2. Rapidity Gaps.

Consider the production of two jets at large rapidity separation in a $p\bar{p}$ collision. At lowest order in QCD perturbation theory one contribution to this arises from quark-quark scattering via gluon exchange. Before the scattering each quark forms a color singlet state with the rest of the quarks and gluons from its parent (anti-)proton. After scattering, this is no longer the case since the gluon exchange causes color charge to be transferred between the quarks. As the parton system hadronizes into jets, color must be exchanged between the outgoing jets. This color exchange manifests itself as soft (low transverse) momentum particles that fill the rapidity interval between the jets. Contrast this with the situation if a colorless object (such as a photon) were exchanged. Now the struck quark and the remnant of its parent are still in a color singlet and can hadronize without communication with the other quark. There is no necessity for color exchange and hence no need for particle production in the rapidity interval between the jets. Both the CDF[53] and D0[54] collaborations have searched for events with rapidity gaps. CDF uses the charged particle multiplicity while D0 uses the energy flow as measured by the calorimeter.

In the D0 case, events with two jets each of transverse energy of at least 30 GeV are selected. The jets are separated by rapidity η. Events are determined to have a gap if there are no calorimeter towers in the region between the two jets with an electromagnetic energy deposit of more than 200 MeV. Figure 6 shows the fraction (f) of events that have such gaps as a function of the rapidity separation of the jets. If all of the events are due to jet production involving color exchange, one expects that f will fall rapidly with increasing η. While this behaviour is observed at small η there is clear evidence for a plateau in $f = f_0$ at large values of η indicating the presence of color singlet exchange.

CDF tags two jets of rapidity η_1 and η_2 (η_2 is assumed to be greater than η_1). They then look at the the multiplicity of charged tracks in region G defined by $\eta_1 < \eta_g < \eta_2$ and region N defined by the remainder of the rapidity range. N then covers the rapidity range between each jet and its parent (anti)proton. If color singlet exchange is contributing to the jet production, then one should expect events with zero multiplicity in region G. Region N always has color flow across it and can therefore be used as a control region. A KNO type multiplicity plot is made for the G and N regions, see figure 7. The shapes of the distributions in the G and N regions are the same except for an excess in the zero multiplicity bin in the G region. This provides clear evidence for events with a rapidity gap at a rate $f = (0.86 \pm 0.12)\%$. There is no evidence for any dependence of f on either the transverse momentum of the jets (E_t) or the width of the gap $(\eta_2 - \eta_1)$. The rate is too large to be due to photon exchange and must represent the exchange of another color singlet object. The obvious candidate is the pomeron.

These data leave several questions unanswered. f cannot be directly interpreted in terms of the strength of the coupling of the pomeron to quarks and gluons since, once two jets are produced by this mechanism, we do not know how often particles are emitted into the gap region by the rest of the event and hence what fraction of these events survive to be detected by the experiments. (Bjorken[55] uses the term survival probability S for this.) Hence $f = S \frac{pomeron-rate}{perturbativeQCD-rate}$. More data are needed on the E_t dependence of f. If, for example, the pomeron couples to quarks and gluons with a form factor as opposed to a hard coupling, then one would expect f to fall as E_t increases. A constant f would indicate that it coupled in a similar way to gluons.

A similar phenomenon has been observed at HERA. In the usual picture of deep-inelastic scattering a quark is struck by the virtual photon and ejected from the target proton. This quark then hadronizes into a jet (the current jet) and since its color is compensated by the target remnant, particles are produced in the rapidity region between the current jet and the beam proton. η_{max} is defined as the rapidity of the particle with the largest rapidity in a particular event. (The proton is initially moving in the positive rapidity direction.) One would expect that there are always particles produced near the initial proton and so the η_{max} distribution would have a peak at large positive value. Figure 8 shows the distribution as measured by ZEUS.[56] The data show, in addition to the expected peak, a large number of events where η_{max} is very small. Approximately 8% of the events have no hadrons in the direction of the initial proton; the fraction is independent of the mass Q^2 of the virtual photon. Similar phenomena have been observed by the H1 collaboration[57]

The rate of events in this region of $\eta_{max} \sim 0$ is much larger than expected from a Monte-Carlo based on this picture of Deep inelastic scat-

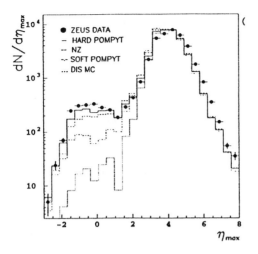

Fig. 8. The η_{max} distribution (see text) as measured by the ZEUS collaboration.[56]

tering. The excess of events can be explained if there is some color neutral component of the proton which itself can be disassociated by the virtual photon. This component will have some fraction of the proton's momentum. After interaction with the virtual photon the hadronization and color neutralization need only take place among the fragments of this color neutral system. There is no necessity for particle production in the rapidity region between the object and its parent proton. A second peak at smaller values of η_{max} will then appear. One candidate for this object is the pomeron,[58] which can be though of as an object similar to other hadrons with quark constituents. A model of this type where the object (pomeron) has a structure function of the form $f(x) \sim x(1 - x)$ is compatible with the ZEUS data.[56]

The simplest candidate for this object in QCD is a two-gluon object[59] as shown in figure 9. This simple picture has been extended to and builds up the BFKL pomeron.[60] This picture is also in qualitative agreement with the data. Note that, as in the case of events with rapidity gaps at hadron colliders, the relative normalization of these color singlet pieces is difficult to extract from the data.

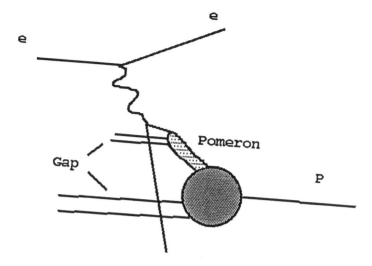

Fig. 9. The simplest contribution to deep inelastic scattering in QCD with the possibility to produce an event with low η_{max}. The simplest object in QCD to play the role of the pomeron is a two gluon system.

6. Particle Multiplicity in Heavy Quark Jets.

While the particle multiplicity cannot be calculated in perturbative QCD, its growth with energy can be. Consider the radiation of a gluon off a quark produced of mass M say in e^+e^- annihilation. If the gluon has energy E and is emitted at angle θ with respect to the quark, then the emission probability behaves as

$$d\sigma = \frac{\theta^2 d\theta^2}{\theta^2 + \delta^2} \frac{dE}{E}$$

This has two consequences, radiation at $\theta < \delta$ is suppressed resulting a what is called a "dead-cone" and a heavy quark radiates less than a light quark.[61] We should expect the particle multiplicity (N) from a heavy quark pair (Q) to be less than that from a light quark pair (q)

$$N(Q\overline{Q}, \sqrt{s}) = N(q\overline{q}, \sqrt{s}) - N(q\overline{q}, M)$$

Note that the difference in multiplicities is independent of $\sqrt{s}$. The naive expectation based on the available phase space

$$N(Q\overline{Q}, \sqrt{s}) = N(q\overline{q}, \sqrt{s} - 2M)$$

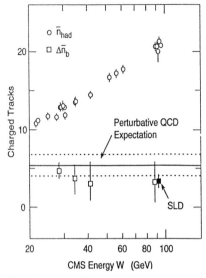

Fig. 10. The behaviour of the charged particle multiplicity in e^+e^- events as a function of $\sqrt{s}$.[62] The plot shows the total multiplicity n_{had} as well as the difference in multiplicity between events tagged as being from or not from the production of a $b\bar{b}$ pair (Δn_b)

predicts a difference that is not independent of $\sqrt{s}$. Figure 10 shows data at various energies.[62] While the total charged particle multiplicity rises with $\sqrt{s}$, the difference between multiplicity in b–quark events (tagged using the finite b lifetime) and average events does not.

7.   Conclusions.

The past few years has seen a continuing development in our understanding of QCD. The strong coupling constant is now known to a precision of order 5% . Its precision can only be expected to improve slowly in the near future since most of the measurements are now limited by various theoretical uncertainties. An improved measurement of the hadronic width of the Z from LEP is on of the few areas where a more precise measurement of a physical quantity will yield a more accurate value of α_s. The recent developments involving lattice gauge theory calculations with propagating light quarks is another area where one can hope for increased precision.

There are still some experimental results that, while accommodated by perturbative QCD, are not entirely satisfactorily explained. While the long standing problem of the prompt photon rate in $p\bar{p}$ collisions may now by going away, the production rate of bottom quarks is still not fully digested. Interesting data on ψ and Υ production from CDF are yet to be fully understood.

Much interest, both theoretical and experimental, has occurred in the area of semi-hard (or small-x) QCD. Diffractive phenomena, for a long time dismissed as incalculable and hence uninteresting, are finally being given the attention that they deserve. There are many "predictions" for phenomena in this region of phase space. However, a systematic procedure for calculating the subleading corrections to the BFKL equation is lacking. Such a procedure is badly needed, for, until we can determine the size of these terms, we cannot say how accurate predictions using BFKL can be expected to be.

The preparation of this talk took place while I was a visitor in the FERMILAB theory group. I am grateful to Keith Ellis and the other members of the group for their hospitality.

References

1. Review of Particle Properties *Phys. Rev.* **D50**, 1174 (1994).
2. A.X. El-Khadra *et al.*, *Phys. Rev. Lett.* **69**, 729 (1992). A. X. El-Khadra, presented at 1993 Lattice conference, OHSTPY-HEP-T-93-020, A.X. El-Khadra *et al.*, FNAL 94-091/T (1994)
3. Martin Luscher, Rainer Sommer,Peter Weisz and Ulli Wolff *Nucl. Phys.* **B413**, 481 (1994).
4. C.T.H. Davies *et al.* OHSTPY-HEP-T-94-013, Aug 1994.
5. G.P. Lepage, and J. Sloan presented at 1993 Lattice conference, hep-lat/9312070
6. R. Balest, *et al.* CLEO-CONF-94-28, Jul 1994.
7. For a recent review see S. Bethke PITHA-94-29, Aug 1994.
8. K. Abe, *et al. Phys. Rev. Lett.* **71**, 2528 (1994).
9. D.A. Bower *et al.* LBL-35812, M. Aolki *et al.* Submitted to the XXVII International Conference on High Energy Physics (Glasgow 1994), D. Decamp *et al. Phys. Lett.* **B284**, 163 (1992).
10. S. Soldner-Rembold, submitted to Int. Conf. on High Energy Physics, Glasgow, Scotland, Jul 20-27, 1994.
11. G. Ingelman,J. Rathsman, TSL-ISV-94-0096, May 1994.

12. CLEO collaboration, reported at the QCD94 Conference in Montpellier.

13. G.Cowan, submitted to Int. Conf. on High Energy Physics, Glasgow, Scotland, Jul 20-27, 1994.

14. P. Abreu, *et al.* *Phys. Lett.* **B311**, 408 (1993).

15. S. Narison and A. Pich *Phys. Lett.* **B211**, 183 (1988), E. Braaten, S. Narison and A. Pich, *Nucl. Phys.* **B373**, 581 (1992)

16. M.A. Shifman, A.I Vainshtein, and V.I. Zakharov *Nucl. Phys.* **B147**, 385 (1979).

17. J. Alexander, *et al.*CLEO-CONF-94-26, Jul 1994, Submitted to Int. Conf. on High Energy Physics, Glasgow, Scotland, Jul 20-27, 1994.

18. D. Buskulic,*et al.* *Phys. Lett.* **B307**, 209 (1993).

19. ALEPH collaboration reported at the QCD94 Conference in Montpellier.

20. F. Abe, *et al.* *Phys. Rev. Lett.* **71**, 2396 (1993), *Phys. Rev. Lett.* **71**, 500 (1993) and CDF collaboration these proceedings.

21. G. Alves and T. Huen for the D0 collaboration, these proceedings.

22. P. Nason, S. Dawson and R.K. Ellis, *Nucl. Phys.* **B303**, 607 (1988).

23. F. Abe, *et al.* *Phys. Rev. Lett.* **73**, 225 (1994).

24. E. Laenen J. Smith, and W.L. van Neerven, *Phys. Lett.* **B321**, 254 (1994).

25. S. Abachi, *et al.* *Phys. Rev. Lett.* **72**, 138 (1994).

26. R. Baier and R. Ruckl, *Zeit. fur Physik* **c19**, 251 (1983), F. Halzen *et al.* *Phys. Rev.* **D30**, 700 (1984).

27. E. Braaten and T.-C. Yuan *Phys. Rev. Lett.* **71**, 1673 (1993).

28. E. Braaten *et al.* *Phys. Lett.* **B333**, 548 (1994).

29. D.P. Roy and K. Sridhar, CERN-TH-7434-94.

30. Vaia Papadimitriou, *et al.* FERMILAB-CONF-94-221-E, Aug 1994.

31. M. Gluck, *et al.* DO-TH-94-02-REV, Feb 1994.

32. H. Baer, J. Ohnemus and J. Owens *Phys. Rev.* **D42**, 61 (1990)

33. M. Gluck, E. Reya, and A. Vogt *Phys. Lett.* **B306**, 391 (1993).

34. F. Abe, *et al.* FERMILAB-PUB-94-208-E, 1994.

35. S. Fahey for the D0 collaboration, these proceedings.

36. James Botts, *et al.* *Phys. Lett.* **B304**, 159 (1993).

37. For a review see, for example, Gunter Wolf DESY-94-178,1994.

38. M. Gluck, E. Reya and M.Voit, *et al.* *Zeit. fur Physik* **C53**, 127 (1992).

39. A.D. Martin, R.G. Roberts and W.J. Stirling HEPPH-9409257, Jul 1994.

40. D0Collaboration, these proceedings.

41. R. Blair, *et al.* FERMILAB-CONF-94-269-E.

42. For a recent review see, for example, J.F. Gunion, UCD-94-24, (1994).

43. B. Bailey, J.F. Owens and J. Ohnemus *Phys. Rev.* **D46**, 2018 (1992).

44. V. N. Gribov and L.N. Lipatov *Sov. J. Nucl. Phys.* 15, 438 (1972), L.N. Lipatov, *Sov. J. Nucl. Phys.* 20, 181 (1974), Yu. L. Dokshitser, *Sov. Phys. JETP* **46**, 641 (1977), G. Altarelli and G. Parisi *Nucl. Phys.* **B126**, 298 (1977).

45. E.A. Kurayev, L.N. Lipatov and V.S. Fadin, *Sov. Phys. JETP* **45**, 119 (1977), Ya. Ya. Bailitsky and L.N. Lipatov, *Sov. J. Nucl. Phys.* 28, 882 (1978).

46. By R.K. Ellis, Z. Kunszt and E.M. Levin, *Nucl. Phys.* **B420**, 517 (1994).

47. L.N. Lipatov *Sov.Phys.JETP* **63**,904 (1986).

48. A.H. Mueller and B. Patel *Nucl. Phys.* **B425**, 471 (1994), A.H. Mueller *Nucl. Phys.* **B415**, 373 (1994), N.N. Nikolaev, B.G. Zakharov and V.R. Zoller J.Exp.Theor.Phys. **78**, 806 (1994).

49. A.H. Mueller, H. Navelet *Nucl. Phys.* **B282**, 87 (19.)

50. Vittorio Del Duca and Carl R. Schmidt, DESY-94-114. W.J. Stirling *Nucl. Phys.* **B423**, 56 (1994).

51. D0 Collaboration, these proceedings.

52. G. Marchesini and B.R. Webber, *Nucl. Phys.* **B310**, 461 (1988).

53. F. Abe, *et al.* FERMILAB-PUB-194-94E, *Phys. Rev. Lett.* (to appear).

54. S. Abachi, *et al.* *Phys. Rev. Lett.* **72**, 2332 (1994).

55. For a recent review see, J.D. Bjorken, SLAC-PUB-6463, (1994),

56. M. Derrick, *et al.* *Phys. Lett.* **B332**, 228 (1994).

57. T. Ahmed, *et al.* DESY-94-133 (1994).

58. G. Ingelmann and P.E. Shlein, *Phys. Lett.* **152B**, 256 (1985).

59. S. Nussinov, *Phys. Rev.* **d41**, 246 (1976), S. Donnachie and P.V. Landshoff, *Nucl. Phys.* **B244**, 322 (1984).

60. N. Nikolaev and B.G. Zakharov, *Zeit. fur Physik* **C53**, 331 (199)2

61. B. A. Schumm, L. Dokshitser and V, A. Khoze *Phys. Rev. Lett.* **69**, 3025 (1992).

62. K. Abe, *et al.* *Phys. Rev. Lett.* **72**, 3145 (1994).

Status of the Standard Model

JONATHAN L. ROSNER, *Chicago, Enrico Fermi Institute*

STATUS OF THE STANDARD MODEL

JONATHAN L. ROSNER

Enrico Fermi Institute and Department of Physics, University of Chicago
5640 S. Ellis Ave., Chicago, IL 60637, USA

ABSTRACT

The standard model of electroweak interactions is reviewed, stressing the top quark's impact on precision tests and on determination of parameters of the Cabibbo-Kobayashi-Maskawa (CKM) matrix. Some opportunities for the study of CP violation in the decays of b-flavored mesons are mentioned, and the possibility of a new "standard model" sector involving neutrino masses is discussed.

1. Introduction

Precision tests of the electroweak theory[1] have reached a mature stage since their beginnings more than twenty years ago. We can now successfully combine weak and electromagnetic interactions in a description which also parametrizes CP violation through phases in the Cabibbo-Kobayashi-Maskawa (CKM)[2,3] matrix. The mass quoted recently by the CDF Collaboration[4] for the top quark is one with which this whole structure is quite comfortable. Since this is the first DPF Meeting at which we can celebrate the existence of more than a dozen top quark candidates rather than just one or two, it is appropriate to review the impact of the top quark's observation in the context of a wide range of other phenomena. While the evidence for the top quark could certainly benefit from a factor of four greater statistics, it seems safe to say that the top is here to stay. Looking beyond it for the next aspects of "standard model physics," we shall propose that the study of neutrinos is a key element in this program.

We begin in Section 2 with a brief review of aspects of the top quark, covered more fully in Mel Shochet's plenary talk[5] and in parallel sessions.[6-8] Section 3 is devoted to electroweak physics, while Section 4 describes the present status of information about the CKM matrix. Some aspects of the study of CP violation in B decays are mentioned in Section 5. We devote Section 6 to a brief overview of neutrino masses and Section 7 to an even briefer treatment of electroweak symmetry breaking. Section 8 concludes.

2. The top quark

2.1. Cross section and mass

The CDF Collaboration[4-6] has reported $m_t = 174 \pm 10^{+13}_{-12}$ GeV/c^2. The production cross section $\sigma(\bar{p}p \to t\bar{t} + \ldots) = 13.9^{+6.1}_{-4.8}$ pb at $\sqrt{s} = 1.8$ TeV is on the high side of the QCD prediction (3 to 10 pb, depending on m_t). The D0 Collaboration[7] does not claim evidence for the top, but if its seven candidates (with a background of 3.2 ± 1.1) are ascribed to top, the cross section for a 174 GeV/c^2 top quark is about 7 ± 5 pb. A cross section in excess of QCD predictions could be a signature for new strongly interacting

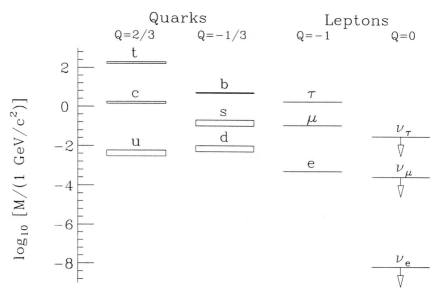

Fig. 1. Masses of quarks and leptons on a logarithmic scale. Widths of bars denote uncertainties in quark masses.

behavior in the electroweak symmetry breaking sector[9,10] or for the production of new quarks.[11] As we shall see, the mass quoted by CDF is just fine to account for loop effects in electroweak processes (through W and Z self-energies) and in giving rise to $B^0 - \overline{B}^0$ and CP-violating $K^0 - \overline{K}^0$ mixing.

2.2. Family structure.

The top quark is the last quark to fit into a set of three families of quarks and leptons, whose masses are shown in Fig. 1:

$$\begin{pmatrix} u \\ d \end{pmatrix} \; ; \; \begin{pmatrix} c \\ s \end{pmatrix} \; ; \; \begin{pmatrix} t \\ b \end{pmatrix} \; ; \tag{1}$$

$$\begin{pmatrix} \nu_e \\ e \end{pmatrix} \; ; \; \begin{pmatrix} \nu_\mu \\ \mu \end{pmatrix} \; ; \; \begin{pmatrix} \nu_\tau \\ \tau \end{pmatrix} \; . \tag{2}$$

Only the ν_τ has not yet been directly observed. If there are any more quarks and leptons, the pattern must change, since the width of the Z implies there are only three light neutrinos.[12]

The question everyone asks, for which we have no answer is: "Why is the top so heavy?" In Section 6 we shall return to this question in another form suggested by

Fig. 1, namely: "Why are the neutrinos so light?" Althought the top quark is by far the heaviest, its separation from the charmed quark (on a logarithmic scale) is no more than the $c - u$ separation. (Amusing exercises on systematics of quark mass ratios have been performed.[13,14]) The fractional errors on the masses of the heavy quarks t, b, c are actually smaller than those on the masses of the light quarks s, d, u.

3.   Electroweak physics

3.1.   Electroweak unification

In contrast to the electromagnetic interaction (involving photon exchange), the four-fermion form of the weak interaction is unsuitable for incorporation into a theory which makes sense to higher orders in perturbation theory. Already in the mid-1930's, Yukawa proposed a particle-exchange model of the weak interactions. At momentum transfers small compared with the mass M_W of the exchanged particle, one identifies

$$\frac{G_F}{\sqrt{2}} = \frac{g^2}{8M_W^2} \quad , \tag{3}$$

where $G_F = 1.11639(2) \times 10^{-5}$ GeV^{-2} is the Fermi coupling, and g is a dimensionless constant.

The simplest version of such a theory[1] predicted not only the existence of a charged $W^\pm$, but also a massive neutral boson Z^0, both of which were discovered in 1983. The exchange of a Z^0 implied the existence of new weak charge-preserving interactions, identified a decade earlier.

The theory involves the gauge group SU(2) × U(1), with respective coupling constants g and g'. Processes involving Z^0 exchange at low momentum transfers can be characterized by a four-fermion interaction with effective coupling

$$\frac{G_F}{\sqrt{2}} = \frac{g^2 + g'^2}{8M_Z^2} \quad . \tag{4}$$

The electric charge is related to g and g' by

$$e = g \sin \theta = g'/cos\theta \quad , \tag{5}$$

where θ is the angle describing the mixtures of the neutral SU(2) boson and U(1) boson in the physical photon and Z^0. These relations can be rearranged to yield

$$M_W^2 = \frac{\pi\alpha}{\sqrt{2}G_F \sin^2 \theta} \quad ; \tag{6}$$

$$M_Z^2 = \frac{\pi\alpha}{\sqrt{2}G_F \sin^2 \theta \cos^2 \theta} \quad . \tag{7}$$

Using the Z mass measured at LEP[12] and a value of the electromagnetic fine structure constant $\alpha(M_Z^2) \simeq 1/128$ evaluated at the appropriate momentum scale, one obtains a value of θ and a consequent prediction for the W mass of about 80 GeV/c^2, which is not too bad. However, one must be careful to define α properly (in one convention it is more like 1/128.9) and to take all vertex and self-energy corrections into account. Crucial contributions are provided by top quarks in W and Z self-energy diagrams.[15] Eq. (4) becomes

$$\frac{G_F}{\sqrt{2}}\hat{\rho} = \frac{g^2 + g'^2}{8M_Z^2} \quad, \tag{8}$$

where

$$\hat{\rho} \simeq 1 + \frac{3G_F m_t^2}{8\pi^2\sqrt{2}} \quad, \tag{9}$$

so that

$$M_Z^2 = \frac{\pi\alpha}{\sqrt{2}G_F\hat{\rho}\sin^2\theta\cos^2\theta} \quad. \tag{10}$$

The angle θ and the mass of the W now acquire implicit dependence on the top quark mass. The quadratic dependence of $\hat{\rho}$ on m_t is a consequence of the chiral nature of the W and Z couplings to quarks; no such dependence occurs in the photon self-energy, which involves purely vector couplings. Small corrections to the right-hand sides of Eqs. (3) and (4), logarithmic in m_t, also arise. We have ignored a QCD correction[16] which replaces m_t^2 by approximately $0.9m_t^2$ in Eq. (9). Taking this into account would increase our quoted m_t values by about 5%.

3.2. The Higgs boson

The electroweak theory requires the existence of something in addition to W's and a Z in order to be self-consistent. For example, W^+W^- scattering would violate probability conservation ("unitarity") at high energy unless a spinless neutral boson H (the "Higgs boson") existed below about 1 TeV.[17] This particle has been searched for in electron-positron collisions with negative results below $M_H = 64$ GeV/c^2.[12]

A Higgs boson contributes to W and Z self-energies and hence to $\hat{\rho}$. We can express the deviation of $\hat{\rho}$ from its value at some nominal top quark and Higgs boson masses $m_t = 175$ GeV/c^2 and $M_H = 300$ GeV/c^2 by means of $\Delta\hat{\rho} = \alpha T$, where

$$T \simeq \frac{3}{16\pi\sin^2\theta}\left[\frac{m_t^2 - (175\text{ GeV})^2}{M_W^2}\right] - \frac{3}{8\pi\cos^2\theta}\ln\frac{M_H}{300\text{ GeV}} \quad. \tag{11}$$

One can also expand $\sin^2\theta$ about its nominal value $x_0 \simeq 0.232$ calculated for the above top and Higgs masses and the Z mass observed at LEP. The angle θ, the W mass, and all other electroweak observables now are functions of both m_t and M_H in the standard model. Additional small corrections to the right-hand sides of (3) and (4) arise which are logarithmic in M_H.

3.3. Electroweak experiments

Direct W mass measurements over the past few years, in GeV/c^2, include 79.92 ± 0.39,[18] 80.35 ± 0.37,[19] 80.37 ± 0.23,[20] 79.86 ± 0.26,[21] with average 80.23 ± 0.18[22]). The ratio $R_\nu \equiv \sigma(\nu N \rightarrow \nu + \ldots)/\sigma(\nu N \rightarrow \mu^- + \ldots)$ depends on $\hat{\rho}$ and $\sin^2\theta$ in such a way that it, too, provides information mainly on M_W. The average of a CCFR Collaboration result presented at this conference[23] and earlier measurements at CERN by the CDHS and CHARM Collaborations[24,25] imply $M_W = 80.27 \pm 0.26$ GeV/c^2.

A number of properties of the Z, as measured at LEP[26] and SLC,[27] are relevant to precise electroweak tests. Global fits to these data have been presented by Steve Olsen at this conference.[12] For our discussion we use the following:

$$M_Z = 91.1888 \pm 0.0044 \text{ GeV}/c^2 \quad , \tag{12}$$

$$\Gamma_Z = 2.4974 \pm 0.0038 \text{ GeV} \quad , \tag{13}$$

$$\sigma_h^0 = 41.49 \pm 0.12 \text{ nb} \quad \text{(hadron production cross section)} \quad , \tag{14}$$

$$R_\ell \equiv \Gamma_{\text{hadrons}}/\Gamma_{\text{leptons}} = 20.795 \pm 0.040 \quad , \tag{15}$$

which may be combined to obtain the Z leptonic width $\Gamma_{\ell\ell}(Z) = 83.96 \pm 0.18$ MeV. Leptonic asymmetries include the forward-backward asymmetry parameter $A_{FB}^\ell = 0.0170 \pm 0.0016$, leading to a value

$$\sin^2 \theta_\ell \equiv \sin^2 \theta_{\text{eff}} = 0.23107 \pm 0.0090 \quad , \tag{16}$$

and independent determinations of $\sin^2 \theta_{\text{eff}} = (1/4)(1 - [g_V^\ell/g_A^\ell])$ from the parameters

$$A_\tau \rightarrow \sin^2 \theta = 0.2320 \pm 0.0013 \quad , \tag{17}$$

$$A_e \rightarrow \sin^2 \theta = 0.2330 \pm 0.0014 \quad . \tag{18}$$

The last three values may be combined to yield

$$\sin^2 \theta = 0.2317 \pm 0.0007 \quad . \tag{19}$$

We do not use values of $\sin^2 \theta$ from forward-backward asymmetries in quark pair production, preferring to discuss them separately. There have been suggestions that the behavior of $Z \rightarrow b\bar{b}$ may be anomalous,[28,29] while the asymmetries in charmed pair production still have little statistical weight and those in light-quark pair production are subject to some model-dependence.

The result of Eq. (19) may be compared with that based on the left-right asymmetry A_{LR} measured with polarized electrons at SLC[27]:

$$\sin^2 \theta = 0.2294 \pm 0.0010 \quad . \tag{20}$$

The results are in conflict with one another at about the level of two standard deviations. This is not a significant discrepancy but we shall use the difference to illustrate the danger of drawing premature conclusions about the impact of electroweak measurements on the Higgs boson sector.

3.4. Dependence of M_W on m_t

We shall illustrate the impact of various electroweak measurements by plotting contours in the M_W vs. m_t plane.[30] A more general language[31] is better for visualizing deviations from the standard model, but space and time limitations prevent its use here. As mentioned, QCD corrections to Eq. (9) are neglected.

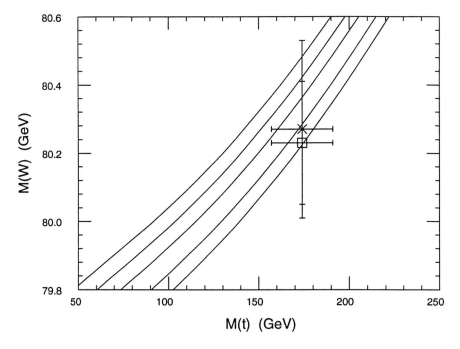

Fig. 2. Dependence of W mass on top quark mass for various values of Higgs boson mass. Curves, from left to right: $M_H = 50, 100, 200, 500, 1000$ GeV/c^2. Horizontal error bars on plotted points correspond to CDF measurement of $m_t = 174 \pm 17$ GeV/c^2. Square: average of direct measurements of W mass; cross: average of determinations based on ratio of neutral-current to charged-current deep inelastic scattering cross sections.

The measurements of M_W via direct observation and via deep inelastic neutrino scattering, together with the CDF top quark mass, are shown as the plotted points in Fig. 2. The results are not yet accurate enough to tell us about the Higgs boson mass, but certainly are consistent with theory. We next ask what information other types of measurements can provide.

The dependence of $\sin^2 \theta_{\text{eff}}$ on m_t and M_H leads to the contours of $\sin^2 \hat{\theta} \approx \sin^2 \theta_{\text{eff}} - 0.0003$ shown in Fig. 3. Here $\sin^2 \hat{\theta}$ is a quantity defined[32] in the $\overline{MS}$ subtraction scheme. Also shown are bands corresponding to the LEP and SLC averages (19) and (20). Taken by itself, the SLC result prefers a high top quark mass. When combined with information on the W mass, however, the main effect of the SLC data is to prefer a lighter Higgs boson mass (indeed, lighter than that already excluded by experiments at LEP).

The observation of parity violation in atomic cesium,[33] together with precise atomic physics calculations,[34] leads to information on the coherent vector coupling of

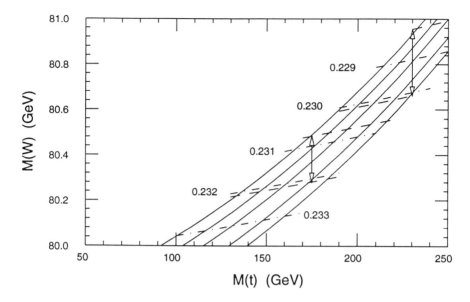

Fig. 3. Dependence of W mass on top quark mass. for various values of Higgs boson mass, together with contours of values of $\sin^2 \hat{\theta} \approx \sin^2 \theta_{\text{eff}} - 0.0003$ predicted by electroweak theory (dot-dashed lines) and measured by LEP (lower region bounded by dashed lines: 1 σ limits) and SLD (upper region).

the Z to the cesium nucleus, encoded in the quantity $Q_W = \rho(Z - N - 4Z \sin^2 \theta)$. Contours of this quantity are shown in Fig. 4. The central value favored by experiment, $Q_W(\text{Cs}) = -71.04 \pm 1.58 \pm 0.88$, lies beyond the upper left-hand corner of the figure, but the present error is large enough to be consistent with predictions. Because of a fortuitous cancellation,[35,36] this quantity is very insensitive to standard-model parameters and very sensitive to effects of new physics (such as exchange of an extra Z boson).

3.5.   Fits to electroweak observables

We now present the results of a fit to the electroweak observables listed in Table 1. The "nominal" values (including[37] $\sin^2 \theta_{\text{eff}} = 0.2320$) are calculated for $m_t = 175$ GeV/c^2 and $M_H = 300$ GeV/c^2. We use $\Gamma_{\ell\ell}(Z)$, even though it is a derived quantity, because it has little correlation with other variables in our fit. It is mainly sensitive to the axial-vector coupling g_A^ℓ, while asymmetries are mainly sensitive to g_V^ℓ. We also omit the total width $\Gamma_{\text{tot}}(Z)$ from the fit, since it is highly correlated with $\Gamma_{\ell\ell}(Z)$ and

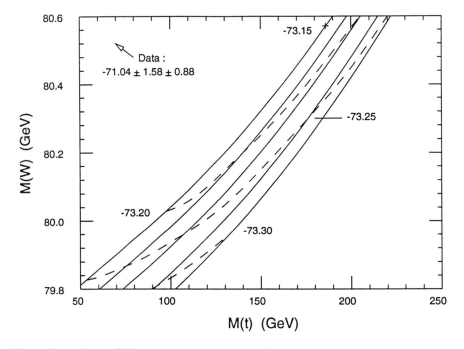

Fig. 4. Dependence of W mass on top quark mass for various values of Higgs boson mass, together with contours of values of weak charge Q_W for cesium as discussed in text.

mainly provides information on the value of the strong fine-structure constant α_s. With $\alpha_s = 0.12 \pm 0.01$, the observed total Z width is consistent with predictions. The partial width $\Gamma(Z \to b\bar{b})$ will be treated separately below.

In addition to the variables in Table 1, we use the constraint $m_t = 174 \pm 17$ GeV/c^2. The results are shown in Fig. 5. To illustrate the impact of the SLD value of $\sin^2 \theta$, we show the effect of omitting it. Conclusions about the Higgs boson mass clearly are premature, especially if they are so sensitive to one input.

3.6.   The decay $Z \to b\bar{b}$

The ratio $R_b \equiv \Gamma(Z \to b\bar{b})/\Gamma(Z \to \text{hadrons})$ has been measured to be slightly above the standard model prediction. In view of the extensive discussion of this process elsewhere at this conference,[12,28,29] we shall be brief.

If one allows R_b and the corresponding quantity for charm, $R_c \equiv \Gamma(Z \to c\bar{c})/\Gamma(Z \to \text{hadrons})$, to be free parameters in a combined fit, the results are[38]

$$R_b = 0.2202 \pm 0.0020 \; ; \quad R_c = 0.1583 \pm 0.0098 \; , \tag{21}$$

to be compared with the standard model predictions $R_b = 0.2156 \pm 0.0006$[39] and

Table 1. Electroweak observables described in fit

Quantity	Experimental value	Nominal value	Experiment/ Nominal
Q_W (Cs)	-71.0 ± 1.8 [a]	-73.2 [b]	0.970 ± 0.025
M_W (GeV/c^2)	80.24 ± 0.15 [c]	80.320 [d]	0.999 ± 0.002
$\Gamma_{\ell\ell}(Z)$ (MeV)	83.96 ± 0.18 [e]	83.90 [f]	1.001 ± 0.002
$\sin^2\theta_{\mathrm{eff}}$	0.2317 ± 0.0007 [f]	0.2320 [g]	0.999 ± 0.003
$\sin^2\theta_{\mathrm{eff}}$	0.2294 ± 0.0010 [h]	0.2320 [g]	0.989 ± 0.004

[a] Weak charge in cesium[33]
[b] Calculation[36] incorporating atomic physics corrections[34]
[c] Average of direct measurements[22] and indirect information
   from neutral/charged current ratio in deep inelastic neutrino scattering[23-25]
[d] Including perturbative QCD corrections[37]
[e] LEP average as of July, 1994[26]
[f] From asymmetries at LEP[26]
[g] As calculated[37] with correction for relation between $\sin^2\theta_{\mathrm{eff}}$ and $\sin^2\hat{\theta}$[32]
[h] From left-right asymmetry in annihilations at SLC[27]

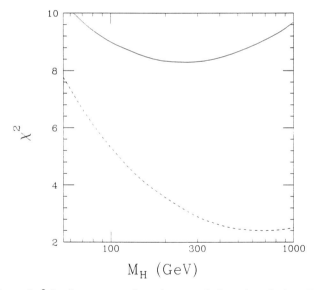

Fig. 5. Values of χ^2 for fits to m_t and to electroweak data described in Table. Solid curve: full data set (5 d. o. f.); dashed curve: without SLD data (4 d. o. f.).

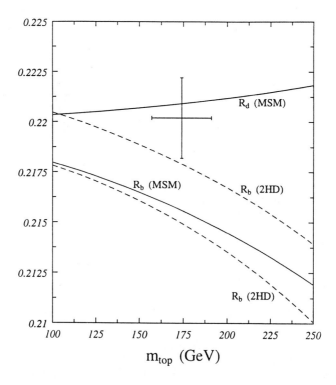

Fig. 6. Dependence of $R_b \equiv \Gamma(Z \to b\bar{b})/\Gamma(Z \to \text{hadrons})$ on top quark mass. Solid curves: predictions of Minimal Standard Model (MSM) for R_b and $R_d \equiv \Gamma(Z \to d\bar{d})/\Gamma(Z \to \text{hadrons})$. Dashed curves: two-Higgs models described in text with $\tan\beta = 70$ (upper) and 1 (lower). Data point: recent LEP and CDF measurements of R_b and m_{top}.

$R_c \approx 0.171$.[38] If one constrains R_c to the standard model prediction, one finds instead $R_b = 0.2192 \pm 0.0018$. The discrepancy is at a level of about 2σ.

Predictions for R_b in the standard model and in two different two-Higgs-doublet models[39] are shown in Fig. 6. With appropriate choices of masses for neutral and charged Higgs bosons, it is possible to reduce the discrepancy between theory and experiment without violating other constraints on the Higgs sector.

A curious item was reported[40] in one of the parallel sessions of this conference. The forward-backward asymmetries in heavy-quark production, $A_{FB}^{0,b}$ and $A_{FB}^{0,c}$, have been measured both on the Z peak and 2 GeV above and below it. All quantities are in accord with standard model expectations except for $A_{FB}^{0,c}$ at $M_Z - 2$ GeV. Off-peak asymmetries can be a hint of extra Z's.[41]

4.   The CKM Matrix

4.1.   Definitions and magnitudes

The CKM matrix for three families of quarks and leptons will have four independent parameters no matter how it is represented. In a parametrization[42] in which the rows of the CKM matrix are labelled by u, c, t and the columns by d, s, b, we may write

$$
V = \begin{pmatrix} V_{ud} & V_{us} & V_{ub} \\ V_{cd} & V_{cs} & V_{cb} \\ V_{td} & V_{ts} & V_{tb} \end{pmatrix} \approx \begin{bmatrix} 1 - \lambda^2/2 & \lambda & A\lambda^3(\rho - i\eta) \\ -\lambda & 1 - \lambda^2/2 & A\lambda^2 \\ A\lambda^3(1 - \rho - i\eta) & -A\lambda^2 & 1 \end{bmatrix} . \tag{22}
$$

Note the phases in the elements V_{ub} and V_{td}. These phases allow the standard $V - A$ interaction to generate CP violation as a higher-order weak effect.

The four parameters are measured as follows:

1. The parameter λ is measured by a comparison of strange particle decays with muon decay and nuclear beta decay, leading to $\lambda \approx \sin\theta \approx 0.22$, where θ is the Cabibbo[2] angle.

2. The dominant decays of b-flavored hadrons occur via the element $V_{cb} = A\lambda^2$. The lifetimes of these hadrons and their semileptonic branching ratios then lead to an estimate $A = 0.79 \pm 0.06$.

3. The decays of b-flavored hadrons to charmless final states allow one to measure the magnitude of the element V_{ub} and thus to conclude that $\sqrt{\rho^2 + \eta^2} = 0.36 \pm 0.09$.

4. The least certain quantity is the phase of V_{ub}: $\mathrm{Arg}\,(V_{ub}^*) = \arctan(\eta/\rho)$. We shall mention ways in which information on this quantity may be improved, in part by indirect information associated with contributions of higher-order diagrams involving the top quark.

The unitarity of V and the fact that V_{ud} and V_{tb} are very close to 1 allows us to write $V_{ub}^* + V_{td} \simeq A\lambda^3$, or, dividing by a common factor of $A\lambda^3$,

$$
\rho + i\eta \; + \; (1 - \rho - i\eta) = 1 \quad . \tag{23}
$$

The point (ρ, η) thus describes in the complex plane one vertex of a triangle whose other two vertices are $(0, 0)$ and $(0, 1)$.

4.2.   Indirect information

Box diagrams involving the quarks with charge 2/3 are responsible for $B^0 - \overline{B}^0$ and CP-violating $K^0 - \overline{K}^0$ mixing in the standard model. Since the top quark provides the dominant contribution, one obtains mainly information on the phase and magnitude of V_{td}.

The evidence for $B^0 - \overline{B}^0$ mixing comes from "wrong-sign" leptons in B meson semileptonic decays and from direct observation of time-dependent oscillations.[43] The

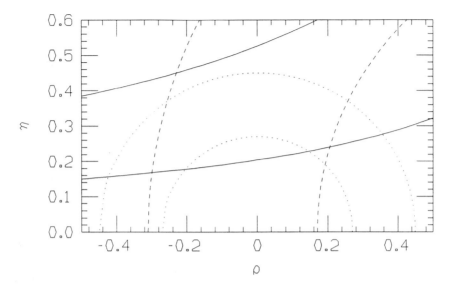

Fig. 7. Region in the (ρ, η) plane allowed by various constraints. Dotted semicircles denote central value and $\pm 1\sigma$ limits implied by $|V_{ub}/V_{cb}| = 0.08 \pm 0.02$. Circular arcs with centers at $(\rho, \eta) = (1, 0)$ denote constraints from $B - \overline{B}$ mixing, while hyperbolae describe region bounded by constraints from CP-violating $K - \overline{K}$ mixing.

splitting Δm between mass eigenstates is proportional to $f_B^2 |V_{td}|^2$ times a function of m_t which can now be considered reasonably well-known. Here f_B is the B meson decay constant, analogous to the pion decay constant $f_\pi = 132$ MeV. Given a range of f_B and the experimental average for B mesons of $\Delta m/\Gamma = 0.71 \pm 0.07$, we can then specify a range of $|V_{td}|$, which is proportional to $|1 - \rho - i\eta|$. We then obtain a band in the (ρ, η) plane bounded by two circles with center $(1,0)$.

The parameter ϵ characterizing CP-violating $K^0 - \overline{K}^0$ mixing arises from an imaginary part in the mass matrix which is dominated by top quark contributions in the loop, with small corrections from charm. In the limit of complete top dominance one would have Im $\mathcal{M} \sim f_K^2$ Im$(V_{td}^2) \sim \eta(1 - \rho)$, so that $\epsilon = (2.26 \pm 0.02) \times 10^{-3}$ would specify a hyperbola in the (ρ, η) plane with focus $(1,0)$. The effect of charm is to shift the focus to about $(1.4, 0)$.

4.3. Constraints on ρ and η

When one combines the indirect information from mixing with the constraint on $(\rho^2 + \eta^2)^{1/2}$ arising from $|V_{ub}/V_{cb}|$, one obtains the allowed region shown in Fig. 7. Here, in addition to parameters mentioned earlier, we have taken $|V_{cb}| = 0.038 \pm 0.003$, the vacuum-saturation factor $B_K = 0.8 \pm 0.2$, and $\eta_B B_B = 0.6 \pm 0.1$, where η_B refers to a

QCD correction. Standard QCD correction factors are taken in the kaon system.[44] We have also assumed $f_B = 180 \pm 30$ MeV, for reasons to be described presently.

The center of the allowed region is near $(\rho, \eta) = (0, 0.35)$, with values of ρ between -0.3 and 0.3 and values of η between 0.2 and 0.45 permitted at the 1σ level. The main error on the constraint from $(\Delta m/\Gamma)_B$ arises from uncertainty in f_B, while the main error on the hyperbolae associated with ϵ comes from uncertainty in the parameter A, which was derived from V_{cb}. Other sources of error have been tabulated by Stone at this conference.[45]

4.4. Improved tests

We can look forward to a number of sources of improved information about CKM matrix elements .[46]

4.4.1 Decay constant information on f_B affects the determination of $|V_{td}|$ (and hence ρ) via $B^0 - \overline{B}^0$ mixing. Lattice gauge theories have become more bold in predicting heavy meson decay constants. For example, one recent calculation obtains the values[47]

$$f_B = 187 \pm 10 \pm 34 \pm 15 \ \text{MeV} \quad ,$$

$$f_{B_s} = 207 \pm 9 \pm 34 \pm 22 \ \text{MeV} \quad ,$$

$$f_D = 208 \pm 9 \pm 35 \pm 12 \ \text{MeV} \quad ,$$

$$f_{D_s} = 230 \pm 7 \pm 30 \pm 18 \ \text{MeV} \quad , \tag{24}$$

where the first errors are statistical, the second are associated with fitting and lattice constant, and the third arise from scaling from the static $(m_Q = \infty)$ limit. An independent lattice calculation[48] finds a similar value of f_B. The spread between these and some other lattice estimates[49] is larger than the errors quoted above, however.

Direct measurements are available for the D_s decay constant. The WA75 collaboration[50] has seen $6 - 7$ $D_s \to \mu\nu$ events and conclude that $f_{D_s} = 232 \pm 69$ MeV. The CLEO Collaboration[51] has a much larger statistical sample; the main errors arise from background subtraction and overall normalization (which relies on the $D_s \to \phi\pi$ branching ratio). Using several methods to estimate this branching ratio, Muheim and Stone[52] estimate $f_{D_s} = 315 \pm 45$ MeV. We average this with the WA75 value to obtain $f_{D_s} = 289 \pm 38$ MeV. A recent value from the BES Collaboration,[53] $f_{D_s} = 434 \pm 160$ MeV (based on one candidate for $D_s \to \mu\nu$ and two for $D_s \to \tau\nu$), and a reanalysis by F. Muheim[54] using the factorization hypothesis,[52] $f_{D_s} = 310 \pm 37$ MeV, should be incorporated in subsequent averages.

Quark models can provide estimates of decay constants and their ratios. In a non-relativistic model,[55] the decay constant f_M of a heavy meson $M = Q\bar{q}$ with mass M_M is related to the square of the $Q\bar{q}$ wave function at the origin by $f_M^2 = 12|\Psi(0)|^2/M_M$. The ratios of squares of wave functions can be estimated from strong hyperfine splittings between vector and pseudoscalar states, $\Delta M_{\text{hfs}} \propto |\Psi(0)|^2/m_Q m_q$. The equality of the $D_s^* - D_s$ and $D^* - D$ splittings then suggests that

$$f_D/f_{D_s} \simeq (m_d/m_s)^{1/2} \simeq 0.8 \simeq f_B/f_{B_s} \quad , \tag{25}$$

where we have assumed that similar dynamics govern the light quarks bound to charmed and b quarks. Using our average for f_{D_s}, we find $f_D = (231 \pm 31)$ MeV. One hopes that the Beijing Electron Synchrotron will be able to find the decay $D \to \mu\nu$ via extended running at the $\Psi(3770)$ resonance, which was the method employed by the Mark III Collaboration to obtain the upper limit[56] $f_D < 290$ MeV (90% c.l.).

An absolute estimate of $|\Psi(0)|^2$ can been obtained using electromagnetic hyperfine splittings,[57] which are probed by comparing isospin splittings in vector and pseudoscalar mesons. On this basis[44] we estimate $f_B = (180 \pm 12)$ MeV. [This is the basis of the value taken above, where we inflated the error arbitrarily.] We also obtain $f_{B_s} = (225 \pm 15)$ MeV from the ratio based on the quark model.

4.4.2 Rates and ratios can constrain $|V_{ub}|$ and possibly $|V_{td}|$. The partial width $\Gamma(B \to \ell\nu)$ is proportional to $f_B^2|V_{ub}|^2$. The expected branching ratios are about $(1/2) \times 10^{-4}$ for $\tau\nu$ and 2×10^{-7} for $\mu\nu$. Another interesting ratio[58] is $\Gamma(B \to \rho\gamma)/\Gamma(B \to K^*\gamma)$, which, aside from phase space corrections, should be $|V_{td}/V_{ts}|^2 \simeq 1/20$. At this conference, however, Soni[59] has argued that there are likely to be long-distance corrections to this relation.

4.4.3 The $K^+ \to \pi^+\nu\bar\nu$ rate is governed by loop diagrams involving the cooperation of charmed and top quark contributions, and lead to constraints which involve circles in the (ρ, η) plane with centers at approximately $(1.4, 0)$.[60] The favored branching ratio is slightly above 10^{-10}, give or take a factor of 2. A low value within this range signifies $\rho > 0$, while a high value signifies $\rho < 0$. The present upper limit[60] is $B(K^+ \to \pi^+\nu\bar\nu) < 3 \times 10^{-9}$ (90% c.l.).

4.4.4 The decays $K_L \to \pi^0 e^+ e^-$ and $K_L \to \pi^0 \mu^+\mu^-$ are expected to be dominated by CP-violating contributions. Two types of CP-violating contributions are expected: "indirect," via the CP-positive component K_1 component of $K_L = K_1 + \epsilon K_2$, and "direct," whose presence would be a detailed verification of the CKM theory of CP violation. These are expected to be of comparable magnitude in most[61,62] but not all[63] calculations, leading to overall branching ratios of order 10^{-11}. The "direct" CP-violating contribution to $K_L \to \pi^0\nu\bar\nu$ is expected to be dominant, making this process an experimentally challenging but theoretically clean source of information on the parameter η.[61]

4.4.5 The ratio ϵ'/ϵ for kaons has long been viewed as one of the most promising ways to disprove a "superweak" theory of CP violation in neutral kaon decays.[61,64] The latest estimates[65] are equivalent (for a top mass of about 170 GeV/c^2) to $[\epsilon'/\epsilon]|_{\text{kaons}} = (6 \pm 3) \times 10^{-4}\eta$, with an additional factor of 2 uncertainty associated with hadronic matrix elements. The Fermilab E731 Collaboration[66] measures $\epsilon'/\epsilon = (7.4 \pm 6) \times 10^{-4}$, consistent with η in the range (0.2 to 0.45) we have already specified. The CERN NA31 Collaboration[67] finds $\epsilon'/\epsilon = (23.0 \pm 6.5) \times 10^{-4}$, which is higher than theoretical expectations. Both groups are preparing new experiments, for which results should be available around 1996.

4.4.6 $B_s - \bar{B}_s$ mixing can probe the ratio $(\Delta m)|_{B_s}/(\Delta m)|_{B_d} = (f_{B_s}/f_{B_d})^2(B_{B_s}/B_{B_d})$ $|V_{ts}/V_{td}|^2$, which should be a very large number (of order 20 or more). Thus, strange B's should undergo many particle-antiparticle oscillations before decaying.

The main uncertainty in an estimate of $x_s \equiv (\Delta m/\Gamma)_{B_s}$ is associated with f_{B_s}. The CKM elements $V_{ts} \simeq -0.04$ and $V_{tb} \simeq 1$ which govern the dominant (top quark)

Table 2. Dependence of mixing parameter x_s on top quark mass and B_s decay constant.

m_t (GeV/c^2)	157	174	191
f_{B_s} (MeV)			
150	7.6	8.9	10.2
200	13.5	15.8	18.2
250	21.1	24.7	28.4

contribution to the mixing are known reasonably well. We show in Table 2 the dependence of x_s on f_{B_s} and m_t. To measure x_s, one must study the time-dependence of decays to specific final states and their charge-conjugates with resolution equal to a small fraction of the B_s lifetime (about 1.5 ps).

The question has been raised: "Can one tell whether $\eta \neq 0$ from $B_s - \overline{B}_s$ mixing?" The ratio of decay constants for strange and nonstrange B mesons is expected to be $f_{B_s}/f_{B_d} \approx 1.19 \pm 0.1$,[44,68] so that

$$\frac{\Delta m_s}{\Delta m_d} = (1.19 \pm 0.10)^2 \left| \frac{V_{ts}}{V_{td}} \right|^2 \quad . \tag{26}$$

At this meeting ALEPH has presented a result[68,69] $x_s = \Delta m_s / \Gamma_s > 9$, leading to a bound $|1 - \rho - i\eta| < 1.7$. However, in order to show that the unitarity triangle has nonzero area, assuming that $|V_{ub}/V_{cb}| > 0.06$, one must show $0.73 < |1 - \rho - i\eta| < 1.27$. With the above expression, taking the B_s and B_d lifetimes to be equal, and assuming $0.64 < x_d < 0.78$, this will be so if $17 < x_s < 29$. An "ideal" measurement would thus be $x_s = 23 \pm 2$.

5.   CP violation and B decays

5.1.   Types of experiments

Soon after the discovery of the Υ states it was realized that CP-violating phenomena in decays of B mesons were expected to be observable and informative.[70,71]

5.1.1 *Decays to CP non-eigenstates* can exhibit rate asymmetries only if there are two different weak decay amplitudes and two different strong phase shifts associated with them. The weak phases change sign under charge conjugation, while the strong phases do not. Thus, the rates for $B^+ \to K^+\pi^0$ and $B^- \to K^-\pi^0$ can differ only if the strong phases differ in the $I = 1/2$ and $I = 3/2$ channels, and interpretation of a rate asymmetry in terms of weak phases requires knowing the difference of strong phases. We shall mention in Sec. 5.3 the results of a recent SU(3) analysis[72] which permits the separation of weak and strong phase shift information without the necessary observation of a CP-violating decay rate asymmetry.

5.1.2 *Decays of neutral B mesons to CP eigenstates f* can exhibit rate asymmetries (or time-dependent asymmetries) as a result of the interference of the direct process $B^0 \to f$ and the two-step process $B^0 \to \bar{B}^0 \to f$ involving mixing.

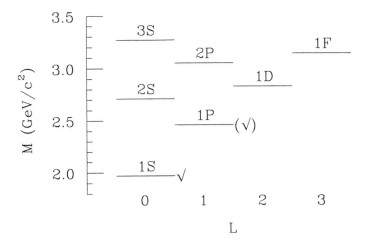

Fig. 8. P-wave nonstrange resonances of a c quark and a light ($\bar{u}$ or $\bar{d}$) antiquark. Check marks with or without parentheses denote observation of some or all predicted states.

Here one does not have to know the strong phase shifts. Decay rate asymmetries directly proble angles of the unitarity triangle. One very promising comparison involves the decays $B^0 \to J/\psi K_S$ and $\overline{B}^0 \to J/\psi K_S$, whose rate asymmetry is sensitive to $\sin[\mathrm{Arg}(V_{td}^2)] \equiv \sin(2\beta)$. It is necessary to know whether the decaying neutral B meson was a B^0 or a $\overline{B}^0$ at some reference time $t = 0$. We now remark briefly on one method[73] for tagging such B^0 mesons using associated pions.

5.2. $\pi - B$ correlations

The correlation of a neutral B meson with a charged pion is easily visualized with the help of quark diagrams. By convention (the same as for kaons), a neutral B meson containing an initially produced $\bar{b}$ is a B^0. It also contains a d quark. The next charged pion down the fragmentation chain must contain a $\bar{d}$, and hence must be a π^+. Similarly, a $\overline{B}^0$ will be correlated with a π^-.

The same conclusion can be drawn by noting that a B^0 can resonate with a positive pion to form an excited B^+, which we shall call B^{**+} (to distinguish it from the B^*, lying less than 50 MeV/c^2 above the B). Similarly, a $\overline{B}^0$ can resonate with a negative pion to form a B^{**-}. The combinations $B^0 \pi^-$ and $\overline{B}^0 \pi^+$ are *exotic*, i.e., they cannot be formed as quark-antiquark states. No evidence for exotic resonances exists. Resonant behavior in the $\pi - B^{(*)}$ system, if discovered, would be very helpful in reducing the combinatorial backgrounds associated with this method.

The lightest states which can decay to $B\pi$ and/or $B^*\pi$ are P-wave resonances of a b quark and a $\bar{u}$ or $\bar{d}$. The expectations for masses of these states may be based on extrapolation from the known D^{**} resonances, for which present data[74] and predictions[75] are summarized in Fig. 8.

The 1S (singlet and triplet) charmed mesons have all been observed, while CLEO[74] has presented at this conference evidence for all six (nonstrange and strange) 1P states in which the light quarks' spins combine with the orbital angular momentum to form a total light-quark angular momentum $j = 3/2$. These states have $J = 1$ and $J = 2$. They are expected to be narrow in the limit of heavy quark symmetry. The strange 1P states are about 110 MeV heavier than the nonstrange ones. In addition, there are expected to be much broader (and probably lower) $j = 1/2$ D^{**} resonances with $J = 0$ and $J = 1$.

For the corresponding B^{**} states, one should add about 3.32 GeV (the difference between b and c quark masses minus a small correction for binding). One then predicts[75] nonstrange B^{**} states with $J = (1, 2)$ at $(5755, 5767)$ MeV. It is surprising that so much progess has been made in identifying D^{**}'s without a corresponding glimmer of hope for the B^{**}'s, especially since we know where to look.

5.3. Decays to pairs of light pseudoscalars

The decays $B \to (\pi\pi, \pi K, K\bar{K})$ are a rich source of information on both weak (CKM) and strong phases, if we are willing to use flavor SU(3) symmetry.

The decays $B \to \pi\pi$ are governed by transitions $b \to dq\bar{q}$ $(q = u, d, \ldots)$ with $\Delta I = 1/2$ and $\Delta I = 3/2$, leading respectively to final states with $I = 0$ and $I = 2$. Since there is a single amplitude for each final isospin but three different charge states in the decays, the amplitudes obey a triangle relation: $A(\pi^+\pi^-) - \sqrt{2}A(\pi^0\pi^0) = \sqrt{2}A(\pi^+\pi^0)$. The triangle may be compared with that for the charge-conjugate processes and combined with information on time-dependent $B \to \pi^+\pi^-$ decays to obtain information on weak phases.[76]

The decays $B \to \pi K$ are governed by transitions $b \to sq\bar{q}$ $(q = u, d, \ldots)$ with $\Delta I = 0$ and $\Delta I = 1$. The $I = 1/2$ final state can be reached by both $\Delta I = 0$ and $\Delta I = 1$ transitions, while only $\Delta I = 1$ contributes to the $I = 3/2$ final state. Consequently, there are three independent amplitudes for four decays, and one quadrangle relation $A(\pi^+K^0) + \sqrt{2}A(\pi^0K^+) = A(\pi^-K^+) + \sqrt{2}A(\pi^0K^0)$. As in the $\pi\pi$ case, this relation may be compared with the charge-conjugate one and the time-dependence of decays to CP eigenstates (in this case π^0K_S) studied to obtain CKM phase information.[77]

We re-examined[72] SU(3) analyses[78] of the decays $B \to PP$ (P = light pseudoscalar). They imply a number of useful relations among $\pi\pi$, πK, and $K\bar{K}$ decays, among which is one relating B^+ amplitudes alone:

$$A(\pi^+K^0) + \sqrt{2}A(\pi^0K^+) = \tilde{r}_u\sqrt{2}A(\pi^+\pi^0) \quad . \tag{27}$$

Here $\tilde{r}_u \equiv (f_K/f_\pi)|V_{us}/V_{ud}|$. This expression relates one side of the $\pi\pi$ amplitude triangle to one of the diagonals of the πK amplitude quadrangle, and thus reduces the quadrangle effectively to two triangles, simplifying previous analyses.[77] Moreover, since one expects the π^+K^0 amplitude to be dominated by a penguin diagram (with expected weak phase π) and the $\pi^+\pi^0$ amplitude to have the phase $\gamma = $ Arg V_{ub}^*, the comparison of this last relation and the corresponding one for charge-conjugate decays can provide information on the weak phase γ. We have estimated[44] that in order to measure γ to $10°$ one needs a sample including about 100 events in the channels $\pi^0 K^\pm$.

Further relations can be obtained[72] by comparing the amplitude triangles involving both charged and neutral B decays to πK. By looking at the amplitude triangles for these decays and their charge conjugates, one can sort out a number of weak and strong phases.

Some combination of the decays $B^0 \to \pi^+\pi^-$ and $B^0 \to \pi^- K^+$ has already been observed,[80] and updated analyses in these and other channels have been presented at this conference.[81]

6. Neutrino masses and new mass scales

6.1. Expected ranges of parameters

Referring back to Fig. 1 in which quark and lepton masses were displayed, we see that the neutrino masses are at least as anomalous as the top quark mass. There are suggestions that the known (direct) upper limits are far above the actual masses, enhancing the puzzle. Why are the neutrinos so light?

A possible answer[82] is that light neutrinos acquire Majorana masses of order $m_M = m_D^2/M_M$, where m_D is a typical Dirac mass and M_M is a large Majorana mass acquired by right-handed neutrinos. One explanation[83] of the apparent deficit of solar neutrinos as observed in various terrestrial experiments invokes matter-induced $\nu_e \to \nu_\mu$ oscillations in the Sun[84] with a muon neutrino mass of a few times 10^{-3} eV. With a Dirac mass of about 0.1 to 1 GeV characterizing the second quark and lepton family, this would correspond to a right-handed Majorana mass $M_M = 10^9 - 10^{12}$ GeV. As stressed by Georgi in his summary talk,[85] nobody really knows what Dirac mass to use for such a calculation, which only enhances the value of experimental information on neutrino masses. However, using the above estimate, and taking a Dirac mass for the third neutrino characteristic of the third quark and lepton family (in the range of 2 to 200 GeV), one is led by the ratios in Fig. 1 to expect the ν_τ to be at least a couple of hundred times as heavy as the ν_μ, and hence to be heavier than 1 eV or so. This begins to be a mass which the cosmologists could use to explain at least part of the missing matter in the Universe.[86]

If $\nu_\mu \leftrightarrow \nu_\tau$ mixing is related to ratios of masses, one might expect the mixing angle to be at least m_μ/m_τ, and hence $\sin^2 2\theta$ to exceed 10^{-2}.

6.2. Present limits and hints

Some limits on neutrino masses and mixings have been summarized at a recent Snowmass workshop.[87] The E531 Collaboration[88] has set limits for $\nu_\mu \to \nu_\tau$ oscillations corresponding to $\Delta m^2 < 1$ eV^2 for large θ and $\sin^2 2\theta < $ (a few) $\times 10^{-3}$ for large Δm^2. The recent measurement of the zenith-angle dependence of the apparent deficit in the ratio of atmospheric ν_μ to ν_e – induced events in the Kamioka detector[89,90] can be interpreted in terms of neutrino oscillations (either $\nu_\mu \to \nu_e$ or $\nu_\mu \to \nu_\tau$), with Δm^2 of order 10^{-2} eV^2. In either case maximal mixing, with $\theta = 45°$, is the most highly favored. We know of at least one other case (the neutral kaon system) where (nearly) maximal mixing occurs; perhaps this will serve as a hint to the pattern not only of neutrino masses but other fermion masses as well. However, it is not possible to fit the Kamioka atmospheric neutrino effect, the apparent solar-neutrino deficit, and a cosmologically significant ν_τ using naive guesses for Dirac masses and a single see-saw scale. Various schemes have been proposed involving near-degeneracies of two or more neutrinos or employing multiple see-saw scales.

6.3. Present and proposed experiments

Opportunities exist and are starting to be realized for filling in a substantial portion of the parameter space for neutrino oscillations. New short-baseline experiments are already in progress at CERN[91,92] and approved at Fermilab.[93] These are capable of pushing the $\nu_\mu \leftrightarrow \nu_\tau$ mixing limits lower for mass differences Δm^2 of at least 1 eV^2. New long-baseline experiments[94] would be sensitive in the same mass range as the Kamioka result to smaller mixing angles. At this conference we have heard a preliminary result from a search for $\bar{\nu}_\mu \rightarrow \bar{\nu}_e$ oscillations using $\bar{\nu}_\mu$ produced in muon decays.[95] An excess of events is seen which, if interpreted in terms of oscillations, would correspond to Δm^2 of several eV^2. (No evidence for oscillations was claimed.) A further look at the solar neutrino problem will be provided by the Sudbury Neutrino Observatory.[96]

We will not understand the pattern of fermion masses until we understand what is going on with the neutrinos. Fortunately this area stands to benefit from much experimental effort in the next few years.

6.4. Electroweak-strong unification

Another potential window on an intermediate mass scale is provided by the pattern of electroweak-strong unification. If the strong and electroweak coupling constants are evolved to high mass scales in accord with the predictions of the renormalization group,[97] as shown in Fig. 9(a), they approach one another in the simplest SU(5) model,[98] but do not really cross at the same point. This "astigmatism" can be cured by invoking supersymmetry,[99] as illustrated in Fig. 9(b). Here the cure is effected not just by the contributions of superpartners, but by the richer Higgs structure in supersymmetric theories. The theory predicts many superpartners below the TeV mass scale, some of which ought to be observable in the next few years.

Alternatively, one can embed SU(5) in an SO(10) model,[100] in which each family of quarks and leptons (together with a right-handed neutrino for each family) fits into a 16-dimensional spinor representation. Fig. 9(c) illustrates one scenario for breaking of SO(10) at two different scales, the lower of which is a comfortable scale for the breaking of left-right symmetry and the generation of right-handed neutrino Majorana masses.

6.5. Baryogenesis

The ratio of baryons to photons in our Universe is a few parts in 10^9. In 1967 Sakharov[101] proposed three ingredients of any theory which sought to explain the preponderance of baryons over antibaryons in our Universe: (1) violation of C and CP; (2) violation of baryon number, and (3) a period in which the Universe was out of thermal equilibrium. Thus our very existence may owe itself to CP violation. However, no consensus exists on a specific implementation of Sakharov's suggestion.

A toy model illustrating Sakharov's idea can be constructed within an SU(5) grand unified theory. The gauge group SU(5) contains "X" bosons which can decay

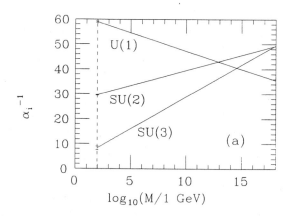

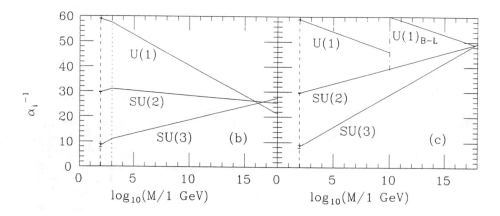

Fig. 9. Behavior of coupling constants predicted by the renormalization group in various grand unified theories. Error bars in plotted points denote uncertainties in coupling constants measured at $M = M_Z$ (dashed vertical line). (a) SU(5); (b) supersymmetric SU(5) with superpartners above 1 TeV (dotted line) (c) example of an SO(10) model with an intermediate mass scale (dot-dashed vertical line).

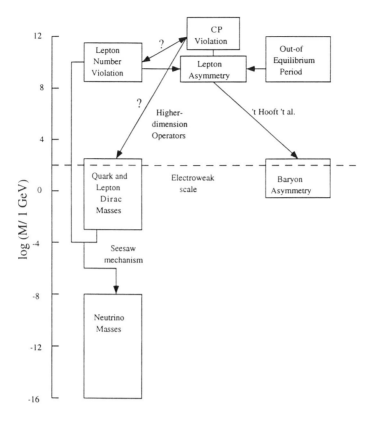

Fig. 10. Mass scales associated with one scenario for baryogenesis.

both to uu and to $e^+\bar{d}$. By CPT, the total decay rates of X and $\bar{X}$ must be equal, but CP-violating rate differences $\Gamma(X \to uu) \neq \Gamma(\bar{X} \to \bar{u}\bar{u})$ and $\Gamma(X \to e^+\bar{d}) \neq \Gamma(\bar{X} \to e^-d)$ are permitted. This example conserves $B - L$, where B is baryon number (1/3 for quarks) and L is lepton number (1 for electrons).

It was pointed out by 't Hooft[102] that the electroweak theory contains an anomaly as a result of nonperturbative effects which conserve $B - L$ but violate $B + L$. If a theory leads to $B - L = 0$ but $B + L \neq 0$ at some primordial temperature T, the anomaly can wipe out any $B + L$ as T sinks below the electroweak scale.[103] Thus, the toy model mentioned above and many others are unsuitable in practice.

One proposed solution is the generation of nonzero $B - L$ at a high temperature, e.g., through the generation of nonzero lepton number L, which is then reprocessed into nonzero baryon number by the 't Hooft anomaly mechanism.[104] We illustrate in Fig. 10 some aspects of the second scenario. The existence of a baryon asymmetry, when combined with information on neutrinos, could provide a window to a new scale of particle physics.

Large Majorana masses acquired by right-handed neutrinos would change lepton number by two units and thus would be ideal for generating a lepton asymmetry if Sakharov's other two conditions are met.

The question of baryogenesis is thus shifted onto the leptons: Do neutrinos indeed have masses? If so, what is their "CKM matrix"? Do the properties of heavy Majorana right-handed neutrinos allow any new and interesting natural mechanisms for violating CP at the same scale where lepton number is violated? Majorana masses for right-handed neutrinos naturally violate left-right symmetry and could be closely connected with the violation of P and C in the weak interactions.[105]

An open question in this scenario, besides the precise form of CP violation at the lepton-number-violating scale, is how this CP violation gets communicated to the lower mass scale at which we see CKM phases. Presumably this occurs through higher-dimension operators which imitate the effect of Higgs boson couplings to quarks and leptons.

7. Electroweak symmetry breaking

A key question facing the standard model of electroweak interactions is the mechanism for breaking $SU(2) \times U(1)$. We discuss two popular alternatives; Nature may turn out to be cleverer than either.

7.1. Fundamental Higgs boson(s)

If there really exists a relatively light fundamental Higgs boson in the context of a grand unified theory, one has to protect its mass from large corrections. Supersymmetry is the popular means for doing so. Then one expects a richer neutral Higgs structure, charged scalar bosons, and superpartners, all below about 1 TeV.

7.2. Strongly interacting Higgs sector

The scattering of longitudinally polarized W and Z bosons becomes strong and violates unitarity above a TeV or two if there does not exist a Higgs boson below this energy.[16] The behavior is similar to what one might expect for pion-pion scattering in the non-linear sigma model above a few hundred MeV. We wouldn't trust such a model above that energy, and perhaps we should not trust the present version of electroweak theory above a TeV. If the theory really has a strongly interacting sector, its $I = J = 0$ boson (like the σ of QCD) may be its least interesting and most elusive feature. Consider, for example, the rich spectrum of resonances in QCD, which we now understand in terms of the interactions of quarks and gluons. Such rich physics in electroweak theory was a prime motivation for the construction of the SSC, and we wish our European colleagues well in their exploration of this energy region via the LHC. [I am also indebted to T. Barklow[106] for reminding me of the merits of TeV e^+e^- colliders in this regard.]

8. Summary

It appears that the top quark, reported by the CDF Collaboration at this meeting, is here to stay. We look forward to its confirmation by the D0 Collaboration and to more precise measurements of its mass and decay properties. Even now, its reported properties are in comfortable accord with standard model expectations based on electroweak physics and mixing effects.

Tests of the electroweak theory continue to achieve greater and greater precision, with occasional excursions into the land of two- and three-standard deviation discrepancies which stimulate our theoretical inventiveness but may be no more than the expected statistical fluctuations. These effects include a low value of $\sin^2 \theta$ from SLD, a high value of R_b from the LEP experiments, and an anomalous forward-backward asymmetry in charmed quark pair production at an energy 2 GeV below the Z.

The Cabibbo-Kobayashi-Maskawa matrix provides an adequate framework for explaining the observed CP violation, which is still confined to a single parameter (ϵ) in the neutral kaon system. We have no deep understanding of the origin of the magnitudes or phases in the CKM matrix, any more than we understand the pattern of quark and lepton masses. Nonetheless, there are many possibilities for testing the present picture, a number of which involve rare kaon and B meson decays.

Numerous opportunities exist for studying CP violation in B decays, and facilities are under construction for doing so. In view of the widespread attention given recently to asymmetric B factories, I have menioned a couple of alternatives which can be pursued at hadron machines and/or symmetric electron-positron colliders.

Neutrino masses may provide us with our next "standard" physics. I have suggested that the mass scale of $10^9 - 10^{12}$ GeV is ripe for exploration not only through the measurement of mass differences in the eV- and sub-eV range, but also through studies of leptogenesis and partial unification of gauge couplings. Searches for axions, which I did not mention, also can shed some light on this mass window.

Alternatives for electroweak symmetry breaking, each with consequences for TeV-scale physics, include fundamental Higgs bosons with masses protected by supersymmetry, a strongly interacting Higgs sector, some new physics which we may have thought of but not learned how to make tractable (like compositeness of Higgs bosons, quarks, and leptons), or even something we have not thought of at all. We will really have solved the problem only when we understand the bewildering question of fermion masses, a signal that while the "standard" model may work very well, it is far from complete.

9. Acknowledgements

I am grateful to Jim Amundson, Aaron Grant, Michael Gronau, Oscar Hernández, Nahmin Horowitz, Mike Kelly, David London, Sheldon Stone, Tatsu Takeuchi, and Mihir Worah for fruitful collaborations on some of the topics mentioned here, to them and to I. Dunietz, M. Goodman, and G. Zapalac for discussions, to the many other speakers in the plenary sessions for sharing their insights and plans so that we could coordinate our efforts, and to the Fermilab Theory Group for hospitality during preparation of the manuscript. This work was supported in part by the United States Department of Energy under Grant No. DE AC02 90ER40560.

References

1. S. L. Glashow, *Nucl. Phys.* **22** (1961) 579; S. Weinberg, *Phys. Rev. Lett.* **19** (1967) 1264; A. Salam, in *Proceedings of the Eighth Nobel Symposium*, ed. by N. Svartholm (Stockholm, Almqvist and Wiksell, and New York, Wiley, 1968) p. 367

2. N. Cabibbo, *Phys. Rev. Lett.* **10** (1963) 531
3. M. Kobayashi and T. Maskawa, *Prog. Theor. Phys.* **49** (1973) 652
4. CDF Collaboration, F. Abe *et al.*, *Phys. Rev.* D **50** (1994) 2966; *Phys. Rev. Lett.* **72** (1994) 225
5. M. Shochet, plenary talk, this conference
6. CDF Collaboration, J. Benlloch, paper no. 446, this conference; G. Watts, papers no. 447, 448, this conference; J. Konigsberg, paper no. 449, this conference; B. Harral, paper no. 450, this conference
7. D0 Collaboration, K. Genser, paper no. 87, this conference; D. Chakraborty, papers no. 80, 89, this conference; W. G. Cobau, paper no. 81, this conference; P. C. Bhat, paper no. 73, this conference. For earlier limits see S. Abachi *et al.*, *Phys. Rev. Lett.* **72** (1994) 2138
8. S. Parke, summary of parallel sessions on top quark, this conference, Fermilab report FERMILAB-CONF-94-322-T, September, 1994
9. C. T. Hill and S. J. Parke, *Phys. Rev.* D **49** (1994) 4454
10. E. Eichten and K. Lane, *Phys. Lett.* B **327** (1994) 129
11. V. Barger and R. J. N. Phillips, Univ. of Wisconsin report MAD-PH-830, May, 1994 (unpublished)
12. S. Olsen, plenary talk, this conference
13. J. D. Bjorken, SLAC report SLAC-PUB-2195, September, 1978 (unpublished)
14. R. Ruchti and M. Wayne, paper no. 319, this conference
15. M. Veltman, *Nucl. Phys.* **B123** (1977) 89
16. A. I. Bochkarev and R. S. Willey, paper no. 423, this conference
17. M. Veltman, *Acta Physica Polonica* **B8** (1977) 475; *Phys. Lett.* **70B** (1977) 253; B. W. Lee, C. Quigg, and H. B. Thacker, *Phys. Rev. Lett.* **38** (1977) 883; *Phys. Rev.* D **16** (1977) 1519
18. CDF Collaboration, F. Abe *et al.*, *Phys. Rev. Lett.* **65** (1990) 2243; *Phys. Rev.* D **43** (1991) 2070
19. UA2 Collaboration, J. Alitti *et al.*, *Phys. Lett.* B **276** (1992) 354
20. CDF Collaboration, F. Abe *et al.*, presented by Y. K. Kim at Moriond Conference on Electroweak Interactions and Unified Theories, Meribel, Savoie, France, Mar. 12–19, 1994
21. D0 Collaboration, S. Abachi *et al.*, Fermilab report FERMILAB-CONF-93-396-E, December, 1993, presented by Q. Zhu at Ninth Topical Workshop on Proton-Antiproton Collider Physics, Oct. 18–22, 1993
22. M. Demarteau *et al.* in CDF note CDF/PHYS/CDF/PUBLIC/2552 and D0 note D0NOTE 2115, May, 1994
23. CCFR Collaboration, C. G. Arroyo *et al.*, *Phys. Rev. Lett.* **72** (1994) 3452; R. Bernstein, paper no. 51, this conference
24. CDHS Collaboration, H. Abramowicz *et al.*, *Phys. Rev. Lett.* **57** (1986) 298; A. Blondel *et al.*, *Zeit. Phys.* C **45** (1990) 361
25. CHARM Collaboration, J. V. Allaby *et al.*, *Phys. Lett.* B **177** (1986) 446; *Zeit. Phys.* C **36** (1987) 611
26. LEP report LEPEWWG/94-02, 12 July 1994, submitted to 27th International Conference on High Energy Physics, Glasgow, Scotland, 20-27 July 1994
27. SLD Collaboration, K. Abe *et al.*, *Phys. Rev. Lett.* **73** (1994) 23
28. S. Chivukula, plenary talk, this conference

29. T. Takeuchi, paper no. 258, this conference; T. Takeuchi, A. K. Grant, and J. L. Rosner, Enrico Fermi Institute Report No. 94-26 (unpublished)
30. J. L. Rosner, *Rev. Mod. Phys.* **64** (1992) 1151. In Eqs. (2.18) – (2.21) of that article the expressions for S_W and S_Z should be interchanged.
31. M. Peskin and T. Takeuchi, *Phys. Rev. Lett.* **65** (1990) 964; *Phys. Rev.* D **46** (1992) 381
32. P. Gambino and A. Sirlin, *Phys. Rev.* D **49** (1994) 1160
33. M. C. Noecker, B. P. Masterson, and C. E. Wieman, *Phys. Rev. Lett.* **61** (1988) 310
34. S. A. Blundell, J. Sapirstein, and W. R. Johnson, *Phys. Rev.* D **45** (1992) 1602
35. P. G. H. Sandars, *J. Phys.* B **23** (1990) L655
36. W. J. Marciano and J. L. Rosner, *Phys. Rev. Lett.* **65** (1990) 2963; *ibid.* **68** (1992) 898(E)
37. G. DeGrassi, B. A. Kniehl, and A. Sirlin, *Phys. Rev.* D **48** (1993) 3963
38. LEP Electroweak Heavy Flavours Working Group, D. Abbaneeo *et al.*, report LEPHF/94-03, 15 July 1994
39. Aaron K. Grant, Enrico Fermi Institute report EFI 94-24, June 1994, to be published in Phys. Rev. D
40. OPAL Collaboration, R. Batley, electroweak parallel session, this conference
41. M. Cvetic and B. W. Lynn, *Phys. Rev.* D **35** (1987) 51; J. L. Rosner, *Phys. Rev.* D **35** (1987) 2244
42. L. Wolfenstein, *Phys. Rev. Lett.* **51** (1983) 1945
43. Y. B. Pan, ALEPH Collaboration, paper no. 476, this conference; T. S. Dai, L3 Collaboration, paper no. 23, this conference; E. Silva, OPAL Collaboration, papers no. 357, 358, 360, this conference; P. Antilogus, DELPHI Collaboration, paper no. 481, this conference
44. J. L. Rosner, Enrico Fermi Institute Report No. 94-25, presented at PASCOS 94 Conference, Syracuse, NY, May 19-24, 1994. Proceedings to be published by World Scientific
45. S. Stone, paper no. 425, this conference
46. J. L. Rosner, in *B Decays*, edited by S. Stone (World Scientific, Singapore, 1994), p. 470
47. C. W. Bernard, J. N. Labrenz, and A. Soni, *Phys. Rev.* D **49** (1994) 2536
48. A. Duncan *et al.*, Fermilab report FERMILAB-PUB-94/164-T, July, 1994 (unpublished)
49. See, e.g., UKQCD Collaboration, R. M. Baxter *et al.*, *Phys. Rev.* D **49** (1994) 1594; P. B. Mackenzie, plenary talk, this conference
50. WA75 Collaboration, S. Aoki *et al.*, *Prog. Theor. Phys.* **89** (1993) 131
51. CLEO Collaboration, D. Acosta *et al.*, *Phys. Rev.* D **49** (1994) 5690
52. F. Muheim and S. Stone, *Phys. Rev.* D **49** (1994) 3767
53. BES Collaboration, M. Kelsey, paper no. 414, this conference
54. F. Muheim, parallel session on *B* physics, this conference
55. E. V. Shuryak, *Nucl. Phys.* **B198** (1982) 83
56. Mark III Collaboration, J. Adler *et al.*, *Phys. Rev. Lett.* **60** (1988) 1375; *ibid.* **63** (1989) 1658(E)

57. J. F. Amundson *et al.*, *Phys. Rev.* D **47** (1993) 3059; J. L. Rosner, *The Fermilab Meeting - DPF 92* (Division of Particles and Fields Meeting, American Physical Society, Fermilab, 10 – 14 November, 1992), ed. by C. H. Albright *et al.* (World Scientific, Singapore, 1993), p. 658

58. A. Ali, *J. Phys.* G **18** (1992) 1065

59. A. Soni, paper no. 282, this conference

60. L. Littenberg, plenary talk, this conference

61. B. Winstein and L. Wolfenstein, *Rev. Mod. Phys.* **65** (1993) 1113

62. C. O. Dib, Ph.D. Thesis, Stanford University, 1990, SLAC Report SLAC-364, April, 1990 (unpublished)

63. P. Ko, *Phys. Rev.* D **44** (1991) 139

64. L. Wolfenstein, *Phys. Rev. Lett.* **13** (1964) 562

65. A. J. Buras, M. Jamin, and M. E. Lautenbacher, *Nucl. Phys.* **B408** (1993) 209

66. E731 Collaboration, L. K. Gibbons *et al.*, *Phys. Rev. Lett.* **70** (1993) 1203

67. NA31 Collaboration, G. D. Barr *et al.*, *Phys. Lett.* B **317** (1993) 233

68. ALEPH Collaboration, Y. B. Pan, paper no. 476, this conference

69. V. Sharma, plenary talk, this conference

70. J. Ellis, M. K. Gaillard, D. V. Nanopoulos, and S. Rudaz, *Nucl. Phys.* **B131** (1977) 285; *ibid.* **B132** (1978) 541(E)

71. See, e.g, I. Dunietz, *Ann. Phys. (N.Y.)* **184** (1988) 350; I. Bigi *et al.*, in *CP Violation,* edited by C. Jarlskog (World Scientific, Singapore, 1989), p. 175; Y. Nir and H. R. Quinn, in *B Decays,* edited by S. Stone (World Scientific, Singapore, 1994), and references therein

72. M. Gronau, J. L. Rosner, and D. London, *Phys. Rev. Lett.* **73** (1994) 21; M. Gronau, O. F. Hernández, D. London, and J. L. Rosner, *Phys. Rev.* D **50** (1994) Oct. 1; O. F. Hernandez, D. London, M. Gronau, and J. L. Rosner, *Phys. Lett.* B **333** (1994) 500

73. M. Gronau, A. Nippe, and J. L. Rosner, *Phys. Rev.* D **47** (1992) 1988; M. Gronau and J. L. Rosner, in *Proceedings of the Workshop on B Physics at Hadron Accelerators (op. cit.),* p. 701; *Phys. Rev. Lett.* **72** (1994) 195; *Phys. Rev.* D **49** (1994) 254

74. CLEO Collaboration, J. Bartelt, parallel session on charm, this conference

75. C. T. Hill, in *Proceedings of the Workshop on B Physics at Hadron Accelerators (op. cit.),* p. 127; C. Quigg, *ibid.,* p. 443; E. Eichten, C. T. Hill, and C. Quigg, *Phys. Rev. Lett.* **71** (1994) 4116, and Fermilab reports FERMILAB-CONF-94-117,118-T, May, 1994, to be published in Proceedings of the the CHARM2000 Workshop, Fermilab, June, 1994

76. M. Gronau and D. London, *Phys. Rev. Lett.* **65** (1990) 3381

77. Y. Nir and H. R. Quinn, *Phys. Rev. Lett.* **67** (1991) 541; M. Gronau, *Phys. Lett.* B **265** (1991) 389; H. J. Lipkin, Y. Nir, H. R. Quinn and A. E. Snyder, *Phys. Rev.* D **44** (1991) 1454; L. Lavoura, *Mod. Phys. Lett.* A **7** (1992) 1553

78. D. Zeppenfeld, *Zeit. Phys.* C **8** (1981) 77; M. Savage and M. Wise, *Phys. Rev.* D **39** (1989) 3346; *ibid.* **40** (1989) 3127(E); J. Silva and L. Wolfenstein, *Phys. Rev.* D **49** (1994) R1151

79. L. Wolfenstein, May, 1994, to be published in *Phys. Rev.* D **50** (1994)

80. CLEO Collaboration, M. Battle *et al.*, *Phys. Rev. Lett.* **71** (1993) 3922

81. CLEO Collaboration, report CLEO CONF 94-3, July 12, 1994; J. Fast, paper no. 169, this conference
82. M. Gell-Mann, R. Slansky, and G. Stephenson (unpublished); M. Gell-Mann, P. Ramond, and R. Slansky in *Supergravity*, edited by P. van Nieuwenhuizen and D. Z. Freedman (Amsterdam, North-Holland, 1979), p. 315; T. Yanagida *Proceedings of the Workshop on Unified Theory and Baryon Number in the Universe*, edited by O. Sawada and A. Sugamoto (Tsukuba, Japan, National Laboratory for High Energy Physics, 1979)
83. S. Bludman, N. Hata, D. C. Kennedy, and P. Langacker, *Phys. Rev.* D **47** (1993) 2220
84. L. Wolfenstein, *Phys. Rev.* D **17** (1978) 2369; S. P. Mikheev and A. Yu. Smirnov, *Yad. Fiz.* **42** (1985) 1441 [Sov. J. Nucl. Phys. **42** (1985) 913]; *Nuovo Cim.* 9C (1986) 17; *Usp. Fiz. Nauk* **153** (1987) 3 [Sov. Phys. - Uspekhi **30** (1987) 759]
85. H. Georgi, summary talk, this conference
86. A. Olinto, plenary talk, this conference
87. M. Goodman, in *Particle and Nuclear Astrophysics and Cosmology in the Next Millenium* (APS Summer Study 94, Snowmass, CO, June 30 – July 14, 1994), to be published
88. Fermilab E531 Collaboration, N. Ushida *et al.*, *Phys. Rev. Lett.* **57** (1986) 2897
89. Kamiokande Collaboration, Y. Fukuda *et al.*, report ICRR-321-94-16, June 1994 (unpublished)
90. F. Halzen, plenary talk, this conference
91. CHORUS Collaboration, M. de Jong *et al.*, CERN report CERN-PPE-93-131, July, 1993 (unpublished)
92. NOMAD Collaboration, CERN report CERN-SPSLC-93-19, 1993 (unpublished)
93. K. Kodama *et al.*, Fermilab proposal FERMILAB-PROPOSAL-P-803, October, 1993 (approved Fermilab experiment)
94. W. W. M. Allison *et al.*, Fermilab proposal FERMILAB-PROPOSAL-P-822-UPD, March, 1994 (unpublished)
95. LSND Collaboration, R. A. Reeder, paper no. 146, this conference
96. E. B. Norman *et al.*, in *The Fermilab Meeting - DPF 92* (Ref. 57), p. 1450
97. H. Georgi, H. Quinn, and S. Weinberg, *Phys. Rev. Lett.* **33** (1974) 451
98. H. Georgi and S. L. Glashow, *Phys. Rev. Lett.* **32** (1974) 438
99. U. Amaldi *et al.*, *Phys. Rev.* D **36** (1987) 1385; U. Amaldi, W. de Boer,and H. Fürstenau, *Phys. Lett.* B **260** (1991) 447; P. Langacker and N. Polonsky, *Phys. Rev.* D **47** (1993) 4028
100. H. Georgi in *Proceedings of the 1974 Williamsburg DPF Meeting*, ed. by C. E. Carlson (New York, AIP, 1975) p. 575; H. Fritzsch and P. Minkowski, *Ann. Phys. (N.Y.)* **93** (1975) 193
101. A. D. Sakharov, *Pis'ma Zh. Eksp. Teor. Fiz.* **5** (1967) 32 [*JETP Lett.* **5** (1967) 24]
102. G. 't Hooft, *Phys. Rev. Lett.* **37** (1976) 8
103. V. A. Kuzmin, V. A. Rubakov, and M. E. Shaposhnikov, *Phys. Lett.* B **155** (1985) 36; *ibid.* **191** (1987) 171
104. M. Fukugita and T. Yanagida, *Phys. Lett.* B **174** (1986) 45; P. Langacker, R. Peccei, and T. Yanagida, *Mod. Phys. Lett.* A **1** (1986) 541
105. B. Kayser, in *CP Violation* (Ref. 71), p. 334
106. T. J. Barklow, paper no. 400, this conference

Summary Talk for DPF'94

HOWARD GEORGI, *Harvard*

SUMMARY TALK FOR DPF94*

HOWARD GEORGI[†]

Lyman Laboratory of Physics, Harvard University

Cambridge, MA   02138, USA

I missed the beginning of this conference because after I had agreed to give the summary talk here, I was invited to a very peculiar conference in Erice — the International Conference on the History of Original Ideas and Basic Discoveries in Particle Physics, which alas, overlapped this one. The Erice conference seemed too entertaining to pass up entirely, so I compromised and attended the beginning of that one, then managed to get from Erice (changing suitcases in Boston's Logan Airport) to Albuquerque in only 25 hours to attend DPF94. The result is that like Billy Pilgrim in Kurt Vonnegut's Slaughterhouse Five — I feel "unstuck in time." Rather than fighting this temporal disorientation, I decided to use it as the theme of my summary talk. Rather than comparing things with the conference 2 years ago, to show progress, I will do serious time-tripping and compare with 25 years ago!

The Erice conference was a bunch of old guys reminiscing about how great we were in the 70's (and even earlier). You couldn't listen to these talks without getting a feeling for how great the decade of the 70's was as a time to do particle physics. Clearly, it wasn't that the young people who participated in the construction of the standard model in the 70's were that much brighter than the young people doing particle physics today — but we were in the right place at the right time.

I rematerialized in Albuquerque on Thursday morning after a mind-numbing trip. My first reaction was to be completely overwhelmed by the parallel sessions. Let me say that the local organizers did a magnificent job of setting up the parallel sessions. The sessions were actually close enough together that it was possible to get from one to another in between talks. But the result was that I tried to set up a detailed schedule to allow me to hit the highlights of each session, and I ended up rather dizzy — with an overload of information to complement my jet-lag.

Nevertheless, the feeling grew as I hurried from talk to talk of how different most of what I was listening to was from what I listened to at conferences 25 years ago, when I started in this business. Most of the major themes in this conference did not even exist at the end of the 60's. I won't have time to talk about all of them, but I'll pick out some that are illustrative of what I think the important issues are to discuss in detail, and just mention the rest. I apologize in advance to those whose contributions I will slight.

Electroweak Physics — This subject today is dominated by the growing body of precise tests of the standard model at LEP, SLC, and FNAL, and in atomic parity violation experiments. These experiments now confront the model at the level of the

---

*Talk presented at DPF-94 in Albuquerque, NM. #HUTP-94/A023.

†Research supported in part by the National Science Foundation under Grant #PHY-9218167.

radiative corrections, the calculation of which was made possible by the theoretical work in the early 70's of 'tHooft and Veltman. Although the experiments are still most sensitive to the gauge structure and the fermion content of the standard model, they are beginning to constrain the mechanism of $SU(2) \times U(1)$ symmetry breaking. More of this later. There is still some room for new physics at modest energies, but it is shrinking fast. One of my favorite comments came from James Branson in his summary of the Electroweak parallel session — that the SLC data supplied the χ^2 that was missing in the LEP data.

Heavy Flavor — The local organizers quite properly divided this subject into four subtopics:

1. τ lepton

2. intermediate quark — charm

3. heavy quark — bottom

4. very heavy quark — top

because these four fermions have such dramatically different properties. Of these, only charm was even a gleam in a theorist's eye in 1970.

The τ lepton — (convener Tomasz Skwarnicki) A complete surprise when Perl found it — even though there were indications for it in $R_{e^+e^-}$, no theorist was really prepared for it — much progress has been made both in pinning down the properties of the τ, and in using it as a probe to test the standard model. The very accurate BES result for the τ mass,

$$m_\tau = 1777.02 \pm 0.25 \text{ MeV}$$

was the first triumph of this fledgling accelerator. The old puzzles with the τ are going away. No problem remains with μ-e universality in τ decay, and we heard from Ian Hinchliffe yesterday that the hadronic decays give a sensible value of α_S. There is still a slight discrepancy for μ-τ universality, measured in the ratio of the partial widths for the electron decay,

$$\frac{G_\tau}{G_\mu} = \frac{\tau_\mu}{\tau_\tau} \left(\frac{m_\mu}{m_\tau}\right)^5 \frac{\text{Br}(\tau \to \nu_\tau e \overline{\nu}_e)}{\text{Br}(\mu \to \nu_\mu e \overline{\nu}_e)}$$

$$\frac{G_\tau}{G_\mu} = 0.987 \pm 0.007 \qquad (1.7\sigma)$$

But it is moving in the right direction — in the 92 Particle Data Book the result was

$$\frac{G_\tau}{G_\mu} = 0.944 \pm 0.023 \qquad (2.4\sigma)$$

The Michel parameters are all known to decent accuracy — all agree with the SM. And almost 98% of the exclusive hadronic decays have been identified. In spite of its large mass, the τ is just a sequential lepton, to a good approximation.

The c quark — (convener Janis McKenna) — I say that it is an intermediate mass quark because its mass is of the order of 1 GeV scale typical of hadronic bound states.

Corrections to spectator approximation in exclusive channels are or order 1 GeV/m_c — that is to say of order 1. This is probably also the size of some corrections to HQET, though we still hope they will be smaller.

I was impressed with how much had been learned about it recently. There is complementary data from fixed target, e^+e^- and $\bar{p}p$ experiments. I will just mention one thing from each type of experiment that caught my attention:

- The observation of the exclusive decay

$$\Omega_c \to \Sigma^+ K^- K^- \pi^+$$

  from E687, $m = 2699.9 \pm 2.5 \pm 1.5$ MeV.

- Production of D_S just above $D_S \overline{D_S}$ threshold at BES — we heard how this allows double tagging —

$$\mathrm{Br}(D_S^\pm \to \phi\pi^\pm) = 4.2\,^{+9.0\,+\,1.7}_{-1.5\,-\,0.0} \pm 0.5\%$$

- Measurements of J/ψ and ψ' production in $\bar{p}p$ — there's a sort of theory, recently developed (Braaten ...) but the corrections aren't well understood and the ψ' doesn't work.

In my view, the intermediate status of the c quark makes it particularly important as a probe of QCD — in the c and the s quarks, nature has provided us with examples on both sides of the boundary between heavy and light. Even more mysterious to me than the lifetime differences between different states are the very large $SU(3)$-breaking effects in the decays — the easiest one to remember is

$$\frac{\mathrm{Br}(D^0 \to K^+K^-)}{\mathrm{Br}(D^0 \to \pi^+\pi^-)}$$

but there are many examples. The current theoretical understanding of this is not very appealing — a bunch of different 30% effects adding up to a factor of 3!

Finally, D^0-$\overline{D^0}$ mixing should be pushed as a signal for new physics.

The b quark — (conveners Marina Artuso and for CP and Rare Decays, German Valencia and George Golin) a huge subject — we heard a lot yesterday in Vivek Sharma's Plenary talk, so I won't repeat it. Let me just note that the lifetime difference between B^+ (for example),

$$\tau(B^+) = 1.652 \pm\,^{+0.035}_{-0.034} \pm 0.06 \text{ ps}$$

and the Λ_b

$$\tau(\Lambda_b) = 1.16 \pm 0.1 \pm 0.05 \text{ ps}$$

doesn't seem crazy, since 1 GeV/m_b is about 0.2.

Clearly, we are going to learn a very great deal about the b quark in the coming years, with major experiments at many different kinds of machines, and one or more dedicated B factories. We can hope to see new physics here and in any event, we will surely beat the standard model to death. I'll come back to this later.

But we can also waste a lot of resources on the B system. Consider, for a moment, the changes in the PDG compilation of particle properties from 25 years ago, when we had only light quarks and leptons. Then there were not a huge number of decay modes for each particle. Now each new weakly interacting state in the c system takes over a page if any significant fraction of its decay modes have been seen. For the B, I hope we will never identify most of the modes! For this system, a list of all the exclusive modes is just not the most interesting way of getting at the physics.

This brings me to the t quark — (convener Steven Parke) for which the message that the exclusive modes are not the most important physics will be emphasized by nature. Because the t decays before it hadronizes, it will not be possible even to define unambiguously define what an exclusive decay mode means. The PDG will have to come up with a different format.

Not quite two months ago, I had the good fortune to spend two weeks on St. Croix at the Advanced Study Institute, a school for young experimenters organized by Tom Ferbel. I was very amused to watch the interaction between the CDF students and the D0 students. It may be too strong to call it a feud, but suffice it to say that the D0 students were not shy about expressing their skepticism about CDF's evidence for the t.

Having witnessed this, I was interested to see how this interaction would be played out at this conference, so I went to the talks of two of the outstanding youngsters from the school, Gordon Watts from CDF and Dhiman Chakraborty from D0, on the t signal in lepton + jets. I was very surprised at how similar the two talks were. Without vertex information, D0 had more background, but the excess events looked, to my naive theorist's eye, very much alike. Something seems to have happened in the last two months to make the D0 people less unhappy about CDF's claims.

There is still a lot of room for (though no evidence for) new physics involving the t — nonstandard properties for the t. Steven Parke talked about a couple of the possibilities in his summary talk this morning. I want to spend a bit of time talking about this because it illustrates a more general moral. The interesting models for producing nonstandard properties of the t involve new strong interactions (topcolor, technicolor, and the like). This is not an accident. With some simple and well understood exceptions involving particle mixing and decay,* new physics involving small couplings shows up first in the production of new particles, while new physics involving a new strong interaction (not QCD — that is already weak at high energies) shows up first in the nonstandard properties of the particles we already know. Something similar happens in the electroweak symmetry breaking sector of the world (it is silly to call it the Higgs sector unless there is actually a Higgs boson, which there may not be). In a weakly interacting model, like the MSSM, the new physics associated with $SU(2) \times U(1)$ breaking shows up first in the production of new narrow particle states

---

*The weak interactions show up in the properties of the quarks and leptons at low energies, even though the couplings are small, only because they change flavor and violate symmetries.

— Higgs's and superpartners. But in a technicolor theory, the new physics shows up first in an modification of the scattering of longitudinal gauge bosons. This may be the tail of a technirho. But actually, even the technirho if such a thing exists, would be an indication of some small parameter in the new interactions. The ρ and the other resonances in QCD are important because they are narrow states, presumably because there is some small parameter (maybe $1/N_c$ but probably involving the dynamics in some more interesting way as well) responsible for Zweig's rule.

I have been thinking about this myself with some of my students, and I think that it is possible and useful to organize our thinking about possible new physics in a way that doesn't depend too much on explicit models. I think this is important, because there are an infinite number of possible models, and none of those that we have found so far look particularly good.

Since it seems likely that any new strongly interacting physics will not show up below the t mass, it should be possible to discuss it in a effective field theory language, like the Fermi theory of weak interactions, without committing ourselves to the precise dynamical details. Thus for example, we could classify the possible new physics of the t in a simple way. Because of the $SU(2) \times U(1)$ structure of the electroweak interactions, we should discuss t_R and the left handed doublet involving the t separately. the right-handed t is easy. If only t_R'' participates in new interactions (where the $''$ indicates possible flavor mixing), the constraints on the new physics from flavor changing neutral current data and precise tests of W and Z properties are not very severe. The flavor mixing can be made small by tuning. The scale of the new physics can be quite low, and the properties of the t can be measurably different from what is expected in the standard model.

The situation is a bit more interesting if it is the doublet, (t_L'', b_L''') that participates in the new physics. Here you cannot tune the mixing to zero because the KM matrix is nontrivial. Nevertheless, you can choose the mixings so that the only GIM violating anomaly is in B_s-$\overline{B_s}$ and t decay.

While I would like to see more model independent analysis, I do actually think that Chris Hill's topcolor model is quite interesting. It is an example of the composite Higgs idea that David Kaplan and I worked out a long time ago, but with more speculative, although still plausible, dynamics. More theoretical work on it is needed.

QCD — (conveners Elizabeth Jenkins and Phillip Burrows) — this is a huge subject, though still not quite 25 years old — I date the beginnings of the subject to the discovery of asymptotic freedom by Gross and Wilczek and Politzer — it was this theoretical development that breathed life into the field-theoretic description of the strong interactions. Combined with the older experimental observations of quarks in deep inelastic lepton-hadron scattering, and the older theoretical idea of dimensional transmutation, this led fairly quickly (though not easily) to the idea of confinement and the modern theory of QCD.

The subject is so huge because it is one thing that we can do with today's technology that is clearly important — clearly has lots of interesting, tough question remaining. We use many different tools to study it. We heard Plenary talks about Lattice QCD from Paul Mackenzie and about Perturbative QCD from Ian Hinchliffe. In both cases, recent progress has been very impressive, but there is still a very long to go before these technologies are exhausted. In addition, chiral perturbation theory and heavy quark effective field theory have developed into large industries, trying to make sense of the enormous range of QCD phenomenology. I can't even begin to summarize all this. Suffice it to say that I spend much of my own time working on QCD, lately on the application of large N_c ideas to baryons.

Beyond the Standard Model — obviously this is completely new from 25 years ago. Unfortunately, I didn't see much of this. It looks from the schedule as if the main subject was supersymmetry. Perhaps one comment is worth making here. What we usually call the MSSM actually has a lot unification in it. Without assumptions about what goes on at very high energies, the number of parameters in a model with softly broken supersymmetry is huge. One possibility is to think of the MSSM as the low energy limit of something like the $SU(5)$ model with softly broken supersymmetry that Savas Dimopoulos and I and independently, Noriske Sakai, constructed in the early 80's. The convergence of the gauge couplings at very high energies that Jon Rosner mentioned yesterday is a test of this idea. The unified theory, in turn, may come from something like string theory, or not — it doesn't really matter for the low energy phenomenology.

Other schemes seem possible as well. For example in the last couple of year, Dine, Nelson and others have revisited the idea of dynamically broken supersymmetry.

Sekhar Chivukula discussed extended technicolor in his Plenary talk, showing clearly how that the measurement of $Z \to b\bar{b}$ is a strong constraint on these models. If ETC has anything to do with the world, and if we were smarter, I suspect that this would be an important clue. But in fact, our ignorance of strongly interacting field theory is so great that I suspect we will have to wait for experiments above the $SU(2) \times U(1)$ breaking scale to see what this clue means, that is if technicolor and ETC have anything to do with the world. The one thing that seems clear is that technicolor, if it exists as a strongly interacting mechanism for $SU(2) \times U(1)$ breaking, is very different from QCD. And QCD is the only strongly interacting field theory we understand. And the reason we understand it is not that theorists are so smart, but that we could look up so many of the answers in the particle data book. For technicolor, without much relevant data, we are still completely at sea. This must be kept in mind in interpreting the data on precise tests of the standard model.

I'll say more about "Beyond the Standard Model" later.

Planck Scale Physics — I missed Jeff Harvey's talk, but I had breakfast with him on Thursday before he left. He talked about the slow "progress" in string theory since 1985 and the current reasons for "optimism." I'm not going to say much about this. Of course, this is the point in the talk where you all expect me to dump on string theory — I might say something like "Planck scale physics is an oxymoron because you can't do physics at the Planck scale." I am, in fact, going to resist the temptation to dump on string theory. I think that there are signs that we are beginning to see some modest returns from the huge intellectual investment in the field in the last 10 years. One example, which I understand only very superficially, is the work by Seiberg and Seiberg and Witten, using some of the fancy mathematics developed to study string theory to analyze an ordinary four dimensional field theory — $N = 2$ supersymmetry. This is a strongly interacting theory with confinement and chiral symmetry breaking.

Just the sort of the thing that we are interested in. They are able to say very detailed things about the dynamics in the model, some of which may teach us more general things about the way strongly interacting field theories work. This is a very hopeful sign. We may or may not care about a theory of everything — but we should certainly care about a deeper understanding of QFT.

Of course, Seiberg and Witten were trained as particle physicists, and made many beautiful and important contributions to particle physics before their apotheosis to the string scale. The real concern about string theory is that too many younger string theorists are completely divorced from particle physics and do not know or care what a ρ meson is. Another hopeful sign is that influential older string theorists are beginning to see this as a problem. I hope that in the next 25 years, string theory will move back closer to the main stream of particle physics. This will take work on both sides.

Finally, there were a few subjects whose roots go back farther than 25 years, but even here, the discussion was dominated by new ideas that are much younger.

Cosmology and Dark Matter — (Kim Griest and Joel Primack)

Alas, I could not get to any of the parallel talks here. I was told that they showed enormous excitement and enthusiasm — new experiments coming on line — speculative theoretical ideas.

I did hear Francis Halzen's excellent Plenary talk on Non-Accelerator Physics, and the talk about the LSND neutrino oscillation search (convener Carl Akerlof). Doing particle physics with things coming down from the sky is hardly new, but the recent developments have been spectacular. Furthermore, I doubt that cosmic ray physics 25 years ago could imagine that we would be now be doing particle physics with things coming **up from the sky** on the other side of the Earth! The only data I plan to show today is one graph of Kamiokande data from Halzen's talk — highly touted by Francis as a possible observation of neutrino oscillations. This shows the angular dependence of the ratio of the observed to expected ratios of μ/e for the high-energy, semi-contained events. It certainly suggests that the ν_μ neutrinos are oscillating into something else when they go through the earth, but not when they come down only through the atmosphere. I am sorry that we did not have a talk by one of the experimenters.

I did want to talk about the interpretation of these events, because it relates to something that I didn't like about Jon Rosner's talk yesterday. It is clear that if this data is a signal of neutrino oscillation, there is large mixing and a mass somewhere in the neighborhood of 0.1 eV. If this is ν_μ-ν_τ mixing, Jon would argue that the see-saw mechanism suggests that it is associated with physics at a high scale of order 10^{10} GeV, because

$$1 \text{ eV} \approx \frac{m_\tau^2}{10^{10} \text{ GeV}}$$

This argument depends, in my view, on too many things that we don't know. First of all, it depends critically on the nature of $SU(2) \times U(1)$ breaking. In a technicolor model, the neutrino mass in a see-saw depends on a much higher power of the large scale, so the large scale is not so large. Secondly, it makes a completely unjustified assumption about how flavor works for the neutrinos. What you actually know, if the symmetry breaking comes from a fundamental Higgs boson, is that the see-saw mass is proportional to two powers of v, the VEV. You might just as well say

$$1 \text{ eV} \approx \frac{M_W^2}{10^{14} \text{ GeV}}$$

or

$$1 \text{ eV} \approx \frac{v^2}{10^{15} \text{ GeV}}$$

The point here is that while the definitive observation of neutrino oscillations would be absolutely wonderful, and while it would indicate the existence of some new physics, most probably at a scale larger than a TeV, we can't really say much more without a much more detailed understanding of both $SU(2) \times U(1)$ and flavor.

Reeder, talking about the LSND experiment at LAMPF, also discussed data that could be interpreted as evidence for neutrino oscillations, $\overline{\nu}_\mu$-$\overline{\nu}_e$, in the coupled reactions

$$\overline{\nu}_e + p \rightarrow e^+ + n$$

$$n + p \rightarrow D + \gamma$$

where the group saw an excess of 8 events. Refusing to let theorists put words into his mouth, he would not make any claims, or even quote a mass, but he told us that the typical energy was 45 MeV and the distance was 30 m, so a mass of the order of few eV is reasonable.

One thing that has changed from 25 years ago is that these days there always seems to be some preliminary indication of a neutrino mass or mixing. Alas, there are not always the same ones!

CP violation/Rare Decays — (conveners German Valencia and George Golin), Instrumentation — (conveners Anna Peisert and Richard Wigmans) and Heavy Ion Physics (conveners Sean Gavin and Richard Seto) also continue much older traditions, but with new twists. I'm not going to say much about any of these beyond what I have said about other related areas.

I did want to briefly discuss Accelerators — (convener John Marriner) — Here I noted a peculiarity of language — despite the words we use accelerator physics is not the opposite of non-accelerator physics! Maybe we should use some different words. It also seemed to me (an ignorant theorist), that in some sense this field seems to change more slowly than anything else we discussed at this conference. At any rate, I was struck by a number of things in Robert Siemann's Plenary talk. One was the incredible amount of art that goes into this business. I mean this in the good sense of really beautiful work — but a less complimentary way of saying this might be to talk about the amount of guesswork and fiddling around with parameters. Since everything so often works beautifully in the end, let's call it art. I found this very scary. What happens to us if the Muse gets disgusted and goes away? Another was his plea that after the foul murder of the SSC (my words not his), we need to find ways of further internationalizing accelerator physics. I hope that this will happen naturally in the next 25 years, but it is not obvious, and it is clearly critical to the long-term survival of the field.

What should we make of all this. The picture, compared to 25 years ago, seems pretty terrific and exciting. There is a lot of physics going on — much of it really interesting. It can, and does, absorb all our energies, and more. Furthermore, the progress made in 25 years is almost beyond description. We study different objects with different tools in different ways than we did 25 years ago. But the more important question is "What will particle physics look like 25 years in the future?" This is the question that worries us. Will things change as much in the next 25 years as they did in the last. I believe that if we are still experimenting below the scale of the physics of $SU(2) \times U(1)$ breaking, they will not — not even if we find neutrino masses, neutrino oscillations, and dark matter. Each of these things would be a wonderful discovery that will open up new field of exploration — but without direct observation of the physics of symmetry breaking, we will still not be able to extrapolate beyond the TeV scale. We will be doomed to plow the same fields forever, and however lush and fertile they are today, they will be exhausted in much less than 25 years. So it goes.

The SSC would have allowed us to begin the study of the TeV scale. Why did the SSC fail? Many reasons — bad political and economic luck not the least. But we need to face up to the internal difficulties in the field — here's my incomplete list — along with what to do about it:

small versus big — Clearly one problem with the SSC was the huge size of the collaborations. Many physicists simply did not want to get involved in such a massive project. We need to find mechanisms to allow people to join big projects without feeling small themselves. Once they get involved, the real work is done in small groups anyway! The exploration of the TeV scale is something that should really involve **everyone!** A related problem is the education of students. As we devote more of our resources to the large projects that will be required to study the TeV scale, we will need to change the way we do it.

possible versus impossible — Clearly, many people were worried about the feasibility of actually doing the experiments. Physics is the art of the possible after all, and if experiments at the TeV scale are impossible, maybe we shouldn't try. This is a real problem, but we don't have any choice. If our tools, which are so terrific for studying what we study today, are inadequate to take the next step, then we have to abandon them and find new ones.

verification versus exploration — I feel particularly bad, as a theorist interested in exploring the TeV scale, that I couldn't even convince all experimenters of how interesting it will be. That may sound silly, but I mean it. It was not just congressmen and newspaper editors who felt "what's the big deal about finding the Higgs." I know that many experimenters felt the same way — that the physics, even if we could do it, would not be that interesting. It is critical to realize that this project is an exploration of the unknown, not the sterile verification of a theory that is almost sure to be right. Whatever the physics of electroweak breaking, whether technicolor or fundamental Higgs and superpartners, or something that we haven't thought of, it will be completely different from anything we have seen in the last 100 years, let alone the last 25. We all need to feel this in our bones!

theory versus experiment — Obviously, it is easier for a theorist to get to a TeV than for an experimenter, because we theorists just have to imagine it. I suspect, that this magnified the natural and healthy mistrust that experimenters should have of theorist and also contributed to uncertainly about how much effort the exploration of the TeV scale was really worth. We need to find ways of improving communication.

present versus future — We clearly spent too much energy worrying about how to preserve and continue the terrific physics that we are doing today. It now seems clear that to get to the TeV scale will require sacrifices. We will have to **not do** some things that we would like to do — things that are worth doing, but which absorb resources that are needed for the study of the TeV scale. We must somehow find the political machinery and the will to make the hard decisions and to make them stick.

lab versus lab — The present system of national laboratories contributes to this difficultly by providing a powerful built-in lobby for the status-quo. We cannot afford a system in which the labs compete with one another and with the rest of the world to do the same physics (B factories, for example). Either we must figure out how to close laboratories, or we must figure out how to expedite the now very slow process of transforming laboratories from particle physics to other fields, or we must figure out how to exercise some higher level control so that the labs can cooperate in the exploration of the TeV scale.

particle physicists versus everyone else — Particle physicists are arrogant! How could it be otherwise? We understand how interesting the questions we study really are. A little bit of arrogance is OK. But snootiness doesn't work. Too often, I think, we let our arrogance hide our enthusiasm for the subject. I am very sorry that I missed the session on outreach, because this is the antidote we need. We need to talk to everyone who will listen about what we do, and we need to find better ways to do it. Consider the success of the astronomers in selling the Shoemaker-Levy comet. It is true that they have pictures. We need to find something equally sexy to the educated populace — I don't know what it is or exactly how to find it, but we must try. We should even be so excited about what we do that it is OK to publish in the New York Times!

If we do all this, and we get lucky at the LHC and beyond, maybe we can transform particle physics in the next 25 years as much as we have transformed it in the last 25 years.

PARALLEL SESSIONS

Electroweak

JAMES G. BRANSON, *California (San Diego)*

Z LINESHAPE MEASUREMENT RESULTS FROM LEP

ZHONG FENG[*],[†]

Physics Department, University of Wisconsin,
1150 University Ave., Madison, Wisconsin 53706 U.S.A.

ABSTRACT

In 1993, LEP ran at three energies around the Z peak. The energy at each point was measured precisely using the resonant depolarization method. Using these data, the Z resonance parameters are measured with reduced statistical and systematic uncertainties. Based on all data taken since 1989, the resonance parameters of Z are determined to be $M_Z = (91.1888 \pm 0.0044)$ GeV, $\Gamma_Z = (2.4974 \pm 0.0038)$ GeV, $\sigma_{had}^0 = (41.49 \pm 0.12)$ nb, and $R_l = 20.795 \pm 0.040$. The corresponding number of light neutrino species $N_\nu = 2.988 \pm 0.023$.

1. Introduction

At LEP, electrons and positrons collide at center-of-mass energies near the Z resonance. The Z then decays into $q\bar{q}$, lepton pairs or neutrinos. There are four detectors, ALEPH, DELPHI, L3 and OPAL taking data at LEP, and the Z lineshape parameters mass M_Z, width Γ_Z and cross sections of the Z decay channels are measured. This report is a summary of the preliminary results of data taken from 1989 to 1993 by all four LEP experiments.

2. LEP Energy Calibration and Luminosity Measurement

Great improvement of the Z lineshape measurements has been made in 1993 when LEP ran at three energies with high luminosity. In addition to the Z mass peak point, LEP scanned on two off peak points with center-of-mass energies roughly 1.8 GeV above and below the Z mass. In 1993, each experiment recorded 15 pb^{-1} at the peak point and 9 pb^{-1} at each of the off peak points. This corresponds to a four fold increase of data at the off peak points.

Since 1992, LEP has been using the spin resonant depolarization method to calibrate the beam energy measurement. In 1993, with better understanding of machine physics, LEP beam energy calibration[1] was performed on each of the energy scan points with reduced systematic errors, leading to a total systematic error of 3.3 MeV for Z mass and 2.2 MeV for Z width, which is a factor of two reduction comparing with 1992.

There were also some efforts made by the four experiments to improve the precision of the luminosity measurement in the past two years. For ALEPH, a solid state (Si-W) luminosity calorimeter has been mounted since September of 1992.

[*]Representing the ALEPH collaboration
[†]Supported by the US Department of Energy, contract DE-AC02-76ER00881

Table 1: Luminosity experimental errors of four LEP experiments

	ALEPH	DELPHI	L3	OPAL
1992	0.15%	0.38%	0.5%	0.41%
1993$_{prelim.}$	0.09%	0.28%	0.16%	0.07%

The improved coverage at low angles of the new luminosity calorimeter increases the Bhabha cross section significantly. The high precision internal alignment of modules and well-defined fiducial regions of silicon pads enable the experiment to reduce the luminosity experimental error to a level of 0.09%. Similar efforts are also made by the other three experiments. Table 1 lists the improvements in experimental luminosity measurement errors since 1992.

The dramatic reduction of the OPAL luminosity measurement experimental error is due to a new solid state luminosity calorimeter mounted in 1993. L3 installed a small angle tracker in 1993 and thus improved the position measurement of the Bhabha events.

3. Combined Z Lineshape Measurement Results

We have used the model independent packages MIZA[2] and ZFITTER[3] to fit the LEP data. For the Z lineshape parameter fit, the inputs are the measured cross sections of the visible Z decay channels at each energy point. Table 2 shows the number of events selected by the four experiments, and Table 3 shows the cross section systematics errors compared to 1992.

Table 2: The LEP statistics in units of 10^3 events used for the analysis of the Z lineshape.

		ALEPH	DELPHI	L3	OPAL	LEP
$q\bar{q}$	'90-'91	451	356	423	454	1684
	'92	680	697	677	733	2787
	'93 prel.	653	677	658	653	2641
	total	1784	1730	1758	1840	7112
$\ell^+\ell^-$	'90-'91	55	37	40	58	190
	'92	82	69	58	88	297
	'93 prel.	79	71	62	81	293
	total	216	177	160	227	780

For each experiment, we fit for a standard set of parameters M_Z, Γ_Z, peak hadronic cross section σ_h^0 and the ratio of hadronic partial width to lepton partial width R_l, since these parameters are only slightly correlated. LEP average is made by using a correlation matrix including the common systematic errors. From the fitted parameters, we can derive some other lineshape parameters. Table 4 summarizes the lineshape fit results and derived parameters.

Assuming lepton universality, we can derive the ratio of invisible width to

Table 3: The experimental systematic errors for the analysis of the Z lineshape measurement. The errors quoted do not include the common uncertainty due to the LEP energy calibration. [a]Only the experimental error including the statistics of small angle Bhabha events is quoted. In addition, there is a theoretical error for the calculation of the small angle Bhabha cross section of 0.25% which is common to all experiments.
[b]The indicated range for the 1993 selection expresses the variation of the systematic error as a function of energy.

	ALEPH		DELPHI		L3		OPAL	
	'92	'93 prel.	'92	'93 prel.	'92	'93 prel.	'92	'93 prel.
$\mathcal{L}^{exp.}$ [a]	0.15%	0.09%	0.38%	0.28%	0.5 %	0.16%	0.41%	0.07%
σ_{had}	0.14%	0.14%	0.13%	0.13%	0.15%	$0.11-0.14\%$ [b]	0.20%	0.20%
σ_e	0.4 %	0.4%	0.59%	1.2%	0.3 %	$0.25-0.76\%$ [b]	0.22%	0.35%
σ_μ	0.5 %	0.5%	0.37%	0.5%	0.5%	$0.45-0.57\%$ [b]	0.19%	0.22%
σ_τ	0.3%	0.3%	0.63%	0.8%	0.7 %	0.54%	0.44%	0.51%

Table 4: Combined Z lineshape fit results from all four experiments

Parameter	Average Value
$M_Z(GeV)$	91.1888 ± 0.0044
$\Gamma_Z(GeV)$	2.4974 ± 0.0038
$\sigma_h^0(GeV)$	41.49 ± 0.12
R_e	20.850 ± 0.067
R_μ	20.824 ± 0.059
R_τ	20.749 ± 0.070
R_l	20.795 ± 0.040
$\Gamma_{ee}(MeV)$	83.35 ± 0.21
$\Gamma_{\mu\mu}(MeV)$	83.95 ± 0.30
$\Gamma_{\tau\tau}(MeV)$	84.26 ± 0.34
$\Gamma_l(MeV)$	83.96 ± 0.18
$\Gamma_{had}(MeV)$	1745.9 ± 4.0
$\Gamma_{inv}(MeV)$	499.8 ± 3.5

lepton partial width: $\Gamma_{inv}/\Gamma_l = 5.953\pm0.046$. By taking the standard model value of the ratio of neutrino partial width to lepton partial width: $\Gamma_\nu/\Gamma_l = 1.992\pm0.003$, and using the formula: $N_\nu = (\frac{\Gamma_{inv}}{\Gamma_l})/(\frac{\Gamma_\nu}{\Gamma_l})$, we can derive the number of neutrinos $N_\nu = 2.988\pm0.023$.

The parameter R_l can be used to determine the strong coupling constant[4] α_s. For $M_Z = 91.1888GeV$, $M_t = 175GeV$ and $M_H = 300GeV$, and taking the lineshape fitted parameter value $R_l = 20.795 \pm 0.040$, we obtain $\alpha_s = 0.126 \pm 0.006$. The error is experimental only.

In the Standard Model, the Z width can be calculated[23] with electroweak loop corrections which depend on M_t^2 and $log(M_H)$. By fitting the lineshape parameters with the LEP lepton and quark asymmetry results[5], the LEP tau polarization measurements and the LEP heavy flavour electroweak results[6], we can constrain $M_t = 173^{+12}_{-13}\,^{+16}_{-20}GeV$. The first error is experimental and second error is due to the variation of the Higgs mass $60 \leq M_H(GeV) \leq 1000$.

4. Conclusions

The 1993 LEP energy scan has greatly reduced the Z lineshape measurement errors. The current relative error on the Z mass has reached 5×10^{-5} approaching the Fermi coupling constant precision of 2×10^{-5}.

5. Acknowledgements

I would like to thank Dr. John Harton and Jim Grahl for help in preparing this talk. The combined LEP results presented here are the teamwork of the LEP Electroweak Working Group.

References

1. L.Arnaudon et al. *CERN SL*/94-71 (August 1994)

2. M.Martinez et al. *Z. Phys.* **C49**(1991)645

3. D.Bardin et al. *CERN-TH* 6443/92 (May 1992)

4. T.Hebbeker et al. *CERN-PPE*/94-44 (March 1994)

5. The LEP Electroweak Working Group, Updated Parameters of the Z lineshape and Lepton Forward-Backward Asymmetries, summary of the combination of preliminary LEP data for the 27th International Conference on High Energy Physics, Glasgow, Scotland, 20-27 July 1994. *CERN Internal Note LEPLINE*/94-01 (July 1994)

6. The LEP Electroweak Working Group, Constraints on Standard Model Parameters from Combined Preliminary Data of the LEP Experiments, summary of the combination of preliminary LEP data for the 27th International Conference on High Energy Physics, Glasgow, Scotland, 20-27 July 1994. *CERN Internal Note LEPEWWG*/94-02 (July 1994)

WEAK COUPLINGS OF THE τ LEPTON

JOHN SWAIN*

Department of Physics, Northeastern University, Dana Research Center, Boston, Mass. 02115, USA

Abstract

The τ lepton is unique among the particles produced at the Z^0 in that its decays allow a complete analysis of the vector and axial vector coupling constants involved in its production. Recent results on these couplings from the LEP experiments, ALEPH, DELPHI, L3, and OPAL are presented. The values obtained for the τ lepton axial vector and vector couplings, a_τ and v_τ, are found to be $a_\tau = -0.5026 \pm 0.0010$ and $v_\tau = -0.0386 \pm 0.0023$. For the electron, $a_e = -0.50093 \pm 0.00064$ and $v_e = -0.0370 \pm 0.0021$, consistent with e$-\tau$ universality. A limit on the weak electric dipole moment, which, if nonzero, would indicate CP violation in τ production, was obtained of $|\tilde{d}_\tau(m_Z^2)| \leq 1.4 \times 10^{-17} e \cdot$ cm at 95% confidence level. Preliminary determination of some of the Michel parameters ρ, δ, and ξ are also given with and without the assumption of e-μ universality.

1.   Introduction

1.1.   Neutral Couplings

The τ lepton interacts with other forms of matter via the exchange of electroweak gauge bosons. The part of that interaction which involves the exchange of electrically uncharged bosons is said to mediated by neutral currents. Neutral current couplings to particles other than the τ have been measured before in a number of experiments [1]. In this summary we consider τ leptons produced in e$^+$e$^-$ collisions at the Large Electron Positron (LEP) collider at CERN. In this case the electromagnetic interactions are extremely weak in comparison to those mediated by the Z^0 and thus we are essentially studying the couplings of the τ to the Z^0. Of course, there is an electromagnetic interaction as well and, at the level of precision attained in the LEP experiments, this must be taken into account. We will consider the *effective* neutral couplings, which are those which appear in the following Feynman diagram describing the process in the improved Born approximation [2]. This procedure absorbs radiative corrections into the definition of these effective couplings.

---

*Presenting results from the LEP Collaborations : ALEPH, DELPHI, L3, and OPAL

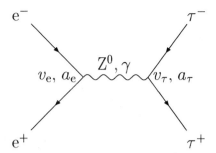

The vector and axial vector couplings v_τ and a_τ of the τ lepton as well as the corresponding couplings v_e and a_e of the electron are both present in this diagram. In general, observable quantities will depend on all four; similar comments can be made for the production of any other fermion from the Z^0. The feature which distinguishes the τ from all other fermions produced in Z^0 decays is the possibility, by a suitable set of observations (some using the decay of the τ itself as a sort of polarimeter) to measure all four coupling constants v_τ, a_τ, v_e, and a_e.

With the unpolarized beams of LEP there are four processes that can be measured for left-handed (L) or right-handed (R) initial and final state fermions. Denoting cross sections by σ, we have

$$
\begin{aligned}
\sigma_1 &= \sigma(e_L^- e_R^+ \longrightarrow \tau_L^- \tau_R^+) \\
\sigma_2 &= \sigma(e_L^- e_R^+ \longrightarrow \tau_R^- \tau_L^+) \\
\sigma_3 &= \sigma(e_R^- e_L^+ \longrightarrow \tau_R^- \tau_L^+) \\
\sigma_4 &= \sigma(e_R^- e_L^+ \longrightarrow \tau_L^- \tau_R^+) \\
\sigma_T &= \sigma_1 + \sigma_2 + \sigma_3 + \sigma_4
\end{aligned}
\tag{1}
$$

While none of these is directly measurable, they can be combined into four quantities which can be measured: the width for Z^0 decay into τ pairs $\Gamma(Z^0 \to \tau^+\tau^-)$, the forward-backward charge asymmetry A_{fb}, the average polarisation P_τ, and the forward-backward polarisation asymmetry A_{pol}^{FB}. These quantities are all calculable in the Standard Model and give rise to the following approximate dependences on the e and τ vector and axial vector coupling constants:

$$
\begin{aligned}
\Gamma(Z^0 \to \tau^+\tau^-) &\propto (a_e^2 + v_e^2)(a_\tau^2 + v_\tau^2) & &\propto \sigma_1 + \sigma_2 + \sigma_3 + \sigma_4 \\
A_{fb} &\simeq a_e v_e a_\tau v_\tau / ((a_e^2 + v_e^2)(a_\tau^2 + v_\tau^2)) & &\simeq (\sigma_1 - \sigma_2 + \sigma_3 - \sigma_4)/\sigma_T \\
P_\tau &\simeq -(2a_\tau v_\tau)/(a_\tau^2 + v_\tau^2) & &\simeq (\sigma_1 - \sigma_2 - \sigma_3 + \sigma_4)/\sigma_T \\
A_{pol}^{FB} &\simeq -(3/2)a_e v_e/(a_e^2 + v_e^2) & &\simeq (\sigma_1 + \sigma_2 - \sigma_3 - \sigma_4)/\sigma_T
\end{aligned}
\tag{2}
$$

All these quantities depend on the centre-of-mass energy, so data taken at energies off the Z^0 peak are corrected accordingly when used to extract a_τ and v_τ.

The Standard Model provides stringent constraints on the possible values of these coupling constants, and tests of the neutral coupling of the τ provide important tests of its basic structure. In the following, we will examine each of these processes and summarize the results from the LEP collaborations. The reader is urged to consult the papers of the relevant LEP experiments for more details.

The measurements of $\Gamma(Z^0 \to \tau^+\tau^-)$, A_{fb}, P_τ, and A_{pol}^{FB} are described in references 3, 4. All numbers cited in this summary are either published values quoted in the references above, or are preliminary numbers provided by the LEP collaborations.

1.2. Charged Couplings

Charged couplings of the τ are also studied at LEP to search for deviations from the V-A structure of the Standard Model. Measurable quantities include the celebrated Michel parameters in $\tau \to$ lepton events, and the chirality parameter, which is essentially the ν_τ polarization in hadronic τ decays.

2. τ selection at LEP and the data sample

$\tau^+\tau^-$ events are selected in different ways by each of the LEP collaborations according the specific strengths of their detectors. All experiments, however, essentially select their events by requiring two collimated almost back-to-back jets with low particle multiplicities (a jet may consist of a single charged particle). This requirement rejects $Z^0 \to$ hadrons. $Z^0 \to \mu^+\mu^-$ and $Z^0 \to e^+e^-$ events are rejected on the basis of their collinearity and very little missing energy. Two-photon events are rejected on the basis of their characteristically low energy and high acollinearity.

τ selections fall into two main classes. For the measurement of $\Gamma(Z^0 \to \tau^+\tau^-)$ and A_{fb} it suffices to have a sample of τ decays where any final state is allowed. For the polarisation measurements P_τ and A_{pol}^{FB} it is important to use only the decay channels for which analyses of the τ polarisation are possible. The branching ratios for the usable channels are [1] BR$(\tau^- \to e^-\nu_\tau\bar{\nu}_e) \sim 18\%$, BR$(\tau^- \to \mu^-\nu_\tau\bar{\nu}_\mu) \sim 18\%$, BR$(\tau^- \to \pi^-\nu_\tau) \sim 12\%$, BR$(\tau^- \to \rho^-\nu_\tau) \sim 24\%$, BR$(\tau^- \to a_1^-\nu_\tau; a_1^- \to \pi^-\pi^-\pi^+$ or $\pi^-\pi^0\pi^0) \sim 16\%$. Identification of these channels in general requires some degree of particle identification. The ρ^- and a_1^- decay modes are complicated by τ^- decays with multiple π^0's in the final state which must be separated by careful use of tracking and calorimeter information. The decay $\tau^- \to K^-\nu_\tau$, with a branching ratio less than 1%, is usually indistinguishable from $\tau^- \to \pi^-\nu_\tau$. For the purpose of polarisation measurements at the level of precision currently attainable this is not important. The branching ratio for $\tau^- \to (K/\pi)^- \geq 2\pi^0\nu_\tau$ is about 13% which makes it a significant background for ρ^- reconstruction.

The data on which the results presented here are based were taken during the LEP runs up to and including 1992, and in some cases also 1993. The present LEP data sample used by each collaboration corresponds to about about 200,000 $\tau^+\tau^-$ events per experiment.

3. $\Gamma(Z^0 \to \tau^+\tau^-)$ from lineshape

Conceptually, the measurement is simple, requiring a knowledge of the integrated luminosity, the number of selected τ pairs, and the efficiency and background content of the selected sample. The cross section varies as a function of centre-of-mass energy and cross sections must be calculated separately for different energies. The cross section as a function of centre-of-mass energy can then be fitted [5] for $\Gamma(Z^0 \rightarrow \tau^+\tau^-)$. The dominant error is due to the determination of luminosity, which is measured using $e^+e^- \rightarrow e^+e^-$ (Bhabha) events at small angles with respect to the beam. Selection uncertainties and uncertainties in the LEP beam energy contribute similar errors. These are smaller than those due to the luminosity determination.

The ratios of the Z^0 width into hadrons divided by its width into $\tau^+\tau^-$ obtained by the four LEP collaborations are shown in table 3.

Table 1. LEP values for R_ℓ.

ALEPH	DELPHI	L3	OPAL
20.69 ± 0.12	20.64 ± 0.16	20.70 ± 0.17	20.91 ± 0.13

Errors given on measured numbers are the combined systematic and statistical errors and the LEP average is made taking into account correlated errors between collaborations. This basic format will be used repeatedly in the presentation of results of various measurements.

For all collaborations preliminary numbers including 1993 data are given. The LEP average is 20.749 ± 0.070.

4. $A_{fb} - \tau$ **Forward-Backward Asymmetry**

The forward-backward (charge) asymmetry is a measure of the probability that a τ^- produced in e^+e^- annihilation will emerge in the same hemisphere as the incoming electron. From this point on, when discussing single τ leptons, we will specialize to the τ^- for clarity, with the understanding that the τ^+ are used in a similar fashion. Defining θ to be the polar angle between the e^- and the τ^-, the forward-backward asymmetry A_{fb} is defined by

$$A_{fb} = \frac{\sigma(\cos\theta > 0) - \sigma(\cos\theta < 0)}{\sigma(\cos\theta > 0) + \sigma(\cos\theta < 0)} \tag{3}$$

and is given by the expression in equation 2. A_{fb} can be extracted by fitting the angular distribution with the theoretical distribution, or by simply counting events. This measurement, like P_τ and A_{pol}^{FB}, has the advantage that many of the systematic errors such as luminosity determination and acceptance cancel in equation 3. Charge determination is a significant concern for this analysis and can be studied with events in which both τ's were assigned the same charge by the reconstruction algorithm. Final state radiative corrections are not negligible, and the use of programs such as ZFITTER [5] and KORALZ [6] is necessary in order to take these properly into account.

A summary of results from LEP is presented in figure 1. The LEP average is 0.0228 ± 0.0026 in good agreement with expectations from the Standard Model, which are also shown in this figure, for a range of top quark, Higgs, and Z^0 masses. For all collaborations preliminary numbers including 1993 data are given.

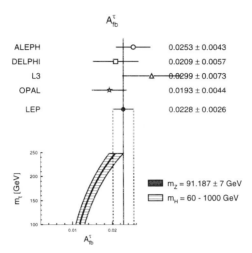

Fig. 1. LEP results on A_{fb}.

5. P_τ – The Average τ Polarisation

Helicity is conserved in τ pair production from Z^0 decays to a level of approximation which is good for the presently attainable precision of electroweak measurements. We can therefore describe the spin of a τ^- in terms of its helicity h. Interference of the vector and axial vector parts of the weak neutral current favours the production of one helicity state of τ^- over the other, and this results in an average polarisation P_τ :

$$P_\tau = \frac{\sigma(h = +1) - \sigma(h = -1)}{\sigma(h = +1) + \sigma(h = -1)} \qquad (4)$$

By convention, P_τ is the τ^- polarisation. The τ^+ is oppositely polarized so that $P_\tau \equiv P_{\tau^-} \equiv -P_{\tau^+}$.

The measurement of τ polarisation assumes that in τ^- decay, the ν_τ is left-handed and thus the decay of the τ tells us something about τ spin direction. Tests of the correctness of this assumption are discussed in section 5.1.

The π^- decay mode is the simplest one to understand and is the most sensitive (since the π^- itself has spin-0). Its branching ratio, however, diminishes its statistical weight in the polarization measurements, so that most of the information comes from $\tau^- \to \rho^- \nu_\tau$ (BR$\sim$ 24%), where the fact that the ρ^- has spin-1 rather than spin-0 is offset by the possibility of analysing its spin from its decay into pions. The other usable channels are $\tau^- \to e^- \nu_\tau \bar{\nu}_e$, $\tau^- \to \mu^- \nu_\tau \bar{\nu}_\mu$, $\tau^- \to \pi^- \nu_\tau$, $\tau^- \to a_1^- \nu_\tau$. The e and μ decay modes suffer from the presence of two undetectable final state particles, however information about polarisation can still be extracted from the laboratory momentum spectra of electrons and muons. The ρ^- decays mainly into $\pi^-\pi^0$ and the a_1^- decays mainly into 3 pions via $\rho\pi$. These two decays provide other variables, such as angles between decay products, in addition to the momentum in the laboratory frame which allow more information to be recovered. P_τ is determined either from analytical fits to detector-corrected distributions or by reweighting samples of reconstructed Monte Carlo τ events with positive and negative τ^- helicities to agree with data.

As most of the information about P_τ comes, in one form or another, from absolute energy or momentum measurements, it is of the utmost importance to know the energy scale of the detector used. Reconstruction of particles of known mass such as the ρ or K_S^0, and electrons and muons of known momenta from Z^0 decays provide valuable means to check this. In all analyses, data have to be corrected for the centre-of-mass energy dependence of P_τ in order to use data taken off the Z^0 peak. Corrections are also applied for radiative effects. The largest sources of systematic error are uncertainties in selection, background, and calibration. Limited Monte Carlo statistics are also a large source of systematic error. Reduction of systematic errors will become an increasingly important concern for future measurements.

P_τ can also be determined by measuring the acollinearity between the charged decay products of the τ^+ and τ^-. The helicities of the two τ leptons are opposite. In the laboratory frame, the energies and angles of the decay products are therefore correlated and the acollinearity angle will contain information on τ polarisation. Care must be taken that the final state particles are correctly identified. Polarisation information can be extracted from a fit to the acollinearity distribution after unfolding the angular resolution. Correction factors must be applied for radiative effects and their uncertainties make up a large part of the systematic error in this method. The other main contributions to the systematic error come from selection, background determination, and uncertainties in the branching ratios of the τ into various final states.

5.1. τ Polarisation and Deviations From V−A

In this section we focus on hadronic τ decays, for which there are experimental results concerning possible deviations from $V - A$ in the charged current coupling. Measurements of P_τ using single τ decays are actually measurements of $\gamma_{AV} P_\tau$ where γ_{AV} is a chirality parameter which measures deviations from $V - A$, being $+1$ for $V - A$ and -1 for $V + A$. γ_{AV} can be described in terms of the vector and axial vector coupling constants $g_v^{\tau\prime}$ and $g_a^{\tau\prime}$ for the charged (rather than the neutral) current interaction via which the τ decays. Note that allowing this possibility is in contradiction with lepton universality, as the $V - A$ structure of the charged current in μ decay is well established [1].

$$\gamma_{AV} = 2\frac{g_v^{\tau\prime} g_a^{\tau\prime}}{g_v^{\tau\prime 2} + g_a^{\tau\prime 2}} \tag{5}$$

The ARGUS collaboration has measured γ_{AV} using an asymmetry in the decay $\tau^- \to a_1^- \nu_\tau$; $a_1^- \to \pi^+\pi^-\pi^-$. They find $\gamma_{AV} = 1.25 \pm 0.23^{+0.15}_{-0.08}$ [7]. While this favours the assignment $\gamma_{AV} = 1$, it does not rule out small deviations from $\gamma_{AV} = 1$.

The ALEPH collaboration has noticed that the arguments made for the indistinguishability, in single τ^- decays, of deviations from $V - A$ from deviations in P_τ from its expected value do not apply to events where both the τ decays are analysed. Using the fact that the τ leptons are produced with opposite helicities, they arrive at a preliminary determination of $|\gamma_{AV}| = 0.951 \pm 0.051 \pm 0.028$ [8] (previous published value was $|\gamma_{AV}| = 0.99 \pm 0.07 \pm 0.04$ [9]) which suggests that the charged current is very nearly pure $V - A$ or very nearly pure $V + A$. Together with the sign from ARGUS, this suggests strongly that the charged current is indeed $V - A$ (assuming only vector and axial vector terms are present).

5.2. *Summary of Average τ Polarisation*

Figure 3 summarizes the measurements of $A_\tau \equiv -P_\tau$ by the LEP collaborations together with expectations from the Standard Model. Also shown is the corresponding value for electrons, described in the following section. Data for ALEPH and DELPHI are preliminary and include data from 1990-1992. L3 numbers are preliminary and include 1993 data. OPAL numbers are final and included data from 1990-1992. The LEP average is $P_\tau = -0.143 \pm 0.010$. It should be noted that the determined values of P_τ are systematically different depending on whether an analytical fit is performed or if Monte Carlo events with differing τ helicities are reweighted to reproduce the data. The difference arises depending on whether diagrams including photons are considered a part of the process or not.

Note that if no assumption is made about the $V - A$ nature of the charged current interaction, then an additional systematic error of about 0.008 due to the uncertainty in γ_{AV} has to be added in quadrature to the relative error on P_τ. This is not negligible compared to the error of 0.010 for the combined LEP result assuming $V - A$ for the τ charged current decays.

6. $A_{pol}^{FB} - \tau$ **Forward-Backward Polarisation Asymmetry**

The τ polarisation P_τ is itself a function of the polar angle θ of the τ^- with respect to the electron beam. The asymmetry in the angular distribution of P_τ is A_{pol}^{FB}. The measurement of A_{pol}^{FB} allows the determination of not just the vector and axial vector coupling constants of the τ, but also those of the electron. This is conveniently described by introducing a quantity A_e defined by $A_e = -(4/3)A_{pol}^{FB}$, where $A_e \simeq 2a_e v_e/(a_e^2 + v_e^2)$. Similarly, $A_\tau = -P_\tau \simeq 2a_\tau v_\tau/(a_\tau^2 + v_\tau^2)$. This facilitates comparisons between the taus and electrons, since $e-\tau$ universality predicts $A_\tau = A_e$.

The ALEPH results (figure 2) show the angular dependence of P_τ together with the fitted curves, both assuming (dashed line) and not assuming (solid line) $e-\tau$ universality. Figure 3 shows the results of the LEP collaborations together with the results expected from the Standard Model, as well as the average τ polarisation values. Data for ALEPH and DELPHI are preliminary and include data from 1990-1992. L3 numbers are preliminary and include 1993 data. OPAL numbers are final and included data from 1990-1992. All of the sources of systematic error which apply to the P_τ, of course, also apply to A_{pol}^{FB}. The averaged results are $A_e = 0.135 \pm 0.011$ and $A_\tau = 0.143 \pm 0.010$, consistent with $e-\tau$ universality.

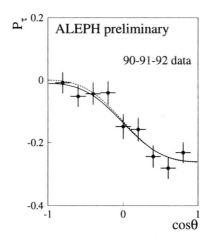

Fig. 2. $P_\tau(\cos\theta)$ from the ALEPH collaboration.

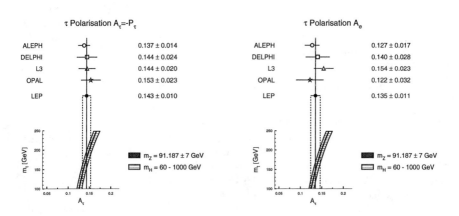

Fig. 3. LEP results for A_τ and A_e.

7. CP violation and Weak Electric Dipole Moments

This section contains the results of analyses which search for anomalous neutral couplings of the τ lepton which cannot be described in terms of v_τ and a_τ. While so far CP violation has only been observed in the kaon system, it is clearly of great interest to search for it in any physical process. Electric dipole coupling is described by a term in the Lagrangian of the form $\bar{\tau}\sigma^{\mu\nu}F_{\mu\nu}\gamma^5\tau$, where $F_{\mu\nu}$ is the electromagnetic field tensor, and $\sigma^{\mu\nu}$ the usual combination of Dirac gamma matrices. One can imagine an analogous term where $F_{\mu\nu}$ is replaced by the weak field tensor $Z_{\mu\nu}$. The proposed CP-violating term to be added to the Standard Model Lagrangian is then

$$L_{\text{CP violation}} = -\frac{1}{2}i\tilde{d}_\tau(q^2)Z_{\mu\nu}\bar{\tau}\sigma^{\mu\nu}\gamma_5\tau \tag{6}$$

where $\tilde{d}_\tau(q^2)$ is the weak-electric dipole moment. The energy scale for this moment is taken to be $q^2 = m_Z^2$, in contrast to the electric and magnetic dipole moments which are usually defined at $q^2 = 0$. For a more detailed description of the theory, see references 10, 12, and 13.

Large values of either $\tilde{d}_\tau(m_Z^2)$ can be ruled out already on the basis of lineshape measurements : $|\tilde{d}_\tau(m_Z^2)| \leq 0.7 \times 10^{-17}e \cdot$ cm at 95% CL. [12] These limits, however, assume no other new physics and it is interesting to derive a limit on $|\tilde{d}_\tau(m_Z^2)|$ using the expected CP violation in observable quantities should it be nonzero.

The basic idea is to look at final states with two charged particles, with momenta $\vec{q}_-$ and $\vec{q}_+$ labelled according to charge. From these a tensor T_{ij} is defined which is CP odd :

$$T_{ij} = (\vec{q}_- - \vec{q}_+)_i(\vec{q}_- \times \vec{q}_+)_j + (i \leftrightarrow j) \tag{7}$$

This tensor can be calculated from theory [10] to be

$$\langle T_{ij}\rangle_{a\bar{b}} = \tilde{d}_\tau(m_Z^2)C_{a\bar{b}}\frac{m_Z}{e}diag(-\frac{1}{6}, -\frac{1}{6}, \frac{1}{3}) \tag{8}$$

where a and $\bar{b}$ are the particles in the final state, and $C_{a\bar{b}}$ is a calculable constant for each final state. The average $\langle\rangle$ is taken over all events used in the analysis. Any deviation of this tensor from zero would indicate CP violation.

The OPAL collaboration obtains $|\tilde{d}_\tau(m_Z^2)| \leq 7.0 \times 10^{-17}e \cdot$ cm at 95% CL [13] where they have taken an average $C_{a\bar{b}}$. ALEPH repeated the analysis taking care to separate decay channels ($C_{a\bar{b}}$ differ in sign and magnitude so this is more sensitive). They obtain a preliminary limit [11] (updating their earlier published value which did not include 1992 data) of $|\tilde{d}_\tau(m_Z^2)| \leq 1.4 \times 10^{-17}e \cdot$ cm at 95% CL [12].

8.   Michel Parameters

The most general τ charged current couplings can include not only V and A, but also scalar (S), pseudoscalar (P), and tensor (T) terms. These are conventionally parametrised by 5 Michel parameters : $\rho, \eta, \xi, \delta, h$ (expected to be 3/4, 0, -1, 3/4, 1). Studying both the τ decays simultaneously, and using the fact that the τ^+ and τ^- must have opposite helicities, allows the determination of some of these parameters. ALEPH [14] has used the e, μ, π, and ρ decay modes, while L3 [15] has used the e, and μ modes to arrive at the preliminary determinations shown in table 2. The Michel parameters were originally defined for the decay $\mu^- \rightarrow e^- \nu_\mu \bar{\nu}_e$, so one may consider measuring potentially different Michel parameters for τ decays into electrons and into muons, or one may assume electron-muon universality. These various options are reflected in the table in parentheses.

Table 2. Preliminary Michel parameter determinations from LEP.

	ρ	δ	ξ
ALEPH (e)	0.838 ± 0.058	0.92 ± 0.29	1.05 ± 0.29
ALEPH (μ)	0.767 ± 0.049	0.66 ± 0.25	1.21 ± 0.25
ALEPH (univ.)	0.798 ± 0.041	0.79 ± 0.18	1.13 ± 0.18
L3 (univ.)		0.38 ± 0.60	1.28 ± 0.58

9.   Combined τ Results from LEP

Table 3 shows the results of a fit including the Z^0 mass and width, the total hadronic cross section, the ratios of e , μ, and τ cross sections to the hadronic cross section, P_τ and the A_{fb} for e, μ and τ, from the 4 LEP experiments [16] taking into account all correlated errors.

Table 3. LEP averages of lepton vector and axial vector coupling constants.

	a_ℓ	v_ℓ
e	-0.50093 ± 0.00064	-0.0370 ± 0.0021
μ	-0.50164 ± 0.00096	-0.0308 ± 0.0051
τ	-0.5026 ± 0.0010	-0.0386 ± 0.0023
ℓ	-0.50128 ± 0.00054	-0.0366 ± 0.0013

The effective weak mixing angle $\sin^2 \theta_w^{\text{eff}}$ is defined for any fermion of charge Q_f by $g_v/g_a = 1 + 4Q_f \sin^2 \theta_w^{\text{eff}}$. This differs by radiative corrections from $\sin^2 \theta_w = 1 - m_w^2/m_z^2$. The LEP average is $\sin^2 \theta_w^{\text{eff}} = .2322 \pm 0.0004^{+0.0001}_{-0.0002}$.

As can be clearly seen, the τ measurements provide some of the best information available on the weak neutral current from LEP. They have the advantage of being much better understood theoretically than the hadronic measurements which despite their large branching ratios, always involve QCD effects and fragmentation models.

10.   Conclusions

Some 10^6 $Z^0 \to \tau^+\tau^-$ events have been recorded by the LEP collaborations to date, and even at the time of writing more data is being collected and analysed. The Standard Model continues to hold up well. Many of the measurements will soon be limited by systematic errors, and the experimental situation becomes more challenging as more data is taken. In the meantime, theory advances with suggestions of new analysis techniques (see, for example, reference 17) and the possibility of searching for effects beyond the Standard Model, such as CP violation.

11.   Acknowledgements

I would like to thank the LEP collaborations, ALEPH, DELPHI, L3 and OPAL for allowing me to present and summarize their data. In addition to the LEP Electroweak Working Group, who did the averaging of the neutral current results, I would like to thank John Harton, Duncan Reid and Michael Roney of ALEPH, DELPHI, and OPAL respectively for their assistance. Special thanks go to Maria-Teresa Dova, Andrei Kunin, Tom Paul, Lucas Taylor, and all my other L3 collaborators in the τ analysis group.

References

1. Particle Data Group, Phys. Rev. **D45** (1992) vol. 11.
2. S. Jadach et al. in "Z Physics at LEP1", CERN Report CERN-89-08. ed. G. Altarelli et al. (CERN, Geneva, 1989) Vol. 1, p. 235.
3. The LEP Electroweak Working Subgroup on Lineshape and Lepton Forward-Backward Asymmetries, Internal Note LEPLINE/94-01, 14 July 1994.
4. Tau Polarisation Subgroup of the LEP Electroweak Working Group, Internal Note LEPTAU/94-02, 14 July 1994.
5. D. Bardin et al., Z. Phys. **C44** (1989) 493, and D. Bardin et al., Nucl. Phys. **B351** (1991) 1.
6. S. Jadach, B.F.L. Ward, and Z. Was, Comp. Phys. Comm. **66** (1991) 276.
7. ARGUS. Collab., Z. Phys. **C58** (1993) 61.
8. ALEPH Collab., D. Buskulic et al., preliminary numbers presented at the 27th International Conference on High Energy Physics, Glasgow, 20-27th July, 1994.
9. ALEPH Collab., Phys. Lett. **B321** (1994) 168.
10. W. Bernreuther, et al., Z. Phys. **C52** (1991) 567.
11. ALEPH Collab., private communication.
12. ALEPH Collab., D. Buskulic et al., Phys. Lett. **B297** (1992) 459.
13. OPAL Collab., P. D. Acton et al., Phys. Lett. **B281** (1992) 405.
14. ALEPH Collab., D. Buskulic et al., preliminary numbers presented at the 27th International Conference on High Energy Physics, Glasgow, 20-27th July, 1994.
15. L3 Collab., O. Adriani et al., preliminary numbers presented at the 27th International Conference on High Energy Physics, Glasgow, 20-27th July, 1994.
16. LEP Electroweak Working Group, Note LEPEWWG/94-02 (1994).
17. M. Davier et al., Phys. Lett. **B306** (1992) 411.

ELECTROWEAK B PHYSICS RESULTS FROM LEP

J.R.BATLEY *

Cavendish Laboratory, Madingley Road, Cambridge CB3 0HE, UK

ABSTRACT

A summary of recent measurements of the partial width ratio $R_b = \Gamma(Z^0 \to b\bar{b})/\Gamma(Z^0 \to \text{hadrons})$ and of the forward-backward asymmetry A^b_{FB} from the ALEPH, DELPHI, L3 and OPAL experiments at the LEP e^+e^- Collider is presented.

1. Introduction

The relative ease with which b quarks can be separated from other quark flavours and the availability of large Z^0 event samples allows precision tests of the Standard Model to be carried out using $Z^0 \to b\bar{b}$ decays at e^+e^- colliders. The quantities of interest at LEP are the partial width ratio $R_b = \Gamma(Z^0 \to b\bar{b})/\Gamma(Z^0 \to \text{hadrons})$, obtained from measurements of the cross section ratio $\sigma(e^+e^- \to b\bar{b})/\sigma(e^+e^- \to \text{hadrons})$ on the peak of the Z^0 resonance, and the forward backward charge asymmetry A^b_{FB}, obtained from measurements of the angular distribution $d\sigma/d\cos\theta \propto 1 + \cos^2\theta + \frac{8}{3}A^b_{FB}\cos\theta$, where θ is the angle of the outgoing b quark with respect to the initial e^- direction. The asymmetry A^b_{FB} arises from differences in the coupling strengths of the Z^0 to left- and right-handed fermions and is proportional to $(g^b_R)^2 - (g^b_L)^2/(g^b_R)^2 + (g^b_L)^2$, while the partial width ratio R_b depends on the combination $(g^b_R)^2 + (g^b_L)^2$.

The decay $Z^0 \to b\bar{b}$ is of particular theoretical interest. In the Standard Model, R_b is predicted to decrease by $\sim 2.5\%$ as the top quark mass m_{top} increases from $100\,\text{GeV}/c^2$ to $250\,\text{GeV}/c^2$. This variation is due largely to virtual top quark corrections to the $Zb\bar{b}$ vertex which are CKM-suppressed for Z^0 decays to other quark flavours. A precision measurement $(\Delta R_b/R_b < 1\%)$ of R_b would therefore represent a unique test of the Standard Model and would provide significant constraints on possible new physics such as additional Higgs bosons or supersymmetry.[1]

By the end of 1993, each of the four LEP experiments (ALEPH, DELPHI, L3 and OPAL) had recorded approximately 1.7×10^6 multihadronic events on and around the Z^0 resonance peak, including nearly 400k $Z^0 \to b\bar{b}$ decays per experiment. Events containing b quarks can be identified either by requiring a high momentum, high transverse momentum electron or muon ("lepton tagging"), by analysing in detail the overall shape and internal structure of the event ("event shape tagging"), or by exploiting the relatively long ($\tau_B \sim 1.5\,\text{ps}$) B hadron lifetime ("lifetime tagging").

2. Measurements of R_b

In inclusive charged lepton analyses, the preferred approach is to fit the two-dimensional p and p_T distributions for single lepton ($\ell^\pm$) and dilepton ($\ell^\pm\ell^\pm$, $\ell^\pm\ell^\mp$)

---

* Representing the ALEPH, DELPHI, L3 and OPAL Collaborations.

events together, and extract simultaneously some or all of the following parameters: Γ_b/Γ_h, Γ_c/Γ_h, $B(b \to \ell)$, $B(c \to \ell)$, $\langle x_E \rangle_b$, $\langle x_E \rangle_c$ and the mixing parameter $\bar{\chi}$.[2-5] This takes into account any correlations between R_b (for example) and the other parameters fitted, and minimises the systematic errors due to uncertainties in these other parameters. Nevertheless, relatively large systematic errors remain due to uncertainties in the modelling of b quark physics (the model used to describe semi-leptonic decays for instance), as well as due to experimental uncertainties in lepton identification efficiencies and backgrounds. The results are summarised in Fig. 1(a). The combined result from all four LEP experiments corresponds to a precision $\Delta R_b/R_b = 2.2\%$.

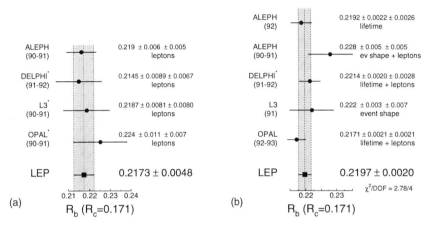

Fig. 1. Summary of LEP results on R_b: (a) from inclusive lepton analyses[2-5] and (b) from lifetime[6,3,7] or event shape[8,9] tagging. An asterisk indicates that the result is preliminary. The experimental numbers quoted are from the original references, while the data points and overall averages are from reference[10]; these assume the Standard Model prediction $R_c = 0.171$ and exclude the systematic error due to uncertainties in R_c.

The most precise measurements of R_b are now provided by "double tagging" analyses involving a lifetime tag, possibly in conjunction with a lepton or event shape tag.[6,3,7] The double tagging technique exploits the fact that the b quark and the $\bar{\text{b}}$ antiquark are typically produced approximately back-to-back, in separate hemispheres as defined by the thrust axis for example. A b quark tag is applied separately to each hemisphere in a sample of N_{had} multihadron events and the total number of tagged hemispheres, N_1, and of double-tagged events (*i.e.* both hemispheres generate a b tag), N_2, is measured. Neglecting background from charm and light quark events we have $N_1 = 2N_{had}R_b\epsilon_b$ and $N_2 = N_{had}R_b\epsilon_b^2$, from which both R_b and the b tagging efficiency, ϵ_b, can be extracted: $R_b = N_1^2/4N_2N_{had}$ and $\epsilon_b = 2N_2/N_1$. Thus the b quark tagging efficiency is determined from the data itself rather than from Monte Carlo, essentially eliminating all systematic errors associated with uncertainties in B hadron production and decay. The statistical precision is limited by the number of double-tagged events available ($\Delta R_b/R_b \sim 1/\sqrt{N_2}$), but is now as low as $\sim 1\%$ per experiment while still maintaining high b purity ($>90\%$).

The ALEPH[6] and DELPHI[3] double tagging analyses employ a lifetime tag derived from charged track impact parameters, requiring a small value of the "hemisphere probability", P_H, defined as the probability that all tracks in a given hemisphere were produced at the primary vertex. The OPAL analysis[7] uses an OR of a secondary vertex tag and a high p, high p_T lepton tag. All four LEP experiments have now installed high spatial precision silicon microvertex detectors close to the interaction point, allowing high purity lifetime tagging at good efficiency; ALEPH for example obtain a hemisphere b purity of 96% with a tagging efficiency of 26%.

Measurements of R_b from analyses involving lifetime or event shape tags (possibly combined with a lepton tag) are summarised in Fig. 1(b). The overall precision from LEP is $\Delta R_b/R_b = 0.9\%$, dominated by the lifetime double tagging analyses. The statistical and systematic errors are comparable, the latter receiving contributions mainly from uncertainties in the modelling of charm and light quark events, from the understanding of the impact parameter resolution of the detector, and from the fact that the tagging efficiencies for the two hemispheres of a given event are in practice slightly correlated.

3. Measurements of A_{FB}^b

The on-peak forward-backward asymmetry for b quarks, A_{FB}^b, is expected to be much larger ($A_{FB}^b \sim 10\%$) than that for leptons ($A_{FB}^\ell \sim 1\%$) and to be more sensitive to $\sin^2\theta_W$ and to m_{top}. However, $Z^0 \to b\bar{b}$ decays are harder to identify than lepton pair events, and it is less straightforward to define the b quark direction (usually approximated by the thrust axis) or to assign the b quark to the forward or backward hemisphere. Two approaches have been used: (i) tag $Z^0 \to b\bar{b}$ events using high p, high p_T leptons and use the lepton charge to identify the b quark hemisphere ($b \to l^-$, $\bar{b} \to l^+$)[11,3,12,5] or (ii) tag $Z^0 \to b\bar{b}$ events using a lifetime tag and use the "jet charge", Q_{jet}, in each hemisphere to flag the b quark.[13,3,14]

The jet charge is a momentum weighted sum over the tracks in each hemisphere: $Q_{\text{jet}} = \sum_i p_i^\kappa Q_i / \sum_i p_i^\kappa$, where p_i is usually the momentum component along the thrust axis, and the free parameter κ (typically $\sim 0.5-0.7$) is chosen to optimise the $b,\bar{b}$ discrimination. Two methods have been used: (i) an event-by-event method in which the b quark is assigned to the hemisphere with lowest Q_{jet}, and a lepton tag in the opposite hemisphere is used to measure the probability that the correct assignment has been made (probabilities $\sim 0.65-0.70$ are typical); (ii) a statistical method whereby A_{FB}^b is derived from measurements of the average charge difference $\langle Q_F - Q_B \rangle$ and charge product $\langle Q_F \cdot Q_B \rangle$.

The lepton and jet charge measurements of A_{FB}^b on the Z^0 peak are summarised in Fig. 2(a). Both approaches are still statistics limited and achieve a similar overall precision. Measurements of A_{FB}^b at energies above and below the resonance peak for the lepton analyses are summarised in Fig. 2(b) and are seen to be in good agreement with the Standard Model prediction.

4. Combined LEP Results

The four LEP experiments have combined their electroweak heavy flavour measurements by performing a simultaneous fit for Γ_b/Γ_h, Γ_c/Γ_h, $B(b \to \ell)$, $B(c \to \ell)$, A_{FB}^b,

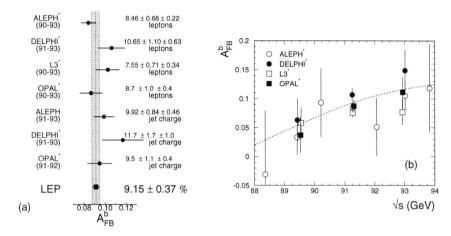

Fig. 2. Summary of LEP results on A_{FB}^b: (a) from on-peak measurements using leptons [11,3,12,5] or jet charge, [13,3,14] and (b) from leptons as a function of $\sqrt{s}$. [11,3,12,5] . An asterisk indicates that the result is preliminary. The combined LEP asymmetry in (a) is from reference[10] and corresponds to the measured asymmetry at $\sqrt{s} = 91.26$ GeV. The dashed curve in (b) is the Standard Model prediction from ZFITTER,[15] assuming $m_{top} = 170$ GeV/c^2 and $m_{Higgs} = 300$ GeV/c^2.

A_{FB}^c and the mixing parameter $\bar{\chi}$.[10] The data from each experiment were adjusted to use a common set of input assumptions and models, and to use common criteria for the estimation of systematic errors. Correlations between different analyses and different experiments were taken fully into account. The combined fit gives $Z^0 \to b\bar{b}$ and $Z^0 \to c\bar{c}$ fractions $R_b = 0.2202 \pm 0.0020$ and $R_c = 0.1583 \pm 0.0098$, and a pole asymmetry (*i.e.* the asymmetry due to Z^0 exchange alone at $\sqrt{s} = M_Z$) of $A_{FB}^{0,b} = 0.0967 \pm 0.0038$. With R_c fixed to the Standard Model value, $R_c = 0.171$, the $Z^0 \to b\bar{b}$ fraction becomes $R_b = 0.2192 \pm 0.0018$. These measurements are compared with the Standard Model predictions in Fig. 3. The result for $A_{FB}^{0,b}$ can be expressed as a measurement of the effective Weinberg angle θ_{eff}: $\sin^2 \theta_{eff} = 0.2327 \pm 0.0007$, a precision slightly better than that achieved from measurements of $e^+e^- \to \ell^+\ell^-$ asymmetries.

5. Summary

Measurements of the $Z^0 \to b\bar{b}$ fraction R_b and the forward backward asymmetry A_{FB}^b at LEP are now precision measurements: $\Delta R_b/R_b < 1\%$ and $\Delta A_{FB}^b \sim 0.4\%$. The R_b measurement is dominated by double tagging analyses involving a lifetime tag, and has comparable statistical and systematic errors. The A_{FB}^b measurement is still statistics limited. Both are already at a level where they are sensitive to the top quark mass via the effects of higher order virtual corrections, and can provide significant constraints on possible extensions of the Standard Model.

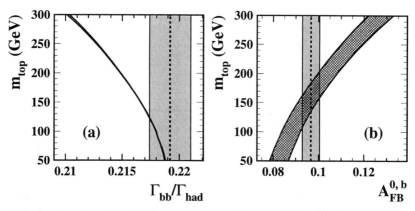

Fig. 3. Comparison of combined LEP measurements of (a) R_b and (b) $A_{FB}^{0,b}$ with Standard Model predictions. The hatched regions represent the effect of varying the Higgs mass from $60-1000\,\mathrm{GeV}/c^2$.

Acknowledgements

I am grateful to I. Brock, A. Halley, S.L. Lloyd, K. Moenig, D. Schaile, P. Wells and, especially, R.W.L. Jones for their help in the preparation of this talk, and to the organisers of the DPF'94 Conference for their hospitality.

References

1. S. Chivukula, these proceedings.
2. ALEPH Collaboration, D. Buskulic et al., Z. Phys. **C62** (1994) 179.
3. DELPHI Collaboration, DELPHI 94-111 PHYS 428, July 1994; G. Borisov, these proceedings; P. Antilogus, these proceedings;
4. L3 Collaboration, L3 Note 1625, July 1994.
5. OPAL Collaboration, OPAL Physics Note PN118, March 1994.
6. ALEPH Collaboration, D. Buskulic et al., Phys. Letters **B313** (1993) 535.
7. OPAL Collaboration, R. Akers et al., CERN PPE/94-106, July 1994.
8. ALEPH Collaboration, D. Buskulic et al., Phys. Letters **B313** (1993) 549.
9. L3 Collaboration, O. Adriani et al., Phys. Letters **B307** (1993) 237.
10. Heavy Flavour Subgroup of the LEP Electroweak Working Group, D. Abbaneo et al., Internal Note LEPHF/94-03, ALEPH 94-119, DELPHI 94-108, L3 Note 1627, OPAL TN242, July 1994.
11. ALEPH Collaboration, ALEPH 94-036, March 1994.
12. L3 Collaboration, L3 Note 1625, July 1994; L3 Collaboration, M. Acciarri et al., CERN PPE/94-89, June 1994.
13. ALEPH Collaboration, D. Buskulic et al., CERN PPE/94-84, July 1994.
14. OPAL Collaboration, OPAL Physics Note PN127, March 1994.
15. D. Bardin et al., Z. Phys. **C44** (1989) 493, Nuc. Phys. **B351** (1991) 1, Phys. Letters **B255** (1991) 290, CERN-TH 6443/92.

PRECISE MEASUREMENT OF THE LEFT-RIGHT CROSS SECTION ASYMMETRY IN Z BOSON PRODUCTION BY e^+e^- COLLISIONS

AMITABH LATH*

Massachussetts Institute of Technology, Cambridge, MA 02139

ABSTRACT

We present a precise measurement of the left-right cross section asymmetry (A_{LR}) for Z boson production by e^+e^- collisions. The measurement was performed at a center-of-mass energy of 91.26 GeV with the SLD detector at the SLAC Linear Collider (SLC). The luminosity-weighted average polarization of the SLC electron beam was $(63.0\pm1.1)\%$. Using a sample of 49,392 Z decays, we measure A_{LR} to be $0.1628\pm0.0071(\text{stat.})\pm0.0028(\text{syst.})$ which determines the effective weak mixing angle to be $\sin^2\theta_W^{\text{eff}} = 0.2292 \pm 0.0009(\text{stat.}) \pm 0.0004(\text{syst.})$. We combine this result with our result obtained in 1992 and obtain $\sin^2\theta_W^{\text{eff}} = 0.2294 \pm 0.0010$, which constitutes the best measurement of $\sin^2\theta_W^{\text{eff}}$ in a single experiment presently available.*

1.   Introduction

The left-right asymmetry at the Z-pole is defined as $A_{LR}^{\text{o}} \equiv (\sigma_L - \sigma_R)/(\sigma_L + \sigma_R)$ where σ_L and σ_R are the e^+e^- production cross sections for Z bosons at the Z pole energy with left-handed and right-handed electrons, respectively. The standard model predicts that this quantity depends strongly on the effective weak mixing angle, $\sin^2\theta_W^{\text{eff}}$.[1] The measurement[2] was performed using the SLD detector, where we counted the number (N_L, N_R) of hadronic decays of the Z boson for each of the two longitudinal polarization states (L,R) of the electron beam. The electron beam polarization was measured precisely with a Compton polarimeter. From these measurements we determined the left-right asymmetry[†],

$$A_{LR}(E_{cm}) = \frac{1}{\mathcal{P}_e^{lum}} \cdot \frac{N_L - N_R}{N_L + N_R}, \tag{1}$$

where E_{cm} is the mean luminosity-weighted collision energy, and $\mathcal{P}_e^{lum}$ is the mean luminosity-weighted polarization.

A_{LR} has several desirable properties. The asymmetry is large in magnitude relative to other asymmetries at the Z pole, and is a sensitive function of $\sin^2\theta_W^{\text{eff}}$, with

---

*Representing the SLD Collaboration.

*Work supported in part by the Department of Energy, contract DE-AC03-76SF00515.

[†]A_{LR}, unlike A_{LR}^{o}, is not corrected for the effects of $\gamma - Z$ interference and initial-state radiation.

$\delta \sin^2 \theta_W^{\text{eff}} \approx \delta A_{LR}/7.9$. It does not depend on the couplings of the Z to final-state fermions, so all final states (except the e^+e^- final state) can be used in the measurement. The measurement does not require knowledge of absolute luminosity, detector acceptance, or detector efficiency. The principal source of systematic uncertainty is in the determination of $\mathcal{P}_e^{lum}$.

2. Production and Transport of Polarized Electrons

The polarized electrons were produced by a strained-lattice photocathode. Such cathodes are capable of delivering electron beams with higher polarizations than conventional GaAs cathodes. Conventional cathodes are limited to a maximum 50% polarization due to the degeneracy of two competing photo-ionization transitions. In strained-lattice cathodes, an epitaxial layer of GaAs is deposited over a GaAsP substrate, which has a smaller lattice spacing than GaAs, and thus creates a mechanical strain which removes the degeneracy.[3] With this technique, polarizations significantly larger than 50% are possible.

After extraction from the cathode, the electrons were injected into the accelerator. The electron spin orientation was longitudinal at the source and remained longitudinal until the transport to the Damping Ring (DR). The Linac-to-Ring transport rotated the electron spin to be vertical in the DR to preserve the polarization during the 8ms storage time. The spin orientation was vertical upon extraction from the DR and remained vertical during injection into the linac and subsequent acceleration to 46 GeV.

The SLC North Arc (NARC) transported the electron beam from the linac to the SLC Interaction Point (IP). The betatron phase advance and the spin-precession in each section of the NARC were close to the same value. The NARC was therefore operating near a spin-tune resonance. We took advantage of this resonance and introduced a pair of vertical betatron oscillations, called *spin bumps*, to adjust the spin direction.[4] The amplitudes of these spin bumps were empirically adjusted to achieve longitudinal polarization at the IP.

3. The Compton Polarimeter

The longitudinal polarization of the electron beam $(\mathcal{P}_e)$ was measured by the Compton polarimeter.[5] This polarimeter detected Compton-scattered electrons from the collision of the longitudinally polarized electron beam with a circularly polarized photon beam; the photon beam was produced from a pulsed Nd:YAG laser operating at 532 nm. The Compton Interaction Point (CIP) was situated 33m past the IP. After the CIP, the electrons passed through a dipole spectrometer; a nine-channel Čerenkov detector (CKV) then measured scattered electrons in the 17 to 30 GeV range.

The counting rates in each channel of the CKV were measured for parallel and anti-parallel combinations of the photon and electron beam helicities. The Compton asymmetry, the difference over the sum of the two combinations, was used to determine the beam polarization. The Compton asymmetry depends on polarization

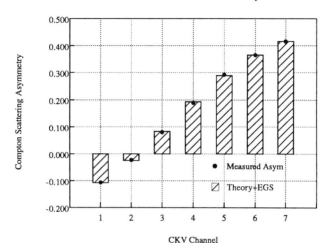

Figure 1: Measured and expected Compton asymmetry.

as $A(E) = \mathcal{P}_e \mathcal{P}_\gamma a_i$, where $A(E)$ is the measured Compton asymmetry, $\mathcal{P}_\gamma$ is the laser polarization, and a_i is the analyzing power for a given channel.

Measurements of $\mathcal{P}_\gamma$ were made before and after the CIP. By monitoring and correcting for small phase shifts in the laser transport line, we were able to achieve $\mathcal{P}_\gamma = (99.2\pm0.6)\%$ for the latter three-quarters of the 1993 data. For the intial quarter of the 1993 data, measurements yielded $\mathcal{P}_\gamma = (97 \pm 2)\%$.

The analyzing power for a given channel was determined by integrating the calculated Compton asymmetry function over the acceptance of the channel, modified by small corrections ($\approx 1\%$) due to electromagnetic shower effects as modeled by the EGS Monte Carlo. Figure 1 shows the good agreement achieved between the measured and expected Compton asymmetry spectrum.

Once the detector was calibrated, we used channel 7 to measure $\mathcal{P}_e$ precisely. The asymmetry spectrum observed in channels 1-6 then served as a cross-check; deviations of the measured asymmetry spectrum from the modeled one are reflected in the inter-channel consistency systematic error.

Polarimeter data were acquired continually during the operation of the SLC. The absolute statistical precision attained in a 3 minute interval was typically $\delta \mathcal{P}_e = 1.0\%$. Averaged over the 1993 run the mean beam polarization was $\mathcal{P}_e = (61.9\pm0.8)\%$. The systematic uncertainties that affect the polarization measurement are summarized in Table 1.

4.   The Chromatic Correction

The Compton polarimeter measured $\mathcal{P}_e$ (the beam polarization weighted by the beam number density), but the quantity of interest in the A_{LR} measurement is the luminosity-weighted beam polarization, $\mathcal{P}_e^{lum}$ (the beam polarization weighted by the beam number density and luminosity).

Off-energy electrons had reduced longitudinal polarization at the IP due to spin precession in the NARC. They also contributed less to the luminosity than on-energy electrons because they did not focus to a small spot at the IP; however they contributed the same as on-energy electrons to the Compton measurement of the beam polarization. Thus, $\mathcal{P}_e^{lum}$ could be greater than $\mathcal{P}_e$. Using dedicated polarization transport studies, the relative difference between $\mathcal{P}_e$ and $\mathcal{P}_e^{lum}$, labelled the chromatic correction, was conservatively estimated to be $(1.7\pm1.1)\%$.

We find the luminosity-weighted polarization for the 1993 run, after the chromatic correction is applied, to be $\mathcal{P}_e^{lum} = (63.0 \pm 1.1)\%$.

5.   Event Selection

The e^+e^- collisions were measured by the SLD detector, which has been described elsewhere.[6] The triggering of the SLD relied on a combination of calorimeter and tracking information, while the event selection was based entirely on the liquid argon calorimeter (LAC).[2]

For each event candidate that passed minimal selection criteria, energy clusters were reconstructed in the LAC. Selected events were required to contain at least 22 GeV of energy in the clusters, and have a normalized energy imbalance, $\sum \vec{E}/\sum E$, of less than 0.6.

Since the left-right asymmetry associated with final state e^+e^- events is diluted by the t-channel photon exchange, we excluded e^+e^- final states by requiring at least nine calorimeter clusters if the event was contained in the central region, ($|\cos\theta| < 0.8$, where θ is the angle the thrust axis makes with the beamline) and at least twelve clusters if the event was in the forward region ($|\cos\theta| \geq 0.8$).

6.   Measurement of A_{LR}

Applying the selection criteria, we find 27,225 (N_L) of the events were produced with the left-polarized electron beam and 22,167 (N_R) were produced with the right-polarized beam. To determine A_{LR}, we use Eq. 1 corrected by some small terms, including terms for backgrounds, luminosity asymmetry, polarization asymmetry, energy asymmetry, efficiency asymmetry and possible positron polarization. These corrections were found to be small, affecting the A_{LR} measurement by $(0.1 \pm 0.08)\%$.

After corrections, we find the left-right asymmetry to be

$$A_{LR}(91.26 \text{ GeV}) = 0.1628 \pm 0.0071(\text{stat.}) \pm 0.0028(\text{syst.}). \qquad (2)$$

Correcting this result to account for photon exchange and for electroweak interference which arises from the deviation of the effective e^+e^- center-of-mass energy from the

Z-pole energy (including the effect of initial-state radiation), we find the effective weak mixing angle and the Z-pole asymmetry to be

$$\sin^2 \theta_W^{\text{eff}} = 0.2292 \pm 0.0009(\text{stat.}) \pm 0.0004(\text{syst.})$$
$$A_{LR}^{\circ} = 0.1656 \pm 0.0071(\text{stat.}) \pm 0.0028(\text{sys.}). \tag{3}$$

Table 1: Systematic Uncertainties

Systematic Uncertainty	$\delta \mathcal{P}_e / \mathcal{P}_e$ (%)	$\delta A_{LR} / A_{LR}$ (%)
Laser Polarization ($\mathcal{P}_\gamma$)	1.0	
Detector Calibration	0.4	
Detector Linearity	0.6	
Inter-channel Consistency	0.5	
Electronic Noise	0.2	
Total Polarimeter Uncertainty	1.3	1.3
Chromatic Correction		1.1
Backgrounds and Other Asymmetries		0.1
Total Systematic Uncertainty		1.7

7.   Conclusions

We note that this is the most precise single determination of $\sin^2 \theta_W^{\text{eff}}$ yet performed. Combining this value of $\sin^2 \theta_W^{\text{eff}}$ with our 1992 measurement[7] at $E_{CM} = 91.55$ GeV, we obtain $\sin^2 \theta_W^{\text{eff}} = 0.2294 \pm 0.0010$. The LEP collaborations combine roughly 30 individual measurements of quark and lepton forward-backward asymmetries, final state τ-polarization, and the forward-backward asymmetry of τ-polarization to give a LEP global average of $\sin^2 \theta_W^{\text{eff}} = 0.2322 \pm 0.0005$.[8]

The 1994 running period for SLD began June 1, 1994 and beam polarizations of over 80% have been achieved. The luminosity goals for the 1994 run call for over 100,000 Z events. By the end of this run, the SLD expects to achieve a statistical precision approaching 0.0004 for $\sin^2 \theta_W^{\text{eff}}$.

References

1. We follow the convention used by the LEP Collaborations in *Phys. Lett.* **B276** (1992) 247.
2. K. Abe *et al. Phys. Rev. Lett.* **73** (1994) 25.
3. T. Maruyama *et al. Phys Rev.* **B46** (1992) 4261
4. T. Limberg, P. Emma, and R. Rossmanith, SLAC-PUB-6210, May 1993.
5. D. Calloway *et al.* in preparation.
6. The SLD Design Report, SLAC Report 273, 1984.
7. K. Abe *et al. Phys. Rev. Lett.* **70**, (1993) 2515
8. The LEP results were presented by B. Pietrzyk at the XXIXth Recontres de Moriond. See also CERN-PPE/93-157, August 1993.

THE LEFT-RIGHT FORWARD-BACKWARD ASYMMETRY OF HEAVY QUARKS MEASURED WITH JET CHARGE AND WITH LEPTONS AT THE SLD*

DAVID C. WILLIAMS

Massachusetts Institute of Technology, Cambridge, MA 02139

Representing the **SLD Collaboration**

ABSTRACT

We present direct measurements of the left-right asymmetry of b- and c-quarks from the decay of Z^0 bosons produced in the annihilation of longitudinally polarized electrons and unpolarized positrons. Two complementary techniques are presented: 1) $Z^0 \to b\bar{b}$ decays are tagged using track impact parameters with $b\bar{b}$ discrimination provided by momentum-weighted track charge; 2) Semileptonic b-decays are tagged using high p and p_T muons and electrons. The preliminary results from our 1993 data sample are: $A_b = 0.93 \pm 0.13 \pm 0.13$ for the jet charge and $A_b = 0.93 \pm 0.14 \pm 0.09$, and $A_c = 0.40 \pm 0.23 \pm 0.20$ for the leptons, where the first error is statistical and the second systematic.

1.   Introduction

Measurements of the fermion asymmetries at the Z^0 provide direct probes of the parity violating left-right asymmetry $A_f = 2v_f a_f / (v_f^2 + a_f^2)$ and hence provide a sensitive test of Standard Model predictions. The left-right forward-backward asymmetry $\tilde{A}_{FB}^b$ isolates the A_f term by taking advantage of the electron beam polarization:[1]

$$\tilde{A}_{FB}^f(y = \cos\theta) = \frac{(\sigma_L(y) - \sigma_L(-y)) - (\sigma_R(y) - \sigma_R(-y))}{(\sigma_L(y) + \sigma_L(-y)) + (\sigma_R(y) + \sigma_R(-y))} = P_e A_f \frac{2y}{1 + y^2} \qquad (1)$$

where σ is the differential cross section for $e^+ e^- \to Z \to f\bar{f}$ and L (R) refers to left (right) incident electron helicity, with $P_e < 0$ ($P_e > 0$).

We present preliminary measurements[2] of A_b and A_c from a sample of 50,000 Z^0 decays observed at $\sqrt{s} = 91.26$ GeV, with an average longitudinal electron polarization of $63.0 \pm 1.1\%$.[3] The SLD detector is described elsewhere.[4,6]

2.   Track-Charge Analysis

Hadronic Z^0 decays are selected by requiring that at least 7 well reconstructed tracks within $|\cos\theta| < 0.80$ carry a large visible energy $E_{vis} > 18$ GeV. The dominant contribution to the residual background, $Z^0 \to \tau\bar{\tau}$ events, is estimated to be less than 0.2% of the sample. Details of the impact parameter b-tag are given in Ref. 5.

The jet charge Q_p can be written:

$$Q_p = \sum_{tracks} Q_i \vec{p}_i \cdot \hat{t} |\vec{p}_i \cdot \hat{t}|^{\kappa - 1} \qquad (2)$$

where $\hat{t}$ is a unit vector in the direction of the thrust axis. To determine A_b, the sign of $\hat{t}$

---

* Work supported by the U.S. Department of Energy under Contract DE-AC03-76SF00515

measured by calorimetry in tagged events is chosen such the Q_p is negative. $\hat{t}$ is then used as the estimate of the b-quark direction where a κ of 0.5 is chosen to maximize the correct-sign probability. The result is shown in Fig. 1.

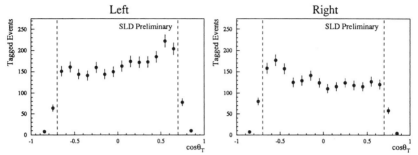

Fig. 1. Signed $\cos\theta_T$ distribution for tagged events produced with left- and right-handed beams, illustrating the significant effect of polarization on the forward-backward asymmetry.

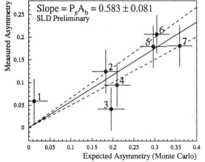

Fig. 2. The observed asymmetry $\tilde{A}^b_{FB}(\cos\theta_T)$, corrected for light-quark contamination, plotted against the Monte Carlo prediction for 100% asymmetry for each bin (numbered points) of $|\cos\theta_T|$. The slope of the fit yields $P_e A_b$.

Table 1. Relative systematic errors on A_b for the track-charge analysis.

Contribution	% Error
Physics	
B-Decay Model	7
b Fragmentation	2
B–$\bar{B}$ mixing	1
Polarization	2
Detector and Tag	
Tracking Efficiency	4
Thrust Axis Resolution	2
Tag Modeling	10
Monte Carlo Statistics	5
Total (Quadrature Sum)	14

To determine A_b, the double asymmetry $\tilde{A}^b_{FB}(\cos\theta)$ is formed for each bin in $|\cos\theta|$ and corrected for bin-dependent tag contamination. The result is compared against Monte Carlo predictions at 100% asymmetry to account for bin-dependent Q_p resolution and thrust axis smearing. In addition a first-order QCD correction[7,8] (2–5%) is applied. A linear fit through points representing each bin plotted as data verses Monte Carlo produces a slope equal to a measurement of $P_e A_b$ (see Fig. 2).

The dominant systematic error arises from a lack of a satisfactory explanation for variation in the measured value of A_b when the tag requirement is varied from 2 tracks at $+2\sigma$ impact parameter to 4 tracks at $+4\sigma$. Table 1 summarizes our estimates of these and other errors, which are combined in quadrature for our preliminary result:

$$A_b = 0.93 \pm 0.13 \text{ (stat)} \pm 0.13 \text{ (sys)} \tag{3}$$

3. Lepton Analysis

Electrons are identified by extrapolating charged tracks to the barrel liquid argon calorimeter (LAC) and analyzing the energy deposited in nearby towers. The energy is required to match the momentum of the track and follow the longitudinal profile of an immediate electromagnetic shower. The efficiency for identifying electrons in hadronic Z^0 events is 55% within acceptance. Above 2 GeV, pion misidentification is less than 0.8%. Ninety percent of electron candidates from gamma conversions are identified and removed by searching for a match with a second track consistent with the decay of a massless particle outside a radius of 2 cm.

Muons are identified by matching charged tracks to hits in the 18 active layers of the warm iron calorimeter. Muons candidates are also required to have at least two hits in the last four active layers, forcing the track to penetrate at least 7.3 interaction lengths. Above 3 GeV efficiency is 85% and pion misidentifcation is below 0.3%.[6]

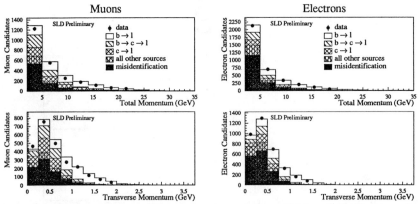

Fig. 3. Spectra of tagged lepton candidates projected onto p and p_T with Monte Carlo estimations of contributions from signal and background sources.

Jets are identified using LAC energy clusters and the JADE[9] algorithm with $y_{cut} = 0.005$. The jet angular resolution is approximately 20 millirad. Event selection is a slightly looser version of the one used for the jet charge.

To determine the best estimates of A_b and A_c, we use the following probability function:

$$p_i(A_b, A_c) = 1 + \cos^2\theta + 2PA\cos\theta$$
$$A = w_b A_b (1 - 2\chi)(1 + \Delta_{QCD}^b(\theta)) - w_c A_c (1 + \Delta_{QCD}^c(\theta)) + w_h A_h \tag{4}$$

where Δ_{QCD} are the θ-dependent first-order QCD corrections[7,8] and A_h is the p and p_T dependent asymmetry of all non-lepton charged tracks. The weights w_b, w_c, and w_h are the

p and p_T dependent amount each lepton is affected by b-quark, c-quark, or hadron asymmetries. Each weight is derived from the Monte Carlo by tabulating the fraction of the leptons at a particular p and p_T that come from one of six relevant sources:

$$w_b(p,p_T) = f_{b\to l} - f_{b\to c\to \bar{l}} + f_{b\to \bar{c}\to l} \qquad w_c(p,p_T) = f_{c\to \bar{l}} \qquad w_h(p,p_T) = f_{hadron} + f_{mis-id} \quad (5)$$

To determine f, the subset of 50 leptons in the Monte Carlo closest in $\ln|p|/2$ and p_T to each lepton in the data is identified and the various sources tallied.

The results of the maximum likelihood fit are:

Muons:	$A_b = 0.94 \pm 0.20 \pm 0.10$	$A_c = 0.42 \pm 0.29 \pm 0.18$
Electrons:	$A_b = 0.92 \pm 0.19 \pm 0.12$	$A_c = 0.37 \pm 0.37 \pm 0.31$
Combined:	$A_b = 0.93 \pm 0.14 \pm 0.09$	$A_c = 0.40 \pm 0.23 \pm 0.20$ (6)

where the first error is statistical and the second systematic (see Table 2). The statistical correlation between A_b and A_c is 0.03 for muons and 0.29 for electrons.

Table 2. The absolute systematic errors for A_b and A_c measured using maximum likelihood.

Source	A_b		A_c	
	Muons	Electrons	Muons	Electrons
Tracking Efficiency	0.013	0.016	0.013	0.004
Jet Axis Simulation	0.060	0.045	0.061	0.130
Background Level	0.019	0.008	0.028	0.014
B mixing	0.027	0.025	—	—
Method of determining w	0.040	0.083	0.082	0.185
$\Gamma (Z^0 \to bb)$	0.005	0.002	0.001	0.003
$\Gamma (Z^0 \to cc)$	0.007	0.002	0.027	0.033
$B^0, B^\pm$ Lepton Spectrum	0.022	0.046	0.116	0.140
B_s Lepton Spectrum	0.032	0.018	0.046	0.054
Λ_b Lepton Spectrum	0.008	0.012	0.023	0.026
D Lepton Spectrum	0.022	0.020	0.037	0.098
b Fragmentation	0.011	0.015	0.012	0.011
c Fragmentation	0.001	0.002	0.001	0.001
Background Asymmetry	0.006	0.011	0.027	0.087
Polarization	0.018	0.018	0.008	0.007
Second Order QCD	0.008	0.008	0.040	0.040
Total (Quadrature Sum)	0.096	0.117	0.179	0.308

References

1 A. Blondel, et al., Nucl. Phys. **B304** (1988) p. 438–450.
2 See also M. King for a report on A_c measured with D^+, these proceedings.
3 A. Lath, these proceedings.
4 SLD Design Report, SLAC Report SLAC-273 (1984); G.D. Agnew, et al., Proceedings of the XXVI Int'l Conf. on High Energy Physics, Dallas (1992); D. Axen, et al., Nucl. Instr. Meth. **A328** (1993) p. 472–494; A. C. Benvenuti, et al., Nucl. Instr. Meth. **A290** (1990) p. 353.
5 K. Abe, et al., SLAC-PUB 6292 (1993), K. Abe, et al., SLAC-PUB 6569 (1994).
6 D. C. Williams, Ph.D. Thesis, SLAC-REPORT-445 (1994).
7 A. Djouadi, J.H. Kühn, and P.M. Zerwas, Z. Phys **C46** (1990) p. 411–417.
8 G. Altarelli and B. Lampe, Nucl. Phys. **B391** (1993) p. 3–22.
9 JADE Collaboration, Z. Phys. **C33** (1986) p. 23.

MEASUREMENT OF THE LEFT–RIGHT FORWARD–BACKWARD ASYMMETRY FOR CHARM QUARKS WITH D^{*+} AND D^+ MESONS

The SLD COLLABORATION [†]

MARY E. KING

Stanford Linear Accelerator Center
Stanford, CA 94309 USA

ABSTRACT

Longitudinal polarization of the SLC e^- beam allows a direct measurement of the left-right forward-backward asymmetry of c-quarks in SLD. Events arising from $Z^0 \rightarrow c\bar{c}$ are tagged using fully and partially reconstructed decays of D^{*+} and fully reconstructed D^+ mesons. We measure $A_c = 0.77 \pm 0.22$ (stat.) ± 0.10 (syst.).

1. Introduction

In the Standard Model (SM) the vector and axial-vector couplings result in forward-backward asymmetries (A_{FB}) in the cross section $d\sigma/d(\cos\theta)$ for $e^+e^- \rightarrow f\bar{f}$, where θ is the angle between the incoming e^- and outgoing fermion f. Asymmetries are related to electroweak coupling parameters, $A_c = 2v_c a_c/(v_c^2 + a_c^2)$, and provide a sensitive test of the SM.[1] At SLC, the selection of incident e^- helicity (left (L) or right (R)) facilitates the cancelation of the e^- coupling such that a *direct* measurement of the final state quark asymmetry A_c can be made from the polarization P_e and A_{FB}: $\tilde{A}_{FB}(P_e) = 3/4 P_e A_c = (\sigma_F^L + \sigma_B^R - \sigma_B^L - \sigma_F^R)/(\sigma_F^L + \sigma_B^R + \sigma_B^L + \sigma_F^R)$. In this double asymmetry any charge or geometric asymmetry in the acceptance will explicitly cancel.

2. Event Selection

The SLD[2] 1993 data sample contains $\sim 50K$ Z^0 produced at 91.26 ± 0.02 GeV. The e^- beam is longitudinally polarized to $63.0 \pm 1.1\%$.[3] For this analysis only the vertex detector,[4] the central drift chamber,[5] and the calorimeter[6] are used. These cover 76%, 85% and 95% of 4π sr, respectively. Multi-hadron events must satisfy: ≥ 5 charged tracks, a charged energy to center-of-mass energy ratio ≥ 0.20, and $|\cos\theta_T| < 0.80$, where θ_T is the thrust direction.

3. Charm Tagging

To isolate $c\bar{c}$ events from $b\bar{b}$ and uds backgrounds, "kinematic" and "vertex" analyses on two D^{*+} decay chains, $D^{*+} \rightarrow D^0 \pi_s^+$, where $D^0 \rightarrow K^- \pi^+$ or $D^0 \rightarrow K^- \pi^+ \pi^0$, (the π^0 is not reconstructed), are performed. Vertexing allows the $D^+ \rightarrow K^- \pi^+ \pi^+$

[†] Work supported by Department of Energy contract DE–AC03–76SF00515 (SLAC).

to be cleanly isolated as well. In the "kinematic" analysis, candidate D^0's are formed cutting on invariant mass $1.765 < m_{D^0 \to K^- \pi^+} < 1.965$ GeV and $1.500 < m_{D^0 \to K^- \pi^+ \pi^0} < 1.700$ GeV (around the "satellite" peak). A helicity angle cut $|\cos \theta^*| > 0.9$ is applied, and the candidates are combined with a "slow" pion having $p_\pi > 1$ GeV/c and correct charge. To reduce both combinatoric background and D^{*+} from $Z^0 \to b\bar{b}$, we require $x_{D^*} = 2E_{D^*}/E_{cm} \geq 0.4$. The mass difference Δm between the D^{*+} and its associated D^0 divides the data into a signal region ($\Delta m < 0.150$ GeV) and a sideband region (0.16 GeV $< \Delta m < 0.20$ GeV.

In the "vertex" analysis the helicity angle and p_π cuts are eliminated and the $x_{D^{*+}}$ cut is reduced to $x_{D^{*+}} > 0.2$. We require D^0 tracks to have a good 3D-vertex fit with a 3D-decay length (L) distinct from the IP by $2.5\sigma_L$. The 2D-impact parameter of the D^0 to the IP is required to be < 20 μm. The vertex and impact cuts strongly reject combinatoric background and the D^{*+} products of beauty cascades.

A "vertex" style analysis is used to isolate $D^+ \to K^- \pi^+ \pi^+$ in $c\bar{c}$ events by cutting on $x_{D^+} \geq 0.4$, the helicity angle of the K^-, $\cos(\theta^*) > -0.8$, and $\Delta m > 0.160$ GeV so that neither $K^- \pi^+ (\pi^+)$ combination lie in the D^{*+} chain. The 3D-decay length (L) is required to lie outside $3.0\sigma_L$ from the IP, and the angle between a line drawn from the D^+ vertex to the IP, and the direction of the D^+, must be < 5 mr in the XY plane and < 20 mr in the RZ plane. This cut is analogous to the impact parameter cut of the D^{*+} analysis. The signal region is defined $1.800 < m_{D^+} < 1.940$ GeV.

Fig. 1a-c shows signals from all modes. To reduce dependence on branching fractions, the shape of the signal and background is taken from Monte Carlo (MC); the data is then fit separately with these two components to determine the number of background events lying in the signal mass region, and the relative number of $c\bar{c}$ and $b\bar{b}$ events. The solid lines in Fig. 1a-c indicate the fitted backgrounds. There are 317 signal events, of which 173 are $c\bar{c}$ signal, 59 are from $b\bar{b}$ events, and 85 are from random combinatoric backgrounds.

Fig. 2 shows the raw asymmetry in $\cos(\theta_D)$ multiplied by the e^- helicity and the charge of the charmed meson. To extract A_c we perform an unbinned maximum likelihood fit to the form of the polarized differential cross section for each contributing flavor.[7] Each event contains three pieces of information: the incident polarization, the D meson direction ($\cos \theta_D$), and x_D. Each event is assigned a probability by its x_D value to arise from $c\bar{c}$, $b\bar{b}$ or random background. Sideband data and MC are used to simulate the parent x_D distributions after cuts, while the relative normalization and errors are fixed by the fitted decomposition of the 317 events, previously described. A mixing diluted $A_b = 0.635 \pm 0.135$ and a sideband asymmetry are input to the fit. The fitted value of A_c is adjusted by $+0.9\%$ ($+6.1\%$) for QED (QCD) radiative corrections (see below) yielding $A_c = 0.77 \pm 0.22$(stat.).

4. Systematic Errors

The largest systematic errors arise from our lack of knowledge of the combinatoric background's asymmetry (4.9%), magnitude (4.4%), x distribution (3.0%), and acceptance (4.7%). The b–asymmetry, feedown, and fragmentation contribute 2.7%, 0.6% and 0.6%, respectively. Charm fragmentation contributes 3.1% while the polarization uncertainty adds 1.3%. The charm asymmetry calculated at tree-level is modified by small QED and QCD radiative corrections. The QED corrections originate from initial and final state radiation, corrections to the $Z^0 e^+ e^-$ and the $Z^0 c\bar{c}$ vertices. Higher order weak processes, such as box diagrams, are negligible. The "oblique" corrections to the Z^0 propagator (seen as changes to the effective $\sin^2 \theta_W$) are not made. The total QED correction is estimated using ZFITTER to be $< 0.85\%$.

Final state QCD corrections have been calculated to $O(\alpha_s^2)$.[8] Mass corrections for the c-quark are negligible. While the lowest order correction is weakly $\cos \theta$ dependent, it has been accounted for by adjusting A_c by 3.6%. The $O(\alpha_s^2)$ correction depends in detail on acceptance cuts for 4-jet events wherein a radiated gluon emerges as a pair of light quarks. We have taken 2.5% for the correction, and include $\pm 1.8\%$ as the total QCD uncertainty. Combining all systematics we find $A_c = 0.77 \pm 0.22(\text{stat.}) \pm 0.10(\text{syst.})$.

5. References

1. G. Burdman, E. Golowich, J. Hewett, and S. Pakvasa, Physics Reports (in preparation).

2. SLD Design Report, SLAC-Report-273 (1984).

3. K. Abe et al., Phys. Rev. Lett. 73, 25 1994.

4. C. Damerell et al., Proceedings of the 1992 Lepton Photon Conf., Dallas, TX (August 1992). M. Strauss, SLAC-PUB-5573, December 1992.

5. W. B. Atwood et al., NIM A252, (1986) 295.

6. D. Axen et al., NIM A328, (1993) 472.

7. J. H. Kühn, P. Zerwas, in Z^0 **Physics at LEP I**, G. Altarelli, et al. (eds), CERN Yellow Report 89-08, Vol 1. (1989).

8. G. Altarelli and B. Lampe, Nucl. Phys. B391 (1993) 3, and references therein.

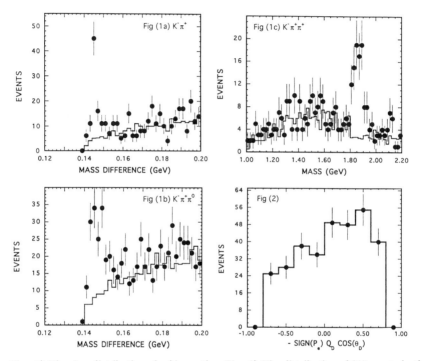

Fig. 1) The Δm distributions for kinematic and vertex analyses for (a) $D^{*+} \rightarrow D^0 \pi_s^+$, $D^0 \rightarrow K^- \pi^+$, (b) $D^{*+} \rightarrow D^0 \pi_s^+$, $D^0 \rightarrow K^- \pi^+ \pi^0$, and (c) the mass distribution for $D^+ \rightarrow K^- \pi^+ \pi^+$

Fig. 2) The distribution of 317 events in the signal mass region region in $-Q_D \cdot SIGN(P_e) \cdot \cos(\theta_D)$, indicating a clear uncorrected asymmetry.

MEASUREMENT OF THE RATIO OF THE INVISIBLE TO LEPTONIC WIDTHS OF THE Z⁰ USING THE OPAL DETECTOR AT LEP

DAVID STROM*

Department of Physics
University of Oregon
Eugene, Oregon 97403, U.S.A.

Abstract

An improved measurement of the ratio of the invisible to leptonic partial widths of the Z^0, $\Gamma_{inv}/\Gamma_{l+l-}$, making use of the new OPAL silicon tungsten luminosity monitor is presented. The experimental error on the luminosity measurement has been reduced to better than 1/1000. Using the new luminosity measurement, $\Gamma_{inv}/\Gamma_{l+l-} = 5.935 \pm 0.068$ is obtained. The error on this quantity is now limited by the theoretical error on the small angle Bhabha cross section and the error on the multihadron acceptance.

1. Introduction

Precision measurements of the Z^0 resonance parameters are a sensitive test of the validity of the Standard Model. The ratio of the invisible partial width to the leptonic partial width, $\Gamma_{inv}/\Gamma_{l+l-}$, is an especially interesting quantity because the radiative correction due to the Higgs and the top mass are similar for both partial widths. As a result, a definitive measurement of this quantity, significantly different from the Standard Model prediction, would imply that the Standard model is wrong or incomplete. New physics such as right handed neutrinos[1] or a heavy Z' boson[2] which mixes with Z^0, could lead to measurable deviations in the ratio from the Standard Model value.

The invisible partial width of the Z^0 results from the decay of Z^0 to particles which do not interact in our detector. Thus, $\Gamma_{inv}/\Gamma_{l+l-}$ must be determined indirectly from

$$\frac{\Gamma_{inv}}{\Gamma_{l+l-}} = \frac{\Gamma_Z - \Gamma_{had} - 3\,\Gamma_{l+l-}}{\Gamma_{l+l-}}$$

where Γ_Z is the total width of the Z^0 resonance, Γ_{had} the total hadronic width and Γ_{l+l-} the partial width due to each of the three charged leptons. Experimentally, one must then measure $\frac{\Gamma_Z}{\Gamma_{l+l-}}$ and $\frac{\Gamma_{had}}{\Gamma_{l+l-}}$. Assuming charged lepton universality, $\frac{\Gamma_Z}{\Gamma_{l+l-}}$ is related to

---

*Representing the OPAL Collaboration

the pole lepton cross section by $\sigma_{ll}^{\text{pole}} = \frac{12\pi}{m_Z^2}(\frac{\Gamma_{l^+l^-}}{\Gamma_Z})^2$. This cross section is measured by counting the number of lepton pairs produced near the Z^0 and normalizing the result to a precision measurement of small angle Bhabha scattering. The quantity $\frac{\Gamma_{\text{had}}}{\Gamma_{l^+l^-}} \equiv R_z$ is simply the ratio of hadronic to leptonic decays of the Z^0.

Since the LEP storage ring now delivers samples on the order of 10^6 Z^0 per year to each experiment, the challenge to the experiments is to measure the number of Z^0 decays and the luminosity with systematic errors of order $1/1000$ so that efficient use of the data collected can be made. In the following, we concentrate on the new OPAL luminosity measurement which is based on data from a precision silicon-tungsten calorimeter. The OPAL selection of mulithadrons and lepton pairs events from Z^0 decay is described in detail elsewhere.[3]

2. Luminosity Measurement

In order to measure the LEP luminosity with a systematic error of $1/1000$, a new luminosity monitor was installed in the OPAL experiment at the beginning of the 1993 run. This device consists of two identical annular position sensitive silicon-tungsten calorimeters located approximately ± 2.5 meters from the OPAL interaction region. These devices register electromagnetic showers from small angle Bhabha scattering in a fiducial range of 28 to 55 mrad in polar angle and 2π in azimuth. The electromagnetic showers are sampled with 19 layers of silicon detectors interleaved between 14 tungsten absorbers of 1 X_0 and 4 tungsten absorbers of 2 X_0. The silicon detectors are finely segmented with a radial pitch of 2.5 mm and an azimuthal pitch of 11.25°. Each layer of both calorimeters is individually readout, giving a total of 38,912 channels.

The small angle Bhabha cross section is proportional to $1/\theta^3$, where θ is the polar angle. Thus the luminosity measurement has a large sensitivity to the position of the inner cut on θ. A simple calculation shows that a bias on this angle of 10 microradians (25 μm at 2.5 m) corresponds to a systematic error of $1/1000$ in the luminosity. As a result great care was taken in the construction of the detector and extensive measurements of the calorimeters were made to establish the position of the average inner radius of the two detectors. The overall uncertainty on the average inner radius of each layer is 9 μ, corresponding to a systematic error on the luminosity of $3.6 \cdot 10^{-4}$. The uncertainty is dominated by the calibration of our metrology instrument against a reference laser interferometer and by measured changes as the detector is moved from the horizontal to vertical position. The effects of the heat load generated by the electronics have been minimized by careful integration of the cooling with the detector structure. The distance between the two calorimeters, the other crucial parameter in the determination of θ, was measured to be 2460.220 ± 0.075 mm, which implies a $6 \cdot 10^{-5}$ uncertainty in the luminosity.

Test beam measurements were used in order to reference the radial coordinate used in the analysis, which combines measurements in the silicon layers between 2 and 10 X_0, to the detector geometry. This reference procedure has three steps. First muons are used in the test beam to establish the position of the pad boundaries relative to an

Definition cuts:	
$N_R(N_L)$	77 mm $<$ $R_R(R_{L)}$ $<$ 127 mm

Isolation cuts:	
Minimum Energy Cut, Right (Left)	E_R (E_L) $>$ $0.5 \cdot E_{Beam}$
Average Energy Cut	$\frac{(E_R+E_L)}{2}$ $>$ $0.75 \cdot E_{Beam}$
Acoplanarity Cut	$||\phi_R - \phi_L| - \pi|$ $<$ 200 mrad
Acollinearity Cut	$|R_R - R_L|$ $<$ 25 mm

Table 1: Definition and isolation cuts used in the Bhabha selection. The energy is calculated from clusters falling with the fiducial radial acceptance of the detector, 67 mm to 137 mm, and the acoplanarity and acollinearity are constructed from the most energetic cluster within this fiducial region. After the definition cuts have been applied, the isolation cuts remove less than 5% of the remaining events.

external tracker. Next the image of the pad boundary in a given layer is reconstructed using the pad in that layer with the maximum energy deposition (pad-max). The small bias in this image from the curved pad boundaries and the finite shower size was measured to be 8 ± 6 μm. Finally the radial coordinate, is referenced to the image of the pad boundary in a particular layer. Using this procedure it is possible to reference the radial coordinate used in the analysis to the detector geometry with a precision of 6 μm which implies a $2.5 \cdot 10^{-4}$ uncertainty in the luminosity.

The high precision with which the calorimeters have been constructed, and the precision with which the radial coordinate can be referenced to the detector geometry, motivated us to define the Bhabha event selection so that its acceptance can be directly related to the geometry of the detector. We define two counters in terms of the reconstructed radial coordinates on the right and left sides of the calorimeter, N_R and N_L (see Table 1).

Fig. 1 shows the radial coordinate on the right and left side of the detector, after the isolation cuts (see Table 1) have been applied. Because of the non-projective geometry of the detector, we can compare the residual bias of the coordinate, as measured at different depths in the calorimeter, with the change in acceptance as the radial cut is varied in data and Monte Carlo. The agreement between the expected and observed shifts as the cut is varied is shown in Fig. 2.

To determine the luminosity we take the average of N_R and N_L. To first order, the effect of longitudinal and transverse offsets of the beam on the luminosity cancel in the average. However, second order effects are important; transverse offsets of order 1 mm can lead to a correction of order 1/1000 to the definition cuts, counter-balanced by an opposite effect in the isolation cuts. The measured luminosity is corrected on a fill by fill basis for shifts in beam positions and tilts.

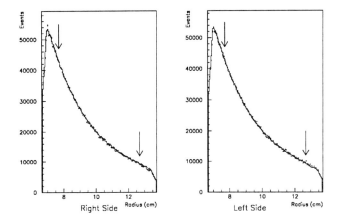

Figure 1: The radial distribution of events on the right and left sides after all other cuts. The solid line is the Monte Carlo prediction, the points with error bars, the data.

The largest potential source of uncertainty on the isolation cuts comes from the energy cuts. In order to determine the acceptance of the energy cuts, the measured energy response of the detector has been used. The shape of the energy response for beam energy electrons (positrons) is measured by selecting collinear events with a tight energy cut on the opposite positron (electron). The extrapolation of the energy response to lower energy has been checked by using acollinear events where a single initial state photon has been radiated in the direction of the beam. Based on a comparison of event kinematics and the measured cluster energies, the nonlinearity of the energy response of our detector was measured to be $1 \pm 2\%$.

To calculate the acceptance of the selection the four vectors from a Monte Carlo of Bhabha scattered are convoluted with the measured energy response of the detector. The change in acceptance from the four vector level to the detector level is $(7.1 \pm 3.8(\text{sys})) \cdot 10^{-4}$, where the systematic error is due to the uncertainties in the measured detector response. As a check on this result, we scale the energy cuts and compare the background subtracted data with Monte Carlo. The measured luminosity changes by less than 10^{-4} as the average energy cut is varied from 0.7 to 0.8 of the beam energy.

The systematic errors in the luminosity analysis are summarized in Table 2 below. The total experimental systematic error is $7.2 \cdot 10^{-4}$. This can be compared with the $25 \cdot 10^{-4}$ theoretical uncertainty of the cross section from the BHLUMI Monte Carlo[4]

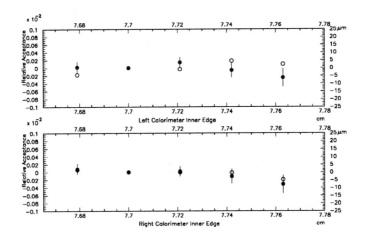

Figure 2: The relative variation in acceptance as the cut on inner right and left radius is varied. The open circles represent the expected change in acceptance based on fixing the radial coordinate to the test beam measurements in layers 4-8 of the calorimeter.

which has been used to calculate the cross section of our selection cuts.

3. Determination of the Invisible Width

In 1993 LEP took data at three energy points in a scan of the Z^0 resonance. The procedure used by OPAL to extract the Z^0 resonance parameters from the measured cross sections is similar to that of reference 3. Including data taken in previous years, values for the Z^0 mass ($M_Z = 91.1862 \pm 0.0054$ GeV), width ($\Gamma_Z = 2.4946 \pm 0.0061$ GeV) and pole hadronic cross section (41.48 ± 0.16 nb) were obtained. Due to the improved luminosity measurement, the pole cross section is determined almost solely from 1993 data. Using the pole cross section, together with ratio of hadronic to leptonic widths ($R_z = 20.864 \pm 0.076$) from all OPAL data, ratio of invisible to leptonic partial widths is determined to be*,

$$\frac{\Gamma_{inv}}{\Gamma_{l+l-}} = 5.935 \pm 0.068$$

agreeing well with the Standard Model prediction of 5.977 ± 0.010 The error on the Standard Model results from Higgs mass variations from 50 to 1000 GeV and top mass

---

*In terms of R_z and σ_{had}^{pole}, $\frac{\Gamma_{inv}}{\Gamma_{l+l-}} = \left(\frac{12\pi}{M_Z{}^2} \frac{R_z}{\sigma_{had}^{pole}} \right)^{\frac{1}{2}} - R_z - 3$ where σ_{had}^{pole} is in GeV^{-2}. In calculating the error on $\frac{\Gamma_{inv}}{\Gamma_{l+l-}}$, a correlation of 0.36 between the error on R_z and σ_{had}^{pole} must be included.

Effect	Correction $\times 10^{-4}$	Systematic $\times 10^{-4}$
SiW radial dimensions ($\pm 9\mu$m)	—	3.6
Radial coordinate bias	—	2.6
Monte Carlo, detector response	-7.3	3.8
Monte Carlo, cross section calculation (stat.)	—	3.7
Detector instability (mechanical + response)	—	0.5
Trigger inefficiency	0	<0.01
LEP Beam parameters (average)	$+3.1$	2.0
Fluctuations in LEP beam parameters	—	0.5
Accidental coincidence background	$+1.0$	0.1
$\gamma\gamma$ background	$+2.0$	0.1
Total	-1.2	7.2

Table 2: Summary of the corrections applied and the corresponding experimental systematic uncertainties on the absolute luminosity measurement.

variations from 50 to 230 GeV. Expressed in terms of a number of neutrino generations, the result is $N_\nu = 2.979 \pm 0.034(\text{exp}) \pm 0.005(\text{M}_t, \text{M}_\text{H})$.

The measurements of N_ν and $\frac{\Gamma_{\text{inv}}}{\Gamma_{l^+l^-}}$ presented here are limited primarily by the theoretical error on the calculation of the cross section of small angle Bhabha scattering. There is considerable activity in the theoretical community aimed at reducing this uncertainty. In addition, it is expected that a considerable improvement can be made in experimental and fragmentation uncertainties on multihadron acceptance. The expected improvement from these two developments would reduce the error $\frac{\Gamma_{\text{inv}}}{\Gamma_{l^+l^-}}$ by more than a factor of two.

4.   Acknowledgments

I wish to thank my OPAL colleagues for their determination and perseverance in obtaining these results.

References

1. C. Jarlskog, *Phys. Lett.* **B241** (1990) 579.
2. See for example, OPAL Collaboration,"Search for Z' Effects on Electroweak Observables",conference paper, DPF 1994.
3. OPAL Collaboration, R. Akers et al., *Z. Phys.* **C61** (1994) 19.
4. S. Jadach et al., *Comp. Phys. Com.* **70** (1992) 305.

IMPROVED LUMINOSITY MEASUREMENT IN L3 AND DETERMINATION OF THE NUMBER OF NEUTRINOS

ELS KOFFEMAN

Nikhef, Kruislaan 409, 1098 SJ Amsterdam
The Netherlands

representing the L3 Collaboration

ABSTRACT

In the L3 experiment the luminosity is determined by measuring small angle Bhabha scattering. In 1993 a silicon tracker was installed in front of the existing BGO calorimeter. With the silicon detector a better measurement of the fiducial volume is obtained. The experimental systematic error on the luminosity measured is 0.12%. One of the results of the improved luminosity measurement is a smaller error on the determination of the number of light neutrino species. Using the 1990−1993 data, the number of light neutrinos is $N_\nu = 2.981 \pm 0.031$.

1. The Detector

At the beginning of 1993 a silicon tracker was installed in L3 at LEP in front of the existing BGO calorimeter in order to improve the luminosity detector. The BGO detector consists of two electromagnetic calorimeters, placed symmetrically at either side of the interaction point at $z = \pm 2.8$ m. Each of these calorimeters is a cylindrical array of 304 BGO crystals with an inner radius of 68.0 mm and an outer radius of 191.4 mm. On each side of the interaction point the silicon tracker contains 4096 silicon strips organised in three layers: an R-measuring layer ($Z = \pm 2610$ mm), a ϕ-measuring layer ($Z = \pm 2650$ mm), and another R-measuring layer ($Z = \pm 2690$ mm). Each of these layers covers 2π in azimuthal angle from a radius of 76 mm to a radius of 154 mm. The R-strips are not equally spaced over the wafer in order to optimize the use of the restricted number of channels. The inner region contains 64 strips of 0.5 mm width, the middle region contains 16 strips of 1.875 mm width and the outer region contains 16 strips of 1 mm width. The ϕ-wafers have 64 strips with a pitch of 0.375 degrees.

The selection of events is based on the measurement of the energy and direction of the scattered particles. To prevent the particles from producing secondary particles, which complicate the direction measurement, a specially shaped beampipe was installed on the $+Z$ side of the interaction point. Installing the same beampipe on the $-Z$ side was not possible due to interference with the installation of the L3 micro-vertex detector. Figure 1 shows the number of radiation lengths as a function of the polar angle of the scattered particles. The advantage of the $+Z$ beampipe is the low number of radiation lengths at small θ-angles where the fiducial volume cut is made.

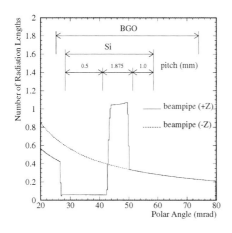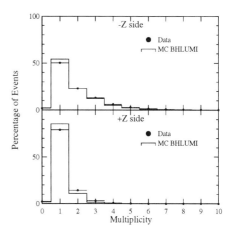

Figure 1: *(left) The profile of the beampipes on the $-Z$ and $+Z$ sides.
(right) The silicon multiplicity distribution for the $-Z$ and $+Z$ sides, along with the
Monte Carlo prediction.*

2.   Event Selection

A Bhabha event is first identified by the BGO calorimeter. In order to accept an event
we require at least $0.8 \cdot E_{\text{beam}}$ on one side of the calorimeter and $0.4 \cdot E_{\text{beam}}$ on the other
side. Subsequently, a window of $(\Delta R, \Delta\phi) = (10\text{mm}, 5°)$ is formed around the impact
point in the BGO. Only silicon hits which lie within this window are considered for the
analysis. In Fig.1 the multiplicity distribution of the number of silicon strips hit inside
the window is shown, compared with the same distribution for Monte Carlo.

   In order to determine the position of the particle, the following strategy is used:

- If there are 0 hits inside the window, we use the BGO impact point for further
  analysis $(-Z : 2.8\%, +Z : 3.5\%)$.

- If there is 1 strip hit, we use the position of this strip $(-Z : 49.6\%, +Z : 76.7\%)$.

- If > 1 strip is hit, we take the average position $(-Z : 47.6\%, +Z : 19.8\%)$.

Using both the detectors requires a precise alignment between the BGO and the silicon.
This alignment is done using the Bhabha events in the data sample. The alignment is
done to a precision of approximately 100 microns.

   The polar angle distribution of the Bhabha events is shown in Fig.2, for both the
$-Z$ and $+Z$ sides, along with the Monte Carlo prediction. The $-Z$ distribution is
characterized by an enhancement of events at the low theta edge.

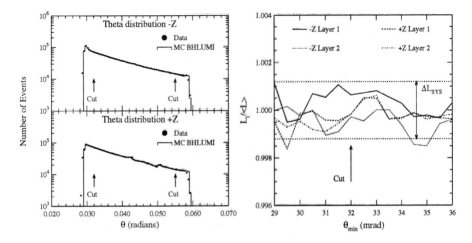

Figure 2: *(left) The polar angle distribution of the reconstructed Bhabha events, along with the Monte Carlo prediction.*
(right) The luminosity as separately measured by each of the 4 R-measuring layers.

This effect is caused by Bhabha events at low theta which radiate a secondary particle into the detector. The effect is reduced on the $+Z$ side, where the beampipe is effectively very thin.

The final luminosity sample is defined by requiring:

- the polar angle to be in the domain: $32 < \theta < 54$ mrad.

- the azimuthal angle to be in the domain: $|\phi - 90°| > 11.25°$ and $|\phi - 270°| > 11.25°$.

In addition the coplanarity angle, $\Delta\phi$, of the two tracks must satisfy: $|\Delta\phi - 180°| < 10°$.

3.   Luminosity Determination

The luminosity is obtained using: $\mathcal{L} = \frac{N_{\text{Bhabha}}}{\sigma_{\text{Bhabha}}}$. The visible cross section, σ_{Bhabha}, of our detector is obtained by a detailed Monte Carlo simulation using a total sample of 2.2 million BHLUMI [1, 2] events and 0.8 million BABAMC [3] events. The visible cross section using the above event selection criteria is 70.46 nb.

The systematic errors on the luminosity determination, for both our previous analysis using only the BGO and the present analysis using both the BGO and the silicon, are listed in table 1. The error due to event selection is estimated by varying each of the event selection cuts over a realistic range and adding the contributions in quadrature. The geometry error in the silicon analysis is mainly due to the wafer positioning precision (0.023%), temperature expansion effects (0.024%) and the accuracy with which the overall distance in Z between the silicon detectors is known (0.060%). The total experimental systematic error is 0.12%. The total systematic error on the luminosity at present is 0.30%, dominated by the contribution of the theory error.

The relative luminosities as determined by each of the 4 R-measuring layers $\mathcal{L}_i/\langle\mathcal{L}\rangle$ are shown in Fig.2. The measurements are seen to agree within the total experimental systematic error.

Source	Contribution to $\Delta\mathcal{L}/\mathcal{L}$ (%) BGO Analysis[4]	Contribution to $\Delta\mathcal{L}/\mathcal{L}$ (%) BGO+Silicon Analysis
Trigger	Negligible	Negligible
Event Selection	0.3	0.09
Background Subt.	Negligible	Negligible
Geometry	0.4	0.07
Total Experimental	0.5	0.12
MC Statistics	0.10	0.10
Theory	0.25	0.25
Total	0.6	0.30

Table 1: *Systematic uncertainties on the luminosity measurement.*

4. Determination of the Number of Neutrinos

Using all the L3 data from 1990 through 1993 the number of light neutrino species can be determined from the mass of the Z boson and the decay widths of the Z into hadrons en leptons. The number of neutrinos is $N_\nu = 2.981 \pm 0.031$.

References

[1] S. Jadach and B.F.L. Ward, *Phys. Rev.* **D 40** (1989) 3582.

[2] S. Jadach *et al.*, *Phys. Lett.* **B 268** (1991) 253;
    S. Jadach *et al.*, *Comp. Phys. Comm.* **70** (1992) 305.

[3] M. Böhm, A. Denner and W. Hollik, *Nucl. Phys.* **B304** (1988) 687;
    F.A. Berends, R. Kleiss and W. Hollik, *Nucl. Phys.* **B304** (1988) 712.

[4] L3 Collab., M. Acciarri *et al.*, *Preprint CERN-PPE/94-45*, CERN, 1994, submitted to *Z. Phys. C*.

TAU POLARIZATION AT ALEPH

MICHAEL WALSH[1,2]
Physics Department, University of Wisconsin
1150 University Ave., Madison, Wisconsin, 53706 U.S.A.

ABSTRACT

Using 41.1 pb^{-1} of data taken near the Z resonance, ALEPH has measured the tau polarization as well as the angular dependence of the tau polarization in the decay modes $\tau \to e\nu\bar{\nu}$, $\tau \to \mu\nu\bar{\nu}$, $\tau \to \pi\nu$, $\tau \to \rho\nu$ and $\tau \to a_1\nu$. The measurement leads to the Z couplings to taus and electrons, A_τ and A_e, where $A_l = 2g_V^l g_A^l / [(g_V^l)^2 + (g_A^l)^2]$. The published results from 1990 and 1991 are preliminarily updated with the data set collected in 1992, yielding $A_\tau = 0.137 \pm 0.014$ and $A_e = 0.127 \pm 0.017$. Assuming lepton universality leads to a value for the effective weak mixing angle of $\sin^2 \theta_W^{\text{eff}} = 0.2334 \pm 0.0014$.

1. Introduction

Tau polarization provides a sensitive measurement of parity violation in the weak neutral current. At LEP, electrons and positrons collide at a center-of-mass energy of 91 GeV on the Z resonance. The Z then decays to a pair of fermions such as taus. The τ^+ and τ^- have correlated helicities, since they are produced from a spin-1 Z. The preference for a lefthanded τ^- is expressed as the polarization P_τ.

The polarization also depends on the tau production angle. It is weakest in the backward direction $\cos\theta = -1$, and strongest in the forward direction $\cos\theta = +1$ where the τ^- continues in the direction of travel of the beam e^-.

The angular dependence of the polarization can be written in terms of the coupling A_τ of the Z to the final state taus, as well as the coupling A_e of the Z to the initial state electrons,

$$P_\tau(\cos\theta) = -\frac{A_\tau(1 + \cos^2\theta) + A_e(2\cos\theta)}{(1 + \cos^2\theta) + A_\tau A_e(2\cos\theta)}, \tag{1}$$

where

$$A_l \equiv 2\frac{g_V^l g_A^l}{[(g_V^l)^2 + (g_A^l)^2]} \approx 2\frac{g_V^l}{g_A^l}. \tag{2}$$

Since these lepton couplings are on the order of 10%, the $A_\tau A_e$ term in equation 1 is small. Consequently, the $\cos\theta$ dependence of the A_τ term nearly cancels. The A_e term becomes an odd function of $\cos\theta$. This means that the average value of the polarization is a measurement of A_τ, while the angular dependence yields A_e. Systematic uncertainties which affect the measurement of A_τ largely cancel for A_e. (See tables 1 and 2).

---

[1]Representing the ALEPH collaboration
[2]Supported by the US Department of Energy, contract DE-AC02-76ER00881

	$e\nu\bar{\nu}$	$\mu\nu\bar{\nu}$	$\pi\nu$	$\rho\nu$	$a_1\nu$
Acceptance	0.003	0.005	0.005	0.014	0.004
Tau backgrground	0.010	0.010	0.005	0.001	0.008
Other background	0.012	0.012	0.003	0.005	0.001
Energy calibration	0.015	0.010	0.001	0.007	0.001
Model dependence	-	-	-	-	0.012
MC statistics	0.025	0.019	0.009	0.008	0.016
Total systematics	0.033	0.027	0.011	0.018	0.022

Table 1: Systematic uncertainties on $\mathcal{A}_\tau$

$\vec{B}$ field	0.001
MC statistics	0.007
Total systematics	0.007

Table 2: Systematic uncertainties on $\mathcal{A}_e$

If the obtained couplings satisfy lepton universality, they can be combined for a single measurement of the effective weak mixing angle

$$\sin^2 \theta_W^{\text{eff}} \equiv \frac{1}{4}(1 - \frac{g_V^l}{g_A^l}). \tag{3}$$

2. Spin Analysis

The tau helicity is analyzed using its decay (assumed to be pure V-A). The simplest case is the decay of the tau to a spin-0 charged pion and a spin-$\frac{1}{2}$ neutrino. The helicity of a lefthanded τ^- will be absorbed by the neutrino, which must be lefthanded, leaving the π^- with a soft energy spectrum in the lab. A righthanded τ^- will give rise to a π^- with a hard energy spectrum. Monte Carlo distributions are fit to the spectrum observed in data, and the polarization is deduced from the relative amount of left- and righthanded contributions. In general, the polarization estimator used in the fit is constructed from track momentum measurements in the ALEPH Time Projection Chamber. The Electromagnetic Calorimeter is used in the case of tau decay to $e\nu\bar{\nu}$ and for photon reconstruction.

Some helicity information is lost to the second neutrino in the tau decays to $e\nu\bar{\nu}$ and $\mu\nu\bar{\nu}$, and these channels have reduced sensitivity to the polarization. Information is also lost to the helicity state of the vector meson in the $\rho\nu$ and $a_1\nu$ decays; however, in these channels part of the information is gained back by analyzing the vector meson decay. (In the case of the $a_1\nu$ channel, only the a_1 decay to three charged pions is used.) The measured momenta of the decay products are

Channel	efficiency	non-tau	prelim 1992 $\overline{P_\tau}$
$e\nu\bar{\nu}$	48%	1.5%	$-0.214 \pm 0.065 \pm 0.033$
$\mu\nu\bar{\nu}$	69%	1.0%	$-0.123 \pm 0.055 \pm 0.027$
$\pi\nu$	60%	1.1%	$-0.148 \pm 0.026 \pm 0.011$
$\rho\nu$	46%	0.4%	$-0.090 \pm 0.024 \pm 0.018$
$a_1\nu$	52%	0.1%	$-0.144 \pm 0.042 \pm 0.022$

Table 3: $\overline{P_\tau}$ by channel, showing overall efficiencies, non-tau backgrounds, and preliminary results for 1992 only. The first error is statistical, the second is systematic.

condensed into a probability density variable, which is used in the fit [1].

3. Selection and Fitting

Two different methods of particle identification are used, one based on a set of conventional cuts, and the other based on a neural net. Both use up to twenty particle identification estimators derived from measurements of track ionization, transverse and longitudinal electromagnetic shower development, hadron calorimeter digital hit pattern, and muon chamber hits. The neural network relies on weights between nodes: when the sum of weighted inputs from the nodes in the previous layer exceeds a threshold, the node fires, *i.e.*, its output goes from zero to one. This nonlinear response allows the net to perform logical operations. The weights are chosen by training on Monte Carlo, and the performance is then checked using tracks in data and in Monte Carlo whose identity is known for other reasons.

The two identification methods lead to separate selections and fits, which are then combined for the final result.

For the angular fit, a global fit to all five channels is performed in each of nine angular bins. Most systematic uncertainties are treated as fully correlated between the bins; uncorrelated systematic uncertainties include Monte Carlo statistics and a small residual distortion of the ALEPH magnetic field.

4. Results

The preliminary average polarization values by individual channel are shown in Table 3, for 1992 data only. The overall selection efficiencies range from 46 to 69%, and non-tau backgrounds are on the order of 1%. A combined angular fit, including the published 1990 and 1991 data [2] as well as preliminary 1992 results, is shown in figure 1. The fit for separate tau and electron couplings (dashed line) yields

$$\mathcal{A}_\tau = 0.137 \pm 0.012 \pm 0.008 \tag{4}$$

and

$$\mathcal{A}_e = 0.127 \pm 0.016 \pm 0.005, \tag{5}$$

confirming electron-tau universality in the weak neutral current. Fitting for a single

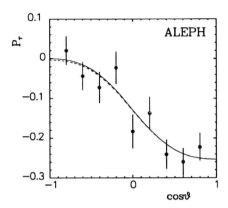

Figure 1: 1990-1992 preliminary fit for $P_\tau(\cos\theta)$ using all five tau decay channels. The dashed line shows the fit for $\mathcal{A}_e$ and $\mathcal{A}_\tau$ separately, the solid line shows the fit for $\mathcal{A}_{e-\tau}$ assuming electron-tau universality.

coupling (solid line), we obtain

$$\mathcal{A}_{e-\tau} = 0.133 \pm 0.010 \pm 0.005, \tag{6}$$

which corresponds to a measurement of the effective weak mixing angle of

$$\sin^2\theta_W^{\text{eff}} = 0.2334 \pm 0.0014. \tag{7}$$

References

1. M. Davier, L. Duflot, F. Le Diberder, A. Rougé, The optimal method for the measurement of tau polarisation, Preprint LAL 92-73 or X-LPNHE 92-22.
2. ALEPH Collab., D Buskulic *et al.*, Z.Phys **C 59** (1993) 369.

MEASUREMENTS OF $\Gamma_{b\bar{b}}/\Gamma_{had}$
WITH THE DELPHI DETECTOR AT LEP

THE DELPHI COLLABORATION

G.BORISOV

Institute for High Energy Physics, Serpukhov P.O. Box 35,
Protvino, Moscow Region, 142284, Russian Federation

ABSTRACT

Several analyses are presented to measure $Z \rightarrow b\bar{b}$ branching ratio with the DELPHI detector at LEP. B hadrons are tagged either by the prompt leptons from their semileptonic decays or using a large impact parameters of their decay products. The branching ratio $Z \rightarrow b\bar{b}$ is obtained from the maximum likelihood fit to single and di-lepton distributions. In another measurement the comparison of single and double hemisphere lifetime tag rates alone or including high transverse momentum lepton tags is preformed. The results of different methods are compared and a combined result is given.

1. Introduction

The measurement of branching ratio of the b quark partial width of the Z^0 to its hadronic width is an important test of the Standard Model. This quantity depends on the top quark mass without residual theoretical uncertainties in Higgs sector, in final state strong interactions or in propagator terms. With a precise measurement of $R_b = \Gamma_{b\bar{b}}/\Gamma_{had}$ the top quark mass can be predicted.[1]

The value of $\Gamma_{b\bar{b}}/\Gamma_{had}$ is measured with the DELPHI detector at LEP using about 1 million hadronic Z^0 decays taken in 1991 and 1992. Different techniques are used for this measurement. The hadrons with b quarks are tagged either by the prompt leptons from their semileptonic decays or by the large impact parameters of their decay products. The obtained results[2-5] are summarized in this note and the combined DELPHI result on $\Gamma_{b\bar{b}}/\Gamma_{had}$ value is given.

2. Tagging of Hadrons with b quark

For tagging of B hadrons through their semileptonic decays both prompt muons and electrons are used.[5] The tagging efficiency for muons with momentum greater than $3GeV/c$, determined in simulation, is found to be 0.76 ± 0.01 in the barrel region $(0.03 < |\cos\Theta_\mu| < 0.62)$ and 0.81 ± 0.02 in the forward region $(0.68 < |\cos\Theta_\mu| < 0.90)$. The hadron misidentification as muon is estimated as 0.007 ± 0.001. The tagging efficiency for isolated electrons with momentum greater than $5GeV/c$ and with $0.03 < |\cos\Theta_e| < 0.70$ is measured in τ decays and Bhabha events as: 0.625 ± 0.006. The average value of hadron misidentification as electron is $(5.91 \pm 0.07) \cdot 10^{-3}$. With the cut on the transverse momentum of the lepton $p_t \geq 1.5 GeV/c$ the purity of b events in the selected sample $P_b = (79.98 \pm 0.80)\%$ is obtained.

For the lifetime tagging of B hadrons the information from the precise silicon microstrip detector[6] is used. The polar angle coverage of this detector for charged particles is from $42.5°$ to $137.5°$. For each charged particle in the event its impact parameter with respect to the primary vertex is determined. The b tagging is performed in each hemisphere of the event separately. For any hemisphere one tagging variable[2,3] P_H is built. It gives the probability for the particles in one hemisphere with their observed impact parameters all to be from the primary vertex. P_H varies from 0 to 1 and its main property is the flat distribution for hemispheres which do not contain the decays of long lived particles. For hemispheres with decays of B hadrons P_H has a very low value corresponding to low probability to obtain the observed values of impact parameters in the primary interaction without secondary decays. Due to this property the tagging variable P_H is named as "hemisphere probability".

B hadrons are tagged by selecting the hemispheres with the value of P_H less than some threshold. For the cut $-\log_{10} P_H > 2.7$ ($P_H < 1.995 \cdot 10^{-3}$) the tagging efficiency inside the acceptance of vertex detector $\epsilon_b = 20\%$ and b-purity of tagged hemispheres $p_b = 90\%$ are obtained.

In the measurement[4] more sophisticated method of tagging is used. A multivariate tagging using lifetime and event shape variables is applied. The tagging efficiency and purity are similar to the pure lifetime tag.

3.   Measurements of $\Gamma_{b\bar{b}}/\Gamma_{had}$

For the measurement of $\Gamma_{b\bar{b}}/\Gamma_{had}$ in a purely lepton tagged events a global fit to single and double lepton spectra is performed. The single lepton spectra are divided in (p, p_t) bins, where p is the momentum of the lepton and p_t is the the transverse momentum of the lepton relative to the direction of the jet. A double lepton spectra are divided in (p_c^{min}, p_c^{max}) bins, where the combined momentum p_c is defined as: $\sqrt{p_t^2 + p^2/100}$.

The data are fitted with a binned maximum likelihood method including the effect of finite simulation statistics. Along with $\Gamma_{b\bar{b}}/\Gamma_{had}$, the following physical parameters are determined: the semileptonic branching ratio $Br(b \to l)$, the cascade branching ratio $Br(b \to c \to l)$, the fraction of $c\bar{c}$ events $\Gamma_{c\bar{c}}/\Gamma_{had}$ and the mean energy of B hadrons $< x_E >$. The final result is:

$$
\begin{aligned}
\Gamma_{b\bar{b}}/\Gamma_{had} &= 0.2145 \pm 0.0089(stat) \pm 0.0063(exp) \pm 0.0023(model) \\
Br(b \to l) &= (11.41 \pm 0.45(stat) \pm 0.50(exp) \pm 0.33(model))\% \\
Br(b \to c \to l) &= (7.26 \pm 0.49(stat) \pm 0.94(exp) \pm 0.55(model))\% \\
\Gamma_{c\bar{c}}/\Gamma_{had} &= 0.1623 \pm 0.0085(stat) \pm 0.0168(exp) \pm 0.0124(model) \\
< x_E > &= 0.702 \pm 0.004(stat) \pm 0.002(exp) \pm 0.011(model)
\end{aligned}
\tag{1}
$$

The accuracy of the measurement of $\Gamma_{b\bar{b}}/\Gamma_{had}$ in purely lepton tagged events is not sufficient, but the results of the fit (see Eq.1) are used for the precise determination of the b purity of tagged lepton sample. It is applied to the mixed tag measurement[5] of $\Gamma_{b\bar{b}}/\Gamma_{had}$, when one jet is tagged by the prompt lepton and the other by the lifetime tag. The result of this analysis is:

$$\Gamma_{b\bar{b}}/\Gamma_{had} = 0.2208 \pm 0.0042(stat) \pm 0.0033(syst) \pm 0.0012(\Gamma_{c\bar{c}}) \qquad (2)$$

The last error in Eq.2 is due to uncertainties in the measurement of $\Gamma(Z^0 \to c\bar{c})$. The LEP average value of $\Gamma_{c\bar{c}} = 0.170 \pm 0.010$ is taken from reference.[7]

Two different analyses are performed for the measurement of $\Gamma_{b\bar{b}}/\Gamma_{had}$ without leptons.[2-4] In the first analysis[2,3] the double hemisphere tagging is used. Two hemispheres of the events are tagged separately by the hemisphere probability, described above. From the comparison of the number of tagged hemispheres and the number of events in which two hemispheres are tagged both the value of $\Gamma_{b\bar{b}}/\Gamma_{had}$ and the b tagging efficiency are determined from the data. The result of this measurement is:

$$\Gamma_{b\bar{b}}/\Gamma_{had} = 0.2217 \pm 0.0022(stat) \pm 0.0032(syst) \pm 0.0018(\Gamma_{c\bar{c}}) \qquad (3)$$

The second analysis[2,4] is more complicated. It is designed to measure the value of $\Gamma_{b\bar{b}}/\Gamma_{had}$ in an as far as possible simulation independent way. The result of this analysis is:

$$\Gamma_{b\bar{b}}/\Gamma_{had} = 0.2196 \pm 0.0044(stat) \pm 0.0029(syst) \pm 0.0005(\Gamma_{c\bar{c}}) \qquad (4)$$

These analyses are combined taking into account common systematic errors and statistical correlations.[8] The combined DELPHI result is:

$$\Gamma_{b\bar{b}}/\Gamma_{had} = 0.2214 \pm 0.0020(stat) \pm 0.0028(syst) \pm 0.0015(\Gamma_{c\bar{c}}) \qquad (5)$$

All separate measurements of $\Gamma_{b\bar{b}}/\Gamma_{had}$ are consistent with the average value given in Eq.5. The average DELPHI value is also consistent with the measurement of $\Gamma_{b\bar{b}}/\Gamma_{had}$ by the other LEP collaborations.

References

1. See, for example, J.H. Kühn, P.M.Zerwas, in: *Z Physics at LEP I*, eds. G. Altarelli *et al.*, CERN 89-08 (1989), Vol.1, p 271.
2. DELPHI Collaboration, *Measurement of $\Gamma_{b\bar{b}}/\Gamma_{had}$ Branching Ratio of the Z^0 by Double Hemisphere Tagging*, DELPHI 94-61 (1994).
3. DELPHI Collaboration, *New Measurement of $\Gamma_{b\bar{b}}/\Gamma_{had}$ with Lifetime Tag Technique*, DELPHI Note 94-90 (1994).
4. DELPHI Collaboration, *New Measurement of $\Gamma_{b\bar{b}}/\Gamma_{had}$ Branching Ratio of the Z^0 with Minimal Model Dependance*, DELPHI Note 94-93 (1994).
5. DELPHI Collaboration, *Measurement of $\Gamma_{b\bar{b}}/\Gamma_{had}$ using micro-vertex and lepton tags*, DELPHI Note 94-91 (1994).
6. N.Bingefors *et al.*, preprint CERN-PPE/92-173 (1992).
7. The LEP Collaborations, *Updated Parameters of the Z^0 Resonance from Combined Preliminary Data of the LEP Experiments*, preprint CERN-PPE/93-157.
8. DELPHI Collaboration, *DELPHI results on Electroweak Physics with Quarks*, DELPHI Note 94-111 (1994).

MEASUREMENT OF THE W MASS AT DØ

ALLEN I. MINCER*

Physics Department, New York University, 4 Washington Place, New York, New York 10003, USA

ABSTRACT

Preliminary results are presented of a precise measurement of the W mass made with the DØ detector using data from Fermilab Tevatron run 1A of August 1992 to May 1993, consisting of $13.3 \pm 1.6 pb^{-1}$ of $\bar{p}p$ collisions with center of mass energy of 1.8 TeV. The transverse mass distribution of produced W's decaying into an electron and neutrino is compared with distributions from a high statistics computer simulation to determine the measured mass. The detector energy scale is determined by measuring the Z mass in electron positron decays and comparing with the LEP value. Detailed studies of data are used in the simulation and to determine systematic offsets and errors.

1. Introduction

The electroweak standard model parameters can be taken to include the fine structure and muon decay constants and the mass of the Z boson, all measured[1,2] to better than 0.01%, and the Higgs particle mass, M_H. Loop order calculations give the mass of the W boson, M_W, and the weak mixing angle, θ_W, as functions of these parameters and of heavy fermion masses.[3] A measurement of M_W thus constrains the top quark mass. In conjunction with a top quark mass measurement, it constrains M_H. Combined with a measurement of $sin^2\theta_W$, it provides a test of the standard model.[4]

The DØ detector has measured $13.3 \pm 1.6 pb^{-1}$ of $\bar{p}p$ collisions with $\sqrt{s} = 1.8$ TeV during Fermilab Tevatron run 1A. Approximately 10% of produced W's decay into an electron,[2] detected by tracking chambers and calorimetry, and a neutrino, measured as event momentum imbalance.

Because much longitudinal momentum is carried down the beam pipe, M_W cannot be measured directly. Instead, the momenta perpendicular to the beam axis, $\vec{P}_t$, are measured for the electron and neutrino to calculate the transverse mass of the W, Mt_W. Mt_W is invariant under longitudinal boosts of the W and is defined by

$$Mt_W^2 = (E_{te} + E_{t\nu})^2 - (\vec{P}_{te} + \vec{P}_{t\nu})^2 \approx 2P_{te}P_{t\nu}(1 - cos(\phi_{e\nu})) \tag{1}$$

where $P_t = | \vec{P}_t |$, the transverse energy E_t is energy times the sine of the angle with the beam axis, and $\phi_{e\nu}$ is the angle between $\vec{P}_{te}$ and $\vec{P}_{t\nu}$. The expected Mt_W distribution

---

*Representing the DØ Collaboration.

is calculated as a function of M_W using a detailed, high statistics computer simulation. This is compared with the measured distribution to determine M_W.

The following sections describe the experiment and data, give details of the simulation, and compare the two to give a preliminary result of the mass measurement.

2. The Experiment and Data

2.1. Data Collection and Analysis

The DØ detector at the Fermilab Tevatron is described fully elsewhere.[5] The Central Detector drift chambers are 85% to 88% efficient at finding electron tracks. Hermetic, fine grained calorimetry (pseudo projective towers mostly of 0.1 in pseudo-rapidity η by 0.1 in azimuthal angle ϕ, with coverage to $\mid \eta \mid \simeq 4.5$) provides electron identification and measures electron energy and missing event momentum. Currently we include only electrons from the Central Calorimeter, the CC, with $\mid \eta \mid < 1.2$. The outermost detector is the solid-iron toroidal magnet and proportional drift tube chamber muon system, used here to include muon energy in measuring $P_{t\nu}$.

Tevatron bunches cross at 3.5 μs intervals causing interactions triggered on by a scintillator hodoscope. The level 1 hardware trigger requires one EM calorimeter trigger tower ($\Delta \eta \times \Delta \phi$ of 0.2×0.2) with $E_T > 10$ GeV for W events. For Z events, two towers are required with $E_T > 7$ GeV each.

For the level 2 software trigger, an electron is defined as a calorimeter energy cluster isolated from other energy deposition with at least 90% of the energy in the electromagnetic (EM) portion of the calorimeter and with shower shape consistent with that of an electron. Missing transverse energy, $\not{E}_t$, is defined as the magnitude of summed $\vec{P}t$ in the calorimeter. The W trigger requires an electron with $E_T > 20$ GeV and $\not{E}_t > 20$ GeV. A Z requires 2 electrons with $E_T > 20$ GeV.

In offline analysis a cluster algorithm identifies candidate electrons. Calorimeter EM energy fraction, shower shape, isolation, track matching, and fiducial cuts are then imposed. A W requires an electron with $E_T > 25$ Gev, $\not{E}_t > 25$ GeV, and reconstructed $P_{tW} < 30$ GeV. A Z requires 2 electrons each with $P_t > 25$ GeV.

In the preliminary analysis there are 5830 W events with an electron in the CC and 313 Z events with both electrons in the CC.

2.2. Determination of the Calorimeter Energy Scale

Pulsers are used to correct electronics channel to channel and time variation at DØ to less than 0.3%. Calorimeter gain is initially determined by calibrating a single CC module in the NW beam at Fermilab.[6] Corrections at DØ , at the test beam, and in comparing the two leaves uncertainties of about 1.5%. A DØ EM energy study of 3.5×10^6 events measured CC module to module variation of about 2%. It is used to correct the DØ variation to less than 0.5%, but cannot be used to compare the DØ and test beam modules.

A more precise calibration is made by measuring the Z mass with DØ using a method similar to that described below for the W, and comparing with the precise LEP result[1] of $91.187 \pm 0.007 GeV/c^2$. Material in front of the calorimeter can, however, result in an energy offset so that true and measured electron energies are related by

$$E_{TRUE} = \alpha E_{MEASURED} + \beta \tag{2}$$

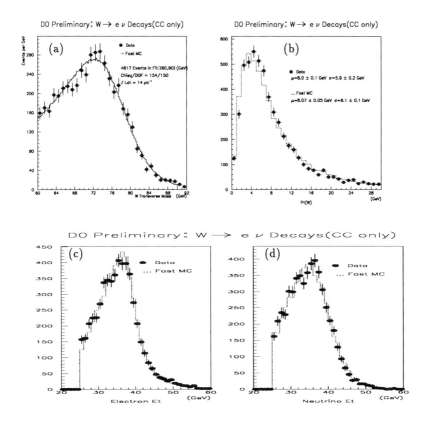

Fig. 1. Simulation compared with data (points), Mt_W (a), P_{tW} (b), P_{te} (c), and $P_{t\nu}$ (d).

A resonance decaying into two electrons of measured energies E_1 and E_2 separated by an angle γ will then have, approximately, a similar relationship between true and measured mass, m_E, with identical slope α but with β replaced by $f\beta$ where f is $2(E_1 + E_2)sin^2(\gamma/2)/m_E$. Three methods are being studied to use this relationship to set limits on β, and therefore on the uncertainty in the energy scale.

In Method I, the measured Z mass dependence on f is fit with a straight line giving $\alpha = 1.050 \pm 0.015$ and $\beta = -320 \pm 780$MeV. Varying these parameters within allowable error in the Mt_W fit gives a 260 MeV variation in M_W. A fit to the Z mass allowing the parameters α and β to vary is in progress and may reduce the error.

Methods II and III use J/Ψ or π^0 signals together with the Z to determine the mass dependence on f with greater sensitivity to β. These methods are expected to give a final uncertainty in α and β which contribute about 200 MeV to the M_W uncertainty.

2.3. Backgrounds

The major background source is QCD two jet events where one jet's calorimeter signal is mostly in the EM layer and energy is missed in the rest of the event. Cutting on electron isolation, $\not{E}_t$, and P_{tW} reduces these to $2 \pm 1\%$ of the event sample.

$W \to \tau\nu$ decays followed by $\tau \to e\bar{\nu}_e\nu_\tau$ are modeled in the simulation. Because the electron from τ decay is typically lower in energy than that from $W \to e\nu$, cuts on P_{te} reduce this background to than 1%, mostly in the low Mt_W region.

Z decays where one electron is missed contribute $0.7 \pm 0.5\%$, mainly in the Mt_W region of 80 to 100 GeV. Fitting Mt_W from 60 GeV to 90 GeV, where the W Jacobian peak gives most of its information, makes this effect negligible.

3. The Simulation

The large number of simulated events required for the experimental precision make a full GEANT-like simulation impossible. Instead, a fast simulation is used which incorporates measured detector resolutions and systematic effects.

A W is generated with momentum $\vec{P}_W$ resulting from structure functions, currently those of Arnold and Kauffman.[7] The W mass is selected using a Breit-Wigner shape folded with a falling exponential calculated from the structure functions. The effect of Bremsstrahlung on measured electron energy is negligible for the DØ calorimeter. Electron energy resolution, determined from the Z width, has the form:

$$\sigma_E/E = \sqrt{(0.003)^2 + \left(0.13\sqrt{GeV}/\sqrt{P_t}\right)^2 + (0.38 GeV/E)^2} \tag{3}$$

A jet is generated with momentum $-\vec{P}_W$ and it energy is scaled as determined from Z events: $\vec{P}_{tZ}$ is calculated from the two electrons, $\vec{P}^{ee}_{tZ}$, and from the energy in the event not including the electrons, $\vec{P}^R_{tZ}$. The smallest error measurement of $\vec{P}^{ee}_{tZ}$ is for its component along the bi-sector of the angle between the two electrons, the $\hat{\eta}$ axis. The difference between the two measurements, $\vec{P}^{ee}_{tZ} \cdot \hat{\eta} - \vec{P}^R_{tZ} \cdot \hat{\eta}$, depends linearly on $\vec{P}^{ee}_{tZ} \cdot \hat{\eta}$. The slope of 0.83 ± 0.06 is the ratio between the electromagnetic and jet energy scales. Jet energy resolution measured in two-jet events is used to smear the jet energy according to

$$\sigma_E/E = \sqrt{(0.04)^2 + \left(0.8\sqrt{GeV}/\sqrt{E}\right)^2 + (1.5 GeV/E)^2} \tag{4}$$

The rest of the $\bar{p}p$ event (called the underlying event) and background from other $\bar{p}p$ interactions are approximated by adding total $\vec{P}_t$ from data minimum bias events to the simulated event. The luminosity distribution of W data is used to generate the luminosity L_W of a simulated W. A minimum bias event is then selected from data at a luminosity such that the number of multiple interactions is equivalent to that in the W event at luminosity L_W plus one for the underlying event. The uncertainty in this method is determined from studies of the width of the P^R_{tZ} distribution.

Energy in the event not from the electron adds some energy to it, found by overlaying W events on electrons to be $91 + 22\eta^2 + 25\eta^4$ MeV. It can also change the

Table 1. Contributions to sytematic error.

Error Source	Error (MeV)	Error Source	Error (MeV)
Trigger efficiency	20	Electron resolution	70
$P_{l\nu}$ resolution	50	Jet energy resolution	20
Energy underlying electron	50	$u_\parallel$ efficiency	10
Hadronic to EM scale	80	QCD background	30
W width	20	Structure functions	70
P_{tW} spectrum	50	Fit systematics	30
Total (quadrature sum)			160

electron shower shape affecting the trigger. This has been studied as a function of the angle between $-\vec{P}_{tW}$, the momentum recoiling against the W, and the electron and has been included in the simulated electron efficiency. The projection of the recoil energy along and perpendicular to the electron direction, $u_\parallel$ and $u_\perp$, depend on this efficiency and are used as a consistency check.

Run by run calibration constants are only used offline, so online energy cuts are not 100% efficient. Efficiency as a function of electron P_{te} and of $\not{E}_t$ are therefore included in the simulation. Also included are the same fiducial cuts as for the data.

4.   Measurement of the W Mass and Comparison of Simulation with Data

The simulation provides the expected Mt_W distribution as a function of M_W, and thus the probability of any event with a particular Mt_W. The data set likelihood can therefore be determined as a function of M_W. Maximization of the log-likelihood gives a W mass of $79.86 \pm 0.16(stat)$ GeV. Figure 1 compares the best fit simulated event characteristics with those of the data. The two are in excellent agreement for P_{te}, $P_{t\nu}$, and Mt_W spectra. The P_{tW} spectra are in good, but not complete, agreement.

Table 1 lists systematic errors. They include those due to effects built into the simulation, errors in modeling the QCD background, and systematic errors in the fit.

The overall result is $M_W = 79.86 \pm 0.16(stat) \pm 0.16(syst) \pm 0.26(scale) GeV/c^2$.

References

1. R. Tanaka, talk given at the XXVI Int. Conf. on High Energy Physics, Dallas, Texas, August 1992.
2. Particle Data Group, *Phys. Rev.* **D45**, part 2 (1992).
3. M. Veltman, *Nucl. Phys.* **B123** (1977) 89; M. Chanowitz, M.A. Furman and I. Hinchliffe, *Phys. Lett.* **78B** (1978), 285; S. Fanchiotti, B. Kniehl and A. Sirlin, *Phys. Rev.* **D48** (1993) 301.
4. A. Sirlin, *Phys. Rev.* **D22** (1980) 971; W. Marciano and Z. Parsa, *Proceedings of the 1982 DPF Summer Study on Elementary Particle Physics and Future Facilities*, p. 155.
5. S. Abachi. et. al., *Nucl. Inst. Meth.* **A338** (1994) 185.
6. M. Abolins et. al., *Nucl. Inst. Meth.* **A280** (1989) 36.
7. P. Arnold and R.P. Kauffman, *W and Z Production at Next-to-Leading Order: from Large q_T to Small*, ANL-HEP-PR-90-70.

A MEASUREMENT OF THE MASS OF THE
W VECTOR BOSON AT CDF

RANDY M. KEUP

Department of Physics, University of Illinois
Urbana, IL 61801, USA
The CDF Collaboration

ABSTRACT

We present a preliminary measurement of the mass of the W vector boson using the decays $W \rightarrow e\nu$ and $W \rightarrow \mu\nu$. The mass of the W is an important parameter of the standard model. Through its relation to both the top quark and the higgs sector, it provides a sensitive probe of the electroweak theory. During the last run of Fermilab's Tevatron, CDF recorded $\bar{p}p$ collisions at $\sqrt{s} = 1.8$ TeV corresponding to an integrated luminosity of 19.3 pb^{-1}. An analysis of both decay channels yields a combined mass measurement of $M_W = 80.38 \pm 0.23$ GeV/c^2.

1.   Introduction

With the precision measurement of the Z° mass from LEP[1] and the recent evidence for the top quark,[2] a precise measurement of the W mass is a crucial step in testing the Standard Model. We present a preliminary measurement of the W mass from data collected by the CDF detector during the 1992 - 1993 run.

2.   Mass Measurement

We determine the mass by fitting the transverse mass distribution given by $m_T = \sqrt{2p_T^l p_T^\nu (1 - \cos \Delta\phi)}$, where transverse is with respect to the beamline. We therefore need to measure the transverse momentum of the lepton, $\vec{p}_T^{\,l}$, and the neutrino, $\vec{p}_T^{\,\nu}$.

2.1.   Lepton Measurement

We measure the energy of the electron using the central electromagnetic calorimeter (CEM) while the muon is measured in the 14 kG magnetic field with the central tracking chamber (CTC).

First the CTC is aligned by requiring electrons and positrons to give the same ratio of energy as measured in the CEM to momentum as measured in the CTC. This works because of the fact that the CEM measurement is independent of the charge of the electron, i.e. whether it is an electron or positron, whereas the alignment of the CTC affects them differently. The absolute momentum scale of the CTC is determined by comparing the mass of the J/ψ as measured by the CTC using muon decays with

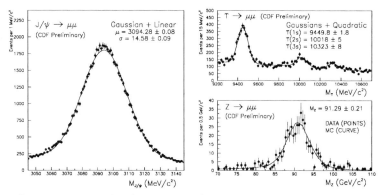

Fig. 1. Left: Invariant mass distribution near J/ψ peak. Points are data, curve is a gaussian plus linear background fit. Upper right: Invariant mass distribution near the Υ peak. The curve is three gaussians plus a quadratic background fit. The fit values have not had the momentum scale shift applied. Lower right: Invariant mass distribution near the Z° pole. The Monte Carlo is a LO generator with radiative corrections.

the world average of 3096.93 ± 0.09 MeV/c². Figure 1 contains a plot of the dimuon mass spectrum near the J/ψ with a fit to a gaussian plus linear background overlaid. The fit value after radiative corrections is 3094.58 ± 2.2 MeV resulting in a momentum scale of 1.00076 ± 0.00071. As further confirmation of the absolute momentum scale, we also check the masses of the three lowest Υ states and the mass of the Z° (Fig. 1).

Next the CEM is calibrated by requiring that the mean E/p be the same everywhere to get relative corrections and by comparing E/p for the entire sample to a radiative Monte Carlo prediction to get the overall scale. Figure 2 is a plot of E/p with the Monte Carlo overlaid. The right-hand tail of the E/p distribution is caused by the emission of both internal and external collinear bremsstrahlung. To verify the scale of the CEM, the Z° mass from electron decays is also measured (Fig. 2).

2.2. Neutrino Momentum Measurement

We determine the neutrino transverse momentum, $\vec{p}_T^{\,\nu}$, by measuring the transverse momentum of the hadrons recoiling from the W, $\vec{P}_T^{\,hadrons}$. We define $\vec{P}_T^{\,hadrons}$ as $\Sigma(E_{tower} \cdot \vec{n}_{tower})_T$, where the T as usual means the transverse components, E is the energy in the ith calorimeter tower, and $\vec{n}$ is a unit vector pointing from the vertex to the ith calorimeter tower. Before constructing this vector, we remove from the calorimeter energy the contributions from the leptons and replace them with an average underlying event energy of 30 MeV. Given this vector, we then have from momentum conservation, $\vec{p}_T^{\,\nu} \;=\; -\vec{p}_T^{\,l} \;-\; \vec{P}_T^{\,hadrons}$.

2.3. Data Selection

We use leptons with $\eta < 1$ and require both $\vec{p}_T^{\,l}$ and $\vec{p}_T^{\,\nu}$ to be greater than 25 GeV and $\vec{P}_T^{\,hadrons}$ less than 20 GeV. We also require no other track greater than 10 GeV

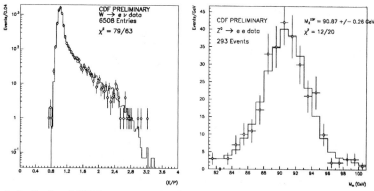

Fig. 2. Left: Ratio of CEM energy to CTC momentum for electrons from the W sample. The points are data and the histogram is a radiative Monte Carlo. Right: Dielectron mass distribution near the Z peak. The points are the data and the histogram is a LO Monte Carlo with radiative corrections.

and no jet greater than 20 GeV. The resultant data samples total 6421 electrons and 4090 muons.

3. Monte Carlo and Fitting

We determine the W mass by fitting the m_T distribution from the data with a set of Monte Carlo m_T distributions generated over a range of mass and width values, which we refer to as templates.

The Monte Carlo we use is a leading order generator using MRSD-$'$ structure functions fed into a parametric detector simulation. The transverse momentum, p_T^W , of the W is added in by hand and constrained by the data. We start with the p_T distribution taken from Z events, which is well measured relative to the p_T^W distribution since we detect both decay leptons, and scale it to account for any differences between the true p_T^W and the measured Z p_T distributions. After p_T^W has been inserted, we simulate the detector response to a vector boson of that p_T by selecting a Z event with the same boson p_T and using the detector response from that event in our Monte Carlo. This means there are no intricate models that need to be constructed to deal with the nonlinear response of the detector at low energies. All resolutions are already contained within the Z data.

The one adjustable parameter in this is the scale factor on the Z p_T distribution. It is constrained by comparing the $u_\parallel$ and $u_\perp$ distributions between Monte Carlo and data. These u distributions are the projections of the recoil vector, $\vec{P}_T^{\,hadrons}$, parallel to ($u_\parallel$) and perpendicular to ($u_\perp$) the lepton direction. We constrain to the widths of these distributions as they are less sensitive to backgrounds than the means.

Once we have accumulated the Monte Carlo templates, we calculate an unbinned negative log likelihood for each m_T template and find the minimum. Figure 3 shows the m_T distributions with the best fit histogram overlaid. The likelihood is calculated from 60 GeV to 100 GeV

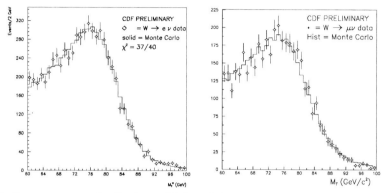

Fig. 3. Transverse mass distributions with best fit histogram overlaid. Left: $W \to e\nu$. Right: $W \to \mu\nu$.

4. Systematic Uncertainties

Lepton Momentum Scale: The momentum scale as determined from the J/ψs contributes 60 MeV common uncertainty. There is an additional 120 MeV uncertainty from the CEM scale for electrons as determined from the E/p fit.

Lepton Momentum Resolution: Both the electron energy resolution and the muon momentum resolution are determined from fitting the Z width. These fits result in resolutions of $(\delta E/E)^2 = (13.5\%/\sqrt{E_T})^2 + (2.1 \pm 1.0)\%^2$ and $\delta P_T/P_T = (0.09 \pm 0.02)\% \cdot P_T$, leading to mass uncertainties of 140 and 120 MeV.

Neutrino Scale: The extent to which we can distinguish a bias in the neutrino measurement is governed by the statistical uncertainty of the mean of the underlying event distribution, $u_{\parallel}$. This contributes 70 and 90 MeV for e and μ respectively.

p_T^W Scale: The scale factor is constrained by the widths of $u_{\parallel}$ and $u_{\perp}$ and results in uncertainties of 40 and 70 MeV for electrons and muons. Because it was constrained independently to e and μ, we take it as an uncorrelated uncertainty.

Detector Model: The detector response was modelled with Z events of which we have only limited statistics and which have an intrinsic smearing from the lepton resolutions. The effect of the limited number of events was studied and found to contribute less than 20 MeV while the smearing of the Z p_T caused a systematic shift of 80 MeV. Both of these are common to e and μ.

Backgrounds: The dominant backgrounds in the muon sample come from Z decays, where one muon is too far forward to have made a track in the CTC (4.35%), $W \to \tau\nu \to \mu\nu\nu$ (1.2%), and cosmic rays where one track was not found (0.8%). Electrons suffer mainly from $W \to \tau\nu \to e\nu\nu$ (also 1.2%). The resulting shifts in mass from backgrounds are 80 ± 50 and 232 ± 50 MeV for e and μ.

Radiative Corrections: Since the Monte Carlo is a leading order generator, we must also apply corrections for the radiative decays of the W. These are simulated with a slower $\mathcal{O}(\alpha)$ generator and result in shifts of 80 and 154 MeV for e and μ.

Structure Functions: To estimate the uncertainty due to not exactly knowing the

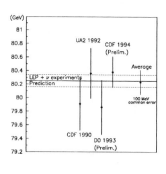 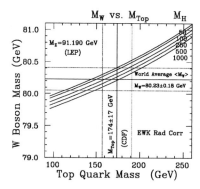

Fig. 4. Left: Comparison of this measurement to others. Right: Standard Model relationship between M_W, M_{TOP}, and M_{HIGGS}.

proton structure, we use a variety of structure functions and take the extreme shifts of 100 MeV as a common uncertainty.

5. Results

For the $W \to e\nu$ channel, we have measured the W mass to be $M_W = 80.47 \pm 0.15(STAT) \pm 0.21(SYST) \pm 0.13(SCALE)$ GeV/c^2 while for $W \to \mu\nu$ we have found it to be $M_W = 80.29 \pm 0.20(STAT) \pm 0.22(SYST) \pm 0.06(SCALE)$ GeV/c^2. Combining the two, taking the structure function and momentum scale uncertainties as common, yields

$$M_W = 80.38 \pm 0.23 \ GeV/c^2.$$

We can also combine this with previous measurements by CDF,[3] UA2,[4] and the current DØ measurement[5] to get $M_W = 80.23 \pm 0.18$ GeV/c^2. In Fig. 4, we compare this measurement of the W mass with other measurements and the current world average as well as the LEP prediction.[6] We also present in Fig. 4 the current state of the Standard Model in the form of the relationship between M_W, M_{TOP}, and M_{HIGGS}.

References

1. L. Arnaudon *et al.* (Working Group on LEP Energy), *Phys. Lett.* **B307** (1993) 187.
2. F. Abe *et al.* (CDF Collaboration), *Phys. Rev. Lett.* **73** (1994) 225, and submitted to Phys. Rev. D.
3. F. Abe *et al.* (CDF Collaboration), *Phys. Rev. Lett.* **65** (1990) 2243.
4. J. Alitti *et al.* (UA2 Collaboration), *Phys. Lett.* **B276** (1992) 354.
5. Q. Zhu (DØ Collaboration), *The Ninth Topical Workshop on Proton-Antiproton Collider Physics,* October 18-22, 1993.
6. M. Swartz. *XVI International Symposium on Lepton-Photon Interactions,* August 10-15, 1993.

UPDATE OF THE SEARCH FOR THE STANDARD MODEL HIGGS BOSON

RASMUS MØLLER*

Niels Bohr Institute, University of Copenhagen,
Blegdamsvej 17, DK–2100 Copenhagen Ø

and

CERN
CH–1211 Geneve 23, Switzerland

ABSTRACT

The search for the Standard Model Higgs boson in the high mass domain at LEP by the DELPHI Collaboration is reviewed. The Higgs boson was searched for through its production in association with neutrinos, electrons or muons. A preliminary update of the search is given to include the data sample of around 730000 hadronic Z^0 decays recorded in 1993. The results restrict the mass of the Higgs boson to be larger than 58.3 GeV/c^2 at 95% confidence level.

1.   Introduction

Identification of the agent responsible for electroweak symmetry breaking is today one of the most pressing questions in particle physics. In the Minimal Standard Model all the particle masses are obtained from one doublet of Higgs fields, which gives rise to just one massive scalar particle, H^0. For a given value of its mass, its production cross section[1] and decay branching ratios[2] can be accurately calculated. The mass of the Higgs particle is, however, theoretically only very loosely constrained. Electroweak precision measurements in combination with the evidence presented by the CDF Collaboration for production of the top quark[3] now start to give indications of the Higgs boson mass through radiative corrections.[4] The preferred solutions have values not far above the limits from direct searches at LEP,[5,6] thus emphasizing the importance of extending these searches as far as possible.

At LEP, the H^0 is expected to be produced mainly in association with a virtual Z^0 boson: $e^+e^- \rightarrow Z^0 \rightarrow H^0Z^{0*}$. In the high mass region, the H^0 decays[2] most often into $b\bar{b}$ pairs and $\tau^+\tau^-$ pairs (9%). The virtual Z^0 is detected through its decay into a $\nu\bar{\nu}$ pair or to e^+e^- or $\mu^+\mu^-$.

In the following the search performed by the DELPHI Collaboration will be briefly discussed and a preliminary update including the data recorded in 1993 will be given. For full details the reader is referred elsewhere.[5,8]

---

*Representing the DELPHI Collaboration.

The 1993 sample corresponds to an integrated luminosity of 36.1 pb^{-1}, or 730000 hadronic Z^0 events, taken at the Z^0 peak and at ± 2 GeV from the peak. The previously recorded samples correspond to about 1.1 million hadronic Z^0 events, thus bringing the total to about 1.8 million events.

2. Neutrino Channel

The channel where the virtual Z^0 decays into neutrinos is the most important in the search because of its large branching ratio (18% of the Higgs cross section). The experimental signature is a high mass unbalanced hadronic system, generally composed of two acollinear and acoplanar b-jets. The main background comes from hadronic Z^0 decays with missing energy and momentum, due to neutrinos or weaknesses in the instrument. In order to reject this background it is necessary to deal with a large number of variables. A traditional analysis based on sequential cuts easily becomes inefficient and precludes the use of variables for which signal and background have a large overlap. We have therefore based the analysis on two methods which attempt in different ways to make use of all available information in a continuous way.

2.1. Neural Network Analysis

The first of these uses a neural network computation. This part of the analysis is still being developed and the results quoted here are preliminary.

The neural network uses 15 variables selected among the many variables characterizing the event and its topology by a step-wise linear discriminant analysis. The network has two hidden layers and one output node.

Before the neural network is applied a preselection is made. It includes cuts on the visible mass and the acollinearity, and an isolation criterion on the missing momentum. The preselection has been tightened with respect to the analysis of the 1991–92 data and uses different cuts. As a consequence, the set of input variables for the network is also different in the new analysis. In the previous analysis, the preselection accepted 81.2% of the signal and rejected 98.5% of the background. The new preselection rejects 99.7% of the background and still accepts 67% of the signal.

The network is trained on simulated samples corresponding to approximately 10^6 $q\bar{q}$ events and 4000 H$^0\nu\bar{\nu}$ events with m(H^0)$=$ 55 GeV/c^2. As demonstrated in Fig. 1, a very good separation between background and signal is achieved. The network output is required to be larger than 0.92, which reduces the $q\bar{q}$ background by a factor 171.

It has, however, to be supplemented by a few extra cuts designed to reduce the remaining background in the $q\bar{q}$ Monte Carlo sample to zero. One is an explicit application of a b-tag, the others are a technical cut against cracks and a cut on the acoplanarity.

The final efficiency is $39.2 \pm 1.1 \pm 1.6\%$ for a Higgs particle at 55 GeV/c^2. One event is left in the data. It has a low reconstructed mass (27.5 ± 3.6 GeV/c^2) and can be neglected in the limit calculation.

2.2. Probabilistic Analysis

In the probabilistic analysis, weights are computed for 14 selected variables. The weights are obtained by integrating the probability density functions of the variables in the simulated $q\bar{q}$ sample. By construction, the weights have values between zero

DELPHI

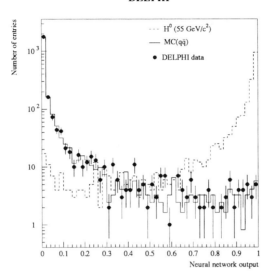

Fig. 1. Distribution of the output of the neural network. The simulated $q\bar{q}$ events (solid line) are normalised to the data (dots). The dashed line shows the expected output for a 55 GeV/c^2 Higgs boson (arbitrarily normalised). The results are preliminary.

and one, and have flat distributions for the background process. For a variable x, the corresponding weight $P(x)$ thus gives the probability to observe a value of the original variable not exceeding x in the background. For the signal events the $P(x)$ distributions are skew and $P(x)$ can, if necessary, be replaced by its complement so that the excess is always below 0.5.

A global variable W is defined as the sum of the weights. For the background the distribution of W should then be approximately Gaussian, if the correlations between the variables can be neglected. Figure 2 demonstrates that this is indeed the case and that a good separation between the expected signal and the background is obtained.

As the shape of the background is predictable (also in the presence of correlations), the expected background for a given cut in W can be calculated. For the Higgs boson search we have choosen a value of the cut, such that the expected background from $q\bar{q}$ events is one.

As in the neural network analysis, a preselection is made before the probabilistic analysis is applied. This reduces the background by 99.8% while keeping 48% of the signal for a 55 GeV/c^2 Higgs particle. In the combined real data sample from 1991–93 it leaves 312 events. The final cut is determined from the same Gaussian fit to the $q\bar{q}$

DELPHI

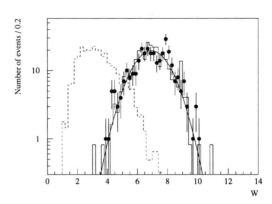

Fig. 2. Distribution of the sum of weights, W, after preselection cuts for the combined 1991–93 real data sample (dots), simulated $q\bar{q}$ events (solid line) and $H^0\nu\bar{\nu}$ events with $m(H^0)=$ 55 GeV/c^2 (dashed line). The $q\bar{q}$ sample has been normalised to the number of hadronic Z^0 events, whereas the normalisation of the signal distribution is arbitrary. The superimposed curve is a fit to the real data by a Gaussian.

simulation as in ref. 5. It was 4.0 in our previous analysis and is now reduced to 3.72, due to the increase in statistics. Below this value there are no events in the real data sample. The expected background is 1.00 ± 0.25 from $q\bar{q}$ events and $0.12 \pm 0.13 \pm 0.03$ from four fermion events with τ leptons in the final state. The efficiency for a 55 GeV/c^2 Higgs particle is $31.0 \pm 0.7 \pm 1.3\%$.

2.3. Combined neutrino channel analysis

Only about 75% of the selected signal events are common to the two analyses of the $H^0\nu\bar{\nu}$ channel. Combining them leads to a final selection efficiency of $44.5 \pm 0.8 \pm 3.1\%$ at 55 GeV/c^2.

3.   Charged lepton channels

The channels in which the virtual Z^0 decays into e^+e^- or $\mu^+\mu^-$ have smaller branching ratios (about 3% of the Higgs cross section each) and contribute less to the limit. On the other hand a better mass resolution can be obtained in these channels by using a constrained fit on the two leptons and the two jets.

The experimental signature consists of a pair of isolated leptons of opposite charge recoiling against a heavy hadronic system. Apart from events with misidentified electrons, the main backgrounds come from semileptonic decays of heavy quarks and four fermion events with leptons in the final state. The requirement of lepton isolation is sufficient to reject almost all of the first background and the other can be much reduced by kinematical cuts. An irreducible background remains, however, and is expected to give rise to candidates at the present level of statistical sensitivity.

The analyses of both channels use sequential cuts and follow the previous publication[5] with only a few additional cuts.[8]

The efficiency for $H^0e^+e^-$ events is $35.2\pm1.5^{+1.3}_{-3.0}\%$ at 55 GeV/c^2. No events are left in the data from 1993. Combined with the full 1990–92 sample, this brings DELPHI to a total of four candidates, all with masses below 40 GeV/c^2. The estimated background is 4.5 ± 1.6 from $q\bar{q}$ events and $2.0 \pm 0.2 \pm 0.3$ from four fermion processes.

The efficiency for $H^0\mu^+\mu^-$ events with $m(H^0) = 55$ GeV/c^2 is $53.6 \pm 1.1^{+0.7}_{-0.5}\%$.

No events survive in the real data sample from 1993. The estimated background from $q\bar{q}$ events is $0.54 \pm 0.54 \pm 0.03$ and from four fermion events $1.15 \pm 0.15 \pm 0.18$. In the 1991–92 data sample one candidate was found, but with a low recoil mass.

4. Limit calculation

Figure 3 shows the efficiencies for selecting Higgs events, and the expected number of observed events, as functions of $m(H^0)$ for the whole sample recorded from 1990–1993. The expected number of events has been conservatively decreased by one standard deviation, and in addition by 4% to take into account the new values for the H^0 branching ratios into $\tau^+\tau^-$ and $b\bar{b}$.[2] Since none of the observed candidates have a mass in the vicinity of the limit, their influence can be neglected. The limit on the mass at 95% confidence level is therefore found where the expected number of events falls below $-\ln 0.05 = 3.0$. The mass of the Standard Model Higgs boson is thus restricted to be larger then 58.3 GeV/c^2 at 95% confidence level.

Similar limits have been obtained by the other LEP experiments.[6] A procedure to combine the limits from LEP is being discussed but has not yet been agreed upon. Simply adding the expected numbers of events, conservatively reduced by 20% to account for the tightening of the cuts due to the increased statistics, leads to a combined limit of 64.5 GeV/c^2.[9]

Data taking at LEP in this and next year is expected to increase the total statistics by a factor 3–4. Because of the steeply falling cross section, this will however only give a modest increase of 4–5 GeV/c^2 in the range of masses that can be investigated. The search for the Standard Model Higgs boson at LEP200 depends crucially on the energy and luminosity which can be achieved. Thus for an energy of 175 GeV and an integrated luminosity of a few hundred pb^{-1}, only masses up to at most 80 GeV/c^2 can be searched for, whereas with higher beam energy and luminosity it might be possible to reach 110–120 GeV/c^2.

References

1. F. A. Berends and R. Kleiss, *Nucl. Phys.* **B260** (1985) 32.
2. E. Gross, B. A. Kniehl and G. Wolf, DESY 94–035, March 1994.
3. CDF Collaboration, F. Abe *et al.*, FERMILAB-PUB-94/097-E, April 1994; FERMILAB-PUB-94/116-E, May 1994.
4. J. Ellis, G. L. Fogli and E. Lisi, CERN-Th. 7261/94.
5. DELPHI Collaboration, P. Abreu *et al.*, CERN-PPE/94-46/Rev., Subm. to Nucl. Phys. B.

DELPHI

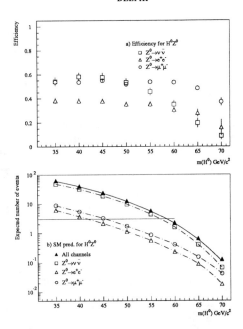

Fig. 3. a) Preliminary selection efficiencies as a function of the Higgs boson mass for the analysis performed on the 1993 data. b) Expected signal in the complete data sample recorded from 1990 to 1993 by DELPHI.

6. ALEPH Collaboration, D. Buskulic *et al.*, *Phys. Lett.* **313B** (1993) 299;
    L3 Collaboration, O. Adriani *et al.*, *Phys. Lett.* **303B** (1993) 391;
    OPAL Collaboration, M. Z. Akrawy *et al.*, CERN-PPE/94-48, Subm. to Phys. Lett. B.
7. DELPHI Collaboration, P. Abreu *et al.*, *Nucl. Phys.* **B373** (1992) 3.
8. DELPHI Collaboration, P. Abreu *et al.*, DELPHI 94-85 PHYS 402, Submitted to the Glasgow Conference, ICHEP94 Ref. gls0304.
9. F. Richard, Talk at the 27th Int. Conf. on High Energy Physics, Glasgow, July 1994.

MEASUREMENT OF W AND Z BOSON PRODUCTION AND EXTRACTION OF THE W WIDTH AND BRANCHING RATIOS

WILLIAM F. BADGETT

Department of Physics, University of Michigan
Ann Arbor, Michigan 48109

For the CDF Collaboration

ABSTRACT

We present results from W and Z boson production in proton-antiproton collisions at $\sqrt{s} = 1.8\ TeV$ using the CDF detector. We measure the W and Z cross sections times lepton branching ratio, $\sigma(p\bar{p} \to W) \cdot B(W \to l\nu_l)$ and $\sigma(p\bar{p} \to Z) \cdot B(Z \to l^+l^-)\ l = e, \mu$. We also measure the ratio: $R_l = \sigma \cdot B(W \to l\nu_l)/\sigma \cdot B(Z \to ll)$. From R_l, we extract the W electron and muon branching ratios, $BR(W \to l\nu_l)$, and the total W width, Γ_W. In addition, high transverse mass $W \to e\nu_e$ candidates provide an alternate, direct measurement of Γ_W.

1. Introduction

The Tevatron $p\bar{p}$ collider, operating at a center of mass energy of $\sqrt{s} = 1.8\ TeV$, provides a unique window into the electroweak and QCD sectors at an energy scale capable of producing real W and Z bosons. Using the CDF detector we measure the W and Z cross sections times branching ratio in the electron and muon channels:

$$\sigma_W^l = \sigma(p\bar{p} \to W^{\pm}) \cdot (W^{\pm} \to l\nu_l) \qquad \sigma_Z^l = \sigma(p\bar{p} \to Z^0) \cdot (Z^0 \to l^+l^-) \qquad l = e, \mu$$

We can measure W and Z production and decay rates with significantly improved precision, with the large 1992-93 data sample and a substantially improved luminosity measurement. We compare our results to recent next-to-next-to-leading order QCD predictions using the latest proton structure functions.

The ratio of cross sections times branching ratios can be expressed as:

$$R_l = \frac{\sigma_W^l}{\sigma_Z^l} = \frac{\sigma(p\bar{p} \to W)}{\sigma(p\bar{p} \to Z)} \cdot \frac{\Gamma(W \to l\nu_l)}{\Gamma(W)} \cdot \frac{\Gamma(Z)}{\Gamma(Z \to l^+l^-)} \qquad l = e, \mu$$

With the theoretical prediction of the production cross sections and the partial and total width measurements from the LEP experiments, we can extract the W leptonic branching ratios and the W total width (Γ_W). The W width is well predicted by the standard model, so the comparison represents a precision test of the consistency of electroweak theory. We also present a direct measurement of the the W total width from the high transverse mass region of the $W \to e\nu$ candidate sample.

2. Measurement of Cross Sections and Ratios

Experimentally, the individual cross sections and their ratio can be expressed as:

$$\sigma_W = \frac{N_W - B_W}{\epsilon_W \cdot A_W \cdot \int \mathcal{L}} \qquad \sigma_Z = \frac{N_Z - B_Z}{\epsilon_Z \cdot A_Z \cdot \int \mathcal{L}} \qquad R = \frac{\sigma_W}{\sigma_Z} = \frac{N_W - B_W}{N_Z - B_Z} \cdot \frac{\epsilon_Z}{\epsilon_W} \cdot \frac{A_Z}{A_W}$$

Where N = number of candidates; B = background; A = geometric/kinematic acceptance; ϵ = trigger and selection efficiencies; $\int \mathcal{L}$ = luminosity. The luminosity

	$W \to \mu\nu$	$Z \to \mu\mu$	$W \to e\nu$	$Z \to ee$
Candidates	6222	423	13796	1312
Background	818 ± 123	1.7 ± 0.8	1700^{+171}_{-163}	21 ± 9
Signal	$5404 \pm 69 \pm 141$	$421 \pm 21 \pm 1.6$	$12096 \pm 117^{+163}_{-171}$	$1291 \pm 36 \pm 9$
A	0.163 ± 0.004	0.159 ± 0.003	0.342 ± 0.008	0.409 ± 0.005
ϵ	0.742 ± 0.027	0.747 ± 0.027	0.754 ± 0.011	0.729 ± 0.016
$\int \mathcal{L}^{vis}$, pb^{-1}	17.99 ± 0.68		19.03 ± 0.72	
$\sigma_W \cdot B$, nb	$2.484 \pm 0.031 \pm 0.129 \pm 0.094$		$2.508 \pm 0.024 \pm 0.072 \pm 0.095$	
$\sigma_Z \cdot B$, nb	$0.2029 \pm 0.0099 \pm 0.0090 \pm 0.0077$		$0.2314 \pm 0.0065 \pm 0.0058 \pm 0.0088$	
R_l	$12.24 \pm 0.62 \pm 0.48$		$10.90 \pm 0.32 \pm 0.29$	
$B(W \to l\nu)$	$0.1237 \pm 0.0062 \pm 0.0051$		$0.1094 \pm 0.0033 \pm 0.0031$	
Γ_W, GeV	$1.825 \pm 0.092 \pm 0.077$		$2.064 \pm 0.061 \pm 0.059$	

Table 1: W and Z production results; errors: statistical, systematic, luminosity

cancels in the ratio, eliminating a large source of uncertainty. Common efficiency terms cancel and the uncertainty in the acceptance is somewhat reduced in R_l.

The event selection maximizes the cancellation of the ratio efficiency terms and minimizes background in the W sample by making tight, identical cuts on the primary lepton in the W and Z samples. We require the inclusive, high P_T electron or muon trigger (all levels) for W and Z candidates. The more important primary muon selection criteria are: track $P_T \geq 20$ GeV; minimum ionizing; muon chamber ($|\eta| < \sim 0.6$) track stub matches to central track; isolation in a surrounding cone $E_T(\Delta R \leq 0.4) \leq 2.0$ GeV [1]. The main primary electron selection criteria are: central calorimeter ($|\eta| < \sim 1.1$) $E_T \geq 20$ GeV; small lateral tower energy sharing; $0.5 < E/p < 2.0$; track to strip chamber matching; isolation $E_T(\Delta R < 0.4)/E_T) < 0.1$. The event vertex must be $|z| \leq 60$ cm from the detector center, e and μ.

The secondary lepton requirements are much looser. W candidates must have at least $\not{E}_T \geq 20$ GeV of unbalanced transverse energy (e and μ), indicating the presence of a neutrino. For Z candidates, the secondary muon must fall anywhere within the central tracking chamber $|\eta| \leq 1.2$ with $P_T \geq 20$ GeV and be minimum ionizing. The secondary electron is allowed to fall in any calorimeter section: central, plug ($1.2 < |\eta| < 2.4$) or forward ($2.4 < |\eta| < 4.0$), with transverse energy cuts of $E_T \geq 20, 15, 10$ GeV, respectively. The Z candidate dilepton invariant mass must be 66 $GeV \leq M_{ll} \leq 116$ GeV.

The geometric/kinematic acceptances, A, come from a simple Monte Carlo model and include the efficiency of the lepton and neutrino P_T cuts and dilepton mass cut. The dominant uncertainties include: PDFs (MRS D'_{-} nominal), boson P_T spectrum and higher order QCD processes, underlying event model and neutrino resolution, and tracking and calorimeter resolution. The other selection cuts and trigger efficiencies, ϵ, are computed using control data samples and tend to be statistics limited.

The dominant backgrounds in the W sample include the processes $W \to \tau \to e$ or μ, $Z \to ll$ (one lepton is lost), and generic QCD jet events. In the muon sample, there are also small contributions from pion decay in flight and cosmic rays. For the $W \to e\nu$ sample, QCD processes dominate the background; for the $W \to \mu\nu$ sample, the $Z \to \mu\mu$ background dominates. The Z sample is almost background free.

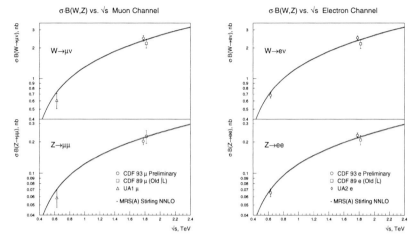

Figure 1: W and Z Cross Sections vs. center of mass energy (μ left, e right)

The $Z \rightarrow ee$ background comes mostly from QCD processes; most of the $Z \rightarrow \mu\mu$ background comes from cosmic rays. In both Z samples, we apply a correction to remove the Drell-Yan continuum (and interference) terms from the Z cross section (1.005 ± 0.002 electrons; 1.03 ± 0.01 muons).

3. Cross Section and Ratio Results

The results for the cross sections, branching ratios and W width are summarized in table 1. Note that a correction factor has been applied to the luminosity to account for the event longitudinal position cut $|z| \leq 60$ cm. Figure 1 plots $\sigma \cdot B$ vs. $\sqrt{s}$ compared with the theoretical prediction using the MRS(A) parton distribution functions from a calculation by Stirling [4]. The former CDF, UA1, and UA2 published values are also plotted, from references [6, 7, 8, 9]. The theory curve uncertainty is due to parton distribution functions and the QCD scale. The error on $\sigma \cdot B$ is still too large to have discriminating power between the different sets of PDFs, but represents an important, independent test of their magnitude at this Q^2 and x scale.

To extract the branching ratio and total width, we take the Z total and partial widths from LEP [5] and the production cross section ratio and ratio of partial widths from references [4, 3]. In figure 2 on the left, we plot the world values of the W width from the UA1, UA2, CDF and DØ published and preliminary numbers [6, 7, 8, 9, 10]. The theory value $\Gamma_W = 2.067 \pm 0.021$ GeV is from reference [2].

4. Direct Γ_W Measurement

We also measure the W total width directly from the high mass region of the W transverse mass spectrum, defined as $M_T \equiv \{2E_T^e E_T^\nu [1 - \cos(\Delta\phi)]\}^{1/2}$. Figure 2 (right) plots the transverse mass distribution for a special $W \rightarrow e\nu$ sample with relaxed quality cuts to avoid biasing against very high E_T electrons; extra cuts are applied to reduce background: $E_T^e \geq 30$ GeV, $P_T^W < 20$ GeV, and more stringent Z

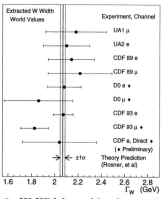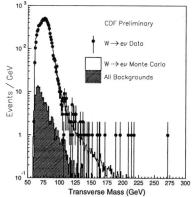

Figure 2: W Width world values and best fit W width to $W \to e\nu$ transverse mass

candidate and dijet removal. Above the point $M_T \geq 110 \ GeV$, the Breit-Wigner line shape starts to dominate over the gaussian resolution, and we perform a binned log-likelihood fit to the transverse mass shape in this region, using Monte Carlo templates of varying Γ_W. The fit yields a result of:

$$\Gamma_W = 2.04 \pm 0.28(stat) \pm 0.16(syst) \quad (direct, \ CDF \ preliminary)$$

The total uncertainty here is not competitive with the indirect value from R_l, but represents the best direct measurement of the W width thus far. The statistical error dominates, which can be easily reduced during the current Tevatron run; the systematic error is dominated by the $W \ P_T$ distribution smearing of M_T; the next largest systematic error includes the neutrino resolution modelling.

We thank James Stirling and Jonathan Rosner for useful discussions and computations. I thank Sacha Kopp and Greg Sullivan for the electron plots and numbers.

References

1. $\Delta R = \sqrt{\Delta \eta^2 + \Delta \phi^2}$; $\eta = -\ln \tan \frac{\theta}{2}$; Polar, azimuthal angles: θ, ϕ.
2. J. Rosner, *et al*, Enrico Fermi Institute Preprint, **EFI-93-40** (1993).
3. A.D. Martin, *et al*, Rutherford Appleton Lab Preprint **RAL-94-055** (1994).
4. W.J. Stirling, *private communication*, June 1994.
5. The LEP Collaborations and Electroweak Working Group, *Proceedings from the Europhysics Conference on High Energy Physics*, Marseille, France (1993).
6. The UA1 Collaboration, Phys. Lett. **B253** (1991) 503-510.
7. The UA2 Collaboration, Phys. Lett. **B276** (1992) 365-374.
8. The CDF Collaboration, Phys. Rev. Lett. **69** (1992) 28-32.
9. The CDF Collaboration, Phys. Rev. D **44** (1991) 29-52.
10. The DØ Collaboration, *Proceedings from this conference* (1994).

MEASUREMENT OF W AND Z PRODUCTION CROSS-SECTIONS IN pp̄ COLLISIONS AT $\sqrt{s} = 1.8$ TeV

PAUL Z. QUINTAS*

Fermi National Accelerator Laboratory
Batavia, IL 60510

ABSTRACT

The cross sections for W and Z production in pp̄ collisions at $\sqrt{s} = 1.8$ TeV are measured using the DØ detector at the Fermilab Tevatron collider. The detected final states are $W \to e\nu_e$, $Z \to e^+e^-$, $W \to \mu\nu_\mu$, and $Z \to \mu^+\mu^-$. In the ratio of these measurements, many common sources of systematic error cancel and we measure $R = \sigma(\text{p}\bar{\text{p}} \to W) \cdot Br(W \to l\nu)/\sigma(\text{p}\bar{\text{p}} \to Z) \cdot Br(Z \to l^+l^-)$. Assuming standard model couplings, this result is used to determine the width of the W boson and to set a limit on the decay $W^+ \to t\bar{b}$.

1. Introduction

The cross-sections times branching ratios for pp̄ $\to W$ and pp̄ $\to Z$ with decays into final states with electrons or muons are measured using the DØ detector. In the ratio of these measurements, many common sources of systematic error cancel and we measure

$$R = \frac{\sigma(\text{p}\bar{\text{p}} \to W + X) \cdot B(W \to l\nu)}{\sigma(\text{p}\bar{\text{p}} \to Z + X) \cdot B(Z \to l^+l^-)}.$$

This ratio is of interest since it can be expressed as the product of calculable or well-measured quantities:

$$R = \frac{\sigma(\text{p}\bar{\text{p}} \to W + X)}{\sigma(\text{p}\bar{\text{p}} \to Z + X)} \frac{\Gamma(W \to l\nu)}{\Gamma(Z \to l^+l^-)} \frac{\Gamma(Z)}{\Gamma(W)}.$$

This gives the most precise measurement of $\Gamma(W)$. A width measurement which exceeds the standard model value might indicate non-standard decays of the W.

2. The DØ Detector

The DØ detector[1] consists of three major subsystems: central tracking detectors, nearly hermetic liquid argon calorimetry, and a muon spectrometer. The central tracking system is used to identify tracks in the psuedorapidity range $|\eta| \leq 3.5$. The calorimeter covers the region up to $|\eta| \leq 4$ with energy resolution for electrons approximately $15\%/\sqrt{E}$. The muon system consists of drift chambers and magnetized iron toroids.

---

*Representing the DØ Collaboration.

3. Electron Channel Cross-sections

The $W \to e\nu_e$ and $Z \to e^+e^-$ candidates were collected on a single trigger consisting of a two levels: the hardware trigger required electromagnetic energy above threshold (usually 10 GeV) in a .2 x .2 (η x ϕ) tower. The software trigger required $E_T \geq 20$ GeV and made loose shower shape and isolation cuts.

Offline electrons were required to have mostly electromagnetic energy, a shower shape consistent transversely and longitudinally with an electron, be isolated from other activity in the calorimeter, and be matched with a central detector track. Fiducial cuts of $|\eta| \leq 1.1$ or $1.5 \leq |\eta| \leq 2.5$ ensured a good trigger, good energy resolution, and low background.

The $W \to e\nu_e$ candidates satisfied $E_T > 25$ GeV and $\not{E}_T > 25$ GeV, resulting in a sample of 10346 candidates. The $Z \to e^+e^-$ candidates had two electrons with $E_T \geq 25$ GeV, one of which satisfied all the $W \to e\nu_e$ cuts, and the other passed all except the track match cut. There was an invariant mass cut of $75 \leq M_{inv} \leq 105$ GeV, creating a sample of 782 candidates.

Electron efficiencies are determined from the $W \to e\nu_e$ sample with a harder $\not{E}_T$ cut and from the $Z \to e^+e^-$ sample. The $W \to e\nu_e$ backgrounds are estimated separately either from data (QCD) or Monte Carlo ($W \to \tau \to e$ and $Z \to e^+e^-$). The $Z \to e^+e^-$ background is estimated by fitting the invariant mass peak to a Breit-Wigner convoluted with the detector resolutions plus a linear background. Luminosity for this trigger was $\int L dt = 12.4 \pm 1.5$ pb^{-1}. Table 1 gives the preliminary values of the cross-sections.

Table 1. Measured W and Z Cross-sections

	$W \to e\nu_e$	$Z \to e^+e^-$	$W \to \mu\nu_\mu$	$Z \to \mu^+\mu^-$
Number of Events	10346	782	1665	77
Acceptance	46.1 ± 0.9 %	36.4 ± 0.5 %	25.1 ± 0.7 %	6.7 ± 0.4 %
Trig + Sel Eff.	73.7 ± 1.8 %	74.6 ± 3.0 %	22.4 ± 2.6 %	53.8 ± 5.0 %
Total Bkgd	5.9 ± 0.7 %	5.2 ± 2.3 %	22.1 ± 1.9 %	10.1 ± 3.7 %
Luminosity	12.4 ± 1.5pb^{-1}		11.1 ± 1.3pb^{-1}	
$\sigma \cdot B$(nb)	2.32	0.220	2.09	0.174
Stat. Err.	± 0.02	± 0.008	± 0.07	± 0.022
Syst. Err.	± 0.07	± 0.011	± 0.22	± 0.018
Lum. Err.	± 0.28	± 0.026	± 0.25	± 0.021

4. Muon Channel Cross-sections

The $W \to \mu\nu_\mu$ and $Z \to \mu^+\mu^-$ trigger consisted of two hits-in-road searches at the hardware level (first a coarse road, then a fine road, with an effective P_T cut of 7 GeV). The software trigger did track finding and reconstruction and required at least one muon with $P_T \geq 15$ GeV.

Offline tracks ("loose" muons) were confirmed by energy in the calorimeter and required to pass through the central iron ($|\eta| \leq 1.0$) and through a minimum magnetic field ($\int B \cdot dl \geq 2\, Tm$). "Tight" muons were defined satisfy the loose requirements plus have a matching track in the central detector, have a good quality fit to the vertex point, central detector track and muon track, be in time with beam crossing, and be isolated in the calorimeter.

The $W \rightarrow \mu\nu_\mu$ candidates have a tight muon and satisfied kinematic cuts of $P_T \geq 20$ GeV and $\not{E}_T \geq 20$ GeV for a sample of 1665 candidates. The $Z \rightarrow \mu^+\mu^-$ candidates have one tight muon and a second loose or tight muon. Kinematic cuts of $P_T^{\mu 1} \geq 20$ and $P_T^{\mu 2} \geq 15$ GeV yielded 77 candidates.

Muon efficiencies are determined from the $Z \rightarrow \mu^+\mu^-$ sample. The $W \rightarrow \mu\nu_\mu$ and the $Z \rightarrow \mu^+\mu^-$ backgrounds are estimated from data for the QCD and cosmic channels, and from Monte Carlo for the $Z \rightarrow \mu^+\mu^-$, $Z \rightarrow \tau^+\tau^-$, $W \rightarrow \tau \rightarrow \mu$, and Drell-Yan backgrounds. The luminosity for this trigger was $11.1 \pm 1.3 \text{pb}^{-1}$ and the preliminary cross-sections are reported in table 1. The electron and muon cross-sections are compared to other measurements in figure 1.

5.   Ratio Measurements and $\Gamma(W)$

The ratio $R = \sigma \cdot B(W \rightarrow l\nu)/\sigma \cdot B(Z \rightarrow l^+l^-)$ is of interest since it can be expressed as the following combination of precisely measurable or calculable quantities:

$$R \equiv \frac{\sigma B(W \rightarrow \mu\nu)}{\sigma B(Z \rightarrow \mu\mu)} = \frac{\sigma(p\bar{p} \rightarrow W + X)}{\sigma(p\bar{p} \rightarrow Z + X)} \frac{\Gamma(W \rightarrow l\nu)}{\Gamma(Z \rightarrow l^+l^-)} \frac{\Gamma(Z)}{\Gamma(W)}. \tag{1}$$

The measured value for the Z width is obtained from the LEP experiments[2]

$$\Gamma(Z) = 2.487 \pm 0.010 \text{ GeV}/c^2.$$

The ratio of the W and Z leptonic decay widths is taken from its theoretical value[3]

$$\frac{\Gamma(W \rightarrow l\nu)}{\Gamma(Z \rightarrow l^+l^-)} = 2.70 \pm 0.01.$$

The ratio of the W to Z production is determined using the complete $O(\alpha_s^2)$ calculation[4] convoluted with various parton distribution functions[5] to obtain

$$\frac{\sigma(p\bar{p} \rightarrow W + X)}{\sigma(p\bar{p} \rightarrow Z + X)} = 3.34 \pm 0.03$$

where the quoted error is dominated by the uncertainty on the W mass and systematic differences in the structure functions.

We take the weighted average of the muon and electron channels' measurements of R:

$$\begin{aligned} R^e &= 10.54 \pm 0.39(stat) \pm 0.55(syst), \\ R^\mu &= 12.0^{+1.8}_{-1.4}(stat) \pm 1.0(syst), \\ R^{e\mu} &= 10.78^{+0.68}_{-0.60}(stat + syst). \end{aligned}$$

Combining this average value with equation 1 yields the total width of the W

$$\Gamma(W) = 2.08 \pm 0.12(stat + syst) \pm 0.02(theory + LEPsyst) \text{ GeV}.$$

This result can be compared with the Standard Model prediction[6] of

$$\Gamma(W) = 2.09 \pm 0.02 \text{ GeV}$$

for $M_t > M_b + M_W$ where M_t, M_b and M_W are the masses of the top quark, bottom quark and W boson, respectively.

We can set a limit on the decay of the W into new quark pairs. If the W couples to a new quark and to the b quark with standard model coupling, a limit on the mass of this quark is set:

$$m_q > 56 \text{ GeV at } 95\% \text{ CL}.$$

This limit applies independent of any assumptions of decay modes of the top quark and is illustrated in figure 2.

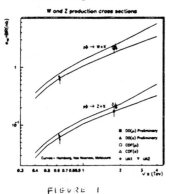

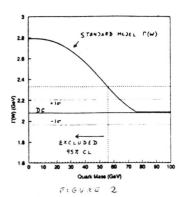

FIGURE 1   FIGURE 2

6.   Summary

DØ has measured the production cross-sections for W and Z in p$\bar{\text{p}}$ at $\sqrt{s} = 1.8$ TeV . We report preliminary results for those cross-sections and for their ratio, $R^{e\mu} = 10.78^{+0.68}_{-0.60}$. This yields a measurement of the width of the W, $\Gamma(W) = 2.08 \pm 0.12(exp) \pm 0.02(thy)$ GeV, and a limit on a new quark $m_q > 56$ GeV at 95% confidence level.

References

1. S. Abachi et al., NIM A **338**, 185 (1994).
2. The LEP collaborations: ALEPH, DELPHI, L3 and OPAL, Phys. Lett. **B276**, 247 (1992)
3. W. F. Hollik, Firtschr. Phys. **38**, 165 (1990)
4. R. Hamberg, W. L. Van Neerven and T. Matsuura, Nucl. Phys. **B359**, 343, (1991).
5. S. Lami and D. Wood, DØ internal Note # 1462, 1992 (Unpublished).
6. T. Alvarez, A. Leites and J. Terrón, Nucl. Phys. **B301**, 1 (1988).

MEASUREMENT OF THE $WW\gamma$, $ZZ\gamma$ AND $Z\gamma\gamma$ COUPLINGS AT THE FERMILAB TEVATRON

JOHN ELLISON

Physics Department, University of California
Riverside, CA 92521, USA

for the DØ Collaboration

ABSTRACT

The processes $p\bar{p} \rightarrow \ell\bar{\nu}\gamma + X$ and $p\bar{p} \rightarrow \ell^+\ell^-\gamma + X$ ($\ell = e, \mu$) have been observed using the DØ detector at the Fermilab Tevatron Collider at $\sqrt{s} = 1.8$ TeV. The observed signals in the electron and muon decay channels are used as a probe of the $WW\gamma$, $ZZ\gamma$ and $Z\gamma\gamma$ vertex couplings. By comparing the event rates and shapes of the photon E_T distributions with theoretical predictions, 95% confidence limits are obtained on the $WW\gamma$, $ZZ\gamma$ and $Z\gamma\gamma$ vertex coupling parameters.

1.   Introduction

In $p\bar{p}$ collisions at the Fermilab Tevatron, the trilinear boson couplings can be probed by diboson production. Measurement of the $WW\gamma$ trilinear coupling, is possible by detecting $W\gamma$ production and measurement of the $ZZ\gamma$ and $Z\gamma\gamma$ trilinear couplings is possible via $Z\gamma$ production.

Requiring Lorentz covariance and gauge invariance of the on-shell photon, the most general $WW\gamma$ vertex can be parametrized by a vertex function which is described by four coupling parameters κ, λ, $\tilde{\kappa}$ and $\tilde{\lambda}$.[1] These parameters are related to the electromagnetic moments of the W. For example, combinations of κ and λ yield the W magnetic dipole moment $\mu_W = \frac{e}{2M_W}(1 + \kappa + \lambda)$ and electric quadrupole moment $Q_W = -\frac{e}{M_W^2}(\kappa - \lambda)$. In the standard model at tree level the $WW\gamma$ coupling is uniquely determined by the $SU(2)_L \otimes U(1)_Y$ gauge symmetry requiring $\kappa = 1$, $\lambda = 0$, $\tilde{\kappa} = 0$, $\tilde{\lambda} = 0$. Therefore, a measurement of the $WW\gamma$ coupling is a powerful test of the Standard Model gauge symmetry.

Anomalous couplings may arise due to radiative loop corrections to the $WW\gamma$ vertex or due to models in which the W is viewed as a composite particle. The experimental signature for anomalous couplings is an increase of the $W\gamma$ production cross section and changes in the kinematic distributions of the final state particles, such as an enhancement of the photon spectrum at high transverse energy.[2]

In contrast to the $WW\gamma$ coupling, the coupling of the photon to the Z boson is forbidden in the SM at tree level. However, to allow for the possibility of anomalous couplings, the vertex functions may be parametrized in terms of four parameters

Table 1. $W\gamma$ and $Z\gamma$ selection trigger requirements and offline cuts.

	$e\nu\gamma$	$\mu\nu\gamma$	$ee\gamma$	$\mu\mu\gamma$
Trigger	$E_T^{em} \geq 20$ GeV	$p_t^\mu \geq 5$ GeV	$2 \times E_T^{em} \geq 10$ GeV	$p_t^\mu \geq 5$ GeV
	$\not{E}_T \geq 20$ GeV	$E_T^{em} \geq 7$ GeV		$E_T^{em} \geq 7$ GeV
Kinematic cuts:	$E_T^e \geq 25$ GeV	$p_T^\mu \geq 15$ GeV	$E_T^{e_1} \geq 25$ GeV	$p_T^{\mu_1} \geq 15$ GeV
	$\not{E}_T \geq 25$ GeV	$\not{E}_T \geq 15$ GeV	$E_T^{e_2} \geq 25$ GeV	$p_T^{\mu_2} \geq 10$ GeV
	$E_T^\gamma \geq 10$ GeV		$E_T^\gamma \geq 10$ GeV	
	$\Delta R \equiv \sqrt{\Delta\phi_{\ell\gamma}^2 + \Delta\eta_{\ell\gamma}^2} \geq 0.7$		$\Delta R \equiv \sqrt{\Delta\phi_{\ell\gamma}^2 + \Delta\eta_{\ell\gamma}^2} \geq 0.7$	

$(h_1^Z, h_2^Z, h_3^Z, h_4^Z)$ for the $ZZ\gamma$ vertex function and four parameters $(h_1^\gamma, h_2^\gamma, h_3^\gamma, h_4^\gamma)$ for the $Z\gamma\gamma$ vertex function.[3]

Tree level unitarity requires that the couplings must be described by form factors which vanish when any of the boson momenta become very large. The form factors used for this study are described in references 2 and 3.

In this paper we present preliminary measurements of the $WW\gamma$, $ZZ\gamma$ and $Z\gamma\gamma$ couplings using $W\gamma$ and $Z\gamma$ events in the electron ($e\bar{\nu}\gamma$ and $e^+e^-\gamma$) and muon ($\mu\bar{\nu}\gamma$ and $\mu^+\mu^-\gamma$) channels observed at DØ during the 1992-93 run, corresponding to an integrated luminosity of approximately 14 pb^{-1}.

2. Event selection

$W\gamma$ candidates were obtained by searching for events containing an isolated high-p_T lepton (electron or muon) and an isolated photon. The events must have large missing transverse energy to signify the presence of a neutrino. Table 1 summarizes the trigger requirements and offline kinematic cuts used.

Since anomalous couplings result in an enhancement of events at large E_T^γ, lowering the E_T^γ cut below 10 GeV does not increase the sensitivity to anomalous couplings. A further requirement was that the photon be well separated from the lepton ($\Delta R_{\ell\gamma} \geq 0.7$). This cut suppresses the contribution of the radiative W decay where the photon originates from bremsstrahlung from the final state lepton.

The $W\gamma$ selection criteria yielded 9 candidate events in the electron channel and 10 candidates in the muon channel.

The $Z\gamma$ candidates were obtained by searching for events containing two isolated electrons or muons and an isolated photon (see table 1). This resulted in 4 candidate events in the electron channel and 2 $Z\gamma$ candidates in the muon channel.

3. Background calculations

The background estimate for the $W\gamma$ events, includes contributions from: $W +$ jets, where a jet is misidentified as a photon; $Z\gamma$, where Z decays to $\ell^+\ell^-$ and one of the leptons is missed or mismeasued by the detector and so contributes to the measured missing E_T; $W\gamma$ with $W \to \tau\nu$ followed by $\tau \to \ell\nu\bar{\nu}$; and $ee +$ jets, where an electron is misidentified as a photon due to tracking inefficiency.

We have estimated the $W +$ jets background using the observed E_T distributions of jets in the inclusive W data sample and the measured probability for a jet to fake a photon. This probability was determined, as a function of E_T of the jet, using samples of

QCD multijet events. The probability was found to be less than 10^{-3} and varied slowly with E_T. The $Z\gamma$ background was estimated using monte carlo with a full GEANT simulation of the DØ detector. The remaining backgrounds were much smaller and were estimated from monte carlo and data.

Subtracting the estimated backgrounds from the observed number of events, we obtain the number of signal events to be $6.5^{+4.0+1.1}_{-2.1-1.2}$ for the electron channel and $7.0^{+4.2+0.5}_{-3.1-1.1}$ for the muon channel, where the first uncertainty is statistical and the second is systematic.

The background estimate for the $Z\gamma$ events includes contributions from $Z +$ jets, where a jet is misidentified as a photon, and $Z\gamma$ with $Z \to \tau\tau$ followed by $\tau \to \ell\nu\bar{\nu}$. These backgrounds were estimated from data and monte carlo respectively, as for the $W\gamma$ background study described above.

Subtracting the estimated backgrounds from the observed number of events, we obtain the number of signal events to be $3.8^{+3.2}_{-1.9} \pm 0.09$ for the electron channel and $1.96^{+2.62}_{-1.29} \pm 0.03$ for the muon channel.

4.   Limits on anomalous couplings

To compare with the SM predictions for $W\gamma$ and $Z\gamma$ production we used the event generator of Baur and Zeppenfeld.[2,3] This program generates 4-vectors for the $W\gamma$ and $Z\gamma$ processes which were then put through a fast DØ detector simulation. This monte carlo simulates the energy and momentum resolutions for electrons, muons and missing E_T and includes fiducial cuts and detector efficiencies.

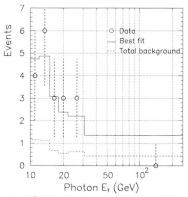
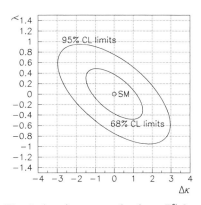

Fig. 1. E_T^γ distribution for $W\gamma$ candidates.   Fig. 2. $\Delta\kappa-\lambda$ contour plot from E_T^γ fit.

The predictions for the number of expected $W\gamma$ events using SM couplings are $7.7\pm0.8\pm0.9$ (e channel) and $6.1\pm1.0\pm0.7$ (μ channel), where the first error is the systematic error and the second error is due to the uncertainly in the integrated luminosity. For $Z\gamma$ the corresponding predictions are $2.3 \pm 0.4 \pm 0.3$ and $1.84 \pm 0.29 \pm 0.22$. The systematic error includes the uncertainty in electron efficiency, muon efficiency, photon efficiency including conversion modelling, $\not{E}_T$ smearing, choice of structure function, and choice of Q^2 scale. Comparing with the observed signal (section 3) we see that the

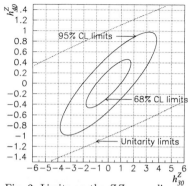

Fig. 3. Limits on the $ZZ\gamma$ couplings.

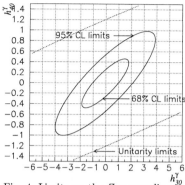

Fig. 4. Limits on the $Z\gamma\gamma$ couplings.

data are in good agreement with the SM within errors. Therefore, we see no evidence for anomalous couplings. However, we are able to set limits on the anomalous couplings by comparing our data with the predictions of the event generator for different anomalous couplings.

To set limits on the anomalous couplings, we fit the E_T^γ spectrum using a binned likelihood method (Fig. 1). The $W\gamma$ 95% confidence level (CL) limits derived from this method are $-2.3 < \Delta\kappa < 2.3$ (for $\lambda = 0$) and $-0.75 < \lambda < 0.75$ (for $\Delta\kappa = 0$) as shown in the $\Delta\kappa - \lambda$ contour plot in Fig. 2. The 95% CL limits on couplings independent of each other, is represented by the furthermost points on the contour: $-3.0 < \Delta\kappa < 3.0$ and $-0.95 < \lambda < 0.95$.

The same method was used to obtain limits for the $ZZ\gamma$ and $Z\gamma\gamma$ couplings. Because of the form of the vertex functions for these couplings our measurement is sensitive to the form factor sacle Λ and the results given here are for $\Lambda = 500$ GeV. For $ZZ\gamma$ (Fig. 3) we obtained the 95% CL limits $-2.1 \leq h_{30}^Z \leq 2.1$ (for $h_{40}^Z = 0$) and $-0.5 \leq h_{40}^Z \leq 0.5$ (for $h_{30}^Z = 0$) and the independent limits $-3.8 \leq h_{30}^Z \leq 3.8$ and $-1.0 \leq h_{40}^Z \leq 1.0$. For $Z\gamma\gamma$ the limits are only slightly weaker as shown in Fig. 4.

5. Conclusions

The processes $p\bar{p} \to \ell\bar{\nu}\gamma + X$ and $p\bar{p} \to \ell^+\ell^-\gamma + X$ ($\ell = e, \mu$) have been observed in DØ and were used to set limits on the $WW\gamma$, $ZZ\gamma$ and $Z\gamma\gamma$ vertex couplings. The limits obtained are a considerable improvement on previous measurements (UA2[4] and the CDF 1988-89 data[5]) and are comparable in sensitivity with the CDF 1992-93 results. Using data from the current Tevatron run (1993-94) and combining results from DØ and CDF, we can expect a significant improvement in sensitivity in the near future.

References

1. K. Hagiwara et al., Nucl. Phys. **B282** (1987) 253.
2. U. Baur and E. L. Berger, Phys. Rev. **D41** (1990) 1476.
3. U. Baur and E. L. Berger, Phys. Rev. **D47** (1993) 4889.
4. UA2 Collaboration, J. Alitti et al., Phys. Lett. **B277** (1992) 194.
5. CDF Collaboration, F. Abe et al., FERMILAB-PUB-94/244-E.

SEARCH FOR W PAIR PRODUCTION WITH DILEPTON DECAY MODES AT D0

NOBUAKI OSHIMA, for the D0 Collaboration

Fermi National Accelerator Laboratory

P.O.Box 500 Batavia, Illinois 60510, USA

ABSTRACT

We present the results of a search of 13.5 pb^{-1} of p$\bar{\text{p}}$ collisions at $\sqrt{s} = 1.8$ TeV for $WW \to$ dileptons $+ X$ using the D0 detector. One $WW \to e\mu\nu\bar{\nu} + X$ candidate was found with an expected background of about 0.96 ± 0.44 events. One $WW \to ee\nu\bar{\nu} + X$ candidate was found with an expected background of about 1.23 ± 0.87 events. The 95% C.L. upper limit for the W pair production cross section is 133 pb without background subtraction. Making the constraint $\kappa = 1$ ($\Delta\kappa = 0$) implies $-2.6 \leq \lambda \leq 2.7(95\%$ C.L.) and making the constraint $\lambda = 0$ implies $-3.3 \leq \Delta\kappa \leq 3.5(95\%$ C.L.) assuming $\kappa_\gamma = \kappa_Z$ and $\lambda_\gamma = \lambda_Z$. We also present the contour of excluded couplings when κ and λ are allowed to simultaneously vary from their standard model values.

1. Introduction

1.1. Physics in W Pair Production

Vector boson pair production is interesting as a probe of many aspects of physics, for example, the standard model decay mode of the Higgs boson,[1] a measurement of possible non-standard model couplings or processes,[2] background for $t\bar{t}$ production which is also background for W^+W^- pair production and a measurement of vector boson self-couplings as test of the standard model.[3] Since the Tevatron experiments D0 and CDF have accumulated data from their 1992-1993 collider runs, the last aspect becomes an interesting study experimentally.

1.2. The D0 Detector

The D0 detector[4] can be divided into three major subsystems; central tracking, calorimeters and the muon system.
- The central tracking system contains a vertex drift chamber, a transition radiation detector, a central drift chamber and forward drift chambers.
- The uranium-liquid argon calorimeters consist of three calorimeters; the central calorimeter and the two mirror-image end calorimeters. These finely segmented calorimeters provide essentially uniform, nearly hermetic coverage over the full range of pseudo-rapidity $|\eta| \leq 4$ and azimuthal angle ϕ. The resolutions of electromagnetic(EM) and

hadronic(HAD) energy measurement are roughly $15\%/\sqrt{E}$ and $50\%/\sqrt{E}$ respectively. - The muon system is divided into the wide angle part which has an acceptance down to pseudorapidity $|\eta| = 1.7$, and the small angle part which has far forward and backward($1.7 < |\eta| < 3.4$) coverage. The wide angle muon system has three large iron toroidal magnets($B \approx 2$ Tesla) and 11434 Proportional Drift Tubes and the resolution is about $(\delta\mathrm{p}/p)^2 = (0.2)^2 + (0.01p)^2$.

2. Event Selection

Both data sets were selected from the "expressline sample" which was created for rapid physics analysis. The corresponding integrated luminosity of both data sets is 13.5 ± 1.6 pb^{-1} after excluding bad runs and correcting for dead time created to mask high loss periods of main ring operation. Requirements to identify electron are the electromagnetic(EM) energy fraction $E_{em}/E_{total} \geq 0.9$, isolation $(E_{total}(\text{cone}=0.4) - E_{em}(\text{cone}=0.2))/E_{em}(\text{cone}=0.2) \leq 0.1$, the η dependant calorimeter cluster shape and the calorimeter cluster and track matching. For muon identification, we required the cosmic ray veto, the track matching between muon chamber and central drift chamber, the confirmation by MIP energy deposition in calorimeter and fiducial cuts to avoid inter-toroid gaps.

2.1. $WW \to e\mu\nu\bar{\nu} + X$

Events in this sample were required to pass the $e\mu$ trigger which was the combination of electron and muon requirements. Those for electron were No. of EM tower ≥ 1 with $E_t > 7$ GeV(Hardware Trigger; Level 1) and No. of EM cluster ≥ 1 with $E_t > 7$ GeV(Software Filter; Level 2) and No. of muon ≥ 1 with $|\eta| \leq 1.7$(Level 1) and No. of muon ≥ 1 with $P_t > 5$ GeV/c(Level 2) for muon. After the loose filtering,[5] we applied the event selection which is described in Table 1.

Table 1. Event Selection for $WW \to e\mu\nu\bar{\nu} + X$.

Event Selection Description	No. of Events Passed
μ and e ID	82
$\Delta R(\mu - \gamma) > 0.25$	80
$|\eta_\mu| \leq 1.0$	70
$\Delta R(\mu - jet) > 0.5$	16
$E_t^e \geq 20$ GeV and $P_t^\mu \geq 15 GeV/c$	5
$\not{E}_t \geq 20$ GeV	1

An expected total background in this channel was 0.96 ± 0.44 events, mostly from $Z \to \tau\tau(0.39 \pm 0.11)$ and Fake electron of W + jets event(0.43 ± 0.43). The efficiency calculation was done using PYTHIA(V5.6)/JETSET(V7.3) $\to$ GEANT(V3.14) $\to$ D0RECO(V11.19) Monte Carlo sample for event selection and using data for triggers.

2.2. $WW \rightarrow ee\nu\bar{\nu} + X$

There are three kinds of trigger for this sample. So called "W" which is $E_t^{EMtower} \geq$ 10 GeV(Level 1) and $E_t^{EM} \geq 20$ GeV with $\not{E}_t \geq 20$ GeV(Level 2), "Z" which is two of $E_t^{EMtower} \geq 7$ GeV(Level 1) and two of $E_t^{EM} \geq 20$ GeV(Level 2), and "W+Jets" which is $E_t^{EMtower} \geq 10$ GeV with two $E_t^{Jet} \geq 3$ GeV(Level 1) and $E_t^{EM} \geq 15$ GeV with $\not{E}_t \geq 20$ GeV plus two of $E_t^{Jet} \geq 16$ GeV(Level 2). The reason for including "W+Jets" trigger is that electrons were counted as jets in the triggers. The event selection for this channel is shown in Table 2.

Table 2. Event Selection for $WW \rightarrow ee\nu\bar{\nu} + X$.

Event Selection Description	No. Events Passed
Two em clusters	2201
Two electrons $E_t^e \geq 20$ GeV	844
$\not{E}_t \geq 20$ GeV	7
$M_{ee} \leq 77$; $M_{ee} \geq 105$ GeV/c^2	5
$M_T(ee, \not{E}_t) \geq 100$ GeV/c^2	1

We expect 1.23 ± 0.87 total background events which is mostly came from Fake electron of W + jets event(0.85 ± 0.85). The efficiency calculation was done by the same way for the $WW \rightarrow e\mu\nu\bar{\nu} + X$.

3. Results and Conclusions

We observe one $e\mu\nu\bar{\nu} + X$ and one $ee\nu\bar{\nu} + X$ as a candidate event of the W pair production. Based on two candidate events and the efficiencies, $\epsilon_{e\mu} = 0.10 \pm 0.01$ and $\epsilon_{ee} = 0.12 \pm 0.01$, we obtained 133 pb as the cross section upper limit for the W pair production at 95% C.L. without background subtraction. In the standard model W pair production[6] we expect about 0.5 events in the two channels combined. We calculated limits on the CP conserving couplings, κ and λ using the event generator for the W pair production[7] and a fast detector simulation. We assumed $\Delta\kappa_\gamma = \Delta\kappa_Z$ and $\lambda_\gamma = \lambda_Z$ in this calculation. Simultaneosly varying κ and λ at compositness scale 800 GeV we obtained limits in the form of a contour. Figure 1 shows the coupling limits with the unitarity limit($\Lambda = 800 GeV$) which allows $\Delta\kappa$ and λ to vary from their standard model values, $\Delta\kappa_V = \kappa_V - 1 = 0$ and $\lambda_V = 0$. Finally limits on the trilinear gauge boson couplings, $-3.3 < \Delta\kappa < 3.5$ with constraint $\lambda = 0$ and $-2.6 < \lambda < 2.7$ with constraint $\Delta\kappa = 0$ are obtained at 95% C.L..

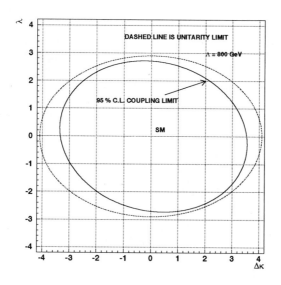

Fig. 1. 95 % C.L. Trilinear Gauge Boson Coupling Limits.

4. Acknowledgements

I would like to thank H. Johari(Northeastern U.) and H.T. Diehl(Fermilab) for their help in preparing material for this talk.

References

1. V. Barger et al, *Finding the Leptonic WW Decay Mode of a Heavy Higgs Boson at Hadron Supercolliders*. **Technical Report MAD/PH/757** Univ. of Wisconsin (1993)
2. G. Valencia, *Anomalous Gauge-Boson Coupling at Hadron Supercollider* **FERMILAB-CONF-92/246-T** Fermilab (1993)
3. E. Eichten, I. Hinchcliffe, K. Lane and C. Quigg, *Rev. Mod. Phys.* **56** (1984) 579.
4. S. Abachi et al, *NIM* **A324** (1993) 53.
5. J. Cochran, *Search for Truth in the eμ channel at D0* **PhD Thesis, State University of New York at Stony Brook** (1993)
6. V. Barger et al, *Phys. Rev.* **D41** (1990) 2782.
7. K. Hagiwara, J. Woodside and D. Zeppenfeld, *Phys. Rev.* **D41** (1990) 2113.

ELECTROWEAK BOSON PAIR PRODUCTION IN $p\bar{p}$ COLLISIONS AT $\sqrt{s} = 1.8$ TeV

THERESA A. FUESS *

High Energy Physics Division, Argonne National Laboratory,
9700 South Cass Avenue, Argonne, Illinois 60439, U.S.A.

ABSTRACT

Results from CDF on W^+W^-, WZ, and $W\gamma$ production in $\sqrt{s} = 1.8$ TeV p̄-p collisions from the 1992-93 collider run are presented. Direct limits on $WW\gamma$ and WWZ anomalous couplings are obtained.

1. Introduction

Since the discovery of the W and Z electroweak force carriers, many predictions of the $SU(2) \times U(1)$ gauge theory have been confirmed and its parameters determined with ever increasing precision. Among the most characteristic and fundamental signatures of non-Abelian symmetry in the theory are the interactions of W, Z, and γ bosons with each other. The interaction between W and γ was previously studied in the process $p\bar{p} \to W\gamma$.[1] Here we report on improved bounds on the $WW\gamma$ and WWZ couplings obtained from production of WW, WZ, and $W\gamma$ in $p\bar{p}$ collisions at $\sqrt{s} = 1.8$ TeV.

The most general $WW\gamma$ and WWZ couplings consistent with Lorentz invariance have been formulated and may be parameterized in terms of fourteen independent constants (or form factors), seven for the $WW\gamma$ vertex and seven for the WWZ vertex.[2] They are g_1^V, g_4^V, g_5^V, λ^V, κ^V, $\tilde{\lambda}^V$, and $\tilde{\kappa}^V$ where V is either γ (for $WW\gamma$) or Z (for WWZ). The standard $SU(2) \times U(1)$ electroweak theory corresponds to the choice $g_1^\gamma = g_1^Z = 1$ and $\kappa^\gamma = \kappa^Z = 1$ with all other couplings set to zero.

In the standard model, the dominant contribution to diboson production in $p\bar{p}$ collisions at $\sqrt{s} = 1.8$ TeV comes from two types of Feynman diagrams (figure 1). There are substantial cancellations between the t- or u-channel diagrams, which involve only the couplings of the bosons to fermions, and the s-channel diagrams which contain the three-boson coupling. To the extent that the fermionic couplings of the W, Z, and γ have been well tested, we may regard diboson production as primarily a test of the three-boson couplings. If any of these couplings differ substantially from the standard model values then the cross section increases. The enhancement is greatest at high boson P_T where the strongest cancellations occur in the standard model. Therefore, this analysis looks for anomalously large cross sections at high boson P_T in order to obtain information on the couplings.

Any couplings, differing from the standard model couplings, that are independent of $\sqrt{\hat{s}}$ cause the diboson production cross section to violate unitarity at some large $\sqrt{\hat{s}}$.

---

*Representing the CDF Collaboration.

To avoid this the couplings are made functions of $\sqrt{\hat{s}}$ and a form factor scale Λ_{FF} in such a way that they approach their standard model values when $\sqrt{\hat{s}}$ is bigger than Λ_{FF}. The sensitivity of the measurement of the couplings can depend on the value of Λ_{FF} used. However, if Λ_{FF} is big enough, there is little effect at lower energies where the measurement is made.

2. $p\bar{p} \rightarrow WW$, WZ, and $W\gamma$ Event Selection

Measurement of WW and WZ production provides information on both $WW\gamma$ and WWZ couplings. $W\gamma$ production provides information on the $WW\gamma$ vertex only.

The standard model cross sections for WW and WZ production are 9.5 and 2.5 pb, respectively.[3] The decay modes $WW \rightarrow l\nu jj$, $WZ \rightarrow l\nu jj$, and $ZW \rightarrow l\bar{l}jj$ are more sensitive to anomalous couplings than purely leptonic decay modes. The leptonic decay modes suffer from top quark background, low branching ratios, and lepton acceptance. The leptons plus jets decay modes have large QCD backgrounds but these background events typically have lower $P_T(V)$ than the events expected from anomalous three-boson couplings. A cut at high $P_T(V)$ eliminates this background so that excess events can be attributed to anomalous three-boson couplings.

Candidate WW and WZ events are selected by looking for events with leptons and missing energy consistent with a W or a Z and with two leading jets with $60 < M(JJ) < 110$ GeV/c^2 and $P_T(JJ) > 130$ or 100 GeV/c for $l\nu jj$ or $l\bar{l}jj$ events. The standard model prediction of the number of diboson events detected, after all cuts, acceptances, efficiencies, and without adding the expected background, is 0.08 in the $l\nu jj$ channel and 0.01 in the $l\bar{l}jj$ channel. One event in the $l\nu jj$ channel passes the cuts. In the $l\bar{l}jj$ channel there are no events that pass.

Candidate $W\gamma$ events are selected from events with a lepton and missing energy consistent with a W and with an isolated, well measured photon with $E_T > 7$ GeV/c. Background in this channel is from QCD $W + j$ production and is calculated from the observed $W + j$ data sample together with a fake photon probability determined from an independent jet data sample. There are 25 $W\gamma$ events selected before background subtraction. Figure 2 shows the cross section times branching ratio for $W\gamma$ production as a function of the minimum photon E_T together with the standard model prediction.

3. Bounds on Three-Boson Couplings

Lack of excess events at high boson P_T leads to bounds on the three-boson couplings. Figures 3a and 3b show the bounds obtained on two of the three-boson couplings. In figure 3a it is assumed that the couplings for the WWZ vertex are the same as that for the $WW\gamma$ vertex. In figure 3b one of the couplings is allowed to be different for the two vertices. Assuming that the couplings for the WWZ vertex are the same as those for the $WW\gamma$ vertex, and using $\Lambda_{FF} = 1.5$ TeV, and holding all other couplings to their standard model values, 95% confidence level limits can be set on $|\Delta\kappa| \equiv |\kappa - 1| < 1.0$ and $|\lambda| < 0.5$. The precision of this measurement is expected to improve by a factor of two with data from the present 1994 CDF data taking run.

4. Acknowledgements

We thank U. Baur and D. Zeppenfeld for providing Monte Carlo programs and for many stimulating discussions. We thank the technical and support staffs of the participating institutions of CDF for their vital contributions. This work supported in part by the U.S. Department of Energy, Division of High Energy Physics, Contract W-31-109-ENG-38.

References

1. J. Alitti, et al., *Phys. Lett.* **B277** (1992) 194.
2. K. Hagiwara, J. Woodside, and D. Zeppenfeld, *Phys. Rev.* **D41** (1990) 2113.
3. J. Ohnemus, *Phys. Rev.* **D44** (1991) 1403; J. Ohnemus, *Phys. Rev.* **D44** (1991) 3477; J. Ohnemus and J.F. Owens, *Phys. Rev.* **D43** (1991)3626.

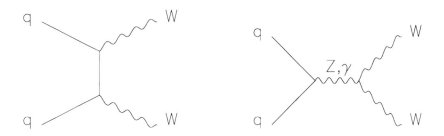

Fig. 1. Feynman diagrams for WW production.

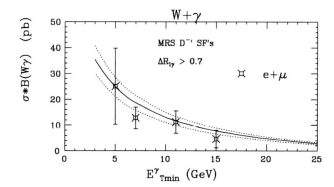

Fig. 2. Cross section times branching ratio for $W\gamma$ production as a function of the minimum photon E_T. The symbols represent the data. The solid line and dashed lines represent the standard model prediction and uncertainty.

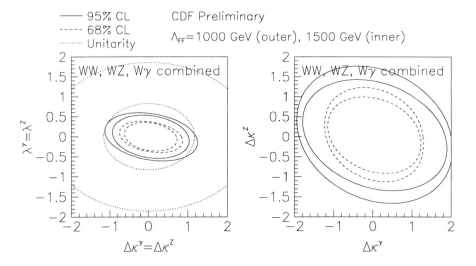

Fig. 3. Bounds on two of the three-boson couplings, λ and $\Delta\kappa \equiv \kappa - 1$. Standard model values are $\lambda = \Delta\kappa = 0$. a) Bounds on the couplings λ versus $\Delta\kappa$ assuming that the couplings are equal for WWZ and $WW\gamma$ vertices. b) Bounds on $\Delta\kappa^Z$ versus $\Delta\kappa^\gamma$ holding $\lambda^Z = \lambda^\gamma = 0$.

MEASUREMENT OF NEUTRAL AND CHARGED CURRENT DEEP INELASTIC SCATTERNING CROSS SECTIONS AT VERY HIGH Q^2

I. ALI*

Physics Department, University of Wisconsin – Madison,
1150 University Ave.
Madison, WI 53706, U.S.A.

ABSTRACT

Neutral Current (NC) and Charged Current(CC) deep inelastic scattering (DIS) are measured using the ZEUS detector at HERA, an ep colliding facility in DESY, Hamburg, using 26.7 GeV electrons colliding with 820 GeV protons. From an integrated luminosity of 0.55 pb^{-1}, 22 CC and 434 NC events with $Q^2 \geq 400\,GeV$ were identified. The differential cross sections, $d\sigma/dQ^2$, for the NC and CC DIS are reported. The results are also compared to theoretical expectations.

1. Introduction

We desribe the measurement of the differential cross sections, $d\sigma/dQ^2$, for Neutral Current (NC) and Charged Current (CC) deep inelastic scatternig (DIS) at very high Q^2. NC DIS, in which the electron and the proton exchange either a photon or a Z^0 particle, are characterized by a scattered isolated electron with hadronic jets balancing it in transverse momentum, P_t. In CC DIS, on the other hand, the electron and the proton exchange a W^- particle resulting in a final state with an undetected neutrino and hadronic jets with large missing transverse momentum, P_{tmiss}.

2. Kinematic Reconstruction

Two methods were used to reconstruct the kinematic variables x and Q^2 from the observed final state particles. One is the *double angle* method (DA) and the other is the *Jacquet-Blondel* method (JB).[1] The DA method, which utilizes the scattered electron and jet angles for reconstruction, is employed to measure the kinematics of NC events. This method is insensitive to the calorimeter energy scales and gives accurate and unbiased results for large x and Q^2 regions.[2] The JB method, however, utilizes only the hadronic part of the event, rendering it well suited for the CC reconstruction. Note that the JB method can also be used to reconstruct the NC event kinematics from the hadronic part only. In contrast to the DA reconstruction, the JB method underestimates the x and Q^2 due to detector effects. Therefore, a set of empirical correction functions

---

*Representing the ZEUS Collaboration.

were determined using an NC $Q^2 \geq 200 \ GeV$ sample that related the JB kinematic variables to the DA variables. These functions were used in the CC analysis.

3. Data Selection

The data were collected during the 1993 running period using the ZEUS detector with a total luminosity of $0.55 \ pb^{-1}$. The cuts applied for CC event selection included a minimum P_{tmiss} of $12 \ GeV$ to remove proton beamgas background interaction. In addtion, P_{tmiss} outside the beampipe region divided by the total P_{tmiss} (P_{tout}/P_{tmiss}) was required to be greater than 0.7. The remaining beamgas background was removed by a vertex cut. Cosmic interactions were removed by requiring 3 or more tracks and agreement between the event times measured in all calorimeter regions. After these cuts, 26 events remained in the sample out of which 23 events were identified as CC events, with 22 events having $Q^2 \geq 400 \ GeV^2$, and 3 as cosmics by a visual scan.

NC data selection included requiring a scattered electron with energy $\geq 10 \ GeV$. The main background came from photoproduction events that could mimic the NC events. To cut those, events were required to have y_e (measured from the scattered electron) ≤ 0.95 and to be fully contained in the calorimeter by selecting events with the total energy of the event minus the longitudinal energy $\geq 35 \ GeV$. Cosmics were removed by requiring the events to be balanced in P_t. 434 NC events remained with $Q^2 \geq 400 \ GeV^2$ after the cuts.

4. Systematics and Background

The data were sorted into four Q^2 bin for this analysis, namely, 400-1000 GeV^2 (bin 1), 1000-2500 GeV^2 (bin 2), 2500-6250 GeV^2 (bin 3), and $\geq 6250 \ GeV^2$ (bin 4). The systematic errors on the CC cross sections arising from the calorimeter energy scale were 15% in bin 1 and 3% in bin 2. There was no effect in the last 2 bins. The number of tracks requirement combined with the P_{tout}/P_{tmiss} cut led to a 12% error in all the bins. The trigger threshold efficiency gave an error of 8% in bin 1 and 5% in the rest of the bins. For the NC cross sections, the main systematic effect came from an uncertainty in the energy scale of 5% which caused a 14% error in all of the bins.

The main background for both NC and CC at $Q^2 \geq 400 \ GeV$ came from photoproduction. This background was estimated to be less than 2% from Monte Carlo studies and was confirmed by visual scanning of the event samples.

5. Results and Conclusions

Table 1 shows the measured total cross sections in each bin. Figure 1 shows $d\sigma/dQ^2$ and overlays the theoretical expectation.[3] From this figure, one can see the effect of the W mass term in the CC propagator. This confirms a result reported by the H1 collaboration.[4] It is also evident from the figure that NC and CC cross sections are comparable at large Q^2.

References

1. J. Feltesse, *Proceedings of the HERA Workshop* **Vol I** (1987) 33.
2. S. Bentvelsen, *Physics at HERA* **Vol I** (1991) 23.
3. G. Ingelman, *Physics at HERA* **Vol III** (1991) 1366.
4. H1 collab., T. Ahmed *et al.*, *Physics Letters* **B324** (1994) 241.

Table 1: Table 1: Measured cross sections fro NC and CC. The numbers in this table are in the form of $\sigma \pm statistical \pm systematic$.

Q^2 bin (GeV^2)	σ_{NC} (pb^{-1})	σ_{CC} (pb^{-1})
400-1000	$766 \pm 44 \pm 107$	$6.3 \pm 3.2 \pm 1.35$
1000-2500	$211 \pm 26 \pm 29.5$	$15.2 \pm 6.2 \pm 2.16$
2500-6250	$47 \pm 12 \pm 6.6$	$12.5 \pm 5.6 \pm 1.74$
≥ 6250	$10 \pm 5.5 \pm 1.4$	$18.2 \pm 6.2 \pm 2.55$

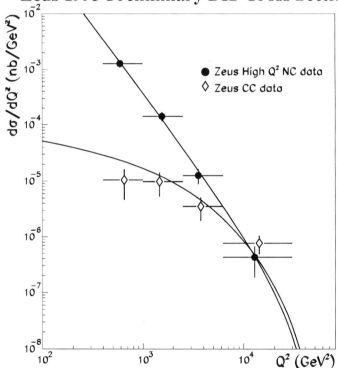

Fig. 1. NC and CC differential cross sections as functions of Q^2. Vertical lines are systematic and statisitical errors added in quadrature. Horizontal lines represent the Q^2 bins.

ELECTROWEAK COUPLING MEASUREMENTS FROM POLARIZED BHABHA SCATTERING AT THE Z^0 RESONANCE

KEVIN T. PITTS*

*Department of Physics, University of Oregon
Eugene, OR 97403†*

ABSTRACT

The cross section for Bhabha scattering ($e^+e^- \to e^+e^-$) with polarized electrons at the center of mass energy of the Z^0 resonance has been measured with the SLD experiment at the SLAC Linear Collider (SLC) during the 1992 and 1993 runs. The first measurement of the left-right asymmetry in Bhabha scattering ($A_{LR}^{e^+e^-}(\theta)$) is presented. From $A_{LR}^{e^+e^-}(\theta)$ the effective weak mixing angle is measured to be $sin^2\theta_W^{\text{eff}} = 0.2245 \pm 0.0049 \pm 0.0010$. The effective electron vector and axial vector couplings to the Z^0 are extracted from a combined analysis of the polarized Bhabha scattering data and and the left-right asymmetry (A_{LR}) previously published by this collaboration. From the combined 1992 and 1993 data the effective electron couplings are measured to be $v_e = -0.0414 \pm 0.0020$ and $a_e = -0.4977 \pm 0.0045$.

The SLD Collaboration has recently performed the most precise single measurement of the effective electroweak mixing angle, $sin^2\theta_W^{\text{eff}}$, by measuring the left-right cross section asymmetry (A_{LR}) in Z boson production at the Z^0 resonance.[1] The left-right cross section asymmetry is a measure of the initial state electron coupling to the Z^0, which allows all visble fermion final states to be included in the measurement. For simplicity, the e^+e^- final state (Bhabha scattering) is omitted in the A_{LR} measurement due to the dilution of the asymmetry from the large QED contribution of the t-channel photon exchange. Here we present two new results: the first measurement of the left-right cross section asymmetry in polarized Bhabha scattering ($A_{LR}^{e^+e^-}(\theta)$), and measurements of the effective electron coupling constants based on a combined analysis of the A_{LR} measurement[1] and the Bhabha cross section and angular distributions. The vector coupling measurement is the most precise yet presented.

In the Standard Model, measuring the left-right asymmetry yields a value for the quantity A_e, a measure of the degree of parity violation in the neutral current, since:

$$A_{LR} = A_e = \frac{2v_e a_e}{v_e^2 + a_e^2} = \frac{2[1 - 4sin^2\theta_W^{\text{eff}}]}{1 + [1 - 4sin^2\theta_W^{\text{eff}}]^2}, \qquad (1)$$

where the effective electroweak mixing parameter is defined[2] as $sin^2\theta_W^{\text{eff}} = \frac{1}{4}(1 - v_e/a_e)$, and v_e and a_e are the effective vector and axial vector electroweak coupling parameters

---

*representing the SLD Collaboration
†current address: Fermilab, P.O. Box 500, Batavia, IL 60510

of the electron. The partial width for Z^0 decaying into e^+e^- is dependent on the coupling parameters:

$$\Gamma_{ee} = \frac{G_F M_Z^3}{6\sqrt{2}\pi}(v_e^2 + a_e^2)(1 + \delta_e), \tag{2}$$

where $\delta_e = \frac{3\alpha}{4\pi}$ is the correction for final state radiation. G_F is the Fermi coupling constant and M_Z is the Z^0 boson mass. By measuring A_e and Γ_{ee}, the above equations can be utilized to extract v_e and a_e.

The data presented at this meeting were collected during the 1992 and 1993 runs of the SLAC Linear Collider (SLC), which collides unpolarized positrons with longitudinally polarized electrons at a center of mass energy near the Z^0 resonance.[3] The luminosity-weighted electron beam polarization ($< \mathcal{P}_e >$) was measured to be $(22.4 \pm 0.7)\%$ for the 1992 run and $(63.0 \pm 1.1)\%$ for the 1993 run.[1,4]

The analysis presented here utilizes the calorimetry systems of the SLD detector.[5] Small angle coverage (28-65 mrad from the beamline) is provided by the finely-segmented silicon-diode/tungsten-radiator luminosity calorimeters (LUM).[6] The LUM measures small angle Bhabha scattering, thereby providing both the absolute luminosity and a measure of the left-right luminosity asymmetry. Events at larger angles from the beamline are measured with the liquid argon calorimeter (LAC).[7]

A detailed description of the systematic error analysis for the luminosity measurement is given elsewhere.[8] The total systematic uncertainty is 0.93%, which is composed of 0.88% experimental and 0.3% theoretical uncertainty. The integrated luminosity is $\mathcal{L} = 385.37 \pm 2.47$ (stat) ± 3.58 (sys) nb^{-1} for the 1992 polarized SLC run and $\mathcal{L} = 1781.1 \pm 5.1$ (stat) ± 16.6 (sys) nb^{-1} for the 1993 SLC run.

The wide angle Bhabha selection algorithm makes use of the distinct topology of the e^+e^- final state. Selected events are required to possess two clusters which contain at least 70% of the center of mass energy and manifest a normalized energy imbalance of less than 0.6. The two largest energy clusters are also required to deposit less than 3.8 GeV of energy in the hadronic calorimeter. The total number of reconstructed clusters found in the event must be less than 9. Collinearity in the final state is controlled by requiring the absolute value of the rapidity sum of the two main clusters to be less than 0.30. The angle-dependent efficiency and contamination is calculated from Monte Carlo simulations. Two small sources of contamination are $e^+e^- \to \gamma\gamma$ (1.25%) and $e^+e^- \to \tau^+\tau^-$ (0.28%). Other sources of contamination were all found to give negligible contributions.

To extract Γ_{ee} and A_e, the data are fit to a calculated differential e^+e^- cross section using the maximum likelihood method. Two programs are used to calculate the differential e^+e^- cross section: EXPOSTAR[9] and DMIBA.[10] To extract the maximal amount of information from the differential polarized Bhabha scattering distribution, the fit is performed over the entire angular region accepted by the LAC ($|cos\theta| < 0.98$). No t-channel subtraction is performed.

The partial width Γ_{ee} is extracted from the data in two ways: (1) using the full fit to the differential cross section to $|cos\,\theta|=0.98$, and (2) measuring the cross section

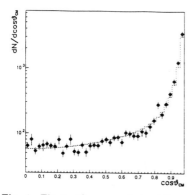

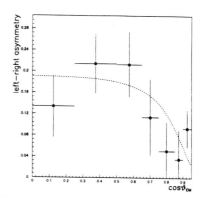

Fig. 1. Fit to the corrected wide angle Bhabha distribution. The points are the corrected data, the curve is the fit.

Fig. 2. EXPOSTAR fit to the wide angle Bhabha left-right asymmetry. The points are the corrected data, the curve is the fit.

in the central region ($|\cos \theta| < 0.6$) where the systematic errors are smaller, yielding a more precise measurement. Figure 1 shows the fit to the full $e^+e^- \rightarrow e^+e^-$ distribution, which yields $\Gamma_{ee} = 83.14 \pm 1.03$ (stat) ± 1.95 (sys) MeV. The 2.4% systematic error is dominated (2.1%) by the uncertainty in the efficiency correction factors in the angular region $0.6 < |\cos\theta| < 0.98$, where the LAC response is difficult to model due to materials from interior detector elements.[8,11]

A more precise determination of Γ_{ee} was performed using only the central region of the LAC and the small angle region in the LUM.[12] The program MIBA[13] is then used to calculate Γ_{ee} based on the total measured cross section within the defined fiducial region. From this method, we find: $\Gamma_{ee} = 82.89 \pm 1.20$ (stat) ± 0.89 (sys) MeV. The loss in statistical precision of the limited fiducial region is more than compensated by the improvement in the systematic uncertainty. The 1.1% systematic uncertainty is dominated by a 1.0% uncertainty in the e^+e^- cross section into the fiducial region arising from the uncertainty in the absolute luminosity and the accuracy of the simulation.

To extract A_e from the Bhabha events, the right- and left-handed differential $e^+e^- \rightarrow e^+e^-$ cross sections are fit directly to v_e and a_e using EXPOSTAR, yielding:

$$A_e = 0.202 \pm 0.038 \text{ (stat)} \pm 0.008 \text{ (sys)}.$$

Figure 2 shows the measured left-right cross section asymmetry for $e^+e^- \rightarrow e^+e^-$ ($A_{LR}^{e^+e^-}(\theta)$) compared to the fit. The measurement of A_e is limited by the statistical uncertainty. The 3.8% systematic is dominated by a 3.2% uncertainty in the angle-dependent response correction factors. The polarization uncertainty contributes 1.7% with other factors contributing less than 1%.[1,8,11]

The results for Γ_{ee} and A_e from above may now be used in equations 1 and 2 to extract the effective vector and axial vector couplings to the Z^0: $v_e = -0.0507 \pm 0.0096$ (stat) ± 0.0020 (sys), and $a_e = -0.4968 \pm 0.0039$ (stat) ± 0.0027 (sys), where

e^+e^- annihilation data have been utilized to assign $|v_e| < |a_e|$, and $\nu_e e$ scattering data have been utilized to establish $v_e < 0$ and $a_e < 0$.[14] Figure 3 shows the one-sigma (68%) contour for these electron vector and axial vector coupling measurements. Most of the sensitivity to the electron vector coupling and, hence, $sin^2\theta_W^{\text{eff}}$ arises from the measurement of A_e, while the sensitivity to the axial vector coupling arises from Γ_{ee}. Also shown are standard model calculations using the program ZFITTER.[15] The effective electroweak mixing angle represented by these vector and axial vector couplings is:

$$sin^2\theta_W^{\text{eff}} = 0.2245 \pm 0.0049 \pm 0.0010,$$

where the first error is statistical, the second systematic.

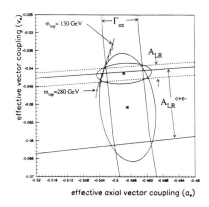

Fig. 3. One standard deviation (68%) contour for v_e and a_e. The points indicate the Standard Model calculation as a function of the mass of the top quark and the Higgs boson.

Combining the Bhabha results with the SLD measurement of A_{LR}[1] gives:

$$v_e = -0.0414 \pm 0.0020 \qquad a_e = -0.4977 \pm 0.0045,$$

the most precise measurement of the electron vector coupling to the Z^0 published to date.[11] The v_e, a_e contour including the A_{LR} measurement is also shown in Figure 3, demonstrating the increased sensitivity in v_e from A_{LR}.

We thank the personnel of the SLAC accelerator department and the technical staffs of our collaborating institutions for their outstanding efforts on our behalf.

References

1. SLD Collaboration, K. Abe *et al.*, Phys. Rev. Lett. **73**, 25 (1994).
2. This follows the convention used by the LEP Collaborations in Phys. Lett. B **276**, 247 (1992).
3. N. Phinney, *Int. J. Mod. Phys. A, Proc. Suppl.* **2A**, 45 (1993).
4. D. Calloway *et al.*, Report No. SLAC-PUB-6423, June 1994.
5. The SLD Design Report, SLAC Report 273, 1984.
6. S.C. Berridge *et al.*, IEEE Trans. Nucl. Sci. **39**, 1242 (1992).
7. D. Axen *et al.*, Nucl. Inst. Meth. **A328**, 472 (1993).
8. K.T. Pitts, Ph.D. Thesis, University of Oregon, SLAC Report 446 (1994).
9. D. Levinthal, F. Bird, R.G. Stuart and B.W. Lynn, Z. Phys. C **53**, 617 (1992).
10. P. Comas and M. Martinez, Z. Phys. C **58**, 15 (1993).
11. SLD Collaboration, K. Abe *et al.*, SLAC-PUB-6605, August 1994.
12. J.M. Yamartino, Ph.D. Thesis, MIT, SLAC Report 426, February 1994.
13. M. Martinez and R. Miquel, Z. Phys. C **53**, 115 (1992).
14. S.L. Wu, Phys. Rep. **107**, 59 (1984).
15. D. Bardin *et al.*, Report No. CERN-TH-6443-92, May 1992.

MEASUREMENT OF THE LEFT-RIGHT CHARGE ASYMMETRY IN HADRONIC Z DECAYS AND A_e DETERMINATION

SLD COLLABORATION

represented by

GREGORY J. BARANKO

University of Colorado, Boulder, CO 80309

ABSTRACT

We present a new method for determining the electron left-right asymmetry factor, A_e, by measuring the forward-backward charge flow in samples of hadronic Z events, produced in e^+e^- collisions, with left-handed and right-handed electron beam polarizations. The raw, left-right charge asymmetry, Q_{LR}^{raw}, is defined. After correction by the measured beam polarization, it provides A_e with greatly reduced dependence on Monte Carlo corrections for acceptances and jet charge measurement. The Q_{LR}^{raw} is also combined with the raw, left-right cross section asymmetry, A_{LR}^{raw}, to obtain A_e without the use of the measured electron beam polarization, and with half the statistical error as that obtained for A_e using the polarization corrected Q_{LR}^{raw} alone. This method gives $\sin^2\theta_W^{eff} = 0.2291 \, ^{+0.0030}_{-0.0053} \, (stat.) \, ^{+0.0011}_{-0.0014} \, (syst.)$. The measurement was performed at a center-of-mass energy of 91.26 GeV with the SLD detector at the SLAC Linear Collider (SLC).

1.   Introduction

SLD has measured the left-right cross section asymmetry, $A_{LR} = (\sigma_L - \sigma_R)/(\sigma_L + \sigma_R)$, in the production of Z bosons by e^+e^- collisions.[1,2] In the Standard Model of the electroweak interactions, this gives the electron left-right asymmetry, $A_e = 2v_e a_e/(v_e^2 + a_e^2)$, from:

$$A_{LR}^{raw} = |\mathcal{P}_e| A_{LR} = |\mathcal{P}_e| \, A_e \tag{1}$$

where $\mathcal{P}_e$ is the electron beam polarization. Since only the total cross sections were observed in this measurement, the forward-backward asymmetries can also be used to provide independent information on the couplings to the Z. At Born level, the forward-backward fermion asymmetry at the Z pole (excluding e^+e^- final states) is given by:

$$A_{FB,f}(\mathcal{P}_e) = -g(a)\frac{\mathcal{P}_e - A_e}{1 - \mathcal{P}_e A_e} \, A_f \tag{2}$$

where $g(a) = a/(1 + \frac{1}{3}a^2)$, $0 < a \leq 1$, $a = |\cos\theta|_{max}$, $\cos\theta$ describes the angle between the outgoing fermion f and the direction of the incident electron, and $A_f = 2v_f a_f/(v_f^2 + a_f^2)$.

Assuming equal polarization magnitudes and luminosities, we can define $A_{FB,L,f} \equiv A_{FB,f}(-|\mathcal{P}_e|)$ and $A_{FB,R,f} \equiv A_{FB,f}(|\mathcal{P}_e|)$ as the forward-backward asymmetries for events produced with left and right beam polarizations.

For a flavor inclusive sample of Z hadronic final states, we must relate the fermion asymmetries above to *charge* asymmetries.[3-5] For a $q_f \bar{q}_f$ final state, at parton level the fermion asymmetries would give non-zero, average forward and backward charges:

$$
\begin{aligned}
< Q^{raw}_{F,L,f} > &= q_f \, A_{FB,L,f} & < Q^{raw}_{F,R,f} > &= q_f \, A_{FB,R,f} \\
< Q^{raw}_{B,L,f} > &= -q_f \, A_{FB,L,f} & < Q^{raw}_{F,R,f} > &= -q_f \, A_{FB,R,f}
\end{aligned}
\tag{3}
$$

where we can separate "left" and "right" events for SLD. The charge magnitudes can then be averaged into the left and right, forward-backward charge flows:

$$
\begin{aligned}
< Q^{raw}_{FB,L,f} > &= < Q^{raw}_{F,L,f} > - < Q^{raw}_{B,L,f} > = 2 \, q_f \, A_{FB,L,f} \\
< Q^{raw}_{FB,R,f} > &= < Q^{raw}_{F,R,f} > - < Q^{raw}_{B,R,f} > = 2 \, q_f \, A_{FB,R,f}.
\end{aligned}
\tag{4}
$$

Finally, the flavor inclusive expressions for the *polarized* ($< \tilde{Q}^{raw}_{FB} >$) and *unpolarized* ($< Q_{FB} >$) forward-backward charge flows can be defined by summing over the flavors, weighting by production, and including modifications ($d_f \epsilon[0,1]$) to account for a reduction in the measured charge magnitudes due to QCD corrections, hadronization effects (including $B^0 \bar{B}^0$ mixing), and decays:[6]

$$
\begin{aligned}
< \tilde{Q}^{raw}_{FB} > &= < Q^{raw}_{FB,L} > f_L - < Q^{raw}_{FB,R} > f_R \\
&= 2 \, g(a) \, |\mathcal{P}_e| \sum_f d_f q_f R_f A_f
\end{aligned}
\tag{5}
$$

$$
\begin{aligned}
< Q_{FB} > &= < Q^{raw}_{FB,L} > f_L + < Q^{raw}_{FB,R} > f_R \\
&= 2 \, g(a) \, A_e \sum_f d_f q_f R_f A_f
\end{aligned}
\tag{6}
$$

where $f_L = \frac{1}{2}(1 + |\mathcal{P}_e|A_e)$ and $f_R = \frac{1}{2}(1 - |\mathcal{P}_e|A_e)$ are the fractions of left and right events, $R_f = \Gamma_f / \Gamma_{had}$, $\Gamma_f \propto v_f^2 + a_f^2$, and $\Gamma_{had} = \sum_f \Gamma_f$. The quantities $< Q^{raw}_{FB,L} >$ and $< Q^{raw}_{FB,R} >$ are now the mean, flavor inclusive, forward-backward charge flows in left and right events. Since $|\mathcal{P}_e|$ and A_e factor out of the sums, the ratio of these charge asymmetries (assuming the sum is not identically zero) has the simple form:

$$
\begin{aligned}
Q^{raw}_{LR} &\equiv \frac{< \tilde{Q}^{raw}_{FB} >}{< Q_{FB} >} \\
&= \frac{< Q^{raw}_{FB,L} > f_L - < Q^{raw}_{FB,R} > f_R}{< Q^{raw}_{FB,L} > f_L + < Q^{raw}_{FB,R} > f_R} \\
&= \frac{|\mathcal{P}_e|}{A_e}
\end{aligned}
\tag{7}
$$

Thus, by measuring the left-right charge asymmetry, $Q_{LR} = Q^{raw}_{LR}/|\mathcal{P}_e|$, A_e can be obtained in a manner largely independent from the A^{raw}_{LR} measurement.[7] At lowest order, the expression for Q_{LR}^{raw} shows that uncertainties in the acceptance and jet charge measurement are cancelled out, thus reducing the dependence on the Monte Carlo for such corrections, and subsequent systematic errors. Many of the remaining errors due to instumental effects are also reduced in the $< \tilde{Q}_{FB} >$ measurement, but not in the $< Q_{FB} >$, and must be considered for corrections and systematic errors to Q_{LR}^{raw}. Comparing A_e obtained from Q_{LR}^{raw} to that obtained from A_{LR}^{raw} provides an inclusive test of our understanding of the charged final states of Z decays. Furthermore, since Q_{LR}^{raw} is formed from the *ratio* of $|\mathcal{P}_e|$ and A_e whereas A_{LR}^{raw} is formed from their *product*, the two measurements may be combined to measure A_e without the use of a direct measurement of the polarization:

$$A_e = \sqrt{\frac{A_{LR}^{raw}}{Q_{LR}^{raw}}}. \tag{8}$$

This method takes advantage of both measurements of A_e and eliminates the systematic error due to the polarization measurement. Alternatively, Q_{LR}^{raw} and A_{LR}^{raw} may be combined to obtain a check on the measured polarization value:

$$|\mathcal{P}_e| = \sqrt{A_{LR}^{raw} Q_{LR}^{raw}}. \tag{9}$$

2. Method of Analysis

Details of the SLAC Linear Collider (SLC), polarized electron source, the measurement of the polarization with the Compton polarimeter, and the SLD experiment are described in Refs. 1 and 2. The results presented here are based on an integrated luminosity of $1.78pb^{-1}$ at $\sqrt{s} = 91.260 GeV$, which SLD obtained during the 1993 run with $|\mathcal{P}_e| \approx 63\%$ at the IP. Hadronic events are selected using track and event selection cuts that were used in a previous measurement of α_s by SLD,[8] with the following additions. Each event is divided by a plane perpendicular to the thrust axis that is determined using all accepted charged tracks in the event. It is required that $|\cos \theta_T| < .7$, $N_{ch}/hemis \geq 3$, $E_{ch}/hemis > .1E_{beam}$, and total $E_{ch} > .2E_{CM}$. Events are rejected if any track has $p_{tot} > 55 GeV/c$. A total of 14723 left events and 12000 right events are obtained. The measured $|\mathcal{P}_e|$ for this sample is 63.5% with a 1.7% systematic error and negligible statistical error as described in Ref. 2.

The forward-backward charge asymmetries are determined as follows. A unit vector along the thrust axis, $\hat{\mathbf{T}}$, is chosen such that $\hat{\mathbf{T}} \cdot \mathbf{p}_{e^-} > 0$. Tracks are then defined as *forward* if $\mathbf{p} \cdot \hat{\mathbf{T}} > 0$, and *backward* otherwise. The weighted charge in the forward hemisphere is then calculated for each event from:

$$Q_F = \frac{\sum_{\mathbf{p}_i \cdot \hat{\mathbf{T}} > 0} |\mathbf{p}_i \cdot \hat{\mathbf{T}}|^\kappa q_i}{\sum_{\mathbf{p}_i \cdot \hat{\mathbf{T}} > 0} |\mathbf{p}_i \cdot \hat{\mathbf{T}}|^\kappa} \tag{10}$$

where q_i is the charge of particle i, and κ is a weighting factor. The results presented here take $\kappa = 1$. The charge in the backward hemisphere, Q_B, is determined in a similar manner for tracks with $\mathbf{p} \cdot \hat{\mathbf{T}} < 0$. The quantity $Q_{FB} = Q_F - Q_B$ is then found for each event, and averaged for all left events as $< Q_{FB,L}^{raw} >$, and all right events as $< Q_{FB,R}^{raw} >$. Finally, $< \tilde{Q}_{FB}^{raw} >$, $< Q_{FB} >$, and Q_{LR}^{raw} are obtained from Eqs. 5, 6, and 7 respectively.

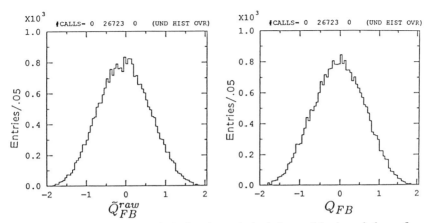

Fig. 1. Distibutions of the polarized and unpolarized, forward-backward charge flows.

3.   Preliminary Results

Distributions for $\tilde{Q}_{FB}^{raw}$ and Q_{FB} are shown in Fig. 1. We obtain the following preliminary results for the measured, or *raw* quantities along with their statistical errors only (a $(-17.5\pm7.3)\%$ correction for a charge dependent, forward-backward bias to the track sagittas has been applied): $< Q_{FB,L}^{raw} >= -0.0467\pm0.0052$, $< Q_{FB,R}^{raw} >= 0.0330\pm 0.0058$, $< \tilde{Q}_{FB}^{raw} >= -0.0405\pm0.0039$, $< Q_{FB} >= -0.0109\pm0.0039$, $Q_{LR}^{raw} = 3.72\pm1.33$, and $A_{LR}^{raw} = 0.1019 \pm 0.0061$.

We have investigated a number of possible systematic errors due to biases of the instrumentation, analysis, and various backgrounds. The relative uncertainties in Q_{LR} due to these errors are summarized in Table 1.

4.   Summary

The value for the left-right charge asymmetry, before radiative corrections but corrected for the measured polarization and including the systematic error is:

$$Q_{LR} = 5.85 \pm 2.10 \; (stat.) \pm 0.66 \; (syst.) \tag{11}$$

Radiative corrections are made to the measured asymmetries using the ZFITTER program.[9] After these corrections, the following values are obtained:

$$Q_{LR}^{0} = 5.93 \pm 2.16 \; (stat.) \pm 0.68 \; (syst.) \tag{12}$$

$$A_e = 0.169 \; {}^{+0.096}_{-0.045} \; (stat.) \; {}^{+0.022}_{-0.017} \; (syst.) \tag{13}$$

$$\sin^2 \theta_W^{\text{eff}} = 0.2288 \; {}^{+0.0058}_{-0.0124} \; (stat.) \; {}^{+0.0022}_{-0.0028} \; (syst.) \tag{14}$$

These results were obtained using $M_Z = 91.187 \; GeV/c^2$, $\alpha_s = .123$, Higgs mass $M_H = 300 \; GeV/c^2$, and top mass $m_t = 250 \pm 170 \; (stat.) \pm 50 \; (syst.) \pm 20 \; (\delta M_H) GeV/c^2$, where the last error on m_t is due to a variation of $M_H \epsilon [70, 10^3 \; GeV/c^2]$. These results are largely independent of those previously obtained by SLD from A_{LR}.

Source of uncertainty	$\delta Q_{LR}/Q_{LR}$ (%)
polarization measurement and chromaticity correction	1.7
q dependent, F-B sagitta bias, $(6.0 \pm 3.3)10^{-4} GeV^{-1}$	8.9
q independent, F-B sagitta bias, $(8.0 \pm 3.2)10^{-4} GeV^{-1}$	0.3
F-B asymmetry of SLD central material, $(.22 \pm .81)\%$	0.5
q independent, F-B track losses from recon. and accept. biases	0.1
unphysical p_{tot} tracks	0.8
dependence of Q_{LR}^{raw} on the value of κ	5.0
polarization effects in hadronization and decays	0.5
SLC track backgrounds, $< Q_{FB} >_{SLC}= (-1.3 \pm 2.0)10^{-4}$	0.0
e^+e^- final state backgrounds, $(1.2 \pm 0.9)10^{-5}$	0.7
two photon backgrounds, $< 1. \times 10^{-4}@95\%C.L.$	1.0
charge dilution factors in calculations of radiative corrections	4.0
Total	11.2

Table 1. Summary of Systematic Errors

From Eq. 8, we can also obtain A_e from Q_{LR}^{raw} and A_{LR}^{raw} alone, without the use of the measured polarization. From the raw results, after radiative corrections we obtain:

$$A_e = 0.1659 \, {}^{+0.0416}_{-0.0238} \, (stat.) \, {}^{+0.0106}_{-0.0090} \, (syst.) \qquad (15)$$

$$\sin^2 \theta_W^{\text{eff}} = 0.2291 \, {}^{+0.0030}_{-0.0053} \, (stat.) \, {}^{+0.0011}_{-0.0014} \, (syst.) \qquad (16)$$

with $m_t = 250\pm80 \, (stat.)\pm25 \, (syst.)\pm20 \, (\delta M_H) \, GeV/c^2$. These results are *alternative*, not independent, of those obtained from A_{LR}^{raw}, and Q_{LR}^{raw}, and using the measured polarization. From Eq. 9, they are also equivalent to $|\mathcal{P}_e| = 0.616 \, {}^{+0.102}_{-0.123} \, (stat.) \, {}^{+0.033}_{-0.035} \, (syst.)$, which agrees well with the measured $|\mathcal{P}_e|$ of the selected data sample.

References

1. SLD Collab., K.Abe et al., *Phys. Rev. Lett.* **70** (1993) 2515.
2. SLD Collab., K.Abe et al., *Phys. Rev. Lett.* **73** (1994) 25.
3. ALEPH Collab., D.Decamp et al., *Phys.Lett.* **B 259** (1991) 377.
4. DELPHI Collab., P.Abreu et al., *Phys.Lett.* **B 277** (1992) 371.
5. OPAL Collab., P.D.Acton et al., *Phys.Lett.* **B 294** (1992) 436.
6. It is assumed here that the d_f factors are the same for both left and right events, or that the QCD corrections and effects of hadronization and decays are the same for both the polarized and unpolarized expressions. Final state polarization effects would require further modifications. SLD has made a preliminary study of jet handedness at the Z. H.Masuda, Report No. SLAC-PUB-6550, 1994 (unpublished).
7. For the measurement presented here, the correlation coefficient between A_{LR}^{raw} and Q_{LR}^{raw} is estimated to be $\approx$ -6% when the charge fluctuations are included in the flavor inclusive sample. M.Swartz (private communication).
8. SLD Collab., K.Abe et al., *Phys. Rev. Lett.* **71** (1993) 2528.
9. D.Bardin et al., CERN-TH. 6443/92, May 1992 (unpublished), and as adapted by M.Swartz (private communication).

MEASUREMENTS OF THE FORWARD-BACKWARD ASYMMETRIES OF
$e^+e^- \to Z \to b\bar{b}$ AND $e^+e^- \to Z \to c\bar{c}$

PIERRE ANTILOGUS

IPN Lyon, 43 bd 11 Novembre 1918, Villeurbanne,
F-69622, France

DELPHI Collaboration

ABSTRACT

The forward-backward asymmetries of the processes $e^+e^- \to Z \to b\bar{b}$ and $e^+e^- \to Z \to c\bar{c}$ were measured using events collected by the DELPHI experiment. The tagging of b and c quarks was performed using three approaches. The first was based on the semileptonic decay channels $b/c \to X + \mu$ and $b/c \to X + e$. The second approach used a lifetime tag with a jet charge reconstruction to extract $A_{\mathrm{FB}}^{b\bar{b}}$. The third approach used the polar angle distribution of reconstructed $D^{*\pm}$. Combining these measurements an effective mixing angle $\sin^2 \theta_{eff}^{lept} = 0.2293 \pm 0.0017$ was obtained.

1.   Introduction

In this paper measurements of $A_{\mathrm{FB}}^{b\bar{b}}$ and $A_{\mathrm{FB}}^{c\bar{c}}$ at LEP with the DELPHI detector using events collected in 1991, 1992 and 1993 are presented. Three independent techniques were used.

The first one[1,2] used the semileptonic decays of the b and c quarks into muons and electrons. The direction of the thrust axis, oriented by the jet containing the lepton, was used as a measurement of the primary quark direction.

The second approach[1] applied a microvertex tag to select an enriched B-sample with efficiency $\epsilon_b = 0.43$ and purity $p_b = 0.77$. The b-quark direction was obtained using the thrust axis and a hemisphere jet-charge algorithm.

The third approach[3] used a sample of $D^{*\pm}$. The sign and the direction of the $D^{*\pm}$ were used to compute the b and c quarks asymmetries.

2. $A_{\mathrm{FB}}^{b\bar{b}}$ and $A_{\mathrm{FB}}^{c\bar{c}}$ measurement using leptons

2.1.   Lepton sample

The muon identification was based on the muon chambers situated in the barrel and in both end-caps. Electron candidates were identified only in the barrel combining the electromagnetic shower information from the High density Projection Chambers

(HPC) with the track ionization measured by the Time Projection Chamber (TPC). Taking into account all selections applied on the muon sample (hadronic selection, tracks selection), a total efficiency of $(46\pm1)\%$ has been estimated for muon comming from direct b semi-leptonic decay. The fraction of misidentified pions was estimated to $(0.92\pm0.16)\%$ in the data. For electrons, the efficiency was found to be $(23\pm1)\%$ and the fraction of misidentified pions was $(0.59\pm0.07)\%$. The main kinematical variable which allows to monitor the flavour composition of the leptonic events is the transverse momentum of the lepton with respect to the closest jet. For the jet reconstruction the JADE algorithm with a scaled invariant mass cut $y_{cut} = m_{ij}^2/E_{vis} = 0.01$ has been used. The jets were reconstructed using well reconstructed charged tracks with $p > 0.2 \text{GeV}/c$ and photons seen by the electromagnetic detectors with $E > 0.8(0.4)\text{GeV}$ in the barrel (end-caps). For the transverse momentum, the p_{Tout} where the lepton is not included in the jet has been chosen.

2.2.   Results

A binned fit of the observed charge asymmetry as a function of θ_{thrust} has been performed. In each bin i of the space $(\cos\theta_{thrust}, p_{Lout}, p_{Tout})$ an asymmetry was measured counting the number of data events of sign $\pm$ in the bin i.

$$A_{FB}^{obs,i} = \frac{N^-(i) - N^+(i)}{N^-(i) + N^+(i)} = \left[\left(f_{b\rightarrow...l^-}^i - f_{\bar{b}\rightarrow...l^-}^i \right) A_{FB}^{b\bar{b},exp} - f_{\bar{c}\rightarrow l^-}^i A_{FB}^{c\bar{c}} + f_{Bkg}^i A_{FB}^{Bkg} \right] W_{\theta_{thurst}}^i$$

where $W_{\theta_{thurst}}^i$ takes into account the θ dependence of the asymmetry and f_x^i are the fractions associated to each channel ($b \rightarrow l^-$,...). To take into account the $B_{s(d)}^0 \bar{B}_{s(d)}^0$ mixing, $A_{FB}^{b\bar{b},exp} = A_{FB}^{b\bar{b}}/(1-2\chi)$ was introduced. The value used in this analysis was $\chi = 0.115 \pm 0.011$. $A_{FB}^{c\bar{c}}$ can be fitted with $A_{FB}^{b\bar{b}}$ or expressed in function of $A_{FB}^{b\bar{b}}$ in the frame work of the Standard Model. A_{FB}^{Bkg}, the background asymmetry, was estimated using the sign correlation between the background track and the initial quark as given by the simulation. A chisquare fit was then performed over the different bins and gave :

at $\sqrt{s} = $ 89.43 GeV (1993 data sample only, it corresponded to $\sim$ 90 000 $q\bar{q}$)
    $A_{FB}^{b\bar{b}} = 0.062 \pm 0.038(stat) \pm 0.002(syst) \pm 0.002(mix)$
at $\sqrt{s} = $ 91.25 GeV (1991-93 data sample, it corresponded to $\sim$ 1 350 000 $q\bar{q}$)
    $A_{FB}^{b\bar{b}} = 0.105 \pm 0.011(stat) \pm 0.004(syst) \pm 0.003(mix)$
at $\sqrt{s} = $ 93.02 GeV (1993 data sample only, it corresponded to $\sim$ 110 000 $q\bar{q}$)
    $A_{FB}^{b\bar{b}} = 0.147 \pm 0.035(stat) \pm 0.006(syst) \pm 0.004(mix)$
at $\sqrt{s} = $ 91.27 GeV (1991-92 data sample,it corresponded to $\sim$ 880 000 $q\bar{q}$)
    $A_{FB}^{c\bar{c}} = 0.080 \pm 0.022(stat) \pm 0.016(syst)$

The correlation between $A_{FB}^{b\bar{b}}$ (at the peak) and $A_{FB}^{c\bar{c}}$ was 0.23. In the 1993 data sample only muons were used. Table 1 contain a list of different contributions to the systematics.

Source of uncertainty	$\delta A_{FB}^{b\bar{b}}$	$\delta A_{FB}^{c\bar{c}}$
production and decay models of b and c quarks.	0.0028	0.0103
knowledge of the background level	0.0024	0.0057
background asymmetry	0.0009	0.0102
p_t reconstruction and fit method	0.0014	0.0046
Total (at $\sqrt{s} \sim M_Z$)	0.004	0.016

Table 1. Contributions to the systematical error in the χ^2 fit of the lepton samples.

Event type	P_f
$u\bar{u},dd,s\bar{s}$	0.04 ± 0.01
$c\bar{c}$	0.11 ± 0.01
$b\bar{b}$	0.77 ± 0.03

Table 2. Composition of the tagged sample for $F_H < 0.01$ ($E_b = 0.43 \pm 0.01$)

3. $A_{FB}^{b\bar{b}}$ measurement using lifetime tag

The b tagging was based on a cumulative probability (F_H) which required that all the tracks of an hemisphere had their impact parameter compatible with the primary vertex. An event was selected if $F_H < 0.01$ in at least one of the two hemispheres. By counting the number of selected events and the number of tagged hemispheres, the efficiency to tag a b event (E_b) and the sample purity in b (P_b) can be extracted from the data with a small input from the simulation (see table 2).

flavour	C_f (%)
u from MC	0.756 ± 0.002
d from MC	0.700 ± 0.002
c from MC	0.652 ± 0.002
s from MC	0.701 ± 0.002
b from MC	0.689 ± 0.002
b from data	$0.673 \pm 0.011(stat)$ $\pm 0.006(syst)$

Table 3. Values of C_f for the different quark species

Source of uncertainty	$\delta A_{FB}^{b\bar{b}}$
uncertainty on C_b	0.0075
B purity	0.0050
hadronization parameters, Λ_{QCD} ...	0.0026
detector acceptance correction	0.0020
Total	0.01

Table 4. Different contributions to the systematical error in the fit of $A_{FB}^{b\bar{b}}$ using the lifetime tag.

The forward (backward) hemisphere charge was defined by $Q_{F(B)} = \sum_i q_i |\vec{p}_i \cdot \vec{T}|^{0.5} / \sum_i |\vec{p}_i \cdot \vec{T}|^{0.5}$, $\vec{p}_i \cdot \vec{T} > 0, (\vec{p}_i \cdot \vec{T} < 0)$. The b quark direction was assumed to be given by the thurst axis oriented toward the hemisphere with the lowest hemisphere charge. C_b, the probability to correctly assign the charge for a b quark, has been extracted from data using a sample of hadronic events with lepton(see table 3). Using the 1992 data sample, corresponding to 690 000 hadronic events at $\sqrt{s} = 91.28$ GeV, the result was

$$A_{FB}^{b\bar{b}} = 0.115 \pm 0.017(stat) \pm 0.010(syst) \pm 0.003(mix)$$

The list of the main systematics is given table 4.

4. $A_{\mathrm{FB}}^{b\bar{b}}$ and $A_{\mathrm{FB}}^{c\bar{c}}$ measurement using $D^{*\pm}$

The D^{*+} was identified through its decay into $D^o\pi^+$, while the D^o was reconstructed in $\mathrm{K}^-\pi^+$, $\mathrm{K}^-\pi^+\pi^o$ or $\mathrm{K}^-\pi^+\pi^-\pi^+$. The charge conjugates were considered as well. The number of candidate after background substraction are given table 5.
The extraction of $A_{\mathrm{FB}}^{b\bar{b}}$ and $A_{\mathrm{FB}}^{c\bar{c}}$ was achieved with an unbinned maximum likelihood fit to the $D^{*\pm}$ sample using the mass difference, scaled energy, D^0 decay distance distributions and $D^{*\pm}$ polar angle. The three first parameters allowed the separation between the different sources of $D^{*\pm}$ ($c\bar{c}$, $b\bar{b}$ or combinatorial background). Using the 1991-92 data sample, the c and b quark asymmetries are determined to be at $\sqrt{s}$ =91.27 GeV

$$A_{\mathrm{FB}}^{b\bar{b}} = 0.046 \pm 0.059(stat) \pm 0.024(syst) \ \ and \ \ A_{\mathrm{FB}}^{c\bar{c}} = 0.081 \pm 0.029(stat) \pm 0.012(syst)$$

The correlation between $A_{\mathrm{FB}}^{b\bar{b}}$ and $A_{\mathrm{FB}}^{c\bar{c}}$ was found to be -0.36.

$K^-\pi^+$	1167 ± 46
$K^-\pi^+\pi^-\pi^+$	899 ± 79
$K^-\pi^+\pi^o$	2691 ± 143

Table 5. $D^{*\pm}$ sample used for the measurement and corresponding D^o decay mode.

Source of uncertainty	$\delta A_{\mathrm{FB}}^{b\bar{b}}$	$\delta A_{\mathrm{FB}}^{c\bar{c}}$
Bkg and reflection fraction	0.009	0.007
b,c fragmentation and $X_E^{D^*}$	0.002	0.002
b mixing, b sample composition, τ_b	0.014	0.004
detector effect / fit method	0.017	0.009
Total	0.024	0.012

Table 6. Different contributions to the systematical errors in the fit of $A_{\mathrm{FB}}^{b\bar{b}}$ and $A_{\mathrm{FB}}^{c\bar{c}}$ using the $D^{*\pm}$ sample.

5. conclusion

The different asymmetries obtained in this note for $\sqrt{s} \sim M_Z$, taking into account their covariance matrix have been combined and yield to :

$$A_{\mathrm{FB}}^{b\bar{b}} = 0.106 \pm 0.010 \ \ (stat+sys) \ \ and \ \ A_{\mathrm{FB}}^{c\bar{c}} = 0.081 \pm 0.021 \ \ (stat+sys)$$

Using all the results presented here and taking $M_Z = 91.187 \ GeV/c^2$, $\alpha_s = 0.12$, $M_{Higgs} = 300^{+700}_{-240} GeV/c$, a value of $m_{top} = 238^{+30}_{-36} \ (expt) \ ^{+12}_{-17} \ (Higgs) \ GeV/c^2$ has been extracted using ZFITTER, it corresponds to $\sin^2\theta_{eff}^{lept} = 0.2293 \ \pm 0.0017$

References

1. DELPHI collaboration, P.Abreu et al., , *DELPHI* **94-62 PHYS 383** (1994)
2. DELPHI collaboration, P.Abreu et al., , *DELPHI* **94-107 PHYS 424** (1994)
3. DELPHI collaboration, P.Abreu et al., , *DELPHI* **94-95 PHYS 412** (1994)

PRECISION DETERMINATION OF THE
Z^0 RESONANCE PARAMETERS

A. L. READ*

*Department of Physics, University of Oslo, P.O. Box 1048, Blindern
0316 Oslo, Norway*

ABSTRACT

During 1993 LEP was run at 3 energies near the Z^0 peak in order to give improved measurements of the mass and width of the resonance. DELPHI accumulated data corresponding to an integrated luminosity of 30.4 pb^{-1} during the dedicated scan with frequent measurements of the LEP energy, and 5.7 pb^{-1} at the Z^0 peak before the start of the scan. Analyses have been carried out on the hadronic and leptonic cross sections, and the leptonic forward–backward asymmetries, using the most accurate evaluations of the LEP energies. Fits to this data and previously published DELPHI lineshape and asymmetry data have been performed. Values of the Z^0 resonance parameters with significantly smaller errors than those previously published are obtained.

1.   Introduction

During 1993 a scan of the Z^0 resonance was performed at LEP with higher statistics than in previous scans (1989-91) and with improved measurements of the LEP beam energies. This note is a brief account of the analysis of the hadronic and leptonic cross-sections and the leptonic forward-backward (FB) asymmetries. Although the results are preliminary, a large reduction of the errors relative to previously published DELPHI results[1] has already been achieved.

2.   LEP Energy Measurement

The scan consisted of datataking at the Z^0 peak and two points at about ± 2 GeV. The beam energy was measured about twice a week at the ends of the ± 2 GeV fills by resonant depolarization of the transversely polarized beams. The evolution of the beam energy between these points was followed by continuous measurements of the magnetic field in a reference magnet with an NMR probe. Additional corrections were made to take into account the deformation of the LEP ring by earth tides, small changes in the magnet currents and the effect of temperature. The LEP experiments were then provided with a table of the estimated average beam energy during 15 minute intervals.

The uncertainty in the interpolations between the precise energy measurements resulted in 3.3 and 2.2 MeV uncertainties on M_Z and Γ_Z. These errors should be reduced to about 2 MeV when a currently unexplained long-term trend in the LEP energy measurements is better understood.

---

*Representing the DELPHI Collaboration.

3.   Luminosity Measurements

The absolute luminosity measurement was performed with the Small Angle Tagger (SAT). The experimental uncertainty on the absolute luminosity of 0.28% is smaller than in previous years due to improved understanding of the data and the construction of a new, more precise system for mounting the acceptance masks.

The relative or point-to-point luminosity measurement was performed with the Very Small Angle Tagger (VSAT). The acceptance of the VSAT is about 20 times larger than that of the SAT and despite the increased sensitivity of the acceptance to LEP beam parameters it gives an overall reduction of about 15% on the experimental uncertainties of the Z^0 mass and width.

4.   Cross Sections and Forward-Backward Asymmetries

Hadronic events were selected by requiring at least four charged tracks and a charged energy greater than 12% of $\sqrt{s}$. A total of 677,000 events were selected at the three energy points of the scan with an efficiency of $95.2 \pm 0.1\%$ and a systematic uncertainty of $\pm 0.13\%$. The e^+e^- event selection was based on two partially independent methods with systematic uncertainties on the normalization ranging from 1 to 2% at the different energy points. The analysis was restricted to $44° < \theta < 136°$. The polar angle range for $\mu^+\mu^-$ events was $20° < \theta < 160°$. A set of three independent selection criteria led to a high efficiency and small systematic errors of $\pm 0.5\%$. The $\tau^+\tau^-$ events were selected by requiring two low-multiplicity back-to-back jets with additional cuts on both energy and momentum based variables. The systematic error on the normalization was roughly 0.8%.

In addition to cross-channel (affecting all four channels) and cosmic ray backgrounds ($\mu^+\mu^-$ and $\tau^+\tau^-$), a flat $\gamma\gamma$ background was subtracted from the hadronic and $\tau^+\tau^-$ samples. A detailed simulation of the detector was used to study features of the background subtractions and efficiency corrections not possible to obtain directly from the data.

The event sample for the e^+e^- FB asymmetry was the same as for the cross-section measurement. The main sources of systematic errors were charge confusion, FB asymmetries in the detector acceptance and the highly asymmetric t-channel subtraction. The $\mu^+\mu^-$ asymmetry measurement extends down to $11° < \theta < 169°$ to take advantage of additional statistics and increased sensitivity to the leptonic couplings. The small systematic error of ± 0.002 is due to *possible* detector asymmetries. The $\tau^+\tau^-$ samples consists of the events in the cross-section sample with one versus N charged tracks. The systematic errors range from ± 0.008 to ± 0.002 and are primarily due to charge confusion and the e^+e^- background subtraction.

The FB asymmetries are calculated with maximum likelihood fits to the resulting charged angular distributions.

An inclusive lepton analysis was also performed as a cross-check of the three exclusive analyses. There is good agreement with the exclusive results.

5.   Results

The program ZFITTER[3] was used to perform fits to the preliminary 1993 cross-sections and leptonic FB asymmetries as well as the previously published DELPHI data. The correlation between the 1991 and '93 LEP energy measurements hasn't been studied properly yet so the Z^0 mass for the '91 scan was left as a free parameter in the fits. A nine parameter fit to the cross-sections and asymmetries gives results consistent with lepton universality. The assumption of lepton universality then results in the following set of fitted resonance parameters: $M_Z = 91.1869 \pm 0.0052$ GeV, $\Gamma_Z = 2.4951 \pm 0.0059$ GeV, $\sigma_0 = 41.26 \pm 0.17$ nb, $R_l = 20.69 \pm 0.09$ and $A_{FB}^0 = 0.0160 \pm 0.0029$. These results are in good agreement with previous DELPHI results and those published by the other LEP collaborations.[4,5,6]

The number of light neutrino species can be extracted by constraining the strong coupling constant $\alpha_s = 0.123 \pm 0.005$ to other DELPHI results[7] and fitting the data to the Standard Model with the number of neutrino species left free: $N_\nu = 3.027 \pm 0.029$.

After constraining the number of neutrino species to three and α_s as before, the DELPHI data can be fitted to the top quark mass: $m_t = 170^{+21}_{-24} \pm 18(Higgs)$ GeV. The second uncertainty corresponds to a variation of the (unobserved) Higgs mass from 60 to 1000 GeV with a central value of 300 GeV. The above value of m_t corresponds to a value of the effective weak mixing angle of $\sin^2 \theta_{eff}^{lept} = 0.2322 \pm 0.0008^{+0.0001}_{-0.0003}(Higgs)$.

6.   Acknowledgements

I would like to thank the organizers for the opportunity to present these results, my DELPHI colleagues for providing them, the CERN-SL division for the superb operation of LEP and the LEP Energy Group for their beautiful energy calibration. This work was supported in part by a grant from the Norwegian Research Council.

References

1. DELPHI Collaboration, P. Abreu et al., Nucl. Phys. **B417** (1994) 3.
2. DELPHI Collaboration, P. Abreu et al., Nucl. Phys. **B418** (1994) 403.
3. "ZFITTER: An Analytical Program for Fermion Pair Production in e^+e^- Annihilation", D. Bardin et al., preprint CERN-TH 6443 (1992) and references therein.
4. ALEPH Collaboration, D. Buskulic et al., "Z Production Cross Sections and Lepton Pair Forward–Backward Asymmetries", preprint CERN-PPE/94-30, Z. Phys **C** to be published.
5. L3 Collaboration, M. Acciarri et al., "Measurement of Cross Sections and Leptonic Forward–Backward Asymmetries at the Z Pole and Determination of Electroweak Parameters", Z. Phys **C** to be published.
6. OPAL Collaboration, R. Akers et al., Z.Phys. **C61** (1994) 19.
7. DELPHI Collaboration, P. Abreu et al., Z. Phys. **C59** (1993) 21.

MEASUREMENTS OF THE τ POLARIZATION
IN Z^0 DECAYS

G. VALENTI[*]

I.N.F.N., Bologna, Italy

ABSTRACT

A sample of $Z^0 \to \tau^+\tau^-$ events observed in the DELPHI detector at LEP in 1991 and 1992, is used to measure the τ polarisation. A fit to the polar angle dependence of the polarisation yields the results for the mean τ polarisation $\langle \mathcal{P}_\tau \rangle = -0.144 \pm 0.024$ and for the Z^0 polarisation $\mathcal{P}_z = -0.140 \pm 0.028$. This leads to the ratio of vector to axial-vector effective couplings for taus $\bar{v}_\tau/\bar{a}_\tau = 0.072 \pm 0.012$ and for electrons $\bar{v}_e/\bar{a}_e = 0.070 \pm 0.014$, compatible with $e - \tau$ universality.

1. Introduction

The results presented are based on a sample $Z^0 \to \tau^+\tau^-$ events corresponding to an integrated luminosity of 33.6 pb^{-1}, selected in the barrel region of DELPHI (polar angle between 43° and 137°). The following exclusive decay channels of the τ have been used as polarimeters: $e\nu\bar{\nu}$ using a combination of the deposited calorimetric energy and the measured momentum; $\mu\nu\bar{\nu}$ using the μ momentum spectrum; $\pi\nu$ and $K\nu$ inclusively using the momentum spectrum of the π/Ks; $\rho\nu$ using the variable ξ described in[1]; $a_1\nu$ where the a_1 decays to 3 charged πs, using moments of various angular distributions sensitive to the τ polarisation.

In addition, an analysis of events where the τ decays inclusively to a charged hadron with or without neutrals has been performed. This has a higher efficiency but lower sensitivity per event than the exclusive hadronic analyses. Corrections for radiation, admixture of photon exchange and for $\sqrt{s} \neq M_Z$ are included in the quoted results.

2. Event sample

The data sample (33.6 pb^{-1}) had the following composition: 22.9 pb^{-1} at $\sqrt{s} = 91.2$ GeV in 1992; 6.7 pb^{-1} at $\sqrt{s} = 91.2$ GeV in 1991; 4.0 pb^{-1} spread across the six centre-of-mass energies $\sqrt{s} = 88.5, 89.5, 90.2, 92.0, 93.0$ and 93.7 GeV in 1991.

Cuts on isolation (minimum angle between any two tracks in different hemisphere greater then 160°), multiplicity (less then or equal to six), vertex pointing (leading track closest approach to the interaction vertex 4.5 cm in z and 1.5 cm in the x-y plane), transverse momentum (greater then 0.4 GeV/c), acollinearity ($\geq 0.5°$), $p_{rad} = \sqrt{p_1^2 + p_2^2}$

---

[*]Representing the DELPHI Collaboration.

($\leq p_{beam}$), $E_{rad} = \sqrt{E_1^2 + E_2^2}$ ($\leq E_{beam}$) were used to remove background from $e^+e^- \to q\bar{q}$, cosmic rays, beam gas and two-photon events.

In all analyses, samples of simulated events were used which had been passed through a detailed simulation of the detector response and reconstructed with the same program as the real data. The Monte Carlo event generators used were: KORALZ[2] for $e^+e^- \to \tau^+\tau^-$ events; DYMU3[3] for $e^+e^- \to \mu^+\mu^-$ events; BABAMC[4] for $e^+e^- \to e^+e^-$ events; Jetset 7.3[5] for $e^+e^- \to q\bar{q}$ events; Berends-Daverveldt-Kleiss[6] for $e^+e^- \to e^+e^-e^+e^-$ events;

3. Combination of results

The fitted values of $\langle \mathcal{P}_\tau \rangle$ and $\mathcal{P}_z$ for each of the individual analyses are shown in Table 1.

Channel	$\langle \mathcal{P}_\tau \rangle$	$\mathcal{P}_z$
$\tau \to e\nu\bar{\nu}$	$-0.130 \pm 0.076 \pm 0.081$	$-0.181 \pm 0.110 \pm 0.003$
$\tau \to \mu\nu\bar{\nu}$	$-0.033 \pm 0.068 \pm 0.041$	$+0.024 \pm 0.099 \pm 0.003$
$\tau \to \pi\nu$	$-0.192 \pm 0.038 \pm 0.040$	$-0.163 \pm 0.063 \pm 0.003$
$\tau \to \rho\nu$	$-0.119 \pm 0.028 \pm 0.031$	$-0.087 \pm 0.050 \pm 0.003$
$\tau \to a_1\nu$	$-0.184 \pm 0.069 \pm 0.059$	$-0.264 \pm 0.103 \pm 0.003$
Inclusive	$-0.147 \pm 0.021 \pm 0.022$	$-0.144 \pm 0.032 \pm 0.003$

Table 1. Values of $\langle \mathcal{P}_\tau \rangle$ and $\mathcal{P}_z$ from the fit of $\mathcal{P}_\tau(\cos\Theta)$ for all channels. Uncertainties are statistical followed by systematic. The systematic error on $\mathcal{P}_z$ is common to all channels.

The correlation coefficient on the statistical uncertainties between the inclusive and $\pi\nu$ channel was estimated to be 0.3 and that between the inclusive and $\rho\nu$ to be 0.4 from the overlap of events in the different samples. A maximal correlation in the systematic errors of the $\pi\nu$ and inclusive hadronic analysis was assumed, while the correlation between the uncertainties in the $\rho\nu$ and inclusive hadronic analyses was low. For each channel the polarisation was also measured in each of six bins in $\cos\Theta$. The six different channels were combined together taking account of the correlations in each of the six bins. The results are shown in Table 2.

The fit gave the results

$$\langle \mathcal{P}_\tau \rangle = -0.144 \pm 0.018 \pm 0.010 \pm 0.012, \qquad \mathcal{P}_z = -0.140 \pm 0.028 \pm 0.001 \pm 0.003.$$

The first error is statistical, the second is due to Monte Carlo statistics and the third due to all other systematic uncertainties. Corrections for propagator effects, radiative effects and the $\sqrt{s}$ spread have also been included. The best fit gave a χ^2 per degree of freedom of 8.5/4. The correlation between $\langle \mathcal{P}_\tau \rangle$ and $\mathcal{P}_z$ was 0.03. A second fit assuming lepton universality gave the result

$$\langle \mathcal{P}_\tau \rangle = \mathcal{P}_z = -0.142 \pm 0.015 \pm 0.007 \pm 0.008.$$

The χ^2 was 8.5 for 5 degrees of freedom.

$\cos\Theta$ range	$\mathcal{P}_\tau$	χ^2/n.d.f.
$[-0.732, -0.488]$	-0.100 ± 0.048	$3.7/5$
$[-0.488, -0.244]$	$+0.002 \pm 0.050$	$3.6/5$
$[-0.244, 0.000]$	-0.065 ± 0.050	$2.3/5$
$[0.000, 0.244]$	-0.116 ± 0.052	$14.6/5$
$[0.244, 0.488]$	-0.286 ± 0.046	$5.4/5$
$[0.488, 0.732]$	-0.248 ± 0.046	$8.7/5$

Table 2. τ polarisation values in bins of $\cos\Theta$ for the combination of all channels.

4.   Summary and Conclusions

The polarisation of the τ and its polar angle dependence has been determined from the study of exclusive channels and a hadronic inclusive analysis. The different measurements were found to be consistent with each other.

A fit to the combined measurement of $\mathcal{P}_\tau$ as a function of of $\cos\Theta$, the charge weighted polar angle of the τ, has been performed. These results yield for the ratios of the effective weak couplings of the τ and electron respectively

$$\bar{v}_\tau/\bar{a}_\tau = 0.072 \pm 0.012, \qquad \bar{v}_e/\bar{a}_e = 0.070 \pm 0.014$$

supporting the hypothesis of lepton universality. A second fit assuming $e-\tau$ universality yields the result

$$\bar{v}_l/\bar{a}_l = 0.071 \pm 0.009.$$

The effective weak mixing angle $\sin^2\theta_{\text{eff}}^{\text{lept}}$ is defined by the relation $\bar{v}_l/\bar{a}_l = 1 - 4\sin^2\theta_{\text{eff}}^{\text{lept}}$, leading to the result

$$\sin^2\theta_{\text{eff}}^{\text{lept}} = 0.2322 \pm 0.0023.$$

References

1. A. Rougé, LPNHE preprint X-LPNHE 92/20 (1992).
   M. Davier et. al., Phys. Lett. B306 (1993) 411.
2. S.Jadach and Z.Was, Comp. Phys. Com. 36 (1985) 191.
   S.Jadach, B.F.L.Ward and Z.Was, Comp. Phys. Com. 66 (1991) 276.
3. J. E. Campagne and R. Zitoun, Z. Phys. C43 (1989) 469.
   Proc. Brighton Workshop on Radiative Corrections (Sussex, July 1989)
4. F. A. Berends, W. Hollik and R. Kleiss, Nucl. Phys. B304 (1988) 712.
5. T.Sjöstrand, Comp. Phys. Comm. 27 (1982) 243, ibid.28 (1983) 229.
   T.Sjöstrand and M. Bengtsson, Comp. Phys. Comm. 43 (1987) 367.
   T.Sjöstrand, "PYTHIA 5.6 JETSET 7.3 Physics and manual", preprint CERN-TH 6488/92 (1992).
6. F.A.Berends, P.H.Daverveldt, R.Kleiss, Phys. Lett. B148 (1984) 489,
   Comp. Phys. Comm. 40 (1986) 271.

Z POLARIZATION IN $pp \to ZZ \to \ell^+\ell^-\nu\bar{\nu}$ AT THE LHC

M. J. DUNCAN and M. H. RENO*

Department of Physics and Astronomy, University of Iowa, Iowa City, Iowa 52242, USA

ABSTRACT

We evaluate the feasibility of measuring the Z polarization from a heavy Higgs signal in the decay channel $H \to ZZ \to \ell^+\ell^-\nu\bar{\nu}$. Including gluon fusion production of the Higgs, continuum production of Z pairs and the QCD background of single Z production with a missing jet, we find that the average value of a new variable to measure the Z polarization, as a function of transverse mass, will demonstrate the existence of a Higgs boson for heavy Higgs masses up to 800 GeV at the LHC with an integrated luminosity of 10^5 pb^{-1}.

1. Introduction

The 'gold-plated' heavy Higgs decay mode $H \to ZZ \to \ell_1^+\ell_1^-\ell_2^+\ell_2^-$ has received much attention in the literature because of the opportunity for a fully reconstructed Higgs signal.[1] For Higgs masses near the unitarity limit of 800 GeV,[2] however, the event rates are low for pp production of the Higgs boson at the Large Hadron Collider (LHC) center of mass energy $\sqrt{S} = 14$ TeV. To augment the Higgs search in the high mass region, the Higgs decay mode $H \to ZZ \to \ell^+\ell^-\nu\bar{\nu}$ should also be considered as it is enhanced relative to the gold-plated decay mode by a factor of six.[3] This mode is easily identifiable by the large missing transverse momentum in the event and a pair of charged leptons that reconstruct to the Z mass. In addition to searching for an enhancement in the event rate, best identified in the transverse mass distribution,[4] one can also use polarization information to identify a Higgs signal. The Higgs boson decays preferentially into longitudinally polarized Z's, (Z_L) while the irreducible and reducible backgrounds to $pp \to \ell^+\ell^-$+missing p_T involve primarily transversely polarized Z's (Z_T). Polarization methods have been adopted for the gold-plated mode[5] and the WW final state[6], and we have suggested that this be applied to the case of $pp \to ZZ \to \ell^+\ell^-\nu\bar{\nu}$.[7] Here, we demonstrate that even with the QCD background included, the LHC with an integrated luminosity of $\mathcal{L} = 10^5$ pb^{-1} has a capability of distinguishing Higgs boson production in the $\ell^+\ell^-\nu\bar{\nu}$ decay mode from the background signal for a range of heavy Higgs masses m_H up to 800 GeV. We consider here masses between 400 GeV and 800 GeV.

2. Signal and Backgrounds

The Higgs coupling to other particles is proportional to the particle's mass, so the main contributions to Higgs production involve top quarks and weak gauge bosons.

*Presenter. Work supported in part by National Science Foundation Grant No. PHY-9307213.

Gunion *et al.*[8] have shown that production of the Higgs, with $t\bar{t}$ in the final state is very small. The main Higgs production processes are gluon fusion into a top quark loop, which couples to the Higgs, and vector boson fusion into a Higgs. For a top quark mass of 175 GeV, the gluon fusion mechanism dominates vector boson fusion for the full range of Higgs masses of interest: $m_H = 400 - 800$ GeV. The vector boson fusion cross section $qq \rightarrow qqZZ$ with an s-channel Higgs, for Higgs masses between 400 GeV and 800 GeV, lies below the gluon fusion cross section by a factor of ~ 10 at the low mass end, to a factor of ~ 3 at the high mass end. Baur and Glover[9] have done the full calculation of $ZZ+2$ parton production and have found that the s-channel approximation overestimates the contribution of $qq \rightarrow qqZZ$ to a polarization measurement because of additional t-channel and non-resonant contributions. To be on the conservative side in our analysis, we include only the gluon fusion mechanism.

Higgs decay into two Z's goes primarily into longitudinal Z's. It is this effect which we exploit in our analysis of the signal. The large irreducible background to $pp \rightarrow H \rightarrow ZZ$ is continuum production of Z pairs via $q\bar{q} \rightarrow ZZ$. The dominant contribution to this background rate comes from production of a pair of transverse Z's.

The main reducible background is from QCD corrections to single Z production: a Z is produced at large transverse momentum, which decays into an $\ell^+\ell^-$ pair, and is accompanied by jets that are missed in the detector because they go down the beam pipe. To model this background, we use the matrix elements for $q\bar{q} \rightarrow Zg \rightarrow e^+e^-g$ and crossed diagrams. We count in the background rate only the part of the cross section where the charged leptons have rapidity $|y_\ell| < y_\ell^c$, so that they are in the central region, and a large final state parton rapidity $|y_p| > y_p^c$.

In Fig. 1, we show the transverse mass distribution for the three contributions above. Here, the transverse mass m_T is defined to be

$$m_T^2 = [(\vec{p}_T{}^2 + m_Z^2)^{1/2} + (\vec{\not{p}}_T{}^2 + m_Z^2)^{1/2}]^2 - (\vec{p}_T + \vec{\not{p}}_T)^2$$

where $\vec{p}_T$ is the Z boson transverse momentum reconstructed from $\ell^+\ell^-$ momenta and $\vec{\not{p}}_T$ is the missing transverse momentum. In our Monte Carlo, with these three processes, all contributions have balanced p_T: $\vec{p}_T = -\vec{\not{p}}_T$. The dashed line in the figure indicates the gluon fusion production of a 500 GeV Higgs, the solid line shows the continuum production of ZZ and the dot-dashed line shows the rate from QCD production of Z+missing jet. Here the rapidity cuts of $y_p^c = y_\ell^c = 3$ are used, and in all the figures, a charged lepton transverse momentum cut of $p_T > 20$ GeV is applied. The heavy solid line is the total of the three contributions.

A longitudinal polarization vector for a massive particle, subjected to an arbitrary boost, will not, in general, retain its longitudinal character, so the statement that the $Z_L Z_L$ production rate is enhanced by a specific amount in Higgs production is a frame dependent statement. For a sufficiently heavy Higgs, the parton center of mass frame (the Higgs rest frame) and the hadron center of mass frame coincide.

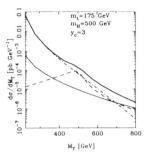

Fig. 1. The transverse mass distributions for $gg \to H \to ZZ \to \ell^+\ell^-\nu\bar{\nu}$ (dashed), $q\bar{q} \to ZZ \to \ell^+\ell^-\nu\bar{\nu}$ (solid) and $q\bar{q} \to Zg \to \ell^+\ell^- g$, with crossed diagrams, (dot-dashed) for pp collisions at $\sqrt{S} = 14$ TeV. Here, the lepton rapidity is $|y_\ell| < 3$ and the final state parton rapidity is $|y_p| > 3$ to mimic a missing jet.

The decay distributions of the lepton from Z_L and Z_T decays, in the Z rest frame, are

$$\phi_L(z) = \frac{3}{4}(1 - z^2) \quad \phi_T(z) = \frac{3}{8}(1 + z^2) \tag{1}$$

where $z = \cos\theta$ for θ, the angle between the lepton and the axis defined by the Z momentum in the parton center of mass frame. The shape of the decay distribution as well as the average value of $|z|$ (for longitudinally polarized Z's, a value of $3/8$, and for transversely polarized Z's, a value of $9/16$) characterize the production mechanism. These decay distributions for the gold plated modes have been discussed elsewhere.[5] For the Higgs decay into ZZ where one Z decays into neutrinos, one loses the information required to reconstruct the parton center of mass.

Note that in the parton center of mass frame (pcm), $z = -2p_\ell^{pcm} \cdot \epsilon_L^{pcm}/M_Z$, where we have a four-vector dot product which involves the longitudinal polarization vector ϵ_L^{pcm}. The parton center of mass frame differs from the hadron center of mass frame by a boost, however, the Lorentz boost of ϵ_L^{pcm} converts the four vector into a linear combination of longitudinal and transverse polarization vectors. If the Higgs is heavy enough, the boost will not be large. To approximate $|z|$ in the hadron center of mass, we introduce the variable z^* where $z^* = 2|p_\ell \cdot \epsilon_L|/M_Z$. Here, all of the vectors are in the hadron center of mass, and $\epsilon_L = (|\vec{p}_Z|/M_Z, E_z\vec{p}_Z/(|\vec{p}_Z|M_Z))$ is the longitudinal polarization vector determined from the reconstructed Z momentum. In the results presented below, we plot the average value of z^* as a function of transverse mass.

3.   Results

We now show our results for several values of the Higgs mass. In Fig. 2, we show $< z^* >$ as a function of m_T for $m_H = 600$ GeV, where the input top quark mass is taken at $m_t = 175$ GeV. The triangles show the $< z^* >$ value for the combination of ZZ continuum production and the QCD missing jet background, and the squares are for the gluon fusion Higgs signal. For each of the separate cross sections σ_i, $i = c$

(continuum), m (missing jet background) and h (Higgs signal), the overall $< z^* >$ value is determined by

$$< z^* >= \frac{< z^* >_h \sigma_h + \kappa(< z^* >_m \sigma_m + < z^* >_c \sigma_c)}{\sigma_h + \kappa(\sigma_m + \sigma_c)} \qquad (2)$$

with $\kappa = 1$. In order to estimate the error on the theoretical prediction for $< z^* >$, we also take $\kappa = 1 \pm 0.3$. This error from the normalization uncertainty from non-Higgs cross sections is combined in quadrature with a statistical error. The statistical error is estimated by $< z^* > /\sqrt{N_{evt}}$, where the number of events assumes an integrated luminosity of 10^5 pb^{-1}. The combined error is the outer error bar, with only the statistical error indicated by the inner vertical error bar on the figures. We take a transverse mass bin width of 50 GeV, except for the two last bins, where the widths are 100 GeV and 250 GeV, respectively, to include a reasonable number of events. The statistical error dominates the overall error. This figure and Figs. 3a and 3b below use $y_\ell^c = y_p^c = 3.0$. Figure 3a shows $< z^* >$ versus m_T for $m_H = 400$ GeV, while Fig. 3b has $m_H = 800$ GeV. In all of these figures, only one charged lepton family is included in the rate.

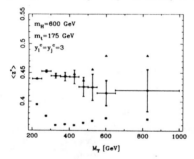

Fig. 2. The value of $< z^* >$ as a function of m_T for $m_H = 600$ GeV and $m_t = 175$ GeV. Rapidity cuts applied are $y_\ell^c = y_p^c = 3.0$. The symbols are described in the text.

At the LHC, a more realistic set of rapidity cuts is to include charged leptons in a more central region, with $y_\ell^c = 2.5$, and to put the "missing jet" cut at $y_p^c = 4$ for the parton rapidity. With these cuts, the signal event rate is not reduced very much, but the QCD missing jet background is greatly reduced. Figures 4a and 4b show $< z^* >$ versus m_T for $m_H = 400$ GeV and $m_H = 800$ GeV, respectively, with these new rapidity cuts. Again, the error bars indicate the combined normalization and statistical error bars as described above.

Figures 2-4 indicate that the value of $< z^* >$ will be a useful tool to characterize high transverse mass data at the LHC. With an integrated luminosity of 10^5 pb^{-1},

Higgs masses up to 800 GeV are accessible using this method.

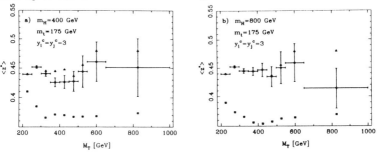

Fig. 3. Value of $< z^* >$ for m_T bins with a) $m_H = 400$ GeV and b) $m_H = 800$ GeV, with
$$y_\ell^c = y_p^c = 3.$$

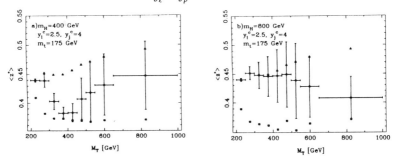

Fig. 4. As in Fig. 3, but with $y_\ell^c = 2.5$ and $y_p^c = 4$.

References

1. R. N. Cahn, *et al.*, in *Experiments, Detectors and Experimental Areas for the SSC*, proceedings of the Workshop, Berkeley, CA, 1987, *eds.* R. Donaldson and M. Gilchriese (World Scientific, Singapore, 1988).
2. B. W. Lee, C. Quigg and H. B. Thacker, *Phys. Rev. Lett.* **38** (1977) 883; *Phys. Rev.* **D16** (1977) 1519.
3. R. N. Cahn and M. S. Chanowitz, *Phys. Rev. Lett.* **56** (1986) 1327.
4. V. Barger, T. Han and R. J. N. Phillips, *Phys. Rev.* **D36** (1987) 295.
5. M. J. Duncan, *Phys. Lett.* **B179** (1986) 393; T. Matsuura and J. J. van der Bij, *Z. Phys.* **C51** (1991) 259.
6. M. J. Duncan, G. L. Kane and W. W. Repko, *Nucl. Phys.* **B272** (1986) 517; G. L. Kane and C. P. Yuan, *Phys. Rev.* **D40** (1989) 2231.
7. M. J. Duncan and M. H. Reno, in *Proceedings of the Workshop on Physics at Current Accelerators and Supercolliders*, Argonne National Laboratory, 1993, *eds.* J. L. Hewett, A. R. White and D. Zeppenfeld.
8. J. F. Gunion, H. E. Haber, F. E. Paige, Wu-Ki Tung and S. S. D. Willenbrock, *Nucl. Phys.* **B294** (1987) 621.
9. U. Baur and E. W. N. Glover, *Nucl. Phys.* **B347** (1990) 12.

AMPLITUDE ZEROS AND RAPIDITY CORRELATIONS IN $W\gamma$ AND WZ PRODUCTION IN HADRONIC COLLISIONS

U. BAUR

Physics Department, Florida State University, Tallahassee, FL 32306, USA

ABSTRACT

A comparative study of amplitude zeros in $W\gamma$ and WZ production in hadronic collisions is presented. The Standard Model amplitude for $q_1\bar{q}_2 \to W^{\pm}Z$ at the Born-level is shown to exhibit an approximate zero located at $\cos\theta = (g_-^{q_1} + g_-^{q_2})/(g_-^{q_1} - g_-^{q_2})$ at high energies, where the $g_-^{q_i}$ $(i = 1, 2)$ are the left-handed couplings of the Z-boson to quarks and θ is the center of mass scattering angle of the W-boson. This approximate zero is similar to the well-known radiation zero in $W\gamma$ production. Prospects to observe the amplitude zeros using rapidity correlations between the final state particles are explored.

1. Introduction

Although the electroweak Standard Model (SM) based on an $SU_L(2) \otimes U_Y(1)$ gauge theory has been very successful in describing contemporary high energy physics experiments, the three vector-boson couplings predicted by this non-Abelian gauge theory remain largely untested experimentally. Careful studies of these couplings, for example in di-boson production in e^+e^- or hadronic collisions, may allow us to test the non-Abelian gauge structure of the SM.[1] Constraints on anomalous $WW\gamma$, WWZ, $ZZ\gamma$ and $Z\gamma\gamma$ couplings have been reported at this conference by the CDF,[2] DØ[3] and L3 collaboration.[4]

2. Amplitude Zeros

The reactions $p\overset{(-)}{p} \to W^{\pm}\gamma$ and $p\overset{(-)}{p} \to W^{\pm}Z$ are of special interest due to the presence of amplitude zeros. It is well known that all SM helicity amplitudes of the parton-level subprocess $q_1\bar{q}_2 \to W^{\pm}\gamma$ vanish for $\cos\theta = (Q_1 + Q_2)/(Q_1 - Q_2)$,[5] where θ is the scattering angle of the W-boson with respect to the quark (q_1) direction, and Q_i $(i = 1, 2)$ are the quark charges in units of the proton electric charge e. This zero is a consequence of the factorizability[6] of the amplitudes in gauge theories into one factor which contains the gauge coupling dependence and another which contains spin information. Although the factorization holds for any four-particle Born-level amplitude in which one or more of the four particles is a gauge-field quantum, the amplitudes for most processes may not necessarily develop a kinematical zero in the physical region. The amplitude zero in the $W^{\pm}\gamma$ process has been further shown to correspond to the absence of dipole radiation by colliding particles with the same charge-to-mass ratio,[7] a realization of classical radiation interference.

Recently, it was found[8] that the SM amplitude of the process $q_1 \bar{q}_2 \to W^{\pm} Z$ also exhibits an approximate zero at high energies. The $(\pm, \mp)$ amplitudes $\mathcal{M}(\pm, \mp)$ vanish for

$$\frac{g_-^{q_1}}{\hat{u}} + \frac{g_-^{q_2}}{\hat{t}} = 0, \tag{1}$$

where $g_-^{q_i}$ is the coupling of the Z boson to left-handed quarks, and $\hat{s}$, $\hat{u}$ and $\hat{t}$ are Mandelstam variables in the parton center of mass frame. For $\hat{s} \gg M_Z^2$, the zero in the $(\pm, \mp)$ amplitudes is located at $\cos \theta_0 = (g_-^{q_1} + g_-^{q_2})/(g_-^{q_1} - g_-^{q_2})$, or

$$\cos \theta_0 \simeq \begin{cases} +\frac{1}{3} \tan^2 \theta_{\rm w} \simeq +0.1 & \text{for } d\bar{u} \to W^- Z, \\ -\frac{1}{3} \tan^2 \theta_{\rm w} \simeq -0.1 & \text{for } u\bar{d} \to W^+ Z. \end{cases}$$

The existence of the zero in $\mathcal{M}(\pm, \mp)$ at $\cos \theta_0$ is a direct consequence of the contributing Feynman diagrams and the left-handed coupling of the W-boson to fermions.

At high energies, strong cancellations occur, and, besides $\mathcal{M}(\pm, \mp)$, only the $(0, 0)$ amplitude remains non-zero. The combined effect of the zero in $\mathcal{M}(\pm, \mp)$ and the gauge cancellations at high energies in the remaining helicity amplitudes results in an approximate zero for the $q_1 \bar{q}_2 \to W^{\pm} Z$ differential cross section at $\cos \theta \approx \cos \theta_0$. This is illustrated in Fig. 1a where we show the differential cross sections for $d\bar{u} \to W^- Z$ for $(\lambda_{\rm w}, \lambda_{\rm z}) = (\pm, \mp)$ and $(0, 0)$, as well as the unpolarized cross section, which is obtained by summing over all W and Z boson helicity combinations (solid line). Although the matrix elements are calculated at $\sqrt{\hat{s}} = 2$ TeV, the results differ little from those obtained in the high energy limit. The total differential cross section displays a pronounced minimum at the location of the zero in $\mathcal{M}(\pm, \mp)$.

Figure 1b illustrates the energy dependence of the differential cross section. At $\sqrt{\hat{s}} = 0.2$ TeV, $i.e.$ close to the threshold, contributions from the $(\pm, \pm)$, $(\pm, 0)$, and $(0, \pm)$ amplitudes are important. Above threshold, these contributions rapidly diminish, as exemplified by the curves for $\sqrt{\hat{s}} = 0.5$ TeV (dashed lines) and $\sqrt{\hat{s}} = 2$ TeV (solid lines). Note that the location of the minimum varies only slightly with energy.

3. Rapidity Correlations

The radiation zero in $q_1 \bar{q}_2 \to W\gamma$ and the approximate amplitude zero in $q_1 \bar{q}_2 \to WZ$ are not easy to observe in the $\cos \theta$ distribution in pp or $p\bar{p}$ collider experiments. Structure function effects transform the zero in the $W\gamma$ case into a dip in the $\cos \theta$ distribution. The approximate zero in WZ production is only slightly affected by structure function effects. Higher order QCD corrections and finite W width effects tend to fill in the dip. In $W\gamma$ production photon radiation from the final state lepton line also diminishes the significance of the dip. Finally, finite detector resolution effects and ambiguities in reconstructing the parton center of mass frame represent an additional major complication in the extraction of the $\cos \theta$ distribution, and further dilute the signal of the amplitude zeros. The ambiguities are associated with the nonobservation of the neutrino arising from W decay. Identifying the missing transverse momentum with the transverse momentum of the neutrino of a given $W\gamma$ or WZ event, the unobservable longitudinal neutrino momentum, $p_L(\nu)$, and thus the parton center of mass

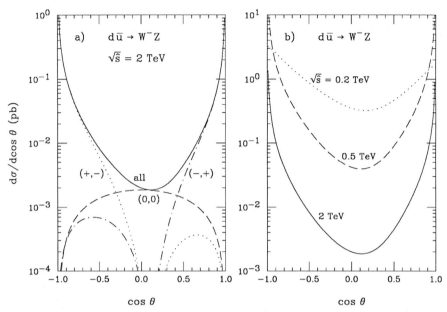

Figure 1: Differential cross section $d\sigma/d\cos\theta$ versus the W^- scattering angle θ in the center of mass frame for the Born-level process $d\bar{u} \to W^-Z$. a) The contributions of the individual helicity amplitudes, together with the total differential cross section (solid line), are shown for $\sqrt{\hat{s}} = 2$ TeV. b) The differential cross section for three different parton center of mass energies.

frame, can be reconstructed by imposing the constraint that the neutrino and charged lepton four momenta combine to form the W rest mass.[9] The resulting quadratic equation, in general, has two solutions. In the approximation of a zero W decay width, one of the two solutions coincides with the true $p_L(\nu)$. On an event to event basis, however, it is impossible to tell which of the two solutions is the correct one. This ambiguity considerably smears out the dip caused by the amplitude zeros.

Instead of trying to reconstruct the parton center of mass frame and measure the $\cos\theta$ or the equivalent rapidity distribution in the center of mass frame, one can study rapidity correlations between the observable final state particles in the laboratory frame.[10] Knowledge of the neutrino longitudinal momentum is not required in determining these correlations. Event mis-reconstruction problems originating from the two possible solutions for $p_L(\nu)$ are thus automatically avoided.

In $2 \to 2$ reactions differences of rapidities are invariant under boosts. One therefore expects that the (lab. frame) rapidity difference distributions $d\sigma/d\Delta y(V,W)$, $V = \gamma$, Z, where $\Delta y(V,W) = y(V) - y(W)$, exhibit a dip signaling the SM amplitude zeros.[10] In $W^{\pm}\gamma$ production, the dominant W helicity is $\lambda_W = \pm 1$,[11] implying that

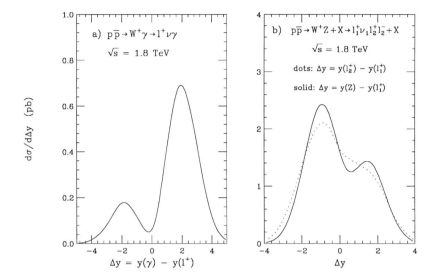

Figure 2: Rapidity difference distributions in the SM at the Tevatron. a) The photon lepton rapidity difference spectrum in $p\bar{p} \to \ell^+ \not{p}_T \gamma$. b) The $y(Z) - y(\ell_1^+)$ and $y(\ell_2^+) - y(\ell_1^+)$ distributions in $p\bar{p} \to W^+ Z$.

the charged lepton, $\ell = e, \mu$, from $W \to \ell\nu$ tends to be emitted in the direction of the parent W, and thus reflects most of its kinematic properties. As a result, the dip signaling the SM radiation zero should manifest itself in the $\Delta y(\gamma, \ell) = y(\gamma) - y(\ell)$ distribution.

The $\Delta y(\gamma, \ell)$ differential cross section for $p\bar{p} \to \ell^+ \not{p}_T \gamma$ at the Tevatron is shown in Fig. 2a. To simulate detector response, transverse momentum cuts of $p_T(\gamma) > 5$ GeV, $p_T(\ell) > 20$ GeV and $\not{p}_T > 20$ GeV, rapidity cuts of $|y(\gamma)| < 3$ and $|y(\ell)| < 3.5$, a cluster transverse mass cut of $M_T(\ell\gamma; \not{p}_T) > 90$ GeV and a lepton photon separation cut of $\Delta R(\gamma, \ell) > 0.7$ have been imposed. The SM radiation zero is seen to lead to a strong dip in the $\Delta y(\gamma, \ell)$ distribution at $\Delta y(\gamma, \ell) \approx -0.3$.

In contrast to the situation in $W\gamma$ production, none of the W helicities dominates in WZ production.[11] The charged lepton originating from the W decay, $W \to \ell_1\nu_1$, thus only partly reflects the kinematical properties of the parent W boson. As a result, a significant part of the correlation present in the $y(Z) - y(W)$ spectrum is lost, and only a slight dip survives in the $y(Z) - y(\ell_1)$ distribution, which is shown for the $W^+ Z$ case in Fig. 2b. Experimentally, the Z boson rapidity, $y(Z)$, can readily be reconstructed from the four momenta of the lepton pair $\ell_2^+ \ell_2^-$ originating from the Z decay. The cuts used in Fig. 2b are the same as those in Fig. 2a except for the lepton rapidity cut which has been replaced by $|y(\ell_{1,2})| < 2.5$.

Instead of the $y(Z) - y(\ell_1)$ distribution, one could also think of studying the rapidity correlations between the charged leptons originating from the $Z \rightarrow \ell_2^+ \ell_2^-$ and $W \rightarrow \ell_1 \nu_1$ decays. The dotted line in Fig. 2b shows the $y(\ell_2^+) - y(\ell_1^+)$ distribution for $W^+ Z$ production at the Tevatron. The $y(\ell_2^-) - y(\ell_1^+)$ spectrum almost coincides with the $y(\ell_2^+) - y(\ell_1^+)$ distribution. Since also none of the Z boson helicities dominates[11] in $q_1 \bar{q}_2 \rightarrow W Z$, the rapidities of the leptons from W and Z decays are almost completely uncorrelated, and essentially no trace of the dip signaling the approximate amplitude zero is left in the $y(\ell_2^+) - y(\ell_1^+)$ distribution.

4. Conclusions

$W\gamma$ and WZ production in hadronic collisions are of special interest due to the presence of amplitude zeros in the physical region. These amplitude zeros are best observed in rapidity correlations between the final state particles. In $W\gamma$ production at the Tevatron, the radiation zero leads to a pronounced dip in the $y(\gamma) - y(\ell)$ distribution. In WZ production, the approximate amplitude zero is signalled by a dip in the analogous $y(Z) - y(\ell_1)$ distribution where ℓ_1 is the charged lepton originating from the W decay. However, the dip in the $y(Z) - y(\ell_1)$ distribution is much less pronounced than that in the $y(\gamma) - y(\ell)$ spectrum in $W\gamma$ production, and thus more difficult to observe experimentally.

5. Acknowledgement

This research was supported by the U.S. Department of Energy under Contract No. DE-FG05-87ER40319.

References

1. K. Gaemers and G. Gounaris, *Z. Phys.* **C1** (1979) 259; K. Hagiwara *et al.*, *Nucl. Phys.* **B282** (1987) 253.
2. T. Fuess, (CDF Collaboration), these proceedings.
3. J. Ellison (DØ Collaboration), these proceedings; N. Oshima (DØ Collaboration), these proceedings.
4. J. Busenitz (L3 Collaboration), these proceedings.
5. R. W. Brown *et al.*, *Phys. Rev.* **D20** (1979) 1164; K. O. Mikaelian *et al.*, *Phys. Rev. Lett.* **43** (1979) 746.
6. D. Zhu, *Phys. Rev.* **D22** (1980) 2266; C. J. Goebel *et al.*, *Phys. Rev.* **D23** (1981) 2682.
7. S. J. Brodsky and R. W. Brown, *Phys. Rev. Lett.* **49** (1982) 966; R. W. Brown *et al.*, *Phys. Rev.* **D28** (1983) 624; R. W. Brown and K. L. Kowalski, *Phys. Rev.* **D29** (1984) 2100.
8. U. Baur, T. Han, and J. Ohnemus, *Phys. Rev. Lett.* **72** (1994) 3941.
9. J. Gunion, Z. Kunszt, and M. Soldate, *Phys. Lett.* **B163** (1985) 389; J. Gunion and M. Soldate, *Phys. Rev.* **D34** (1986) 826; W. J. Stirling *et al.*, *Phys. Lett.* **B163** (1985) 261.
10. U. Baur, S. Errede and G. Landsberg, FSU-HEP-940214 (preprint), to appear in *Phys. Rev.* **D**; U. Baur, T. Han and J. Ohnemus, in preparation.
11. C. Bilchak, R. Brown and J. Stroughair, *Phys. Rev.* **D29** (1984) 375.

STANDARD MODEL HIGGS PHYSICS AT AN UPGRADED TEVATRON

JACK F. GUNION and TAO HAN*

*Davis Institute for High Energy Physics, Department of Physics,
University of California, Davis, CA 95616, USA*

ABSTRACT

We compute an array of Standard Model Higgs boson signals and back-grounds for a possible upgrade of the Tevatron to $\sqrt{s} = 4$ TeV. Assuming a total accumulated luminosity of $L = 30$ fb^{-1}, we find that a Standard Model Higgs boson with $m_H \lesssim 110$ GeV could be detected using the $W^\pm H \to l\nu b\bar{b}$ mode. A Higgs boson with mass between 120 GeV and 140 GeV or above $230 - 250$ GeV would not be seen. A Higgs boson with $m_H \sim 150$ GeV or $200 \lesssim m_H \lesssim 230 - 250$ GeV has a decent chance of being detected in the $ZZ^{(*)} \to 4l$ mode. A comparison with a high luminosity Tevatron ($\sqrt{s} = 2$ TeV) is made.

1. Introduction

Given the cancellation of the SSC, it is important to reassess the possibilities for exploring the Electroweak Symmetry Breaking (EWSB) sector at possibly available machines. It is generally believed that the LHC can find the Standard Model Higgs boson (H) for all masses between about 80 GeV and ~ 800 GeV,[1] and possibly explore a strongly interacting EWSB sector that would arise for very large effective m_H values.[2] However, the time scale for construction of the LHC may be rather long without significant U.S. participation. The question arises as to whether the U.S. should consider an alternative investment in existing U.S. laboratories. A Tevatron upgrade to the $p\bar{p}$ center of mass energy $\sqrt{s} = 4$ TeV with yearly luminosity of $L = 10$ fb^{-1} has emerged as a subject of discussion in this context.[3] The possibility of a pp collider with $L = 100$ fb^{-1} is even being considered. Here, we present a first-level examination of the ability of such upgraded Tevatrons to probe the SM Higgs sector. In particular, we wish to establish the extent to which an upgraded Tevatron can search for Higgs bosons with mass beyond the reach of LEP-200 (*i. e.* LEP-II operated at $\sqrt{s} = 200$ GeV). We compute signals and those backgrounds that are not highly detector-dependent for most of conceivably useful channels. Of these processes, we find[4] that four of them have some real chance of being useful. They are:

- associated $W^\pm H$ (or ZH) production followed by $W \to l\nu$ ($Z \to l\bar{l}$) and $H \to b\bar{b}$, leading to an $lb\bar{b}$ ($l\bar{l}b\bar{b}$) final state;

---

*Talk presented by T. Han.

- inclusive H production followed by $H \to ZZ^{(*)} \to 4l$: the gold-plated events;

- inclusive H production followed by $H \to W^+W^- \to l\nu jj$;

- inclusive H production followed by $H \to W^+W^- \to 2l2\nu$.

After discussing our assumptions for the detector performance and acceptance cuts, we briefly present our results for those channels in the next section. It is interesting to know how crucial the upgrade in energy to 4 TeV is for extending the Tevatron's ability to search for the SM Higgs, as compared to simply increasing the luminosity at the standard 2 TeV Tevatron energy. We comment on this comparison for the $W^{\pm}H \to l\nu b\bar{b}$ channel, and present results for the $ZZ^{(*)} \to 4l$ channel.

2. Feasibility for SM Higgs Boson Searches at an Upgraded Tevatron

2.1. Detector Characteristics and Acceptance Cuts

Our approach[4] with regard to the detector is to consider generic resolutions and acceptance for: i) one very much like the current CDF and D0 detectors; and ii) a much more optimized detector, significantly upgraded from the current CDF and D0 detector characteristics. Here we only present results with the optimized option:

$$\frac{\Delta E}{E} = \begin{cases} \frac{0.2}{\sqrt{E}} \oplus 0.01 & \text{for } l = e, \mu \\ \frac{0.5}{\sqrt{E}} \oplus 0.03 & \text{for jets}. \end{cases} \qquad (1)$$

The kinematical cuts we choose are for leptons: $p_T^l > 10$ GeV; $|y_l| < 2.5$; $\Delta R_l > 0.3$; $p_T^{miss} > 15$ GeV. Non-b jets that are specifically utilized are required to have: $p_T^j \geq 15$ GeV; $|y_j| \leq 2.5$; $\Delta R_j \geq 0.7$, where ΔR_j is the separation from other jets. We assume that any given b-jet can be tagged with 30% efficiency and 99% purity for $p_T^b \geq 15$ GeV; $|y_b| < 2$, provided the b is separated by an appropriate $\Delta R_b > 0.5$ from neighboring b and light quark/gluon jets.

In our analysis b quarks are allowed to decay semi-leptonically to $cl\nu$ according to the measured branching ratio. The result is a moderate broadening of the $b\bar{b}$ mass distribution compared to that obtained if semi-leptonic decays are not included.

Table 1. We tabulate S, B, and $S/\sqrt{B}$ for $WH \to lb\bar{b}X$ with a double b-tagging, at $p\bar{p}$ and pp colliders for $L = 30$ fb^{-1} and $\sqrt{s} = 4$ TeV, for a series of m_H values, using the tabulated mass bins. Also given is the L (in fb^{-1}) required for $S/\sqrt{B} = 5$.

m_H	ΔM	$p\bar{p}$				pp			
(GeV)	(GeV)	S	B	$S/\sqrt{B}$	$L(5\sigma)$	S	B	$S/\sqrt{B}$	$L(5\sigma)$
60	10	467	505	20.8	1.7	352	375	18.2	2.3
80	10	229	320	12.8	4.6	163	228	10.8	6.4
100	20	167	372	8.7	10.0	112	263	6.9	15.6
110	20	122	278	7.3	14.0	80	196	5.7	23.2
120	20	85	217	5.8	22.4	54	155	4.3	40.0
130	20	55	182	4.0	46.1	34	127	3.0	84.5

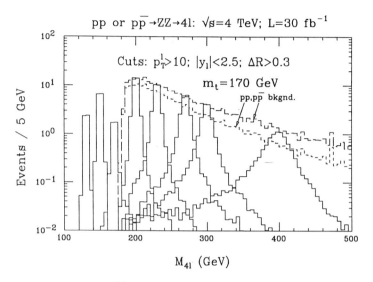

Figure 1. The $gg \to H \to ZZ^{(*)} \to 4l$ ($l = e, \mu$) signal and $q\bar{q} \to ZZ^{(*)} \to 4l$ background as a function of the four-lepton mass, M_{4l}, in 5 GeV bins. Higgs boson signals for $m_H = 130$, 150, 170, 200, 230, 270, 300 and 400 GeV are illustrated (including a K factor of 1.5).

2.2. $W^{\pm}H \to l\nu b\bar{b}$ and $ZH \to l\bar{l}b\bar{b}$ Channels

The most promising channel for H detection in the mass region $m_H \lesssim 120$ GeV is $W^{\pm}H \to l\nu b\bar{b}$ associated production, leading to the $lb\bar{b}X$ final state. To suppress the QCD background from Wjj, we select events with 2 tagged b's. The dominant background* then is $Wb\bar{b}$. The statistical significances that could be achieved after optimizing the signal mass intervals appear in Table 1, which gives signal and background rates, S and B, along with $S/\sqrt{B}$ for $L = 30$ fb^{-1} at both a $p\bar{p}$ and a pp collider. Also shown is the L required to achieve $S/\sqrt{B} = 5$. Based on this table, we conclude that $m_H \lesssim 110$ GeV could be probed in the $WH \to lb\bar{b}X$ mode with larger than 5σ effects, reaching possibly as high as $m_H \sim 120$ GeV if the optimal acceptance cuts and the double b-tagging can be effectively implemented.

The analogous process of the $ZH \to 2lb\bar{b}$ channel may also provide clean signal. Unfortunately, the overall event rates are much lower — roughly by a factor of 9–10 in the case of the signal. As a result, it is not very useful to extend the LEP-II coverage.

2.3. $gg \to H \to ZZ^{(*)} \to 4l$ Channel

It is well recognized that the $H \to ZZ^{(*)} \to 4l$ channel provides the most signal, but the event rate is rather low. It is nevertheless important to evaluate the feasibility

---

*We have not included the backgrounds from a single top quark production $Wg \to t\bar{b}$ and $q\bar{q} \to t\bar{b}$. Those were found to be not negligible[5] for $m_H \gtrsim 120$ GeV at $\sqrt{s} = 3.5$ TeV.

Table 2. For $ZZ^* \to 4l$ channel, we tabulate, as a function of Higgs boson mass, the signal and background rates S, B (summed over a 10 GeV interval); the associated $S/\sqrt{B}$ values; and the L (in fb^{-1}) required for a $S/\sqrt{B} = 5$ or 3 signal level. The background rate for $M_{4l} < 2m_Z$ is negligible.

m_H	$p\bar{p}(pp)$:	4 TeV,	30 fb^{-1}		$p\bar{p}$:	2 TeV,	100 fb^{-1}	
(GeV)	S	B	$S/\sqrt{B}$	$L(5\sigma)$	S	B	$S/\sqrt{B}$	$L(3\sigma)$
130	5	-	-	-	3	-	-	-
150	13	-	-	-	9	-	-	-
170	4	-	-	-	2	-	-	-
200	28	28(21)	5.3(6.2)	27(20)	14	43	2.1	204
230	20	21(15)	4.3(5.2)	41(28)	9	34	1.5	400
270	12	12(8)	3.5(4.3)	61(41)	4	17	1.0	900
300	8	8(5)	2.8(3.6)	96(58)	2	12	0.6	2500

of this channel at the upgraded Tevatron. The signal rate for $gg \to H \to ZZ^{(*)} \to 4l$ as a function of the $4l$ mass is plotted in 5 GeV bins in Fig. 1, for several m_H values. For $m_H < 2m_Z$ there is no significant background so long as the two leptons that do not reconstruct to an on-shell Z are constrained to have significant mass.[6] Meanwhile, the continuum ZZ^* background, included in Fig. 1, quickly falls to a negligible level below $2m_Z$. Note that we have required that the charged leptons be isolated from each other and from other hadrons, so that the backgrounds from heavy quark decays, such as $Zb\bar{b}, Zc\bar{c}$, can be effectively removed. Table 2 presents the event rates for the $4l$ signal and the ZZ background, as well as the statistical significance. We see immediately that this clean channel suffers from the low event rate. For the best case of $m_H \sim 150$ GeV, there are only about 13 events altogether for $L = 30$ fb^{-1}. For another optimum choice of $m_H \sim 200$ GeV, $S/\sqrt{B} = 28/\sqrt{28} \sim 5$ is achieved for $L = 30$ fb^{-1}.

2.4. $gg \to H \to W^+W^- \to l\nu jj$ and $2l2\nu$ Channels

These two channels have sizeable event rates, typically of a few hundred to thousand events for 30 fb^{-1}, and may result in some statistical significance. However, due to the rather small signal-to-background ratio, typically of 1% to 10%, systematic uncertainties may prevent these channels from being useful. Better understanding of background normalization and distribution shapes is crucial in these cases.

3.   Discussions and Conclusions

We have seen that at low m_H, the $Wb\bar{b}$ mode provides clear signals, especially if double b-tagging is employed. However, this mode cannot be pushed beyond $m_H \sim 120$ GeV.

In the $ZZ^{(*)} \to 4l$ mode, the feasibility of detecting the H is clearly limited by the small event rate. It may be possible to detect the H in this mode with $m_H \sim 150$ GeV or $200 \lesssim m_H \lesssim 230 - 250$ GeV, given the cleanliness of this mode. The $H \to ZZ \to l\bar{l}\nu\bar{\nu}$ channel has a larger rate than the $4l$ mode by about a factor of 6, and the H signals exhibit a decent ($\gtrsim 5\sigma$) nominal statistical significance with respect to the ZZ

continuum background for $2m_Z \lesssim m_H \lesssim 250$ GeV. However, we have not evaluated the potentially large, but detector-dependent background, the QCD Z+hadrons. No conclusion can been drawn for this channel without a careful study of this background with full detector simulation.

It is instructive to compare our results at 4 TeV with those from a high luminosity Tevatron at 2 TeV. For the channel $W^{\pm}H \rightarrow l\nu b\bar{b}$, the signal significance at 4 TeV is only slightly larger than that at 2 TeV for a given integrated luminosity.[5] This is because there are gg initiated backgrounds that rise faster with increasing energy than the $q\bar{q}$ initiated signal. In contrast, for the gluon initiated signal processes, such as $gg \rightarrow H \rightarrow ZZ^{(*)}, WW$, the signal significance declines substantially in going from 4 TeV to 2 TeV (at fixed L). This is because the primary background sources are proportional to the $q\bar{q}$ luminosity functions which decrease less rapidly with decreasing energy than does the gg luminosity function to which the signals are proportional. Table 2 also presents the $L = 100$ fb^{-1} signal and background rates, the corresponding statistical significance, and the accumulated luminosity needed to observe 3σ effects at 2 TeV. Clearly very large integrated luminosities (requiring multiple years of operation at high instantaneous luminosity) would be needed to establish a SM Higgs signal in the $4l$ channel.

Two recent papers[7] have discussed the possibility of making use of τ-mode from $W^{\pm}H \rightarrow jj, \tau^+\tau^-$, but are not in agreement. Since the τ-pair events have not been experimentally studied in hadronic colliders, it would need some more work done, both theoretical and experimental, on this channel before one can draw further conclusion.

We have studied the ability of an upgraded Tevatron with $\sqrt{s} = 4$ TeV to search for a Standard Model Higgs boson with mass beyond the reach of LEP-200. Since such an upgrade would be most useful if detection were possible prior to the full luminosity operation of the LHC, we have employed an integrated luminosity of $L = 30$ fb^{-1} in our evaluations of discovery potential. To obtain potential signals for Higgs boson masses beyond the reach of LEP-200 will require multiple years of running at $L = 10$ fb^{-1} annual luminosity, even in the more favored $\lesssim 110 - 120$ GeV and $150 - 230$ GeV mass ranges. Overall, it is difficult to carry out a comprehensive search for the Higgs boson at a 4 TeV upgrade of the Tevatron.

References

1. For a recent review, see *e. g.*, A. Rubbia, p. 601, in *Physics and Experiments with Linear e^+e^- Colliders*, Waikoloa, Hawaii, eds. F. A. Harris *et al.* (1993).
2. J. Bagger, V. Barger, K. Cheung, J.F. Gunion, T. Han, G. Ladinsky, R. Rosenfeld, and C.P. Yuan, Phys. Rev. **D49**, 1246 (1994); M. Chanowitz and W. Kilgore, Phys. Lett. **B322**, 147 (1994).
3. G. Jackson, talk presented at the Workshop on Electroweak Symmetry Breaking and TeV-Scale Physics, UC–Santa Barbara, Feb., 1994.
4. J. F. Gunion and T. Han, UCD-94-10, to appear in Phys. Rev. D (1994).
5. A. Stange, W. Marciano, and S. Willenbrock, ILL-(TH)-94-8.
6. J. F. Gunion, G. Kane, and J. Wudka, Nucl. Phys. **B299**, 231 (1988).
7. S. Mrenna and G. L. Kane, CALT-68-1938; T. Kamon, J. L. Lopez, P. McIntyre, and J. T. White, CTP-TAMU-19/94.

DETECTING THE INTERMEDIATE-MASS NEUTRAL HIGGS BOSON AT THE LHC THROUGH $pp \to WH$

PANKAJ AGRAWAL and DAVID BOWSER-CHAO.[*]
Department of Physics & Astronomy, Michigan State University
East Lansing, Michigan 48824, USA

KINGMAN CHEUNG
Department of Physics & Astronomy, Northwestern University
Evanston, Illinois 60208, USA

and

DUANE A. DICUS
Center for Particle Physics, University of Texas at Austin
Austin, Texas 78712, USA

ABSTRACT

We examine the exclusive signature $pp \to WH \to \ell\nu\, b\bar{b}$ at the LHC. Although the backgrounds, principally arising from top production and Wjj, are quite severe, it is shown that judicious application of phase-space cuts and the use of b-tagging can in fact greatly enhance the detectability of this channel.

1. Introduction

Discovery of the intermediate mass Higgs boson is covered by LEP II from below (through $e^+e^- \to ZH$), and from above by the LHC (through $pp \to H \to ZZ^*$). The LHC strategy for the intermediate region ($100 \lesssim m_H \lesssim 120$ GeV) is to use the the rare decay mode $H \to \gamma\gamma$ along with very high di-photon mass resolution to beat down both the very large irreducible background as well as the even larger reducible background $pp \to \gamma + \text{jet}$. The signal/background ratios are quite low, so a thorough understanding of the systematic error in mass-resolution and jet/γ rejection is crucial.

As an alternative/complement to this search mode, we consider the primary decay mode of the Higgs boson, $H \to b\bar{b}$ produced in the process $pp \to WH \to \ell\nu b\bar{b}$. Even with tagging of the lepton, however, the QCD background $pp \to Wjj$ completely overwhelms the signal. With an efficient means of tagging b-quark jets and effectively rejecting light quark and gluon jets, this background could be cut down to the level of the much smaller subprocess $pp \to Wb\bar{b}$; after Higgs boson mass reconstruction, the latter is comparable to the signal.[1]

---

*Talk presented by David Bowser-Chao.

A potentially more serious threat comes from the expected copious production of the top quark, whose current lower mass limit[2] ($m_t > 131\,\text{GeV} > m_b + m_W$) implies a large source of both b-quarks and leptonically decaying W bosons. For this reason, we consider the *exclusive* production of $pp \to WH$ — by rejecting "extra" jets and leptons, we can greatly suppress backgrounds from single and pair production of top quarks[3]*.

2.   Backgrounds

The backgrounds considered include:

$$W jj \;\to\; \ell\nu jj \,, \tag{1}$$
$$W b\bar{b} \;\to\; \ell\nu b\bar{b} \,, \tag{2}$$
$$W Z \;\to\; \ell\nu b\bar{b} \,, \tag{3}$$
$$t\bar{t} \;\to\; (\ell\nu\ell\nu, \ell\nu jj) + b\bar{b} \,, \tag{4}$$
$$tbq \;\to\; \ell\nu b\bar{b} j \,, \tag{5}$$
$$tb \;\to\; \ell\nu b\bar{b} \,, \tag{6}$$
$$tq \;\to\; \ell\nu bj \tag{7}$$

An overview of the cuts made and their effect on the backgrounds are as follows (with ATLAS-inspired parameters):

Our primary weapon, of course, is b-tagging (through displaced vertices) and light quark/gluon rejection, which we depend on to beat down backgrounds (1) and (7). Reconstruction of the Higgs boson will reduce these and the rest of the backgrounds. This is especially true of background (3), whose significance at the upper end of the m_H range considered here will strongly depend on the detector resolution.

To reduce the top backgrounds (4–7), we essentially veto events with any "extra" particles, including jets or a second lepton. The second neutrino of $t\bar{t} \to WW b\bar{b} \to \ell\nu\ell\nu b\bar{b}$ is partially vetoed by a transverse mass cut, to require consistency with the decay of a W boson. A second e or jet is vetoed in the same way in the forward region. A second μ in the forward region cannot be vetoed, and in fact contributes to the apparent $\not{E}_T$.

3.   Results

For detector acceptance and resolution, we have modelled the ATLAS detector[4] on a parton level†. The $\not{E}_T$ for each process was constructed assuming a hermetic calorimeter, and adding the p_T of all jets and leptons (except for muons beyond the μ-detector) smeared as indicated:

$$\Delta E_{\text{had}}/E_{\text{had}} \;=\; \frac{50\%}{\sqrt{E_{\text{had}}}} \oplus 3\% \tag{8}$$

---

*This reference examines the same process considered here, but at the Tevatron (and its proposed upgrades in luminosity and/or energy) and at the LHC, assuming Tevatron-type acceptance and resolution.

†Except that the constant term in lepton resolution was taken to be twice as large.

$$\Delta E_\ell / E_\ell \;=\; \frac{10\%}{\sqrt{E_\ell}} \oplus 2\% \tag{9}$$

The detailed cuts are:

$$
\begin{aligned}
p_T(l) &> 20\,\text{GeV}, & |\eta(l)| &< 2.5 & e,\mu \text{ tagging}, & \tag{10}\\
p_T(b) &> 30\,\text{GeV}, & |\eta(b)| &< 2.0 & b\text{-tagging}, & \tag{11}\\
p_T(\mu) &> 10\,\text{GeV}, & |\eta(\mu)| &< 3.0 & \text{extra } \mu \text{ veto}, & \tag{12}\\
p_T(e) &> 10\,\text{GeV}, & |\eta(e)| &< 4.5 & \text{extra } e \text{ veto}, & \tag{13}\\
p_T(j) &> 10\,\text{GeV}, & |\eta(j)| &< 4.5 & \text{extra jet veto}, & \tag{14}\\
|m(b\bar b) - m_H| &< 7.5\,\text{GeV} & & & m_H \text{ reconstruction}, & \tag{15}\\
m_T(\ell, \not{p}_T) &< 80\,\text{GeV} & & & W\text{-boson consistency cut} & \tag{16}
\end{aligned}
$$

The b-tagging efficiency is assumed to be 30%; the mistagging rate for light quarks/gluons is taken to be 1%.

Total event rates with all the above cuts (for 10 fb^{-1} of integrated luminosity) are provided in Table 1, for a range of m_t and m_H. The last row gives the number of years at $\int \mathcal{L}\,dt = 10$ fb^{-1} per year for 5σ detection. Because of the light quark/gluon jet rejection, the principal background is process (2), while the combined top quark backgrounds account for roughly half. The signal falls from 67.2 ($m_H = 100$ GeV) to 40 events ($m_H = 120$ GeV) principally because of the falling BR($H \to b\bar b$). For $m_t = 175$ GeV, approximately the central value of CDF's search for the top quark,[5] detection at the 5σ level would require around 2-4 years for the range of m_H considered here.

As noted above, without a factor of 10^{-4} suppression of background (1), this process would have swamped the signal. Increasing the mistagging rate would quadruple and double processes (1) and (5) respectively. Similarly, if jet reconstruction has an appreciably lower efficiency around 10 GeV for the forward compared to the central region, backgrounds (4–5) will grow in relative importance.

4.   Improving the S/B ratio

We note that it is possible, by employing certain cuts, to increase the S/B ratio at the cost of a small decrease in significance. This may be a price worth paying, since significance is calculated here using only statistical uncertainty, and in the end, results may well be dominated by systematic errors in measuring the $m(b\bar b)$ distribution.

Top reconstruction (and vetoing) is an obvious strategy to try, given that processes (4–7) make up around 1/3 to 1/2 of the total background. Another strategy is to consider topological cuts. In figures 1 and 2 are shown two distributions that can be helpful in increasing the S/B ratio by specifically reducing the top backgrounds.

Backgrounds (1–3) and backgrounds (4–7), respectively, fall in the same way for large $\sum |p_T(b)|$ and small $\Delta R(b, \bar b)$. For example, for $m_t = 150$ GeV, $m_H = 100$ GeV, a cut of $\Delta R(b, \bar b) < 1.6$ yields about 17 signal events (for 10 fb^{-1}), a total of 30 events from the first set of backgrounds, and 5 from the top-quark production background; the S/B ratio is increased from 1/5 to 1/2, while the significance falls from 3.4 to 2.9. A more complete appraisal of this and other issues raised above is given elsewhere.[6]

5. Acknowlegements

We would like to thank A. Stange and S. Willenbrock for discussion of their work. This work was supported in part by the National Science Foundation under Grant No. PHY-9307980

Processes	$m_t = 150$ GeV		$m_t = 175$ GeV		$m_t = 200$ GeV	
	m_H 100 GeV	m_H 120 GeV	m_H 100 GeV	m_H 120 GeV	m_H 100 GeV	m_H 120 GeV
WH	67.2	40.0	67.2	40.0	67.2	40.0
Wjj	44.4	41.1	44.4	41.1	44.4	41.1
Wbb	113.6	82.2	113.6	82.2	113.6	82.2
WZ	45.8	1.6	45.8	1.6	45.8	1.6
tt	73.6	72.3	31.9	33.2	12.9	14.8
tbq	44.6	43.6	39.6	36.6	34.6	35.4
tb	44.8	43.4	23.6	25.8	12.6	15.0
tq	22.7	24.7	16.0	19	10.3	13.3
S/B	67/389	40/309	67/315	40/240	67/274	40/203
Signif.	3.4	2.3	3.8	2.6	4.0	2.8
$\int \mathcal{L}dt$ (10 fb^{-1}) for 5σ signif.	2.2	4.7	1.7	3.7	1.6	3.2

Table 1. Event rates (for 10 fb^{-1}) at the LHC, $\sqrt{s} = 14$ TeV with the kinematic cuts specified in the text. The b-tagging efficiency/mistagging rates are taken to be 30% and 1% respectively.

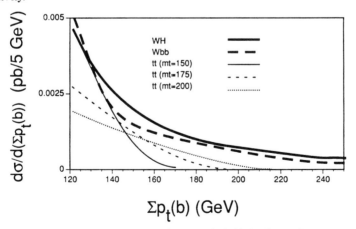

Fig. 1. Differential cross-section $d\sigma/d\sum_{i=b,\bar{b}} |p_t(b_i)|$ (pb/5 GeV).

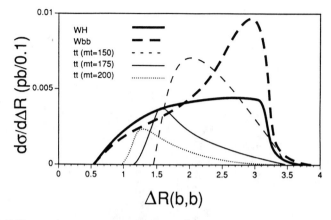

Fig. 2. Differential cross-section $d\sigma/d\Delta R(b,\bar{b})$ (pb/0.1).

References

1. P. Agrawal and S.D. Ellis, *Phys. Lett.* **B229** (1989) 2015.
2. S. Abachi, et al., D0 Collaboration, Phys.Rev.Lett.**72**, (1994) 2138.
3. A. Stange, W. Marciano, and S. Willenbrock, ILL-TH-94-8, and S. Willenbrock (in these proceedings).
4. ATLAS Letter of Intent, CERN/LHCC/92-4 (1992).
5. F. Abe et al., CDF Collaboration, FERMILAB Pub-94/097-E
6. P. Agrawal, D. Bowser-Chao, K. Cheung, and D. Dicus, MSU-940829.

HIGGS DECAY TO BOTTOM QUARKS
AT FUTURE HADRON COLLIDERS

A. STANGE and W. MARCIANO

Physics Department, Brookhaven National Laboratory, Upton, NY 11973

and

S. WILLENBROCK*

Department of Physics, University of Illinois, 1110 W. Green St., Urbana, IL 61801

ABSTRACT

We consider the search for the Higgs boson at a high-luminosity Fermilab Tevatron ($\sqrt{s} = 2$ TeV), an upgraded Tevatron of energy $\sqrt{s} = 3.5$ TeV, and the CERN Large Hadron Collider (LHC, $\sqrt{s} = 14$ TeV), via WH production followed by $H \to b\bar{b}$ and leptonic decay of the weak vector boson. We show that each of these colliders can potentially observe the standard Higgs boson in the intermediate-mass range, 80 GeV $< m_H <$ 120 GeV. This mode complements the search for and the study of the intermediate-mass Higgs boson via $H \to \gamma\gamma$ at the LHC. In addition, it can potentially be used to observe the lightest Higgs scalar of the minimal supersymmetric model, h, in a region of parameter space not accessible to CERN LEP II or the LHC (using $h \to \gamma\gamma, ZZ^*$).

The intermediate-mass Higgs boson, 80 GeV $< m_H <$ 120 GeV, is elusive. It is too heavy to be produced at LEP II, and can be discovered at the LHC using the rare two-photon decay mode only if excellent photon energy and angular resolution can be maintained while running at full luminosity ($\mathcal{L} \approx 10^{34}/cm^2/s$). In a recent paper, we have shown that Higgs-boson production in association with a weak vector boson, followed by $H \to b\bar{b}$ and leptonic decay of the weak vector boson, can potentially be used to observe the standard Higgs boson in the intermediate-mass range at future hadron colliders.[1] The machines we considered are the Fermilab Tevatron ($\sqrt{s} = 2$ TeV) with high luminosity ($\mathcal{L} \geq 10^{33}/cm^2/s$), an upgraded Tevatron of energy $\sqrt{s} = 3.5$ TeV with high luminosity, and the LHC ($\sqrt{s} = 14$ TeV). This mode has also recently been studied at a 4 TeV $p\bar{p}$ collider in Ref. 2, and at the Tevatron, a 4 TeV pp collider, and the LHC in Ref. 3. These analyses are in agreement with ours. Other work in progress on this mode at the LHC was report at this conference.[4]

Associated production of Higgs and weak vector bosons, with $H \to b\bar{b}$, was considered at the Tevatron by us in a previous study.[5] The analysis in our recent paper closely follows that work. The weak bosons are detected via their leptonic decays ($W \to \ell\nu$).

---

*Presenter

The cuts made to simulate the acceptance of the detector are listed in Table 1. The jet energy resolution is taken to be $\Delta E_j/E_j = 0.80/\sqrt{E_j} \oplus 0.05$, which corresponds to a two-jet invariant-mass resolution of about $\Delta M_{jj}/M_{jj} = 0.80/\sqrt{M_{jj}} \oplus 0.03$ (added in quadrature).* We integrate the background over an invariant-mass bin of $\pm 2\Delta M_{jj}$; for $M_{jj} = 100$ GeV, this amounts to a bin width of 34 GeV. A change from our previous analysis is that we reduce the p_T threshold for observing charged leptons to $p_T > 10$ GeV (although we continue to trigger on leptons of $p_T > 20$ GeV). This is important for rejecting the top-quark background to the WH signal. We also extend the coverage for jets out to a rapidity of 4, for the same reason.

Table 1. Acceptance cuts used to simulate the detector. The $p_{T\ell}$ threshold is greater for charged leptons which are used as triggers (in parentheses).

$$
\begin{array}{ll}
|\eta_b| < 2 & p_{Tb} > 15 \text{ GeV} \\
|\eta_\ell| < 2.5 & p_{T\ell} > 10 \text{ GeV (20 GeV)} \\
|\eta_j| < 4 & p_{Tj} > 15 \text{ GeV} \\
|\Delta R_{b\bar{b}}| > 0.7 & |\Delta R_{b\ell}| > 0.7 \\
\not{p}_T > 20 \text{ GeV} & (\text{for } W \to \ell\bar{\nu})
\end{array}
$$

The cross sections for the signals and backgrounds at the Tevatron, the 3.5 TeV $p\bar{p}$ collider, and the LHC, including all branching ratios and acceptances, are shown in Fig. 1. We list in Table 2 the number of signal and background events per 10 fb^{-1} of integrated luminosity, with a double b tag, at the Tevatron, the 3.5 TeV $p\bar{p}$ collider[†], and the LHC, for a variety of Higgs-boson masses. The statistical significance of the signal, $S/\sqrt{B}$, is listed in the last column. If we define discovery of the WH signal by the criterion of a 5σ significance, then with 30 fb^{-1} of integrated luminosity the reach of the Tevatron is about $m_H = 95$ GeV, the reach of the 3.5 TeV $p\bar{p}$ collider about 100 GeV, and the reach of the LHC also about 100 GeV. The largest background at the Fermilab machines is $Wb\bar{b}$, while at the LHC it is $t\bar{t}$ and $Wg \to t\bar{b}$. The $t\bar{t}$ cross sections in Fig. 1 represent a reduction of the $t\bar{t}$ cross section, before rejecting the additional W boson but with acceptance cuts, by a factor of about 1/35 at the Fermilab machines, and about 1/25 at the LHC. If the rejection could be improved at the LHC, it would increase the significance of the signal. The corresponding rejection factor for the extra jet in the $Wg \to t\bar{b}$ process is only a factor of about 1/3. If the misidentification probability of a light-quark jet as a b jet is increased from 1% to 2%, the Wjj background quadruples, but since it is not the largest background at any of the machines, the statistical significance of the WH signal drops by only about 15%.

---

*The two-jet invariant-mass resolution may be degraded somewhat due to semileptonic b decays, which occur in 40% of the events.

[†]At a 4 TeV $p\bar{p}$ collider, the WH signal and the backgrounds $Wb\bar{b}$ and WZ increase by about 15% from the 3.5 TeV machine, increasing the significance of the signal by roughly 7%.

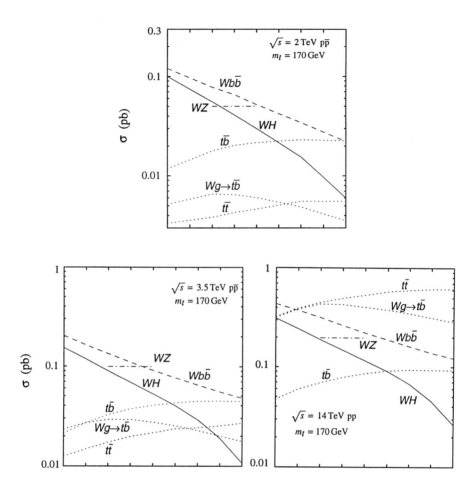

Fig. 1. Cross sections and backgrounds at the Tevatron ($\sqrt{s} = 2$ TeV), the $\sqrt{s} = 3.5$ TeV $p\bar{p}$ collider, and the LHC ($\sqrt{s} = 14$ TeV pp), for WH production followed by $H \to b\bar{b}$ and $W \to \ell\bar{\nu}$, versus the Higgs-boson mass. The cuts made to simulate the acceptance of the detector are listed in Table 1. The backgrounds are from $Wb\bar{b}$, WZ followed by $Z \to b\bar{b}$, $t\bar{t} \to W^{+}W^{-}b\bar{b}$ with one W missed, $Wg \to t\bar{b} \to Wb\bar{b}$, and $q\bar{q} \to t\bar{b} \to Wb\bar{b}$. The top-quark mass is taken to be 170 GeV.

Table 2. Number of signal and background events at the Tevatron, the $\sqrt{s} = 3.5$ TeV $p\bar{p}$ collider, and the LHC, per 10 fb^{-1} of integrated luminosity, for production of the Higgs boson in association with a weak vector boson, followed by $H \rightarrow b\bar{b}$ and $W \rightarrow \ell\bar{\nu}$. The statistical significance of the signal, $S/\sqrt{B}$, is listed in the last column. The cuts made to simulate the acceptance of the detector are listed in Table 1. We assume a 30% efficiency for detecting a secondary vertex per b jet, within the rapidity coverage of the vertex detector and with $p_{Tb} > 15$ GeV. We demand that both b jets be identified. We also assume a 1% misidentification of light-quark and gluon jets as a b jet. The $t\bar{t} \rightarrow b\bar{b}W^+W^-$ background to the WH signal is reduced by rejecting events with an additional W boson, which is identified either via a charged lepton, a jet of $p_T > 30$ GeV, or two jets with $p_T > 15$ GeV. The $Wg \rightarrow t\bar{b}$ background is reduced by rejecting events with an additional jet of $p_T > 30$ GeV.

	Tevatron	2 TeV	$p\bar{p}$					
m_H (GeV)	WH	$Wb\bar{b}$	WZ	Wjj	$t\bar{t}$	$Wg \rightarrow t\bar{b}$	$q\bar{q} \rightarrow t\bar{b}$	Signif.
60	90	108	-	17	3	4.7	11	7.5
80	50	70	30	13	3	5.9	16	4.3
90	37	59	45	11	4	5.8	18	3.1
100	27	45	37	9	4	5.4	20	2.5
120	14	31	3	6.5	5	4.4	21	1.7
140	5.4	20	-	4.5	5	3.2	20	0.74

		3.5 TeV	$p\bar{p}$					
60	138	183	-	43	11	21	19	8.3
80	81	121	59	33	16	26	29	4.8
90	61	99	89	30	18	26	33	3.6
100	47	82	73	26	19	25	36	2.9
120	26	58	6	19	22	20	40	2.0
140	10	43	-	15	24	16	40	0.83

	LHC	14 TeV	pp					
60	281	394	-	186	291	286	42	8.5
80	170	285	119	169	407	392	62	4.6
90	132	237	178	159	448	376	70	3.4
100	104	199	146	142	481	364	77	2.8
120	60	145	12	115	529	312	84	1.7
140	24	110	-	94	552	257	83	0.72

If integrated luminosities in excess of 30 fb^{-1} can be achieved, one can observe Higgs bosons of greater mass. With 100 fb^{-1}, the reach of the Tevatron is about 120 GeV, the 3.5 TeV $p\bar{p}$ collider about 125 GeV, and the LHC about 120 GeV. Thus each machine is potentially capable of covering the intermediate-mass region. However, b-tagging with high efficiency and purity, which is essential to the extraction of the signal, may prove to be difficult at higher instantaneous luminosities. With a bunch spacing of 20 ns, a luminosity of $10^{33}/cm^2/s$ yields about 1.5 interactions per bunch crossing, which is acceptable for tagging secondary vertices with the SVX. However, with the same bunch spacing, a luminosity of $10^{34}/cm^2/s$ yields about 15 interactions per bunch crossing. It is not known how successful b-tagging with the SVX will be in an environment with many interactions per bunch crossing.

If the coupling of the Higgs boson to bottom quarks is enhanced, the decay of the intermediate-mass Higgs boson to two photons can be suppressed such that it is unobservable at the LHC. This can occur for the light Higgs scalar, h, of the minimal supersymmetric model, whose mass lies within or below the intermediate-mass region if the top quark and the top squarks are not too heavy. The enhanced coupling of h to bottom quarks increases the mass at which the branching ratio of $h \to b\bar{b}$ begins to fall off due to the competition from $h \to WW^{(*)}$. We showed in Ref. 1 that the mode Wh, with $h \to b\bar{b}$, can potentially cover part of the parameter space of the minimal supersymmetric model not accessible to LEP II or the LHC (using $h \to \gamma\gamma, ZZ^*$). It is also possible that the heavy Higgs scalar of the minimal supersymmetric model, H, can be observed via WH production over part of the parameter space.

References

1. A. Stange, W. Marciano, and S. Willenbrock, Phys. Rev. D **50**, 4491 (1994).
2. J. Gunion and T. Han, UCD-94-10 (1994).
3. S. Mrenna and G. Kane, CALT-68-1938 (1994).
4. P. Agrawal, D. Bowser-Chao, K. Cheung, and D. Dicus, these proceedings.
5. A. Stange, W. Marciano, and S. Willenbrock, Phys. Rev. D **49**, 1354 (1994).

STRUCTURE FUNCTIONS AND DISTRIBUTIONS IN SEMILEPTONIC TAU DECAYS

J. H. KÜHN

Institut für Theoretische Teilchenphysik, Universität Karlsruhe, Kaiserstr.12, 76128 Karlsruhe, Germany.

and

E. MIRKES*

Physics Department, University of Wisconsin, Madison, WI 53706, USA

ABSTRACT

Semileptonic decays of polarized τ leptons are investigated. The most general angular distribution of three meson final states ($\tau \to \pi\pi\nu$, $K\pi\nu$, $\pi\pi\pi\nu$, $K\pi\pi\nu$, $KK\pi\nu$, $KKK\nu$, $\eta\pi\pi\nu$, ...) is discussed. It is shown, that the most general distribution can be characterized by 16 structure functions, most of which can be determined in currently ongoing high statistics experiments. Emphasis is put on τ decays in e^+e^- experiments where the neutrino escapes detection and the τ rest frame cannot be reconstructed. The structure of the hadronic matrix elements, based on CVC and chiral lagrangians, is discussed.

With the experimental progress in τ-decays an ideal tool for studying strong interaction physics has been developed. In this paper we show, that detailed informations about the hadronic charged current for the decay into three pseudoscalar mesons can be derived from the study of angular distributions. Consider the semileptonic τ-decay

$$\tau(l,s) \to \nu(l',s') + h_1(q_1,m_1) + h_2(q_2,m_2) + h_3(q_3,m_3) , \tag{1}$$

where $h_i(q_i,m_i)$ are pseudoscalar mesons. The matrix element reads as

$$\mathcal{M} = \frac{G}{\sqrt{2}} \left(_{\sin\theta_c}^{\cos\theta_c}\right) M_\mu J^\mu , \tag{2}$$

with G the Fermi-coupling constant. The cosine and the sine of the Cabibbo angle (θ_C) in (2) have to be used for Cabibbo allowed $\Delta S = 0$ and Cabibbo suppressed $|\Delta S| = 1$ decays, respectively. The leptonic (M_μ) and hadronic (J^μ) currents are given by $M_\mu = \bar{u}(l',s')\gamma_\mu(g_V - g_A\gamma_5)u(l,s)$ and $J^\mu(q_1,q_2,q_3) = \langle h_1(q_1)h_2(q_2)h_3(q_3)|V^\mu(0) - A^\mu(0)|0\rangle$. V^μ and A^μ are the vector and axial vector quark currents, respectively. The most

---

*Presented by E. Mirkes.

general ansatz for the matrix element of the quark current J^μ is characterized by four formfactors[1]

$$J^\mu(q_1, q_2, q_3) = V_1^\mu F_1 + V_2^\mu F_2 + i V_3^\mu F_3 + V_4^\mu F_4 , \tag{3}$$

with

$$
\begin{aligned}
V_1^\mu &= q_1^\mu - q_3^\mu - Q^\mu \frac{Q(q_1 - q_3)}{Q^2} , \\
V_2^\mu &= q_2^\mu - q_3^\mu - Q^\mu \frac{Q(q_2 - q_3)}{Q^2} , \\
V_3^\mu &= \epsilon^{\mu\alpha\beta\gamma} q_{1\alpha} q_{2\beta} q_{3\gamma} , \\
V_4^\mu &= q_1^\mu + q_2^\mu + q_3^\mu = Q^\mu .
\end{aligned}
\tag{4}
$$

The formfactors F_1 and $F_2(F_3)$ originate from the axial vector hadronic current (vector current) and correspond to spin 1, whereas F_4 is due to the spin zero part of the axial current matrix element. The formfactors F_1 and F_2 can be predicted by chiral lagrangians, supplemented by informations about resonance parameters. Parametrizations for the 3π final states can be found in refs. [1-3]. In this case, only the axial vector current formfactors F_1 and F_2 contribute due to the G parity of the pions. The 3π decay mode offers a unique tool for the study of ρ, ρ' resonance parameters competing well with low energy $e^+ e^-$ colliders with energies in the region below 1.7 GeV. As we will see later, the two body (ρ and ρ') resonances can be fixed by taking ratios of hadronic structure functions, whereas the measurement of four structure functions can be used to put constraints on the a_1 parameters. The decay modes involving different mesons (for example $K\pi\pi$, $KK\pi$ or $\eta\pi\pi$) allow for axial and vector current contributions at the same time. Explicit parametrizations for the form factors in these decay modes are presented in refs. [4,5]. The vector formfactor F_3 is related to the Wess-Zumino anomaly,[6] whereas the axial vector form factors are again predicted by chiral lagrangians. The latter decay modes allow also for the study of $J^{PC} = 0^{-+}$ and $J^{PC} = 1^{++}$ resonances which are not directly accessible from other experiments.

Let us now introduce the formalism of the hadronic structure functions. The differential decay rate is obtained from

$$d\Gamma(\tau \to \nu_\tau 3h) = \frac{1}{2m_\tau} \frac{G^2}{2} \binom{\cos^2 \theta_c}{\sin^2 \theta_c} \{ L_{\mu\nu} H^{\mu\nu} \} \, dPS^{(4)} , \tag{5}$$

where $L_{\mu\nu} = M_\mu (M_\nu)^\dagger$ and $H^{\mu\nu} \equiv J^\mu (J^\nu)^\dagger$. The considered decays (1) are most easily analyzed in the hadronic rest frame $\vec{q}_1 + \vec{q}_2 + \vec{q}_3 = \vec{Q} = 0$. The orientation of the hadronic system is characterized by three Euler angles (α, β and γ) as introduced in refs. [1,3]. Performing the analysis of $\tau \to \nu_\tau + 3$ mesons in the hadronic rest frame has the advantage that the product of the hadronic and the leptonic tensors reduce to a sum[1]

$$L^{\mu\nu} H_{\mu\nu} = \sum_X L_X W_X . \tag{6}$$

In fact in this system the hadronic tensor $H^{\mu\nu}$ is decomposed into 16 hadronic structure functions W_X corresponding to 16 density matrix elements for a hadronic system in a spin one $[V_1^\mu, V_2^\mu, V_3^\mu]$ and spin zero state $[V_4^\mu]$ (nine of them originate from a pure spin one and the remaining are pure spin zero or interference terms). The 16 structure functions contain the dynamics of the three meson decay and depend only on the hadronic invariants Q^2 and the Dalitz plot variables s_i. The leptonic factors L_X contain the dependence on the Euler angles, (which determine the orientation of the hadronic system), on the τ polarization, on the chirality parameter γ_{VA} and on the total energy of the hadrons in the laboratory frame as well. Analytical expressions for the 16 coefficients L_X are first presented in ref. [1]. They can also be applied to the case, where the τ rest frame cannot be reconstructed because of the unknown neutrino momentum. The dependence of these coefficients on the τ polarization allow for an improved measurement of the τ polarization at LEP [see for example refs. [7,8] and references therein].

The hadronic structure functions W_X on the other hand contain the full dynamics of the hadronic decay and a measurement of these structure functions provide a unique tool for low enery hadronic physics. They can be calculated from a decomposition of the hadronic matrix element J^μ and can be expressed in terms of the form factors F_i. We list here only the result for the pure spin one state.

$$
\begin{aligned}
W_A &= (x_1^2 + x_3^2)\,|F_1|^2 + (x_2^2 + x_3^2)\,|F_2|^2 + 2(x_1 x_2 - x_3^2)\,\mathrm{Re}\,(F_1 F_2^*)\ , \\[4pt]
W_B &= x_4^2|F_3|^2\ , \\[4pt]
W_C &= (x_1^2 - x_3^2)\,|F_1|^2 + (x_2^2 - x_3^2)\,|F_2|^2 + 2(x_1 x_2 + x_3^2)\,\mathrm{Re}\,(F_1 F_2^*)\ , \\[4pt]
W_D &= 2\left[x_1 x_3\,|F_1|^2 - x_2 x_3\,|F_2|^2 + x_3(x_2 - x_1)\,\mathrm{Re}\,(F_1 F_2^*)\right]\ , \\[4pt]
W_E &= -2x_3(x_1 + x_2)\,\mathrm{Im}\,(F_1 F_2^*)\ , \qquad\qquad\qquad\qquad\qquad\quad (7) \\[4pt]
W_F &= 2x_4\left[x_1\,\mathrm{Im}\,(F_1 F_3^*) + x_2\,\mathrm{Im}\,(F_2 F_3^*)\right]\ , \\[4pt]
W_G &= -2x_4\left[x_1\,\mathrm{Re}\,(F_1 F_3^*) + x_2\,\mathrm{Re}\,(F_2 F_3^*)\right]]\ , \\[4pt]
W_H &= 2x_3 x_4\left[\,\mathrm{Im}\,(F_1 F_3^*) - \mathrm{Im}\,(F_2 F_3^*)\right]\ , \\[4pt]
W_I &= -2x_3 x_4\left[\,\mathrm{Re}\,(F_1 F_3^*) - \mathrm{Re}\,(F_2 F_3^*)\right].
\end{aligned}
$$

The remaining structure functions originating from a possible (small) contribution from a spin zero state are presented in [1]. The variables x_i are defined by $x_1 = V_1^x = q_1^x - q_3^x$, $x_2 = V_2^x = q_2^x - q_3^x$, $x_3 = V_1^y = q_1^y = -q_2^y$, $x_4 = V_3^z = \sqrt{Q^2} x_3 q_3^x$, where q_i^x (q_i^y) denotes the x (y) component of the momentum of meson i in the hadronic rest frame. The structure functions can be extracted by taking suitable moments with respect to an appropriate product of two Euler angles.[1] An alternative method to extract the structure functions can be achieved by a direct fit to the expressions (7).[8]

As an example, we will now present numerical results for the non vanishing structure functions W_A, W_C, W_D and W_E in the 3π decay mode, which originates from the spin one part of the hadronic current. Figure 1 shows predictions for the structure function ratios w_C/w_A, w_D/w_A and w_E/w_A as a function of Q^2, where we have integrated over the Dalitz plot variables s_i [The integrated structure functions are denoted

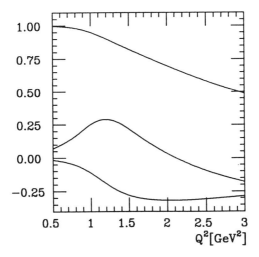

Fig. 1. Ratio of the spin one hadronic structure functions $w_C/w_A, w_D/w_A, w_E/w_A$ (from top to bottom) for $\tau \to \nu\pi\pi\pi$ as a function of Q^2.

by lower case letter w_X]. The results are based on the same parametrization of the formfactors as used in [1]. Although we have lost informations on the resonance parameters in the two body decays by integrating over s_1 and s_2, we observe interesting structures*. One observes that all normalized structure functions are sizable. w_C/w_A approaches its maximal value 1 for small Q^2. Note, that the dependence on the a_1 mass and width parameters cancel in the ratio w_X/w_A in fig. 1. On the other hand, the Q^2 distributions of the structure functions $w_{A,C,D,E}$ presented in fig. 2 are very sensitive to the a_1 parameters. As an example, fig. 2 shows predictions for the structure functions, where two different values for the a_1 width has been used. Therefore, the ratios in fig. 1 can be used to fix the model dependence in the two body resonances, whereas the structure functions itself put then rigid contraints on the a_1 parameters. It is therefore possible to test the hadronic physics in much more detail than it is possible by rate measurements alone.

The technique of structure functions also allow for a model independent test of the presence of spin zero components in the hadronic current. Such a contribution would lead to additional structure functions to (7) originating from the interference with the (large) spin one contributions.[1]

A detailed discussion of the matrix elements for the decay modes involving different pseudo scalar mesons [$K\pi\pi\nu$, $KKK\nu$, $\eta\pi\pi\nu$] together with predictions for the corresponding structure functions and angular distributions is presented in ref. [4]. In

---

*More contsraints on the two body resonances can be obtained by analyzing the full dependence on Q^2 and s_i, which should be accessible with the present high statistic experiments.

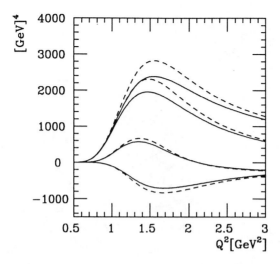

Fig. 2. Spin one hadronic structure functions w_A, w_C, w_D, w_E (from top to bottom) for $\tau \to \nu\pi\pi\pi$ as a function of Q^2. Results are shown for two sets of a_1 parameters: $m_{a_1} = 1.251$ GeV, $\Gamma_{a_1} = 0.599$ GeV (solid) and $m_{a_1} = 1.251$ GeV, $\Gamma_{a_1} = 0.550$ GeV (dashed)

this case, all 9 structure functions in (7) are nonvanishing because of the interference of the anomaly with the axial vector contributions. An analyses of these distributions would allow to test the underlying hadronic physics [like the different contributions from the axial and vector (Wess-Zumino anomaly) current] in detail. It is thus possible to confirm (not only qualitatively) the presence of the Wess-Zumino anomaly in the decays modes $\tau \to \nu K\pi\pi$ and $\tau \to \nu K\pi\pi$.

In ref. [9], the technique of the structure functions has been extended to the $\tau \to \nu\omega\pi$ decay mode. This allows to test the model for the hadronic matrix element in this decay mode, which involves both a vector and a second class axial vector current.

References

1. J.H. Kühn and E. Mirkes, Z.Phys.**C56**, 661, (1992).
2. J.H. Kühn and A. Santamaria, Z. Phys. **C48**, 445, (1990).
3. J.H. Kühn and E. Mirkes, Phys. Lett. **B286**, 381, (1992).
4. R. Decker and E. Mirkes, Phys. Rev. **D47**, 4012 (1993).
5. R. Decker, E. Mirkes, R. Sauer and Z. Was, Z. Phys. **C58**, 445, (1993).
6. J. Wess and B. Zumino, Phys. Lett. **B37**, 95, (1971).
7. P. Privitera, Phys. Lett. **B318**, 249, (1993).
8. J. Shukla, these proceedings.
9. R. Decker and E. Mirkes, Z.Phys. **C57**, 495, (1993).

EXPONENTIATION OF SOFT PHOTONS IN THE PROCESS
$e^+e^- \to \mu^+\mu^- + n$ hard photons

DAVID J. SUMMERS

Department of Physics, University of Wisconsin – Madison,
1150 University Avenue, Madison, WI 53706, U.S.A.

ABSTRACT

We present a method of exponentiating soft photons in a process that involves hard photon emission, and apply this to the process $e^+e^- \to \mu^+\mu^- + n$ hard photons. In this we arrive at a consistent description of both soft and hard photon radiation simultaneously. The effect of exponentiation is to remove the singularities associated with soft photon emission, and as such exponentiation suppresses the soft photon cross-section over the leading order prediction. This can be of important experimentally where often the theoretical predictions for hard photon emission are normalised to the experimental results through the soft photon event rates.

At leading order (LO) in perturbation theory radiation off charged particles shows divergences as the radiation becomes soft. This means that for low energy radiation LO cross-sections grow without limit; and this means that at some low energy a purely LO description breaks down. For example, if we consider n soft photons $(k_1, \ldots, k_n)$ radiated off a $\mu^-(p_1)\mu^+(p_2)$ pair both the matrix element and the phase space factorise into a product of 1 photon factors multiplied by the lowest order matrix element, this gives the cross-section to produce n photons with energy between E and $E_{\min}$ as,

$$d\sigma_n = \frac{1}{\text{Flux}} |\mathcal{M}_0|^2 \, d(\text{LIPS})_2 \; \frac{1}{n!} \prod_{i=1}^{n} -e^2 \left(\frac{p_1}{p_1 \cdot k_i} - \frac{p_2}{p_2 \cdot k_i} \right)^2 \; \prod_{i=1}^{n} \frac{E_i^2 \, dE_i \, d\Omega_i}{2E_i(2\pi)^3} \quad (1)$$

$$= d\sigma_0 \frac{1}{n!} \left(\frac{\alpha}{4\pi^2} \ln\left(\frac{E}{E_{\min}}\right) g(\Omega_c) \right)^n \quad (2)$$

where $g(\Omega_c) = -\int_{\Omega_c} E^2 (p_1/p_1 \cdot k - p_2/p_2 \cdot k)^2$. The $1/n!$ term is the symmetry factor arising from integrating the n photons over the same phase space. It is clear that as $E_{\min}$ tends to zero the cross-section diverges to $+\infty$ logrithmically. Now this divergence is canceled by the virtual diagrams, up to finite corrections the cross-section to produce a photon with energy less the E is given by the same equation 1, replacing $E_{\min}$ with E_{reg}, where E_{reg} is constrained to values given by $\ln(E_{\max}/E_{\text{reg}}) = \mathcal{O}(1)$. In this form the total n photon cross-section is finite, however the cross-section for soft photons is still badly behaved. For photon energies less than E_{reg} the cross-section goes negative, and eventually diverges to $-\infty$.

Now these problems with various cross-sections going to $\pm\infty$ can not be solved at any finite order in perturbation theory, however for soft photons we can resum the n photon cross-sections to infinite order in perturbation theory. If we ask for the inclusive cross-section for $\mu^-\mu^+$ production where we observe no photons with energy greater than E we find,[1]

$$d\sigma^{\mathrm{exp}} = d\sigma_0 + d\sigma_1 + d\sigma_2 + \cdots \quad = d\sigma_0 \left(\frac{E}{E_{\mathrm{reg}}}\right)^{(\alpha g(\Omega_c)/4\pi^2)} \tag{3}$$

To form the more exclusive cross-section where we observe 1 or more photons we can differentiate 1 or more times to obtain the photon spectrum. This description is well behaved as the photon energy goes to zero and is the most basic form of exponentiation.

Now for exponentiation to work we require both the matrix element and the phase space to factorise into a product of 1 photon terms. Unfortunatly this does not happen for hard photons, and this means that hard photons can not be exponentiated like soft photons. However if we take a process involving hard photons then additional soft photons do still factorise both the hard photon matrix element and phase space exactly as in the case with no hard photons. This means that we can exponentiate just the soft photons, and calculate the hard photons using exact hard matrix elements and phase space. As before the differential cross-section for n hard photons, and m soft photons can be written as the differential cross-section for n hard photons multiplied by the factor $\frac{1}{m!}\left(\frac{\alpha}{4\pi^2}\ln\left(\frac{E}{E_{\mathrm{reg}}}\right)g(\Omega_c)\right)^m$; where we have the symmetry factor of $1/m!$ if and only if the soft photons are distinguishable from the hard photons. This can be imposed by forcing the soft and hard photons into mutually exclusive areas of phase space, we do this by insisting that the hard photons have an energy larger than E_{cut} and the soft photons have energy less than E_{cut}, i.e. $E_{\mathrm{hard}} > E_{\mathrm{cut}} > E_{\mathrm{soft}}$. If we now sum over the unobserved soft photons we arrive at an exponentiated hard photon cross-section,

$$d\sigma_n^{\mathrm{exp}} = d\sigma_n \left(\frac{E_{\mathrm{cut}}}{E_{\mathrm{reg}}}\right)^{(\alpha g(\Omega_c)/4\pi^2)} \tag{4}$$

In this form we can exponentiate soft photon radiation, while still treating hard photon radiation exactly. First we generate n hard photons with energy larger than E_{cut} and calculate the exact hard matrix element, then we weight this by $(E_{\mathrm{cut}}/E_{\mathrm{reg}})^{(\alpha g(\Omega_c)/4\pi^2)}$ and integrate over the hard photon phase space.

By using this prescription we have apparently introduced two extra parameters, E_{reg} and E_{cut}, over the LO calculation. Now E_{reg} is introduced to regulate the soft photon singularity, and as without calculating the virtual diagrams we do not know how much of the singularity cancels, the value of E_{reg} is unknown beyond $\ln(E_{\mathrm{max}}/E_{\mathrm{reg}}) = \mathcal{O}(1)$. Thus E_{reg} is a reflection of our uncertainty from working at LO.

The dependence upon E_{cut} is more worrying, as E_{cut} was introduced purely as a parameter to distinguish "hard" photons from "soft" photons, however there no physical meaning to these words. Hence if this procedure is meaningful our results should be independent of the value of E_{cut}. Consider what happens when we decrease E_{cut}, the hard cross-section grows as we approach the soft singularity, however this is

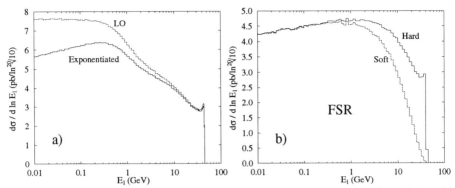

Fig.1 Comparisons between the exponentiated hard cross-section and the LO prediction (a) and exponentiated soft approximation (b). In both graphs we set $\sqrt{s} = M_Z$ and $E_{\text{reg}} = 10$ GeV.

compensated by the factor $(E_{\text{cut}}/E_{\text{reg}})^{(\alpha g(\Omega_c)/4\pi^2)}$ which decreases. However note that as the exponential factor sums up contributions from arbitarily large numbers of photons we are required to sum over an arbitrary number of hard photons.

We have applied this procedure to the process $e^+e^- \to \mu^+\mu^- + n\gamma$ including initial state radiation, final state radiation, all interference effects, both Z^0 and photon exchange, and including the mass effects of the muons.[2] All distributions that we have tested are independent of E_{cut} for values of E_{cut} smaller than 1 GeV, for larger values the approximation that have gone into calculating the exponential factor break down.

As an illustration of this method we show in Fig.1a the differential cross-section $d\sigma/d\ln E_1$ where E_1 is the energy of the most energetic photon. We also show the results from working strictly at LO. In Fig.1b we show the the contributions just arising from final state radiation (FSR), and also show the cross-section you would obtain from using the soft photon approximation over the whole of phase space.

In conclusion, we can see that exponentiation is important for an accurate description of soft photons; whereas exact matrix elements are important for an accurate description of hard photons. It is not unusual when analysing experimental data for hard photon emission to normalise the theoretical predictions to the experimental measurement through the event rates for soft photons, however if we are to do this we become both sensitive to the exact matrix elements (to describe the hard photons) and exponentiation (to normalise the predictions).

References

1. D.R. Yennie, S. Frautschi, and H. Suura, *Ann. Phys.* **13** (1961) 379.
   K.T. Mahanthappa, *Phys. Rev.* **126** (1961) 329.
   S. Jadach and B.F.L Ward, *Phys. Lett.B* **274** (1992) 470.
   S. Jadach and B.F.L Ward, *Comp.Phys.Comm.* **56** (1990) 351.
2. D.J. Summers, Univ. of Wisconsin – Madison preprint, MAD/PH/833.

Charm Physics

JANIS A. McKENNA, *British Columbia*

OBSERVATION OF $D^{*\pm}$ PRODUCTION AND STUDY OF STRANGE PARTICLE PRODUCTION WITH THE ZEUS DETECTOR AT HERA

URI KARSHON[*]

*Particle Physics Department, Weizmann Institute,
Rehovot, Israel*

ABSTRACT

We report the first observation in ep collisions of charmed mesons in the decay channel $D^{*+} \to D^o \pi^+ \to (K^- \pi^+)\pi^+$ (+c.c), and the first measurements on K^o_S, and Λ, $\bar{\Lambda}$ production with the ZEUS detector at HERA. The 1993 data sample, corresponding to an integrated luminosity of 0.5 pb^{-1}, was analyzed with the central tracking detectors. The evidence for D^* production is a peak observed at 145.5 MeV in $\Delta M = M(K\pi\pi) - M(K\pi)$, where $M(K\pi) = M(D^o)$. Cross sections for the charged particle decays of D^o, K^o_S and Λ, $\bar{\Lambda}$ are determined within restricted kinematic regions, for which the detector acceptance is large and well determined.

1. Introduction and Experiment

The study of the reactions $ep \to K^o_S(\Lambda) + X$ is important for the understanding of parton fragmentation and the nature of the strange sea of the proton. The reaction $ep \to c\bar{c} + X$ is predicted[1] to be mainly produced by a direct photon interacting with a gluon in the proton (Boson Gluon Fusion, BGF), contrary to light quark production, which is dominated by resolved photon interactions[2]. If BGF is the dominant mechanism in $c\bar{c}$ production, its measurement is a clean way to determine[1] the gluon density in the proton at low values of Bjorken x, which is still unknown.

The data was taken in 1993 at the ep collider HERA by the ZEUS detector. Electrons with 26.7 GeV were colliding with 820 GeV protons, and the integrated luminosity collected was about 0.5 pb^{-1}. Charged tracks were measured by a vertex detector with a spatial resolution of 50-150 micron and by a Central Tracking Detector (CTD) with about 260 micron resolution. The transverse momentum resolution for full length tracks is $\sigma(p_T)/p_T \approx \sqrt{(0.005p_T)^2 + (0.016)^2}$, where p_T is in GeV. Reliable tracking is obtained with $p_T \geq 0.15$ GeV and track polar angle of $25 \leq \theta \leq 155$ degrees, or $|\eta| \leq 1.5$, where $\eta = -\ln\tan(\theta/2)$ is pseudorapidity.

2. K^o_S and Λ production

A preliminary analysis of K^o_S and Λ production has been performed in Deep Inelastic Scattering (DIS) events with $10 \leq Q^2 \leq 640 GeV^2, 0.0003 \leq x \leq 0.01$, and $y \geq 0.04$, where Q^2, x and y are the usual DIS kinematic variables, and for K^o_S (Λ) kinematic range of $0.5 \leq p_T \leq 4.0(3.5)$ GeV and $|\eta| \leq 1.3$, where the CTD has a high and well determined efficiency. The decay modes investigated were $K^o \to \pi^+\pi^-$ and

---

*Representing the ZEUS COLLABORATION.

$\Lambda \to p\pi$. Since the K° and Λ lifetimes are of the order of 10^{-10} seconds, the average $K^\circ(\Lambda)$ decay length is a few cm. Secondary vertices are identified and background is reduced by applying various selection cuts. In Fig.1 the invariant masses $M(p\pi)$ and $M(\pi\pi)$ are shown, where clear K_S° and Λ signals are seen. Fitting the mass spectra to a gaussian shaped signal and a linear background, we get $971 \pm 35 K_S^\circ$ with a mass of 497.4 ± 0.3 MeV and a width of 7.8 ± 0.3 MeV, and $80 \pm 9\Lambda$ with a mass 1116.2 ± 0.4 MeV and a width 3.0 ± 0.5 MeV.

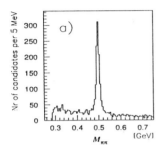 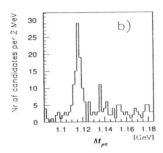

Fig.1.a)Reconstructed K_S° signal and b)reconstructed Λ signal.

Defining the K° multiplicity as the average number of K°'s per event, corrected for the branching ratio into the measured decay channel and by a factor 2 for K° to K_S°, the number of neutral kaons per event in the restricted η, p_T range is $0.279 \pm 0.014 \pm 0.025$ corresponding to a total cross section of $16.2 \pm 0.8 \pm 1.8$ nb. The first error is statistical and the second gives the systematic uncertainty. Defining P_i as the probability to create in the fragmentation process a quark i from the proton sea, we compare the above result with various Monte Carlo simulation programs with $P_s/P_u = 0.3$ (the default value) or with $P_s/P_u = 0.2$. The results of the K° multiplicity are too high with the default value, and are better reproduced with $P_s/P_u = 0.2$. Similar conclusions are obtained for the Λ results, where the measured multiplicity (corrected for the branching ratio into $\Lambda \to p\pi$) in the restricted η, p_T range is $0.038 \pm 0.006 \pm 0.003$ and the cross section is $2.2 \pm 0.3 \pm 0.3$ nb.

3. Observation of D^* production

We have searched for $c\bar{c}$ production at HERA via the production and decay chain $D^{*+} \to D^\circ \pi_L^+ \to (K^-\pi^+)\pi_L^+$ (+c.c.). The method relies on the very low Q-value of the D^{*+} decay, where the momentum of the π_L^+ in the D^* rest frame is only 40 MeV. This yields a prominent signal in the threshold region of the mass difference $\Delta M = M(D^\circ\pi) - M(D^\circ)$(at 145.4 MeV) , which is otherwise a phase space suppressed kinematical region. The procedure used is to select D° candidates in the mass range $1.75 \le M(K\pi) \le 1.95$ GeV and to add a π_L with a charge opposite to that of the kaon. No particle identification is used in this analysis.

The Monte Carlo program PYTHIA[3] was used to model the hadronic final states in $c\bar{c}$ production. The generator includes parton showers in the initial and final states and the fragmentation into hadrons is performed with the LUND string model[4] as implemented in JETSET 7.3[5]. We have generated both direct and resolved photon events, where the photon structure function was parametrized with GRV[6], ACFGP[7] or LAC1[8] and the parton densities in the proton were described by MRSD- or MRSD0[9]. The contribution of the resolved component is about 10-15% in GRV and ACFGP and reaches 50% or more with the LAC1 parametrization. Consequently, the inclusive cross section for open charm production at HERA vary in the wide range between 0.5 and 2.0 μb.

Various cuts were imposed in order to reduce the background and have a high detection efficiency. The main cut is $p_T(D^*) \geq 1.7$ GeV. We also restrict the analysis to the angular coverage $|\eta(D^*)| \leq 1.5$. In Fig.2a the ΔM plot is shown after the above cuts. A clear peak around 145 MeV is observed with a signal-to-noise ratio of about 1, which is improved by increasing the value of the $p_T(D^*)$ cut up to about 3 for $p_T(D^*) \geq 3$ GeV. When we plot the $M(K\pi)$ distribution for events in the ΔM range from 144 to 147 MeV without the constraint $1.75 \leq M(K\pi) \leq 1.95$ GeV, a clear signal is also seen around the nominal value of the D^o mass (not shown). The background shape has been determined by selecting pairs of tracks with the same charge for calculating the $K\pi$ invariant mass ("wrong combination"). The background (Fig.2b) has been fitted with a function $A.(\Delta M - m(\pi))^B$, where A,B are free parameters and $m(\pi)$ is the pion mass. The signal distribution was then fitted, assuming this function for the background plus a gaussian to parametrize the signal. The fit yielded $\Delta M = 145.4 \pm 0.2$ MeV and a width of 1.0 ± 0.2 MeV, consistent with the Monte Carlo simulation. We obtain $78 \pm 15 D^*$ over a background of ≈ 80 events.

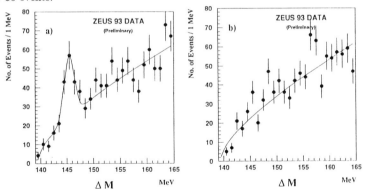

Fig.2.ΔM distribution for a)right combination and b)wrong combination. The curves are fits as described in the text.

In order to determine the total cross section in the selected kinematic region $|\eta| \leq 1.5$ and for different cuts on $p_T(D^*)$, the efficiency was evaluated using Monte Carlo sets with different structure function parametrizations. The typical efficiency, defined as the number of detected over generated $D^* \to K\pi\pi$ decays in the above kinematic range, is about 10% for $p_T(D^*) \geq 1.7$ GeV and increases when the $p_T(D^*)$ cut is increased. The resulting cross sections are decreasing as function of the $p_T(D^*)$ cut. For $p_T(D^*) \geq 1.7$ GeV we get $\sigma(ep \to D^{*+}X + c.c.)B(D^{*+} \to D^0\pi^+ \to (K^-\pi^+)\pi^+) = 1.4 \pm 0.3(stat.) \pm 0.3(syst.)nb$. For comparison, PYTHIA gives 1.36 nb for MRSD-/LAC1 and 1.24 nb for MRSD-/GRV. We also compared with the predictions for different $p_T(D^*)$ cuts for MRSD0/GRV and MRSD0/LAC1. The generated quantities are in better agreement with data using MRSD- for the proton structure function. The $|\eta(D^*)| \leq 1.5$ cut removes most of the resolved contribution, mainly for the LAC1 parametrization. Thus, with the present statistics, the results are insensitive to the photon structure function parametrization. Extrapolating outside the selected kinematic range, assuming a branching ratio[10] $B(c \to D^{*+} \to D^0\pi^+ \to (K^-\pi^+)\pi^+) = 7.1.10^{-3}$, we get a total charm cross section of $\sigma(ep \to c\bar{c}X) \approx 1.7\mu b$ for MRSD-/LAC1 and $\approx 1.0\mu b$ for MRSD-/GRV.

References

1. A. Ali et al., in Proc. of the HERA Workshop, Vol.1 DESY (1987) 395; R. van Woudenberg et al., in Proc. of the Workshop Physics at HERA, Vol.2, DESY (1991) 739.

2. G. Kramer and S. G. Salesch, in Proc. of the Workshop Physics at HERA, DESY (1991) 649; L. E. Gordon and J. K. Storrow, *Phys. Lett.* **B291** (1992) 320; D. Boedeker, *Phys. Lett.* **B292** (1992) 164; M. Greco and A. Vicini, Frascati preprint, LNF-93/017 (April 1993).

3. T. Sjoestrand, *Z. Phys.* **C42** (1989) 301 and in Proc. of the Workshop on Physics at HERA, DESY (1992) 1405.

4. B. Andersson et al., *Phys. Rep.* **97** (1983) 31.

5. T. Sjoestrand, *Comp. Phys. Comm.* **39** (1986) 347; T. Sjoestrand and M. Mengtsson, *Comp. Phys. Comm.* **43** (1987) 367.

6. M. Gluck, E. Reya and A. Vogt, *Phys. Rev.* **D46** (1992) 1973.

7. P. Aurenche et al., *Z. Phys.* **C56** (1992) 589.

8. H. Abramowicz, K. Charchula and A. Levy, *Phys. Lett.* **269B** (1991) 458.

9. A. D. Martin, R. G. Roberts and W. J. Stirling, *Phys. Rev.* **D47** (1993) 867.

10. F. Butler et al. (CLEO Collaboration), *Phys. Rev. Lett.* **69** (1992) 2041; D. S. Akerib et al. (CLEO Collaboration), *Phys. Rev. Lett.* **71** (1993) 3070; H. Albrecht et al. (ARGUS Collaboration), DESY 94-094 preprint.

PRODUCTION ASYMMETRIES IN x_f AND P_t^2 FOR $D^\pm$ MESONS

TOM CARTER*

Fermi National Accelerator Laboratory
Batavia, IL 60510, U.S.A.

ABSTRACT

We present differences in leading and non-leading charged D meson production as a doubly-differential function of both P_t^2 and x_f. Comparisons to specific models are made. This information is from the analysis of half the data from Fermilab experiment E791, taken during the 1991-2 fixed target run with a 500 GeV/c π^- beam incident on a segmented target.

1. Experiment E791

The results given in this paper come from the 1991/92 run of fixed target experiment E791 at Fermilab. The experiment used a 500 GeV/c π^- beam incident on a segmented target. The detector was the Tagged Photon Spectrometer, an open geometry multiparticle spectrometer, and has been described elsewhere.[1,2] The experiment combined an extremely fast data acquisition system with a very open trigger to record the world's largest sample of hadronically produced charm. Over 20 billion events were recorded on 24,000 8-mm tapes. The reconstruction of this large data set will be completed by late summer on parallel processing computer farms in four locations: Kansas State University, University of Mississippi, Fermilab and Centro Brasileiro de Pesquisas Fisicas in Brazil.

2. Physics

Previous experiments[3,4] have seen asymmetries in the hadronic production of charmed mesons. By asymmetries, we mean that a particular charmed meson may have a different production distribution from its anti-particle. The specific case we dicuss in this paper is the one in which the D^- meson has a *harder* Feynman-x (x_f) distribution than the D^+ meson. We will also show initial results on the comparison of their P_t^2 distributions where P_t is the net momentum transverse to the beamline.

There are several possible causes for a difference in production direction for charmed particles and anti-particles. For a π^- beam next-to-leading-order calculations[5]

---

*Representing the E791 Collaboration: Centro Brasileiro de Pesquisas Fisicas, University of California, Santa Cruz, University of Cincinnati, CINVESTAV, Fermilab, Illinois Institute of Technology, Kansas State University, University of Mississippi, Princeton University, Universidad Autonoma de Puebla, Tel Aviv University, Tufts University, University of Wisconsin, Yale University

predict a small enhancement in the number of mesons containing a $\bar{c}$ quark over those containing c quarks in the very forward direction. However, a much larger enhancement is seen in the data.[3]

Another possible explanation of the asymmetry is provided by the Lund "string fragmentation" model that effects the formation of the visible particles. In this model, forward momentum is added to the produced heavy quarks if they combine with the remnant light quarks from the incoming beam particle.[6] This causes charmed mesons with a light quark in common with the incoming beam ("leading particles") to have a harder x_f spectrum than those which do not ("non-leading particles"). In the case of E791, with a $\pi^-(d\bar{u})$ beam, the $D^-(d\bar{c})$ is leading and the $D^+(\bar{d}c)$ is non-leading.

A third possible model that would affect the production distributions for charmed hadrons is that of intrinsic charm.[7] Here, a virtual $c\bar{c}$ pair is formed in the incoming beam particle and is knocked onto its mass shell in a small percentage of interactions. Since these intrinsic quarks would have the same velocity as the original $\bar{u}$ and d quarks in the pion, they are more likely to form a leading particle, enhancing the "beam dragging" effect described in the previous paragraph.

3. Analysis

For the case of a π^- beam in E791, the most copious leading charmed mesons are the D^0, D^- and D^{*-}. One might set out to study the directions in which these particles are produced in comparison with their non-leading counterparts, the $\bar{D}^0$, D^+ and D^{*+}. Unfortunately, a large fraction of the D^0's (typically $1/3$ of those observed) may have been produced by the $D^{*+} \rightarrow D^0\pi^+$ decay process. The original D^{*+} is actually a *non*-leading particle. Therefore the observed D^0's come from a mixture of leading and non-leading processes, making the study more complex. This document will study only the D^+/D^- comparisons although D^{*+}/D^{*-} comparisons will follow soon. The direction of a produced meson may be described by its x_f and P_t values. In order to show small differences in the number of mesons observed over many different x_f and P_t ranges, an asymmetry parameter, A, is calculated for each ranges.

$$A \equiv \frac{N_{D^-} - N_{D^+}}{N_{D^-} + N_{D^+}}$$

where $N_{D^\pm}$ is the number of $D^\pm$ mesons produced within that x_f or P_t range. Note that since the acceptance for D^+ and D^- is the same in our detector, A is independent of the acceptance values. In addition to the x_f distributions, E791 can use its high statistics ($21,469 \pm 146$ candidates used here) to plot the asymmetries as a function of P_t^2 for different regions of x_f. This will allow us to test predictions made for the intrinsic charm model.[7]

In this paper, the asymmetry parameter, A, is calculated directly from data with no correction for detector acceptance. To ensure that no acceptance correction was required, Monte Carlo software simulation of our detector was used. No differences in D^- and D^+ acceptance were found, assuring us that any non-zero values of A came from differences in production, not detector efficiencies.

4. Results

The results presented in this paper come from the analysis of approximately half the total E791 data set. Figure 1 shows the resulting $K\pi\pi$ mass peak from the decay of $D^{\pm} \rightarrow K2\pi$. A much larger sample of $D^{\pm}$ can be produced if the background level is allowed to increase.

In Figure 2, the value of A for the E791 $D^{\pm}$ mesons is plotted as a function of x_f and compared to the predictions of two models. The error bars shown on this and all other plots correspond to statisical uncertainities only. As with previous experimental results, the PYTHIA prediction appears slightly to be systematically higher although the curve shape is similar. The recent prediction involving intrinsic charm[8] is also shown.

Figure 3 shows A as a function of P_t^2 for E791 and for the same two models. There is a surprising prediction of increasing A at low P_t^2 from the PYTHIA model. There are indications of such an increase in our data, but at a much smaller level. The intrinsic charm model predicts an almost constant value of $A \simeq 0.0$.

Figure 4 shows A as function of P_t^2 for the high x_f region. The E791 and PYTHIA values are given for mesons with x_f between 0.3 and 0.8. The intrinsic charm values are given for mesons with x_f between 0.4 and 0.8. The intrinsic charm model predicts an increase in A at very low values of P_t^2. Our data indicates no such increase, although it has large statistical uncertainities.

It should be noted again that the E791 values of A are plotted without correction for acceptance. As our acceptance varies with x_f, this may be a problem when plotting A vs. P_t^2 for limited regions of x_f (as in figure 4). This concern is being addressed now.

5. Future Analysis

In the future, we hope to expand our study of these production asymmetries. First, we will include our entire data set and include decays of the vector mesons $D^{*\pm}$. This will decrease our statistical errors, allowing us to make more definitive statements about various models. At the same time we can increase the range of x_f examined. Second, we hope to include completely different particles such as the D_s or Λ_c into our study. Finally, we will complete our studies of possible systematic errors in our analysis.

References

1. L. Cremaldi, *Proceedings, XXVI International Conference on High Energy Physics, Dallas, Texas August 1992* Vol. 1, p 1058.
2. A. Amato, et al, *Nucl Instr Meth* **A324:535-542** (1993).
3. E769 Collaboration, G.A. Alves *et al*, *Phy Rev Lett* **72,6** (1994).
4. WA82 Collaboration, M Adamovich *et al.*, *Phy Let* **B305**,(1993).
5. P. Nason, S. Dawson, and K. Ellis, *Nucl Phys* **B327, 49** (1989)
6. T. Sjostrand *CERN-TH.6488/92 (PYTHIA 5.6 Manual)*
7. R. Vogt and S.J. Brodsky *SLAC-PUB-6468* (1994)
8. R. Vogt - Private Communication

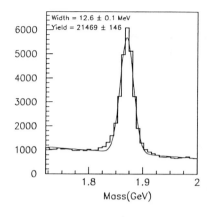

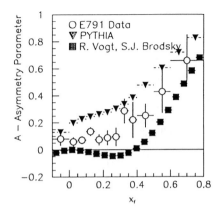

Figure 1: $D^{\pm} \to K\pi\pi$

Figure 2: Asymmetry Parameter vs. x_f

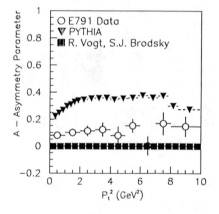

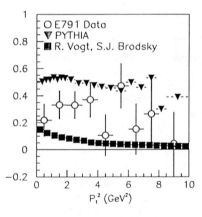

Figure 3: Asymmetry Parameter vs. P_t^2

Figure 4: Asymmetry Paramter vs P_t^2 for high x_f Region

THE PHYSICS OF CHARM LIFETIMES

HARRY W. K. CHEUNG*

*Fermi National Accelerator Laboratory, Wilson Road,
P.O. Box 500, Batavia, IL 60510*

ABSTRACT

Charm particle lifetimes provide a laboratory for the study of inclusive charm decays. Recent charm lifetime results from Fermilab photoproduction experiment E687 are presented and the physics impact of these results is diccussed.

1. Introduction

There is growing interest in the application of Heavy Quark Effective Theory to weak decays of charm and beauty hadrons.[1] This treatment constitutes an expansion in inverse powers of the heavy quark mass, and thus leads to a systematic procedure for including non-perturbative effects if the heavy quark mass is large enough. Although this is not the case for charm, a $1/m_c$ expansion may still be useful. However, rather than using this method to make predictions for weak decays of charm, we can use experimental measurements on charm decays to define the validity of the method and help discriminate between "choices" that arise in its application.[2] Recent charm lifetime results will be presented and it will be shown that we have now entered the stage where these measurements can tell us how QCD should work at the charm mass scale.

2. Lifetime Measurements

Data collected by Fermilab photoproduction experiment E687 have been used to measure the lifetimes of the D^0, D^+, D_s^+, Λ_c^+, Ξ_c^+ and Ξ_c^0 charm hadrons. The E687 detector is described in detail elsewhere.[3]

*E687 Collaboration coauthors: P.L.Frabetti **Bologna**; H.W.K.Cheung, J.P.Cumalat, C.Dallapiccola, J.F.Ginkel, S.V.Greene, W.E.Johns, M.S.Nehring **Colorado**; J.N.Butler, S.Cihangir, I.Gaines, P.H.Garbincius, L.Garren, S.A.Gourlay, D.J.Harding, P.Kasper, A.Kreymer, P.Lebrun, S.Shukla **Fermilab**; S.Bianco, F.L.Fabbri, S.Sarwar, A.Zallo **Frascati**; R.Culbertson, R.Gardner, R.Greene, J.Wiss **Illinois**; G.Alimonti, G.Bellini, B.Caccianiga, L.Cinquini, M.Di Corato, M.Giammarchi, P.Inzani, F.Leveraro, S.Malzezzi, D.Menasce, E.Meroni, L.Moroni, D.Pedrini,L.Perasso, A.Sala, S.Sala, D.Torretta, M. Vittone **Milano**; D.Buchholz, D.Claes, B.Gobbi, B.O'Reilly **Northwestern**; J.M.Bishop, N.M.Cason, C.J.Kennedy, G.N.Kim, T.F.Lin, D.L.Puseljic, R.C.Ruchti, W.D.Shephard, J.A.Swiatek, Z.Y.Wu **Notre Dame**; V.Arena, G.Boca, C.Castoldi, G.Gianini, S.Malvezzi, S.P.Ratti, C.Riccardi L.Viola, P.Vitulo **Pavia**; A.Lopez **Puerto Rico**;G.P.Grim, V.S.Paolone, P.M.Yager **Davis**; J.R.Wilson **South Carolina**; P.D.Sheldon **Vanderbilt**; F.Davenport **North Carolina**; J.F.Filaseta **Northern Kentucky**; G.R.Blackett, M.Pisharody, T.Handler **Tennessee**; B.G.Cheon, J.S.Kang, K.Y.Kim **Korea**

The following decay modes were used in the lifetime analyses: $D^0 \rightarrow K^-\pi^+$; $D^+ \rightarrow K^-\pi^+\pi^+$; $D_s^+ \rightarrow \phi\pi^+$; $\Lambda_c^+ \rightarrow pK^-\pi^+$; $\Xi_c^+ \rightarrow \Xi^-\pi^+\pi^+$; and $\Xi_c^0 \rightarrow \Xi^-\pi^+$. For the D^0 lifetime analysis, both directly produced D^0 mesons and ones produced via $D^{*+} \rightarrow D^0\pi^+$ were used. The decay length was used in the lifetime analyses. To avoid using large acceptance corrections at short proper times, the reduced proper time was used. This amounts to "starting the clock" at a later time, instead of using the absolute time. The lifetime resolution is typically 0.05-0.06 ps depending on the decay mode. The lifetimes were determined using a binned maximum likelihood method, where events from the mass sidebands were used to model the background lifetime distributions. Extensive studies were made to investigate possible systematic effects associated with backgrounds, the fit method and acceptance uncertainties. The lifetime measurements are shown in Table 1. The 1992 world averaged lifetimes are also shown for comparison.[4] Details of the different lifetime measurements are described in detail elsewhere.[5]

Table 1. Lifetime measurements from E687 and Particle Data Group 1992 world averages. The mean of these two are also given together with the total percentage error on the mean.

Particle	Lifetimes in 10^{-12} sec.		
	E687	PDG'92	Mean (% error)
D^0	$0.413\pm0.004\pm0.003$	0.420 ± 0.008	0.415 ± 0.004 (1%)
D^+	$1.048\pm0.015\pm0.011$	1.066 ± 0.023	1.055 ± 0.015 (1.4%)
D_s^+	$0.475\pm0.020\pm0.007$	0.450 ± 0.030	0.466 ± 0.017 (3.6%)
Λ_c^+	$0.215\pm0.016\pm0.008$	$0.191^{+0.015}_{-0.012}$	0.200 ± 0.011 (5.5%)
Ξ_c^+	$0.41^{+0.11}_{-0.08}\pm0.02$	$0.30^{+0.10}_{-0.06}$	0.35 ± 0.06 (17%)
Ξ_c^0	$0.101^{+0.025}_{-0.017}\pm0.005$	$0.082^{+0.059}_{-0.030}$	$0.097^{+0.023}_{-0.015}$ (20%)

It is interesting to compare the lifetimes of the different charm particles. The lifetime ratios $\frac{\tau(D^+)}{\tau(D^0)} = 2.54 \pm 0.04$ and $\frac{\tau(D^0)}{\tau(\Lambda_c^+)} = 2.08 \pm 0.12$ are conclusively different from unity. The ratios $\frac{\tau(D_s^+)}{\tau(D^0)} = 1.12 \pm 0.04$, $\frac{\tau(\Xi_c^+)}{\tau(\Lambda_c^+)} = 1.75 \pm 0.32$, $\frac{\tau(\Lambda_c^+)}{\tau(\Xi_c^0)} = 1.75 \pm 0.32$, and $\frac{\tau(\Xi_c^+)}{\tau(\Xi_c^0)} = 3.61 \pm 0.94$ are different from unity by about 2.5-3.0σ. A definite lifetime hierarchy is emerging but we still require better precision to to make this more conclusive. We also still need an accurate and precise measurement of the Ω_c^0 lifetime to complete the picture. At present the data shows the following lifetime hierarchy:

$$\tau(D^+) > \tau(D_s^+) > \tau(D^0) \sim \tau(\Xi_c^+) > \tau(\Lambda_c^+) > \tau(\Xi_c^0).$$

3. Theoretical Picture

Much progress has been made in the use of "heavy quark expansions" as a systematic procedure for including non-perturbative corrections to inclusive charm decays.[2,6-8]

The decay width of the charm hadron is given by:

$$\Gamma(H_c) = \Gamma_0 \left(1 + \frac{A_2}{m_c^2} + \frac{A_3}{m_c^3} + ...\right).$$

Higher order $(1/m_c^4)$ terms have not been calculated but they are expected to play a role in charm decays. The A_2 term is different between mesons and baryons, and also it causes a 50% correction in the D meson inclusive semileptonic branching ratio (BR_{SL}).[7] In the spectator model with QCD radiative corrections BR_{SL} is expected to be about 15%. With just the $1/m_c^2$ corrections, the D mesons should all have BR_{SL} about 8%. The usual spectator corrections due to W-exchange/W-annhilation (WX/WA) and Pauli Interference (PI) are contained in the $1/m_c^3$ term. This term is as large as the $1/m_c^2$ and normal spectator decay terms and are different between the different mesons. This picture fits in well with the charm meson lifetimes and BR_{SL}. The D$^+$ lifetime and BR_{SL} is increased due to PI with WX in D^0 playing a small part. WX is both helicity and color suppressed and its contribution is difficult to calculate reliably. However the difference between the D^0 and D$_s^+$ lifetimes can give information on the size of WX/WA contributions in charm meson decays.[8]

Table 2. Charm baryon lifetime hierarchies from different authors.

Blok and Shifman[2] using F_D	$\tau(\Xi_c^+) > \tau(\Lambda_c^+) > \tau(\Xi_c^0) > \tau(\Omega_c^0)$
using f_D	$\tau(\Omega_c^0) > \tau(\Xi_c^+)$
Voloshin and Shifman[9]	$\tau(\Xi_c^+) \approx \tau(\Lambda_c^+) > \tau(\Xi_c^0) > \tau(\Omega_c^0)$
Guberina, Rückl and Trampetić[10]	$\tau(\Xi_c^+) > \tau(\Lambda_c^+) > \tau(\Xi_c^0) \approx \tau(\Omega_c^0)$
Cheng[11]	$\tau(\Xi_c^+) > \tau(\Lambda_c^+) > \tau(\Xi_c^0) \geq \tau(\Omega_c^0)$
Gupta and Sarmar[12]	$\tau(\Xi_c^+) > \tau(\Omega_c^0) > \tau(\Xi_c^0) \geq \tau(\Lambda_c^+)$

The lifetime hierarchy for charm baryons has been calculated by a number of authors.[2,9-12] Some of these are shown in Table 2. Gupta and Sarmar[12] use a purely phenomenological analysis without considering PI. Blok and Shifman[2] are the only ones to include $1/m_c^2$ terms and not drop non-leading $1/N_c$ terms. Only Blok and Shifman, and Voloshin and Shifman[9] take hybrid logs into account. These partly take into account soft gluon radiative corrections.[2] Voloshin and Shifman use values of C_+, C_- calculated at $\Lambda_{QCD} \sim 100$ MeV, whereas the others use $\Lambda_{QCD} \sim$250-300 MeV. In calculations of the WX/WA and PI terms using the NQM, $|\psi(0)|$ can be related to the decay constant of the D meson, f_D. However, since f_D can also be expanded in $1/m_c$, $f_D = F_D[1 + (a_1/m_c) + ...]$, one should use F_D to be consistent. In fact Blok and Shifman have to use f_D for mesons and F_D for baryons in order that their lifetimes match data. The Ω_c^0 has a spin correlation with the spectator quarks like the D mesons but unlike the other charm baryons. This gives rise to a spin $1/m_c^2$ term which has been included only by Blok and Shifman. Because of the spin correlation for Ω_c^0, Blok and Shifman also try to use f_D for the Ω_c^0. Guberina, Rückl and Trampetić[10] use the "real world values" of f_D and $(M_{\Sigma_c^+} - M_{\Lambda_c^+})$ that appear in $|\psi(0)|$. Cheng[11] includes

Cabibbo suppressed WX terms for Ξ_c^+ and Ω_c^0, and uses a semileptonic width of half that used by Guberina, Rückl and Trampetić.

It can be seen that although a heavy quark expansion provides a systematic procedure for the inclusion of non-perturbative corrections, the realm of its validity and the exact implementation are not yet clear. However it appears that experimental measurements of the charm lifetimes can provide information on this.

4. Summary and Conclusions

Recent measurements of the lifetimes of D^0, D^+, D_s^+, Λ_c^+, Ξ_c^+ and Ξ_c^0 charm hadrons have been presented. All the measurements are comparable in precision and accuracy with the Particle Data Group 1992 world averages. Taken together, a definite lifetime hierarchy is emergying:

$$\tau(D^+) > \tau(D_s^+) > \tau(D^0) \sim \tau(\Xi_c^+) > \tau(\Lambda_c^+) > \tau(\Xi_c^0).$$

This is still at the 2.5–3.0σ level, so more precise measurements are needed to determine the hierarchy more conclusively. There exists a systematic procedure for the inclusion of non-perturbative corrections to inclusive charm decays. However its realm of validity and the exact implementation are not yet clear. A conclusive lifetime hierarchy based on experimental measurements should help define the correct implementation and hopefully in the process educate us on applications of QCD at the charm scale.

References

1. B. Grinstein, *Heavy Quark Effective Theory: Applications to Weak Decays*, in proceedings of the 26th Int. Conf. on High Energy Physics, Dallas TX 1992, ed. J. R. Sanford (AIP, 1993) p. 408.
2. B. Blok and M. Shifman, *Lifetimes of Charmed Hadrons Revisited. Facts and Fancy*, in proceedings of the 3rd Workshop on the Tau-Charm Factory, Marbella, Spain, June 1993.
3. P. L. Frabetti et al., *Nucl. Instrumm. Methods Phys. Res., Sect. A* **320** (1992) 519.
4. Particle Data Group, K. Hikasa et al., *Phys. Rev.* **D45** (1992), S1.
5. P. L. Frabetti et al., *Phys. Rev. Lett.* **70** (1993) 1381; *Phys. Rev. Lett.* **70** (1993) 1755; *Phys. Rev. Lett.* **70** (1993) 2058; *Phys. Rev. Lett.* **71** (1993) 827; and *Phys. Lett.* **B232** (1994) 429.
6. I. I. Bigi, *Inclusive Decays of Beauty (& Charm): QCD vs. Phenomenological Models*, in proceedings of the 26th Int. Conf. on High Energy Physics, Dallas TX 1992, ed. J. R. Sanford (AIP, 1993) p. 402
7. I. I. Bigi, N. G. Uraltsev and A. I. Vainshtein, *Nonperturbative Corrections to Inclusive Beauty and Charm Decays: QCD versus Phenomenological Models*, Fermilab Preprint FERMILAB-PUB-92/158-T.
8. I. I. Bigi and N. G. Uraltsev, D_s *Lifetime, m_b, m_c and $|V_{cb}|$ in the Heavy Quark Expansion*, CERN Preprint CERN-TH.7063/93.
9. M. B. Voloshin and M. A. Shifman, *Sov. Phys. JETP*, **64** (1986) 698.
10. B. Guberina, R. Rückl and J. Trampetić, *Z. Phys.* **C33** (1986) 297.
11. H. Y. Cheng, *Phys. Lett.* **B289** (1992) 455.
12. V. Gupta and K. V. L. Sarma, *Int. Jour. Mod. Phys.* **A5** (1990) 879.

PHOTOPRODUCTION OF CHARMED HADRONS

ROB GARDNER*

Department of Physics
University of Illinois at Urbana-Champaign
1110 W. Green St., Urbana, IL 61801, USA

Representing the E687 Collaboration

ABSTRACT

Photoproduction data can be used to test QCD production mechanisms. In this paper we present results on the single-inclusive p_t^2 distributions of charm mesons and baryons, correlations between fully reconstructed charm pairs, and production asymmetries between charm and anticharm particles.

1. Introduction

Distributions such as the semi-inclusive p_t^2 spectra of photoproduced charmed hadrons as well as correlations between fully reconstructed $D\overline{D}$ pairs can be used to test QCD production mechanisms. In particular, the data may be compared to the next-to-leading order (NLO) single-inclusive cross sections[1] and the more recent NLO calculation for doubly-differential distributions.[2] Nonperturbative effects such as fragmentation also play an important role; one gauge of such effects is the production asymmetry between charm and anticharm particles.

The results presented here are from data collected by the Fermilab high energy photoproduction experiment E687 (average photon energy $\approx$ 200 GeV) during the 1990-91 fixed-target run at Fermilab. The data sample for this analysis consists of approximately 55000 fully reconstructed charm particles in the following decay topologies: $D^+ \to K^- \pi^+ \pi^+$, $D^0 \to K^- \pi^+$ and $D^0 \to K^- \pi^+ \pi^+ \pi^-$ (both D^0 modes reconstructed with and without a D^{*+} tag), $D_s^+ \to K^- K^+ \pi^+$, $\Lambda_c^+ \to p K^- \pi^+$, together with their charged conjugates.

2. Inclusive p_t^2 Distribution

In figure 1 are shown the observed p_t^2 spectra for the D^+ and Λ_c^+ states. The distribution was fit to the the form $dN/dp_t^2 = A \exp(a p_t^2 + b p_t^4)$. In Table 1 the data are compared to a Monte Carlo simulation in which the charm quarks are produced through the photon-gluon fusion mechanism at leading order in QCD and are decayed using the Lund model of string fragmentation.[4] Also shown is the prediction from the NLO calculation of FMNR[2] for photoproduced charm quarks at our beam energy. The

---

*For a complete list of co-authors, see H.W.K. Cheung, *"The Physics of Charm Lifetimes"*, these proceedings.

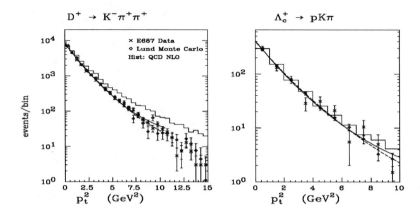

Fig. 1. The inclusive p_t^2 distribution from the E687 experiment (crosses) compared to Monte Carlo (diamonds) and to the NLO calculation for charm quarks from FMNR[2] (solid histogram).

significant disagreement with our data indicates that nonperturbative effects play an important role for D decay but less so for Λ_c decays. The FMNR authors have studied the inclusion of nonperturbative effects by supplementing their NLO result with a fragmentation function (which softens the p_t^2 distribution considerably) and by adding an intrinsic k_t component to the incoming partons. After inclusion of these effects the agreement with our data (Table 1) is considerably improved.

Table 1. Fits to p_t^2 distribution for photoproduced charm decays: NLO QCD theory[2], E687 data, and Monte Carlo simulations based on LO QCD and the Lund fragmentation model.

	a	b
QCD NLO	-0.656	0.021
$+$ frag $+ < k_t^2 >= 1$ GeV^2	-0.947	0.035
$D^+ \to K\pi\pi$	-0.84 ± 0.02	0.030 ± 0.002
Lund Monte Carlo	-0.79 ± 0.02	0.023 ± 0.002
$\Lambda_c^+ \to pK\pi$	-0.75 ± 0.09	0.025 ± 0.011
Lund Monte Carlo	-0.70 ± 0.03	0.018 ± 0.004

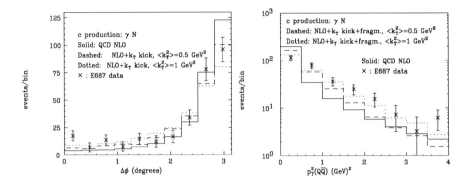

Fig. 2. The $\Delta\phi$ and p_t^2 distributions of the $D\overline{D}$ pair (E687 data) compared to the NLO prediction of FMNR[2] with and without the inclusion of nonperturbative effects. (Figure from Ref. 2.)

3. Charm Pairs

Correlations between two fully reconstructed D's can also be used to test QCD production models. The acoplanarity angle $\Delta\phi$ is the azimuthal angle between the D and $\overline{D}$ momentum vectors in the plane transverse to the photon direction. At leading order it is π radians. The p_t^2 of the $D\overline{D}$ pair is expected to be 0 at leading order. In Fig. 2 our sample[3] of 325 ± 23 fully reconstructed $D\overline{D}$ pairs is compared to the NLO QCD predictions.[2] Again, when the NLO result is supplemented with a fragmentation model and an intrinsic k_t kick the agreement with our data is considerably improved.

4. Production Asymmetries

At leading order in QCD, charm and anticharm quarks are produced symmetrically through the photon-gluon fusion mechanism. As the charm quarks fragment into colorless charmed hadrons, soft gluons are exchanged and thus nonperturbative effects may induce an asymmetry between charmed and anticharmed species. For example, in the Lund string fragmentation model[4] the struck gluon leaves the target nucleon in a color octet state which can be divided into a color antitriplet diquark and triplet quark, both of which are color connected to the charm quarks participating in the hard interaction. Since the color triplet charm quark must on average be closer in rapidity to the diquark, a condition which would favor charm baryon production as well as soften the D momentum spectrum, the model predicts a small enhancement of $\overline{D}$ meson production over D which should decrease with increasing beam energy.

In Fig. 3 we plot preliminary measurements of the production asymmetry, defined

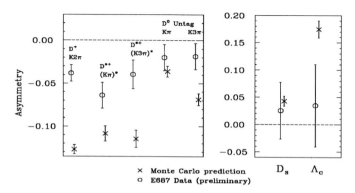

Fig. 3. The production asymmetry α (see text for definition) for D, D_s and Λ_c.

as $\alpha = \frac{N_D - N_{\overline{D}}}{N_D + N_{\overline{D}}}$. Also shown are Monte Carlo predictions based on the Lund Model. We observe small negative asymmetries for each of the D decay modes and small (but not statistically significant) positive asymmetries for both D_s and Λ_c decays. In general the measured asymmetries are much smaller than the Lund Model predictions.

5. Conclusions

We find that NLO calculations combined with nonperturbative models can yield a satisfactory description of our data. The $\Delta\phi$ ($D\overline{D}$ acoplanarity) is consistent with NLO supplemented with dressing and an intrinsic k_t kick of $\approx .5$ GeV^2. The inclusive p_t^2 distribution requires a larger k_t kick ($\approx$ 1-2 GeV^2) to overcome fragmentation effects. E687 observes a small enhancement of $\overline{D}$ over D production in the kinematic region $E_\gamma \approx 200$ GeV. Our preliminary result for the asymmetry averaged over D meson states is $\alpha = -3.7 \pm 0.6\%$. We note that no asymmetry is expected in leading order perturbative QCD. The asymmetry is smaller than Lund fragmentation model predictions and exhibits a much less severe kinematic dependence. Finally, at our energies we observe no significant excess of D_s^+ over D_s^- or Λ_c^+ over Λ_c^-.

References

1. R.K. Ellis and P. Nason, *Nucl. Phys.* **B312** (1989) 551.
2. S.Frixione, M.Mangano, P.Nason, G.Ridolfi, Nucl. Phys. **B412** (1994) 225;
   Charm and Bottom Production: Theoretical Results Versus Experimental Data, CERN-TH.7292 (1994).
3. E687 Collab., P.L.Frabetti *et al.*, *Phys. Lett.* **B308** (1993) 194.
4. The Monte Carlo program consisted of simulation algorithms for the E687 apparatus combined with the Lund model packages JETSET 7.3 and PYTHIA 5.6:
   T.Sjostrand, *Computer Phys. Comm.* **39** (1986) 347;
   T.Sjostrand, H.-U.Bengtsson, *Computer Phys. Comm.* **43** (1987) 367;
   H.-U.Bengtsson, T.Sjostrand, *Computer Phys. Comm.* **46** (1987) 43.

CHARMED MESONS

SANDRA MALVEZZI*

*Dipartimento di Fisica Nucleare e Teorica dell'Università and Sezione INFN
via Bassi 6, 27100 Pavia, ITALY*

ABSTRACT

Charmed meson decays into hadronic final states have been extensively studied in the E687 photoproduction experiment at Fermilab. Multi-meson-decay
modes offer a remarkable chance to address some of the main issues in charm
physics. Results on the D^0, D^+ and D_s 3-body amplitude analysis have direct
implications on the role of the different decay mechanisms. In addition a study
of 4 and 5 body decays into kaons and pions has been carried out; the relative
branching ratios are presented.

1.   Three-body decays

A detailed Dalitz plot study has been performed on the decays of charmed mesons
into three-body final states. The phenomenological amplitude used to fit the data is a
sum of Breit-Wigner terms with appropriate angular factors and a constant term to
represent the assumed uniform non-resonant contribution.

1.1. $K\pi\pi$ results

The $K\pi\pi$ Dalitz plot analysis has already been completed.[1] The $I = 1/2$ and
$3/2$ amplitudes have been computed using the branching fractions obtained from
this analysis in the modes $D^0 \rightarrow K^{*-}\pi^+, D^0 \rightarrow \bar{K}^{*0}\pi^0$ and $D^+ \rightarrow \bar{K}^{*0}\pi^+$. A ratio
$|A_{1/2}|/|A_{3/2}| = 5.9 \pm .3 \pm .3$ and a phase shift $\delta_{1/2} - \delta_{3/2} = 95 \pm 16 \pm 21°$ have been
measured; this angular difference indicates the importance of final-state interactions in
charm decays.

1.2. $KK\pi$ preliminary results

Fig. 1 shows the D^+ and $D_s^+ \rightarrow K^-K^+\pi^+$ signals along with their Dalitz plots.
In the D_s case the reflection from the $D^+ \rightarrow K^-\pi^+\pi^+$ has been removed by rejecting
events if they are consistent with a D^+ when reconstructed as $K^-\pi^+\pi^+$. The resulting
D_s Dalitz plot (Fig. 1) shows clearly that the $KK\pi$ decay is dominated by the $\phi\pi$ and
the $\bar{K}^{*0}K$ channels. The non-ϕ and non-$\bar{K}^{*0}$ events may be attributed either to the

Coauthors: P.L.Frabetti Bologna; J.P.Cumalat, C.Dallapiccola, J.F.Ginkel, S.V.Greene, W.E.Johns, M.S.Nehring Colorado; J.N.Butler, H.W.K.Cheung, S.Cihangir, I.Gaines, P.H.Garbincius, L.Garren, S.A.Gourlay, D.J.Harding, P.Kasper, A.Kreymer, P.Lebrun, S.Shukla, M.Vittone Fermilab; S.Bianco, F.L.Fabbri, S.Sarwar, A.Zallo Frascati; R.Culbertson, R.Gardner, R.Greene, J.Wiss Illinois; G.Alimonti, G.Bellini, B.Caccianiga, L.Cinquini, M.Di Corato, M.Giammarchi, P.Inzani, F.Leveraro, D.Menasce, E.Meroni, L.Moroni, D.Pedrini, L.Perasso, A.Sala, S.Sala, D.Torretta Milano; D.Buchholz, D.Claes, B.Gobbi, B.O'Reilly Northwestern; J.M.Bishop, N.M.Cason, C.J.Kennedy, G.N.Kim, T.F.Lin, D.L.Puseljic, R.C.Ruchti, W.D.Shephard, J.A.Swiatek, Z.Y.Wu Notre Dame; V.Arena, G.Boca, C.Castoldi, G.Gianini, S.P.Ratti, C.Riccardi, L.Viola, P.Vitulo Pavia; A.Lopes Puerto Rico;G.P.Grim, V.S.Paolone, P.M.Yager Davis; J.R.Wilson South Carolina; P.D.Sheldon Vanderbilt; F.Davenport North Carolina; J.F.Filaseta Northern Kentucky; G.R.Blackett, M.Pisharody, T.Handler Tennessee; B.G.Cheon, S.Kang, K.Y.Kim Korea

non-resonant or to the $f_0(975)\pi$ channel. Indeed the D_s into 3-pion data (Fig. 3) reveal the presence of the scalar $f_0(975)$ also decaying into a KK state.

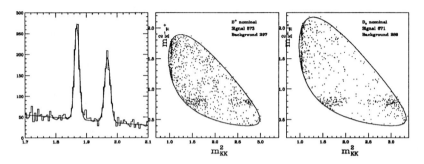

Fig. 1: D^+ and $D_s^+ \rightarrow K^- K^+ \pi^+$ signals and Dalitz plots.

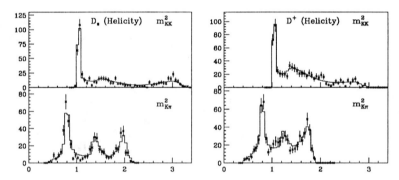

Fig. 2: D_s^+ and D^+ Dalitz plot projections and fit.

The parametrization and the effect of the $f_0(975)$ amplitude in the analysis is still problematic. On the other hand the "out-of-target"* sample shows a blur of events near the ϕ Zemach zero, which might be due to the $f_0(975)$. For the time being, preliminary results are reported as obtained with a 3-amplitude fit consisting only of ϕ, K^{*0} and non-resonant amplitudes. Two differently selected samples of data (skims) have been analyzed with two different formalisms (Zemach and Helicity); in Fig. 2 the Dalitz projections and their fit are shown as obtained via the helicity approach. The quoted results are evaluated as an average of the analyses and should be considered very preliminary. The systematic errors have been estimated from fluctuations of the results of the different analyses. Results for the D_s fit fractions and phases are shown in Table

---

*A sample of events has been selected requiring the decay to be downstream of the experimental target. This very clean sample serves as an important check of the systematic errors caused by background fluctuations and model assumptions

I. An analogous model has been applied to D^+ (Fig. 2) and the corresponding results reported in the same Table I. In this case all the different analyses give very consistent results, with the exception of the $\phi/\bar{K}^{*0}$ phase that shows a certain shift in the out-of-target sample. This uncertainty is quoted as a systematic error. [†]

Table I: D_s and D^+ preliminary results

fit fractions and Phases	D_s	D^+
non-res	$.082 \pm .041 \pm .073$	$.401 \pm .025 \pm .018$
ϕ	$.437 \pm .039 \pm .062$	$.334 \pm .018 \pm .027$
$\bar{K}^{*0}$	$.584 \pm .040 \pm .060$	$.302 \pm .024 \pm .018$
non-resonant Phase	$-177 \pm 20 \pm 11°$	$96 \pm 7 \pm 11°$
ϕ Phase	$121 \pm 11 \pm 17°$	$162 \pm 10 \pm 41°$
$\bar{K}^{*0}$ Phase	0 (fixed)	0 (fixed)

1.3. $\pi\pi\pi$ preliminary results

The $\pi\pi\pi$ data sample was selected using the out-of-target cut already mentioned because of the huge background. Although the statistics is low,

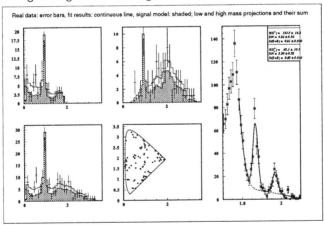

Fig. 3: D_s Dalitz plot and projections and $\pi\pi\pi$ invariant mass distribution.

the very good signal-to-noise ratio makes the analysis promising (Fig. 3). In this case, because of Bose symmetry, the data are represented by a folded Dalitz plot in $(m_{\pi\pi}^2)_{high}$ and $(m_{\pi\pi}^2)_{low}$. The model includes the $\rho(770)$, scalar $f_0(975)$, spin-2 $f_2(1270)$ and a flat non-resonant term. Again, as in the $KK\pi$ decay, there are remarkable differences in the fit predictions among the various $f_0(975)$ models. The analysis is still

---

[†]In this $K^-K^+\pi^+$ analysis, the helicity angle is defined to be the angle between the K^- and the resonance in the resonance rest frame.

in progress and only a very preliminary fit for the D_s is shown is Fig. 3. There is an indication that the decay is dominated by the $f_0(975)$ and f_2 channels.

2. Four and Five-body decays

The $D^0 \to 4\pi$, $D^0 \to 2K2\pi$ and $D^0 \to 3K\pi$ decays have been studied. The branching ratios relative to $D^0 \to K3\pi$ are reported in Table II.

Table II: Four-body decays of D^0

B.R.	E687	Previous results
$\frac{\Gamma(D^0 \to 4\pi)}{\Gamma(D^0 \to K3\pi)}$	$.097 \pm .006 \pm .002$	$.096 \pm .018 \pm .007 (E691)$ $.102 \pm .013 (CLEO)$
$\frac{\Gamma(D^0 \to 2K2\pi)}{\Gamma(D^0 \to K3\pi)}$	$.034 \pm .004 \pm .002$	$.028^{+.008}_{-.007}$ (E691) $.0314 \pm .010$ (CLEO) $.041 \pm .007 \pm .005$ (ARGUS)
$\frac{\Gamma(D^0 \to 3K\pi)}{\Gamma(D^0 \to K3\pi)}$	$.0028 \pm .007 \pm .002$	

These results constitute a considerable improvement over previous measurements.

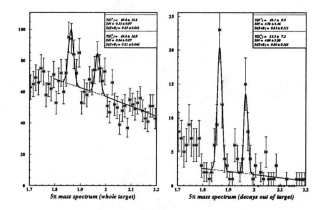

Fig. 4: D^+ and $D_s \to 5\pi$ signals with and without the out-of-target cut.

The first significant evidence of the $D^+, D_s \to 5\pi$ has been observed. The signals are shown in Fig. 4 with and without the out-of-target cut.

References

1. P.L. Frabetti et al., Phys. Lett. **B331** (1994) 217.

OBSERVATION OF THE DECAY $\Omega_c^0 \to \Sigma^+ K^- K^- \pi^+$

CRISTINA RICCARDI*

*Dipartimento di Fisica Nucleare e Teorica dell'Università and Sezione INFN
via Bassi 6, 27100 Pavia, ITALY*

ABSTRACT

We report on the observation of a new decay mode of the Ω_c^0, $\Omega_c^0 \to \Sigma^+ K^- K^- \pi^+$, where Σ^+ decays into either $p\pi^0$ and $n\pi^+$. We observe a clear signal of 42.4 ± 9.0 events and we give a new measurement of the mass.

1. Introduction

At present the published evidence for the doubly strange charmed baryon Ω_c^0 is not strong. Three Ω_c^0 events were reported by WA62[1] in the decay channel $\Xi^- K^- \pi^+ \pi^+$ and 6.5 ± 3.2 events were reported by ARGUS[2] for $\Omega_c^0 \to \Omega^- \pi^+ \pi^+ \pi^-$. The best published evidence so far are results from ARGUS[3] with 12.2 ± 4.5 events of $\Omega_c^0 \to \Xi^- K^- \pi^+ \pi^+$ at a mass of $2719 \pm 7.0 \pm 2.5$ MeV/c^2 and our results published in an earlier paper on $\Omega_c^0 \to \Omega^- \pi^{+4}$ with with 10.3 ± 3.9 events at a mass of $2705.9 \pm 3.3 \pm 2.0$ MeV/c^2. Both these signals are about 2.7 standard deviations in statistical significance. In a recent preliminary analysis,[5] the CLEO collaboration finds no evidence for $\Omega_c^0 \to \Xi^- K^- \pi^+ \pi^+$, and reports a 90% confidence level upper limit on the cross-section $\times$ branching ratio of 0.40 pb, which is much lower than the value of $2.41 \pm 0.90 \pm 0.30$ pb measured by ARGUS. There is clearly a need for further confirmation of the Ω_c^0 and an improvement in the determination of its mass. In this paper we present further evidence for the Ω_c^0 in a new decay mode $\Omega_c^0 \to \Sigma^+ K^- K^- \pi^+$ (the charge conjugate state is implied when a decay mode of a specific charge is stated) and give a new measurement of its mass. The data have been collected by Fermilab high energy photoproduction experiment E687.[6]

2. $\Sigma^+ K^- K^- \pi^+$ candidates selection

The Σ^+ hyperons are reconstructed through both the decays modes $\Sigma^+ \to p\pi^0$, and $\Sigma^+ \to n\pi^+$.[7] The Ω_c^0 decays were reconstructed using the standard E687 candidate driven vertex algorithm.[6] The decay secondaries, other than the Σ^+ hyperons, must be reconstructed in both the SSD and MWPC, and the two sets of track parameters have to agree within measurement errors. The K^- must be identified by the Čerenkov counters as kaon definite or K/p ambiguous, and the pion must not be identified as electron, kaon or proton definite or K/p ambiguous. The four microstrip tracks of the $\Sigma^+ K^- K^- \pi^+$ combination are required to form a (secondary decay) vertex with

Coauthors: P.L.Frabetti Bologna; J.P.Cumalat, C.Dallapiccola, J.F.Ginkel, S.V.Greene, W.E.Johns, M.S.Nehring Colorado; J.N.Butler, H.W.K.Cheung, S.Cihangir, I.Gaines, P.H.Garbincius, L.Garren, S.A.Gourlay, D.J.Harding, P.Kasper, A.Kreymer, P.Lebrun, S.Shukla, M.Vittone Fermilab; S.Bianco, F.L.Fabbri, S.Sarwar, A.Zallo Frascati; R.Culbertson, R.Gardner, R.Greene, J.Wiss Illinois; G.Alimonti, G.Bellini, B.Caccianiga, L.Cinquini, M.Di Corato, M.Giammarchi, P.Inzani, F.Leveraro, D.Menasce, E.Meroni, L.Moroni, D.Pedrini, L.Perasso, A.Sala, S.Sala, D.Torretta Milano; D.Buchholz, D.Claes, B.Gobbi, B.O'Reilly Northwestern; J.M.Bishop, N.M.Cason, C.J.Kennedy, G.N.Kim, T.F.Lin, D.L.Puseljic, R.C.Ruchti, W.D.Shephard, J.A.Swiatek, Z.Y.Wu Notre Dame; V.Arena, G.Boca, C.Castoldi, G.Gianini, S.Malvezzi, S.P.Ratti, L.Viola, P.Vitulo Pavia; A.Lopez Puerto Rico; G.P.Grim, V.S.Paolone, P.M.Yager Davis; J.R.Wilson South Carolina; P.D.Sheldon Vanderbilt; F.Davenport North Carolina; J.F.Filaseta Northern Kentucky; G.R.Blackett, M.Pisharody, T.Handler Tennessee; B.G.Cheon, S.Kang, K.Y.Kim Korea

a confidence level (CLD) greater than 3%. The candidate Ω_c^0 "track" must form a primary vertex with at least one other microstrip track with a confidence level (CLP) larger than 3%. We eliminate fake decay secondaries that were actually produced in the primary interaction vertex by requiring that the confidence level for any of the four $\Sigma^+ K^- K^- \pi^+$ tracks to individually extrapolate back to the primary vertex is less than 85%. In addition, the confidence level that other SSD tracks, not already assigned to the primary or secondary vertices, point back to the secondary vertex is required to be less than 0.1%. This cut is designed to eliminate higher multiplicity decays and fake decay vertices caused by secondary interactions. Since it is possible for the secondary vertex to be reconstructed upstream of the primary vertex because of finite resolution and due to fake vertices, we require that the secondary decay vertex be reconstructed downstream of the primary vertex. Based on Monte Carlo studies of the momentum spectrum of this Ω_c^0 decay, the momentum of the $\Sigma^+ K^- K^- \pi^+$ combination is required to be greater than 50 GeV/c.

Figures 1(a) and 1(b) show separately the $\Sigma^+ K^- K^- \pi^+$ invariant mass plots for the two decay modes of the Σ^+, $\Sigma^+ \to p\pi^0$ and $\Sigma^+ \to n\pi^+$ respectively. The cuts used were as described above. A peak in both plots at a mass of about 2.7 GeV/c^2 is clearly evident and remains even with a large variety of different event selection criteria.

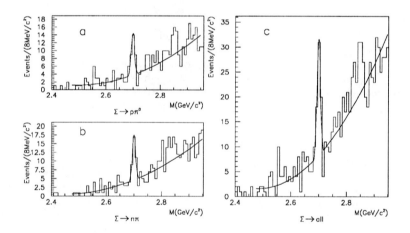

Fig. 1: $\Sigma^+ K^- K^- \pi^+$ invariant mass: (a) for $\Sigma^+ \to p\pi^0$ decay mode, (b) for $\Sigma^+ \to n\pi^+$ decay mode, (c) for both decay modes of Σ^+.

The fits shown are to a Gaussian peak over a quadratic background. The fitted mass and width of the peak for the $p\pi^0$ mode are 2699.7±2.5 MeV/c^2 and 5.5±1.3 MeV/c^2 respectively, and for the $n\pi^+$ mode they are 2700.4±2.0 MeV/c^2 and 5.7±1.3 MeV/c^2. The two masses agree well with each other. Figure 1(c) shows the total $\Sigma^+ K^- K^- \pi^+$ invariant mass plot using both Σ^+ decay modes. The fit to this mass distribution gives a yield of 42.4 ± 9.0 events at a mass of 2699.9 ± 1.5 MeV/c^2. The fitted width is 5.9 ± 1.0 MeV/c^2, which is consistent with the experimental mass resolution. For the mass measurement we have investigated the effect of events containing a two-fold ambiguity in the Σ^+ momentum[7]: these events make up less than about 30% of the total. Any two-fold ambiguity leads to two candidate combinations with different $\Sigma^+K^-K^-\pi^+$ invariant masses. This was studied in a Monte Carlo analysis and also in data by using different methods of resolving the two-fold ambiguity. No significant shift in the measured mass was seen and we assign an uncertainty of 0.2 MeV/c^2 associated with the two-fold ambiguity. We have also investigated the overestimate of the true number of signal events due to the two-fold ambiguity. We estimated this effect to be less than 8%. According to Monte Carlo analyses we assign a systematic uncertainty of 2.0 MeV/c^2 on our absolute mass scale for this mass measurement. No systematic shifts associated with our choice of binning, fitting function, fitting method and choice of selection criteria have been observed, but, since their study was limited by statistics, we conservatively assign an upper limit of 1.5 MeV/c^2 in the systematic uncertainty in the mass. Adding the different systematic uncertainties incoherently we obtain a final mass measurement for the Ω_c^0 of $2699.9\pm1.5(\text{stat})\pm2.5(\text{syst})$ MeV/c^2.

3.   Conclusions

In summary, we report an observation of 42.4 ± 9.0 events in a new decay mode of the $\Omega_c^0 \to \Sigma^+K^-K^-\pi^+$ at a mass of $2699.9\pm1.5(\text{stat.})\pm2.5(\text{syst.})$ MeV/c^2. This strengthens the evidence for the existence of the Ω_c^0, and at a mass lower than that given by the ARGUS measurements but consistent with our previous measurement using the decay mode $\Omega_c^0 \to \Omega^-\pi^+$.

References

1. S. F. Biagi *et al.*, *Z. Phys* **C28** (1985) 175.
2. J. Stiewe, ARGUS collaboration, *Recent ARGUS results on charm baryon physics*, in *Proceedings of the 26th Int. Conf. on High Energy Physics* **Vol. 1**, (Dallas, TX 6–12 August, 1992) pp 1076–108.
3. H. Albrecht *et al.*, *Phys. Lett.* **B288** (1992) 367.
4. P. L. Frabetti *et al.*, *Phys. Lett.* **B300** (1993) 190.
5. M. Battle *et al.*, *Search for the Ω_c^0 in e^+e^- Annihilations*, contributed paper submitted to *Int. Symp. on Lepton and Photon Interactions* (10–15 August 1993, Ithaca, New York).
6. P. L. Frabetti *et al.*, *Nucl. Instr. and Meth. in Phys. Res. Sect.* **A320** (1992) 519.
7. P. L. Frabetti *et al.*, *Phys. Lett.* **B328** (1994) 193.

PHOTON - CHARM PRODUCTION IN $P\bar{P}$
COLLISIONS AT $\sqrt{S} = 1.8$ TeV

ROWAN T. HAMILTON*

Harvard University, Cambridge, MA 02138

ABSTRACT

We present preliminary results of a measurement of inclusive photon-charm production in p̄p collisions. Using data collected by the CDF experiment during the 1992-1993 Tevatron collider run, we have measured the inclusive photon-muon cross section, and estimated the decay-in-flight and punch-through backgrounds. We compare our measured photon-muon cross section and photon E_T spectrum to Pythia predictions for the charm Compton contribution.

1. Introduction

The measurement of the photon-charm process is interesting because it is believed that the charm Compton process $gc \rightarrow \gamma c$ is the dominant contribution, and that it may thus be possible to directly measure the charm density of the proton at LO accuracy. The charm structure function is important for current and future hadron colliders, but at present is poorly understood. For instance, at $Q = 80$GeV and $x < 0.005$, CTEQ2M predicts that more than 20% of available quarks are charm, while there are to date no measurements of the charm structure function for the x ranges probed by either the LHC or the Tevatron.

CDF has several analyses in progress attempting to measure associated photon-charm production.Here we will discuss a measurement of the photon-muon cross section, and show why we believe this cross section to be dominated by the charm Compton process $gc \rightarrow \gamma c \rightarrow \gamma \mu X$. Theoretical estimates lead us to believe that the bottom-to-charm ratio in direct photon production is about 1:8,[1] and that the gluon-splitting contribution to the photon-charm process is only about 10%.

2. Data

A data sample was selected using the 16 GeV photon trigger from the CDF 1992-1993 run. The events were required to have an isolated electromagnetic energy cluster with $|\eta| < 1.0$ and $E_T > 16.0$, and no track pointing at the cluster. The offline photon cuts used in this analysis are identical to those used in the CDF inclusive photon analysis.[2] This sample corresponded to an integrated luminosity of 15.0pb^{-1}.

Photon-muon events were selected by additionally requiring a good offline muon in the central region. For this analysis, a good muon was defined by a coincidence in

---

*Representing the CDF Collaboration

the inner and outer muon chambers, and a good track-stub match for both the inner and outer stubs. These requirements left us with 127 events.

3. Backgrounds

The backgrounds for a photon - muon analysis are fairly easy to classify. A photon can be faked by a neutral meson like a π^o or an η, and a muon can be faked by a decay-in-flight of a kaon or a pion or by a punch-through (a jet which isn't completely absorbed by the calorimeter).

The photon backgrounds were statistically subtracted using the profile method described in the CDF inclusive photon analysis.[2] Briefly, a χ^2 was assigned to the transverse shower profile of the photon candidate, and on this basis the event was assigned a photon weight determined from Monte Carlo.

The muon backgrounds were estimated analytically and subtracted from the data. The punch-through rate was found to be negligible when we required inner/outer co-incidence. The decay-in-flight rate we calculated from the charged particle spectrum of the photon sample.

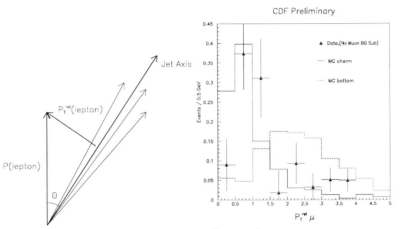

Fig. 1. **a)** Diagram explaining the calculation of P_T^{rel}. **b)** P_T^{rel} distribution of the data, compared to MC of photon - charm and photon - bottom.

Finally, in order to ensure that our cuts are efficient for charm, we used P_T^{rel} to estimate the relative charm/bottom content of our sample. The P_T^{rel} of the muon is defined to be the P_T of the muon relative to the nearest jet, as shown in Fig. 1a. Figure 1b shows the P_T^{rel} distribution for our data, compared to Pythia predictions for both photon-charm and photon-bottom, and our data is clearly more consistent with charm than with bottom.

4. Conclusions

After background subtractions, we are left with $80 \pm 22\text{pb(stat.)} \pm 20\text{pb(syst.)}$ events. The muon P_T spectrum is shown in Fig. 2a, with the decay-in-flight and punch-through estimates superimposed. Figure 2b shows the measured differential cross section for associated photon-muon production. If one assumes that the bottom contamination of our sample is small, and that the gluon-splitting contribution is negligible in comparison to the charm Compton signal, then one can compare our data to a Monte Carlo of the charm Compton contribution to the photon-muon cross section. In Fig. 2b the Pythia prediction for the charm Compton process is superimposed on the data.

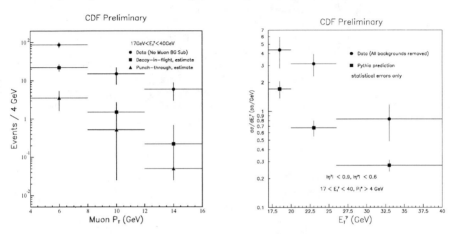

Fig. 2. **a)** P_T^μ spectrum for the data, with decay-in-flight and punch-through estimates superimposed. **b)** Measured photon - muon differential cross section.

If one integrates the differential cross sections in Fig. 2b one finds a measured cross section of $43 \pm 12\text{pb(stat.)} \pm 11\text{pb(syst.)}$ for $17 < E_T^\gamma < 40\text{GeV}$, $P_T^\mu > 4\text{GeV}, |\eta^\gamma| < 0.9$ and $|\eta^\mu| < 0.6$. The corresponding Pythia prediction for the charm Compton process alone is $15 \pm 3\text{pb}$, for MRSD0 structure functions with $q^2 = \hat{s}$, where the error includes only the statistical error due to the number of events simulated. We emphasize that there are potentially large systematic uncertainties in the Monte Carlo cross section. Structure function choice, q^2 choice, and modeling of initial and final state parton shower result in roughly a factor of 2 systematic uncertainty. Work is in progress to better quantify these issues.

References

1. Michelangelo Mangano, private communication.
2. A. Maghakian these proceedings and F Abe,*et. al.,Prompt Photon Cross Section Measurement in $\overline{p}p$ Collisions at $\sqrt{s} = 1.8\text{TeV}$*, Phys.Rev.D48:2998-3025,1993.

BEAUTY AND CHARM PRODUCTION FROM FERMILAB EXPERIMENT E789

DOUGLAS M. JANSEN*

Los Alamos National Laboratory, Los Alamos, NM, 87545

ABSTRACT

Experiment 789 is a fixed-target experiment at Fermilab designed to study low-multiplicity decays of charm and beauty. During the 1991 run, E789 collected $\approx 10^9$ events using an 800 GeV proton beam incident upon gold and beryllium targets. Preliminary results on the beauty cross section $\sigma(b\bar{b})$, determined using $b \rightarrow J/\psi + X$ inclusive decays, and on the A-dependence of neutral D-meson production are presented.

1. Introduction

Experiment 789 at Fermilab was designed to measure low-multiplicity decays of B and D-mesons. The existing E605/E772[1] spectrometer, used in previous experiments to detect hadron and lepton pairs, was significantly upgraded for E789. In particular, a silicon microstrip vertex spectrometer and a vertex trigger processor[2] were installed. The main physics goals of experiment 789 are 1) to measure $\sigma(b\bar{b})$ at 800 GeV via the detection of $b \rightarrow J/\psi + X$ inclusive decays, and 2) to measure the A-dependence of neutral D-meson production. Results from analysis of data obtained by E789 during the 1991 fixed-target run are presented in this paper.

2. Summary of the 1991 Run

During the 1991 run, an 800 GeV proton beam was incident upon thin wire targets, 0.1 mm to 0.2 mm high and 0.8 mm to 3 mm thick, constructed of gold or beryllium. Sixteen silicon microstrip detectors were positioned 40 – 115 cm downstream of the target and covered the angular range 20 – 60 mr above and below the beam axis. The silicon detectors, type 'B' from Micron Semiconductor, had 5×5 cm^2 area, 300 μm thickness, and 50 μm pitch. They were oriented to measure either the Y (vertical) or the U,V coordinates ($\pm 5°$ stereo angles). Signals from 8,544 silicon strips were individually read out via Fermilab 128-channel amplifier cards[3] and LBL discriminators[4] synchronized to the accelerator RF.

The magnetic field of the SM12 magnet was varied in order to optimize the acceptance for either charm or beauty decays. For the charm data, the current of the SM12 magnet was set at 1000 and 900 A. Data were taken at 1×10^{10} protons per spill on 1.5 mm thick targets constructed of gold and beryllium in sequential, but separate, running periods. Approximately 380 million events were written onto 240 8-mm tapes during the charm running period.

For the beauty data, the current of the SM12 magnet was set to 1500 A. In this configuration, the spectrometer had acceptance for both $B \rightarrow J/\psi + X$ and $B \rightarrow h^+ h^-$

---

*Representing the E789 Collaboration

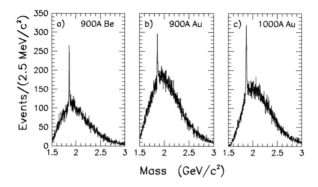

Fig. 1. Dihadron invariant mass spectra: a) Be target, 900 A, b) Au target, 900 A, c) Au target, 1000 A.

decays. Data were taken at 3×10^{10} protons per spill on a 3 mm thick gold target giving a 50 MHz interaction rate. Approximately 860 million events were written onto 700 8-mm tapes during the beauty running period.

3. Results from E789

A full description of E789's charm results is given in Reference.[5] Figure 1 shows the dihadron invariant mass spectra obtained from the charm data sample. The various target materials and SM12 currents used to obtain the data are indicated in the figure. To reject the dihadron background originating from the target, we cut on the variable τ/σ where τ is the proper lifetime of the dihadron and σ is the calculated error on τ.

Figure 2 shows the J/ψ A-dependence as a function of x_F. The high x_F data were obtained using the E789 beam dump[6] and the low x_F data were obtained during the 1990 E789 test run.[7] The data obtained by E772 have been described previously.[8] Analysis of the beryllium and gold data described above results in $\alpha = 1.02\pm0.03\pm0.02$ for the A-dependence of neutral D-meson production. This value is also indicated in Fig. 2. The lack of nuclear dependence in the production of neutral D-mesons suggests that the strong dependence seen in the production of J/ψ is the result of final-state effects.

The dimuon invariant mass spectrum from a preliminary analysis of the beauty data is shown in Fig. 3a. For events shown in Fig. 3a, good silicon tracks are required and vertex cuts have not yet been imposed. Approximately 85,000 J/ψ and 1600 ψ' events are observed. To search for $b \rightarrow J/\psi + X$ events contained in this event sample, impact parameter and Z_{vtx} cuts are applied. Fig. 3b shows the dimuon invariant mass spectrum for events in which both muon tracks have target impact parameters $\delta > 150\mu$m and the event has a decay vertex at least 7 mm downstream of the target (7 mm $< Z_{vtx} <$ 5 cm). The number of J/ψ events above the continuum is 17 ± 4. To estimate the background which might be caused by silicon tracking errors or from the tails of the target distribution, we also select events with decay vertices upstream of the target. Figure 3c shows the dimuon invariant mass for events with $\delta > 150\mu$m and -5 cm $< Z_{vtx} <$ -7 mm. This figure contains no events near the J/ψ region.

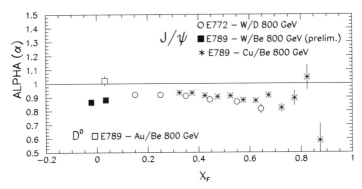

Fig. 2. J/ψ A-dependence as a function of x_F. The neutral D-meson value of α is also indicated in the figure.

Using the 17 ± 4 events seen in Fig. 3c and a detailed Monte-Carlo program to determine spectrometer acceptance and efficiency, we obtain a preliminary doubly-differential cross section for J/ψ mesons originating from b-decays equal to $d^2\sigma/dx_F\,dp_T^2 = 77 \pm 17 \pm 15$ pb/(GeV/c)^2/nucleon at $x_F = 0.05$ and $p_T = 1$ GeV/c, where the first error is statistical and the second one is systematic. We have assumed A^1 nuclear dependence. In Fig. 4 we show our measurement with two predictions by Mangano et al.[9] Using the shapes of the theoretical curves shown in Fig. 4 to extrapolate over all x_F and p_T, and using $BR(b\bar{b} \to J/\psi) = (2.24 \pm 0.32)\%$ which is twice the inclusive branching ratio for $B^+ \to J/\psi + X$,[10] we derive a preliminary total cross section $\sigma(b\bar{b}) = 4.9 \pm 1.4 \pm 1.2$ nb/nucleon.

4. Summary

Experiment 789 explores the feasibility of studying beauty and charm physics in a high-rate fixed-target environment. Over 3000 neutral $D \to \pi K$ events have been observed from gold and beryllium targets. The analysis of the neutral D-meson A-dependence gives $\alpha = 1.02 \pm 0.03 \pm 0.02$. A preliminary analysis of the dimuon beauty data yields a doubly-differential cross section for J/ψ mesons originating from b-decays equal to $d^2\sigma/dx_F\,dp_T^2 = 77 \pm 17 \pm 15$ pb/(GeV/c)^2/nucleon at $x_F = 0.05$ and $p_T = 1$ GeV/c. Based on this measurement, we derive a total cross section equal to $\sigma(b\bar{b}) = 4.9 \pm 1.4 \pm 1.2$ nb/nucleon.

References
1. J. A. Crittenden et al., Phys. Rev. D **34**, 2584 (1986).
2. C. Lee et al., IEEE Trans. Nucl. Sci. **38**, 461 (1989).
3. D. Christian et al., IEEE Trans. Nucl. Sci. **36**, 507 (1989).
4. B. T. Turko et al., IEEE Trans. Nucl. Sci. **39**, 758 (1992).
5. M. J. Leitch et al., Phys. Rev. Lett. **72**, 2542 (1994).
6. M. S. Kowitt et al., Phys. Rev. Lett. 72 1318 (1994).
7. M. J. Leitch, personal communication.
8. D. M. Alde et al., Phys. Rev. Lett. **66**, 133 (1991).
9. M. Mangano et al., Nucl. Phys. B. **405**, 507 (1993).
10. K. Hikasa et al., Phys. Rev. D **45**, Part 2 (June 1992).

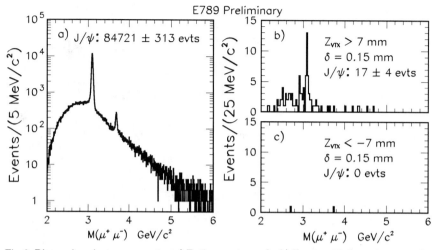

Fig. 3. Dimuon invariant mass spectra: a) Entire event sample, b) Events with downstream vertices, c) Events with upstream vertices.

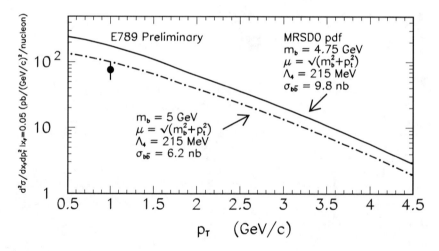

Fig. 4. $d^2\sigma/dx_F\,dp_T^2$ for J/ψ mesons originating from b-decays compared to theoretical predictions. The theoretical predictions, which include our simulation of $b \to J/\psi$ decays, are based upon the $b\bar{b}$ calculations described in Reference [8].

OBSERVATION OF QUASI-ELASTIC ELECTROPRODUCTION OF J/Ψ AT HERA

G. POPE*

*Physics Department, University of California, Davis,
Davis, CA 95616*

ABSTRACT

A signal for J/ψ production in the dileptonic decay channels has been observed in the data taken by the H1 experiment during the 1993 run. The extraction was done using criteria designed for selecting elastic-like event topology. A preliminary value for elastic J/ψ production cross section is presented. A comparison with predictions for the cross section is made. Based on a Monte Carlo study, an estimate of the contamination of this sample from inelastic J/ψ production is presented.

1. Introduction

The HERA collider offers a unique opportunity to study J/ψ electroproduction. With a proton beam of 820 GeV and an electron beam of 26.7 GeV, a very large center-of-mass energy of about 300 GeV is available. This allows the investigation of low-x physics. Furthermore, the energy scale, which is set by the mass of the charm quark, lies in a region of transition from perturbative to non-perturbative QCD. Of particular interest is elastic production of the J/ψ meson, as elastic events can be easily tagged using the dileptonic decay modes of the J/ψ. However, the problem of contamination from inelastic processes, in which the proton remnants do not enter the detector, must be understood and accounted for. This paper presents a method of using independent hardware triggers to estimate the amount of contamination in the elastic data sample.

2. Theory

Models of elastic production of the J/ψ involve the exchange of a Pomeron between the proton and a photon from the electron. Three such models will be discussed: the Vector Meson Dominance model,[1] the Two Gluon Pomeron model,[2] and a Photon Gluon Fusion model.[3] All of these processes will have an inelastic contribution due to cases in which the proton breaks up. Also, the Color Singlet model[4] describes a hard inelastic J/ψ production mechanism.

2.1. The Vector Meson Dominance model

This model takes its characterization of the Pomeron from elastic proton-proton scattering. The photon has some probability of fluctuating into a J/ψ meson, which

---

*Representing the H1 Collaboration.

has the same quantum numbers as the photon. The Pomeron then couples to one of the charm quarks. This model predicts a cross section which rises slowly with increasing energy.

2.2. The Two Gluon Pomeron model

This model, advanced by Jung, describes the production of a color-neutral $c\bar{c}g$ state via the hard scattering of the photon with a gluon. In this case the gluons originate from the Pomeron, so the corresponding gluon density must be input. In this model, the resulting cross section is constant with increasing energy.

2.3. The Photon Gluon Fusion model

Ryskin has proposed an alternate hard scattering model of elastic J/ψ production, in which the Pomeron is modeled as a two-gluon state which can be evolved via perturbative QCD. Since the gluons come from the proton, the square of the gluon density of the proton directly enters the cross section, which shows a dramatic rise with increasing energy.

2.4. The Color Singlet model

The dominant contribution to inelastic processes which can mimic elastic production is given by this model. The hard scattering of the photon off of a gluon from the proton results in a color singlet $c\bar{c}$ state. Color neutrality is preserved by the emission of an additional hard gluon. In cases where the initial photon energy is close to that of the final J/ψ (as measured in the proton rest frame), the process appears elastic. As with Ryskin's Photon Gluon Fusion model, the gluon density of the proton directly enters the cross section, which rises with increasing energy.

3. Triggers

Raw data from the H1 detector come from a wide variety of low-level hardware triggers. After a level of software trigger verification, the data are saved for a subsequent physics class selection routine. All events that were assigned some physics class were taken for this analysis. Then the data were filtered to extract events that were triggered by one of three hardware triggers. These triggers are a BEMC trigger, a Tracking trigger, and an Iron Endcap trigger. The triggers are independent of one another, and have overlapping regions of acceptance in the detector. The BEMC trigger is a single-electron trigger designed to tag events in which the electron scatters into the Backward ElectroMagnetic Calorimeter and deposits all of its energy in a localized cluster. The decay electrons from the J/ψ can also fire this trigger. The Tracking trigger combines z-vertex information from the central multiwire proportional chambers with a fast track finder that operates in the r-ϕ plane. Within the coarse-binned z-vertex histogram that is produced, a central peak with no contents in the sidebands is required. The trigger takes advantage of the low efficiency of the r-ϕ trigger by requiring exactly one track. This gives the highest acceptance for events with exactly two tracks. In addition, there is a veto on energy deposited in the inner forward region of the liquid argon calorimeter. The Iron Endcap trigger uses the instrumented iron subdetector to tag events with high momentum muons. It requires a signal in at least three out of five streamer tubes in any one of the 16 modules in the endcap. The trigger also includes the drift chamber r-ϕ subtrigger and requires at least one track found.

4. Data Selection

In order to select J/ψ events, a program was used which identifies candidate electron and muon clusters in the liquid argon calorimeter, as well as tracks in the instrumented iron. Electron clusters were required to have a minimum energy of 1.25 GeV, and muons that stopped in the calorimeter were required to deposit no more than 2 GeV. The cluster parameter of ratio of length over width was required to be at least 1.2 for electrons, and 1.8 for muons. These candidates were then linked to tracks in the central and forward trackers. Only tracks with a minimum momentum of 0.75 GeV and a minimum transverse momentum of 0.2 GeV were considered for linking. Central tracks had the additional requirement of a minimum radial length of 10 cm, and forward tracks were required to be long enough for an accurate momentum measurement. All tracks had to originate within 50 cm along the z-axis of the nominal vertex, and within 2 cm in the r-ϕ plane. For a valid link, electron cluster energy had to be at least one half that of the track momentum, and both muon and electron and cluster energies could not be more than twice the track momenta. Masses and momenta of J/ψ candidates were calculated using the tracker information. All events that had at least two tracks, one of which was linked to a lepton candidate, were passed to the final J/ψ selection. This selection required 2 or 3 tracks, two of which were link candidates, and a minimum transverse momentum of 1.1 GeV for each track. Cosmics were excluded by a cutting out track pairs that were exactly back-to-back. A further cut on event T0 from the central tracking chamber removed poorly reconstructed cosmics which passed the first cosmic cut. In order to have a well-defined kinematic range, a cut of $Q^2 < 4$ GeV^2 was also applied. Figure 1 shows the resulting mass spectrum from 1993 data corresponding to a luminosity of 320 nb^{-1}. Of the events in the mass peak, 10 were triggered by the Iron Endcap trigger and 9 were triggered by the Tracking trigger. The cross-hatched histogram is a Monte Carlo prediction of the background due to QED lepton pair production.

5. Method

A major problem in the study of elastic J/ψ production is that the amount of contamination from inelastic processes is unknown. However, it is expected that trigger efficiencies and selection efficiencies will be different for different physics processes. These efficiencies can be determined for each trigger type, process, and decay mode. The largest uncertainty in this determination will be due to the simulation of the hardware triggers. For each trigger type, we can vary our assumption of R (ratio of inelastic to elastic events) resulting in a curve $\sigma(R)$. Applying this procedure for several independent triggers will result in different curves. If these curves all intersect, then both R and the total cross section σ will be determined.

6. Results

This method was tested using the PYTHIA MC program, in conjunction with the IJRAY generator which provided the correct photon flux. A full detector simulation was performed using H1SIM, which is a GEANT-based package. A Monte Carlo sample of single diffractive and elastic events in an *apriori* ratio of R=0.7 was prepared and passed through the entire data processing chain. The resulting events were used to

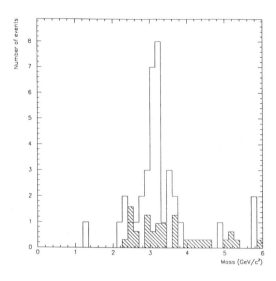

Fig. 1. Mass of dilepton pairs for 1993 data (solid line) and MC estimate of QED background (hatched).

calculate the cross section and a curve $\sigma(R)$ was generated for the three triggers of interest. The value of R=0.8 ± 0.1 obtained from the intersection of these curves is consistent with the input value. This procedure was then repeated for the events in the J/ψ mass peak shown in Fig. 1. While the resulting curves did not intersect in the region where R runs from 0.5 (elastic and inelastic production equal) to 4.0 (large contamination due to inelastic production), there was a region of overlap within their statistical bands. In order to reduce statistical errors, the curves were averaged, and preliminary values for the ep cross section of 13±3 nb for R=0.5 and 17±4 nb for R=4.0 were obtained. This result is consistent with the preliminary value of 12.0 ± 2.5 ± 3.0 nb obtained by the H1 collaboration[5] in a study where all triggers are taken and an average trigger efficiency is used. The photon flux factor can be calculated for the kinematical range given by the lower limit of $W_{\gamma P} > 20$ GeV and the upper limit of $Q^2 < 4$ GeV^2 and is equal to 0.21.

References

1. M. Gell-Mann, F. Zachariasen, *Phys. Rev.* 24, 953 (1961)
2. H. Jung et al., *Z. f. Phys.* C60, 721 (1993)
3. M.G. Ryskin, *Z. f. Phys.* C57, 89 (1993)
4. E.L. Berger and D. Jones, *Phys. Rev.* D23, 1521 (1981)
5. H1 Collaboration, in preparation

BOTTOM AND CHARMONIUM PRODUCTION
IN PION-NUCLEON INTERACTIONS AT 515 GeV/c

RICHARD JESIK*

*Department of Physics, University of Illinois at Chicago,
Chicago, IL 60607, USA*

ABSTRACT

We report on a sample of twelve thousand reconstructed $J/\psi \to \mu^+\mu^-$ events produced in π^--nucleon collisions at 515 GeV/c. Higher mass charmonium states, χ_{c1}, χ_{c2}, and $\psi(2S)$, are observed by their decays into J/ψ. A χ signal in the $J/\psi\gamma$, $\gamma \to e^+e^-$ mode has been reconstructed with a mass resolution sufficient to resolve the χ_{c1} and χ_{c2} states. Bottom-quark states are identified by J/ψs originating from secondary vertices. A few events in the exclusive decay modes $B \to J/\psi K$ and $J/\psi K^*$ have been observed. The inclusive bottom cross section, as well as the relative rates for the observed charmonium states, are presented and compared with recent QCD predictions.

1. Introduction

Hadroproduction of heavy-quark states can provide important information on both perturbative and non-perturbative QCD. Subprocess diagrams for heavy-quark production are known to next-to-leading order, $O(\alpha_s^3)$, in the perturbative expansion, and predictions have been made for the $b\bar{b}$ cross section.[1] Initially, $q\bar{q}$ pairs are produced in states of definite spin, parity, charge conjugation, and color depending upon the number of gluon-quark couplings. One of the problems in charmonium physics is trying to understand how the $c\bar{c}$ pairs form the observed mesons which must be in a color-singlet state. Does the meson retain the quantum numbers of the $c\bar{c}$ pair with each J^{PC} charmonium state coupling only to specific color-singlet subprocesses (color-singlet model), or is the colorless state achieved by emission of soft gluons (color-evaporation model)? These different methods of formation lead to very different rates for the relative production of charmonium states. This paper reports on an experimental measurement of these rates, as well as a measurement of the inclusive b-quark cross section.

The experiment was carried out in the Fermilab Meson West beamline using a large-aperture, open-geometry spectrometer with the capability of studying high-mass muon pairs. The muon detector, located 20 m downstream of the target, consisted of two muon PWC stations, a beam dump, a toroid magnet with an average p_T impulse of 1.3 GeV/c, four more muon PWC stations separated by iron and concrete shielding, and two muon hodoscopes. The upstream part of the spectrometer consisted of 16 planes

---
*Representing the E672 and E706 collaborations. Work supported by the NSF and DOE.

of silicon-strip detectors (SSDs), a dipole magnet with a p_T impulse of 0.45 GeV/c, 16 planes of PWCs and 16 planes of straw drift tubes, a liquid-argon calorimeter with electromagnetic and hadronic sections, and a forward calorimeter. Data were collected using a 515 GeV/c π^- beam incident on Be and Cu targets. Approximately 5×10^6 dimuon triggers were recorded (corresponding to a luminosity of 8 pb^{-1} per nucleon).

2. Charmonium Hadroproduction

A fit was performed to link muon tracks through the entire detector. This fit included multiple-scattering considerations and a consistency requirement for the track momentum as measured via the dipole magnet and by the toroid. The remaining segments in the SSDs and upstream PWCs were used to find the hadronic tracks and event vertices. The reconstructed opposite-sign dimuon mass in the J/ψ region is shown in Fig. 1. A fit to this sample yields $12,640 \pm 150$ J/ψs and approximately 300 ψ(2S)s.

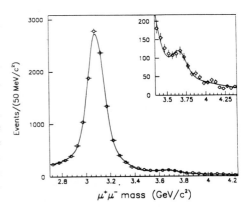

Fig. 1. $\mu^+\mu^-$ invariant mass.

In addition to the dimuon mode, we have detected ψ(2S) in the decay channel into $J/\psi\pi^+\pi^-$. We looked for opposite sign hadrons from the same vertex as the J/ψ and required $M_{\pi\pi} > 0.8(M_{J/\psi\pi\pi} - M_{J/\psi})$, where the latter condition follows from the chiral symmetry of the transition matrix. The resulting $M_{J/\psi\pi^+\pi^-}$ mass plot, shown in Fig. 2, shows a clear ψ(2S) signal. The number of events over background is 300 ± 33.

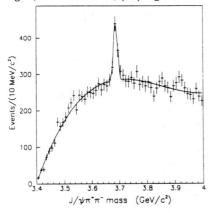

Fig. 2. $J/\psi\pi^+\pi^-$ invariant mass.

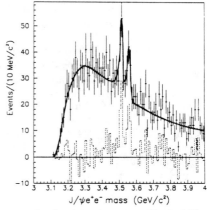

Fig. 3. $J/\psi e^+e^-$ invariant mass (dashed histogram is background-subtracted signal).

To observe $\chi_c \rightarrow J/\psi\gamma$ decays, we looked for J/ψ events with e^+e^- pairs resulting from photon conversions. Due to its small opening angle, a conversion pair appears as a single track (or two closely spaced tracks) until they reach the dipole magnet. After the magnet they are separated in the bend-plane, but remain together in the orthogonal one. After cuts based on this geometry, we obtain the $J/\psi e^+e^-$ mass distribution shown in Fig. 3. Two peaks corresponding to χ_{c1} and χ_{c2} are clearly visible. A fit to the background plus two signal shapes determined by Monte Carlo, gives 59 ± 12 χ_{c1} and 49 ± 11 χ_{c2} events.

Using these signals, we can determine the fractions of J/ψs coming from decays of higher mass states, and from direct production. We assume that all charmonium states are produced with the same x_F distribution as that measured for inclusive J/ψs. After corrections for acceptance, reconstruction efficiencies, and conversion probabilities, we attribute $(44 \pm 8)\%$, $(26 \pm 6)\%$, $(22 \pm 5)\%$, and $(7 \pm 2)\%$ of the J/ψ yield to direct J/ψ production, χ_{c1}, χ_{c2}, and $\psi(2S)$, respectively. This is consistent with predictions of a recent color-singlet model[2] which finds central values for the relative fractions of 43%, 25%, 24%, and 7%. This model, however, predicts a lower ratio of χ_{c1} to χ_{c2} production than is seen in our data.

3.   Bottom-quark Hadroproduction

We have used our J/ψ sample to study bottom-quark production as well. $b \rightarrow J/\psi + X$ decays are uniquely tagged by J/ψs emerging from secondary vertices. To increase the efficiency and precision for finding a secondary vertex, the origin of the J/ψ was found by a muon-oriented vertex finding procedure. Opposite-sign muon pairs in the mass range $2.85 \text{ GeV}/c^2 < M_{\mu\mu} < 3.35 \text{ GeV}/c^2$ with consistent intersections in the $x-z$ and $y-z$ planes were fit with simultaneous vertex and J/ψ mass constraints. The resulting vertex served as a seed for an iterative fitting procedure which associated other tracks to this vertex. After several cuts on vertex quality and separation, we were left with a sample of 121 events in which a J/ψ emerges from a secondary vertex. Events which had more than 4 other tracks associated with the J/ψ vertex were discarded in order to reduce the background from secondary production of J/ψ in our target, leaving 73 events.

We searched this sample for the exclusive channels $B^\pm \rightarrow J/\psi + K^\pm$ and $B^0 \rightarrow J/\psi + K^*$. Events with three-prong secondaries (2 muons plus another track) were selected as candidate $B^\pm \rightarrow J/\psi + K^\pm$ decays. To reduce background, only combinations with a secondary hadron having $p_T > 0.5 \text{ GeV}/c$ were considered. This track was assumed to be a kaon, and its momentum was added to that of the J/ψ to form a candidate B momentum vector, which was required to point back to primary vertex to within 80 μm. Since tracks from the underlying event can be coincidentally associated with a secondary vertex, 4-prong secondary vertices were also included in the selection. None of the 4-prong vertex events in the sample had both secondary hadron tracks pass the p_T cut. The secondary vertices for these events were refit omitting the low p_T track, and the events were analyzed under the hypothesis of 3-prong B decays. To search for $B^0 \rightarrow J/\psi + K^*$ decays, K^*s were observed by their decays into $K^\pm\pi^\mp$ pairs and combined with the J/ψ in a similar manner as above. The combined $J/\psi K^\pm, J/\psi K^*$ invariant-mass distribution obtained by this analysis, shown in Fig. 4, shows an excess of events at the nominal B-meson mass.

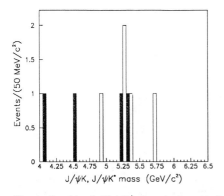

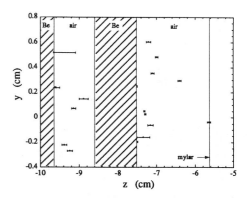

Fig. 4. Combined $J/\psi K^{\pm}$ (hatched entries), $J/\psi K^*$ invariant mass.

Fig. 5. Secondary-vertex J/ψ positions in the $y - z$ plane showing air gap events.

The background due to secondary interactions is negligible for vertices occurring in the air-gap region of our target-SSD system. In ten events there is a secondary-vertex J/ψ that occurs at least three standard deviations away from the Be target elements. The positions of these vertices in the $y - z$ plane (with z error bars) are shown in Fig. 5. The background to this signal, estimated using Monte Carlo simulations, was found to be 2 ± 1 events, leaving 8 ± 3.3 events attributed to b-decays.

Based upon this signal we obtain a cross section in the forward direction ($x_F > 0$) of $\sigma_{b\bar{b}} = (47 \pm 19 \pm 14)$ nb/nucleon (assuming A^1 production). Extrapolating to all values of x_F, using predictions by MNR,[1] we obtain a total cross section of $\sigma_{b\bar{b}} = (75 \pm 31 \pm 26)$ nb/nucleon. A comparison of our measurement to predictions for bottom cross sections in πN collisions[1] and to those of previous experiments,[3] is shown in Fig. 6. Our measurement favors predictions with smaller b-mass values and renormalization scales that are a smaller fraction of the mass.

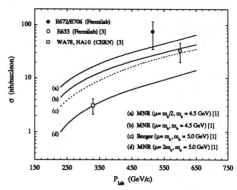

Fig. 6. $b\bar{b}$ cross section in πN collisions.

References

1. P. Nason, S. Dawson, and R. K. Ellis, *Nucl. Phys.* **B303** (1988) 607, **B327** (1989) 49, **B335** (1990) 260; E. L. Berger, *Phys. Rev.* **D37** (1988) 1810; M. Mangano, P. Nason, G. Ridolfi, *Nucl. Phys.* **B405** (1993) 507; **B373** (1992) 295.
2. G. Schuler, CERN-TH.7170 (1994).
3. K. Kodama *et al.*, *Phys. Lett.* **B303** (1993) 359; M. Catanesi *et al.*, *Phys. Lett.* **B187** (1987) 431; P. Bordalo *et al.*, *Z. Phys.* **C39** (1988) 7.

A SEARCH FOR NEW CHARMONIUM STATES
DECAYING INTO J/Ψ PLUS CHARGED PIONS

A. MCMANUS[3]

The E-771 collaboration

T. Alexopoulos[1], L. Antoniazzi[2], M. Arenton[3], H.C. Ballagh[4], H. Bingham[4],
A. Blankman[5], M. Block[6], A. Boden[7], G. Bonomi[2], S.V. Borodin[5], J. Budagov[8],
Z.L. Cao[3], G. Cataldi[9], T.Y. Chen[10], K. Clark[11], D. Cline[7], S. Conetti[3], M. Cooper[12],
G. Corti[3], B. Cox[3], P. Creti[9], C. Dukes[3], C. Durandet[1], V. Elia[9], A.R. Erwin[1],
E. Evangelista[9], L. Fortney[13], V. Golovatyuk[3], E. Gorini[9], F. Grancagnolo[9],
K. Hagan-Ingram[3], M. Haire[14], P. Hanlet[3], M. He[15], G. Introzzi[2], M. Jenkins[11],
J. Jennings[1], D. Judd[14], W. Kononenko[5], W. Kowald[13], K. Lau[16], T. Lawry[3],
A. Ledovskoy[3], G. Liguori[2], J. Lys[4], P.O. Mazur[17], S. Misawa[4], G.H. Mo[16],
C.T. Murphy[17], K. Nelson[3], M. Panareo[9], V. Pogosian[3], S. Ramachandran[7],
M. Recagni[3], J. Rhoades[7], J. Segal[3], W. Selove[5], R.P. Smith[17], L. Spiegel[17],
J.G. Sun[3], S. Tokar[18], P. Torre[2], J. Trischuk[19], L. Turnbull[14], I. Tzamouranis[3],
D.E. Wagoner[14], C.R. Wang[15], C. Wei[13], W. Yang[17], N. Yao[10], N.J. Zhang[15],
S.N. Zhang[5] and B.T. Zou[13]

1. University of Wisconsin, Madison, Wisconsin, USA
2. Pavia INFN and University, Pavia, Italy
3. University of Virginia, Charlottesville, Virginia, USA
4. University of California at Berkeley, Berkeley, California, USA
5. University of Pennsylvania, Philadelphia, Pennsylvania, USA
6. Northwestern University, Evanston, Illinois, USA
7. University of California at Los Angeles, Los Angeles, California, USA
8. JINR, Dubna, Russia
9. Lecce INFN and University, Lecce, Italy
10. Nanjing University, Nanjing, People's Republic of China
11. University of South Alabama, Mobile, Alabama, USA
12. Vanier College, Montreal, Quebec, Canada
13. Duke University, Durham, North Carolina, USA
14. Prairie View A&M, Prairie View, Texas, USA
15. Shandong University, Jinan, Shandong, People's Republic of China
16. University of Houston, Houston, Texas, USA
17. Fermilab, Batavia, Illinois, USA
18. Comenius University, Bratislava, Slovakia
19. McGill University, Montreal, Quebec, Canada

ABSTRACT

The FNAL E771 data was examined to search for new states of hidden charm decaying into J/Ψ plus pions. The search was motivated by the previously reported observations of a possible $^3D_2$ state decaying into $J/\Psi\pi^+\pi^-$. Preliminary results, based on the analysis of a sample of about 15,000 J/Ψ's produced in 800 GeV/c pSi interactions, are reported with a tentative confirmation of a state decaying into $J/\Psi\pi^+\pi^-$ at $3.833 \pm 0.006 GeV/c^2$.

1. Introduction

Fermilab experiment E771 was designed to study the production of particles decaying into final states containing one or two muons. The experiment used the an 800 GeV/c proton beam extracted from the Fermilab Tevatron and transported to the Fermilab High Intensity Laboratory. The experiment ran for approximately one month in the 1991 fixed target run at about 2×10^6 interactions/sec and collected 127 million dimuon and 62 million single muon triggers.

2. E771 Spectrometer

The E771 spectrometer[1] consisted of a silicon foil target, a silicon microvertex detector and 31 planes of wire chambers upstream of an analyzing magnet. Downstream of the analyzing magnet were 20 drift chambers, 4 pad chamber planes and 4 strip (Y) planes. Downstream of the wire chambers was an electromagnetic calorimeter and a muon detector consisting of three planes each of Resistive Plate Counters (RPC) and scintillator counters embedded in layers of concrete and steel which provide muon shielding equivalent to an integrated $\frac{dE}{dx}$ of 10 GeV/c.

3. Data Analysis

For this analysis, the 127 million dimuon triggers, which corresponded to a live beam of $(1.65 \pm 0.17) \times 10^{13}$ protons, were processed first by a muon tracking program followed by a full hadronic tracking program. The muon tracking program reconstructed only tracks which had rear track segments consistent with at least five out of six RPC or scintillator counters hit in the muon detector. The rear track segments were required to match at the magnet with a front track that pointed at the target. Only wire chambers upstrea of the magnet were used to construct front tracks. The silicon detector was not used in this analysis.

The reconstructed dimuon mass is presented in Fig 1. In a given event the dimuon pair which had the smallest χ^2 when fit to the assumption that the muons originated from a common point in space was chosen for the analysis. The total number of events in the J/Ψ mass region was approximately 21000. There are approximately 15000 events in the J/Ψ peak above background. Dimuon events with the best muon pair having a reconstructed mass within a window of 150 MeV/c around the J/Ψ mass were then tracked by a full tracking program. All charged tracks not identified as a muon in the first pass were assumed to be pions.

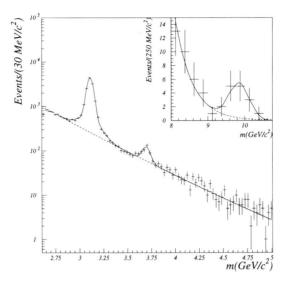

Fig. 1: Invariant dimuon mass.

After full tracking the J/Ψ region events, the two muons were assumed to have come from the decay of the J/Ψ and their momenta were rescaled using the J/Ψ mass as a constraint. Additional track quality cuts were made on the pion track front segments: track $\chi^2/DOF \leq 2.5$, no track was allowed to share more than 45% hits with another track, and the number of x projection hits on a track must be ≥ 5. On a rear track segment, no track was allowed to share more than 15% of the hits with another track, track $\chi^2/DOF \leq 4.0$, and the number of x projection hits on a track had to be ≥ 4. In addition the front and rear track segments had to match at the magnet with a matching $\chi^2/DOF \leq 25$.

4. States With a J/Ψ Plus two charged pions

For the J/Ψ events used in the search of $J/\Psi\pi\pi$ final states, three additional criteria were imposed on the two pion tracks: pions with momenta greater than 100 GeV/c were eliminated, the pions were constrained to come from a common vertex with the J/Ψ, and finally the pion pairs were limited to those that had a dipion mass in the range $0.40GeV/c^2 \leq M_{\pi\pi} \leq 0.75GeV/c^2$. The dipion mass cut was chosen based on predictions that the phase space for decays like $\Psi(2s) \to J/\Psi\pi\pi$ is modified by chiral symmetry such that the dipion mass is skewed toward high values[2]. The resulting $J/\Psi\pi^+\pi^-$ mass spectrum is shown in Fig 2a.

The $J/\Psi\pi^+\pi^+$ and $J/\Psi\pi^-\pi^-$ mass spectrum which was used to construct the background function is shown in Fig 2b. The same sign pion plus J/Ψ mass spectrum

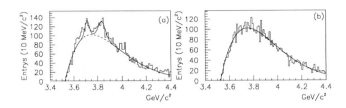

Fig. 2: a) Invariant $J/\Psi\pi^+\pi^-$ mass. b) Invariant $J/\Psi\pi^\pm\pi^\pm$ mass.

was subjected to the same cuts as the $J/\Psi\pi^+\pi^-$ mass spectrum. It is evident that the same sign spectrum is quite different from the opposite sign spectrum.

The $J/\Psi\pi^+\pi^-$ mass spectrum shows two enhancements. The enhancement at a mass of 3.695 ± 0.013 GeV/c^2 is identified as the $\Psi(2s)$. From the cross section measured by E771[3], the branching ratio[4] of the $\Psi(2s) \to \mu\mu$, and an acceptance times reconstruction efficiency for reconstruction of the dipions of 14% (after reconstructing the J/Ψ), the branching ratio of the $\Psi(2s) \to J/\Psi\pi^+\pi^-$ is found to be $47 \pm 17\%$ which agrees within errors with the established branching ratio[4] of $32.4 \pm 2.6\%$. The second peak at a mass of 3.833 ± 0.006 GeV/c^2 is consistent with an enhancement seen by E705[5] and is tentatively identified as the $^3D_2$ state of charmonium. There are 145 ± 28 events above background at a mass of 3.695 GeV/c^2 and 248 ± 35 events above background at a mass of 3.833 GeV/c^2, which indicates that the enhancement at 3.833 GeV/c^2 has a branching ratio times cross section of 20 ± 7 nb/nucleon.

While it is impossible at this time to completely rule out the possibility that the enhancement at $3.833 GeV/c^2$ is an artifact of the E771 spectrometer or reconstruction program the tests made to date tend to rule out that explanation. The events at $3.833 GeV/c^2$ are not common to the events containing the $\Psi(2s)$. If the events at $3.833 GeV/c^2$ were $\Psi(2s)$ events distorted or displaced by some mechanism to $3.833 GeV/c^2$ then the branching ratio of the $\Psi(2s)$ would be more than twice the accepted value and well outside the errors for this measurement. No comparable enhancements are observed in the $J/\Psi\pi^+\pi^+$, $J/\Psi\pi^-\pi^-$, and $J/\Psi\pi^\pm$ mass spectra. We tentatively identify the enhancement at $3.833 GeV/c^2$ as the $^3D_2$ state of charmonium since such a state although above the $D\bar{D}$ threshold would be prevented from decay into $D\bar{D}$ by quantum numbers. The expected mass[6,7] for the $^3D_2$ is in the range from 3.810 to 3.840 GeV/c^2 which is consistent with the observed mass of the enhancement at 3.833 GeV/c^2.

5. Conclusions

We have observed the decay $\Psi(2s) \to J/\Psi\pi^+\pi^-$ and a structure at 3.833 GeV/c^2 that is consistent in position and magnitude with a state reported by E705 and tentatively assigned as the $^3D_2(2^{--})$ state of charmonium.

6. Acknowledgments

We thank the Fermilab staff including Research and Computing Divisions. This work is supported by DOE, NSF, Natural science and Engineering Research Council of Canada, and Instituto Nazionale Di Fisica Nucleare (INFN).

References

1. A. McManus *et al. I.E.E.E Transactions on Nuclear Science* **Vol 39, No. 5** (1992) 1249.
2. L.S. Brown and R.S. Cahn, *Phys. Rev. Lett.* **35** (1975) 1.
3. G. Corti *et al.* "High Mass Dimuon states Produced in 800 GeV/c pN interactions", these (DPF) proceedings (1994)
4. L.S. Brown *et al.* "Review of particle properties", *Phys. Rev. D*, **Vol 45, No 11** (1992) VII.179
5. L. Antoniazzi *et al.* "Search for Hidden Charm States Decaying into J/Ψ or Ψ' plus pions", *accepted by Phys. Rev. D*, 1994
6. S. Godfrey and N. Isgur. *Rhys. Rev.* **D32** (1985) 189.
7. W. Kwong, J Rosner, and C. Quigg. *Ann. Rev. of Nucl. and Part. Phys.* **37** (1987) 343.

J/ψ, $\psi' \to \mu^+\mu^-$ and $B \to J/\psi$, ψ' CROSS SECTIONS

TROY DANIELS*

Fermilab, Batavia, IL 60510

ABSTRACT

This paper presents a measurement of $J/\psi,\psi'$ differential cross sections in $p\bar{p}$ collisions at $\sqrt{s} = 1.8$ TeV. The cross sections are measured above 4 GeV/c in the central region ($|\eta| < 0.6$) using the dimuon decay channel. The fraction of events from B decays is measured, and used to calculate b quark cross sections and direct $J/\psi,\psi'$ cross sections. The direct cross sections for the ψ' are found to be more than an order of magnitude above theoretical expectations.

1. Introduction

Charmonium production is currently the best way to study b quark production at CDF at the lowest transverse momentum of the b. While the signal-to-noise for the J/ψ is excellent, conclusions regarding b production are dependent on the fraction of the sample due to b decays. One way of determining this fraction is to use the decay distance of the J/ψ state. In contrast, the fraction of ψ''s from b decays was thought to be close to one.[1] In using ψ' events to study b quarks, one should not have to worry about the fraction due to b decays. However, a very large zero-lifetime component for the observed ψ' signal is seen. Therefore, one still must make use of the lifetime information.

The CDF detector has been described in detail elsewhere.[2] We mention here briefly the components relevant to this analysis. A solenoidal magnet generating a 1.4 T magnetic field surrounds the two tracking chambers used. The Central Tracking Chamber (CTC) is a cylindrical drift chamber surrounding the beam line. The Silicon Vertex Detector (SVX) is a silicon microvertex detector that provides an impact parameter resolution of $(13+40/P_T)\mu$m. Because the SVX is shorter than the interaction region, only about 60% of the CTC tracks pass through the SVX. Outside the CTC are electromagnetic and hadronic calorimeters, which provide five absorption lengths of material before the Central Muon Chambers (CMU).

2. Event Selection

Events are selected from a dimuon trigger. Both muons are required to have $P_T^\mu > 2.0$ GeV/c and one muon must have $P_T^\mu > 2.8$ GeV/c. The dimuon is required to have $|\eta^{\mu\mu}| < 0.6$ and $P_T^{\mu\mu} > 4$ GeV/c. The tracks are constrained to improve the mass resolution, and the resulting invariant mass is used to define signal regions of $3.0441 < m_{\mu\mu} < 3.1443$ GeV/c^2 for the J/ψ and $3.636 < m_{\mu\mu} < 3.736$ GeV/c^2 for the ψ'. Sideband regions from $2.9606 < m_{\mu\mu} < 3.0274$ GeV/c^2 or $3.1610 < m_{\mu\mu} < 3.2278$ GeV/c^2 for the J/ψ and of $3.52 < m_{\mu\mu} < 3.62$ GeV/c^2 or $3.75 < m_{\mu\mu} < 3.85$ GeV/c^2 for the ψ' are used.

---

*Representing the CDF collaboration

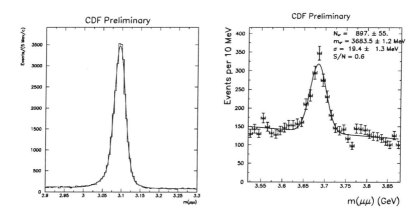

Fig. 1. Mass distribution $(J/\psi, \psi')$

3. b Fraction

For events where both muons have well-measured SVX tracks, we vertex constrain the tracks to measure the decay length. This is converted into the proper lifetime of the parent using the $J/\psi,\psi'$ transverse momentum and a Monte Carlo correction factor. The background shape is measured from the sidebands. The signal region is fit to a resolution function, an exponential convoluted with the resolution function, and the background shape. We fix the b lifetime to 438 μm, as found by the CDF inclusive b lifetime measurement.[3]

Separate binned fits are done on the J/ψ signal and sideband regions. A log-likelihood fit is performed on the ψ' data. The background fraction is varied within Poisson statistics, but the resolution function is limited to a Gaussian, as measured errors are used. Varying the fitting methods changes the b fraction by 7%, which we assign as the systematic uncertainty.

3.1. P_T dependence

It is expected that the b fraction in the $J/\psi,\psi'$ samples rises with P_T. To measure this, we divide the ψ' sample into three P_T bins from $4-6$ GeV, $6-9$ GeV, and $9-20$ GeV. We repeat the above fitting procedure in each of the P_T regions. In the J/ψ sample, the spectrum of events with $c\tau > 250\mu$m is found. This is divided by the J/ψ spectrum, normalized so that the ratio of the two areas is the inclusive b fraction.

4.   Cross Section

4.1. $J/\psi,\psi'$ Cross Section

After correcting the measured P_T distributions for acceptance and efficiencies, we find

CDF Preliminary

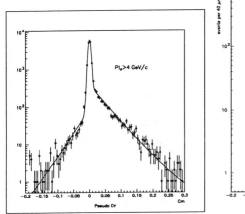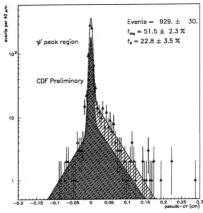

Fig. 2. J/ψ and ψ' Lifetime Distributions. In the ψ' plot, the dark region is the background shape. The slashed region is the B component plus the background shape.

$\sigma(J/\psi) \cdot Br(J/\psi \to \mu^+\mu^-) = 29.10 \pm 0.19(stat) \, {}^{+3.05}_{-2.84}(syst)$ nb and $\sigma(\psi') \cdot Br(\psi' \to \mu^+\mu^-) = 0.721 \pm 0.058(stat) \pm 0.072(syst)$ nb, where $\sigma(\psi) = \sigma(p\bar{p} \to \psi X, P_T^\psi > 4$ GeV/$c, |\eta| < 0.6)$. The systematic uncertainties are dominated by the trigger efficiency (7-8%), the polarization (5%), and the luminosity (4%). The cross section from b decays is extracted by multiplying the differential b fraction by the cross section. We also obtain a prompt cross section by multiplying the cross section by one minus the b fraction. The prompt theory curves are from reference.[4]

4.2. Inclusive b Cross Section

The cross sections from b decays are combined with b quark acceptances to produce integrated b quark cross sections. The P_T of the b quark is described by P_T^{min}, the P_T such that 90% of the b quarks in our sample have $P_T^b > P_T^{min}$. We use branching ratios of $Br(b \to J/\psi X) = 1.16 \pm 0.09\%$,[5] $Br(b \to \psi'X) = 0.30 \pm 0.06\%$,[6] $Br(J/\psi \to \mu\mu) = 6.27 \pm 0.20\%$[7] and $Br(\psi' \to \mu\mu) = 0.88 \pm 0.13\%$.[8] Using the J/ψ, we find $\sigma(6.0$ GeV/$c) = 12.16 \pm 2.07$ μb, $\sigma(7.3$ GeV/$c) = 8.24 \pm 1.34$ μb and $\sigma(8.7$ GeV/$c) = 5.20 \pm 0.83$ μb. The ψ' results are $\sigma(5.9$ GeV/$c) = 6.12 \pm 2.04$ μb and $\sigma(8.3$ GeV/$c) = 3.85 \pm 1.23$ μb, where $\sigma(P_T^{min}) = \sigma(p\bar{p} \to bX, |y^b| < 1.0, P_T^b > P_T^{min})$. The errors are statistical and systematic, added in quadrature. The systematic uncertainties are dominated by the $J/\psi, \psi'$ momentum distribution (5%), Peterson ϵ (5%), and the values of μ and Λ in the NDE spectrum (4.6%). A 4% uncertainty is added to the ψ' values because of the large bin size used in calculating the b fraction.

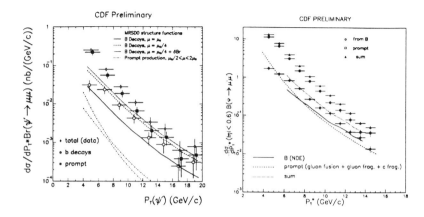

Fig. 3. ψ' and J/ψ Differential Cross Sections. The b(prompt) values have been artificially offset by 200(-200) MeV/c for clarity in the ψ'.

5.   Acknowledgements

We thank the Fermilab staff and the technical staffs of the participating institutions for their vital contributions. This work was supported by the U.S. Department of Energy and National Science Foundation; the Italian Istituto Nazionale di Fisica Nucleare; the Ministry of Education, Science and Culture of Japan; the Natural Sciences and Engineering Research Council of Canada; the National Science Council of the Republic of China; the A. P. Sloan Foundation; and the Alexander von Humboldt-Stiftung.

References

1. S.D. Ellis, et al., Phys. Rev. Lett. **36**, 1263 (1976); C.E. Carlson and R. Suaya, Phys. Rev. D **15**, 1416 (1977)
2. F.Abe et al., CDF Collab., Nucl. Instrum. Methods Phys. Res., Sect. A, **271**, 387 (1988)
3. F. Abe et al., Published Proceedings Advanced Study Conference On Heavy Flavours, Pavia, Italy, September 3-7, 1993. FERMILAB-CONF-93/319-E
4. E. Braaten et al., Fermilab Preprint, FERMILAB-PUB-94/135-T (1994)
5. Y. Kubota et al., "Inclusive Decays of B Mesons to J/ψ and ψ'," Contributed paper at XVI International Symposium on Lepton-Photon Interactions: Cornell University, Ithaca, N.Y., U.S.A., August 10-15, 1993
6. T. Browder et al., "A Review of Hadronic and Rare B Decays", To appear in B Decays, 2nd edition, Ed. by S. Stone, World Scientific
7. Particle Data Group, M.Aguillar-Benitez et al., Phys. Rev. **D45**, (1992)
8. G.J. Feldman and M.L. Perl, Physics Letters, **33C**, 285 (1977)

FRAGMENTATION PRODUCTION OF J/ψ AND ψ' AT THE TEVATRON*

MICHAEL A. DONCHESKI

Department of Physics, Carleton University, Ottawa, Ontario K1S 5B6, Canada

ERIC BRAATEN†

Fermi National Accelerator Laboratory, P.O. Box 500, Batavia, IL 60510

SEAN FLEMING

Department of Physics and Astronomy, Northwestern University, Evanston, IL 60208
and

MICHELANGELO L. MANGANO

INFN, Scuola Normale Superiore and Dipartimento di Fisica, Pisa, Italy

ABSTRACT

We present a calculation of the charm and gluon fragmentation contributions to inclusive J/ψ and ψ' production at large transverse momentum at the Tevatron. For ψ production, we include both fragmentation directly into ψ and fragmentation into χ_c followed by the radiative decay to ψ. We find that fragmentation overwhelms the leading-order mechanisms for prompt ψ production at large p_T, and that the dominant contributions come from fragmentation into χ_c. Our results are consistent with recent data on ψ production from the Tevatron. In the case of prompt ψ' production, the dominant mechanism at large p_T is charm fragmentation into ψ'. We find serious disagreement between our theoretical predictions and recent ψ' data from CDF.

The study of charmonium production in high energy hadronic collisions provides an important testing ground for perturbative quantum chromodynamics (QCD). The J/ψ and ψ' states are of particular interest since they are produced in abundance and are relatively easy to detect at a collider such as the Tevatron. In earlier calculations of direct charmonium production at large transverse momentum (p_T) in $p\bar{p}$ collisions[1] , it was assumed that the leading-order diagrams give the dominant contributions to the cross section. These calculations did not reproduce all aspects of the available data,[2-4] suggesting that there are other important production mechanisms. It was pointed out by Braaten and Yuan[5] that fragmentation processes, while formally of higher order

---

*Talk presented by M. A. Donchseki
†on leave from the Dept. of Physics and Astronomy, Northwestern University, Evanston, IL 60208

in strong coupling (α_s), will dominate at sufficiently large p_T. Explicit calculations of the contribution to ψ production at the Tevatron from fragmentation revealed that fragmentation dominates over leading-order gluon-gluon fusion for p_T greater than about 6 GeV.[6] In addition to being directly produced, the ψ signal is also fed by the radiative decay of the χ_c states. Full details of our calculation can be found in Ref. [7]. Similar results are obtained by Cacciari and Greco[8] and by Roy and Sridhar.[9]

The differential cross section for ψ production (to be specific) can be written as:

$$d\sigma(p\bar{p} \to \psi(p_T, y) + X) = \sum_i \int_0^1 dz \, d\sigma(p\bar{p} \to i(\frac{p_T}{z}, y) + X, \mu_{\text{frag}}^2) \, D_{i \to \psi}(z, \mu_{\text{frag}}^2), \quad (1)$$

where z is the longitudinal momentum fraction of the ψ relative to parton i, y is the rapidity of the ψ, and $D_{i \to \psi}(z, \mu_{\text{frag}}^2)$ is the fragmentation function. The dependence on the fragmentation scale μ_{frag} cancels between the two factors only after inclusion of all orders of the perturbative expansion. For the charm quark mass, we use $m_c = 1.5$ GeV. The S-wave fragmentation functions depend on the wavefunction at the origin, which we take to be $|R_\psi(0)|^2 = 0.7$ GeV^3. This value is obtained from the electronic width of the ψ, including the effect of the NLO perturbative correction (which is about 50%) but not relativistic corrections. The P-wave fragmentation functions depend on two nonperturbative parameters. For the derivative of the radial wavefunction at the origin, we use the value $|R'_{\chi_c}(0)|^2 = 0.053$ GeV^5. This value is determined from the annihilation rates of the χ_c states,[11] neglecting the as-yet-uncalculated NLO perturbative corrections as well as relativistic corrections. The least well-determined parameter in the P-wave fragmentation functions is a parameter H'_8 associated with the color-octet mechanism for P-wave production.[12] This parameter is poorly constrained, lying in the range $2.2 < H'_8 < 25$ MeV.[10,13] We use the value $H'_8 = 3$ MeV, which is consistent with measured branching fractions for B mesons into the χ_c states.[12] Considering all the theoretical uncertainties, we believe that the error in our fragmentation calculations can easily be larger than a factor of 2, but it is definitely less than an order of magnitude. One should of course keep in mind that in addition to fragmentation, which must dominate at sufficiently large p_T, there are other contributions suppressed by factors of m_c^2/p_T^2 that may be important at the values of p_T that are currently accessible.

In fig. 1, the sum of the fragmentation contributions (solid curves) and the sum of the leading-order contributions (dashed curves)* are compared with preliminary CDF data for prompt ψ production.[14] We used the MRSD0 parton distribution set. In order to compare with available data from the Tevatron, we imposed a pseudorapidity cut of $|\eta| < 0.6$ on the ψ. The contribution to ψ production from b-hadron decays has been removed from the data via detection of the secondary vertex from which the ψ's originate.[14] The upper and lower curves in fig. 1 were obtained by varying the renormalization (μ_R), factorization (μ_F) and fragmentation (μ_{frag}) scales used in the calculation, in order to provide an estimate of the systematic uncertainty associated with the LO calculation. The upper curve corresponds to $\mu_R = \mu_F = P_T/2$ and $\mu_{\text{frag}} = \max(P_T/2, \mu_0)$, while the lower curve is obtained for $\mu_R = \mu_F = \mu_{\text{frag}} = 2P_T$ (here, P_T is the transverse momentum of the fragmenting parton). The cross-over of the curves at small p_T is

---

*The results for direct ψ inclusive p_T distributions given in ref. [4] are incorrect, due to a coding error.

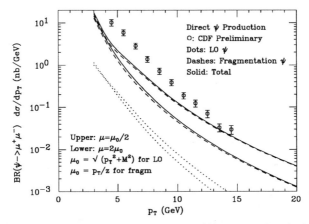

Fig. 1. Preliminary CDF data for prompt ψ production (O) compared with theoretical predictions of the total fragmentation contribution (solid curves) and the total leading-order contribution (dashed curves).

due to the rapid growth of the parton distribution functions with increasing scale, and should be considered an artificial reduction of scale sensitivity. While the shapes of the leading-order curve and the fragmentation curve are both consistent with the data over the available range of p_T, the leading-order contribution is too small by over an order of magnitude. The fragmentation contribution has the correct normalization to within a factor of 2 or 3, which can be easily accounted for by theoretical uncertainties. We conclude that the fragmentation calculation is not inconsistent with the CDF data.

We next consider the production of ψ', which should not receive any contributions from higher charmonium states. The ψ' fragmentation contribution can be obtained from the ψ fragmentation contribution simply by multiplying by the ratio of the electronic widths of the ψ' and ψ. The total fragmentation contribution (solid curves) and the leading-order contribution (dashed curves) are shown in fig. 2, along with the preliminary CDF data.[15] Again the contribution from b–hadron decays has been subtracted. The pairs of curves correspond to the same choices of scales as in fig. 1. The dominant production mechanisms are gluon-gluon fusion for p_T below about 5 GeV, and c-quark fragmentation into ψ' for larger p_T. The leading-order curve falls much too rapidly with p_T to explain the data, but the fragmentation curve has the correct shape. However, in striking contrast to the case of ψ production, the normalization of the fragmentation contribution to ψ' production is too small by more than an order of magnitude. That there is such a large discrepancy between theory and experiment in the case of ψ', but not for ψ, is extremely interesting. It suggests that there are other important mechanisms for production of S-wave states at large p_T beyond those that have presently been calculated. While such processes would certainly affect ψ production as well, their effect may not be as dramatic because of the large contribution from χ_c-production in the case of the ψ. Whether the normalization agrees can only

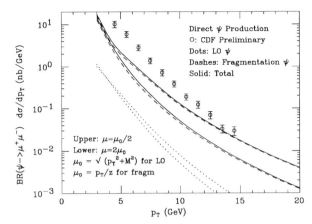

Fig. 2. Preliminary CDF data for prompt ψ' production (O) compared with theoretical predictions of the total fragmentation contribution (solid curves) and the total leading-order contribution (dashed curves).

be determined by explicit calculation, and such a calculation is in progress.

References

1. R. Baier and R. Rückl, *Z. Phys.* **C19** (1983) 251; F. Halzen et al., *Phys. Rev.* **D30** (1984) 700; B. van Eijk and R. Kinnunen, *Z. Phys.* **C41** (1988) 489; E.W.N. Glover, A.D. Martin, W.J. Stirling, *Z. Phys.* **C38** (1988) 473.
2. C. Albajar et al., UA1 Coll., *Phys. Lett.* **256B** (1991) 112.
3. F. Abe et al., CDF Coll., *Phys. Rev. Lett.* **69** (1992) 3704.
4. M. L. Mangano, *Z. Phys.* **C58** (1993) 651.
5. E. Braaten and T.C. Yuan, *Phys. Rev. Lett.* **71** (1993) 1673.
6. M.A. Doncheski, S. Fleming and M.L. Mangano, in Proceedings of the Workshop on Physics at Current Accelerators and the Supercollider, ANL, 1993, p. 481.
7. E. Braaten, et al., FERMILAB-PUB-94/135-T, to appear in Phys. Lett. **B**.
8. M. Cacciari and M. Greco, INFN preprint, FNT/T-94/13, hep-ph/9405241.
9. D.P. Roy and K. Sridhar, CERN-TH.7329/94.
10. E. Braaten and T.C. Yuan, Fermilab preprint FERMILAB-PUB-94/040-T (1994).
11. G.T. Bodwin, E. Braaten, and G.P. Lepage, *Phys. Rev.* **D46** (1992) 1914.
12. G.T. Bodwin, et al., *Phys. Rev.* **D46** (1992) R3703.
13. H. Trottier, *Phys. Lett.* **B320** (1994) 145.
14. V. Papadimitriou, CDF Coll., presented at the "Rencontres de la Vallee d'Aoste", La Thuile, March 1994.
15. T. Daniels, CDF Coll., Fermilab-Conf-94/136-E.

PRELIMINARY MEASUREMENT OF THE D^{*+} BRANCHING FRACTIONS AND THE $D\overline{D}^*$ PRODUCTION CROSS SECTIONS AT 4.03 GEV

OLIVER BARDON[*]

Massachusetts Institute of Technology, Cambridge, MA 02139, USA

ABSTRACT

The Beijing Spectrometer experiment has collected a data sample corresponding to about 22 pb^{-1} at a center-of-mass energy of 4.03 GeV. At this energy $D^*\overline{D}^*$ and $D\overline{D}^*$ events are distinguishable from one another using the distinct momentum spectra of the D mesons produced in each type of event. The number of D candidates observed in the $D^0 \to K^-\pi^+$ and $D^+ \to K^-\pi^+\pi^+$ modes in $D\overline{D}^*$ events have been counted, and from these numbers preliminary values for the D^{*+} branching ratios and the $D\overline{D}^*$ production cross sections at 4.03 GeV have been extracted.

1. Motivation

There are several interesting D^* physics topics which can be addressed by the Beijing Spectrometer experiment (BES) (the Beijing Spectrometer has been described in detail elsewhere[1]) using some 22 pb^{-1} of e^+e^- annihilations accumulated at $\sqrt{s} = 4.03$ GeV at the Beijing Electron-Positron Collider (BEPC).

a) BES can perform a direct measurement of $B(D^{*+} \to D^0\pi^+)$, from which the other D^{*+} decay fractions can be obtained.[†] This measurement is largely motivated by the ≈ 2 standard deviation difference between the two most recent and statistically significant reported measurements (Table 1), done by the Mark III[2] and CLEO II[3] experiments.[4] The Mark III measurement dominates the 1992 Particle Data Group (PDG) D^{*+} decay values,[5] while the single CLEO II result now dominates the 1994 PDG values.[6]

Table 1. D^{*+} decay branching fraction measurements.

Decay Mode	Mark III (1988)	PDG (1992)	CLEO II (1992)	PDG (1994)
$B(D^{*+} \to D^0\pi^+)$	$57 \pm 4 \pm 4$	55 ± 4	$68.1 \pm 1.0 \pm 1.3$	68.1 ± 1.3
$B(D^{*+} \to D^+\pi^0)$	$26 \pm 2 \pm 2$	27.2 ± 2.5	$30.8 \pm 0.4 \pm 0.8$	30.8 ± 0.8
$B(D^{*+} \to D^+\gamma)$	$17 \pm 5 \pm 5$	18 ± 4	$1.1 \pm 1.4 \pm 1.6$	$1.1^{+1.4}_{-0.7}$

---

[*]Representing the Beijing Spectrometer Collaboration.

[†]Reference to a specific particle or decay also implies the charge conjugate particle or decay.

b) An anomalously high ratio of the $D^*\overline{D}^*$ production cross section to the $D\overline{D}^*$ production cross section has been observed in this energy region, near the $D^*\overline{D}^*$ production threshold.[7] Several mechanisms have been suggested to explain this situation, including the existence of a $D^*\overline{D}^*$ bound state.[8] BES can measure these cross sections and further study $D^*\overline{D}^*$ production.

c) The process $D^0\overline{D}^{*0} \rightarrow D^0\overline{D}^0\gamma$ provides an opportunity to search for indirect CP violation using a time-integrated measurement of a CP eigenstate.[9] BES can perform prototype studies for future high-statistic experiments at a τ-charm factory.

2. D^* Physics at 4.03 GeV

At 4.03 GeV, D^* mesons are produced only in the reactions $e^+e^- \rightarrow D\overline{D}^*$ and $e^+e^- \rightarrow D^*\overline{D}^*$. At this energy, the momentum spectra of D mesons, both primary and secondary (from $D^* \rightarrow DX$), from these events and from $e^+e^- \rightarrow D\overline{D}$ events are all distinct from one another (Table 2). This fact allows the separation of the three types of events, and the selection of a pure $D\overline{D}^*$ sample.

Table 2. D momenta from production at $\sqrt{s}$=4.03 GeV.

$D^*\overline{D}^*$	.003 GeV $< P_D < $ .308 GeV
$D\overline{D}^*$	.368 GeV $< P_D < $ .665 GeV
$D\overline{D}$	$P_{D^0} = .752$ GeV, $P_{D^+} = .765$ GeV

The D^{*+} decays via three modes, to $D^+\gamma$, $D^+\pi^0$, and $D^0\pi^+$, while the D^{*0} decays only to $D^0\gamma$ and $D^0\pi^0$. Requiring that the D^{*+} decay branching fractions sum to 1, the D^+ and D^0 production cross sections for DD^* events can be expressed as a function of $B(D^{*+} \rightarrow D^0\pi^+)$ $(\equiv B_{\pi+})$ and the $D\overline{D}^*$ cross sections:

$$\sigma_{D^0} = 2 \cdot \sigma_{D^0\overline{D}^{*0}} + B_{\pi+} \cdot \sigma_{D^+D^{*-}}, \tag{1}$$

$$\sigma_{D^+} = (2 - B_{\pi+}) \cdot \sigma_{D^+D^{*-}}. \tag{2}$$

Assuming that charged and neutral $D\overline{D}^*$ pairs are produced at equal rates except for p-wave phase space factors,

$$r = \frac{\sigma_{D^0\overline{D}^{*0}}}{\sigma_{D^+D^{*-}}} = \frac{P_{D^{*0}}^3}{P_{D^{*+}}^3}, \tag{3}$$

where P_{D^*} is the D^* momentum.

The ratio of the D^+ and D^0 cross sections can be expressed as

$$R = \frac{\sigma_{D^+}}{\sigma_{D^0}} = \frac{N_{D^+}^{obs} \cdot B(D^0) \cdot \epsilon_{D^0}}{N_{D^0}^{obs} \cdot B(D^+) \cdot \epsilon_{D^+}} \tag{4}$$

where N_D^{obs} is the number of observed D's, $B(D)$ is the D decay branching fraction, and ϵ_D is the D reconstruction efficiency. Combining Eq. 1, Eq. 2, Eq. 3, and Eq. 4,

$$B_{\pi+} = \frac{2(1 - rR)}{1 + R}. \tag{5}$$

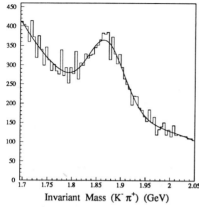

Fig. 1. Invariant mass $D^0 \to K^-\pi^+$

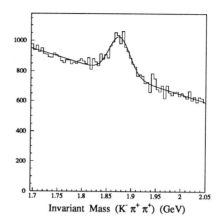

Fig. 2. Invariant mass $D^+ \to K^-\pi^+\pi^+$

3. D^{*+} Branching Ratio Measurement

D mesons are reconstructed in the $D^0 \to K^-\pi^+$ and $D^+ \to K^-\pi^+\pi^+$ decay modes. Each K candidate must pass a consistency check, using its measured time-of-flight and dE/dx information combined in the form of a confidence level. In addition, for the $D^0 \to K^-\pi^+$ mode there are two candidates for every pair of charged tracks; of these, only the candidate with the greater K confidence level is chosen, to prevent double counting. A momentum cut of 340 MeV $< P_D <$ 700 MeV is applied to the D candidates to select only D mesons from $D\overline{D}^*$ events. Using these selection criteria, fits to the D invariant mass plots (Fig. 1, Fig. 2) give 3258 ± 155 (stat.) ± 385 (sys.) D^0 candidates, and 2644 ± 174 (stat.) ± 342 (sys.) D^+ candidates. The systematic errors on these numbers are dominated by disagreement between the observed D signal widths in the data and in Monte Carlo simulations. The errors were estimated by comparing unconstrained fits to the data with ones in which the widths were fixed to those found in fits to Monte Carlo. Other contributions to the systematic errors include those of the D decay mode branching fractions and of the D reconstruction efficiencies.

Applying these numbers in Eq. 5, a result of $B(D^{*+} \to D^0\pi^+) = (65 \pm 8 \pm 18)\%$ is obtained. The other two D^{*+} branching fractions can be obtained using the unitarity constraint that the three branching fractions sum to 1, and the relationship between the $D^{*+} \to D\pi$ fractions,

$$\frac{B(D^{*+} \to D^+\pi^0)}{B(D^{*+} \to D^0\pi^+)} = \frac{P_{D+}^3}{P_{D^0}^3} \qquad (6)$$

due to isospin conservation and p-wave phase space dependence, giving a result of $B(D^{*+} \to D^+\pi^0) = (26 \pm 3 \pm 7)\%$ and $B(D^{*+} \to D^+\gamma) = (9 \pm 11 \pm 25)\%$.

4. $D\overline{D}^*$ Cross Section Measurement

The D^0 production cross section can be expressed as a function of $\sigma_{D^+D^{*-}}$ and B_{π^+} by combining Eq. 1 and Eq. 3:

$$\sigma_{D^0} = (2 \cdot r + B_{\pi^+}) \cdot \sigma_{D^+D^{*-}}. \tag{7}$$

Using this equation together with Eq. 2,

$$\sigma_{D^+D^{*-}} = \frac{\sigma_{D^0} + \sigma_{D^+}}{2(1+r)}, \tag{8}$$

where

$$N_D^{obs} = \sigma_D \cdot B(D) \cdot \epsilon(D) \cdot L, \tag{9}$$

and $L = 21.6 \ pb^{-1}$ is the integrated luminosity. Inserting the measured values into this equation, the result is $\sigma_{D^+D^{*-}} = (2.35 \pm .11 \pm .23) \ nb$. The $D^0 \overline{D}^{*0}$ cross section can be related to $\sigma_{D^+D^{*-}}$ by Eq. 3, giving $\sigma_{D^0\overline{D}^{*0}} = (2.55 \pm .12 \pm .25) \ nb$.

5. Conclusion

The preliminary measurement of D^{*+} branching fractions obtained by BES is consistent with the different values obtained by Mark III and CLEO II. Further reduction of systematic errors and reconstruction efficiencies should allow BES to distinguish between these previous results. A preliminary measurement of $\sigma_{D^+D^{*-}}$ and $\sigma_{D^0\overline{D}^{*0}}$ has also been made. Work to reduce the uncertainties on all these measurements will continue. Other future plans include application of these analyses to $D^*\overline{D}^*$ events, in order to obtain a statistically independent D^{*+} branching ratio measurement and to measure the relative $D\overline{D}^*$ and $D^*\overline{D}^*$ cross sections.

The BES collaboration wishes to thank the staff of the BEPC, and the staff of the Parallel Distributed Systems Facility of the Superconducting Super Collider Laboratory. This work was supported in part by the National Science Foundation and the U.S. Department of Energy.

References

1. BES Collaboration, J. Bai et al., Phys. Rev. Lett. 69 (1992) 3021.
2. Mark III Collaboration, J. Adler et al., Phys. Lett. B208 (1988) 152.
3. CLEO Collaboration, F. Butler et al., Phys. Rev. Lett. 69 (1992) 2041.
4. Since the completion of this manuscript, the Argus Collaboration has also reported a D^{*+} branching ratio measurement consistent with the CLEO II result (H. Albrecht et al., Preprint DESY 94-111, July 1994).
5. Particle Data Group, K. Hikasa et al., Phys. Rev. D45 (1992) S1.
6. Particle Data Group, 1994 Summary Tables of the Review of Particle Properties by the Particle Data Group (to appear in Phys. Rev. D50 (1994) 1173).
7. Mark II Collaboration, M. W. Coles et al., Phys. Rev. D26 (1982) 2190.
8. E. Eichten et al., Phys. Rev. D21 (1980) 203; A. De Rujula et al., Phys. Rev. Lett. 38 (1977) 317.
9. G. Gladding, Proceedings of the Tau-Charm Factory Workshop, (1989) 152; I. I. Bigi, Proceedings of the Tau-Charm Factory Workshop, (1989) 169.

OBSERVATION OF $D_1(2430)^+$ AND $D_2^*(2470)^+$ AND OTHER RECENT CLEO D^{**} RESULTS

JOHN BARTELT

Department of Physics & Astronomy, Vanderbilt University,
Nashville, Tennessee USA 37235

(Representing the CLEO collaboration)

ABSTRACT

Six excited charmed mesons have been observed by CLEO: two neutral, two charged and two strange. The latest results, concerning the two charged mesons are presented: new mass and width measurements, plus angular distributions which support the spin-parity assignments of these states. We also present a summary of the published results on the other four states.

1. Introduction

Besides the six pseudoscalar and vector charmed mesons, six heavier charmed mesons, commonly called D^{**}'s, have been observed. They are believed to be states with orbital angular momentum $L = 1$. There should be a total of 12 such states. Heavy Quark Effective Theory[1,2] describes mesons with one heavy quark in terms of the total light angular momentum ($j_\ell = L + s_\ell$) plus the spin of the heavy quark ($m_{s_h} = \pm 1/2$). This leads to the prediction of two doublets of mesons for each each flavor of light anti-quark (up, down or strange).

The members of the $j_\ell = 1/2$ doublet, $J^P = 0^+$, 1^+ states, are predicted to be broad ($\Gamma \sim 200$ MeV/c^2). They should decay exclusively through the S-wave. The members of the $j_\ell = 3/2$ doublet, the $J^P = 1^+$, 2^+ states, should be narrow ($\Gamma \sim 20$ MeV/c^2). Their HQET-favored decay is through the D-wave. It is these narrow states which (presumably) have been observed.

Over the last two years, CLEO has made new measurements of all six narrow states. This includes the first observation of one mesons, the $D_{s2}^*(2573)^+$, and the first complete reconstruction of another, the $D_1(2430)^+$. Our measurements of the $D_{s1}(2536)^+$, the $D_{s2}^*(2573)^+$, and the neutral states, $D_1(2420)^0$ and $D_2^*(2460)^0$, have been published.[3-5] Our results on the charged states, the $D_1(2430)^+$ and $D_2^*(2470)^+$, are presented here and will be submitted for publication later this year. This paper will summarize the measurements of all six states.

The data were taken at the Cornell Electron Storage Ring (CESR), using the CLEO II detector. CLEO II is a general purpose detector with both excellent charged particle tracking and photon energy measurement. We also use dE/dx measurements from tracking chambers, and time-of-flight measurements, for particle identification.

The events were produced in e^+e^- annihilation in the energy region at and near the $\Upsilon(4S)$. Only continuum production of the charmed mesons was included in each analysis; no B decay products were used. In general, a cut of $x \geq 0.6$, or higher, was used on the reconstructed mesons. We define the scaled momentum $x \equiv p/p_{max}$, where $p_{max} \equiv [(E_b^2 - M^2)]^{1/2}$. The purpose of the x-cut is to reduce combinatoric background; it also eliminates any charmed mesons coming from bottom decay.

2. Data Analysis

2.1. Decay Modes

Angular momentum and parity conservation impose the following constraints on possible decay modes:

$$1^+ \rightarrow 1^-0^- \qquad \text{S- or D-wave}$$
$$2^+ \rightarrow 0^-0^- \text{ or } 1^-0^- \qquad \text{D-wave only.}$$

Thus, for example, the $D_1(2430)^+$ was reconstructed in the $D^{*0}\pi^+$ mode, while the $D_2^*(2470)^+$ was reconstructed in both the $D^0\pi^+$ and $D^{*0}\pi^+$ modes.

2.2. Masses and Widths

In Table 1, we present the CLEO measurements of the masses and widths of the D^{**}'s, along with statistical and systematic errors. We also show the intradoublet splitting derived from the mass measurements.

Table 1. CLEO Measurements of Masses and Widths (MeV/c^2)

Meson	Mass	Width	ΔM
$D_1(2430)^+$	$2425 \pm 2 \pm 2$	$26^{+8}_{-7} \pm 4$	$\left.\right\}38 \pm 5$
$D_2^*(2470)^+$	$2463 \pm 3 \pm 3$	$27^{+11}_{-8} \pm 5$	
$D_1(2420)^0$	$2421^{+1}_{-2} \pm 2$	$20^{+6}_{-5} \pm 3$	$\left.\right\}44 \pm 5$
$D_2^*(2460)^0$	$2465 \pm 3 \pm 3$	$28^{+8}_{-7} \pm 6$	
$D_{s1}(2536)^+$	$2535.1 \pm 0.2 \pm 0.5$	< 2.3 (90% C.L.)	$\left.\right\}38.1 \pm 1.9$
$D_{s2}^*(2573)^+$	$2573.2^{+1.7}_{-1.6} \pm 0.5$	$16^{+5}_{-4} \pm 3$	

The only previously reported observation of the $D_1(2430)^+$ was by E691,[6] which used only partial reconstruction to determine a mass of $2443 \pm 7 \pm 5$ MeV/c^2 and a width of $41 \pm 19 \pm 8$ MeV/c^2. The $D_2^*(2470)^+$ was first reported by ARGUS[7] with a mass of $2469 \pm 4 \pm 6$ MeV/c^2. More recently E687[8] measured the mass to be $2453 \pm 3 \pm 2$ MeV/c^2, and the width to be $23 \pm 9 \pm 5$ MeV/c^2.

We determine the isospin mass splittings between the charged and neutral states to be $M(D_1^+) - M(D_1^0) = +4^{+2}_{-3} \pm 3$ MeV/c^2 and $M(D_2^{*+}) - M(D_2^{*0}) = -2 \pm 4 \pm 4$ MeV/c^2. E687 measured the latter to be 0 ± 4 MeV/c^2.[6]

2.3. Helicity Angle Distributions: Spin-Parity and Partial Waves

In decays of a D^{**} to a D^* and a pseudoscalar, the subsequent decay of the D^* yields information about the helicity of the D^* and hence about its parent D^{**}. In the decay sequence $D_1(2430)^+ \to D^{*0}\pi^+$, $D^{*0} \to D^0\pi^0$, the helicity angle α is defined to be the angle between the π^+ and the π^0, measured in the D^{*0} rest-frame. Thus measuring the helicity angle distribution can provide information on the spin-parity of the D^{**} and the relative sizes of partial waves.

For example, if the parent D^{**} is in the "natural" sequence (0^+, 1^-, 2^+ ...), it must produce a $\sin^2\alpha$ distribution. Any particle that decays to two pseudoscalars must be natural; thus the higher mass member of each doublet (the D_2^{*+}, D_2^{*0}, and D_{s2}^{*+}) must exhibit the $\sin^2\alpha$ distribution. If it is also seen to decay to a vector and a pseudoscalar, 0^+ is eliminated as a possibility.

If a D^{**} had $J^P = 0^-$, the D^* decays would have a helicity angle distribution proportional to $\cos^2\alpha$. The pure S-wave decay of a 1^+ would produce a flat distribution, while a pure D-wave decay would produce a $1 + 3\cos^2\alpha$ distribution.

Table 2. Testing Helicity Angle Distribution Hypotheses

Meson	Angular Dist.	χ^2/N_{dof}	C.L.
$D_1(2430)^+$	$1 + 3\cos^2\alpha$	1.7/3	63.7%
	$\cos^2\alpha$	6.7/3	8.2%
	isotropic	8.7/3	3.4%
	$\sin^2\alpha$	27.8/3	4.0×10^{-6}
$D_2^*(2470)^+$	$\sin^2\alpha$	1.3/3	72.9%
	isotropic	6.1/3	10.7%
	$\cos^2\alpha$	29.2/3	2.0×10^{-6}
$D_1(2420)^0$	$1 + 3\cos^2\alpha$	2.3/4	68.1%
	$\cos^2\alpha$	28.2/4	11.4×10^{-6}
	isotropic	23.9/4	83.6×10^{-6}
	$\sin^2\alpha$	93.2/4	27.5×10^{-20}
$D_2^*(2460)^0$	$\sin^2\alpha$	1.2/4	87.8%
	isotropic	2.5/4	64.5%
$D_{s1}(2536)^+$	isotropic	7.6/4	10.7%
	$\cos^2\alpha$	68.3/4	5.2×10^{-14}
	$\sin^2\alpha$	16.4/4	2.5×10^{-3}

For the $D_2^*(2470)^+$, the data are well fit by $\sin^2\alpha$ as expected, though a flat function is also acceptable. For the $D_1(2430)^+$, $1 + 3\cos^2\alpha$ gives the best fit, indicating that the decay is consistent with being a pure D-wave. These results are similar to those found for the neutral states; see Table 2 for a summary. The $D_{s1}(2536)^+$ appears unusual in that its helicity distribution is consistent with a pure S-wave decay. The D-wave decay is apparently highly supressed by the very small phase space available. Note, however, that the S-wave width of the D_{s1}^+ is less than 2.3 MeV/c^2, which is very small for an S-wave decay. Compare, for example, the decay of the $a_1(1260) \to \rho\pi$.

2.4. Branching Ratios

We have measured the following branching ratios for the 2^+ states. Godfrey & Kokoski's predictions[2] for these ratios are given in brackets at the end of each line.

$$\frac{B[D_2^{*+} \to D^{*0}\pi^+]}{B[D_2^{*+} \to D^0\pi^+]} = 0.53 \pm 0.30 \pm 0.16 \quad [0.43\text{--}0.61]$$

$$\frac{B[D_2^{*0} \to D^{*+}\pi^-]}{B[D_2^{*0} \to D^+\pi^-]} = 0.45 \pm 0.14 \pm 0.12 \quad [0.43\text{--}0.61]$$

$$\frac{B[D_{s2}^{*+} \to D^{*0}K^+]}{B[D_{s2}^{*+} \to D^0K^+]} < 0.33 \ (90\%\text{C.L.}) \quad [0.10\text{--}0.16]$$

For a 1^+ state, decay to two pseudoscalars is forbidden by angular momentum and parity conservation. We set 90% C.L. upper limits on these decays:

$$\frac{B[D_1(2430)^+ \to D^0\pi^+]}{B[D_1(2430)^+ \to D^{*0}\pi^+]} < 0.18$$

$$\frac{B[D_{s1}(2536)^+ \to D^0K^+]}{B[D_{s1}(2536)^+ \to D^{*0}K^+]} < 0.12$$

2.5. Fragmentation

The momentum spectra for five of the D^{**}'s were fit with the Peterson fragmentation function.[9] We find $\epsilon_P = 0.013 \pm 0.005 \pm 0.004$ and $0.020^{+0.011}_{-0.006} \pm 0.003$ for the $D_1(2430)^+$ and $D_2^*(2470)^+$, respectively. These values are close to those found for the other D^{**}'s; thus they have much harder spectra than the D's and D^*'s.

3. Conclusions

The $D_1(2430)^+$ and $D_2^*(2470)^+$ are now well established, with good mass and width measurements. Branching ratios and helicity angle distributions are completely consistent with the spin-parity assignments of 1^+ and 2^+, respectively, though other assignments are not absolutely ruled out. Their status as isospin partners of the $D_1(2420)^0$ and $D_2^*(2460)^0$ seems obvious, and the spin-parity assignments of the neutral mesons is on even firmer ground. The $D_{s1}(2536)^+$ is almost certainly a 1^+ state, and it seems reasonable to believe that the $D_{s2}^*(2573)^+$ is its doublet partner. Thus all six narrow $L = 1$ charmed mesons (those with $j_\ell = 3/2$) are firmly established and their properties are in good agreement with those predicted by HQET models.

References

1. N. Isgur and M.B. Wise, *Phys. Rev. Lett.* **66** (1991) 1130.
2. S. Godfrey and R. Kokoski, *Phys. Rev.* **D43** (1991) 1679.
3. J. Alexander *et al.*, *Phys. Lett.* **B303** (1993) 377.
4. Y. Kubota *et al.*, *Phys. Rev. Lett.* **72** (1994) 1972.
5. P. Avery *et al.*, *Phys. Lett.* **B331** (1994) 236.
6. J.C. Anjos *et al.*, *Phys. Rev. Lett.* **62** (1989) 1717.
7. H. Albrecht *et al.*, *Phys. Lett.* **2331** (1989) 208.
8. P.L. Frabetti *et al.*, *Phys. Rev. Lett.* **72**, (1994) 324.
9. C. Peterson *Phys. Rev.* **D27** (1983) 105.

DIRECT MEASUREMENT OF B($D_s \to \phi\pi$) BY THE BEIJING SPECTROMETER COLLABORATION

BRUCE LOWERY [*]

ABSTRACT

The Beijing Spectrometer has made a preliminary measurement of the absolute $D_s \to \phi\pi$ branching fraction using approximately 22 pb^{-1} at center-of-mass energy 4.03 GeV which is 92 MeV above threshold for $e^+e^- \to D_s^+ D_s^-$. We observe events with one or both D_s mesons decaying in the modes $D_s+ \to \phi\pi^+$, $\overline{K}^{o*}K$, or $\overline{K}^o K$, and these events are used to measure B($D_s \to \phi\pi$) model independently.

1. Introduction

The D_s meson has been observed in e^+e^- annihilations at $\sqrt{s} = 4.03$ GeV by the Beijing Spectrometer collaboration (BES). The outstanding problems in D_s meson physics include measurement of leptonic decay modes and a direct measurement of the branching fraction for $D_s \to \phi\pi$. Previous mesurements of B($D_s \to \phi\pi$) have relied on SU(3) flavor symmetry based models in relating $\Gamma(D^+ \to K^{o*}e^+\nu_e)$ to $\Gamma(D_s^+ \to \phi e^+\nu_e)$,[1] but the ability of the Beijing Electron Positron Collider (BEPC) to operate with high luminosity just above D_s threshold gives BES a key advantage to measure B($D_s \to \phi\pi$) model independently as well as directly observe leptonic D_s decays.

The BES detector is a solenoidal detector similar in design to the Mark III. A detailed description of the detector may be found elsewhere.[2]

2. D_s physics at 4.03 GeV

At 4.03 GeV D_s mesons are produced *only* in pairs in the reaction, $e^+e^- \to D_s^+ D_s^-$. Events with only one D_s reconstructed are called single tags whereas those with both D_s reconstructed are called double tags. From the measured number of $\phi\pi$ single tags, $N_{\phi\pi}$, the product of cross section times $B_{\phi\pi}$ can be measured according to $\sigma_{D_s \overline{D}_s} \cdot B_{\phi\pi} = \frac{N_{\phi\pi}}{2 \cdot L \cdot \epsilon_{\phi\pi}}$ where $L = \int \mathcal{L}dt$ is the integrated luminosity, and $\epsilon_{\phi\pi}$ is the reconstruction efficiency. Using Eq. 1 the expected number of double tags, N_{DT}, is

$$N_{DT} = \sigma \cdot L \cdot \sum_{ij} B_i B_j \epsilon_{ij} = \frac{1}{2} B_{\phi\pi} \cdot N_{\phi\pi} \cdot \frac{\sum_{ij} b_i b_j \epsilon_{ij}}{\epsilon_{\phi\pi}} \tag{1}$$

where the B_i are D_s branching fractions, $b_i = B_i/B_{\phi\pi}$, ϵ_{ij} is the efficiency for reconstructing modes i and j simultaneously, and the double sum in the numerator extends only over the double tag modes. Once $N_{\phi\pi}$ and N_{DT} have been measured a maximum likelihood estimation may be made of $B_{\phi\pi}$.

---

*representing the Beijing Spectrometer Collaboration

BES has observed the decays $D_s^+ \to \phi\pi^+, \overline{K}^{o*}K^+$, and $\overline{K}^o K^+$ with $\phi \to K^+K^-$, $\overline{K}^{o*} \to K^-\pi^+$, and $\overline{K}^o \to K_s \to \pi^+\pi^-$.[3] The $\phi\pi^+$ mode is the cleanest due to the narrow width of the ϕ and is the only mode included in the single tag search, but all 3 modes are used in the double tag search since the double tag backgrounds are very small.

3. Measurement of $\sigma(e^+e^- \to D_s^+D_s^-) \cdot B_{\phi\pi}$

The $D_s \to \phi\pi$ final state is isolated by combining K^+K^- pairs which have an invariant mass with 18 MeV of the ϕ mass with an additional pion. The energy of the $KK\pi$ combination must be within 50 MeV of the beam energy (2.015 GeV). Each track is required to be consistent with it's assigned mass hypothesis at the 1% level by combining it's measured time-of-flight and dE/dx into the form of a confidence level, and tracks assigned to be kaons are required to have a normalized likelihood greater than 0.5 where the likelihood ratio is constructed to be near one for genuine kaons and near zero for background pions. Figure 1 shows the $\phi\pi$ single tag signal in

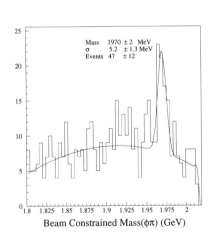

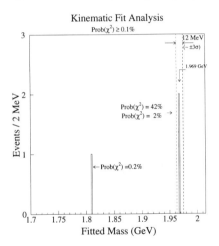

Fig. 1. Beam constrained mass $D_s \to \phi\pi$ Fig. 2. Kinematic fit double tag analysis.

beam constrained mass defined by $M_{bc} = \sqrt{2.015^2 - |\sum_i \vec{p}_i|^2}$, where $\vec{p}_i$ is the vector momentum of the ith track in the combination. The beam constrained mass is simply the invariant mass with the energy replaced by the beam energy (the "true" D_s energy). The number of $\phi\pi$ single tags are measured from a fit to Fig. 1 giving 47 ± 12 $\phi\pi$ single tags. The $\phi\pi$ reconstruction efficiency is 11.3%, and the integrated luminosity of the data is 21.6 pb^{-1}. Using Eq. 1 we measure $\sigma \cdot B_{\phi\pi} = (9.6 \pm 2.4 \pm 0.1)$ pb where the first error is statistica,l and the second error is a systematic due to uncertainty in the shape of the single tag background.

Table 1. $D_s^+ D_s^-$ double tags

Mode	Tag 1			Tag 2		
	Mass (GeV)	P (D_s)		Mass(GeV)	P (D_s)	
	D_s	ϕ/K^{o*}	(GeV)	D_s	ϕ/K^{o*}	(GeV)
$\phi\pi$ vs $\overline{K}^{o*}K$	1.952	1.013	.422	2.013	0.855	0.487
$\overline{K}^{o*}K$ vs $\overline{K}^{o*}K$	1.988	0.879	0.447	1.952	0.913	0.458

4.   Measurement of Absolute B($D_s \to \phi\pi$)

To measure B($D_s \to \phi\pi$) we search for events in which both D_s are reconstructed to one of the decay modes $\phi\pi$, $\overline{K}^{o*}K^+$, and $\overline{K}^o K^+$. Each double tag candidate is subjected to a kinematic fit in which the invariant masses of each D_s are constrained to be equal but unspecified. Candidates for which the fit probability is less than 0.1% are rejected. The remaining candidates are required to have all tracks consistent with the assigned mass hypothesis at the 1% confidence level, and all ϕ, K^{o*}, and K^o candidates are required to have invariant masses within 18 MeV, 50 MeV, and 20 MeV respectively of the nominal values. Additionally, the pion from the decay of the K^{o*} must have a helicity angle in the K^{o*} rest frame satisfying $|\cos\theta_\pi| \geq 0.4$. The contribution from nonresonant $D_s \to KK\pi$ which pass the ϕ and K^{o*} resonance cuts is accounted for in the analysis. We find 2 D_s double tag events as shown in Fig. 2 which shows the fitted mass for the events passing the cuts. Information for the two events is given in table 1 including the raw invariant mass ($\sigma(M_{KK\pi}) = 20~MeV$), the raw reconstructed momenta and the masses of the two body resonances.

To check the kinematic fit analysis, we subject the same data to two more analyses that differ from the original analysis by (1) applying momentum cuts, 348 MeV < $|\Sigma P|$ < 498 MeV on both D_s candidate tracks and looking for pairs of invariant masses that are near the D_s mass and (2) applying energy cuts, $|\Sigma E - 2.015~GeV|$ < 0.075 GeV on both D_s candidate tracks and looking for pairs of beam constrained masses near the D_s mass. All three double tag analyses find the same two events with very small backgrounds.

The sidebands in each analysis are used to estimate the number of background events. From the sideband analysis and a Monte Carlo background study (in which 1.6 million nonstrange charm events are passed through the kinematic fit analysis) we estimate 0.2 background events in the D_s signal region. Using the value of the $D^{+*}D^-$ cross section measured by BES we estimate 0.1 background event from the Monte Carlo study. From the several background estimates we estimate the background to the D_s double tag signal to be 0.2.

To estimate $B(D_s \to \phi\pi)$, we build a likelihood function $\mathcal{L} = \frac{<\tilde{N}_{DT}>^{N_{DT}} \cdot e^{-<\tilde{N}_{DT}>}}{N_{DT}!}$ where $< \tilde{N}_{DT} >$ is the expected number of double tag signal plus background events,

$$< \tilde{N}_{DT} >= \frac{1}{2} B_{\phi\pi} \cdot \tilde{N}_{\phi\pi} \cdot \frac{\sum_{ij} b_i b_j \epsilon_{ij}}{\epsilon_{\phi\pi}} + 0.2 \qquad (2)$$

and N_{DT} is the number of observed double tag events in the data. The marginal likelihood obtained by integrating the product of $\mathcal{L}$ and the likelihood for $\tilde{N}_{\phi\pi}$ (assumed to be gaussian). We neglect the errors on b_i and the Monte Carlo reconstruction efficiencies. Maximizing the marginalized likelihood gives the most likely value of $B_{\phi\pi}$ corresponding to our measured double and single tag sample and yields $B_{\phi\pi} = (4.2^{+9.0}_{-1.5})\%$ (preliminary). where the error is statistical and corresponds to those values of $B(\phi\pi)$ above and below which lie 15.8% of the area of the likelihood function. From the central value of $\sigma \cdot B_{\phi\pi}$ and the preliminary measurement of absolute $B_{\phi\pi}$ the D_s production cross section is estimated to be $\sigma(e^+e^- \to D_s^+ D_s^-) = 229$ pb.

5. Systematics

We estimate the systematic error on $B_{\phi\pi}$ by varying the double tag cuts and the background estimate. Varying the background estimate by ± 0.2 leads to a systematic uncertainty of ± 0.5 in the $\phi\pi$ branching fraction measurement. Opening up the double tag cuts gives an additional double tag event which passes the less restrictive K^{o*} resonance mass cut which leads to a systematic uncertainty $+1.7 - 0.0$ in $B_{\phi\pi}$. The final value for $B_{\phi\pi}$ including systematic effects is

$$B_{\phi\pi} = 4.2^{+9.0+0.5+1.7}_{-1.5-0.5-0.0} \quad (preliminary). \tag{3}$$

6. Conclusion

Threshold production of D_s mesons provides the most favorable kinematics for measuring D_s leptonic decays and measuring the absolute value of $B(D_s \to \phi\pi)$ model independently. However, the large statistical error is dominated by the small number of observed double tag events resulting from small branching ratios and a modest production cross section. We can improve this measurement by reconstructing additional modes such as $D_s \to \eta\pi$, $\phi\rho$, $\eta'\pi$, nonresonant $KK\pi$, $K^{*o}K^{*-}$, and $f_o\pi$.

In summary, BES observes 47 ± 12 single tagged $D_s \to \phi\pi$ decays and two fully reconstructed events in which both D_s mesons are reconstructed over an estimated 0.2 event background. From these events we extract the absolute $\phi\pi$ branching fraction to be $B_{\phi\pi} = 4.2^{+9.0+0.5+1.7}_{-1.5-0.5-0.0}$, and the D_s production cross section at 4.03 GeV is estimated to be $\sigma = 229$ pb.

The BES collaboration wishes to thank the staff of the BEPC, and the staff of the Physics Detector Simulation Facility of the Superconducting Super Collider Laboratory. This work was supported in part by the National Science Foundation and the U.S. Department of Energy.

References

1. CLEO Collaboration, M. Daoudi, et al., Phys. Rev. D **45**, 3972 (1992)
2. BES Collaboration, J. Bai, et al., Phys. Rev. Lett. **69**, 3021 (1992)
3. BES Collaboration, C. Zhang, et al., Direct Measurement of the Pseudoscalar Decay Constant, f_{D_s}, from BES, Contributed paper to ICHEP 94, Glasgow, Scotland

A MEASUREMENT OF THE PSEUDOSCALAR
DECAY CONSTANT, f_{D_s}*

Michael KELSEY†

California Institute of Technology, Pasadena, California 91125

ABSTRACT

The Beijing Spectrometer (BES) experiment has collected a data sample corresponding to about 20 pb^{-1} at center of mass energy 4.03 GeV, which is just above the threshold for $D_s^+ D_s^-$ production. This permits D_s reconstruction with excellent mass resolution by applying the beam-energy constraint to the D_s system. By investigating the system recoiling against a reconstructed D_s, a few candidates for D_s leptonic decay are identified. From these events, the value of the pseudoscalar decay constant f_{D_s}, is estimated.

1. Introduction

The pseudoscalar decay constants, f_p, for mesons containing a heavy quark are sensitive in testing the the quark content of the meson.[1,2] Estimates of mixing and CP violation in the B meson system require knowledge of f_B, which may be computed via lattice gauge theory[3-8] if f_D or f_{D_s} are known. For several years, theorists have made efforts to calculate these decay constants; the results have some uncertainties, and large discrepancies exist between various approaches.[9-21] Some measurements of f_{D_s} have been made from $\mathcal{B}\left(D_s \to \mu\nu_\mu\right)$ with theoretical or experimental assumptions.[22-24] f_{D_s} appears in the partial leptonic decay width of D_s[25]

$$\Gamma(D_s \to \ell\nu_\ell) = \frac{G_F^2 |V_{cs}|^2}{8\pi} f_{D_s}^2 m_{D_s} m_\ell^2 \left(1 - \frac{m_\ell^2}{m_{D_s}^2}\right)^2 \tag{1}$$

where m_{D_s} is the D_s mass, m_ℓ the lepton mass, $V_{cs} = 0.974$ is the $c \to s$ CKM matrix element, and G_F is the Fermi constant. Measuring $\Gamma(D_s \to \ell\nu_\ell)$ gives a direct determination of f_{D_s}.

The BES experiment[26,27] has collected a data sample of 22.3 pb^{-1} at a center of mass energy 4.03 GeV, which is just above the threshold for $D_s^+ D_s^-$ production.[28,29] Since no D_s^* mesons are produced at this energy,[28] a beam-energy constraint can be used to select $D_s^+ D_s^-$ events and to obtain very narrow mass resolution in fully-reconstructed decay channels. From a sample of tagged D_s events, a search for candidates of $D_s \to \mu\nu_\mu$ or $\tau\nu_\tau$ in the system recoiling against a reconstructed D_s will give a direct, model-independent measurement of the pseudoscalar decay constant f_{D_s}.

---

*Work supported in part by the National Natural Science Foundation of China and the U.S. Department of Energy

†Representing the Beijing Spectrometer Collaboration

2. D_s Event Selection

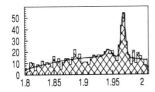

Fig. 1. $D_s \to \phi\pi$, $K^{*0}K$, K^0K masses (GeV/c^2) after kinematic fit.

A D_s signal is searched for in three decay channels, $D_s^+ \to \phi\pi^+$ ($\phi \to K^+K^-$), $D_s^+ \to \overline{K^{*0}}K^+$ ($\overline{K^{*0}} \to K^-\pi^+$), and $D_s^+ \to \overline{K}^0 K^+$ ($\overline{K}^0 \to \pi^+\pi^-$) (and charge-conjugates). The mass distribution of candidate events, after kinematic fit, is shown at left. There is an expected background to the D_s from DD^* events (observed in the $D_s \to K^0K$ channel), which appears as an enhancement near 1.94 GeV/c^2 in a mass plot. We fit the mass distribution to a Gaussion signal plus background combining a polynomial and DD^* enhancement. The total number of observed D_s tags from $\phi\pi$, $K^{*0}K$, and K^0K is $N_{\text{tag}} = 95.1 \pm 13.5$. The D_s mass obtained from the fit is $m(D_s) = 1969.0 \pm 0.6\,(\text{stat}) \pm 0.5\,(\text{sys})$ MeV$/c^2$.

3. Search for D_s Leptonic Decays

The single-tag sample of D_s events described in Section 2 is searched for single identified leptons in the recoil. Possible backgrounds from $e^+e^- \to \tau^+\tau^-$ can be suppressed with a polar angle cut: $e^+e^- \to D_s^+D_s^- \sim \sin^2\theta_{D_s}$, while $e^+e^- \to \tau^+\tau^- \sim 1 + A\cos^2\theta_\tau$. We require $|\cos\theta_{D_s}| \leq 0.7$. D_s decay modes that contain photons and π^0's are not considered in this analysis. Events in the single-tag sample containing one or more isolated photons are rejected.

The recoil to the reconstructed D_s tag should be a single track with charge opposite the tag, and must be consistent with a muon or electron hypothesis. Visual scans eliminate Cosmic rays superimposed on D_ℓ events, and events with TOF or energy hits in the endcap associated with untracked drift chamber hits.

Candidate lepton tracks are required to have a good helix fit, and to be close to the primary vertex: $|r_{xy}| \leq 2$ cm, $|z| \leq 20$ cm. Muons are selected using information from the muon counters, and with momenta $0.55 < p_\mu < 1.25$ GeV$/c$. The acceptance for muons is about 85% within the fiducial volume of the counters. $\pi \to \mu$ misidentification is about 4%. Electrons are selected with momenta $p_e \geq 400$ MeV$/c$ and $|\cos\theta| \leq 0.75$. The TOF and dE/dx measurements should be consistent with electrons. Hits in each calorimeter layer are weighted to give e/π separation based on longitudinal shower development as well as total energy. These cuts give an average acceptance of over 80%, and more than 90% for $p > 1$ GeV$/c$. $\pi \to e$ misidentification is less than 5%.

We define the efficiency ε_i as the probability of finding the decay mode i in the recoil, including all detector acceptance effects, and tau decay branching fractions where appropriate. From Monte Carlo studies, we find that the backgrounds, $n_{\text{bkg},i}$, mostly arise from processes such as $D_s \to K_L^0K$ and $D_s \to \tau\nu_\tau$, $\tau \to \pi\nu$. Backgrounds from D decays are negligible. The results are summarized in Table 1.

We find a total of three candidate leptonic decays in the D_s single tag sample (Table 2). There are two leptonic decay candidates with a muon track, and one with

Table 1. Leptonic decay selection efficiency and backgrounds.

Recoil mode	ε	n_{bkg}	Acceptance cuts
$D_s \to \mu\nu_\mu$	51.4%	0.04	$p_\mu \geq 750$ MeV/c, $\lvert\cos\theta\rvert \leq 0.65$
$D_s \to \tau\nu_\tau$	6.28%		$p_\mu \geq 550$ MeV/c, $\lvert\cos\theta\rvert \leq 0.65$
$\tau \to \mu\nu\nu$		0.6	$\mathcal{B}(\tau \to \mu\nu\nu) = 17.58\%$
$D_s \to \tau\nu_\tau$	8.23%	(both)	$p_e \geq 400$ MeV/c, $\lvert\cos\theta\rvert \leq 0.75$
$\tau \to e\nu\nu$			$\mathcal{B}(\tau \to e\nu\nu) = 17.93\%$

an electron. The missing momentum for these events are pointing to the barrel part of the detector. No photons or extra charged tracks are found in the detector.

Table 2. Observed candidates for Ds leptonic decay.

Tag Mode	$m(res)$ (MeV/c^2)	$m_{bc}(D_s)$ (MeV/c^2)	Lepton ID	p_ℓ (MeV/c)	m_{miss}^2 (GeV^2/c^4)	Recoil channel	τ decay
$\phi\pi^+$	1019.3	1970.2	μ^-	751	0.763	$\tau^-\bar\nu_\tau$	$\mu\nu\nu$
$\overline{K^{*0}}K^+$	873.4	1970.9	μ^-	1216	$-.112$	$\mu^-\bar\nu_\mu$	
$K_S^0 K^+$	491.5	1969.0	e^-	489	1.627	$\tau^-\bar\nu_\tau$	$e\nu\nu$

4. Measuring f_{D_s}

From Eq. 1 and $B_\ell \equiv \mathcal{B}(D_s \to \ell\nu_\ell) = \tau_{D_s}\Gamma(D_s \to \ell\nu_\ell)$, where τ_{D_s} is the D_s lifetime, with a measured $\mathcal{B}(D_s \to \ell\nu_\ell)$ we can estimate f_{D_s}. We use a maximum-likelihood technique to determine the branching fractions B_μ and B_τ. The likelihood function $\mathcal{L} = \mathcal{L}_\mu \cdot \mathcal{L}_e \cdot \mathcal{L}_{\text{cand}} \cdot \mathcal{L}_{\text{tag}}$, where $\mathcal{L}_{\text{cand}}$ is a Poission distribution for the number of candidates, and $\mathcal{L}_{\text{tag}}$ is a Gaussian distribution for the number of single-tag events. $\mathcal{L}_\mu$ and $\mathcal{L}_e$ are likelihood functions for the individual lepton recoils:

$$\mathcal{L}_\mu(m_{\text{miss}}^2) = \frac{n_\mu \cdot f_\mu(m_{\text{miss}}^2) + n_{\tau\mu} \cdot f_{\tau\mu}(m_{\text{miss}}^2) + n_{\text{bkg},\mu} \cdot f_{\text{bkg},\mu}(m_{\text{miss}}^2)}{n_\mu + n_{\tau\mu} + n_{\text{bkg},\mu}} \qquad (2)$$

$$\mathcal{L}_e(m_{\text{miss}}^2) = \frac{n_{\tau e} \cdot f_{\tau e}(m_{\text{miss}}^2) + n_{\text{bkg},e} \cdot f_{\text{bkg},e}(m_{\text{miss}}^2)}{n_{\tau e} + n_{\text{bkg},e}}. \qquad (3)$$

The functions $f_i(m_{\text{miss}}^2)$ are kinematic distributions for missing mass in events of type i ($\mu \equiv D_s \to \mu\nu_\mu$, $\tau\mu \equiv D_s \to \tau\nu_\tau(\tau \to \mu\nu\nu)$, and $\tau e \equiv D_s \to \tau\nu_\tau(\tau \to e\nu\nu)$). The likelihood parameters n_i represent the number of observed events in each mode, weighted for branching ratio and efficiency. The total number of observed events $N_{\text{cand}} = n_\mu + n_{\tau\mu} + n_{\tau e} + n_{\text{bkg},\mu} + n_{\text{bkg},e}$. The $n_i = N_{\text{tag}} \cdot B_i \cdot \varepsilon_i$ for each mode $i \in \{\mu, \tau\mu, \tau e\}$. ε_i is the detection efficiency for the channel (see section 3).

Taking B_μ and B_τ as independent parameters, we maximize $\mathcal{L}$ and obtain $B_\mu = 1.8^{+5.2}_{-0.5}$ % and $B_\tau = 14^{+20}_{-5}$ %. If we assume lepton universality, $B_\tau/B_\mu = 9.78$, and fitting for B_μ alone we obtain the consistent result $\mathcal{B}(D_s \to \mu\nu_\mu) = 1.4^{+1.2}_{-0.7}$ %. Finally, using Eq. 1 to write $\mathcal{L}$ in terms of f_{D_s} we obtain

$$f_{D_s} = 434^{+153+35}_{-133-33} \text{ MeV} .\tag{4}$$

The first error is statistical, and the second is the combined systematic uncertainty from tagging efficiency, background estimation and D_s lifetime.

This measurement of f_{D_s} is absolute and model-independent. It is not normalized to other D_s branching ratios, nor to assumptions about D_s production rates. In Fig. 2 we present a variety of theoretical and experimental determinations of f_{D_s}, along with our absolute measurement.

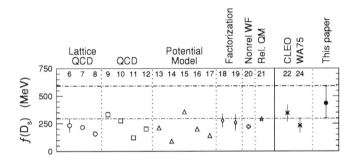

Fig. 2. Various calculations and measurements of f_{D_s}. References are indicated by the numbers above each point.

Our determination is higher than the normal range of theoretical predictions (150 to 360 MeV), but is consistent within errors. Studies are ongoing[30] to determine f_{D_s} more precisely, using additional tag modes and semi-inclusive methods that require normalziation to specific D_s branching ratios.

5. Conclusion

From a sample of tagged D_s events produced at BES, three D_s leptonic decay events with very low background are observed. The first direct measurment of f_{D_s}, using the leptonic decays $D_s \to \mu\nu_\mu$ and $D_s \to \tau\nu_\tau$ ($\tau \to \mu\nu\nu$, $e\nu\nu$), is higher than current theoretical expectations.

Acknowledgements

We acknowledge the excellent support of BEPC machine physicists and the efforts of all the engineers and technicians who have participated in the construction and maintenance of BEPC machine and BES detector.

References

1. J.L. Rosner, "Determining Heavy Meson Decay Constants," presented at Snowmass 90, EFI 90-80 November 1990.
2. C.H. Chang and Y.Q. Chen, *Phys. Rev.* **D46** (1992) 3845, and *Phys. Rev.* **D49** (1994) 3399.
3. M. Witherell, "Charm Weak Decay," presented at XVI International Symposium on Photon-Lepton Physics, Cornell U., Ithaca, New York, August 1993.
4. I. Claudio *et al.*, *Phys. Rev.* **D41** (1990) 1522.
5. F.J. Gilman and M.B. Wise, *Phys. Rev.* **D27** (1983) 1128.
6. C. Bernard *et al.*, *Phys. Rev.* **D38** (1988) 3540.
7. M.B. Gavela *et al.*, *Phys. Lett.* **206B** (1988).
8. T.A. De Grand and R.D. Loft, *Phys. Rev.* **D38** (1988) 954.
9. J.G. Bian and T. Huang, *Modern Phys. Lett.* **8** (1993) 635.
10. C. Dominguez and N. Paver, *Phys. Lett.* **197B** (1987) 423.
11. S. Narison, *Phys. Lett.* **198B** (1987) 104.
12. M.A. Shifman, *Ups. Fiz. Nauk* **151** (1987) 193.
13. H. Krasemann, *Phys. Lett.* **96B** (1980) 397.
14. M. Suzuki, *Phys. Lett.* **162B** (1985) 391.
15. S.N. Sinha, *Phys. Lett.* **178B** (1986) 110.
16. P. Cea *et al.*, *Phys. Lett.* **206B** (1988) 691.
17. P. Colagelo, G. Nardulli and M. Pietroni, *Phys. Rev.* **D43** (1991) 3002.
18. D. Bortoletto snd S. Stone, *Phys. Rev. Lett.* **65** (1990) 2951.
19. J.L. Rosner, *Phys. Rev.* **D42** (1990) 3732.
20. E.V. Shuryak, *Nucl. Phys.* **B198** (1982) 83.
21. S. Capstick and S. Godfrey, *Phys. Rev.* **D41** (1988) 2856.
22. Y. Kubota *et al.* (CLEO), presented at XVI International Symposium on Photon-Lepton Physics, Cornell U., Ithaca, New York, August 1993.
23. F. Butler *et al.*, "A measurement of $\mathcal{B}(D_s^+ \to \phi\ell^+\nu)/\mathcal{B}(D_s^+ \to \phi\pi^+)$," CLNS 94/1272, CLEO 94-7.
24. S. Aoki *et al.* (WA75), *Progress of Theor. Phys.* **89** (1993) 131.
25. K. Hikasa *et al.* (Particle Data Group), *Review of Particle Properties, section VII.i*, *Phys. Rev.* **D45** (1992) 1.
26. J.Z. Bai *et al.*, *Nucl. Instr. and Methods* **A344** (1994) 319.
27. J.Z. Bai *et al.*, *Phys. Rev. Lett.* **69** (1992) 3021.
28. C.C. Zhang and Z.X. Zhang, "A proposal on D_s physics run at BES," National Workshop on Charm Physics, Beijing, China, October 1990.
29. M. Fero snd W. Toki, "D_s physics at BES/BEPC," BES internal report, January 1991.
30. J.H. Gu, presented at the BES Annual Meeting, Hangzhou, China, July 1994.

A MEASUREMENT OF THE PRODUCTION OF D*± MESONS ON THE Z⁰ RESONANCE

ALFRED M. LEE IV[†]

Department of Physics, Duke University,
P.O.Box 90305, Durham, North Carolina 27708, USA

ABSTRACT

In an analysis of 1.25 million multihadronic decays of the Z^0, we have identified 2001 candidate D*± mesons. Using a combination of four bottom tagging methods, D*± mesons arising from primary c quarks are separated from those of b quarks, and are used to measure $\Gamma_{c\bar{c}}/\Gamma_{had} = 0.141 \pm 0.008(stat) \pm 0.009(sys) \pm 0.011(ext)$. Studying D*± production at low x_{D^*}, we find indications for $c\bar{c}$ production from gluon splitting and measure the mean multiplicity of this process in multihadronic Z^0 decays to be $\bar{n}_{g \to c\bar{c}} = (4.4 \pm 1.6(stat) \pm 1.7(sys))\%$.

1. Introduction

The analysis presented here combines four different methods for separating D* [‡] mesons arising from primary charm quarks from those in bottom quark decays. By tagging separately c and b quarks, $\Gamma_{c\bar{c}}/\Gamma_{had}$ and production methods of charm quarks can be studied. We use the D* decay chain, $D^{*+} \to D^0\pi^+$ (68%); $D^0 \to K^-\pi^+$ (3.7%), which provides a clean sample of events and direct information about the shape of the fragmentation function and the average D* energy. Measuring the production of D* mesons in the full range of scaled energy, $x_{D^*} \equiv 2E_{D^*}/E_{cm}$, the contribution to charm meson production from gluon splitting can be disentangled from that of $Z^0 \to c\bar{c}$ production.

The data presented here were collected with the OPAL detector, a multi-purpose particle detector at the electron-positron collider LEP. Details of its construction and performance are described elsewhere.[1,2]

2. D* Event Selection and Background Subtraction

The method used to select D* events has been previously described,[3,4] so only a few relevant details will be given here. Track quality, invariant mass, dE/dx, and D^0 decay kinematics are used to isolate a pure sample of events. Candidates are selected by trying all combinations of tracks which pass track quality cuts and charge assignments for D* and D^0 mesons. D* candidates have D^0 candidates with masses of $1790 \, \text{MeV}/c^2 < M_{D^0}^{cand} < 1940 \, \text{MeV}/c^2$, and a kaon candidate with $\mathcal{P}_K > 0.2$ for $x_{D^*} < 0.2$ or $\mathcal{P}_K > 0.1$ for $0.2 < x_{D^*} < 0.5$, where $\mathcal{P}_K$ is the probability that the measured dE/dx is consistent with that expected for a real kaon. Since the D^0 decays isotropically, true

---

[*]This paper is an abridgment of OPAL Physics Note PN–148. **All results are preliminary.**
[†]Representing the OPAL Collaboration.
[‡]Charge conjugate modes are always implied, and D*, K, etc. refer to the charged particles only.

Fig. 1. Distributions of the invariant mass difference between the D*candidate and the D⁰candidate, in four ranges of x_{D^*} corresponding to regions of different selection cuts. Contrary to convention, the line histogram shows the signal sample, and the points the distribution of the background estimator obtained from wrong charge, reflected pion and reflected pion wrong charge combinations. The shape of the background in ΔM is fit with an empirical function for the four different x_{D^*} ranges, which is then normalized to the signal distribution for $155 < \Delta M < 200\,\text{MeV}/c^2$ in each bin of x_{D^*} when subtracting the background to find the number of signal events.

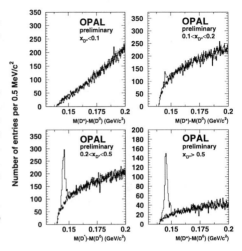

D^0 events have a flat distribution in the angle between the kaon direction in the D^0 rest frame and the D^0 direction in the laboratory frame, while the background events have pronounced peaks around $\cos\theta^* = \pm 1$. Cuts are applied to remove these events.

The D^* signal region is defined as $142\,\text{MeV}/c^2 < \Delta M < 149\,\text{MeV}/c^2$, where $\Delta M \equiv M_{D^*}^{cand} - M_{D^0}^{cand}$. The D^* selection was found to have an efficiency of between 20 and 40% for $0.1 < x_{D^*} < 1.0$, with the efficiency dropping to 7.8% for $x_{D^*} < 0.1$.

The dominant source of background comes from random combinations of tracks. The shape and rate are determined from the data using three special event samples: *wrong charge candidates,* where the charges of both tracks of the D^0 candidate decay products are equal, and the candidate D^* pion has opposite charge; *reflected pion candidates,* where a candidate D^* pion track from the hemisphere opposite a normal D^0 candidate, reflected about the origin, is added to the event; and *reflected pion wrong charge candidates,* where both a wrong charge candidate and a reflected pion are used. These events are analyzed in the same way as the signal events. Monte Carlo studies indicate that all three samples describe the background well, so they are combined for maximum statistical sensitivity (see figure 1).

3. Separating c → D* from b → D*

A clean separation between c → D* and b → D* production can be achieved by identifying b quark events using additional information in the event. We use four different techniques to tag b quark events: leptons with high momenta and transverse momenta; an artificial neural network of several jet shape variables which distinguish types of jets; the separation of tracks from the Z^0 decay vertex in the hemisphere opposite from the D^*; and the apparent decay length significance of the D^0 candidate in the D^* candidate decay. Each method provides a statistical separation of the event sample into b → D* events and those not originating from a b decay. Aside from the possibility of g → c̄c (which will be discussed later), the number of D* candidates

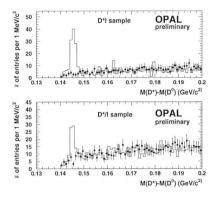

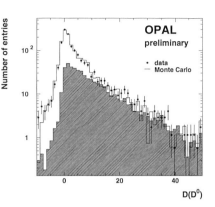

Fig. 2. ΔM for the D*ℓ and D*/ℓ candidate samples. The line histogram shows the signal sample, and the points show the background estimator combined with leptons of all charge.

Fig. 3. Distribution of decay length significance for D* candidates. The points are data and the line histogram Monte Carlo. The hatched histogram is the predicted contribution from b quarks.

tagged as coming from b decay, N_{b-tag}, in a sample of N_{cand} events, is given by

$$\frac{N_{b-tag}}{N_{cand}} = (1 - f_{BG}) \cdot [f_b \cdot \mathcal{P}_b + (1 - f_b) \cdot \mathcal{P}_c] + f_{BG} \cdot \mathcal{P}_{BG}. \qquad (1)$$

where f_{BG} is the fraction of background candidates in the D* sample and f_b is the fraction of true D* mesons from b-flavored events; the fraction of D*mesons from c-flavored events is $f_c = 1 - f_b$. The variables $\mathcal{P}_c$ and $\mathcal{P}_b$ describe the probability that c → D* and b → D* events will be tagged as being bottom-flavored events respectively, and $\mathcal{P}_{BG}$ is the probability that a background D* candidate is thus tagged. Solving equation 1 for f_b, the fraction of D* events from b → D* decays can be calculated.

Detailed descriptions of the four methods will not be given. See figures 2 and 3 for an indication of the effectiveness of two of these methods, and figure 4 for the c → D* and b → D* fractions found by each method.

4. Evidence for Gluon Splitting to cc̄ and Measurement of $\Gamma_{c\bar{c}}/\Gamma_{had}$

After subtracting the contribution from b → D* from the x_{D*} distribution, we fit c → D* data with a function which has the g → D* component, Z^0 → cc̄ → D* component, and average x_{D*} as free parameters (the shapes are determined from Monte Carlo). Using the standard model prediction of $F_{c\bar{c}} = 0.171$, and the assumption that the product branching ratio of c → D* is equal for directly produced charm quarks and for charm quarks produced in gluon splitting, we find $\bar{n}_{g \to c\bar{c}} = (4.4 \pm 1.6 \pm 1.7)\%$. The largest systematic error (1.3%) comes from uncertainty in the b and c quark event separation; uncertainty from fragmentation modeling is negligible (0.4%). The differential yield of D* mesons in the c-tagged sample is shown in figure 5, together

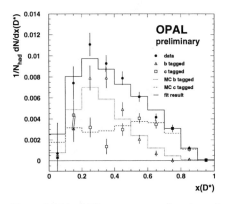 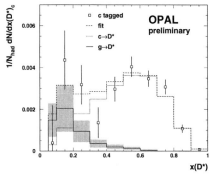

Fig. 4. Yield of D* mesons as a function of x_{D*}. The filled dots are the total measured yield, open squares c → D*, open triangles b → D*. The histograms are Monte Carlo predictions for c → D* (dashed line) and b → D* (dotted line).

Fig. 5. Yield of D* mesons in c-tagged events. The dashed curve is the result of the fit (see text), with the dotted and solid line giving the direct c → D* and g → D* components respectively. The hashed area indicates the errors in g → D*.

with the component from primary quark production and from gluon splitting. If we repeat the fit with the gluon component set to zero, the χ^2 increases from 8.3 for seven degrees of freedom to 12.5 for 8 degrees of freedom (including systematic errors). Our data seem to favour a non-zero g → c̄c rate, representing the first experimental signs of this process in e^+e^- collisions.

To convert the production rate for D* mesons from primary c̄c into the partial width, $\Gamma_{c\bar{c}}/\Gamma_{had}$, we need the branching ratio of c → D*. No published measurement of this exists for LEP energies, so we use an average of measurements at lower energy e^+e^- accelerators, similar to the procedure followed in an earlier publication.[3] The experiments at lower energies quote directly the product branching ratio into the $\pi K \pi$ final state, making the calculation independent of the errors on the individual branching ratios. The validity of this procedure depends on the assumption that the sources of D* mesons in charm decays are the same at LEP energies as they are at lower energies. With this caveat, we measure $\frac{\Gamma_{c\bar{c}}}{\Gamma_{had}} = 0.141 \pm 0.008 \pm 0.009 \pm 0.011$. The errors quoted are statistical, systematic from this analysis, and systematic due to the uncertainty in the branching ratios used to extract the result. This result is 1.8 standard deviations below the expectation in the standard model of $\Gamma_{c\bar{c}}/\Gamma_{had} = 0.171$.

References

1. OPAL Collaboration, K. Ahmet *et al.*, *Nucl. Instr. Meth.* **A 305** (1991) 275.
2. P.P. Allport *et al.*, *Nucl. Instr. Meth.* **A 324** (1993) 34.
3. OPAL Collaboration, G. Alexander *et al.*, *Phys. Lett.* **B 262** (1991) 341.
4. OPAL Collaboration, R. Akers *et al.*, *Z. Phys.* **C 60** (1993) 601.

Λ_c^+ PRODUCTION IN Z^0 DECAYS

JANIS A. McKENNA *

*Department of Physics, University of British Columbia,
Vancouver, British Columbia, V6T 1Z1, Canada*

ABSTRACT

The production of Λ_c^+ baryons in Z^0 decays was measured using 1.72 million multihadronic events collected during 1991-93 with the OPAL detector at LEP. A sample of 423±71 Λ_c^+ candidates was obtained. By examining the energy spectrum of Λ_c^+ candidates, separation of Λ_c^+ from $Z^0 \rightarrow \bar{b}b \rightarrow \Lambda_c^+ X$ and $Z^0 \rightarrow \bar{c}c \rightarrow \Lambda_c^+ X$ sources was achieved, yielding production fraction times branching ratio:

$$f(b \rightarrow \Lambda_c^+ X) \cdot Br(\Lambda_c^+ \rightarrow pK^-\pi^+) = (0.48 \pm 0.12 \pm 0.05)\%$$
$$f(c \rightarrow \Lambda_c^+ X) \cdot Br(\Lambda_c^+ \rightarrow pK^-\pi^+) = (0.33 \pm 0.15 \pm 0.08)\%$$

and the inclusive rate from Z^0 decays:

$$\left[\Gamma\left(Z^0 \rightarrow \Lambda_c^+ X\right)/\Gamma_{had}\right] \cdot Br\left(\Lambda_c^+ \rightarrow pK^-\pi^+\right) = (0.322 \pm 0.049 \pm 0.038)\%$$

An excess of Λ_c^+ in $Z^0 \rightarrow \bar{b}b$ events is seen when compared with the rate measured at $\Upsilon(4S)$ energies, indicating the presence of b baryons decaying to Λ_c^+.

1. Introduction – Charmed Baryons at LEP

There are many measurements * of Λ_c^+ production from $c\bar{c}$ continuum and B meson decays (CLEO and ARGUS),[1] yet very few measurements have been performed at higher energies.[2] At LEP, it is interesting to examine inclusive Λ_c^+ production, as Λ_c^+ could also be produced in the decays of b baryons, which are not accessible at the lower energy $\Upsilon(4S)$ experiments.

2. The OPAL Detector

A complete description of the OPAL detector may be found elsewhere.[3] The crucial component of the detector in this analysis is the tracking system, which is immersed in a uniform .435 T magnetic field. The main tracking chamber is a jet-cell drift chamber, which provides particle identification via specific ionization (dE/dx) measurements, and together with the precision vertex drift chamber, tracks charged particles. The Silicon Microvertex detector[4] consists of 2 barrels of silicon wafers providing $r - \phi$ space point resolution of $9\mu m$ and an impact parameter resolution of $16\mu m$ for 45 GeV lepton pairs. In 1993, the silicon detector was upgraded and presently provides $r - \phi$ and z coordinate measurements.

3. Event Selection

This analysis used only events with Silicon Microvertex detector information available. The OPAL hadron selection[5] was applied and events which passed data quality and detector status were selected, leaving a sample of 1.72 million events.

---

*Representing The OPAL Collaboration
*Throughout this note, whenever a specific particle or decay chain is specified, the conjugate particle or process is also implied.

At this point, three prong vertices were formed. Only tracks with $|\cos\theta| < 0.85$ were considered. Silicon $r - \phi$ hits were matched to tracks when possible. Candidate Λ_c^+ vertices were formed in the r-ϕ plane, using all possible 3 track combinations for the decay $\Lambda_c^+ \rightarrow pK^-\pi^+$. A signed decay length was constructed using the position of the reconstructed vertex to average e^+e^- interaction point. At least 2 of the 3 tracks were required to have associated silicon hits, and the χ^2 probability of the vertex fit was required to be $> 1\%$. Particle identification cuts were then applied:

- $\pi^\pm$ candidates have $p > 1\,\mathrm{GeV}/c$ and dE/dx π probability $>1\%$

- p candidates have $p > 2\,\mathrm{GeV}/c$ and dE/dx proton probability $>1\%$ ($>3\%$) if below (above) the proton dE/dx expectation.
  The probability of the proton candidate being a pion must be $< 1\%$.

- K candidates have $p > 2\,\mathrm{GeV}/c$ and dE/dx K probability $>1\%$ ($>3\%$) if below (above) the kaon dE/dx expectation.
  The probability of the kaon candidate being a pion must be $< 1\%$.

- reject event if $K^- - p$ combination is consistent with being a ϕ when kaon mass hypotheses are assumed.

Finally, the large combinatorial background at low x_E was rejected by considering only Λ_c^+ candidates with $x_E > .15$ and applying the following cuts on decay length significance (L/σ) :

$$
(L/\sigma) \quad
\begin{array}{l}
>4 \text{ for } .15 < x_E < .3 \\
>1 \text{ for } .3 < x_E < .6 \\
>0 \text{ for } .6 < x_E < 1.0
\end{array}
$$

4. The $\Lambda_c^+ \rightarrow pK^-\pi^+$ Signal

The $pK^-\pi^+$ invariant mass distribution of all track combinations satisfying the selection criteria is shown in Figure 1. The line is a χ^2 fit to the Λ_c^+ spectrum, with a Gaussian signal and polynomial background.

Efficiencies were determined using a full simulation of the OPAL detector.[7] The JETSET 7.3 event generator was used, with heavy flavour branching ratios from EU-RODEC 205. This combination reproduces the CLEO Br($\overline{B} \rightarrow \Lambda_c^+ X$) very well. The Peterson fragmentation scheme in JETSET was used for heavy charm and beauty quarks, with parameters $\epsilon_b = 0.0055$ and $\epsilon_c = 0.050$. The OPAL measurements for b hadron lifetimes were used,[6] and the Z^0 widths used were $\Gamma_{\overline{b}b} : \Gamma_{\overline{c}c} = 17.2 : 21.7$.[8] Extra smearing on track parameters in the Monte Carlo was applied in order to get good agreement between data and Monte Carlo for tracking resolution.

The Λ_c^+ energy spectrum from Monte Carlo generator is shown in Figure 2. It is evident that the Λ_c^+ spectra from primary $Z^0 \rightarrow \overline{b}b$ and primary $Z^0 \rightarrow \overline{c}c$ quarks differ sufficiently to allow separation of Λ_c^+ into the two sources. The full Monte Carlo was then used to obtain acceptances as function of Λ_c^+ decay length and Λ_c^+ energy. The acceptances are low, due to the severity of the dE/dx and decay length significance cuts for Λ_c^+'s at low x_E.

5. Measurement of $b \rightarrow \Lambda_c^+ X$, $c \rightarrow \Lambda_c^+ X$ rates

The Λ_c^+ candidate events were divided into four ranges of the scaled energy variable x_E. Signal and backgrounds were fit as previously, and each x_E range was corrected for efficiencies.

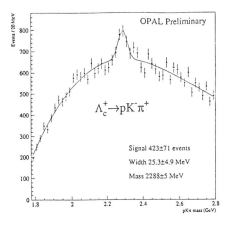

Figure 1. Invariant mass of $pK^-\pi^+$ combinations passing selection criteria described in Section 3.

Figure 2. Monte Carlo simulation of the Λ_c^+ energy spectrum, showing contributions from $c \to \Lambda_c^+ X$ and $b \to \Lambda_c^+ X$.

5.1.   Systematic Uncertainties

The dominant systematic uncertainties arise from the physics input to Monte Carlo generators and from the simulation of the detector.

The b hadron lifetimes from OPAL were varied over a $1\,\sigma$ range of the OPAL measurements,[6] and the variation in the Λ_c^+ lifetime is taken from the Particle Data Group, over a 1σ range. The Peterson fragmentation parameters were varied over the ranges: $\epsilon_b = 0.0055^{+0.0045}_{-0.0035}$, $\epsilon_c = 0.050 \pm 0.020$, corresponding to the ranges for x_E: $\langle x_E \rangle_b = 0.70 \mp 0.02$, $\langle x_E \rangle_c = 0.51 \mp 0.02$.

The simulation of the detector response is another source of systematic uncertainty. Track impact parameter and angular resolutions were varied by $\pm 15\%$.

The efficiency of matching silicon hits to tracks was studied using the decay $D^{*+} \to D^0\pi^+$, with $D^0 \to K^-\pi^+$, from which the single track matching efficiency was determined to be $90.6 \pm 0.9\%$, in excellent agreement with the Monte Carlo prediction.

The Monte Carlo simulation of the dE/dx response of the detector was studied using a sample of Λ^0 particles, decaying via $\Lambda^0 \to p\pi^-$. The dE/dx efficiency was measured as a function of proton momentum in both the data and the Monte Carlo simulation. The two were found to be in excellent agreement. Additionally, the kaon dE/dx simulation was checked using the decay chain $D^{*+} \to D^0\pi^+$, with $D^0 \to K^-\pi^+$. The measured efficiency of the dE/dx kaon identification cuts were found to agree excellently in the Monte Carlo and the data. Statistical uncertainties in the ratio of the data and Monte Carlo efficiencies were used to obtain the systematic uncertainty.

All systematic uncertainties are summarized in Table 1. A more detailed description of this analysis may be found elsewhere.[9]

6.   Results and Conclusions

The OPAL detector was used to obtain the first inclusive measurements of Λ_c^+ in Z^0 decays:

$$\left[\Gamma\left(Z^0 \to \Lambda_c^+ X\right)/\Gamma_{had}\right] \cdot Br\left(\Lambda_c^+ \to pK^-\pi^+\right) = (0.322 \pm 0.049 \pm 0.038)\%$$

Quantity	$\Delta f(c \to \Lambda_c^+ X) \cdot$ Br	$\Delta f(b \to \Lambda_c^+ X) \cdot$ Br	$\Delta(\Gamma(Z^0 \to \Lambda_c^+ X)/\Gamma_{had}) \cdot$ Br
τ_b	$\pm 1.00\%$	$\mp 4.62\ \%$	$\mp 2.68\ \%$
τ_{Λ_c}	$\mp 4.30\%$	$\pm 1.44\ \%$	$\mp 2.44\ \%$
ϵ_b	$\pm 11.2\%$	$\mp 2.94\ \%$	$\pm 5.80\ \%$
ϵ_c	$\pm 16.5\%$	$\mp 4.83\ \%$	$\pm 2.55\ \%$
$\Gamma_{q\bar{q}}/\Gamma_{had}$	$\mp 8.19\ \%$	$\mp 1.23\%$	$-$
Track Resolution	$\pm 1.72\%$	$\pm 1.56\ \%$	$\pm 1.10\ \%$
p dE/dx	12.8%	6.24%	8.11%
K dE/dx		4.4%	
Silicon match		0.5%	
Total Syst. Err.	25.9%	10.9%	11.8%

Table 1. Summary of systematic uncertainties.

which may be separated into charm and beauty components:

$$f(b \to \Lambda_c^+ X) \cdot Br(\Lambda_c^+ \to pK^-\pi^+) = (0.48 \pm 0.12 \pm 0.05)\%$$
$$f(c \to \Lambda_c^+ X) \cdot Br(\Lambda_c^+ \to pK^-\pi^+) = (0.33 \pm 0.15 \pm 0.08)\%$$

These measurements may be combined with the ARGUS/CLEO average for $Br(\Lambda_c^+ \to pK^-\pi^+)$:

$$f(b \to \Lambda_c^+ X) = (11.1 \pm 4.0 \pm 1.2 \pm 2.8)\%$$
$$f(c \to \Lambda_c^+ X) = (7.6 \pm 4.1 \pm 2.0 \pm 1.9)\%$$
$$\Gamma(Z^0 \to \Lambda_c^+ X)/\Gamma_{had} = (7.5 \pm 2.2 \pm 0.9 \pm 1.9)\%$$

In conclusion, the $f(c \to \Lambda_c^+ X) \cdot Br(\Lambda_c^+ \to pK^-\pi^+)$ has been measured at $\sqrt{s} \sim$ 90 GeV, and is compatible with measurements at $\sqrt{s} \sim 10$ GeV. b baryon decays to Λ_c^+ could produce an excess of Λ_c^+ over the number expected from B meson decays alone, and using the CLEO measurement of $Br(\overline{B} \to \Lambda_c^+ X) \cdot Br(\Lambda_c^+ \to pK^-\pi^+) =$ $(0.27 \pm 0.06)\%$, and the results of this analysis for $f(b \to \Lambda_c^+ X)$, one obtains

$$f(b \to Y) \cdot Br(Y \to \Lambda_c^+ X) \cdot Br(\Lambda_c^+ \to pK^-\pi^+) = (0.21 \pm 0.14)\%$$

where Y is a source of Λ_c^+ in $Z^0 \to b\bar{b}$ events other than B mesons. This may be interpreted as a 1.4σ excess of Λ_c^+ production due to b baryon decays.

References

1. G.Crawford et al., CLEO Collab., Phys. Rev. **D45**, 752 1992.
   H. Albrecht et al., ARGUS Collab., Phys. Lett. **B210** , 263 1988.
2. S.R.Klein et al., Mark II Collab., Phys. Rev. Lett. **62**, 2444 1989.
3. K.Ahmet et al., OPAL Collab., Nucl. Instrum. and Meth. **A305**, 275 1991.
4. P.P.Allport et al., Nucl. Instrum. and Meth. **A324**, 34 1993.
   P.P.Allport et al., CERN-PPE-94-016 1994.
5. G.Alexander et al., OPAL Collab., Z. Phys. **C52**, 175 1991.
6. P.Acton et al., OPAL Collab., Phys. Lett. **B312**, 501 1993.
   R.Akers et al., OPAL Collab., Phys. Lett. **B316**, 435 1993.
   OPAL Physics Note 106, July 1993.
7. J.Allison et al., Nucl. Instrum. and Meth. **A317**, 47 1992.
8. D.Bardin et al., CERN-TH-6443-92 1992.
9. OPAL Physics Note 141, July 1994.

THE Λ_c - NOT YOUR AVERAGE BEARYON

DAVE Z. BESSON*

*Physics Department, University of Kansas,
Lawrence, KS USA 66045*

(Representing the CLEO collaboration)

ABSTRACT

We report on studies of hadronic Λ_c decay modes which are expected to proceed through diagrams other than the simple external W-emission diagram. Signals are reported for the decay modes: $pK_s^0\eta$, $\Lambda\eta\pi^+$, $\Sigma^+\eta$, and $\Lambda K_s^0 K^+$. Branching fractions are calculated relative to the normalizing mode: $\Lambda_c \to pK\pi$.

1. Introduction

Since its discovery almost 15 years ago, the Λ_c has been a source of important information on weak decays of heavy baryons. The last five years have witnessed increasingly precise measurements of the Λ_c lifetime,[1,2] as well as probes of the rare decay sector. The Λ_c is typically modeled as a system containing a heavy charm quark surrounded by an inert (ud) diquark, bound in an S=0 state, with L=0 relative to the heavy charm quark. The (ud) system is in an isospin=0 state, unlike the Σ_c, which has I=1. The simplest picture of Λ_c decays then is as the weak decay of the charm quark to the strange quark by external W-emission, such that the (ud) diquark is entirely decoupled from the $c \to s$ transition. In such a naive picture, $\mathcal{B}(\Lambda_c \to \Lambda X)$=100%. However, the tabulated value for this inclusive decay is 27±9%[7]; moreover, the simplest spectator decay $\mathcal{B}(\Lambda_c \to \Lambda\pi^+) = 0.58 \pm 0.16\%$[7]; by comparison the comparable decay in the charmed meson sector: $D^0 \to K^-\pi^+$ has a branching ratio about 8 times as large. This, coupled with the measurement of a Λ_c lifetime half that of the D^0 and the D_s, indicates that there is a large hadronic width proceeding by diagrams other than the simple external W-emission diagram. To some extent, this is to be expected in the baryonic sector, where exchange decays ($\Lambda_c \to \Sigma^+\phi$, e.g.) are not subject to the same helicity suppression operative in charmed meson decays. The fact that $\Lambda_c \to pK\pi$ is so large also indicates that internal W-emission could play a substantial role in charmed baryon decays. In this paper, we report on an analysis of Λ_c's to rare final states which are candidates for non-external spectator decays.

---

*With inspiration from Yogi Berra and Don Fujino

2. Data Sample and Event Selection

The data were collected with the CLEO II detector at the Cornell e^+e^- storage ring CESR, which operated on, and just below, the $\Upsilon(4S)$ resonance. The CLEO II detector[8] is a large solenoidal detector with 67 tracking layers and a CsI electromagnetic calorimeter that provides efficient π^0 and η reconstruction. We have used a total integrated luminosity of 3.25 fb^{-1}, which corresponds to roughly 4 million $c\bar{c}$ events. Events with three charged tracks originating from the known primary vertex which satisfy a minimum energy cut in the crystal calorimeter constitute the candidate hadronic event sample.

K_s^o and Λ candidates are selected through their decays $K_s^o \to \pi^+\pi^-$ and $\Lambda \to p\pi^-$ by reconstructing a secondary decay vertex. Protons, kaons, and pions are identified using dE/dx and time-of-flight information, when available. η and π^0 candidates are selected through their decays to $\gamma\gamma$ from pairs of well-defined showers in the CsI calorimeter. We kinematically fit the photon momenta to the nominal parent mass in order to improve the parent momentum estimate. The Σ^+ particle is identified through its decay into $p\pi^0$.[6] The Σ^+ momentum measurement is refined by computing the proton and π^0 momenta at the location of the Σ^+ decay vertex. Charged tracks that do not arise from long-lived particles (Λ, K_s^o, and Σ^+) must have a distance of closest approach in the $r - \phi$ plane of less than 5 mm to the interaction point, and be within 5 cm in $\hat{z}$.

3. Observation of new Λ_c decay modes

We search for the decay modes: $\Lambda_c \to pK^+\eta$, $\Lambda_c \to \Lambda K^oK^+$, $\Lambda_c \to \Sigma^+\eta$, and $\Lambda_c \to \Lambda\eta\pi^+$ by combining the candidate daughter particles and requiring that the scaled momentum $x_p(\Lambda_c^+) > 0.5$, where $x_p(\Lambda_c^+) = P_{\Lambda_c^+}/\sqrt{E_{beam}^2 - M_{\Lambda_c^+}^2}$. Invariant mass spectra for the four decay modes of interest are displayed in Figures 1.

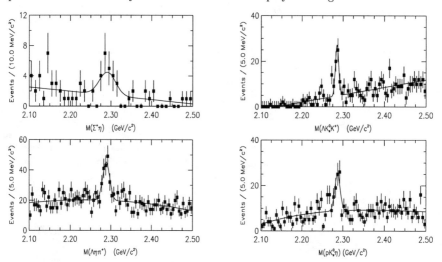

Figs. 1. Candidate Λ_c invariant mass spectra for the modes indicated.

4. Results and discussion

Event yields, calculated efficiencies based on a GEANT-based full CLEO detector simulation, and the extracted ratio of branching ratios relative to the decay $\Lambda_c \to pK\pi$ are presented in Table 1. Table 2 gives the breakdown of our systematic errors. Table 3 presents our results in comparison with theoretical predictions.

Table 1. Summary of event yields for new Λ_c modes.

Λ_c^+ decay mode	Number of events	Efficiency (%)	$\mathcal{B}/\mathcal{B}(\Lambda_c^+ \to pK^-\pi^+)$
$p\bar{K}^0\eta$	53 ± 10	7.3	$0.25 \pm 0.05 \pm 0.04$
$\Lambda\eta\pi^+$	109 ± 16	8.5	$0.36 \pm 0.06 \pm 0.05$
$\Sigma^+\eta$	25 ± 7	5.2	$0.10 \pm 0.03 \pm 0.02$
$\Lambda\bar{K}^0 K^+$	46 ± 8	7.9	$0.11 \pm 0.02 \pm 0.02$

Table 2. Summary of systematic errors

Source	Fractional Error (%)			
	$pK_s^0\eta$	$\Lambda\eta\pi^+$	$\Sigma^+\eta$	$\Lambda K_s^0 K^+$
K_s^0, Λ, and Σ^+ finding	10	10	10	14
η and π^0 finding	5	5	7	—
Tracking efficiency	—	—	4	4
Proton and kaon identification	8	2	2	—
$pK^-\pi^+$ substructure	7	2	2	5
Λ_c^+ substructure	—	6	—	—
Uncertainties in $\sigma_{\Lambda_c^+}$	5	5	5	5
Monte Carlo statistics	3	3	3	3
TOTAL	16	14	14	16

Table 3. Results for $\Lambda_c \to \Sigma^+\pi^o$ compared with theoretical predictions.

	$\frac{\mathcal{B}(\Lambda_c^+ \to \Sigma^+\pi^0)}{B(pK^-\pi^+)}$	$\frac{\mathcal{B}(\Lambda_c^+ \to \Sigma^+\eta)}{B(pK^-\pi^+)}$
CLEO	$0.20 \pm 0.03 \pm 0.03$	$0.10 \pm 0.03 \pm 0.02$
Körner and Krämer[9]	0.10	0.05
Zenczykowski[10]	0.13	0.08

5. Comment on Results and $\mathcal{B}(\Lambda_c \to pK\pi)$

The branching ratios relative to $\Lambda_c \to pK\pi$ are shown in Figure 2. We note the following:

1. $\frac{\mathcal{B}(\Lambda_c \to pK^o\eta)}{\mathcal{B}(\Lambda_c \to pK^o\pi^o)} \sim 0.5$. If the modes $\Lambda_c \to pKX$ occur through the internal spectator diagram, with $W_{int} \to u\bar{d}$ and $d\bar{d}$ or $u\bar{u}$ popping between the $\bar{d}$ and the s-quark (as shown in the diagram below), then, since $\pi^0 = \frac{u\bar{u}-d\bar{d}}{\sqrt{2}}$, $\pi^+ = u\bar{d}$, and $\eta^o = \frac{u\bar{u}+d\bar{d}-2s\bar{s}}{\sqrt{6}}$,

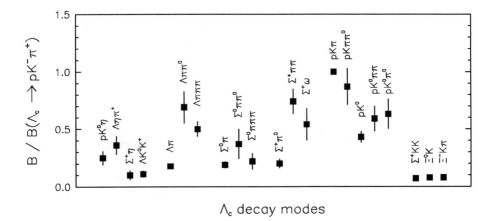

Fig. 1. Figure 2. Branching fractions of Λ_c^+ decays relative to the normalizing decay mode $\Lambda_c^+ \to pK^-\pi^+$ from recent CLEO measurements.

we would expect $\mathcal{B}(\Lambda_c \to pK\pi) : \mathcal{B}(\Lambda_c \to p\bar{K}^o\pi^o) : \mathcal{B}(\Lambda_c \to pK\eta) = 1{:}2{:}6$. Our data are in agreement with this.

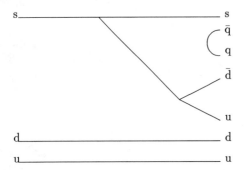

2. $\frac{\mathcal{B}(\Lambda_c \to \Lambda\eta\pi^+)}{\mathcal{B}(\Lambda_c \to \Lambda\pi^o\pi^+)} \sim 0.5$. By the same argument above, if there is no resonant enhancement of $\Lambda_c \to \Lambda\pi^o\pi^+$ through the external W-diagram: $\Lambda_c \to \Lambda\rho^+; \rho^+ \to \pi^o\pi^+$, then the ratio of these branching ratios should be again 1:3. In fact, examination of the $\pi^o\pi^+$ invariant mass spectrum in the $\Lambda\pi^+\pi^o$ final state is entirely consistent with phase space; our data are therefore consistent with a simple quark popping model.

3. $\frac{\mathcal{B}(\Lambda_c \to \Lambda K^o K^+)}{\mathcal{B}(\Lambda_c \to \Lambda\pi^o\pi^+)} \sim 0.15$. The ratio of inclusive kaon to inclusive pion yields at $\sqrt{s}=10$ GeV suggests that the fraction of times an s$\bar{s}$ pair is popped from the vacuum compared with the sum of u$\bar{u}$ and d$\bar{d}$ is about $\frac{1}{6}$. Our data are therefore once again consistent with the simple quark-popping model given above.

4. $\frac{\mathcal{B}(\Lambda_c \to \Sigma^+ \eta)}{\mathcal{B}(\Lambda_c \to \Sigma^+ \pi^o)} \sim 0.5$, and $\frac{\mathcal{B}(\Lambda_c \to \Sigma^+ \eta)}{\mathcal{B}(\Lambda_c \to \Sigma^+ \phi)} \sim 1.0$. The $\Lambda_c \to \Sigma^+ \phi$ decay mode is expected to proceed dominantly through a W-exchange diagram, the $\Lambda_c \to \Sigma^+ \pi^o$ mode through an internal spectator diagram. The $\Lambda_c \to \Sigma^+ \eta$ mode is accessible through both of these diagrams, with some unknown fraction of each contributing to the final state. Non-observation of the $\Sigma^+ \eta$ final state would then be somewhat surprising; given our large errors, we can only state here that this mode appears to have approximately the branching ratio one might expect by inference from the other $\Sigma^+ X$ modes.

Taken together, these modes constitute new evidence for diagrams other than the external W-diagram contributing to the total hadronic Λ_c decay width. This is consistent with the observed shortness of the Λ_c lifetime and also the fact that the inclusive $\Lambda_c \to \Lambda + X$ rate is not large.

Finally, it should be remarked here that the normalizing mode, $\Lambda_c \to pK\pi$ was derived[11] using a model of $B \to baryons$ which has recently been shown to be incomplete.[12] The impact of this on the derived value of $\mathcal{B}(\Lambda_c \to pK\pi)$ has yet to be evaluated; it could result in a 20-30% shift in the branching fraction.

6. Acknowledgements

I thank D. Fujino for doing the analysis contained herein.

References

1. P.L. Frabetti (E687), Phys. Rev. Lett. **70** 1381 (1993) and P.L. Frabetti (E697), Phys. Rev. Lett. **70** 2058 (1993).
2. S. Barlag, (NA32), Phys. Lett. **B233** 522 (1989), and S. Barlag, (NA32), Phys. Lett **B236** 495 (1990).
3. CLEO Collaboration, D. Cinabro et al., CLEO Conf 93-27, contrib. paper in Lepton Photon Conf. (1993).
4. CLEO Collaboration, P. Avery et al., Phys. Rev. Lett. **71**, 2391(1993).
5. CLEO Collaboration, P. Avery et al., Phys. Lett. B **325**, 257(1994). No significant $\Lambda \rho^+$ or $\Sigma^*(1385)\pi$ resonant substructure was evident in the $\Lambda_c^+ \to \Lambda \pi^+ \pi^0$ decay mode.
6. CLEO Collaboration, Y. Kubota et al., Phys. Rev. Lett. **71**, 3255(1993).
7. Particle Data Group, K. Hikasa et al., Phys. Rev. D **45**, 1(1992).
8. CLEO Collaboration, Y. Kubota et al., Nucl. Inst. and Meth. **A320**, 66(1992).
9. J.G. Körner and M. Krämer, Z. Phys. C **55**, 659(1992).
10. P. Zenczykowski, Phys. Rev. D **50**, 402(1994).
11. G. Crawford et al. (The CLEO Collaboration) Phys. Rev. **D45**, 752 1992, and M. Procario et al., CLNS 93/1264, submitted to PRL.
12. D. Cinabro et al. "A Study of Baryon-Production in B-decay: Search for Semileptonic decays of B mesons to Charmed Baryons and the First Observation of Ξ_c Production in B-decay", (the CLEO collaboration), submitted to the XXVII International Conference on High Energy Physics, Glasgow, Scotland.

RECENT RESULTS ON THE SEMILEPTONIC DECAY $D^0 \to K^- \mu^+ \nu_\mu$

WILL E. JOHNS*

Department of Physics, University of Colorado, Campus Box 390
Boulder, Co. 80309, USA

ABSTRACT

High statistics results on the decay $D^0 \to K^- \mu^+ \nu_\mu$(+c.c.) will be presented. The results include the relative branching fraction of $D^0 \to K^- \mu^+ \nu_\mu$ to $D^0 \to K^- \pi^+$, a measurement of the pole dependence of the f_+ form factor and a determination of $f_+(0)$.

1. Introduction

The next generation of fixed target charm experiments will have high precision, high statistics, low background measurements of semileptonic D decays. To take full advantage of this bounty, the analysis of the decay $D^0 \to K^- \mu^+ \nu_\mu$ from data taken during the 1990 and 1991 runs of Fermilab experiment E687 is used to test the possibility of relaxing the D^* tag requirement to increase statistics. The result is a better measurement of the ratio $\frac{BR(D^0 \to K^- \mu^+ \nu_\mu)}{BR(D^0 \to K^- \pi^+)}$ and the pole mass(from the single pole form of the f_+) than was realized when the analysis utilized a D^* tag.

2. Motivation

Using the expression for the full rate,[1]

$$\frac{d\Gamma}{dE_K} = \frac{G_F^2}{4\pi^3} |V_{cs}|^2 \left| f_+(q^2) \right|^2 P_K \left(\frac{W_0 - E_K}{F_0} \right)^2 \left[\frac{1}{3} m_D P_K^2 + \frac{m_l^2}{8 m_D} (m_D^2 + m_K^2 + 2 m_D E_K) \right.$$

$$\left. + \frac{1}{3} m_l^2 \frac{P_K^2}{F_0} + \frac{1}{4} m_l^2 \frac{m_D^2 - m_K^2}{m_D} Re \left(\frac{f_-(q^2)}{f_+(q^2)} \right) + \frac{1}{4} m_l^2 F_0 \left| \frac{f_-(q^2)}{f_+(q^2)} \right|^2 \right],$$

$$W_0 = \frac{m_D^2 + m_K^2 - m_l^2}{2 m_D}, F_0 = W_0 - E_K + \frac{m_l^2}{m_D}, \tag{1}$$

and the single pole representation of the form factor, (other forms are possible)

$$f_{+,-}(q^2) = f_{+,-}(0)/(1 - q^2/m_{pole}^2) \tag{2}$$

we see that the semi–muonic channel must be measured if f_- is large. The best measurement presently available of the $\frac{BR(D^0 \to K^- l^+ \nu_l)}{BR(D^0 \to K^- \pi^+)}$ is from CLEO[2] and is dominated by the semi–electronic mode. Our measurement will also roughly double the available statistics now available for measurements of the ratio $\frac{BR(D^0 \to K^- l^+ \nu_l)}{BR(D^0 \to K^- \pi^+)}$, m_{pole} and $f_+(0)$.

---

*Representing the E687 Collaboration

3.   Data and Analysis

3.1.   Data Selection

The E687 spectrometer is described elsewhere.[3] To find a D^0, we select Kμ candidates by choosing kaons that the Čerenkov system identifies as a definite kaon or kaon/proton ambiguous. This kaon is combined with an opposite charge muon identified as a muon by the inner muon system of the E687 spectrometer. Typically, $\sim 1.0\%$ of all pions or kaons are identified as muons due to pattern recognition failures, the decay of the pion or the decay of the kaon. The kaon and the muon must form a vertex with a confidence level of 5% or better and have an invariant mass less than $1.855\ GeV/c^2$ to reduce contamination from $D^0 \to K^- \pi^+$. The remaining tracks in the event are used to find primary vertex candidates. The highest multiplicity primary vertex candidate with a significance of separation (l/σ) from the secondary Kμ candidate of 4.5 or greater and confidence level of fit 1% or greater is retained. To estimate the D^0 momentum, following Ref. 4, we assume the D^0 travels from the primary to the secondary; then boost the Kμ candidate to a frame where $(\bar{P}_\mu + \bar{P}_K) \cdot \check{l} = 0$ ($\check{l}$ is the D^0 decay direction) and determine the amount of momentum parallel to the D^0 carried by the missing neutrino. In the boosted frame, we require that $E_\nu > 0$, $(\bar{P}_\nu \cdot \check{l})^2 = (E_\nu^2 - P_{K\mu}^2) > -.7$ $(GeV/c^2)^2$, and for those events with $-.7 < (E_\nu^2 - P_{K\mu}^2) < 0$ we set $(E_\nu^2 - P_{K\mu}^2) = 0$. Hence, $(\bar{P}_\nu \cdot \check{l}) = \pm\sqrt{E_\nu^2 - P_{K\mu}^2} = \pm\bar{P}_D$, and one simply boosts the solutions into the lab frame to estimate the D^0 momentum. We resolve the kinematic ambiguity by choosing the lower D^0 momentum solution since it has the best E_k(hence q^2) resolution in the region of highest m_{pole} information density. A fit is then performed that takes microvertex tracks not assigned to the primary or the secondary vertex and gives a confidence level that any of these tracks are consistent with being in the secondary vertex. We require this confidence level to be $< 1\%$. To reduce backgrounds from muon misidentification, we further require that the $K\mu$ invariant mass be larger than $0.95\ GeV/c^2$, the $K\mu$ momentum be greater than $35\ GeV$ and the muon momentum be greater than $15\ GeV$.

3.2.   Fit

The emphasis in the fit is to include as much of the background as we understand. The largest source of background, muon misidentification, is estimated by redoing the $k\mu$ analysis without the requirement of muon identification and reweighting the event based on the probability that the particle could be misidentified. The backgrounds $D^0 \to K^{*-}\mu^+\nu_\mu$, $D^+ \to K^0\mu^+\nu_\mu$ and $D_s^+ \to \phi\mu^+\nu_\mu$ are generated with our Monte Carlo and given weights relative to $BR(D^0 \to K^-\mu^+\nu_\mu)$ based on production and measured branching ratios. The remainder of the background is simulated using events from the data where the K and the μ have the same charge. Of the components in the signal: $D^0 \to K^-\mu^+\nu_\mu$ accounts for 64.3% of the signal, muon misidentification 18.5%, $D^0 \to K^{*-}\mu^+\nu_\mu$ 7.6%, $D^+ \to K^0\mu^+\nu_\mu$ 7.1%, $D_s^+ \to \phi\mu^+\nu_\mu$ 0.8% and other backgrounds(from $Q(K) = Q(\mu)$) 1.7%.

To fit the data, we construct a likelihood function,

$$\mathcal{L} = \prod_{bins_i} \frac{n_i^{s_i} e^{-n_i}}{s_i!} \tag{3}$$

where

$$s_i = number\ of\ events\ in\ bin_i \tag{4}$$

$$
\begin{aligned}
n_i \;=\; Yield \times & \frac{[(K^-\mu^+\nu_\mu MC_i)+f_1\times(K^{*-}\mu^+\nu_\mu MC_i)+f_2\times((D^+\to)K^{*0}\mu^+MC_i)+f_3\times((D_s^+\to)\phi\mu^+MC_i)]}{(1+f_1+f_2+f_3)} \\
& + number\ of\ events\ from\ misid\ in\ bin_i \\
& + BKYield \times (background(from\ same\ sign\ data)\ in\ bin_i)
\end{aligned} \tag{5}
$$

and each fraction(f_i) multiplying a normalized Monte Carlo shape is determined by

$$f_1 = \frac{Efficiency(D^0 \to K^{*-}(\to K^-\pi^0)\mu^+\nu_\mu)}{Efficiency(D^0 \to K^-\mu^+\nu_\mu)} \times \frac{BR(D^0 \to K^{*-}\mu^+\nu_\mu)}{BR(D^0 \to K^-\mu^+\nu_\mu)} \tag{6}$$

$$f_2 = \frac{Efficiency(D^+ \to K^{*0}(\to K^-\pi^+)\mu^+\nu_\mu)}{Efficiency(D^0 \to K^-\mu^+\nu_\mu)} \times \frac{BR(D^0 \to K^{*-}\mu^+\nu_\mu)}{BR(D^0 \to K^-\mu^+\nu_\mu)} \times \frac{\Gamma(D^+ \to K^{*0}\mu^+\nu_\mu)}{\Gamma(D^0 \to K^{*+}\mu^+\nu_\mu)} \times \frac{\sigma_{D^+}}{\sigma_{D^0}} \times \frac{\tau_{D^+}}{\tau_{D^0}} \tag{7}$$

$$f_3 = \frac{Efficiency(C\bar{C} \to D_s^+ \to \phi(\to K^-K^+)\mu^+\nu_\mu)}{Efficiency(C\bar{C} \to D^0 \to K^-\mu^+\nu_\mu)} \tag{8}$$

where we account for the change in efficiency for different pole masses via,

$$Efficiency(D^0 \to K^-\mu^+\nu_\mu)_{bin_i} \Rightarrow Efficiency(D^0 \to K^-\mu^+\nu_\mu)_{bin_i} \times$$

$$\left[\sum_{in\ bin_i}^{\#events} \left(\frac{M_{fit}^2(q^2,E_\mu)}{M_{gen}^2(q^2,E_\mu)} \right) \right] \times \left(\frac{\int_{q_{low}^2}^{q_{hi}^2} \int_{E_{\mu low}}^{E_{\mu hi}} f_+^2(q^2, m_{pole_0}) M_{gen}^2(q^2, E_\mu) dq^2 dE_{mu}}{\int_{q_{low}^2}^{q_{hi}^2} \int_{E_{\mu low}}^{E_{\mu hi}} f_+^2(q^2, m_{pole}) M_{fit}^2(q^2, E_\mu) dq^2 dE_{mu}} \right) \tag{9}$$

and we allow the yield, background and pole mass to vary in the fit.

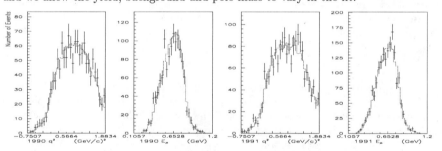

Figure 1. Fit overlayed to projections of the 1990 and 1991 signals.

The fit yields 823.6 ± 31.2 $D^0 \to K^-\mu^+\nu_\mu$ events for 1990 data, 1038.2 ± 37.4 $D^0 \to K^-\mu^+\nu_\mu$ events for 1991 data and a pole mass of $1.98^{+0.12}_{-0.10}$. Subsequent studies using simulated data sets show that the fit underestimates the error on all the parameters by 4%, and we inflate the returned errors by this amount in the final result. We also find that the likelihood from the fit to the data lies within 2σ of the centroid of the spread of likelihoods returned from fitting the simulated data sets. The result of the fit in projections for the 1990 and 1991 data is shown in Fig. 1.

3.3. Systematic Studies

Systematic error in the final result is due to uncertainty in the muon (mis)identification, the ratio $BR(D^0 \to K^{*-}\mu^+\nu_\mu)/BR(D^0 \to K^-\mu^+\nu_\mu)$, the ratio of production between D^0 and D^+ and the amount of Monte Carlo used for establishing efficiencies. To investigate the stability of the signal, we look at the variation with l/σ, $M(K\mu)$, confidence level of the fit to the $K\mu$ vertex, momentum of the $K\mu$ pair, momentum of the μ, choice of D momentum solution and 1990 and 1991 data sets. We also take all microvertex tracks not in the $K\mu$ vertex and vary the confidence level that any of these tracks are consistent with being in the $K\mu$ vertex.(note that this cut is more powerful than that used for the data selection process). Of these stability tests, we find the largest contributions to the systematic error coming from the choice of D momentum solution and the momentum behavior of the $K\mu$ pair(1991 data yield only). We also analyzed a subset of the non–tagged signal for the case where the D^0 is consistent with the hypothesis that it was produced from a D^{*+}. The results from this tagged data indicate statistical departures of less than 1σ for all parameters we measure with the non–tagged data. All sources of systematic error are added in quadrature.

3.4. Preliminary Results

In Table 1. we combine the results of an analysis of the mode $D^0 \to K^-\pi^+$ (where the π was limited to the acceptance of the muon system) for the ratio of branching ratios with the results of the fit and the systematic studies.

Table 1. Preliminary results from the non–tagged signal.

Measured Quantity	Value	Statistical Error	Systematic Error
$\frac{BR(D^0 \to K^-\mu^+\nu_\mu)}{BR(D^0 \to K^-\pi^+)}$	0.860	0.028	$+0.042$ -0.039
Pole Mass M_{pole}	1.98	$+0.13$ -0.10	$+0.04$ -0.10
$f_+(0)$	0.730	$+0.020$ -0.021	$+0.029$ -0.033

Our results agree to within 2σ with the recent CLEO[2] measurement, but we are unable to rule out a significant f_- contribution to the semi–muonic branching fraction. Since $\left(\frac{E687\ BR(D^0 \to K^-\mu^+\nu_\mu)}{CLEO\ BR(D^0 \to K^-l^+\nu_l)}\right) = \left(\frac{0.86 \pm 0.05}{0.98 \pm 0.05}\right) = 0.88 \pm 0.07$, it will take the next generation of charm experiments to show whether there is a significant difference between the semi–electronic and the semi–muonic modes.

References

1. L. Jauneau, in *Methods in Subnuclear Physics*, ed. M. Nicolic (Gordon and Brach, New York, 1969), p. 125.
2. A. Bean *et al.*, *Phys. Lett. B 62* (1993) 647
3. P. L. Frabetti *et al.*, *Nucl. Instr. and Meth. in Phys. Res., sect. A 320* (1992) 519.
4. J. C. Anjos *et al.*, *Phys. Rev. Lett. 62* (1989) 1587

NEW CLEO RESULTS ON
CHARMED BARYON SEMILEPTONIC DECAYS

TING MIAO

Physics Department, Purdue University,
West Lafayette, Indiana 47907, USA

(Representing the CLEO collaboration)

ABSTRACT

We report new results from CLEO on charmed baryon semileptonic decays, including the first observation of the decay modes $\Xi_c^+ \rightarrow \Xi^0 e^+ \nu_e$, the first absolute upper bounds on Ξ_c hadronic branching ratios, a measurement of the lifetime ratio of Ξ_c^+/Ξ_c^0 and the preliminary result on form factor ratio measurement in $\Lambda_c^+ \rightarrow \Lambda e^+ \nu_e$.

Charm semileptonic decays are important to test theoretical models as V_{cs} is known from unitarity. Within Heavy Quark Effective Theory (HQET), Λ-type baryons are more straightforward to treat than mesons as they consist of a heavy quark and a spin zero light diquark. The semileptonic decays of charmed baryons provide an interesting opportunity to test many of the predictions of HQET. We report here two new results from CLEO II, Ξ_c semileptonic decays[1] and form factor ratio measurement in $\Lambda_c^+ \rightarrow \Lambda e^+ \nu_e$.[3]

1. Ξ_c semileptonic decays and a measurement of the Ξ_c^+/Ξ_c^0 lifetime ratio

The naive spectator model of heavy quark decay predicts that the lifetimes of the charmed hadrons are equal. Experimentally this is not the case, as $\frac{\tau(D^+)}{\tau(D^0)} = 2.54 \pm 0.07$.[4] It is believed that the source of the lifetime difference is the hadronic width and destructive interference between the internal and external Cabibbo allowed spectator graphs is sufficient to decrease the hadronic width of the D^+ relative to the D^0 by about the amount required to explain the lifetime ratio.[5]

In the charmed baryon sector, destructive interference between the external and internal spectator graphs can occur when a spectator is a u quark. In addition, when a spectator quark is an s, constructive interference between two internal spectator graphs can occur. Finally, the W-exchange diagram is not helicity suppressed for baryons. The relative importance of these three effects leads to different predictions for the lifetime hierarchy of charm baryons[6],[7]

1.1. Search $\Xi_c^+ \rightarrow \Xi^0 e^+ \nu_e$ and $\Xi_c^0 \rightarrow \Xi^- e^+ \nu_e$

The data sample used in this study contains about 3 million $e^+ e^- \rightarrow c \bar{c}$ events. We search for the decays $\Xi_c \rightarrow \Xi e^+ \nu_e$ by detecting a Ξe^+ (right sign) pair with $m_\Xi <$

$m_{\Xi e^+} < m_{\Xi_c}$. The Λ is reconstructed through its decay to $p\pi$. We reconstruct Ξ^0 and Ξ^- through the decay modes $\Lambda\pi^0$ and $\Lambda\pi^-$ respectively. The minimum allowed momentum for positrons is 0.5 GeV/c. The combined momentum of Ξ and positrons is required to be greater than 1.4 GeV/c to reduce background. To determine the number of events passing our cuts we fit the $\Lambda\pi$ invariant mass distributions, shown in Fig.1, and the fit results are given in Table 1.

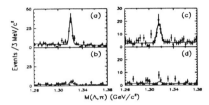

Fig. 1. The $\Lambda\pi$ invariant mass for right sign and wrong sign Ξe combinations;(a)$\Lambda\pi^-$ RS, (b)$\Lambda\pi^-$ WS;(c)$\Lambda\pi^0$ RS, (d)$\Lambda\pi^0$ WS.

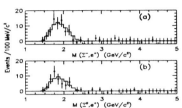

Fig. 2. The Ξe^+ invariant mass for: (a) $\Xi^- e^+$, (b)$\Xi^0 e^+$. The points are data and the solid lines are MC predictions.

Background from the decays $\Xi_c \to \Xi(1530)e^+\nu_e$ with $\Xi(1530) \to \Xi\pi$ is estimated to be 0.4 ± 3.3 events. Backgrounds produced from fake electrons, generic continuum and B decay are expected to have approximately equal number of Ξe^+ as Ξe^- (wrong sign) pairs. We therefore use the number of Ξe^- pairs in the data to estimate these backgrounds. The yield given in Table 1 is then obtained by subtracting the number of wrong sign events from the right sign events. Comparisons between data and Monte Carlo (MC) predictions for Ξe^+ invariant mass distributions are shown in Fig.2. Our results are: $\mathcal{B}(\Xi_c^+ \to \Xi^0 e^+\nu_e) \cdot \sigma(e^+e^- \to \Xi_c^+ X) = (1.55 \pm 0.33 \pm 0.25)$pb and $\mathcal{B}(\Xi_c^0 \to \Xi^- e^+\nu_e) \cdot \sigma(e^+e^- \to \Xi_c^0 X) = (0.63 \pm 0.12 \pm 0.10)$pb. This is the first observation of $\Xi_c^+ \to \Xi^0 e^+\nu_e$.

Table 1. Signals and backgrounds

mode	$\Xi_c^+ \to \Xi^0 e^+\nu_e$	$\Xi_c^0 \to \Xi^- e^+\nu_e$
$N_{\Xi e^+}$ (right sign)	47 ± 8	62 ± 9
$N_{\Xi e^-}$ (wrong sign)	6 ± 3	8 ± 4
corrected yield	41 ± 9	54 ± 10
efficiency (%)	1.17 ± 0.02	3.80 ± 0.05
$\sigma \cdot \mathcal{B}$ (pb)	$1.55 \pm 0.33 \pm 0.25$	$0.63 \pm 0.12 \pm 0.10$

1.2. First model independent upper bounds on the absolute branching ratios of Ξ_c

Under the assumption that the semileptonic widths of all charmed particles are equal, we use the weighted average of the inclusive semileptonic widths of the D^0 and D^+, $\langle\Gamma_{SL}\rangle$[4] and the weighted average of the Ξ_c lifetime measurements of NA32[8] and E687[9] to setimate $\mathcal{B}_{\ell X}^{+(0)} = \mathcal{B}(\Xi_c^{+(0)} \to \ell^+ X) = \langle\Gamma_{SL}\rangle \cdot \tau(\Xi_c^{+(0)})$. We find $\mathcal{B}_{\ell X}^+ = (4.8^{+1.3}_{-1.0})\%$ and $\mathcal{B}_{\ell X}^0 = (1.6^{+0.4}_{-0.3})\%$. There exists no theoretical relationship between $\mathcal{B}(\Xi_c \to \Xi\ell^+\nu_l)$ and $\mathcal{B}_{\ell X}$, however $\mathcal{B}(\Xi_c \to \Xi\ell^+\nu_l) \leq \mathcal{B}_{\ell X}$.

We measure $R_0 = \frac{\mathcal{B}(\Xi_c^0 \to \Xi^- \pi^+)}{\mathcal{B}(\Xi_c^0 \to \Xi^- e^+ \nu_e)} = 0.32 \pm 0.10^{+0.05}_{-0.03}$ and $R_+ = \frac{\mathcal{B}(\Xi_c^+ \to \Xi^- \pi^+ \pi^+)}{\mathcal{B}(\Xi_c^+ \to \Xi^0 e^+ \nu_e)} = 0.44 \pm 0.11^{+0.11}_{-0.06}$. Therefore $\mathcal{B}(\Xi_c^0 \to \Xi^- \pi^+) \leq (5.2 \pm 1.6^{+1.4}_{-1.2}) \times 10^{-3}$ and $\mathcal{B}(\Xi_c^+ \to \Xi^- \pi^+ \pi^+) \leq (2.1 \pm 0.5 \pm 0.7) \times 10^{-2}$ where the first error is from the determination of $R_{0(+)}$ and the second error is from our estimate of the $\mathcal{B}_{\ell X}$. This is the first model independent absolute normalization of the hadronic decay mode of the charmed cascades.

1.3. Lifetime ratio of Ξ_c^+/Ξ_c^0

Since we measure both Ξ_c^+ and Ξ_c^0 semileptonic decays using similar cuts in one experiment, the lifetime ratio $R_\tau = \tau_{\Xi_c^+}/\tau_{\Xi_c^0}$ can be extracted from the ratio of our measurements under the following assumptions. We assume that semileptonic decay width (Γ_{sl}) of Ξ_c^+ and Ξ_c^0 are equal and $\Gamma(\Xi_c^+ \to \Xi^0 e^+ \nu_e) = \Gamma(\Xi_c^0 \to \Xi^- e^+ \nu_e)$. Both of these assumptions follow from isospin invariance. We also assume that the Ξ_c^+ and Ξ_c^0 are equally produced from e^+e^- annihilations at 10 GeV. This is reasonable because the Ξ_c^*'s are expected to decay to the ground state via pion emission, and to be sufficiently heavy that no channel related by isospin is excluded by phase space, and the Ξ_c' decays electromagnetically to a Ξ_c.[10] The lifetime ratio is then related to the ratio of branching ratios:

$$R_\tau = \frac{\mathcal{B}(\Xi_c^+ \to \Xi^0 e^+ \nu_e)}{\mathcal{B}(\Xi_c^0 \to \Xi^- e^+ \nu_e)} = \frac{\Gamma_{sl}(\Xi_c^+) \cdot \tau(\Xi_c^+)}{\Gamma_{sl}(\Xi_c^0) \cdot \tau(\Xi_c^0)} = \frac{\tau(\Xi_c^+)}{\tau(\Xi_c^0)} = 2.46 \pm 0.70^{+0.33}_{-0.23}$$

Agreement among the existing lifetime ratio measurements, 4.06 ± 1.26 (E687),[9] 2.44 ± 1.68 (NA32)[8] and this result, is good. As our result is not a direct measurement of the lifetimes, it has entirely different systematic errors than E687 and NA32 and therefore serves as valuable independent confirmation of the fixed target results.

2. Form factor ratio measurement in $\Lambda_c^+ \to \Lambda e^+ \nu_e$

In the limit of negligible lepton mass, the semileptonic decay $\Lambda_c^+ \to \Lambda e^+ \nu_e$ is usually parametrized in terms of four form factors: two axial form factors F_1^A and F_2^A, and two vector form factors F_1^V and F_2^V. Within the framework of HQET the heavy flavor and spin symmetries imply relations among the form factors which reduce their number to one when the decay involves only heavy quarks.[11],[12] Treating the s quark as a light quark, two independent form factors f_1 and f_2 to $O(\bar{\Lambda}/m_s)$[11],[13] are required to describe the hadronic current. In general f_2 is expected to be less than f_1. The relationship between these form factors and the standard form factors is: $F_1^V(q^2) = -F_1^A(q^2) = f_1(q^2) + \frac{M_\Lambda}{M_{\Lambda_c}} f_2(q^2)$ and $F_2^V(q^2) = -F_2^A(q^2) = \frac{1}{M_{\Lambda_c}} f_2(q^2)$. Here q^2 is the invariant mass squared of the virtual W. In order to extract the form factor ratio $R = f_2/f_1$ using all the data we follow the model of Körner and Krämer (KK)[14] and assume the dipole form of q^2 dependence of f_1 and f_2: $f(q^2) = \frac{f(q^2_{max})}{(1 - q^2/m^2_{ff})^2}(1 - q^2_{max}/m^2_{ff})^2$.

where the pole mass is chosen to be $m_{ff} = 2.11 \text{GeV}/c^2$.

The data sample used in this study contains about 4 million $e^+e^- \to c\bar{c}$ events. We search for the decay $\Lambda_c^+ \to \Lambda e^+ \nu_e$ by detecting a Λe^+ pair with $m_\Lambda < m_{\Lambda e^+} < m_{\Lambda_c}$. The minimum allowed momentum for positrons is 0.7 GeV/c. The Λ is reconstructed through its decay to $p\pi$. We require the momentum of the $p\pi$ pair to be greater than 0.8 GeV/c. The sum of Λ and e^+ momentum is required to be greater than 1.4 GeV/c. The number of events passing the cuts described above is 1101, of which 135^{+22}_{-14} are consistent with fake Λ background, 133 ± 40 are consistent with fake electron background and 116 ± 23 are $\Xi_c \to \Xi e^+ \nu_e$ feedthrough. In Fig. 3, we show the $m_{\Lambda e^+}$ distribution after fake Λ subtraction and comparisons between the data and MC predictions.

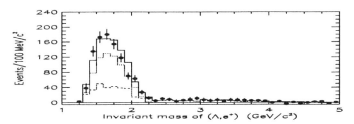

Fig. 3. Invariant Λe^+ mass for right sign combinations. The points are data after subtraction of fake Λs. The dashed line shows the backgrounds. The dotted line shows the MC prediction for the signal. The solid line shows the sum of the MC prediction and the backgrounds.

We estimate the direction of the Λ_c from the event thrust axis. The magnitude of the Λ_c momentum is then obtained by solving the equation $\vec{P}_{\Lambda_c}^2 = (\vec{P}_\Lambda + \vec{P}_e + \vec{P}_{\nu_e})^2$. After the Λ_c momentum is estimated, the four kinematic variables are obtained by working in Λ_c center-of-mass frame. The resolutions (RMS) on $t = q^2/q_{max}^2$, $\cos\Theta_\Lambda$, $\cos\Theta_W$ and χ determined by MC are 0.25, 0.25, 0.2 and 45^0 respectively. Due to poor resolution on χ, this angle is not used here to extract the form factor ratio.

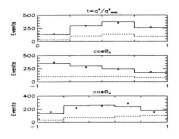

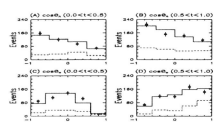

Fig. 4. Projections of the data (points) and the fit (solid lines) for t, $\cos\Theta_\Lambda$ and $\cos\Theta_W$. The dashed lines show the background.

Fig. 5. Projections of the data and the fit onto $\cos\Theta_\Lambda$ and $\cos\Theta_W$ for two t regions. (A) and (C) for $0.0 < t < 0.5$, and (B) and (D) for $0.5 < t < 1.0$. The dashed lines show the background.

Using t, $\cos\Theta_\Lambda$ and $\cos\Theta_W$, we perform a 3-dimensional unbinned maximum likelihood fit in a manner similar to that developed by E691.[15] This technique enables a multi-dimensional likelihood fit to be performed to variables modified by experimental acceptance and resolution, and is necessary for this analysis due to the substantial smearing of the kinematic variables. Fig. 4 and Fig. 5 show the projections of the t, $\cos\Theta_\Lambda$ and $\cos\Theta_W$ distributions for the data and for the fit. The confidence level of the fit is determined to be 23% by the K nearest neighbors method[16] and by comparing the likelihood of the fit with the distribution of likelihoods obtained by fitting many MC samples of the same number of events as the data sample. The systematic error on R arises from MC statistics, volume size used in the likelihood fit, uncertainties in the background normalization and the method of background incorporation into the fit. Our preliminary result is:

$$R = -0.33 \pm 0.15 \pm 0.16$$

3. Summary

CLEO II is carrying out a systematic study of the semileptonic decays of charmed baryons. New results on Ξ_c semileptonic decays and form factor ratio in $\Lambda_c^+ \rightarrow \Lambda e^+ \nu_e$ are reported here. We have measured $\mathcal{B}(\Xi_c^+ \rightarrow \Xi^0 e^+ \nu_e) \cdot \sigma(e^+ e^- \rightarrow \Xi_c^+ X) = (1.55 \pm 0.33 \pm 0.25)$ pb and $\mathcal{B}(\Xi_c^0 \rightarrow \Xi^- e^+ \nu_e) \cdot \sigma(e^+ e^- \rightarrow \Xi_c^0 X) = (0.63 \pm 0.12 \pm 0.10)$ pb. We set upper bounds on absolute branching ratios of Ξ_c decays, $\mathcal{B}(\Xi_c^0 \rightarrow \Xi^- \pi^+) \leq (5.2 \pm 1.6^{+1.4}_{-1.2}) \times 10^{-3}$ and $\mathcal{B}(\Xi_c^+ \rightarrow \Xi^- \pi^+ \pi^+) \leq (2.1 \pm 0.5 \pm 0.7) \times 10^{-2}$. Assuming the Ξ_c^+ and Ξ_c^0 are equally produced in $e^+ e^-$ annihilation events at 10 GeV, the lifetime ratio of Ξ_c^+ / Ξ_c^0 is found to be $2.46 \pm 0.70^{+0.33}_{-0.23}$. From a 3-dimensional unbinned maximum likelihood fit we obtain a preliminary result on form factor ratio in $\Lambda_c^+ \rightarrow \Lambda e^+ \nu_e$, $R = f_2/f_1 = -0.33 \pm 0.15 \pm 0.16$.

References

1. J. Alexander et al. (CLEO), CLNS 94/1288 (submitted to Phys. Rev. Lett.).
2. J. Dominick et al. (CLEO), CLEO CONF 94-19.
3. Throughout this paper charge conjugate modes are implied.
4. Particle Data Group, Phys. Rev. Lett. **45** 1 (1992).
5. M. Bauer, B. Stech and M. Wirbel, Zeit. Phys. C **34**, 103 (1987).
6. B. Guberina, R. Rückl, and J. Trampetić, Zeit. Phys. C **33** 297 (1986).
7. M.B. Voloshin and M.A. Shifman, Sov. Phys. JETP 64, 698 (1986).
8. S. Barlag, (NA32), Phys. Lett. **B233** 522 (1989), and S. Barlag, (NA32), Phys. Lett **B236** 495 (1990).
9. P.L. Frabetti (E687), Phys. Rev. Lett. **70** 1381 (1993) and P.L. Frabetti (E697), Phys. Rev. Lett. **70** 2058 (1993).
10. J.M. Richard and P. Taxil, Phys. Lett **B128** 453 (1983). S. Fleck and J.M. Richard, Particle. World Vol. 1 No.3 67 (1990). J.G. Körner and H. Siebert, Ann. Rev. Nucl. Part. Sci.**41** 511 (1991). R.E. Cutkosky and P. Geiger, Phys. Rev. **D48** 1315 (1993)
11. T. Mannel, W. Roberts and Z. Ryzak, Nucl. Phys. **B355** 38 (1991).
12. H. Georgi, B. Grinstein and M.B. Wise, Phys. Lett. **B252** 456 (1990); N. Isgur and M.B. Wise, Nucl. Phys.**B348** 276 (1991); H. Georgi, Nucl. Phys. **B348** 293 (1991).
13. T. Mannel, W. Roberts and Z. Ryzak, Phys. Lett. **B255** 593 (1991); F. Hussain, J.G. Körner, M. Krämer and G. Thompson; Z. Phys. **C51** 321 (1991).
14. J.G. Körner and M. Krämer, Phys. Lett. **B275** 495 (1992).
15. D.M.Schmidt, R.M.Morrison and M.S.Witherell, Nucl. Instr. and Meth. **A328** 547 (1993).
16. J.H. Friedman, CERN 74-23 271(1974).

SEMILEPTONIC CHARM MESON DECAY RESULTS FROM CLEO-II

HIRO TAJIMA

Physics Department, University of California
Santa Barbara, California 93106, USA
(Representing the CLEO collaboration)

ABSTRACT

Using a sample of over 2 million $e^+e^- \to c\bar{c}$ continuum events collected by the CLEO-II detector at the Cornell Electron Storage Ring, we have measured branching fractions and form factors for semileptonic charm meson decays. We report a new measurement of $\mathcal{B}(D^0 \to Xe^+\nu) = (6.97 \pm 0.18 \pm 0.30)\%$. We present a first measurement of ratios $\mathcal{B}(D_s^+ \to \eta e^+\nu)/\mathcal{B}(D_s^+ \to \phi e^+\nu) = 1.74 \pm 0.34 \pm 0.24$ and $\mathcal{B}(D_s^+ \to \eta'e^+\nu)/\mathcal{B}(D_s^+ \to \phi e^+\nu) = 0.71 \pm 0.19 \pm 0.09$. Using these results, we estimate the absolute branching fraction for $D_s^+ \to \phi e^+\nu$ and $D_s^+ \to \phi\pi^+$ to be $(2.06 \pm 0.33)\%$ and $(3.82 \pm 0.74)\%$, respectively. Also presented is a new measurement of form factor ratios for $D_s^+ \to \phi e^+\nu$ decay, $R_2 = 1.4 \pm 0.5 \pm 0.3$ and $R_V = 0.9 \pm 0.6 \pm 0.3$.

1. Introduction

In B semileptonic decays, we measure $|V_{cb}|$ and $|V_{ub}|$ using the theoretical predictions on form factors. It is important to test and improve those theoretical predictions. In charm semileptonic decays, the absolute scale of form factors can be measured, since Cabibbo-Kobayashi-Masukawa (CKM) matrix elements, $|V_{cs}|$ and $|V_{cd}|$ are well known. In particular, theoretical predictions for $b \to u\ell^-\bar{\nu}$ must be improved in order to determine $|V_{ub}|$ and studies of charm semileptonic decay will provide theoretical information.

One of the outstanding problems in charm semileptonic decay is disagreement on vector-to-pseudoscalar (V/P) ratio, $\mathcal{B}(D \to \overline{K^*}e^+\nu)/\mathcal{B}(D \to \overline{K}e^+\nu)$.[1] The world average is 0.56 ± 0.06, while theory predicts $0.8 - 1.2$. While D^0 and D^+ semileptonic form factors have been studied in detail, in the D_s^+ sector only the $D_s^+ \to \phi\ell^+\nu$ has been studied. Fermilab experiment E653 has also seen evidence for $D_s^+ \to (\eta+\eta')\mu^+\nu$.[2]

In this report, we present a new measurement of $\mathcal{B}(D^0 \to Xe^+\nu)$ and a first measurement of the ratios $\mathcal{B}(D_s^+ \to \eta e^+\nu)/\mathcal{B}(D_s^+ \to \phi e^+\nu)$ and $\mathcal{B}(D_s^+ \to \eta'e^+\nu)/\mathcal{B}(D_s^+ \to \phi e^+\nu)$. Using these results, we estimate the absolute branching fraction for $D_s^+ \to \phi e^+\nu$. We also report a new measurement of the form factor ratios in $D_s^+ \to \phi e^+\nu$.

These analyses are based on a data sample of over 2 million $e^+e^- \to c\bar{c}$ continuum events at $\sqrt{s} \sim 10.6$ GeV collected by the CLEO-II detector[3] at the Cornell Electron Storage Ring.

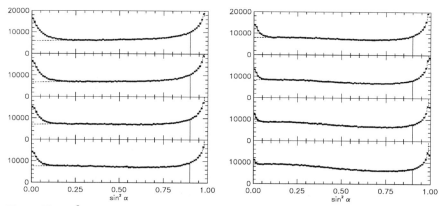

Fig. 1. The $\sin^2 \alpha$ distribution for the pions accompanied by the right sign electron for the eight 25 MeV/c momentum slices. The solid histogram is the result of a fit which includes signal and background functions. The dashed histogram shows the background contribution.

2.   Measurement of $\mathcal{B}(D^0 \to Xe^+\nu)$

In order to measure an absolute branching fraction for D^0 decay, we must first determine the number of D^0 mesons in the data sample. We use the decay $D^{*+} \to D^0\pi^+$ to achieve this. We exploit the fact that transverse momentum (p_T) of π^+ relative to D^{*+} direction is very small, less than 40 MeV/c, due to the small Q-value of the decay. This low transverse momentum provides the $D^{*+} \to D^0\pi^+$ signature. The D^0 is not reconstructed, so the D^{*+} direction is not measured. It is well approximated, however, by the thrust axis of the event. The resolution on the D^{*+} direction using this technique has been measured with data to be 0.08 rad FWHM. We use the $\sin^2 \alpha = (p_T/p_\pi)^2$ distribution in eight 25 MeV/c slices in p_π to determine the total number of $D^{*+} \to D^0\pi^+$, $N(D^{*+} \to D^0\pi^+)$, where α is the angle between the pion and the thrust axis. This is the technique used to obtain the absolute branching fraction for $D^0 \to K^-\pi^+$.[4] We then require an electron with the same charge as the tagged pion in the same jet (right-sign combination) to obtain the total number of $D^{*+} \to D^0\pi^+$, $D^0 \to Xe^+\nu$, $N(D^{*+} \to D^0\pi^+, D^0 \to Xe^+\nu)$. The wrong-sign combination is used to model the background shape in the right-sign combination. Figure 1 shows the $\sin^2 \alpha$ distributions for the pions accompanied by a right-sign electron in eight momentum slices from 225 to 425 MeV/c. A clear peak from $D^{*+} \to D^0\pi^+$ decays is observed near $\sin^2 \alpha = 0$. The histogram is the result of a fit to a signal shape obtained from a Monte Carlo simulation and a background shape obtained from the wrong sign combination.

Dividing $N(D^{*+} \to D^0\pi^+, D^0 \to Xe^+\nu)$ by $N(D^{*+} \to D^0\pi^+)$, we obtain $\mathcal{B}(D^0 \to Xe^+\nu) = (6.97 \pm 0.18 \pm 0.30)\%$, where the first error is statistical and the second error is systematic. The dominant source of the systematic error is the uncertainty in the electron identification efficiency (3%).

CLEO-II has measured three exclusive modes, $\mathcal{B}(D \to \overline{K}e^+\nu) = (3.77 \pm 0.28)\%$,

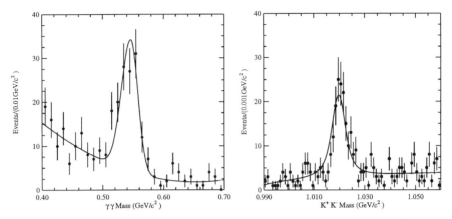

Fig. 2. The $\gamma\gamma$ and K^+K^- invariant-mass distributions for $D_s^+ \to \eta\ell^+\nu$ and $D_s^+ \to \phi\ell^+\nu$ candidates. The curve is the fit to the invariant-mass distribution.

$\mathcal{B}(D \to \overline{K}^* e^+\nu) = (2.36 \pm 0.29)\%$,[7] and $\mathcal{B}(D \to \pi e^+\nu) = (0.64 \pm 0.22)\%$.[8] The sum of those exclusive modes is $(6.77 \pm 0.45)\%$, which is consistent with expected small contributions from unobserved modes.

3. Study of $D_s^+ \to \phi\ell^+\nu$ and $D_s^+ \to \eta\ell^+\nu$ and $D_s^+ \to \eta'\ell^+\nu$

3.1. $D_s^+ \to \eta\ell^+\nu$ analysis

Due to the undetected neutrino, we cannot fully reconstruct $D_s^+ \to X_{s\bar{s}}\ell^+\nu$ ($X_{s\bar{s}} = \phi, \eta, \eta'$) decays. However, there are few processes which produce both a ϕ, η or η' meson and a lepton contained in the same jet. Consequently, this correlation can be used to extract a clean $D_s^+ \to X_{s\bar{s}}\ell^+\nu$ signal. The backgrounds due to misidentified leptons and from random $X_{s\bar{s}}\ell^+$ combinations can be reliably estimated, and the possible contamination from other decay modes can be shown to be small. We extract the yields of $D_s^+ \to X_{s\bar{s}}\ell^+\nu$ using a D_s^{*+} tag in order to reduce the random $X_{s\bar{s}}\ell^+$ background and normalize $D_s^+ \to \eta\ell^+\nu$ yield to $D_s^+ \to \phi\ell^+\nu$ yield. A detailed description of this analysis can be found elsewhere.[5]

We identify ϕ and η candidates by using the decay modes $\phi \to K^+K^-$ and $\eta \to \gamma\gamma$. After selecting $X_{s\bar{s}}\ell^+$ candidates which pass certain criteria, we then combine them with a photon which lies in the same hemisphere. To select $X_{s\bar{s}}\ell^+\gamma$ candidates which come from the $D_s^{*+} \to D_s^+\gamma$, $D_s^+ \to X_{s\bar{s}}\ell^+\nu$ decay chain, we require that $\Delta M \equiv M_{X_{s\bar{s}}\ell\gamma} - M_{X_{s\bar{s}}\ell}$ be between 0.1 and 0.2 GeV/c^2.

Figure 2 shows the $\gamma\gamma$ and K^+K^- invariant-mass distributions for all $\eta\ell^+\gamma$ and $\phi\ell^+\gamma$ combinations which pass our selection criteria. The fits yield 103.0 ± 11.9 $D_s^+ \to \eta\ell^+\nu$ candidates and 167.9 ± 15.8 $D_s^+ \to \phi\ell^+\nu$ candidates.

There are three main sources of background in the sample: $X_{s\bar{s}}$'s accompanied by fake leptons, random $X_{s\bar{s}}\ell^+$ combinations, and random photon combinations with true $X_{s\bar{s}}\ell^+$. The background due to fake leptons is estimated by using the data. The

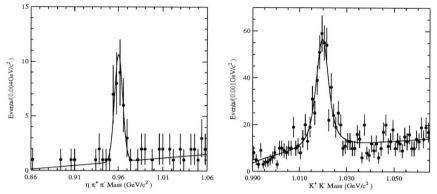

Fig. 3. The $\eta\pi^+\pi^-$ and K^+K^- invariant-mass distributions for $D_s^+ \to \eta'\ell^+\nu$ and $D_s^+ \to \phi\ell^+\nu$ candidates. The curve is the fit to the invariant-mass distribution.

backgrounds from random $X_{s\bar{s}}\ell^+$ combinations and random photon combinations are estimated using the Monte Carlo simulation. We scale the Monte Carlo prediction to account for the $X_{s\bar{s}}$ and γ production rate observed in the data.

We estimate 1.2 ± 0.4 background events for $D_s^+ \to \eta\ell^+\nu$ from $D_s^+ \to \eta'\ell^+\nu$ feed down. The contamination from the decays $D^+ \to \eta\ell^+\nu$ and $D^+ \to \eta'\ell^+\nu$ is negligible since these decay modes are Cabibbo suppressed and $\mathcal{B}(D^{*+} \to D^+\gamma)$ is small.

After subtracting all backgrounds, we find 60.9 ± 9.7 $D_s^+ \to \eta\ell^+\nu$ events and 105.3 ± 11.8 $D_s^+ \to \phi\ell^+\nu$ events. After correcting for the detection efficiency and for the $\eta \to \gamma\gamma$ and $\phi \to K^+K^-$ branching fractions,[6] the efficiency-corrected yields for $D_s^+ \to \eta\ell^+\nu$ and $D_s^+ \to \phi\ell^+\nu$ are 8320 ± 1320 and 4770 ± 530 events, respectively.

The ratio of branching fractions is

$$R_\eta = \frac{\mathcal{B}(D_s^+ \to \eta e^+\nu)}{\mathcal{B}(D_s^+ \to \phi e^+\nu)} = \frac{8320 \pm 1320}{4770 \pm 530} = 1.74 \pm 0.34 \pm 0.24, \qquad (1)$$

where the first error is statistical, and the second is an estimate of systematic effects. The dominant systematic error is the uncertainty in the level of continuum charm background (8.2%). Note that systematics associated with the rate of fake leptons and the scaling of the random photo background mostly cancel since they are correlated in both analyses.

3.2. $D_s^+ \to \eta'\ell^+\nu$ analysis

We reconstruct η' with the decay chain $\eta' \to \eta\pi^+\pi^-$, $\eta \to \gamma\gamma$. We do not use the D_s^{*+} tag for the $D_s^+ \to \eta'\ell^+\nu$ analysis since the sensitivity for the η' decay chain is $1/5$ of that for the $\eta \to \gamma\gamma$ decay. To compensate for the absence of the D_s^{*+} constraint, we apply tighter cuts than those used for $D_s^+ \to \eta\ell^+\nu$ analysis.

Figure 3 shows the $\eta\pi^+\pi^-$ and K^+K^- invariant-mass distributions for all $\eta'\ell^+$ and $\phi\ell^+$ combinations which pass the our selection criteria. The fits yield $29.1^{+6.2}_{-5.6}$ $D_s^+ \to \eta'\ell^+\nu$ candidates and 419 ± 24 $D_s^+ \to \phi\ell^+\nu$ candidates.

The backgrounds are estimated in the same manner as the $D_s^+ \to \eta \ell^+ \nu$ analysis. After subtracting all backgrounds, the efficiency corrected yields for $D_s^+ \to \eta' \ell^+ \nu$ and $D_s^+ \to \phi \ell^+ \nu$ decays are 6100 ± 1600 and 8580 ± 600 events, respectively.

The ratio of branching fractions is

$$R_{\eta'} = \frac{\mathcal{B}(D_s^+ \to \eta' e^+ \nu)}{\mathcal{B}(D_s^+ \to \phi e^+ \nu)} = \frac{6100 \pm 1600}{8580 \pm 600} = 0.71 \pm 0.19 \pm 0.09, \quad (2)$$

3.3. Implication of $D_s^+ \to \eta \ell^+ \nu$ and $D_s^+ \to \eta' \ell^+ \nu$ results

The V/P ratio, $\mathcal{B}(D_s^+ \to \phi e^+ \nu)/\mathcal{B}(D_s^+ \to (\eta + \eta') e^+ \nu)$ is predicted to be 1.19 by a modified ISGW model.[9] This value is considerably higher than our measurement of $1/(R_\eta + R_{\eta'}) = 0.41 \pm 0.06 \pm 0.04$. Our result is consistent with E653 result of $0.26^{+1.7}_{-0.08}$.[2] Our result is also comparable with the V/P ratio for D^0 and D^+ decay, $\mathcal{B}(D \to \overline{K}^* e^+ \nu)/\mathcal{B}(D \to \overline{K} e^+ \nu) = 0.56 \pm 0.06$.

The factorization hypothesis implies $\mathcal{B}(D_s^+ \to \eta' \rho^+)/\mathcal{B}(D_s^+ \to \eta \rho^+) = \mathcal{B}(D_s^+ \to \eta' e^+ \nu)/\mathcal{B}(D_s^+ \to \eta e^+ \nu)$. CLEO-II has measured $\mathcal{B}(D_s^+ \to \eta' \rho^+)/\mathcal{B}(D_s^+ \to \eta \rho^+) = 1.20 \pm 0.33$, which is inconsistent with our measurement of $\mathcal{B}(D_s^+ \to \eta' e^+ \nu)/\mathcal{B}(D_s^+ \to \eta e^+ \nu) = 0.41 \pm 0.13 \pm 0.05$. This disagreement suggests that the factorization hypothesis doesn't hold for D_s^+ decay.

Having measured dominant modes of D_s^+ semileptonic decay, we calculate the absolute branching fraction for $D_s^+ \to \phi e^+ \nu$ as,

$$\Gamma(D_s^+ \to \phi e^+ \nu) = \frac{\Gamma(D_s^+ \to \phi e^+ \nu)}{\Gamma(D_s^+ \to (\phi + \eta + \eta') e^+ \nu)} \frac{\Gamma(D_s^+ \to (\phi + \eta + \eta') e^+ \nu)}{\Gamma(D_s^+ \to X e^+ \nu)} \Gamma(D_s^+ \to X e^+ \nu)$$

$$(3)$$

We assume that $\Gamma(D_s^+ \to (\phi + \eta + \eta') e^+ \nu)/\Gamma(D_s^+ \to X e^+ \nu) = 0.89 \pm 0.04$ and $\Gamma(D_s^+ \to X e^+ \nu) = \Gamma(D \to X e^+ \nu)$. Using our measurement of $\Gamma(D_s^+ \to \phi e^+ \nu)/\Gamma(D_s^+ \to (\phi + \eta + \eta') e^+ \nu) = 1/(1 + R_\eta + R_{\eta'}) = 0.292 \pm 0.040$ and $\Gamma(D^0 \to X e^+ \nu) = (16.9 \pm 0.9) \times 10^{10} \ s^{-1}$, we obtain $\Gamma(D_s^+ \to \phi e^+ \nu) = (4.35 \pm 0.66) \times 10^{10} \ s^{-1}$. Comparing this result with $\Gamma(D \to \overline{K}^* e^+ \nu)$, we find $\Gamma(D_s^+ \to \phi e^+ \nu)/\Gamma(D \to \overline{K}^* e^+ \nu) = 0.91 \pm 0.16$, which is consistent with prediction of 1.02 by ISGW model.[9] We also find $\mathcal{B}(D_s^+ \to \phi e^+ \nu) = (2.06 \pm 0.33)\%$, which is combined with our measurement of $\mathcal{B}(D_s^+ \to \phi e^+ \nu)/\mathcal{B}(D_s^+ \to \phi \pi^+) = 0.54 \pm 0.05 \pm 0.04$[10] to obtain $\mathcal{B}(D_s^+ \to \phi \pi^+) = (3.82 \pm 0.74)\%$. The branching fraction for $D_s^+ \to \phi \pi^+$ sets the scale for all D_s^+ hadronic branching fractions.

4. Measurement of $D_s^+ \to \phi e^+ \nu$ Form Factor Ratios

We measure $R_2 = A_2(0)/A_1(0)$ and $R_V = V(0)/A_1(0)$ by performing a two-parameter likelihood fit to the decay rate using the three kinematic variables, $|\cos \theta_V|$, $\cos \theta_e$ and q^2/q_{max}^2. A detailed description of this analysis can be found elsewhere.[11]

In order to perform the fit, the functional form for the form factors is assumed to be a single pole form, $F(q^2) = F(0)/(1 - q^2/M_p^2)$. Since the q^2 range in the decay is from 0 to 0.9 GeV/c^2, the results are rather insensitive to the particular parametrization chosen.

The fitting technique follows that developed by the E691 experiment for the determination of the $D^+ \to \overline{K}^{*0} e^+ \nu$ form factor ratios.[12] The essence of the method is to determine the probability density function by using the population of appropriately weighted Monte Carlo events in the three-dimensional kinematic space.

Background is incorporated into the fit by constructing the probability $P = N_s p_s + N_b p_b$ where N_s is the number of signal events in the data sample, N_b is the number of background events in this sample, p_s and p_b are the probability densities for the signal and background distributions, respectively. Since the level of background in this analysis is greater than that for any previous analysis, the inclusion of background is studied using two other approaches. The difference in final result due to different approaches is taken into account as a systematic error.

The 474 events of the final data set are fit to the expression for the decay rate in terms of the three kinematic variables. We find: $R_2 = 1.4 \pm 0.5 \pm 0.3$ and $R_V = 0.9 \pm 0.6 \pm 0.3$. Our results are consistent with the previous measurements of the $D_s^+ \to \phi e^+ \nu$ form factor ratios.[13,14] Assuming $SU(3)$ symmetry, R_2 and R_V for the decays $D_s^+ \to \phi e^+ \nu_e$ and $D^+ \to \overline{K^{*0}} e^+ \nu$, should be approximately equal. Further, equality of the form factors for the two decays would imply equal partial widths. The latter point has been the basic assumption for determining the absolute branching ratio $D_s^+ \to \phi \pi^+$ from the ratio of branching ratios $\mathcal{B}(D_s^+ \to \phi e^+ \nu)/\mathcal{B}(D_s^+ \to \phi \pi^+)$. The new world averages including our results, $R_2 = 1.6 \pm 0.4$ and $R_V = 1.4 \pm 0.5$ are similar to those for $D \to \overline{K^*} e^+ \nu$, $R_2 = 0.73 \pm 0.15$ and $R_V = 1.89 \pm 0.25$.

5.   Conclusions

In conclusion, we report a new measurement of $\mathcal{B}(D^0 \to X e^+ \nu) = (6.97 \pm 0.18 \pm 0.30)\%$. In D_s^+ sector we still see the disagreement on the V/P ratio between the experimental result and the theoretical prediction. We find similar characteristics for the V/P ratio and form factors between D_s^+ and D semileptonic decays. We obtain the absolute branching fraction of $D_s^+ \to \phi \pi^+$ decay to be $(3.82 \pm 0.74)\%$.

References

1. For all states described, the charge conjugate state is also implied.
2. E653 Collaboration, K. Kodama et al., Phys. Lett. **B309** (1993) 483.
3. CLEO Collaboration, Y. Kubota et al., Nucl. Inst. and Meth. **A320** (1992) 66.
4. CLEO Collaboration, D. S. Akerib et al., Phys. Rev. Lett. **71** (1993) 3070.
5. CLEO Collaboration, M. Battle et al., preprint CLEO CONF 94-18.
6. Particle Data Group, K. Hikasa et al., Review of Particle Properties, Phys. Rev. **D45** (1992) 1.
7. CLEO Collaboration, J. Alexander et al., Phys. Lett. **B317** (1993) 647.
8. CLEO Collaboration, M. S. Alam et al., Phys. Rev. Lett. **71** (1993) 1311.
9. D. Scora, Ph.D. thesis, University of Toronto, (1993). We take the predictions for $\theta_P = -20°$.
10. CLEO Collaboration, F. Butler et al., Phys. Lett. **B324** (1994) 255.
11. CLEO Collaboration, P. Avery et al., submitted to Phys. Lett. B.
12. J.C. Anjos et al., Phys. Rev. Lett. **65** (1990) 2630; D.M. Schmidt, R.M. Morrison, and M. Witherell, Nucl. Inst. and Meth. **A328** (1993) 547; D.M. Schmidt, Ph. D. Thesis, Univ. of California at Santa Barbara (1992).
13. K. Kodama et al., Phys. Lett. **B274** (1992) 246.
14. P.L. Frabetti et al., Phys. Lett. **B307** (1993) 262.

SEMI-LEPTONIC FORM-FACTORS FROM LATTICE QCD*

TANMOY BHATTACHARYA and RAJAN GUPTA

T-8 Group, MS B285, Los Alamos National Laboratory, Los Alamos, New Mexico 87545, U. S. A.

ABSTRACT

We present results for semi-leptonic form-factors obtained on a statistical sample of 66 $32^3 \times 64$ lattices at $\beta = 6.0$ using quenched Wilson fermions. We find $f_+^{D \to Kl\nu}(q^2 = 0) = 0.73 \pm 0.06$, $A_2/A_1(D \to K^*l\nu) = 0.72 \pm 0.22$, $V/A_1(D_s \to \phi l\nu) = 1.91 \pm 0.04$, and $A_2/A_1(D_s \to \phi l\nu) = 0.68 \pm 0.09$, where the error estimate includes statistical errors and errors due to extrapolation to $q^2 = 0$ and to physical values of $(m_u + m_d)/2$ and m_s. The remaining sources of systematic errors are those due to $O(a)$ discretization errors and those due to quenching, which our results indicate may be small. We also comment on the validity of pole-dominance in these form-factors.

1. INTRODUCTION

Exclusive semi-leptonic decays of D and B mesons provide the cleanest measurements of the CKM quark mixing matrix. For example, the decay rate for $D \to Kl\nu$,

$$\frac{d\Gamma^{D \to Kl\nu}}{dq^2} = \frac{G_F^2}{24\pi^3} |V_{cs}|^2 \, p_K^3 \, f_+^2(q^2), \tag{1}$$

depends on kinematic factors, a single CKM matrix element V_{cs}, and the form-factor $f_+(q^2)$. To extract CKM matrix elements from such processes requires non-perturbative calculations of the form-factors as they encapsulate all strong interaction effects. In this talk we report on results obtained from numerical simulations of lattice QCD.

2. LATTICE PARAMETERS

The results presented here have been obtained using the following lattice parameters. The $32^3 \times 64$ gauge lattices were generated at $\beta = 6.0$ using the combination 5 over-relaxed (OR) sweeps followed by 1 Metropolis sweep. Quark propagators are calculated on lattices separated by 2000 OR sweeps using the simple Wilson action. Periodic boundary conditions are used in all 4 directions, both during lattice update and propagator calculation. Quark propagators have been calculated using one version of Wuppertal smeared sources at $\kappa = 0.135$ (C), 0.153 (S), 0.155 (U_1), 0.1558 (U_2), and 0.1563 (U_3). These quark masses correspond to pseudoscalar mesons of mass 2800, 980, 700, 550 and 440 MeV respectively using $1/a = 2.25(10)GeV$ set by m_ρ. On each of the 66 configurations we make two independent measurements of the form-factors, which we average before doing the statistical analysis using the jackknife method.

---

*Talk presented by R. Gupta at DPF94.

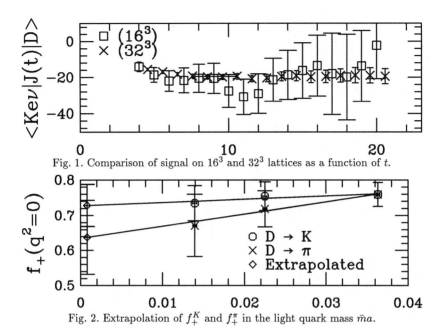

Fig. 1. Comparison of signal on 16^3 and 32^3 lattices as a function of t.

Fig. 2. Extrapolation of f_+^K and f_+^π in the light quark mass $\bar{m}a$.

Our procedure for extracting form-factors is very similar to that proposed by Lubicz *el al.*,[1] and a detailed paper is under preparation. The D meson is created at $\vec{p} = (0,0,0)$ and the momentum inserted by the current is carried by the final kaon. The five values of momenta analyzed are $\vec{p} = (0,0,0)$, $\vec{p} = (1,0,0)$, $\vec{p} = (1,1,0)$, $\vec{p} = (1,1,1)$, and $\vec{p} = (2,0,0)$ in units of $\pi/16a$. These correspond to roughly 0, 440, 625, 765, and 880 MeV respectively.

The use of large lattices to study form-factors leads to a dramatic improvement in reliability. In Fig. 1 we show a comparison of the signal in $\langle K \mid V_i \mid D \rangle$ with $\vec{p} = (\pi/8a, 0, 0)$ for our current data set (132 measurements) with a previous study using 35 16^3 lattices. The reduction in errors by a factor of ≈ 5 is consistent with the increase in statistics and lattice volume. In addition, the larger lattice allows measurements at three smaller values of non-zero momentum transfer, for which the signal is even better. These points bracket $q^2 = 0$ and allow a reliable extraction of $f(q^2 = 0)$, which we do in two ways. Our best fit uses a two parameter fit to the pole-dominance ansatz $f(q^2) = f(0)/(1 - q^2/\mathcal{M}^2)$. In the second method we fix the pole mass $\mathcal{M}$ to its lattice measured value. The relative merits of the two methods are discussed below.

We take $\kappa = 0.135$ as the physical charm quark. The ratio m_π^2/m_K^2 fixes the strange quark at $\kappa = 0.1550(2)$. The three light quarks $U_1 - U_3$ are used to extrapolate the $q^2 = 0$ data to the physical value of $\bar{m} \equiv (m_u + m_d)/2$ (fixed by the experimental ratio m_π^2/m_ρ^2) assuming that the form-factors depend linearly on the light quark mass. For example, the extrapolation of f_+ is shown in Fig. 2. Also, to calculate A_2/A_1, V/A_1 etc, the ratios of form-factors are taken at the very beginning of the jackknife process.

| | $f_+(q^2 = 0)$ | | | $f_0(q^2 = 0)$ | |
	(a)	(b)	EXPT.	(a)	(b)
$D \to Kl\nu$	0.72(5)	0.81(3)	0.77(4)	0.73(4)	0.73(2)
$D \to \pi l\nu$	0.64(9)	0.75(4)		0.63(6)	0.65(3)
$(D \to \pi l\nu)/(D \to Kl\nu)$	0.88(6)	0.93(2)	$1.29 \pm 0.21 \pm 0.11$	0.88(4)	0.89(2)

Table 1. form-factors, f_+ and f_0, extracted using (a) best fit and (b) lattice pole masses.

| | V | | A_1 | | A_2 | |
		exp.		exp.		exp.
$D \to K^*l\nu$	1.24(8)	1.16(16)	0.66(3)	0.61(5)	0.45(19)	0.45(9)
$D \to \rho l\nu$	1.08(12)		0.56(4)		0.19(24)	
$D_s \to \phi l\nu$	1.29(5)		0.66(1)		0.46(8)	

Table 2. Estimates for vector form-factors using fits with lattice pole masses.

3. FINAL RESULTS

The results for the decay $D \to Kl\nu$ are given in Table 1. Present errors preclude a serious test of the pole-dominance hypothesis even though the best fit value for $\mathcal{M}_{1^-}$ is about $10-20\%$ below the mass measured on the lattice and $20-30\%$ below the known experimental values. Since f_+ is known from experiments,[2] one can regard the lattice measurements as providing a measure of systematic errors due to quenching and lattice discretization that we cannot otherwise estimate. The data for f_+ in Table 1 suggest that these are small, i.e. at the 10% level. Our best estimate for $(D \to \pi l\nu)/(D \to Kl\nu) = 0.88(6)$ lies at the lower end of the range of experimental values.[2]

In the case of the vector final states we find that the estimates for A_2 are not very stable for $\vec{p} = (1,1,1)$ and $(2,0,0)$, making a two free parameter fit (best fit method) unreliable. However, the point $\vec{p} = (1,1,0)$ lies very close to the desired limit $q^2 = 0$, and can be used as a consistency check. With this criterion we find that the pole fits give reasonable estimates. For V and A_1 the two kinds of fits give consistent estimates, therefore in Table 2 we give pole fit results as our best estimates for all three form-factors. The results for $D \to K^*l\nu$ are in surprisingly good agreement with the averaged experimental values.[2] The form-factors for $D \to \rho l\nu$ are consistently smaller and we find little difference, qualitatively or quantitatively, between the two final states K^* and ϕ. The experimental errors in $D_s \to \phi l\nu$ are too large (see summary talk by Janis McKenna in these proceedings) to make a meaningful comparison.

We gratefully acknowledge the tremendous support provided by the ACL at Los Alamos, and NCSA at Urbana-Champaign. These calculations have been done on CM5 parallel supercomputers as part of the DOE HPCC Grand Challenges program (at ACL) and an NSF Metacenter allocation (at NCSA).

References

1 V. Lubicz, G. Martinelli, C. Sachrajda, *Nucl. Phys.* **B356** (1991) 301.
2 M. Witherell, Proceedings of the International Symposium on *Lepton and Photon Interactions at High Energies*, Cornell University, 1993.

CHARM PHYSICS: Summary and Outlook

JANIS A. McKENNA

Department of Physics, University of British Columbia,
Vancouver, British Columbia, V6T 1Z1, Canada

ABSTRACT

Highlights of recent results in charm physics are reviewed and summarized.

1. Introduction – Charm

The era of high precision, high statistics studies in weak decays has finally arrived. As its twentieth birthday approaches, the charm quark has gracefully matured from the uncertain and trying times of its teenage years into solid adulthood. But despite its age, new physics is continually emerging in our studies of charm. Because the mass of the charm quark is not too much larger than Λ_{QCD}, (considering QCD scales logarithmically) it is not truly a 'heavy' quark, and therefore QCD effects and final state interactions play a much larger role in weak charm decays than, for example, in weak beauty decays.

A weak decay is depicted in Figure 1. The short-range weak interaction occurs in the small inner circle, while the confinement region is the large outer circle. Both electroweak and strong QCD effects play a role in weak decays. On the short distance scale, the weak interaction and QCD hard gluon radiation may both be exactly calculated. The long distance QCD effects, which include final state interactions, soft gluon radiation, confinement of quarks to form hadrons, and bound state effects, cannot be calculated perturbatively in QCD. Decoupling or factorization simplifies the weak decay problem into two distance scales and lets us calculate at least some of the QCD corrections, the short range dynamics part, along with the weak interaction:

- Weak Interaction – Small scale   Exactly Calculable
- Hard Gluon Radiation – Small scale   Exactly Calculable
- QCD Confinement – Larger scale   Non-Perturbative

Phenomenological models are however necessary to calculate the long-range, non-perturbative QCD effects.

Thanks to excellent emulsion and silicon detector vertex resolution, extremely high statistics in charm particle production, and the charm quark's long lifetime, huge, clean samples of charm have been collected, making precision studies of charm, and studies of the interplay of the strong and electroweak interactions possible.

We've come far beyond the naïve spectator model of weak decays shown in Figure 2a. In the case where the virtual W decays to leptons, the amplitude for such an external spectator semileptonic decay is factorizable into two parts: the leptonic part,

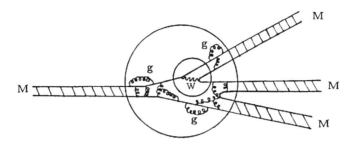

Fig. 1. The weak decay of a meson M involves both electroweak and strong QCD effects. The weak interaction occurs in the small inner circle, with the W intermediate vector boson, while the QCD confinement region is the larger outer circle. Initial and final state mesons are denoted M, and some soft gluons are denoted g.

which is exactly calculable, and the hadronic part, which describes how the quarks form hadrons. The hadronic part of the amplitude may be specified in terms of form factors, which unfortunately are not exactly known, so we rely on phenomenologically-inspired q^2 dependent functions for the form factors.

In Figure 2b, an internal spectator decay is depicted. Note that gluon radiation and final state interactions may occur between the spectator quark and the quarks resulting from the decay of the virtual W. (Semileptonic decays cannot occur via this diagram, so they have no strong interaction complications from the W decay products.) This diagram is sometimes referred to as colour-suppressed, as the colour of the quarks resulting from the virtual W^+ decay must match the colour of the spectator quark to form a hadron. Figures 2c and 2d depict non-spectator W-exchange and annihilation diagrams. These amplitudes are in principle helicity-suppressed, although in the case of W-exchange, soft gluonic radiation may greatly reduce any any helicity-suppression. In the case of charmed baryon decays, it is expected that there is very little helicity suppression. There are many detailed reviews of weak decays in the charm quark system.[1,2]

2. New Experiments in Charm Physics: BES and HERA

The measurement of absolute branching fractions of charmed hadrons is essential for both the understanding of details of non-perturbative QCD contributions to weak decay rates, and in order to do B physics, which usually requires complete reconstruction of charmed hadrons, and knowledge of their branching fractions to specific decay modes.

For some time now, all D_S branching fractions have been measured relative to the $D_S \to \phi\pi$ decay mode. Theory, models, and many assumptions are then used to deduce and extract absolute branching fractions for D_S decays. BES has recently completed a run just above threshold for $e^+e^- \to D_S^+ D_S^-$ production (below D_S^*

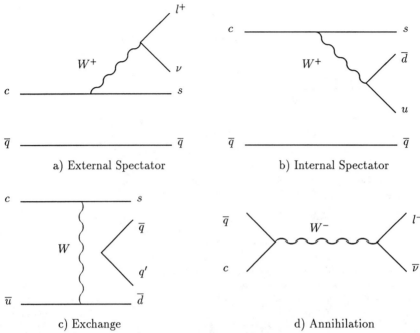

Fig. 2. Charm quark decay diagrams.

threshold), so that only $D_S^+ D_S^-$ pairs are produced, each with the beam energy. By reconstructing events in which either one or both D_S are reconstructed, the absolute branching fraction $D_S \to \phi\pi$ may be measured model-independently:

$$Br(D_S^\pm \to \phi\pi^\pm) = (4.2^{+9.0+1.7}_{-1.5-0.0} \pm 0.5)\%$$

The excellent particle identification capabilities of the dE/dx and Time-of-Flight systems, and the fact that the D_S energy is constrained to the beam energy, result in excellent signal/background enhancement. BES has also taken advantage of doubly-tagged D_S candidates to measure the D_S pseudoscalar decay constant via the decay $D_S \to \ell\bar{\nu}_l$. Measurement of this decay constant gives the coupling of the weak hadronic current to the vacuum, a measure of the c and $\bar{s}$ quark wave function overlap, which provides critical tests of QCD calculations involving sum rules, lattice gauge calculations, and potential models. Measurements of f_{D_S} from CLEO, WA75, and BES are listed in Figure 3, alongside theoretical predictions. Additionally, the BES group has obtained a sample of doubly-tagged $D^*\bar{D}$ and $D^*\bar{D}^*$ events, which has been used to confirm CLEO's relatively new D^* branching fraction results.[3]

Charmed D^* mesons and J/ψ mesons have been reconstructed at the ZEUS

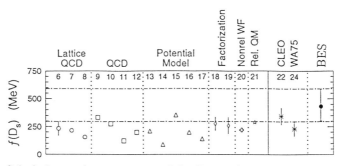

Fig. 3. Calculations and measurements of the D_S pseudoscalar coupling constant.

and H1 experiments at the HERA $e - p$ collider, and provide a probe of QCD. The charm cross section and its Q^2 dependence provides a test of our understanding of QCD charm production mechanisms such as photoproduction via low $p_\perp$ γ−proton diffractive scattering or high $p_\perp$ γ−gluon fusion. Elastic and quasi-elastic electroproduction via γ−gluon fusion probes the gluon density of the proton or pomeron and tests vector meson dominance models. Prompt μ from charm and beauty decays have been observed and the charm photoproduction cross-section $\sigma(\gamma p \rightarrow c\bar{c}X)$ measured. The data compare well with theory and with extrapolations from lower energy experiments. It is clear that critical tests of QCD in the charm sector have only just begun at HERA, but preliminary charm results to date are well described by QCD.

3. High Precision Studies in Charm

Detailed high precision understanding of charm decays continues at fixed target experiments, e^+e^- colliders and hadron colliders.

3.1. Semileptonic Decays of Charmed Hadrons

Semileptonic decays may proceed only through the external spectator diagram, depicted in Figure 2a. Since the virtual W decays to leptons, there is only one diagram which contributes to the amplitude, and any QCD effects are confined to the hadronic part of the decay. While the leptonic part of the weak decay is exactly calculable, the hadronic part is typically described by form-factors which are q^2 dependent functions which are deduced from phenomenological models. (q^2 is the invariant mass of the lepton and neutrino.) For a decay of a pseudoscalar meson into leptons plus a pseudoscalar meson, (for example, the decay $D^0 \rightarrow K^-\ell^+\nu_\ell$) only one form factor is necessary to describe the decay (in the limit the leptons have zero mass). For the semileptonic decay of a vector meson to a pseudoscalar meson plus leptons, three form factors are needed to describe the decay. Models differ in the form of the form factors: of two current popular models, the ISGW model uses exponential form factors,[5] while WBS model has single-pole type form factors.[6]

CLEO has recently constructed decays of the charmed baryons $\Xi_c^+ \rightarrow \Xi^0 e^+ \nu$ and $\Xi_c^0 \rightarrow \Xi^- e^+ \nu$. These decays are particularly interesting: if one assumes that the production rates of Ξ_c^+ and Ξ_c^0 are identical, and that their semileptonic widths are identical, then the ratio of their semileptonic branching ratios is the ratio of their lifetimes. Recall that this is the case in D^0 and D^+ decays, where differences in lifetimes originate from the hadronic sector, in which non-spectator decays are responsible for lifetime differences. CLEO measures $\tau(\Xi_c^+)/\tau(\Xi_c^0) = 2.46 \pm 0.70^{+0.33}_{-0.23}$, which suggests that the W-exchange amplitude is significant in decays of charmed baryons. Using theoretical predictions that the interference in the case of the Ξ_C^+ has both destructive and constructive components, while the Ξ_c^0 has components from W-exchange and constructive interference, each of these components can be deduced if all charmed baryon amplitudes/lifetimes are fit simultaneously. This result agrees well with direct measurements at fixed target experiments.

CLEO has also presented results from semileptonic decays of charmed hadrons, and has deduced form factors at $q^2 = 0$. In the case of the decay $D_S^+ \to \phi \ell^+ \nu$, the three form factors may be denoted $A_1(q^2)$, $A_2(q^2)$ and $V(q^2)$, and from the angular distributions of the decay the ratios $\frac{V(q^2)}{A_1(q^2)}$ and $\frac{A_2(q^2)}{A_1(q^2)}$ may be deduced. These ratios should be the same as the ratios measured in the decay $D^+ \to \overline{K^{*0}} \ell^+ \nu$, if SU(3) flavour symmetry holds. CLEO has measured these ratios for the decay $D_S^+ \to \phi \ell^+ \nu$ and results are consistent with world average results for the decay $D^+ \to \overline{K^{*0}} \ell^+ \nu$: $\frac{A_2(q^2)}{A_1(q^2)} = 1.4 \pm 0.5 \pm 0.3$ and $\frac{V(q^2)}{A_1(q^2)} = 0.9 \pm 0.6 \pm 0.3$, which agree with predictions from the theory of ISGW.[5]

There are however problems in understanding semileptonic D_S decays. CLEO has measured the ratio of pseudoscalar to vector rates in D_S semileptonic decays, a quantity which many otherwise successful theories fail to predict correctly in analogous D decays. Not only do the branching ratios $Br(D_S^+ \to \eta' \ell^+ \nu)/Br(D_S^+ \to \phi \ell^+ \nu) = 0.71 \pm ^{+0.19}_{-0.18}{}^{+0.08}_{-0.10}$ and $Br(D_S^+ \to \eta \ell^+ \nu)/Br(D_S^+ \to \phi \ell^+ \nu) = 1.74 \pm 0.34 \pm 0.24$ disagree with theory, but the ratio of these branching ratios does not agree with expectations from the factorization hypothesis. Clearly there is a need for further work in understanding semileptonic decays.

Finally, CLEO has studied the decay $\Lambda_c^+ \to \Lambda \ell^+ \nu$ in order to study the form factors in Heavy Quark Effective Theory.[7] Decays of charmed baryons are characterized by form factors, two axial, two vector. But in the limit of the charmed quark being infinitely heavy, the spin and flavour symmetries reduce the number of independent form factors to one, which may be measured via studying the angular distribution of the Λ_c^+ semileptonic decays.

3.2.   Spectroscopy

There are four $\ell = 1$ p-wave mesons with spin-parity $J^P = 0^+$, 1^+, 1^+, and 2^+ for each the D^0, D_S^+, and D^+. The approximate spin-flavour symmetry for heavy quarks[7] results in the charmed p-wave mesons being arranged in two doublets:
- 0^+ 1^+ : very broad resonance ($j = S_q + \ell = \frac{1}{2}$)
- 1^+ 2^+ : narrow resonance ($j = S_q + \ell = \frac{3}{2}$)

where S_q is the spin of the light quark in the charmed meson. CLEO has observed both narrow p-wave states for each the D^0, D_S and D^+ mesons, verifying their spin-parity assignments through helicity analyses. The masses and branching fractions agree well with predictions of potential models.[8]

CLEO has reported new measurements of branching ratios of the Λ_c^+. While only slightly more than 25% of all Λ_c^+ decay modes are known, some decay modes which can only occur via $W-$exchange have been observed (along with decays which can occur via both spectator and exchange diagrams). These contributions to the total Λ_c^+ width leads one to conclude that the lifetimes of charmed baryons could be much smaller than for charmed mesons–a fact which has been observed. Additionally, when Pauli interference and final state interactions have been considered, the complete picture of hierarchy in charmed hadron lifetimes may be deduced.

From Fermilab, there is possible evidence (preliminary, pending full helicity analysis) from E-771 for the $^1P_1$ and $^3D_2$ charmonium states. The observed masses agree well with potential model predictions of these charmonium states.

3.3. Charm at LEP

At LEP, separated primary charm quark and primary beauty data samples are used to test electroweak theory such as the coupling of the Z^0 to $b\bar{b}$ and $c\bar{c}$ quarks, and to test QCD, in studying, for example, fragmentation processes of each the b and c quarks. In contributions to this conference, OPAL has performed detailed studies of D^* mesons and Λ_c baryons including measurement of total production rates from the Z^0, and in particular, separating out contributions from primary b quarks from primary c quarks, either based on momentum distributions, or if momentum distributions themselves are to be studied to gain insight into QCD hadronization. This separation may be achieved by several techniques:

- the presence of a high p, high p_T lepton, characteristic in semileptonic B decay

- global event topology and jet shape variables input to a neural network

- the displacement of tracks from the Z^0 vertex in the opposite thrust hemisphere to a D^* candidate is used to tag $Z^0 \to b\bar{b}$ events.

- the reconstruction of the decay vertex of the D^0 candidate produced in D^* decay, taking advantage of the fact that the B_d^0 lifetime is about four times that of the D^0, so that D^0s produced from D^*s from B decay have significantly larger displacements from the primary Z^0 decay vertex than those produced from primary charm quarks.

Fragmentation studies are approaching the precision necessary to rule out some fragmentation function models. The Peterson fragmentation function is only slightly favoured over several other popular models. Characteristics of charm fragmentation near 90 GeV may be compared to previous studies at lower energies to study scaling violations as an interesting test of QCD.

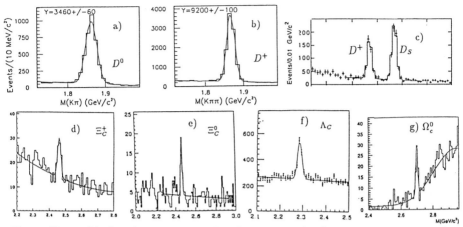

Fig. 4. Charmed hadron invariant mass plots from E-687. a) D^0, b) D^+, c) D_S, d) Ξ_C^+, e) Ξ_C^0, f) Λ_C, g) Ω_c^0

Additionally, completely reconstructed charmed hadrons are crucial building blocks in the reconstruction of B hadrons. Semileptonic B decays to D, D^* and D^{**} mesons accompanied by a lepton have been reconstructed using kinematic and vertex information. OPAL has made the first observation of semileptonic B meson decays to charged p-wave D mesons, and has determined that p-wave D mesons account for $(34\pm7)\%$ of charmed semileptonic B meson decays, in agreement with earlier results from the CLEO[9] and ARGUS[10] collaborations.

3.4. Charm at Fixed Target Experiments

Fixed target experiments have contributed immensely to our knowledge of charm production, lifetimes, and spectroscopy, noted the large number of contributed talks from Fermilab experiments E-672, E-687, E-706, E-771, E-789 and E-791.

Powerful vertexing techniques have been used to tag very clean charmed hadron samples, from which high precision, high statistics charmed hadron lifetimes have been measured. Detailed studies involving Dalitz analysis of multi-body charm meson decays been performed, including 5-body decay modes and D_S decay modes involving the f_0 and f_2, and the observation of $\Omega_c^0 \to \Sigma^+ K^- K^+ \pi^+$, a new decay mode, (shown in Figure 4g), with statistics sufficient (over 40 events) to soon measure its lifetime.

Measurements of 1% to a few % on charm hadron lifetimes have been been performed with the E-687 spectrometer at Fermilab. The clean charmed hadron signals from which lifetimes are extracted are shown in Figure 4. In the naïve spectator model, a free charm quark would have lifetime given by analogy to muon decay, approximately of $1\text{-}2\times10^{-12}$s, and so the spectator model predicts all charmed hadrons to have the same lifetime. Experimentally, a lifetime hierarchy is observed, with lifetimes varying over an order of magnitude:[11]

$$\tau(D^+) > \tau(D_S^+) > \tau(D^0) \approx \tau(\Xi_C^+) > \tau(\Lambda_C^+) > \tau(\Xi_C^0)$$

$\tau(ps)$	1.057	0.467	0.414	0.35	0.202	0.098
	± 0.015	± 0.017	± 0.004	$^{+0.07}_{-0.05}$	± 0.012	± 0.023

The D_S and D^0 have lifetimes which now differ significantly, by 3σ. More notably, in the charmed baryon sector, final state interactions are large, and helicity suppression in exchange diagrams is not very significant, due to the presence of an extra spectator quark, which can exchange soft gluons, overcoming helicity suppression.

Production asymmetries in Feynman x distributions of leading and non-leading charmed mesons produced in $\gamma - g$ fusion have been observed. Production mechanisms and models may be studied by investigating the $p_T{}^2$ and x_F dependences of charmed meson production, and such QCD studies are underway at several FNAL fixed target experiments.

At fixed target experiments at Fermilab it has been determined that not all $J\psi$'s come from B decays; in fact, at E-672/706, it was determined that less than half of J/ψ's detected are direct or from B decays. The remaining $J\psi$'s come from decays of $\psi(2S)$ and χ_c charmonium states. Additionally, the fixed target experiments have reconstructed a handful of B decays to $J/\psi K$ and $J/\psi K^*$.

3.5.   Charm at Hadron Colliders

At the TeVatron collider, a few $\times 10^4$ J/ψ candidates have been reconstructed and the CDF precision vertex detector has enabled the determination that most J/ψ's are prompt, the result of gluon fusion, gluon fragmentation and charm fragmentation processes, and not from B decays, previously thought to be the dominant source of J/ψ. Recent advances on the theoretical front in perturbative calculations of fragmentation now correctly predict prompt J/ψ production and J/ψ production from B decays. There remains however a problem with the theory being an order of magnitude away from the measured ψ' production rates. CDF has also measured Compton production of charm and beauty via the production of D^*'s and prompt muons.

4.   Conclusions and Outlook

Much progress has recently been made in the field of charm physics. The hierarchy of charmed hadron lifetimes has emerged and measurements may be used to study the effects of spectator and non-spectator decay amplitudes in charmed hadron decays, and interference effects in weak decays. We anxiously await a measurement of the Ω_c lifetime, which will help us fill in more details on non-spectator contributions to heavy quark decay.

CLEO-II, with its superb photon resolution, will continue to measure new decay modes of charm in modes involving neutral particles, and reconstructing B mesons from charmed hadrons to study weak decays in the B system. The BES detector, although no longer running just above $D_S^+ D_S^-$ threshold, will include more decay modes in its double tagging analyses, so we can expect better statistical precision on absolute D_S^+ branching fractions. At HERA, we are just beginning to explore the charm content in the proton and pomeron, performing critical tests of QCD.

In semileptonic decays, form factors are measured, but in order to gain detailed insight into the decays, more data are needed. It will be interesting when statistics are sufficient to study q^2 dependences of form factors in order to distinguish various models and thoroughly test lattice QCD calculations.

The TeVatron is a copious producer of heavy quarks, and detailed studies of charmed quark production from B decays, gluon fusion, and gluon fragmentation provide tests of QCD. Theoretical advances in the understanding of charmonium production (specifically ψ') are necessary, as present calculations fail to reproduce experimental measurements.

LEP, with high charm and beauty statistics, and high precision vertex detectors is contributing much to the field of weak heavy quark decays, both in detailed studies of charm, and the development of charm as a tool for tagging and reconstructing B decays.

Finally, all the theories formulated, and insight and experience gained in studying charm decays will be directly applicable and testable in the B system soon.

References

1. M.S.Witherell, *Proceedings of the International Symposium on Lepton and Photon Interactions at High-Energies*, ed. P.Drell, D.Rubin, Ithaca, NY, A.I.P., 1994.
2. J.A.McKenna, *Proceedings of the 8th Lake Louise Winter Institute: Collider Physics*, ed. A.Astbury *et al.*, World Scientific, 1993.
3. F.Butler *et al.*(CLEO Collaboration), Phys. Rev. Lett. **69**, 2041 1992.
4. M.Kelsey, contribution to this conference.
5. N.Isgur, D.Scora, B.Grinstein, and M.B.Wise, Phys. Rev. **D39**, 799 (1989).
6. M.Wirbel, B.Stech, M.Bauer, Z. Phys. **C29**, 637(1985).
7. N.Isgur and M.B.Wise, Phys. Lett. **B232**, 113 (1989),
   N.Isgur and M.B.Wise, Phys. Lett. **B237**, 527 (1990).
8. S.Godfrey, R.Kokoski, Phys. Rev. **D43**, 1679 1991,
   S.Godfrey, N.Isgur, Phys. Rev. **D32**, 189 1985.
9. S.Henderson *et al.*CLEO Collaboration, Phys. Rev. **D45**, 2212 1992.
10. H.Albrecht *et al.*(ARGUS Collaboration) Z. Phys. **C57**, 533 1993.
11. V.Sharma, compiled for plenary talk at this conference.

Tau Physics

TOMASZ SKWARNICKI, *Southern Methodist*

NEW RESULTS FROM CLEO-II
ON HADRONIC DECAYS OF THE TAU LEPTON

JON URHEIM

California Institute of Technology
Pasadena, California USA 91125

(Representing the CLEO collaboration)

ABSTRACT

Results on semi-hadronic decays of the τ lepton are presented, from studies of e^+e^- annihilation data obtained at the Cornell Electron Storage Ring with the CLEO-II detector. Branching fractions have been measured for decays to two, three and four hadrons, namely $\tau^- \to \nu_\tau h^- \pi^0$, $\tau^- \to \nu_\tau h^- h^+ h^-$, and $\tau^- \to \nu_\tau h^- h^+ h^- \pi^0$, where $h^\pm$ represents a charged pion or kaon. CLEO-II has also observed decays with charged and/or neutral kaons; preliminary results for branching ratios and structure arising from the decay dynamics are given. Connections are made with predictions derived from theoretical models, the Conserved Vector Current theorem, isospin constraints and sum rules.

1. Introduction

Semi-hadronic decays of the τ lepton provide a unique opportunity to explore the hadronic charged weak current. The leptonic current is well-understood in electroweak theory, and the τ is the only lepton which can decay semi-hadronically. Thus, one can probe the weak and strong dynamics governing the formation of hadronic final states.

A cartoon of semi-hadronic τ decay is shown in Figure 1. The values of q^2 carried by the W are small, so the strong physics is non-perturbative, dominated by resonance formation, and gives rise to final states consisting of light pseudoscalar mesons.

Spin, parity, isospin and charge conjugation symmetries constrain the hadronic states which can form.[1] The spin and parity quantum numbers of a given state select out either the vector or axial vector part of the weak V–A current. Isospin and charge conjugation for Cabibbo-favored $(\bar{u}d)$ currents require that vector states produced be G-parity even and axial vector states be G-parity odd, and further restrict them to decay to even and odd numbers of pions respectively. While there is little other theoretical input regarding τ decays to axial vector states, production of non-strange vector states can be related to low energy e^+e^- cross-sections by way of the Conserved Vector Current (CVC) theorem. Also sum-rules can be invoked to relate the vector and axial-vector (or strange and non-strange) hadronic spectral functions (of q^2) which describe the resonant content of the hadronic system.

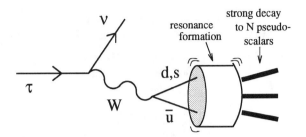

Fig. 1. Simplified picture of semi-hadronic τ lepton decay.

The present experimental situation is somewhat murky however. Only a few τ decays have been studied with sufficient precision to test the elements of the hadronic weak and strong physics described above. Inconsistencies are present, both in regard to the overall picture of τ decay and within the data for several specific decays. Below, I present results from CLEO-II on semi-hadronic τ decay branching fractions. Comments on resonant structure are made where appropriate. This talk is organized as follows: A brief description of the CLEO-II detector is given in Section 2. In Section 3, results are given for decays containing from 2 to 5 hadrons. Results on decays containing one or two kaons are discussed in Section 4.

2.   The CLEO-II Detector

CLEO-II[2] is a solenoidal spectrometer and calorimeter operating at the south interaction region of the Cornell Electron Storage Ring (CESR), where τ leptons are produced in pairs through e^+e^- annihilation at $\sqrt{s} \sim 10.6$ GeV, with a ~ 0.91 nb cross section. Since the turn-on of CLEO-II in late 1989, CESR has delivered in excess of 4 fb^{-1} of integrated luminosity, corresponding to ~ 3.6 million τ-pairs. The components of CLEO-II most relevant to the analyses presented here are the 67-layer tracking system and the 7800-element CsI(Tl) crystal calorimeter which are contained within a 1.5T magnetic field, and cover $> 96\%$ of 4π in solid angle. The segmentation and energy resolution of the calorimeter permit reconstruction of $\pi^0 \rightarrow \gamma\gamma$ decays with high efficiency for π^0's of all energies: typical π^0 mass resolutions range from 5 to 7 MeV/c^2, and we have observed τ decays containing as many as four fully reconstructed π^0's.[5] Muons above ~ 1 GeV/c are tracked in Iarocci tubes deployed at three depths in the flux return steel. Limited π/K separation is achieved below 1 GeV/c and above 2 GeV/c using dE/dx information from the 51-layer main drift chamber and/or time of flight measurements from scintillation counters located in front of the calorimeter.

3.   Decays to Two, Three, Four and Five Hadrons

CLEO-II has directed much effort to precisely measure branching fractions for decays to charged hadrons with and without accompanying π^0's. This focus was motivated by the persistence of the one-prong deficit.[3,4] Earlier work includes studies of τ decays to one charged hadron plus two or more π^0's[5] as well as decays containing η

mesons.[6] We observed for the first time the decay $\tau^- \to \nu_\tau 2\pi^- \pi^+ 2\pi^0$,[7] finding a surprisingly large branching fraction $(0.48 \pm 0.04 \pm 0.04\%)$. Below I present results for the decays $\tau^- \to \nu_\tau h^- \pi^0$, $\tau^- \to \nu_\tau 2h^- h^+ [\pi^0]$ and $\tau^- \to \nu_\tau 3h^- 2h^+ [\pi^0]$. In these analyses, $h^\pm$ represents a charged hadron $(\pi$ or $K)$ which we do not attempt to identify.

3.1. The Decay $\tau^- \to \nu_\tau h^- \pi^0$

The primary semi-hadronic τ decay, $\tau^- \to \nu_\tau \pi^- \pi^0$, occurs through the vector portion of the hadronic charged weak current. The $\rho(770)$ meson dominates the spectral function responsible for this decay, but contributions from excited ρ's and non-resonant $\pi^- \pi^0$ formation may also be present.

The CVC theorem allows one to relate the $\tau^- \to \nu_\tau \pi^- \pi^0$ decay width to the iso-vector contribution to the $e^+ e^- \to \pi^+ \pi^-$ cross section. A branching fraction of $24.58 \pm 0.93 \pm 0.27 \pm 0.50\%$ is predicted,[8,9] where the errors are from uncertainties in the $e^+ e^-$ data, τ lifetime, and knowledge of the radiative corrections. Experimental values range from 22% to 26%,[4] limited in precision by difficulties in π^0-finding and rejection of decays with multiple neutrals (themselves poorly measured until recently).

We have studied the decay $\tau^- \to \nu_\tau h^- \pi^0$,[10] $h^- \equiv \pi^-, K^-$, using a 1.58 fb^{-1} data sample $(1.44 \times 10^6 \ \tau$-pairs$)$. We employ three statistically independent methods, determining the branching fraction, $B_{h\pi^0}$, according the following forms:

$$B_{h\pi^0} = \sqrt{\frac{N_{e-h\pi^0} N_{\mu-h\pi^0}}{2N_{e-\mu}N_{\tau\tau}}}, \quad \sqrt{\frac{N_{h\pi^0 - h\pi^0}}{N_{\tau\tau}}}, \quad \text{or} \quad \frac{N_{3h-h\pi^0}}{N_{3h-1}} B_1. \tag{1}$$

The N's denote the measured background-subtracted, efficiency-corrected event yields for the given final states $(e$ denotes $\nu e \bar{\nu}$, etc.$)$. $N_{\tau\tau}$ is the number of τ-pairs, as determined from the integrated luminosity (known to 1% from studies of Bhabha scattering, $e^+ e^- \to \gamma\gamma$, and $e^+ e^- \to \mu^+ \mu^-)$, and the theoretical cross-section (also known to 1%). B_1 is the world average topological one-prong branching fraction.

Due to space constraints, we merely give the results below, and refer the reader elsewhere[10] for experimental details. The three methods agree; combining them we find:

$$B_{h\pi^0} = 25.87 \pm 0.12 \,\text{(stat.)} \pm 0.42 \,\text{(syst.)} \,\%, \tag{2}$$

based on samples totalling $\sim 44,000$ reconstructed $\tau^- \to \nu_\tau h^- \pi^0$ decays. The dominant systematic errors (common to all three methods) are those due to π^0-finding efficiency (0.9%) and the veto on events with additional energy deposition in the calorimeter applied to reject backgrounds (0.9%).

This result is higher than most other measurements,[4] but is consistent with recent values obtained at LEP by OPAL,[11] L3,[12] and Aleph[13] which also tend to be high. Subtracting the measured contribution from $\tau^- \to \nu_\tau K^- \pi^0$,[14] we find $B_{\pi^\pm \pi^0} = 25.36 \pm 0.44\%$, in good agreement with the CVC prediction.

3.2. The Decays $\tau^- \to \nu_\tau h^- h^+ h^- [\pi^0]$

As with the decay described above, our understanding of τ decay into final states with three charged particles has been confused by conflicting experimental data.[4] The CLEO-II observation of decays containing two π^0's with a branching fraction around

$0.5\%^7$ has implications for 3-prong results from other experiments. Below, I present preliminary results from CLEO-II on branching fractions for the decays $\tau^- \to \nu_\tau h^- h^+ h^-$ and $\tau^- \to \nu_\tau h^- h^+ h^- \pi^0$.

The decay $\tau^- \to \nu_\tau h^- h^+ h^-$ is expected to occur through the axial-vector current. Hence theoretical constraints are limited. The hadronic system is expected to be dominated by the $a_1(1260)$, which gives rise to both $\pi^- \pi^+ \pi^-$ and $\pi^- \pi^0 \pi^0$ final states. However channels with charged kaons are also known to contribute.

Because the decay $\tau^- \to \nu_\tau h^- h^+ h^- \pi^0$ has an even number of final state mesons, the vector current is expected to play the primary role. Due to severe Cabibbo and phase-space suppression for modes containing kaons, the four-pion final state is expected to dominate. Thus, predictions can be made from $e^+ e^- \to 4\pi$ data through application of the CVC theorem. Two such calculations give the branching fraction to be $4.3 \pm 0.3\%^{15}$ and $4.8 \pm 0.7\%.^{16}$ Experimental results cluster around the 5% level.

At CLEO-II, branching fractions are obtained from analyses of the following event types: $3h^\pm - 3h^\pm$, $\ell - 3h^\pm$ and $\ell - 3h^\pm \pi^0$, where ℓ represents an electron or muon.[17] The branching fractions are determined from measured quantities as follows:

$$B_{3h^\pm} = \sqrt{\frac{N_{3h^\pm - 3h^\pm}}{N_{\tau\tau}}} \quad \text{or} \quad \frac{N_{\ell - 3h^\pm}}{N_{\tau\tau} B_\ell}, \qquad B_{3h^\pm \pi^0} = \frac{N_{\ell - 3h^\pm \pi^0}}{N_{\tau\tau} B_\ell}, \qquad (3)$$

where as before the N's represent background-subtracted, efficiency-corrected event yields, $N_{\tau\tau}$ is the number of produced tau-pairs (1.44×10^6), and B_ℓ represents the leptonic branching fraction. The much smaller $3h^\pm - 3h^\pm$ sample (2,030 events as opposed to 17,479 in the $\ell - 3h^\pm$ sample) gives a nearly comparable statistical error because of the square root. Also systematic errors associated with event-related efficiencies are smaller by a factor of 2 than in the $\ell - 3h^\pm$ analysis.

In Figure 2(a,b), we plot the invariant mass spectra for the hadronic system in the $\tau \to \nu_\tau h^- h^+ h^-$ event samples, taking the pion mass for the hadrons. These samples are relatively background-free as demonstrated by the scarcity of events with $M_{3h} > M_\tau$: non-τ backgrounds are estimated to comprise $4.7 \pm 0.9\%$ in the $3h^\pm - 3h^\pm$ sample and only $0.2 \pm 0.2\%$ in the $\ell - 3h^\pm$ sample. In Figure 2(c), the presence of π^0's is demonstrated in events containing two energy clusters unassociated with the charged tracks.

The combined results obtained from these event samples are:

$$B_{3h^\pm} = 9.82 \pm 0.09 \,(\text{stat.}) \pm 0.34 \,(\text{syst.}) \,\% \qquad (4)$$
$$B_{3h^\pm \pi^0} = 4.25 \pm 0.09 \,(\text{stat.}) \pm 0.26 \,(\text{syst.}) \,\%. \qquad (5)$$

While the statistical errors are significantly smaller than those previously attained, the systematic errors are comparable to or slightly worse than recent reports from LEP.[20] Uncertainties in modelling track-finding efficiency (2–2.5%), in effects of cuts on unaccounted-for energy deposition applied to suppress backgrounds (3–4%), and in the π^0-reconstruction efficiency for $3h^\pm \pi^0$ events (3%) dominate the systematic errors. These errors are preliminary, and they should shrink after further study.

The $\tau \to \nu_\tau h^- h^+ h^-$ result is significantly higher than the PDG '94 average value of $8.0 \pm 0.4\%,^4$ supporting the recent preliminary value reported by Aleph $(9.57 \pm 0.24 \pm 0.22\%)$. After correcting for modes containing kaons, CLEO-II results in both channels[5,10] indicate $B_{3\pi^\pm} \approx B_{\pi^\pm 2\pi^0}$, as expected.

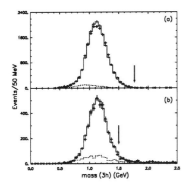
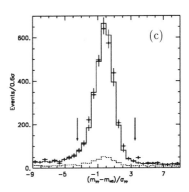

Fig. 2. Invariant mass spectra for $\tau^- \to \nu_\tau h^- h^+ h^-$ decays, taking $M_h = M_\pi$, for lepton-tagged events (a) and $3h^\pm$–$3h^\pm$ (b) event samples. In (c) is shown the quantity $(M_{\gamma\gamma} - M_{\pi^0})/\sigma_{M_{\gamma\gamma}}$ for events in the lepton tagged $\tau^- \to \nu_\tau h^- h^+ h^- \pi^0$ candidate sample. The data (points) are shown along with the expectations for signal plus background from the Monte Carlo[18,19] (solid histogram). Also shown are the expected contributions from backgrounds (dashed histogram). Arrows indicate the location of cuts.

Correspondingly, the $\tau^- \to \nu_\tau h^- h^+ h^- \pi^0$ result is somewhat lower than previous results, although still agreeing with the CVC predictions noted earlier. Summing the two results along with the result for $B_{3h^\pm 2\pi^0}$[7] and accounting for common errors, we obtain a total 3-prong τ branching fraction of $14.55 \pm 0.13 \pm 0.59\%$. This is consistent with the PDG average value of $14.32 \pm 0.27\%$[4] for the topological branching fraction.

3.3. Five-Prong Tau Decays

CLEO-II has measured branching fractions for decays into final states with five charged particles.[21] The results are

$$B(\tau^- \to \nu_\tau 3\pi^- 2\pi^+ + \geq 0\,\text{neutrals}) = 0.097 \pm 0.005 \pm 0.011\,\% \qquad (6)$$
$$B(\tau^- \to \nu_\tau 3\pi^- 2\pi^+) = 0.077 \pm 0.005 \pm 0.009\,\% \qquad (7)$$
$$B(\tau^- \to \nu_\tau 3\pi^- 2\pi^+ \pi^0) = 0.019 \pm 0.004 \pm 0.004\,\%, \qquad (8)$$

the most precise to date.

4. Decays to Final States Containing Kaons

The production of final states containing kaons represents a largely unexplored sector of τ lepton decay. In the following sections, I present results[14,23] from CLEO-II on decays with one, two, or three final state mesons of which at least one is a kaon.

4.1. The Decay $\tau^- \to \nu_\tau K^-$

The decay $\tau^- \to \nu_\tau K^-$ is completely specified in electroweak theory through analogy with kaon decay. The branching ratio, accounting for radiative effects, is predicted to be $0.74 \pm 0.01\%$.[9] CLEO-II has measured it to be $0.66 \pm 0.07 \pm 0.09\%$.[14]

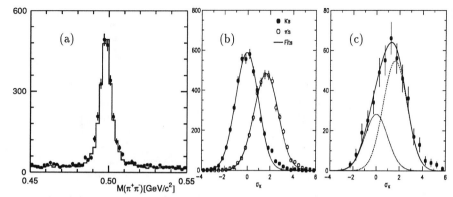

Fig. 3. (a) Invariant mass of $\pi^+\pi^-$ pairs forming detached vertices in the $\tau^- \to \nu_\tau h^- K^0_S$ candidate data (points) and Monte Carlo (histogram) samples. (b) Plots of σ_K (see text) for π's and K's with momenta above 2 GeV/c, identified from kinematics in $D^{*+} \to D^0 \pi^+ \to$ $(K^- \pi^+)\pi^+$ decays. (c) Plot of σ_K for the charged hadron accompanying the K^0_S in $\tau^- \to$ $\nu_\tau h^- K^0_S$, where $p_h > 2$ GeV/c, fit to a sum (solid curve) of Gaussians parameterizing the dE/dx response to pions (dashed) and kaons (dotted) taken from the D^* data shown in (b).

4.2. Decays of the Type $\tau^- \to \nu_\tau [Kh]^-$

Like the decay $\tau^- \to \nu_\tau \pi^- \pi^0$, decays of the type $\tau^- \to \nu_\tau [Kh]^-$ are expected to occur through the vector part of the weak current. The $K^- K^0$ final state has zero net strangeness, and CVC predictions based on $e^+ e^- \to \pi^+ \pi^-$ data and SU(3) symmetry give branching fractions of 0.11±0.03%[15] and 0.16±0.02.[16] A recent analysis by Aleph[22] gives $B_{K-K^0} = 0.29 \pm 0.12 \pm 0.03\%$. The $K^- \pi^0$ and $\pi^- \bar{K}^0$ final states are expected to appear in the ratio 1 : 2 from isospin, with a spectral function dominated by the $K^*(892)$ resonance. In this case, the Das-Mathur-Okubo (DMO) sum rule[24] allows one to predict $B_{[\pi K]^-}$ from the measured value of $B_{\pi^- \pi^0}$ discussed earlier.

At CLEO-II, all three final states $(K^- K^0, K^- \pi^0$ and $\pi^- \bar{K}^0)$ have been observed, using dE/dX (and time-of-flight for $K^- \pi^0$) for charged kaon identification, and detection of $K^0_S \to \pi^+ \pi^-$ decay for neutral kaons. The $K^- \pi^0$ decay branching fraction has been measured with limited statistics to be $0.51 \pm 0.10 \pm 0.07\%$, in the analysis[14] described in the previous section. A higher-statistics study of the $\pi^- \bar{K}^0$ decay based on 2.96 fb^{-1} of data $(2.71 \times 10^6$ τ-pairs) is described below, in which the $K^- K^0$ decay is also extracted.

We select events of a 1–3 charged-track topology in which two oppositely charged tracks of the '3' form a vertex separated from the $e^+ e^-$ interaction point by at least 5 mm. In Figure 3(a), the presence of $K^0_S \to \pi^+ \pi^-$ decays is demonstrated by plotting the invariant mass of the detached tracks as pions. After background subtraction, we have $1,938 \pm 126$ candidates for the decay $\tau^- \to \nu_\tau h^- K^0_S$.

For momenta above 2 GeV/c, dE/dx gives $\sim 2\sigma$ π/K separation, as shown in 3(b) for kinematically selected π's and K's from $D^0 \to K^- \pi^+$ data. Selecting $\tau^- \to \nu_\tau h^- K^0_S$

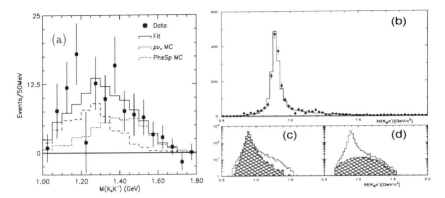

Fig. 4. (a) Invariant mass of $K^- K_S^0$ system for $\tau^- \to \nu_\tau K^- K_S^0$ candidate events in the data (points), along with a fit (solid histogram) to ρ^- (dashed) and phase space (dotted) contributions as determined through Monte Carlo simulation. (b) Plot of $M_{\pi^- K^0}$ for $\tau^- \to \nu_\tau h^- K_S^0$ candidates in the data (points), along with a sum (solid histogram) of distributions obtained from $\tau^- \to \nu_\tau K^{*-}(892)$ and $\tau^- \to \nu_\tau K^- K_S^0$ Monte Carlo samples (shown separately in (c) and (d), respectively), where the latter is determined according to the mixture of ρ and phase-space constributions shown in (a).

events with $p_h > 2$ GeV/c, we plot in Figure 3(c) the difference between the measured dE/dx and that expected for kaons, divided by experimental resolution, (denoted σ_K), and fit to a linear combination of response functions for π's and K's.

We obtain the following preliminary branching fractions from these studies:

$$B(\tau^- \to \nu_\tau h^- K^0) \quad = \quad 0.977 \pm 0.023 \pm 0.108\,\% \tag{9}$$

$$B(\tau^- \to \nu_\tau K^- K^0) \quad = \quad 0.123 \pm 0.023 \pm 0.023\,\%, \tag{10}$$

The result for B_{K-K^0} is lower than that from Aleph, and consistent with expectations from CVC. In Figure 4, we plot the invariant masses of the $K^- K_S^0$ and $h^- K_S^0$ systems in these event samples. The $K^- K_S^0$ system is softer than that predicted by a phase-space model, and may have some resonant content such as in the 'ρ' model, based on the $\tau^- \to \nu_\tau \pi^- \pi^0$ spectral function. The $h^- K_S^0$ system, after accounting for the admixture of $K^- K_S^0$ events, is consistent with saturation by $\tau^- \to \nu_\tau K^{*-}(892)$, as shown in Fig. 4(b), permitting a test of the DMO sum rule. Using the CLEO-II results on $h^- \pi^0$, $K^- \pi^0$, $h^- K^0$ and $K^- K^0$ final states we find $B_{K^*}/B_\rho = (0.98 \pm 0.03 \pm 0.13) \times \tan^2 \theta_C$, where the DMO sum rule prediction is $0.93 \tan^2 \theta_C$ in the narrow-width approximation.

4.3.   Decays of the Type $\tau^- \to \nu_\tau [K h \pi]^-$

Space constraints forbid a detailed discussion of the very interesting 3-meson final states containing one or two kaons, and only the results can be presented below. CLEO-II has observed the decays $\tau^- \to \nu_\tau h^- K_S^0 \pi^0$, $\tau^- \to \nu_\tau K^- K_S^0 \pi^0$, and $\tau^- \to \nu_\tau \pi^- K_S^0 K_S^0$, the first two being studied in a fashion similar to what was described in the previous

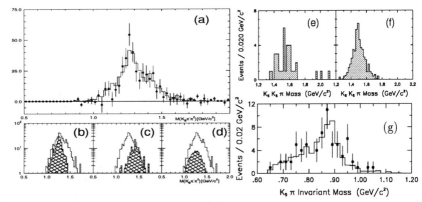

Fig. 5. (a) Plot of $M_{\pi^- K^0 \pi^0}$ for $\tau^- \to \nu_\tau h^- K_S^0 \pi^0$ candidates in the data (points), along with a fit (solid histogram) to distributions obtained from $\tau^- \to \nu_\tau K_1^-(1270)$, $\tau^- \to \nu_\tau K_1^-(1400)$, and $\tau^- \to \nu_\tau K^- K^0 \pi^0$ Monte Carlo samples (shown separately in (b), (c), (d) respectively), with the latter fixed according to the measured branching fraction. In (e) and (f) the $K_S^0 K_S^0 \pi^-$ invariant mass is plotted for $\tau^- \to \nu_\tau \pi^- K_S K_S$ candidate events in the data and Monte Carlo respectively, where the latter is based on the model of Decker et al.[26] In (g), we plot the $K_S \pi^-$ submass (two entries per event) for data (points) and Monte Carlo.

section. We obtain the following preliminary results:

$$B(\tau^- \to \nu_\tau h^- K^0 \pi^0) = 0.519 \pm 0.035 \pm 0.062\,\% \qquad (11)$$

$$B(\tau^- \to \nu_\tau K^- K^0 \pi^0) = 0.129 \pm 0.050 \pm 0.032\,\% \qquad (12)$$

$$B(\tau^- \to \nu_\tau \pi^- K^0 \bar{K}^0) = 0.083 \pm 0.017 \pm 0.017\,\%, \qquad (13)$$

where the last result assumes that the $\pi^- K^0 \bar{K}^0$ system manifests itself as $\pi^- K_S K_S$ 25% of the time. The decay $\tau^- \to \nu_\tau \pi^- \bar{K}^0 \pi^0$ is thought to proceed via the strange axial vector $K_1(1270)$ and $K_1(1400)$ resonances. As shown in Figure 5(a-d), we find that both contribute, with a preliminary fit favoring $K_1(1270)$ production, contrary to indications from TPC/2γ[25] in the all-charged channel, (although the admixture of $K^- K_S \pi^0$ events in our $h^- K_S \pi^0$ sample makes this difficult to quantify).

The resonant structure for the non-strange $[KK\pi]^-$ system is unknown. Our results support crudely a recent model based on chiral perturbation theory,[26] as shown in Figure 5(e-g) for the $\pi^- K_S K_S$ event sample. However the branching fraction for the $K^- K^0 \pi^0$ channel and indications for substructure containing K^{*0} and K^{*-} in this channel are at variance with the predictions of this model.

5. Conclusion

To conclude, CLEO-II has obtained new results for many semi-hadronic τ decay channels. Many aspects of these processes are for the first time being tested through precision measurements of branching fractions and studies of spectral functions and resonant substructure. We anticipate many more results from CLEO in years to come.

6. Acknowledgements

The CLEO collaboration acknowledges the effort of the CESR staff in providing excellent luminosity and running conditions. This work was supported by the U.S. Dept. of Energy and the National Science Foundation.

References

1. Y.-S. Tsai, *Phys. Rev. D* **4**, 2821 (1971).
2. Y. Kubota *et al.* (CLEO Collaboration), *Nucl. Instrum. Methods A* **320**, 66 (1992).
3. F.J. Gilman and S.H. Rhie, *Phys. Rev. D* **31**, 1066 (1985).
4. L. Montanet *et al.* (Particle Data Group), *Phys. Rev. D* **50**, 1173 (1994).
5. M. Procario *et al.* (CLEO Collaboration), *Phys. Rev. Lett.* **71**, 1791 (1993).
6. M. Artuso *et al.* (CLEO Collaboration), *Phys. Rev. Lett.* **69**, 3278 (1992).
7. D. Bortoletto *et al.* (CLEO Collaboration), *Phys. Rev. Lett.* **71**, 1791 (1993).
8. J. Kühn and A. Santamaria, *Z. Phys. C* **48**, 443 (1990).
9. W. Marciano, in *Proceedings of the Second Workshop on Tau Lepton Physics*, September 1992, ed. K.K. Gan, World Scientific (1993).
10. M. Artuso *et al.* (CLEO Collaboration), *Phys. Rev. Lett.* **72**, 3762 (1994).
11. R. Akers *et al.* (OPAL Collaboration), *Phys. Lett. B* **328**, 207 (1994).
12. A. Kounine, presentation at XXIX Rencontres de Moriond, Electroweak Interactions and Unified Theories, Meribel, France, March 1994.
13. S. Snow, in *Proceedings of the Second Workshop on Tau Lepton Physics*, September 1992, ed. K.K. Gan, World Scientific (1993).
14. M. Battle *et al.* (CLEO Collaboration), Cornell Preprint CLNS 94/1273 (1994), to appear in *Phys. Rev. Lett.*
15. S.I. Eidelman and V.N. Ivanchenko, *Phys. Lett. B* **257**, 437 (1991).
16. S. Narison and A. Pich, *Phys. Lett. B* **304**, 359 (1993).
17. For further details, see B. Barish *et al.*, GLS0249, CLEO-CONF 94-22, contributed to the XXVII International Conference on High Energy Physics, Glasgow, Scotland, July 1994.
18. KORALB 2.1/ TAUOLA 1.5: S. Jadach and Z. Was, *Comput. Phys. Commun.* **36**, 191 (1985) and *ibid*, **64**, 267 (1991); S. Jadach J.H. Kühn, and Z. Was, *Comput. Phys. Commun.* **64**, 275 (1991), *ibid*, **70**, 69 (1992), *ibid*, **76**, 361 (1993).
19. R. Brun *et al.*, GEANT 3.15, CERN DD/EE/84-1.
20. See contribution from Tomasz Skwarnicki in these proceedings.
21. D. Gibaut *et al.* (CLEO Collaboration), Cornell Preprint CLNS 94/1284 (1994), submitted to *Phys. Rev. Lett.*
22. D. Buskulic *et al.* (ALEPH Collaboration), *Phys. Lett. B* **332** 219, (1994).
23. For further details, please see the following papers contributed to the XXVII International Conference on High Energy Physics, Glasgow, Scotland, July 1994: M. Athanas *et al.*, GLS0460, CLEO-CONF 94-23; J. Gronberg *et al.*, GLS0242, CLEO-CONF 94-24; and R. Balest *et al.*, GLS0455, CLEO-CONF 94-25.
24. T. Das, V.S. Mathur, and S. Okubo, *Phys. Rev. Lett.* **18**, 761 (1967).
25. D.A. Bauer *et al.* (TPC/2γ Collaboration), *Phys. Rev. D* **50**, R13 (1994).
26. R. Decker, E. Mirkes, R. Sauer, and Z. Was, *Z. Phys. C* **58**, 445 (1993).

A MEASUREMENT OF EXCLUSIVE BRANCHING FRACTIONS
OF HADRONIC ONE-PRONG TAU DECAY MODES

JAYANT SHUKLA *

Department of Physics, Carnegie Mellon University, 5000 Forbes Ave.,
Pittsburgh, PA 15213, U.S.A.

ABSTRACT

We have measured the branching fractions for the hadronic τ decays, $\tau \to \pi/K\, n\pi^\circ\, \nu (0 \le n \le 3)$, with the L3 detector at LEP. The results are: $\mathrm{BR}(\tau \to \pi/K\, \nu) = (12.18 \pm 0.26 \pm 0.42)\%$, $\mathrm{BR}(\tau \to \pi/K\, \pi^\circ\, \nu) = (25.20 \pm 0.35 \pm 0.50)\%$, $\mathrm{BR}(\tau \to \pi/K\, 2\pi^\circ\, \nu) = (8.88 \pm 0.37 \pm 0.42)\%$, $\mathrm{BR}(\tau \to \pi/K\, 3\pi^\circ\, \nu) = (1.70 \pm 0.24 \pm 0.40)\%$, where the first error is statistical and the second is systematic.

1. Introduction

Almost 20 years after the discovery of the τ lepton[1] its hadronic decay modes are still a subject of debate.[2] It is not clear whether all hadronic τ decays have actually been observed experimentally or whether the measured branching fractions allow non-standard decay modes.

LEP experiments are well suited to study τ decays. The high center-of-mass energy facilitates the selection of τ-pair events and the rejection of background from hadron events. In addition, the high luminosity of LEP and the large τ-pair production cross section at the Z° pole make a high statistics study possible. In particular, the high resolution and fine granularity of the L3 electromagnetic calorimeter provide a clean separation of multiphoton decay channels. The sample corresponds to an integrated luminosity of 21.68 pb^{-1} collected at $\sqrt{s} = M_Z$.

2. Selection

To select the one-prong hadronic tau decay mode, we first select l^+l^- events by requiring low multiplicity and back-to-back topology. Every event is then divided into two hemispheres by a plane perpendicular to the thrust axis. All hemispheres with lepton(e or μ) signatures are rejected for further analysis.

2.1. neutral reconstruction

The shower caused by the tau decay in the electromagnetic calorimeter is then analyzed in terms of one charged hadron and a variable number of neutrals. First, the average hadronic shower energy is subtracted from each crystal in the cluster, as predicted on the basis of the impact point given by the central tracking chamber and the energy measured in the impact crystal. The rest of the shower is then distributed

---

*Representing the L3 collaboration.

to five or fewer electromagnetic showers, with a profile as predicted by the L3 data and energies corresponding to local maxima. Leakage from electromagnetic showers into hadronic showers and vice versa are then recalculated. The procedure is iterated, varying the energies and centers of the shower components, until convergence is reached with an optimum description of the total cluster. There are 12757 tau decays without any neutrals, 6801 with one neutral, 3778 with 2 neutrals 1034 with 3 neutrals, 309 with 4 neutrals and 97 with 5 neutrals in the data sample.

2.2. selection using neural networks(NN)

The identification of the decay modes with one or more neutrals was carried out by three layered feed-forward and backward-propagation neural networks.[3]

The input variables to the networks were:

- energy variables describing the whole hemisphere.

- variables describing the shower profile:

- the measured invariant masses between all pairs of particles, charged and neutral:

Each of the four networks was trained on Monte Carlo samples of definite τ-decay channels. The training sample of 250k $\tau^+\tau^-$ events was generated with the KORALZ program.

As examples of the neural network input, Fig. 1 shows the distributions in the energy of the charged particle, the neutral particles and the energy in the electromagnetic calorimeter transverse to the charged particle direction. The Monte Carlo description of details of the shower shape is in good agreement with data. In the four experimental categories h, h π°, h $2\pi^\circ$ and h $3\pi^\circ$, we find 3109, 6613, 1060 and 293 decays respectively. The selection efficiencies are shown in Table. 1. The distribution of the total hadronic invariant mass in the last three categories is shown in Fig. 2 together with the respective Monte Carlo result for signal and background after adjustment of the branching fractions.

3. Branching ratio calculation

Branching fractions are determined using the relation

$$N_{exp}^{i} = N_\tau \sum_{j=1,9} \mathrm{BR}_j \epsilon_j^i + \sum_{k=1,2} N_k^{\mathrm{bg}} \epsilon_k^i. \tag{1}$$

The index i runs over the four decay catagories, j over the nine τ decay channels considered as specified in Table. 1 and k over the main background channels. ϵ is the respective detection efficiency as specified in Table. 1. N_{exp} is the expected number of events in an experimental category; N_τ is the total number of τ decays. N^{bg} is the number of background events after the initial selection. The branching fractions BR_j are determined with a maximum likelihood fit comparing the predicted numbers to the

Source	Classification Efficiency (%)			
	h	h $\pi°$	h $2\pi°$	h $3\pi°$
$\pi/K\ \nu$	69.05	3.96	0.36	0.04
$\pi/K\ \pi°\ \nu$	3.64	78.64	1.49	0.18
$\pi/K\ 2\pi°\ \nu$	0.37	12.33	33.13	4.79
$\pi/K\ 3\pi°\ \nu$	0.15	2.81	12.38	28.58
$\pi/K\ 4\pi°\ \nu$	-	1.42	6.03	40.78
$\pi/K\ \eta\pi°\ \nu$	-	2.61	7.48	34.70
$e\ \nu\ \nu$	1.18	0.93	0.24	-
$\mu\ \nu\ \nu$	0.80	0.61	0.02	-
3 prong	0.08	0.21	0.06	-
e^+e^-	0.03	0.004	-	-
$\mu^+\mu^-$	0.26	0.04	-	-

Table 1. Efficiencies to classify a given final state. These efficiencies are determined within the fiducial region by a Monte Carlo sample of events.

observed numbers in the four experimental categories. No constraint was put on the branching fraction sum. The results are:

$$BR(\pi/K\ \nu) = (12.18 \pm 0.26 \pm 0.42)\%$$
$$BR(\pi/K\ \pi°\ \nu) = (25.20 \pm 0.35 \pm 0.50)\%$$
$$BR(\pi/K\ 2\pi°\ \nu) = (8.88 \pm 0.37 \pm 0.42)\%$$
$$BR(\pi/K\ 3\pi°\ \nu) = (1.70 \pm 0.24 \pm 0.40)\%$$

The first error is statistical and the second is systematic.

4. Errors

Contributions to the systematic error for each decay channel are listed in Table 2. We vary all cuts by $\pm10\%$ and take the maximum observed change in the branching fraction as the contribution to the systematic error.

The dominant contribution to the systematic errors come from the procedure identifying the final states and from the efficiency determination. A first part in this error is taken by the statistical uncertainties of the efficiencies predicted by the Monte Carlo program. The lower limit on the energy of neutral culsters (400 - 500 MeV) used by the shower analysis program adds a small uncertainty.

The systematic uncertainty due to the neural network selection is estimated by varying the cut on each neural network's output by 10%. In addition, we identify the input variables which have the largest influence on the neural net's decisions by applying a shift or a scale factor to the input quantities one by one.

Systematic errors coming from the $\rho°$ substructure to the $\pi/K\ 3\pi°\ \nu$ modes not modelled correctly by the Monte Carlo event generator, and the uncertainity in the branching fraction of h $4\pi°$ and h $\eta\ \pi°$, where η decays into neutrals, are also taken into account.

Source	$\Delta BR(\pi/K\ \nu)$	$\Delta BR(\pi/K\ \pi^\circ\ \nu)$	$\Delta BR(\pi/K\ 2\pi^\circ\ \nu)$	$\Delta BR(\pi/K\ 3\pi^\circ\ \nu)$
$\tau\tau$ selection	0.14	0.20	0.11	0.17
N_τ	0.10	0.22	0.08	0.02
MC statistics	0.18	0.26	0.18	0.15
E_n threshold	0.032	0.065	0.065	0.095
NN output	0.04	0.13	0.22	0.18
NN input	0.29	0.28	0.22	0.20
h selection cuts	0.16	-	-	-
h $4\pi^\circ$	-	-	0.04	0.15
h $\eta\ \pi^\circ$	-	-	0.02	0.04
Form factor	-	-	0.18	0.085
total	0.42	0.50	0.42	0.40

Table 2. Breakdown of the systematic error (in %) on the four measured branching fractions. The sources of systematic error are explained in the text.

5. Conclusions

The branching fractions for the hadronic τ decays, $\tau \to \pi/Kn\pi^\circ\nu$ with n between zero and three, were measured with the L3 detector at LEP. Multiphoton final states are identified by a neural network method. Results are in good agreement with the current world averages.[5]

References

1. M. Perl *et al.*, Phys. Rev. Lett. **35** (1975) 1489.
2. For a comprehensive review see: K.G. Hayes in Particle Data Group, K. Hikasa *et al.*, Phys. Rev. **D 45** (1992) VI.19, and references therein.
3. V. Innocente, Y.F. Wang and Z.P. Zhang, Nucl. Instr. and Meth. **A323** (1992) 647.
4. L3 collab.. B. Adeva *et al.*, Z. Phys. C51 (1991) 179;
   L3 collab.. M. Acciarri *et al.*, to be published in Phys. Lett. B. CERN preprint CERN-PPE/94-45; March 14, 1994.
5. Particle Data Group, K. Hikasa *et al.*, Phys. Rev. **D 45** (1992).

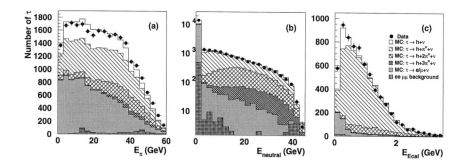

Fig. 1. Some of the input variables to the neural networks.

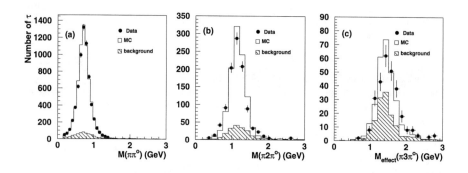

Fig. 2. The invariant mass distributions for selected events.

SUMMARY OF τ BRANCHING RATIO RESULTS FROM DELPHI

D. M. EDSALL*

Ames HEP Group, Iowa State Univeristy,
12 Physics, Ames,Iowa 50011, USA

ABSTRACT

On a sample of 25000 $Z \to \tau^+\tau^-$ observed in the DELPHI detector at LEP in the period 1991-1992, measurements are made of the τ decay branching ratios for the exclusive channels $\tau \to \pi\nu, K\nu, \rho\nu, e\nu\bar{\nu}, \mu\nu\bar{\nu}$ and $a_1\nu$ and the inclusive channels $\tau \to K\nu \geq 0$ *neutrals* and $\tau \to K\nu \geq 1$ *neutrals*. In addition, tau decay into K^{*-} was studied in the channel $K^{*-} \to K^-\pi^0$. The charged kaons were identified over a large momentum range by the DELPHI barrel Ring Imaging Cherenkov detector.

1.   Introduction

At $\sqrt{s} \simeq 91$ GeV, the reaction e^+e^- to a fermion-antifermion pair is dominated by Z^0 exchange. The clean signature of the $Z \to \tau^+\tau^-$ decay at LEP makes this reaction well suited to do precise measurements of both exclusive and topological branching ratios due to the low background levels and good particle identification of the DELPHI subdetectors.

In addition, the τ decays involving kaons probe the W coupling to the weak hadronic current. While these decays are strongly suppressed by the Cabibbo angle in weak interactions, this is generally not expected for new superweak interactions. The branching ratio BR($\tau \to K\nu_\tau$) would be sensitive to extensions of the Standard Model that violate lepton universality. The present measurements exploit the charged kaon identification capability of the Ring Imaging Cherenkov (RICH) detector over a large momentum range.

2.   The DELPHI Detector

The DELPHI detector has been described in detail elsewhere.[1] All of the analyses contained in this paper were done in the polar region defined by $43° < \theta < 137°$ and known as the *barrel* region. Tracking was performed by a set of cylindrical drift chambers known as the microvertex detector (VD), the inner detector (ID), the Time Projection Chamber (TPC), the outer detector (OD) and the muon chambers. The momentum resolution was measured to be $\sigma_p/p = 0.0008 \times p(GeV/c)$. The charged kaons were identified by the barrel Ring Imaging Cherenkov (barrel RICH) detector which surrounds the TPC.

---

*representing the DELPHI Collaboration

Electromagnetic showers were reconstructed in the High Density Projection Chamber (HPC) in the barrel region. The energy resolution was studied using electromagnetic showers from Bhabha events, Compton electrons and $\mu^+\mu^-\gamma$ radiative events. The measured energy resolution for photons of energy E is $\sigma_E/E = 0.29/\sqrt{E(GeV)} \oplus 0.04$. Hadronic energies were measured with the hadron calorimeter (HCAL), which surrounds the coil of the magnet.

3.   Event Selection

The following sections describe the criteria imposed to select candidates for each tau decay mode discussed here. All of the channels studied (with the exception of the exclusive kaon channel) made a selection of candidates from an 87% pure $\tau^+\tau^-$ sample. A description of how the $\tau^+\tau^-$ selection was done has been described in.[2] The a1 channel[3] and the exclusive and inclusive charged kaon channels[4] have been described elsewhere and thus only the results are included here.

In all analyses, samples of simulated events were used which had been passed through a detailed simulation of the detector response and reconstructed with the same program as the real data. The Monte Carlo event generators used were: KORALZ[5] for $e^+e^- \to \tau^+\tau^-$ events; DYMU3[6] for $e^+e^- \to \mu^+\mu^-$ events; BABAMC[7] for $e^+e^- \to e^+e^-$ events; Jetset 7.3[8] for $e^+e^- \to q\bar{q}$ events; Berends-Daverveldt-Kleiss[9] for $e^+e^- \to e^+e^-e^+e^-$ events; the generator described in[10] for $e^+e^- \to e^+e^-\mu^+\mu^-$, $e^+e^- \to e^+e^-\tau^+\tau^-$ events.

3.1. $\tau \to e\nu\bar{\nu}$

An electron candidate consisted of an isolated charged particle track with momentum greater than $0.01 \times p_{beam}$. To ensure a optimal performance of the HPC it was required that the track lie in the polar angle region $45° < \theta < 88°$ or $92° < \theta < 135°$, and that its extrapolation to the HPC be further than $1°$ from the centre of an HPC Φ boundary region.

It was required that the TPC dE/dx measurement be compatible with that of an electron by demanding that it lie within two standard deviations of the electron hypothesis from the negative side: $P_{dE/dx}(e) > -2$. This reduced the background from hadrons and muons at low momentum, with a low loss of signal. This was complemented by an **OR** of two independent sets of selection criteria based on the HPC and the TPC dE/dx respectively, ensuring a high identification efficiency over the full momentum range. The first required that for particles with a measured momentum greater than $0.05 \times p_{beam}$ the associated HPC energy had to be compatible with the momentum measurement within two standard deviations. The second required that the TPC dE/dx signal lie more than three standard deviations above that expected for a pion: $P_{dE/dx}(\pi) > 3$. This was applied only to tracks with a momentum less than $0.5 \times p_{beam}$.

In order to remove a residual background from hadronic tau decays it was required that the particle have no muon chamber hits and no associated energy in the HCAL beyond the first layer. Furthermore there should be no neutral HPC shower in a $18°$ cone about the track with an energy greater than 4 GeV. Neutral showers compatible with a bremsstrahlung photon were not included in this cut.

These cuts had an identification efficiency of 95% within the angular and momentum acceptance. This efficiency was constant within 2% over the full electron energy range. The background from other τ decays, primarily the $\tau \to \rho\nu$ channel, was found to be $2.2 \pm 0.5\%$

Most $e^+e^- \to e^+e^-$ events were rejected already with the event acollinearity cut. Remaining Bhabha contamination was removed using cuts on the opposite hemisphere to the identified decay. These cuts were dependent on the value of a variable X_{el}, constructed from the a combination of the momentum measurement of the charged track and the electromagnetic associated energy of the identified electron. If X_{el} was less than $0.7 \times E_{beam}$ for the identified electron, the total energy in a cone of half-angle $30°$ about the track had to be less than $0.8 \times E_{beam}$; otherwise the cone energy had to be less than $0.7 \times E_{beam}$ and the momentum of the highest momentum track had to be less than $0.7 \times p_{beam}$.

The selection efficiency for $\tau \to e\nu\bar{\nu}$ decays after the Bhabha rejection cuts was 85%, with a background of $1.5\pm0.5\%$ from Bhabha events and $1.5\pm0.3\%$ from $e^+e^- \to e^+e^-e^+e^-$ events. The selected sample consisted of 5437 candidate decays.

3.2. $\tau \to \mu\nu\bar{\nu}$

In order to identify such a decay it was required that there be only one charged particle track in a hemisphere and that it be able to penetrate to the outside of the DELPHI magnet iron. Thus the charged particle had to have a momentum greater than $0.067 \times p_{beam}$ and lie in the polar angle region $43° < \theta < 88°$ or $92° < \theta < 137°$. To positively identify the particle as a muon it was required that it have an associated hit in the muon chambers or deposited energy in the outer layer of the HCAL. Further rejection of τ decays containing a high energy hadron whose showers penetrated deep into the HCAL was performed by demanding that average energy deposited per layer of the hadron calorimeter be less than 3 GeV. Decays containing a hadron which was associated with a large hadronic shower in the HPC were rejected by the cut $E_{ass} < 3$ GeV, which was very efficient for muons. By demanding that the neutral electromagnetic energy in a cone of half-angle $30°$ about the track be less than 1 GeV, both the $\tau \to \rho\nu$ contamination and that from $\mu^+\mu^-\gamma$ events were reduced.

The efficiency of the muon identification in the angular and momentum acceptance was 95%.

Contamination from cosmic ray events were reduced by using tightening the standard impact parameter cuts.

Background from $\mu^+\mu^-$ events remaining after the event acollinearity cut were rejected by demanding that p_{rad}, as defined in,[2] be less than $1.2 \times p_{beam}$. Where the highest momentum track in the other hemisphere had muon chamber hits or energy deposition in the outer layers of the HCAL it was required that the maximum momentum of any charged track in the event be less than $0.7 \times p_{beam}$. This gave a step in the efficiency at $0.7 \times p_{beam}$ which was well controlled with data test samples.

The number of candidate τ decays remaining after these cuts was 6617. The overall efficiency to identify a $\tau \to \mu\nu\bar{\nu}$ decay inside the angular and momentum acceptance was 88%. The background was composed of $3.4\pm0.3\%$ from other τ decays, $0.5\pm0.1\%$ from $\mu^+\mu^-$ events, $0.6\pm0.1\%$ from $e^+e^- \to e^+e^-\mu^+\mu^-$ events and $0.4\pm0.1\%$ from cosmic rays.

3.3. $\tau \rightarrow \pi(K)\nu$

In order to perform efficient muon rejection it was required that the isolated charged track lie in the polar angle region $53° < \theta < 88°$ or $92° < \theta < 127°$ and that it have a momentum greater than $0.05 \times E_{beam}$.

Muon rejection was then performed using the muon chambers and cuts on the energy deposition in the HCAL which were optimised to account for the fluctuations in the penetration of the pion in the HCAL.

Rejection of electrons was performed with the HPC and with the TPC dE/dx. It was required that the energy deposition in first 4 layers of the HPC be less than 350 MeV. This cut was highly efficient for electron rejection except in the cracks of the HPC.

In order to remove background from other hadronic decays of the τ containing π^0s, it was required that there be no unassociated energy deposition in the HPC within $18°$ of the particle track.

The remaining backgrounds due to $\mu^+\mu^-$ and e^+e^- events were removed using cuts on the opposite hemisphere to the identified candidate decay. The momentum of the highest momentum charged particle in the opposite hemisphere had to be less than $0.75 \times p_{beam}$ and its associated energy in the HPC had to be less than $0.75 \times E_{beam}$.

The estimated efficiency within the fiducial region was 59%. The selected sample contained 2503 candidate decays with a total background from other τ decays of $8.6 \pm 2.5\%$. This included 2.5% of $\nu\pi K_L^0$ events from $\nu K^*(892)$ decays which were subtracted from the distribution, as were the other backgrounds. Background from other sources, primarily $e^+e^- \rightarrow e^+e^-$ events, was $1.4 \pm 0.4\%$.

3.4. $\tau \rightarrow \rho\nu$

The τ decay to a charged rho was selected by requesting an isolated charged track with an associated π^0 candidate. The charged track had to be incompatible with being an electron.

Candidate π^0s were subdivided in four different classes dependent on their energy and the number of separate showers due to photons expected in the calorimeter. The first class consisted of events with two showers of energy E_1 and E_2 with

$$2.5 \ GeV < E_1 + E_2 < 10 \ GeV$$

and an angle greater than $1°$ between the photons and the charged track. The reconstructed invariant mass had to lie in the range 0.04 GeV/c^2 to 0.25 GeV/c^2. The second class contained events having one shower with energy greater than 5 GeV and with an angle to the charged track of less than $1°$. The third class selected events having two showers with

$$E_1 + E_2 > 10 \ GeV.$$

In this case the second shower is generally from either a hadronic interaction of the charged pion or due to a secondary associated with the main shower. Only the highest energy shower is used in the calculation of the ρ invariant mass and ξ variable and an additional cut is used to reduce contamination from other channels:

$$\frac{E_1}{E_1 + E_2} > 0.85.$$

The fourth class contained events with two neutrals, including at least one converted photon, having a reconstructed invariant mass in the range 0.04 GeV/c² to 0.25 GeV/c².

To reduce background it was required that the reconstructed ρ invariant mass lie in the range 0.48 GeV/c² to 1.20 GeV/c². The sample remaining after the cuts contained 5903 τ decays. The overall efficiency inside the acceptance was 37%. The background consisted of 16% originating from other τ decays consisting primarily of: $\pi 2\pi^0 \nu$ (10.7%); $\pi n\pi^0 \nu, n > 2$ (2.0%); $K\pi^0 \nu$ (1.8%); $\pi\nu$ (1.1%). Contamination from non-τ sources was 1.1%.

4. τ Branching Ratios

Table 1 gives a list of the different exclusive and inclusive τ decay Branching Ratios:

Table 1. τ exclusive branching ratio measurements. The first error is statistical, the second is systematic. For the Kaon channels, the statistical and systematic errors have been added in quadrature.

	Branching Ratio (%)	Events	Data
$e\nu\nu$	$17.61 \pm 0.22 \pm 0.40$	6011	1990+91+92
$\mu\nu\nu$	$17.47 \pm 0.19 \pm 0.27$	7154	1990+91+92
$\pi/K\nu$	$11.85 \pm 0.26 \pm 0.49$	2090	1990+92
$\rho\nu$	$23.82 \pm 0.32 \pm 0.61$	6597	1990+91+92
$3\pi\nu$	$8.35 \pm 0.35 \pm 0.24$	585	1991
$K\nu$	0.85 ± 0.18		1992
$K\nu X$	1.54 ± 0.24		1992
$K\nu \geq 1neutral$	0.69 ± 0.25		1992
$K^* \to K^- \pi^0 \nu$	0.57 ± 0.23		1992
$K_s^0 X$	0.99 ± 0.19		1991+92
$K^* \to K_s^0 \pi$	0.58 ± 0.14		1991+92

The decay to $\pi\nu$ includes both charged pions and kaons.

In the decay to $\rho\nu$ the contribution from final states with charged or neutral kaons are treated as background and have been subtracted from the branching ratio. The background from $K^*\nu$ was evaluated with KORALZ and amounted to 1.8% of the $h^-\pi^0\nu$ decays.

5. Summary

The branching ratios presented for the charged kaon channels are final. All other results are **preliminary**. All results appear to be consistent with Standard Model predictions. In addition, the use of the barrel RICH subdetector in DELPHI has allowed a more precise measurement to be made of the inclusive and exclusive charged kaon than the previous world average.

References

1. DELPHI Collaboration, *Nucl. Instr. and Meth.* **A303** (1991) 233.
2. DELPHI Collaboration,*Nucl. Phys.* **B367** (1991) 511-574.
3. P. Privitera, Doctoral thesis,
   preprint *IEKP-KA* **93-01**, University of Karlsruhe, January 1993.
4. DELPHI Collaboration, P.Abreu et al., *CERN PPE* **94-88**
5. S.Jadach and Z.Was, *Comp. Phys. Com.* **36** (1985) 191.
   S.Jadach, B.F.L.Ward and Z.Was, *Comp. Phys. Com.* **66** (1991) 276.
6. J. E. Campagne and R. Zitoun, *Z. Phys.* **C43** (1989) 469.
   Proc. Brighton Workshop on Radiative Corrections, Sussex, (July 1989)
7. F. A. Berends, W. Hollik and R. Kleiss, *Nucl. Phys.* **B304** (1988) 712.
8. T.Sjöstrand, *Comp. Phys. Comm.* **27** (1982) 243,
   ibid. **28** (1983) 229.
   T.Sjöstrand and M. Bengtsson, *Comp. Phys. Comm.* **43** (1987) 367.
   T.Sjöstrand, PYTHIA 5.6 JETSET 7.3 Physics and manual,
   preprint *CERN-TH* **6488/92** (1992).
9. F.A.Berends, P.H.Daverveldt, R.Kleiss, *Phys. Lett.* **B148** (1984) 489,
   Comp. Phys. Comm. **40** (1986) 271.
10. T. Todorov, doctoral thesis, CRN Strasbourg, preprint *CRN/HE* **94-21**, 1994.
11. J. H. Kühn, A. Santamaria, Z. Phys. C48 (1990) 445.

FINAL RESULT ON THE MASS OF THE TAU LEPTON FROM THE BES COLLABORATION

ERIC SODERSTROM[*†]

Stanford Linear Accelerator Center
Stanford, CA 94309, USA

ABSTRACT

A data-driven energy scan in the immediate vicinity of τ pair production threshold has been performed using the Beijing Spectrometer (BES) at the Beijing Electron-Positron Collider (BEPC). Approximately 5 pb^{-1} of data, distributed over twelve scan points, have been collected. An initial τ lepton mass value, obtained using only the $e\mu$ final state, has been published. In this paper, the final BES result on the mass measurement, based on the combined data samples of ee, $e\mu$, $e\pi$, eK, $\mu\mu$, $\mu\pi$, μK, $\pi\pi$ and πK events, is presented. A maximum likelihood fit yields the value $m_\tau = 1776.96 \, ^{+0.18}_{-0.19} \, ^{+0.20}_{-0.16}$ MeV.

The measurement of the mass of the tau lepton by BES is based on a data-driven energy scan within 25 MeV of the threshold for $e^+e^- \to \tau^+\tau^-$ production. During the scan, $\tau^+\tau^-$ events were identified by means of the $e\mu$ topology in which one tau decays via $\tau \to e\nu\bar{\nu}$ and the other via $\tau \to \mu\nu\bar{\nu}$, and the mass value obtained from a likelihood fit to the 14 observed $e\mu$ events was $m_\tau = 1776.9 \, ^{+0.4}_{-0.5} \pm 0.2$ MeV.[1] This paper presents a new result based on a much larger event sample which combines events from the ee, $e\mu$, $e\pi$, eK, $\mu\mu$, $\mu\pi$, μK, $\pi\pi$ and πK channels, where the $\pi(K)$ result from the decay $\tau \to \pi(K)\nu_\tau$.

The experimental details are described in Ref. 1. The total data sample consists of 10^7 event triggers corresponding to 5 pb^{-1} of integrated luminosity. The initial selection of $\tau^+\tau^-$ candidates is effected by means of simple topological and particle identification criteria.

It is required that exactly two charged tracks be well reconstructed, without regard to net charge. For each track, the point of closest approach to the beam line should have radius ≤ 1.5 cm and $|z| \leq 15$ cm where z is measured along the beam line from the nominal beam crossing point; in addition $|z_1 - z_2|$ must be less than 5 cm. Furthermore, each track is required to satisfy $|cos\theta| \leq 0.75$, where θ is the polar angle, to ensure it is contained within the cylindrical barrel region of the detector. These criteria reduce the data sample by a factor of ~ 20.

Next it is required that the transverse momentum of each track be above the 100 MeV/c minimum needed to traverse the barrel time-of-flight counter and reach

---

[*]Representing the Beijing Spectrometer Collaboration.
[†]Work supported by the Department of Energy, contract DE-AC03-76SF00515.

the outer radius of the barrel shower counter in the 0.4 Tesla axial magnetic field. In addition, the magnitude of the momentum must be less than the maximum expected in any tau decay at the given center-of-mass energy within a tolerance of 3 standard deviations in momentum resolution. These constraints on momentum together reduce the data sample by over an order of magnitude leaving $\sim 40,000$ events. Most of this reduction is due to the removal of Bhabha and μ pair events.

A further requirement is that there be no isolated photon present with measured energy greater than 60 MeV and making an angle of greater than $12°$ with the original direction of each of the charged tracks. This reduces the data sample to ~ 33000 events.

The particle identification procedure is applied to the remaining events. For each mass hypothesis (e, μ, π, K, or p) for each track, the measured momentum is used to predict the expected values of dE/dx, time-of-flight and shower counter energy. The corresponding measured quantities and resolutions are then used to create an overall χ^2 value, which is converted to a confidence level using the number of contributing devices as the number of degrees of freedom. Events are rejected for which a positively charged track has $CL(p) \geq 5\%$ (this removes 3441 events due mainly to beam-gas interactions). Next, the μ hypothesis is assigned to a track if $CL(\mu) \geq 5\ \%$, $|\vec{P}| \geq 0.5\ GeV/c$ and there are corroborating muon counter hits. Failing the muon requirement, a track is assigned the one of the e, π, and K hypotheses with the highest confidence level, provided it is at least 5%. For the π or K assignments it is further required that $|\vec{P}|$ be consistent with 2-body tau decay at the three sigma level; for the calculation of the relevant momentum limits, m_τ is taken 1.0 MeV $below$ the previous measurement[1] to make the $\tau \to \pi$, K selection efficiencies independent of center-of-mass energy and to avoid biasing the new mass measurement.

After the particle identification procedures there remain 12571 ee, 1340 $\mu\mu$ and 127 events in the other channels. The ee sample results predominantly from two-photon $e^+e^- \to e^+(e^-e^+)e^-$ events for which the leading e^+ and e^- in the final state are undetected. The $\mu\mu$ sample similarly results from $e^+e^- \to e^+(\mu^+\mu^-)e^-$ events with additional contributions from cosmic rays. These background events are characterized by small observed net transverse momentum and, for the QED events, large missing energy. The $\tau^+\tau^-$ final states are separated from these sources of background by requiring that the ratio of total transverse momentum to the maximum possible missing energy be greater than 0.4 for ee, 0.1 for $e\mu$, and 0.2 for all other channels. The final event sample consists of 64 events distributed as 4 ee, 18 $e\mu$, 19 $e\pi$, 2 eK, 3 $\mu\mu$, 5 $\mu\pi$, 3 μK, 4 $\pi\pi$ and 6 πK.

A maximum likelihood fit is performed to find the values of the mass, m_τ, of the τ lepton, the overall absolute efficiency, ϵ, for identifying $\tau^+\tau^-$ events, and the effective background cross section, σ_B, which is assumed constant over the limited range of center-of-mass energy involved. By fitting the parameter, ϵ, uncertainties in the luminosity scale and trigger and detector efficiencies are implicitly taken into account. The likelihood function is the product over the twelve scan points of the Poisson probability for the observed number of events at each point, where the expected number of events, μ, is given by

$$\mu = (\epsilon \times \sigma(W, m_\tau) + \sigma_B) \times L(W)\ ;$$

here L is the integrated luminosity, and the cross section, $\sigma(W, m_\tau)$, for $\tau^+\tau^-$ production is corrected for Coulomb interaction, initial and final state radiation, vacuum polarization, and the spread ($\sigma \cong 1.4 MeV$) in center-of-mass energy, W.

The parameter values resulting from the fit are $m_\tau = 1776.96$, $\epsilon = 4.14\%$, and $\sigma_B = 0$. The statistical uncertainty of m_τ is found by setting $\epsilon = 4.14\%$, $\sigma_B = 0$, and integrating the likelihood function to find the 68.3% confidence level interval, with result $m_\tau = 1776.96^{+0.18}_{-0.19}$ MeV. The cross section curve which results from this mass value is plotted over the data in Figs. 1(a), and 1(b); the corresponding likelihood function is shown in Fig. 1(c).

The uncertainty in the efficiency, $\Delta\epsilon = ^{+0.54}_{-0.50}$, contributes systematic errors $\Delta m_\tau = ^{+0.08}_{-0.09}$ MeV to the mass of the τ, and that in the background cross section, $\Delta\sigma_B = +0.61$ pb (which corresponds to a 1σ background level of 3.1 events), contributes systematic error $\Delta m_\tau = +0.12$ MeV; possible bias in the scanning procedure contributes $\Delta m_\tau = \pm 0.10$ MeV, as estimated from Monte Carlo simulations; and uncertainties in the center-of-mass energy scale and energy spread yield $\Delta m_\tau = \pm 0.09$ MeV and $\Delta m_\tau = \pm 0.02$ MeV respectively.

Combining the systematic errors in quadrature, the final BES result on the mass of the τ lepton is

$$m_\tau = 1776.96\ ^{+0.18}_{-0.19}\ ^{+0.20}_{-0.16}\ \text{MeV}$$

where the first errors are statistical, and the second systematic.

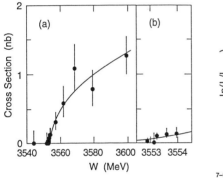 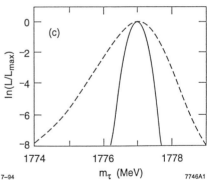

Fig. 1. (a) The center-of-mass energy dependence of the $\tau^+\tau^-$ cross section resulting from the likelihood fit (curve), compared to the data (Poisson errors). It should be emphasized that the curve does not result from a direct fit to these data points. (b) An expanded version of (a), in the immediate vicinity of $\tau^+\tau^-$ threshold. (c) The solid curve shows the dependence of the logarithm of the likelihood function on m_τ, with the efficiency and background parameters fixed at their most likely values; the dashed curve shows the likelihood function from Ref. 1.

References

1. J.Z. Bai et al., Phys. Rev. Lett. **69** (1992) 3021.

PRELIMINARY MEASUREMENT OF THE TAU LIFETIME AT SLD

SARAH HEDGES

Physics Department, Brunel University, Kingston Lane, Uxbridge,
Middlesex, UB83PH, U.K.

REPRESENTING THE SLD COLLABORATION

ABSTRACT

A measurement of the tau lifetime has been made using a sample of 1638 $Z^0 \to \tau^+\tau^-$ decays collected by the SLD at SLC during 1992 and 1993. The measurement benefits from the small stable beam spot of SLC and the 3-D Vertex Detector of SLD. Two methods have been used, the vertex method for 3-prong tau decays and the impact parameter method for events with 1-1 topology. The measured tau lifetime is $285 \pm 10 \pm 4 fs$.

1.   Motivation

Over the last few years improved measurements of the tau lifetime have reduced the discrepancy between the experimentally determined value and that predicted by theory. These improvements have been mainly due to the use of micro-vertex detectors.[1]

The unique high resolution 3-d silicon Vertex Detector (VXD) of SLD, in combination with the small stable beam spot of SLC, reduce the uncertainties in the production and decay points of the tau particle, providing an excellent environment for measuring the tau lifetime.

2.   Detector and beam position resolution

The SLD is a 4π hermetic detector at the interaction region of SLC. This analysis makes use of the VXD, the Central Drift Chamber (CDC) and the Liquid Argon Calorimeter. Ninety-six percent of all well measured CDC tracks which fall within the VXD acceptance are associated to ≥ 1 hit in the VXD. The local errors $\sigma_{r-\phi}$ and σ_z for the VXD clusters are 7μm and 8μm respectively; when combined with the CDC angular errors, an impact parameter resolution of 11μm in the plane orthogonal to the beam and 38μm in the plane parallel to the beam is obtained for infinite momentum tracks.

During 1993 the $\langle rms \rangle$ xyz beam profile at the interaction point (IP) was $2 \times 0.8 \times 650\mu m^3$. The IP was tracked using hadronic Z^o events. Small impact parameter tracks from ~ 30 consecutive events were fitted to a common vertex to obtain the average transverse IP position.[2] Muon pairs were not used in the IP determination but were used as a check of the error on the transverse position, resulting in a σ_{xy} of $7\pm2\mu$m.

3.   Event selection

The tau-pair sample used for this analysis was obtained using a standard set of cuts based on event topology, calorimetry and track quality. The cuts were designed to reduce background from multihadrons, bhabhas, muon pairs and two- photon events. The efficiencies for event selection and background rejection were studied using Monte Carlo. The Monte Carlo generator KoralZ (version 3.8) was used to generate tau pairs which were then subjected to Geant-based SLD detector simulation. The efficiency for event selection is estimated to be 61% and the trigger efficiency for tau pairs to be 95%, giving a total efficiency of 58%. The major source of inefficiency results from the limit of coverage of the central tracking systems, 73% of the total solid angle. A sample of 1638 tau pair was obtained from the 60K Z^o events collected in 1992 and 1993. The purity of the sample is estimated to be 96%.

4.   Measurement techniques

4.1.   The vertex method

The vertex method uses the events in which one tau decays to 3 prongs and the other to 1 prong. The common vertex formed by the three tracks in one hemisphere determines the decay point of the tau and the nominal beam position is used as the production point. The flight distance, l, is then related to the lifetime by: $l=\beta\gamma c\tau_\tau$. Events in which there are tracks consistent with originating from a photon conversion are rejected. All tracks are required to have a link with the VXD. After all cuts 271 events remain. The only significant background in the sample is from multihadrons and is estimated to be 0.2%.

The average decay length is determined by performing a maximum likelihood fit to the data distribution (see Fig.1a), where the fit function is a convolution of an exponential with a Gaussian resolution function. The parameters of the fit are the decay length and a scale factor on the decay length error. The scale factor allows for uncertainties in the calculation of the error and ideally should be 1. The decay length error is calculated on an event-by-event basis and is a function of the beam position error and vertex resolution; the latter is the dominant source of error at SLD. The average decay length error is 494±11μm. From the fit the average decay length is 0.205cm and the scale factor is 0.93±0.12. After correction for backgrounds the decay length method gives a lifetime of 272±17±3 fs. Systematic errors due to biases in the method, reconstruction, backgrounds, detector alignment, initial and final state radiation and resolution have been studied and the total systematic error is conservatively estimated to be 1%.

4.2.   The impact parameter method

The impact parameter method has been used to determine the lifetime for events in which both taus decay to a 1 prong. The impact parameter, d, is the distance of closest approach of the track to the IP in the xy plane. It is related to the decay length l_{xy}: $d=l_{xy}\sin\Phi$, where Φ is the angle of the track with respect to the tau direction, in this case taken to be the thrust axis determined from charged tracks and unassociated neutral clusters. The impact parameter is assigned a positive (negative) value if the track crosses the tau direction before (after) the point of closest approach. The negative

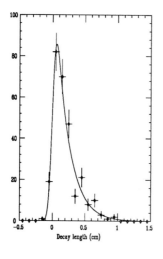
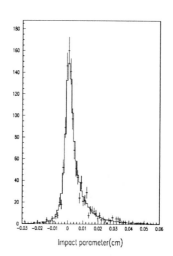

Fig. 1(a) The distribution of measured tau decay lengths in the data, over-layed with a maximum likelihood fit curve. (b) Data and Monte Carlo impact parameters for tracks from 1 prong tau decays.

impact parameters result from the finite resolution and uncertainties in both the beam position and the reconstruction of the tau direction. Additional track quality cuts are made: for example CDC tracks must have ≥ 50 hits and the momentum of the track must be ≥ 3GeV. The distance of closest approach in z is required to be ≤ 0.25cm. Following the cuts the number of tracks remaining is 1399. The purity of the sample is 96%. The dominant source of background is from muon pairs, 1.3%.

A multinodal log likelihood fit is performed to the data, in which the fit function is determined from the normalised Monte Carlo distribution, corrected for background, see Fig 1b. The effects of uncertainties in the thrust axis, background, fit procedure, detector alignment and beam position have been studied. The systematic error due to these effects is 1.7%. A tau lifetime of 292 $\pm 12 \pm$ 5 fs has been obtained with the impact parameter technique.

5. Combined result

The samples used for the two methods are independent and the results may be combined assuming no correlation to give a preliminary measurement of 285 $\pm 10 \pm 4 fs$.

References

1. E. Fernandez, *Proceedings of the International Europhysics Conference on High Energy Physics, Marseille, France, Jul 1993.*
2. SLD Collaboration, *SLAC-Pub 6569, Jul 1994.*

MEASUREMENT OF TAU LEPTON LIFETIME FROM ALEPH [†]

ALBERTO LUSIANI [*]

INFN & Scuola Normale Superiore, Pisa, Italy

ABSTRACT

The tau lepton lifetime has been measured from data collected by ALEPH in 1992. The lifetime is determined from the sum and the difference of impact parameters in events where both tau leptons decay in one charged prong; furthermore, the decay lenght is measured by reconstructing the decay vertex for the tau leptons that decay in three charged prongs. Including the previous ALEPH measurements from 1989 through 1991, the preliminary result is:

$$\tau_\tau = 292.5 \pm 2.8 \,(\text{stat.}) \pm 1.5 \,(\text{sys.}) \text{ fs} .$$

1.   Introduction

The ratio of the weak charged-current couplings for τ^- and μ^- may be derived from the reaction rates for $\tau^- \to e^- \nu \bar{\nu}$ and $\mu^- \to e^- \nu \bar{\nu}$ through the following relation:

$$\left(\frac{g_\tau}{g_\mu}\right)^2 = \left(\frac{m_\mu}{m_\tau}\right)^5 \left(\frac{\tau_\mu}{\tau_\tau}\right) \, B(\tau \to e\bar{\nu}_e \nu_\tau) * \delta \, , \tag{1}$$

where m_τ is the mass of the τ, m_μ and τ_μ are the muon mass and lifetime; $\delta = 0.9996$ is a small adjustment due to radiative corrections. The assumption of charged-current universality predicts therefore a relation between the electronic branching ratio of the τ and its lifetime, in terms of other more precisely measured quantities.

A few years ago, the above formula resulted in a coupling constant ratio two sigmas away from unity,[1] while more recent results are consistent with one.[2,3] The τ lifetime measurement contributes the largest uncertainty.

A new preliminary measurement is presented in the following, based on the 1992 ALEPH data sample and on several different analysis techniques.

2.   The ALEPH Detector

The ALEPH detector is described in detail elsewhere.[4,5] The tracking system consists of a silicon strip vertex detector (VDET) with double sided readout, a cylindrical drift chamber (the inner tracking chamber or ITC), and a large time-projection chamber (TPC). The VDET features two layers of 300 μm thick silicon wafers. Each layer provides measurements in both the r-ϕ and r-z views at average radii of 6.3 and 10.8 cm. The spatial resolution for r-ϕ coordinates is 12 μm and varies between 12

[†] Invited talk given at DPF94, University of New Mexico, Albuquerque, August 1994.
[*] Representing the ALEPH Collaboration.

and $22\,\mu$m for z coordinates, depending on track polar angle. The angular coverage is $|\cos\theta| < 0.85$ for the inner layer and $|\cos\theta| < 0.69$ for the outer layer. The ITC has eight axial wire layers at radii of 16 to 26 cm. The TPC provides up to 21 three-dimensional coordinates per track at radii between 40 and 171 cm. A superconducting solenoid produces a magnetic field of 1.5 T.

Charged tracks measured in the VDET-ITC-TPC system are reconstructed with a momentum resolution of $\Delta p/p = 6 \times 10^{-4} p\,(\text{GeV}/c)^{-1} \oplus 0.005$. An impact parameter resolution of $28\,\mu$m in the r-ϕ plane is achieved for muons from $Z^0 \to \mu^+\mu^-$ having at least one VDET r-ϕ hit.

3.   The $\tau^+\tau^-$ Data and Monte Carlo Samples

The data sample used in this analysis was collected in 1992 at $\sqrt{s} = 91.3$ GeV and corresponds to an integrated luminosity of $21.5\,\text{pb}^{-1}$, or 32100 produced τ pairs. Candidate tau pair events are selected according to an algorithm described elsewhere.[6] The overall efficiency for this selection is 78%, with an expected background contamination of 1.6%. The $\tau^+\tau^-$ sample contains 25679 candidate events to which further cuts are applied for the different analyses.

The KORALZ Monte Carlo program, version 3.8,[7] was used for the simulation of $\tau^+\tau^-$ events at $\sqrt{s} = 91.25$ GeV. 300,000 events were generated.

4.   Impact Parameter Sum Analysis

The tau lepton lifetime is derived by fitting the distribution of the sum of the impact parameters of the decay products in $Z^0 \to \tau^+\tau^-$ events where both tau leptons decay in one charged prong. The impact parameters are measured with respect to the beam axis and are signed like the z component of the particle's momentum about the same axis. The sum of the impact parameters of the two daughter tracks, $\delta = d_+ + d_-$, corresponds roughly to the distance between the two tracks at their closest approach to the beam axis and is relatively insensitive to the smearing associated with the beam size. In the ALEPH analysis, the daughter track directions and the event sphericity axis (as reconstructed from all charged and neutral particles) are used to approximate the τ decay angles in a maximum likelihood fit.

This analysis method has been used for previously published results,[8,9] and has been recently improved.[10] In particular, special care has been devoted to the understanding of the impact parameter sum measurement resolution. Real data events like $e^+e^- \to e^+e^-$, and $\gamma\gamma \to e^+e^-$ are now used in addition to $Z^0 \to \mu^+\mu^-$ and $\gamma\gamma \to \mu^+\mu^-$. A dedicated study of the Monte Carlo simulation accuracy for the non-gaussian resolution tails has been accomplished with $Z^0 \to q\bar{q}$ events.

Starting from the basic $\tau^+\tau^-$ sample, events of 1-1 topology are selected. A maximum likelihood fit is performed on the remaining 10464 events. The complete analysis is repeated on Monte Carlo events, including the resolution studies, to determine a number of systematic biases.

Several sources of systematic errors have been investigated. The uncertainty associated with the parametrization of the core of the impact parameter resolution in real data events is $\pm0.37\%$, while the uncertainty associated with the simulation of the non-Gaussian tails is $\pm1.28\%$. The transverse τ polarization correlation gives a bias and uncertainty of $-0.22 \pm 0.44\%$. The τ branching fractions contribute $\pm 0.28\%$ and a minor systematic contribution comes from the effect of background events. After corrections, the mean τ lifetime obtained in the impact parameter sum analysis is:

$$\tau_\tau = 295.0 \pm 3.9\,(\text{stat}) \pm 4.4\,(\text{syst})\,\text{fs}. \tag{2}$$

5. Momentum-dependent Impact Parameter Sum Analysis

There is a strong correlation between a daughter track's momentum and the τ decay angle. Such correlation becomes a one-to-one relationship for monoenergetic τ's decaying into two bodies. It is therefore possible to derive the τ lifetime from a fit of the impact parameter sum distribution as a function of the two track momenta for 1-1 decay topology $\tau^+\tau^-$ events.

This new analysis method has been described in detail elsewhere.[11] There are several similarities with the impact parameter sum analysis described in the previous section, especially for what concerns the parametrization of the impact parameter sum resolution.

The resulting mean lifetime is:

$$\tau_\tau = 298.5 \pm 3.4\,(\text{stat}) \pm 4.5\,(\text{syst})\,\text{fs}. \tag{3}$$

6. Impact Parameter Difference Analysis

The 1-1 topology events were also analyzed with the impact parameter difference method. The averaged difference of the two impact parameters depends linearly from the difference of the two azymuthal decay angles. The mean lifetime is determined by means of an unbinned, weighted least-squares fit with an iterative trimming procedure to remove poorly measured events. The analysis is described in full detail elsewhere.[8,9] The result is:

$$\tau_\tau = 288.1 \pm 5.4\,(\text{stat}) \pm 1.2\,(\text{syst})\,\text{fs}. \tag{4}$$

The statistical error is inflated by the fact that the impact parameter difference is doubly smeared by the size of the interaction region. On the other hand, the systematic error is particularly small: its largest contributions come from resolution and trimming (0.25%) and the uncertainty on the $\tau^+\tau^-$ acollinarity (0.18%). The averaging and trimming procedures effectively reduce the effects due to the resolution, and the use of the difference of the decay angles is insensitive to the unknown τ direction. This also means that the correlation of the transverse tau polarization does not infuence this analysis.

7. Decay Length Analysis

The classical vertex method is used to measure the mean lifetime of τ's decaying into three charged tracks. The mean decay length is extracted from the decay length distribution by means of a maximum likelihood fit, as described elsewhere.[8,9] The lifetime, determined from 2794 τ decays, is:

$$\tau_\tau = 290.0 \pm 5.8\,(\text{stat}) \pm 2.1\,(\text{syst})\,\text{fs}. \tag{5}$$

The systematic error comes from the estimated contamination of hadronic events (0.3%), from the contamination of one-prong decays with photon conversions (0.2%) and from the understanding of pattern recognition errors (0.6%).

8. Combined Result

The momentum-dependent impact parameter sum (MIPS), impact parameter sum (IPS), and impact parameter difference (IPD) analyses are all based on the 1-1 topology events. According to a Monte Carlo simulation, the statistical correlation coefficients are 0.84 ± 0.02 between MIPS and IPS, 0.46 ± 0.04 between MIPS and IPD, and 0.48 ± 0.04 between IPS and IPD.

The statistical and systematic errors in the MIPS and IPS analyses are strongly correlated; these two measurements are averaged first, by means of a procedure described elsewhere,[12] applied to the statistical errors only. The result is $\tau_\tau = 298.2 \pm 3.4 \pm 4.6$ fs.

The MIPS+IPS average is then combined with the IPD and decay length results, according to the respective statistical and systematic errors, following a prescription given elsewhere,[12] and accounting for some small common sistematic contributions. The preliminary combined result for the 1992 data is

$$\tau_\tau = 291.9 \pm 3.2 \,(\text{stat}) \pm 1.8 \,(\text{syst}) \,\text{fs}, \quad (\chi^2 = 2.6 \text{ for 2 d.o.f., CL} = 27\%) \quad (6)$$

Including the previous ALEPH measurements,[8,9] the combined result is

$$\tau_\tau = 292.5 \pm 2.8 \,(\text{stat}) \pm 1.5 \,(\text{syst}) \,\text{fs}, \quad (\chi^2 = 7.0 \text{ for 9 d.o.f., CL} = 64\%) \quad (7)$$

$m_\tau = 1777.1 \,\text{MeV}/\text{c}^2$ has been assumed.

References

1. Particle Data Group, M. Aguilar-Benitez et al., Phys. Rev. **D45** (1992) 1.
2. E. Fernandez, *Review of Tau Physics*, Proceedings of the International Europhysics Conference on High Energy Physics, Marseille, France 22-28 July 1993.
3. Particle Data Group, M. Aguilar-Benitez et al., Phys. Rev. **D50** (1994) 1173.
4. D. Decamp *et al.,* Nucl. Instrum. Methods A 294 (1990) 121.
5. G. Batignani *et al.,* conference record of the 1991 IEEE Nuclear Science Symposium (November 1991, Sante Fe, NM, USA), Vol. 1, p. 438.
6. D. Buskulic *et al.* (ALEPH Collaboration), preprint CERN-PPE/94-30 (1994), to be published in Z. Phys. C.
7. S. Jadach et al., Computer Physics Commun. **66** (1991) 276.
8. D. Buskulic *et al.* (ALEPH Collaboration), Phys. Lett. B **297** (1992) 432.
9. D. Decamp *et al.* (ALEPH Collaboration), Phys. Lett. B **279** (1992) 411.
10. F. Fidecaro, A. Lusiani, A. Messineo, A. Sciabà, internal ALEPH note 94-115 (1994).
11. The ALEPH Collaboration, *Measurement of the tau lepton lifetime by ALEPH*, ICHEP94 Ref. 0574, contribution submitted to the 27th International Conference on High Energy Physicsm Glasgow, Scotland, 20-27 July 1994.
12. L. Lyons, D. Gibaut, and P. Clifford, Nucl. Instrum. Methods A **270** (1988) 110.

LIMIT ON m_{ν_τ}, τ LEPTONIC BRANCHING FRACTIONS, AND τ LIFETIME FROM OPAL [†]

PAUL WEBER [*]

Physics Department, Carleton University, Ottawa, Ontario K1S 5B6, Canada

ABSTRACT

The results of three measurements made with the OPAL detector are presented. Firstly, using the 1992 data sample we have determined an upper limit for the mass of the τ neutrino, using a novel technique based on both missing energy and mass. Secondly, the leptonic branching ratios of the τ lepton have been determined using the 1990 and 1991 data. Thirdly, the τ lifetime has been updated to include all OPAL data from 1990-1993. The results of these measurements are consistent with the relationship predicted by lepton universality, assuming zero τ neutrino mass.

1. Introduction

The assumption of charged-current universality leads to a strong prediction on the relation of the electronic branching ratio of the τ and its lifetime:

$$\left(\frac{g_\tau}{g_\mu}\right)^2 = \left(\frac{\tau_\mu}{\tau_\tau}\right)\left(\frac{m_\mu}{m_\tau}\right)^5 B(\tau \to e\bar{\nu}_e\nu_\tau) * \delta * f\left(\frac{m_{\nu_\tau}}{m_\tau}\right) , \tag{1}$$

where m_τ is the mass of the τ, m_μ and τ_μ are the muon mass and lifetime; $\delta = 0.9996$ is a small adjustment due to radiative corrections; and where

$$f(x) = 1 - 8x^2 + 8x^3 - x^4 - 12x^2 \ln x \tag{2}$$

is the usual phase space correction factor for non-zero neutrino mass (a similar correction, due to a potential non-zero μ neutrino mass, is negligibly small due to the more stringent limit on m_{ν_μ} in comparison to the μ mass). The electroweak couplings, g_μ and g_τ, are equal under the universality assumption and a significant difference between them could be a signal for new physics.

Over the last few years, the value of the coupling constant ratio has changed considerably, from what used to be a two-sigma discrepancy from unity[1] to values which are essentially consistent with one.[2,3] In this report measurements of the τ leptonic branching ratios and lifetime are presented, which are the least precisely determined inputs to the universality relation. We also present a determination of an upper limit to the τ neutrino mass, which is currently constrained to values[3] small enough so as to have an almost negligible effect on the ratio of g_τ/g_μ. However, as will be seen in the conclusions, the τ lifetime and electronic branching ratios may soon reach a level of precision where it may be possible to place limits on the neutrino mass which rival those obtained through direct measurement.

---

[*] Representing the OPAL Collaboration.

2.   The OPAL detector

The analyses described here rely on the high-precision tracking of charged particles, dE/dx particle ID information, and calorimetry of the OPAL detector[4] which are described here only briefly. The portion of the central tracking detector common to all the data sets includes a vertex drift chamber, a large-volume central jet chamber, and drift chambers for more accurate measurement of charged-track z-coordinates. These drift chambers operate in a common 4–bar pressure vessel which is immersed in a solenoid field of 0.435 Tesla. The vertex drift chamber provides up to 18 radial measurements (including 6 stereo layers) at distances from 10.3 to 21.3 cm. The 159-layer central jet chamber provides both accurate position and dE/dx information from radii of 25.5 to 183.5 cm. Beyond this, the z-chambers provide six additional coordinate measurements from wires arranged perpendicular to the beam axis. The impact parameter resolution for 45 GeV leptons is about 40 μm using the drift chambers alone.

For the latter part of the 1991 run onwards, a high-precision silicon microvertex detector[5] was operational inside the central drift chamber pressure vessel. This detector provided two additional high-precision tracking points (12 μm resolution) at radii of 6 and 7.5 cm, over the acceptance $|\cos\theta| \leq 0.8$. With the addition of these hits the extrapolated track resolution of high-momentum tracks is improved to 18 μm. This silicon detector was replaced for the 1993 run, by a device that includes z-readout.[6] The xy-configuration was unchanged, however, from that of the previous detector.

Calorimetry in the central part of the detector is composed of a shower detector containing an array of 9440 lead-glass blocks in a pointing geometry; together with a hadron calorimeter consisting of 9 layers of drift tubes operated in limited streamer mode, with 10 cm iron absorber between layers. The barrel calorimeter covers the acceptance $|\cos\theta| \leq 0.81$. Beyond this are found four layers of drift cells for muon detection, over the slightly smaller acceptance of $|\cos\theta| \leq 0.68$ (for all four layers).

3.   The $\tau^+\tau^-$ Data and Monte Carlo Samples

The OPAL detector integrated luminosities for the run years from 1990 through 1993 were, respectively, 6.8 pb^{-1}, 14.0 pb^{-1}, 24 pb^{-1}, and 34 pb^{-1}. The basic τ event selection[7] required six or fewer charged tracks, and less than 15 total charged tracks and electromagnetic calorimeter clusters in exactly two cones. Electron-pair events were rejected by requiring the event visible energy to be less than 80% of the center-of-mass energy. Muon-pair events were rejected by requiring an acollinearity angle of at least 18° when two muons were identified. Events were required to have $|\cos\theta_{thrust}| < 0.9$. The efficiency of these cuts is about 75%, with a residual background of 2.0% coming from μ-pairs (1.0%), two-photon (0.5%), multihadron (0.4%), e-pair (0.2%), and cosmic ray/beam gas (0.2%) events. The τ sample sizes for the 1990-1993 run years were, respectively, 5 206, 12 347, 28 792, and 26 894 events.*

The KORALZ Monte Carlo program, version 3.8,[8] was used for the simulation of $\tau^+\tau^-$ events at $\sqrt{s} = 91.160$ GeV, for the 1990–1992 detector configurations. Sample sizes of 50K, 100K, and 300K events, respectively, were generated for those years. For the 1993 detector configuration, 300K events were generated using KORALZ 4.0.[10]

---

* Energy scans were made during all but the 1992 run years, giving rise to numbers of events different than those obtained from a simple scaling of the integrated luminosity.

4. An Upper Limit for the τ Neutrino Mass

A new method for determining an upper limit for the τ neutrino, which includes both missing mass and missing energy information, has been used with the 1992 OPAL data.[10] Decays of the τ to five charged tracks are used, and only events in which the other τ decayed to a single charged track are considered because of the non-negligible background from multihadronic Z^0 decays which remains in candidate "5-3" $\tau^+\tau^-$ events. In the decay $\tau^- \rightarrow \nu_\tau + X^-$, the missing mass and energy must satisfy the relations: $m_{\nu_\tau} \leq m_\tau - m_X$, and $E_{\nu_\tau} \leq E_\tau - E_X$. The advantage of performing this measurement at LEP energies is that the τ samples tend to be of extraordinarily high purity, especially with regard to the multihadron background that afflicts samples obtained from running at low energies. The disadvantage, however, is the large boost of the τ's which leads to degraded mass and energy resolutions.

The most important backgrounds to the analysis are multihadrons and three-prong τ decays with a Dalitz π^0 decay, with one of the photons then undergoing pair-conversion. To remove the latter, electron rejection from the candidate five-prong decay was effected by using both the shower calorimeter and dE/dx information. The remaining background from such conversions is less than 0.14 events. The multihadron contamination was estimated by using a Monte Carlo which was calibrated to the data using 5-2 and 5-4 τ decays, and was found to contribute less than 0.09 events. The five-prong decay vertex fit probability of at least 5%, and an xy decay length less than 2 cm. A total of five candidates were found in the 1992 data.

The candidate five-prong decays are shown in Figure 1, together with their mass and energy resolutions (shown as error ellipses). Also shown is the predicted population of events from a Monte Carlo with input zero neutrino mass. The most probable value of the neutrino mass is found from the likelihood function:

$$P_i(m_i, E_i; m_\nu) = \frac{\int dm \int dE \frac{d^2\Gamma}{dmdE} R(m_i, E_i, \sigma_{m_i}, \sigma_{E_i}; m, E)\epsilon(m, E)}{\int dm \int dE \frac{d^2\Gamma}{dmdE}\epsilon(m, E)} \ , \tag{3}$$

where $d^2\Gamma(m_\nu, m, E)/dmdE$ is the theoretical distribution for a neutrino mass m_ν that is obtained from the Monte Carlo; $\epsilon(m, E)$ is an energy- and mass-dependent detection efficiency; σ_{m_i} and σ_{E_i} are the mass and energy resolutions; and R is the experimental resolution function, determined for each event individually through Monte Carlo studies. From the likelihood function we find a most probable neutrino mass of zero, and obtain an upper limit, at 95% confidence level, of:

$$m_{\nu_\tau} < 74 \text{ MeV}/c^2 \ . \tag{4}$$

5. The τ Leptonic Branching Ratios

The leptonic branching ratio measurements benefit from the low backgrounds at LEP, and good particle identification of the OPAL detector. For this analysis, only the 1990 and 1991 data were used, and the acceptance was restricted to the barrel calorimeter, $|\cos\theta| < 0.68$, in order to avoid regions of non-uniform calorimeter response. The resulting event selection efficiency was $54.1 \pm 0.2\%$, with a background of $1.9 \pm 0.6\%$. The total number of events selected was $11\,381$.

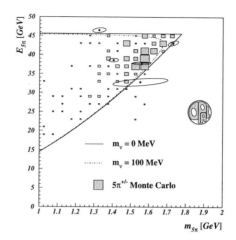

Fig. 1. The selected $\tau \to 5\pi^{\pm}\nu_\tau$ events from the data together with their mass and energy resolutions (points with ellipses). Also shown is the predicted population of these decays from the Monte Carlo (shaded boxes, with the size at each grid point being proportional to the number of decays), and the predicted outline of the boundary region for neutrino masses of zero, and 100 MeV/c^2.

Candidate electronic τ decays were selected by requiring E/p to be in the range $[0.7, 2.0]$, where E is the track energy measured by the shower calorimeter, and p is the momentum obtained from tracking. The background from $Z^0 \to e^+e^-$ was reduced by vetoing candidates with a nearly collinear track in the opposite hemisphere, that showed significant energy deposit in the calorimeter. The final decay selection efficiency was $78.4 \pm 0.9\%$, with a background of $6.1 \pm 1.2\%$ which is mostly misidentified hadrons.

Muonic τ decays were identified by requiring any two of the following three criteria: minimum ionization in the electromagnetic, or shower calorimeters, and signals in the outer muon detector. Events selected by the $Z^0 \to \mu^+\mu^-$ identification[9] were removed. The decay selection efficiency was $86.2 \pm 0.9\%$, with a background of $4.0 \pm 0.8\%$, which again is mostly misidentified hadrons.

From the total of 3 327 electronic and 3 238 muonic τ decays selected, the measured branching ratios were found:

$$
\begin{aligned}
B(\tau \to e\bar{\nu}_e\nu_\tau) &= 17.5 \pm 0.3 \text{ (stat.)} \pm 0.3 \text{ (sys.)} \% \,, \\
B(\tau \to \mu\bar{\nu}_\mu\nu_\tau) &= 16.8 \pm 0.3 \text{ (stat.)} \pm 0.3 \text{ (sys.)} \% \,.
\end{aligned} \tag{5}
$$

6.   Updated Measurement of the τ Lifetime

We have updated[11] our published value of the τ lifetime,[12] which was based on the 1990-1991 data samples, to include the 1992 and 1993 data. The measurement techniques remain unchanged from the previous analysis. They involve the 10% trimmed mean of the impact parameter distribution from decays of the τ into single charged particles, in a comparison to the value obtained from a Monte Carlo calibration generated with a known input lifetime; and a maximum likelihood fit for the decay length of vertices in decays of the τ to three charged particles.

The one-prong lifetime was determined with the 35 168/32 703 tracks selected from the 1992/1993 data which contain at least one silicon detector hit. The impact parameter was signed by using the beam position to approximate the τ production point, and thrust axis (determined from all the charged tracks in the event) to approximate the τ flight direction.[11] The impact parameter distribution is shown in Figure 2. Because of the oblate LEP beamsize, the data and Monte Carlo were split into six equally-sized azimuthal divisions, according to the amount of beamsize subtended,[12] and the lifetimes were measured separately in these bins. A small correction of $(+0.82 \pm 0.28)\%$ was applied for residual (zero-lifetime) backgrounds in the sample. The systematic error is dominated by the uncertainty with which the resolution of the Monte Carlo matches that of the data $(\pm 0.5\%)$, the beamsize uncertainty $(\pm 0.4\%)$, and Monte Carlo statistics (0.55%). The value obtained from the 1992 plus 1993 data was $\tau_\tau = 287.2 \pm 3.4(\text{stat.}) \pm 2.0$ (sys.) fs.

The three-prong lifetime was based on 4 671/4 417 vertices selected from the 1992/1993 data. As before,[12] at least two of the three tracks were required to have either at least one silicon detector hit, or a majority of first hits in the vertex drift chamber.* The vertex fit probability was required to be better than 1%. The thrust axis was used in a best fit with the vertex and beam positions and error ellipses[12] to improve the precision of determination on the decay length in the xy plane. The decay lengths in three dimensions were determined by projection using the thrust axis. A maximum likelihood fit determined the best average decay length of the ensembles for each of the two years, which was then converted to a lifetime using a boost factor which included initial state radiation effects. The lifetime was corrected for the small background of multihadronic events $(+0.28 \pm 0.28)\%$. The systematic errors were dominated by silicon detector alignment uncertainty $(\pm 0.4\%)$, and the uncertainty on residual biases remaining in the method $(\pm 0.6\%)$ as determined from the Monte Carlo sample. The value of the lifetime obtained from 1992 and 1993 data was $\tau_\tau = 288.9 \pm 3.6(\text{stat.}) \pm 1.7$ (sys.) fs. The distribution of decay lengths is shown in Figure 2.

The combined value of the lifetime from both methods, and in combination with the published measurement gives the final result for all OPAL data from 1990 through 1993 inclusive of:

$$\tau_\tau = 288.8 \pm 2.2 \text{ (stat.)} \pm 1.4 \text{ (sys.) fs} . \tag{6}$$

7. Conclusions

The OPAL values of the leptonic branching ratios and lifetime, when inserted into Equation (1), give the coupling constant ratio:

$$\frac{g_\tau}{g_\mu} = 0.990 \pm 0.014 , \tag{7}$$

---

* I.e., hits which are the first ones recorded on a given wire.

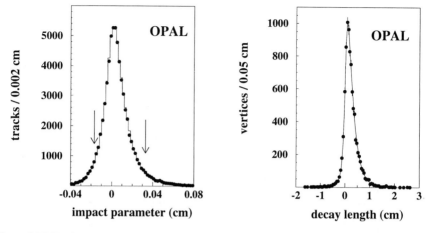

Fig. 2. LEFT: The impact parameter distribution for one-prong tracks from the data (points) together with the predicted distribution from the Monte Carlo (histogram). The arrows indicate the cut points for the 10% trimmed mean. RIGHT: The decay length distribution from the data (points), together with the result of the maximum likelihood fit (curve).

which is in good agreement with the lepton universality hypothesis. If universality is assumed in Equation (1), then a modest limit for the mass of the τ neutrino is obtained: $m_{\nu_\tau} < 173$ MeV/c^2, at 95% c.l. As the precisions of determination of the lifetime and leptonic branching ratios are improved, however, improves, this universality test may eventually provide the most stringent limits for m_{ν_τ}.

References

1. Particle Data Group, M. Aguilar-Benitez et al., *Phys. Rev.* **D45** (1992) 1.
2. E. Fernandez, *Review of Tau Physics*, Proceedings of the International Europhysics Conference on High Energy Physics, Marseille, France, 22-28 July 1993.
3. Particle Data Group, M. Aguilar-Benitez et al., *Phys. Rev.* **D50** (1994) 1173.
4. K. Ahmet et al., OPAL Collaboration, *Nucl. Instr. and Meth.* **A305** (1991) 275.
5. P.P. Allport et al., *Nucl. Instr. and Meth.* **A324** (1993) 34.
6. P.P. Allport et al., *The OPAL Silicon Strip Microvertex Detector with Two Coordinate Readout*, submitted to Nucl. Instr. and Meth.
7. M.Z.Akrawy et al., OPAL Collaboration, *Z. Phys.* **C52** (1991) 175.
8. S. Jadach et al., *Comp. Phys. Commun.* **66** (1991) 276.
9. S. Jadach et al., *Comp. Phys. Commun.* **79** (1994) 503.
10. R. Akers et al., OPAL Collaboration, *Determination of an Upper Limit for the Mass of the τ Neutrino at LEP*, CERN-PPE/94-107, 15 July 1994 (submitted to Phys. Lett.). This technique was originally suggested by Johannes Raab.
11. R. Akers et al., OPAL Collaboration, *Updated Measurement of the τ Lifetime*, CERN-PPE/94-129, 28 July 1994 (submitted to Phys. Lett.).
12. P. D. Acton et al., OPAL Collaboration, *Z. Phys.* **C59** (1993) 183.

SOME RECENT τ RESULTS FROM CLEO-II

RICHARD KASS *

Department of Physics, Ohio State University, Columbus, OH, 43210

ABSTRACT

We report new results on the lifetime of the τ-lepton (τ_τ) and the magnitude of the helicity of the τ neutrino ($|h_\nu|$). The data were collected with the CLEO-II detector at CESR and consist of ≈ 2 fb^{-1} taken at energies near the $\Upsilon(4S)$. Using $\tau^+\tau^-$ pairs in which one (both) of the τ's decays to three charged particles we measure $\tau_\tau = (2.91 \pm 0.04 \pm 0.07) \times 10^{-13}$ ($(2.85 \pm 0.13 \pm 0.10) \times 10^{-13}$)s. We measure $|h_\nu| = 0.99 \pm 0.06 \pm 0.10$ using events where both τ's decay via $\tau \to \pi\nu$.

1. Introduction

In this paper we present new measurements of the τ lifetime based on a high statistics sample of 1 vs 3 and, for the first time, 3 vs 3 τ decays. We also present a new measurement of the magnitude of the helicity of ν_τ ($|h_\nu|$) using the reaction $e^+e^- \to \tau^+\tau^-$, $\tau^- \to \pi^-\nu_\tau$, $\tau^+ \to \pi^+\bar{\nu}_\tau$.

2. The Data Set

The data used in this analysis was taken with the CLEO detector at the Cornell Electron-Positron Storage Ring (CESR). It corresponds to a total integrated luminosity of 2 fb^{-1}, with $\approx$ two thirds of the data collected at the $\Upsilon(4S)$ ($\sqrt{s} =$10.58 GeV), and the rest at energies slightly below the $\Upsilon(4S)$. This luminosity corresponds to the production of $\approx$1.8x10^6 τ-pairs. The CLEO-II detector emphasizes precision charged particle tracking and high resolution electromagnetic calorimetry. A detailed description of the detector can be found in Ref. [1].

3. τ Lifetime

The decay of the τ-lepton provides a useful testing ground for the Standard Model of electroweak interactions.[2] If the τ is a sequential lepton, then its coupling to the W is the same as that of the μ and its lifetime is directly related to the μ mass and lifetime . To lowest order[3] and neglecting the electron and neutrino masses, the Standard Model predicts:

$$\tau_\tau = \tau_\mu (m_\mu/m_\tau)^5 B(\tau^- \to e^- \nu_\tau \bar{\nu}_e). \tag{1}$$

The calculated τ lifetime depends directly on experimental measurements of the μ mass and lifetime and the τ mass and electronic branching ratio. Using the world average values for these quantities[4] the predicted lifetime becomes:

$$\tau_\tau = (1.63 \pm 0.002) \times 10^{-12} B(\tau^- \to e^- \nu_\tau \bar{\nu}_e) = (2.92 \pm 0.03) \times 10^{-13} \text{ s}. \tag{2}$$

---

*For the CLEO Collaboration.

3.1. Event Selection

The selection of 1 vs 3 tau events follows the procedure of Ref. [5]. Selection criteria include charge conservation, event topology, momentum and energy balance, and quality of charged track reconstruction. A total of 37471 events pass all of the 1 vs 3 event selection criteria.

The 3 vs 3 analysis is based on a sample of events in which both τ's decay into three charged hadrons and no π^o's. The selection of double 3-prong events is discussed in detail in Ref. [6]. A total of 1447 events pass all selection criteria.

3.2. Background Estimates

The hadronic background in the 1 vs 3 (3 vs 3) sample is estimated using both data and Monte Carlo * calculations to be 1.0% (4.7%). From a Monte Carlo calculation we estimate the two photon background in both samples to be < 0.2%, dominated by hadronic final states. By varying the selection criteria and studying the data and Monte Carlo we estimate the remaining background from Bhabha and mu-pair events with a photon to be <0.5% for the 1 vs 3 analysis and negligible for the 3 vs 3 analysis. We have also investigated possible contamination from beam gas interactions and $\Upsilon(4S) \to B\bar{B}$ decays and found these sources have a negligible effect on both analyses.

3.3. Lifetime Determination Procedure

The tau proper flight distance, $c\tau$, is calculated using:

$$c\tau = \frac{1}{\gamma\beta}\frac{L_{xy}}{\sin\theta} = \frac{m_\tau}{p_\tau}\frac{L_{xy}}{\sin\theta} \qquad (3)$$

where γ, β, and p_τ, the magnitude of the τ's momentum, are calculated using the beam energy. L_{xy} is the flight path in the plane transverse to the z axis. This decay distance is converted to the full decay distance (L) using $\sin\theta$, with θ the polar angle. The τ polar angle is approximated using the vector momentum of the 3 charged tracks.

Initial state radiation reduces the τ energy to less than the beam energy. Consequently, the τ's average flight path $(\approx 250\mu m)$ is less than that predicted by using the beam energy to calculate p_τ. Although on an event by event basis the true τ production energy is not known, the average correction to the τ energy can be determined from a Monte Carlo simulation.[7] We find the average decay distance of the selected τ sample is reduced by $f_{rad}=3.1\%$ from this effect.

3.4. Vertex Reconstruction

For the 1 vs 3 analysis the most probable decay length in the transverse plane, L_{xy}, is determined via the equations:

$$L_{xy} = \frac{Xt_x\sigma_y^2 + Yt_y\sigma_x^2 - (Xt_y + Yt_x)\sigma_{xy}^2}{t_y^2\sigma_x^2 + t_x^2\sigma_y^2 - 2t_xt_y\sigma_{xy}^2} \qquad (4)$$

---

*Monte Carlo events have been generated using a $c\tau$ of 126 μm for the D^0 , 133 μm for the D_s, and 318 μm for D^+.

$$X = X_v - X_b \text{ and } Y = Y_v - Y_b \qquad (5)$$

In the above equations, X_v and Y_v are the horizontal and vertical decay coordinates of the τ and X_b and Y_b are the horizontal and vertical production points of the τ. Here $\sigma_x^2(\sigma_y^2)$ is the variance of $X(Y)$ and σ_{xy}^2 is the correlation term for the X and Y vertex errors.[†] Beam positions in the xy plane, X_b and Y_b, are determined for each run. The uncertainties in the beam position per run (σ) for x and y are typically 35 μm and 15 μm, respectively. The full error on the production point also includes a contribution from the finite extent of the beams (350 μm in x, 10 μm in y). Finally, t_x and t_y are direction cosines calculated from the momentum vector of the three charged tracks.

An event-by-event estimate of X_v and Y_v is obtained using a χ^2 minimization algorithm.[8] Events with a χ^2 per degree of freedom > 24 or reconstructed $|c\tau| > 0.42$ cm (≈ 50 τ lifetimes) are eliminated from further consideration.

In the double vertex method we take 3 vs 3 τ-pairs, reconstruct both vertices, and compute the projection of the difference of these decay lengths along the estimated line of flight of the τ's. This measurement is independent of the beam position. For the separation λ between two exponential decays the distribution is

$$f(\lambda) = \frac{\lambda}{d^2} e^{-\lambda/d} \qquad (\lambda \geq 0). \qquad (6)$$

Note that $f(\lambda)$ vanishes at $\lambda = 0$, and $\langle\lambda\rangle/\sigma_\lambda = \sqrt{2}$, compared with unity for the exponential distribution. Identifying the transverse projection λ_{xy} we have $L_{xy} = \frac{1}{2}\lambda_{xy}$ given by Eq. 4 with the replacements

$$X = 1/2(X_{v2} - X_{v1}), \quad Y = 1/2(Y_{v2} - Y_{v1}), \quad \sigma^{-1} = (\sigma_1^{-1} + \sigma_2^{-1}),$$

and $t_{x,y} = $ the direction cosines of the momentum-difference vector of the three-prong jets. The positions X_v and Y_v and their associated error matrices σ are obtained from χ^2 fits for each vertex.[9]

3.5.   The Lifetime

We determine the lifetime of our data sample using a weighted average of the proper flight distances calculated according to Eq. 3. The weight of each event is calculated using the error matrix from the vertex fit. The mean τ lifetime, τ_τ, and the mean lifetime of the event sample, τ_{meas}, are related by:

$$\tau_{meas} = (1 - f_{bg})(1 - f_{rad})\tau_\tau + f_{bg}\tau_{bg} \qquad (7)$$

where f_{bg} is the fraction of non-τ events in the sample, τ_{bg} is the mean lifetime of these events, and f_{rad} is the correction due to initial-state radiation. The lifetime of

---

[†]We expect the uncertainty in X to be larger than in Y since the error in X is dominated by the error in X_b which is dominated by the beam spread.

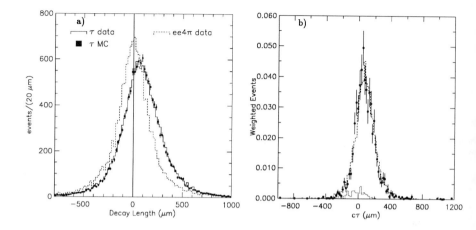

Fig. 1. Decay length distribution for Monte Carlo and Data for 1 vs 3 events (a) and for 3 vs 3 events (b). For 3 vs 3 sample data are indicated by points with error bars, simulation (signal plus background) by the dashed histogram, and background simulation by the dotted histogram.

the sample, τ_{meas}, is the average decay length found with the vertex-fitting process described above.

The lifetime of the background sample is computed using a Monte Carlo simulation and confirmed using data. For the hadronic background in the 1 vs 3 (3 vs 3) sample we find $c\tau_{bg} = 7 \pm 22 \ \mu$m (-21 $\pm$ 13 μm).

The measured $c\tau$ distributions for 1 vs 3 and 3 vs 3 events are shown in Fig. 1a) and 1b) respectively. Also displayed is a Monte Carlo calculation for this distribution (including contributions from the backgrounds) showing good agreement with the shape of the data distribution. We measure $c\tau_{meas}$ to be 83.7±1.1 μm for the 1 vs 3 sample and 79.8±3.6 μm for the 3 vs 3 sample. After correcting for backgrounds and initial state radiation we measure $c\tau_\tau$ to be 87.2±1.2 μm for the 1 vs 3 sample and 85.3±3.8μm for the 3 vs 3 sample.

3.6.   Systematic Errors

We have considered systematic errors from several sources: charged particle tracking and subsequent vertex reconstruction, uncertainty in the beam position, uncertainty in the amount of background in the sample, and uncertainty in the lifetime of the background.

For the 1 vs 3 analysis we used a large sample of two-photon events with 4 charged tracks in the final state as a systematic check of the vertexing code and method. Here we expect the event sample to have a mean decay length of 0 μm. After correcting for background events in this sample we measure an average decay length of -1.8 ± 1.8 μm, consistent with no bias. Thus we conservatively assign a systematic error of $\pm 1.8 \mu$m to account for effects due to tracking, vertexing and the lifetime method.

Systematic errors for the two analyses are summarized in Table 1. The total systematic error for each analysis is obtained by combining each term in quadrature. Thus for the 1 vs 3 sample we measure $c\tau_\tau = 87.2 \pm 1.2 \pm 2.1$ μm corresponding to $\tau_\tau = (2.91 \pm 0.04 \pm 0.07) \times 10^{-13}$ s, with the first error statistical and the second systematic. For the 3 vs 3 sample we measure $c\tau_\tau = 85.3 \pm 3.8 \pm 3.0$ μm corresponding to $\tau_\tau = (2.85 \pm 0.13 \pm 0.10) \times 10^{-13}$ s.

Table 1: Systematic errors for both analyses in μm.

Source	1 vs 3	3 vs 3
Tracking and Vertexing	1.8	2.8
Beam Position + Size	0.3	0.0
Background Fraction	1.0	1.0
Background Lifetime	0.2	0.6
Total	2.1	3.0

3.7. Discussion

The results presented here are consistent with other measurements of τ_τ and the prediction of the Standard Model based on the world average of the τ electron branching ratio and m_τ.

4. Helicity of ν_τ

In the Standard Model neutrinos are left-handed leptons, tau decay is a V-A process, and the ν_τ helicity is $h_\nu = -1$.[10] In the rest frame of a $\tau^{-(+)}$ that decays to $\pi\nu$, the parent spin is maximally correlated (anticorrelated) to the daughter meson direction. Tau pairs have preferentially aligned spins when produced via the QED process $e^+e^- \to \gamma^* \to \tau^+\tau^-$. At center-of-mass energies above $\tau^+\tau^-$ threshold, each τ has a relativistic boost, and the meson directions in the center-of-mass systems are partially Lorentz-transformed into hadron energies. Thus for a left-handed ν_τ, the energies of the π^+ and the π^- are correlated. This correlation is also present if ν_τ was instead right handed with $h_\nu = +1$. However, if ν_τ has no preferred handedness there will be no correlations between the daughter energies.

For the decay of the τ to a pseudoscalar ($e.g.$, π) the neutrino helicity h_ν is related to to the parameters γ_{av} and ξ and the charged current couplings g_a and g_v by[11]

$$h_\nu = \xi_\pi = -\gamma_{av} = \frac{-2g_a g_v}{g_a^2 + g_v^2}. \tag{8}$$

The double differential cross section in terms of the scaled laboratory energies ($x = E/E_{beam}$) of the hadrons is given by

$$\frac{d^2\sigma}{dx_+ dx_-} = F_1(x_+, x_-) + h_\nu^2 \cdot F_2(x_+, x_-), \tag{9}$$

with F_1 and F_2 being known kinematic functions. For either $V - A$ or $V + A$ interactions the correlations will be maximal, whereas for pure V or pure A interactions the correlations will vanish.

4.1. Event Selection

Events consistent with $\tau^+\tau^- \to (\pi^+\bar{\nu})(\pi^-\nu)$ are extracted for this analysis. The event selection procedure is also designed to minimize backgrounds from non-τ QED processes such as $e^+e^-\gamma$, $\mu^+\mu^-\gamma$ and $\gamma\gamma$ interactions as well as τ backgrounds such as $\tau \to \rho\nu$, $\rho \to \pi^-\pi^o$. A total of 1972 events pass all selection criteria.

4.2. Determination of $|h_\nu|$

We use a maximum likelihood procedure to extract $|h_\nu|$. A large sample of $\tau^+\tau^-$ events were generated using KORALB[7] and GEANT[12] to assess which decay modes were significant sources of background for the analysis. It was found that 97% of all such events that passed the selection criteria came from $\pi\pi$, πK, $\pi\rho$, $\pi\mu$, and πK^* final states. Samples corresponding to several times the actual data are then generated for these five modes both with the $V - A$ and with the V hypotheses. The samples are scaled to the appropriate luminosity and binned as x_+ vs. x_-. For each bin (i, j) the predicted number of events, denoted n_{ij} for *any* value of h_ν is given by[13]

$$n_{ij}(h_\nu) = h_\nu^2 \cdot N_{ij}^{V-A} + (1 - h_\nu^2) \cdot N_{ij}^V. \tag{10}$$

A binned Poisson likelihood is then taken as a function of h_ν as

$$\log \mathcal{L}(h_\nu) = \sum_{i,j} [d_{ij} \log n_{ij} - n_{ij} - \log(d_{ij}!)]. \tag{11}$$

The result of the maximum likelihood analysis is:

$$|h_\nu| = 0.99 \pm 0.06,$$

with the uncertainty being statistical.

4.3. Systematic Errors

We have made many systematic checks which can be divided into four broad categories: (i) dependence on the values of variables used in event selection, (ii) possible biases in the methodology, (iii) possible biases in the simulations, and (iv) our estimate of backgrounds. The systematic errors are summarized in the Table 2. Taking all the effects in quadrature we obtain an overall systematic uncertainty of $\delta|h_\nu| = \pm 0.10$.

4.4. Discussion

From a sample of ≈ 2000 events of the type $\tau^+\tau^- \to (\pi^+\bar{\nu})(\pi^-\nu)$ taken at $\sqrt{s} \sim$ 10.6 GeV CLEO has extracted the magnitude of the helicity of the tau neutrino. Given that this helicity is known to be negative,[14] the result is $h_\nu = -0.99 \pm 0.06 \pm 0.10$. This is consistent with the $V - A$ nature of the charged weak current mediating τ decay and with other measures of the ν_τ helicity.[15,16]

Table 2: Summary of systematic errors.

Parameter	$\Delta\xi$
Kinematic criteria	0.02
$\gamma\gamma$ veto	0.02
Isolated shower veto	0.03
Hadron identification	0.05
Trigger and LEVEL3	0.03
Simulation comparisons	0.04
Normalizations	0.01
$\log \mathcal{L}$ vs χ^2	0.04
Data-based corrections	0.03
Tau branching ratios	0.03
$\mu\mu\gamma$ backgrounds	0.01
Total syst uncertainty	0.10

References

1. Y. Kubota et al., Nucl Inst. Meth. **320** (1992) 66.
2. S. L. Glashow, A Salam, S. Weinberg, Rev. Mod. Phys, **52**, (1980) 515.
3. W. J. Marciano, Proc. DPF91 (Vancouver, August 1991), Vol. 1, ed. D. Axen, D. Bryman, and M. Comyn (World Scientific Publishing Co. Pte. Ltd., Singapore, 1992) p. 461. The radiative corrections to the lifetime are of the order of α and increase the average decay length by $< 1\%$.
4. Particle Data Group, L. Montanet et al., Phys. Rev. D **50**, 1173 (1994).
5. D. Bortoletto et al., Phys. Rev. Lett. **71** (1993) 1791.
6. B. Barish et al., contribution GLS0249 to the 27th Inter. Conf. on High Energy Physics, Glasgow Scotland.
7. KORALB 2.1 S. Jadach and Z. Was, Comp. Phys. Commun. **64** (1991) 267, and TAUOLA1.5, S. Jadach, J. Kuhn, Z. Was, Comp. Phys. Commun. **64** (1991) 275. The size of the correction to the lifetime is insensitive to the photon energy cutoff used in the simulation.
8. J. Whitmore, Ph.D. thesis, Ohio State University (1992) unpublished.
9. W. T. Ford, CLEO Software Note CSN 94/329 (1994).
10. We use the convention that $h = \vec{s} \cdot \vec{p}/sp$, such that $-1 < h < 1$.
11. See for example, J.H. Kühn and F. Wagner, Nuclear Physics **B236**, 16 (1984), C. Nelson, Physical Review **D43**, 1465 (1991), W. Fetscher, Physical Review **D42**, 1544 (1990).
12. R. Brun et al., CERN Report No. CERN-DD/EE/84-1, 1987 (unpublished).
13. This form is *exact* for the four final states involving only charged hadrons. For the final state $(\pi\nu)(\mu\nu\bar{\nu})$ the expression has negligible corrections near $h_\nu = -1$.
14. H. Albrecht et al. (ARGUS Collaboration), Phys. Lett. **B250**, 164 (1990); H. Albrecht et al. (ARGUS Collaboration), Z. Phys. **C58**, 61 (1993). The signed result is $h_\nu = -1.25 \pm 0.23 \, {}^{+0.15}_{-0.08}$.
15. C. Hast, "Selected ARGUS Results on τ-physics", XXIX Recontres de Moriond, March 1994. Their preliminary value is $|h_\nu| = 0.98 \pm 0.04 \pm 0.03$.
16. D. Buskulic et al. ($\aleph$ Collab.), Phys. Lett. **B321**, 168 (1994). They report $|h_\nu| = 0.99 \pm 0.07 \pm 0.04$.

SUMMARY OF TAU LEPTON RESULTS

TOMASZ SKWARNICKI

Department of Physics, Southern Methodist University, Dallas
TX 75275

ABSTRACT

We summarize τ lepton results presented at the heavy flavor physics session.

In the Standard Model the tau is a heavy electron-like particle, which comes with its massless partner, the left-handed tau neutrino. From the measurements of the Z° width at SLC and LEP we know that the tau lepton is the heaviest sequential lepton, i.e. the last charged lepton with a light neutrino. Most of the experimental activity is concerned with the verification that the tau is just a heavier duplicate of the electron and the muon.

1. Lepton universality tests

1.1. Purely leptonic decays

The Standard Model predicts that the couplings of leptons to charged and neutral currents are flavor independent. A test of this assumption can be obtained by comparing the electronic decay of the tau, $\tau \to \nu_\tau e \bar{\nu}_e$, to muon decay, $\mu \to \nu_\mu e \bar{\nu}_e$. These two processes differ only by different phase-space (dependence on m_τ and m_μ) and possibly different couplings of τ and μ (G_τ and G_μ). In comparison of the measured decay branching fractions, one also has to correct for different lifetimes (τ_τ and τ_μ):

$$\frac{G_\tau}{G_\mu} = \frac{\tau_\mu}{\tau_\tau} \left(\frac{m_\mu}{m_\tau}\right)^5 \frac{B(\tau \to \nu_\tau e \bar{\nu}_e)}{B(\mu \to \nu_\mu e \bar{\nu}_e)}$$

Since each of the μ parameters is known with accuracy roughly three orders of magnitude better than the corresponding τ parameters, the above test is totally dominated by experimental errors for the τ lepton. Theoretical uncertainty is only of the order of 0.01%.[1]

Many results presented at this conference refer to the parameters used in testing lepton universality. The BES collaboration presented[2] the final result on the τ mass with the astonishing accuracy of about 1 part in 10,000 which totally dominates the world average: $m_\tau = 1777.02 \pm 0.25$ MeV. New results on the τ lifetime were presented by OPAL,[3] ALEPH,[4] CLEO-II,[5] and SLD.[6] The new world average, $\tau_\tau = 291.9 \pm 1.7$ fs, is dominated by the LEP experiments. ALEPH[4] and DELPHI[7] presented new determinations of leptonic branching fractions. The ALEPH results dominate the world average values (see Table 2). Using the assumption of $e - \mu$ universality, which was experimentally established at the 0.3% level from pion decays,[8] we can average muonic and electronic branching fractions to obtain, $B(\tau \to \nu_\tau e \bar{\nu}_e)_{e,\mu} = (17.65 \pm 0.08)\%$.

The $\tau - \mu$ universality constraint can be illustrated on a graph of τ lifetime plotted as a function of $B(\tau \rightarrow \nu_\tau e \bar{\nu}_e)$ (see Fig.1). If lepton universality holds, measurements of these parameters must produce a point falling within the band defined by the measurement of the τ mass. An impressive improvement in the measurements of τ_τ, $B(\tau \rightarrow \nu_\tau e \bar{\nu}_e)$ and m_τ over the past two years is shown in Fig. 1. From these measurements we obtain, $G_\tau/G_\mu = 0.987 \pm 0.007$, which is very close to the universality prediction, $G_\tau/G_\mu = 1$. At present, accuracy of the test is limited by the uncertainity in τ_τ and $B(\tau \rightarrow \nu_\tau e \bar{\nu}_e)$.

1.2. Hadronic decays

An independent test of $\tau - \mu$ universality can be obtained from comparison of decays $\tau \rightarrow \nu_\tau \pi$, K to π, $K \rightarrow \mu \bar{\nu}_\mu$. At present, this test is limited to a precision of about 2% due to an error on $B(\tau \rightarrow \nu_\tau \pi, K)$. The result is $G_\tau/G_\mu = 0.999 \pm 0.018$.

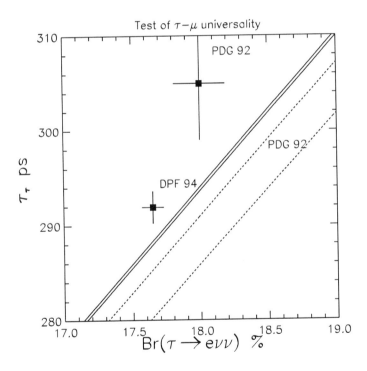

Figure 1. Test of $\tau - \mu$ universality. The bands illustrate the relations between τ_τ and $B(\tau \rightarrow \nu_\tau e \bar{\nu}_e)$ imposed by lepton universality and the measurements of m_τ and of the muon parameters. Widths of these bands indicate $\pm 1\sigma$ uncertainties in the τ mass measurements. Actual measurements of the lifetime and electronic branching ratio are shown by crosses. The present status and the 1992 status[9] (the upper cross, the lower band) are shown.

2.   Lorentz structure of the weak currents

In the Standard Model the $\tau - W - \nu_\tau$ vertex has exactly a "V-A" structure. This can be tested either in purely leptonic or in semi-hadronic decays of the τ.

2.1.   Purely leptonic decays

The Lorentz structure of four-fermion interactions in τ decays can be studied in full analogy with μ decays. The charged lepton momentum spectrum and angular distribution in the τ rest frame can be written in the most general form which allows for scalar, vector and tensor interactions with the help of Michel parameters $(\rho, \eta, \xi, \delta)$:

$$
\begin{aligned}
\frac{d\Gamma}{dx\, d\cos\vartheta} \propto \; & \frac{x^2}{1+4\eta(m_l/m_\tau)} \left[12(1-x) + \rho\tfrac{4}{3}(8x-6) + \eta(m_l/m_\tau)\tfrac{1-x}{x} \right] \\
& - \quad \xi \mathcal{P}_\tau \cos\vartheta\, x^2 \left[4(1-x) + \delta\tfrac{4}{3}(8x-6) \right]
\end{aligned}
$$

where $\mathcal{P}_\tau$ is the τ lepton polarization, $x = P_l/P_{max}$ is a fraction of τ momentum carried by the final state lepton, and ϑ is the angle between the τ spin and $\vec{P}_l$. The Standard Model ("V-A") predicts: $\rho = \tfrac{3}{4}$, $\eta = 0$, $\xi = 1$, and $\delta = \tfrac{3}{4}$. Values of these parameters are sensitive to small admixtures of scalar or right-handed intermediate bosons, and thus their measurement allows for a search for new physics.

The spectral shape parameters, ρ and η, are measured from fits to the observed electron and muon spectra in τ decays. New results were presented by ARGUS and ALEPH at the Glasgow conference.[10] Sensitivity to the η parameter is better in $\tau \to \nu_\tau \mu \bar{\nu}_\mu$ because this parameter enters multiplied by the ratio m_l/m_τ. This year ARGUS presented the first direct measurement of η from the muon spectrum, $\eta = 0.03 \pm 0.18 \pm 0.12$. Since the η parameter also changes the overall rate, the world average values of B_e and B_μ allow a better determination of η: $\eta = -0.01 \pm 0.04$. The decay asymmetry parameters, ξ and δ, are more difficult to measure because they depend on tau polarization which averages out to zero when integrated over all angles in $\gamma \to \tau^+\tau^-$ production. Thus, measurements at lower energy e^+e^- colliders must study correlations in leptonic decays of the two τ's in the event. Measurements using $Z^\circ \to \tau^+\tau^-$ have an advantage that the average $\mathcal{P}_\tau \neq 0$. ALEPH and L3 presented new results on ξ and δ at the Glasgow conference.[10]

2.2.   Hadronic decays

Semi-hadronic decays offer independent opportunity to test the V-A coupling at the $\tau - W - \nu_\tau$ vertex. Studies of energy and angle correlations among final state hadrons from decays of both τ's in the event allow one to measure the tau neutrino helicity (-1 in the Standard Model). Measurements with an accuracy of a few percent are now available from the ARGUS, CLEO-II[5] and ALEPH experiments.

The status of Michel parameter determination in τ and μ decays is shown in Table 1.

Table 1. Michel parameters measured in τ and μ^{11} decays.

Michel parameter	τ	μ	SM
ρ	0.752 ± 0.022	0.752 ± 0.003	0.75
η	$-0.01 \;\; \pm 0.04$	-0.007 ± 0.013	0
δ	$0.75 \;\; \pm 0.17$	0.749 ± 0.004	0.75
ξ	$1.03 \;\; \pm 0.12$	1.003 ± 0.008	1
ξ_h	$-1.00 \;\; \pm 0.03$	—	-1

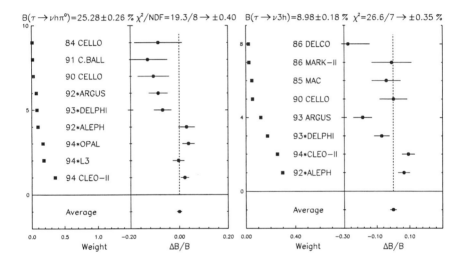

Figure 2. Measurements of $B(\tau \to \nu_\tau h\pi^\circ)$. The left part of the picture shows a weight in which each experiment contributes to the world average. The right part of the picture shows a relative deviation of each measurement from the world average value. Error on the world average included in Table 2 has been scaled up due to inconsistency in the measurements.

Figure 3. Measurements of $B(\tau \to \nu_\tau 3h)$. The left part of the picture shows a weight in which each experiment contributes to the world average. The right part of the picture shows a relative deviation of each measurement from the world average value. Error on the world average included in Table 2 has been scaled up due to inconsistency in the measurements.

3.   Hadronic branching fractions

Thanks to its large mass, the tau is the only lepton which can decay to hadrons (i.e. $\bar{u}d$ and $\bar{u}s$ quark pairs). This provides an opportunity to study strong interaction phenomena in weak charged current decays in particularly clean experimental and theoretical environments.

When approached in an inclusive way, hadronic decays of the tau provide interesting tests of perturbative QCD and give a determination of the scale of strong interactions. New results on this subject were presented by the CLEO-II experiment at the QCD session at this conference.[12]

The results presented at the Heavy Flavor session were concerned mostly with determination of branching fractions for individual hadronic decay modes. CLEO-II,[13] OPAL,[14] L3,[15] and DEL-PHI[7] presented new measurements of $B(\tau \rightarrow \nu_\tau h\pi^\circ)$ ($h = \pi^\pm$ or $K^\pm$). All new measurements point to a higher value of this largest exclusive branching fraction of τ than less precise determinations by the older experiments (see Fig.2). Similar change is observed for the most common decay in 3-prong topology, where the measurements of $B(\tau \rightarrow \nu_\tau 3h)$ by CLEO-II[13] and ALEPH are significantly higher than the previous determinations (see Fig.3). Table 2 shows world average values for exclusive branching fractions sorted according to charge track multiplicity. Their sum is close to 100%. Inclusive measurements of branching fraction for 1,3, and 5 charged particles produced in τ decay roughly agree with the sum over exclusively measured branching fractions in each topology.

Table 2. World average branching fractions for τ decays.

Decay Mode	Branching Ratio %
$\nu_\tau e \bar{\nu}_e$	17.67 ± 0.11
$\nu_\tau \mu \bar{\nu}_\mu$	17.15 ± 0.12
$\nu_\tau h$	12.29 ± 0.20
$\nu_\tau h\pi^\circ$	25.28 ± 0.40
$\nu_\tau h2\pi^\circ$	8.89 ± 0.26
$\nu_\tau h3\pi^\circ$	1.22 ± 0.13
$\nu_\tau h4\pi^\circ$	0.15 ± 0.07
$\nu_\tau h\pi^\circ \eta,\ \eta \rightarrow \gamma\gamma$	0.07 ± 0.01
$\nu_\tau h\omega,\ \omega \rightarrow \gamma\pi^\circ$	0.15 ± 0.02
$\nu_\tau h\pi^\circ \omega,\ \omega \rightarrow \gamma\pi^\circ$	0.033 ± 0.005
$\nu_\tau h K_L^0$	0.51 ± 0.03
$\nu_\tau h\pi^\circ K_L^0$	0.26 ± 0.04
Sum	83.67 ± 0.56
Topological 1−prong	85.28 ± 0.20
Difference	-1.61 ± 0.59
$\nu_\tau 3h$	8.98 ± 0.18
$\nu_\tau 3h\pi^\circ$	4.36 ± 0.14
$\nu_\tau 3h2\pi^\circ$	0.48 ± 0.06
Sum	13.82 ± 0.24
Topological 3−prong	14.57 ± 0.20
Difference	-0.75 ± 0.31
$\nu_\tau 5h$	0.071 ± 0.009
$\nu_\tau 5h\pi^\circ$	0.021 ± 0.005
Sum	0.092 ± 0.010
Topological 5−prong	0.105 ± 0.010
Difference	-0.013 ± 0.014
Total sum	97.58 ± 0.67
Missing	-2.42 ± 0.67

4.   Conclusions

Tau physics is a very active experimental field. There have been about 50 new measurements presented this year, many of them presented at this conference. New generation of experiments at low (BEPC), intermediate (CESR) and high (LEP) e^+e^- energies has reached a new level of precision. Inclusive and exclusive measurements of τ decay branching fractions are now consistent. The data are in full agreement with the Standard Model:

- Tau-muon universality has been tested at a $\sim 1\%$ level.

- V–A structure of the decay current has been established at a few percent level.

- Tau lepton number is conserved at least at a few times 10^{-6} level.[16]

- No evidence has been found for a τ neutrino mass ($m_{\nu_\tau} < 31$ MeV at 90% C.L.)[17]

- No evidence for τ neutrino oscillations has been found.[11]

However, experimental uncertainities are much larger than similar tests for the muon and the electron. Thus, searches for deviations from the Standard Model prediction in τ physics will continue.

References

1. W.J. Marciano, A. Sirlin, Phys. Rev. Lett. **61** (1988) 1815.
2. E. Sonderstrom (BES collaboration), paper contributed to the Heavy Flavor Physics (Tau) session at this conference.
3. P. Weber (OPAL collaboration), paper contributed to the Heavy Flavor Physics (Tau) session at this conference.
4. A. Lusiani (ALEPH collaboration), paper contributed to the Heavy Flavor Physics (Tau) session at this conference.
5. R. Kass (CLEO-II collaboration), paper contributed to the Heavy Flavor Physics (Tau) session at this conference.
6. S. Hedges (SLD collaboration), paper contributed to the Heavy Flavor Physics (Tau) session at this conference.
7. D. Edsall (DELPHI collaboration), paper contributed to the Heavy Flavor Physics (Tau) session at this conference.
8. D. Britton et al., Phys. Rev. Lett. **68** (1992) 3000; G. Czapek et al., Phys. Rev. Lett. **70** (1993) 17.
9. Particle Data Group, Phys. Rev. **D45** (1992) 1.
10. D. Wegener, Talk presented at the parallel session Pa-16a at the 27th International Conference on High Energy Physics, Glasgow, 20-27 July 1994.
11. Particle Data Group, Phys. Rev. **D50** (1994) 1173.
12. Sukhpal Sanghera (CLEO-II collaboration), paper contributed to the QCD session at this conference, SMU-HEP-94-17.
13. J. Urheim (CLEO-II collaboration), paper contributed to the Heavy Flavor Physics (Tau) session at this conference.
14. K. Anderson (OPAL collaboration), paper contributed to the Heavy Flavor Physics (Tau) session at this conference.
15. J. Shukla (L3 collaboration), paper contributed to the Heavy Flavor Physics (Tau) session at this conference.
16. R. Kass (CLEO-II collaboration), paper contributed to the Rare Decay session at this conference.
17. H. Albrecht et al. (ARGUS collaboration), Phys. Lett. **B202** (1988) 149.

Top Physics

STEPHEN J. PARKE, *Fermilab*

CDF TOP RESULTS IN THE DILEPTON CHANNEL

JOSE M. BENLLOCH *

*Massachusetts Institute of Technology,
Cambridge, Massachusetts 02139, USA*

ABSTRACT

The current status of the top quark search at CDF in the dilepton channel is presented. In the 1992-93 run (Run Ia), with 19.3 pb^{-1} collected, two $e\mu$ events survived all the cuts, including a two-jet cut for high mass top, with a total estimated background of $0.56^{+0.25}_{-0.13}$ events. With approximately 9 pb^{-1} of data analyzed from the 1993-94 run (Run Ib) a new $e\mu$ event passing all the cuts has been detected.

1. Top Quark Production and Decay at the Tevatron. The Dilepton Channel

The process $q\bar{q} \to t\bar{t}$ is the dominant one for top quark production at the Fermilab Tevatron (center of mass energy of 1.8 TeV). The cross section is a steeply falling function of the top quark mass varying from 30 pb at 120 GeV to 5 pb at 190 GeV. The top quark is expected to decay into a W boson and a b-quark assuming no deviations from the Standard Model. So, the produced top quark pairs have three possible decay channels, giving three different search methods:

1. The dilepton, electrons or muons (taus are not considered) and two jets (the b-quark jets) in the final state. This is the cleanest channel but has the lowest branching ratio (4/81).

2. The single electron or muon (taus are not considered) plus three or four jets (the b-quark jets and the two jets coming from the hadronic decay of a W) in the final state, having a branching ratio of 24/81.

3. The six jets final state, having a branching ratio of 36/81.

In this contribution we report on the results from the dilepton channel search mainly from the 19.3 ± 0.7 pb^{-1} collected by CDF during the 1992-93 run (Run Ia).

---

*Representing the CDF Collaboration.

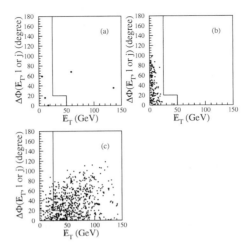

Fig. 1. Distributions of the azimuthal angle between missing E_T and the closest lepton or jet versus E_T. a) $e\mu$ data. b) Dielectron and dimuon data after the invariant mass cut. c) Monte Carlo (unnormalized) events for $M_{top} = 160 \ GeV/c^2$.

2. Dilepton Event Selection

Two high p_T $(P_T > 20 \ GeV)$ leptons (electrons or muons) with opposite charges are required in this channel. At least one lepton has to be central $(|\eta| < 1.0)$ and pass tight identification requirements and a track isolation cut (with an efficiency of $87 \pm 1\%$). The second lepton may pass looser cuts (with an efficiency of $94 \pm 1\%$).

An invariant mass cut is applied to remove $Z's$ in ee, $\mu\mu$: $75 \ GeV/c^2 < M_{ll} < 105 \ GeV/c^2$. For $M_{top} = 160 \ GeV/c^2$, 80 % of dielectron and dimuon events pass the invariant mass cut.

Due to the presence of two high P_T neutrinos in the top dilepton channel a missing E_T cut is applied on its magnitude: $\not E_T > 25 \ GeV$. At this point, the background is dominated by Drell-Yan in the ee, $\mu\mu$ channels and by $Z \to \tau\tau$ in the $e\mu$ channel. For events with $\not E_T < 50 GeV$ it is also required that the azimuthal angle between the $\not E_T$ direction and the direction of the closest jet be $\Delta\phi(\not E_T, jet) > 20^0$ to reject the Drell-Yan continuum background, which would pass the missing $\not E_T$ magnitude cut due to mis-measured jet energies from cracks, etc. A similar cut is applied to the direction of $\not E_T$ with respect to leptons to reject the $Z \to \tau\tau$ background $(\Delta\phi(\not E_T, l) > 20^0$ when $\not E_T < 50 \ GeV/c^2)$, since the missing $\not E_T$ produced by neutrinos from $\tau's$ goes primarily along the lepton direction. For $M_{top} = 160 \ GeV/c^2$, 76 % of the dilepton events pass all the missing $\not E_T$ cuts.

After all these cuts, there are two $e\mu$ events (see figure 1) and no dielectron nor dimuon events in the signal region.

For a high mass top ($M_{top} > 120 \ GeV/c^2$) the two b-quarks will form high E_T jets with high efficiency. So, we require two or more jets with tranverse energy greater than 10 GeV. The two-jet cut reduces the background by a factor 4 while being 84 % efficient for $M_{top} = 160 \ GeV/c^2$. The remaining two events also pass the two-jet cut.

The number of data events surviving the sequence of cuts is shown in table 1.

Cut	$e\mu$	ee	$\mu\mu$
P_T	8	702	588
Opposite-Charge	6	695	583
Isolation	5	685	571
Invariant Mass	5	58	62
$\not{E}_T$ magnitude	2	0	1
$\not{E}_T$ direction	2	0	0
Two-jet	2	0	0

Table 1. Number of data events surviving consecutive requirements.

3. Dilepton Acceptance

The efficiency has been studied mainly with the Isajet Monte Carlo program[1]. The detection efficiencies, the predicted central value of the $t\bar{t}$ production cross section from the NNLO theoretical cross section[2] and the number of expected events in $19.3pb^{-1}$ is given in table 2. After all the cuts, the contribution from the different channels is: $e\mu : 59\%; \ ee : 21\%; \ \mu\mu : 20\%$.

M_{top} GeV/c^2	ϵ_{DIL}	$\sigma_{t\bar{t}}$ pb	$N_{e\mu}$ events	$N_{e\mu,ee,\mu\mu}$ events
120	0.0049	38.9	2.2	3.7
140	0.0066	16.9	1.3	2.2
160	0.0078	8.2	0.8	1.3
180	0.0086	4.2	0.4	0.7

Table 2. Detection efficiencies, $\epsilon_{DIL} = Br \cdot \epsilon_{total}$, the predicted central value of $t\bar{t}$ production cross section and the number of events expected in 19.3 pb^{-1}, as functions of top mass.

The dilepton acceptance uncertainty is dominated by the two-jet cut which varies from 36% to 3% ($M_{top} \ 100 \ - \ 180 \ GeV/c^2$). The other cuts have uncertainties which do not depend strongly on the value of the top mass.

4. Dilepton Backgrounds

The most important backgrounds have been estimated to be WW, Drell-Yan $(\gamma/Z \rightarrow ee, \mu\mu)$, $Z \rightarrow \tau\tau$, $b\bar{b}$ processes and QCD or W+jets processes with at least one misidentified lepton.

The WW background is estimated using the Isajet Monte Carlo program, normalized to a total cross section of 9.5 pb, and the CDF detector simulation. Before the two-jet cut we expect 1.17 ± 0.37 events, where the error is dominated by the theoretical uncertainty in the cross section. Only 13% of the events pass the two-jet cut. The effects of initial state radiation for the two-jet cut are checked using the Drell-Yan process. The total WW background after the two-jet cut is predicted to be 0.16 ± 0.06 events.

The $\tau\tau$ background is determined using the hybrid technique of taking the $Z \rightarrow ee$ events and substituting Monte Carlo generated taus in place of the electrons. We expect 0.13 ± 0.04 events after the two-jet cut.

The $b\bar{b}$ background is estimated using the Isajet Monte Carlo for the production and the CLEO Monte Carlo[3] for b decays. To check Isajet, the lepton P_T cut is lowered to 15 GeV/c and 5 GeV/c (for the first and second lepton, respectively), where heavy flavor is dominant, and the distributions found to be in good agreement with the $e\mu$ data. We predict 0.10 ± 0.06 events from this background.

The lepton misidentification background (QCD or W+jets processes with at least one fake lepton) is determined using jet data samples to assign a probability per track to pass the electron and muon identification cuts. These fake probabilities are applied to events with a good lepton plus additional tracks to estimate a total background of 0.07 ± 0.05 events.

The non-resonant Drell-Yan backgrounds for ee, $\mu\mu$ are determined using Z data to get scale factors for cuts (0.1% of Z events pass all the cuts, based on one $Z \rightarrow \mu\mu$ event), and apply the same scale factors to the number of non-resonant ee $\mu\mu$ events. We expect $0.10^{+0.23}_{-0.08}$ events from this background.

The total expected background is 0.24 ± 0.06 events in the $e\mu$ channel and $0.31^{+0.24}_{-0.10}$ in the ee, $\mu\mu$ channels. As a cross check of the background estimation in the $e\mu$ channel the lepton P_T is lowered to 15 GeV and the number of predicted events (25 ± 3) is found to be in good agreement with the number of events observed (18).

5. Conclusions

There are two events passing all the top selection cuts in the $e\mu$ channel from Run Ia $(19.3 \pm pb^{-1})$, while 3.7 (0.7) events are expected for $M_{top} = 120$ (180) GeV/c^2. The total expected background is $0.56^{+0.25}_{-0.13}$. One of the two top candidate events has a jet that has been b-tagged by both the SVX detector and the soft muon tagging algorithms. From the analysis of the new data from Run Ib $(\approx 9\ pb^{-1})$ one more $e\mu$ event has appeared passing all cuts.

References

1. F. Paige and S.D. Protopopescu, BNL Report No. 38034 (1986).
2. E. Laenen, J. Smith, and W.L. Van Neerven, *Phys. Lett.* **321B** (1994) 254.
3. P. Avery, K. Read, and G. Trahern, Cornell Internal Note CSN-212 (1985).

CDF TOP RESULTS IN THE LEPTON + JETS CHANNEL

GORDON WATTS*

Department of Physics and Astronomy,
University of Rochester,
Rochester, New York 14627, USA

ABSTRACT

Results from the 1992-1993 Tevatron run (run IA) top search at CDF in the lepton+jets channel are reported. The jet-vertexing algorithm tags 6 events on a background of 2.3 ± 0.3 events in the $W+ \geq 3$ jets data sample, and the soft lepton analysis finds 7 events on a background of 3.1 ± 0.3 events in the same data sample. Jet-vertex and soft-lepton tag correlations are described. Also given are the expected and observed number of tags in the Z+mulitjet control sample.

1. The Lepton+Jets Channel

In the minimal standard model, top decays almost exclusively by $t\bar{t} \to W^+ b W^- \bar{b}$. The top search in lepton+jets looks for a lepton from one of the Ws and jets from the other ($W^+ \to e\nu$ or $\mu\nu$ and $W^- \to$ jets, or the charge conjugate). Top decays to this final state about 30% of the time. Events selected for the lepton+jets search are required to have an isolated lepton with E_T (P_T for muons) > 20 GeV and $|\eta| < 1.0$, and to have $\not{E}_T > 20$ GeV.[1] Events containing Z bosons, with an ee or $\mu\mu$ invariant mass between 70 and 110 GeV/c^2, are removed. The dominant background in the lepton+jets top search is the direct production of W+jets. The ratio of the $t\bar{t}$ signal to W+jets background can be greatly improved by requiring $N_{jet} \geq 3$. This requirement has a rejection factor of $\approx$400 against inclusive W production while keeping approximately 75% of the $t\bar{t}$ signal for $M_{top} = 160$ GeV/c^2. In the $W+ \geq$3-jet sample, we expect 12±2 (6.6±0.7) $t\bar{t}$ events for $M_{top} = 160$ (180) GeV/c^2 using the acceptance discussed below and the theoretical cross section.[2] We observe 52 events with $N_{jet} \geq 3$ in the W sample.

The VECBOS[3] Monte Carlo program is used to make an estimate of the direct W+jets background, and predicts 46 events with ≥ 3 jets in Run 1A's 19.3pb^{-1} of data. There are, however, uncertainties of about a factor of two due to the choice in the Q^2 scale used in the calculation. We have, therefore, developed a technique for estimating backgrounds in the lepton+jets search directly from the data, described below. Other backgrounds (direct $b\bar{b}$, Z bosons, W pairs, and hadrons misidentified as leptons) contribute an additional 12.2±3.1 events.[1] Additional background rejection is needed to isolate a possible $t\bar{t}$ signal. Requiring the presence of a b-quark jet, tagged either by a secondary vertex or by a semileptonic decay, provides such a rejection.

---

*Representing the CDF Collaboration at Fermilab.

1.1. Jet Vertex Search

The lifetime of b hadrons can cause the b-decay vertex to be measurably displaced from the $\bar{p}p$ interaction vertex. CDF's new Silicon Vertex Detector (SVX)[4] is a high resolution $r - \phi$ tracker capable of resolving a displaced B decay vertex.

Found high quality SVX tracks $(P_T \geq 2\text{GeV/c}, |d|/\sigma_d \geq 3.0$, and with clean SVX reconstruction)[1] are associated with jets that have $E_T > 15$ GeV and $|\eta| < 2.0$ if $\cos(\text{jetaxis} - \text{track}) > 0.8$. For each jet a transverse decay length (L_{xy}) and its uncertainty (typically $\sigma_{L_{xy}} \approx 130$ μm) are calculated using a three-dimensional fit, with the jet's tracks constrained to originate from a common vertex. Jets that have a secondary vertex displaced in the direction of the jet, with significance $L_{xy}/\sigma_{L_{xy}} \geq 3.0$ are positively tagged $(+L_{xy})$, and those with a secondary vertex behind the primary, with significance $L_{xy}/\sigma_{L_{xy}} \leq 3.0$ are negatively tagged $(-L_{xy})$. The largest contributions to the negative tags are resolution smearing and mistakes in the tracking reconstruction.

We use a control sample of inclusive electrons, which are enriched in b-decays, $(E_T > 10$ GeV) to measure an efficiency for SVX-tagging of a semileptonic b jet.[1] We compare this efficiency with that predicted by the ISAJET + CLEO[5] $b\bar{b}$ Monte Carlo and find our measured efficiency to be lower than the Monte Carlo prediction by a factor of 0.72 ± 0.21. To verify that our tags are due to heavy flavor, the transverse decay length, L_{xy}, is converted into an estimate of the effective proper decay length $(c\tau_{eff})$ using the expression $c\tau_{eff} = L_{xy}\frac{M}{P_T}F$, where M is the invariant mass of the tracks associated with the displaced vertex, P_T is their total vector transverse momentum, and F is a scale factor, determined from a b Monte Carlo sample. The scale factor F accounts for b-hadron decay products that are not attached to the secondary vertex. Figure 1 is the $c\tau_{eff}$ of both the tags in the inclusive electron sample (points) and the tags in $b\bar{b}$ Monte Carlo (histogram); the two agree.

We then determine the efficiency for tagging at least one b jet in a $t\bar{t}$ event with three or more observed jets, ϵ_{tag}, from $t\bar{t}$ Monte Carlo. The efficiency must be rescaled by 0.72, as determined above. We find $\epsilon_{tag} = 22 \pm 6\%$ independent of top mass for $M_{top} > 120$ GeV/c^2. The efficiency, ϵ_{SVX}, for inclusive $t\bar{t}$ events to pass the lepton-identification, kinematic, and SVX b-tag requirements is shown in Table 1. The number of expected SVX-tagged $t\bar{t}$ events with $N_{jet} \geq 3$ is shown in the same table. Six SVX-tagged events are observed in the 52-event $W+ \geq 3$-jet sample.

Rather than rely on Monte Carlo predictions, we estimate directly from our data how many tags we would expect in the 52-event sample if it were entirely background. We assume that the heavy-quark (b and c) content of jets in $W+$jets background events is the same as in an inclusive-jet sample.[1] This assumption is expected to be conservative, since the inclusive-jet sample contains heavy-quark contributions from direct production (e.g. $gg \rightarrow b\bar{b}$), gluon splitting (where a final-state gluon branches into a heavy-quark pair), and flavor excitation (where an initial-state gluon excites a heavy quark in the proton or antiproton sea), while heavy quarks in $W+$jets background $(Wb\bar{b}, Wc\bar{c})$ events are expected to be produced almost entirely from gluon splitting.[6]

We apply the tag rates measured in the inclusive-jet sample, parameterized by the E_T, track multiplicity and η of each jet, to the jets in the 52 events to yield the total expected number of SVX-tagged events from $Wb\bar{b}$, $Wc\bar{c}$, and fake tags due to track mismeasurement. We have tested this technique in a number of control samples

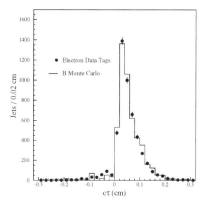

Fig. 1. The $c\tau_{eff}$ distribution for jets with a secondary vertex in the inclusive electron data (points with errors) compared to a Monte Carlo simulation (histogram) with the world average B lifetime.

M_{top}	120 GeV/c^2	140 GeV/c^2	160 GeV/c^2	180 GeV/c^2
ϵ_{SVX}	$1.0 \pm 0.3\%$	$1.5 \pm 0.4\%$	$1.7 \pm 0.5\%$	$1.8 \pm 0.6\%$
ϵ_{SLT}	$0.84 \pm 0.17\%$	$1.1 \pm 0.2\%$	$1.2 \pm 0.2\%$	$1.3 \pm 0.2\%$
$\sigma_{t\bar{t}}^{Theor}$ (pb)	$38.9^{+10.8}_{-5.2}$	$16.9^{+3.6}_{-1.8}$	$8.2^{+1.4}_{-0.8}$	$4.2^{+0.6}_{-0.4}$
$N_{expected}(SVX)$	7.7 ± 2.5	4.8 ± 1.7	2.7 ± 0.9	1.4 ± 0.4
$N_{expected}(SLT)$	6.3 ± 1.3	3.5 ± 0.7	1.9 ± 0.3	1.1 ± 0.2

Table 1. Summary of top acceptance (including branching ratios) and the theoretical cross section.[2] The middle line gives the $t\bar{t}$ production cross section obtained from this measurement. The last two lines are the expected event yield for $t\bar{t}$ in Run 1A's $19.3pb^{-1}$.

and use the level of agreement with the number of observed tags to determine the systematic uncertainty on the predicted tag rate.

The backgrounds from non-W sources (direct $b\bar{b}$ production and hadrons misidentified as leptons) are also determined from the data[1] by studying the isolation of the lepton candidates in the low $\not{E}_T(\not{E}_T < 15\text{GeV}/c)$ and high $\not{E}_T(\not{E}_T > 20\ \text{GeV}/c)$ regions.[7] The total number of non-W background events in the signal region (high $\not{E}_T$ and isolated lepton) is estimated as the number of non-isolated lepton candidates in the high $\not{E}_T$ region scaled by the ratio isolated to non-isolated lepton candidates in the low $\not{E}_T$ region (dominated by backgrounds). To predict the number of tagged non-W events in the signal region we scale by the tagging rate in the low $\not{E}_T$, low isolation region.

The small contributions from Wc, WW, WZ production, and $Z \to \tau\tau$ are estimated with Monte Carlo. The total estimated background, including the inclusive jet prediction, to SVX tags in the 52-event sample is 2.3 ± 0.3 events. Approximately 85% of this background is from heavy flavor and fake tags.

An alternate background estimate, using Monte Carlo calculations of the heavy-quark processes in W+jets events and a fake-tag estimate from jet data, predicts a heavy-quark content per jet approximately a factor of three lower than in inclusive-jet events and gives an overall background estimate a factor of 1.6 lower than the number presented above, supporting the conservative nature of our background estimate.

1.2. Soft Lepton Search

A second technique for tagging b quarks is to search for leptons arising from the decays $b \to \ell\nu X$ ($\ell = e$ or μ), or $b \to c \to \ell\nu X$. Because these leptons typically have lower P_T than leptons from W decays, we refer to them as "soft lepton tags", or SLT tags. We require lepton $P_T > 2$ GeV/c. To keep this analysis statistically independent of the dilepton search, leptons that pass the dilepton requirements are not considered as SLT candidates.

In searching for electrons from b and c decays, each CTC track is extrapolated to the calorimeter, and a match is sought to an electromagnetic cluster consistent in size, shape, and position with expectations for electron showers. The efficiency of the electron selection criteria, excluding isolation cuts, is determined from a sample of electron pairs from photon conversions, where the first electron is identified in the calorimeter and the second, unbiased, electron is selected using a track-pairing algorithm. Combined with electron isolation efficiencies determined from $t\bar{t}$ Monte Carlo, the total efficiencies are $(53\pm3)\%$ and $(23\pm3)\%$ (statistical uncertainties only) for electrons from b and sequential c decays respectively. To identify muons, track segments in the muon chambers are matched to tracks in the CTC. The efficiency for reconstructing track segments in the muon chambers is measured to be 96% using $J/\psi \to \mu^+\mu^-$ and $Z \to \mu^+\mu^-$ decays. This number is combined with the P_T-dependent efficiency of the track-matching requirements to give an overall efficiency of approximately 85% for muons from a combination of both b and c decays.

The acceptance of the SLT analysis for $t\bar{t}$ events is calculated using the ISAJET + CLEO Monte Carlo programs. The efficiency for tagging at least one jet in a $t\bar{t}$ event by detecting an additional lepton with $P_T > 2$ GeV/c is $\epsilon_{tag} = 16 \pm 2\%$, approximately independent of M_{top}. The efficiency, ϵ_{SLT}, for inclusive $t\bar{t}$ events to pass the lepton-identification, kinematic, and SLT b-tag requirements is shown in Table 1 along with

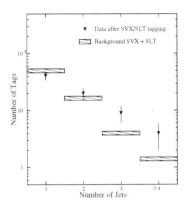

Fig. 2. The sum of SVX and SLT tags observed in the W+jets data (solid triangles). Events tagged by both algorithms are counted twice. The shaded area is the sum of the background estimates for SVX and SLT, with its uncertainty. The 3-jet and $\geq$ 4-jet bins are the $t\bar{t}$ signal region.

the number of expected SLT-tagged $t\bar{t}$ events. We find seven SLT-tagged events with $N_{jet} \geq 3$. Three of the seven also have SVX tags.

The main backgrounds to the SLT search are hadrons misidentified as leptons, and $Wb\bar{b}$, $Wc\bar{c}$ production. As in the SVX analysis, we estimate these backgrounds from the data by conservatively assuming that the heavy-quark content per jet in W+jets events is the same as in inclusive-jet events. By studying tracks in such events, we measure the probability of misidentifying a hadron as an electron or muon, or of tagging a true semileptonic decay. We use these probabilities to predict the number of tags in a variety of control samples, and obtain good agreement with the number observed. We expect 2.70±0.27 tags in the W+ $\geq$3 jet sample from these sources. Other sources (direct $b\bar{b}$, W/Z pairs, $Z \rightarrow \tau\tau$, Wc, and Drell-Yan) contribute 0.36±0.09 events, for a total SLT background of 3.1±0.3 events. The number of SLT tags in the W+1 and W+2-jet samples, which should have only a small contribution from $t\bar{t}$, agrees with the background expectation (45 events tagged, 44±3.4 predicted). Figure 2 shows the combined number of SVX and SLT tags, together with the estimated background, as a function of jet multiplicity.

2.   Jet-Vertex and Soft Lepton Correlations

The correlations between the SVX and SLT taggers must be understood before we can understand their combined significance. We do this by looking at a large sample of jets. Because the jet vertexing's negative decay $(-L_{xy})$ length tags are not expected to be enriched in heavy flavor (as the positive tags are), they are not expected to be correlated with soft lepton tags. Applying the soft lepton tagger and its prediction to the $-L_{xy}$ tags yields a observed-minus-predicted rate that differs from zero by $+1.3\sigma$

for double tagged jets and $+1.0\sigma$ for double tagged events. Note that an exact match means only that correlations in the two taggers are accounted for in the background prediction. In the case of the $-L_{xy}$ sample, the correlations are accounted for by the parameterization.

The SVX $+L_{xy}$ sample is enriched in heavy flavor, and thus the SLT prediction is much smaller than the observed number of SVX-SLT double tags. If we assume the excess jet vertex tags, $(+L_{xy}) - (-L_{xy})$, are due to heavy flavor, we can calculate a second prediction which closely matches the number of observed double tags. We conclude that the correlation in the $-L_{xy}$ sample is properly modeled by the tag rate probabilities, and the correlation in the $+L_{xy}$ sample is understood as resulting from the heavy flavor content of the sample. Monte Carlo calculations predict 1.8 ± 1.3 double tagged events from $t\bar{t}$ and background, thus the three double tags in the 52 signal events are consistent with the $t\bar{t}$ hypothesis.

3.   Tags in the Zs

The Z+multijet sample is, in principle, a good cross check of the heavy flavor content of the W+multijet sample. The standard model production mechanisms are similar in Ws and Zs, but there should be no top in the Z sample. In practice, however, $\sigma \times BR$ for $Z \to ee, \mu\mu$ is an order of magnitude less than for $W \to e\nu, \mu\nu$.

Events with oppositely charged ee or $\mu\mu$ pairs with 75 GeV/c^2 $< m_{ee}, m_{\mu\mu} < 105$ GeV/c^2 are selected. The first lepton is required to pass the same cuts as the high P_T leptons, and the second lepton has a E_T cut of 10 GeV (with a tighter calorimeter isolation cut to remove background).

Backgrounds are calculated using the same method we used in the W+jets background calculation, but no backgrounds from non-Z sources are included ($t\bar{t}$, $b\bar{b}$, etc.). 8 tags are observed, and 5.8 ± 0.4 are predicted. If we look in only the $Z+ \geq 3$ jets region, equivalent to the signal sample in the top search, we find 2 events with a background of .64 events. This is a low statistics check and we are currently taking more data so that we can study the Z+multijets sample further.

4.   Conclusion

The jet vertex search finds 6 tags in the $W+ \geq 3$ sample with a background of 2.3 ± 0.3 events. Using a background estimation technique that relies more on Monte Carlo, we see a background of 1.4 ± 0.5. The soft lepton search finds seven tags with a background of 3.1 ± 0.3 events. Correlations between the two taggers are understood. A low statistics check of the tagging algorithms has been made in the Z+multijet sample; more data is required to provide conclusive results.

References

1. F Abe et el., Fermilab-Pub-94/097-E, submitted to *Phys. Rev.* **D**.
2. E. Laenen, J. Smith, and W.L. Van Neerven, *Phys. Lett.* **B321** (1994) 254.
3. F.A. Berends, W.T. Giele, H. Kuijf, and B. Tausk, *Nucl. Phys.* **B357** (1991) 32.
4. Fermilab-94/024-E, submitted to *Nucl. Instrum. Methods Phys. Res.*.
5. E. Paige and S.D. Protopopescu, *BNL Report* No. 38034, (1986).
6. M.L. Mangano, *Nucl. Phys.* **B405** (1993) 536.
7. F. Abe et al., *Phys. Rev.* **D44** (1991) 29.

EVIDENCE FOR $t\bar{t}$ PRODUCTION AT THE TEVATRON: STATISTICAL SIGNIFICANCE AND CROSS-SECTION

JACOBO KONIGSBERG*

Physics Department, Harvard University, 42 Oxford St., Cambridge, Mass. 02138

ABSTRACT

We summarize here the results of the "counting experiments" by the CDF Collaboration in the search for $t\bar{t}$ production in $p\bar{p}$ collisions at $\sqrt{s} = 1800\ TeV$ at the Tevatron.[1] We analyze their statistical significance by calculating the probability that the observed excess is a fluctuation of the expected backgrounds and, assuming the excess is from top events, extract a measurement of the $t\bar{t}$ production cross-section.

1. Statistical Significance of the Counting Experiments

The counting experiments that search for the top quark in the CDF experiment all yield an excess of events over the estimated background. The reader is referred to the talks in these proceedings that summarize the searches.[1] In this Section we test the "Null Hypothesis"; we find the probability that the number of observed events is consistent with a fluctuation of the estimated background. We first do this for each of the experiments separately and then we combine the experiments. The detailed account of these derivations can be found in the PRD article published by the CDF collaboration[2]

1.1. Significance of the Individual Analyses

If n_B^i is the estimated background for each of the analyses and σ_B^i is its uncertainty on n_B^i, then the probability $P_i(n)$ that we observe n events is formed by smearing a Poisson distribution with mean n_B^i by a Gaussian distribution with mean n_B^i and width σ_B^i.

$$P_i(n) \sim Gaussian(\mathbf{n_B^i}, \sigma_\mathbf{B}^\mathbf{i}) \times Poisson(\mathbf{n_B^i}; n)$$

From this distribution we find the probability for having $n \geq n_{obs}$. Table 1 summarizes the results for the individual experiments. We notice that although the individual probabilities are small they are not negligible. The SVX tag refers to tagging heavy flavour with the CDF silicon vertex detector. The SLT algorithm tags heavy flavor by looking for soft leptons.

---

*Representing the CDF Collaboration

Analysis	n_B^i	σ_B^i	n_{obs}	$P_i(n \geq n_{obs})$
Dileptons	0.56	$^{+0.25}_{-0.13}$	2	12%
Lepton + jets + SVX tag	2.3	± 0.3	6	3.2%
Lepton + jets + SLT tag	3.1	± 0.3	7	3.8%

Table 1. Probability of a background fluctuation for the individual experiments

1.2. Significance of the Combined Analyses

1.2.1. Significance by counting events

In the sets of 6 and 7 events found by the independent SVX and SLT analyses respectively, 3 events are in common. Thus 3 events are double-tagged (the two tags are not necessarily in the same jet). If we combine the analyses by simply counting events that are tagged we need to subtract the overlap between the SLT and the SVX analyses when we add the number of events found and the background expected. We have therefore 10 events in the lepton + jets channels and 2 events in the dilepton channel for a total of 12. The total background is found by adding the third column in Table 1 and subtracting the expected overlap in background between the SLT and SVX analyses which amounts to 0.26 events.[2] This results in $5.7^{+0.47}_{-0.44}$ background events.

The probability that 5.7 events fluctuate to 12 or more is again found by smearing a Poisson distribution with a mean of 5.7 with a Gaussian distribution with the same mean and a sigma of 0.47. We find this probability to be 1.6%.

In combining the experiments this way we've ignored the fact that 3 of our lepton + jets events are double-tagged.

1.2.2. Significance including double-tags

Given that a tag in an event tagged by both SVX and SLT is approximately six times more likely to be from heavy flavor than from a mistag,[2] we combine the SVX and SLT analyses by simply counting the number of tags (both from SLT and SVX) in the lepton + jets sample. If an event is double-tagged (either in the same or in a different jet) we count twice. We therefore have 13 "counts" from the lepton + jets analyses and 2 from the dilepton analysis, for a total of 15 "counts". Note that for dileptons we do not count the b-tags in the events given that a-priori we did not request a b-tag for the selection. The acceptance for top events would be too small. However one of the dilepton candidate events is double-tagged.

The total background, without taking into account correlations between the tagging methods, is $5.96^{+0.49}_{-0.44}$, found by adding the fourth column in Table 1. The probability for 5.96 counts to fluctuate to 15 or more is 0.16%.

In order to properly take into account the background correlations between SVX and SLT taggers we estimate the combined probability of a background fluctuation using a Monte Carlo method. The correlations are studied in a sample of generic QCD jets. We find that the number of expected SLT tags in those jets with a negative L_{xy} is

17 ± 2 which is consistent with the 22 observed. (L_{xy} is the secondary vertex distance in the transverse plane. Negative values are obtained due to tracking errors and resolution effects). The probability for these mistags is found to be twice that of the independent probabilities. This correlation is due to the preference by both algorithms to tag jets with many tracks. For jets with positive L_{xy}, after subtracting the mistag content from resolution effects and leaving only heavy flavor, we expect 77 ± 9 SLT tags, consistent with the 66 observed. The probabilities for SLT and SVX to tag heavy flavor are found to be uncorrelated.

In the Monte Carlo method used for estimating the combined probability of a background fluctuation we take the estimates for the mean number of events and the uncertainty for each background type in the sample of 52 $W+ \geq 3$ jets. This includes $Wb\bar{b}$, $Wc\bar{c}$, Wc, W+no heavy flavor (which are estimated from Monte Carlo but scaled up to the amounts indicated by background method I[1,2]), and $b\bar{b}$, WW and $Z \rightarrow \tau\tau$. We perform a large number of Monte Carlo "background experiments". In each we sample from these populations with the constraint that $\sum n_i = 52$. We then "apply" the tagging algorithms including the correlations for mistags. We add the dilepton background independently.

We find that the probability for the background to fluctuate to 15 or more counts is:

$$P(n \geq 15) = 0.26\%$$

Had we used the method II background estimates[1,2] we would have obtained that $P(n \geq 15) = 0.036\%$.

We've tested the robustness of the excess by changing the jet E_T thresholds by $+5$ GeV(-5 GeV) in the event selection. In this case we expect 3.6 (12.7) counts in background events, 5.8 (7.9) from top events (for $m_{top} = 160$ GeV/c^2) and we observe 11 (19). We've also doubled the correlations between the tagging algorithms in the Monte Carlo method described above and we obtain $P(n \geq 15) = 0.37\%$.

We conclude that the excess has a small probability of coming from a background fluctuation and seems robust. However the limited statistics do not allow us to firmly establish the existence of the top quark. If we interpret this excess as due to $t\bar{t}$ production we can measure the cross-section, $\sigma_{t\bar{t}}$.

2.   Cross-Section Measurement for $p\bar{p} \rightarrow t\bar{t} + X$

In order to calculate the top cross-section we need to re-estimate the amount of background in the 52 $W+ \geq 3$ jets event sample. (The estimates in the previous Section were done assuming the "Null Hypothesis"). We do this by an iterative method[2] and obtain that for the SVX (SLT) analysis we expect 1.6 ± 0.7 (1.5 ± 0.7) counts from backgrounds.

The cross sections are calculated by maximizing the following likelihood function:

$$L = e^{-\frac{(\int \mathcal{L}dt - \overline{\int \mathcal{L}dt})^2}{2\sigma_{\mathcal{L}}^2}} L_{DIL} \cdot L_{SVX} \cdot L_{SLT}$$

where each of the individual likelihoods is of the form

$$L_i = G(\epsilon_i, \overline{\epsilon_i}, \sigma_{\epsilon_i}) G(b_i, \overline{b_i}, \sigma_{b_i}) P(\{\epsilon_i \cdot \sigma_{t\bar{t}} \cdot \int \mathcal{L}dt + b_i\}, n_i).$$

Here $G(x, \overline{x}, \sigma)$ is a Gaussian in x, with mean $\overline{x}$ and variance σ^2, and $P(\mu, n)$ is a Poisson probability for n with mean μ. In each of these likelihoods, $\overline{\epsilon_i}$ and σ_{ϵ_i} are the total acceptance and its uncertainty, $\overline{b_i}$ and σ_{b_i} are the expected background and its uncertainty, n_i is the number of observed candidate events, $\sigma_{t\bar{t}}$ is the $t\bar{t}$ production cross section, and $\int \mathcal{L}dt = 19.3 \ pb^{-1}$ is the integrated luminosity, with $\sigma_{\mathcal{L}}$ its 3.6% uncertainty.

To calculate the cross section from an individual analysis, the individual likelihood functions are used, in which case the maximum likelihood solution for $\sigma_{t\bar{t}}$ is just

$$\sigma_{t\bar{t}} = \frac{n - \overline{b}}{\overline{\epsilon} \cdot \int \mathcal{L}dt}$$

The uncertainties on the measured cross section values are calculated as the $\Delta \log L = \frac{1}{2}$ points of the likelihood function.

The calculated cross sections from the individual SVX, SLT and dilepton results, as well as the combined result (labeled $\sigma_{t\bar{t}}^{ALL}$) are shown in Table 2. In Figure 1 we plot the calculated combined cross section as a function of mass, and the theoretical expectation from Reference.[3] Because the acceptance depends on m_{top}, four points are shown corresponding to measured values of the acceptance. Had we chosen to use the method II background estimate for SVX, and an equivalent estimate for SLT, the $t\bar{t}$ cross section measurement would have shifted upward by 11%. We also note that an alternate method of calculating the cross section, based on the total number of observed events, gives a result approximately 12% lower with comparable uncertainties. From a dataset of $lepton+ \geq 4jets$ we have estimated[4] the top mass to be $m_{top} = 174 \pm 16 \ GeV/c^2$. This mass yields:

$$\sigma_{t\bar{t}} = 13.9^{+6.1}_{-4.8} \ pb$$

This cross section is somewhat higher than the theoretical calculation[3] for the same mass. We address the mutual compatibility with a χ^2 analysis on our measured mass, our cross section as a function of mass, the theoretical cross section versus mass, and their respective uncertainties. We find that the three results are compatible at a confidence level of 13%. We note, however, that the QCD uncertainties on the top cross sections can be larger[5] than those reported in Reference.[3]

3.   Conclusions

We have calculated the probability that the excess of events in our search for the top quark be due to a fluctuation in the estimated backgrounds. We find this probability to be 0.26%. The limited statistics do not allow us to definitely establish the existence of the top quark. Under the assumption of $t\bar{t}$ production we measure the mass to be $m_{top} = 174 \pm 16 \ GeV/c^2$ and the corresponding cross-section for $p\bar{p} \rightarrow t\bar{t} + X$ to be $\sigma_{t\bar{t}} = 13.9^{+6.1}_{-4.8} \ pb$.

M_{top}	120 GeV/c^2	140 GeV/c^2	160 GeV/c^2	180 GeV/c^2
$\sigma_{t\bar{t}}^{SVX}$(pb)	$21.5^{+17.2}_{-11.3}$	$14.9^{+12.2}_{-7.9}$	$13.0^{+10.6}_{-6.9}$	$12.4^{+10.0}_{-6.5}$
$\sigma_{t\bar{t}}^{SLT}$(pb)	$28.9^{+18.5}_{-13.3}$	$22.7^{+14.3}_{-10.4}$	$20.4^{+12.7}_{-9.3}$	$18.8^{+11.7}_{-8.6}$
$\sigma_{t\bar{t}}^{DIL}$(pb)	$15.2^{+19.5}_{-12.2}$	$11.3^{+14.2}_{-9.0}$	$9.6^{+12.0}_{-7.6}$	$8.8^{+11.0}_{-7.0}$
$\sigma_{t\bar{t}}^{ALL}$(pb)	$22.7^{+10.0}_{-7.9}$	$16.8^{+7.4}_{-5.9}$	$14.7^{+6.5}_{-5.1}$	$13.7^{+6.0}_{-4.7}$

Table 2. $t\bar{t}$ production cross sections calculated from the individual analyses and from the combination of the three analyses.

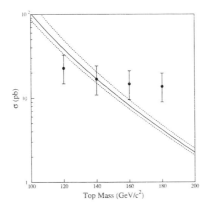

Fig. 1. Combined $t\bar{t}$ production cross section vs. M_{top} from data (points) and theory.[3] The dashed lines are estimates of the theoretical uncertainty quoted in Reference.[3]

References

1. These analyses are described in the talks by J. Benlloch and G. Watts in these proceedings.
2. F. Abe *el al*, Fermilab-Pub-94/097-E, submitted to *Phys. Rev.* **D**.
3. E. Laenen, J. Smith, and W. L. van Neerven, Nucl. Phys. **B369**, 543 (1992).
4. Talk by B. Harral in these proceedings.
5. R.K. Ellis, Phys. Letters, B259, 492(1991).

STUDIES OF KINEMATIC DISTRIBUTIONS AND MASS RECONSTRUCTION OF CDF TOP CANDIDATE EVENTS

BRIAN D. HARRAL*

*Department of Physics, University of Pennsylvania,
209 S 33rd St, Philadelphia PA 19104-6396, USA*

ABSTRACT

We present analyses of CDF top-candidate events which study the separation of signal from background using easily-defined kinematic variables, and which study invariant masses of three-body systems to fully reconstruct the top quark decay products.

1. Introduction

To check whether the CDF data are consistent with a significant top content, various techniques involving the event kinematics have been attempted. One method to be discussed here is to find simple kinematic quantities which can provide sufficient background rejection. We find that using various combinations of jet energies provides the needed discrimination. We also discuss our method of fully reconstructing events under the hypothesis of top decays, which provides us with our estimate of the top mass. Both methods use the "lepton+jets" event sample described in these proceedings[3] as opposed to the "dilepton" events[4] to take advantage of the increased branching ratio and the less-ambiguous event structure.

2. Kinematics

We expect that the most significant background to the top signal in lepton+jets is standard production of W bosons with associated QCD production of jets; our estimates of other backgrounds (due to lepton misidentification, Z decays, dibosons, and others) are at least an order of magnitude lower.[1] The jets produced by QCD processes are produced, on average, with less transverse energy than jets from $t\bar{t}$ events. This leads reasonably to comparisons between jet energies in $t\bar{t}$ events and in W+jet events from Monte Carlo.

To generate $t\bar{t}$ events, we have used both the ISAJET Monte Carlo[5] (using the CLEO fragmentation scheme for b hadrons[6]) and the HERWIG Monte Carlo[7] for purposes of comparison. We use the VECBOS Monte Carlo[8] to generate W+jet events.

Jets are defined by clusters of energy found in the calorimeters using a cone of fixed size in η-ϕ space.[9] In both Monte Carlo and detector data, the calorimeter energy has a correction applied to take into account detector nonlinearities, energy lost in

---

*Representing the CDF Collaboration.

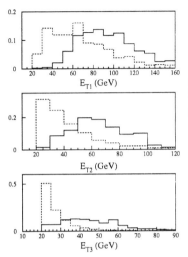

Fig. 1. Comparison of first-, second-, and third-highest E_T jets in ISAJET (solid line) and VECBOS (dashed line) events. Curves are normalized to unit area.

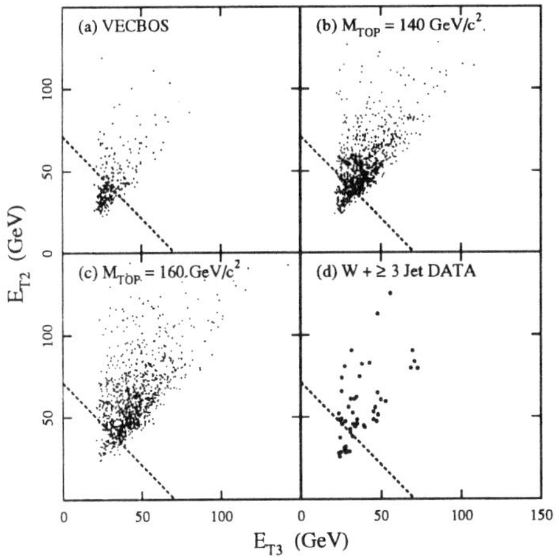

Fig. 2. Plot of second-highest E_T jet $vs.$ third-highest for background MC, $m_{top} = 140$ MC, $m_{top} = 160$ MC, and detector data. The diagonal line ($E_{T2} + E_{T3} = 71$ GeV) is chosen such that half the background events lie below it.

gaps between detectors, energy lost due to particles falling outside the fixed-size cone, and other effects.[10] The jet energy transverse to the beam line (E_T) is compared between VECBOS and $t\bar{t}$ Monte Carlos after this correction has been applied. Figure 1 shows this comparison for VECBOS and ISAJET ($m_{top} = 170$ GeV) for the highest, second-highest, and third-highest E_T jets in events which pass the W+$\geq$ 3-jet criteria.[3] The difference between background and signal becomes more evident when scatter plots of the second-highest $vs.$ the third-highest E_T jets are viewed (Fig. 2). The 52 events of the detector data are also shown in Fig. 2. Top events are expected to populate the high-E_T region of this plot much more heavily than W+jet background; of the 52 events, 39 lie in the high-E_T region defined in Fig. 2, and of the 10 tagged events, eight are in this region. Work is currently in progress to reduce systematic uncertainties (mainly jet energy scale and jet energy corrections), to improve understanding of the background, and to make the analysis more quantitative.

3. Mass Analysis

The mass analysis adds a loose fourth-jet requirement to the W+3-jet event selection; forms appropriate invariant masses with the four highest-E_T jets, the lepton,

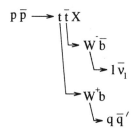

$$p\,\bar{p} \longrightarrow t\,\bar{t}\,X$$

	known	unknown
p's	8	0
t's	1	7
X	2	2
W's	2	6
b's	8	0
q's	8	0
l	4	0
ν	1	3
	34	18

5 vertices — 20 equations
13 4-vectors — 52 variables   18 unknowns — "2C" fit

Fig. 3. Counting constraints in W+4-jet event. See Ref. 1 for more details.

and a vector representing the assumed neutrino; and performs a constrained fit under the hypothesis of top decay. Rather than taking the attitude that an invariant mass analysis can separate signal from background, we use the heavy-flavor tagging to provide both background reduction and reduction in combinatorics. In the hypothesized decay system as we specify it (see Fig. 3), there are two more constraints than unknowns, thus allowing a fit. Note that the dilepton system would then have one more unknown than constraints; work is in progress in deciding the best method of removing this degree of freedom.

In order to completely reconstruct the decay products of the hypothesized top decays, we must therefore require at least four jets in the events. Because of our small amount of data, we loosen the cluster requirements on the fourth jet to $E_T >$ 8 GeV, $|\eta| < 2.4$ to improve our statistics. We also require the presence of an SVX or SLT tagged jet in the event; of the ten tagged events observed in the 1992-93 CDF

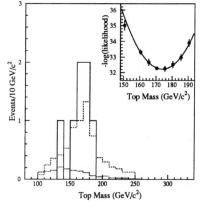

Fig. 4. Top mass distribution for data (solid histogram), W+jets background (dots), and the sum of background and Monte Carlo with m_{top} = 175 GeV (dashes). Background is normalized to 1.4 events. Inset shows likelihood fit for top mass.

run, seven pass the loose fourth-jet requirement. Of these, we estimate a contribution of $1.4^{+2.0}_{-1.1}$ events from non-$t\bar{t}$ processes.

In forming invariant masses (under the assumptions of Fig. 3), we consider only the four highest-E_T jets in each event. Of the 12 ways to associate these four jets with the final-state quarks, we consider only those where the tagged jet is assigned to a b quark, leaving six possibilities (reducing combinatoric background and improving mass resolution); however, there are in general two possible solutions for the z-momentum of the neutrino satisfying the W mass constraint, giving 12 potential event configurations.

The main difficulty (and source of systematic error) is making the connection between energy measured in the calorimeter and quark momentum; the correction function referred to in Section 2 above is not sufficient, mainly due to the possible presence of muons and neutrinos in b-quark jets. We have therefore developed corrections specifically for jets from W decays, and for b jets with and without semileptonic decays, and applied these to correct the jets in the candidate events before mass fitting. Details of these corrections can be found in Ref. 1.

A χ^2 is formed using the uncertainties in measurement of the leptons and the uncertainty in making the jet-quark connection mentioned above. This χ^2 is minimized subject to the mass constraints of the problem for each of the 12 event configurations. We choose the lowest-χ^2 configuration with $m_{top} < 260$ GeV as giving our best estimate of the mass for that event.

To get a best mass for the experiment, we begin by computing mass spectra for $t\bar{t}$ Monte Carlo events at various values of m_{top}, and for W+jet Monte Carlo. We then fit our observed mass distribution to a linear combination of signal and background, subject to our background estimate given above. The likelihood for these fits can then be plotted as a function of m_{top} to find the best overall mass and its statistical error. This information is summarized in Fig. 4, with the inset plot showing the likelihood fit and the main plot displaying the mixture of signal and background spectra superimposed on the data. Details of systematic errors can be found in Ref. 1; our result is $m_{top} = 174 \pm 10^{+13}_{-12}$ GeV/c^2.

We thank the Fermilab staff and the technical staffs of the participating institutions for their vital contributions. This work was supported by the U.S. Department of Energy and National Science Foundation; the Italian Istituto Nazionale di Fisica Nucleare; the Ministry of Education, Science and Culture of Japan; the Natural Sciences and Engineering Research Council of Canada; the National Science Council of the Republic of China; the A.P. Sloan Foundation; and the Alexander von Humboldt-Stiftung.

References

1. F. Abe *et al.*, FERMILAB-PUB-94-097-E (1994), to be published in Phys. Rev. D.
2. F. Abe *et al.*, Phys. Rev. Lett. **73** (1994) 225.
3. G. Watts, these proceedings.
4. J. Benlloch, these proceedings.
5. F. Paige and S. Protopopescu, BNL Report No. 38034 (1986).
6. P. Avery, K. Read, G. Trahern, Cornell Internal Note CSN-212 (1985).
7. G. Marchesini *et al.*, Comput. Phys. Comm. **67** (1992) 465.
8. F.A. Berends *et al.*, Nucl. Phys. **B357** (1991) 32.
9. F. Abe *et al.*, Phys. Rev. **D45** (1992) 1448.
10. F. Abe *et al.*, Phys. Rev. **D47** (1993) 4857.

SEARCH FOR TOP WITH DØ DETECTOR IN DILEPTON CHANNEL

KRZYSZTOF GENSER

Fermi National Accelerator Laboratory
Batavia, IL 60510-0500, USA

Representing the DØ Collaboration

ABSTRACT

Preliminary results from a search for high mass $t\bar{t}$ quark pair production in $p\bar{p}$ collisions at $\sqrt{s} = 1.8$ TeV with the DØ detector in the ee+jets, $e\mu$+jets, and $\mu\mu$+jets decay channels are presented. No conclusive evidence for top quark production for an integrated luminosity of 13.5 ± 1.6 pb^{-1} is observed.

1. Introduction

At the Fermilab Tevatron $p\bar{p}$ center of mass energy the Standard Model top quark t with the mass m_t greater than about 90 GeV/c^2 is expected to be produced mainly in $t\bar{t}$ pairs through the quark-quark annihilation with some contribution from gluon-gluon fusion. The gluon-gluon contribution decreases as m_t increases.[1] Once m_t is greater than the mass of the W-boson the top quark is expected to decay via the weak charged current $(t \rightarrow W^+ b)$ and the event signatures follow from the W branching fractions.

DØ has published a lower limit for m_t of 131 GeV/c^2 at 95% confidence level.[2] The prediction for the top mass based on the precision electroweak LEP measurements[3] is $m_t = 172^{+13+18}_{-14-20}$ GeV/c^2. The CDF evidence for the top quark indicates its mass to be[4] $m_t = 174 \pm 10^{+13}_{-12}$ GeV/c^2

The analysis presented in this paper is focused on the top mass above 120 GeV/c^2 and includes a new channel $\mu\mu$ which was not used in the previous analysis.[2] The dilepton channels considered: ee, $e\mu$, and $\mu\mu$ constitute 1/81, 2/81, and 1/81 of all $t\bar{t}$ events respectively. The contributions from $e\tau$, $\mu\tau$, and $\tau\tau$ channels having the same signatures as ee, $e\mu$, and $\mu\mu$ are also taken into account in the event yields. The lepton plus jets channels are discussed in complementary DØ papers,[5,6] where also the cross-section results are summarized.[6]

2. DØ Detector and Data Sample

The DØ detector[7] is based on a hermetic, high granularity, high resolution, compensating liquid Argon – depleted Uranium calorimeter. The calorimeter is surrounded by a hermetic muon system which has one super layer of proportional drift chambers before and two super layers of chambers behind magnetized iron toroids. The central tracking system placed between the beam pipe and the calorimeter consists of a vertex proportional chamber, transition radiation detector and central, and forward drift chambers. There is no central magnetic field. The calorimeter energy resolutions can be parametrized as $\sigma(E)/E \approx A/\sqrt{E}$ (E in GeV), where $A=0.15$ for electrons, and $A=0.80$ for jets. The muon momentum resolution is $\sigma(1/p) \approx 0.2/p + 0.01$ GeV/c^{-1}. For minimum bias events, the resolution for each component of the missing transverse energy $\not{E}_T$ is about 1.1 GeV+0.02×(ΣE_T), where ΣE_T is the scalar sum of all the transverse energy E_T in the calorimeter.

The data sample analyzed was collected during the Fermilab Tevatron Collider run 1992/93 and has a corresponding luminosity of 13.5 ± 1.6 pb^{-1} for the ee and $e\mu$ channels, and 9.8 ± 1.2 pb^{-1} for the $\mu\mu$ channel.

3. Lepton Identification

All the $t\bar{t}$ dilepton decay channels are characterized by the presence of two high transverse momentum isolated leptons, large missing transverse momentum and significant jet activity. Once appropriate lepton identification and kinematic cuts are applied one can expect a good signal to background ratio and a good efficiency.

3.1. Electron Identification

Electrons are identified as energy clusters in the electromagnetic calorimeter within the pseudorapidity region of $|\eta| < 2.5$. The clusters are required to have at least 90% of their total energy in the electromagnetic part of the calorimeter. The longitudinal and transverse cluster profile has to be consistent with the shape of the electromagnetic shower initiated by an electron.[8] The isolation requirement is $(E_{TOT}^{0.4} - E_{EM}^{0.2})/E_{EM}^{0.2} < 0.1$ where $E_{TOT}^{0.4}$ and $E_{EM}^{0.2}$ are the total and electromagnetic cluster energy contained within the cone of $\mathcal{R} < 0.4$ and $\mathcal{R} < 0.2$ respectively and $\mathcal{R}$ is defined as $\mathcal{R} = \sqrt{(\Delta\phi)^2 + (\Delta\eta)^2}$. It is also required that there is a central detector track coming from the interaction vertex which is matched with each cluster. In the case of the ee channel, the track has to have the corresponding energy deposition in the drift chambers which is consistent with that of a single charged particle.

3.2. Muon Identification

Muons are identified as tracks in the muon drift chambers after penetrating through 13 to 19 interaction lengths of calorimeter and muon toroids. It is required that there is an associated minimum ionizing deposition of at least 0.5 GeV in the calorimeter along the path of the muon if there is a central track matching the muon track, and a deposition of at least 1.5 GeV if there is not a central detector track which matches the muon track. To reject cosmic rays, no track or hits consistent with a track back-to-back in η and ϕ to the considered muon are allowed. It is required that the magnetic field integrated over the path of the muon is greater than 1.83 Tm in order to assure good momentum determination. In case of the $e\mu$ channel, the isolation is imposed by rejecting muons for which there is a jet with the transverse energy E_T greater than 8 GeV within the cone of $\mathcal{R} < 0.5$ or for which there is an energy deposit of 4 GeV or more in an annular cone of $0.2 < \mathcal{R} < 0.4$ around the muon direction. For the $\mu\mu$ channel, to impose isolation, muons with the transverse momentum relative to the nearest jet smaller than 5 GeV/c are rejected. For the $e\mu$ channel, all muons within pseudorapidity region of $|\eta| < 1.7$ are considered. For the $\mu\mu$ channel, central muons ($|\eta| < 1.1$) are used.

4. Analysis of Dilepton Channels and Background Processes

4.1. μμ channel

This is a new channel which was not included in the previous analysis.[2] The trigger for this channel requires a muon candidate with the transverse momentum p_T greater than 14 GeV/c and a jet candidate with $E_T > 15$ GeV. The offline cuts demand two muons, each with $p_T > 15$ GeV/c. To reject background processes, two jets with $E_T > 15$ GeV are required. Dimuon invariant mass $M_{\mu\mu}$ is required to be greater than 10 GeV/c^2 to remove J/Ψ or Ψ' events. $Z^0 \to \mu\mu$ events are rejected by demanding that the angle between $\not{E}_T$ and the leading muon to be smaller than 165° (165°) for two (three) layer muon tracks. Since the calorimeter missing transverse energy $\not{E}_T^{cal}$ is an independent measure of the transverse momentum of the muon pair, the events are rejected if $\Delta\phi(\not{E}_T^{cal}, \vec{p}_T^{\mu\mu}) > 30°$ where $\Delta\phi$ is the azimuthal opening angle. Further rejection of the $Z^0 \to \mu\mu$ events is achieved by the cut $\not{E}_T > 40$ GeV when $\Delta\phi(\vec{p}_T^{\mu 1}, \vec{p}_T^{\mu 2}) > 140°$. No events remain after all the cuts. Very good agreement between the data and the dominant background process $Z^0 \to \mu\mu$ is seen at all the stages of the analysis. The predicted number of $Z^0 \to \mu\mu$ events is 0.28 ± 0.05. The event yields for this and for the other channels, for the data, $t\bar{t}$ Monte Carlo and for the background processes are summarized in table 1. The event yields for the $t\bar{t}$ production were calculated using central value of the $t\bar{t}$ cross-section.[1] The number of background events coming from processes of the same signature as the process under study was predicted using Monte Carlo simulations. The errors quoted are statistical and systematic. In addition to the quoted errors there is an additional normalization error due to the 12% uncertainty in the value of the integrated luminosity.

4.2. ee channel

The trigger used in this channel is a logical OR of triggers demanding one electron candidate with $E_T > 20$ GeV, or two electron candidates with $E_T > 10$ GeV, or an electron candidate with $E_T > 20$ GeV and $\not{E}_T > 20$ GeV, or one electron candidate with $E_T > 15$ GeV and $\not{E}_T > 20$ GeV and two jets with $E_T > 16$ GeV. The offline selection cuts require two electrons, each with $E_T > 20$ GeV, $\not{E}_T > 25$ GeV and two jets with $E_T > 15$ GeV. $Z^0 \to ee$ events are rejected by demanding $|M_Z - M_{ee}| > 12$ GeV/c^2 if $\not{E}_T < 40$ GeV. No events remain after all the cuts. The complete event yields can be found in table 1. After all cuts, the remaining dominant background processes are $Z^0 \to \tau\tau \to ee$ and $Z^0 \to ee$. There is also an instrumental background when a jet is misidentified as an electron in the W+jets events. The probability of a jet faking an electron was estimated using an independent jet data sample.

4.3. eμ channel

The trigger used in this channel is also a logical OR of triggers requiring one electron candidate with $E_T > 7$ GeV and one muon candidate with $p_T > 5$ GeV/c, or a muon candidate with $p_T > 14$ GeV/c and a jet candidate with $E_T > 15$ GeV, or an electron candidate with $E_T > 15$ GeV and $\not{E}_T > 20$ GeV and two jets with $E_T > 16$ GeV. The offline kinematic cuts require muon $p_T > 12$ GeV/c, electron $E_T > 15$ GeV, $\not{E}_T > 10$ GeV, and two jets with $E_T > 15$ GeV. To reject W($\mu\nu_\mu$)+jets events it is required that $\not{E}_T^{cal} > 20$ GeV. The $\Delta R_{e\mu} > 0.25$ cut is used to remove muon bremsstrahlung events. One event survives all the cuts. This is the same remarkable event which was found in the previous analysis. The event is characterized by two very high p_T leptons. Figure 1a shows muon $1/p_T$ versus electron E_T for the data before the two jet cut is applied. The surviving event is marked by a $\star$. Figure 1b shows the corresponding Monte Carlo distribution. The event yields are summarized in table 1. The remaining dominant background processes are $Z^0 \to \tau\tau \to e\mu$ and $W(\mu\nu_\mu)$+jets with one jet misidentified as an electron.

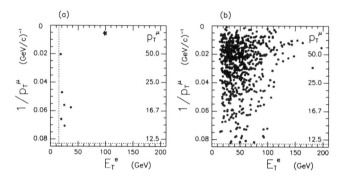

Fig. 1. Muon $1/p_T$ versus electron E_T for the $e\mu$ channel; a) Data 13.5 pb^{-1}, b) 170 GeV/c^2 $t\bar{t}$ Monte Carlo 21.3 fb^{-1}.

Table 1. Summary of event yields. The errors quoted are statistical and systematic. In addition to the quoted errors there is an additional normalization error due to the 12% uncertainty in the value of the integrated luminosity.

	$e\mu$	ee	$\mu\mu$
Luminosity [pb^{-1}]	13.5 ± 1.6	13.5 ± 1.6	9.8 ± 1.2
$t\bar{t}$ MC			
140 GeV/c^2	0.72 ± 0.16	0.41 ± 0.07	0.24 ± 0.05
160 GeV/c^2	0.40 ± 0.09	0.22 ± 0.04	0.12 ± 0.02
180 GeV/c^2	0.23 ± 0.05	0.12 ± 0.02	0.06 ± 0.01
Data	1	0	0
Background	0.27 ± 0.09	0.16 ± 0.07	0.33 ± 0.06

5. Conclusions

DØ has performed a search for high mass $t\bar{t}$ decays into the three dilepton decay modes: $e\mu$+jets, ee+jets, and $\mu\mu$+jets. One event survives in the $e\mu$ channel. This is the same event as found in the original DØ analysis.[2] No candidates are found in the ee and $\mu\mu$ channels. Since the number of expected background events is consistent with the one event seen, it is concluded that no significant $t\bar{t}$ signal is observed. The results are preliminary.

References

1. E. Laenen et al., Nucl. Phys. **B369** (1992) 543, Phys. Lett. **321B** (1994) 254.
2. DØ Collaboration, S. Abachi et al., Phys. Rev. Lett. **72** (1994) 2138.
3. B. Pietrzyk, in Proceedings of XXIXth, Recontres de Moriond, Méribel, 1994.
4. CDF Collaboration, F. Abe et al., Phys. Rev. Lett. **73** (1994) 225, FERMILAB-PUB-94/097-E, 1994, submited to Phys. Rev. D.
5. D. Chakraborty, Search for Top in the Lepton + Jet Channels at DØ, these Proceedings.
6. W. G. Cobau, Search for the Top Quark in Electron + Jets + Bottom Quark Tag using the DØ Detector, these Proceedings.
7. DØ Collaboration, S. Abachi et al., The DØ Detector, Nucl. Instr. and Meth. in Phys. Res. **A338** (1994) 185.
8. DØ Collaboration, M. Narain, in Proceedings of the 7th Meeting of the American Physical Society Division of Particles and Fields, Fermilab (1992) p. 1678, eds. R. Raja and J. Yoh; R. Engelmann et al., Nucl. Instr. Methods **A216** (1983) 45.

A SEARCH FOR $t\bar{t} \to$ *lepton* $+ \not{E}_T + $ *jets* SIGNATURE IN $p\bar{p}$ COLLISIONS AT $\sqrt{s} = 1.8$ *TeV* WITH THE DØ DETECTOR

DHIMAN CHAKRABORTY

Department of Physics, State University of New York at Stony Brook
Stony Brook, NY 11794, USA
for the DØ collaboration

ABSTRACT

We report the results from a search for $t\bar{t}$ production in $p\bar{p}$ collisions at $\sqrt{s}$ =1.8 TeV with the DØ detector at Fermilab in the final states consisting of one isolated high p_T lepton (e or μ), multiple jets and a substantial missing transverse energy ($\not{E}_T$), excluding the events that have μ-tagged jets. Two independent analysis approaches lead us to similar estimates for the cross-section. We see no conclusive evidence for top production in the 13.5 pb^{-1} of data taken during the 1992-1993 run of the Tevatron.

1. Introduction

According to the minimal Standard Model, a t quark should decay almost exclusively into a b quark and a real W boson. Our search focuses on the monoleptonic channels where one of the two W's decays into $l\nu_l$ ("l" stands for an electron or a muon) while the other decays into $q_1\bar{q}_2$. The high mass of top drives the expectation values of the transverse momenta (p_T) of the final state partons pretty high, typically to $\frac{m_W}{2}$. In the detector, we expect to see an isolated high-p_T e or μ, a large $\not{E}_T$ arising from the ν and about 4 high-E_T jets: two from the W decaying hadronically and the two b jets. The number of jets reconstructed is often larger due to initial or final state radiation and sometimes smaller due to jet merging. The total branching ratio of $t\bar{t}$ into such a final state is approximately 0.30. However, in this analysis, we reject any event that has a μ-tagged jet in order to avoid a possible overlap with another analysis that deals specifically with such events only. This results in a reduction in the effective branching ratio within the scope of our analysis to about 0.24.

Having established a new lower limit of 131 GeV (95% CL) from a previous analysis [1], our interest has since moved to the region above it. A vast improvement in our understanding of the detector performance and the background processes, combined with the shifting of the region of interest, has resulted in a significant modification in our analysis approaches. Here we present the new techniques and the results.

2. Signal and Backgrounds

Our signal can be viewed as a cascade production of $W + jets$ through high-mass resonance states. The dominant physical source of background for either channel

(e and μ) is the continuum QCD production of $W + jets$. We shall refer to this as the "W background". The complementary set, i.e. the "non-W background" is due to measurement fluctuations.

To model the kinematics of our signal, we use the ISAJET Monte Carlo [2]. The cross-sections necessary for the calculation of limits are taken from the NNLO calculations by Laenen et al [3]. For the calculation of backgrounds, we rely entirely on our own data except for one of our two analysis approaches where we employ the VECBOS parton generator [4] with ISAJET fragmentation to model the W-background. DØ GEANT is used for Monte Carlo detector simulation.

3.   Event selection

We use very loose sets of requirements for triggering on the events of our interest online. The following offline event selection criteria help us improve the S/B ratio:

- $E_T(j) > E_T^{min}(j) = 15$ GeV. The leptons from W decay are expected to be fairly energetic. A cut $E_T(l) > E_T^{min}(l)$ helps reduce the non-W background. We use $E_T^{min}(e) = 20$ GeV and $p_T^{min}(\mu) = 15$ GeV.

- $\mid \eta(i) \mid < \mid \eta(i) \mid^{max}$ where "i" stands for e, μ, j. We use $\mid \eta(e) \mid^{max} = 2.0$, $\mid \eta(\mu) \mid^{max} = 1.7$ and $\mid \eta(j) \mid^{max} = 2.0$.

- The non-W backgrounds acquire lepton-id and $\not{E}_T$ by statistical fluctuations. Therefore, we impose very tight lepton quality cuts and require $\not{E}_T > \not{E}_T^{min}$. For $e + jets$, $\not{E}_T^{min} = 25$ GeV; for $\mu + jets$, $\not{E}_T^{min} = 20$ GeV. For $\mu + jets$, we further require $\not{E}_{T\,cal}^{min} = 20$ GeV where $\not{E}_{T\,cal}^{min}$ is the $\not{E}_T$ measured by the calorimeter only.

- From a global point of view, the jets tend to be more isotropically distributed in the signal than in the backgrounds. In one of our two analysis approaches, we exploit this feature by requiring a minimum "spherical shape" of the events.

4.   Analysis

We take two independent approaches to analyze our data:

A. Jet multiplicity analysis

In this approach, we assume the well-known exponential scaling law [5] to be valid for all non-resonant QCD multijet processes i.e.

$$\frac{\sigma(X + nj)}{\sigma(X + (n+1)j)} = \rho \tag{1}$$

where ρ, a constant determined by α_s and $E_T^{min}(j)$, is independent of n (at least for $n < \sim 6$). Knowing the contribution of our signal to different values of n, we can make a fit to the observed data to estimate the amount of signal in it. Henceforth, whenever we talk about the number of jets, "n_m" will stand for the inclusive count, meaning "the number of events with m or more jets". Figure 1a shows the multiplicity of jets in the W candidate events after subtracting the non-W backgrounds shown in Figure 1b (not to scale, see Figure caption) for two different values of $E_T^{min}(j)$. The exponential fits are made through the points n_2 and n_3 only .

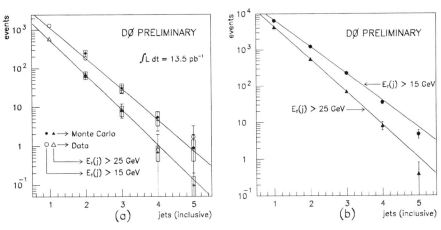

Fig. 1. Jet multiplicity distribution in the electron channel (a) W events: DØ data and Monte Carlo (VECBOS); (b) non-W background corresponding to ~ 67 times our data sample.

m	n_m			
	data	non-W bkg.	W bkg.	$t\bar{t}$
1	1686	$142. \pm 20.$	$1656. \pm 96.$	5.7 ± 10.1
2	335	50.6 ± 7.1	278.8 ± 16.7	5.6 ± 9.9
3	57	10.3 ± 1.5	42.2 ± 7.1	4.5 ± 8.2
4	12	2.4 ± 0.4	6.6 ± 3.8	3.0 ± 4.6

Table 1. The inclusive jet multiplicities in DØ data, estimated non-W background and the results of fitting for the W background and signal ($t\bar{t}$), e and μ channels combined

Let n_{ml} stand for the contribution of $t\bar{t}$ events to the n_m sample in our data. Then, if the background processes honor the scaling law, we have

$$n_m - n_{ml} = \frac{\left(n_{m-1} - a_{(m-1),m}.n_{ml}\right)^2}{n_{m-2} - a_{(m-2),m}.n_{ml}} \qquad (2)$$

where $a_{k,l} \equiv n_{kl}/n_{ll}$ are ratios that we calculate using Monte Carlo. For $m = 4$, we can safely ignore the term containing n_{ml}^2 whereby eq (3) yields the simple solution for n_{4l}:

$$n_{4l} = \frac{n_2.n_4 - n_3^2}{n_2 + a_{2,4}.n_4 - 2a_{3,4}.n_3} \qquad (3)$$

Table 1 summarizes the results thus obtained. We find 12 events where we expect 9.0 ± 3.8 from background processes. The 3.0 ± 4.6 signal events translate to a cross-section obtained solely from the monoleptonic channels:

$$\sigma_{t\bar{t}} = 6.4 \pm 9.8(stat.) \pm 4.0(sys.) \quad pb \qquad (4)$$

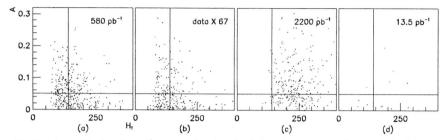

Fig. 2. A vs. H_T in $e/\mu + \not{E}_T + \geq 4jets$ events: (a) W background (VECBOS); (b) non-W background; (c) $t\bar{t}$ (ISAJET, $m_{top} = 180$ GeV); (d) DØ data.

Quadrants →	$A > A_0$ $H_T > H_{T0}$	$A > A_0$ $H_T < H_{T0}$	$A < A_0$ $H_T < H_{T0}$	$A < A_0$ $H_T > H_{T0}$
Fractions →	ϵ^1	ϵ^2	ϵ^3	ϵ^4
W bkg (VECBOS)	0.21 ± 0.03	0.27 ± 0.03	0.21 ± 0.03	0.31 ± 0.03
Non-W bkg	0.19 ± 0.04	0.25 ± 0.05	0.28 ± 0.05	0.28 ± 0.05
$t\bar{t}$ ($m_t = 180$ GeV)	0.60 ± 0.05	0.02 ± 0.01	0.02 ± 0.01	0.36 ± 0.04
Number of events →	N^1	N^2	N^3	N^4
Data	4	1	3	4
Best fit	4.1	2.4	2.0	3.5
No $t\bar{t}$	2.4	3.2	2.9	3.5

Table 2. The split of backgrounds, signal ($t\bar{t}$), DØ data and the results of fitting among the four quadrants of the A vs. H_T plane. The origin is at $A_0 = 0.05$, $H_{T0} = 140$ GeV

B. Event shape analysis

Here we examine the global characteristics such as the shape and the size of events which have at least 4 jets. To this end, we use a single variable called *"Aplanarity" (A)* [6], calculated using only the jets in an event, to quantify its shape. The higher the minimum one requires of A, which is bounded in the range $0 \leq A \leq 0.5$, the more spherical an event has to be in order to meet that requirement. We use H_T, defined as the scalar sum of the E_T's of all the jets in an event, to quantify its global size. Figure 2 shows the distribution of the backgrounds, signal and our data in the A vs. H_T plane. Table 2 shows the splits when we divide the plane into four quadrants with the origin at $A = 0.05$, $H_T = 140$ GeV. The origin was chosen so as to split the backgrounds more or less evenly among the four quadrants. This minimizes the statistical error in the procedure when, using Poisson distributions, we perform a maximum likelihood fit to the number of events observed in the four quadrants with respect to f, the unknown fraction of $t\bar{t}$ events in our sample. The results from the best fit, shown in the Table, corresponds to

$$f = 0.32 \pm 0.30 \tag{5}$$

This gives
$$N_{tt} = 3.8 \pm 3.6(stat) \pm 1.5(sys) \tag{6}$$
which translates into
$$\sigma_{tt} = 8.1 \pm 7.6(stat.) \pm 3.8(sys.) \quad pb \tag{7}$$
The last row in the Table shows the split expected if we force $f = 0$.

Finally, we apply the cuts $A > 0.05, H_T > 140$ GeV on the backgrounds. The two methods, A and B give independent estimates for the number of events to be expected from the background processes in the first quadrant of the A vs H_T plane which turn out to be very close:
$$N_B^A = \epsilon_W^1 . N_W + \epsilon_{non-W}^1 . N_{non-W} = 1.8 \pm 0.8 \pm 0.4 \tag{8}$$
and
$$N_B^B = \epsilon_B^1 .(1 - f).N = 1.7 \pm 0.8 \pm 0.5 \tag{9}$$
Subtracting this background from the observed number of events (i.e. 4), we get yet another estimate for the cross-section:
$$\sigma_{tt} = 7.5 \pm 7.3(stat.) \pm 3.3(sys.) \quad pb \tag{10}$$
This number is valid in the range 160 GeV$\leq m_{top} \leq 180$ GeV. For $m_{top} = 140$ GeV, it would be $\sim 25\%$ higher.

5.   Conclusions

Two independent ways of analyzing the *lepton + jets (sans μ-tag)* data collected from the first collider run of DØ in 1992-93 result in similar estimates for the $t\bar{t}$ production cross-section. Owing to the large branching ratio, the monoleptonic channels play a key role in the overall estimation of the cross-section that combines all the available channels. At this point, the uncertainties are dominated by statistical limitation. With the systematic uncertainties under control, we hope to be able to make a stronger statement based on a ~ 5 times larger set of data that is expected by the end of the ongoing run.

References

[1] DØ collaboration, S. Abachi *et al* Phys. Rev. Lett. **72**, 2138 (1994)

[2] F. Paige and S. Protopopescu, BNL Report no. BNL38034, 1986 (unpublished), release v 6.49.

[3] E. Laenen, J. Smith, and W. van Neerven, Phys. Lett. B **321**, 254 (1994).

[4] W. Giele, E. Glover, and D. Kosower, Nucl. Phys. B **403**, 633 (1993).

[5] F. Berends, H. Kuijf, B. Tausk, and W. Giele, Nucl. Phys. B **357**, 32 (1991).

[6] V. Barger and R. J. N. Phillips, *"Collider Physics"* (Addison-Wesley, Reading, MA, 1987), p. 281

SEARCH FOR THE TOP QUARK
IN THE ELECTRON + JETS + MUON TAG CHANNEL AT DØ

WILLIAM G. COBAU[†]

Department of Physics,
University of Maryland,
College Park, Maryland, 20742

ABSTRACT

Results are presented from the search for the top quark in the electron + jets + soft muon tag channel using the DØ detector. This search is based on data taken during the 1992–93 running of the Tevatron collider. In addition, the results from the DØ searches for the top quark in all analyzed channels are combined and summarized. All the results presented in this paper are preliminary.

1. Introduction

The top quark is the only standard model quark which has not been observed. At the present time the mass of the top quark is found to be greater than 131 GeV/c^2 at 95% confidence level.[1] In addition, the CDF collaboration has reported an excess above background in their top search which, when attributed to standard model top results in a $t\bar{t}$ production cross section of $13.9^{+6.1}_{-4.8}$ pb.[2] CDF then preforms a constrained fit to individual events which results in a measured top mass of $174 \pm 10^{+13}_{-12}$.

At the Tevatron, top quarks are produced dominantly in pairs. Each top quark then decays to a real W boson and a b quark. The cleanest signature for top production occurs when both W's decay leptonically and one has the signature of 2 high p_T leptons and 2 jets (from the b quarks) but the branching ratio for these signatures is small (<3%). When only one W decays leptonically and the other W decays hadronically, the branching ratio is much more favorable (~15%) but the backgrounds are larger. If in addition to the high p_T lepton + multiple jets, one "tags" a b quark jet, there is the possibility of taking advantage of the large lepton + jets branching ratio while having a relatively background free sample. One is then able to loosen cuts used to reject backgrounds in the non-tagged analyses. At DØ, we use soft muons to tag b quark jets. From branching ratios, one expects that 40% of $t\bar{t}$ events should have a soft muon from either the semi-leptonic decay or the cascade decay, b → c → μ, of the b quark. In the analysis presented here, the muon tag efficiency is estimated from Monte Carlo to be ~20% per $t\bar{t}$ → e + jets event.

This paper reports on a preliminary search for the $t\bar{t}$ production at $\sqrt{s}$ = 1.8 TeV in the electron + jets + muon tag channel. In addition, we summarize the DØ searches for top quark production and combine the results of all the analyses presented at this conference.

2. Data Analysis

The DØ detector is a large hermetic detector with an inner tracking volume surrounded by a liquid Argon/Uranium calorimeter and a toroidal muon system.[3] The detector has no central magnetic field. The calorimeter provides hermetic coverage up to $|\eta| \cong 4$ with fine segmentation. The muon system covers the range $|\eta| \leq 3.3$. During the 1992–93 collider run of the Tevatron, a data set representing an integrated luminosity of 13.5±1.6 pb^{-1} was collected.

The search for $t\bar{t}$ production in the electron + jets + muon tag channel starts by reconstructing the W → ev decay. A high p_T electron is required. The electron identification cuts

---

[†] *Representing the DØ Collaboration*

for this analysis require the EM fraction of the electron cluster in the calorimeter be greater than 90%, the shape of the cluster be consistent with an electron, the cluster be isolated, there be a track in the tracking volume matching the calorimeter cluster and the track not be consistent with a conversion of a γ. The electron id for this analysis is very strict to eliminate backgrounds from QCD multi-jet events where one jet fakes an electron. Once an electron is found, we require that it satisfy the kinematic requirements of $E_T > 20$ GeV and $|\eta_{\text{detector}}| < 2.0$. For the neutrino, we require that the missing transverse energy, $\not{E}_T$, exceed 20 GeV. Once the W candidate is identified, 3 jets ($\eta\phi$ cone of 0.5) with a $E_T > 20$ GeV and $|\eta| < 2.0$ are required. This leaves 34 candidates in the data sample.

To tag a b quark jet, we require a high quality muon with $p_T > 4$ GeV and $|\eta_{\text{detector}}| < 1.7$. In addition, if the muon has $p_T > 12$ GeV, we require the muon be non-isolated (to make this analysis orthogonal to the eμ dilepton search) and if $\not{E}_T < 35$ GeV that $|\phi_\mu - \phi_\nu| > 25°$. The last cut on the difference between the ϕ angles of the muon and the missing transverse energy significantly reduces the background from QCD multi-jet fake events. After all cuts, 2 events are observed.

3. Background Calculations

The two dominant backgrounds in this channel are the QCD production of W $+ \geq 3$ jets + a muon (we label this background, "W + jets") and QCD multi-jet events where one jet passes the electron id cuts, event mis-measurement results in missing transverse energy and a muon is found (we label this background "fakes"). To estimate the background to $t\bar{t}$ production, we treat these two backgrounds separately but with a common philosophy: backgrounds are estimated before muon tag from the data and then the measured probability of finding a muon associated with a jet in the events is applied to get the final background estimate.

3.1. W + jets Background

The W+jets background estimate starts with the 34 events observed before muon tag less the background from fakes before muon tag as calculated in the next section. We measure the muon rate per jet as a function of jet E_T using an inclusive electron trigger sample which is predominately QCD background. Applying the measured rate, we find the background due to W + jets is 0.43±0.14.

We have double checked this method by applying the muon tag rate per jet to a number of different samples and comparing the observed numbers of muon tags to the predicted. For all samples, we find that the observed and predicted numbers of events are consistent.

3.2. Fake Background

The fake background is estimated starting with a inclusive electron trigger sample. From the inclusive electron trigger, two samples are culled: a sample with the same cuts as the final data sample except for the $\not{E}_T$ cut and the muon tag and a sample with the same kinematic cuts as the first sample but with an EM cluster that fails the electron id. The second sample can be thought of as a sample of "bad electron" events which have the same kinematic properties as the fake background. The background before muon tag is estimated using the $\not{E}_T$ distribution of the two samples. The "bad electron" sample is normalized to the good electron distribution below the $\not{E}_T$ cut and the "bad electron" distribution above the cut gives the background before muon tag. This measurement gives a background from fakes before muon tag of 6.3±1.7 events.

The muon tagging rate per jet is then calculated from the inclusive electron trigger sample for events that pass the final data sample $\not{E}_T$ cut. One must be careful to include the $\not{E}_T$ cut since the presence of a muon implies that there should be real $\not{E}_T$, thus the muon tag rate increases as a function of the $\not{E}_T$ cut. After applying the measured muon tag rate, the background from fakes is found to be 0.12±0.05 events.

3.3 Final Background Cross-Check

The complete background calculation is subjected to final cross-checks. A sample of W events with exactly one additional jet and looser electron id is examined. This sample, because of the looser electron id, has a greater fraction of events due to fakes than the sample with tight electron id. The number of muon tagged events was calculated using the above methods and compared to the observed number of events as a function of the jet E_T threshold. There is good agreement between the observation and prediction.

The W + jets sample was also examined as a function of jet multiplicity. Figure 1 shows the observed number of events and the estimated backgrounds as a function of inclusive jet multiplicity. The predicted background for W + ≥1 and ≥2 jets with a muon tag is consistent with the number of events observed.

4. Results for Electron + Jets + Muon Tag Channel

The preliminary results from a search for $t\bar{t} \rightarrow$ e + jets + muon tag are two events are observed passing all cuts with a total background of 0.55±0.15 events. Efficiencies and expected number of top events have been calculated using the theoretical $t\bar{t}$ production cross section of Laenen *et al.*[4] as a function of top mass. These numbers are presented as part of Table 1. For a top mass of 160 (180) GeV/c^2, the expected number of events is 1.0±0.2 (0.6±0.2) and the product of the efficiency and branching ratio is 0.9±0.2% (1.1±0.2%).

5. Summary of DØ Top Quark Searches

The DØ collaboration has presented preliminary results from a new set of searches for $t\bar{t}$ production in the following channels: ee, eμ, μμ, e + jets, μ + jets and e + jets + muon tag.[5,6] The channels μμ and e + jet + muon tag are new channels that were not included in

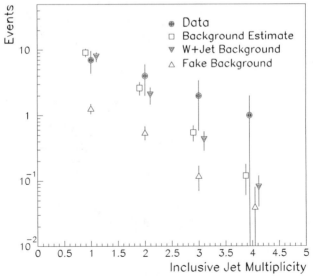

Figure 1. Events with muon tag vs. Inclusive Jet Multiplicity. Shown are the number of events with a muon tag for data, W + Jets background, Fake background and the total background estimate.

Table 1.
Summary of DØ Top Searches

M_{top} (GeV/c^2)	eμ	ee	μμ	e + jets	μ+ jets	e + jets (μ)	ALL
New Channel?	No	No	Yes	No	No	Yes	
ε×BR (%)	0.32±0.06	0.18±0.02	0.11±0.02	1.2±0.3	0.8±0.2	0.6±0.2	
140 $\langle N_{t\bar{t}} \rangle$	0.72±0.16	0.41±0.07	0.24±0.05	2.8±0.7	1.3±0.4	1.3±0.4	6.7±1.2
ε×BR (%)	0.36±0.07	0.20±0.03	0.11±0.01	1.6±0.4	1.1±0.3	0.9±0.2	
160 $\langle N_{t\bar{t}} \rangle$	0.40±0.09	0.22±0.04	0.12±0.02	1.8±0.5	0.9±0.3	1.0±0.2	4.4±0.7
ε×BR (%)	0.41±0.07	0.21±0.03	0.11±0.01	1.7±0.4	1.2±0.3	1.1±0.2	
180 $\langle N_{t\bar{t}} \rangle$	0.23±0.05	0.12±0.02	0.06±0.01	1.0±0.2	0.5±0.2	0.6±0.2	2.5±0.4
Background	0.27±0.09	0.16±0.07	0.33±0.06	1.2±0.7	0.6±0.5	0.6±0.2	3.2±1.1
$\int \mathcal{L} dt$ (pb^{-1})	13.5±1.6	13.5±1.6	9.8±1.2	13.5±1.6	9.8±1.2	13.5±1.6	
DATA	1	0	0	2	2	2	7

Table 1. This table summarizes the DØ top searches. Each channel and the combined result is represented by a column. The rows indicate the channel, whether this channel was included in Ref. 1, the product of branching ratio and efficiency and the expected number of top events for top masses of 140, 160 and 180 GeV/c², the estimated background, the integrated luminosity and the number of observed events.

reference 1. For the channels included in reference 1, the searches have been re-optimized to search for top above the present top mass limit of 131 GeV/c². When all channels are combined, a total of 7 events are observed with an estimated background of 3.2±1.1. Table 1 summarizes the results from all channels and the combination of the channels. We calculate a 7.2% probability that the estimated background of 3.2 events fluctuates to give the observed signal of 7 events using Gaussian statistics for the acceptance, luminosity and background errors and Poisson statistics for the number of events observed. We conclude that we do not observe a statistically significant signal.

It is, of course, natural to note that we do have an excess number of events above our background estimates. If we make the assumption that the excess number of events is due to standard model top production, we can then calculate a cross section for $t\bar{t}$ production as a function of top mass. The result of this calculation is shown in Table 2 and in Fig. 2. Figure 2 shows the ±1 sigma band of our cross section calculation along with the cross section prediction of Laenen *et al.* and the CDF published result from reference 2. As one can see, the DØ result is consistent with a null top hypothesis, the CDF result and the theoretical expectations for any top mass above the top mass limit.

Table 2
Measured Top Cross Section

M_{top} (GeV/c²)	σ (pb)
140	9.6±7.2
160	7.2±5.4
180	6.5±4.9

Table 2. Shown are calculated top cross sections as a function of top mass. Errors are combined statistical and systematic errors.

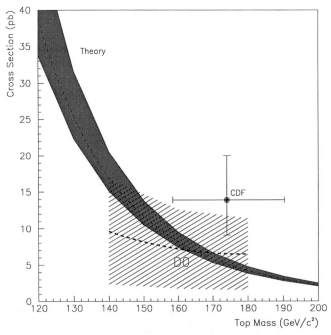

Figure 2. Top Search Results. Plotted are the results from the DØ top quark search. The band labeled DØ is the cross section result of assuming that the excess events observed above background are from Standard Model Top production. Also shown are the recent result from CDF and the calculation of Laenen *et al.*

6. Conclusion

We have presented preliminary results from the DØ searches for $t\bar{t}$ production at $\sqrt{s} =$ 1.8 TeV. These results have been optimized for the search for top above 131 GeV/c^2 and include 2 new channels. We observe no significant signal for top quark production in our data sample representing an integrated luminosity of 13.5 pb^{-1}. If we make the assumption that the events unaccounted for by our background calculations are due to standard model top production, we find that for M_{top} = 160 (180) GeV/c^2, the $t\bar{t}$ production cross section is 7.2±5.4 (6.5±4.9) pb. This result is consistent with a null top result and the published CDF result.

References

1. S. Abachi *et al.*, *Phys. Rev. Lett.* **72**, 2138 (1994).
2. F. Abe *et al.*, *Phys. Rev. Lett.* **73**, 225 (1994) and references there in.
3. S. Abachi *et al.*, *Nucl. Instrum. Methods Phys Res, Sect A,* **338**, 185 (1994).
4. E. Laenen, J. Smith and W.L. Van Neerven, *Phys. Lett.* **B321**, 254 (1994).
5. K.Genser, *In these proceedings.*
6. D. Chakraborty, *In these proceedings.*

SEARCH FOR THE TOP QUARK USING MULTIVARIATE ANALYSIS TECHNIQUES

PUSHPALATHA C. BHAT [*]

Fermi National Accelerator Laboratory,
Batavia, IL 60510, U.S.A.

ABSTRACT

The DØ collaboration is developing top search strategies using multivariate analysis techniques. We report here on applications of the H-matrix method to the $e\mu$ channel and neural networks to the e+jets channel.

1. Introduction

Top quark events are being searched for in the di-lepton, lepton+jets and all-jets channels at the Collider detectors at Fermilab, in $p\bar{p}$ collisions at $\sqrt{(s)}$ =1.8 TeV. The DØ collaboration has been applying multivariate techniques such as the Covariance matrix (H-matrix) method, Probability Density Estimation (PDE) method and Neural Networks, in the search for the top quark. In this paper, we present a brief discussion of these techniques and report on some aspects of the on-going analyses.

2. Multivariate Techniques

Multivariate classifiers provide a discriminating boundary between the signal and background in multidimensional space that can yield discrimination close to the theoretical maximum (Bayes' limit[1]). If $P(s|x)$ $(P(b|x))$ is the probability that a given event with feature vector x is a signal (background) event, then the optimal way to partition the feature space is to cut on the ratio of these probabilities. This ratio is the Bayes discriminant function,

$$R(x) = \frac{P(s|x)}{P(b|x)} = \frac{P(x|s)P(s)}{P(x|b)P(b)}. \tag{1}$$

$P(s|x), P(b|x)$ are also known as Bayesian probabilities. The quantities $P(x|s), P(x|b)$ are the likelihood functions for signal and background, respectively (hereafter referred to as f(x)). The ratio of the prior probabilities $\frac{P(s)}{P(b)}$ is the ratio of signal and background cross-sections. Different multivariate classifiers approximate the likelihood functions with different functional forms and attempt to arrive at the Bayes discriminant. The three classifiers being used at DØ are briefly described below.

[*]Representing the DØ Collaboration

2.1. H-Matrix Method

This is the familiar covariance matrix method which is also known as the Gaussian Classifier. The likelihood function is taken to be gaussian,

$$f(x) = A.exp\{-\frac{1}{2}\sum_{i,j}(x_i - \bar{x}_i)^T M^{-1}(x_j - \bar{x}_j)\} = A.exp(-\chi^2) \tag{2}$$

where $H = M^{-1}$ is the covariance matrix. We use Fisher's formulation of the discriminant, which is $F = \frac{1}{2}(\chi_b^2 - \chi_s^2)$, where χ_b^2 and χ_s^2 are the χ^2 terms of a sample calculated using background and signal H-matrices respectively. The Bayes discriminant in terms of the Fisher variable F can be shown to be $R(x) = exp(F)$, when P(s)=P(b).

2.2. Probability Density Estimation (PDE) Method

The likelihood function is approximated as,

$$f(x) = \frac{1}{N_{events}h} \sum_{i=1}^{N_{events}} \prod_{j=1}^{d} K(\frac{x_i - x_{ij}}{h_j}) \tag{3}$$

where K is a kernel which we take to be a multivariate gaussian centered at each data point x_{ij} with variance h_j^2 (for the jth variable). The Bayes discriminant is $R(x) = \frac{f_s(x)}{f_b(x)}$.

2.3. Neural Networks

It has been shown[2] that neural networks do not calculate the likelihood function for each class separately, but arrive at the Bayesian probability for the signal directly. The discriminant in this case is the output of the network

$$O(x) = g(\sum_j w_{kj}g(\sum_i w_{ji}x_i)) = P(s|x) \tag{4}$$

(assuming a three layer feed-forward neural network). The x_i's are the input variables, g represents a non-linear function $(e.g., \frac{1}{(1+e^{-2x})})$, w_{kj} and w_{ji} are the weights that are adjusted during the "learning" process. Descriptions of the neural network approach and details of the training algorithms are available in many articles and books. The Bayes discriminant in terms of the network output will be $R(x) = \frac{O(x)}{(1-O(x))}$.

3. H-matrix Analysis of $e\mu$ data

From the conventional analysis,[3] DØ has one top candidate event in the $e\mu$ channel. The dominant backgrounds are $Z \to \tau\tau$, WW and instrumental fake events. We have applied the H-matrix method to enhance the signal to background particularly w.r.t. $Z \to \tau\tau$. We have built H-matrices using the variables E_T^e, P_T^μ, E_T^{jet1}, E_T^{jet2}, $\not{E}_T^{cal}(\not{E}_T$ in the calorimeter), $H_T(\Sigma E_T$ of jets), $M_{e\mu}$, $\Delta\phi_{e\mu}$ for $t\bar{t}$, $Z \to \tau\tau$ Monte Carlo (MC) and data events, after applying loose electron identification (ID) criteria and requiring $P_T^e > 11$ GeV and $P_T^\mu > 11$ GeV. We use data to represent one set of backgrounds ($b\bar{b} \to e\mu$ and fakes -bkg1) and $Z \to \tau\tau$ as the other background (bkg2). These H-matrices are then applied to data, $t\bar{t}$ (top mass 140, 160 and 180 GeV), $Z \to \tau\tau$ and WW samples and χ^2 values calculated. The lego plots of χ_{Top}^2 vs χ_{bkg1}^2 are shown in

Fig.1. It can be seen that the data and $Z \to \tau\tau$ events have small χ^2 values with both signal and background H-matrices whereas top events have small χ^2_{Top} but large χ^2_{bkg1}. We define two Fisher discriminants $F_1 = \frac{1}{2}(\chi^2_{bkg1} - \chi^2_{Top})$ and $F_2 = \frac{1}{2}(\chi^2_{bkg2} - \chi^2_{Top})$. In Fig. 2, are shown F_1 for various event samples. We search for a combination of F_1 and F_2 to keep the same efficiency as in the conventional analysis for 140 GeV top events and to maximize background rejection. By applying $F_1 > 15$ and $F_2 > 3$ we have 16%, 22% and 25% efficiency for top events with top mass of 140, 160 and 180 GeV respectively. The signal to background ratio (S/B) is about 18 w.r.t. to $Z \to \tau\tau$ and 10 w.r.t. WW events for 180 GeV top mass. For lower masses the S/B is higher.

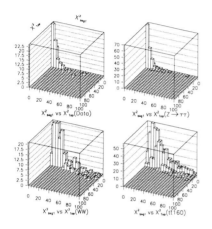

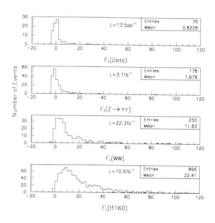

Fig. 1. χ^2_{Top} vs χ^2_{bkg1} for DØ data, $Z \to \tau\tau$, WW and $t\bar{t}160$ samples

Fig. 2. Fisher variable distributions

4. Neural Networks Analysis of $e +$ jets data

Details of our conventional analyses in the $e +$ jets channel can be found elsewhere in these proceedings.[4,5] Here, we have analysed $e + 4$jets events using neural networks after applying the following kinematic cuts: $E^e_T \geq 20$ GeV (+tight electron ID), $\not{E}_T \geq 20$ GeV and $E_T(jet4) \geq 15$ GeV. We use two separate neural networks to handle the dominant backgrounds to this channel viz., $W +$ jets and QCD events where one of the jets gives a false electron ID (QCD fake).

Figure 3 shows a schematic of using networks in parallel. Their outputs can be fed into another network to make higher level decisions. We train the first network with $t\bar{t}160$ and $W +$ jets samples and the second network with $t\bar{t}160$ and QCD fake data sample (obtained from multi-jet triggers at DØ). We have carried out analyses with different sets of input variables. In a 2-variable analysis we use network with 2 input nodes, 3 hidden nodes (one hidden layer) and 1 output node. We use the H_T (sum of E_T of jets with $\eta_{jet} \leq 1.7$) and the aplanarity[4] (A) of the event as input variables. In Fig. 4, we show the distributions of the output for $t\bar{t}160$ and $W +$ jets from network 1.

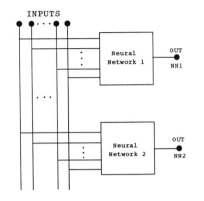

Fig. 3. Processing with many Neural Networks

Fig. 4. Distributions of output from network-1 trained on $t\bar{t}160$ and W + jets events

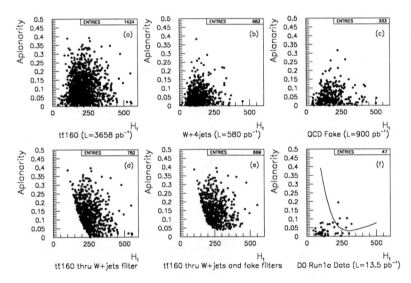

Fig. 5. H_t vs Aplanarity plots from the 2-variable analysis for (a)$t\bar{t}160$, (b)W+4jets (VECBOS MC), (c)QCD fakes, (d)$t\bar{t}160$ after cut on network 1 (NN1>.8), (e)$t\bar{t}160$ after cuts on networks 1 & 2 (NN1>.8, NN2>.6) and (f)D0 data with contour from the cuts NN1>.8,NN2>.6.

From the plots shown in Fig. 5, it can be seen that a combination of neural networks can provide good rejection to individual backgrounds in different parts of the multi-dimensional phase space. The DØ data with the decision boundary generated by the neural networks (NN1> 0.8,NN2>.6) is shown in Fig. 5(f). Figure 6 shows the compound probability (signal probability surface) for the two networks together as a function of H_T and A. When the analysis is extended to many dimensions, it is still possible to examine the distributions of the variables and their correlations before and after the network selection to understand the decision boundary. Figure 7 shows such distributions for a 6 variable analysis done using H_T, A, E_T^e, $\not{E}_T$, $E_T(jet4)$, $\theta(e)$ (6 inputs, 6 hidden nodes and 1 output node used).

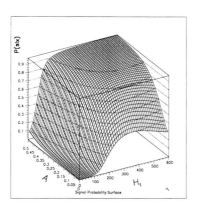

Fig. 6. Signal probability surface for the two networks together as a function of H_T and A

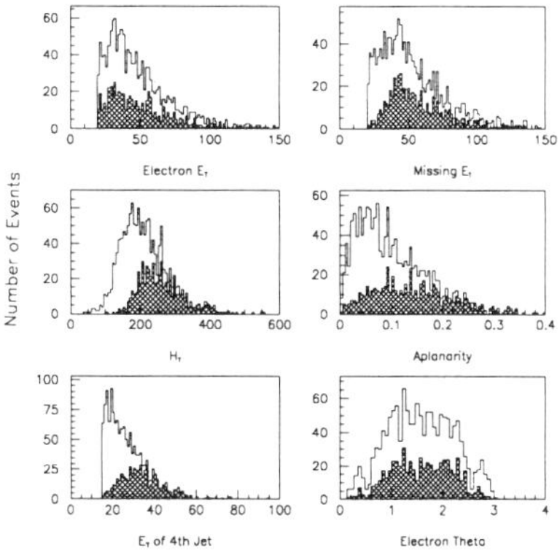

Fig. 7. Distributions of variables in 6-variable analysis before (open histograms) and after(hatched histograms) neural network cuts

5. Summary

A discussion of the multivariate techniques being used in the top search at DØ and some preliminary results have been presented. For example, the H-matrix applied to the $e\mu$ channel yields a S/B=18 for top(180 GeV) to $Z \to \tau\tau$. This is a significant improvement over conventional analyses.[3] Preliminary results from the neural networks analysis of the lepton+jets channels show promise for better background rejection and higher efficiency than conventional analysis techniques.

This work is supported in part by the U.S. Department of Energy.

References

1. R.O. Duda and P.E. Hart, *Pattern Classification and Scene Analysis* (Wiley, New York, 1973).
2. D.W. Ruck *et al.*, "The multilayer perceptron as an approximation to a Bayes optimal discriminant function", IEEE Trans. Neural Networks 1 (4), 296-298(1990).
3. K. Genser, DØ Collaboration, these proceedings.
4. D. Chakraborty, DØ Collaboration, these proceedings.
5. W. Cobau, DØ Collaboration, these proceedings.

TOP MOMENTUM RECONSTRUCTION AND EXTRA SOFT JETS*

LYNNE H. ORR

Department of Physics and Astronomy, University of Rochester,
Rochester, NY 14627, USA

and

W.J. STIRLING

Departments of Physics and Mathematical Sciences, University of Durham
Durham DH1 3LE, England

ABSTRACT

Top pairs produced at the Tevatron may be accompanied by extra jets due to radiation of soft gluons. Whether these jets should be included in top momentum reconstruction depends on the source of the gluons, which can be radiated off the initial partons, the final state b quarks, or the intermediate top quarks. We compute soft gluon distributions and discuss their implications for top mass measurements.

Top quark pairs produced at the Tevatron may be accompanied by additional soft jets due to radiation of soft gluons. In attempts to measure the top mass m_t by reconstructing the top quark's four–momentum from the momenta of its decay products, the question of how to deal with additional soft jets must be addressed. In particular, should such jets be combined with the W and b in reconstructing m_t? Obviously, if the gluon was radiated off an initial state quark, the answer is "no," and if it was radiated off one of the final b's, the answer is "yes." Correspondingly, we would guess that soft jets near the beam come from initial state radiation, and thus should be ignored, and that central jets are due to final state radiation from b quarks, and should be included. But what about gluons radiated off the top quarks themselves? Although top decays before hadronizing, it does have time to radiate gluons. Do such gluons belong to the initial or final state?

Clearly, an initial/final state interpretation is too naïve, and we must consider top production and decay simultaneously in our treatment of soft gluon radiation; *i.e.*, we must include all possible diagrams. Fortunately, the result can be decomposed into contributions associated with $t\bar{t}$ production, t and $\bar{t}$ decay, and their interference. (See Ref. [1] for a discussion of the associated formalism.) This is exactly the decomposition we need for purposes of momentum reconstruction.

Consider the process $q\bar{q} \to t\bar{t} \to bW^+\bar{b}W^-$ with emission of a gluon. In the soft approximation, the lowest order cross section and gluon distribution factorize, and we

---

*Presented by L.H. Orr.

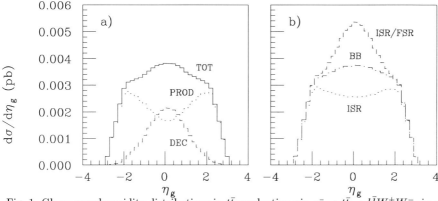

Fig. 1. Gluon pseudorapidity distributions in $t\bar{t}$ production via $q\bar{q} \to t\bar{t} \to b\bar{b}W^+W^-$, in $p\bar{p}$ collisions at $\sqrt{s} = 1.8$ TeV. (a) Net distribution and contributions from production and decay. (b) Distributions arising from ISR, ISR/FSR, and BB models described in the text.

have $(1/d\sigma_0)\, d\sigma/(dE_g\, d\cos\theta_g\, d\phi_g) = \frac{\alpha_s}{4\pi^2}\, E_g\, (\mathcal{F}_{\mathrm{PROD}} + \mathcal{F}_{\mathrm{DEC}} + \mathcal{F}_{\mathrm{INT}})$. $\mathcal{F}_{\mathrm{PROD}}$ corresponds to gluons radiated in association with $t\bar{t}$ production, *i.e.*, radiated before the t or $\bar{t}$ quark goes on shell. Similarly, $\mathcal{F}_{\mathrm{DEC}}$ corrseponds to gluons radiated in the decay of the t or $\bar{t}$. $\mathcal{F}_{\mathrm{INT}}$ represents the interferences between the two and depends on the top width Γ_t. Expressions for $\mathcal{F}_{\mathrm{PROD}}$, $\mathcal{F}_{\mathrm{DEC}}$, and $\mathcal{F}_{\mathrm{INT}}$ can be found in Ref. [2]. This production–decay-interference decomposition determines for us whether the gluon momentum should be combined with those of the W and b to form the top momentum. Gluons from the production piece are not to be included, while gluons from the decay piece are. For the interference term, whether to add the gluon is undetermined, but this contribution is negligible in the case of interest here.

Let us now obtain the full soft gluon distribution for $t\bar{t}$ production at the Tevatron. The results shown here are from Ref. [3], to which the reader is referred for a more complete discussion. We consider angular distributions since we want to know where the soft jets are expected to appear in the detector. We use $m_t = 174$ GeV, work at the parton level, and keep kinematic cuts to a minimum. The cuts are $|\eta_b|, |\eta_{\bar{b}}| \leq 1.5$; $|\eta_g| \leq 3.5$; 10 GeV$/c \leq p_T^g \leq 25$ GeV$/c$; $E_g \leq 100$ GeV; and $\Delta R_{bg}, \Delta R_{\bar{b}g} \geq 0.5$. The resulting distribution in the gluon pseudorapidity is shown in Fig. 1(a) as a solid line, along with the contributions from production (dotted line) and decay (dashed line). We see that, as we might expect, the decay piece is strongly peaked in the center, and the production piece contributes most of the gluons in the forward and backward directions. Note, however, that gluons in the central region are almost as likely to have come from production as from decay, hence we cannot simply assign central soft jets to the decay piece.

We can hope to get a better handle on how to assign soft jets by using proximity to the b quark. Figure 2(a) shows the distribution in cosine of the angle between the gluon and the b quark with the same decomposition as in Fig. 1(a). We see a slight enhancement near the b in the decay piece compared to production, but the difference

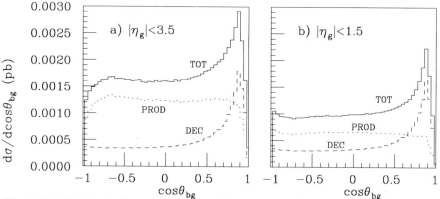

Fig. 2. Distribution in the cosine of the angle between the gluon and the b-quark, (a) with cuts described in the text and (b) with the additional cut $|\eta_g| \leq 1.5$.

is not terribly dramatic and is very sensitive to the cut on ΔR_{bg}. In addition, it must be kept in mind that no fragmentation or detector effects have been included. The situation improves a little if we consider only central gluons, as shown in Fig. 2(b), where we take $|\eta_g| \leq 1.5$. The excess in the decay piece is now more pronounced.

Finally, we compare the correct pseudorapidity distribution in Fig. 1(a) to those in Fig. 1(b), obtained from some simpler models that are intuitively appealing and easily implemented in Monte Carlo simulations. The ISR model (dotted line) includes radiation off the initial $q\bar{q}$ state only, as if the q and $\bar{q}$ formed a color singlet. We might expect this to correspond to the contribution associated with production, but we see by comparing to the dotted line in Fig. 1(a) that the ISR model overestimates radiation in the central region. In the ISR/FSR model (dashed line) we add to the ISR model radiation from the final $b\bar{b}$ pair as if they too formed a color singlet. This model corresponds roughly to the naïve expectation mentioned in the introductory paragraph. Figure 1(b) shows that this model overestimates the total radiation and gets the shape wrong. In the BB model (dot-dashed line) we use the correct color structure but ignore radiation off the top quark. This model approximately reproduces the correct pseudorapidity distribution. However, it does not give the correct azimuthal distribution,[3] and, more important for m_t reconstruction, does not permit a production–decay decomposition.

In conclusion, gluon radiation in $t\bar{t}$ production and decay has a structure which gives rise to interesting and sometimes subtle effects, a few of which we have illustrated here. We have seen that there is no simple prescription for treating extra soft jets in m_t reconstruction, but the production–decay decomposition provides some guidance. We have also shown that simpler, intuitively appealing models either do not reproduce the correct distributions, or do not allow for the production–decay decomposition.

References

1. V.A. Khoze, L.H. Orr and W.J. Stirling, Nucl. Phys. **B378** (1992) 413.
2. V.A. Khoze, J. Ohnemus and W.J. Stirling, Phys. Rev. **D49** (1994) 1237.
3. L.H. Orr and W.J. Stirling, DTP/94/60, UR-1365, July 1994.

LEADING WEAK CORRECTIONS
TO TOP QUARK PAIR PRODUCTION
AT HADRON COLLIDERS

CHUNG KAO* †

*Department of Physics, Florida State University,
Tallahassee, FL 32306, USA*

ABSTRACT

The weak corrections to the top quark pair production via $q\bar{q}$ annihilation at hadron colliders are evaluated. The effects of parity violation are studied in the Standard Model (SM) and the Minimal Supersymmetric Model. The parity violation effect from the SM weak corrections is small. Any large parity violation appearing in this process will indicate new physics.

1. Introduction

If the top quark is heavy, as now it appears to be,[1] the Yukawa couplings between the top quark and spin-0 bosons will be interestingly large. Many processes involving the third generation quarks and the Higgs bosons might provide opportunities to test the Standard Model (SM) and lead to signals of new physics. At the Tevatron, $q\bar{q}$ annihilation dominates the production of top quark pairs. At the Large Hadron Collider (LHC), although gluon fusion is the major source of $t\bar{t}$ in the SM, it might be possible that some large new physics effects[2] will appear in the $q\bar{q}$ annihilation process, $q\bar{q} \to t\bar{t}$.

The QCD corrections[3] to the production rate of $q\bar{q} \to t\bar{t}$ are larger than the weak corrections.[4,5] However, the parity violation manifestation in the electroweak interactions might produce effects unobtainable from the QCD that conserves parity (P) and charge conjugation (C) symmetries. We have employed the on shell renormalization scheme in the 't Hooft-Feynman gauge to evaluate weak corrections to $q\bar{q} \to g \to t\bar{t}$ with $M_W = 80.0$ GeV, $M_Z = 91.17$ GeV, $m_b = 5.0$ GeV, $m_t = 170$ GeV, and $\sin^2 \theta_W = 0.230$. The parity violating chromoanapole moment of the top quark is studied in the SM and the Minimal Supersymmetric Model (MSSM). The quark mixing from the CKM matrix is not included, thus, CP is conserved.

In the SM, the weak radiative corrections to the $gt\bar{t}$ vertex have contributions from the Higgs boson (H^0), the Nambu-Goldstone bosons $(G^0, G^\pm)$, the Z boson (Z), and the W bosons $(W^\pm)$. In the MSSM, there are two Higgs doublets and five Higgs bosons: two neutral CP-even, H (heavier) and h (lighter); one neutral CP-odd, A; and a pair of singly charged Higgs bosons, $H^\pm$.

---

*Research supported in part by the U. S. Department of Energy under contract DE-FG05-87ER40319.
†I am grateful to Glenn Ladinsky and Chien-Peng Yuan for a fruitful and enjoyable collaboration.

The Yukawa Lagrange density with the top quark in the MSSM is

$$
\begin{aligned}
\mathcal{L}_I &= -[\frac{\sin\alpha}{\sin\beta}]\frac{m_t}{v}\bar{t}tH - [\frac{\cos\alpha}{\sin\beta}]\frac{m_t}{v}\bar{t}th + i[\cot\beta]\frac{m_t}{v}\bar{t}\gamma_5 tA \\
&\quad +\frac{m_t}{\sqrt{2}v}[\bar{t}(\lambda_s^H - \lambda_p^H\gamma_5)bH^+ + \bar{b}(\lambda_s^H + \lambda_p^H\gamma_5)tH^-]
\end{aligned}
\tag{1}
$$

where $\tan\beta \equiv v_2/v_1$ = ratio of the vacuum expectation values of the two Higgs doublets, $\lambda_s^H = \cot\beta + \frac{m_b}{m_t}\tan\beta$, $\lambda_p^H = \cot\beta - \frac{m_b}{m_t}\tan\beta$, and α is the mixing angle between the neutral CP-even Higgs bosons.

2. The Top Quark Chromoanapole Moment

Let's write the $gt\bar{t}$ vertex as $-ig_s\bar{u}(p)T^a\Gamma^\mu v(q)$, where g_s = the strong coupling, T^a = the $SU(3)$ matrices, $u(p)$ and $v(q)$ are the Dirac spinors of t and $\bar{t}$ with outgoing momenta p and q. At the tree level, $\Gamma_0^\mu = \gamma^\mu$. The 1-loop vertex function can be expressed as

$$
\begin{aligned}
\Gamma_1^\mu &= \gamma^\mu[A(k^2) - B(k^2)\gamma_5] \\
&\quad +(p-q)^\mu[C(k^2) - D(k^2)\gamma_5] + (p+q)^\mu[E(k^2) - F(k^2)\gamma_5] \\
&= F_1(k^2)\gamma_\mu + F_2(k^2)i\sigma^{\mu\nu}k_\nu \\
&\quad +a(k^2)\gamma_\nu\gamma_5(k^2g^{\mu\nu} - k^\mu k^\nu) + d(k^2)i\sigma^{\mu\nu}k_\nu\gamma_5
\end{aligned}
\tag{2}
$$

where $k = p + q$ and $k^2 = \hat{s}$ in the $t\bar{t}$ center of mass frame. The conservation of vector current demands that $E = 0$ and $B = -k^2F/(2m_t)$. Applying the Gordon identities, we obtain $F_1 = A + 2m_tC$, $F_1(0)$ = the chromocharge; $F_2 = C$, $F_2(0)$ = the anomalous chromomagnetic moment; $a = -B/k^2 = F/(2m_t)$, $a(0)$ = chromoanapole moment; and $d = D$, $d(0)$ = the chromoelectric dipole moment*.

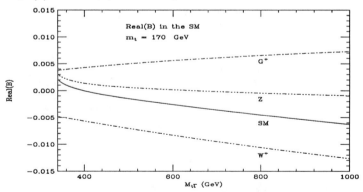

Fig. 1. The form factors B_{G^+}, B_{W^+}, B_Z and B_{SM} versus $M_{t\bar{t}}$.

*A nonzero $d(k^2)$ will lead to CP violation.

It is convenient to choose the form factor $B(k^2) = -k^2 a(k^2)$ to study parity violation effects in the $t\bar{t}$ production. A large chromoanapole form factor of the top quark might produce interesting signals of parity violation. In the SM, parity violation appears only in diagrams with the Z, the W^+, and the G^+. Figure 1 shows the form factor B as a function of the $t\bar{t}$ invariant mass $(M_{t\bar{t}})$, from diagrams with the charged Nambu-Goldstone boson alone (B_{G^+}), the W boson alone (B_{W^+}), the Z boson alone (B_Z), and the total in the SM (B_{SM}). The form factor B_Z is found to be very small. The $|Re(B_{G^+})|$ is larger than the full $|Re(B_{SM})|$, due to a cancellation between the the pure $O(\frac{m_t^2}{v^2})$ and $O(g^2)$ contributions. In the 't Hooft–Feynman gauge, this cancellation occurs because the diagrams with the G^+ and the W^+ interfere destructively.[5]

In the MSSM, the charged Higgs boson makes an additional contribution to parity violation. The form factor B from the charged Higgs boson, B_{H^+}, is a function of M_{H^+} and $\tan\beta$, and is proportional to $\lambda^H = \lambda_s^H \lambda_p^H = \cot^2\beta\{1 - [(m_b/m_t)\tan^2\beta]^2\}$. The total B in the MSSM, B_{MSSM} is the sum of B_{H^+} and B_{SM}. For $M_{t\bar{t}}$ larger than about 400 GeV, if $\tan\beta > \sqrt{m_t/m_b}$, the B_{H^+} and the B_{SM} interfere constructively, therefore, $|Re(B_{MSSM})| > |Re(B_{SM})|$; while if $\tan\beta < \sqrt{m_t/m_b}$, the B_{H^+} and the B_{SM} interfere destructively. The reverse occurs when $M_{t\bar{t}}$ is close to the threshold of $2m_t$. Figure 2 shows the form factor B_{H^+}, and Fig. 3 shows the form factor B_{MSSM} for various values of M_{H^+} and $\tan\beta$. The B_{SM} is also presented.

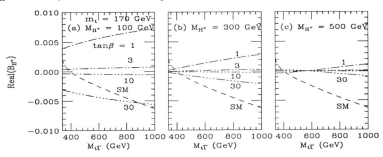

Fig. 2. The form factors B_{SM} and B_{H^+} versus $M_{t\bar{t}}$ for various values of M_{H^+} and $\tan\beta$.

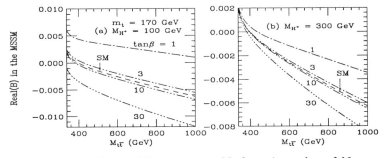

Fig. 3. The form factors B_{SM} and B_{MSSM} versus $M_{t\bar{t}}$ for various values of M_{H^+} and $\tan\beta$.

3. Asymmetries and Parity Violation

The parity violation from weak corrections appears in the production rate of polarized $t\bar{t}$. In the SM with $M_{H^0} = 100$ GeV, the $t_L\bar{t}_R$ (LR) states generally receive a slightly larger enhancement near the threshold than that for the $t_R\bar{t}_L$ (RL) states, while the suppression the LR states receive in the TeV region is greater than that for the RL states.[5] The effects of parity violation appear in the ratio

$$\mathcal{A}(M_{t\bar{t}}) = \frac{\hat{\sigma}(RL) - \hat{\sigma}(LR)}{\hat{\sigma}(RL) + \hat{\sigma}(LR)}$$
$$\sim 2Re[B(M_{t\bar{t}})], \qquad (3)$$

which is zero for parity conserving interactions. For the corrections we have computed, $\mathcal{A}$ is approximately $1 - 2\%$ at $M(t\bar{t}) \approx 1$ TeV.

4. Conclusions

In the SM, for heavy top quarks produced near the threshold region the Yukawa corrections to the production rate dominate only if the Higgs boson is light. The complete 1-loop weak corrections in the final state makes the suppression of the event rate at large $M_{t\bar{t}}$ about a factor of two different from that predicted by the Yukawa corrections alone.[5] The parity violation effects derived from the SM weak corrections are a few percent in the asymmetry $\mathcal{A}$. Therefore, any observation of large parity violation in this process will indicate new physics.

In the MSSM, the effects of parity violation can be enhanced in two regions of the parameter space: (i) $\tan\beta > \sqrt{m_t/m_b}$, and (ii) $\tan\beta$ is close to 1 and M_{H^+} is close to 100 GeV. If $M_{H^+} > 300$ GeV, the parity violation effect from the charged Higgs boson will be very small unless $\tan\beta \gg \sqrt{m_t/m_b}$.

References

1. S. Abachi et al., the D0 collaboration, Phys. Rev. Lett. **72** (1994) 2138; F. Abe, et al., the CDF collaboration, FERMILAB-PUB-94-097-E (1994); FERMILAB-PUB-94/116-E (1994).
2. C. Hill and S. Parke, Fermilab-Pub-93/397-T (1993); E. Eichten and K. Lane, Phys. Lett. **B327** (1994) 129.
3. P. Nason, S. Dawson and R. K. Ellis, Nucl. Phys. **B303** (1988) 607; **B327** (1989) 49; W. Beenakker, H. Kuijf, W. van Neerven and J. Smith, Phys. Rev. **D40** (1989) 54; W. Beenakker, W. van Neerven, R. Meng, G. Schuler and J. Smith, Nucl. Phys. **B351** (1991) 507; R. Meng, G. Schuler, J. Smith and W. van Neerven, *ibid.* **B339** (90) 325.
4. A. Stange and S. Willenbrock Phys. Rev. **D48** (1993) 2054; C. Kao, G. Ladinsky, and C.-P. Yuan, FSU-HEP-930508, MSUHEP-93/04; W. Beenakker, A. Denner, W. Hollik, R. Mertig, T. Sack, et al. Nucl. Phys. **B411** (1994) 343.
5. C. Kao, G. Ladinsky, and C.-P. Yuan, FSU-HEP-940508, MSUHEP-94/04, and references therein.

SIGNALS FOR TOP QUARK ANOMALOUS CHROMOMAGNETIC MOMENTS AT COLLIDERS

THOMAS G. RIZZO

Stanford Linear Accelerator Center, Stanford University, Stanford, CA 94309, USA

ABSTRACT

The Tevatron and the Next Linear Collider(NLC) will be excellent tools for probing the detailed nature of the top quark. We perform a preliminary examination of the influence of an anomalous chromomagnetic moment for the top, κ, on the characteristics of $t\bar{t}$ production at the Tevatron and on the spectrum of gluon radiation associated with $t\bar{t}$ production at the NLC. In particular, we analyze the sensitivity of future data to non-zero values of κ and estimate the limits that can be placed on this parameter at the Tevatron and at the NLC with center of mass energies of $\sqrt{s} = 500$ and 1000 GeV. Constraints on κ from low energy processes, such as $b \to s\gamma$, are briefly discussed.

The probable discovery of the top quark at the Tevatron[1] has renewed thinking about what may be learned from a detailed study of its properties. It is believed that the details of top quark physics may shed some light on new physics beyond the Standard Model(SM). Amongst others, one set of the top's properties which deserve study is its couplings to the various gauge bosons; up until now such analyses[2] have concentrated on the electroweak couplings of the top. In this work we consider the possible existence of an anomalous chromomagnetic moment, dimension-5 coupling, κ, at the $t\bar{t}g$ vertex and explore the capability of the Tevatron and NLC to probe this kind of new physics. Such interactions may arise in extended technicolor or compositeness scenarios. At present, only rather weak limits on κ (of order 10) exist, in particular, from operator mixing contributions to the $b \to s\gamma$ decay. (See the last paper in Ref. 2). For details of the analysis presented below, see Ref. 3.

At the Tevatron, both $gg, q\bar{q} \to t\bar{t}$ subprocesses are modified by the existence of $\kappa \neq 0$ with the $q\bar{q}(gg)$ case displaying a quadratic(quartic) κ dependence. In the results below only the SM NLO and gluon resummation corrections[4] are incorporated by way of 'K-factors'. For $\kappa \neq 0$ the relative weights of the gg and $q\bar{q}$ subprocesses can be drastically altered as can be seen in Fig.1a. We also see that the total σ can be dramatically increased or decreased via $\kappa \neq 0$; in particular, the CDF σ result can be reproduced if $\kappa \simeq 0.25$. If we assume that in the future the $t\bar{t}$ σ settles down to its SM value, we can estimate the constraints that this would impose on κ including uncertainties(which we estimate using Refs. 1 and 5) due to (i)scale ambiguities, (ii)parton density variations, (iii)NNLO QCD corrections, and (iv)the machine luminosity, as well as statistics. For $\mathcal{L} = 100(250, 500, 1000)pb^{-1}$ we obtain the 95% CL ranges of

$-0.14 \leq \kappa \leq 0.15$, $-0.11 \leq \kappa \leq 0.12$, $-0.09 \leq \kappa \leq 0.11$, and $-0.08 \leq \kappa \leq 0.11$, respectively.

One might ask if the top pair p_t-, rapidity(y-), or invariant mass(M-) distributions can be used to increase the sensitivity to $\kappa \neq 0$; we find that the by far dominant effect on these observables (for small values of κ) is an approximate overall rescaling of the observable by the ratio of the κ-dependent to the SM cross sections. (This results from the fact that the $t\bar{t}$ threshold region is found to dominate in the evaluation of σ's.) Almost all deviations from this simple rescaling occur at very large M or p_t values where statistics will always remain quite meager for interesting values of κ. Fig. 1b shows this situation explicitly for the $t\bar{t}$ p_t-distribution. We conclude that the total $t\bar{t}$ cross section provides the best probe of κ at the Tevatron.

At the NLC, the $t\bar{t}g$ vertex can only be directly explored via the QCD radiative process $e^+e^- \rightarrow t\bar{t}g$[5]. Relative to the Tevatron, this results in a substantial loss in statistics which can be compensated for by the cleanliness of the environment as well as a reduction in the associated theoretical uncertainties. Since the new κ-dependent interaction is proportional to the gluon 4-momentum, we are thus lead to a study of the gluon energy distribution associated with $t\bar{t}$. The dominant effect of $\kappa \neq 0$ is to induce an increase in the high energy tail of this distribution. This same energy dependence leads to the observation that the finite κ contributions grow rapidly with increasing $\sqrt{s}/2m_t$, implying increased sensitivity at an NLC with $\sqrt{s} = 1$ TeV instead of 500 GeV. In this first study, we ignore effects from top decay(except in the statistics) and perform a LO analysis following the work in Ref. 5. To reduce scale ambiguities and contributions from higher orders, we employ the scheme of Brodsky et al.(BLM) in Ref. 6. Estimates of contributions from these higher order are lumped into the uncertainties when obtaining limits. Fig. 2a shows this distribution for the case of $\sqrt{s} = 1$ TeV for $\alpha_s = 0.10$ while Fig. 2b shows the result of integrating this distribution for values of $z = 2E_{glu}/\sqrt{s} > 0.4$. Assuming that the SM results are realized, bounds on κ may be obtainable by either (i)counting excess events with high energy gluon jets or (ii)by a fit to the gluon energy distribution via a Monte Carlo analysis. Events are selected with at least one b-tag as well as one high p_t lepton and gluon jet energies larger than 200 GeV. Such large jet energies will allow a clean separation from the top decays and will simultaneously place us in the region of greatest κ sensitivity. For a luminosity of 200 fb^{-1} the resulting 95% CL allowed range is found to be $-1.0 \leq \kappa \leq 0.25$. Substantial improvement is obtained by fitting the spectrum itself; Fig. 2c shows the Monte Carlo generated spectrum and best fit($\kappa = 0.06$) assuming that the SM is realized. At 95% CL, one now obtains the allowed range of $-0.12 \leq \kappa \leq 0.21$ for the same luminosity as above. For a $\sqrt{s} = 500$ GeV machine with an integrated luminosity of $30fb^{-1}$, these limits are significantly loosened; following the same procedure yields the corresponding bounds $-1.98 \leq \kappa \leq 0.44$. A full analysis including the effect of top decay and NLO corrections should be performed to confirm these 'first pass' results.

Both the Tevatron and the NLC provide complementary windows on the possible anomalous chromomagnetic couplings of the top with different systematics. If such a coupling were observed, it would provide a unique signature for new physics beyond the Standard Model.

References

1. F. Abe *et al.*, CDF Collaboration, Fermilab report Fermilab-PUB-94/097-E, 1994; S. Abachi *et al.*, D0 Collaboration, Phys. Rev. Lett. **72**, 2138 (1994).
2. G. Kane, G.A. Ladinsky, and C.P. Yuan, Phys. Rev. **D45**, 124 (1992); C.P. Yuan, Phys. Rev. **D45**, 782 (1992); D. Atwood, A. Aeppli, and A. Soni, Phys. Rev. Lett. **69**, 2754 (1992); M. Peskin, talk presented at the *Second International Workshop on Physics and Experiments at Linear e^+e^- Collider*, Waikoloa, HI, April 1993; M. Peskin and C.R. Schmidt, talk presented at the *First Workshop on Linear Colliders*, Saariselkä, Finland, September 1991; P. Zerwas, *ibid.*; W. Bernreuther *et al.*, in *Proceedings of the Workshop on e^+e^- Collisions at 500 GeV, The Physics Potential*, (DESY, Hamburg) ed. by P. Igo-Kemenes and J.H. Kühn, 1992; A. Djouadi, ENSLAPP-A-365-92 (1992); D.O. Carlson, E. Malkawi, and C.-P. Yuan, Michigan State report MSUHEP-94/05, 1994; J.L. Hewett and T.G. Rizzo, Phys. Rev. **D49**, 319 (1994).
3. D. Atwood, A. Kagan and T. Rizzo, SLAC-PUB-6580, 1994; T.G. Rizzo, SLAC-PUB-6512, 1994.
4. E. Laenen, J. Smith, and W.L. van Neerven, Phys. Lett. **B321**, 254 (1994). A.D. Martin, W.J. Stirling and R.G. Roberts, Rutherford Laboratory report RAL-94-055, 1994.
5. For a recent study, see Yu.L. Dokshitzer, V.A. Khoze and W.J. Sterling, Durham report DTP/94/14, 1994.
6. S. Brodsky, P. LePage and P.B. Mackenzie, Phys. Rev. **D28**, 228 (1983).

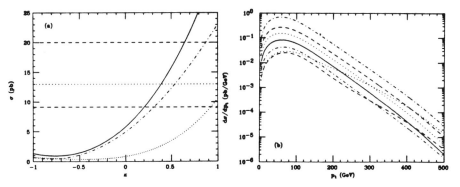

Fig. 1: (a)NLO cross sections for the $q\bar{q} \to t\bar{t}$(dash-dotted) and $gg \to t\bar{t}$(dotted) subprocesses as well as the total cross section(solid) at the Tevatron as functions of κ for $m_t = 170$ GeV using the CTEQ parton distribution functions. The horizontal dashed lines provide the $\pm 1\sigma$ CDF cross section determination while the horizontal dotted line is the D0 95% CL upper limit. (b)p_t distribution for top quark pairs produced at the Tevatron assuming $m_t = 170$ GeV and CTEQ parton densities. The solid curve is the SM prediction and the upper(lower) dash-dotted, dashed, and dotted curves correspond to $\kappa = 1$, 0.5, 0.25(-1, -0.5, -0.25), respectively.

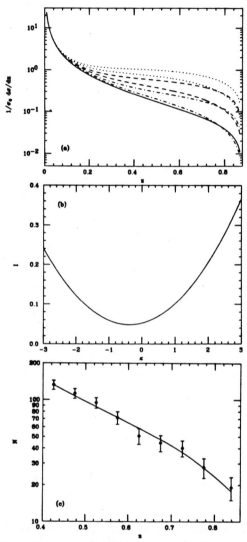

Fig. 2: (a)Gluon jet energy spectrum assuming $\alpha_s = 0.10$ for $m_t = 175$ GeV at a $\sqrt{s} = 1$ TeV NLC. The upper(lower) dotted, dashed, and dot-dashed curves correspond to κ values of 3(-3), 2(-2), and 1(-1) respectively while the solid curve is conventional QCD with $\kappa = 0$. (b)Integrated gluon energy spectrum for the same input parameters and labelings as in Fig. 3 as a function of κ assuming $z_{cut} = 0.4$. (c)Best fit gluon spectrum through the points generated by the Monte Carlo analysis corresponding to $\kappa = 0.06$.

PROBING THE TOP QUARK PRODUCTION AND DECAY VERTICES AT A 400 GEV e^+e^- COLLIDER

TIMOTHY L. BARKLOW

Stanford Linear Accelerator Center
Stanford University, Stanford, CA 94309, USA

and

CARL R. SCHMIDT*

Santa Cruz Institute for Particle Physics
University of California, Santa Cruz, CA 95064, USA

ABSTRACT

A high energy e^+e^- linear collider (NLC) is an ideal tool for probing the structure of the $\gamma t\bar{t}$, $Zt\bar{t}$, and $Wt\bar{b}$ vertices to look for effects both within and beyond the standard model. In this talk we discuss how to use the high level of top quark polarization at the NLC to disentangle its production and decay vertices, and we make some estimates of limits on anomalous couplings that can be reached for realistic NLC parameters.

The CDF evidence for a top quark of mass around 174 GeV [1] indicates that the top quark couples more strongly to the electroweak symmetry breaking sector than any other known particle, and much more strongly than any of the other quarks. This suggests that the top may play a special role in the generation of the quark masses, or at least it may offer hints to this mass generation. Thus, we should try to study its properties as precisely as possible. Due to the clean event environment and the possibility of initial-state polarization, a high energy e^+e^- linear collider would be an ideal tool for extracting the properties of the top quark. There have already been detailed analyses of the top quark production at threshold at the Next Linear Collider (NLC), where the mass and width can be measured to considerable accuracy [2]. In this talk we will give some initial results of an analysis of the top quark above threshold. In particular we will concentrate on extracting the couplings of the top quark at the NLC [3].

The most general top quark couplings can be written in terms of form factors in the production and decay vertices. The $t \to W^+b$ decay vertex is

$$i\mathcal{M}^{W\mu} = i\frac{g}{\sqrt{2}}\left\{\gamma^\mu[F_{1L}^W P_L + F_{1R}^W P_R] + \frac{i\sigma^{\mu\nu}q_\nu}{2m_t}[F_{2L}^W P_R + F_{2R}^W P_L]\right\}. \tag{1}$$

At tree level in the standard model $F_{1L}^W = 1$ and all other form factors are zero. The antitop form factors are identical to these in the limit of CP invariance. Similarly,

---
*Talk presented by C.R.S.

the $\gamma, Z \to t\bar{t}$ production vertices are

$$i\mathcal{M}^{i\mu} = ie\left\{\gamma^\mu [Q_V^i F_{1V}^i + Q_A^i F_{1A}^i \gamma_5] + \frac{i\sigma^{\mu\nu} q_\nu}{2m_t}[Q_V^i F_{2V}^i + Q_A^i F_{2A}^i \gamma_5]\right\}, \qquad (2)$$

where the superscript is $i = \gamma, Z$. In this formula $Q_V^\gamma = Q_A^\gamma = \frac{2}{3}$, $Q_V^Z = (\frac{1}{4} - \frac{2}{3}s^2)/sc$, and $Q_A^Z = (-\frac{1}{4})/sc$, so that at tree level in the standard model $F_{1V}^\gamma = F_{1V}^Z = F_{1A}^Z = 1$ and all the others are zero. In the limit of CP invariance $F_{2A}^i = F_{3A}^i = 0$. The form factors are typically corrected by a few percent or less from QCD and electroweak loops. Thus, any large deviations would indicate new physics, which may occur if the top quark is strongly coupled to the symmetry breaking sector [4].

In the present analysis we will study the process $e^+e^- \to t\bar{t} \to b\ell^+\nu\bar{b}q\bar{q}$ at the parton level without detector resolution effects, hadronization, or initial-state radiation. These will be included in the future. We use the complete helicity amplitudes [5] for the event in the zero-width approximation for the top quark and the W and with $m_b = 0$. The top quark mass is set to its nominal value of $m_t = 174$ GeV. The form factors can be obtained from a maximum likelihood fit using all of the information available in the helicity angles of the top quarks and their decay products in the sample of $t\bar{t}$ events. Here we estimate the sensitivity possible in such an analysis at the NLC.

We will present our results for an integrated luminosity of 100 fb^{-1}. This is an optimistic value for one year of running and is more likely characteristic of a few years worth of data. Of course, the estimated errors will simply scale as $(\int \mathcal{L}dt)^{-1/2}$. For simplicity in this analysis we will just keep events where one of the W's decays hadronically and the other leptonically to an electron or a muon. This allows complete reconstruction of the event with little background. The resulting branching fraction is then $2 \times (2/9) \times (6/9) = 30\%$. In addition, as a crude model of the detector efficiency we require that all of the partons except the neutrino must lie within $|\cos\theta_{\text{lab}}| < 0.8$. The bottom quarks are assumed to be tagged and signed, but the ambiguity in the sign of the decay angle of the $W \to q\bar{q}$ decay has been included. At the NLC the backgrounds to this process are quite small; the dominant one is radiative W pair production.

Table 1: Number of $t\bar{t}$ pairs.

	50 fb^{-1} e_L	50 fb^{-1} e_R	100 fb^{-1} unpol.
Total	42403	20488	62891
× BR	12564	6070	18634
× cuts	3981	1891	5872

In table 1 we show the number of $t\bar{t}$ pairs produced at 400 GeV center-of-mass energy and 100 fb^{-1} integrated luminosity, along with the effects of the branching fraction and the angular cuts. In addition we show the breakdown if the collider is run with 50 fb^{-1} each of left-handed and right-handed electrons at 90%

polarization. The table shows that the acceptance cut reduces the event sample by a factor of about 0.3 beyond that of the branching fraction. To explain this we plot the cross section in units of R as a function of the top production angle for 90% left-polarized electrons, including the subprocess cross sections for top quarks of specific helicities. From this figure we see that the electron helicity typically carries through to the top quark, with the top quark produced strongly peaked in the forward direction. Thus, good forward detector coverage is very important for obtaining the most complete event information. However, we must emphasize that events that are neglected here, such as ones with a missing quark jet, may still be used in a less complete analysis for constraining the form factors. In addition, the purely hadronic $t\bar{t}$ mode may also be usable in the e^+e^- event environment.

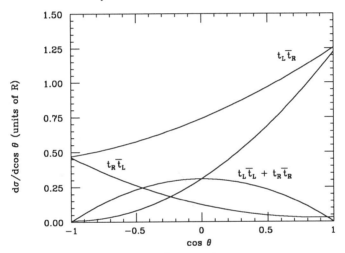

Figure 1: $t\bar{t}$ cross section.

Figure 1 also shows that electron polarization will be an important tool in studying top quarks at the NLC. We exhibit this in figure 2 where we plot the 95% confidence level contours for F_{1V}^Z and F_{1A}^Z with all other form factors set to their standard model values. The dotted (dotdashed) curves are obtained from the 50 fb^{-1} of events with 90% L (R) polarized electrons, while the solid band is obtained from the measurement of $A_{LR} = (\sigma_L - \sigma_R)/(\sigma_L + \sigma_R)$. Note that the A_{LR} measurement is very important for constraining F_{1V}^Z, gaining almost a factor of two in sensitivity over an unpolarized event sample. On the other hand, the absolute total cross section is quite insensitive to these form factors[†].

[†]The total cross section is, however, important for constraining the $t\bar{t}$ production magnetic moment form factors.

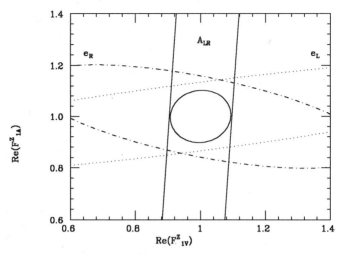

Figure 2: 95% Confidence level contours.

The polarization of the top quarks can also be used in constraining the decay form factors. The standard method for obtaining F_{2L}^W is by measuring the ratio $W_{\text{long.}}/W_{\text{trans.}}$ of polarized W's in top quark decays [6]. This simply requires fitting to the polar angle of the decay products of the W in its rest frame. However, for polarized top quarks, the top decay angles are also correlated with the W polarization, offering further constraints. To gauge the importance of these effects we made a one parameter fit to F_{2L}^W, using 1) 50 fb^{-1} each of L and R polarized electrons with a complete event analysis; 2) 100 fb^{-1} of unpolarized electrons with a complete event analysis; and 3) the same events, but analyzing only the W-decay polar angle information. The 95% confidence limits are 1) $\pm.080$, 2) ±0.098, 3) ±0.114. Thus, a complete likelihood analysis can increase the sensitivity by a factor of 1.4. However, the W-decay angle analysis should still be useful for events where complete reconstruction is impossible.

A likelihood fit can also be used to search for transverse polarization effects. Transverse polarization can arise within the standard model at the one-loop level from absorptive QCD and electroweak diagrams [7]. In the formalism used here these loop effects contribute to the imaginary parts of the form factors. In figure 3 we show the sensitivity to the imaginary parts of F_{2V}^Z and F_{2V}^γ around their one-loop QCD values for 50 fb^{-1} each of L and R polarized electrons. This plot suggests that a factor of 4 in statistics is still needed to confidently infer a nonzero value for the form factors.

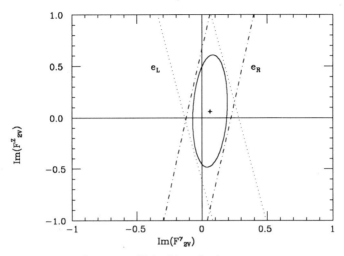

Figure 3: 95% Confidence level contours.

In this talk we have investigated the sensitivity to the top production and decay vertices of a likelihood analysis using the complete event reconstruction at the NLC. Further work will include experimental resolution effects, non-$t\bar{t}$ backgrounds, and more efficient ways to use all of the $t\bar{t}$ events. Implications for detector design are important because simply increasing coverage to $|\cos\theta_{lab}| < 0.9$ from 0.8 increases the statistics by a factor of ~ 2.

We would like to thank Michael Peskin for collaboration on the event generator used in this analysis and for offering many useful suggestions along the way.

References

1. F. Abe *et al.* (CDF Collaboration), *Phys. Rev. Lett.* **73** (1994) 225.

2. K. Fujii, T. Matsui, and Y. Sumino, KEK Preprint 93-125 (1993).

3. A previous related work on this subject is G.A. Ladinsky and C.-P. Yuan, *Phys. Rev.* **D49** (1994) 4415.

4. R.S. Chivukula, S.B. Selipsy, and E.H. Simmons, *Phys. Rev. Lett.* **69** (1992) 575; R.S. Chivukula, E.H. Simmons, and J. Terning, *Phys. Lett.* **B331** (1994) 383.

5. G. Kane, G.A. Ladinsky, and C.-P. Yuan, *Phys. Rev.* **D45**, (1992) 124.

6. K. Fujii, in *Physics and Experiments at Linear Colliders*, R. Orava, P. Eerola, and M. Nordberg, eds. (World Scientific, 1992).

7. C.-P. Yuan, *Phys. Rev.* **D45** (1992) 782.

SUMMARY OF TOP QUARK PHYSICS

STEPHEN PARKE

parke @ fnal.gov
Department of Theoretical Physics
Fermi National Accelerator Laboratory
Batavia, Illinois 60510, USA

ABSTRACT

I briefly review standard top quark physics at hadron colliders and summarize the contributions to this conference. The possibility of new mechanisms for $t\bar{t}$ production are also discussed.

1. Standard

In hadron colliders the dominant mode of top quark production is via quark-antiquark annihilation or gluon-gluon fusion,

$$q\,\bar{q} \;\rightarrow\; t\,\bar{t}$$
$$g\,g \;\rightarrow\; t\,\bar{t}.$$

However there are other modes,

$$W^+\,g \;\rightarrow\; t\,\bar{b}$$
$$W^{+*} \;\rightarrow\; t\,\bar{b}$$
$$(\gamma, Z)\,g \;\rightarrow\; t\,\bar{t}$$
$$(\gamma, Z)^* \;\rightarrow\; t\,\bar{t}$$
$$\cdots \;.$$

In this list I have not included processes which pick a b-quark out of the hadron.

These processes are approximately ordered according to their rates in hadron colliders. Fig. 1(a) has the rates for the first three processes at the Tevatron assuming that the dominant decay model for the top quark is $W^+\,b$. The channel, positron plus jets, was chosen so that the final state for all three processes is positron, $b\bar{b}$ plus jets. The QCD, W-gluon and W^* processes have two, one and zero non-b-quark jets, respectively.

Fig. 1(b) contains the QCD cross section for $m_t = 175\ GeV$ verses $\sqrt{s}$ for both proton-proton colliders and for proton-antiproton colliders. At $\sqrt{s} = 1.8\ TeV$ the gluon-gluon fusion is only 10% of the cross section for a proton-antiproton collider, the Tavatron, whereas at a $14\ TeV$ proton-proton collider, the LHC, gluon-gluon fusion is 90% of the cross section.

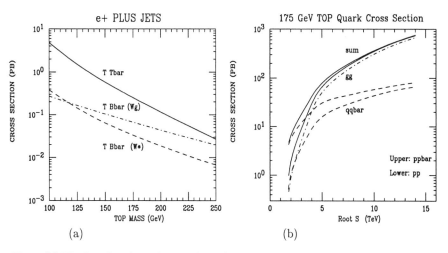

Fig. 1. (a) The Top Quark Production Cross Section at the Tevatron. The three curves are for quark-antiquark annihilation plus gluon-gluon fusion (solid), W-gluon fusion (dot-dash) and through an off-mass shell W-Boson (dashes). (b) QCD Top quark Production cross section as a function of $\sqrt{s}$, for quark-antiquark annihilation (dashes), gluon-gluon fusion (dot-dash) and the sum (solid) for both proton-antiproton (upper) and proton-proton (lower) colliders.

For an accurate determination of the QCD top cross section, we need to consider the next to leading order calculations and the soft gluon resummation of the next to leading order calculations.[1] In Fig. 2, the results of these calculations using the same structure functions are shown. At high top quark masses the difference between these two calculations is at the 20% level.

The standard model decays of the top are

$$
\begin{aligned}
t &\rightarrow W^+ b \\
t &\rightarrow W^+(\ s\ or\ d) \\
t &\rightarrow g\ W^+(\ b,\ s\ or\ d) \\
t &\rightarrow \gamma\ W^+(\ b,\ s\ or\ d) \\
t &\rightarrow Z\ W^+(\ b,\ s\ or\ d) \\
t &\rightarrow \phi\ W^+(\ b,\ s\ or\ d) \\
&\cdots
\end{aligned}
$$

For $m_t = 175\ GeV$, the total width of the top quark is approximately 1.5 GeV. The CKM suppressed decays are expected to be less than 0.1% of the non-suppressed decays. Whereas the decays including a Z or Higgs will be extremely small unless the on-mass shell decay is kinematically allowed.

Flavor changing neutral currents,

$$
t \rightarrow (\gamma,\ g,\ Z\ or\ \phi^0) + (c\ or\ u),
$$

have branching ratios less than 10^{-10} in the Standard Model.

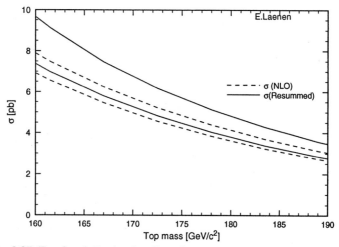

Fig. 2. The QCD Top Quark Production Cross Section at the Tevatron. The solid curves give the range of values using the resummed next to leading order calculations and the dashed curves are the range for the next to leading oder calculations. The same structure functions were used.

2. Searches

Both CDF[2] and D0[3] presented detailed results on the search for top at the Tevatron. The data presented included both the dilepton and the lepton plus jets mode for the decay of the $t\bar{t}$ pair. However, neither collaboration presented data on the six jet mode. Theoretical calculations suggest that with an efficient b-quark tag, that this mode will be accessible at the Tevatron. A detailed summary of the experimental results present can be found in the review of hadron collider physics by Shochet.[4]

CDF observes a 2.8 σ (0.26%) effect which is not sufficient to firmly establish the existence of top but which, if interpreted as top, yields $m_t = 174 \pm 10^{+13}_{-12} \, GeV/c^2$ and $\sigma_{t\bar{t}} = 13.9^{+6.9}_{-4.8} \, pb$.

D0 does not observe a significant excess of events due to $t\bar{t}$ production. The probability for the background to fluctuate to give greater than or equal to the observed number of events is 7.2% (1.5 σ). If $m_t = 180 \, GeV$ then $\sigma_{t\bar{t}} = 6.5 \pm 4.9 \, pb$.

Orr[5] presented the results of a study on the effects of soft gluon radiation in the determination of the top quark momentum. The results of this study will be important for precision measurements of the top quark mass at hadron colliders.

3. Surprises

In the Standard Model the couplings of the top quark to each of the gauge bosons, g, γ, Z and $W^{\pm}$, are determined, including radiative corrections. Therefore potential new physics could show up as deviations of these vertices from Standard Model expectations. Kao[6] and Rizzo[7] discussed corrections to the QCD coupling, $gt\bar{t}$. Kao's paper concentrated on the one-loop weak corrections in both the Standard Model and the Minimal Supersymmetric Standard Model. Whereas Rizzo considered the effects of an

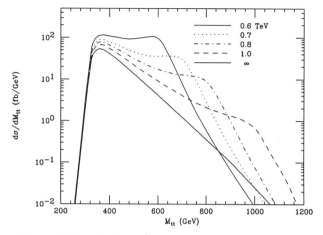

Fig. 3. The differential distribution, $\frac{d\sigma}{dM_{t\bar{t}}}$ verses $M_{t\bar{t}}$ for top-antitop production with the curves labelled by the mass of the octet top-color vector boson.

anomalous chromomagnetic moment to this coupling.

Schmidt[8] summarized top quark physics at e^+e^- colliders and in particular discussed the signatures of deviations to the $\gamma t\bar{t}$, $Zt\bar{t}$ and $Wt\bar{b}$ vertices at such machines.

Of course the top quark could present more dramatic surprises such as charged Higgs decays, $t \to H^+ b$, large flavor changing neutral current decays, $t \to (g, Z \; or \; \gamma) + (c \; or \; d)$ or enhanced production through a new resonance. This latest possibility could occur either through the quark-antiquark production mode,[9] as expected in Top-color models of electroweak symmetry breaking or via the gluon-gluon fusion[10] as suggested by some Technicolor models. In both cases the $t\bar{t}$ pair is produced by the decay of a heavy new particle, Top-color Boson or Techni-eta, which distorts the P_T, $\cos\theta^*$ and $M_{t\bar{t}}$ distributions from the Standard Model expectation. Fig. 3 is the change in the shape of the $M_{t\bar{t}}$ distribution in the Top-color Model discussed by Hill and Parke.[9] We should think of top quark production as a new Drell-Yan process probing extremely high mass scales, greater than 500 GeV.

4.   Conclusion

The top quark is an exciting new window on very high mass scale physics. While exploring the vista from this window we should be on the lookout for any deviation from the Standard Model which will provide us with information about that elusive beast, the mechanism of electro-weak symmetry breaking. Because the mass of the top quark is very heavy, this quark is the particle most strongly coupled to the electro-weak symmetry breaking sector. Therefore the deviations could be seen at zeroth order or may require more subtle measurements.

What is needed is hundreds of top-antitop pairs as soon as possible. Then, watch out for surprises at DPF'96!

References

1. R. K. Ellis, Phys. Lett. **B259** 492, (1991);
   E. Laenen, J. Smith and W. L. Van Neerven, Phys. Lett. **B321** 254, (1994).
2. J. Benlloch, *'Results from the CDF Top Search in the Dilepton Channel,'*
   G. Watts, *'CDF Top Results in the Lepton Plus Jets Channel,'*
   J. Konigsberg, *'Discussion of Results of the CDF Top Searches,'*
   B. Harral, *'Kinematics Properties of Lepton Plus Multijet Events at CDF,'*
   these proceedings.
3. K. Genser, *'Search for Top with D0 Detector in the Dilepton Channel,'*
   D. Chakraborty, *'Search for Top in the Lepton plus Jets Channel at D0,'*
   W. Cobau, *'Search for the Top Quark in Electron plus Jets plus B-quark Tag Using the D0 Detector,'*
   P. Bhat, *'Search for the Top Quark Using Multi-Variate Analysis Techniques,'*
   these proceedings.
4. M. Shochet, *'Physics at the Fermilab Collider,'* plenary session, these proceedings.
5. L. Orr and W. Stirling, *'Top Momentum Reconstruction and Extra Soft Jets,'* these proceedings.
6. C. Kao, *'Leading Weak Corrections to the Production of Top Quark Pairs at Hadron Colliders,'* these proceedings.
7. T. Rizzo, *'Signals for Top Quark Anomalous Chromomagnetic Moments,'* these proceedings.
8. T. Barklow and C. Schmidt, *'Probing the Top Quark Production and Decay Vertices at a 500 GeV e^+e^- Collider,'* these proceedings.
9. C. Hill and S. Parke, Phys. Rev. **D49** 4454, (1994).
10. E. Eichten and K. Lane, Phys. Lett. **B327** 129, (1994);
    K. Lane, BUHEP-94-12 (1994).

B Physics

MARINA ARTUSO, *Syracuse*

SEMILEPTONIC DECAYS OF B MESONS

MICHAEL S. SAULNIER

High Energy Physics Laboratory, Harvard University,
42 Oxford Street, Cambridge, MA 02138, USA

(Representing the CLEO Collaboration)

ABSTRACT

CLEO has measured the inclusive B meson semileptonic branching fraction using a number of methods. We measure the average B semileptonic branching fraction from fits to the single lepton spectrum and from a model-independent method using lepton tags. We also measure the B^+ and B^0 semileptonic branching fractions using four tagging methods. Assuming equal semileptonic partial widths, we obtain a value for the ratio of lifetimes.

1.   Introduction

The B semileptonic branching fraction[*] is an interesting quantity because measured values have traditionally been below theoretical expectations. Naive counting of quarks and leptons lead to a branching fraction of $\sim 17\%$. More careful second order QCD calculations predict $\mathrm{Br}(B \to X\ell\nu) \geq 12.5\%$, while the measurements consistently lie below 11%.[1,2]

It is also predicted that the B^+ has a slightly longer lifetime than the B^0 at the 10% level.[2] Assuming that the semileptonic partial widths are equal, this implies that $\mathrm{Br}(B^+ \to X\ell\nu)$ should be larger than $\mathrm{Br}(B^0 \to X\ell\nu)$ to the same degree.

We report two measurements of the average B semileptonic branching fraction in Sections 2 and 3, followed by measurements of the separate charged and neutral B branching fractions and their ratio in Sections 4, 5 and 6.

2. $\mathrm{Br}(B \to X\ell\nu)$ From Single Lepton Spectrum

Fits to the shape of the single lepton spectrum have been the standard method for measuring $\mathrm{Br}(B \to X\ell\nu)$ at the $\Upsilon(4S)$. Figure 1 shows the lepton spectrum observed with 0.94 fb^{-1} of on-resonance data collected by the CLEOII detector at the Cornell Electron Storage Ring. The spectrum is a sum of a primary $B \to X\ell\nu$ and a secondary $b \to c \to Y\ell\nu$ component. A fit to the model of Isgur *et al.*[3] (ISGW) with a floating D^{**} fraction yields $\mathrm{Br}(B \to X\ell\nu) = (10.98 \pm 0.10 \pm 0.33)\%$ for a D^{**} fraction of 21%. With the model of Altarelli *et al.*[4] (ACCMM), a branching fraction of $(10.65 \pm 0.05 \pm 0.33)\%$ is found for $p_f = 230$ MeV/c, $m_c = 1700$ MeV/c^2 and m_{spec} fixed at 150 MeV/c^2.

---

[*]Throughout, "semileptonic" is understood to mean "semielectronic" or "semimuonic".

These results are systematics-limited from the uncertainty in the choice of model to extrapolate the primary spectrum to low momentum. Additionally, this method is sensitive to possible non-$B\bar{B}$ decays of the $\Upsilon(4S)$ since the lepton yield is normalized to the number of $\Upsilon(4S)$ decays. Both of these concerns are addressed by using a method employing lepton tags.

3. Br($B \to X\ell\nu$) From Tagged Lepton Spectrum

The semileptonic branching fraction can be measured with little model dependence by using a fast ($p_\ell > 1.4$ GeV/c) lepton as a B tag. Remaining electrons in the event can be separated into primary and secondary spectra by exploiting charge and angular correlations. Only electrons are used so that the primary spectrum can be probed at low momentum. Assuming that the tag lepton is primary, the semileptonic decay of the opposite B will produce an oppositely charged electron, except when $B^0 - \bar{B^0}$ mixing occurs. If the electron is from the semileptonic decay of a charmed particle produced in the decay of the opposite B, then it will be like-sign (again, except for mixing). Finally, electrons from charm decay originating from the same B as the tag lepton will always be unlike-sign. This last contribution can be suppressed using kinematic correlations.

At the $\Upsilon(4S)$, the B's are nearly at rest and there is no angular correlation between particles from opposite B's. However, there is an angular correlation between the tag lepton and electrons from the same B. Unlike-sign electrons are required to satisfy $p_e + \cos\theta_{\ell e} > 1$, where p_e is the electron momentum in GeV/c, and $\theta_{\ell e}$ is the angle between the tag lepton and the electron. This "diagonal" cut in the $p_e - \cos\theta_{\ell e}$ plane suppresses same-B electrons by a factor of 25, while keeping 67% of opposite-B electrons.

Figure 2 shows the unlike and like-sign spectra obtained before a number of backgrounds are subtracted. This analysis was performed using 2.07 fb^{-1} of on-resonance data. Corrections are made for continuum, for either tag leptons or electrons which were misidentified hadrons (fakes), for either tag leptons or electrons from J/ψ, π^0, τ, Λ_c or D_s decays or from γ conversions, for tag leptons from semileptonic charm decay, and for the residual contribution of electrons from the same B. The corrected spectra $N(\ell^\pm e^\mp)$ and $N(\ell^\pm e^\pm)$ are related to the primary $\mathcal{B}(b)$ and secondary $\mathcal{B}(c)$ branching fractions as

$$\frac{dN(\ell^\pm e^\mp)}{dp} = N_\ell \eta(p) \left[\frac{d\mathcal{B}(b)}{dp}(1 - \chi) + \frac{d\mathcal{B}(c)}{dp}\chi \right] \epsilon(p), \qquad (1)$$

$$\frac{dN(\ell^\pm e^\pm)}{dp} = N_\ell \eta(p) \left[\frac{d\mathcal{B}(b)}{dp}\chi + \frac{d\mathcal{B}(c)}{dp}(1 - \chi) \right], \qquad (2)$$

where $\eta(p)$ is the efficiency of electron identification and $\epsilon(p)$ is the efficiency of the diagonal cut. N_ℓ is the number of tag leptons and $\chi = f_0\chi_0$ represents the fraction of all $\Upsilon(4S)$ decays with $B^0 - \bar{B^0}$ mixing, approximately 8%.

Figure 3 shows primary and secondary spectra obtained after solving equations (1) and (2). Integrating the primary spectrum from 0.6 to 2.5 GeV/c, and correcting for the $(5.8 \pm 0.5)\%$ of the spectrum expected to fall below 0.6 GeV/c, we find Br($B \to Xe\nu$) = $(10.36 \pm 0.17 \pm 0.40)\%$. The solid curve is the fit to the ACCMM model with

p_f, m_c and m_{spec} fixed to the values found in the first method, giving a χ^2 of 28 for 19 d.o.f. For the ISGW model we find a best fit with $\chi^2 = 36$ for 19 d.o.f. for the default D^{**} fraction, and $\chi^2 = 20$ for 19 d.o.f. when the D^{**} is fixed at 21%. Finally, using the ACCMM model to extrapolate the secondary spectrum (dashed curve), we find $Br(b \to c \to Y e\nu) = (7.7 \pm 0.3 \pm 1.2)\%$ with a χ^2 of 27 for 13 d.o.f.

4. $Br(B^+ \to X\ell\nu)$

To measure the B^+ semileptonic branching fraction, we examine the yield of fast leptons across from a sample of fully reconstructed B^- decays (charge conjugation implied). CLEO reconstructs B^- and $\overline{B^0}$ in eight different modes: $D^{(*)}\pi^-$, $D^{(*)}\rho^-$, $D^{(*)}a_1^-$, and $K^{(*)}J/\psi$. The summed momenta and energies, $\vec{p}_B$ and E_B, of the decay products are used to construct the beam-constrained mass $M_B = \sqrt{E_{\text{beam}}^2 - |\vec{p}_B|^2}$ and the normalized energy difference between the B and the beam energy $\delta(\Delta E) = (E_{\text{beam}} - E_B)/\sigma(\Delta E)$, where $\sigma(\Delta E)$ is the expected resolution on $E_{\text{beam}} - E_B$. For signal, M_B peaks at the B mass, and $\delta(\Delta E)$ is a gaussian of zero mean and unit width. The $\delta(\Delta E)$ sidebands are used to estimate the background in the M_B distribution. Figure 4(a) shows the M_B distribution for the B^- tags in the $\delta(\Delta E)$ signal (points) and sideband (histogram) regions. We find 834 ± 42 tags using 1.35 fb^{-1} of on-resonance data.

Next, a search is made for a fast ($p > 1.4$ GeV/c) electron or muon elsewhere in the event as a signature of semileptonic decay of the remaining B^+. Only leptons of the expected charge are considered since the flavor of the remaining B is known. Two M_B distributions are formed for events satisfying this additional criterion – one for tags with muons, and one for tags with electrons. Their sum is shown in Fig. 4(b).

The lepton yields must be efficiency-corrected, including the correction for the portion of the primary spectrum below 1.4 GeV/c. We use the ISGW model with 21%D^{**}. Assuming lepton universality, we average the semielectronic and semimuonic branching fractions to obtain $Br(B^+ \to X\ell\nu) = (10.1 \pm 1.8 \pm 1.4)\%$. This is the first measurement of the charged B semileptonic branching fraction.

5. $Br(B^0 \to X\ell\nu)$

Three measurements are made of $Br(B^0 \to X\ell\nu)$. The approach is similar to the B^+ measurement, except that additional leptons of either sign are accepted. This eliminates the need for a mixing correction, but requires a small correction for secondary leptons from charm. As before, the electron and muon results are averaged. These results are all obtained with 1.35 fb^{-1} of on-resonance data.

The first measurement uses a sample of fully reconstructed $\overline{B^0}$ tags. The technique has already been described in Section 4. We find 515 ± 31 tags.

Next, partially reconstructed $\overline{B^0} \to D^{*+}\pi^-$, $D^{*+} \to D^0\pi^+$ tags are used. Here, the fast π^- recoils against a slow π^+ and an unseen D^0. The energy of the D^0 is calculated from energy conservation. This fixes the opening angle between the π^+ and the D^0 in the laboratory frame. The azimuthal angle of the D^0 about the slow pion axis is chosen to maximize the apparent B mass. This "pseudomass" must lie in the narrow window between M_{B^0} and E_{beam}. A fit to the pseudomass distribution yields 822 ± 53 tags.

The third measurement utilizes the partially reconstructed tag $\overline{B^0} \to D^{*+}\ell^-\bar{\nu}$, $D^{*+} \to D^0\pi^+$. The ℓ^- must have momentum > 1.8 GeV/c to suppress feed-down from $B^- \to D^{**0}\ell^-\bar{\nu}$. An approximate D^{*+} 4-momentum is calculated from the soft π^+ momentum only. This is possible because the decay of the D^{*+} is slightly above kinematic threshold. With the D^{*+} and ℓ^- 4-momentum, the square of the neutrino mass $\widetilde{M}_\nu^2$ is calculated assuming that the B is at rest, but with energy E_{beam}. Signal populates $\widetilde{M}_\nu^2 = 0$. Figure 5(a) shows the $\widetilde{M}_\nu^2$ distribution, which gives 7119 ± 139 tags. Figure 5(b) is the subset with an additional fast electron or muon.

The results for $\text{Br}(B^0 \to X\ell\nu)$ are $(13.5 \pm 2.6 \pm 1.9)\%, (10.2 \pm 1.9 \pm 1.3)\%$ and $(10.5 \pm 0.8 \pm 1.3)\%$ for the fully reconstructed, the $D^{*+}\pi^-$, and the $D^{*+}\ell^-\bar{\nu}$ tags, respectively. We average the three measurements weighting each by the statistical and uncorrelated systematic errors to obtain $\text{Br}(B^0 \to X\ell\nu) = (10.9 \pm 0.7 \pm 1.1)\%$, using the same model as the B^+ measurement.

6. τ_+/τ_0

If the B^+ and B^0 semileptonic partial widths are assumed to be equal, then the ratio of branching fractions $\text{Br}(B^+ \to X\ell\nu)/\text{Br}(B^0 \to X\ell\nu)$ equals the ratio of lifetimes. Many correlated systematic errors cancel when taking the ratio of branching fractions, giving $\tau_+/\tau_0 = 0.93 \pm 0.18 \pm 0.12$.

7. Conclusions

CLEO has measured the inclusive B meson semileptonic branching fraction using a number of methods. From fits to the single lepton spectrum, we find $\text{Br}(B \to X\ell\nu) = (10.98 \pm 0.10 \pm 0.33)\%$ for the model of Isgur *et al.* where the $B \to D^{**}\ell\nu$ fraction is allowed to float. Using a lepton tag technique with very little model dependence, we find $\text{Br}(B \to X\ell\nu) = (10.36 \pm 0.17 \pm 0.40)\%$. The separate B^+ and B^0 semileptonic branching fractions have also been measured using four tagging methods. We find $\text{Br}(B^+ \to X\ell\nu) = (10.1 \pm 1.8 \pm 1.4)\%$ and $\text{Br}(B^0 \to X\ell\nu) = (10.9 \pm 0.7 \pm 1.1)\%$. This is the first such measurement for the B^+. If the semileptonic partial widths are assumed to be equal, we obtain a lifetime ratio $\tau_+/\tau_0 = 0.93 \pm 0.18 \pm 0.12$.

Although the central value of the lifetime ratio is on the unexpected side of unity, it is consistent with theoretical predictions within errors. However, the measured values for the average B semileptonic branching fraction remain well below 12.5%. In particular, the lepton-tagged result indicates that the discrepancy between theory and experiment is unlikely to be due to uncertainty in the shape of the primary lepton spectrum at intermediate momenta, or from possibilities of non-$B\bar{B}$ decays of the $\Upsilon(4S)$.

References

1. G. Altarelli and S. Petrarca, *Phys. Lett.* **B261** (1991), 303.
2. I. Bigi, B. Blok, M. Shifman and A. Vainshtein, *Phys. Lett.* **B323** (1994) 408.
3. N. Isgur, D. Scora, B. Grinstein and M. Wise, *Phys. Rev.* **D39** (1989) 799.
4. G. Altarelli, N. Cabibbo, G. Corbo, L. Maiani and G. Martinelli, *Nucl. Phys.* **B208** (1982) 365.

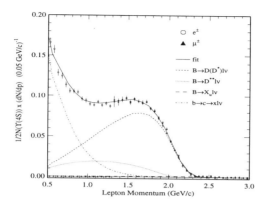

Fig. 1. Fit to the lepton spectrum of the Isgur *et al.* model with a floating D^{**} fraction.

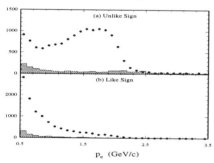

Fig. 2. Raw spectra of electrons of (a) unlike-sign and (b) like-sign relative to tag lepton. The points represent $\Upsilon(4S)$ data and the shaded histograms nearby continuum.

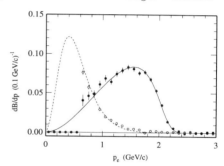

Fig. 3. Spectra of electrons from primary decays (filled circles) and secondary decays (open circles). The curves show fits to the model of Altarelli *et al.*

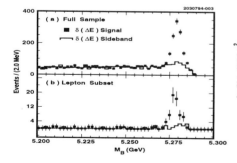

Fig. 4. $B^- \to$ hadrons tags

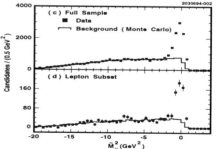

Fig. 5. $\overline{B^0} \to D^{*+}\ell^-\overline{\nu}$ tags

CHARM MESON PRODUCTION IN SEMILEPTONIC B DECAYS

A. J. MARTIN*

*Physics Department, Queen Mary and Westfield College,
University of London, London, E1 4NS, U.K.*

Abstract

New OPAL measurements of D^0, D^+ and D^{*+} production in semileptonic b hadron decays are reported. These measurements are used to infer the fraction of b quarks which fragment to B^0 and B^+ mesons in Z^0 decays. The results of a search for P-wave charmed mesons in semileptonic B decays are also reported. The observed signals of $44 \pm 8^{+3}_{-7}$ D^0_J and $48 \pm 10^{+3}_{-6}$ D^+_J events provide evidence for the production of both pseudovector and tensor states.

1. Introduction

It is well established that a large fraction of charmed semileptonic B decays are not accounted for by the exclusive decays $B \rightarrow D\ell\nu$ and $B \rightarrow D^*\ell\nu$. Given that several theoretical models predict that these exclusive modes should be dominant[1] and the lack of theoretical understanding of the total semileptonic rate[2] it is of interest to search for and understand the other exclusive decays. The remaining modes may contain P-wave mesons, higher spin states or be non-resonant $B \rightarrow D^{(*)}n(\pi)\ell X$ decays.

One method of understanding the contributions from the various processes is to measure the relative rates of the different charmed mesons in semileptonic b decays. At LEP these measurements are complicated by the production of B_s and b-baryons as well as B^0 and B^+ mesons. However, by comparing the rates with those at $\Upsilon(4S)$ energies the fraction of b quarks which fragment to B^0 and B^+ mesons may be determined.

In addition it is possible to search directly for the decays involving excited charmed states. Evidence for the production of neutral P-wave mesons in semileptonic B decay has been reported recently.[3] The study of these decays is difficult at low energies because of high combinatorial backgrounds. However, at LEP the boost of the B mesons, combined with precision vertex reconstruction is expected to allow relatively pure samples to be isolated.

2. Event Selection

This analysis is based on a data sample of 1.72 million hadronic Z^0 decays collected by OPAL between 1991 and 1993. A complete description of the OPAL detector may be found elsewhere.[4] The features of most importance to this analysis are the precise tracking system, including a silicon microvertex detector and the good particle identification provided by dE/dx measurements in the central jet chamber.

---

*Representing the OPAL Collaboration.

The selection of $b \to D^{(*)}\ell X$ decays uses both the kinematic and vertex information of the decays. The selection criteria are similar to those used in a previous publication.[5] D mesons are reconstructed in the decay modes * $D^0 \to K^-\pi^+$, $D^0 \to K^-\pi^-\pi^+\pi^+$ and $D^+ \to K^-\pi^+\pi^+$. D^{*+} mesons are selected in the decay mode $D^{*+} \to D^0\pi^+$. Electron and muon candidates are selected using criteria similar to those used previously.[6] The leptons are required to have momentum greater than 2 GeV. To suppress random combinations the mass $(m_{D^{(*)}\ell})$ and energy $(E_{D^{(*)}\ell})$ of the $D^{(*)}\ell^-$ candidates are required to satisfy certain minimum criteria. In addition, to suppress $D^{(*)}\ell$ combinations from different b hadrons, all candidates are required to have $M_{D^{(*)}\ell} < 5.35$ GeV.

To facilitate precise vertex reconstruction, the lepton candidate, and at least two of the $D^{(*)}$ decay tracks are required to be associated with hits in the microvertex detector. The vertices are reconstructed by preforming a simultaneous fit in two dimensions. The probablity of the vertex fit is required to be greater than 1% in order to suppress random track combinations. The decay length of the B candidate is calculated with respect to the average e^+e^- interaction point and converted in three dimensions using the direction cosines of the $D^{(*)}\ell$ combination. The decay length resolution is typically 300 μm, compared with the average decay length of about 3 mm. For the inclusive D^0 and D^+ samples loose selection criteria are applied to both B and D decay lengths to further reduce the combinatorial backgrounds.

Figure 1 shows the observed D signals for two of the decay modes. The number of events are determined by fitting these mass distributions using functions consisting of Gaussians and second order polynomials to describe the combinatorial backgrounds. The enhancements due to partially reconstructed D decays are included in the fits.

3.   Measurement of D^+, D^0 and D^{*+} production in semileptonic b decays

In addition to the decays $b \to D^{(*)}\ell^-X$, several other processes may contribute to the observed signals. The background due to $D^{(*)}$ mesons combined with fake leptons has been studied by searching for 'wrong sign' $D^{(*)}\ell^+$ combinations with the same selection criteria. No signals are observed, indicating the background from this source is negligible. Contributions to the signals from the decays $B \to \overline{D}^{(*)}D_s^{(*)}$ and $B \to \overline{D}^{(*)}\tau^-X$ have been studied using simulations and known branching ratios. These are estimated to form between 1.7 % and 3.0 % of the signals and have been subtracted.

The reconstruction efficiences for the $b \to D^{(*)}\ell^-X$ were determined using simulated events produced using the JETSET 7.3 program. These were processed using a detailed simulation of the OPAL detector. The exclusive semileptonic branching ratios used in this simulation were choosen to have 36 % of decays containing a P-wave charmed meson, and the P-wave mesons to decay via $D^*\pi$ (as opposed $D\pi$) 54 % of the time. The b quarks were fragmented using a Peterson function with $\epsilon_b = 0.0057$. Variations in these parameters were considered as systematic errors. In order to verify that the simulation provides an adequate description of the data, the main kinematic

---

*Charge conjugation is implied throughout this paper.

variables have been compared. Examples are shown in figure 2. There is good agreement in the kinematic regions in which we select events.

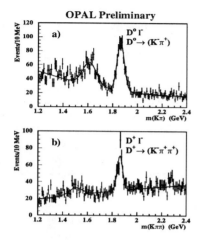

Fig. 1: Observed signals for $D^0 \to K^- \pi^+$ and $D^+ \to K^- \pi^+ \pi^+$ associated with leptons.

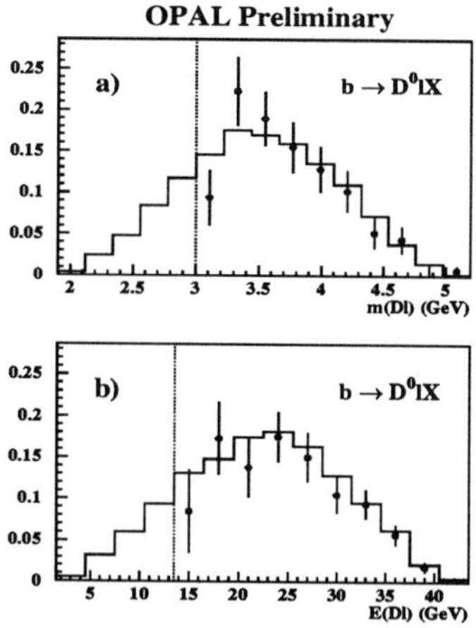

Fig. 2: Comparison of $m(D^0 \ell^-)$ and $E(D^0 \ell^-)$ for data (points) and simulation.

Systematic source	$(b \to D^+ \ell X)$ %	$(b \to D^0 \ell X)$ %	$(b \to D^{*+} \ell X)$ %
f^{**} 0.36 ± 0.12	4.0	3.2	3.0
p_v 0.54 ± 0.3	2.7	1.0	3.7
b fragmentation	2.9	2.9	2.9
Monte Carlo statistics	2.5	2.6	4.1
Background subtraction	0.7	0.8	0.9
e ID	2.8	2.8	2.8
μ ID	1.1	1.1	1.1
Si matching efficiency	1.1	2.0	1.1
dE/dx (Kaon ID)	3.3	3.3	2.7
$\Gamma_{b\bar{b}}/\Gamma_{had}$	1.2	1.2	1.2
Relative D^0 B.R.'s	-	1.7	2.0
Total	7.5	7.1	8.2

Table 1: Summary of Systematic errors.

The rates for the different $b \to D^{(*)} \ell^- X$ were determined using the reconstruction efficiencies determined from the simulation. Good agreement was found for the electron and muon channels and the results were averaged. Similarly the two D^0 decay modes were found to give consistent results, after allowing for their relative branching ratios.

Table 1 summarizes the systematic errors. The main uncertainties arise from the decay modelling. Combining all systematic errors in quadrature the following results are obtained:

$$B(b \to D^+\ell X) \cdot B(D^+ \to K^-\pi^+\pi^+) = (1.93 \pm 0.22 \pm 0.14) \times 10^{-3},$$

$$B(b \to D^0\ell X) \cdot B(D^0 \to K^-\pi^+) = (2.49 \pm 0.14 \pm 0.18) \times 10^{-3},$$

$$B(b \to D^{*+}\ell X) \cdot B(D^{*+} \to D^0\pi^+) \cdot B(D^0 \to K^-\pi^+) = (7.41 \pm 0.46 \pm 0.61) \times 10^{-4}.$$

It should be noted that all the systematic errors, other than those due to Monte Carlo statistics are correlated between the different samples.

Using the D^0 and D^+ branching ratios given by the PDG,[7] the semileptonic rate to modes containing a D^0 or D^+ meson is found to be $(9.23 \pm 0.48 \pm 0.69 \pm 0.47)\%$, where the third error is due to the charm branching ratios. This should be compared with our measurement of the total b semileptonic rate of $(10.5 \pm 0.6 \pm 0.5)\%$.[6]

By comparison with CLEO results[8] it is possible to estimate the fraction of b quarks which fragment to B^0 and B^+ mesons. After subtraction of an estimated contribution of $2 \pm 1\%$ from the decays $B_s \to D_{sJ}^-\ell^+X$, $D_{sJ} \to D^{(*)}K$ we obtain:

$$f(b \to B^0) + f(b \to B^+) = 0.83 \pm 0.07 \pm 0.10,$$

where the first error is our combined uncertainty and the second is from CLEO.

4.   Search for P-wave charmed mesons in semileptonic b decays

The excited charm P-wave mesons are expected to consist of four different charged and neutral states with spin-parity $(0^+, 1^+, 1^+, 2^+)$. The doublets $(1^+, 2^+)$ are narrow with natural widths of about 20 MeV and have been identified with the experimentally observed states[9] $D_1^0(2420)$, $D_2^{*0}(2460)$, $D_1^+(2430)$ and $D_2^{*+}(2470)$. The remaining states are expected to be broad and have not been observed.

Decay Mode	B.R.	Δ_m (MeV)	Decay Mode	B.R.	Δ_m (MeV)
$D_2^{*0}(2460) \to \underline{D}^+\pi^-$	48%	589	$D_2^{*+}(2470) \to \underline{D}^0\pi^+$	48%	591
$D_2^{*0}(2460) \to D^{*+}\pi^-$			$D_2^{*+}(2470) \to D^{*0}\pi^+$		
$\hookrightarrow \underline{D}^+X$	6%	448	$\hookrightarrow \underline{D}^0X$	18%	449
$D_2^{*0}(2460) \to D^{*+}\pi^-$					
$\hookrightarrow \underline{D}^0\pi^+$	12%	448			
$D_1^0(2420) \to D^{*+}\pi^-$			$D_1^+(2430) \to D^{*0}\pi^+$		
$\hookrightarrow \underline{D}^+X$	22%	414	$\hookrightarrow \underline{D}^0X$	67%	436
$D_1^0(2420) \to D^{*+}\pi^-$					
$\hookrightarrow \underline{D}^0\pi^+$	44%	414			

Table 2: A summary of D_J decay modes to which this analysis is sensitive. Detected particles are indicated by an underscore.

Using a combination of the known D* branching ratios, isospin and the measured ratio of decay modes for the $D_2^{*0}(2460)$ meson: $\frac{B(D_2^{*0}(2460) \to D^+\pi^-)}{B(D_2^{*0}(2460) \to D^{*+}\pi^-)} = 2.6 \pm 0.6$, it is possible to estimate the decay modes of the D_J mesons to different $D^{(*)}\pi$ final states, assuming these saturate the D_J decays. These predictions are listed in table 2. The expected mean values of $\Delta_m = m(D^{(*)}\pi) - m(D^{(*)})$ are also shown. In analogy to the familar $D^{*+} \to D^0\pi^+$ decays, Δ_m has a better experimental resolution than the raw $m(D^{(*)}\pi)$ distribution. We have simulated all of these decays. The near degeneracy of the pseudovector and tensor states results in peaks in the $D^{*+}\pi^-$, $D^+\pi^-$ and $D^0\pi^+$ spectra at ~ 430 MeV which are difficult to resolve experimentally. However, the direct decays of the tensor mesons to $D\pi$ final states provide a distinct signature with peaks at $\Delta_m \sim 590$ MeV.

The inclusive $b \to D^{(*)}\ell^- X$ samples are used to search for these decays. Events are selected from mass regions of approximately 2σ around the D masses and slightly tighter selection criteria are applied. To increase the available statistics $b \to D^{*+}\ell^- X$ candidates are also selected in the $D^0 \to K^-\pi^+X$ mass region 1.54-1.70 GeV. These partially reconstructed decays are largely due to $D^0 \to K^-\rho^+$. To avoid overlap, decays consistent with $D^{*+} \to D^0\pi^+$ are rejected from the $D^0\ell^-$ sample.

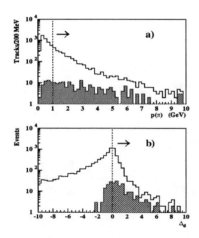

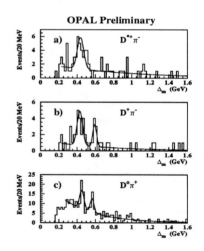

Fig. 3: Selection power of $p(\pi)$ and Δ_d discriminants. The hatched area shows the signal contribution to the total.

Fig. 4: Observed Δ_m distributions for different $D^{(*)}\pi$ combinations.

The $D^{(*)}$ mesons are combined with transition pion candidates to search for D_J decays. The invariant mass of the $D_J\ell^-$ system is required to be less than 5 GeV. The isolation of the signals remains difficult because of the presence of many soft tracks not related to the B decays. To reduce this background two further criteria are used. These are the momentum of the transition pion candidate $p(\pi)$ and $\Delta_d = |d_{bm}/\sigma_{bm}| - |d_{vx}/\sigma_{vx}|$, where $d_{bm}(d_{vx})$ is the impact parameter with respect to the beamspot (B decay vertex) and $\sigma_{bm}(\sigma_{vx})$ its error. This makes use of the fact that the background tracks arise from the primary vertex, whilst the signal pions originate from the B decay vertex. In addition it is required that $|d_{vx}/\sigma_{vx}| < 2$. The selection powers of these discriminants are illustrated in figure 3.

Applying these selection criteria to the data, the Δ_m distributions shown in figure 4 are obtained. Signals are observed in all cases close to the expected values of Δ_m. Studies of the corresponding samples of $D^{(*)}$ mesons combined with pions of the wrong charge and background events selected using the Δ_d variable show no enhancements in the signal regions. The number of signal events were determined by fitting the Δ_m distributions using functions consisting of Gaussians plus exponential background terms. The signals consist of $44 \pm 8^{+3}_{-7}$ D^0_J and $48 \pm 10^{+3}_{-6}$ D^+_J decays.

Decay mode	Br. Ratio (10^{-3})
$(b \to D^0_J \ell^- X) \cdot (D^0_J \to D^{*+}\pi^-)$	$6.9 \pm 1.5 \pm 1.4$
$(b \to D^0_J \ell^- X) \cdot (D^0_J \to D^+\pi^-)$	$1.8 \pm 0.7 \pm 0.3$
$(b \to D^+_J \ell^- X) \cdot (D^+_J \to D^{*0}\pi^+)$	$7.5 \pm 2.0^{+1.1}_{-1.2}$
$(b \to D^+_J \ell^- X) \cdot (D^+_J \to D^0\pi^+)$	$4.5 \pm 1.4^{+0.7}_{-1.3}$

Table 3: Observed b quark product B.R.'s

Decay mode	Br. Ratio $(\%)$
$B^- \to D^0_1(2420)\ell^-\nu$	2.3 ± 0.8
$B^- \to D^{*0}_2(2460)\ell^-\nu$	0.9 ± 0.4
$B^0 \to D^+_1(2430)\ell^-\nu$	2.1 ± 1.0
$B^0 \to D^{*+}_2(2470)\ell^-\nu$	2.3 ± 0.9

Table 4: Inferred B branching ratios.

Using these signals the product branching ratios listed in table 3 are obtained. Here the two signals resulting from D^{*+} have been averaged. Assuming that these signals are due solely to the presence of the narrow P-wave mesons and are predominantly due to exclusive B^0 and B^+ decays then one can estimate these rates using the D_J branching ratios given in table 2. These are listed in table 4. No additional errors have been added for the uncertainties involved in deriving these results. By comparison with the inclusive $b \to D\ell^- X$ rates these decays are estimated to form 34 ± 7 % of charmed semileptonic B decays.

5.   Summary

Precise measurements of the production rates of D^0, D^+ and D^{*+} mesons in semileptonic b decays have been presented. These results been used to infer the probability that a b quark produced in Z^0 decay fragments to a B^0 or B^+ meson. The results of a search for P-wave charmed mesons in semileptonic B decays has also been reported. The observed signals provide the first evidence for the production of charged P-wave states and evidence for both pseudovector and tensor production. The branching ratios to these decay modes are found to be somewhat larger than the predictions of several theoretical models.[1]

References

1. N. Isgur, D. Scora, B. Grinstein and M. B. Wise, *Phys. Rev.* **D39** (1989) 175; P. Colangelo, G. Nardulli and N. Paver, *Phys. Lett.* **B293** (1992) 207.
2. I. Bigi *et al. CERN-TH.7082/93*, (1993).
3. H. Albrecht *et al.*, *Z. Phys.* **C57** (1993) 533.
4. K. Ahmet *et al.*, *Nucl. Instrum. Methods* **A305** (1991) 275; P. P. Allport *et al.*, *Nucl. Instrum. Methods* **A324** (1993) 34; P. P. Allport *et al.*, *Nucl. Instrum. Methods* **A346** (1994) 476; M. Hauschild *et al.*, *Nucl. Instrum. Methods* **A314** (1992) 74.
5. P. D. Acton *et al.*, *Phys. Lett.* **B307** (1993) 247;
6. P. D. Acton *et al.*, *Z. Phys.* **C58** (1993) 523; R. Akers *et al.*, *Z. Phys.* **C60** (1993) 199.
7. K. Hikasa *et al.*, *Phys. Rev.* **D45** (1992).
8. R. Fulton *et al.*,*Phys. Rev.* **D43** (1991) 651.
9. H. Albrecht *et al.*, *Phys. Lett.* **B221** (1989) 422; H. Albrecht *et al.*, *Phys. Lett.* **B231** (1989) 208; H. Albrecht *et al.*, *Phys. Lett.* **B232** (1989) 398; J. C. Anjos *et al.*, *Phys. Rev. Lett.* **62** (1989) 1717; P. Avery *et al.*, *Phys. Rev.* **D41** (1990) 774; P. L. Frabetti *et al.*, *Phys. Rev. Lett.* **72** (1994) 324; P. Avery *et al.*, *CLNS 94/1280*.

OBSERVATION OF B⁻ → D**⁰ℓ⁻ν̄ WITH THE

ALEPH DETECTOR

LEO BELLANTONI*

Department of Physics, University of Wisconsin
Madison, WI 53706

of the

ALEPH Collaboration

ABSTRACT

In a sample of 1.5 million hadronic decays of the Z collected by the ALEPH detector, a search for the decay B→D**⁰(2420)ℓ⁻ν̄X and B→D**⁰(2460)ℓ⁻ν̄X is carried out. A topological search, sensitive to the production of the process above, to wide D** resonances that decay to D*⁺π⁻, and to the non-resonant decay B → D*πℓ⁻ν̄X is carried out. Excluding the events with masses close to observed resonances, there are nine events for an estimated background of two events.

1. Introduction

Experimentally, 36±12% of all semileptonic B decays are not accounted for[1] by the decays B → Dℓ⁻ν̄ and B →D*ℓ⁻ν̄, contrary to expectations[2]. There are several possible explanations. There may be four-body decays, such as B → D*πℓ⁻ν̄. There may be decays to ℓ = 1 charm mesons; four such states are expected[3]. Two of the states are wide resonances which can not be distinguished from four-body decays with the available experimental statistics.

The decays of the Z at LEP provides a source of boosted B mesons ($\gamma\beta \sim 6$) and the tracking resolution of the ALEPH detector permits the measurement of previously unseen semileptonic decays of the B. Figure 1 shows an event with such a decay mode which is unambiguously identified with its vertex topology. The decay point of the B can be found with a typical resolution of 280 μm, compared with a typical B decay length of 3.0 mm, permitting the differentiation of tracks originating at the primary interaction point due to background processes from a tracks originating at the B decay point due to interesting processes.

This paper describes briefly two measurements which are more fully described[4] elsewhere.

2. B→D**⁰ℓ⁻ν̄X

The decay B→D**⁰ℓ⁻ν̄X is identified in events with a D*⁺ (reconstructed in the channel D⁰π⁺) and a lepton. Tracks of the same sign as the lepton are assigned

---

0* Supported by the US Department of Energy, contract DE-AC02-76ER00881.

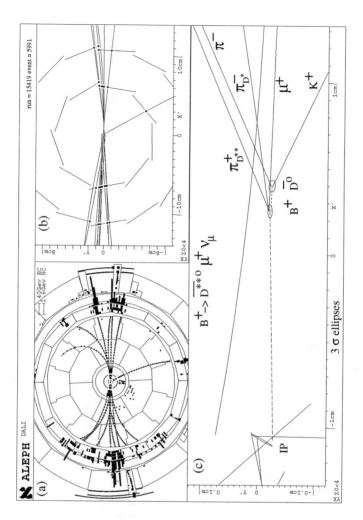

Figure 1: A reconstructed $B^+ \rightarrow \bar{D}^{**0}(2420)\mu^+\nu$ event. (a) A fisheye r-ϕ view showing all the tracking detectors, the calorimeters and the muon chambers. (b) An r-ϕ view with the silicon vertex detector's two planes of 3D information, and the inner tracking chamber's hits. (c) A close-up view of the region around the interaction point showing the decay vertices of the B^+ and the $\bar{D}^0$. The K^+ and π^- form a 3D vertex with an error ellipse of $125\,\mu$m length. The flight of the $\bar{D}^0$ is extrapolated to the point where it intersects the μ^+; that point is known to $166\,\mu$m. Two pions, corresponding to the decays of a $D^{**0}(2420)$ and a D^{*+}, come from this point, which is the decay point of the B^+.

the mass of a pion and the mass of the D^{*+}-π^- system is calculated. The distribution of $\Delta m^* = M(D^{*+}$-$\pi^-) - M(D^{*+})$ then shows an excess of events at values of Δm^* that correspond to the narrow D^{**0} resonances. The resolution of Δm^* is 3 to $4\,\mathrm{MeV/c^2}$.

Two complementary approaches to event selection are taken. In one approach, a large sample of D^{*+}-lepton events is chosen, using the D^0 decay channels $K^-\pi^+$, $K^-\pi^+\pi^-\pi^+$, $K^0_S\pi^+\pi^-$, and $K^-\pi^+\pi^0$. Stringent requirements are applied to ensure that the π^- candidate does not pass through the interaction point. The overall efficiency of this event selection is 0.77 ± 0.09 %, with a background level of about 1 event per $20\,\mathrm{MeV/c^2}$ bin in Δm^*. In the second, a cleaner sample of D^{*+}-lepton events is chosen, using the D^0 decay channels $K^-\pi^+$, and $K^-\pi^+\pi^-\pi^+$. Stringent requirements are then made on the quality of the D^{*+}-lepton vertex and the measurement of the π^- position relative to that vertex, but the requirement that the π^- candidate does not pass through the interaction point is relaxed. The overall efficiency of this event selection is 0.75 ± 0.11 %, with a background level of about 2 events per $20\,\mathrm{MeV/c^2}$ bin in Δm^*. The results from the two selection procedures are consistent. The complementary approaches to the event selection procedure and the low overall efficiencies result in two event samples with little overlap; in the region of the resonances, only six events are found by both selection procedures. The results are combined, yielding

$$\mathrm{Br}(b{\to}B) \times \mathrm{Br}(B{\to}D^{**0}(2420)\ell^-\bar{\nu}X) \times \mathrm{Br}(D^{**0}(2420){\to}D^{*+}\pi^-)$$

$$= (2.08 \pm 0.59(stat.) \pm 0.34(syst.)) \times 10^{-3}.$$

where the systematic uncertainties are detailed in table 1. No significant $D^{**0}(2460)$ signal is seen.

Source	Uncertainty in units of 10^{-3} 1st selection	Uncertainty in units of 10^{-3} 2nd selection
B^- Vertex Efficiency	± 0.18	± 0.26
Background Function	± 0.23	± 0.15
M,Γ of D^{**0}	± 0.24	± 0.25
Monte Carlo Statistics	± 0.08	± 0.20
D^{*+}, D^0 Branching Ratios	± 0.06	± 0.13
Lepton ID Efficiency	± 0.09	± 0.13
D^0 ID Efficiency	± 0.04	± 0.08
b fragmentation	± 0.03	± 0.03

Table 1: Systematic uncertainties in $\mathrm{Br}(b{\to}B)$ $\times$ $\mathrm{Br}(B{\to}D^{**0}(2420)\ell^-\bar{\nu}X)$ $\times$ $\mathrm{Br}(D^{**0}(2420){\to}D^{*+}\pi^-)$.

Isospin arguments imply that if three body decays of $D^{**0}(2420)$ can be neglected, $\mathrm{Br}(D^{**0}(2420){\to}D^{*+}\pi^-)$ is 2/3. Using also the branching ratio $\mathrm{Br}(b{\to}B^-)$

$= 0.37 \pm 0.03$ from reference[5], $\mathrm{Br}(\mathrm{B} \rightarrow \mathrm{D}^{**0}(2420)\, \ell^- \bar{\nu} X) \geq 0.84 \pm 0.28$ %. which explains at least $19 \pm 8\%$ of unidentified semileptonic branching ratio of the B meson.

3. $\mathrm{B} \rightarrow \mathrm{D}^{*+} \pi^- \ell^- \bar{\nu} X$

For $\mathrm{B} \rightarrow \mathrm{D}^{*+} \pi^- \ell^- \bar{\nu} X$, it is not possible to identify resonance structures in a narrow mass window, but the unique topology of this decay mode is sufficient for its identification.

The events are selected using essentially the tight D^{*+}-lepton procedure previously described, in conjunction with the stringent requirement that the π^- candidate does not pass through the interaction point. Additionally, the candidate π^- must have an error on the distance from the D^{*+}-lepton vertex of less than $150\,\mu m$.

For every candidate π^- the distance from the D^{*+}-lepton vertex is measured. This impact parameter, divided by its resolution as evaluated on a track by track basis, is used to calculate an integrated probability that the event is consistent with being a signal event.

The distribution of the integrated probability in the data, figure 2, shows a large peak in the first bin due to background events, in both the same and opposite sign samples. Eighteen events are found with $P_\delta > 0.2$ in the same sign sample; only one is found in the wrong sign sample, confirming a Monte Carlo based estimate of 2.2 ± 0.7 background events. Excluding the mass region $0.38 < \Delta m^* < 0.44$, nine events are found in the same sign sample; the probability that the background, estimated to be 2.0 events in this case, will fluctuate up to this level is 2.4×10^{-4}.

Source	Uncertainty in units of 10^{-3}
B^- Vertex Efficiency	± 0.40
Wide and non res. Efficiency	± 0.40
Monte Carlo Statistics	± 0.41
D^{*+}, D^0 Branching Ratios	± 0.19
Lepton ID Efficiency	± 0.20
Probability function	± 0.12
b fragmentation	± 0.03

Table 2: Systematic uncertainties in $\mathrm{Br}(b \rightarrow B) \times \mathrm{Br}(\mathrm{B} \rightarrow \mathrm{D}^{*+} \pi^- \ell^- \bar{\nu} X)$

The branching ratio, calculated using an efficiency of 0.45 ± 0.05 %, is

$$\mathrm{Br}(b \rightarrow B) \times \mathrm{Br}(\mathrm{B} \rightarrow \mathrm{D}^{*+} \pi^- \ell^- \bar{\nu} X) = (4.0 \pm 1.1(stat) \pm 0.8(syst)) \times 10^{-3}.$$

The systematic uncertainties are described in table 2.

References

1. S. Stone, "Semileptonic B Decays", *to be published in* **B Decays,**

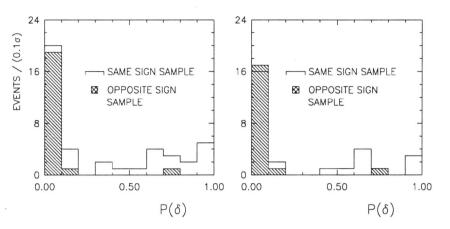

Figure 2: Probability distribution for D** candidates; the plot on the left is for the entire range of Δm^* values; on the left is the distribution for Δm^* inconsistent with the narrow D**^{0} resonances.

Second Edition World Scientific, Singapore.

2. N. Isgur, D. Scora, B. Grinstein, and M. B. Wise, Phys. Rev. **D39** (1989) 799

3. E. J. Eichten, C. T. Hill, and C. Quigg, Phys. Rev. Lett. **71** (1993) 4116
   S. Godfrey and R. Kokoski, Phys. Rev. **D43** (1991) 43
   T. B. Suzuki, T. Ito, S. Sawada, and M. Matsuda, DPNU-93-35, AUE-04-93
   CLEO Coll., P. Avery *et al.* , CLEO 94-10, CLNS 94/1280

4. "A Study of B$^-$ → (D*$^+$$\pi^-$)$\ell^-$$\bar\nu$ at LEP", *Contribution to the 27th International Conference on High Energy Physics, Glasgow, Scotland, 20-27 July 1994*
   A Phys. Lett. note is in preparation.

5. ALEPH Coll., D. Buskulic *et al.* , CERN-PPE/93-221

OPAL MEASUREMENT OF THE B^0 AND B^+ LIFETIMES

DAVID L. REES

CERN, CH-1211 Geneva 23, Switzerland

ABSTRACT

From a data sample of 1.72 million hadronic Z^0 decays recorded by the OPAL detector at LEP, approximately 1000 semileptonic B hadron decays with a reconstructed D^0, D^+ or D^{*+} and lepton were found. Using information from the microvertex detector, the decay length of the B hadron was reconstructed and used to extract preliminary lifetimes of 1.62 ± 0.10(stat.) ± 0.10(syst.) ps and 1.53 ± 0.14(stat.) ± 0.11(syst.) ps for the B^0 and B^+ mesons respectively. This corresponds to a ratio of B^+ to B^0 lifetime of 0.94 ± 0.12(stat.) ± 0.07(syst.).

1.   Introduction

In the spectator model the light quark(s) within a heavy hadron play no part in its decay and so all heavy hadrons are predicted to have the same lifetime. However, it is known that the charm hadrons display large non-spectator variations in lifetime:

$$\tau(D^+) : \tau(D^0) : \tau(D_s^+) : \tau(\Lambda_c^+) \sim 2.5 : 1 : 1 : 0.5.$$

Similar differences are also expected among the B hadrons although greatly reduced because of the greater mass of the b-quark compared to the c-quark.[1] The B^+ lifetime is expected to be 0-7% larger than the B^0 lifetime and the Λ_b^0 lifetime is expected to be about 10% smaller than the B^0 lifetime.

The average B hadron lifetime measured in Z^0 decays is used to measure the CKM matrix element governing the coupling between the b and c quarks. This has been measured to a precision of a few percent[2] and so the expected small differences between B hadron lifetimes are becoming more important. The measurement of the individual lifetimes is therefore of interest and OPAL's measurement of the B^0 and B^+ lifetimes is described here. The preliminary results presented are an update to those previously published[3] to include data collected during 1992 and 1993 and an improved lifetime fitting method.

2.   Identification of B^0 and B^+ Mesons

B^0 and B^+ mesons were tagged in the semileptonic decay modes $B \to \overline{D}^0 l^+ X$, $B \to D^- l^+ X$ and $B \to D^{*-} l^+ X$ where only the lepton and $D^{(*)}$ meson were identified. The $\overline{D}^0 l^+$ combination is mainly produced in B^+ decays while the $D^{(*)-} l^+$ combinations come largely from B^0 decays. The important exception to this rule occurs when a D^{**} is produced. For example, a B^+ meson gives a $D^- l^+$ combination in the decay

Table 1. Decay mode dependent kinematic cuts. All values are in GeV.

Decay mode	p_l	p_π	p_K	$E_{D^{(*)}}$	$E_{D^{(*)}l}$	$m_{D^{(*)}l}$
$D^0 \to K^-\pi^+$	> 2	> 0.15	> 2	> 6	> 13.5	3.0 − 5.35
$D^+ \to K^-\pi^+\pi^+$	> 2	> 0.15	> 2	> 7	> 13.5	3.0 − 5.35
$D^{*+} \to D^0\pi^+$						
$\hookrightarrow K^-\pi^+$	> 2	> 0.15	> 0.15	> 5	> 9.0	2.8 − 5.35
$D^{*+} \to D^0\pi^+$						
$\hookrightarrow K^-\pi^+\pi^+\pi^-$	> 2	> 0.15	> 2	> 6	> 13.5	3.0 − 5.35

$B^+ \to \overline{D}^{**0}l^+\nu_l$, $\overline{D}^{**0} \to D^-\pi^+$. In the presence of D^{**} production it is estimated that about 75% of the decays giving $D^{(*)-}l^+$ come from B^0 and about 70% of the decays giving $\overline{D}^0l^+$ are from B^+ mesons.

2.1. Lepton Identification and $D^{(*)}$ Meson Reconstruction

Electrons were identified by requiring that the energy loss measured along a central detector track (dE/dx) is consistent with an electron and that the ratio of the electromagnetic calorimeter energy and the track momentum, $E_{ECAL}/p \sim 1$. Muons were identified by matching a track in the outer muon chambers with a central detector track and cuts on dE/dx were made to reject protons and kaons.

$D^{(*)}$ mesons were reconstructed in the four decays modes given in Table 1. Pions and kaons were identified using dE/dx information. D^{*+} candidates were found by looking for a π^+ in addition to a D^0 candidate with a mass difference between the D^{*+} and D^0 in the range 0.142 to 0.148 GeV. For the D^0 and D^+ modes it was also required that there not be a $D^{*+} \to D^0\pi^+$ candidate with a mass difference less than 0.16 GeV.

Once a lepton and $D^{(*)}$ meson were identified any 'wrong sign' $D^{(*)}$-lepton combinations were rejected. Various kinematic cuts were then applied to the lepton, pion and kaon momentum $(p_l, p_\pi$ and $p_K)$ the energy of the $D^{(*)}$ candidate $(E_{D^{(*)}})$ and the energy and mass of the $D^{(*)}$-lepton pair $(E_{D^{(*)}l}$ and $m_{D^{(*)}l})$ in order to reduce background. The cuts change slightly with each decay mode and are summarised in Table 1. For the D^0 and D^+ modes the cut $|\cos\theta^*| < 0.7$ was also made where θ^* is the angle between the K^- and the D boost direction in the D rest frame.

2.2. Vertex Finding

The decay length of the B meson, l_B, was measured by reconstructing the B decay vertex from the lepton track and D meson momentum vector and using the average beam spot position. The D decay vertex was also found from its decay tracks. This vertex fitting was done in two dimensions and three dimensional decay lengths for the B and D mesons were calculated using the lepton and D meson momentum vectors.

To ensure good vertex resolution the lepton and at least two of the D decay tracks were required to have hits in the silicon microvertex detector. The vertex fit was

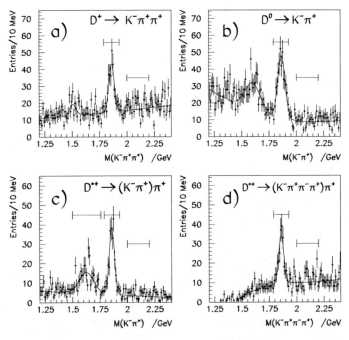

Fig. 1. Reconstructed D meson mass for (a) $D^+ \to K^-\pi^+\pi^+$ (b) $D^0 \to K^-\pi^+$ (c) $D^{*+} \to D^0\pi^+$, $D^0 \to K^-\pi^+$ and (d) $D^{*+} \to D^0\pi^+$, $D^0 \to K^-\pi^+\pi^+\pi^-$.

required to give a good χ^2 and a small error on the B decay length, σ_{l_B}, helping to reject background further and ensure the decay was well reconstructed. For the $D^0 \to K^-\pi^+$ mode the D decay length significance, l_D/σ_{l_D}, was also required to be greater than -1 and for the $D^+ \to K^-\pi^+\pi^+$ mode it was required to be greater than zero. A typical B decay length was 3 mm with an error of 300 μm and typical D decay lengths were 1 mm with an 800 μm uncertainty.

Figure 1 shows the reconstructed D mass in each of the four decay modes after making all selection cuts. In addition to the clear peak at the D meson mass there are also significant satellite peaks at lower mass in plots b) and c). These are due to partially reconstructed decays where a π^0 is not seen $e.g.$ $D^0 \to K^-\rho^+$, $\rho^+ \to \pi^+\pi^0$. In c) the satellite peak contains little background and was used in addition to the four main mass peaks. The five signal regions within ± 3 sigma of the fitted peak mass contain approximately 1000 signal events when combined. Sideband data in the mass range 2.0 to 2.2 GeV were used to describe the background in the signal regions.

3.   Lifetime Fitting

The proper decay time of the B meson is needed to measure the lifetime. This can be calculated using the decay length and the boost of the B. However, since some energy is always carried away by a neutrino, the B energy, E_B, is not known precisely and the boost cannot be calculated exactly. It is usually necessary to estimate E_B and this can be done using the kinematics of the D-lepton pair. To avoid such an estimation and its associated uncertainty on the proper time, E_{Dl} and m_{Dl} were used as *constraints* on E_B and all allowed values of E_B were considered. The probability function that describes the likelihood of a signal event with a given decay length is

$$\mathcal{L}_{\text{sig}}(l_B) = \int_0^{E_{\text{BEAM}}} P(E_B|E_{Dl}, m_{Dl}) \left[\mathcal{E}(\tau_B) \otimes \mathcal{G}(\sigma_{l_B}) \right] dE_B \qquad (1)$$

where $P(E_B|E_{Dl}, m_{Dl})$ is the probability of having E_B given E_{Dl} and m_{Dl}, and the part in square brackets represents an exponential decay convoluted over decay length with a gaussian resolution function. $P(E_B|E_{Dl}, m_{Dl})$ was calculated using simple kinematics and assuming Peterson fragmentation and is only non-zero for values of E_B allowed by E_{Dl} and m_{Dl}.

An unbinned maximum likelihood fit was performed using the likelihood

$$\mathcal{L} = f(m_D)\mathcal{L}_{\text{sig}} + [1 - f(m_D)]\mathcal{L}_{\text{back}} \qquad (2)$$

where $f(m_D)$ is the probability an event is signal given its reconstructed D mass and was taken from the fit to the mass distributions shown in Fig. 1, and $\mathcal{L}(l_B)_{\text{back}}$ is a function that describes the background decay length distribution. The result of a fit to the signal and sideband regions of all decay modes is shown in Fig. 2.

3.1.   Extracting the B^0 and B^+ Lifetimes

To extract the B^0 and B^+ lifetimes the relative contribution of B^0 and B^+ decays to each of the $D^{(*)}$-lepton combinations is needed. The main source of uncertainty comes from the relatively little that is known about D^{**} production and decays. The following parameters and assumptions were made;

- semileptonic branching ratios are proportional to the B meson lifetime,

- the production rates for B^0 and B^+ in Z^0 decays are equal,

- the fraction of B decays that produce a D^{**} is 0.36 ± 0.12,[4]

- the fraction of D^{**} that decay to a D^* is 0.54 ± 0.30 and

- D^{**} decays are dominated by $D^{**} \to D^{(*)}\pi$ and follow isospin.

A fit was performed to all channels using a signal likelihood containing two terms like that of Eq. 1, each with a different lifetime and weighted by the probability the $D^{(*)}$-lepton comes from a B^0 and B^+ decay respectively. Preliminary results from the fit were a B^0 lifetime of $1.62 \pm 0.10(\text{stat.}) \pm 0.10(\text{syst.})$ ps and a B^+ lifetime of $1.53 \pm 0.14(\text{stat.}) \pm 0.11(\text{syst.})$ ps.

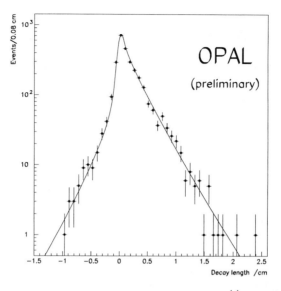

Fig. 2. Fit to decay length distribution for all B → $\overline{\mathrm{D}}^{(*)}l^{+}$X candidates.

4.   Summary

OPAL has measured the lifetime of the B^0 and B^+ mesons using events containing a $D^{(*)}$ meson and lepton. The ratio of B^+ to B^0 lifetime was found to be 0.94 ± 0.12(stat.) ± 0.07(syst.), which is in agreement with the theoretical expectation of a 0-7% difference in lifetime. When this measurement is combined with other recent measurements the ratio becomes 1.003 ± 0.069.[5] With more data we will see the first meaningful test of the theoretical expectation.

Acknowledgments

I would like to thank the OPAL Collaboration for providing me with the opportunity of presenting this talk. In particular, I thank Alex Martin and Tara Shears for doing the analysis and for very useful discussions.

References

1. I. I. Bigi and N. G. Uraltsev, *Phys. Lett.* **B280** (1992) 271.
2. P. Roudeau, Plenary talk, Glasgow Conference 1994.
3. OPAL Collaboration, P. D. Acton *et al.*, *Phys. Lett.* **B307** (1993) 247.
4. CLEO Collaboration, R. Fulton *et al.*, *Phys. Rev.* **D43** (1991) 651.
5. F. Dejong, Parallel session talk, Glasgow Conference 1994.

OPAL MEASUREMENTS OF THE B_s^0
AND B BARYON LIFETIMES

DEAN KARLEN

Physics Department, Carleton University
Ottawa, Canada, K1S 5B6

representing the OPAL collaboration

ABSTRACT

This talk presents the preliminary updates of the B_s^0 and b baryon lifetime measurements using data collected with the OPAL detector between 1990 and 1993. By including the 1993 data and increasing the selection efficiency, both measurements are significantly improved over our previously published results.

1. Introduction

Before measurements were made, all the b hadrons were predicted to have similar lifetimes. Because of the large mass of the b quark, the other quarks in the hadron were not expected to play a large role in the decay (and hence these quarks were called spectator quarks). In this model, known as the spectator model, the CKM mixing parameter V_{cb} can be directly determined from lifetimes of b-flavoured hadrons.

The amount that the b hadron lifetimes differ indicates the relative strength of decay amplitudes that involve the light quarks. For example, the B_s^0 and Λ_b^0 can decay as a result of W exchange. The effect is expected to be small in B meson decays due to helicity suppression.[1] The W exchange amplitude is not helicity suppressed in b baryon decays; it is estimated[2] that the neutral b baryon lifetimes are reduced by about 10% due to this effect. To verify the predictions for the strength of non-spectator decay amplitudes, it is important to measure individual b hadron lifetimes.

2. B_s^0 Lifetime Measurement

As in the published result[3], this update of the B_s^0 lifetime uses events containing the decay channels

$$
\begin{array}{ll}
\begin{aligned}
B_s^0 &\to D_s^- \, \ell^+ \, \nu \, X \\
&\hookrightarrow K^{*0} \, K^- \\
&\quad\hookrightarrow K^+ \, \pi^-
\end{aligned}
\qquad \text{and} \qquad
\begin{aligned}
B_s^0 &\to D_s^- \, \ell^+ \, \nu \, X \\
&\hookrightarrow \phi \, \pi^- \\
&\quad\hookrightarrow K^+ \, K^-
\end{aligned}
\end{array}
\qquad (1)
$$

Charge conjugate modes are implied. Changes to the analysis from the publication are, inclusion of the 1993 data sample; improved electron, pion, and kaon selections; looser requirements on the lepton momentum; and modifications to the decay length fitting procedure.

For a $D_s^- \rightarrow K^+K^-\pi^-$ candidate to be selected, all three particles are required to have dE/dx consistent with the particle hypothesis and the D_s^- energy must be greater than 20% of the beam energy. The K^{*0} and ϕ candidates are required to have a mass within 50 MeV of their PDG values. Cuts are made on angles between particles in the D_s^-, K^{*0}, and ϕ frames. The mass of the D_s^- lepton system is required to be between 3.2 and 5.5 GeV and the total momentum above 17 GeV. To further reduce combinatorial backgrounds a vertex quality cut is made for the $D_s^- \rightarrow K^+K^-\pi^-$ candidate. Also, the proper decay time of the D_s^- candidate is required to be well measured and consistent with the nominal D_s^- lifetime.

The invariant mass of the D_s^- candidates are shown in figure 1a. The signal region, within 50 MeV of the D_s^- mass, contains 55 ± 10 candidates above background. To determine the B_s^0 lifetime, a likelihood function for the decay length for B_s^0 decays is defined by a convolution of an exponential proper time distribution convoluted with a boost distribution (which is a function of the mass and momentum of the $D_s^-\ell$ system) and with a Gaussian resolution function.

The candidates from the high mass sideband, 50 to 200 MeV above the D_s^- mass, are used to define the decay length distribution from the combinatorial background. Physics backgrounds containing a real D_s^- and lepton from a B^+ or B^0 decay are expected to comprise 8 ± 6 % of the signal. This decay length distribution from this source is taken to be the same form as the signal, except with the lifetime fixed to 1.5 ps.

The result of the fit to the signal and combinatorial background regions are shown in figure 1b. The B_s^0 lifetime is found to be

$$\tau_{B_s^0} = 1.33 \; {}^{+0.26}_{-0.21} \; (\text{stat}) \; \pm 0.06 \; (\text{syst}) \; \text{ps}.$$

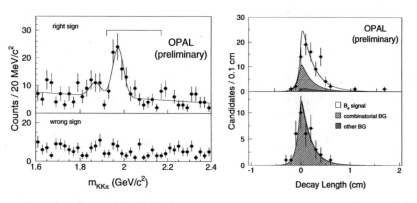

Figure 1. a) The $K^+K^-\pi^-$ invariant mass distribution for is shown for events with a right sign (upper) and wrong sign (lower) lepton. b) The decay length distribution is shown for the signal region (upper) and the sideband region (lower). The curves show the result of the fits to these distributions.

3. B Baryon Lifetime Measurement

As in the published result[4], the average b baryon lifetime is determined from events which contain decay chains of the form,

$$
\begin{aligned}
\Lambda_b^0 &\to \Lambda_c^+ \, \ell^- \, \bar{\nu} \, X & & & \Xi_b^0 &\to \Xi_c^+ \, \ell^- \, \bar{\nu} \, X \\
&\hookrightarrow \Lambda \, X & \text{and} & & &\hookrightarrow \Lambda \, X \\
&\hookrightarrow p \, \pi^- & & & &\hookrightarrow p \, \pi^-
\end{aligned}
\tag{2}
$$

Changes to the analysis in the publication are, inclusion of the 1993 data sample; improved electron and Λ selections; and inclusion of silicon vertex information. An analysis of the lepton impact parameters in these events has also been performed.

To be selected as a $\Lambda \to p\pi^-$ candidate, the proton must have a consistent dE/dx measurement, the momentum must be above 4 GeV (to reduce fragmentation background), the mass must be within 7.8 MeV of the Λ mass, the $\pi^+\pi^-$ mass must be away from the K_s^0 mass, and additional topological requirements must be satisfied. The Λ lepton system must have a mass greater than 2.2 GeV (to reduce Λ_c background) and momentum greater than 9 GeV. A total of 449 right sign and 173 wrong sign Λ lepton combinations are selected by these criteria.

Two analyses are performed on the data sample to determine the lifetime. In both cases the background (Λ's from fragmentation, and misidentified Λ's and leptons) is primarily estimated from the wrong sign events. A simultaneous fit to the decay length distribution of right sign and wrong sign candidates is performed as described in the publication. The result of this fit is shown in figure 2a and the b baryon lifetime is found to be $\tau_{\Lambda_b} = 1.25 \, ^{+0.17}_{-0.16}$ (stat) ± 0.07 (syst) ps.

A fit of the impact parameter distributions for events in which the lepton has a hit associated in the silicon microvertex detector is also performed. The physics function is found from Monte Carlo studies and the resolution is allowed to float in the fit. This fit is shown in figure 2b with the result $\tau_{\Lambda_b} = 1.28 \, ^{+0.18}_{-0.17}$ (stat) ± 0.11 (syst) ps.

The statistical correlation coefficient between the two methods is determined using Monte Carlo studies to be 0.59. The two results are combined, taking into account the correlated statistical and systematic errors, to produce the final result,

$$
\tau_{\Lambda_b} = 1.26 \, ^{+0.16}_{-0.15} \text{ (stat)} \pm 0.07 \text{ (syst) ps.}
$$

4. Summary

A summary of recent measurements of the B_s^0 and b baryon lifetimes are shown in figure 3. It is seen that these lifetimes are closer to being equal than in the charm system, in accordance with expectations. The b baryon lifetime is measured to be shorter than the other b hadrons, also in agreement with predictions. The OPAL results will continue to improve as more data is collected in the coming years.

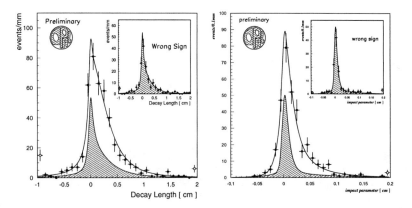

Figure 2. a) Result of decay length fit. b) Result of impact parameter fit. In both figures the right sign contribution (signal) is shown by the open region and the wrong sign events (background) are shown by the shaded region.

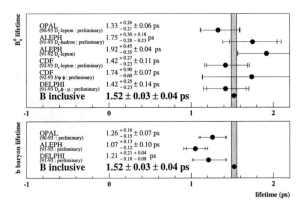

Figure 3. Summary of recent B_s^0 and b baryon lifetime measurements. The B inclusive value shown is the OPAL measurement of the average b hadron lifetime[5].

REFERENCES

1. R. Rückl, in *Z physics at LEP*, CERN 89-08, 1989, p. 311;
   G. Altarelli, S. Petrarca, Phys. Lett. **B261** (1991) 303;
   I. I. Bigi and N. G. Uraltsev, Phys. Lett. **B280** (1992) 271.
2. H.-Y. Cheng, Phys. Lett. **B289** (1992) 455.
3. OPAL Collaboration, P. D. Acton *et al.*, Phys. Lett. **B312** (1993) 501.
4. OPAL Collaboration, R. Akers *et al.*, Phys. Lett. **B316** (1993) 435.
5. OPAL Collaboration, P. D. Acton *et al.*, Zeit. Phys. **C60** (1993) 217.

A MEASUREMENT OF THE b BARYON LIFETIME

Yongsheng GAO*

Physics Dept., University of Wisconsin, Madison, Wisconsin 53706, USA

ABSTRACT

In about 1,500,000 hadronic Z^0 decays, recorded with the ALEPH detector in 1991, 1992 and 1993, the yields of $\Lambda\ell^-$ and $\Lambda\ell^+$ combinations are measured. Semileptonic decays of b baryons result in a signal of $290 \pm 35(stat)^{+38}_{-45}(syst)$ $\Lambda\ell^-$ combinations from b baryon decays, which corresponds to a product branching ratio

$$Br(b \to \Lambda_b) \cdot Br(\Lambda_b \to \Lambda_c^+ \ell^- \bar{\nu} X) \cdot Br(\Lambda_c^+ \to \Lambda X) = (0.61 \pm 0.07_{stat} \pm 0.10_{syst})\%.$$

From a fit to the impact parameter distribution of the leptons in the $\Lambda\ell^-$ sample, the lifetime of b baryons is measured to be:

$$\tau_{\Lambda_b} = 1.07^{+0.13}_{-0.12}(stat) \pm 0.10(syst) \ ps.$$

1. Introduction

The copious production of b hadrons at the Z^0 resonance has allowed many measurements of the individual b hadron lifetimes. Of all the b hadron species, the lifetime of the b baryons are the least precisely measured. More precise b baryon lifetime measurements are therefore essential for the understanding of the dynamics in the b hadron system.

Evidence for b baryons in Z^0 decays via correlation between a Λ and a high transverse momentum prompt lepton has been reported previously by the ALEPH OPAL and DELPHI collaborations.[1] The signal results from the decay* $\Lambda_b \to \Lambda_c^+ \ell^- \bar{\nu}$ followed by the decay $\Lambda_c^+ \to \Lambda X$, with the Λ decaying to $p\pi^-$. The correlation $\Lambda\ell^-$, as opposed to $\Lambda\ell^+$, is a distinctive signature of semileptonic b baryon decay.

In this analysis the b baryon are selected using $\Lambda\ell^-$ correlations and the lifetime is extracted from a maximum likelihood fit to the impact parameter distribution of the lepton tracks.

2. Isolation of the b Baryon Signal

Not all $\Lambda\ell^-$ are from b baryon decays. There are five possible sources of $\Lambda\ell$ correlations:

$$\Lambda_b \to \Lambda_c^+ \ell^- \bar{\nu}, \quad \Lambda_c^+ \to \Lambda X \tag{1}$$

---

*Representing the ALEPH Collaboration
*Charge conjugate decays are implied throughout this paper.

$$\overline{B} \to \Lambda_c^+ X \ell^- \overline{\nu}, \quad \Lambda_c^+ \to \Lambda X \tag{2}$$

$$b \to \Lambda_c^+ X, \quad \Lambda_c^+ \to \Lambda \ell^+ X \tag{3}$$

$$c \to \Lambda_c^+ X, \quad \Lambda_c^+ \to \Lambda \ell^+ X \tag{4}$$

$$\text{Accidental Combinations} \tag{5}$$

By requiring a lepton candidate to have at least 1 GeV of $p_\perp$ and 3 GeV of momentum, about 90% of the $\Lambda\ell$ correlations from process (2) and more than 95% of those from processes (3) and (4) are removed, while keeping over half of the $\Lambda\ell^-$ correlations from b baryon decay. The remaining $\Lambda\ell^-$ combinations originate mostly from either b baryon semileptonic decays (1) or accidental combinations (5), while the $\Lambda\ell^+$ combinations are mostly accidental combinations (5) where the leptons are mainly from b decays plus real Λ from fragmentation. The ratio of $\Lambda\ell^-$ to $\Lambda\ell^+$ accidental combinations is taken to be 0.8 ± 0.2 from Monte Carlo study. The excess of $\Lambda\ell^-$ combinations over $\Lambda\ell^+$ is then taken to be due to semileptonic decay of b baryons.

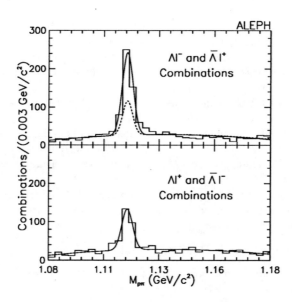

Fig. 1. *The $p\pi^-$ invariant mass distribution of the $\Lambda\ell^-$ and $\Lambda\ell^+$.*

This analysis is based on data taken from 1991 to 1993, consisting of about 1,500,000 hadronic Z^0 decays. The yields of $\Lambda\ell^-$ and $\Lambda\ell^+$ combinations, after application of the Λ and lepton selection criteria described above, are shown in Fig. 1, where the $p\pi^-$ invariant mass is plotted.

To estimate the b baryon contribution to the observed excess, the residual contributions of the background processes are evaluated as follows. Based on a simulation of process (2) and the 90% confidence level upper limit of $Br(\overline{B} \to pe^-X) \leq 0.16\%$ set by the ARGUS collaboration,[2] less than 23 $\Lambda \ell^-$ combinations are expected from this process. This possible contribution is included in the evaluation of the systematic error. From a study of approximately 1.6×10^6 $Z^0 \to q\overline{q}$ events generated with the JETSET 7.3 Monte Carlo, the contribution to the $\Lambda \ell^+$ sample of processes (3) and (4) is estimated to be 28 ± 5 combinations, which is subtracted from the wrong sign peak as a correction to the accidental combinations in the wrong sign peak. These considerations together imply a b baryon signal of 290 ± 35 (stat) $^{+38}_{-45}$ (syst) events which corresponds a product branching ratio:

$$Br(b \to \Lambda_b) \cdot Br(\Lambda_b \to \Lambda_c^+ \ell^- \overline{\nu}X) \cdot Br(\Lambda_c^+ \to \Lambda X) = (0.61 \pm 0.07_{stat} \pm 0.10_{syst})\%.$$

3. The Lifetime Fit

The lifetime of the Λ_b is extracted by a maximum likelihood fit to the observed impact parameter distribution of the lepton candidates belonging to the Λl^- sample.

The lepton track is required to have at least 1 $r\phi$ associated coordinate in the VDET, 5 hits in the TPC and 2 in the ITC and a $\chi^2/$d.o.f for the track fit to be less than 4. These cuts, together with the Λ selection and the lepton identification requirements, described in the previous sections, as well as a cut of 6 MeV around the Λ mass to select the $p\pi^-$ combinations, yield a final sample of 519 Λl^- candidates in the data when the vertex detector was operational.

The various sources of lepton candidates which are considered in the fit are:

1. semileptonic decay of b baryon ($\Lambda_b \to l$)

2. semileptonic decay of generic b hadron ($b \to l$)

3. cascade decays ($b \to c(\tau) \to l$)

4. c hadron decay in $c\overline{c}$ events ($c \to l$)

5. hadron misidentification (misid)

6. π, K decay in flight or γ conversion (decay)

The first component is the signal while the remainder are the background. The fraction of signal relative to the background is taken from the ratio between right and wrong sign events, corrected for the production asymmetry for the accidental Λl pairs.

The physics function for the signal depends on the unknown Λ_b lifetime, the only parameter of the fit, while the background physics functions depend on the inclusive b lifetime. The value used in the fit is a world average of all the b lifetime measurements, $\tau_b = 1.52 \pm 0.02$ ps.[3]

The maximum likelihood fit to the impact parameter distribution of the lepton candidates in the 519 Λl^- candidates yields a b baryon lifetime:

$$\tau_{\Lambda_b} = 1.02 ^{+0.12}_{-0.11} \text{ ps}$$

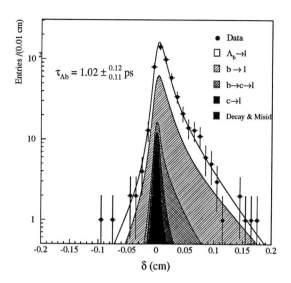

Fig. 2. *Impact parameter distribution of the selected* Λl^- *candidates. The solid line is the probability function at the fitted value of the lifetime.*

4. Consistency Checks

Several checks have been performed in order to look for possible systematic effect which could affect the fitted b baryon lifetime value. First, in order to check the fitting procedure, the analysis has been applied to a Monte Carlo sample of about 3,000,000 equivalent $q\bar{q}$ events generated with an input Λ_b lifetime of 1.5 ps. From 987 Λl^- and 569 Λl^+ selected events, the fit yields a result of $\tau_{\Lambda_b} = 1.48 \pm 0.13$ which is consistent with the input value of 1.50 ps.

A further check is performed repeating the fit on various samples of data selected with different cuts. The lepton kinematic cuts on p and $p_\perp$ and the Λ momentum are varied. No significant variation is observed.

5. Systematic Errors

Sources of systematic error and their contributions are listed in Table 1.

References

1. ALEPH Collab., D. Decamp et al., Phys. Lett. **B278** (1992), 209;
   DELPHI Collab., P. Abreu et al., Phys. Lett. **B311** (1993), 379;
   OPAL Collab., P.D. Acton et al., Phys. Lett. **B281** (1992), 394.
2. ARGUS Collab., H. Albrecht et al., Phys. Lett. **B249** (1990), 359.
3. V. Sharma and F. Weber, Recent measurements of the lifetimes of b hadron (in *"B decays", second edition, editted by S. Stone, to be published by World Scientific, Singapore*)

Source of systematic error	σ_τ^{sys} (ps)
$\Lambda_b \to l$ physics function	± 0.03
$b \to l$ physics functions	± 0.03
$b \to c \to l$ and $c \to l$ physics functions	± 0.01
Decay background and misid function	± 0.02
Resolution function	± 0.02
Fraction of Λ_b	$^{+0.04}_{-0.05}$
Combinatorial Λ background	± 0.01
Lepton fractions	± 0.03
Background lifetime	$^{+0.03}_{-0.01}$
Fragmentation and decay model	± 0.03
Fragmentation Λ spectrum	± 0.02
Λ_b polarization	0.05 ± 0.05
Total	± 0.10

Table 1. *Contributions to the systematic error.*

MEASUREMENT OF THE B_s LIFETIME WITH D_s + HADRON EVENTS

SARAH D. JOHNSON*

*Niels Bohr Institute, University of Copenhagen
Blegdamsvej 17, 2100 Copenhagen, Denmark*

ABSTRACT

The B_s lifetime has been measured using a sample of D_s + *hadron* candidates selected in the decay channels $D_s^+ \to \varphi \pi^+$ and $D_s^+ \to \bar{K}^{*0} K^+$. An unbinned maximum likelihood fit to the proper time that takes into account the various sources of D_s + *hadron* events has been used to extract the B_s lifetime.

1. Introduction

The B_s lifetime has been measured by ALEPH using the decay $\bar{B}_s \to D_s^+ \ell^- \bar{\nu}$.[1] The B_s vertex is reconstructed by extrapolating the D_s candidate backwards to form a vertex with the lepton. This method has low backgrounds, but suffers from lack of statistics.

The present analysis uses a similar technique, but instead of a lepton, uses a charged hadron to vertex the B_s. This has the advantage that about 68% of B_s mesons decay into a D_s accompanied by at least one hadron, but there are several non-B_s sources of D_s + *hadron* candidates.

2. The ALEPH Detector

The ALEPH detector is described in detail elsewhere.[2] The high resolution silicon vertex detector (VDET) was important for this analysis. It provides measurements in the $r\phi$ and z directions with $\simeq 12$ μm precision.[3] Outside VDET particles traverse the inner tracking chamber (ITC) and the time projection chamber (TPC). The ALEPH detector also includes a lead/wire-chamber sandwich electromagnetic calorimeter and a hadron calorimeter consisting of iron and streamer tubes.

3. Sources of D_s + Hadron Events

D_s mesons in Z decays arise from three different sources:

1) decay of B_s mesons $\left(\text{not via } b \to W^- \to D_s^- \right)$,

2) decay of b hadrons via $b \to W^- \to D_s^-$,

3) direct production in $Z^0 \to c\bar{c}$ events.

---

*Representing the ALEPH Collaboration

The rates of sources 1) and 3) have been measured in ALEPH[4] and have been updated for this analysis. The rate of source 2), measured in B^0 and B^+ decays,[5,6] was assumed to be valid for B_s and b baryon decays.

Additionally, the hadron added to the D_s candidate can originate from a) the B decay vertex, b) a D cascade decay vertex, or c) the primary vertex. Information on τ_{B_s} comes from sources 1a) and 1b) (where 1a) corresponds to sources 1)+a) listed above). Sources 2a) and 2b) have essentially the b hadron inclusive lifetime, while sources with a hadron originating from the primary vertex 1c), 2c) and 3c), have essentially zero lifetime.

4.   Event Selection

The analysis is based on about 1.5M hadronic events collected between 1991 and 1993. D_s mesons have been reconstructed using two decay channels: $D_s^+ \to \varphi \pi^+$ and $D_s^+ \to \bar{K}^{*0} K^+$ with a selection similar to that described in reference [1].

To reconstruct the B_s decay vertex a hadron has been added to the D_s. Cuts are made on the hadron momentum and the $D_s + hadron$ combined momentum. Other cuts include requiring the D_s and the hadron have opposite charge and that the hadron not be identified as a lepton. Cuts on various decay angles and on the χ^2 of the $D_s + hadron$ vertex were also applied. If for a given D_s candidate several hadrons pass the selection criteria, the hadron candidate that makes the vertex with the lowest χ^2 value is chosen.

Figure 1 shows the D_s mass distribution for the $D_s + hadron$ candidates selected in the $\phi\pi$ decay channel. The purity of D_s for the events selected in a mass window of two sigma is approximately 55% for both D_s decay channels.

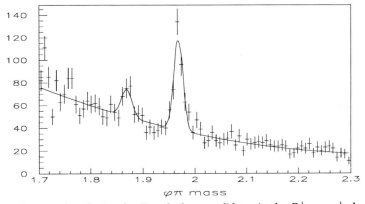

$\varphi\pi$ mass

Fig. 1. The mass distribution for $D_s + hadron$ candidates in the $D_s^+ \to \varphi\pi^+$ channel.

5.   Proper Time Measurement

The decay length δ of a $D_s + hadron$ candidate is calculated as the distance between the primary vertex and the secondary vertex projected onto the jet axis.

The decay length *pull*, that is $(\delta - \delta_{true})/\sigma_\delta$, was fitted to two gaussians for B_s signal events and for B_{had} background events. The error on the proper time is scaled up according to the widths of the Gaussians determined in the fit. For the B_s signal one finds $s_1 = 1.04$, $s_2 = 2.94$ and the fraction of events in the second Gaussian $a_2 = 0.04$. For the B_{had} background one obtains $s_1 = 1.38$, $s_2 = 3.14$ and $a_2 = 0.24$.

The proper time of a $D_s + hadron$ candidate is given by

$$t = \frac{\delta}{\beta \gamma c} = \frac{\delta m_{B_s}}{p_{B_s} c} \tag{1}$$

An estimate of p/m for the B_s meson using p/m of the B_s jet is made, as described in reference [7].

6.   Fitting Function

The B_s lifetime is determined by performing a maximum likelihood fit to the proper time distribution of the $D_s + hadron$ candidates. The proper time shape of the combinatorial background is taken into account by simultaneous fitting the proper time distribution of the sideband region of the D_s mass spectrum $(2 < m(D_s) < 2.3 \text{ GeV})$. The likelihood function is given by

$$\mathcal{L} = \prod_{i=1}^{n_{peak}} \mathcal{P}_{peak}(t_i, \sigma_{ti}) \prod_{i=1}^{n_{side}} \mathcal{P}_{comb}(t_i) \tag{2}$$

where t_i and σ_{ti} are the measured proper time and error for event i. The probability function for events in the peak consists of four components which correspond to $D_s + hadron$ source 1a), source 2b), the PV background and the combinatorial background.

The probability function for the B_s signal is given by an exponential convoluted with the vertex resolution function and the momentum resolution function. The probability function for the B_{had} background is similar to the one for the B_s signal, but the lifetime of the exponential is the inclusive b hadron lifetime, $\tau_b = 1.49 \pm 0.04$ ps. The proper time distribution for the primary vertex background is taken from Monte Carlo. The combinatorial background probability function is fitted to a sum of a Gaussian and two exponentials. The free parameter τ_{B_s} is determined by the fit.

7.   Fit Results

The fit was performed on 397 peak events and 1039 sideband events for the $D_s^+ \rightarrow \varphi \pi^+$ channel, and 160 peak events and 401 sideband events for the $D_s^+ \rightarrow \bar{K}^{*0} K^+$ channel. The result obtained is

$$\tau_{B_s} = 1.75^{+0.30}_{-0.28}(\text{stat}) \text{ ps}.$$

The fit result is shown in Fig. 2.

A sample of Monte Carlo events, generated with a lifetime of 1.5 ps, that reflected the correct mixture of B_s signal events and B_{had} background events was fitted to a lifetime of $\tau_{B_s} = 1.41 \pm 0.17$ ps.

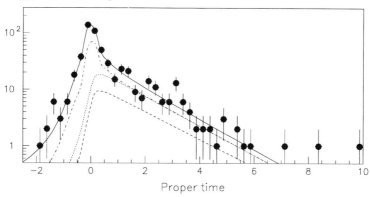

Fig. 2. The proper time distributions for $D_s + hadron$ candidates in the peak region. The solid curve represents the total fit, while the dashed, dotted, dash-dotted show the contributions from B_s signal, B_{had} background, and PV background, respectively.

8. Systematic Uncertainties

The largest contribution to the total systematic error arises from the uncertainties on the selection efficiencies for the various $D_s + hadron$ combinations: $\Delta\tau_{B_s}(ps) = ^{+0.08}_{-0.17}$. To consider these uncertainties the efficiency for the B_s signal as a function of the charged multiplicity of the B_s decay was studied in the Monte Carlo.

The second largest contribution to the systematic error comes from the statistical error on $f_s^b \cdot Br(B_s \to D_s)$ and f_s^c: $\Delta\tau_{B_s}(ps) = ^{+0.12}_{-0.12}$

9. Conclusion

The B_s lifetime has been measured using a sample of $D_s + hadron$ candidates. The preliminary result obtained is

$$\tau_{B_s} = 1.75^{+0.30}_{-0.28}(\text{stat.})^{+0.18}_{-0.23}(\text{syst.}) \text{ ps.}$$

References

1. D. Buskulic et al., (ALEPH Collaboration) *"Measurement of the B_s Lifetime"*, *Physics Letters* **B322** (1994) 275.
2. D. Decamp et al. (ALEPH Collaboration), *Nucl. Instrum. Methods* **A294** (1990) 121.
3. G. Batignani et al., *"Recent Results and Running Experience of the New ALEPH Vertex Detector"*, *IEEE Transactions on Nuclear Science* **39** (1992) 701.
4. G. Rizzo et al. *"Measurement of the B_s and D_s production rates in Z decays,"* *ALEPH-NOTE 93-120.*
5. H. Albrecht et al., (ARGUS Collaboration), *Phys. Lett.* **B187** (1987) 425.
6. D. Bortoletto et al., (CLEO Collaboration), *Phys. Rev.* **D45** (1992) 21.
7. D. Buskulic et al., (ALEPH Collaboration), *Physics Letters* **B295** (1992) 396.

MEASUREMENTS OF LIFETIMES AND MASSES OF B-HADRONS

JAVIER CUEVAS*

Departamento de Física, Universidad de Oviedo,
C/ Julián Clavería S/N, 33006-Oviedo, Spain

ABSTRACT

Experimental results on lifetimes and masses of B-hadrons obtained with the DELPHI detector at LEP are presented.

They are based on approximately 1.79 Million hadronic Z^0 decays recorded between 1991 and 1993.

1. B_u^- and B_d^0 Lifetimes

The lifetimes of B_u^- and B_d^0 hadrons have been measured using a method based on the inclusive reconstruction of secondary vertices, the charge of the B meson being determined summing the charges of the observed decay products.[1] Using a sample of approximately 1 564 000 multihadronic Z^0 decays, 1816 B hadron candidates with well measured charge were obtained. The B purity was estimated to be $99^{+0.5}_{-1.0}\%$, and the charge was correctly estimated 70% of the time. The B lifetime is fitted from the proper time distribution of the reconstructed decays, which requires a knowledge of the decay length (found from the positions of the primary and secondary vertices) and the B-hadron velocity (calculated from the measured momentum and invariant mass). Two different fits were performed. The assumption that all charged B species had one lifetime and all neutral ones another was assumed in the first fit. The (preliminary) results for this mean charged and neutral lifetimes were:

$$< \tau_{charged} > \ = \ 1.72 \pm 0.08 \ (stat) \pm 0.06 \ (syst) \ \text{ps}$$
$$< \tau_{neutral} > \ = \ 1.63 \pm 0.11 \ (stat) \pm 0.07 \ (syst) \ \text{ps}$$
$$< \tau_{charged} > / < \tau_{neutral} > \ = \ 1.06^{+0.11}_{-0.09} \ (stat) \pm 0.07 \ (syst) \tag{1}$$

In the second fit the data were interpreted in terms of the lifetimes of the B^+ and B^0 mesons, by assuming that the selected neutral B hadrons are composed of B^0, B_s^0 and Λ_b^0. The (preliminary) values for the B^+ and B^0 meson lifetimes were:

$$\tau_{B^+} \ = \ 1.72 \pm 0.08 \ (stat) \pm 0.06 \ (syst) \ ps$$
$$\tau_{B^0} \ = \ 1.68 \pm 0.15 \ (stat)^{+0.13}_{-0.17} \ (syst) \ ps$$
$$\tau_{B^+}/\tau_{B^0} \ = \ 1.02^{+0.13}_{-0.10} \ (stat)^{+0.13}_{-0.10} \ (syst) \tag{2}$$

---

*Representing the DELPHI Collaboration.

2. B_s^0 Lifetime

The lifetime of the B_s^0 meson was measured using the inclusive decay into D_s and the semi-leptonic decay into the $D_s - l$ final state.

In the inclusive D_s meson analysis, the D_s meson has been reconstructed in the mode[2]: $D_s \to \phi\pi$, $\phi \to K^+K^-$.

410 events were selected from the mass interval $1930 < m < 2010$ MeV/c^2, corresponding to ± 2 standard deviations around the D_s fitted mass. The expected combinatorial background fraction is (53 ± 6) %. The B_s lifetime after fitting the proper lifetime distribution using an unbinned maximum likelihood method was: $\tau(B \to D_sX) = 1.55 \, {}^{+0.23}_{-0.19} \, (stat) \pm 0.14 \, (syst) \, ps$. The purity of the sample was estimated to be (60 ± 7)%. Using the average lifetime of B hadrons[3] $\tau(B) = 1.536 \pm 0.034 \, ps$ resulted in:

$$\tau(B_s) = 1.56 \, {}^{+0.38}_{-0.32} \, (stat) \pm 0.23 \, (syst) \, ps.$$

In the second analysis, the B_s is searched for by looking at the correlation between a D_s meson and a lepton of opposite charge present in the same hemisphere. The D_s meson has been reconstructed using two different decay modes: $D_s^+ \to \phi\pi$, $\phi \to K^+K^-$, and $D_s^+ \to \bar{K}^{*0}K^+$, $\bar{K}^{*0} \to K^+\pi^-$. A signal of 37 ± 8 events is observed, centered at the expected D_s mass. The lifetime of the sample was found to be: $\tau(B \to D_s^\pm l^\mp) = 1.34 \, {}^{+0.37}_{-0.29}(stat) \pm 0.16(syst) \, ps$. and considering that the $D_s l$ sample is 90 % pure in B_s^0 and using the afore mentioned average B lifetime, the B_s^0 was found to be:

$$\tau(B_s) = 1.32 \, {}^{+0.41}_{-0.32}(stat) \pm 0.18(syst) \, ps.$$

Taking into account the B_s purities in these two samples and combining the two measurements with the one obtained with the 1991 data using ϕl events[2] the B_s lifetime was (preliminary) measured to be:

$$\tau(B_s) = 1.42 \, {}^{+0.25}_{-0.23}(stat) \pm 0.14(syst) \, ps.$$

3. B baryon Lifetime and Production Rates

Beauty baryon signals were obtained using three complementary methods: a excess of Λl^- pairs with respect to Λl^+ pairs, when the lepton has high transverse momentum, presence of a fully reconstructed Λ_c baryon associated with a high p_T lepton, or presence of a secondary vertex formed by a fast proton, identified by the RICH,[4] and a muon of opposite sign.

In all the cases, an excess of events with *right sign* pairs i.e. with a baryon having baryon number opposite to the lepton charge is observed, with respect to events with *wrong sign*, pairs i.e. with identical baryon number and lepton charge. Reconstructed vertices from the three data samples were used to determine average b-baryon lifetimes.

The Λl sample was obtained using the selection criteria described in.[5] Assuming lepton universality the production rate was measured to be :

$$f(b \to b - baryon) \times Br(b - baryon \to \Lambda l\nu X) = (0.32 \pm 0.07(stat) \pm 0.04(syst))\%$$

The measurement of the lifetime is based on the muon sample only.[5] Secondary vertices are reconstructed using the Λ, the correlated high p_T muon and an oppositely charged particle (supposed to be a pion). A maximum likelihood fit was performed simultaneously to the lifetime distribution of the 63 events of the signal sample and to the one of the background sample, giving the result:

$$\tau(b - baryon) = 1.13^{+0.30}_{-0.23}(stat)^{+0.05}_{-0.08}(syst) \ ps \ (63 \ decays, \ \Lambda\mu\nu X \ channel)$$

In the second analysis, based on 1991-1992 data, the Λ_c is reconstructed via the decay $\Lambda_c \rightarrow pK\pi$, which takes adavantage of the good particle identification capability of the DELPHI detector to tag the proton and the kaon.

Assuming lepton universality the measured production rate was :

$$f(b \rightarrow b - baryon) \times Br(b - baryon \rightarrow \Lambda_c l\nu X) = (1.30 \pm 0.32(stat)^{+0.42}_{-0.35}(syst))\%$$

The proper time distribution of the $\Lambda_c\mu\nu X$ has been fitted with the same technique used before. The result was:

$$\tau(b - baryon) = 1.32^{+0.71}_{-0.42}(stat)^{+0.08}_{-0.09}(syst) \ ps \ (\Lambda_c\mu\nu X \ channel, \ 28 \ decays)$$

In the third analysis the two essential ingredients are the hadron identification (dE/dx and RICH) and the muon-proton vertex reconstruction. A maximum likelihood fit was used to estimate the number of muon-proton pairs from b-baryon decays and the average lifetime of b-baryons, giving the results:

$$f(b \rightarrow b - baryon) \times Br(b - baryon \rightarrow p\mu\nu X) = (0.36 \pm 0.08(stat)^{+.15}_{-.07}(syst))\%$$

$$\tau(b - baryon) = 1.28^{+0.35}_{-0.29}(stat) \pm 0.09(syst) \ ps \ (47 \ decays, \ p\mu\nu X \ channel)$$

Averaging the three (independent) results, the mean b-baryon lifetime was (preliminary) measured to be:

$$\tau(b - baryon) = 1.21^{+0.21}_{-0.18}(stat) \pm 0.04(exp.syst)^{+0.02}_{-0.07}(theory) \ ps$$

4. B^0_s Meson Mass

Using data collected in 1992 by the DELPHI experiment, strange beauty mesons B^0_s have been reconstructed through the following exclusive hadronic final states[6]: $D^-_s \ \pi^+$, $D^-_s \ a^+_1(1260)$, and $J/\psi \ \phi$. It is important to notice that all the selected channels lead to a final state formed by two kaons and either two or four lighter particles. The charged K identification performed by the combined use of the RICH and the dE/dx measurements allows clean reconstruction of these states free from the effect of kinematical reflections. The combinatorial background is then further suppressed by cutting on the intermediate state masses and by requiring a well reconstructed secondary decay vertex.

In the $D^-_s \ \pi^+$ channel, the D^-_s was reconstructed in the two modes $\phi\pi$ or $K^*(892)^0$ K and in the $D^-_s \ a^+_1(1260)$ only the $\phi\pi$ mode was used because of the larger combinatorial background predicted by the simulation. One event was found in each of the two channels. In the $J/\psi \ \phi$ channel, the J/ψ was reconstructed through decay into muon pairs. One candidate was found in this channel.

Several sources of error for the B_s^0 mass measurement have been studied: possible reflection of the more abundantly produced B_d^0 meson when a misidentified pion is attributed the kaon mass, the combinatorial background, and possible incomplete reconstruction of decays through a D_s^{*}. In order to take into account the possible distortions of the measured mass and the contribution to the error from these sources, a global likelihood fit was performed. An additional source of systematic error comes from the absolute mass scale calibration. The final result was:

$$m_{B_s} = (5374 \pm 16 \pm 2) \ \text{MeV}/c^2$$

5. Λ_b^0 mass

The Λ_b^0 beauty baryon was searched for in the following channels: $J/\psi \, \Lambda^0$, $\Lambda_c^+ \, \pi^-$, $\Lambda_c^+ \, a_1(1260)$ and $D^0 \, p\pi^-$ using the data recorded by DELPHI in 1992. The $J/\psi \, \Lambda^0$ channel allows a clean background free reconstruction of the Λ_b^0 by tagging the decay $J/\psi \rightarrow \mu^+\mu^-$ combined with a long flying Λ^0 . The other selected channels involve either a Λ_c^+ or a D^0 intermediate state, leading to a final configuration consisting of a proton, a kaon and either two or four pions. For these channels both the proton and the kaon had to be identified in the RICH.

Two Λ_b^0 candidates were observed in the $\Lambda_c^+ \, \pi^-$ and $D^0 \, p\pi^-$ decay modes. Their masses are compatible within errors. Their weighted average gives: ($5630 \pm 25 \ (stat)$) MeV/c^2

After considering the different sources of systematic errors (as in the B_s^0 case), a maximum likelihood fit was performed using the measured masses and errors, the estimated combinatorial background and incomplete final state reconstruction.

The preliminary measurement using the two fully reconstructed Λ_b^0 candidates correspond to a mass of:

$$(\ 5635^{+38}_{-29} \ \pm 4 \ (mass \ scale) \ ) \ \ \text{MeV}/c^2$$

6. Conclusions

The DELPHI experiment at LEP has measured the lifetimes of different B hadron states: B^+, B^0, B_s^0, and Λ_b^0 with high precision, using the approximately 1.79 Million hadronic Z^0 decays collected during the years 1991-1993.

The B_s^0 meson mass has been measured within a few Mev/c^2 using three fully reconstructed events found in the data recorded during 1992. The mass of the Λ_b^0 beauty baryon has been preliminary measured using the data collected during 1992.

References

1. P. Abreu et al., *Phys Lett.* **B312** (1993) 533.
2. P. Abreu et al., *Z Phys.* **C61** (1994) 407.
3. T. Hessing: *Measurement of exclusive B lifetimes and masses at LEP* Rencontres de Moriond, March 12-19, 1994.
4. C. Bourdarios, *Contribution to this conference*
5. P. Abreu et al., *Phys Lett.* **B311** (1993) 379.
6. P. Abreu et al., *Phys Lett.* **B324** (1994) 500.

MEASUREMENT OF THE AVERAGE LIFETIME
OF B HADRONS AT SLD

GREG PUNKAR *

*Stanford Linear Accelerator Center, Stanford University,
Stanford, California 94309, USA*

ABSTRACT

We present preliminary measurements of the average B hadron lifetime using a sample of 50,000 Z^0 events collected by SLD at the SLC in 1993. Our first technique uses the impact parameter of tracks in jets opposite tagged b jets. We obtain $\tau_B(\delta) = 1.617 \pm 0.048 \pm 0.086$ ps, and $\tau_B(\sum \delta) = 1.627 \pm 0.054 \pm 0.132$ ps from single and summed impact parameter distributions. The second technique uses inclusive vertices reconstructed in three dimensions. From the decay length distribution, we extract $\tau_B = 1.577 \pm 0.032 \pm 0.046$ ps.

1.   General Description and Experimental Procedure

The precise measurement of the average lifetime of B hadrons, τ_B, is important for the study of the b quark and its weak couplings to u and c quarks. Furthermore, recent measurements have shown a marked departure from the 1992 world average.[1]

The preliminary results presented here use the pixel-based Vertex Detector (VXD) and the Central Drift Chamber (CDC) for tracking, and the Liquid Argon Calorimeter (LAC)[2] for triggering and determining event shape properties. Within the VXD solid angle, 96% of all CDC tracks correctly link to one or more pixel-clusters. Angular errors in the extrapolated tracks combined with local errors $\sigma_{r\phi}$ and σ_{rz} of ~ 6 μm for the VXD clusters, lead to xy (orthogonal to the e^- beam) and rz (plane containing the e^- axis) impact parameter resolutions of $(\alpha,\beta)_{xy} = (11$ μm, 70 μm) and $(\alpha,\beta)_{rz} = (38$ μm, 70 μm).[3] During the 1993 SLD run the $\langle rms \rangle_{xyz}$ profile of SLC beams was $2.4 \times 0.8 \times 700$ μm^3 at the interaction point (IP). The IP x and y positions are tracked by SLD using reconstructed tracks from hadronic Z^0 events. Muon pairs (not used in the average IP determination) are used to check the IP xy position, giving $\sigma_{xy}^{IP} = 7 \pm 2$ μm. The z position of the IP is measured event by event with $\sigma_z \simeq 35$ μm as determined by simulation.

The Monte Carlo (MC) physics simulation models Z^0 and heavy flavor decays with the LUND JETSET (version 6.3) Monte Carlo generator, which has been adjusted to reflect current knowledge of the B and D decay spectra. The lifetime for B mesons (baryons) is set to 1.55 ps (1.10 ps). The MC detector simulation is based on GEANT and produces raw hits that are superimposed on randomly triggered events from the

---

*Representing the SLD Collaboration.

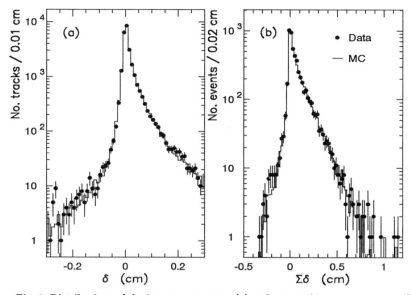

Fig. 1. Distributions of the impact parameter (a) and summed impact parameter (b).

data, to simulate the unique SLC backgrounds. A difference in track finding efficiency between data and MC is observed and corrected for as a function of momentum, θ and ϕ (polar and azimuthal angles), and ξ (angle to jet direction), by randomly removing MC tracks. Approximately 7% of the MC tracks passing all track selection cuts are removed.

Standard hadronic event selection cuts[4] are applied, resulting in a sample of 29,400 Z^0 events. Jet axes are determined from energy clusters in the LAC using the JADE algorithm with $y_{cut} = 0.02$. For each track passing selection criteria,[4] we form the xy impact parameter (δ) relative to the IP and $\delta_{norm} \equiv \delta/\sigma$, where σ is obtained from σ_δ and σ_{xy}^{IP} added in quadrature. The impact parameter is signed with respect to the nearest jet axis following the standard convention.[4] An event is b-tagged by requiring a minimum number of 3 tracks with $\delta_{norm} > 3$. A jet is b-tagged by applying the same criteria to tracks in a given jet. The corresponding efficiencies and purities are $\epsilon_{event} = 69\%$, $\epsilon_{jet} = 30\%$, $\Pi_{event} = 82\%$, and $\Pi_{jet} = 93\%$ for event- and jet-tagging, respectively.

2. Lifetime Analyses

The **impact parameter** lifetime measurements use 2- and 3-jet events and take advantage of the fact that tracks resulting from heavy quark decays have large positive δ while those resulting from fragmentation and light quark decays have small positive or negative δ. In the first method, the lifetime is determined from the impact parameter distribution for *all* quality tracks in the jet(s) opposite the tagged jet (in the case of double tagged events, for both analyses, all jets are used). Since all tracks

Detector Modeling	δ (%)	$\sum \delta$ (%)	Physics Modeling	δ (%)	$\sum \delta$ (%)
Track Resolution	1.4	3.6	R_b (0.218 ± 0.015)	1.9	2.3
Track./Link. Eff.	0.9	3.8	R_c (0.181 ± 0.030)	2.0	2.8
IP Position Tails	0.2	0.2	b fragmentation	2.9	2.9
Subtotal	1.7	5.2	$(\langle x_E \rangle = 0.700 \pm 0.011)$		
			c fragmentation	0.4	1.4
			$(\langle x_E \rangle = 0.49 \pm 0.03)$		
			B multiplicity (5.5 ± 0.2)	1.9	3.1
			B baryon fraction	2.0	2.0
			(0.088 ± 0.050)		
			Charm content of B decay	1.3	1.3
			Subtotal	5.0	6.2
TOTAL				5.3	8.1

Table 1. Systematic errors in the average B hadron lifetime for impact parameter methods.

are used directly in the lifetime measurement, the method provides a high analyzing power. In the summed impact parameter ($\sum \delta$) method,[5] a scalar sum of the signed impact parameters from quality tracks in the jet opposite the tagged jet is formed. The advantage of this method is in its enhancement of the lifetime signal since jets from light quarks will have $\sum \delta \simeq 0$.

To extract τ_B, the δ and $\sum \delta$ distributions are fit to their corresponding MC distributions. Since the b MC sample is generated at fixed lifetime, the τ_B dependence is introduced to the fitting function through a weighting procedure. Jets containing a B hadron are given a weight which represents the probability of its being generated with a new lifetime, relative to the probability for its generated lifetime. Each entry in the δ and $\sum \delta$ MC distributions is weighted by the product of the weight for the un-tagged jet to which it belongs and the weight for the jet tagging the event. This accounts for the b-jet tagging purity and efficiency as a function of B hadron lifetime. MC δ and $\sum \delta$ distributions are formed ranging from $\tau_B = 0.5$ to 2.5 ps in 0.02 ps steps. The data distributions are fit to each of the MC distributions using a Maximum Likelihood procedure utilizing a multinomial probability function. The resulting δ and $\sum \delta$ distributions for data and best fit MC are shown in Fig. 1.

Table 1 summarizes the fractional systematic errors for the δ and $\sum \delta$ lifetime measurements. The detector errors contribute 1.7% (5.2%) for the δ ($\sum \delta$) method; these are expected to improve in the near future as our understanding of the tracking systems continues to improve. The systematic errors are dominated by the uncertainties in b-quark fragmentation and modeling. The physics modeling errors contribute 5.0% (6.2%) for the δ ($\sum \delta$) method. Added in quadrature, the net systematic error is 5.3% (8.1%) for the δ ($\sum \delta$) method. The resulting values for the lifetime are $\tau_B(\delta) = 1.617 \pm 0.048(stat.) \pm 0.086(syst.)$ ps, and $\tau_B(\sum \delta) = 1.627 \pm 0.054(stat.) \pm 0.132(syst.)$ ps.

The fit results are stable with respect to the following: different tag requirement, maximum number of jets allowed in the event (2 or 3), average lifetime of generated MC (1.55 ps or 2.00 ps for B mesons), fit method, range and binning.

The lifetime is also extracted using **3-D reconstructed vertices**. In this technique we use quality tracks from b-tagged events to reconstruct all possible geometrical vertices in three dimensions. The number of vertices is then reduced by looking at the global event topology and by choosing the set of independent (i.e. not sharing any track) vertices that maximizes the product of the vertex fit probablities. Further cuts are applied to remove vertices containing tracks originating from the IP. In particular, the decay length is required to be greater than 1 mm. Only the vertex closest to the IP is kept in each event hemisphere. The final sample consists of 4294 b-tagged events with 5427 selected vertices. Monte Carlo studies indicate that 88% of selected vertices carry B hadron lifetime information.

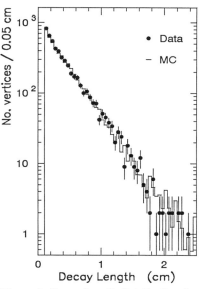

Figure 2. Decay length distribution for selected vertices.

The lifetime is extracted from the decay length distribution (Fig. 2) by following a similar weighting and fit procedure as described above. The lifetime is measured to be $\tau_B = 1.577 \pm 0.032(stat.) \pm 0.046(syst.)$ ps, where the systematic error is dominated by the uncertainty in the b-quark fragmentation. A complete account of this analysis technique and results is given elsewhere.[6]

3. Conclusions

We have made preliminary measurements of the average B hadron lifetime using 2-D impact parameters and 3-D vertices. Each of these measurements is currently limited by systematic errors which should decrease as we continue to improve our understanding of the detector and the physics modeling. In addition, we hope to be able to extend the 3-D vertexing technique to study exclusive B decay lifetimes in the near future.

References

1. See for example, W. Venus, Talk Presented at the 1993 Lepton-Photon Interactions Conference, August 10-15, 1993, Ithaca, New York.
2. SLD Design Report, SLAC-REPORT 273, 1984.
3. The impact parameter resolution function is parametrized as $\alpha \oplus \beta/p\sqrt{sin^3\theta}$.
4. K. Abe *et al.*, SLAC-PUB-6569, Aug. 1994.
5. D. Fujino, SLAC-PUB-5635, Nov. 1991.
6. K. Abe *et al.*, SLAC-PUB-6586, July 1994.

B MESON LIFETIMES AT CDF

JOHN E. SKARHA*

Department of Physics and Astronomy

The Johns Hopkins University

Baltimore, Maryland 21218, USA

ABSTRACT

Measurements of the B_u, B_d, and B_s meson lifetime using semileptonic $B_u \rightarrow e\nu D^0 X, B_d \rightarrow e\nu D^* X, B_s \rightarrow l\nu D_s X$ events and exclusive $B_u \rightarrow \psi^{(l)} K^{(*)}, B_d \rightarrow \psi^{(l)} K_{(s)}^{(*)}, B_s \rightarrow \psi\phi$ events are presented. These results used the precise position measurements of the CDF SVX silicon vertex detector and were obtained from a 19.3 pb^{-1} sample of 1.8 TeV $\bar{p}p$ collisions collected in 1992-93 at the Fermilab Tevatron collider. Comparisons with previous measurements will be shown.

1. Introduction

During the 1992-93 Tevatron collider Run Ia, the Collider Detector at Fermilab (CDF)[1] collected a data sample of $\bar{p}p$ collisions at $\sqrt{s} = 1.8$ TeV with an integrated luminosity of 19.3 pb^{-1}. This data sample, in combination with improvements to the data acquisition system, the muon coverage, and most importantly, the installation of the CDF SVX silicon vertex detector,[2] has allowed the first measurements of inclusive and exclusive B meson lifetimes at a hadron collider. In this paper we report results on the B_u, B_d, and B_s meson lifetime using semileptonic $B_u \rightarrow e\nu D^0 X, B_d \rightarrow e\nu D^* X, B_s \rightarrow l\nu D_s X$ events and exclusive $B_u \rightarrow \psi^{(l)} K^{(*)}, B_d \rightarrow \psi^{(l)} K_{(s)}^{(*)}, B_s \rightarrow \psi\phi$ events.

2. Charged and Neutral B Meson Lifetimes

Measuring the lifetime differences of the individual B mesons is a direct probe to possible non-spectator contributions in B meson decay. Only small lifetime differences are expected among the different B mesons (possibly as low as $\sim 5\%$[3]) and experiments are now approaching this precision.

At CDF, the measurement of the charged and neutral B meson lifetimes was performed using fully reconstructed B decays in the following modes[4]:

$$B^+ \rightarrow J/\psi K^+ \quad \rightarrow \mu^+\mu^- K^+; \qquad B^+ \rightarrow J/\psi K^{*+} \quad \rightarrow \mu^+\mu^- K_s^0 \pi^+$$
$$B^+ \rightarrow \psi(2S)K^+ \quad \rightarrow \mu^+\mu^-\pi^+\pi^- K^+; \quad B^+ \rightarrow \psi(2S)K^{*+} \quad \rightarrow \mu^+\mu^-\pi^+\pi^- K_s^0\pi^+$$
$$B^0 \rightarrow \psi K_s^0 \quad \rightarrow \mu^+\mu^- K_s^0; \qquad B^0 \rightarrow \psi K^{*0} \quad \rightarrow \mu^+\mu^- K^+\pi^-$$
$$B^0 \rightarrow \psi(2S)K_s^0 \quad \rightarrow \mu^+\mu^-\pi^+\pi^- K_s^0; \quad B^0 \rightarrow \psi(2S)K^{*0} \quad \rightarrow \mu^+\mu^-\pi^+\pi^- K^+\pi^-$$

---

*For the CDF Collaboration; Currently Guest Scientist at Fermi National Accelerator Laboratory, P.O. Box 500, M.S. 318, Batavia, IL 60510, USA

These measurements using exclusive decay modes are rather unique, provide a statistical precision of $10 - 12\%$, and are now published.[5] We quote only the results here:

$$\tau_{exc}^{+} = 1.61 \pm 0.16(stat) \pm 0.05(syst)ps$$
$$\tau_{exc}^{0} = 1.57 \pm 0.18(stat) \pm 0.08(syst)ps$$
$$\tau_{exc}^{+}/\tau_{exc}^{0} = 1.02 \pm 0.16(stat) \pm 0.05(syst)$$

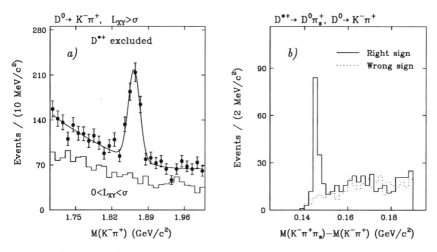

Fig. 1: a) The $K^-\pi^+$ invariant mass distribution in the electron sample. Events from D^{*+} decay are excluded. b) The distribution of the mass difference, $\Delta m = m(K^-\pi^+\pi_s) - m(K^-\pi^+)$, for $D^{*+} \to D^0\pi_s^+, D^0 \to K^-\pi^+$ candidates.

We now turn to a new, preliminary measurement of the charged and neutral B meson lifetimes using semileptonic decays,[6] as have been previously done by the LEP experiments.[7] Partially reconstructed semileptonic decays of B mesons, namely a lepton in association with a charm D^0 or D^{*+} meson will provide *nearly orthogonal* samples of charged and neutral B mesons and thus enable a determination of their individual lifetimes.

The present measurement uses only the single electron sample and makes electron identification cuts which have been described previously.[8] After applying these cuts, approximately 400,000 electron candidates remain. The next step is the reconstruction of the D^0 meson in the decay mode $D^0 \to K^-\pi^+$. This is done by using all oppositely charged track pairs, assigning kaon and pion masses, and considering only tracks within a cone of $\Delta R = \sqrt{(\Delta\eta)^2 + (\Delta\phi)^2} < 0.7$ around the electron. In addition, the momentum of the kaon and pion are required to satisfy $p(K) > 1.0$ GeV/c, $p(\pi) > 0.5$ GeV/c and all three tracks, the electron, kaon, and pion candidates must be well reconstructed within the SVX detector. Three sources of D^0 mesons are considered in this analysis: 1) $B^- \to e^-\bar{\nu}D^0X, D^0 \to K^-\pi^+$, where the D^0 is *not* from

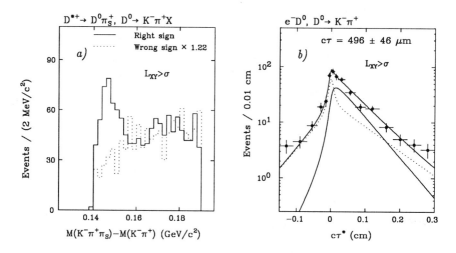

Fig. 2: a) The mass difference distribution for $D^0 \to K^-\pi^+\pi^0$ candidates. b) The combined, signal, and background $c\tau^*$ distributions in the D^0 signal sample.

$D^{*+} \to D^0\pi_s^+$ ("D^0 sample"); 2) $\overline{B}^0 \to e^-\overline{\nu}D^{*+}X, D^{*+} \to D^0\pi_s^+, D^0 \to K^-\pi^+$ ("D^{*+} sample"); and 3) $\overline{B}^0 \to e^-\overline{\nu}D^{*+}X, D^{*+} \to D^0\pi_s^+, D^0 \to K^-\pi^+\pi^0$ ("satellite sample") . The decay length of the D^0 in the plane transverse to the colliding beams, L_{XY}, must satisfy $L_{XY} > \sigma(L_{XY})$, where $\sigma(L_{XY})$ is the calculated error on the transverse decay length, for the D^0 and satellite samples. Figure 1a shows the resulting $K^-\pi^+$ mass distribution, containing 389 ± 31 events in the D^0 peak, for the D^0 sample. Figure 1b gives the mass difference distribution for the D^{*+} sample, and Fig. 2a gives the Δm distribution for the satellite sample where a cut is made at $\Delta m < 0.155$ GeV/c^2. These figures show the production of $e - D^0$ combinations in the expected ("right sign") charge combinations and little evidence above combinatoric background in the "wrong sign" combinations.

The electron and D^0 tracks are then intersected to determine the B decay vertex position and decay length from the primary vertex. Since the B is only partially reconstructed, the $e - D^0$ system transverse momentum can be used to determine a "pseudo-$c\tau$" value $c\tau^* = L_B m_B/p_T(e + D^0) = c\tau/K$, where K is a momentum correction factor determined from Monte Carlo. It is determined separately for each D^0 signal sample.

The signal $c\tau^*$ distributions are fit with an exponential lifetime term convoluted with a Gaussian resolution function and the momentum correction distribution. The lifetime of the background under the signal peak is determined from the wrong sign and signal sideband distributions and is modeled by a Gaussian resolution function plus two exponential tails. Figure 2b shows the result of the lifetime fit for the D^0 signal sample. The fit quality and results for the D^{*+} and satellite samples are similar. The fraction of B^- and $\overline{B}^0$ contributing to each of the D^0, D^{*+}, and satellite samples is determined and includes the effects of cross-talk due to: 1) the π_s^+ reconstruction efficiency, $\epsilon(\pi_s^+) = 0.93 \pm 0.21$. A missed spectator pion from D^{*+} decay can cause a D^0 to be associated with B^- rather than $\overline{B}^0$; 2) the D^{**} fraction, $f^{**} = \text{BR}(\overline{B} \to l^-\overline{\nu}D^{**})/\text{BR}(\overline{B} \to l^-\overline{\nu}X) = 0.36 \pm 0.12^9$; 3) the fraction of D^{**} decaying to D^*, from the QQ Monte Carlo is found to be $\text{BR}(D^{**} \to D^*\pi)/(\text{BR}(D^{**} \to D^*\pi) + \text{BR}(D^{**} \to D\pi)) = 0.78$; and 4) the charged-to-neutral lifetime ratio can affect the event mixture, $\text{BR}(B^- \to l^-\overline{\nu}X)/\text{BR}(\overline{B}^0 \to l^-\overline{\nu}X) = \tau(B^-)/\tau(\overline{B}^0)$. In spite of these effects, we find that the D^0 and D^{*+} signals provide *nearly orthogonal* samples of B^- and $\overline{B}^0$ mesons. A combined likelihood function is used to simultaneously fit the signal samples for the B^- and $\overline{B}^0$ meson lifetimes. Variations in the sample composition due to the above effects are included in the systematic uncertainty. The results are:

$$\tau_{semi}^+ = 1.63 \pm 0.20(\text{stat})^{+0.15}_{-0.16}(\text{syst})\text{ps}$$
$$\tau_{semi}^0 = 1.62 \pm 0.16(\text{stat})^{+0.14}_{-0.15}(\text{syst})\text{ps}$$
$$\tau_{semi}^+/\tau_{semi}^0 = 1.01 \pm 0.19(\text{stat}) \pm 0.17(\text{syst})$$

Tables 1 and 2 show a comparison of the latest[10] τ^+, τ^0, and τ^+/τ^0 values from CDF, LEP, and CLEO. Averaging asymmetric errors and computing a weighted average, we find that the error on the world value for τ^+ and τ^0 is at 5% and the lifetime ratio has an uncertainty of 7%. Clearly, with some additional statistics and work on systematic errors, non-spectator contributions to B meson decay will soon be tested.

Table 1: Comparison of charged and neutral B meson lifetime measurements.

Experiment	τ^+ (ps)	τ^0 (ps)	Reference
Delphi	1.30 ± 0.35	1.17 ± 0.31	Z. Phys. C57, 181 (1993)
Delphi	1.72 ± 0.10	1.68 ± 0.21	DELPHI 94-97 PHYS 414
Opal	1.53 ± 0.18	1.62 ± 0.14	OPAL Note PN149
Aleph	1.47 ± 0.25	1.52 ± 0.21	Phys. Lett. B307, 194 (1993)
Aleph	1.30 ± 0.23	1.17 ± 0.22	ALEPH Note 94-100
CDF	1.61 ± 0.17	1.57 ± 0.20	PRL 72, 3456 (1994)
CDF	1.63 ± 0.26	1.62 ± 0.22	CDF Note 2598
World Ave.	1.60 ± 0.07	1.53 ± 0.08	DPF 1994

3. B_s Meson Lifetime

A similar technique has been used to measure the B_s meson lifetime using semileptonic $B_s \to l\overline{\nu}D_s$, $D_s \to \phi\pi$, $\phi \to K^+K^-$ decays.[12] Figure 3a shows the $K^+K^-\pi^+$ invariant mass spectrum after all cuts for the combined electron and muon samples. Some 76 ± 8 events are found in the right-sign mass peak and a hint of the Cabbibo

Table 2: Comparison of charged-to-neutral B meson lifetime ratio results.

Experiment	τ^+/τ^0	Reference
Delphi	1.11 ± 0.46	Z. Phys. C57, 181 (1993)
Delphi	1.02 ± 0.16	DELPHI 94-97 PHYS 414
Opal	0.94 ± 0.14	OPAL Note PN149
Aleph	0.96 ± 0.23	Phys. Lett. B307, 194 (1993)
Cleo	0.93 ± 0.22	CLNS 94/1286[11]
CDF	1.02 ± 0.17	PRL 72, 3456 (1994)
CDF	1.01 ± 0.25	CDF Note 2598
World Ave.	0.98 ± 0.07	DPF 1994

Table 3: B_s meson lifetime results from CDF and LEP.

Experiment	τ_s (ps)	Reference
Delphi	1.42 ± 0.24	Z. Phys. C61, 407 (1994)
Opal	1.33 ± 0.24	OPAL Note PN113
Aleph	1.75 ± 0.36	ALEPH Note 94-044
Aleph	1.92 ± 0.40	Phys. Lett. B322, 275 (1994)
CDF	1.42 ± 0.27	CDF Note 2472
CDF	1.74 ± 0.75	CDF Note 2515
World Ave.	1.49 ± 0.12	DPF 1994

suppressed $D^+ \rightarrow \phi\pi^+$ decay is seen. Following the same procedure as above, the B_s lifetime from semileptonic $B_s \rightarrow l\bar{\nu}D_s$ is measured to be (Fig. 3b):

$$\tau_s^{semi} = 1.42^{+0.27}_{-0.23}(\text{stat}) \pm 0.11(\text{syst})\text{ps}$$

Finally, there is a new low-statistics measurement of the B_s lifetime using fully reconstructed $B_s \rightarrow J/\psi\phi, J/\psi \rightarrow \mu^+\mu^-, \phi \rightarrow K^+K^-$ decays.[13] At least two of the four daughter tracks are required to be reconstructed in the SVX. Based on a sample of 10 events, the B_s lifetime using exclusive $B_s \rightarrow J/\psi\phi$ is measured to be:

$$\tau_s^{exc} = 1.74^{+0.90}_{-0.60}(\text{stat}) \pm 0.07(\text{syst})\text{ps} \qquad (1)$$

Table 3 shows a comparison of the latest B_s lifetime measurements at CDF and LEP.[14] Calculated similar to above, the world average value has only an 8% uncertainty.

4. Conclusions

The first measurements of the B_u, B_d, B_s meson lifetimes and the B_u/B_d lifetime ratio by CDF are comparable to the latest results from other experiments. With the installation of a rad-hard SVX and possibly x5 more data from the present Run Ib, the prospects for precision measurements of the B meson lifetimes in both inclusive and exclusive modes in the near future is very promising.

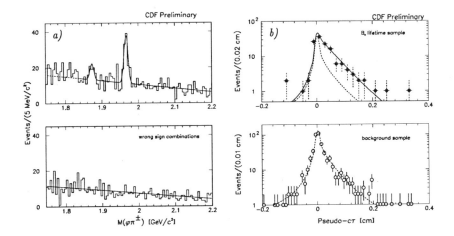

Fig. 3: a) The $\phi\pi^-$ mass distribution for "right sign" and "wrong sign" lepton-D_s combinations. b) Pseudo-$c\tau$ distribution of the $l^-D_s^+$ signal sample showing the lifetime fits of the combined (signal plus background) and background distributions separately.

References

1. F. Abe et al. (CDF Collaboration), *Nucl. Instrum. Methods Phys. Res., Sect. A* **271** (1988) 387, and references therein.
2. D. Amidei et al., preprint FERMILAB-PUB-94/024-E, submitted to *Nucl. Instrum. Methods Phys. Res.*
3. M. B. Voloshin and M. A. Shifman, *Sov. Phys. JETP* **64**, 698 (1986), and more recently V. Chernyak, preprint BudkerINP 94-69.
4. Throughout this paper, references to a specific charge state imply the charge-conjugate state as well.
5. F. Abe et al., *Phys. Rev. Lett.* **72** 3456 (1994).
6. CDF Coll., *CDF internal note 2598*.
7. ALEPH Coll., *Phys. Lett.* **B307**, 194 (1993); OPAL Coll., *OPAL internal note PN149*; DELPHI Coll., *Z. Phys.* **C57**, 181 (1993).
8. F. Abe et al., *Phys. Rev. Lett.* **71** 500 (1993).
9. CLEO Coll., *Phys. Rev.* **D43**, 651 (1991).
10. T. Hessing (DELPHI), private communication.
11. M. Saulnier (CLEO), private communication.
12. CDF Coll., *CDF internal note 2472*.
13. CDF Coll., *CDF internal note 2515*.
14. D. Karlen (OPAL), private communication.

LIFETIME RESOLUTION FOR BEAUTY PHYSICS AT HADRON COLLIDERS

CHRISTOPHER J. KENNEDY, ROBERT F. HARR, PAUL E. KARCHIN

Physics Department, Yale University, P.O. Box 208121,
New Haven, Connecticut 06520-8121

ABSTRACT

Monte Carlo and analytic techniques are used to estimate the lifetime resolution of $B_d^0 \rightarrow J/\psi K_s$, for model forward and central detectors at the Tevatron.

The large cross section for beauty production in hadron interactions leads to unique opportunities to perform precision b decay measurements. A hadronic B experiment will complement the capabilities of an e^+e^- B factory. This has prompted proposals for b-physics experiments at RHIC, the Tevatron, LHC, and SSC.

We report here the results of studies on the expected performance of detectors at the Tevatron. These studies examine the effect of detector configuration on the lifetime resolution for $B_d^0 \rightarrow J/\psi K_s$. We consider two different geometries, forward and central, representative of the large number of potential detector configurations. We use detailed models of vertexing and tracking detectors, which take into account spatial resolution and scattering material. We investigate the effects of varying the important vertex detector parameters of inner radius and spatial resolution.

We use PYTHIA to generate events and then analytically calculate the track error matrices. These matrices are used to calculate the primary and secondary vertex error matrices, from which we calculate the significance of detachment, $\tau/\delta\tau$. This technique gives an accurate estimate of the lifetime resolution without a time consuming hit level simulation, allowing a large number of detector parameters to be analyzed quickly.

The model forward detector is a single arm spectrometer. It contains a silicon vertex detector with 10 μm resolution and 10 planes spaced every 4 cm from 2 to 38 cm downstream from the interaction point. The downstream spectrometer has a geometric acceptance of $\eta > 1.5$ and a momentum resolution that is based on current fixed target experiments. The central detector model is based on an upgraded CDF detector. It includes the silicon vertex detector (SVX), the vertex time projection chamber, and the central tracking chamber. The SVX has resolution of 13 μm in r-ϕ and 30 μm in z.

The value of $\langle\tau/\delta\tau\rangle$ versus η_B is shown in Fig. 1 for both detectors. The forward detector has much more favorable values of $\langle\tau/\delta\tau\rangle$ over the entire range of η_B. This is due to a variety of reasons. The forward vertex detector has a smaller inner radius, and better resolution. There is less scattering material at high η_B due to the orientation of

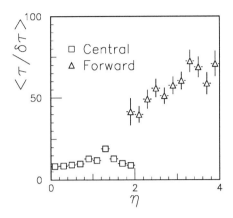

Figure 1: $\langle \tau/\delta\tau \rangle$ vs η_B for the forward and central detector models.

the detector planes and the momentum increases with η_B, so the multiple scattering contribution is reduced.

We study the effects of varying the inner radius and resolution of the two vertex detectors. The nominal value of inner radius, R_{min}, of the forward detector is 0.5 cm and of the central detector is 3.0 cm. Radii in this range are examined. The values of $\langle \tau/\delta\tau \rangle$ versus η_B for the central detector are presented in Fig. 2a. Four values of R_{min} are studied. There is a factor of two improvement in $\langle \tau/\delta\tau \rangle$ for a factor of six reduction in inner radius. The naive scaling law that $\langle \tau/\delta\tau \rangle$ is proportional to $1/R_{min}$ is not a good approximation in this range. There is little improvement in $\langle \tau/\delta\tau \rangle$ for $R_{min} < 1$ cm.

For the forward detector, $\langle \tau/\delta\tau \rangle$ vs η_B is presented in Fig. 2b. The values of R_{min} used are the same as in 2a. Both the average value and the shape of the distribution change with R_{min}. At low η_B ($2 < \eta_B < 2.5$) there is a factor of about 4 improvement in $\langle \tau/\delta\tau \rangle$ over this range of radii. For $\eta_B > 3.5$, there is a factor of 6 to 8 improvement in $\langle \tau/\delta\tau \rangle$ for a factor of 6 reduction in R_{min}. As with the central detector, little improvement is gained from reducing R_{min} below 1 cm.

For a central detector, the addition of stereo strips allows for three dimensional measurement of vertices. These stereo strips are usually at small angle or at $90°$ (z-strips). For this study, the r-ϕ resolution is held constant at 13 μm and the z-strip resolution is varied from 5 to 500 μm. The values of $\langle \tau/\delta\tau \rangle$ versus η_B for various z-strip resolutions at the Tevatron are shown in Fig. 3. The factor of 100 variation in resolution yields a factor of between 3 and 5 in $\langle \tau/\delta\tau \rangle$. There is little improvement in $\langle \tau/\delta\tau \rangle$ for improvement of z-strip resolution below 50 μm.

For the forward detector, the x and y resolutions are varied together from 5 to 30 μm. The values of $\langle \tau/\delta\tau \rangle$ vs η_B for these resolutions are presented in Fig. 3. There

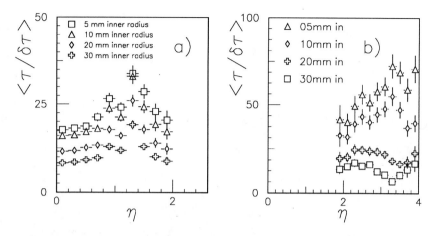

Figure 2: $\langle \tau/\delta\tau \rangle$ vs η_B for various inner radii for a) the central and b) the forward detector.

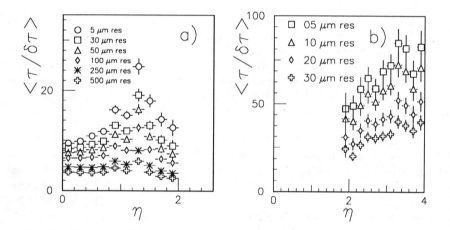

Figure 3: Same as Fig. 2 but varying detector resolution.

is a factor of 2 improvement in $\langle \tau/\delta\tau \rangle$ for the factor of 6 improvement in resolution. The shape of the distribution is similar for each resolution. There is only modest improvement in $\langle \tau/\delta\tau \rangle$ from reducing the resolution below 10 μm.

We thank the Texas National Research Laboratory Commission and the U.S. Department of Energy for financial support.

RECENT PROGRESS IN THE THEORY OF HEAVY QUARK PRODUCTION IN HADRONIC COLLISIONS

*M.L. MANGANO[a], S. FRIXIONE[b], P. NASON[c] and G. RIDOLFI[c]

[a] INFN Pisa, [b] INFN Genova, [c] CERN

ABSTRACT

We review heavy quark production in high energy hadronic collisions. We discuss the status of the theoretical uncertainties and compare the most recent measurements from the Tevatron Collider experiments with the results of next-to-leading order QCD calculations for single and double inclusive distributions.

1. Introduction

Heavy quark production in high energy hadronic collisions consitutes a benchmark process for the study of perturbative QCD. The comparison of experimental data with the predictions of QCD provides a necessary check that the ingredients entering the evaluation of hadronic processes (partonic distribution functions and higher order corrections) are under control and can be used to evaluate the rates for more exotic phenomena or to extrapolate the calculations to even higher energies. The estimates of production rates for the *top* quark rely on the understanding of heavy quark production properties within QCD.

In this presentation we review the current status of theoretical calculations, and discuss the implications of the most recent experimental measurements of single and double inclusive distributions of b quarks performed at the Tevatron $p\bar{p}$ Collider. For a more complete review, including a discussion of heavy quark production at fixed target energies, see Ref. 1.

2. Theory Overview

To start with, we briefly report on the current status of the theoretical calculations. One has to distinguish between calculations performed at a complete but fixed order in perturbation theory (PT), and those performed resumming classes of potentially large logarithmic contributions which arise at any order in PT. The exact matrix elements squared for heavy quark production in hadronic collisions are fully known up to the $\mathcal{O}(\alpha_s^3)$, both for real and virtual processes. These matrix elements have been used to evaluate at NLO the total production cross section,[2] single particle inclusive distributions[3] and two particle inclusive distributions (a.k.a. correlations).[4]

Three classes of large logarithms can appear in the perturbative expansion for heavy quark production:

---

*Presenting author.

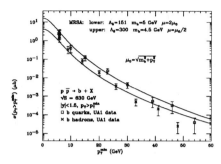

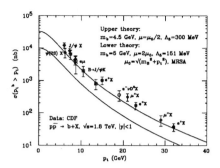

Fig. 1. Bottom cross sections at UA1.   Fig. 2. Bottom cross sections at CDF.

1. $[\alpha_s \log(S/m_Q^2)]^n \sim [\alpha_s \log(1/x_{Bj})]^n$ terms, where S is the hadronic CM energy squared. These small x effects are possibly relevant for the production of charm or bottom quarks at the current energies, while should have no effect on the determination of the *top* cross section, given the large t mass. Several theoretical studies have been performed,[5] and the indications are that b production cross sections should not increase by more than 30-50% at Tevatron energies due to these effects.

2. $[\alpha_s \log(m_Q/p_{QQ}^T)]^n$ terms, where p_{QQ}^T is the transverse momentum of the heavy quark pair. These contributions come from the multiple emission of initial state soft gluons, similarly to standard Drell Yan corrections. These corrections have been studied in detail in the case of top production, where the effect is potentially large due to the heavy top mass.[6] They are not relevant for the redefinition of the total cross section of b quarks, but affect the kinematical distributions of pairs produced just above threshold,[7] or in regions at the edge of phase space, such as $\Delta\phi = \pi$.

3. $[\alpha_s \log(p_T/m_Q)]^n$ terms, where p_T is the transverse momentum of the heavy quark. These terms arise from multiple collinear gluons emitted by a heavy quark produced at large transverse momentum, or from almost collinear branching of gluons into heavy quark pairs. Again these corrections are not expected to affect the total production rates, but will contribute to the large p_T distributions of c and b quarks. No effect is expected for the top at current energies. These logarithms can be resummed using a fragmentation function formalism. A first step in this direction was taken by Cacciari and Greco,[8] who convoluted the NLO fragmentation functions for heavy quarks[9] with the NLO parton level cross section for production of massless partons.[10] A significant improvement in the stability w.r.t. scale changes has been observed for $p_T > 50$ GeV.

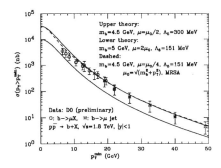

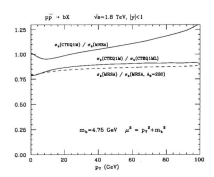

Fig. 3. Bottom cross sections at D0.

Fig. 4. Dependence of predicted bottom cross section on different choices of structure functions and values of Λ_5. See the text for a complete discussion.

3. Single Inclusive Bottom Production

The status of b production at hadron colliders has been quite puzzling for some time. Data collected by UA1[11] at the CERN Collider ($\sqrt{S}$=630 GeV) were in good agreement with theoretical expectations based on the NLO QCD calculations[3] (see Fig.1). On the contrary, the first measurements performed at 1.8 TeV by the CDF[12] experiment at the Fermilab Collider showed a significant discrepancy with the same calculation.

Owing to recent progress, the situation has considerably clarified. The latest results from the Fermilab 1.8 TeV $p\bar{p}$ Collider have been presented at this Conference by CDF[13] and by the new experiment, D0.[14] The current situation is summarized in Figs. 2 and 3, showing a comparison of the theoretical expectations with the results from CDF and D0 for integrated p_T distributions of b quarks.

The theoretical curves require some explanation. First of all, they do not differ much from the original prediction[3] using the DFLM structure functions. New structure function fits, including the first results from HERA, have recently become available. We use in our prediction one of these sets, namely MRSA.[15] Since the values of x probed by b production at the Tevatron in the currently measured p_T range only cover the region $x > 5 \times 10^{-3}$, we observe no significant change relative to the results obtained using older fits.

The second important point is the choice of a range for Λ_5. Deep inelastic scattering results tend to favour smaller values of Λ. For example, the set MRSA uses $\Lambda_5 = 151$ MeV. On the other hand, LEP data favour a higher value: the central value of Λ_5 at LEP is around 300 MeV. This value is also supported by other lower-energy results, such as the τ hadronic width (for a review of Λ_5 determinations, see the work by Catani[16]). It is therefore sensible to use the range from 151 to 300 MeV for Λ_5.

The upper curves in Fig.2 and Fig.3 correspond to the PDF set MRSA,[15] $\Lambda_5 = 300$ MeV, $m_b = 4.5$ GeV and $\mu_R = \mu_F = \mu_0/2$ ($\mu_0 = \sqrt{p_T^2 + m_b^2}$). The lower curves correspond to $\Lambda_5 = 151$ MeV, $m_b = 5$ GeV and $\mu_R = \mu_F = 2\mu_0$. In the absence of fits with Λ_5 frozen to the desired values we chose to simply change the value of Λ_5 in the partonic cross section. A priori, one would expect that this amounts to an overestimate of the variation due to the uncertainty in Λ_5. As a test, we used the structure-function fits performed by the CTEQ collaboration,[17] which presents results both for the best DIS fit of Λ_5 ($\Lambda_5 = 152$ MeV, set CTEQ1M) and for a value artificially frozen to a number closer to the LEP measurement (set CTEQ1ML, $\Lambda_5 = 220$ MeV). In Fig.4 we plot the ratio of the b cross sections obtained using the two sets CTEQ1M and CTEQ1ML, together with the ratio between the result of the default MRSA set and of MRSA with Λ_5 fixed at 220 MeV. We used $m_b = 4.75$ GeV and $\mu_R = \mu_F = \mu_0$. As is clear from the figure, the two results are equal to within a few per cent, indicating that varying Λ_5 within a limited range, without refitting the parton distributions, is a reasonable way to estimate the effect of the Λ_5 uncertainty on the b cross section.

In Fig.3 we also show as a dashed line the theoretical prediction obtained using a rather extreme choice of scale, namely $\mu_R = \mu_F = \mu_0/4$, toghether with $m_b = 4.5$ GeV and MRSA PDF ($\Lambda_5 = 151$ MeV). It is interesting to notice that this choice results in rates almost equal to those obtained using $\mu_R = \mu_F = \mu_0/2$ and the extreme value of $\Lambda_5 = 300$ MeV. This suggests that indeed most of the variation in the cross section is simply due to the changes in α_s, as $\alpha_s = \alpha_s(\mu_R/\Lambda)$; the argument is the same for the two choices $\mu_R = \mu_0/2$, $\Lambda_5 = 300$ MeV, and $\mu_R = \mu_0/4$, $\Lambda_5 = 151$ MeV.

For completeness, we include in Fig.4 the ratio between the predictions obtained using the two default sets, CTEQ1M and MRSA. This number is equal to 1 to within 5% for a large range of p_T. While such a stability is partly artificial, being related to the large overlap of correlated measurements entering the determination of the parton distribution fits, it however suggests that by now the uncertainty in the structure functions does not leave much room by itself for significant changes in the expected b cross section at Tevatron energies.

Coming back to the comparison of theory and data, from Fig.2 we see that the CDF data points are now consistent with the fixed-order theoretical prediction, although on the high side. The D0 points, instead, comfortably sit within the theoretical range. In order to better compare data among themselves and with theory, we plot the ratio between data points and the upper theoretical predition on a linear scale (Fig.5). From this figure we see that the UA1 and D0 data are well consistent with the upper theoretical curve, while CDF points are slightly above. Until the apparent discrepancy between D0 and CDF will be resolved, it is therefore appropriate to conclude that at present no significant discrepancy between theory and data or between data at different energies is being observed. Once the experimental statistics and systematics will be further reduced, it will be reasonable to assume that residual discrepancies of the same order as those currently observed may be explained in terms of small-x effects. Additional theoretical studies of these effects, such as a better understanding of the matching with the fixed-order next-to-leading-order calculations, should therefore be pursued.

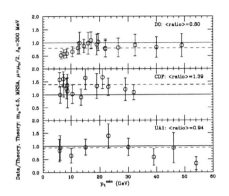

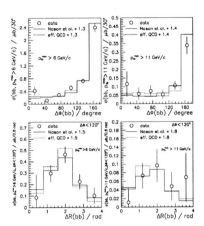

Fig. 5. Ratio of data and theory for the integrated b p_T distribution at UA1, CDF and D0.

Fig. 6. Bottom pair correlations at UA1.

4. Correlations and Heavy Quark Jets

Another important test of the dynamics of heavy quark production is the study of correlations. Aside from providing interesting tests of QCD, spatial and p_T correlations between b and $\bar{b}$ are also important to establish relevant detector parameters, such as acceptances and tagging efficiencies, that will play a critical role in the design of future hadron collider experiments aimed at the measurement of CP violation in the b system. At the leading order in PT heavy quarks are produced back-to-back in ϕ and with equal p_T. Radiative corrections modify this topology, via emission of radiation both from the initial and the final state. Measurements of the $\Delta\phi = |\phi(b) - \phi(\bar{b})|$ distribution have been reported by UA1[18] and more recently by CDF.[19] We include them here, compared with the available NLO QCD predictions.[4] The agreement, within the large errors, is good.

From the previous plots, one is tempted to conclude that the largest fraction of b and $\bar{b}$ pairs are produced back-to-back, as expected from the leading order production mechanisms. This observation seems to contradict the common belief that higher order processes, and in particular the so-called final state gluon splitting mechanism which would lead to b pairs very close in ϕ, dominate b production. The solution to this apparent puzzle, which I will discuss briefly now, will introduce us to the last item of this presentation, namely the discussion of b-jets.

The sample of events used for the measurement of the $b\bar{b}$ correlations in Fig.6-7 is triggered by the presence of a large p_T lepton. This is equivalent to selecting events with a b quark above a given p_T threshold. This can be contrasted with another possible choice, which is to select events with a b-jet above a given E_T threshold. A b-jet can be defined as a standard calorimetric jet containing a b in it. Fig.8 shows the predicted $\Delta\phi$ correlations obtained using the two different selection criteria (b-p_T or b-jet E_T above a

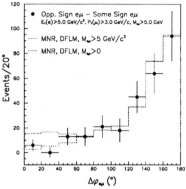

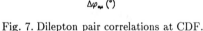

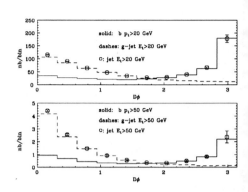

Fig. 7. Dilepton pair correlations at CDF.

Fig. 8. Azimuthal correlations between b pairs, as a function of different triggers.

threshold), for two different thresholds, 20 and 50 GeV. The calculation was performed using the double differential NLO cross section,[4] which allows to define b-jets using the standard cone algorithms applied to inclusive jets in hadronic collisions. The difference between the two shapes can be easily understood in terms of the production dynamics and of differences in the available phase space. The peak which develops at $\Delta\phi = 0$ indicates that the presence of a high E_T b-jet in the event enhances the contribution from gluon splitting processes. The results indicate that different mechanisms will dominate production of b-hadrons at a given E_T and production of b-flavoured jets of the same E_T.

5. Conclusions

Significant progress has taken place in this field over the past few years, both in theory and experiments. The latest measurements at 1.8 TeV indicate an acceptable agreement between the data and NLO QCD predictions for the b inclusive p_T spectrum, and the presence now of two experiments will hopefully reduce experimental uncertanties. Previously detected discrepancies, observed in the inclusive ψ final states, are now attributed to large sources of ψ direct production. The measurement of $b\text{-}\bar{b}$ correlations shows good agreement between data and theory, and suggests extending the measurements to events selected according to the presence of a b-jet.

Acknowledgements The work of MLM is supported in part by the EEC Programme "Human Capital and Mobility", Network "Physics at High Energy Colliders", contract CHRX-CT93-0357 (DG 12 COMA).

References

1. S. Frixione, M. Mangano, P. Nason and G. Ridolfi, CERN-TH.7292/94 (1994).
2. P. Nason, S. Dawson and R. K. Ellis, *Nucl. Phys.* **B303** (1988), 607; W. Beenakker, H. Kuijf, W.L. van Neerven and J. Smith, *Phys. Rev.* **D40** (1989), 54.
3. P. Nason, S. Dawson and R. K. Ellis, *Nucl. Phys.* **B327** (1988), 49 ; W. Beenakker et al., *Nucl. Phys.* **B351** (1991), 507.
4. M. Mangano, P. Nason and G. Ridolfi, *Nucl. Phys.* **B373** (1992), 295.
5. J.C. Collins and R.K. Ellis, *Nucl. Phys.* **B360** (1991), 3; S. Catani, M. Ciafaloni and F. Hautmann, *Nucl. Phys.* **B366** (1991), 135; E.M. Levin, M.G. Ryskin and Yu.M. Shabelsky, *Phys. Lett.* **260B** (1991), 429.
6. E. Laenen, J. Smith and W.L. van Neerven, *Nucl. Phys.* **B369** (1992), 543.
7. E. Berger and R. Meng, *Phys. Rev.* **D49** (1994), 3248.
8. M. Cacciari and M. Greco, Univ. of Pavia FNT/T-93/43, hep-ph/9311260.
9. B. Mele and P. Nason, *Nucl. Phys.* **B361** (1991), 626
10. F. Aversa et al., *Phys. Lett.* **210B** (1988), 225; *Z. Phys.* **C49** (459), 1991; S. Ellis, Z. Kunszt, D. Soper, *Phys. Rev. Lett.* **62** (1989), 2188; **64** (1990), 2121.
11. C. Albajar et al., UA1 Coll., *Phys. Lett.* **256B** (1991), 121.
12. F. Abe et al., CDF Coll., *Phys. Rev. Lett.* **68** (1992), 3403; **69**(1992)3704; **71**(1993)500, 2396 and 2537.
13. I. Yu, J.D. Lewis, M.W. Bailey and D. Gerdes, CDF Collaboration, presented at this Conference.
14. G. Alves and T. Huehn, D0 Collaboration, presented at this Conference.
15. A.D. Martin, W.J. Stirling and R.G. Roberts, Rutherford Lab preprint RAL-94-055, DTP/94/34 (1994).
16. S. Catani, preprint DFF 194/11/93, to appear in the *Proceedings of the EPS conference*, Marseille, 1993.
17. J. Botts et al., *Phys. Lett.* **304B** (1993), 159.
18. C. Albajar et al., UA1 Coll., *Z. Phys.* **C61** (1994), 41.
19. F. Abe et al., CDF Coll., Fermilab-Pub-94/131-E; Fermilab-Conf-94/129-E.

A MEASUREMENT OF THE $b\bar{b}$ CROSS SECTION AT CDF

I. YU*

Physics Department, Yale University,
New Haven, CT, 06520

ABSTRACT

We report on a measurement of the $b\bar{b}$ cross section at CDF from the 1992-1993 run of the Tevatron Collider. Dimuon events from inclusive $b \to \mu$ decays of $b\bar{b}$ pairs are used to obtain the cross section as a function of $P_T(b_1)$ and $P_T(b_2)$. The results are compared to the predictions of next-to-leading order QCD and are found to be consistent.

1. Introduction

In $p\bar{p}$ collisions at $\sqrt{s} = 1.8$ TeV, the strong coupling constant α_s becomes small for heavy quark production processes and perturbative QCD can provide reliable predictions. Studies of $b\bar{b}$ production provide an opportunity to check pertubative QCD at next-to-leading order.[1] In these proceedings we present a preliminary measurement of the $b\bar{b}$ cross section as a function of $P_T(b_1)$ and $P_T(b_2)$ using dimuon data with an integrated luminosity of $16.7 \pm 0.60pb^{-1}$.

2. Data selection

Dimuon events are studied using inclusive $b \to \mu$ decays of $b\bar{b}$ pairs. At CDF, the muon identification is done by associating a track in the central tracking chamber (CTC) with a track in the central muon chamber (CMU).[2] The trigger requires at least 2 muons in the central region ($|\eta| < 0.6$). We also require P_T to be greater than 3 GeV/c for both muons.

The dimuon data come from $b\bar{b}$ production, $c\bar{c}$ production, Drell Yan, J/ψ, Υ, and $fakes$, where $fakes$ are due to hadronic punchthroughs or decay muons from pions and kaons. The cascade decays of single b quarks and J/ψ dimuons are removed by requiring the dimuon invariant mass to be greater than 5 GeV/c^2. The $b\bar{b}$ production produces like-sign (LS) dimuons through $B^0\bar{B}^0$ mixing and the semileptonic decay from the daughter charm quark as well as opposite sign (OS) dimuons. $Fake$ dimuon events with at least one $fake$ contribute equally to LS and OS dimuons as there is no sign correlation between muons. The dimuons from all the other sources can be removed by requiring the sign of dimuons to be of like-sign as they only generate OS dimuons. The $fake$ background fraction may be different in $\mu^-\mu^-$ and $\mu^+\mu^+$ sample and they must be treated seperately. In this analysis we use $\mu^-\mu^-$ events.

---

*Representing the CDF Collaboration.

3. The Method

After the central muon chamber (CMU), there is a steel absorber followed by the central muon upgrade chamber (CMP).[2] The $b\bar{b}$ fraction in LS dimuon events is determined by measuring the CMP efficiency, the probability of muons having additional stubs in the CMP chamber. Most of the $b\bar{b}$ muons or decay muons from pions or kaons travel to the CMP while most of the hardronic punchthroughs are absorbed within the steel. With the CMP efficiency ϵ_μ for the $b\bar{b}$ muons and the CMP efficiency ϵ_f for $fakes$, we contruct three equations for the three different types of dimuon events (CMU-CMU, CMU-CMP, and CMP-CMP).

$$N_{CMUCMU} = M + F_1 + F_2 \tag{1}$$

$$N_{CMUCMP} = 2\epsilon_\mu(1 - \epsilon_\mu)M + \{\epsilon_\mu(1 - \epsilon_f) + \epsilon_f(1 - \epsilon_\mu)\}F_1 + 2\epsilon_f(1 - \epsilon_f)F_2 \tag{2}$$

$$N_{CMPCMP} = \epsilon_\mu^2 M + \epsilon_\mu\epsilon_f F_1 + \epsilon_f^2 F_2 \tag{3}$$

where the N's are the number of dimuon events of each type from the data and the subscript of N denotes the type of dimuon event. The number of the $b\bar{b}$ dimuon events is represented by M and the number of $fake$ dimuon events by F. The subscript of F denotes the number of $fakes$ in a event.

The CMP efficiency ϵ_μ is measured to be 0.94 ± 0.01 using J/ψ dimuons. The CMP efficiency ϵ_f is bounded from above analytically (0.53 ± 0.04) from those equations given the numbers available. The punchthrough probability[3] of a K^- is almost equal to that of a π^- while the decay probability of a K^- to a CMU muon is greater than that of a π^- to a CMU muon. We set a lower limit on ϵ_f by determining the CMP efficiency of $fakes$ from negative pions as the fraction of decay muons in π^- $fakes$ is higher. This efficiency is measured to be 0.49 ± 0.04 from K_S^0 sample reconstructed with a negatively charged muon track and a positively charged track. From the equations and the CMP efficiencies, we calculate the number of $b\bar{b}$ dimuons, M, for different P_T thresholds for the second muon (3GeV/c, 4GeV/c, and 5GeV/c).

4. $b\bar{b}$ Cross Section

To obtain the $b\bar{b}$ cross section, the number of the $b\bar{b}$ dimuon events $N_{b\bar{b}}$ is divided by the dimuon reconstruction efficiency ϵ, the acceptance A, the branching ratio of the dual semimuonic decay of a $b\bar{b}$ pair,[3] and the integrated luminosity L.

$$\sigma^{b\bar{b}}(P_T(b_1) > P_T^{min,1}, P_T(b_2) > P_T^{min,2}, |y_{1,2}^b| < 1) = \frac{N_{b\bar{b}}}{\epsilon \cdot A \cdot Br(b\bar{b} \to \mu_1\mu_2 X) \cdot L} \tag{4}$$

The dimuon reconstruction efficiency is measured using J/ψ muons. The acceptance is determined using the $b\bar{b}$ generator based on next-to-leading order QCD calculations[1] and the CDF detector simulation. The b quark P_T^{min} is chosen such that 90% of the muons with $P_T > P_T^{thres}$ also have $P_T^b > P_T^{min}$. The corresponding P_T^{min} values are 6.5, 7.5, 8.75 GeV/c for $P_T^{thres} = 3.0, 4.0, 5.0$ GeV/c. We also use the value of $B^0\bar{B}^0$ mixing parameter from CDF 1988-89 measurement[4] for the acceptance calculations.

Table 1 shows the measurements of the $b\bar{b}$ cross section. The preliminary results are in good agreement with the CDF measurement using $e\mu$ data[5] and consistent with next-to-leading order (NLO) QCD predictions as shown in Figure 1a.

Table 1. $b\bar{b}$ cross section

$P_T^{min,1}$	$P_T^{min,2}$	$\sigma^{b\bar{b}}$
6.5 GeV/c	6.5 GeV/c	$2.00 \pm 0.38(stat) \pm 0.37(sys)$ μb
6.5 GeV/c	7.5 GeV/c	$1.71 \pm 0.42(stat) \pm 0.31(sys)$ μb
6.5 GeV/c	8.75 GeV/c	$1.20 \pm 0.42(stat) \pm 0.24(sys)$ μb

We also investigate the P_T correlations between the $b\bar{b}$ muons. Figure 1b shows the differential cross section of the observed dimuons as a function of $P_T(\mu_2)$ for $P_T(\mu_1) > 3$ GeV/c. We use the same method as in the $b\bar{b}$ cross section measurements to determine the number of $b\bar{b}$ dimuons for each P_T bin. The result is consistent with the prediction from NLO QCD and the detector simulation (CDFSIM).

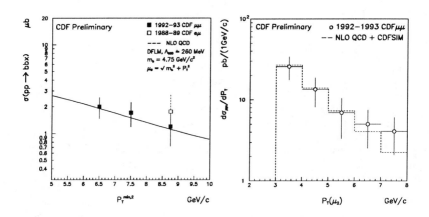

Fig. 1. a) $\sigma^{b\bar{b}}(P_T(b_1) > 6.5 GeV/c, P_T(b_2) > P_T^{min,2}, |y_{1,2}^b| < 1)$ and b) $d\sigma_{\mu\mu}/dP_T(\mu_2)$ for $P_T(\mu_1) > 3$ GeV/c.

References

1. M. Mangano *et al.*, *Nucl.Phys.* B373 (1992).
2. F Abe *et al.*, *Nucl. Inst. and Meth.* **A271** (1988) 387 and references therein.
3. *Review of Particle Properties*, *Phys.Rev.* D45 (1992).
4. F. Abe *et al.*, *Phys. Rev. Lett.* <u>67</u>, 3351 (1991).
5. F. Abe *et al.*, submitted to *Physical Review Letters*.

MEASUREMENT OF THE B CROSS SECTION AT CDF VIA B SEMILEPTONIC DECAYS

JONATHAN LEWIS*

Fermi National Accelerator Laboratory
Box 500, Batavia, IL   60510, USA

ABSTRACT

Using data collected during the 1992-1993 collider run at Fermilab, CDF has reconstructed several hundred charmed mesons (D^0, D^+, D^{*+} and D_s) in association with leptons from B semileptonic decays. We report on a measurement of the cross section of B and B_s mesons as a function of transverse momentum using this sample. The observation of a charmed meson eliminates many systematic uncertainties in the background subtraction inherent in previous measurements from inclusive lepton samples, and allows the backgrounds to be measured from the data. The B meson p_T range probed by the lepton+charm technique is 18 GeV and above, and thus these measurements complement similar measurements at lower p_T in the fully exclusive channels $B \to J/\psi K$ and $B \to J/\psi K^*$. Results are compared to other Tevatron measurements and Next-To-Leading-Order QCD predictions.

1. Introduction

One source of leptons in $\bar{p}p$ interactions is semileptonic B decays. The requirement of an identified charm particle in the final state provides a narrow peak signature for B's and low backgrounds. Further, backgrounds can be constrained from particle combinations inconsistent with coming from a B decay. This paper describes a measurement of B and B_s production using 17.9 ± 0.6 pb^{-1} of $\bar{p}p$ collisions at $\sqrt{s} = 1.8$ TeV collected with the CDF detector between August 1992 and May 1993. The B cross section is measured from the number of D^0's[†] and D^{*+}'s reconstructed in the inclusive muon sample. The B_s fraction is measured from the ratio of D_s to D^+ events in a combined electron and muon sample. The CDF detector is described in detail elsewhere[1] Muons are identified[2] as central tracking chamber (CTC) tracks that extrapolate to track segments in both the central muon detector (CMU) located behind the central calorimeter the central muon upgrade chambers (CMP) behind an additional 60 cm of steel. The combined efficiency for muons in the geometric acceptance of the detector is $> 95\%$ The electrons are identified by CTC tracks that extrapolate to electromagnetic showers in the central calorimeter. They must pass requirements on shower shape and

---

*Representing the CDF Collaboration

[†]Reference to any particular state implies the charge conjugate state as well.

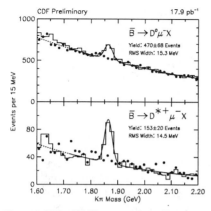

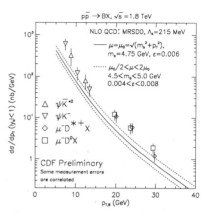

Figure 1: The solid histogram shows $K^-\pi^+$ mass for $D^0\mu^-$ and $D^{*+}\mu^-$ candidates. The dotted histograms are the background distributions described in the text. The curve indicates the results of the fit.

Figure 2: B meson cross sections determined from $D\mu$ and $J/\psi K$ yields. Curves show predictions of next-to-leading-order QCD.

depth profile and on the matching of CTC tracks to clusters in the wire chambers located near shower maximum in the calorimeter.[3]

2. B Cross section with $\mu + D^0$ and $\mu + D^{*+}$

In events with identified muons with $p_T > 7.5$ GeV, we assign to charged particle tracks the kaon and pion masses and compute $(K^-\pi^+)$ and $(K^-\pi^+\pi^+)$ invariant masses. A combination with a $K\pi$ mass between 1.55 and 2.25 GeV and a $K\pi\mu$ mass less than 5.3 GeV is considered a D^0 candidate, and the mass is recalculated applying the constraint that the K and π originate from a common point. If difference in mass of the $(K^-\pi^+\pi^+)$ and $(K^-\pi^+)$ systems is less than 153 MeV, the event is identified as containing a D^{*+} candidate as well.

To obtain a $B \rightarrow \mu D^0$ sample, we make the following requirements on the $K\pi$ combination: the charge is consistent with B decay; $\min(p_T(K), p_T(\pi)) \geq 1.5$ GeV; $\max(p_T(K), p_T(\pi)) \geq 3.0$ GeV; $m(\mu K\pi) \leq 5.3$ GeV; and $|\cos\theta^*| \leq 0.8$, where θ^* is the angle between the kaon and the muon in the $K\pi$ rest frame. In the μD^0 search, we do not reject identified D^{*+}'s: this is a measurement of the B cross section via the inclusive decay $B \rightarrow \mu D^0 X$.

We make a separate search for $B \rightarrow \mu D^{*+} X$, where $D^{*+} \rightarrow D^0\pi^+$, applying the D^{*+} pion tag. The $K\pi$ signal to noise ratio is much larger, resulting in comparable statistical uncertainties despite the smaller number of events. We apply the same requirements to the $K\pi$ combination as above with the $\cos\theta^*$ cut removed and the additional requirement that $p_T \geq 450$ MeV for the soft pion from the D^{*+} decay. This sample is not statistically independent of the D^0 sample: 80% of the events in the D^{*+}

sample are also included in the D^0 sample. The D^0 mass distributions with the cuts applied are shown in Fig. 1.

Because $B \to \mu D$ decays always include an undetected particles, we must estimate the B meson p_T from kinematic properties of detected particles. We parameterize this correction in terms of m_{visible} and $p_{T,\text{visible}}$, the mass and transverse momentum of the lepton$+D^0$ or D^* system. The correction is applied in two steps. We multiply $p_{T,\text{visible}}$ by a fourth-order polynomial in the missing mass fraction, $(m_B - m_{\text{visible}})/m_B$. We correct the intermediate value, p'_T, to arrive at the estimate of the B momentum: $p_T = b_1 p'_T/(1 - b_2 p'_T)$. The parameters of the correction functions are obtained by comparing the reconstructed m_{visible} and $p_{T,\text{visible}}$ from a sample of two million Monte Carlo events with the generated B meson p_T. Because we can determine the B meson momentum only to about 15% on an event-by-event basis, the acceptance has some sensitivity to the generated B meson spectrum. Replacing the generated spectrum with one that is a factor two smaller for every 4 GeV increase in p_T changes the measured cross section changes by less than 4% per bin. We also use the Monte Carlo sample to calculate the reconstruction efficiency and geometric acceptance.

We fit the $K\pi$ mass spectra in the D^0 and D^* searches separately to a Gaussian on a quadratic background. The mean of the Gaussian is fixed to the known D^0 mass[4]. A background $K\pi$ sample is simultaneously fitted. For the D^0 search the wrong lepton charge is chosen, and for the D^* search wrong-charge leptons and soft pions are used so that both signal and background distributions are for neutral objects. The backgrounds from the right- and wrong-charge distributions are constrained to be the same, and the width of the peak in the wrong sign distribution is determined from width of the D^0 in the Monte Carlo sample when the K and π assignments are reversed. We exclude the region below 1.72 GeV to avoid the D^0 decay to $K^-\pi^+\pi^0$. We then divide the sample into three bins of B-meson p_T and repeat the process with the widths of the Gaussians constrained to the value returned by the Monte Carlo, scaled by the ratio of the inclusive D^0 width to the Monte Carlo prediction.

The sources of systematic uncertainty are listed in Table 1. The numbers of reconstructed D^0's and the single-species B cross sections derived from these numbers are shown in Table 2. We assume that $\sigma(B^0) = \sigma(B^+)$. We reduce the number of D^* tagged D^0's by $6 \pm 6\%$ to account for the possibility of accidental tags: finding a π that is not from a D^* that happens to pass all the D^* tagging requirements. This value comes from counting the number of D^* tags in a different mass difference window, one that has the same number of pions as in the D^* mass window.

Source of Uncertainty	$D^0\mu^-$	$D^{*+}\mu^-$
Luminosity	4 %	4 %
Acceptance	6 %	6 %
Tracking efficiency	2 %	7 %
Trigger efficiency	3 %	3 %
K,π decays in flight	2 %	2 %
B p_T resolution	4 %	4 %
Fitting method and background shape	8 %	8 %
Br($B \to \mu X$)	10 %	10 %
Br($B \to \mu D^{**}$)	2 %	2 %
Br($D^{*+} \to \pi^+ D$)		2 %
Br($D \to K\pi$)	6 %	6 %
Total	17 %	18 %

Table 1: Systematic errors in measurement of B cross section.

The results are compared in Fig. 2 to the lower p_T CDF measurements from $B^+ \to J/\psi K^+$ and $B^0 \to J\psi K^{*0}$ decays as well as the Next-to-Leading Order QCD prediction[5] using the MRSD0 parton distribution functions[6] and Peterson fragmentation[7] using $\epsilon = 0.006 \pm 0.002$.[8]

3. B_s Fraction

The same technique could be used to measure the B_s cross section by reconstructing D_s mesons produced in association with muons. Because the B_s production rate is smaller than the non-strange B production rate, and because the combined branching ratio $\mathrm{Br}(D_s \to \phi\pi) \times \mathrm{Br}(\phi \to K^+K^-)$ is smaller than $\mathrm{Br}(D^0 \to K\pi)$, there are not nearly as many reconstructed μD_s events as μD^0 events. We expand the sample to include electrons as well and apply rather tight isolation cuts on the lepton to improve the signal to noise ratio. In the $\phi\pi$ spectrum, the D_s and D^+ peaks are visible. By measuring the ratio of observed D_s's (coming predominantly from B_s decays) to D^+'s (coming predominantly from B^- or $\overline{B}^0$ decays), we can extract the ratio of produced B_s's relative to B^0 or B^+'s.

We relax the lepton p_T cut to 6 GeV and require that the leptons originate within 30 cm of the center of the detector along the beamline. We consider tracks within a cone of $R(\equiv \sqrt{(\Delta\phi)^2 + (\Delta\eta)^2}) < 0.8$ around the lepton. Two tracks from oppositely charged particles are assigned the kaon mass, and if their invariant mass is within ± 8 MeV of the ϕ mass of 1019 MeV, the combination is classified as a ϕ candidate and is combined with another track that is assigned the pion mass. The three-track combination is fitted with the constraint that all three tracks intersect at a common point, requiring the probability $P(\chi^2) \geq 1\%$. We require candidates to satisfy the following conditions: $p_T(K) \geq 1$ GeV; $p_T(\pi) \geq 800$ MeV; $p_T(\phi) \geq 2$ GeV; the lepton is isolated such that the transverse energy in a cone of $R \leq 0.4$ around the lepton does not exceed the $\phi\pi$ transverse momentum by more than 20%; and the displacement of the $\phi\pi$ vertex projected along the momentum direction is positive. Figure 3 shows the $\phi\pi$ mass distribution where the associated lepton had a charge consistent with B decay, (i.e. $B_s^0 \to D_s^{(*)-}\ell^+\nu$). Peaks from both $D_s^+ \to \phi\pi^+$ and $D^+ \to \phi\pi^+$ are visible. We fit the ratio of the number of D_s mesons observed to the number of D^+'s and find it to be $3.5^{+2.3}_{-1.1}$ for the combined muons and electron sample, where the ratio of the widths of the two peaks has been determined in Monte Carlo simulations.

p_T range	D^0		D^{*+}	
	Number of Events	$d\sigma/dp_T$ ($|y| < 1$) (nb/GeV)	Number of Events	$d\sigma/dp_T$ ($|y| < 1$) (nb/GeV)
All	459 ± 69		153 ± 20	
18-22 GeV	112 ± 26	$12.1 \pm 2.8 \pm 2.0$	39 ± 9	$10.8 \pm 2.6 \pm 1.9$
22-26 GeV	98 ± 22	$5.7 \pm 1.3 \pm 1.0$	43 ± 8	$5.9 \pm 1.2 \pm 1.1$
26-34 GeV	98 ± 23	$2.0 \pm 0.5 \pm 0.3$	27 ± 7	$1.2 \pm 0.3 \pm 0.2$

Table 2: $D\mu$ yields and B cross sections.

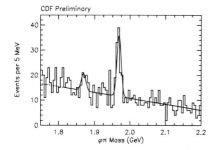

Figure 3: The solid histogram shows $\phi\pi^+$ mass for $D^+\mu^-$ and $D_s^+\mu^-$ candidates. The curve indicates the results of the fit.

Source of Uncertainty	Value
$\mathrm{Br}(\overline{B} \to D, D^*l^- X$	10 %
$\mathrm{Br}(\overline{B} \to D^{**}l^- X$	10 %
$\mathrm{Br}(D_s^+ \to \phi\pi^+$	18 %
$\mathrm{Br}(D^+ \to \phi\pi^+$	13 %
Relative acceptance and efficiency for D_s^+ and D^+	10 %
$\tau(B_s)/\tau(B_{u,d})$	12 %
Total	31 %

Table 3: Fractional uncertainties in B_s^0 production fraction.

The relationship between the cross section and the number of reconstructed D^+ and D_s^+ mesons is given by:

$$\frac{\sigma(B_s)}{\sigma(B_{u,d})} = \frac{\Gamma(B_{u,d} \to D^+ lX)}{\Gamma(B_s \to D_s lX)} \frac{\tau(B_{u,d})}{\tau(B_s)} \frac{BR(D^+ \to \phi\pi)}{BR(D_s \to \phi\pi)} \frac{N(D_s \to \phi\pi)}{N(D^+ \to \phi\pi)}$$

where the B semileptonic branching fractions have been replaced by partial widths and lifetimes. The acceptance and reconstruction efficiency of the two states is equal to within 10%. We assume that the semileptonic partial widths for all B mesons are the same and that the lifetimes are equal to within 12%. Additionally, we assume that the D^{**} fraction is the same ($20 \pm 10\%$) in all decays and that all D_s^{**}'s decay to DK. From these assumptions and the branching fractions $D^+ \to \phi\pi^+$ of $0.57 \pm 0.11\%$[4] and $D_s \to \phi\pi^+$ of $5.1 \pm 0.9\%$,[9] we can calculate the relative fraction of B_s mesons: $\sigma(B_s)/(\sigma(B_u) + \sigma(B_d)) = .26^{+.17}_{-.08} \pm .08$ The systematic uncertainties are dominated by branching ratios and are enumerated in the Table 3. This measurement is consistent with the common assumption that 15% of b's fragment into B_s and 75% into B_u or B_d. It is slightly larger than the LEP average[10] of $.112 \pm .024 \pm .020$ when one includes the new CLEO $D_s^+ \to \phi\pi^+$ branching ratio.

References

1. F. Abe, et al., Nucl. Inst. Meth. **A271**, 387 (1988).
2. F. Abe, et al., "Evidence for Top Quark Production in $\bar{p}p$ Collisions at $\sqrt{s} = 1.8$ TeV," FERMILAB-PUB-94-097-E, Submitted to Physical Review D.
3. Y.Cen, "B Decays at CDF." To be published in the proceedings of the XXIXth Rencontres de Moriond, QCD and High Energy Hadronic Interactions (1994), FERMILAB-CONF-94/113-E.
4. K. Hikasa et al., "Review of Particle Properties," Phys. Rev. **D45**, S1 (1992).
5. P. Dawson et al., Nucl. Phys. **B327**, 49 (1988). M. Mangano, et al., Nucl. Phys. **B373**, 295 (1992).
6. A. Martin et al. "New Information on the Parton Distributions," RAL-92-021, DTP/92/16 (1992).
7. C. Peterson et al., Phys. Rev. **D27**, 105 (1983).
8. J. Chirin, Z. Phys. **C36**, 163 (1987).
9. F. Butler et al., Phys. Lett. **B324**, 255 (1994).
10. D. Buskulic et al., Phys. Lett. **B294**, 145 (1993). P. Abreu et al., Phys. Lett. **B289**, 199 (1992). P.D. Acton et al., Phys. Lett. **B295**, 357 (1993).

MEASUREMENT OF THE B MESON DIFFERENTIAL CROSS-SECTIONS IN $p\bar{p}$ COLLISIONS AT $\sqrt{s} = 1.8$ TeV USING THE EXCLUSIVE DECAYS $B^{\pm} \to J/\psi K^{\pm}$ AND $B^0 \to J/\psi K^{*0}$

MARK W. BAILEY*

Dept. of Physics and Astronomy, University of New Mexico,
800 Yale Blvd. NE, Albuquerque, NM, 87131

ABSTRACT

This paper presents the first measurement of the differential $B^{\pm}$ and B^0 cross-sections, $d\sigma/dp_T$, in $p\bar{p}$ collisions at $\sqrt{s} = 1.8$ TeV. The data sample used represents an integrated luminosity of 19.3 ± 0.8 pb^{-1} accumulated by the Collider Detector at Fermilab(CDF). The cross-sections are measured over the p_T range 6-15 GeV/c in the central pseudorapidity region $|\eta| < 1$ by fully reconstructing the B meson decays $B^{\pm} \to J/\psi K^{\pm}$ and $B^0 \to J/\psi K^{*0}$, where the J/ψ is required to decay to two muons, and the K^{*0} is required to decay to $K^{\pm}\pi^{\mp}$. The results are compared to the theoretical QCD prediction calculated at next-to-leading order.

We present the first direct measurement of the differential B meson cross-section in transverse momentum(p_T) in $p\bar{p}$ collisions at $\sqrt{s}= 1.8$ TeV by measuring the mass and momentum of fully reconstructed B mesons decays. Such a measurement is free of the model-dependant procedures used to infer the b quark p_T from an inclusive lepton sample, for example, and thus provides a better test of the QCD prediction. The data sample used represents 19.3 ± 0.8 pb^{-1} collected by CDF during the 1992-93 run.

B mesons are reconstructed via the decays $B^{\pm} \to J/\psi K^{\pm}$ and $B^0 \to J/\psi K^{*0}$, with $J/\psi \to \mu^{+}\mu^{-}$ and $K^{*0} \to K^{+}\pi^{-}$, and their charge conjugates. By triggering on dimuons, we obtain a sample of J/ψ of which more than 15% comes from B decays.[1] Each muon is required to have $p_T \geq 1.8$ GeV/c, and at least one muon is required to have $p_T \geq 2.8$ GeV/c, to match the trigger requirements. The $K^{\pm}$ candidates from $B^{\pm} \to J/\psi K^{\pm}$ and the K^{*0} candidates from $B^0 \to J/\psi K^{*0}$ are required to have $p_T > 1.25$ GeV/c. The B meson transverse momentum is required to be greater than 6.0 GeV/c. The $K\pi$ invariant mass is required to be within 50 MeV/c² of the K^{*0} mass(896.1 MeV/c²). The decay tracks are vertex constrained, and the muons are mass constrained to the J/ψ mass. The confidence level of the fit χ^2 is required to be greater than 0.5%, and the proper decay length of the B candidate is required to be greater than 100 μm.

The $B^{\pm}$ candidates are divided into subsamples in B p_T ranges 6-9, 9-12, 12-15, and > 15 GeV/c, while the B^0 candidates are divided into p_T ranges 7-11, 11-15, and

*Representing the CDF Collaboration

Table 1. Differential B meson cross-sections, $d\sigma(|y| < 1.0)/dp_T$ (nb/GeV/c), for the p_T range 6-15 GeV/c and integrated cross-sections (nb) for $p_T > 15$ GeV/c.

	$p_T(B)$ GeV/c	$\langle p_T \rangle$ GeV/c	Acceptance %	No. of Events	Cross-section
B^+	6-9	7.4	1.29 ± 0.01	53 ± 12	$610 \pm 138 \pm 141$
	9-12	10.4	3.58 ± 0.04	29 ± 8	$121 \pm 33 \pm 28$
	12-15	13.4	5.71 ± 0.07	19 ± 5	$49 \pm 13 \pm 11$
	>15	19.7	9.03 ± 0.08	25 ± 6	$12 \pm 3 \pm 3$
B^0	7-11	8.8	1.18 ± 0.01	31 ± 11	$324 \pm 115 \pm 99$
	11-15	12.8	3.46 ± 0.04	22 ± 6	$79 \pm 22 \pm 24$
	>15	20.5	6.54 ± 0.06	8 ± 4	$15 \pm 8 \pm 5$

> 15 GeV/c. For each of these subsamples, the invariant mass distribution is fitted to a Gaussian plus a linear background over the signal region 5.2-5.6 GeV/c^2. The fitted numbers of $B^\pm$ and B^0 mesons are given in Table 1.

The differential cross-section is calculated from the following equation:

$$\frac{d\sigma}{dp_T} = \frac{N/2}{\mathcal{L} \cdot A \cdot e \cdot F \cdot \Delta p_T} \tag{1}$$

where N is the number of events observed, $\mathcal{L}$ is the integrated luminosity, A is the detector acceptance and selection cut efficiency, e is the combined tracking efficiency, F is the branching fraction, and Δp_T is the width of the p_T bin. The factor of $1/2$ is included because decays involving both B and $\overline{B}$ mesons have been reconstructed, but the quoted cross-sections are for B mesons only. Integrated cross-sections are calculated for B $p_T > 15$ GeV/c.

A Monte Carlo employing the next-to-leading order QCD calculation[2,3], the MRS$D\_'$[4] proton structure functions, and the Peterson parameterization[5,6] for fragmentation was used to determine the acceptance. After simulating the detector response and the J/ψ trigger efficiency, the selection requirements were applied, with the resulting acceptance shown in Table 1.

The online tracking efficiency for the muon pairs was determined to be $93 \pm 2\%$, while the offline tracking efficiency was found to be $98.9 \pm 1\%$. The efficiency of the matching requirement between the CTC track and the muon chamber track segment was measured to be $98.7 \pm 1\%$. Branching fractions of $(1.12 \pm 0.17) \times 10^{-3}$, $(1.53 \pm 0.37) \times 10^{-3}$,[7] 0.0597 ± 0.0025,[8] and $2/3$ were used for the $B^\pm \to J/\psi K^\pm$, $B^0 \to J/\psi K^{*0}$, $J/\psi \to \mu^+\mu^-$, and $K^{*0} \to K^+\pi^-$ decays, respectively.

Varying the b production and fragmentation parameters used in the Monte Carlo simulation indicates an uncertainty in the acceptance calculation of 9%. The systematic uncertainty in the J/ψ trigger efficiency parameterization was determined to be $\pm 4\%$. Additionally, a systematic uncertainty of $\pm 4\%$ is associated with the reconstruction of kaons which decay inside the CTC volume. A 10% uncertainty is associated with the pseudorapidity dependence of the tracking efficiency. The uncertainty in the $c\tau$ cut was determined to be 4%, and the vertex fit χ^2 cut had an additional 1% uncertainty. Varying the polarization of the $B^0 \to J/\psi K^{*0}$ decay products in the range $(84 \pm 10)\%$[7] changes the calculated acceptance by $\pm 7\%$. Combining these uncertainties in quadrature, the reconstruction efficiency has overall systematic errors of 17.6% for the B^+ decay and 19.0% for the B^0 decay. The branching fractions contribute additional uncertainties of 15% and 24%, respectively.

The differential B meson cross-section measurements are listed in Table 1 and plotted in Fig. 1. The integrated cross-sections for $p_T > 15$ GeV/c are given in the table but not plotted. Each measurement is quoted for the momentum $\langle p_T \rangle$, at which the cross-section integrated over the bin divided by the bin width is estimated to equal the differential cross-section. The solid curve shows the differential B meson cross-sections predicted by the Monte Carlo, which includes the assumption that 75%[9] of $\bar{b}$ quarks fragment in equal amounts to B^+ and B^0 mesons. While this theory correctly predicts the shape, as measured here, the predicted rate using the natural choice for the renormalization scale, $\mu = \sqrt{m_b^2 + p_T^2}$, remains low.

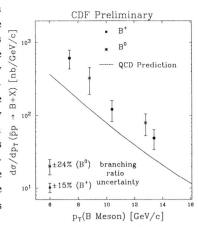

Figure 1: B meson differential cross-sections compared to the QCD prediction.

We thank the Fermilab staff and the technical staffs of the participating institutions for their vital contributions. This work was supported by the U.S. Department of Energy and National Science Foundation; the Italian Istituto Nazionale di Fisica Nucleare; the Ministry of Education, Science and Culture of Japan; the Natural Sciences and Engineering Research Council of Canada; the National Science Council of the Republic of China; the A. P. Sloan Foundation; and the Alexander von Humboldt-Stiftung.

References

1. F. Abe *et al.*, CDF Collab., *Phys. Rev. Lett.* **71** (1993) 3421.
2. P. Dawson *et al.*, *Nucl. Phys.* **B327** (1988) 49.
3. M. Mangano *et al.*, *Nucl. Phys.* **B373** (1992) 295.
4. A. Martin, R. Roberts and J. Stirling, "New Information on the Parton Distributions", **RAL-92-021, DTP/92/16**, (1992).
5. C. Peterson *et al.*, *Phys. Rev. D* **27** (1983) 105.
6. J. Chrin, *Z. Phys. C* **36** (1987) 163.
7. M.S. Alam *et al.*, CLEO Collab., 1993 Lepton-Photon Conference, Cornell, August 10-15, 1993.
8. M. Aguilar-Benitez *et al.*, *Phys. Rev. D* **45**, Part 2 (1992).
9. B. Adeva *et al.*, L3 Collab., *Phys. Lett. B* **252** (1990) 703.

MEASUREMENT OF CORRELATED b QUARK CROSS SECTIONS AT CDF

DAN AMIDEI, PAUL DERWENT, DAVID GERDES, and TAE SONG*

Randall Laboratory of Physics, University of Michigan,

500 E. University Avenue, Ann Arbor, MI 48109

ABSTRACT

Using data collected during the 1992-93 collider run at Fermilab, CDF has made measurements of correlated b quark cross section where one b is detected from a muon from semileptonic decay and the second b is detected with secondary vertex techniques. We report on measurements of the cross section as a function of the momentum of the second b and as a function of the azimuthal separation of the two b quarks, for transverse momentum of the initial b quark greater than 15 GeV. Results are compared to QCD predictions.

Studies of b production in $p\bar{p}$ collisions provide quantitative tests of perturbative QCD. Measurements of the cross section for $p\bar{p} \to bX$ have been made at CDF[1] and UA1.[2] In this analysis we extend the range of comparisons between theory and experiment by performing a measurement of b-$\bar{b}$ correlations in the process $p\bar{p} \to b\bar{b}X$. We identify the first b via its semileptonic decay to a muon, and the other b (referred to for simplicity as the $\bar{b}$, though we do not perform explicit flavor identification for either b) by using precision track reconstruction to measure the displaced tracks from $\bar{b}$ decay. Identification of the b and $\bar{b}$ permits a measurement of the cross section as a function of the transverse momentum of the $\bar{b}$, $\frac{d\sigma_{\bar{b}}}{dE_T}$, and as a function of the azimuthal separation of the two b quarks, $\frac{d\sigma_{\bar{b}}}{d\delta\phi}$, for transverse momentum of the initial b quark greater than 15 GeV. The data used here were collected by the Collider Detector at Fermilab (CDF) during the 1992-93 Tevatron collider run, and correspond to an integrated luminosity of 15.1 ± 0.5 pb^{-1}.

The CDF has been described in detail elsewhere.[3] The tracking systems used for this analysis are the silicon vertex detector[4] (SVX), the central tracking chamber (CTC), and the muon system. The central muon system consists of two detector elements. The Central Muon chambers (CMU) provide muon identification over 85% of ϕ in the pseudorapidity range $|\eta| \leq 0.6$, where $\eta = -\ln[\tan(\theta/2)]$. This η region is further instrumented by the Central Muon Upgrade chambers (CMP), located behind the CMU after an additional ≈ 3 absorption lengths. The calorimeter systems used for this analysis are the central and plug systems, which give 2π azimuthal coverage in the range $|\eta| < 1.1$ and $1.1 < |\eta| < 2.4$ respectively.

From events that pass an inclusive muon trigger, we select good-quality CMU muons with $P_T > 9$ GeV that have an associated track segment in the CMP. We

---

*Presented by David Gerdes, representing the CDF Collaboration.

further require the muon track to fall within the fiducial region of the SVX. This sample contains 145,784 events. An independent analysis has measured the fraction of muons from b-decay in this sample to be $36.0\pm2.4\pm2.5\%$,[5] making this an excellent sample to look for the presence of additional $\bar{b}$ jets.

The long lifetime of b quarks causes the tracks from b-decay to be displaced relative to the primary $p\bar{p}$ interaction point. The high-precision track reconstruction made possible by the SVX allows us to identify these tracks with good efficiency. We use the "jet-probability" algorithm,[6] which compares the impact parameters of the tracks in a jet to the measured resolution of the SVX and determines an overall probability that the jet is primary. This probability is flat between 0 and 1 for jets from zero-lifetime particles, and has a peak at low values for jets from b and c decays.

To identify the $\bar{b}$ jets in the inclusive muon sample, we require that the event contain at least one jet with $E_T > 10$ GeV, $|\eta| < 1.5$, and at least 2 good tracks. The jet is required to be separated from the muon in $\eta - \phi$ space by $\Delta R \geq 1.0$, so that the tracks clustered around the jet axis are separated from the μ direction. There are 17810 events in this sample. We then fit the jet probability distribution of these jets to a sum of Monte Carlo templates for b, c, and primary jets, thereby obtaining the fraction of $\bar{b}$ jets in the sample.

The Monte Carlo samples for b and c jets are produced using the ISAJET event generator[7] and a full detector simulation. The CLEO Monte Carlo program[8] is used to model the decay of B mesons, using an average b lifetime of $c\tau = 420$ μm. The input jet probability templates for b, c, and zero-lifetime jets are shown in Figure 1a. The fit is actually done using the variable $\log_{10}$(jet probability), which magnifies the interesting region of low jet probability. Tests of the fitter in Monte Carlo samples with known admixtures of b, c, and primary jets show that the fitter returns the correct number of jets of each type to within the uncertainties. The data are shown in Figure 1b together with the results of the fit. The fit agrees very well with the data over ten orders of magnitude in jet probability, and predicts 2620 ± 97 $\bar{b}$ jets, 2085 ± 180 c jets, and 13103 ± 161 primary jets for a total of 17808.

To convert the results of the above fit into a measurement of the cross section, we calculate the acceptance for both $b \to \mu$ and $\bar{b}$ jets. The μ acceptance and efficiency has three parts: (1) the fiducial acceptance for muons coming from b's with $|\eta| < 1$, (2) the fraction of b's, $P_T^b > P_T^{min}$, that decay to muons with $P_T > 9$ GeV, and (3) the trigger and identification efficiencies for 9 GeV muons. The first two factors are studied using the Monte Carlo sample described above. The P_T^{min} value, chosen such that 90% of muons with $P_T > 9$ GeV come from b quarks with $P_T > P_T^{min}$, is 15 GeV. The trigger and identification efficiencies for muons are determined from J/ψ and Z° samples. The overall acceptance for $b \to \mu(P_T^\mu > 9\text{GeV}, P_T^b > 15$ GeV$, |y^b| < 1)$, including the semileptonic branching ratio,[10] is $0.239^{+0.030}_{-0.018}\%$.

The $\bar{b}$ jet acceptance represents the fraction of $\bar{b}$ quarks that produce jets with $E_T > 10$ GeV, $|\eta| < 1.5$ and at least 2 good tracks inside a cone of 0.4 around the jet axis, in events where there is also a b quark which decays to a μ with $P_T > 9$ GeV within the CMU-CMP acceptance. The $\bar{b}$ jet acceptance is calculated separately as a function of the jet E_T and azimuthal opening angle between the two quarks, using the Monte Carlo sample described above. The average acceptance for the $\bar{b}$ is $\approx 40\%$, and

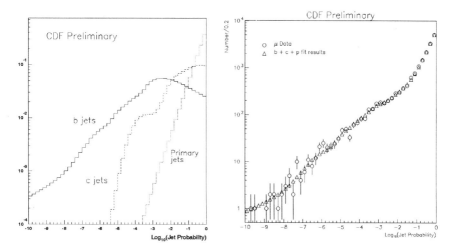

Fig. 1. (a) Jet probability distributions for b, c, and primary jets. (b) Observed jet probability distribution for jets in the inclusive muon sample, together with the results of the fit.

ranges from $32.9 \pm 1.9\%$ (statistical error only) for $10 < E_T < 15$ GeV to $49.8 \pm 7.3\%$ for $40 < E_T < 50$ GeV. For $\delta\phi < \frac{\pi}{8}$ radians, the acceptance is $7.3 \pm 2.2\%$, while for $\frac{7\pi}{8} < \delta\phi < \pi$, the acceptance is $51.4 \pm 0.8\%$.

We have compared the values for the $\bar{b}$ jet acceptance from ISAJET samples to the acceptance from HERWIG samples. The acceptance agrees within the statistical error in the samples as a function of E_T, differing at the 5% level. We take this as an additional systematic uncertainty on the acceptance. In combination with a 10% uncertainty due to the vertex distribution for events in the SVX fiducial volume, we have a common 11.2% systematic uncertainty in all the jet acceptance numbers.

We use the jet probability fit to determine the number of $\bar{b}$ jets as a function of (1) the azimuthal separation $\delta\phi$ between the jet and the muon, and (2) the E_T of the jet, and convert these numbers into the differential cross section using the aceptances calculated above. Figure 2a shows the measured distribution of $d\sigma_{\bar{b}}/d(\delta\phi)$, together with a prediction from the Mangano-Nason-Ridolfi (MNR) calculation.[11] There is a large change in the acceptance for $\delta\phi < \frac{3\pi}{8}$ due to the ΔR separation requirement on the μ–jet system. The shapes of the theoretical prediction and the experimental data agree well, especially for $\delta\phi > \frac{\pi}{2}$, but the overall normalization of the data is about a factor of 1.3 higher than predicted.

We also divide the jets into six E_T bins between 10 and 50 GeV, and fit the jet probability distribution for each bin to determine $d\sigma_{\bar{b}}/dE_T$. This cross section is shown in Figure 2b. Work to compare this measurement to the MNR prediction, which requires a bin-by-bin understanding of the detector response, is in progess.

We thank the Fermilab staff and the technical staffs of the participating institutions for their vital contributions. This work was supported by the U.S. Department of

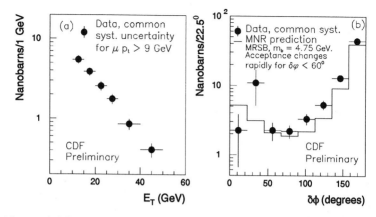

Fig. 2. Measured differential cross section as a function of (a) the E_T of the $\bar{b}$ jet, and (b) the azimuthal angle between the jet and the muon, for jets with $E_T > 10$ GeV and $|\eta| < 1.5$. There is a common systematic uncertainty of $^{+18.3}_{-15.4}\%$ not shown in the experimental points. Also shown in (b) is the prediction from the MNR calculation.

Energy and National Science Foundation, the Italian Istituto Nazionale di Fisica Nucleare, the Ministry of Education, Science and Culture of Japan, the Natural Sciences and Engineering Research Council of Canada, the National Science Council of the Republic of China, the A. P. Sloan Foundation, and the Alexander von Humboldt-Stiftung.

References

1. F. Abe et al., Phys. Rev. Lett. **71** 2396 (1993).
2. C. Albajar et al., Phys. Lett. **B256** (1991).
3. F. Abe et al., Nucl. Instrum. Methods Phys. Res. **A271** 387 (1988) and references therein.
4. D. Amidei et al., Fermilab-Pub-94/024-E, to appear in Nucl. Instrum. Methods Phys. Res.
5. T.Y. Song et al., CDF Internal Note 2004 (unpublished).
6. D. Buskulic et al., Phys. Lett. **313B**, 535 (1993); F. Abe et al., Fermilab-Pub-94/097-E, to appear in Phys. Rev. D.
7. F. Paige and S.D. Protopopescu, BNL Report No. 38034, 1986 (unpublished).
8. P. Avery, K. Read, G. Trahern, Cornell Internal Note CSN-212, March 25, 1985, (unpublished).
9. G. Marchesini and B.R. Webber, Nucl. Phys. **B310**, (1988), 461; G. Marchesini, et al., Comput. Phys. Comm. **67** 465 (1992).
10. S. Henderson et al., Phys. Rev. **D45**, 2212 (1992).
11. M. Mangano, P. Nason, and G. Ridolfi, Nucl. Phys. **B373**, 295 (1992).

INCLUSIVE MUON AND B-QUARK CROSS SECTION AT DØ

GILVAN A. ALVES*

Laboratório de Cosmologia e Física Experimental de Altas Energias (LAFEX), CBPF
Rua Xavier Sigaud, 150 5° andar
Rio de Janeiro, RJ 22290-180, Brazil

ABSTRACT

We report a measurement of the inclusive muon and b-quark production cross sections in $p\bar{p}$ collisions at $\sqrt{s}$ =1.8 TeV using the DØ detector at the Fermilab Tevatron collider. The inclusive muon spectrum, in the kinematic range $|\eta_\mu| < 0.8$ and $3.5 < p_T^\mu < 60.0$ GeV/c, is well described by the sum of all expected sources of muons. The extracted b-quark cross section is in good agreement with next-to-leading order QCD predictions for $|y^b| < 1.0$ and $p_T^b > 7 \; GeV/c$.

1. Introduction

The study of b-quark production in high energy hadronic collisions is important for testing the perturbative QCD description of heavy quark production.[1] The fact that next-to-leading order (NLO) terms are as large as the leading order contribution and are very sensitive to the input normalization scale μ suggests that even higher order terms may be significant. A comparison with experimental data should reveal how reliable NLO predictions are. The b-quark integrated p_T cross section measured at $\sqrt{s}$ =0.63 TeV by UA1[2] is in agreement with the theoretical predictions. The corresponding data from CDF[3] at $\sqrt{s}$ =1.8 TeV are consistent with the upper extreme of the theoretical band but a discrepancy is observed for values of transverse momentum (p_T) less than 10 GeV/c. A measurement by DØ experiment provides an independent test of the reliability of the description of heavy quark production within the QCD framework.

This paper presents a measurement of the inclusive muon and b-quark production cross sections at $\sqrt{s}$ =1.8 TeV. The data, which correspond to an integrated luminosity $\int \mathcal{L} \, dt = 73 \pm 8 \; nb^{-1}$, were taken in dedicated runs during the 1992-93 Tevatron collider run using the DØ detector.

2. The DØ Detector

DØ is a large multi-purpose detector[4] operating at the Fermilab $p\bar{p}$ collider. It features an inner tracking system, calorimetry for the detection of electrons, jets and missing transverse energy, and a muon system. The inner tracking covers the region $|\eta| < 3$ with wire drift chambers to detect charged tracks. The calorimeter is a uranium-liquid argon sampling device with a thickness of 7 - 9 interaction lengths (λ) and a

---

*Representing the DØ Collaboration.

measured fractional energy resolution for pions of $50\%/\sqrt{E}$. The muon system consists of three layers of chambers and 1.9 T magnetized iron toroids located between the first and second layers. The thickness of the calorimeter plus iron toroids varies from 14 λ in the central region to 19 λ in the forward region.

3. Data Sample

The data sample used in this analysis was streamed from two hardware trigger levels (level 0 and 1), and a software level 2. The level 0 trigger used scintillator hodoscopes to select inelastic collisions which occurred within 1 meter of the nominal interaction point along the beam direction. The level 1 muon trigger[5] required that hits in 60 cm wide roads from 2 or 3 layers of muon chambers be consistent with a track pointing to the interaction region. At level 2 at least one muon track must be reconstructed and have a minimum p_T of 3 GeV/c. Events were fully reconstructed offline and were retained for further analysis if they contained at least one muon track within the range $|\eta^\mu| < 0.8$ and $3.5 < p_T^\mu < 60.0$ GeV/c. Tracks were required to have a minimum of 1 GeV/c of associated energy in the calorimeter, and a matching track in the central tracking chambers. To ensure a better momentum measurement only tracks with hits in all three layers of chambers were kept and a minimum field integral of 0.6 GeV/c in the toroids was required. Furthermore, to reject cosmic ray muons, the reconstructed time of passage (T_0) through the muon chambers was required to be within 100 ns of the beam crossing.

4. Inclusive Muon Cross Section

The inclusive muon differential cross section was extracted according to the following expression:

$$\frac{d\sigma^\mu}{\Delta\eta^\mu \, dp_T^\mu} = \left(\frac{dN^\mu}{dp_T^\mu}\right) \frac{[1 - R_c]}{\int \mathcal{L} \, dt \, . \, \epsilon \, . \, \Delta\eta} \tag{1}$$

where N^μ is the number of muons passing the offline selection , $\Delta\eta = 1.6$ units of covered η range and R_c is the cosmic ray contamination, estimated to be $R_c = 13 \pm 4\%$. The resulting cross section as a function of the measured muon p_T (p_T^μ) is shown in Figure 1. The curves represent the expected different sources of muons using the ISAJET Monte Carlo package[6] convoluted with the muon momentum resolution. The solid curve, which represents the sum of all expected contributions, describes the data very well.

To determine the b-quark fraction (f_b) in the data, muon events from W decays were first removed on an event by event basis by requiring a maximum missing transverse energy ($\not{E}_T$) of 20 GeV/c. This cut rejects about 97% of W and 50% of Z events with an efficiency of $95 \pm 3\%$. Residual W and Z events were estimated and subtracted using Monte Carlo events run through the same offline selection as the data. The remaining events are from b-quark, c-quark, and π/K decays.

The fraction (f_b) was then determined from the Monte Carlo calculation of the spectra shown in Figure 1. Furthermore, it was cross checked using the transverse momentum of the muon (p_T^{rel}) with respect to the associated jet axis (which also includes the muon momentum) for that subset of events which contained a reconstructed jet

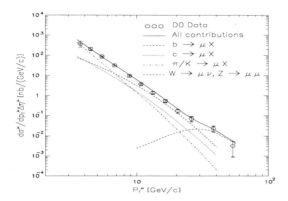

Fig. 1. Comparison between the expected and measured inclusive muon cross section. The errors on the data points are statistical only (inner bars) and total of statistical and systematic added in quadrature (outer bars).

near the muon with $E_T^{jet} > 10$ GeV/c. The different p_T^{rel} shapes for different flavours were taken from Monte Carlo distributions. They were then fit to the data for different momentum ranges to determine f_b as a function of p_T^μ.

To convert the muon spectrum to a b-cross section, we followed the method previously used by UA1 and CDF.

The integrated b-quark cross section is determined from the following expression:

$$\sigma^b = \sigma^\mu \cdot f_b \cdot f_{\mu \to b} \qquad (2)$$

where σ^μ is the measured muon cross section integrated over the studied p_T^μ range, and $f_{\mu \to b}$ is given by:

$$f_{\mu \to b} = \frac{\sigma_{MC}^b \left(p_T^b > p_T^{min}, |y^b| < 1.0 \right)}{\sigma_{MC}^\mu \left(p_T^{\mu 1} < p_T^\mu < p_T^{\mu 2}, |\eta^\mu| < 0.8 \right)} \qquad (3)$$

In the above expression, p_T^{min} is the b-quark transverse momentum for which 90% of muons within the range $p_T^{\mu 1}$ - $p_T^{\mu 2}$ come from b-quarks with $p_T^b > p_T^{min}$, σ_{MC}^b is the b-quark production cross section and σ_{MC}^μ is the cross section for producing b-quarks which decay to muons within the concerned range.

The extracted cross section for b-quark production ($\bar{b}$ not included) as a function of p_T^{min}, for $|y^b| < 1.0$, is shown in Figure 2. The data are in good agreement with NLO QCD predictions over the entire measured p_T range. In the plot, both statistic and systematic erros are included. The main sources of systematic errors are the uncertainties in the muon detection efficiency, muon momentum resolution and the model dependence of equation(3).

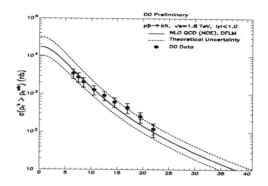

Fig. 2. b-quark production cross section extracted from the inclusive muon data compared to NLO QCD predictions.

5.   Conclusions and Prospects

We have measured the beauty quark production cross section using the inclusive muon spectrum at DØ . We find good agreement between the data and theoretical calculations based on NLO QCD. In this sense the data indicates that the perturbative QCD description of the heavy quark production is reliable in our kinematic range.

We are presently using a trigger which includes a jet requirement,[7] in addition to the muon, to enhance the heavy quark content of the data being taken. We have also added scintillator coverage to the top and sides of the detector, to help in the cosmic ray rejection, and installed a new level of trigger(level 1.5) in the forward region to further reduce the rates. We expect to have about 10 times more data and extend the rapidity coverage down to $|\eta| < 3.3$ in the present collider run.

References

1. P. Nason, S. Dawson and R.K. Ellis, Nucl. Phys. **B327** (1989) 49.
2. C. Albajar $et.$ $al.$, Phys. Lett. **B213** (1988) 405.
3. F. Abe $et.$ $al.$ Phys. Rev. Lett. **71** (1993) 2396.
4. S. Abachi $et.$ $al.$, Nucl. Instr. Meth. **A338** (1994) 185.
5. M. Fortner et $al.$, IEEE Trans. on Nucl. Science **38, No. 2, Part 1** (1991) 480.
6. F. Paige and S.D. Protopopescu, BNL (1986) 38034. Version 7.0
7. T. Huehn, these proceedings.

BOTTOM-QUARK PRODUCTION FROM MUON-JET AND DIMUON EVENTS IN $p\bar{p}$-INTERACTIONS AT $\sqrt{s}$ =1.8 TeV

THORSTEN HUEHN*

Department of Physics, University of California, Riverside
CA 92521

ABSTRACT

Bottom quark production in $p\bar{p}$-interactions has been measured in the rapidity range $| \, y^b \, | < 1$ with the DØ detector at the Fermilab Tevatron collider. The cross section is determined from events containing a muon and jets as well as from dimuon events. Preliminary results are presented based on 197 nb^{-1} and 6.4 pb^{-1} of data for the muon-jets and dimuon analysis, respectively, and are compared to next-to-leading order QCD predictions. The measurements are consistent within errors and are in reasonable agreement with QCD predictions.

1. Introduction

We report measurements of the bottom quark production cross section in proton-antiproton collisions at a center of mass energy of 1.8 TeV. The data were collected at the Fermilab Tevatron Collider during the 1992/93 run. One measurement is based upon data recorded with a trigger requiring a muon-jet coincidence which selects the signal expected from hadronization of a b-quark, followed by a muonic decay of one of the B-hadrons. The second measurement is based on data collected with a dimuon trigger.

Bottom quark production was measured at 0.6 TeV by the UA1 collaboration.[1] The result is in good agreement with the $\mathcal{O}(\alpha_s^3)$ QCD prediction by Nason, Dawson and Ellis,[2] while the CDF experiment has measured b-quark production at the Tevatron collider[3] with results in general above the QCD predictions.

2. Experimental Details

The DØ detector is described in detail elsewhere.[4] Briefly, it is a hermetic large acceptance collider detector with successive layers of tracking, calorimetry, and muon detection. The calorimeter is a uranium-liquid argon sampling device with segmentation 0.1×0.1 in $\Delta\eta \times \Delta\phi$ and $\Delta E/E$ of $0.8/\sqrt{E}$ for jets.

The muon system is a toroidal magnetic spectrometer with 3 sets of proportional drift tubes, a four layer assembly of drift tubes inside and two sets of 3-layer modules separated by 1 m outside the toroid. The momentum of muons is determined from the bending of the track in the 1m thick magnetic field of 2T. The muon momentum

---

*Representing the DØ Collaboration.

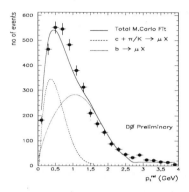

Fig. 1. p_t^{rel}-distribution fit to data

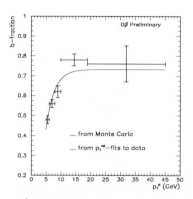

Fig. 2. b-fraction from μ-jets data

resolution is $\Delta p/p = 0.18(p-2)/p \oplus 0.008p$. Muon coverage extends down to $\mid \eta \mid = 3.3$

3. b-Production from Muon-Jets Events

A total of 197 nb^{-1} was collected in dedicated physics runs. The level 1 hardware trigger required 1) a hit pattern in the muon chambers $\mid \eta \mid < 1.7$ consistent with a track pointing back to the interaction region, 2) transverse energy E_t above 3 GeV in a trigger tower of 0.2 in both $\Delta\eta$ and $\Delta\phi$. The level 2 software trigger required at least one good muon track with $p_t^\mu > 3\,\mathrm{GeV/c}$, $\mid \eta^\mu \mid < 1.7$ and a jet with $E_t > 10$ GeV, using a cone algorithm of $\Delta R = \sqrt{(\Delta\eta)^2 + (\Delta\phi)^2} = 0.7$.

Offline selection criteria were chosen to provide good momentum measurement and to reduce cosmic ray background. Muon tracks are restricted to $\mid \eta^\mu \mid < 0.8$, must have hits in all three layers of the muon system and traverse a minimum field integral of 1.9 Tm.

The muon track must be matched, within expected multiple Coulomb scattering deviation, by a track in the central tracking system. Similarly, the impact parameter relative to the interaction vertex must be compatible with a vertex origin for the track. Muon tracks were also required to have matching energy deposition in the calorimeter. To suppress cosmic rays, muons must traverse the muon system within $\Delta T_0 = 100$ ns of the beam crossing. A minimum p_t-cut of $p_t^\mu > 6\,\mathrm{GeV/c}$ was imposed to be above the muon trigger threshold and to ensure a reasonable b-fraction in the data. Standard jet quality cuts reject fake jets due to hot cells and noise in the calorimeter. The jet E_t-threshold was set to 12 GeV.

The extraction of the b-cross section requires knowledge of the efficiency, the fraction of b decays in the data, and a procedure to convert the muon spectrum into the parent b-quark cross-section.

Estimation of efficiencies for trigger, reconstruction and cuts on the data relies mainly on a Monte Carlo simulation which includes the ISAJET event generator,[5] the GEANT detector simulation, trigger simulation and the reconstruction process. The

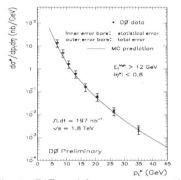

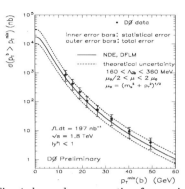

Fig. 3. Differential muon cross section from $b \to \mu X$ processes.

Fig. 4. b-quark cross section from μ-jets production.

triggering efficiencies were checked against appropriate data samples.

The transverse momentum of the muon relative to the jet axis, p_t^{rel}, is proportional to the Q-value of the c or b-decay and can be used to estimate the relative production rates for b and c quarks in the data. Figure 1 shows the measured p_t^{rel} for events with $\Delta R^{\mu-jet} < 1$. The curves show Monte Carlo calculations of p_t^{rel} for b-production and for c and π/K-production combined. The sum of these distributions was fit to the data, giving an overall b-fraction of about 60 %. Muons from W and Z^0-decays are suppressed by the jet requirement.

The b-fraction was taken from the Monte Carlo simulations of b, c, π/K and W-production and is shown in Fig. 2 (smooth line). The fraction was checked against fits of Monte Carlo b and $c + \pi/K$-decays to the data p_t^{rel}-distribution, obtained as in Fig. 1. The data points from the fits are in satisfactory agreement with the Monte Carlo estimate.

Figure 3 shows the differential cross section for muons from bottom quark decay, $d\sigma/dp_t d\eta(p\bar{p} \to bX \to \mu + jet + X)$, with $E_t^{jet} > 12 GeV$. The non-b contribution has been subtracted using the information in Fig. 2, and the momentum resolution has been unfolded using a Monte Carlo calculation. It is compared to an ISAJET prediction, where the b-quark cross-section has been scaled to an NLO calculation.[2]

The measured differential cross section shown in Fig. 3 is transformed into the integrated cross section for b-quark production, $\sigma(p_t^b > p_t^{min})$, for $|y^b| < 1$ according to the UA1 prescription:

$$\sigma(p_t^b > p_t^{min}, |y^b| < 1.0) = \sigma^{Data}(p_t^\mu > X, |\eta^\mu| < 0.8, E_t^{high} > 12\,\mathrm{GeV}) \cdot f_{\mu \to b}^{MC}$$

$$\text{where} \quad f_{\mu \to b}^{MC} = \frac{\sigma^{MC}(p_t^b > p_t^{min}, |y^b| < 1.0)}{\sigma^{MC}(p_t^\mu > X, |\eta^\mu| < 0.8, E_t^{high} > 12\,\mathrm{GeV})}.$$

Here p_t^{min} is the minimum momentum of b-quarks such that 90% of the decay muons satisfy the jet and muon kinematic cuts. The resulting b-quark cross section is shown in Fig. 4, along with a NLO QCD prediction by Nason, Dawson and Ellis.[2]

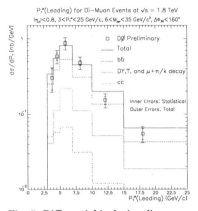

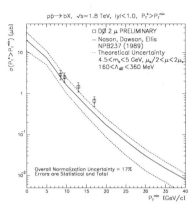

Fig. 5. Differential inclusive dimuon spectrum

Fig. 6. b-quark production cross-section from dimuon events

Contributions to the systematic error come from uncertainties in the muon efficiency (11%), jet efficiency (30%), cosmic ray subtraction (4%), b-fraction (10%) and integrated luminosity (12%), totalling 36%. The b-cross section is subject to additional uncertainties in the p_t-shape of the b-spectrum in the Monte Carlo (12%), the Peterson fragmentation parameter(15%), the $b \to \mu$ branching ratio (5%), and the muon decay spectrum (10%), totalling 42%.

4. b-Quark Production from Dimuon Events

The dimuon selection criteria are similar to those applied to the muon-jets analysis, except that no jet requirement was imposed, and that two muons were required at both trigger levels and in the offline selection cuts. The dimuon analysis admits muons which have hits in 2 out of the 3 layers of the muon-system.

Kinematic cuts require both muons to have $3 < p_t^\mu < 25$ GeV/c. The upper p_t^μ-cut excludes muons with degraded momentum resolution. A dimuon invariant mass requirement of $6 < M_{\mu\mu} < 35$ GeV/c^2 suppresses J/ψ as well as Z^0-decays. Cosmic ray background has been suppressed by requiring a dimuon opening angle of less that $160°$. A total integrated luminosity of 6.4 pb^{-1} yielded 562 events after all cuts. Residual cosmic background was estimated using fits to the ΔT_0-distribution.

The muon efficiencies were estimated in a similar way as described for the μ-jets analysis.

The resulting dimuon cross-section as a function of the leading muon p_t has been plotted in Fig. 5. Also shown in Fig. 5 are the ISAJET predictions for the predominant contributions to dimuon production. The muon momentum resolution in the data has been unfolded. The systematic errors on the muon spectrum are 17% due to uncertainties in the muon efficiency (12%), luminosity (12%) and cosmic subtraction (4%).

The b-quark cross section (Fig. 6) was obtained from the muon spectrum by the same method as for the μ-jets analysis. Systematic errors in the muon spectrum are due to the uncertainty in the muon cross-section, b-fraction (20%), b-quark p_t-shape (12%), fragmentation parameterization (15%), branching ratio for $b \to \mu$ (5%), and the spec-

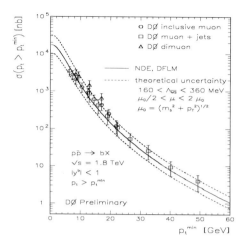

Fig. 7. b-cross section from inclusive muon, μ-jets and inclusive dimuon events

trum of the $b \to \mu$ decay (10%) in the Monte Carlo. The total systematic error is 34%.

5.   Conclusion and Outlook

We have presented preliminary measurements of the b-quark production cross-section from muon-jets and inclusive dimuon events. Both cross-sections, as well as the one derived from DØ -inclusive muon data[6] are shown in Fig. 7. The measurements agree with each other within errors and are in reasonable agreement with NLO theory predictions.

The muon-jets analysis shows good agreement with theory predictions over a large range in p_t^b. The uncertainties are dominated by the error in the jet-efficiency at low E_t^{jet}, further understanding of which will reduce the systematic errors.

The b quark cross section obtained from dimuon events is in reasonable agreement with results from the inclusive muon and muon-jets analyses. Dimuon events can also be used to measure $p\bar{p}$-correlations and to directly separate LO and NLO contributions to be the b-quark cross section.[7]

References

1. C.Albajar *et. al.*, Phys. Lett. **B186** (1987) 237.
2. P. Nason, S. Dawson and R.K. Ellis, Nucl.Phys. **303** (1988); Nucl. Phys. **B327** (1989) 49.
3. F. Abe *et. al.*, Phys. Rev. Lett. **71** (1993) 2396.
4. S. Abachi *et. al.*, Nucl. Instr. Meth. **A338** (1994) 185.
5. F. Paige and S.D. Protopopescu, BNL (1986) 38034.
6. G. Alves, these proceedings.
7. L. Markosky, Proceedings, 27th International Conference on High Energy Physics, Glasgow, July 1994, in preparation.

Υ PRODUCTION AT CDF

VAIA PAPADIMITRIOU*

*Fermi National Accelerator Laboratory, PO Box 500, Batavia,
Illinois 60510, U.S.A.*

ABSTRACT

We report on measurements of the $\Upsilon(1S)$, $\Upsilon(2S)$ and $\Upsilon(3S)$ differential and integrated cross sections in $p\bar{p}$ collisions at $\sqrt{s} = 1.8$ TeV. The three resonances were reconstucted through the decay $\Upsilon \rightarrow \mu^+\mu^-$. The cross section measurements are compared to theoretical models of direct bottomonium production.

1.   Introduction

We report a study of the reaction $p\bar{p} \rightarrow \Upsilon X \rightarrow \mu^+\mu^- X$ at $\sqrt{s} = 1.8$ TeV. This study yields the P_T dependence of the production cross sections for the $\Upsilon(1S)$, $\Upsilon(2S)$, and $\Upsilon(3S)$ states and also the integrated over P_t cross sections. This is the first measurement of the individual Υ cross sections at the Tevatron energies and it is important for the investigation of bottomonium production mechanisms in $p\bar{p}$ collisions.[1,2] Although, due to triggering constraints, we cannot extend our charmonia production cross section measurements below 4 GeV/c, with the Υ's we can go as low as $P_t = 0$, and therefore, besides the differential cross sections, we can measure the total cross sections as well.

2.   Event Selection, Acceptance, Other Efficiencies

The measurements reported here are based on a data sample of opposite sign dimuons collected with a multilevel trigger and with total integrated luminosity of 16.6 ± 0.6 pb^{-1}. The muons from the $\Upsilon \rightarrow \mu^+\mu^-$ decay were required to satisfy the following criteria: Both muons are identified by the central muon system[3]; $P_t(\mu) > 2.0$ GeV/c for each muon; $P_t(\mu) > 2.8$ GeV/c for at least one of the two muons; less than a $(3-4)\sigma$ difference in position between each muon chamber track and its associated, extrapolated CTC track, where σ is the calculated uncertainty due to multiple scattering, energy loss, and measurement uncertainties; a common vertex along the beam axis for the two muons; region of rapidity of the dimuon pair $|y| < 0.4$. In addition the Level 1, Level 2 trigger had to be satisfied by the dimuon pair. The transverse momentum P_t of the dimuon pairs had to be greater than 0.5 GeV/c and less than 20 GeV/c. We observe 1248 $\Upsilon(1S)$, 300 $\Upsilon(2S)$ and 203 $\Upsilon(3S)$ events.

The geometric and kinematic acceptances for $\Upsilon(1S), \Upsilon(2S)$, $\Upsilon(3S) \rightarrow \mu^+\mu^-$ were evaluated by a dimuon event generator that produces Υ's with flat P_t and y distributions. The generated events were processed with the full detector simulation and the

---

*Representing the CDF Collaboration.

same reconstruction used for the data. The acceptances are P_t and y dependent and they are very similar for the three different states. To verify that the acceptance is independent of the kinematic distribution of generated events, the acceptance calculation was repeated for the $\Upsilon(1S)$ state using a parton level generator[1] which provides on an event by event basis the four momentum of various bottomonium states decaying to $\Upsilon(1S)$. This generator together with the CLEO decay table was used to obtain the theoretical predictions shown in Figures 1 and 2.

Efficiency corrections required for the Υ cross section calculations are as following: the Level 1 and Level 2 trigger efficiencies for each muon are increasing functions of P_t, reaching a plateau of 92% above 3.1 GeV/c. The Level 3 tracking efficiency is $92 \pm 2\%$. The total trigger efficiency for each dimuon event is taken to be the product of the Level 1 and Level 2 efficiencies for each muon and the Level 3 efficiency for the event. The offline CTC track reconstruction efficiency is $(97.8 \pm 1.4)\%$ and the muon reconstruction efficiency is $(96 \pm 1.4)\%$ for each dimuon event. The matching cut on the difference between the muon chamber track and the extrapolated CTC track is $(98.7 \pm 0.2)\%$ efficient.

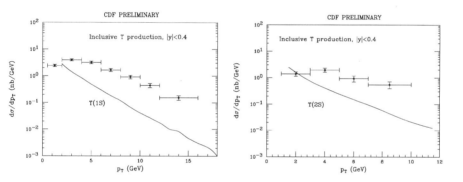

Fig. 1. $\Upsilon(1S)$ and $\Upsilon(2S)$ differential cross section for $|y| < 0.4$ compared to the theoretical prediction. Error bars include statistical error and P_T dependent systematic error. There is also a common systematic error of 15%(22%) for the $\Upsilon(1S)(\Upsilon(2S))$ not shown. The theoretical curves are leading order and were generated using MRSDO PDF with scale $\mu^2 = P_t^2 + m_\Upsilon^2$. Production and decay of higher bottomonium states are included.

3. Differential and total cross sections

The acceptance and efficiency corrected Υ cross sections are displayed in Figures 1 and 2 as functions of P_t. The vertical error bars are from statistical fluctuations in the number of counts (background fluctuations included) and the P_t-dependent systematic uncertainties added in quadrature. The signal and background contributions have been determined independently in the P_t bins shown in the differential cross section plots, by fitting the dimuon mass distribution in each bin to a gaussian plus a first degree polynomial. The integrated cross section results are:

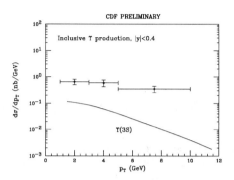

Fig. 2. $\Upsilon(3S)$ differential cross section for $|y| < 0.4$ compared to the theoretical prediction. Error bars include statistical error and P_T dependent systematic error. There is also a common systematic error of 18% not shown. The theoretical curves are leading order and were generated using MRSD0 PDF with scale $\mu^2 = P_t^2 + m_\Upsilon^2$.

$\sigma(\bar{p}p \to \Upsilon(1S), |y| < 0.4, P_t > 0.5\,\text{GeV/c}) = 23.48 \pm 0.99\,(\text{stat}) \pm 2.80\,(\text{sys})\,\text{nb}, \sigma(\bar{p}p \to \Upsilon(2S), |y| < 0.4, P_t > 1.0\,\text{GeV/c}) = 10.07 \pm 1.01\,(\text{stat}) \pm 1.99\,(\text{sys})\,\text{nb}, \sigma(\bar{p}p \to \Upsilon(3S), |y| < 0.4, P_t > 1.0\,\text{GeV/c}) = 4.79 \pm 0.64\,(\text{stat}) \pm 0.72\,(\text{sys})\,\text{nb}$. The systematic errors for the total cross section measurements are coming from the trigger efficiency (4-5%), from the acceptance model (10%), from the luminosity determination (3.6%) and from the various reconstruction efficiencies (2.8%). We have an additional systematic coming from the uncertainty of the branching ratios of $\Upsilon \to \mu^+\mu^-$, which is 2.4%, 16% and 9.4% for the $\Upsilon(1S)$, $\Upsilon(2S)$ and $\Upsilon(3S)$ respectively. Total cross sections in the range $|y| < 0.4$ can be calculated at LO QCD.[2] Use of MRSD0 PDF with scale $\mu^2 = m_\Upsilon^2$ yields 4.2, 2.5 and 0.12 nb for the $\Upsilon(1S)$, $\Upsilon(2S)$ and $\Upsilon(3S)$ respectively.[1] These results include the production and decay of $\chi_b(1P)$ and $\chi_b(2P)$ states, which are found to dominate the rate of $\Upsilon(1S)$ and $\Upsilon(2S)$ respectively. No $\chi_b(3P)$ state has been observed at present, and therefore its possible contribution to $\Upsilon(3S)$ is not included. Our measurements are a factor of 4-6 higher than the theoretical prediction for the $\Upsilon(1S)$ and $\Upsilon(2S)$ states, and by a factor of 40 for the $\Upsilon(3S)$ state. While a factor 4-6 can be partly accomodated by the inclusion of higher order corrections and possible new production mechanisms, the big discrepancy for the $\Upsilon(3S)$ state suggests that there are additional χ_b states below the $B\bar{B}$ threshold that contribute significantly to the $\Upsilon(3S)$ production.

References

1. M. Mangano, private communication.
2. R. Baier and R. Rückl, Z. Phys. C **19**, 251 (1983).
3. F. Abe et al., Nucl. Instrum. Methods A **271**, 387 (1988).

MEASUREMENT OF THE Υ CROSS SECTION
AT DØ USING DI-MUONS

ALEX SMITH*

Department of Physics, The University of Arizona
Tucson, Arizona, 85721

ABSTRACT

The Υ production cross section in $p\bar{p}$ collisions at $\sqrt{s} = 1.8\ TeV$ has been measured by the DØ experiment at the Fermilab Tevatron. Using di-muon events collected from 6.1 pb^{-1} of data, the Υ cross section is measured to be $\sigma_{\Upsilon} = (3.8 \pm 0.4(stat) \pm 1.9(sys)) \times 10^5\ pb$ which appears to be above theoretical predictions.

1. Introduction

At $p\bar{p}$ colliders, production of $b\bar{b}$ bound states, the Υ, Υ', and Υ'', proceeds either through gluon-gluon fusion or through the production of a χ_b, which then decays to an Υ.[1] The decay of an Υ to di-muons has a branching fraction of 2.49%.[2] Generally, it is expected that the muons from the Υ decays will be isolated compared with decays of heavy quarks.

The sample used for this analysis represents $6.1 \pm 0.7\ pb^{-1}$ of data collected in the 1992-93 collider run of the Fermilab Tevatron at $\sqrt{s} = 1.8\ TeV$.

2. Data Selection Cuts and Efficiencies

Events were collected using a di-muon trigger,[3] which requires two muon tracks with $p_T^{\mu} \geq 3\ GeV$ and $|\eta^{\mu}| \leq 1.7$. The efficiency of this trigger (including chamber efficiencies) was determined from Monte Carlo simulations to be $\varepsilon_{trig} = (8.8 \pm 1.0)\%$.

Cuts were first applied to select two high quality muons (Table 1). Kinematic cuts of $|\eta^{\mu}| \leq 0.8$ and $p_T^{\mu} \geq 3\ GeV$ ($\varepsilon_{\eta^{\mu}} \otimes \varepsilon_{p_T^{\mu}} = (6.5 \pm 1.2)\%$) were applied and tracks were required to have opening angle $\leq 165°$ ($\varepsilon_{opening} = (88 \pm 4)\%$) to reduce the cosmic background. Finally, Υ candidates were selected by requiring opposite signed muons ($\varepsilon_{sign} = (93 \pm 4)\%$), and isolation of one muon, $E_{expected}^{2NN} - E_{observed}^{2NN} \leq 3\sigma$ ($\varepsilon_{isolation} = (96 \pm 4)\%$), where E^{2NN} refers to the energy in calorimeter cells hit by the track plus two nearest neighbors, and σ is the uncertainty of the expected energy. The total acceptance of the offline selection cuts is $A_{\Upsilon} = (1.6 \pm 0.4)\%$.

3. Determination of Signal and Background Contributions

The signal and background contributions were resolved using a maximum likelihood fit of the data. As shown in Fig. 1, a simultaneous fit was made to the di-muon invariant mass, halo of the highest p_T^{μ} muon ($E_{cal}^{\Delta R=0.6} - E_{cal}^{\Delta R=0.2}$), and time offset from beam crossing (t_0^f) distributions. The value of t_0^f was calculated by allowing the beam

---

*Representing the DØ Collaboration

Table 1. Muon selection cuts and efficiencies.

Muon Selection Efficiency(%)	
$\int Bdl \geq 0.5\ GeV$	96 ± 3
CD Match $(\Delta\theta, \Delta\varphi \leq 0.45\ \text{radians})$	82 ± 4
Global Fit $\chi^2 \leq 100$	83 ± 4
MIP Energy $(E_{cal}^{1NN} \geq 1\ GeV)$	90 ± 3
Muon Reconstruction	95 ± 3
Total Muon Selection	56 ± 5

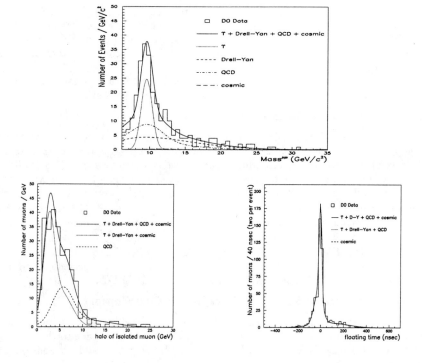

Fig. 1. Result of simultaneous fit of the mass (top), halo (bottom left) and floating time (bottom right) data distributions.

crossing time to be a free parameter and fitting the drift times of the hits on the track. The results of this fit are 81^{+14}_{-13} Υ , 101 ± 14 QCD, 89^{+14}_{-13} Drell-Yan, and 2^{+8}_{-2} cosmic ray events.

4. Results and Conclusions

The cross section is calculated using $\sigma \cdot Br(\Upsilon \rightarrow \mu\mu) = \frac{N_\Upsilon}{(\epsilon_{trig} \cdot A_\Upsilon \cdot lum)}$ to be $\sigma \cdot Br(\Upsilon \rightarrow \mu\mu) = (9.4 \pm 1.0(stat)^{+4.7}_{-4.6}(sys)) \times 10^3$ pb. Using $2.49\%^2$ for the branching fraction, the total Υ production cross section was found to be $\sigma_\Upsilon = (3.8 \pm 0.4(stat) \pm 1.9(sys)) \times 10^5$ pb. Figure 2 shows the DØ preliminary point with the UA1 data and theoretical predictions.

We expect to reduce the systematic errors associated with the trigger and event selection criteria. Sources of systematic error include uncertainties in the Υ Monte Carlo (25%), uncertainties in the input distributions to the fit (25%), and uncertainties from the fit itself (17%). We expect to reduce each of these by further analyses. Finally, the statistical error will be reduced when the full data sample is analyzed.

References

1. V. Barger and A. Martin. *Phys. Rev.* **D31** (1985) 1051.
2. *Review of Particle Properties. Phys. Rev.* **D45** (1992).
3. S. Abachi et. al. *Nucl. Instr. and Meth.*. **A338** (1994) 185.

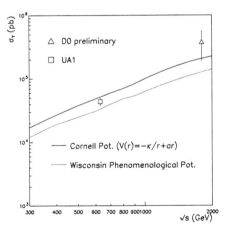

Fig. 2. Preliminary D0 and UA1 Υ cross sections. The theoretical predictions are from models using the "Cornell" potential, $V(r) = -\kappa/r + ar$ (solid line), and a phenomenological potential used by the Wisconsin group (dotted line).

S-WAVE AND P-WAVE B_c MESON PRODUCTION AT HADRON COLLIDERS BY HEAVY QUARK FRAGMENTATION

KINGMAN CHEUNG

Department of Physics, Northwestern University, Evantson, Illinois 60208, U.S.A.

and

TZU CHIANG YUAN

Davis Institute for High Energy Physics, Department of Physics,
University of California, Davis CA 95616, U.S.A.

ABSTRACT

We compute model-independently the production rates and transverse momentum spectra for the B_c mesons in various spin-orbital states ($n\,^1S_0$, $n\,^3S_1$, $n\,^1P_1$, and $n\,^3P_J$ ($J = 0, 1, 2$)) at hadron colliders via the direct fragmentation of the bottom antiquark and via the Altarelli-Parisi-induced gluon fragmentation. Since all the radially and orbitally excited states below the BD flavor threshold will decay, either electromagnetically, hadronically, or a combination of both, into the pseudoscalar ground state $1\,^1S_0$, they all contribute significantly to the inclusive B_c meson production.

The next and the last family of B mesons to be observed will be the $B_c(\bar{b}c)$ made up of one charm quark and one bottom antiquark. Like the J/ψ and Υ quarkonia, dynamical properties of B_c can be predicted reliably by using perturbative QCD, in contrast to the heavy-light mesons. In the limit $m_c/m_b \to 0$, the B_c system enables us to test the heavy quark symmetry and to understand the next-to-leading terms in the heavy quark effective theory in the applications to the heavy-light B mesons. In addition, the production rates for different spin-orbital states also help us to understand the spin symmetry breaking effects. Phenomenologically, B_c mesons can be used to analyze the mixing of the $B_s^0 - \overline{B_s^0}$ without ambiguity by tagging the charge of the lepton in the decay $B_c^+ \to B_s^0 + \ell^+ \nu$ or $B_c^- \to \overline{B_s^0} + \ell^- \bar{\nu}$

Calculations on the production of B_c mesons at e^+e^- colliders were previously performed. But the calculation for the production at hadronic colliders is rather tedious until Braaten and Yuan[1] pointed out that the heavy quarkonium production at the large transverse momentum region is dominated by heavy quark fragmentation. The fragmentation of a heavy quark into a heavy-heavy-quark bound state essentially involves the creation of a heavy quark-antiquark pair, which tells us that the process should be hard enough to be calculable in perturbative QCD (PQCD). Explicit calculation of the production rates and the transverse momentum spectra was performed in Ref.,[2] which was based on the direct $\bar{b}$ antiquark fragmentation functions $D_{\bar{b} \to B_c}(z)$

and $D_{\bar{b} \to B_c^*}(z)$ calculated in Ref.[3] It was found[4] that the Altarelli-Parisi-induced gluon fragmentation $g \to B_c$ also contribute significantly to the total production. Since the P-wave fragmentation functions have just been completed,[5] it is natural to include all the S-wave and P-wave contributions to calculate the inclusive production rate.

The direct $\bar{b} \to B_c$ fragmentation function and the induced $g \to B_c$ fragmentation function are coupled by the following Altarelli-Parisi equations:

$$\mu \frac{\partial}{\partial \mu} D_{\bar{b} \to B_c}(z, \mu) = \int_z^1 \frac{dy}{y} P_{\bar{b} \to \bar{b}}(z/y, \mu) \, D_{\bar{b} \to B_c}(y, \mu) + \int_z^1 \frac{dy}{y} P_{\bar{b} \to g}(z/y, \mu) \, D_{g \to B_c}(y, \mu) \,,$$
(1)

$$\mu \frac{\partial}{\partial \mu} D_{g \to B_c}(z, \mu) = \int_z^1 \frac{dy}{y} P_{g \to \bar{b}}(z/y, \mu) \, D_{\bar{b} \to B_c}(y, \mu) + \int_z^1 \frac{dy}{y} P_{g \to g}(z/y, \mu) \, D_{g \to B_c}(y, \mu) \,.$$
(2)

where $P_{i \to j}$ can be approximated by the usual massless Altarelli-Parisi splitting functions. Similar equations can be written down for the $^3S_1$, $^1P_1$, and ^3P_J ($J = 0, 1, 2$) states. The boundary conditions for the coupled equations are $D_{g \to B_c}(z, \mu) = 0$ for $\mu \leq 2(m_b + m_c)$ and $D_{\bar{b} \to B_c}(z, \mu_0 = m_b + 2m_c)$, which is the heavy quark fragmentation function calculated to the order of α_s^2 at the initial scale μ_0 by PQCD. Expressions for the initial fragmentation functions in S-wave states can be found in Ref.[3] and those for P-wave states can be found in Ref.[5]

Numerically integrating the coupled equations with the above boundary conditions, we obtain the direct $\bar{b}$ antiquark fragmentation functions and the induced gluon fragmentation functions for the S-wave and P-wave states at any arbitrary scale $\mu \geq \mu_0$. The inputs to these fragmentation functions are the quark masses m_b and m_c and the nonperturbative parameters associated with the wavefunction of the bound state. These nonperturbative parameters can be calculated within the framework of the potential models.[6] For the two S-wave states there is only one nonperturbative parameter, which is the radial wavefunction $R(0)$ at the origin. The P-wave fragmentation functions have two nonperturbative parameters associated with the color-singlet and color-octet mechanisms. Two of the P-wave states ($^1P_1$ and $^3P_1$) mix to form two physical states because they have the same quantum numbers. The two physical states are denoted by $|1+\rangle$ and $|1+'\rangle$. The mixing and further details can be found in Ref.[5]

The calculation of B_c meson production is simplified by factorizing the whole process into a short-distance process of producing the heavy quark and a long-distance process, which is the fragmentation of the heavy quark into the B_c meson. The differential cross-section for the B_c meson in the $^1S_0$ state is given by

$$d\sigma\left(B_c(p_T)\right) = \sum_{ij} \int f_{i/p}(x_1, \mu) f_{j/p}(x_2, \mu) \left[d\hat{\sigma}(ij \to \bar{b}(p_T/z)X, \mu) \, D_{\bar{b} \to B_c}(z, \mu) \right.$$
$$\left. + d\hat{\sigma}(ij \to g(p_T/z)X, \mu) \, D_{g \to B_c}(z, \mu) \right],$$
(3)

where i, j denote all the possible initial partons. The first term in the square bracket is the direct $\bar{b}$ fragmentation contribution and the second term is the induced gluon fragmentation contribution. Similar expressions can be written down for the $^3S_1$ and the P-wave states. In the above equation, the factorization scale μ is chosen to be of the order of p_T/z to avoid large logarithms in $d\hat{\sigma}$, while the large logarithms in $D(z)$

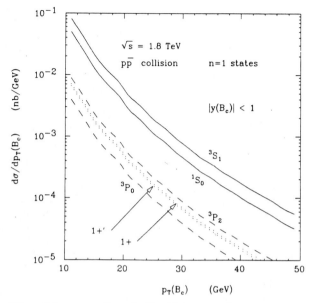

Fig. 1. The differential cross sections $d\sigma/dp_T(B_c)$ versus the transverse momentum $p_T(B_c)$ of the B_c meson for different spin-orbital states at the Tevatron.

can be summed up by evolving the $D(z)$ according to the coupled equations in Eqs. (1) and (2).

In our calculation the factorization scale μ in Eq. (3) is chosen to be

$$\mu = \sqrt{p_{T_{b,g}}^2 + m_b^2}\,, \tag{4}$$

where $p_{T_{b,g}}$ is the transverse momentum of the fragmenting parton. $p_{T_{b,g}}$ is related to $p_T(B_c)$ by $p_{T_{b,g}} = p_T(B_c)/z$. This scale is also used for the parton distributions and the running coupling constant α_s. Explicitly, we used $m_b = 4.9$ GeV, $m_c = 1.5$ GeV, and CTEQ2[7] for the parton distributions. The $\alpha_s(Q)$ is evaluated at 1-loop by evolving from the experimental value $\alpha_s(m_Z) = 0.118$ by $\alpha_s(Q) = \alpha_s(m_Z)/(1 + ((33 - 2n_f)/6\pi)\alpha_s(m_Z)\log(Q/m_Z))$, where n_f is the number of active flavors at the scale Q. We included $gg \to b\bar{b}$, $g\bar{b} \to g\bar{b}$, and $q\bar{q} \to b\bar{b}$ as the hard subprocesses for the inclusive $\bar{b}$ production, and $gg \to gg$, $gq(\bar{q}) \to gq(\bar{q})$, and $q\bar{q} \to gg$ for the inclusive g production. The fragmentation functions at the scale μ are obtained by solving Eqs. (1) and (2) with the boundary conditions mentioned above.

The transverse momentum spectra for different spin-orbital states are shown in Fig. 1, in which the contributions from the direct $\bar{b}$ fragmentation and the induced gluon fragmentation have been added, with the acceptance cuts

$$p_T(B_c) > 10 \text{ GeV} \qquad \text{and} \qquad |y(B_c)| < 1 \tag{5}$$

Table 1. The integrated cross sections in pb for the B_c mesons in various spin-orbital states, with the acceptance cuts in Eq. (5), at the Tevatron. $|1+\rangle$ and $|1+'\rangle$ are the two physical P-wave states resulted from the mixing of the $^1P_1$ and $^3P_1$ states.

	$n = 1$	$n = 2$
$^1S_0$	210	130
$^3S_1$	350	210
$^3P_0$	17	24
$^3P_2$	38	54
$|1+\rangle$	31	29
$|1+'\rangle$	28	54

on the B_c mesons. The curves in Fig. 1 are for $n = 1$ states. The radially excited $n = 2$ states can be calculated similarly with the corresponding nonperturbative parameters. We present the integrated cross sections for $n = 1$ and $n = 2$ states in Table 1.

Since the annihilation channel for the decay of the excited B_c meson states is highly suppressed relative to the electromagnetic and hadronic transitions, all the excited states below the BD threshold will decay into the ground states by emitting photons or pions. Thus, they all contribute to the inclusive production. Adding all the contributions shown in Table 1, we have a total cross section of 1.2 nb, which implies about 1.2×10^5 B_c mesons for 100 pb^{-1} at the Tevatron. This number should almost represent the total inclusive rate, except for a small contribution from the D-wave states.

Thus, we have presented the so far most complete B_c meson production via heavy quark fragmentation, including S-wave and P-wave contributions, at the Tevatron. The B_c meson can be detected via the decays into $J/\psi + X$, in which the J/ψ is fully reconstructed by the leptonic decay. If X is a charged lepton, then the event has a very distinct signature of three charged leptons coming out from the same displaced vertex. If X can be fully reconstructed, together with the reconstructed J/ψ the B_c meson can be fully reconstructed.

This work was supported by the U. S. Department of Energy, Division of High Energy Physics, under Grants DE-FG02-91-ER40684 and DE-FG03-91ER40674.

References

1. E. Braaten and T.C. Yuan, Phys. Rev. Lett. **71**, 1673 (1993).
2. K. Cheung, Phys. Rev. Lett **71**, 3413 (1993).
3. E. Braaten, K. Cheung, and T.C. Yuan, Phys. Rev. **D48**, R5049 (1993).
4. K. Cheung and T.C. Yuan, Phys. Lett. **B325**, 481 (1994).
5. T.C. Yuan, preprint UCD-94-2 (May 1994).
6. E. Eichten and C. Quigg, Phys. Rev. **D49**, 5845 (1994).
7. CTEQ Collaboration, J. Botts *et al.*, Phys. Lett. **B304**, 159 (1993).

HEAVY MESON DECAY CONSTANTS
WITH PHYSICAL HEAVY-QUARK MASSES*

T. ONOGI[†] and J.N. SIMONE

*Theory Group, FERMILAB, Box 500, Batavia,
IL 60510, USA*

ABSTRACT

We report preliminary numerical results for weak matrix elements of heavy mesons computed in the quenched approximation of Lattice QCD using a discretization of the heavy-quark action that allows calculations to be performed directly for the charm and bottom quark masses.

1. Introduction

The era of B factories will see B meson decay rates, mixing rates and asymmetries measured to unprecedented levels of precision. These data will lead to better determinations of the least well known CKM matrix elements such as V_{td}, V_{ub}, and V_{cb} that describe the electroweak couplings of the heavy quarks. Accurate values for these CKM matrix elements will greatly improve the sensitivity of tests for consistency of the Standard Model which may provide valuable clues to the origin of CP violation and the physics beyond the Standard Model.

The precision to which CKM matrix elements can be found depends upon the reliability of form factors that occur in combination with the elements of the CKM matrix in electroweak amplitudes. Consider the extraction of $|V_{td}|$ from B_d-$\overline{B}_d$ mixing. Within the Standard Model, the splitting between eigenstates, Δm_d, arises from box diagrams. The experimentally measurable ratio $x_d \equiv \Delta m_d / \Gamma$ is proportional to $|V_{td}^* V_{tb}|^2$ times the product of the B meson bag parameter, $\hat{B}_d$, and the *square* of the B meson leptonic decay constant, f_{B_d}. Currently, the theoretical uncertainty in f_{B_d} is the dominant source of uncertainty in $|V_{td}|$ determined from B_d-$\overline{B}_d$ mixing.[1]

Electroweak form factors cannot be calculated reliably in perturbation theory due to long distance QCD effects. Lattice QCD is the only nonperturbative, first principles, systematically improvable method available for calculating such form factors.

Many lattice studies of weak matrix elements are underway.[2] Studies, such as those calculating the leptonic decay constants[2-5] f_{B_d} and f_{B_s} are already having an impact upon B phenomenology. Building upon these studies requires understanding the effect of the quenched approximation and continued careful attention to sources of systematic error on the lattice such as finite volume errors, discretization errors, and uncertainties in extrapolations of lattice light quarks to physical light-quark masses.

---

*Presented by J. Simone
[†]On leave from: Dept. of Physics, Hiroshima University

Weak matrix elements for B mesons have typically been obtained in lattice QCD by computing matrix elements using Wilson[7] or Sheikholeslami-Wohlert[8] * (SW) fermions for the heavy quarks, keeping heavy-quark masses around the charm quark mass, and with infinitely heavy quarks using Static fermions.[9] The heavy-quark mass dependence of these results yields, by interpolation, matrix elements for the b quark. On current lattices, directly simulating the b quark with the standard definition of the field renormalization, appropriate for light Wilson and light SW fermions, would introduce large discretization errors since the b-quark mass is large compared to the lattice spacing, a.

This study presents preliminary numerical results for heavy-light mesons using a general formalism for fermions with arbitrary quark mass. This general fermion action[10] has the nonrelativistic heavy-quark action,[11] the Wilson fermion action, and the SW fermion action as well defined limits. In particular, the infinite quark mass limit corresponds to the Static approximation. A fermion action is written as a series expansion of the continuum action in powers of a times local discrete gauge-invariant lattice operators with coefficients that are functions of g^2 and the quark mass in lattice units, am_Q. The coefficients can be expanded order-by-order in g^2. Such expansions allows discretization errors to be estimated and systematically reduced by including in the discrete action terms of higher order in a.[12]

With suitably defined massive fermions, matrix elements can be directly computed for heavy quarks with masses near the physical c-quark and b-quark masses. This methodology for heavy quarks has been used to study Charmonium and Bottomonium spectroscopy.[14] Some of the techniques for treating massive Wilson fermions, needed to remove the largest discretization errors, have already been used in to compute decay constants.[3] This study aims to apply this formalism systematically to $o(a)$.

This study uses the mean field improved[13] SW action for both heavy quarks and light quarks. For massive fermions, the fermion fields that appear in the Wilson and SW actions differ from the canonically normalized fermion fields that must be used to form properly normalized current operators. To $o(a)$, the relation between the fields in the action and the fermion fields in currents is given by the "rotation"[10]

$$\psi^{current} = \sqrt{2\tilde{\kappa}}\, e^{am_Q/2} \left[1 + ad_1(am_Q, g^2)\vec{\gamma} \cdot \vec{D}\right] \psi^{action} \qquad (1)$$

where $\tilde{\kappa}$ is the mean field improved quark hopping parameter.[13] The effect of the d_1 term for large quark masses is an additional $O(1/m_Q)$ correction to matrix elements. The coefficient d_1 has a perturbative expansion in the QCD coupling parameter g^2. This study uses the tree level value for d_1. With tree level coefficients, these $o(a)$ corrections are about 10% or less for typical lattice spacings.

2. Results

Results are reported for a spacetime lattice of 16^3 spatial sites by 32 time sites and a sample of 60 gauge configurations for which effects of the disconnected fermion loops upon the gauge fields have been neglected (quenched approximation). The gauge

---

*The SW fermion action differs from the Wilson fermion action by an additional $o(a)$ improvement term.

configurations were generated at $\beta \equiv 6/g_0^2 = 5.9$, where g_0 is the bare gauge coupling parameter, giving an inverse lattice spacing of $a^{-1} = 1.75\,\text{GeV}$ as determined from the $1P$-$1S$ mass splitting in Charmonium.[14]

Leptonic decay constants are obtained for mesons composed of one light valence quark of mass m_q and one heavy valence quark of mass m_Q. Results are presented for the light-quark mass *fixed* near the strange quark mass, $m_q \approx m_s$, and variable heavy-quark mass. Heavy quarks with masses of approximately 1.5, 1.8, 2.6, 4.6, and 6.1 GeV are used.

Two-point correlators of the lattice axial current, A^μ, and the operator $\chi = \overline{q}\gamma_5 Q$ having nonzero overlap, Z_5, with the pseudoscalar channel become,

$$\sum_{\vec{x}} e^{-i\vec{p}\cdot\vec{x}} \left\langle \chi(x)\,\chi(0)^\dagger \right\rangle \quad \underset{0 \ll t \ll T}{\longrightarrow} \quad Z_5^2 \cosh\left(E(T/2 - t) \right) \tag{2}$$

$$\sum_{\vec{x}} e^{-i\vec{p}\cdot\vec{x}} \left\langle A^\mu(x)\,\chi(0)^\dagger \right\rangle \quad \longrightarrow \quad Z_5 \frac{Z_a^{-1} f p^\mu}{\sqrt{2E}} \sinh\left(E(T/2 - t) \right) \tag{3}$$

for "large" Euclidean times, $0 \ll t \ll T$, on a periodic lattice of extent T. The energy $E(\vec{p})$, $Z_a^{-1} f p^\mu / \sqrt{2E}$, and Z_5 are found by a minimal χ^2 fit of the lattice correlators to the functional forms in Eqns. 2 and 3. Statistical errors are determined by a bootstrap procedure.[16]

The perturbatively calculable renormalization constant, Z_a, relates the lattice axial current to the continuum current,

$$A_\mu^{cont} = Z_a \overline{\psi}^{current} \gamma_\mu \gamma_5 \psi^{current} \tag{4}$$

in order that continuum decay constants may be obtained. Perturbative calculations are underway.[15]

In the limit $m \to \infty$, the quantity $f\sqrt{m}\,[\alpha_s(m)]^{2/\beta_0}$ tends to a constant. It is then convenient to define

$$\hat{\Phi}(m) \equiv \left[\frac{\alpha_s(m)}{\alpha_s(m_B)} \right]^{2/\beta_0} Z_a^{-1} f \sqrt{m} \tag{5}$$

where α_s is the QCD coupling parameter and $\beta_0 = 11 - \frac{2}{3}n_f$, and study the dependence of $\hat{\Phi}(m)$ upon $1/m$.

Figure 1 shows $\hat{\Phi}$ plotted as a function of $1/m$ for the values of m_Q used in this study. Also shown is the value for $\hat{\Phi}(\infty)$, with $m_q = m_s$, obtained by the Fermilab Static Collaboration,[4] using static heavy quarks and the same lattice but with a superset of 100 gauge configurations. The solid curve is a fit of the five points in this study to a quadratic function of $1/m$. The two dashed curves above and below the solid curve are the 68% confidence limits to the fit. Note that the extrapolation of the finite mass results of this study to the infinite mass limit agrees well with the static result. In the future, we plan to include the static result in fits.

Figure 1 indicates large deviations in $\hat{\Phi}(m)$ from the scaling limit, $\hat{\Phi}(\infty)$, for $m = m_{B_s}$ and $m = m_{D_s}$. Studies of the infinite quark mass limit have shown a large

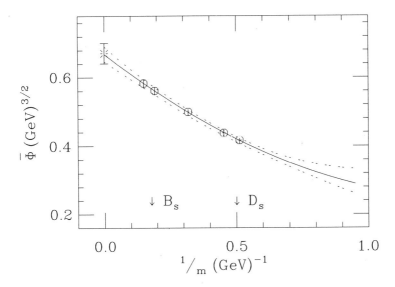

Fig. 1. The scaling quantity $\hat{\Phi}$ plotted as a function of the reciprocal of the pseudoscalar mass. The light-quark mass is approximately equal to the strange quark mass. These preliminary results are for $\beta = 5.9$ corresponding to a lattice spacing of $a = 0.11\,\mathrm{fm}$. Octagons are results from this study. The burst is the static result in reference 4 for this lattice.

dependence of $\hat{\Phi}(\infty)$ upon the lattice spacing.[4] This then calls into question the extent to which these large scaling violations at *finite* lattice spacing will persist after extrapolation to zero lattice spacing.

3. Conclusion

This study presents some first results for leptonic decay constants from our investigation of weak matrix elements using an $o(a)$-improved heavy quark action. The important feature of this study is that it aims to systematically correct matrix elements for massive fermions through $o(a)$ whereas, some previous studies[3,5] have included only some of the $o(a)$ corrections for heavy quarks. Differences in $o(a)$ effects between this study and these other studies for a given typical lattice spacing are expected to be on the order of 10%.

The cost of numerical simulation at fixed physical volume is roughly proportional to a^{-6}. "Brute force" reduction of discretization errors by greatly reducing the lattice spacing is prohibitively expensive. It is then cost effective to systematically improve the action, through analytic effort, and perform simulations with errors of $o(a^2)$ or less. Then, less expensive simulations, performed on lattices coarser than otherwise would be necessary, can be used to extrapolate results to zero lattice spacing.

With the massive quark formulation, quantities should approach the Static result in the infinitely massive heavy quark limit. We test this by extrapolating the scaling quantity, $\hat{\Phi}(m)$, to the infinitely massive heavy quark limit and find good agreement with the result in the Static approximation.

We have selected a rather coarse lattice spacing compared to lattice spacings that are considered "state of the art" for weak matrix element calculations with heavy quarks in order to test the effect of improvement. In this preliminary study, we do not yet report values for decay constants. We note that a thorough study of leptonic decay constants must include the following goals: a) Study the light-quark mass dependence of $\hat{\Phi}(m)$ and determine decay constants in the chiral limit, $m_q \to m_d$. b) Determine the axial current renormalization constant Z_a. c) Improve upon the tree level calculation of coefficient d_1 in the renormalization of $\psi^{current}$. d) Reduce residual $o(a)$ errors by extending the study to additional lattice spacings and extrapolating results to zero lattice spacing. e) Estimate finite volume errors by studying the volume dependence of decay constants. f) Estimate the effects due to the quenched approximation.

Acknowledgements

We wish to thank our collaborators A. El-Khadra, G. Hockney, A. Kronfeld, and P. Mackenzie. We also thank E. Eichten and H. Thacker for valuable discussions. T. O. acknowledges the support of the Nishina Foundation. These calculations were performed on the Fermilab ACPMAPS supercomputer.

References

1. J. Rosner, these proceedings.
2. reviewed by C. Bernard in *Nucl. Phys.* **B (Proc. Suppl.)** (1994) 47.
3. C. Bernard, *et al.*, *Phys. Rev.* **D49** (1994) 2536.
4. A. Duncan, *et al.*, FERMILAB-PUB-94/164-T.
5. UKQCD Collaboration, R. Baxter, *et al.*, *Phys. Rev.* **D49** (1994) 1594.
6. P. Mackenzie, these proceedings.
7. K. Wilson, in *New Phenomena in Subnuclear Physics*, ed. by A. Zichichi (Plenum, New York, 1977).
8. B. Sheikholeslami and R. Wohlert, *Nucl. Phys.* **B259** (1985) 572.
9. E. Eichten and F. Feinberg, *Phys. Rev.* **D23** (1981) 2724.
10. A. El-Khadra, *et al.*, in progress, FERMILAB-PUB-93/195-T.
11. B. Thacker and G. Lepage, *Phys. Rev.* **D43** (1991) 196.
12. K. Symanzik, in *Recent Developments in Gauge Theories*, ed. by G. 't Hooft, *et al.*(Plenum, New York, 1980).
13. G. Lepage and P. Mackenzie, *Phys. Rev.* **D48** (1993) 2250.
14. A. El-Khadra, these proceedings.
15. A. Kronfeld and B. Mertens, *Nucl. Phys.* **B (Proc. Suppl.)** (1994) 495.
16. B. Efron, *SIAM Review* **21** (1979) 460

ESTIMATES OF HEAVY QUARK SYMMETRY-BREAKING EFFECTS IN MESON DECAY CONSTANTS USING POTENTIAL MODELS

JAMES F. AMUNDSON

Department of Physics, University of Wisconsin–Madison,
1150 University Avenue, Madison, Wisconsin, 53706, USA

ABSTRACT

Heavy quark symmetry predicts the ratio of D- and B-meson decay constants in the limit of infinite charm and bottom quark masses. Results from both lattice and QCD sum rule calculations indicate there may be significant finite-mass corrections to this ratio. I discuss a potential model calculation of these decay constants that is consistent with heavy quark symmetry at both leading and subleading order in the heavy quark mass. This calculation gives insight into both the probable size of heavy quark symmetry-breaking effects and the limitations of potential models.

1.   Decay Constants in Heavy Quark Effective Theory

The scaling of heavy-light decay constants with heavy quark (meson) mass in the $m_Q \to \infty$ limit

$$f_M \sqrt{m_M} = F = \text{const.} \tag{1}$$

has been known since the sixties. The logarithmic radiative corrections[1] to the above relation

$$\frac{f_B}{f_D} = \left[\frac{\alpha_s(m_c)}{\alpha_s(m_b)}\right]^{6/25} \sqrt{\frac{m_D}{m_B}} \tag{2}$$

can be considered first result of Heavy Quark Effective Theory (HQET). In the limit of infinite charm (and bottom) quark mass Eq. 2 allows the experimentally feasible measurement of f_D to tell us the experimentally elusive value of f_B.

Unfortunately, the charm quark mass is not infinite, so finite-mass corrections to Eq. 2 must be investigated. Including the leading finite-mass corrections (while neglecting the radiative corrections for simplicity) gives

$$f_M \sqrt{m_M} = F \left[1 + \frac{1}{m_Q}(G_1 + 2d_M G_2) - d_M \frac{\tilde{\Lambda}}{6m_Q}\right], \tag{3}$$

where $d_M = -3$ $(+1)$ for pseudoscalar (vector) mesons. To this order there are $3 + 1$ constants in the theory. F gives the normalization of decays constants in the infinite-mass limit. G_1 and G_2 represent the effects of insertion of the heavy quark kinetic energy operator and the chromomagnetic operator, respectively. The remaining ("+1") constant, $\tilde{\Lambda} = \overline{\Lambda} - m_{\text{light}} \approx \overline{\Lambda} \equiv m_M - m_Q$, also appears in other HQET calculations. These constants are not calculable using perturbative QCD, so they must be estimated in models or in lattice QCD.

QCD sum rules and lattice calculations estimates of these quantities roughly agree[3] that $F(\mu = m_b) \approx (0.4\text{--}0.6)$ GeV$^{3/2}$. Unfortunately, at least for phenomenological applications of HQET, estimates also agree that the $1/m_Q$-suppressed corrections are large even for f_B, of order 10–20%, so they may be prohibitively large for f_D.

2. Results from Potential Models

In the nonrelativistic quark model, meson decay constants are given by

$$f_M \sqrt{m_M} = \sqrt{12}|\psi_M(\boldsymbol{r} = 0)|, \tag{4}$$

where ψ_M is the solution to the Schrödinger equation with a phenomenological potential. The Hamiltonian used in this work is

$$\mathcal{H} = \frac{p^2}{2m_l} + \frac{p^2}{2m_Q} - \frac{4\alpha_s}{3r} + ar + \frac{2\boldsymbol{\sigma_1} \cdot \boldsymbol{\sigma_2}}{3m_l m_Q}\delta^3(\boldsymbol{r}), \tag{5}$$

where m_l is the light quark constituent mass. In the heavy quark limit the two terms explicitly involving the heavy quark mass, i.e., the heavy quark kinetic energy and the hyperfine interaction, must be ignored. Doing so gives a result consistent with Eq. 3 at leading order in $1/m_Q$. (Of course, the scaling of decay constants was deduced from potential models long before the advent of HQET.) Inserting the resultant wave function in Eq. 4 gives $F \approx 0.55$ GeV$^{3/2}$.

Including the heavy quark kinetic energy and the hyperfine interaction in the Hamiltonian gives a result consistent with Eq. 3 to subleading order in the heavy quark mass, except that the term involving $\tilde{\Lambda}$ is absent. The discrepancy arises from the lack of distinction between the constituent and current quark masses in the nonrelativistic quark model. Since $\overline{\Lambda} \equiv m_l$ in the model, $\tilde{\Lambda}$ vanishes. In reality, the relevant mass for a process with $q^2 = m_B^2$ is undoubtedly not the constituent quark mass. The NRQM is simply wrong in this case. Fortunately, however, adding in the $\tilde{\Lambda}$ term by hand corrects the problem at this order.

In the model calculation, G_1 is proportional to the effect of the perturbation due to the heavy quark kinetic energy term. Numerically, it comes out to 400 MeV, substantially smaller than other estimates.[3] G_2 comes from the perturbation due to the hyperfine interaction. Unfortunately, the resultant wave function at the origin diverges. Reasonable schemes for regulating this divergence give a magnitude of G_2 that is substantially smaller than G_1. Nonetheless, the result is unsatisfactory in that G_2 depends on scheme introduced to regularize it.

In an attempt to understand the robustness of the results obtained with the nonrelativistic quark model, we can consider the effect of replacing the nonrelativistic kinetic energy of the light quark, $p^2/2m_l$, with the fully relativistic expression $\sqrt{p^2 + m_l^2}$. A subtlety immediately arises: the resultant wave function diverges at the origin. This is a feature common to relativistic wave equations. The problem can be resolved by considering both the running of α_s in the potential and the radiative corrections to the decay constant in HQET. Then the small-r divergence of the wave function

$$\Psi(r \to 0) \propto [\ln(r_0/r)]^{4/3\beta_0} \tag{6}$$

is exactly canceled by the μ-dependence of the vertex correction in HQET. The result is therefore finite and unambiguous as long as the radiative correction to the decay constant is included. The result gives $F(\mu = m_b) \approx 460$ MeV. This is only about 20% different from the result using the NRQM.

Results for G_1 and G_2 using this model do differ substantially from those using the NRQM, however. G_2, divergent in the fully nonrelativistic model, is finite in the relativistic model as long as the radiative corrections to the chromomagnetic operator are included. The numerical calculation gives $G_2 \approx -50$ MeV, however the calculation proves to be very sensitive and should be considered preliminary. Nonetheless, the answer is qualitatively consistent with the NRQM result being much smaller than the other $1/m_Q$ effects. The important difference is in the G_1 term. G_1 in the model with the relativistic kinetic energy comes out to be -1.1 GeV. This is nearly three times larger than the result in the NRQM.

Why do the calculations of G_1 differ so much between the two models while the calculation of F differs so little? The answer can be seen by considering ordinary nonrelativistic quantum mechanical perturbation theory:

$$G_1 \propto \sum_{n \neq 0} \frac{\phi_n(0)}{E_n - E_0} \langle \phi_n | p^2 | \phi_0 \rangle. \tag{7}$$

Here ϕ_n are the complete set of S-wave bound states which are the solution to unperturbed (infinite-mass) problem. The explicit factor p^2 makes the result sensitive to the large-p dependence of the wave functions. This is precisely where the difference between the nonrelativistic and relativistic kinetic energies becomes apparent. The NRQM fails to get the large-p dependence of the wave functions. This result has implications for other estimates using the NRQM of processes which are sensitive to the large-p dependence of the wave functions, $e.g.$ $B \to K^*\gamma$.

3. Conclusions

The NRQM estimates of $1/m_Q$ corrections to the heavy quark limit for decay constants are substantially smaller than other estimates. Including the relativistic kinetic energy of the light quark brings the results into qualitative agreement with other estimates. The $1/m_Q$ effects turn out to be sensitive probes of heavy-light wave functions. Unfortunately, the calculation indicates that the $1/m_c$ effects may be prohibitively large for f_D.

References

1. M.B. Voloshin and M.A. Shifman, Yad. Fiz. **45**, 463 (1987); H.D. Politzer and M.B. Wise, Phys. Lett. B **206**, 681 (1988); **208**, 504 (1988).
2. M. Neubert, Phys. Rev. D **46**, 1076 (1992).
3. See, $e.g.$, P. Ball, Nucl. Phys. **B421**, 593 (1994) and references therein.

$\frac{1}{M}$ CHIRAL CORRECTIONS FOR B DECAYS

C. GLENN BOYD

Dept. of Physics 0319, University of California at San Diego
La Jolla, CA 92093

ABSTRACT

The leading nonanalytic heavy quark and chiral symmetry violating corrections to heavy meson decay constants are computed in heavy quark chiral perturbation theory, and implications to B decays are discussed; measurements of the experimentally accessible form factors for $D \to \pi e \nu$ and $B \to \pi e \nu$, along with knowledge of the $D^* D \pi$ coupling, determine the eight decay constants $f_D, f_{D_s}, f_{D^*}, f_{D_s^*}$, f_B, f_{B_s}, f_{B^*}, and $f_{B_s^*}$, as well as one relation among the $B \to \pi e \nu$ form factors, to $\mathcal{O}(\frac{1}{M})$. The ratio $R_1 = \frac{f_{B_s}}{f_B} / \frac{f_{D_s}}{f_D}$ is expressed in terms of two dimensionful couplings.

1.Introduction

The incorporation of chiral and heavy quark symmetries into a Lagrangian governing the interactions of heavy mesons with pseudo-Nambu-Goldstone bosons leads to new relations among heavy meson leptonic and semi-leptonic form factors[1]. Because the charm quark mass is not large, and the strange mass not small, one expects significant deviations from the heavy quark and chiral limits in many processes. It is therefore important to compute these corrections both to improve the accuracy of theoretical relations and to test the validity of the perturbative expansion.

Of particular interest is the ratio [2] $R_1 = \frac{f_{B_s}}{f_B} / \frac{f_{D_s}}{f_D}$, which is relevent to the extraction from $B - \bar{B}$ mixing of the CKM angle V_{ts}. To see how R_1 deviates from unity requires the heavy quark chiral Lagrangian to order $\mathcal{O}(\frac{1}{M})$. We present both the Lagrangian and the heavy-light current to this order, and discuss implications for $B \to \pi e \nu$ decays. We will see that all the unknown parameters involved in heavy decay constants can be determined, in principle, from $B \to \pi l \nu$ and $D \to \pi l \nu$ form factors. The leading heavy quark and chiral symmetry violating corrections to heavy meson decay constants will then be computed using a chiral log approximation.

The low momentum strong interactions of B and B^* mesons are governed, at $\mathcal{O}(\frac{1}{M})$, by the chiral Lagrangian

$$\mathcal{L} = -\operatorname{Tr}\left[\overline{H}_a(v) i v \cdot D_{ba} H_b(v)\right] + \tilde{g}_{\overline{HH}} \operatorname{Tr}\left[\overline{H}_a(v) H_b(v) \, \slashed{A}_{ba} \gamma_5\right] \qquad (1)$$

where $\tilde{g}$ is $\tilde{g}_{B^*} = g + \frac{1}{M}(g_1 + g_2)$ for $B^* B^*$ couplings and $\tilde{g}_B = g + \frac{1}{M}(g_1 - g_2)$ for $B^* B$, and M is the heavy quark mass. The field ξ contains the octet of pseudo-Nambu-Goldstone bosons, while the field A represents the $SU(3)$ axial current. The B and B^* heavy meson fields are incorporated into the 4×4 matrix $H_a = \frac{1}{2}(1 + \slashed{v})\left[\overline{B}_a^{*\mu} \gamma_\mu - \overline{B}_a \gamma_5\right]$, where a is an isospin index and v is the four-velocity of the heavy meson. To the order we will be working, the propagators for heavy mesons should include the effects of hyperfine splitting $\Delta = M_{B^*} - M_B$ and $SU(3)$ splitting $\delta = M_{D_s} - M_D$. Details are presented in reference [3].

The left handed current is represented by

$$J_a^\lambda = \frac{i\alpha}{2}(1 + \frac{\rho_1}{M}) \operatorname{Tr}[\Gamma H_b(v) \xi_{ba}^\dagger] + \frac{i\alpha}{2} \frac{\rho_2}{M} \operatorname{Tr}[\gamma^\mu \Gamma \gamma_\mu H_b(v) \xi_{ba}^\dagger], \qquad (2)$$

where $\Gamma = \gamma^\lambda (1 - \gamma_5)$.

All formulas hold for the D meson as well, after the substitution $M_B, f_B \to M_D, f_D$. Higher derivative operators have been shown to be irrelevent to the processes and order considered here[3].

2. $B \to \pi l \nu$ Form Factors

To go beyond leading order, we make some approximations based on the formal hierarchy $m_\pi \ll m_K \ll \bar\Lambda \sim \Lambda_\chi \ll M$ and $\ln \frac{\mu^2}{m_K^2} \gg 1$, where $\bar\Lambda = M_B - M_b$ and the subtraction scale μ is of order Λ_χ, the chiral symmetry breaking scale. Theoretically this hierarchy holds arbitrarily well in the $m_{u,d,s} \to 0$ and $M \to \infty$ limits. Numerically, there may be large $\mathcal{O}(m_K^2)$ corrections.

The decay constants defined by

$$\langle 0 \,|\, \bar{q}_a \gamma^\mu \gamma^5 b \,|\, \overline{B}_a(p) \rangle = i f_{B_a} p^\mu, \qquad \langle 0 \,|\, \bar{q}_a \gamma^\mu b \,|\, \overline{B^*_a}(p, \epsilon) \rangle = i f_{B^*_a} \epsilon^\mu \qquad (3)$$

are altered only by current corrections at $\mathcal{O}(\frac{1}{M})$:

$$\sqrt{M_D} f_D = \alpha[1 + \frac{\rho_1 + 2\rho_2}{M_D}] \qquad \frac{1}{\sqrt{M_D}} f_{D^*} = \alpha[1 + \frac{\rho_1 - 2\rho_2}{M_D}] \qquad (4)$$

We can also compute the leading SU(3) symmetric but heavy quark spin/flavor violating corrections to semileptonic $\overline{B} \to K$. The relevent matrix elements are

$$\langle \overline{K}(p_K) \,|\, \bar{s} \gamma^\mu b \,|\, \overline{B}(p_B) \rangle = f_+ (p_B + p_K)^\mu + f_- (p_B - p_K)^\mu \,,$$
$$\langle \overline{K}(p_K) \,|\, \bar{s} \sigma^{\mu\nu} b \,|\, \overline{B}(p_B) \rangle = 2ih \,[p_K^\mu p_B^\nu - p_K^\nu p_B^\mu] \,. \qquad (5)$$

Only the form factors $h(p_K \cdot p_B)$ and $f_+(p_K \cdot p_B)$ enter into the differential partial decay rate[4].

The operators which match onto the heavy-light currents above are linear combinations of the left and right handed heavy to light current. A B meson can emit a pion and turn into a B^* which is then annihilated by the current, or it can interact with the current directly. Computation of the two corresponding graphs using the current eq. (2) gives

$$f_+ = -\frac{1}{2f_\pi}[\frac{\tilde{g}_B f_{B^*}}{v \cdot k - \Delta} - \frac{\tilde{g}_B f_{B^*}}{M_B} + f_B], \qquad h = -\frac{1}{2f_\pi M_B}[\frac{\tilde{g}_B f_{B^*}}{v \cdot k - \Delta}(1 + \frac{2\rho_2}{M_B})], \quad (6)$$

where k is the pion momentum. In principle, measurements of $f_+^{(B)}$, $f_+^{(D)}$ and either $h^{(B)}$ or $h^{(D)}$, coupled with a precise measurement of $D^* \to D\pi$, would determine all six unknown constants α, ρ_1, ρ_2, g, g_1 and g_2.

3. Loop Corrections To Decay Constants

The leading SU(3) and heavy quark symmetry violating contributions to f_B come from one loop diagrams involving both virtual kaons and $\mathcal{O}(\frac{\bar\Lambda}{M})$ heavy quark violating contact terms (with coefficients ρ_1, ρ_2, g_1, g_2). The leading chiral SU(3) violation arises from non-analytic dependence on the strange quark mass m_s. Since this nonanalytic dependence is only generated by chiral loops, we can compute chiral and heavy quark symmetry violation, at leading order, in terms of the six parameters α, ρ_1, ρ_2, g, g_1 and g_2.

Wavefunction renormalization is computed using the Gell-Mann-Okubo formula $m_\eta^2 = \frac{4}{3}m_K^2$. Vertex corrections consist of one graph in which a B converts to a pion loop via the current. The final results for the decay constants, valid to $\mathcal{O}(\frac{\bar{\Lambda}}{M}, \frac{m_K^2}{M})$, are found by combining the wavefunction and vertex corrections:

$$FB = \hat{\alpha}\left\{1 - \hat{C}_1 \frac{m_K^2 \ln\frac{m_K^2}{\mu^2}}{16\pi^2 f_K^2}\left(\hat{g}^2 + \frac{1}{3}\right) + \hat{C}_2 \frac{x^2 \hat{g}^2}{\pi^2 f_K^2}\left[\ln\frac{m_K^2}{\mu^2} + 2F(\frac{m_K}{x})\right]\right\}, \qquad (7)$$

where FB may be either $\sqrt{M}f_B$ or $\frac{1}{\sqrt{M}}f_{B^*}$, $\hat{\alpha} = \alpha(1 + \frac{\rho_1+2\rho_2}{M})$ for the B and B_s, $\hat{\alpha} = \alpha(1 + \frac{\rho_1-2\rho_2}{M})$ for the B^* and B_s^*, $C_1 = \frac{11}{6}$ for B, B^* and $C_1 = \frac{13}{3}$ for B_s, B_s^*, $C_2 = \frac{3}{16}, \frac{3}{8}, \frac{1}{16}, \frac{1}{8}$ for the B, B_s, B^*, B_s^*, respectively, $x = \delta + \Delta$ for B, B_s^* and $x = \delta - \Delta$ for B^*, B_s, $\hat{g} = \tilde{g}_B$ for B, B_s, and $\hat{g} = g + \frac{g_1+\frac{1}{3}g_2}{M}$ for the B^*, B_s^*. The function F may be found in [3].

Substituting physical values and a subtraction scale of $\mu = 1$ GeV, into these equations, it follows that the ratio of decay constants R_1 is

$$R_1 - 1 = -.10g^2 - .14\text{GeV}^{-1}g(g_1 - g_2)] \qquad (8)$$

If we take $g^2 = .5$ and $g_1 = g_2$, we get a -5% correction to R_1. A better determination of R_1 requires information about g_1 and g_2, which we have shown may be determined from $B, D \to \pi l\nu$ form factors. These parameters also enter into relations for $\overline{B}_s \to D_s l\nu$ decays, and hyperfine mass splittings. These relations are worth investigating.

References

[1] M. Wise, Phys. Rev. D45 (1992) 2188;

    G. Burdman and J. F. Donoghue, Phys. Lett. 280B (1992) 287;

    T. M. Yan et al. , Phys. Rev. D46 (1992) 1148

[2] B. Grinstein, Phys. Rev. Lett. 71 (3067) 1993.

[3] G. Boyd and B. Grinstein, UCSD/PTH 93-46 and SMU-Hep/94-03 (hep-ph/9402340)

[4] N. Isgur and M. Wise, Phys. Rev. D42 (1990) 2388.

QCD POTENTIAL MODEL FOR LIGHT-HEAVY QUARKONIA AND THE HEAVY QUARK EFFECTIVE THEORY

SURAJ N. GUPTA and JAMES M. JOHNSON*

Department of Physics, Wayne State University, Detroit, Michigan, 48202

ABSTRACT

We have investigated the spectra of light-heavy quarkonia with the use of a quantum-chromodynamic potential model which is similar to that used earlier for the heavy quarkonia. An essential feature of our treatment is the inclusion of the one-loop radiative corrections to the quark-antiquark potential, which contribute significantly to the spin-splittings among the quarkonium energy levels. Unlike $c\bar{c}$ and $b\bar{b}$, the potential for a light-heavy system has a complicated dependence on the light and heavy quark masses m and M, and it contains a spin-orbit mixing term. We have obtained excellent results for the observed energy levels of D^0, D_s, B^0, and B_s, and we are able to provide predicted results for many unobserved energy levels.

We have also used our investigation to test the accuracy of the heavy quark effective theory. We find that the heavy quark expansion yields generally good results for the B^0 and B_s energy levels provided that M^{-1} and $M^{-1}\ln M$ corrections are taken into account in the quark-antiquark interactions. It does not, however, provide equally good results for the energy levels of D^0 and D_s, which shows that the effective theory can be applied more accurately to the b quark than the c quark.

1.   Introduction

The light-heavy quarkonia D, D_s, B, and B_s are at present of much experimental and theoretical interest, and their exploration is necessary for our understanding of the strong as well as the electroweak interactions. We shall here investigate the spectra of light-heavy quarkonia with the use of a quantum chromodynamic model similar to the highly successful model used earlier for the heavy quarkonia $c\bar{c}$ and $b\bar{b}$.[1] The complexity of the model is necessarily enhanced for a light-heavy system because the potential has a complicated dependence on the light and heavy quark masses m and M, and it contains a spin-orbit mixing term.

2.   Light-heavy Quarkonium Spectra

Our treatment for the light-heavy quarkonia is similar to that for $c\bar{c}$ and $b\bar{b}$ except for the complications arising from the difference in the quark and antiquark masses.

---

*Presented by James M. Johnson

Thus, our model is based on the Hamiltonian

$$H = H_0 + V_p + V_c, \quad H_0 = (m^2 + \mathbf{p}^2)^{1/2} + (M^2 + \mathbf{p}^2)^{1/2} \quad (1)$$

where V_p and V_c are nonsingular quasistatic perturbative and confining potentials. Since our potentials are nonsingular, we are able to avoid the use of an illegitimate perturbative treatment.

The experimental and theoretical results for the energy levels of the light-heavy quarkonia D^0, B^0, D_s, and B_s, together with the $^3P_1'$-$^1P_1'$ mixing angles arising from the spin-orbit mixing terms, are given in Table 1. For experimental data we have relied on the Particle Data Group[2] except that we have used the more recent results from the CLEO collaboration[3,4] for D_1^0, D_2^{*0}, and D_{s2} and from the CDF collaboration[5] for B_s. In this table, one set of theoretical results corresponds to the direct use of our model, while the other two sets are obtained by means of heavy quark expansions of our potentials to test the accuracy of the heavy quark effective theory with the inclusion of the M^{-1} and $M^{-1} \ln M$ corrections as well as without these corrections.

We expect the dynamics of a light-heavy system to be primarily dependent on the light quark. Therefore, our potential parameters for D^0 and B^0 are the same except for the difference in the c and b quark masses. We have also ensured that the parameters for D_s and B_s are related to those for D^0 and B^0 through quantum chromodynamic transformation relations.

3. Conclusion

We have obtained excellent results for the observed energy levels of D^0, B^0, D_s, and B_s, and provided predicted results for many unobserved energy levels in Table 1. Although the use of a semirelativistic model may seem questionable for a system containing a light quark, ultimately such an approach should be judged on the basis of its predictions. Additional experimental data on the light-heavy quarkonia should be available in the near future.

We have also used our results to test the accuracy of the heavy quark effective theory. According to Table 1, the heavy quark expansion with the inclusion of the M^{-1} and $M^{-1} \ln M$ corrections yields generally good results for the B^0 and B_s energy levels. It does not, however, provide equally good results for the energy levels of D^0 and D_s, which indicates that the effective theory can be applied more accurately to the b quark than the c quark. We further find that the results for the energy levels in the limit $M \to \infty$ are unacceptable.

This work was supported in part by the U.S. Department of Energy under Grant No. DE-FG02-85ER40209.

References

1. S. N. Gupta, J. M. Johnson, W. W. Repko, and C. J. Suchyta III, Phys. Rev D **49**, 1551 (1994). See also earlier related papers cited therein.
2. Particle Data Group, K. Hikasa et al., Phys. Rev. D **45**, S1 (1992).
3. CLEO Collaboration, Y. Kubota et al., Phys. Rev. Lett. **72**, 1972 (1994).
4. CLEO Collaboration, P. Avery et al., Phys. Lett. B **331**, 236 (1994).
5. CDF Collaboration, F. Abe et al., Phys. Rev. Lett. **71**, 1685 (1993).

Table 1. D^0, D_s, B^0 and B_s energy levels in MeV. Effective theory results are given with the M^{-1} and $M^{-1}\ln M$ corrections as well as in the limit of $M \to \infty$.

	Expt.	Theory	Effective theory	$M \to \infty$
$1\,^1S_0\ (D^0)$	1864.5±0.5	1864.5	1864.5	1864.5
$1\,^3S_1\ (D^{*0})$	2007±1.4	2007.0	2010.9	1864.5
$2\,^1S_0$		2547.7	2566.5	2431.9
$2\,^3S_1$		2647.0	2662.1	2431.9
$1\,^3P_0$		2278.6	2310.2	2244.8
$1\,^3P_1'$		2407.3	2414.6	2244.8
$1\,^3P_2\ (D_2^0)$	2465±4.2	2465.0	2474.0	2287.2
$1\,^1P_1'\ (D_1^0)$	2421±2.8	2421.0	2438.2	2287.2
θ		29.0°	30.9°	35.6°
$1\,^1S_0\ (D_s)$	1968.8±0.7	1968.8	1968.8	1968.8
$1\,^3S_1\ (D_s^*)$	2110.3±2.0	2110.5	2113.1	1968.8
$2\,^1S_0$		2656.5	2678.8	2536.5
$2\,^3S_1$		2757.8	2774.3	2536.5
$1\,^3P_0$		2387.8	2422.2	2382.2
$1\,^3P_1'$		2521.2	2528.8	2382.2
$1\,^3P_2\ (D_{s2})$	2573.2±1.9	2573.1	2582.8	2402.8
$1\,^1P_1'\ (D_{s1})$	2536.5±0.8	2536.5	2552.1	2402.8
θ		26.0°	31.8°	35.6°
$1\,^1S_0\ (B^0)$	5278.7±2.1	5278.7	5278.7	5278.7
$1\,^3S_1\ (B^{*0})$	5324.6±2.1	5324.0	5325.8	5278.7
$2\,^1S_0$		5892.1	5893.9	5846.3
$2\,^3S_1$		5924.3	5927.1	5846.3
$1\,^3P_0$		5689.5	5692.5	5659.1
$1\,^3P_1'$		5730.8	5734.1	5659.1
$1\,^3P_2$		5759.1	5761.4	5701.5
$1\,^1P_1'$		5743.6	5745.4	5701.5
θ		31.7°	31.3°	35.6°
$1\,^1S_0\ (B_s)$	5383.3±6.7	5383.3	5383.3	5383.3
$1\,^3S_1\ (B_s^*)$	5430.5±2.6	5431.9	5434.1	5383.3
$2\,^1S_0$		6000.9	6003.1	5950.9
$2\,^3S_1$		6035.8	6039.1	5950.9
$1\,^3P_0$		5810.1	5814.2	5796.7
$1\,^3P_1'$		5855.0	5857.9	5796.7
$1\,^3P_2$		5875.2	5878.1	5817.1
$1\,^1P_1'$		5860.2	5863.2	5817.1
θ		27.3°	27.1°	35.6°

TESTS OF FACTORIZATION IN B DECAYS

PATRICK SKUBIC[*][†]

Department of Physics and Astronomy, University of Oklahoma, 440 W. Brooks, Norman, Oklahoma 73019, USA

ABSTRACT

Predictions using the factorization hypothesis are compared to B meson decays from data taken with the CLEO II detector. Within experimental errors, reasonable agreement between theory and data is obtained. We conclude that the factorization hypothesis is valid over the kinematical range tested. A positive value for the sign of the ratio of BSW parameters a_2/a_1 was obtained from a fit to CLEO II B-decay data in contrast to charm decay where a negative sign is obtained. This result indicates that constructive interference occurs between internal and external spectator diagrams in nonleptonic B-decays.

1. Introduction

The factorization hypothesis, which is the basis of most theoretical treatments of hadronic B decays, assumes that two body decays of B mesons which occur via the spectator process may be expressed theoretically as the product of two independent hadronic currents in analogy to semileptonic decays. One current describes the formation of a charm meson and the other the hadronization of the $\bar{u}d$ (or $\bar{c}s$) system from the virtual W^-. Consider the case of $\bar{B}^0 \to D^{*+}\pi^-$. The amplitude for this reaction is

$$A = G_F/\sqrt{2}\ V_{cb}V_{ud}^*\langle\pi^-(p)|(\bar{d}u)|0\rangle\langle D^{*+}|(\bar{c}b)|\bar{B}^0\rangle \tag{1}$$

where V_{ud} is the well measured CKM factor from the $W^- \to \bar{u}d$ vertex. The first hadron current, which creates the π^- from the vacuum, is related to the decay constant f_π, and is known for π and ρ. We have

$$\langle\pi^-(p)|(\bar{d}u)|0\rangle = -if_\pi p_\mu, \tag{2}$$

where p_μ is the π^- four momentum. The other hadron current can be found from semileptonic $\bar{B}^0 \to D^{*+}\ell^-\bar{\nu}_\ell$ decays.

---

[*]This work is supported by the National Science Foundation and the U.S. Department of Energy.
[†]Representing the CLEO Collaboration.

Table 1. Comparison of R_{exp} and R_{theor}. The first error in R_{exp} is statistical and the second is systematic.

	R_{exp} (GeV^2)	R_{theor} (GeV^2)
$\bar{B}^0 \to D^{*+}\pi^-$	$1.1 \pm 0.1 \pm 0.2$	1.2 ± 0.2
$\bar{B}^0 \to D^{*+}\rho^-$	$3.0 \pm 0.4 \pm 0.6$	3.3 ± 0.5
$\bar{B}^0 \to D^{*+}a_1^-$	$4.0 \pm 0.6 \pm 0.5$	3.0 ± 0.5

Table 2. Ratios of B decay widths.

	Exp.	Factorization	RI Model	BSW Model
$\mathcal{B}(\bar{B}^0 \to D^{*+}\rho^-)/\mathcal{B}(\bar{B}^0 \to D^{*+}\pi^-)$	$2.9 \pm 0.5 \pm 0.5$	2.9 ± 0.05	$2.2 - 2.3$	2.8
$\mathcal{B}(\bar{B}^0 \to D^{*+}a_1^-)/\mathcal{B}(\bar{B}^0 \to D^{*+}\pi^-)$	$5.0 \pm 1.0 \pm 0.6$	3.4 ± 0.3	$2.0 - 2.1$	3.4

2. Branching Ratio Tests

Factorization can be tested experimentally by verifying whether the relation

$$R_{exp} \equiv \frac{\Gamma\left(\bar{B}^0 \to D^{*+}h^-\right)}{\frac{d\Gamma}{dq^2}\left(\bar{B}^0 \to D^{*+}l^-\bar{\nu}_l\right)\Big|_{q^2=m_h^2}} = 6\pi^2 c_1^2 f_h^2 |V_{ud}|^2 \equiv R_{theor}, \qquad (3)$$

is satisfied, where q^2 is the four momentum transfer from the B meson to the D^* meson. Since q^2 is also the mass of the lepton-neutrino system, by setting $q^2 = m_h^2$ we are simply requiring that the lepton-neutrino system has the same kinematic properties as the h^- in the hadronic decay. The c_1^2 term accounts for hard gluon corrections.

To derive numerical predictions for branching ratios, we must use models to interpolate the differential q^2 distribution for $\bar{B} \to D^*\ell\,\nu$ to $q^2 = m_\pi^2$, m_ρ^2, and $m_{a_1}^2$, respectively. Fortunately, the spread in the theoretical models which describe $\bar{B} \to D^*\ell\,\nu$ is small.[1] We now have all the required ingredients[2] for the test with decay rates and obtain the results given in Table 1.

If we form ratios of branching fractions some of the systematic uncertainties on R_{exp} will cancel, as does the QCD correction c_1 in R_{theor}. For example in the case of $D^{*+}\rho^-/D^{*+}\pi^-$, the expectation from factorization is given by $R_{theor}(\rho)/R_{theor}(\pi)$ times the ratio of the semileptonic branching ratios evaluated at the appropriate q^2 values. In Table 2 we show the comparison of the data, the expectation from factorization as defined above and two theoretical predictions of Bauer, Stech and Wirbel (BSW),[3] and Reader and Isgur (RI).[4]

A two-body decay of the type $B \to D_s^{(*)+}D^{(*)}$ is produced from the fragmentation of the W^+ in the external spectator diagram. While there are 50% differences between model predictions for any particular branching ratio, the variations in the predictions for the ratios of widths due to the different parameterizations of the form factors are much smaller.[5] Predictions of these ratios of widths are given in Table 3.

Table 3. Predictions for ratios of widths in $B \to D_s^{(*)+} D^{(*)}$ decays.

Model	$\frac{\Gamma(B \to D_s^{*+} D)}{\Gamma(B \to D_s^+ D^*)}$	$\frac{\Gamma(B \to D_s^+ D)}{\Gamma(B \to D_s^+ D^*)}$	$\frac{\Gamma(B \to D_s^{*+} D^*)}{\Gamma(B \to D_s^{*+} D)}$
BSW[3]	$1.31 \left(f_{D_s^*}/f_{D_s} \right)^2$	2.24	2.77
Rosner[6]	$1.00 \left(f_{D_s^*}/f_{D_s} \right)^2$	1.43	2.59
Neubert *et al.*[10]	$1.04 \left(f_{D_s^*}/f_{D_s} \right)^2$	1.47	2.56
Du & Liu[7]	$0.91 \left(f_{D_s^*}/f_{D_s} \right)^2$	1.35	2.72
Deandrea *et al.*[8]	$0.97 \left(f_{D_s^*}/f_{D_s} \right)^2$	1.42	3.84
Mannel *et al.*[9]	$1.00 \left(f_{D_s^*}/f_{D_s} \right)^2$	1.56	3.01
CLEO-II	0.75 ± 0.26	1.16 ± 0.32	2.46 ± 0.84

3.   Determination of $|a_1|$, $|a_2|$ and the relative sign of (a_2/a_1)

In the BSW model,[10,3] the branching fractions of the B^0 normalization modes are proportional to a_1^2 while the branching fractions of the $B \to \psi$ decay modes depend on a_2^2. We can relate a_1 and a_2 to the QCD coefficients c_1 and c_2 by $a_1 = c_1 + \xi c_2$ and $a_2 = c_2 + \xi c_1$ where $\xi = 1/N_{\text{color}}$. A fit to the branching ratios that we have measured for the modes $\bar{B}^0 \to D^+\pi^-$, $D^+\rho^-$, $D^{*+}\pi^-$ and $D^{*+}\rho^-$ yields

$$|a_1| = 1.15 \pm 0.04 \pm 0.10 \tag{4}$$

and a fit to the modes with ψ mesons in the final state gives

$$|a_2| = 0.26 \pm 0.01 \pm 0.02 \tag{5}$$

The comparison of B^- and $\bar{B}^0$ modes can be used to distinguish between the two possible choices for the sign of a_2 relative to a_1. The BSW model,[10] predicts the following ratios:

$$R_1 = \frac{\mathcal{B}(B^- \to D^0\pi^-)}{\mathcal{B}(\bar{B}^0 \to D^+\pi^-)} = (1 + 1.23 a_2/a_1)^2 \tag{6}$$

$$R_2 = \frac{\mathcal{B}(B^- \to D^0\rho^-)}{\mathcal{B}(\bar{B}^0 \to D^+\rho^-)} = (1 + 0.66 a_2/a_1)^2 \tag{7}$$

Table 4. Ratios of normalization modes to determine the sign of a_2/a_1. The magnitude of a_2/a_1 is the value in the BSW model which agrees with our result from $B \to \psi$ modes.

Ratio	$a_2/a_1 = -0.24$	$a_2/a_1 = 0.24$	CLEO II	RI model
R_1	0.50	1.68	$1.89 \pm 0.26 \pm 0.32$	$1.20 - 1.28$
R_2	0.71	1.34	$1.67 \pm 0.27 \pm 0.30$	$1.09 - 1.12$
R_3	0.48	1.72	$2.00 \pm 0.37 \pm 0.28$	$1.19 - 1.27$
R_4	0.41	1.85	$2.27 \pm 0.41 \pm 0.41$	$1.10 - 1.36$

$$R_3 = \frac{\mathcal{B}(B^- \to D^{*0}\pi^-)}{\mathcal{B}(\bar{B}^0 \to D^{*+}\pi^-)} = (1 + 1.29 a_2/a_1)^2 \tag{8}$$

$$R_4 = \frac{\mathcal{B}(B^- \to D^{*0}\rho^-)}{\mathcal{B}(\bar{B}^0 \to D^{*+}\rho^-)} \approx (1 + 0.75 a_2/a_1)^2 \tag{9}$$

Table 4 shows a comparison between the experimental results and the two allowed solutions in the BSW model indicating disagreement with the theoretical extrapolation from data on charmed meson decay which predicts a negative value for a_2/a_1.[3]

4. Conclusions

We have presented results on B meson non-leptonic decays and compared them with predictions using the factorization hypothesis. Reasonable agreement between theory and data is obtained. A positive value for the sign of the ratio of BSW parameters a_2/a_1 was obtained from a fit to CLEO II B-decay data in contrast to charm decay where a negative sign is obtained. This result may indicate that QCD corrections are of a perturbative nature in nonleptonic B-decays, where α_s would be small, and $a_1 \approx c_1 \approx 1$ and $a_2 \approx c_1/3 \approx 1/3$.[10]

References

1. CLEO Collab., M. S. Alam et al., Cornell Preprint CLNS 94-1270.
2. The value of $c_1 = 1.12 \pm 0.1$ is the Wilson coefficient evaluated at the b quark mass scale.[10] The value of $f_\rho = (215 \pm 4)$ MeV is from $e^+e^- \to \rho^0$ data as determined by Pham and Vu and a value of $f_{a_1} = (205 \pm 16)$ MeV was used.[4] The value of $V_{ud} = 0.975 \pm 0.001$ is taken from J. Rosner, in B decays, edited by S. Stone, (World Scientific, Singapore, 1992).
3. M. Bauer, B. Stech, and M. Wirbel, Z. Phys. C29, (1985) 637; Z. Phys. C34, (1987) 103 and Z. Phys. C42, (1989) 671.
4. C. Reader and N. Isgur, Phys. Rev. D47, (1993) 1007.
5. CLEO collab., T. Bergfeld et al., CLEO CONF 94-9.
6. J. Rosner Phys. Rev. D42 (1990) 3732.
7. D. Du and C. Liu, BIHEP-TH-92-58.
8. A. Deandrea et al., Phys. Lett. B318 (1993) 549.
9. T. Mannel et al., Phys. Lett. B259 (1991) 359.
10. M. Neubert, V. Riekert, Q. P. Xu and B. Stech in Heavy Flavours, edited by A. J. Buras and H. Lindner (World Scientific, Singapore, 1992).

NON-FACTORIZATION AND THE DECAYS $B \to J/\psi + K^{(*)}$

CARL E. CARLSON*

Physics Department, College of William and Mary, Williamsburg, VA 23187, USA

and

J. MILANA

Physics Department, University of Maryland, College Park, MD 20742, USA

ABSTRACT

Many known models, which generally use a factorization hypothesis, give a poor account of the decays $B \to J/\psi + K^{(*)}$. Usually there is a free overall factor, which is fit to the data, so that tests of the models rely upon ratios. The models tend to give too much K^* compared to K and too much transverse polarization compared to longitudinal. Our microscopic calculations, which use perturbative QCD, do well for both ratios. A microscopic calculation allows us to see how well factorization, heavy quark symmetry, and other features of various models are working. In the present case, agreement with the experimental ratios stems from a breakdown of factorization for one of the amplitudes.

1. Factorization and the data

Gourdin, Kamal, and Pham[1] and Aleksan et al.[2] point out that many known models,[3-6] all of which use the factorization hypothesis, give a poor account of the decays $B \to J/\psi + K^{(*)}$.

Since in most models there is an overall factor, generally called a_2,[3] which is fit to the data, the tests of the models rely upon ratios of K^* and K decays and upon ratios of longitudinal and tranverse polarization in the $J/\psi + K^*$ decays. The models tend to give too much K^* compared to K and too much transverse polarization compared to longitudinal.

Our microscopic calculations, using perturbative QCD, do well for both ratios. Although the end results have been published in a fuller context,[7-9] charmonium B decays deserve further comment because of current interest in them. We will try to make our remarks self-contained.

The ratios under study and their experimental values are,[10]

$$R \equiv \frac{\mathrm{Br}(B \to J/\psi + K^*)}{\mathrm{Br}(B \to J/\psi + K)} = 1.64 \pm 0.34 \quad \text{and} \quad \frac{\Gamma_{LL}}{\Gamma_{LL} + \Gamma_{TT}} = 0.84 \pm 0.06 \pm 0.08. \quad (1)$$

where Γ_{LL} and Γ_{TT} represent the longitudinal and transverse parts of the K^* rate. Our own results for the two ratios are 1.76 and 0.65, respectively (using Table IV of[7]).

Factorization implies that the decays depend upon a set of form factors for a current connecting B to $K^{(*)}$. As a benchmark—yes, we know the $K^{(*)}$ is light—the relations that heavy quark symmetry[11-13] implies among the form factors lead to

$$R = \frac{m_B^2 + 4m_{J/\psi}^2}{m_B^2} \approx 2.38 \quad \text{and} \quad \frac{\Gamma_{LL}}{\Gamma_{LL} + \Gamma_{TT}} = \frac{m_B^2}{m_B^2 + 4m_{J/\psi}^2} \approx 0.42, \quad (2)$$

*presenter

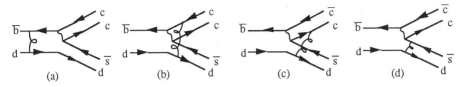

Fig. 1. Lowest order perturbation theory diagrams for B decays involving charmonium. (a) and (d) are factorizable, (b) and (c) are nonfactorizable.

which do not agree with the data but are not bad as a representation of many of the models. For information, in each term $m_B^2 + 4m_{J/\psi}^2$, the m_B^2 is from Γ_{LL} and the $4m_{J/\psi}^2$ from Γ_{TT}.

Why does our calculation work for the ratios when others do not? Most importantly, because the factorization hypothesis fails. It does not fail always. Its failure is significant only for the transverse polarization final state of $B \to J/\psi + K^*$. In this amplitude the nonfactorizable contributions are about half the size and opposite in sign to the factorizable ones, which has roughly the effect of turning the "4" into a "1" in the previous equations, and giving decent agreement with the $\Gamma_{LL}/(\Gamma_{LL} + \Gamma_{TT})$ data.

Also, surprisingly in this context, we find the heavy quark symmetry symmetry predictions for the form factors of the factorizable parts of the amplitude work well; small corrections and nonfactorizable contributions keep the two ratios from just being inverses of each other.

2. More detailed discussion

2.1. Factorization

The factorization hypothesis for $B \to J/\psi + K^{(*)}$ is that after the charmed quarks which go into the J/ψ are created, they are unconnected to other quarks in the process.

The matrix element we want is $M = \langle X, J/\psi | (G_F/\sqrt{2}) V_{cb} V_{cs}\, \bar{s}\gamma_{\mu L} c\, \bar{c}\gamma_L^\mu b | B \rangle$. If factorization is valid, one can reorganize this into

$$M = (-1/N_c)(G_F/\sqrt{2}) V_{cb} V_{cs} \langle J/\psi | \bar{c}\gamma_{\mu L} c | 0 \rangle \langle X | \bar{s}\gamma_L^\mu b | B \rangle, \tag{3}$$

that is, the matrix element is a product of two hadronic factors. Quantity N_c is the number of colors, and to allow for mixing effects one usually replaces $(-1/N_c)$ with a constant a_2.

2.2. Non-factorization

Fig. 1 shows factorizable contributions in (a) and (d) and nonfactorizable ones in (b) and (c).

Upon first view, it is easy to believe that the nonfactorizable contributions are small. The gluon couples to two oppositely colored quarks that are nearly at the same point because of the W-exchange. Indeed, the largest parts of diagrams (b) and (c) cancel each other and the subleading $O(q_G)$, where q_G is the gluon momentum, terms give the surviving result. However, the gluon momentum is not so small; in fact we argue that it is large enough that a perturbative calculation is plausibly valid. It supplies the momentum transfer needed by the light quark, which is of order $\bar{\Lambda}_B$, the part of the mass of the B meson carried by the light quark, which is about 500 MeV or a few times Λ_{QCD}.

However, for $B \rightarrow J/\psi + K$ and longitudinal $B \rightarrow J/\psi + K^*$, the subleading parts of the two nonfactorizable diagrams themselves partially cancel. In contrast, they add for the transverse decay, so this nonfactorizable amplitude can get large. Transverse $B \rightarrow J/\psi + K^*$ does require chirality violations, but the ensuing suppression of $O(m_{J/\psi}/m_B)$ is not decisive.

3. Closing Remarks

We have seen, in one explicit calculation, how nonfactorizable contributions to $B \rightarrow J/\psi + K^{(*)}$ are significant and are crucial to giving agreement with data for the K^* to K ratio and the tranverse to longitudinal polarization ratio.

Ratios are used to test the models because in most models the overall rate is determined by a constant that is fit to the data. Our calculations also have trouble with the overall rate. At present, with the parameters we choose, we do well for non-color-suppressed decays such as $B \rightarrow D\pi$ but the rates are rather low compared to data for $B \rightarrow J/\psi + K^{(*)}$.

We believe the use of pQCD is valid for high recoil decays of the B. This has been given better support by Akhoury, Sterman, and Yao[14] who show how Sudakov effects suppress contributions from end point regions where use of pQCD would be questionable.

It is possible that perturbative contributions are part but not all of what gives $B \rightarrow J/\psi + K^{(*)}$ decay. In this case, the details of our remarks will only be part of something larger, but the scenario can well stand: the factorizable contributions to $B \rightarrow J/\psi + K^{(*)}$ will not suffice to explain those decays, and nonfactorizable contributions will be crucial.

References

1. M. Gourdin, A.N. Kamal, and X.Y. Pham, Report PAR/LPTHE/ 94-19 (hep-ph/9405318).
2. R. Aleksan, A. Le Yaouanc, L. Oliver, O. Pène, and J.-C. Raynal, Report LPTHE-Orsay 94/54 (hep-ph/9406334).
3. M. Bauer, B. Stech, and M. Wirbel, Z. Phys. C **34**, 103 (1987).
4. R. Casalbuoni, A. Deandrea, N. Di Bartolomeo, R. Gatto, F. Feruglio, and G. Nardulli, Phys. Lett. B **292**, 371 (1992); A. Deandrea, N. Di Bartolomeo, R. Gatto, and G. Nardulli, Phys. Lett. B **318**, 549 (1993).
5. A. Ali and T. Mannel, Phys. Lett. B **264**, 447 (1991) and **274**, 256 (1992); A. Ali, T. Ohland, and T. Mannel, Phys. Lett. B **298**,195 (1993); M. Neubert and V. Rieckert, Nucl. Phys. B **285**, 97 (1992).
6. W. Jaus, Phys. Rev. D **41**, 3394 (1990); W. Jaus and D. Wyler, Phys. Rev. D **41**, 3405 (1990).
7. C. E. Carlson and J. Milana, Phys. Rev. D **49**, 5908 (1994).
8. C. E. Carlson and J. Milana, Phys. Lett. B **301**, 237 (1993).
9. A. Szczepaniak, E.M. Henley, and S.J. Brodsky, Phys. Lett. B **243**, 287 (1990).
10. T.E. Browder, K. Honscheid, and S. Playfer, Report CLNS 93/1261 (1994) [to appear in *B Decays*, 2nd edition, Ed. S. Stone, World Scientific, Singapore].
11. S. Nussinov and W. Wetzel, Phys. Rev. D **36**, 130 (1987).
12. N.B. Voloshin and M.A. Shifman, Sov. J. Nucl. Phys. **47**, 511 (1988).
13. N. Isgur and M. Wise, Phys. Lett. B**232**, 113 (1989); *ibid*, **237**, 527 (1990)
14. R. Akhoury, G. Sterman, and Y.-P. Yao, Phys. Rev. D **50**, 358 (1994).

MEASUREMENT OF THE FORM FACTORS FOR $\bar{B}^0 \to D^{*+}\ell^-\bar{\nu}$

ANDERS RYD

Physics Department, University of California
Santa Barbara, CA 93106, USA

Representing the CLEO Collaboration.

ABSTRACT

Using a sample of 2.2×10^6 $B\bar{B}$ pairs collected by the CLEO-II detector at the Cornell Electron Storage Ring, we have studied the form factors for the process $\bar{B}^0 \to D^{*+}\ell^-\bar{\nu}$. We have obtained the preliminary result for the form factor ratios $R_1 = 1.30 \pm 0.36 \pm 0.16$ and $R_2 = 0.64 \pm 0.26 \pm 0.12$ and the form factor slope $\rho^2 = 1.01 \pm 0.15 \pm 0.09$.

The decay $B \to D^*\ell\nu$ is a key process for measuring the magnitude of the CKM matrix element V_{cb} and for testing theoretical predictions of semileptonic decay form factors. This decay has a large branching fraction and very little background, which makes it well-suited experimentally for the detailed studies of the kinematic distributions that are required for the measurement of the form factors. This analysis is described in more detail elsewhere.[1]

Recently, there has been important progress in the theoretical understanding of the process $\bar{B}^0 \to D^{*+}\ell^-\bar{\nu}$ with the development of heavy quark effective theory (HQET).[2] HQET relates the three form factors $A_1(q^2)$, $V(q^2)$, and $A_2(q^2)$, that describe the decay $\bar{B}^0 \to D^{*+}\ell^-\bar{\nu}$, to one form factor, the Isgur-Wise function, $\xi(q^2)$:

$$V(q^2) = A_2(q^2) = A_1(q^2) \left[1 - \frac{q^2}{(m_B + m_{D^*})^2} \right]^{-1} = R^{*-1}\xi(q^2), \qquad (1)$$

where $R^* = 2\sqrt{m_B m_{D^*}}/(m_B + m_{D^*}) \approx 0.89$. In HQET, ξ is normally written in terms of $w = (m_B^2 + m_{D^*}^2 - q^2)/2m_B m_{D^*} = E_{D^*}/m_{D^*}$. HQET cannot predict $\xi(w)$ over the full w range, but it gives us an absolute normalization in the zero-recoil configuration: $\xi(w = 1) = 1$. In our fits we assume $\xi(w) = 1 - \rho^2(w - 1)$, where ρ^2 is called the slope of the Isgur-Wise function or simply the form-factor slope. The linear form is expected to be a good approximation over the range of w available in the decay $B \to D^*\ell\nu$.

Following Neubert, [3] we define the form factor ratios

$$R_1 \equiv \left[1 - \frac{q^2}{(m_B + m_{D^*})^2} \right] \frac{V(q^2)}{A_1(q^2)} \qquad R_2 \equiv \left[1 - \frac{q^2}{(m_B + m_{D^*})^2} \right] \frac{A_2(q^2)}{A_1(q^2)}. \qquad (2)$$

These form factor ratios, which are both predicted to be unity in the heavy-quark symmetry limit, are slightly different from those used in studies of $D \to K^*\ell\nu$. They are defined in such a way that they are constant in the heavy-quark symmetry limit. However, the b and c quarks are not infinitely heavy, and corrections have been estimated by Neubert[3] who finds $R_1 \approx 1.3$ and $R_2 \approx 0.8$. Lattice QCD and QCD sum rules predict ρ^2 in the range from 0.5 to 2.

We perform a fit to the joint four-dimensional distribution for the decay $\bar{B}^0 \to D^{*+}\ell^-\bar{\nu}$, $D^{*+} \to D^0\pi^+$ for the three parameters R_1, R_2, and ρ^2. The four kinematic variables are q^2 and the three angles $\cos\theta_\ell$, $\cos\theta_V$, and χ, where θ_ℓ is the lepton decay angle in the W rest frame, θ_V is the D^0 decay angle in the D^{*+} rest frame, and χ is the azimuthal angle in the $\bar{B}^0$ rest frame between the decay planes of the D^{*+} and the W.

The D^* is reconstructed in the mode $D^{*+} \to D^0\pi^+$ with $D^0 \to K^-\pi^+$ or $D^0 \to K^-\pi^+\pi^0$. Lepton candidates are either electrons or muons. The D^*–lepton candidates are required to be kinematically consistent with coming from the decay $B \to D^*\ell\nu$, where the B is slow ($\beta \approx 0.06$) and a neutrino is missing. The cuts used in the analysis are designed to suppress the background strongly to minimize the uncertainty in the modeling of the background in the form factor fit. Figure 1 shows the $\delta m = m_{D^*} - m_{D^0}$

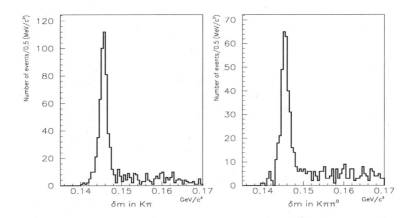

Fig. 1. The distribution of $\delta m = M(D^{*+}) - M(D^0)$ for $D^0 \to K^-\pi^+$ candidates (left) and $D^0 \to K^-\pi^+\pi^0$ candidates (right) after all analysis cuts except the δm cut.

distribution for the two modes in which the D^0's are reconstructed, after all cuts except that on δm. The signal region in δm is ± 2 MeV/c^2 around the nominal value; this is approximately a 2σ cut. To study the background from fake D^*'s, which is the dominant background in both $D^0 \to K^-\pi^+$ and $D^0 \to K^-\pi^+\pi^0$, we use the sideband region $0.15 < \delta m < 0.17$ GeV/c^2. After all event selection criteria are applied we have 783 events with a 16% background contamination.

We then fit the joint four-dimensional distribution using an unbinned maximum likelihood fit. We use a method that was originally developed by the E691 collaboration in their measurement of the $D \to K^* \ell \nu$ form factor ratios. This method uses a re-weighting technique of MC events to numerically evaluate the likelihood function. The advantage of this method is that it directly incorporates the effects of smearing and acceptance. In Fig. 2, the fit is shown in the four 1-dimensional projections. The plots

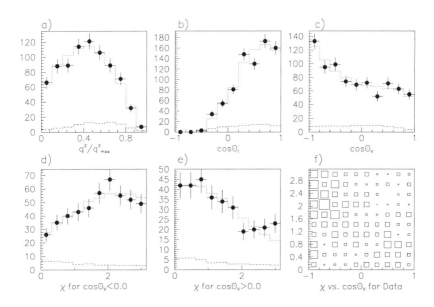

Fig. 2. In plots a) through e), the data are shown with error bars, the fit as the dotted line, and the background as the dashed line. In a) through c), the one dimensional projections on the q^2, $\cos \theta_\ell$, and $\cos \theta_V$ axes are shown. Plots d) and e) show the projection on the χ axis in a low and high $\cos \theta_V$ bin. The two-dimensional projection onto the plane χ vs. $\cos \theta_V$ is shown in f) for the data . The density of points is proportional to the size of the boxes.

show that the fit describes the data well in all four projections. The last plot, f), shows an example of a 2-dimensional projection, χ vs. $\cos \theta_V$. The structure in this plot — the enhancement in the upper-left and lower-right corner and the depletion in the opposite corners — is an effect of a parity-violating interference term. The observed structure is in agreement with the expectations from the standard model. If the $b \to c \ell \nu$ vertex had a $V + A$ coupling, we would expect that the pattern of enhanced and depleted regions to be the opposite, which is in disagreement with the data. Table 1 shows the result of the fit for R_1, R_2, and ρ^2. Since the errors on these quantities are highly correlated, we also present the off-diagonal elements of the correlation matrix. Several sources of systematic errors have been investigated. The dominant sources of systematic errors are from the uncertainty in the background modeling and the fitting method itself. Our value of $\rho^2 = 1.01 \pm 0.15 \pm 0.09$ is in agreement with the measurement of $\tilde{\rho}^2 = 0.84 \pm 0.13 \pm 0.08$

in the CLEO measurement[4] of $|V_{cb}|$. Neubert[3] estimates that $\hat{\rho}^2 \approx \rho^2 - 0.09$.

Table 2 shows the results from a series of fits for R_1 and R_2 in which the slope ρ^2 was fixed to a series of values. If theoretical calculations of ρ^2 become sufficiently precise we would gain significantly in precision, especially for R_2.

In conclusion, we have performed the first simultaneous measurement of R_1, R_2, and ρ^2 for the decay $\bar{B}^0 \to D^{*+}\ell^-\bar{\nu}$. We make a precise measurement of ρ^2 even with R_1 and R_2 floating. The measurement is in good agreement with estimated corrections to heavy-quark symmetry for the values of R_1 and R_2 and is in the range of predictions from lattice QCD and QCD sum rules for ρ^2.

Table 1. The preliminary result of the three parameter fit. The first error is statistical and the second is systematic.

Fit Parameters		Correlation Coefficients	
R_1	$1.30 \pm 0.36 \pm 0.16$	$C(R_1 R_2)$	-0.83
R_2	$0.64 \pm 0.26 \pm 0.12$	$C(R_1 \rho^2)$	0.63
ρ^2	$1.01 \pm 0.15 \pm 0.09$	$C(R_2 \rho^2)$	-0.82

Table 2. The preliminary results of the two parameter fits for different slopes, ρ^2. The first error is statistical and the second is systematic. The correlation, $C(R_1 R_2)$, between the errors on R_1 and R_2 is strong.

ρ^2	R_1	R_2	$C(R_1 R_2)$
0.50	$0.77 \pm 0.21 \pm 0.16$	$1.11 \pm 0.10 \pm 0.10$	-0.67
0.75	$0.99 \pm 0.24 \pm 0.16$	$0.93 \pm 0.12 \pm 0.10$	-0.68
1.00	$1.29 \pm 0.27 \pm 0.16$	$0.65 \pm 0.15 \pm 0.10$	-0.73
1.25	$1.75 \pm 0.34 \pm 0.17$	$0.20 \pm 0.20 \pm 0.11$	-0.77

References

1. P. Avery *et al.* (CLEO), *Measurement of the Form Factors for* $\bar{B}^0 \to D^{*+}\ell^-\bar{\nu}$, ICHEP94 Ref. 0144.

2. N. Isgur and M.B. Wise, Phys. Lett. B **232**, 113 (1989); **237**, 527 (1990); H. Georgi, Phys. Lett. B **240**, 447 (1990).

3. M. Neubert, *Heavy Quark Symmetry*, SLAC–PUB–6263, June 1993, to appear in Physics Reports.

4. B. Barish *et al.* (CLEO), *Measurement of the* $B \to D^*\ell\nu$ *Branching Fractions and* $|V_{cb}|$, ICHEP94 Ref. 0251, CLNS 94/1285.

CHARM PRODUCTION IN B MESON DECAYS

FRANZ MUHEIM

Physics Department, Syracuse University,
Syracuse, New York, 13244-1130

(*Representing the CLEO collaboration*)

ABSTRACT

We present new preliminary results from CLEO of inclusive branching fractions $\mathcal{B}(B \to D^0 X) = (63.8 \pm 1.1 \pm 2.0 \pm 1.7)\%$, and $\mathcal{B}(B \to D_s^+ X) = (11.8 \pm 0.4 \pm 0.9 \pm 2.9)\%$, and of the exclusive two-body decays $B \to D_s^{+(*)} \bar{D}^{(*)}$. These results are used to extract the D_s^+ decay constant f_{D_s} and the total charm production rate in B meson decays.

1. Introduction

Most B mesons decay via either the external or color-suppressed internal spectator diagram into charmed mesons or charmed baryons. Due to the large phase space available in these decays many exclusive final states are formed. The calculation of rates into final states is model-dependent whereas an attempt has been made to calculate the rates into inclusive final states based on QCD.[1-3] Thus it is important to measure inclusive decay rates into charmed particles. In this talk we present new results from CLEO on the inclusive decays $B \to D^0 X$ and $B \to D_s^+ X$ (charge-conjugate states are always implied), respectively. We also present measurements of the exclusive branchings of the two-body decays where the W fragments into $\bar{c}s$ quark pair, $B \to D_s^{+(*)} \bar{D}^{(*)}$. These results are used to extract the D_s^+ meson decay constant f_{D_s} by applying factorization in B meson decays. The new measurements of inclusive branching fractions are combined with other measurements from CLEO and ARGUS to obtain a total charm production rate from B mesons.

2. New CLEO Results

2.1. Measurement of $B \to D^0 X$ Decays

The data sample consists of 2020 pb^{-1} and 960 pb^{-1} of $e^+ e^-$ annihilations taken at CESR with the CLEO II detector[4] at the $\Upsilon(4S)$ resonance and in the nearby continuum, respectively. The $\Upsilon(4S)$ data correspond to about 2.15 million $B\bar{B}$ events.

Hadronic events are selected based on measured charged and neutral energy, at least three charged tracks and an event vertex consistent with the nominal interaction point. To suppress events from continuum production it is required that $R_2 < 0.5$ where R_2, the ratio of the second to the zeroth Fox-Wolfram moment,[5] exploits the different event shapes of "spherical" $B\bar{B}$ events as opposed to "jetty" continuum events.

The $D^0 \to K^- \pi^+$ channel is used. Each track candidate has to be consistent with originating from the event vertex. We also require that the specific ionization loss (dE/dx) is consistent with the K or π hypothesis. The $K^- \pi^+$ invariant mass distribution is used to extract the yield in the following way. First, the data is subdivided in 20 bins in reduced momentum x where $x = p_{D^0}/p_{\max}$. The continuum production which is estimated by the continuum data, scaled for luminosity and energy differences, is subtracted for each bin. The subtracted distribution is then fitted in each x-bin to a double Gaussian signal and background contributions. The combinatorial background is parameterized with a 2nd order Chebyshev polynomial. Additional sources of background are due to switched particle mass assignments from the decay $D^0 \to K^- \pi^+$ itself and from the Cabibbo suppressed decays $D^0 \to K^- K^+$ and $D^0 \to \pi^- \pi^+$, which peak underneath or near the signal region, respectively. The shape and normalization of these backgrounds is determined with D^0 decays stemming from $D^{*+} \to D^0 \pi^+$ where the charge of the slow π uniquely identifies the mass assignments of the D^0 daughters. The efficiency is determined by Monte Carlo simulations. The signal shape parameters and the efficiency have been smoothed as a function of x. The raw yield is 61050 reconstructed $D^0 \to K^- \pi^+$ decays. In Fig. 1 we show the obtained D^0 momentum spectrum corrected for efficiency.

By summing all bins with $0 < x < 0.5$ we obtain a branching fraction $\mathcal{B}(B \to D^0 X) = (63.8 \pm 1.1 \pm 2.0 \pm 1.7)\%$ where the first error is statistical, and the second error accounts for the systematic uncertainties which have been estimated by varying the R_2 cut and redoing the analysis by a single fit to the integral $(0 < x < 0.5)$ $K^- \pi^+$ mass distribution. The systematic error also comprises contributions which arise when signal shape parameters are changed within their errors and from the total number of B meson pairs. The third error is due to the branching ratio of $D^0 \to K^- \pi^+$.[6] We combine this result with earlier measurements of ARGUS and CLEO (Table 1) which have been rescaled to the new CLEO result[6] of $\mathcal{B}(D^0 \to K^- \pi^+) = (3.91 \pm 0.08 \pm 0.17)\%$ and obtain a new world average of $\mathcal{B}(B \to D^0 X) = (62.1 \pm 2.6)\%$.

2.2. Measurement of $B \to D_s^+ X$ Decays

The $D_s^+ \to \phi \pi^+, \phi \to K^+ K^-$ decay channel is chosen. The event and track selection has been described in the preceding section. Here we require $R_2 < 0.35$. The invariant mass of the $K^+ K^-$ pairs has to be within 10 MeV of the ϕ mass and we make angular cuts using the spin-parity structure of the decay. As in the $B \to D^0 X$ analysis we divide the data sample in x bins and fit the $\phi \pi^+$ invariant mass distribution in each bin to a Gaussian signal and polynomial background shape. To increase the statistical precision of the continuum subtraction we fit the continuum data before subtraction. The raw yield of reconstructed D_s^+ mesons from B decay is 1978 ± 78 events. The detector efficiency as a function of x is determined with Monte Carlo simulations. We plot the obtained D_s^+ momentum spectrum, corrected for efficiency, in Fig. 2.

Combining all bins with $x < 0.5$ we obtain $\mathcal{B}(B \to D_s^+ X) = (11.81 \pm 0.43 \pm 0.94 \pm 2.9)\%$ where the first error is statistical and the second is the systematic error dominated by the uncertainty in the tracking efficiency. The third error is due to $\mathcal{B}(D_s^+ \to \phi \pi^+) = (3.7 \pm 0.9)\%$[7] which is not directly measured and has a large uncertainty. Previous CLEO and ARGUS measurements for $\mathcal{B}(B \to D_s^+ X)$ are given in Table 1. The new world average is $\mathcal{B}(B \to D_s^+ X) = (10.0 \pm 2.5)\%$.

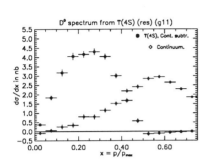

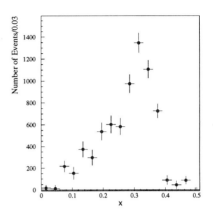

Fig. 1. The continuum-subtracted, efficiency-corrected yield of D^0 mesons versus x.

Fig. 2. The continuum-subtracted, efficiency-corrected yield of D_s^+ mesons versus x.

Table 1. Branching fractions $\mathcal{B}(B \to D^0 X)$ and $\mathcal{B}(B \to D_s^+ X)$.

Experiment	$\mathcal{B}(B \to D^0 X)[\%]$	$\mathcal{B}(B \to D_s^+ X)[\%]\frac{3.7\%}{\mathcal{B}(D_s^+ \to \phi\pi^+)}$
CLEO (preliminary)	$63.8 \pm 1.1 \pm 2.0 \pm 1.7$	$11.81 \pm 0.43 \pm 0.94$
ARGUS[8]	$49.6 \pm 3.8 \pm 6.4 \pm 2.4$	$7.89 \pm 1.05 \pm 0.84$
CLEO[8]	$59.6 \pm 3.1 \pm 3.6 \pm 2.9$	$8.27 \pm 1.24 \pm 0.81$
ARGUS (87)[8]		11.4 ± 3.2
CLEO (86)[8]		10.3 ± 2.7
New world average	62.1 ± 2.6	$10.0 \pm 0.7 \pm 2.4$

We have also measured all eight two-body decay modes $B \to D_s^{+(*)} \bar{D}^{(*)}$. The following decay channels are used: $D_s^{*+} \to \gamma D_s^+$, $D^{*+} \to \pi^+ D^0$, $D^{*0} \to \pi^0 D^0$, $D_s^+ \to \phi\pi^+, \bar{K}^0 K^+, \bar{K}^{*0} K^+, \phi\rho^+, \eta\pi^+, \eta\rho^+, D^0 \to K^-\pi^+, K^-\pi^+\pi^0, K^-\pi^+\pi^-\pi^+, D^+ \to K^-\pi^+\pi^+$ for the charmed mesons and $\pi^0 \to \gamma\gamma, \eta \to \gamma\gamma, \bar{K}^0 \to K_s \to \pi^+\pi^-, \rho^+ \to \pi^+\pi^0, \bar{K}^{*0} \to K^-\pi^+, \phi \to K^+ K^-$ for the resonances. Photons have to have energies larger than 30 MeV. The summed energy of the candidate has to be within 25 MeV of the beam energy E_{beam}. In Fig. 3 we show the beam constrained mass $m_B = \sqrt{E_{\text{beam}}^2 - p_B^2}$ where p_B is the summed momentum of the candidate. The fit to a Gaussian signal plus a background shape yields 134 ± 15 signal events. The branching ratios, averaged over charged and neutral B meson decays, are given in Table 2.

3. D_s^+ Decay Constant f_{D_s}

Factorization is the assumption that hadronic two-body B decays $(B \to \bar{D}^* h^+)$ can be expressed as the product of two amplitudes, one that describes $B \to \bar{D}^*$ and can be measured in the semileptonic decay $B \to \bar{D}^* \ell^+ \nu$, and another one that couples the h^+ to the virtual W^+. This assumption has been shown to be valid at the 20 % level for

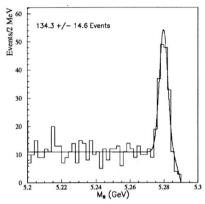

Fig. 3. The beam-constrained mass for the sum of the eight $B \to D_s^{+(*)}\bar{D}^{(*)}$ modes.

Table 2. Branching fractions $\mathcal{B}(B \to D_s^{+(*)}\bar{D}^{(*)})$.

Mode	$\mathcal{B}[\%]\frac{3.7\%}{\mathcal{B}(D_s^+ \to \phi\pi^+)}$
$B \to D_s^+\bar{D}$	$1.24 \pm 0.19 \pm 0.29$
$B \to D_s^+\bar{D}^*$	$1.07 \pm 0.23\pm 0.23$
$B \to D_s^{*+}\bar{D}$	$0.81 \pm 0.21 \pm 0.19$
$B \to D_s^{*+}\bar{D}^*$	$1.98 \pm 0.42\pm 0.45$

light hadrons.[9] Factorization can be applied to the case where the h^+ is a $D_s^{+(*)}$. This leads to the relation

$$\frac{\Gamma(B \to \bar{D}^*D_s^{+(*)})}{\frac{d\Gamma}{dq^2}(B \to \bar{D}^*\ell^+\nu)|_{q^2=m_{D_s}^2}} = \delta 6\pi^2 |V_{cs}|^2 f_{D_s}^2 \tag{1}$$

where $\delta = 0.41$ (1.0) for $D_s^{+(*)}$ and we assume $f_{D_s^*} = f_{D_s}$. Using the presented CLEO measurements for $\mathcal{B}(B \to D_s^{+(*)}\bar{D}^*)$ (Table 2) and for $\frac{dB}{dq^2}(B \to \bar{D}^*\ell^+\nu)|_{q^2=m_{D_s}^2} = (5.0 \pm 0.7) \cdot 10^{-3}$ GeV^{-2},[10] averaged over charged and neutral B mesons, we obtain

$$f_{D_s} = (280 \pm 43)\sqrt{\frac{3.7\%}{\mathcal{B}(D_s^+ \to \phi\pi^+)}} \text{ MeV} \quad . \tag{2}$$

CLEO has also observed[7] the leptonic decay mode of the D_s^+ into a muon and a neutrino which measures the decay constant f_{D_s} as

$$f_{D_s} = (344 \pm 64)\sqrt{\frac{\mathcal{B}(D_s \to \phi\pi^+)}{3.7\%}} \text{ MeV} \quad . \tag{3}$$

Equations (3) and (2) both depend on $\mathcal{B}(D_s^+ \to \phi\pi^+)$, but in an opposite manner. Solving the two equations gives

$$f_{D_s} = (310 \pm 37) \text{ MeV}, \text{ and } \mathcal{B}(D_s \to \phi\pi^+) = (3.0 \pm 0.7)\% \quad . \tag{4}$$

This value of f_{D_s} is higher than most theoretical predictions, eg. lattice gauge calculations which give $f_{D_s} = (230 \pm 35)$ MeV.[11]

4. Total Charm Production

B meson decays proceed via $b \to c$ transitions most of the time. From the parton model we expect about 115% charm where the additional 15% arises when the W fragments as a $\bar{c}s$ quark pair. The small contributions from $b \to u$ and Penguin transitions are neglected. The various charm contributions are listed in Table 3. We use the new world averages for the D^0 and D_s^+ branching fractions, and we have averaged previous ARGUS and CLEO measurements of $\mathcal{B}(B \to D^+X)$ rescaled to $\mathcal{B}(D^+ \to K^-\pi^+\pi^+) = (9.23 \pm 0.81)\%$ which is an average of the MARK III[8] and the new CLEO measurement.[12] The charmonium production is estimated by multiplying $\mathcal{B}(B \to \psi X)$[13] by $2f$ where the factor of 2 accounts for the two charm quarks and $f = 2.0 \pm 0.5$ allows for charmonium states not decaying via a ψ. Decays into baryons mostly proceed via a Λ_c^+. We use the average of previous ARGUS and CLEO measurements.[8] Finally, we add a small contribution of $b \to c\bar{c}s$ transitions into $\Xi_c\Lambda_c^-$ baryon pairs for which CLEO has some evidence.[14] Thus we obtain a value of $(108 \pm 6)\%$ for the number of charm quarks produced in B decays and $(15.6 \pm 2.9)\%$ for the sum of all $b \to c\bar{c}s$ transitions. These measurements are in good agreement with the naive parton model prediction and do not support the larger estimates of 130%.[3]

Table 3. $b \to c$ branching fractions.

$\mathcal{B}(B \to D^0X)$	$(62.1 \pm 2.6)\,\%$	$\mathcal{B}(B \to \psi X)$	$(1.1 \pm 0.1)\,\%$
$\mathcal{B}(B \to D^+X)$	$(23.9 \pm 3.7)\,\%$	$\mathcal{B}(B \to \Lambda_c^+X)$	$(6.4 \pm 2.3)\,\%$
$\mathcal{B}(B \to D_s^+X)$	$(10.0 \pm 2.5)\,\%$	$\mathcal{B}(B \to \Xi_c\Lambda_c^-X)$	$(1.1 \pm 0.8)\,\%$
$\Sigma(\text{Charm})$	$(108 \pm 6)\,\%$	$\Sigma(b \to c\bar{c}s)$	$15.6 \pm 2.9\,\%$

5. Conclusions

We have presented new CLEO results on inclusive $b \to c$ decays of $\mathcal{B}(B \to D^0X) = (63.8 \pm 1.1 \pm 2.0 \pm 1.7)\%$, and $\mathcal{B}(B \to D_s^+X) = (11.8 \pm 0.4 \pm 0.9 \pm 2.9)\%$. New results of exclusive two-body decays $B \to D_s^{+(*)}\bar{D}^{(*)}$ combined with the CLEO measurement of $D_s^+ \to \mu^+\nu$ give $f_{D_s} = (310 \pm 37)$ MeV if factorization holds in B meson decays. The total charm production rate in B meson decays is $(108 \pm 6)\%$ and the $b \to c\bar{c}s$ rate is $(15.6 \pm 2.9)\,\%$.

I thank Scott Menary, Gian-Carlo Moneti, and Sheldon Stone for help in preparing this talk. This work was supported by the National Science Foundation.

References

1. W.F. Palmer and B. Stech, *Phys. Rev.* **D48** (1993) 4174.
2. I.I. Bigi, B. Blok, M.A. Shifman and A. Vainshtein, *Phys. Lett.* **B323** (1994) 408.
3. A.F. Falk, M.B. Wise, and I. Dunietz, preprint hep-ph-9405346.
4. Y. Kubota et al., *Nucl. Inst. Meth.* **A320** (1992) 66.
5. G.C. Fox and S. Wolfram, *Phys. Rev. Lett.* **41** (1978) 1581.
6. D.S. Akerib et al., *Phys. Rev. Lett.* **71** (1993) 3070.
7. D. Acosta et al., *Phys. Rev.* **D49** (1994) 5690.
8. K. Hikasa et al., *Phys. Rev.* **D45** (1992) 1.
9. P. Skubic, in *these proceedings*.
10. B. Barish et al., CLNS preprint 94/1285, submitted to *Phys. Rev.* **D**.
11. C.T. Sachrajda, in *B Decays, Revised 2nd Edition*, ed. S. Stone (World Scientific, Singapore, 1994) p. 602.
12. R. Balest et al., *Phys. Rev. Lett.* **72** (1994) 2328.
13. S. Schrenk, in *these proceedings*.
14. P. Baringer, in *these proceedings*.

DECAYS OF THE B MESON INVOLVING CHARMONIUM

S. SCHRENK

University of Minnesota, Minneapolis, MN 55455

(Representing the CLEO collaboration)

ABSTRACT

We have used the CLEO-II detector at the Cornell Electron Storage Ring to measure the branching fractions for B-meson decays into charmonium mesons (J/ψ, ψ', χ_{c1}, χ_{c2} and η_c) in a sample of 2.15 million $B\overline{B}$ events. Momentum spectra for inclusive J/ψ and ψ' production are presented. We also find an exclusive branching fraction for the Cabibbo suppressed decay $B \to J/\psi \pi^-$.

1. Introduction

Inclusive decays of B mesons to charmonium states provide a testing ground for QCD calculations of quark dynamics. The dominant mechanism for production of charmonium is the color-suppressed internal spectator diagram. Virtual gluon interactions complicate this picture.[1,2] These are difficult to handle in QCD, and alternative approaches[3,4] result in significantly different predictions.

The data used in this analysis were recorded with the CLEO-II detector. An integrated luminosity of 2.02 fb^{-1} was accumulated at the $\Upsilon(4S)$ resonance, and an additional 0.99 fb^{-1} was collected at energies just below that resonance. Only the first half of this sample was used to study $B \to J/\psi X$, for which systematic effects were the primary limitation. With event selection optimized for B decays to charmonium we found $(2.15 \pm 0.04) \times 10^6$ $B\overline{B}$ events.

The CLEO-II detector[5] consists of three concentric cylindrical wire drift chambers, a time-of-flight system, and an electromagnetic calorimeter of 7800 thallium-doped cesium iodide crystals. Outside the calorimeter are a 1.5-Tesla superconducting solenoidal magnet and muon detectors embedded in steel.

2. $B \to J/\psi X$

We reconstructed J/ψ's through their decays to e^+e^- and $\mu^+\mu^-$. The invariant mass distributions for dielectrons and dimuons are shown in Fig. 1. The signal shapes were derived from a Monte Carlo simulation of $B \to J/\psi X$, $J/\psi \to e^+e^-$ or $\mu^+\mu^-$, that included bremsstrahlung in the detector material and final-state radiation. Continuum background was determined with the below $\Upsilon(4S)$ data. Fig. 2 shows the continuum dilepton mass spectra for momenta below 2.0 GeV/c. Using the MARK-III[6] branching fractions, $\mathcal{B}(J/\psi \to e^+e^-)=(5.92\pm0.25)\%$, and $\mathcal{B}(J/\psi \to \mu^+\mu^-)=(5.90\pm0.25)\%$, we find $\mathcal{B}(B \to J/\psi X)=(1.14\pm0.06)\%$ using dielectrons and $\mathcal{B}(B \to J/\psi X)=(1.13\pm0.05)\%$ using dimuons, where the errors are statistical only. There are significant systematic uncertainties, principally in lepton identification and the J/ψ to dilepton branching fractions. Combining the two modes, we find $\mathcal{B}(B \to J/\psi X)=(1.13\pm0.04\pm0.06)\%$.

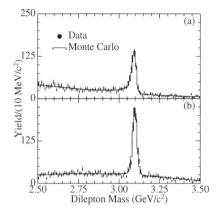

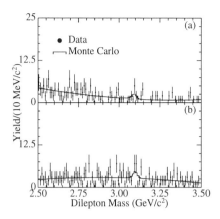

Fig. 1. Mass distributions for (a) dielectron and (b) dimuon modes from B decays.

Fig. 2. Mass distributions from continuum data for (a) dielectrons and (b) dimuons, with $P_{l+l-}<2$ GeV/c and Fox-Wolfram $R_2<0.5$.

3. $B \rightarrow \psi' X$

We measured B-meson decays to ψ' using two decay channels, $\psi' \rightarrow \ell^+ \ell^-$ and $\psi' \rightarrow J/\psi \pi^+ \pi^-$. The distribution of dilepton invariant masses in the ψ' region is shown in Fig. 3. The distribution of the difference between the masses of reconstructed ψ' and J/ψ candidates for $\psi' \rightarrow J/\psi \pi^+ \pi^-$ is shown in Fig. 4. Using the mass difference reduces the effect of the error in the J/ψ mass measurement. We find the $B \rightarrow \psi' X$ branching fraction for the modes $\psi' \rightarrow \ell^+ \ell^-$ and $\psi' \rightarrow J/\psi \pi^+ \pi^-$ to be $(0.30 \pm 0.05 \pm 0.04)\%$ and $(0.38 \pm 0.05 \pm 0.05)\%$, respectively. Since the samples are statistically independent they can be combined into a single result: $(0.34 \pm 0.04 \pm 0.03)\%$.

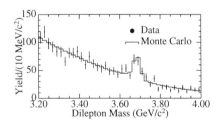

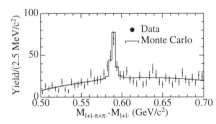

Fig. 3. Dilepton mass distribution.

Fig. 4. Distribution of the difference between the $\ell^+ \ell^- \pi^+ \pi^-$ and $\ell^+ \ell^-$ masses.

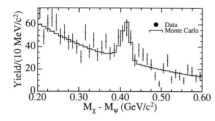
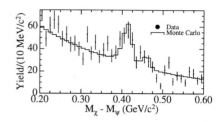

Fig. 5. Mass-difference distribution for $J/\psi\gamma$; fit to $B\to\chi_{c1}X$ Monte Carlo.

Fig. 6. Mass-difference distribution for $J/\psi\gamma$; fit to $B\to\chi_{c1}X$ and $B\to\chi_{c2}X$ Monte Carlo.

4. $B\to\chi_c X$

Inclusive χ_c events were reconstructed by combining detected photons with any accompanying J/ψ candidates. The largest background is random combinations of photons from π^0's with real J/ψ's from B-meson decays. Any candidate photon which could be combined with another photon to produce an effective mass near the π^0 mass was rejected. Fig. 5 shows a fit to the resulting mass-difference distribution with only the χ_{c1} allowed, which gives a signal of 112 ± 17 events. A better fit is obtained by allowing for both χ_{c1} and χ_{c2} (Fig. 6). The χ_{c1} signal is unchanged while there are 35 ± 13 entries in the χ_{c2} region. We find $\mathcal{B}(B\to\chi_{c1}X)=(0.40\pm0.06\pm0.04)\%$. The marginal excess in the χ_{c2} region corresponds to $\mathcal{B}(B\to\chi_{c2}X)=(0.25\pm0.10\pm0.03)\%$, or a 90% confidence lever upper limit of 0.38%.

5. $B\to\eta_c X$

We searched for the decay $\eta_c \to \phi\phi$, $\phi \to K^+K^-$. The signal region was defined to be the interval 2960 to 3010 MeV, allowing for the uncertainty in the η_c mass.[7,8] The dominant systematic error is the 40% uncertainty in $\mathcal{B}(\eta_c \to \phi\phi)$. We derive an upper limit on the branching ratio $\mathcal{B}(B \to \eta_c X) < 1.0\%$ at 90% confidence level.

6.   Momentum Spectra of J/ψ and ψ' From B Decay

The J/ψ momentum spectrum is shown in Fig. 7 with the expected contributions from the feed-down modes $B\to\psi'X$ and $B\to\chi_c X$ subtracted. The momentum spectrum of the ψ''s is shown in Fig. 8. In both cases, the observed shapes include soft components attributable to $B\to J/\psi(\psi')$ X with higher K* resonances or multiple hadrons.

7.   The Exclusive Decay $B\to J/\psi\pi$

We combine π^- candidates with detected J/ψ's to reconstruct the B meson. Figure 9 shows the energy difference (ΔE) between the B and beam energies, plotted against the reconstructed B mass, where the beam energy has been substituted for the reconstructed energy. The background is dominated by $B^-\to J/\psi K^-$, where the K^- is misidentified as a pion (Fig. 10). We find $\mathcal{B}(B^-\to J/\psi\pi^-)=(1.10\pm0.15\pm0.09)\times10^{-3}$, in agreement with expectations relative to $B^-\to J/\psi K^-$ for a Cabibbo suppressed decay. We also find $\mathcal{B}(\overline{B}^0\to J/\psi\pi^0)<6.9\times10^{-5}$ at 90% confidence limit.

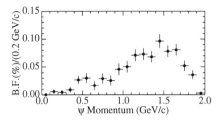

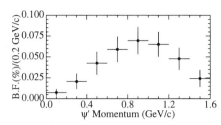

Fig. 7. Momentum spectrum for direct inclusive J/ψ production from B decays, see text.

Fig. 8. Momentum spectrum for inclusive ψ' production from B decays.

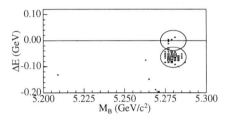

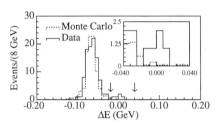

Fig. 9. Correlation between ΔE and M_B in the expected $B^- \to J/\psi\pi^-$ signal region. The ellipses show the 3 standard deviation contours expected for $B \to J/\psi\pi^-$ and $B \to J/\psi K^-$.

Fig. 10. The ΔE projection for events with reconstructed mass within 3σ of the B mass. The arrows point to the $B \to J/\psi\pi^-$ signal region.

8. Summary of Results and Comparison to Theory

Table 1 summarizes our inclusive measurements with the corresponding Particle Data Group (PDG) values. (The PDG values for $B \to J/\psi X$ and $B \to \psi' X$ have been corrected to use the Mark III $J/\psi \to \ell^+\ell^-$ branching fractions.) The measured branching fraction for $B \to J/\psi X$ and $B \to \chi_c X$ are composed of "direct" production, and "feed down" from charmonium modes, such as $B \to \psi' X$, $\psi' \to J/\psi \pi^+\pi^-$ or $\psi' \to \chi_{c1}\gamma$. To obtain the direct rate for comparison with theoretical predictions, we correct the inclusive branching fraction for the feed-down component. The direct branching fractions for J/ψ and χ_{c1} are included in Table 1.

The interaction Hamiltonian for charmonium production in B-meson decay is

$$H_{Effective} = \frac{G_F}{\sqrt{2}}V_{cb}V_{cs}^* \left[\left(\frac{1}{3}c_1(\mu) + c_2(\mu)\right)(\bar{c}c)(\bar{s}b) + \frac{1}{2}c_1(\mu)(\bar{s}\lambda_i b)(\bar{c}\lambda_i c) \right]. \quad (1)$$

The first part of Eq. (1) transforms as a color singlet and contributes only to J/ψ, ψ', χ_{c1} and η_c production. The second part transforms as a color octet and contribute to χ_{c0}, χ_{c1}, χ_{c2} and h_c production.[4] If the $\frac{1}{3}$ (reflecting color suppression) is replaced by $1/N_c$ then the coefficient of the color singlet part is equivalent to the a_2 term in the

Charmonium	Yield	Branching Fraction (%)		
		Measured	PDG (%)[7]†	Direct
η_c		<1.0		
J/ψ	1455±49	1.13±0.04±0.06	1.31±0.19	0.81±0.08
$\psi'\to\ell^+\ell^-$	127±21	0.30±0.05±0.04		
$\psi'\to J/\psi\pi^+\pi^-$	113±16	0.38±0.05±0.05		
ψ' combined		0.34±0.04±0.03	0.53±0.23	
χ_{c1}	112±17	0.40±0.06±0.04	1.05±0.35±0.25‡	0.37±0.07
χ_{c2}	35±13	<0.38		

†Corrected for recently updated $J/\psi\to\ell^+\ell^-$ branching ratios. ‡ARGUS[9]

Table 1. Inclusive $B\to$Charmonium + X. For η_c and χ_{c2} the limits are at the 90% C.L.

factorization model of Wirbel, Stech and Bauer for exclusive decays.[10] The difference between $1/N_c$ and $\frac{1}{3}$ parameterizes non-factorizable contributions to B-meson decay.[11]

Predictions for $\mathcal{B}(B\to J/\psi X)$ range from 0.2% to 2.0%, depending on the Wilson coefficients c_1 and c_2, and $1/N_c$.[3,4] With $1/N_c=3$, contributions of c_1 and c_2 almost cancel in the singlet term. Following the method of Bodwin et al.,[4] and using the values $c_1 = 1.13$ and $c_2 = -0.29$,[11] and a b mass of 5.0 GeV/c^2, we find the prediction for $\mathcal{B}(B\to J/\psi X)$ direct to be 0.10%. If one replaces $(\frac{1}{3}c_1 + c_2)$ by the measured value of a_2 from exclusive decays $(0.23\pm0.01\pm0.01^{12})$ then one predicts the direct branching fraction for $B\to J/\psi X$ to be 0.75%, in agreement with our measurement.

The success of using a_2 in this way prompts us to examine the predictions of Bodwin et al. for χ_{c2} production.[4] The difference in the predicted χ_{c1} branching fraction from the color-singlet term and our measured branching fraction gives a measure of color-octet production. Following the method of Bodwin et al., but with the values of c_1, c_2 and b mass given above and using the branching fraction for χ_{c1}, we calculate $\mathcal{B}(B\to\chi_{c2}X)=(0.56\pm0.15)$%, larger than our measured value. They have assumed $N_c=3$, implying a small contribution from the color-singlet mode. When we replace $(\frac{1}{3}c_1 + c_2)$ by a_2, we predict $\mathcal{B}(B\to\chi_{c2}X)=(0.30\pm0.11)$%, in good agreement with the measured value of (0.25 ± 0.10)% we obtain if we assume the χ_{c2} signal is real.

References

1. M.K. Gaillard, B.W. Lee, Phys. Rev. Lett. **33** (1974) 108.
2. G. Altarelli, L. Maiani, Phys. Lett. **52B** (1974) 351.
3. J.H. Kühn, S. Nussinov, and R. Rückl, Z. Physik **C5** (1980) 117.
4. G.T. Bodwin, et al., Phys. Rev. **D46** (1992) 3703.
5. Y. Kubota, et al., Nucl. Instrum. Meth. **A320** (1992) 66–113.
6. D. Coffman et al., Phys. Rev. Lett. **68** (1992) 282.
7. K. Hikasa, et al. (Particle Data Group), Phys. Rev. **D45** (1992) 1.
8. T.A. Armstrong et al. (E760 Collaboration), Phys. Rev. Lett. **68** (1992) 1468.
9. H. Albrecht, et al. (ARGUS Collaboration), Phys. Let. **B277** (1992) 209.
10. M. Bauer, B. Stech, and M. Wirbel, Z. Physik **C34** (1987) 103.
11. I. Bigi et al., Preprint UMN-TH-1234/94 (1994).
12. T. Browder, K. Honscheid and S. Playfer, Preprint CLNS 93/1261 (1993).

MEASUREMENT OF THE POLARIZATION IN THE DECAY
$B^0 \to J/\psi K^{*0}$ IN $\bar{p}p$ COLLISIONS AT $\sqrt{s} = 1.8$ TeV

KAREN E. OHL*

Physics Department, Yale University,
New Haven, CT, 06520

ABSTRACT

We report on a measurement of the polarization in the decay $B^0 \to J/\psi K^{*0}$ using data collected at the Collider Detector at Fermilab in $\bar{p}p$ collisions at $\sqrt{s} = 1.8$ TeV. B^0 mesons were reconstructed through the decay chain $B^0 \to J/\psi K^{*0}, J/\psi \to \mu^+\mu^-, K^{*0} \to K^+\pi^-$. The result, based on a sample of 60 ± 11 events, is $\Gamma_L/\Gamma = 0.66 \pm 0.10$ (stat) $^{+0.08}_{-0.10}$ (sys).

The pseudoscalar to vector-vector decay $B^0 \to J/\psi K^{*0}$ allows different polarizations in the final state. The measurement of this polarization tests the factorization hypothesis for hadronic decays and also helps determine if the decay is useful for studies of CP violation. Using the form factor calculated with the model of Bauer, Stech, and Wirbel,[1] Kramer and Palmer predict Γ_L/Γ =0.57. If instead heavy quark symmetries are used to relate the experimental results for $D \to K^* l \nu$ to $B^0 \to J/\psi K^{*0}$, the prediction is $\Gamma_L/\Gamma = 0.73$.[2] Measurements from ARGUS suggest that the decay is completely longitudinally polarized.[3] A recent result from CLEO gives the value $\Gamma_L/\Gamma = 0.80 \pm 0.08 \pm 0.05$.[4] This paper describes a preliminary polarization measurement performed by the CDF collaboration using 19 pb^{-1} of $\bar{p}p$ collisions collected during the 1992–93 run at the Fermilab Tevatron.

The CDF detector has been described in detail elsewhere.[5] B^0 mesons were reconstructed through the decay chain $B^0 \to J/\psi K^{*0}$, $J/\psi \to \mu^+\mu^-$, $K^{*0} \to K^+\pi^-$ using a data sample selected online by dimuon triggers in the CDF three level trigger system. At Level 1 two charged track segments were required in the muon chambers. The Level 2 trigger required that at least one of the muon segments was matched in azimuthal angle to a track in the Central Tracking Chamber (CTC). At Level 3 the trigger, using online track reconstruction software, required a pair of oppositely charged muons with an invariant mass between 2.8 and 3.4 GeV/c^2.

In order to isolate the J/ψ signal, and to keep the systematic effects from the trigger under control, additional offline requirements were placed on the muons. The match between the extrapolated CTC track and the segment in the muon chamber was required to be less than 3σ, where σ is the expected multiple scattering error combined in quadrature with the measurement errors. Information from the Silicon Vertex Detector (SVX) was added to the CTC tracks when it was available. Both

---

*Representing the CDF Collaboration.

muons were required to have a transverse momentum greater than 2.0 GeV/c, and at least one had to have a transverse momentum greater than 2.8 GeV/c. The invariant mass of the dimuon pair was formed while constraining the muon tracks to come from a common vertex. After all of the above requirements were applied there were approximately 41000 J/ψ candidates remaining, with a signal width of about 20 MeV.

Those dimuon pairs within 80 MeV/c^2 of the world average J/ψ mass were combined with other charged tracks to search for B^0 mesons. K^{*0} candidates were formed by selecting pairs of oppositely charged tracks. Both $K^{\pm}\pi^{\mp}$ mass assignments were tried. The assignment closer to the K^{*0} mass was used, and the candidate was kept if the $K\pi$ invariant mass was within 80 MeV/c^2 of the mass of the K^{*0}. All of the tracks were constrained to come from a common vertex, and the dimuons were mass constrained to the world average J/ψ mass. Combinatoric backgrounds were reduced by applying requirements on the proper decay distance $c\tau$ and transverse momentum of the B^0 candidate, and on the transverse momentum of the K^{*0}. The requirements were $c\tau > 100$ μm, $P_{T_B} > 8.0$ GeV/c and $P_{T_{K^{*0}}} > 2.0$ GeV/c. After all the constraints were applied there were 60 ± 11 B^0 candidates remaining in the signal region defined by $|m_{\mu\mu K\pi} - m_B| < 0.03$ GeV/c^2.

The polarization was measured using a helicity angle analysis. The differential decay distribution for $B^0 \to J/\psi K^{*0}$, $J/\psi \to \mu^+\mu^-$, $K^{*0} \to K^+\pi^-$ can be written in terms of the helicity amplitudes H_λ as (e.g. see Ref. 2)

$$\frac{d^2\Gamma}{d\cos\theta_K \cdot d\cos\theta_\psi} \propto \frac{1}{4}\sin^2\theta_{K^*}(1 + \cos^2\theta_\psi)(|H_{+1}|^2 + |H_{-1}|^2) + \cos^2\theta_{K^*}\sin^2\theta_\psi|H_0|^2 \quad (1)$$

where the helicity angle θ_{K^*} is the decay angle of the kaon in the K^{*0} rest frame with respect to the K^{*0} direction in the B^0 rest frame. Similarly, θ_ψ is the decay angle of the muon in the J/ψ rest frame with respect to the J/ψ direction in the B^0 rest frame. Integrating separately over θ_{K^*} and θ_ψ gives the relations

$$\frac{d\Gamma}{d\cos\theta_{K^*}} \propto \frac{1}{2} - \alpha\left(-\frac{1}{2} + \cos^2\theta_{K^*}\right) ; \qquad \frac{d\Gamma}{d\cos\theta_\psi} \propto 1 + \alpha\cos^2\theta_\psi \quad (2)$$

where $\alpha = (1 - 3\Gamma_L/\Gamma)/(1 + \Gamma_L/\Gamma)$. The ratio Γ_L/Γ measures the amount of longitudinally polarized K^{*0} or J/ψ.

The helicity angles were calculated for events in the B^0 signal region. An unbinned likelihood fit was performed using information from both the K^{*0} and the J/ψ by forming an event by event product of likelihoods

$$L = \prod_{i=1}^{n} L_{K_i^*} L_{\psi_i} \quad (3)$$

where the main components of $L_{K_i^*}$ and L_{ψ_i} are the theoretical distributions in equation 2, the acceptance functions, and the background shapes. The acceptance functions were calculated using Monte Carlo and the background shapes were determined by studying events in the sidebands of the B^0 invariant mass distribution. The fit results in a polarization measurement of $\Gamma_L/\Gamma = 0.66 \pm 0.10$ (stat) $^{+0.08}_{-0.10}$ (sys). The fit is projected onto background subtracted, acceptance corrected, data plots in Fig. 1. The systematic error is dominated by the uncertainty associated with the background estimation.

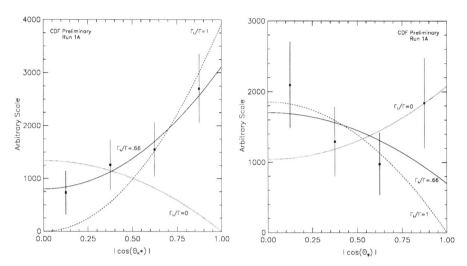

Fig. 1. a) K^{*0} helicity angle distribution and b) J/ψ helicity angle distribution. The fit value of $\Gamma_L/\Gamma = 0.66$ is shown along with the extremes.

This preliminary result is in good agreement with the recent CLEO measurement. Although not fully polarized, this decay mode still appears to be useful for CP violation studies at B factories. This result also demonstrates the feasibility of studying the dynamics of B meson decays in a hadron collider environment.

References

1. M. Bauer, B. Stech, and M. Wirbel, *Z. Phys.* **C34** (1987) 103; M. Wirbel, B. Stech, and M. Bauer, *Z. Phys.* **C29** (1985) 637.
2. G. Kramer and W.F. Palmer, *Phys. Rev.* **D46** (1992) 2969; G. Kramer and W.F. Palmer, *Phys. Lett.* **B279** (1992) 181.
3. M.V. Danilov, in *Proceedings of the Joint International Lepton-Photon Symposium and Europhysics Conference on High Energy Physics*, Geneva, Switzerland, (1991) 331.
4. M.S. Alam *et al.*, *Phys. Rev.* **D50** (1994) 43.
5. F Abe *et al.*, *Nucl. Inst. and Meth.* **A271** (1988) 387 and references therein.

NEW RESULTS ON BARYON PRODUCTION IN B MESON DECAYS FROM THE CLEO II DETECTOR

P. BARINGER

Department of Physics and Astronomy, University of Kansas,
Lawrence, Kansas USA 66045

(Representing the CLEO collaboration)

ABSTRACT

The CLEO II collaboration has made preliminary studies exploring baryon production in B meson decay using a data sample containing over two million events where $e^+e^- \to \Upsilon(4S) \to B\bar{B}$. Baryon-lepton correlation data imply that external W-emission does not saturate charmed baryon production in B meson decay. Further, these data and the observation of Ξ_c production indicate a contribution from $b \to cW^-$, $W^- \to \bar{c}s$. Finally, we measure $B(\bar{B}^0 \to \Lambda_c^+ \bar{p}\pi^+\pi^-) = (0.187 \pm 0.059 \pm 0.056 \pm 0.045)\%$.

1. Introduction

The B meson has been previously observed to decay into charmed baryons, specifically the Λ_c[1] and Σ_c.[2] This is the only meson system observed to decay into these baryons and little is understood about the mechanism. The simplest picture is the "external spectator", shown in Fig. 1(a) below. From such a picture one would expect a semileptonic contribution of about 10%, as is the case in decays to charmed mesons. Strange charmed baryons would result only from $s\bar{s}$ quark popping from the vacuum. The "internal spectator" of Fig. 1(b) would not result in semileptonic decays.

One puzzling feature of the previous data is the softness of the momentum spectrum of the charmed baryons from B decay. Figure 2 shows the spectrum for Λ_c's and the spectrum for Σ_c's is similarly soft. In Fig. 1, the W^- is shown as going to $\bar{u}d$, but it may also go to $\bar{c}s$ (and to Cabibbo suppressed combinations, which we ignore here). It has been suggested recently that the soft momentum spectrum could be explained if this $b \to c\bar{c}s$ transition dominates the B to baryons decay width.[3]

Here we present preliminary data bearing on the B to baryons mechanism using a high statistics sample taken with the CLEO II detector.[4] These results have been presented in more detail elsewhere.[5] Data were taken at the energy of the $\Upsilon(4S)$ resonance (2036 pb^{-1}) and just below that energy (967 pb^{-1}). This latter data sample is referred to as the "continuum" and is used to subtract non-$B\bar{B}$ contributions from the on-resonance data. There are $(2.19 \pm 0.04) \times 10^6$ $B\bar{B}$ pairs produced via $\Upsilon(4S)$ decay in our dataset.

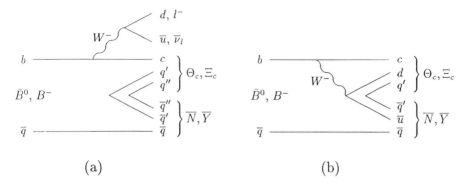

Fig. 1. Graphs for (a) "external spectator" and (b) "internal spectator" decays of the B meson into charmed baryons. Θ_c represents a non-strange charmed baryon, Ξ_c a strange charmed baryon.

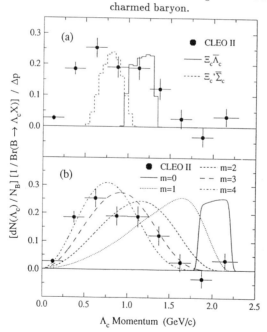

Fig. 2. Measured Λ_c momentum spectrum from B Decays compared to the shape expected from (a) $B \to \Xi_c \bar{\Lambda}_c$, $B \to \Xi'_c \bar{\Sigma}_c$, (b) $B \to \Lambda_c \bar{N}(m\pi)$ for $m = 0...4$.

2.   Baryon-Lepton correlations

We observe 3154 ± 160 Λ_c's coming from B decay in our dataset using four different Λ_c decay modes: $pK^-\pi^+$, $\Lambda\pi^+$, $p\bar{K}^0$, and $\Sigma^0\pi^+$. A high momentum lepton ($p > 1.4$ GeV/c) can be used to tag the flavor of the *other* B meson in the event, shedding light on the production mechanism for the charmed baryon. Electrons are identified using dE/dx measurements in the drift chamber and measurements of energy deposited in the electromagnetic calorimeter. Muons are identified using muon chambers to detect charged particles which penetrate the magnet iron.

2.1. Λ_c-Lepton correlations

If we observe a Λ_c^+ coming from a b quark decay (hence from a $\bar{B}^0$ or B^- meson), then it must contain the c quark from $b \to cW^-$. If instead we observe a Λ_c^-, that must arise from the $\bar{c}$ antiquark from $W^- \to \bar{c}s$. The high momentum lepton coming from the other B meson in the event would be positively charged ($\bar{b} \to \bar{c}W^+; W^+ \to l^+\nu$). Thus like-sign Λ_c-lepton correlations point to the $b \to c\bar{u}d$ mechanism, and opposite sign to $b \to c\bar{c}s$. The data yield 141 ± 16 like-sign combinations and 43 ± 16 opposite-sign combinations. We now remove backgrounds from misidentified leptons, and secondary leptons (that is, those produced via the decay chain $b \to cX, c \to sW^+, W^+ \to l^+\nu$). A correction is applied for the effect of $B^0\bar{B}^0$ mixing and we obtain 148 ± 19 like-sign events and 29 ± 19 opposite-sign events. These numbers would indicate that in B decays to charmed baryons the ratio of events produced via the $b \to c\bar{c}s$ mechanism to those from $b \to c\bar{u}d$ is $(20 \pm 13 \pm 4)\%$. The former mechanism does not dominate, but does contribute.

2.2.   Search for Semileptonic B Decays with Λ_c

We now try to determine the semileptonic contribution to charmed baryon production in B meson decay. This contribution must come from an external spectator process. The lepton coming from the same B as the charmed baryon will be softer than the lepton we used previously to tag the flavor of the other B in the event. At most 2% of these leptons would pass our cut of $p > 1.4$ GeV/c. We use three techniques to try to observe the semileptonic contribution.

First we look for events with a Λ, a soft lepton which could come from the same B, and a hard lepton which could come from the opposite B. We observe only four such events and find that backgrounds from continuum processes, fake leptons, and secondary leptons can account for this number. We obtain an upper limit[*] of: $\frac{B \to \Lambda_c \bar{N} X l\nu}{B \to \Lambda_c X} < 5.7\%$.

Next we take our fully reconstructed Λ_c's and match them with a soft electron. We observe 95 ± 20 events where the Λ_c and electron have opposite charges (as expected when they come from the same B in a semileptonic decay) and 74 ± 16 events with the same charge. After subtracting backgrounds and correcting for $B^0\bar{B}^0$ mixing we have a signal excess of 35 ± 26 events in the opposite charge channel, and find that the backgrounds account for all of the events in the same charge channel (as they should). This yields an upper limit of: $\frac{B \to \Lambda_c \bar{N} l\nu}{B \to \Lambda_c X} < 6\%$.

---

[*]Upper limits in this paper are at the 90% confidence level.

Third, a search was made for the exclusive decay $B \to \Lambda_c \bar{p} l^- \nu$ using a missing mass technique to account for the undetected neutrino. We obtained an upper limit on this decay of: $\frac{B \to \Lambda_c \bar{p} l^- \nu}{B \to \Lambda_c X} < 10\%$.

3. Ξ_c Production in B Decays

We observe here for the first time, Ξ_c's from B meson decay. Figure 3 shows the continuum subtracted signals of 71.2 ± 21.1 events for the Ξ_c^+ and 51.8 ± 18.7 events for the Ξ_c^0. Using reasonable estimates of the Ξ_c branching ratios, we get $B(B \to \Xi_c^+ X) = (1.5 \pm 0.7)\%$ and $B(B \to \Xi_c^0 X) = (2.4 \pm 1.3)\%$. These can be compared to $B(B \to \Lambda_c^+ X) = (6.3 \pm 2.4)\%$; so the sum of the Ξ_c yields is roughly half the Λ_c yield. The momentum spectrum of the Ξ_c's is soft, as one would expect for the $b \to c\bar{c}s$ mechanism. However, we do see a small number of events above $p = 1.2$ GeV/c, which is the kinematic limit for a Ξ_c from $B \to \Xi_c \bar{\Lambda}_c X$. This suggests that some Ξ_c's are produced from $b \to c\bar{u}d$ with $s\bar{s}$ popping.

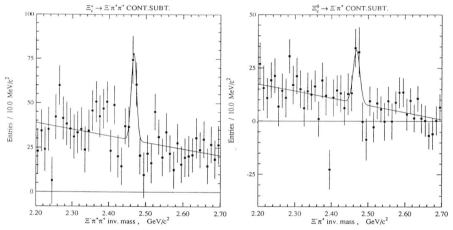

Fig. 3. Signals observed after continuum subtraction for $B \to \Xi_c^+ X, \Xi_c^+ \to \Xi^- \pi^+ \pi^+$ and $B \to \Xi_c^0 X, \Xi_c^0 \to \Xi^- \pi^+$.

4. Exclusive B Decays with a Λ_c

Now we try to fully reconstruct some previously unobserved B meson decays. Instead of an invariant mass, a beam constrained mass is used since the energy of the B meson must equal E_{beam}. We plot $M_B = \sqrt{E_{beam}^2 - (\Sigma \vec{p})^2}$ and select events where $|\Sigma E - E_{beam}| < 25$ MeV. The Monte Carlo predicts the width in $|\Sigma E - E_{beam}|$ to be 12 to 16 MeV depending upon the signal being considered. Nine different decay channels were studied and a signal of 15.0 ± 4.7 events, shown in Fig. 4, was seen in the $\bar{B}^0 \to \Lambda_c^+ \bar{p} \pi^- \pi^+$ channel. No events are seen in the continuum, the energy sidebands or the Λ_c sidebands. We obtain a branching fraction $B(\bar{B}^0 \to \Lambda_c^+ \bar{p} \pi^+ \pi^-) = (0.187 \pm 0.059 \pm 0.056 \pm 0.045)\%$, where the third error reflects the uncertainty in the Λ_c branching fractions used. No statistically significant signals were seen in the other eight channels and the resulting upper limits are tabulated below.

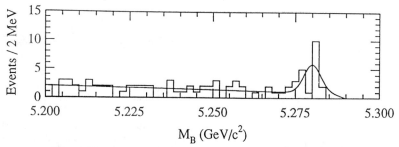

Fig. 4. Beam constrained mass distribution for $\bar{B}^0 \to \Lambda_c^+ \bar{p} \pi^+ \pi^-$.

channel	upper limit (%)	channel	upper limit (%)
$\bar{B}^0 \to \Lambda_c^+ \bar{p}$	.044	$B^- \to \Lambda_c^+ \bar{p} \pi^- \pi^+ \pi^-$	.553
$B^- \to \Lambda_c^+ \bar{p} \pi^-$	.084	$\bar{B}^0 \to \Lambda_c^+ \bar{p} \pi^- \pi^+ \pi^0$	.763
$\bar{B}^0 \to \Lambda_c^+ \bar{p} \pi^0$	.076	$\bar{B}^0 \to \Lambda_c^+ \bar{p} \pi^+ \pi^- \pi^+ \pi^-$	.341
$B^- \to \Lambda_c^+ \bar{p} \pi^- \pi^0$	.364	$B^- \to \Lambda_c^+ \bar{p} \pi^- \pi^+ \pi^- \pi^0$	2.165

5.   Conclusions

We have explored the production of charmed baryons in B meson decay with a high statistics data sample. Baryon-lepton correlation data imply that external W-emission does not saturate charmed baryon production in B meson decay. The observation of Ξ_c production indicates a contribution from $b \to cW^-$, $W^- \to \bar{c}s$ and the correlations data indicate this contribution is $(20 \pm 13 \pm 4)\%$ of $b \to cW^-$, $W^- \to \bar{u}d$. Finally, we have observed the first fully reconstructed B meson decays involving charmed baryons.

References

1. ARGUS Collab., H. Albrecht *et al.*, *Phys. Lett.* **B210**, (1988) 263; CLEO Collab., G. Crawford *et al.*, Phys. Rev. **D45**, (1992) 752.
2. CLEO Collab., M. Procario *et al.*, CLNS 93/1264, 1994 submitted to PRL.
3. Dunietz, Cooper, Falk and Wise, FERMILAB-PUB-94-132-T, 1994.
4. Y. Kubota *et al.*, *Nucl. Instr. and Meth.* **A320**, (1992) 66.
5. CLEO Collab., D. Cinabro *et al.*, CLEO CONF 94-8, 1994, submitted to ICHEP94, Glasgow; CLEO Collab., Y. Kubota *et al.*, CLEO CONF 94-13, 1994, submitted to ICHEP94, Glasgow.

CP Violation, Rare Decays, and B Physics

MARINA ARTUSO (L), *Syracuse,* and GEORGE G. GOLLIN (R), *Illinois (Urbana-Champaign)*

PROBING THE CKM MATRIX WITH b DECAYS

SHELDON STONE

Physics Department, Syracuse University
Syracuse, N.Y. 13244, U. S. A.

ABSTRACT

Best estimates of the CKM matrix elements V_{cb} and V_{ub} are extracted from data on semileptonic b decays. Three independent techniques are averaged to determine V_{cb}= 0.038±0.003, while the endpoint of the lepton spectrum gives $V_{ub} = 0.0030$±0.0008. This information is combined with results on $B - \bar{B}$ mixing and CP violation in K_L^0 decay in order to determine constraints on standard model parameters. Expectations of CP violating angles in the B system and B_s mixing are given.

1. Introduction

Quark mixing in the standard model is implemented via the CKM matrix, which is a function of 4 parameters. A useful first order expansion, due to Wolfenstein,[1] is shown in Fig. 1.

$$V_{ij} = \begin{pmatrix} 1 - \lambda^2/2 & \lambda & A\lambda^3(\rho - i\eta) \\ -\lambda & 1 - \lambda^2/2 & A\lambda^2 \\ A\lambda^3(1 - \rho - i\eta) & -A\lambda^2 & 1 \end{pmatrix}$$

Fig. 1. The Wolfenstein parameterization of the CKM matrix

The parameter λ is determined from charged current decays by measuring the decay rate of kaon and hyperon semileptonic decays. After applying suitable corrections, a value of

$$\lambda = V_{us} = 0.2205 \pm 0.0018 \tag{1}$$

is found.[3] Similarly, A is determined from $b \to c\ell\nu$ decays. Constraints on ρ and η are found from other measurements.

The fact that the CKM matrix is complex allows CP violation for 3 or more generations, as first shown by Kobayashi and Maskawa. Examples of CP violation have been found in the K^0 system.[4] In the Standard Model, CP violation results from the interference between "tree" decay diagram and the "box" decay mixing diagram. *If we could find CP violation in the B system we could see if the standard model works or perhaps go beyond the model.* Speculation has it that CP violation is responsible for the baryon-antibaryon asymmetry in our section of the Universe. *If so, understanding the mechanism of CP violation is critical in our conjectures of why we may exist.*

2. CKM Elements from B decays

2.1. V_{cb}

2.1.1. Introduction

The charged current semileptonic B decay diagram, used for the study of V_{ub} and V_{cb}, is shown in Fig. 2. Either inclusive or exclusive decays can be used to extract the CKM elements. It is informative to consider the fraction of semileptonic decays of heavy mesons to the lowest lying exclusive finals states, those with a pseudoscalar or vector meson in the final state, see Table 1.

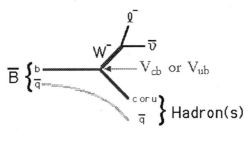

Fig. 2. Diagram discribing semileptonic B decays.

Table 1. Fraction of $Q \to q\ell\nu$ to lowest lying states

quark	percentage	process
s	100%	$K \to \pi\ell\nu$
c	>90%	$D \to (K + K^*)\ell\nu$
	?	$D \to (\pi + \rho)\ell\nu$
b	≈66%	$B \to (D + D^*)\ell\nu$
	?	$B \to (\pi + \rho)\ell\nu$
t	0%	t does not form hadrons

Whereas strange or charm decays must use exclusive final states to ascertain the value of λ, bottom decays can use both exclusive and inclusive decays to determine V_{cb} and possibly V_{ub}. I will discuss three different ways of determining V_{cb}, all with comparable accuracy.

2.1.2. Measurements of $\mathcal{B}(B \to D^*\ell\bar{\nu})$

After selecting candidate D^* and candidate leptons, the missing mass squared is used to find the signal. The missing mass squared is calculated as:

$$MM^2 = (E_B - (E_{D^*} + E_\ell))^2 - (\overrightarrow{p}_B - (\overrightarrow{p}_{D^*} + \overrightarrow{p}_\ell))^2. \qquad (2)$$

The B meson energy, E_B, is set equal to the beam energy, E_{beam}, and the B momentum, p_B, which is 325 MeV/c, is approximated as zero because the direction is unknown. Signal events will have a missing mass consistent with zero. The approximation of setting $p_B = 0$ causes the MM^2 distribution to be widened significantly; this is much larger than any widening caused by detector mismeasurements.

This technique has been used for isolating exclusive decays into both $D^{*+}\ell^-\bar{\nu}_\ell$ and $D^{*0}\ell^-\bar{\nu}_\ell$. Let us consider first the case of the D^{*+}. Data from ARGUS, which pioneered this technique, is shown in Fig. 3. A clear signal is evident as well backround from $B^0 - \bar{B}^0$ mixing and $B \to D^{**}\ell^-\bar{\nu}_\ell$.

The branching ratio measurements are given in Table 2.

Table 2. $\mathcal{B}(B \to D^*\ell^-\bar{\nu}) \left(\frac{0.5}{f_i}\right) (\%)$

Mode	CLEO[6]	ARGUS[7,8]	CLEO II[9]	Average
$D^{*+}\ell^-\bar{\nu}$	4.1±0.5±0.7	4.7±0.6±0.6	4.49±0.32±0.39	4.46±0.39
$D^{*0}\ell^-\bar{\nu}$		6.8±1.6±1.5	5.13±0.54±0.64	5.3±0.8

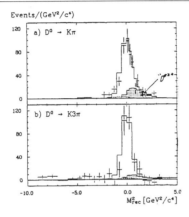

Fig. 3. Missing mass squared data from ARGUS for the $D^* + \ell^-\bar{\nu}$ final state, for two different D^0 decay modes

To extract $|V_{cb}|$, I use the average branching ratio for $\mathcal{B}(\bar{B}^0 \to D^{*+}\ell^-\bar{\nu})$ and $\mathcal{B}(B^- \to D^{*0}\ell^-\bar{\nu})$ from CLEO II only, since this is the only experiment which has measured both of these rates accurately. In this average branching ratio the poorly known fractions of neutral, f_0 and charged, f_-, $B's$ from $\Upsilon(4S)$ decay cancel and do not add to the uncertainty. Explicitly,

$$< B >= f_0 \mathcal{B}(\bar{B}^0 \to D^{*+}\ell^-\bar{\nu}) \left(\frac{0.5}{f_0}\right) + f_- \mathcal{B}(B^- \to D^{*0}\ell^-\bar{\nu}) \left(\frac{0.5}{f_-}\right) = (4.72 \pm 0.52)\% \tag{3}$$

These branching ratios can be used directly to find V_{cb} when combined with lifetime measurements from other experiments. I use 1.53±0.09 ps, 1.68±0.12 ps, and 1.58±0.07 ps, for the lifetimes of $\bar{B}^0$, B^- and their average, respectively.[10] Using these values, the experimental value of the width for $\Gamma(B \to D^*\ell^-\bar{\nu})$ is $(29.9 \pm 2.3 \pm 2.7)$ ns^{-1}. The resulting values for V_{cb} are given in Table 3, along with the predicted values for the width.

I take an average value, in the center of the model predictions, and include an error due to the range of the model predictions. This gives a value

$$|V_{cb}| = 0.0356 \pm 0.0022 \pm 0.0015. \tag{4}$$

Table 3. Values of $|V_{cb}|$ from $\Gamma(B \to D^*\ell^-\bar\nu)$

Model	$\Gamma(B \to D^*\ell^-\bar\nu)$ps	$|V_{cb}|$
ISGW[11]	$25.2|V_{cb}|^2$	0.0344 ± 0.0021
KS[12]	$25.7|V_{cb}|^2$	0.0341 ± 0.0020
WBS[13]	$21.9|V_{cb}|^2$	0.0369 ± 0.0022
Jaus[14]	$21.7|V_{cb}|^2$	0.0371 ± 0.0022

The first error is formed from the errors on the branching ratio ($\pm3.0\%$) and the lifetime ($\pm2.3\%$), while the second error arises from the model dependence ($\pm4.2\%$).

2.1.3. V_{cb} using the "Universal" form factor

One theory based on QCD, called Heavy Quark Effective Theory (HQET),[15] assumes very heavy quarks. Then the spin degrees of freedom decouple, and there is only one form factor function $\xi(y)$ which is a function of the Lorentz invariant 4-velocity transfer y

$$y = \frac{M_B^2 + M_{D^*}^2 - q^2}{2M_B M_D^*}.$$

(5)

The point y equals 1 corresponds to the situation where the B decays to a D^* which is at rest in the B frame. At this point the "universal" form factor function $\xi(y)$ has the value $\xi(1) = 1$ in lowest order. There are, however, corrections even at $y = 1$. These are due to hard gluons, which cause a first order correction, and the finite values of the b and c quark masses, which enter only in second order. Neubert estimates the correction factor as 0.97 ± 0.04 for $\xi(1)$.[16] This value has been challanged by a QCD sum rule calculation of Shifman et al.;[17] they set an upper limit of < 0.94 and give an "educated guess" of 0.89 ± 0.03.

In order to find the experimental value of the cross section at y of one, the data need to be fit to an "unknown" functional form. The curvature is expected to be positive, since there is a pole as y approaches 1, outside of the physical region, and $\xi(y) \to 0$ as y increases.

CLEO[9] assumes the form

$$\xi(y) = \xi(1) - \rho^2(y-1) + b(y-1)^2,$$

(6)

which represents a second order expansion in the vicinity of y of one. The CLEO data plotted as function of y are shown in Fig. 4. The resulting values are shown in Table 4. The b parameter is found to be consistent with zero and the second row reflects the result of the fit with b constrained to be zero. The first error is due to the statistical uncertainty, including the uncertainty due to the background, while the second gives the systematic uncertainty.

CLEO uses the latter value to extract a value of $|V_{cb}|\xi(1)$, although it would be more conservative to use the unconstrained b parameter fit. Another way of seeing what the error is, due to various shapes of $\xi(y)$, is to fit the data points to different forms that have been predicted theoretically. I have done this using the CLEO data points. The fits are shown in Fig. 5 and the results given shown in Table 5.

The values extracted for V_{cb} using either the Neubert or Shifman estimates for the QCD corrections are shown in Table 6. The first two entries, for the linear and

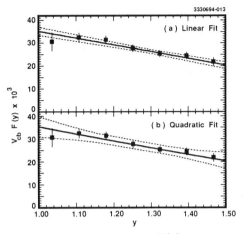

Fig. 4. Linear and quadratic fits to the CLEO data. $F(y)$ is equivalent to $\xi(y)$ as used in this paper.

Table 4. Values for $|V_{cb}|\xi(1)$ from CLEO data

$|V_{cb}|\xi(1)$	ρ^2	b
$0.0353\pm\ 0.0032\pm0.0030$	$0.92\pm0.64\pm0.40$	$0.15\pm1.24\pm0.90$
$0.0351\pm\ 0.0019\pm0.0019$	$0.84\pm0.13\pm0.08$	0

quadratic fits give the CLEO values from Table 4, while the last three entries are derived from Table 5. The quoted errors are the quadrature of the errors given in the above mentioned tables.

Using the average of the values derived using the Neubert and Shifman values for $\xi(1)$,[18] and the exponential fit, I derive a value of

$$|V_{cb}| = 0.0387 \pm 0.0030 \pm 0.0020, \tag{7}$$

where the last error results from the spread in functional forms and theoretical values for $\xi(1)$.

2.1.4. V_{cb} from Inclusive Decays

What is actually measured here is the semileptonic branching ratio $\mathcal{B}(B \to X e^- \bar{\nu})$. While this has traditionally been done by measuring the inclusive lepton momemtum spectrum using only single lepton data, recently dilepton data have been used. The inclusive lepton spectrum from the latest CLEO II data is shown in Fig. 6. Both electrons and muons are shown. Leptons which arise from the continuum have been statistically subtracted using the below resonance sample. The peak at low momentum is due to the decay chain $\bar{B} \to DX$, $D \to Y\ell^+\nu$. The data are fit to two shapes whose normalizations are allowed to float. The first shape is taken from models of B decay while the second comes from the measured shape of leptons from D mesons

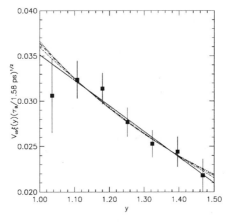

Fig. 5. Fits to the CLEO data with the functions listed in Table 5.

Table 5. Values for $|V_{cb}|\xi(1)$ from fits to different shapes

$\xi(y)$	name	ρ	$V_{cb}\xi(1)$
$1 - \rho^2(y - 1)$	linear	0.90 ± 0.07	$0.0351\pm0.0018\pm0.0018$
$\frac{2}{y+1}exp\left[-(2\rho^2 - 1)\frac{y-1}{y+1}\right]$	NR exp	0.90 ± 0.12	$0.0366\pm0.0024\pm0.0018$
$\left(\frac{2}{y+1}\right)^{2\rho^2}$	pole	1.07 ± 0.11	$0.0364\pm0.0023\pm0.0018$
$exp\left[-\rho^2(y - 1)\right]$	exp	1.01 ± 0.10	$0.0360\pm0.0022\pm0.0018$

produced nearly at rest at the ψ'', which is then smeared using the measured momentum distribution of $D's$ produced in B decay. CLEO finds $\mathcal{B}_{sl}$ of 10.5±0.2% and 11.1±0.3% in the ACM[19] and ISGW* models, respectively.[20]

Next, I discuss how to use dilepton events to eliminate the secondary leptons at low momentum. Consider the sign of the lepton charges for the four leptons in the following decay sequence: $\Upsilon(4S) \rightarrow B^-B^+$; $B^- \rightarrow D\ell_1^-\bar{\nu}$, $B^+ \rightarrow \bar{D}\ell_3^+\nu$; $D \rightarrow Y\ell_2^+\nu$, $\bar{D} \rightarrow Y'\ell_4^-\bar{\nu}$. If a high momentum negative lepton (ℓ_1^-) is found, then if the second lepton is also negative it must come from the cascade decay of the B^+ (i.e. it must be ℓ_4^-). On the other hand the second lepton being positive shows that it must be either the primary lepton from the opposite B^+, (ℓ_3^+), or the cascade from the same B^-, (ℓ_2^+). However the cascades from the same B^- can be greatly reduced by insisting that the cosine of the opening angle between the two leptons be greater than zero as they tend to be aligned. The same arguments are applicable to $\Upsilon(4S) \rightarrow B^0\bar{B}^0$, except that an additional correction must be made to account for $B\bar{B}$ mixing.

The CLEO II data are shown in Fig. 7. The data fit nicely to either the ACM or ISGW model. They find that the semileptonic branching ratio, $\mathcal{B}_{sl}$, equals (10.36 ± 0.17 ± 0.40)% with a negligible dependence on the model.[22] This result confirms that the B model shapes are appropriate down to lepton momenta of 0.6 GeV/c. ARGUS[21]

Table 6. Values for $|V_{cb}|$

$\xi(y)$	name	V_{cb} (Neubert)	V_{cb} (Shifman)
$1 - \rho^2(y-1) + b(y-1)^2$	quadratic	0.0364 ± 0.0045	0.0397 ± 0.0049
$1 - \rho^2(y-1)$	linear	0.0362 ± 0.0030	0.0395 ± 0.0033
$\frac{2}{y+1} exp\left[-(2\rho^2-1)\frac{y-1}{y+1}\right]$	NR exp	0.0377 ± 0.0031	0.0411 ± 0.0034
$\left(\frac{2}{y+1}\right)^{2\rho^2}$	pole	0.0375 ± 0.0030	0.0409 ± 0.0033
$exp\left[-\rho^2(y-1)\right]$	exp	0.0371 ± 0.0029	0.0404 ± 0.0032

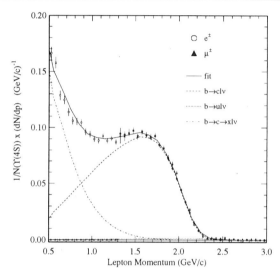

Fig. 6. Fit to the CLEO inclusive lepton spectrum with the ACM model.

did the first analyis using this technique and found $\mathcal{B}_{sl} = (9.6 \pm 0.5 \pm 0.4)\%$.

To extract $|V_{cb}|$ we can use the ACM model and the ISGW model because it includes final states beyond D and D^*. CLEO lets the "extra" component float it the fit. This model is denoted ISGW*. The resulting values are given in Table 7.

The representative value of $|V_{cb}|$ found from this analysis alone is $0.0395 \pm 0.0010 \pm 0.0040$.

2.1.5. Average of all three methods

The values of V_{cb} derived using all three methods are shown in Fig. 8. Consistent results are found. An average value is derived using all three results, but adding the statistical and systematic errors for each method linearly. These three numbers are then used in a weighted average to extract

$$|V_{cb}| = 0.0378 \pm 0.0026, \quad \text{and} \quad A = 0.777 \pm 0.053. \tag{8}$$

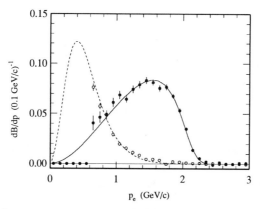

Fig. 7. The lepton momentum spectrum in dilepton events from CLEO. The solid points are for opposite sign leptons, while the open circles indicate like sign lepton pairs. The fit is to the ACM model.

Table 7. V_{cb} Values from Inclusive leptons

Model	Experiment	V_{cb}
ACM	CLEO I	$0.042\pm0.002\pm0.004$
ACM	ARGUS	$0.039\pm0.001\pm0.003$
ACM	CLEO II	$0.040\pm0.001\pm0.004$
ISGW	CLEO I	$0.039\pm0.002\pm0.004$
ISGW	ARGUS	$0.039\pm0.001\pm0.005$
ISGW	CLEO II	$0.040\pm0.001\pm0.004$
ISGW*	CLEO I	$0.037\pm0.002\pm0.004$
ISGW*	CLEO II	$0.040\pm0.002\pm0.004$

2.2. The $b \to u$ transistion

2.2.1. Introduction

The only direct experimental evidence for the $b \to u$ transistion is from inclusive $b \to u\ell\nu$ decays, where e^- and μ^- are found beyond the endpoint for $B \to D\ell\nu$ decays. The branching ratios are small. CLEO II finds that the rate in the lepton momentum interval $2.6 > p_\ell > 2.4$ GeV/c, $\mathcal{B}_u(p)$, is $(1.5 \pm 0.2 \pm 0.2) \times 10^{-4}$. To extract V_{ub} from this measurement we need to use theoretical models. It is convienent to define: $\Gamma(b \to u\ell\nu) = \gamma_u|V_{ub}|^2$, and $\Gamma(b \to c\ell\nu) = \gamma_c|V_{cb}|^2$. In addition, $f_u(p)$ is the fraction of the spectrum predicted in the end point region by different models, and $\mathcal{B}_{sl}$ is the semileptonic branching ratio. Then:

$$\frac{|V_{ub}|^2}{|V_{cb}|^2} = \frac{\mathcal{B}_u(p)}{\mathcal{B}_{sl}} \cdot \frac{\gamma_c}{f_u(p)\gamma_u}. \tag{9}$$

These models disagree as to which final states populate endpoint region. Most

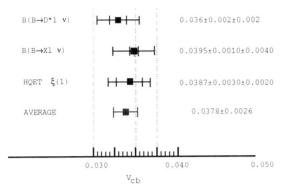

Fig. 8. Summary of $|V_{cb}|$ values found using different methods.

models agree roughly on values of γ_c. However, models differ greatly in the value of the product $\gamma_u \cdot f_u(p)$. There are two important reasons for these differences. First of all, different authors disagree as to the importance of the specific exclusive final states such as $\pi\ell^-\bar{\nu}_\ell$, $\rho\ell^-\bar{\nu}_\ell$ in the lepton endpoint region. For example, the Altarelli et al. model doesn't consider individual final states and thus can be seriously misleading if the endpoint region is dominated by only one or two final states. In fact, several inventors of exclusive models have claimed that the endpoint is dominated by only a few final states.[11,13] Secondly, even among the exclusive form-factor models there are large differences in the absolute decay rate predictions. The differences in the exclusive models are much larger in $b \rightarrow u$ transitions than in $b \rightarrow c$ transitions because the q^2 range is much larger. Ramirez, Donoghue and Burdman[24] claim that the lepton endpoint region is comprised both of exclusive final states and inclusive ones with multiple pions. T

2.2.2. Value of V_{ub}/V_{cb}

Fig. 9 shows V_{ub}/V_{cb} for different models from an average of data reported by CLEO I,[25] ARGUS[26] and CLEO II[23]. The differences among the models dominates the uncertainty. The best estimate is that $V_{ub}/V_{cb}=0.08\pm0.02$.

2.2.3. Limits on exclusive charmless final states

There isn't any convincing evidence for the exclusive final states $\pi\ell\nu$, $\rho\ell\nu$, or $\omega\ell\nu$. The CLEO II upper limits, in WSB model are $\mathcal{B}(B^o \rightarrow \pi^-\ell^+\nu) < 4.5 \times 10^{-4}$ and $\mathcal{B}(B^o \rightarrow \rho^-\ell^+\nu) < 2.7 \times 10^{-4}$ at 90% confidence level. These give upper limits on V_{ub}/V_{cb} of < 0.18 and < 0.10, respectively.

2.2.4. Constraints on ρ and η

In terms of the Wolfenstein parameters,

$$\left|\frac{V_{ub}}{V_{cb}}\right|^2 = \lambda^2 \left(\rho^2 + \eta^2\right), \qquad (10)$$

which describes a circle centered at zero in the $\rho - \eta$ plane. From CP violation

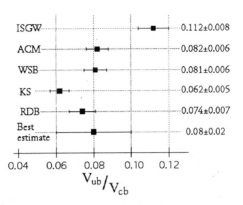

Fig. 9. $|V_{ub}/V_{cb}|$ from an average of CLEO I, ARGUS and CLEO II measurements using different models.

measurements in neutral kaon decay $\eta > 0$, which lets us describe this constraint as a semiannular region, of radius 0.36 ± 0.09.

2.3. V_{td}, Information from $B_d^0 - \bar{B}_d^0$ mixing
For $x \equiv \Delta M/\Gamma$, the CKM elements are related to x via

$$ x = \frac{G_F^2}{6\pi^2} B_B f_B^2 m_b \tau_B |V_{tb}^* V_{td}|^2 F\left(\frac{m_t^2}{M_W^2}\right)\eta_{QCD}, \tag{11} $$

where G_F is the Fermi constant and f_B is the decay constant of the B meson, which has been calculated theoretically, albeit with very large uncertainty. Since

$$ |V_{tb}^* V_{td}|^2 \propto |(1 - \rho - i\eta)|^2 = (\rho - 1)^2 + \eta^2, \tag{12} $$

the mixing measurement gives a circle centered at $(1,0)$ in the $\rho - \eta$ plane.
    The width of the band is caused primarily by the uncertainty in f_B. To measure x experiments have measured the ratio of mixed events to total events either integrating over time, as done by ARGUS and CLEO or recently measuring the explict time dependence, as done by ALEPH and OPAL. One such measurement is shown in Fig. 10. The extracted x values are shown in Table 8.

3. What is learned from CP Violation measurments in K_L^0 decay.
    The CP violating parameter ϵ is well measured. The constraint equation arising from this measurement is

$$ \eta\left[(1 - \rho)A^2 + 0.20\right] A^2 \frac{B_K}{0.65} = 0.15, \tag{13} $$

where B_K is a parameter which is related to the probability that the s and d quarks form a neutral K meson. This parameter must be calculated from theory. (Note also, that the numbers 0.20 and 0.15 depend somewhat on the charm quark and top

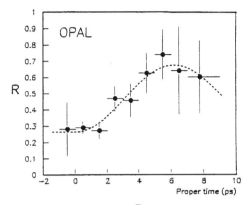

Fig. 10. The ratio R of mixed B events as a function of time.

Table 8. $x = \Delta M / \Gamma$ Values from B_d^0 mixing measurements

Experiment	x
CLEO [27]	0.65±0.10
ALEPH [28]	0.76±0.12
OPAL [29]	0.73±0.14
ARGUS [30]	0.75±0.15
AVERAGE	0.71±0.06

quark masses.) Following Buras and Harlander,[31] I take $B_K = 0.65 \pm 0.15$. The constraints in the $\rho - \eta$ plane from $|V_{ub}/V_{cb}|$, B mixing and ϵ are shown in Fig. 11. I use $240 > f_B > 160$ MeV, which gives a range consistent with most calculations, and will be discussed later.

The bands are shown with one standard deviation error bars. The consistency between all three measurements is remarkable. While it is of utmost importance to reduce the errors on all of these measurements, presently the Standard Model is spot on.

The error sources for the ϵ constraint are shown in Fig. 12. The largest error arises from the error on the A parameter, which results from the A^4 dependence. The error on B_K is also important. It is interesting that the uncertainty due to the error on m_t already is smaller than these other sources.

4. The unitarity triangle

4.1. Introduction

Since the CKM matrix is unitary we can multiply any row or any column by the complex conjugate of another row or column. The most useful such relationship is:

$$V_{td} \cdot V_{ud}^* + V_{ts} \cdot V_{us}^* + V_{tb} \cdot V_{ub}^* = 0$$

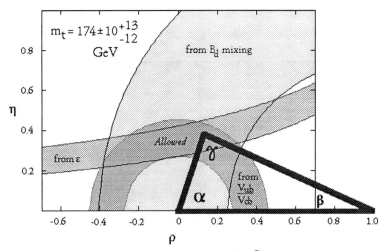

Fig. 11. Constraints in the $\rho - \eta$ plane from $|V_{ub}/V_{cb}|$, $B_d^0 - \bar{B}_d^0$ mixing and CP violation in the K_L^0 system (ϵ). The bands are $\pm 1\sigma$. Also shown is the "unitary" triangle.

$$\frac{1}{\lambda}V_{td}/V_{ts} + \frac{1}{\lambda}V_{ub}/V_{cb} = 1 \qquad (14)$$

Think of this as a vector equation with each "vector" representing the sides of a triangle. A triangle consistent with the data is shown in Fig. 11, where the angles α, β and γ are defined.

4.2. To test the Standard Model

We can measure all 3 sides AND all 3 angles. If we see consistency between all of these measurements we have defined the parameters of the Standard Model. If we see inconsistency the breakdown can point us to a more complete theory.

We know two sides already. The base is defined as 1, and the leftmost side comes from $|V_{ub}/V_{cb}|$. The righmost side can be found using $B^0 - \bar{B}^0$ mixing. As we have seen this introduces a large error due to f_B uncertainty. There are two solutions to this problem.

One solution is to measure B_s^0 mixing. The ratio between x_s and x is given by

$$\frac{x_s}{x} = \left(\frac{B_s}{B}\right)\left(\frac{f_{B_s}}{f_B}\right)^2 \left(\frac{\tau_{B_s}}{\tau_B}\right)\left|\frac{V_{ts}}{V_{td}}\right|^2, \qquad (15)$$

The ratios of the B parameters and decay constants are much better known than the absolute values. The allowed values for x_s as a function of ρ are given in Fig. ??.

The second method for finding the rightmost side of the triangle is to use the measurement of V_{td}/V_{ts} from "Penguin" diagrams. CLEO found the first unambiguous evidence for such graphs by finding the decay $B \rightarrow K^*\gamma$.[33] The diagram for this process is shown in Fig. 13. Recently they have also measured the inclusive rate $\mathcal{B}(b \rightarrow s\gamma) = (2.3 \pm 0.5 \pm 0.4) \times 10^{-4}$.[34]

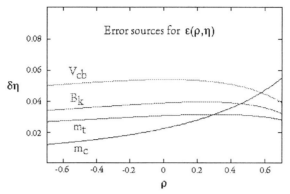

Fig. 12. The contributions to the error on the ϵ constraint as a function of ρ.

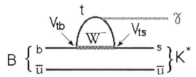

Fig. 13. Loop diagram for $B \rightarrow K^*\gamma$.

If the ts vertex in Fig. 13 where to be replaced with a td vertex, the final state would be $\rho\gamma$. Therefore a measurement of the relative rates

$$R_p = \mathcal{B}(B \rightarrow \rho\gamma)/\mathcal{B}(\rightarrow K^*\gamma) = \zeta|V_{td}/V_{ts}|^2, \tag{16}$$

where ζ is a model dependent correction due to different form-factors for the K^* and the ρ. Current models predict $\zeta = 0.58, 0.77, 0.81$. The CLEO II data[35] are shown in Fig. 14, from which it is found that $R_p < 0.34$ @ 90 % c. l. This is far from the range suggested from our allowed region in the $\rho - \eta$ plane, which is $0.07 > R_p > 0.02$ for ζ of 0.7. The upper value is for ρ of -0.040, while the lower value is for ρ of 0.30. A. Soni has claimed that "long distance" effects may pollute this measurement.[37]

4.3. Measure angles using CP violation

There are several ways of measuring CP violation in B decays. All of them rely on the interference between two amplitudes. In the "classic" case, CP violation via $B^0 - \overline{B}^0$ mixing, we choose a final state which is accessible to both B^0 and $\overline{B}^0$ decays. The second amplitude necessary for interference is provided by mixing as depicted in Fig. 15.

At the $\Upsilon(4S)$, $e^+e^- \rightarrow \gamma \rightarrow B^0\overline{B}^0$, which is a state of negative charge conjugation. For final states which are CP eigenstates an asymmetry exists, but its time integral is zero. Therefore we need to make time dependent measurments, which can be done by using asymmetric beam energies in order to get the B's moving. An alternative is to measure the process $e^+e^- \rightarrow \gamma \rightarrow B^0\overline{B}^0\gamma$, which is a C=+1 final state.[38] However,

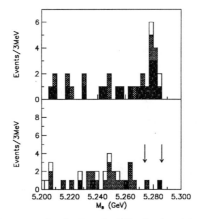

Fig. 14. The B candidate mass distributions for $K^*\gamma$ (top) and $(\rho + \omega)\gamma$ (bottom). The bottom plot includes 50% more integrated luminosity than the top. The arrows indicate the signal region. The dark entries for the upper (lower) plot are the $K^-\pi^+$ $(\pi^-\pi^+)$ events, the cross-hatched $K^-\pi^0$ $(\pi^-\pi^0)$, and the white $K_s^0\pi^+$ $(\pi^-\pi^+\pi^0)$.

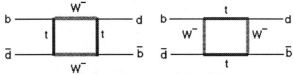

Fig. 15. Quark level diagrams for $B^0 - \bar{B}^0$ mixing.

the measured cross section is down by factor of 7 with respect to the $\Upsilon(4S)$.[39] Another alternative is to use a hadron collider. The Main Injector at FNAL will produce 1000 times as many B's as an e^+e^- machine.

Examples of final states most discussed to measure CP violation via mixing are ψK_s which measures the angle 2β, and $\pi^+\pi^-$ which measures the angle 2α.

Interference can also arise between "Penguin" and "Tree" graphs. In this case we have two distinct processes which yield the same final state so they interfere. This can lead to a rate asymmetry between B^- and B^+. An example is the decay $B^- \to K^-\pi^0$, depicted in Fig. 16.

Gronau, Rosner and London have[40] shown using the assumption of SU(3) and isospin relations that this mode coupled with a study of $B^- \to \pi^-\pi^0$ and $B^- \to \pi^-K^0$ can be used to measure the angle γ independent of any hadronic matrix elements. There is a similar plan using the decays $B^- \to D^0K^-$, $\overline{D}^0K^-$ and $D_{CP}^0K^-$, where $D_C^0{}_P$ indicates the decay into a CP eigenstate.[41]

4.4. Predictions of CP violating angles

The preferred range of CP violating angles as a function of the CKM parameter ρ is shown in Figs. 17, 18, 19. It should be kept in mind that ρ is closely related to f_B, so that negative ρ corresponds to small values of f_B, while positive ρ corresponds to large values. I have plotted the variables $sin(2\alpha)$, $sin(2\beta)$, and $sin(\gamma)$, since these are

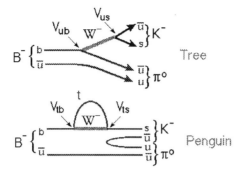

Fig. 16. Tree level $b \to u$ diagram and loop diagram for the final state $K\pi$.

directly related to the asymmetries that can be measured in the methods mentioned here. Large asymmetries are expected for the last two $(2\beta, \ \gamma)$, while the first, (2α) can have almost any asymmetry.

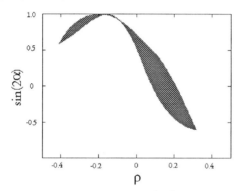

Fig. 17. Allowed region of $sin(2\alpha)$ versus ρ.

5.   Conclusions

The CKM model is consistent with data on V_{ub}/V_{cb}, $B^0 - \overline{B}^0$ mixing and the value of ϵ from CP violation in the kaon system, for a heavy top quark. Specifically,

- $|V_{cb}| = 0.038 \pm 0.003 \ \Rightarrow \ A = 0.78 \pm 0.06$

- $\left| \frac{V_{ub}}{V_{cb}} \right| = 0.08 \pm 0.02, \ \Rightarrow \ |V_{ub}| = 0.0030 \pm 0.0008$

- Large CP violating asymmetries are expected in B decays

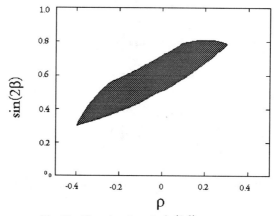

Fig. 18. Allowed region of $sin(2\beta)$ versus ρ.

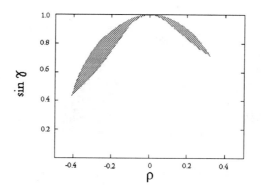

Fig. 19. Allowed region of $sin(\gamma)$ versus ρ.

6. Acknowledgements

I thank A. Ali, J. Rosner, A. Soni, M. Artuso, F. Muheim and S. Playfer for useful discussions. This work was supported by the National Science Foundation.

7. References

[1] L. Wolfenstein, Phys. Rev. D **31**, 2381 (1985).

[2] A. Ali, "*B* Decays - Introduction and Overview," in *B Decays 2nd edition*, ed. S. Stone, World Scientific, Singapore (1994), and references contained therein.

[3] L. Montanet et al., (Particle Data Group) Phys. Rev. D **50**, 1173 (1994).

[4] K. Kleinknecht in *CP Violation*, ed. C. Jarlskog, World Scientific, Singapore (1989), p 41.

[5] See F. J. Gilman and R. L. Singleton,Phys. Rev. D **41**, 142 (1990);K. Hagiwara, A. D. Martin and M. F. Wade Nucl. Phys. **B327**, 569 (1989); S. Stone "Semileptonic *B* Decays," in *B Decays 2nd edition*, ed. S. Stone, World Scientific, Singapore (1994).

[6] D. Bortoletto et al., (CLEO) Phys. Rev. Lett. **16**, 1667 (1989).

[7] H. Albrecht et al., (ARGUS) Zeit. Phys. C **57**, 533 (1993).

[8] H. Albrecht et al., (ARGUS) Phys. Lett. B **275**, 195 (1992).

[9] B. Barish et al., (CLEO II) CLNS 94/1285 (1994).

[10] W. Venus, "*b* Weak Interaction Physics at High Energies," in proc. of XVI Int. Symp. on Lepton-Photon Interactions, ed. P. Drell and D. Rubin, AIP Press, NY (1993).

[11] N. Isgur, D. Scora, B. Grinstein, and M. B. Wise, Phys. Rev. D **39**, 799 (1989).

[12] J. G. Korner and G. A. Schuler, Zeit. Phys. C **38**, 511 (1988); ibid, (erratum) **C41**, (1989), 690.

[13] M. Wirbel, B. Stech and M. Bauer, Zeit. Phys. C **29**, 637 (1985); M. Bauer and M. Wirbel, Z. Phys. **C42**, 671(1989).

[14] W. Jaus, Phys. Rev. D **41**, 3394 (1990).

[15] N. Isgur and M. B. Wise, Phys. Lett. B **232**, 113 (1989), ibid Phys. Lett. B **237**, 527 (1990).

[16] M. Neubert, SLAC-PUB-6263, to appear in Physics Reports; A. Falk, M. Neubert and M. Luke, Nucl. Phys. **B388**, 3363 (1992); A. Falk and M. Neubert, Phys. Rev. D **47**, 2965 (1993); M. Neubert, Phys. Rev. D **47**, 4063 (1993); M. Neubert and V. Rieckert, Phys. Lett. B **382**, 97 (1992); M. E. Luke, Phys. Lett. B **252**, 447 (1990).

[17] M. Shifman, N. Uraltsev and A. Vainshtein, Univ. of Minn. preprint TPI-MINN-94/13-T (1994).

[18] A new value of $\xi(1)$ from Neubert of 0.93 ± 0.03 is consistent with the average value used here. M. Neubert, "Theoretical Update on the Model-Independent Determination of $|V_{cb}|$ Using Heavy Quark Symmetry," CERN-TH.7395/94 (1994).

[19] G. Altarelli, N. Cabibbo, G. Corbo and L. Maiani, Nucl. Phys. **B207**, 365 (1982).

[20] J. Bartelt et al. (CLEO), " Inclusive Measurements of *B*-meson Semileptonic Branching Fractions," submitted to Lepton Photon conf., Cornell (1993), CLEO-CONF 93-19.

[21] H. Albrecht et al., (ARGUS) Phys. Lett. B **318**, 397 (1993).

[22] J. Gronbert et al., (CLEO) "Measurement of the Branching Ratio $B \to X e\nu$ with Lepton Tags," CONF 94-6.

[23] J. Bartelt et al., (CLEO) Phys. Rev. Lett. **71**, 4111 (1993).

[24] C. Ramirez, J. F. Donoghue and G. Burdman, Phys. Rev. D **41**, 1496 (1990).

[25] R. Fulton et al., (CLEO) Phys. Rev. Lett. **64**, 16 (1990).

[26] H. Albrecht et al., (ARGUS) Phys. Lett. B **234**, 409 (1990).

[27] J. Bartelt et al., (CLEO) Phys. Rev. Lett. **71**, 1680 (1993).

[28] D. Buskulic et al., (ALEPH) Phys. Lett. B **322**, 441 (1994).

[29] R. Akers et al., (OPAL) Zeit. Phys. C **60**, 199 (1993).

[30] H. Albrecht et al., (ARGUS) Zeit. Phys. C **55**, 357 (1992).

[31] A. Buras and M. Harlander, "A Top Quark Story: Quark Mixing, CP Violation and Rare Decays in the Standard Model," in *Heavy Flavours*, ed. A. Buras and M. Linder, World Scientific, Singapore (1992).

[32] J. Alexander et al., (CLEO) "A Search for $B \to \tau\nu$," CONF 94-5

[33] R. Ammar et al., (CLEO) Phys. Rev. Lett. **71**, 674 (1993).

[34] B. Barish et al., (CLEO), "First Measuremnt of the Inclusive Rate for the Radiative Penguin Decay $b \to s\gamma$," CONF94-1 (1994).

[35] M. Athanas et al., (CLEO), "A Constraint on $|V_{td}/V_{ts}|$ from $B \to \rho(\omega)\gamma/B \to K^*\gamma$," CONF94-2 (1994).

[36] A. Ali et al., CERN-Th.7118/93 (1993); J. M. Soares, Phys. Rev. D **49**, 283 (1994); S. Narison, Phys. Lett. B **327**, 354 (1994).

[37] D. Atwood, B. Blok andA. Soni, "Feasibility of Extracting V_{td} from Radiative $B(B_s)$ Decays, SLAC-PUB-6635 (1994); see also E. Golowich and S. Pakvasa, "Uncertainties from Long Range Effects in $B \to K^*\gamma$," Univ. of Mass. UMHEP-411 (1994).

[38] S. Stone, Mod. Phys. Lett. **A3**, 541, (1988).

[39] D. S. Akerib et al., (CLEO) Phys. Rev. Lett. **67**, 1692 (1991).

[40] M. Gronau, J. Rosner and D. London, Phys. Rev. Lett. **73**, 21 (1994).

[41] M. Gronau and D. Wyler, Phys. Lett. B **265**, 172 (1991); S. Stone, Nucl. Instr. & Meth. **A333**, 15, (1993).

MEASUREMENT OF THE $\bar{B} \to D^* \ell^- \bar{\nu}$ BRANCHING FRACTIONS AND $|V_{cb}|$

XU FU[*]

Department of Physics and Astronomy, University of Oklahoma
Norman, OK, 73019, USA

ABSTRACT

Using the CLEO II detector, we have made new measurements on the $\bar{B}^0 \to D^{*+} \ell^- \bar{\nu}$ and $B^- \to D^{*0} \ell^- \bar{\nu}$ branching fractions. The product of the CKM matrix element $|V_{cb}|$ times the normalization of the decay form factor at the point of no recoil of the D^* meson is determined. Using theoretical calculations of the form factor normalization we extract a value for $|V_{cb}|$.

1. Introduction

In the framework of the Standard Model of weak interactions the elements of the 3×3 Cabibbo-Kobayashi-Maskawa (CKM) mixing matrix must be determined experimentally. The recent development of Heavy Quark Effective Theory (HQET)[1] yields an expression for the $\bar{B} \to D^* \ell^- \bar{\nu}$ decay rate in terms of a single unknown form factor which, at the point of no recoil of the D^* meson, is absolutely normalized up to corrections of order $1/m_Q^2$ (where m_Q is the b or c quark mass). It is currently believed that these corrections can be calculated with less than 5% uncertainty,[2,3] which would permit a precise determination of $|V_{cb}|$ from the study of $\bar{B} \to D^* \ell^- \bar{\nu}$ as a function of the recoil of the D^* meson.

We report on new measurements[4] of the branching fractions and differential decay rate for the decays, $\bar{B}^0 \to D^{*+} \ell^- \bar{\nu}$ and $B^- \to D^{*0} \ell^- \bar{\nu}$. The CKM matrix element $|V_{cb}|$ is extracted from fits to the differential decay rates.

2. Event Reconstruction

The data used in this analysis were produced at the Cornell Electron-positron Storage Ring (CESR) and recorded with the CLEO II detector.[5] The signal comes from an integrated luminosity of 1.55 fb^{-1} collected at the $\Upsilon(4S)$ center-of-mass energy. An additional 0.69 fb^{-1} of data collected below the $B\bar{B}$ production threshold are used for continuum background determination.

For this analysis we select hadronic events that have at least one track identified as a lepton(e or μ) with momentum $1.4 \leq |\mathbf{p}_\ell| \leq 2.4$ GeV. All identified electrons must spatially match a cluster of energy in the CsI calorimeter of approximately equal energy to the measured momentum. All identified muons must penetrate at least 5 nuclear

---

*Representing the CLEO collaboration

absorption lengths of the muon chamber. Both electrons and muons must lie within the angular region $|\cos\theta_l| < 0.71$, where θ_l is the polar angle from the beam axis. The ratio of the second to the zeroth Fox-Wolfram moments[6] of the event is required to be less than 0.4 to suppress background from continuum events. For each lepton in these events we search for D^0 candidates in the decay mode $D^0 \rightarrow K^-\pi^+$ using charge correlation with the lepton to make unambiguous mass assignments (the lepton and kaon charges must be the same).

D^0 candidates are combined with pion candidates to fully reconstruct D^* mesons in the two modes $D^{*+} \rightarrow D^0\pi_s^+$ and $D^{*0} \rightarrow D^0\pi_s^0$. Charged pions emitted from D^{*+} are accepted if they lie within the polar angle region $|\cos\theta_{\pi_s^+}| < 0.71$ and have momentum above 65 MeV. Neutral pions emitted from D^{*0} are constructed from pairs of showers in the electromagnetic calorimeter which do not match the projection of any drift chamber track and have an invariant mass within 3σ of the measured π_s^0 mass ($\sigma = 5$ to 8 MeV, depending on shower energies and polar angles). The momentum of D^* candidates must satisfy $|\mathbf{p}_{D^*}|/\sqrt{E_B^2 - m_{D^*}^2} < 0.5$ to be consistent with B decay.

The 2-D 'missing mass squared' vs. $|\mathbf{p}_{D^*} + \mathbf{p}_\ell|$ distribution is investigated. A signal region is chosen to suppress the $\bar{B} \rightarrow D^{**} \ell \bar{\nu}$ background. The $\bar{B} \rightarrow D^{**} \ell \bar{\nu}$ background that falls within the signal region is estimated using Monte Carlo events generated according to ISGW model. Uncorrelated D^*, ℓ pairs from different B's are corrected from the study of inclusive $\Upsilon(4S) \rightarrow D^*$ rate and inclusive $B \rightarrow \ell$ rate.

The raw yield of $\bar{B} \rightarrow D^* \ell \bar{\nu}$ events is obtained by fitting the $M_{K\pi}$ distribution after cutting on δ_m and doing a δ_m sideband subtraction, where $\delta_m \equiv M_{K\pi\pi_s} - M_{K\pi}$. We find $376 \pm 27 \pm 16$ * $(302 \pm 32 \pm 13)$ $\bar{B}^0 \rightarrow D^{*+} \ell^- \bar{\nu}(B^- \rightarrow D^{*0} \ell^- \bar{\nu})$ raw signal events, as shown in Fig. 1, where the dashed lines are scaled δ_m sidebands.

The detection efficiencies for the $\bar{B}^0 \rightarrow D^{*+} \ell^- \bar{\nu}$ and $B^- \rightarrow D^{*0} \ell^- \bar{\nu}$ events are predicted by Monte Carlo simulation to be $[9.54 \pm 0.77(sys.)]\%$ and $[7.18 \pm 0.74(sys.)]\%$

3. Branching Fraction Results

The $\bar{B} \rightarrow D^* \ell \bar{\nu}$ branching fractions can be written as the following,

$$\mathcal{B}(\bar{B}^0 \rightarrow D^{*+} \ell^- \bar{\nu}) = \frac{N_0}{4N_{\Upsilon(4S)}f_{00}\mathcal{B}_{D^{*+}}\mathcal{B}_{D^0}} \tag{1}$$

$$\mathcal{B}(B^0 \rightarrow D^{*0} \ell^- \bar{\nu}) = \frac{N_-}{4N_{\Upsilon(4S)}f_{+-}\mathcal{B}_{D^{*0}}\mathcal{B}_{D^0}} \tag{2}$$

where $\mathcal{B}_{D^{*+}} \equiv \mathcal{B}(D^{*+} \rightarrow D^0\pi^+) = [68.1 \pm 1.0 \pm 1.3]\%$ [7], $\mathcal{B}_{D^{*0}} \equiv \mathcal{B}(D^{*0} \rightarrow D^0\pi^0) = [63.6 \pm 2.3 \pm 3.3]\%$ [7], $\mathcal{B}_{D^0} \equiv \mathcal{B}(D^0 \rightarrow K^-\pi^+) = [3.91 \pm 0.08 \pm 0.17]\%$,[8] $N_0(N_-)$ is efficiency-corrected yield for the $\bar{B}^0 \rightarrow D^{*+} \ell^- \bar{\nu}(B^- \rightarrow D^{*0} \ell^- \bar{\nu})$ events. $N_{\Upsilon(4S)} = (1.65 \pm 0.01) \times 10^6$.

If we assume equal production of charged and neutral meson pairs at the $\Upsilon(4S)$, $f_{+-} = f_{00} = 0.5$, we obtain,

$$\mathcal{B}_{0.5}(\bar{B}^0 \rightarrow D^{*+} \ell^- \bar{\nu}) = [4.49 \pm 0.32 \pm 0.39]\%, \tag{3}$$

$$\mathcal{B}_{0.5}(B^- \rightarrow D^{*0} \ell^- \bar{\nu}) = [5.13 \pm 0.54 \pm 0.64]\%, \tag{4}$$

---

*Through out this paper, the first error is statistical, the second systematic, the third due to B lifetime. Except otherwise specified.

On the other hand, by assuming $\Gamma(\bar{B}^0 \to D^{*+} \ell^- \bar{\nu}) = \Gamma(B^- \to D^{*0} \ell^- \bar{\nu})$, which follows from isospin invariance, and $f_{+-} + f_{00} = 1$, Eqs. (1) and (2) can be combined to give

$$\Gamma(\bar{B} \to D^* \ell \, \bar{\nu}) = \frac{1}{4N_{\Upsilon(4S)}\mathcal{B}_{D^0}} \left[\frac{N_0}{\tau_{\bar{B}^0}\mathcal{B}_{D^{*+}}} + \frac{N_-}{\tau_{B^-}\mathcal{B}_{D^{*0}}} \right]. \tag{5}$$

Using $\tau_{B^0} = [1.53 \pm 0.09]$ps and $\tau_{B^+} = [1.68 \pm 0.12]$ps,[9] yields,

$$\Gamma(\bar{B} \to D^* \ell \, \bar{\nu}) = [29.9 \pm 1.9 \pm 2.7 \pm 2.0] \text{ ns}^{-1}, \tag{6}$$

independent of f_{+-}/f_{00}.

The systematic errors on all the above results are dominated by the uncertainties in reconstructing the π_s^+ from D^{*+}(5%) and and the π_s^0 from D^{*0}(8.6%).

4.   Determination of $|V_{cb}|$

HQET predicts the differential decay rate for $\bar{B} \to D^* \ell^- \bar{\nu}$ at heavy quark symmetry limit in term of the Isgur-Wise function $\xi(y)$ with normalization $\xi(1) = 1$, where $y = (m_B^2 + m_{D^*}^2 - q^2)/(2m_B m_{D^*})$. For finite b and c quark masses the corrections necessary to account for the deviation from the heavy quark symmetry limit are absorbed into $\mathcal{F}(y)$. Following Neubert,[10] we write,

$$\frac{d\Gamma}{dy} = \frac{G_F^2}{48\pi^3} m_{D^*}^3 (m_B - m_D^*)^2 |V_{cb}|^2 \mathcal{F}^2(y) \times \sqrt{y^2 - 1} \left[4y(y+1)\frac{1 - 2yr + r^2}{(1-r)^2} + (y+1)^2 \right], \tag{7}$$

where $r = m_{D^*}/m_B$. The function $\mathcal{F}(y)$[11] can be related to the Isgur-Wise function and correction terms that vanish in the infinite b and c mass limit, $\mathcal{F}(1) = \eta_A \xi(1) + \mathcal{O}(\Lambda_{QCD}/m_Q)^2$, where η_A is a perturbatively calculable QCD radiative correction and Λ_{QCD} is the QCD scale parameter. Since we want to determine the product $|V_{cb}|\mathcal{F}(y)$,we approximate the unknown function $\mathcal{F}(y)$ with an linear expansion about $y = 1$, $\mathcal{F}(y) = \mathcal{F}(1)\left[1 - \hat{\rho}^2(y-1)\right]$.

The unbinned maximum likelihood method developed by E691[12] in the $D \to K^* l \bar{\nu}$ study is utilized. The likelihood function also contains a Poisson probability factor to determine the normalization for $|V_{cb}|$. The detector acceptance and smearing are taken into account appropriately.

We fit the differential decay distribution of the two decay modes $\bar{B}^0 \to D^{*+} \ell^- \bar{\nu}$ and $B^- \to D^{*0} \ell^- \bar{\nu}$ simultaneously by calculating the likelihood of the two modes separately and maximizing their product. The result of the simultaneous fit is shown in Fig. 2, along with the combined data and background level for the $\bar{B}^0 \to D^{*+} \ell^- \bar{\nu}$ and $B^- \to D^{*0} \ell^- \bar{\nu}$ modes. From a fit to these data we obtain the most precise measurement to date of $|V_{cb}|\mathcal{F}(1)$ and $\hat{\rho}^2$,

$$|V_{cb}|\mathcal{F}(1) = 0.0351 \pm 0.0019 \pm 0.0018 \pm 0.0008 \tag{8}$$
$$\hat{\rho}^2 = 0.84 \pm 0.12 \pm 0.08, \tag{9}$$

independent of f_{+-}/f_{00}.

To extract a value of $|V_{cb}|$ one requires a prediction for $\mathcal{F}(1)$. Using $\mathcal{F}(1) = 0.97 \pm 0.04$[14] we obtain, $|V_{cb}| = 0.0362 \pm 0.0019(stat.) \pm 0.0020(sys., lifetime) \pm 0.0014(theo.)$.

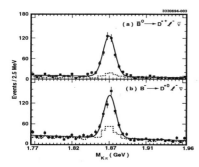

Fig. 1. Raw yield for $\bar{B} \to D^* \ell^- \bar{\nu}$ represented by $K^-\pi^+$ invariant mass spectrum.

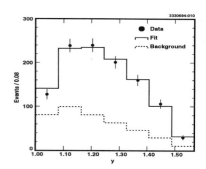

Fig. 2. Raw differential yield for $\bar{B} \to D^* \ell^- \bar{\nu}$ overlayed with fits.

The systematic errors propagate in a similar fashion as in the branching fraction measurement. The uncertainties due to the fitting procedure are estimated by comparing the results from the unbinned method with that from a binned method using an analytic smearing form.[13]

References

1. N. Isgur and M. Wise, Phys. Lett. B **232**, 113 (1989); N. Isgur and M. Wise, Phys. Lett. B **237**, 527 (1990); E. Eichten and B. Hill, Phys. Lett. B **234** 511 (1990); H. Georgi Phys. Lett. B **240** 447 (1990).
2. A. Falk, M. Neubert and M. Luke, Nucl. Phys B **388**, 3363 (1992); A. F. Falk and M. Neubert, Phys. Rev. D **47**, 2965 (1993); M. Neubert, Phys. Rev. D **47**, 4063 (1993).
3. T. Mannel, CERN Report No. CERN-TH.7162/94 (unpublished).
4. For a more detailed report see **CLNS** 94/1285 (to appear in Phys. Rev. D).
5. CLEO collaboration, Y. Kubota *et.al.*, Nucl. Instr. and Meth. A **320**, 66(1992).
6. G. Fox and S. Wolfram, Phys. Rev. Lett. **41**, 1581 (1978).
7. CLEO collaboration, F. Butler *et.al.*, Phys. Rev. Lett **69**, 2041 (1992).
8. CLEO collaboration, D. S. Akerib *et.al.*, Phys. Rev. Lett **71**, 3070 (1993).
9. W. Venus, in: Lepton and Photon Interactions, XVI International Symposium, eds. Persis Drell and David Rubin, AIP Press, New York (1994).
10. M. Neubert, SLAC Report No. SLAC-PUB-6263 (to appear in Physics Reports).
11. Neubert writes the form factor $\mathcal{F}(y)$ as $\eta_A \hat{\xi}(y)$ [10], where η_A is a calculable radiative correction for hard gluon processes and $\hat{\xi}(y)$ is the non perturbative part.
12. D. M. Schmidt *et.al.*, Nucl. Instr. and Meth. A **328**, 547 (1993).
13. M. Garcia-Sciveres, Ph.D. thesis, Cornell University, 1994 (unpublished).
14. M. Neubert, private communication. Ref. 10 gives $\hat{\xi}(1) = 1.00 \pm 0.04$ and $\eta_A = 0.99 \pm 0.006$, but the final version, to appear in Physics Reports, will have $\hat{\xi}(1) = 0.98 \pm 0.04$ and $\eta_A = 0.986 \pm 0.006$, leading to $\mathcal{F}(1) = 0.97 \pm 0.04$.

A MEASUREMENT OF $|V_{cb}|$ FROM $\overline{B}^0 \to D^{*+} \ell^- \overline{\nu}_\ell$

IAN J. SCOTT*

Physics Dept., University of Wisconsin, Madison, Wisconsin 53706, USA

ABSTRACT

In a sample of approximately 1.6 million hadronic decays of Z bosons recorded with the ALEPH detector at LEP, $\overline{B} \to D^{*+}\ell^- X$ candidates have been identified. From this sample, the differential width $d\Gamma(\overline{B}^0 \to D^{*+}\ell^-\overline{\nu}_\ell)/dq^2$ has been measured. From a fit to this spectrum the product of the CKM matrix element $|V_{cb}|$ and the normalization of the decay form factor at the point of zero recoil of the D^* meson $\mathcal{F}(1)$ has been measured to be

$$\mathcal{F}(1)|V_{cb}| = (36.4 \pm 4.2_{\text{stat}} \pm 3.1_{\text{syst}}) \times 10^{-3}.$$

A value for $|V_{cb}|$ has been extracted using theoretical calculations of the form factor normalization.

1. Introduction

Recent developments in Heavy Quark Effective Theory (HQET) have raised hopes for a precise and a model-independent measurement of $|V_{cb}|$ using exclusive decays such as $\overline{B}^0 \to D^{*+}\ell^-\overline{\nu}_\ell$ (For a review, see for example[2] and references therein). The expression for the differential partial width is[3]:

$$
\begin{aligned}
\frac{d\Gamma}{dy} &= \frac{1}{\tau_{B^0}} \frac{d\mathcal{B}(\overline{B}^0 \to D^{*+}\ell^-\overline{\nu}_\ell)}{dy} \\
&= \frac{G_F^2}{48\pi^3\hbar} m_{D^{*+}}^3 (m_{B^0} - m_{D^{*+}})^2 \mathcal{F}^2(y)|V_{cb}|^2 \\
&\quad \times \sqrt{y^2 - 1}\left[4y(y+1)\frac{1-2yr+r^2}{(1-r)^2} + (y+1)^2\right],
\end{aligned}
\tag{1}
$$

where $r = m_{D^{*+}}/m_{B^0}$ and $y = (m_{B^0}^2 + m_{D^{*+}}^2 - q^2)/2m_{B^0}m_{D^{*+}}$. The variable q^2 is the square of the four-momentum transfer in the decay. The unknown quantities in this expression are $|V_{cb}|$ and $\mathcal{F}(y)$. The function $\mathcal{F}(y)$ is not specified by HQET but its magnitude at maximum q^2 ($y = 1$) is normalized to one in the heavy quark limit. While the QCD corrections can be estimated at this point, the experimental data are statistically deficient due to the vanishing rate. Consequently, $|V_{cb}|$ is presently measured from the $d\Gamma/dy$ spectrum by an extrapolation to $y = 1$.

So far all such measurements of $|V_{cb}|$ have come from the ARGUS and CLEO experiments at the $\Upsilon(4S)$ resonance where the B mesons are produced with momentum of 325 MeV.[3,4] At maximum q^2 the D^{*+} are produced at rest and

---

*Representing the ALEPH Collaboration

subsequently decay into a D meson and a pion with momentum of about 40 MeV and can only be reconstructed with substantially reduced efficiency. In contrast, at the Z resonance, the B hadrons are produced with a large boost ($\beta\gamma \approx 6$). Consequently, the pion from the D*+ decay has a typical momentum of about 1 GeV. This feature of B meson production at the Z resonance allows access to the entire $d\Gamma/dy$ spectrum in $\overline{\mathrm{B}}^0 \to \mathrm{D}^{*+}\ell^-\overline{\nu}_\ell$ decays with approximately equal (and high) efficiency.

2. q^2 Reconstruction

Equation 2 expresses the q^2 in terms of the neutrino energy E_ν and the angle ϕ between the planes formed by the D* and the lepton and by the B and the neutrino:

$$
\begin{aligned}
q^2 &= m_\ell^2 + p_\ell \cos\theta_\ell (M_\mathrm{B}^2 - m^2)/P \\
&+ 2E_\nu \left(E_\ell - p_\ell \cos\theta_\ell E/P\right) \\
&+ 2p_\ell \sin\theta_\ell E_\nu \sin\theta_\nu \cos\phi \,,
\end{aligned}
\tag{2}
$$

where m, P, and E are the mass, momentum and energy of the reconstructed D*+ℓ^- and M_B is the B meson mass. The variables θ_ℓ and θ_ν are respectively the opening angles of the lepton and the neutrino with respect to the axis defined by the D*+ℓ^- system. The variable θ_ν can be expressed as a function of E_ν, which is estimated from the center of mass energy and the visible energy in the hemisphere containing the D*+ℓ^- candidate. The angle ϕ is calculated from the B-meson direction as measured by the production and decay vertices. The resolution is approximately $1.4\,\mathrm{GeV}^2$, which corresponds to about 13% of the allowed q^2 range.

The opening angle θ_B between the reconstructed B meson direction and the D*+ℓ^- direction can be used in conjunction with the measured neutrino energy to reconstruct the missing squared mass MM^2 of the system recoiling against the D*+ℓ^- :

$$
MM^2 = M_\mathrm{B}^2 + m^2 - 2\left(E + E_\nu\right)\left(E - \beta P \cos\theta_\mathrm{B}\right) \,,
$$

where $\beta = \sqrt{1 - 1/\gamma^2}$ and $\gamma = (E + E_\nu)/M_\mathrm{B}$. This is useful in rejecting background with additional particles coming from the B decay vertex.

3.   Event Selection and Sample Composition

The decay $\overline{\mathrm{B}} \to \mathrm{D}^{*+}\ell^-\overline{\nu}X$ was identified in events where a D*+ and a lepton were found in the same hemisphere of a hadronic Z decay. The D*+ candidates were reconstructed in the channel $\mathrm{D}^{*+} \to \mathrm{D}^0\pi^+$. The D^0 candidates were reconstructed in three decay modes; $\mathrm{D}^0 \to \mathrm{K}^-\pi^+$, $\mathrm{D}^0 \to \mathrm{K}^-\pi^+\pi^-\pi^+$ and $\mathrm{D}^0 \to \mathrm{K}^0_\mathrm{S}\pi^+\pi^-$.

In order to suppress background events with extra particles originating from the B-hadron decay vertex (e.g. $\overline{\mathrm{B}} \to \mathrm{D}^{*+}\pi\ell^-\overline{\nu}$), additional event selection criteria were applied. In candidate events, tracks consistent within three standard

Channel	Branching Ratio	Reference
$\mathcal{B}(b \to \overline{B}{}^0)$	37±3%	6
$\mathcal{B}(\ B^- \to D^{*+}\pi^-\ell^-\overline{\nu})$	0.36±0.12%	1
$\mathcal{B}(\ \overline{B} \to D^{*+}X_c, X_c \to \ell^-\overline{\nu}Y)$	0.32±0.06%	1
$\mathcal{B}(\ \overline{B}{}^0 \to D^{*+}\tau^-\overline{\nu}_\tau)$	2.34±0.39%	1
$\mathcal{B}(D^{*+} \to D^0\pi^+)$	68.1±1.3%	7
$\mathcal{B}(D^0 \to K^-\pi^+)$	4.01±0.14%	7
$\mathcal{B}(D^0 \to K^-\pi^+\pi^-\pi^+)$	8.1±0.5%	7
$\mathcal{B}(D^0 \to K_S^0\pi^+\pi^-)$ $\times\mathcal{B}(K_S^0 \to \pi^+\pi^-)$	1.8±0.2%	7

Table 1. Summary of the branching fractions used in this analysis.

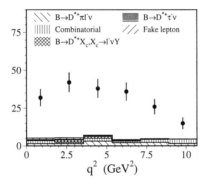

Fig. 1. The reconstructed q^2 of events with estimated background.

deviations with the reconstructed $D^{*+}\ell^-$ vertex and having the same charge as the lepton candidate were identified. Events containing at least one such track were rejected. Furthermore, the reconstructed missing mass squared in the event was required to be less than $1\,\text{GeV}^2$. Finally, as the measurement precision of the B meson direction improves with the decay length, candidates were rejected if the $D^{*+}\ell^-$ vertex was less than 1 mm away from the interaction point. The selection results in a sample of 190±14 $D^{*+}\ell^-$ candidates with an estimated background of 35±6.

Table 3 lists the branching ratios used in this analysis, and Fig. 1 shows the reconstructed q^2 spectrum and the expected background.

4.   Measurement of $|V_{cb}|$

A binned χ^2 fit was performed to the background subtracted q^2 spectrum. The fitting function is given in Eq. 1, with the following assumed functional form

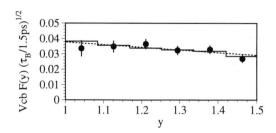

Fig. 2. $|V_{cb}|\mathcal{F}(y)\,(\tau_{B^0}/1.5\,\mathrm{ps})^{1/2}$ vs. the reconstructed y. See text for details. The solid curve is the fit. The dashed curve is the fitted function without taking into account resolution.

for $\mathcal{F}(y)$[3]:

$$\mathcal{F}(y) = \mathcal{F}(1)(1 + a^2(1 - y))\,.$$

The two free parameters in the fit are $\mathcal{F}(1)|V_{cb}|$ and a^2. The fit takes into account both resolution and efficiency as determined from a Monte Carlo simulation.[5]

The fit gives a result of

$$|V_{cb}|\mathcal{F}(1)\sqrt{\tau_{B^0}/1.5\,\mathrm{ps}} = (37.7 \pm 4.3) \times 10^{-3}$$
$$a^2 = 0.46 \pm 0.30\,,$$

where the errors are statistical only. The χ^2 of the fit is 1.6 for 4 degrees of freedom.

The measured B^0 lifetime of $\tau_{B^0} = 1.61 \pm 0.09\mathrm{ps}$[8] is used to obtain

$$|V_{cb}|\mathcal{F}(1) = (36.4 \pm 4.2_{\mathrm{stat}} \pm 1.0_{\mathrm{lifetime}}) \times 10^{-3}\,,$$

where the first error is statistical and the second error is due to the lifetime uncertainty.

Using $\mathcal{F}(1) = 0.93 \pm 0.03$[9] we obtain

$$|V_{cb}| = (39.1 \pm 4.5_{\mathrm{stat}} \pm 1.1_{\mathrm{lifetime}} \pm 1.3_{\mathrm{theory}}) \times 10^{-3}\,,$$

where the the third error is the quoted theoretical uncertainty in $\mathcal{F}(1)$.

Plotting the reconstructed q^2 spectrum as a function of y and factoring out known q^2 dependent terms, a graph can be made where the intercept at $y = 1$ is given by $\mathcal{F}(1)|V_{cb}|$ and the slope is a^2. In practice finite q^2 resolution distorts the shape, destroying the simple interpretation. The graph does provide a qualitative description of the data, as is shown in Fig. 2.

The background-subtracted data sample can also be used to extract a measurement of $\mathcal{B}(\overline{B}^0 \to D^{*+}\ell^-\overline{\nu}_\ell)$:

$$\mathcal{B}(\overline{B}^0 \to D^{*+}\ell^-\overline{\nu}_\ell) = (5.07 \pm 0.48)\%\,,$$

where the quoted error is statistical only.

Source	$\Delta\mathcal{B}/\mathcal{B}$	$\Delta|V_{cb}|/|V_{cb}|$	Δa^2
$\mathcal{B}(b \to B^0)$	8.2%	4.1%	–
$\mathcal{B}(\overline{B} \to D^{*+}\pi\ell^-\overline{\nu})$	3.0%	2.4%	0.03
$\mathcal{B}(D^{*+} \to D^0\pi^+)$	1.9%	1.0%	–
$\mathcal{B}(D^0 \to K^-\pi^+)$	1.4%	0.7%	–
$\mathcal{B}(D^0 \to K^-\pi^+\pi^-\pi^+)$	3.4%	1.7%	–
$\mathcal{B}(D^0 \to K_S^0\pi^+\pi^-)$	0.6%	0.3%	–
τ_{B^0}	–	2.8%	–
Absolute efficiency	10%	5%	–
Efficiency shape	–	3.6%	0.13
Total	13.9%	8.5%	0.13

Table 2. Summary of systematic uncertainties.

5. Systematic Uncertainties and Summary

Uncertainties due to branching ratios and the B^0 lifetime are determined from the effect of varying the values within quoted errors. The error on the absolute efficiency is the estimated from the statistical uncertainty due to the finite Monte Carlo sample. The systematic uncertainties are summarized in Table 2.

Including the systematic uncertainties, the following quantities have been measured:

$$\mathcal{B}(\overline{B}^0 \to D^{*+}\ell^-\overline{\nu}_\ell) =$$
$$(5.07 \pm 0.48_{\text{stat}} \pm 0.70_{\text{syst}})\%$$
$$\mathcal{F}(1)|V_{cb}| = (36.4 \pm 4.2_{\text{stat}} \pm 3.1_{\text{syst}}) \times 10^{-3}$$
$$|V_{cb}| = (39.1 \pm 4.5_{\text{stat}} \pm 3.3_{\text{syst}} \pm 1.3_{\text{theo}})$$
$$\times 10^{-3}$$
$$a^2 = 0.46 \pm 0.30_{\text{stat}} \pm 0.13_{\text{syst}}.$$

References

1. ALEPH Collab.: D. Decamp *et al.*, **GLS0605**. Submitted paper to ICHEP94.
2. M. Neubert, SLAC preprint: SLAC-PUB-6263 (1993).
3. CLEO Collab.: B. Barish *et al.*, preprint CLNS 94/1285.
4. ARGUS Collab.: H. Albrecht *et al.*, Z. Phys. C **57** (1993) 533.
5. Throughout this analysis, JETSET 7.3 program was used to generate $Z \to q\overline{q}$ events. T. Sjöstrand and M. Bengtsson, Comp. Phys. Com. **46**, (1987) 43.
6. ALEPH Collab.: D. Decamp *et al.*, Phys. Lett. B **322** (1992) 441.
7. Particle Data Group: K. Hikasa *et al.* Phys. Rev. D **50** (1994).
8. P. Roudeau, these proceedings.
9. M. Neubert, CERN preprint CERN-TH.7395/94

TIME DEPENDENT $B^0 \leftrightarrow \overline{B^0}$ MIXING

EDUARDO DO COUTO E SILVA*

Department of Physics, Indiana University, Bloomington, Indiana, 47405, USA

ABSTRACT

Time dependent $B^0 \leftrightarrow \overline{B^0}$ mixing is observed by extracting the parameter Δm_d in three different analyses. A limit on Δm_s is also determined. These measurements were obtained using the hadronic decays of the Z^0 collected in the OPAL detector from 1990 to 1993.

1. Introduction

Within the framework of the Standard Model, B_d^0 and B_s^0 mesons are expected to oscillate between particle and antiparticle states. The mixing mechanism occurs via box diagrams which include W exchange and are dominated by the top quark contributions.[1] The oscillation frequency is proportional to the mass difference (Δm_d or Δm_s) between the mass eigenstates. To study the time dependent mixing B_d^0 needs to distinguished from $\overline{B_d^0}$ at production and decay time, and also the proper decay time has to be reconstructed. A complication arises due to the mistagging of a B_d^0 as a $\overline{B_d^0}$ and vice-versa, which will reduce the amplitude of the observed oscillations. Finally one looks for an oscillatory behavior in the fraction of mixed events over the total number of events as a function of the estimated proper decay time, and extracts a value for the mass difference Δm_d.

OPAL measures Δm_d in three different ways. This paper discusses separately these three different analysis and presents the values for Δm_d and the limit on Δm_s.

2. The D* and Lepton Analysis

Each event is divided into two hemispheres.[2] In one hemisphere one looks for inclusive decays of the type $B_d^0 \to D^{*-} + X$ where the D^* is reconstructed in the channel $D^{*-} \to \overline{D^0} + \pi^-$. The charge of the D^* determines whether the decaying hadron was a B_d^0 or a $\overline{B_d^0}$. There is background due to D^* mesons from $B^+ \to D^{*-} + X$ and produced in $Z^0 \to c\bar{c}$ decays.

In the hemisphere opposite to the reconstructed D^*, a lepton with high transverse momentum with respect to the jet axis is identified. This tags the production state of the B hadron. Cascade leptons ($b \to c \to \ell$), leptons from semileptonic charm decays and particles misidentified as leptons cause wrong production time tags.

Figure 1 shows the D^* - D^0 mass difference, which is used to suppress and estimate the combinatorial background. Because of the slow pion from the D^* decay, the B_d^0 decay

---

*Representing the OPAL Collaboration.

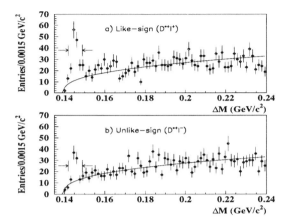

Fig. 1. D^* - D^0 mass difference. a) Like-sign combinations ($D^{*+}\ell^+$); b) Unlike-sign combinations ($D^{*+}\ell^-$). The curves represent the background. The arrows indicate the region used for the Δm_d fit .

length cannot be well measured. Therefore, only the D^0 decay vertex is reconstructed. The boost is estimated on a statistical basis by convoluting the fragmentation function in the Peterson parametrization with the decay length distribution. The uncertainty in the fragmentation function is the dominant source of systematic error in this analysis.

The sample is divided into like sign ($D^{*+}\ell^+$) and unlike-sign ($D^{*+}\ell^-$) events. An unbinned likelihood fit is performed to the asymmetry in mixed versus unmixed events as a function of decay length. This one paremeter fit is shown in Fig. 2 and gives

$$\Delta m_d = 0.57 \pm 0.11 \text{ (stat) } \pm 0.02 \text{ (syst) ps}^{-1}. \qquad (1)$$

3.   The $D^*\ell$ and Jet Charge Analysis

The B_d^0 meson is identified in the decay mode $B_d^0 \to D^{*-}\ell^+\nu$, by the presence of a $D^{*-}\ell^+$ pair in the same jet.[3] The lepton is used for flavor tagging of the B_d^0 at its decay time. The D^* is identified by: $D^{*-} \to \overline{D^0}\,\pi^-$ followed by the decay $D^0 \to K^-\pi^+$ or $\overline{D^0} \to K^+\pi^-(\pi^0)$ (called the satellite channel) where the π^0 is not reconstructed. Contamination from secondary leptons and from charm decays is negligible. Background processes such as $B \to \overline{D^*}D_s^+$ and $D_s^+ \to \ell^+X$ are also found to be negligible. Contamination from $B^+ \to D^{**0}\ell\nu$, where the D^{**0} decays into a D^* and a charged pion, is accounted for in the fit for Δm_d.

A jet charge technique is used to obtain the information about the production flavor of the B_d^0 meson. The jet charge is defined as

$$Q_{jet} = \frac{1}{(E_{beam})^\kappa} \cdot \sum_{i=1}^{n} q_i \cdot (p_i^l)^\kappa. \tag{2}$$

where E_{beam} is the energy of the beam and q_i and p_i^l are the charge and the longitudinal along the jet direction momentum of each track, respectively. In the jet where a B_d^0 is tagged, κ is set equal to zero. Therefore, the jet charge is the sum of the charges of all tracks and is mainly sensitive to the fragmentation tracks. For the other jet, the value of κ is set equal to one. In this case, particles with higher momentum (probably decay products of b hadrons) contribute more to the value of the jet charge. The variable that correlates the jet charge information between both sides is

$$Q_{2jet} = Q_{jet}^{\kappa=0}(B_d^0) - 10 \cdot Q_{jet}^{\kappa=1}(opp), \tag{3}$$

where $Q_{jet}(B_d^0)$ and $Q_{jet}(opp)$ are the jet charges of the B_d^0 and of the most energetic other jet, respectively. Figure 3 shows the Q_{2jet} distribution for simulated events. Note that Fig. 3 is used only for illustration and the analysis does not rely on the simulation of Q_{2jet}.

The boost of the B_d^0 is determined by its parametrization as a function of the momentum and the invariant mass of the $D^*\ell$ pair. The decay length is first measured in two dimensions and converted into three dimensions by using the direction of the reconstructed $D^*\ell$ pair.

A fit is carried out to the analytical expression which represents the fraction of events in which the jet charge, Q_{2jet}, and the lepton charge have opposite signs. The flavor mistag probability (f) and Δm_d are free parameters and the fraction of events coming from charged B mesons is assumed to be 0.16 ± 0.09. The resulting fitted values are

$$\Delta m_d = 0.508 \pm 0.075 \text{ (stat)} \pm 0.025 \text{ (syst) ps}^{-1} \tag{4}$$

and

$$f = 0.263 \pm 0.033 \text{ (stat)}, \tag{5}$$

in good agreement with the mistag fraction of 0.28 predicted by a Monte Carlo simulation. In Fig.4a the fitted curve is shown. Systematic uncertainties are dominated by the knowledge on the charged B fraction and differences in lifetimes between neutral and charged B's.

4.   The Dilepton Analysis and the Δm_s Limit

Events are separated into two thrust hemispheres.[4] Leptons with high transverse momentum relative to the jet axis are used to identify b hadrons on both sides. Samples of like-sign $(\ell^\pm\ell^\pm)$ and unlike-sign $(\ell^\pm\ell^\mp)$ lepton pairs are selected by choosing events with one lepton in each thrust hemisphere. The momentum of the particle, its transverse component with respect to the jet axis and the energy near the lepton candidate are combined into a feed forward type neural network to minimize the background from cascade leptons. Secondary vertices which include the lepton are reconstructed. For the boost estimate, each event is divided into two parts, each of which is considered

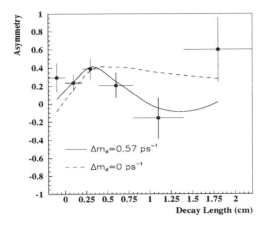

Fig. 2. The asymmetry between mixed and unmixed events as a function of decay length. The solid line corresponds to the fit to Δm_d, and the dashed line represents the case with no mixing .

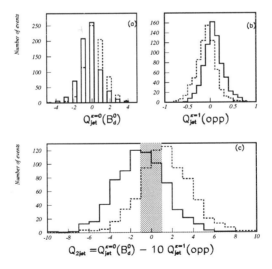

Fig. 3. The jet charge distribution for (a) B_d^0 jets, (b) opposite jets, and (c) the combined jet charge measure. The hatched area is excluded by the cut $|Q_{2jet}| > 1$. The solid (dashed) lines are the distributions for simulated $B_d^0(\overline{B_d^0})$ events .

to be a result of a two-body decay: the lepton jet (ν included) and the rest of the event. Using this information together with the energy of the beam, the energy of the b hadron is estimated and consequently its boost. An unbinned maximum likelihood fit to the fraction of mixed events is peformed. Fixing the value for Δm_s to 10 ps^{-1} (varying Δm_s from 2 to 20 ps^{-1} corresponds to a variation of $^{+0.011}_{-0.002}$ ps^{-1} in Δm_d), the fitted value for Δm_d is

$$\Delta m_d = 0.50 \pm 0.04 \text{ (stat)} \pm 0.09 \text{ (syst) ps}^{-1}. \tag{6}$$

The fraction of mixed over total number of events is shown in Fig.4b. Dominant systematic uncertainties come from the knowledge of B_s^0 and cascade decay fractions.

To study the sensitivity of the experiment to Δm_s, 140 Monte Carlo samples were generated each with the same number of events as in the data. For each of the samples Δm_d was fixed at 0.5 ps^{-1} and Δm_s was set to 1, 2, 4 and 8 ps^{-1}. The difference between the log-likelihood at the generated Δm_s and that of the mimimum, is shown in the first column of Fig. 5, where 95% of the trials fall below the limit of 1.92. In the middle column the fitted minimum value is plotted for each trial. In the third column the 95 % C.L. statistical limit on Δm_s for each trial is displayed. These studies indicate that there is no sensitivity for values above $\Delta m_s \doteq 4$ ps^{-1}. The statistical limit is set to $\Delta m_s > 1.9$ ps^{-1}. By doing a multidimensional constrained fit sytematics are included and a limit of $\Delta m_s > 1.3$ ps^{-1} is derived.

5.   Summary

The value of Δm_d was measured using three different analyses. Ignoring the small statistical correlation and assuming systematic correlations between the three measurements, the average value is

$$\Delta m_d = 0.520 \pm 0.054 \text{ ps}^{-1}. \tag{7}$$

The difference in the log-likelihood from the minimum in the Δm_s-Δm_d plane is used to derive the limit $\Delta m_s > 1.3$ps^{-1} at 95 % confidence level.

References

1. P. J. Franzini, *Physics Reports* **173** (1989) 1.
2. OPAL Collaboration, R. Akers *et al.*, CERN-PPE/94-90 (13 June 94), submitted to *Phys. Lett.* **B**.
3. OPAL Collaboration, R. Akers *et al.*,*Phys. Lett.* **B327** (1994) 411.
4. H. Fukui, M. Jimack, S. Komamiya and R. Kowalewski, Internal OPAL Physics Note 152.

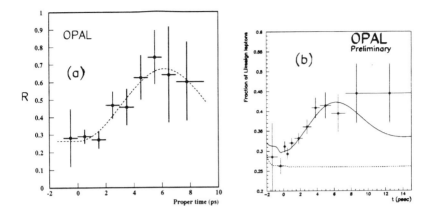

Fig. 4. a) Proper time dependence of fraction of mixed over total number of events, R, and the fit to the data as described in the jet charge analysis; b) The fraction of like-sign leptons as a function of the proper decay time. The solid line represents the situation with $\Delta m_d = 0.50$ ps^{-1} and $\Delta m_s = 10$ ps^{-1},and the dashed line corresponds to no mixing .

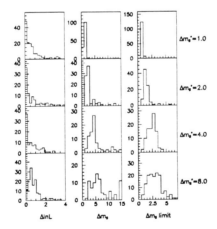

Fig. 5. .The results of fits to 140 toy Monte Carlo datasets are shown. The Δm_s value indicates the generated value of Δm_s. The three quantities plotted for each Δm_s value are the difference between the log-likelihood values at the fitted minimum and at the generated value, the fitted value of Δm_s, and the statistical lower limit one would derive for Δm_s on each trial.

STUDIES OF TIME DEPENDENT $B_d^o - \bar{B}_d^o$ MIXING

PIERRE ANTILOGUS

IPN Lyon, 43 bd 11 Novembre 1918, Villeurbanne,
F-69622, France

DELPHI Collaboration

ABSTRACT

Using events collected by DELPHI at LEP in 1991-1993, the time dependent mixing of the B_d^o meson has been observed throught D^*–hemisphere charge correlation and D^*lepton–hemispere charge correlation. The last approach provided a purer sample of B_d^o mesons.

Further studies have been made by selecting events where the sign is given by a lepton, charged kaon or the jet charge on one side of the event and in the other one, where the B flight distance is evaluated, a lepton or charged kaon is used.

The average value of these measurements gave for the mass difference between the two B_d^o mass eigenstates the following result

$$\Delta m_d c^2 = 0.52 \pm 0.06 (stat + syst) \hbar/ps$$

1.   Introduction

To measure the $B_d^o - \bar{B}_d^o$ time dependent mixing two types of information are needed. The first one is related to the detection of oscillating events. This is done by measuring the charge of the b quark in both hemispheres. The second one is the decay time of the B_d^o .

In this note, three different approaches are presented to perform the measurement of Δm_d. Only the main lines of these measurements will be presented. The references to more detailed notes will be given.

2. Δm_d measurement using $D^{*\pm}$/hem. charge

In this analysis the tagging of the flavor and decay time of the B_d^o is done throught the decay chain $\bar{B}_d^o \to D^{*+}$ followed by $D^{*+} \to D^o \pi^+$ with the D^o reconstructed in $K^- \pi^+$ (1816 events) or $K^- \pi^+ \pi^o$ (1853 events). The decay time is identified as $t \sim m_B.d/P_B$ with $P_B \sim 0.7 E_{beam}$ and d = measured D^o decay distance. A resolution of $\sim 0.4 ps$ is reached on t.

To be sensitive to oscillating event, the charge of the $D^{*\pm}$ is compared with the weighted mean charge difference of the 2 hemispheres, $\Delta C_H = C_H^{opposite\ to\ D^*\ side} -$

$C_H^{D^* \ side}$ with $C_H = \sum_i \left(\overrightarrow{p_i}\,\overrightarrow{e_s}\right)^{0.6} q_i / \sum_i \left(\overrightarrow{p_i}\,\overrightarrow{e_s}\right)^{0.6}$ ($\overrightarrow{e_s}$ = sphericity axis direction and p_i, q_i = momentum and charge of track "i").

An event is unlike sign when ΔC_H is opposite in sign to the D^* charge (~ the initial B_d^o has mixed). From this, $\epsilon^{unlike} \left(= 1 - \epsilon^{like}\right)$ is introduced as the probability of classifying a reconstructed D^* event in the unlike sign category. An unbinned maximum likelihood fit is performed on the D^* sample attributing a probability to each like sign or unlike sign event in function of the sample composition for a given value of t and Δm_d. In this approach many parameters needed to define the likelihood function can be determined directly from the data as shown table 1. The result of the fit is

$$\Delta m_d c^2 = 0.470 \pm 0.086(stat) \pm 0.061(syst)\hbar/ps$$

, the detail of the systematical errors can be found table 2. In the same fit the value of $\epsilon^{unlike}_{B_d^o \to D^{*+}} = 0.289 \pm 0.028$ is obtained (= 0.275 in the simulation).

fitted parameter	using	$D^o \to K^- \pi^+$	$D^o \to K^- \pi^+ \pi^o$
$r_{c\bar{c}} = \frac{f_{c\bar{c}}}{f_{b\bar{b}} + f_{c\bar{c}}}$	$X_E = \frac{E_{D^*}}{E_{beam}}$ distribution	0.36 ± 0.06	0.42 ± 0.08
ϵ^{unlike}_{Bkg}	$M_{D^*} - M_{D^o} > 0.16$ GeV/c^2 events	0.481 ± 0.008	0.480 ± 0.011
frac. of Bkg	mass spectra distribution	0.318 ± 0.020	0.425 ± 0.029
τ_B	t distribution	1.63 ± 0.10 ps	

Table 1. Some of the parameters fitted using different distribution of the D^* sample events

Source of uncertainty	$\delta \Delta m$
time param. and resolution	0.030
B momentum	0.030
B lifetime	0.010
fraction of $c\bar{c}, B^\pm, Bkg$	0.032
$\epsilon_{mix}, \epsilon_{c\bar{c}}, \epsilon_{B^\pm}$	0.027

Table 2. Contribution to the systematics in the measurement of Δm_d with the D^* sample

Source of uncertainty	$\delta \Delta m$
B momentum	0.020
B lifetime	0.010
fraction of fake lepton	0.005
param&fraction of Bkg	0.010
fraction of $B^\pm$	0.020

Table 3. Contribution to the systematics in the measurement of Δm_d with the $D^{*\pm} l^\mp$ sample

3. Δm_d measurement using $D^{*\pm} l^\mp$/hem. charge

To increase the contribution from B_d^o a lepton with the charge opposite to the $D^{*\pm}$ is required in the $D^{*\pm}$ hemisphere. As the sample is purer the parametrisation of the time distribution is easier to control and the $c\bar{c}$ contribution becomes negligible. The result of the fit is

$$\Delta m_d c^2 = 0.44 \pm 0.10(stat) \pm 0.03(syst)\hbar/ps$$

, the detail of the systematical errors can be found table 3.

4. Δm_d measurement using $l/K/$jet charge

In this analysis to determine the b flavor, three tag technics are used: lepton , kaon or jet charge. Different requirements are applied for each tag. They give different probabilities to correctly determine the initial quark sign (N_{right}/N_{wrong}) and have different tagging efficiency:

- Leptons ($b \rightarrow l$), to enrich in lepton from B : $p > 3$ GeV/c and $p_T^{out} > 1$ GeV/c
  $\rightarrow N_{right}/N_{wrong} \sim 6$ and B_d^o tag efficiency $\sim 8\%$

- Kaons ($b \rightarrow c \rightarrow s$) identifed by the RICH, to enrich in B : cut on $\frac{\delta_K}{\sigma_{\delta_K}} > 1.5$
  $\rightarrow N_{right}/N_{wrong} \sim 2.5$ and B_d^o tag efficiency $\sim 18\%$

- Jet sign : $Q_{jet} = \frac{\sum_i q_i p_i^{0.6}}{\sum_i p_i^{0.6}}$, to enrich in B : $\sum_i \left(\frac{\delta_i}{\sigma_{\delta_i}}\right) / (N-1) > 2$ and $|Q_{jet}| > 0.1$
  $\rightarrow N_{right}/N_{wrong} \sim 2$ and B_d^o tag efficiency $\sim 32\%$

tag	discri. power
l+jet	0.21 $\sqrt{N}$
K+jet	0.19 $\sqrt{N}$
jet only	0.13 $\sqrt{N}$

Table 4. $b\bar{b}$ discriminating power of the different tags used in the tagging hemisphere.

Source of uncertainty	$\delta\Delta m$
B lifetimes, b fragmentation	0.030
B_d, B_s fraction	0.023
K multiplicity in B decay	0.022
lepton eff. and Bkg	0.011
K eff. and Bkg	0.038
parametrization, δ_f min	0.028

Table 5. Systematical erros in the Δm_d measurement using $l/K/$jet charge.

For well measured tracks within a cone with an opening 1/2 angle of 25 deg around the jet direction a "pseudo-secondary vertex" is fitted. It is a mixture of primary and secondary tracks. As there is no cut on χ^2, the vertex position (δ_f) depends linearly on the B flight. 7 "types" of hemispheres are considered : $u/d, s, c, B^{\pm}, B_d^o, B_s^o, b-$baryon. For each of them the MC parametrizes the probability ($P_{mes,cat}^{right}(\delta_f)$) to identify correctly the charge of the initial quark in function of δ_f, the flavor tagging itself is done using the lepton or kaon tag (with no cut on impact parameter) as described previously. The probability to determine correctly the initial quark charge ($P_{tag,cat}^{right}$) in the opposite hemisphere is estimated using three tagging indicators (l+jet,K+jet or jet alone). They have more or less the same $b\bar{b}$ discriminating power $= (N_{right} - N_{wrong})/(N_{right} + N_{wrong})/error$ as shown table 4. The probabilities ($P_{tag,cat}$ and $P_{mes,cat}$) are combined to give the probability to have a like/unlike sign correlation between the hemispheres as a function of δ_f and Δm_d. An unbinned likelihood fit give :

$$\Delta m_d c^2 = 0.586 \pm 0.049(stat) \pm 0.062(syst)\hbar/ps$$

A detail of the systematical errors can be found table 5.

5.   Conclusion

A clear time dependent $B_d^o - \bar{B}_d^o$ mixing has been observed using a $D^{*\pm}/D^{*\pm}l^{\mp}$-hemisphere charge tagging[1,2]:

$$\Delta m_d c^2 = 0.456 \pm 0.068(stat) \pm 0.043(syst)\hbar/ps$$

An inclusive technique[3,4] gave compatible results :

$$\Delta m_d c^2 = 0.586 \pm 0.049(stat) \pm 0.062(syst)\hbar/ps$$

with better statistical precision than the $D^{*\pm}$ approach but as it relied more on the simulation it had bigger systematical errors than the exclusive approach.
The average value of these two measurements, taking into account their common systematic, gave

$$\Delta m_d c^2 = 0.52 \pm 0.06(stat + syst)\hbar/ps$$

For the future the Δm_d measurement based on a $D^{*\pm}l^{\mp}$ sample seem promising as this method had the lowest systematical error ($\delta\Delta m_d = 0.030$).

References

1. DELPHI collaboration, P.Abreu et al., , *DELPHI* **94-100 PHYS 417** (1994)
2. DELPHI collaboration, P.Abreu et al., , *DELPHI* **94-101 PHYS 418** (1994)
3. DELPHI collaboration, P.Abreu et al., , *DELPHI* **94-63 PHYS 384** (1994)
4. DELPHI collaboration, P.Abreu et al., , *DELPHI* **94-118 PHYS 435** (1994)

MEASUREMENT OF $B° - \bar{B}°$ MIXING USING DIMUONS AT DØ

ERIC JAMES*

Department of Physics, University of Arizona
Tucson, Arizona 85721

ABSTRACT

The DØ experiment at Fermilab has determined the $B° - \bar{B}°$ mixing probability χ using dimuon events produced in $p\bar{p}$ collisions at $\sqrt{s} = 1.8$ TeV. Using a sample of 174 dimuon events, we have determined the time and flavor averaged mixing probability χ to be $0.13 \pm 0.05(stat) \pm 0.04(sys)$ in good agreement with the world average.

1. Introduction

Mixing between $B°$ and its anti-particle can occur in the Standard Model via well-known box diagrams which result from the non-conservation of quark flavor in the weak interaction. The time averaged mixing probability χ is given in terms of the mixing parameter x as

$$\chi = \frac{P(B° \to \bar{B}°)}{P(B° \to B°) + P(B° \to \bar{B}°)} \approx \frac{x^2}{2 + 2x^2}, \tag{1}$$

where x is the mass difference of the mass eigenstates divided by their average decay width. The mixing parameters x_d and x_s are of interest because they can be written in terms of parameters of the Standard Model. In particular, x_d and x_s depend on the CKM matrix elements V_{td} and V_{ts}. An accurate measurement of χ (or χ_s) can be used to set a lower limit on x_s and thus help constrain elements of the CKM matrix.

In the case of the semileptonic decay of B mesons into muons, the combined mixing probability χ is redefined as

$$\chi = \frac{BR(b \to B^0 \to \bar{B}^0 \to \mu^+)}{BR(b \to \mu^\pm)}, \tag{2}$$

which is an average over both $B_d°$ and $B_s°$ mesons which can mix as well as charged B mesons which can not. The sign of the muon produced via the semi-leptonic decay of the B^0 or $\bar{B}^0$ can be used to tag events in which mixing has occurred.

Experimentally, one measures the ratio R of like to unlike sign dimuons. In order to extract χ from R it is necessary to model the relative contributions of all processes which contribute to dimuon production. Using the redefined mixing probability χ, the fraction of like and unlike sign dimuons in the presence of mixing are given in Table 1 for the different production processes. Once the relative fractions of the contributing processes are modeled using Monte Carlo, χ can be extracted from R as the solution to a quadratic equation.

---

*Representing the DØ Collaboration

Process	Type	Like Sign	Unlike Sign
P1	$b \to \mu^-,\, \bar{b} \to \mu^+$	$2\chi(1-\chi)$	$(1-\chi)^2 + \chi^2$
P2	$b \to \mu^-,\, \bar{b} \to \bar{c} \to \mu^-$	$(1-\chi)^2 + \chi^2$	$2\chi(1-\chi)$
P3	$b \to c \to \mu^+,\, \bar{b} \to \bar{c} \to \mu^-$	$2\chi(1-\chi)$	$(1-\chi)^2 + \chi^2$
P4	$b \to c\mu^-,\, c \to \mu^+$	0%	100%
P5	$c \to \mu^+,\, \bar{c} \to \mu^-$	0%	100%
P6	Drell-Yan, J/ψ, Υ	0%	100%
P7	decay background	50%	50%

Table 1. Fraction of like and unlike sign dimuons from contributing processes

2. Trigger and Selection Cuts

The DØ detector has been described in detail elsewhere.[1] The data set used in this preliminary analysis was collected during the FNAL 1992-93 collider run using both a dimuon and a single muon plus jet trigger. The corresponding integrated luminosity for these triggering conditions in our data sample was 6.4 pb^{-1} and 3.4 pb^{-1} respectively. The offline cuts for the mixing analysis required two high quality muon tracks in the region $|\eta| < 0.8$. For each track energy deposition of at least 1 GeV was required in the calorimeter cells along the muon track plus their nearest neighbors. Matching central detector tracks were also required. The transverse momentum of each muon was constrained to the range 4 GeV $< p_T^\mu < 25$ GeV with the added requirement that $\int B \cdot d\ell > 0.5$ GeV along the muon track to help ensure a good momentum measurement.

Additional criteria were imposed to help remove specific backgrounds from the data sample. The dimuon opening angle was required to be less than 165° to help lower the cosmic ray background. The dimuon effective mass was required to be between 6 GeV and 40 GeV to remove events from J/ψ and Z^0 decay. In addition, each event was required to have at least one associated jet where an associated jet was defined as a jet with $E_T^{jet} > 12$ GeV within $\Delta R = \sqrt{\Delta\eta^2 + \Delta\phi^2} = 0.8$ of the muon. Further, all muons having associated jets in the event were required to satisfy $p_T^{rel} > 1.0$ GeV where p_T^{rel} is the transverse momentum of the muon relative to the jet axis. Here the jet axis was determined using the vector addition of the muon and jet momenta. These cuts were applied to help remove events from isolated processes such as Drell-Yan and Υ production as well as to enhance the fraction of dimuons coming directly from $b\bar{b}$ decay.

3. Data Analysis

Using these cuts we find a total of 61 like sign and 113 unlike sign dimuon events. The fraction of cosmic rays in these events is estimated by performing a maximum likelihood fit to the *floating t0* distributions for the muon tracks in the data sample. The *floating t0* parameter for a given muon track is determined via a refit of the track which leaves the beam crossing time as a free parameter in the fit. The best beam crossing time from this fit is the *floating t0*, the distribution of which is peaked sharply at zero for beam produced events and flatter for randomly dispersed cosmic rays. Correcting for the estimated cosmic ray background, one finds that the ratio of

Process	Type	Fraction
P1	$b \to \mu^-,\, \bar{b} \to \mu^+$	0.688 ± 0.234
P2	$b \to \mu^-,\, \bar{b} \to \bar{c} \to \mu^-$	0.141 ± 0.051
P3	$b \to c \to \mu^+,\, \bar{b} \to \bar{c} \to \mu^-$	0.022 ± 0.011
P4-P6	$b \to c\mu^- + c \to \mu^+,\, c\bar{c},\, J/\psi,\, \Upsilon$	0.025 ± 0.016
P7	decay background	0.123 ± 0.065

Table 2. Fraction of contributing processes to dimuon events

like to unlike-sign dimuons to be

$$R = \frac{like}{unlike} = 0.49 \pm 0.08(stat) \pm 0.05(sys) \tag{3}$$

where the systematic error reflects the uncertainties associated with the fits used to estimate the background fraction of cosmic rays. The relative contributions of the different dimuon production processes were determined using ISAJET Monte Carlo plus the full DØ detector and trigger simulations. The relative fractions of the contributing processes to the data sample are shown in Table 2. The errors in the relative fractions are attributable to both the Monte Carlo statistics and systematic uncertainties inherent in the simulations. The most prominent contributions to the systematic error are due to uncertainties in the production cross sections ($b\bar{b}$ -25%, $c\bar{c}$ -50%, π -50%), fragmentation functions (14%), muon decay spectrum shape (14%), and muon branching ratios (7%).

4. Results and Conclusions

Assuming the relative fractions of contributing processes given in Table 2 and using the measured value of R from equation (3) we find the time and flavor averaged mixing parameter χ to be

$$\chi = 0.13 \pm 0.05(stat) \pm 0.04(sys) \qquad (Preliminary), \tag{4}$$

where the systematic error is dominated by the uncertainties in the estimation of the fractions of contributing processes. Our preliminary value of χ is in good agreement with earlier results from UA1,[2] CDF,[3] and LEP[4,5,6].[7] Prospects for reducing both the statistical and systematic errors in this measurement are good as the backgrounds from the non-$b\bar{b}$ dimuon production processes are more fully understood.

References

1. S. Abachi et al., *Nucl. Instr. and Meth.* **A338** (1994) 185.
2. C. Albajar et al., *Phys. Lett.* **B262** (1991) 171.
3. F. Abe et al., *Phys. Rev. Lett.* **67** (1991) 3351.
4. B. Adeva et al., *Phys. Lett.* **B288** (1992) 395.
5. D. Decamp et al.,*Phys. Lett.* **B258** (1991) 236.
6. P. Acton et al., *Phys. Lett.* **B276** (1992) 379.
7. P. Abreu et al., *Phys. Lett.* **B301** (1993) 145.

FIRST MEASUREMENT OF $b \to s\gamma$

JESSE ERNST *

Dept. of Physics and Astronomy, University of Rochester
Rochester, NY 14627, USA

ABSTRACT

The CLEO Collaboration has measured the inclusive branching ratio for $b \to s\gamma$ and finds $\mathcal{B}\,(b \to s\gamma) = (2.32 \pm 0.51 \pm 0.29 \pm 0.32) \times 10^{-4}$. This article summarizes the techniques and results of the measurement.

I also report CLEO's upper limits on two other radiative penguin decays; B $\to \rho(\omega)\gamma$ and B $\to$ K(K*)l+l− .

1. Introduction

The April 1993, discovery of $B \to K^*(890)\gamma$ by the CLEO collaboration[1] verified the existence of One-Loop Flavor Changing Neutral Current decays (penguins). While measurement of the branching ratio for this exclusive mode demonstrates penguins' existence, the sensitivity of $b \to s\gamma$ to physics beyond the Standard Model[2,3] is realized only with a determination of the inclusive rate. Unfortunately, theoretical predictions for the fraction of $b \to s\gamma$ that hadronizes into K*(890) range from 5 to 40%; this uncertainty prevents us from converting the branching ratio for $B \to K^*(890)\gamma$ into an inclusive branching ratio. Here, I discuss the techniques and results of the fully inclusive measurement. $b \to s\gamma$ is an ideal decay channel in which to measure penguin contribution because 1) the GIM mechanism prevents Flavor Changing Neutral Current (FCNC) through tree diagrams, 2) QCD corrections significantly enhance the one-loop rate, and 3) contributions from non-penguin diagrams are expected to be at least an order of magnitude smaller than that from the penguin.

2. Detector and Dataset

CLEO II is a general purpose solenoidal detector. It measures charged particles over 90% of 4π steradians with three nested cylindrical drift chambers. Its barrel and endcap CsI electromagnetic calorimeters cover 98% of 4π. The region used for detecting the high energy photons for this analysis ($\cos(\theta) < .7$) has an energy resolution σ_E/E of 2% near 2.5 GeV. A more complete description of CLEO II is available.[4]

CLEO data are collected at two center of mass energies. The "ON" data are taken at E_{cm} = 10.58 GeV, which is on the $\Upsilon(4S)$ resonance. At this energy, roughly 25% of hadronic events are B meson pairs; the rest are continuum ($u\bar{u}$, $d\bar{d}$, $c\bar{c}$, or $s\bar{s}$). The "OFF" data are taken at E_{cm} = 10.52 GeV. This energy is below the B meson production threshold, so the events are strictly continuum. Within small corrections, the difference between an ON and an OFF spectrum represents the spectrum from B decays. The dataset used here is 2.01 fb^{-1} ON (2.15×10^6 B$\overline{\text{B}}$ events and 6.6×10^6 continuum events) and .96 fb^{-1} OFF.

---

*Representing the CLEO Collaboration

3. Inclusive $b \to s\gamma$

Our general strategy is simply to count the number of photons from B decays that have energy between 2.2 and 2.7 GeV. $b \to s\gamma$ will contribute photons between 2 and 2.8 GeV, while contributions from most other B decay channels lie below 2.4 GeV. The background from B decays between 2.2 and 2.7 GeV is small and can be estimated. The background from continuum events is large and must be suppressed. We chose 2.2–2.7 GeV rather than 2.0–3.0 GeV as the signal range because 1) B backgrounds become significant below 2.2 GeV and thus including this range would have required an unacceptable reliance on Monte Carlo, and 2) there is little signal above 2.7 Gev, yet the continuum background remains large. Model calculations indicate that 75–90% of $b \to s\gamma$ photons lie between 2.2 and 2.7 GeV. In selecting photon candidates, we require that the shower be well isolated (not too close to a charged track), that it be in the highest resolution portion of the calorimeter ($\cos(\theta) < .7$), and that the energy distribution among the crystals in the shower be consistent with that from a photon. If the photon can be paired with another photon to form a π^0 or an η then the candidate photon is discarded.

Two distinct event types contribute to the continuum background; generic $q\bar{q}$ events ($q\bar{q}$), and $q\bar{q}$ events that undergo Initial State Radiation (ISR). In $q\bar{q}$ events, 83% of the photon background is from π^0 or η decay. In ISR events, the background is the radiated photon. The size of these continuum backgrounds is such that a signal for $b \to s\gamma$ would be lost in statistical noise if we measured the difference between the ON and OFF photon spectra without continuum suppression. We have measured $b \to s\gamma$ with two independent analyses. Both analyses follow the same general strategy, but use different techniques to suppress the continuum background.

3.1. Neural Net Analysis

This analysis uses eight event shape variables to separate signal from the two types of background.

Since the ISR events are $q\bar{q}$ events recoiling against the radiated photon, we can restore the $q\bar{q}$ shape of an ISR event by transforming it to this recoiling frame *. We call this the primed frame.

The B mesons at CLEO are produced nearly at rest and decay isotropically; continuum events are more jet-like. We use R_2, R_2', and $S_\perp$ to measure isotropy. R_2 is the normalized second Fox-Wolfram moment. In the lab frame, R_2 separates signal from $q\bar{q}$; in the primed frame it separates signal from ISR. $S_\perp$ is the fraction of the momentum carried by particles more than 45° from the photon; it separates signal from $q\bar{q}$.

In the primed frame, an ISR photon's direction is uncorrelated with the direction of the continuum jet. We define $Cos(\theta')$ as the cosine of the angle between the candidate photon and the thrust axis of the rest of the event. For ISR events, $|Cos(\theta')|$ will be flat; for signal events, it will peak near +1.

To make use of differences in the energy/angular distribution of particles in signal

---

*Having lost the energy of the radiated photon, an ISR event in the primed frame has the shape of a reduced energy continuum event.

and in background events, we define four additional variables. We compute the energy in four cones (20° and 30° along and opposed to the photon direction).

All eight variables have some power to discriminate between signal and one or both of the continuum backgrounds. However, in order to gain enough continuum suppression, we must allow for correlations between the variables. To do this, we create a single variable (r) which is a function of all eight shape variables and allows for linear and non-linear correlations between them. The functional form of r is such that its free parameters can be optimized using a neural network algorithm.[5] The network [†] was trained (i.e., the parameters of r were optimized) on a Monte Carlo sample. Once optimized, r tends toward +1 for signal events and −1 for either type of background. We do not make a cut on r. Instead, we gain additional statistical power by weighting events according to their r value. The improvement in continuum suppression over using uncorrelated cuts on the eight variables is equivalent to having 77% more data.

3.2. *B-Reconstruction Analysis*

In this analysis, we suppress the continuum backgrounds by reconstructing some subset of the particles in the event into a $b \to s\gamma$ candidate. For each event, we consider only one combination of particles; the combination that best satisfies a $B \to X_s\gamma$ decay hypothesis. A fully inclusive analysis is achieved by using a large number of exclusive decay modes. We allow γ with $K^{\pm}$ or K^0_s (reconstructed in $\pi^+\pi^-$ decays), and up to four pions (which may include zero or one π^0).

For each set of candidate particles, we form the total energy E and momentum P. The energy difference ΔE, beam-constrained mass M, and mass difference ΔM are then calculated as $\Delta E = E - E_{beam}$, $M = \sqrt{E^2_{beam} - P^2}$, and $\Delta M = M - 5.279$ GeV. For the correct set of candidate particles, ΔE and ΔM will be zero, within measurement error. Consequently we construct a variable $\chi^2_B = (\Delta M/\sigma_M)^2 + (\Delta E/\sigma_E)^2$, where σ_E and σ_M are the expected measurement errors. χ^2_B reflects how well the kinematic properties of the set of candidate particles represent a B meson.

Once we have the best combination of particles, we make cuts on two variables; χ^2_B and $|\cos(\theta_{tt})|$. χ^2_B is defined above, and $|\cos(\theta_{tt})|$ is the cosine of the angle between the thrust axis of the particles used to form the candidate B and the thrust axis of the rest of the particles in the event. For a $B\overline{B}$ event, there is no angular correlation between the decay products of the two B mesons, so $|\cos(\theta_{tt})|$ should be flat. For continuum events, the two thrust angles will tend to align and hence $|\cos(\theta_{tt})|$ will peak near 1. Based on Monte Carlo studies, we require $\chi^2_B < 6.0$ and $|\cos(\theta_{tt})| < 0.6$.

3.3. *Yield and Background From the* $\Upsilon(4S)$

The $\Upsilon(4S)$ background is dominated by γ's from π^0 and η decays. To determine this background, we treat high momentum π^0's and η's as the candidate photon and compare the data with the Monte Carlo predictions. We then apply a correction to the Monte Carlo. With this procedure, we rely on the Monte Carlo predictions only for the very small backgrounds other than π^0 and η decays. Table 1 lists the yields and the $\Upsilon(4S)$ backgrounds for the two analyses. Figures 1a and 2a show the yield from

---

[†]We use a two layer, fully-connected neural network with 10 nodes in the first layer, and 7 in the second.

ON, OFF (scaled to correct for the smaller OFF luminosity), and OFF+B backgrounds. Figures 1b and 2b show the final yields (the curve gives the shape expected for $b \rightarrow s\gamma$).

3.4.   Combining the Two Analyses

Both analyses use a spectator model from Ali-Greub[6] to determine efficiency. The spectator model gives the X_s mass distribution. We use conventional models of quark hadronization to describe the decay of the X_s.

The results (with statistical errors only) from the neural-net and the B-reconstruction analyses are $\mathcal{B}(b \rightarrow s\gamma) = (1.88 \pm 0.67) \times 10^{-4}$, and $\mathcal{B}(b \rightarrow s\gamma) = (2.76 \pm 0.63) \times 10^{-4}$ respectively. Monte Carlo indicates that the two methods have approximately equal statistical power, so we weight the results equally. To compute the combined error, we use Monte Carlo samples to determine the degree of correlation between the analyses. From this, we determine that the results agree within 1.1σ and that the average has a statistical error of $.51 \times 10^{-4}$.

3.5.   Systematic Errors

The largest sources of additive systematic errors are uncertainties in subtracting the OFF data from the ON data and uncertainty in the π^0 and η corrections to the Monte Carlo when estimating the $\Upsilon(4S)$ background. The ON–OFF subtraction error is from uncertainty in a correction factor which accounts for the lower E_{cm} of the OFF data. The error in the π^0 and η corrections is statistical.

The largest multiplicative systematic error is from model dependence. We use the Ali-Greub spectator model with $< m_b > = 4.87 \pm 0.10$ GeV and $P_f = 270 \pm 40$ MeV. The uncertainties on P_f and m_b correspond to uncertainties in the X_s mass distribution. Because the efficiency is not completely flat across the various X_s masses and decay modes for either analysis, the variations in X_s mass distribution cause a variation in the overall efficiency.

3.6.   Final Results and Implications

We find
$$\mathcal{B}(b \rightarrow s\gamma) = (2.32 \pm 0.51 \pm 0.29 \pm 0.32) \times 10^{-4} \ ,$$
where the first error is statistical, the second is the additive systematic error from uncertainty in yield, and the third is the multiplicative systematic error from uncertainty in efficiency, which includes model dependence. This result is in good agreement with the Standard Model expectations of 2.0×10^{-4} to 3.8×10^{-4}.[7] At 95% CL, we find $1 \times 10^{-4} < \mathcal{B}(b \rightarrow s\gamma) < 4 \times 10^{-4}$

One example of the sensitivity of $b \rightarrow s\gamma$ to new physics is the limit it places on the charged Higgs mass in a Two-Higgs-Doublet-Model II. In this model, the down and up type quarks acquire mass from separate vacuum expectation values ($\tan(\beta) \equiv v_2/v_1$). All points below the curve in figure 3 are excluded by our upper limit (where we have allowed for theoretical uncertainty in renormalization scale, and an additional theoretical uncertainty of 10%). J. Hewett has recently published a review which covers a wide variety of $b \rightarrow s\gamma$ implications.[8]

4.   Upper Limits on $\mathbf{B} \to \rho(\omega)\gamma$ and $\mathbf{B} \to \mathrm{K(K^*)l^+l^-}$

CLEO has recently obtained new upper limits from two other radiative penguin decays.

For $\mathrm{B} \to \rho(\omega)\gamma$ we find (at 90% CL) $\mathcal{B}(B^- \to \rho^-\gamma) < 2.0 \times 10^{-5}$, $\mathcal{B}(\bar{B}^0 \to \rho^0\gamma) < 2.4 \times 10^{-5}$, and $\mathcal{B}(\bar{B}^0 \to \omega\gamma) < 1.1 \times 10^{-5}$. If long-distance effects are negligible, we can combine these results with the measurement of $\mathcal{B}(B \to K^*\gamma)$ to set upper limits on $|V_{td}/V_{ts}|$. Depending on the choice of model, $|V_{td}/V_{ts}| < 0.64 - 0.76$.

For $\mathrm{B} \to \mathrm{K(K^*)l^+l^-}$ modes, we find the following upper limits at 90% CL:
$\mathcal{B}(B^- \to K^- e^+ e^-) < 1.2 \times 10^{-5}$, $\mathcal{B}(\bar{B}^0 \to \bar{K}^{*0} e^+ e^-) < 1.6 \times 10^{-5}$,
$\mathcal{B}(B^- \to K^- \mu^+ \mu^-) < 0.9 \times 10^{-5}$, $\mathcal{B}(\bar{B}^0 \to \bar{K}^{*0} \mu^+ \mu^-) < 3.1 \times 10^{-5}$,
$\mathcal{B}(B^- \to K^- e^\pm \mu^\mp) < 1.2 \times 10^{-5}$, and $\mathcal{B}(\bar{B}^0 \to \bar{K}^{*0} e^\pm \mu^\mp) < 2.7 \times 10^{-5}$. The upper limit for $\mathcal{B}(\bar{B}^0 \to \bar{K}^{*0} e^+ e^-)$ is within a factor of 3 of the Standard Model prediction.

5.   Acknowledgements

I thank Ed Thorndike and Peter C. Kim for assistance in preparing this article. I also thank W.I. for agreeing to delay a clear definition of non-nutritive variables.

References

1. R. Ammar *et al.* (CLEO Collaboration), Phys. Rev. Lett. **71**, 674 (1993).
2. T. G. Rizzo, Phys Rev. **D38**, 820 (1988); W.-S. Hou and R.S. Wiley, Phys Lett. **B202**, 591 (1988); C.Q.Geng and J.N. Ng, Phys Rev. **D38**, 2858 (1988); V. Barger, J.L. Hewett, and R.J.N. Phillips, Phys. Rev. **D41**, 3421 (1990); J.L. Hewett, Phys. Rev. Lett. **70**, 1045 (1993); V. Barger, M. Berger, and R.J.N. Phillips, Phys. Rev. Lett. **70**, 1368 (1993).
3. S.-P. Chia, Phys. Lett. **B240**, 465 (1990); K.A. Peterson, Phys. Lett. **B282**, 207 (1992); T.G. Rizzo, Phys. Lett. **B315**, 471 (1993); U. Baur, in Proceedings of the *Summer Workshop on B Physics*, Snowmass, CO, 1993; X.-G. He and B. McKellar, Phys. Lett. **B320**, 165 (1994).
4. Y. Kubota *et al.* (CLEO Collaboration), Nucl. Instrum. Methods Phys. Res., Sect. **A320**, 66 (1992).
5. Rumelhart D.E. and McClelland J.L., *Parallel Distributed Processing*, Vol. **1–2** (Bradford Books, MIT Press, Cambridge, MA) 1986.
6. A. Ali and C. Greub, Phys. Lett. **B259**, 182 (1991).
7. S. Bertolini *et al.*, Phys. Rev. Lett. **59**, 180 (1987); N. Deshpande *et al.*, Phys. Rev. Lett. **59**, 183 (1987); B. Grinstein *et al.*, Phys. Lett. **B202**,138 (1988); R. Grigjanis *et al.*, Phys. Lett. **B224**, 209 (1989); G. Cella *et al.*, Phys. Lett. **B248**, 181 (1990); A.J. Buras *et al.*, Max Planck Institute preprint MPI-Ph/93-77.
8. J.L. Hewett, SLAC-PUB-6521 (May 1994).

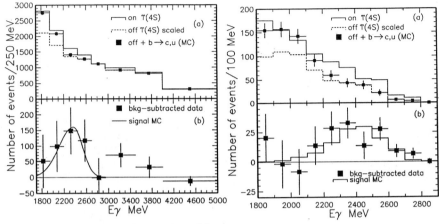

Figure 1: Photon energy spectra from the neural net analysis.

Figure 2: Photon energy spectra from the B reconstruction analysis.

Table 1: Yields for the two analysis procedures.

	Neural Net	B Reconstruction
On	3013 ± 59	281 ± 17
Off(scaled)	2618 ± 73	155 ± 18
$\Upsilon(4S)$ background		
$b \to c$	50.7 ± 5.1	$^-12 \pm 2$
$b \to u$	11.9 ± 4.0	2 ± 1
π^0 correction	50.2 ± 27.7	-0.7 ± 2.3
η correction	16.5 ± 33.7	2.0 ± 8.5
Non-$B\bar{B}$	2.3	
$\Upsilon(4S)$ total	132 ± 44	$^-15 \pm 9$
On$-$Off$-\Upsilon(4S)$	263 ± 94	110 ± 25
	± 49	± 10

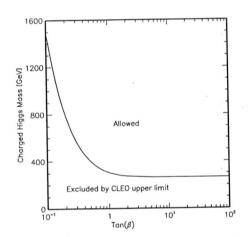

Fig. 3. Lower limit on Two-Higgs-Doublet-Model-II Higgs mass as a function of $\tan(\beta)$. The

LIMITS ON RARE B DECAYS $B \to \mu^+\mu^- K^{\pm}$ AND $B \to \mu^+\mu^- K^*$

CAROL ANWAY-WIESE[*]

Physics Department, University California, Los Angeles
405 Hilgard Ave, Los Angeles, California 90024-1547, USA

ABSTRACT

We report on a search for flavor-changing neutral current decays of B mesons into $\mu\mu K^{\pm}$ and $\mu\mu K^*$ using data obtained in the Collider Detector at Fermilab (CDF) 1992-1993 data taking run. To reduce the amount of background in our data we use precise tracking information from the CDF silicon vertex detector to pinpoint the location of the decay vertex of the B candidate, and accept only events which have a large decay time. We compare this data to a B meson signal obtained in a similar fashion, but where the muon pairs originate from ψ decays, and calculate the relative branching ratios. In absence of any indication of flavor-changing neutral current decays we set an upper limits of BR($B \to \mu\mu K^{\pm}$) $< 3.5 \times 10^{-5}$, and BR($B \to \mu\mu K^*$) $< 5.1 \times 10^{-5}$ at 90% confidence level, which are consistent with Standard Model expectations but leave little room for non-standard physics.

1. Introduction

Rare B decays provide us a way to test the Standard Model against possible effects of different form factors, anomalous magnetic moment of the $W^{\pm}$, and charged Higgs. Several theorists have predicted the rate of these decays. Differences in the form factors used in these calculations give relatively small uncertainties, but deviations from Standard Model physics can dramatically increase the expected rate. For a top quark mass of 150 GeV/c², A. Ali[1] predicts branching ratios of BR($B \to \mu\mu K^{\pm}$) $= 4.4 \times 10^{-7}$ and BR($B \to \mu\mu K^*$) $= 2.3 \times 10^{-6}$ using the hadronic matrix elements of Isgur and Wise.[2] G. Baillie[3] has also used heavy quark effective theory to calculate the ratio of the decay rate in a portion of the non-resonant dimuon mass spectrum to the decay rate of $B \to \psi K, \psi \to \mu\mu$. He finds the ratio of the decay rates to be 2×10^{-3} for each decay mode at a top quark mass of 170 GeV/c². The portion of the non-resonant dimuon mass spectrum he uses goes from $\hat{s} \equiv M(\mu\mu)^2/M_B^2 = 0.35$ to 0.48 and 0.50 to $\hat{s}_{max} = (M_B - M_K)^2/M_B^2$, similar to the region we use here. We also use the non-resonant theoretical differential decay rate as a function of dimuon mass from Isgur and Wise to extrapolate our results from the small dimuon mass region to the overall non-resonant region.

---

[*]Representing the CDF Collaboration MS 318, Fermilab, PO Box 500 Batavia, IL 60510 caw@fnald.fnal.gov

2.   Data and Method

At CDF[4] we have accurate momentum resolution in the central tracking chamber (CTC), further improved by using vertex position information from the silicon vertex detector (SVX).[5] In this analysis we accept only pairs of muons which have traversed the SVX, the CTC, and have muon stubs in the central muon chambers (up to $\eta = 0.6$). Both muons must have $Pt(\mu) > 2.0\,$GeV/c, and one $Pt(\mu) > 2.8$ GeV/c.

We select muon pairs with an invariant mass between 2.8 and 4.5 GeV/c^2, assign the K$^\pm$ and $\pi^\pm$ masses to tracks in the central tracking chamber and use a secondary vertex fit to help reconstruct candidate B's. Tight cuts on the vertex fit quality and the transverse proper B candidate decay time (0.1 mm) help reduce combinatoric background. Background from hadronic punch-through is largely reduced by requiring the B candidate to carry the majority of the momentum in a cone. The transverse momentum cuts on the events are as follows: For $B \to \mu\mu K^\pm$, Pt(B) > 5.0 GeV/c, Pt(K) > 1.0 GeV/c. For $B \to \mu\mu K^*$, Pt(B) > 6.0 GeV/c, Pt(K*) > 2.0 GeV/c. We assign dimuons to the ψ resonance if their invariant mass falls between 3.017 and 3.177 GeV/c^2, and to the (partial) non-resonant region if their mass falls in the range 3.3 to 3.6 or 3.8 to 4.5 GeV/c^2.

3. $B \to \mu\mu K$ Results

After making the cuts above, we see 57 ± 10 $B \to \psi K^\pm, \psi \to \mu\mu$ events above background. We compare this to 10 $B \to \mu\mu K^\pm$ candidate events in our signal region, which is consistent with 13.0 ± 2.5 background events, as estimated from the sidebands. Using BR($B \to \psi K^\pm, \psi \to \mu\mu$) = $6.5 \pm 1.0 \times 10^{-5}$ (ref. 6) we can compare the number of ψ events to the number of non-resonant events to calculate the branching ratio limits according to the method of G. Zech.[9] We extrapolate to the overall dimuon mass region by multiplying the partial branching ratios by 4.4, as calculated using Monte Carlo and the theoretical models of Isgur and Wise.[2,3]

$$\frac{\text{BR}(B \to \mu\mu K, partial)}{\text{BR}(B \to \psi K, \psi \to \mu\mu)} \quad < \quad 0.12 \text{ at } 90\% \text{ CL}, \qquad < \quad 0.15 \text{ at } 95\% \text{ CL}$$

$$\frac{\text{BR}(B \to \mu\mu K)}{\text{BR}(B \to \psi K, \psi \to \mu\mu)} \quad < \quad 0.50 \text{ at } 90\% \text{ CL}, \qquad < \quad 0.66 \text{ at } 95\% \text{ CL}$$

$$\text{BR}(B \to \mu\mu K, \text{partial}) \quad < \quad 0.8 \times 10^{-5} \text{ at } 90\% \text{ CL}, \quad < \quad 1.0 \times 10^{-5} \text{ at } 95\% \text{ CL}$$

$$\text{BR}(B \to \mu\mu K) \quad < \quad 3.5 \times 10^{-5} \text{ at } 90\% \text{ CL}, \quad < \quad 4.4 \times 10^{-5} \text{ at } 95\% \text{ CL}$$

4. $B \to \mu\mu K^*$ Results

We see 33.5 ± 6.7 $B \to \psi K^*, \psi \to \mu\mu$ events above background. We compare this to 7 $B \to \mu\mu K^*$ candidate events in our signal region, which is consistent with 7.5 ± 1.9 background events. Using BR($B \to \psi K^*, \psi \to \mu\mu$) = $7.76 \pm 2.41 \times 10^{-5}$ (ref. 7) we can compare the number of ψ events to the number of non-resonant events to calculate the branching ratio limits. We extrapolate to the overall dimuon mass region by multiplying the partial branching ratios by 3.75.

$$\frac{\text{BR}(B \to \mu\mu K^*, partial)}{\text{BR}(B \to \psi K^*, \psi \to \mu\mu)} \quad < \quad 0.21 \text{ at } 90\% \text{ CL}, \qquad < \quad 0.26 \text{ at } 95\% \text{ CL}$$

$$\frac{\text{BR}(B \to \mu\mu K^*)}{\text{BR}(B \to \psi K^*, \psi \to \mu\mu)} \quad < \quad 0.77 \text{ at } 90\% \text{ CL}, \qquad < \quad 0.96 \text{ at } 95\% \text{ CL}$$

$$\text{BR}(B \to \mu\mu K^*, \text{partial}) \quad < \quad 1.3 \times 10^{-5} \text{ at } 90\% \text{ CL}, \quad < \quad 1.7 \times 10^{-5} \text{ at } 95\% \text{ CL}$$

$$\text{BR}(B \to \mu\mu K^*) \quad < \quad 5.1 \times 10^{-5} \text{ at } 90\% \text{ CL}, \quad < \quad 6.4 \times 10^{-5} \text{ at } 95\% \text{ CL}$$

5. Acknowledgements

We thank Grant Baillie, UCLA, for his help in the theory behind the RareB Monte Carlo event generator and for many valuable discussions on the subject of rare B decays.

We thank the Fermilab staff and the technical staffs of the participating institutions for their vital contributions. This work was supported by the U.S. Department of Energy and National Science Foundation; the Italian Istituto Nazionale di Fisica Nucleare; the Ministry of Education, Science and Culture of Japan; the Natural Sciences and Engineering Research Council of Canada; the National Science Council of the Republic of China; the A. P. Sloan Foundation; and the Alexander von Humboldt-Stiftung.

References

1. A. Ali, Rare Decays of B Mesons, DESY 91-080 (1991), Proceedings of the First International A. D. Sakharov Conference On Physics, Lebedev Physics Institute, Moscow, USSR (1991).
2. N. Isgur and M. B. Wise, Phys. Lett. B 232 (1989) 113 and Phys. Lett. B 237 (1990) 527.
3. G. Baillie, Observing $b \to s\mu^+\mu^-$ Decays at Hadron Colliders, UCLA/93/TEP/26, hep-ph 9307369.
4. CDF detector, F. Abe *et al.* Nucl. Inst. and Meth. A271 (1988) 387, and references therein.
5. Silicon vertex detector, B. Barnett *et al.* Nucl. Inst. and Meth. A315 (1992) 125.
6. BR($B \to \psi K^\pm, \psi \to \mu\mu$), T. Browder, K. Honscheid, S. Playfer, A Review of Hadronic and Rare B Decays, CLNS 93/1261, Jan 15, 1994.
7. Review of Particle Properties, Phys. Rev. D, 45 (1992)
8. M. B. Çakir, Angular distribution and helicity dependence in $B \to K^{(*)}l^+l^-$, Phys. Rev. D, 46 (1992) 2961.
9. G. Zech, Upper Limits in Experiments with Backgound or Measurement Errors, NIM, A 277 (1989) 608.

SEARCH FOR CHARMLESS B-DECAYS

J. FAST

Physics Department, Purdue University,
West Lafayette, Indiana USA 47907

(Representing the CLEO collaboration)

ABSTRACT
We have searched for several two-body charmless hadronic decays of B
mesons. These final states include $K\pi$ and $\pi\pi$ with both charged and
neutral kaons and pions; $K^*\pi$, $K\rho$, and $\pi\rho$; and ϕX_s, where
$X_s = K$, K^*, or ϕ. We have also searched for the decay $B \to \tau\bar{\nu}_\tau$. Using
a dataset which includes 2.2 million $B\overline{B}$ pairs, we place upper limits on
the charmless hadronic decay branching ratios in the range from 10^{-4}
to 10^{-5}, and set a limit $BR(B \to \tau\bar{\nu}_\tau) < 2.2 \times 10^{-3}$.

1.   Introduction

The decays of B-mesons to two charmless hadrons can proceed via a a $b \to u$ tree-level spectator diagram (Fig. 1a), or via a $b \to sg$ one-loop "penguin-diagram" (Fig. 1b) and to a lesser extent, via the CKM-suppressed $b \to dg$ penguin diagram. Although rare decays can also proceed via $b \to u$ internal, color-suppressed diagrams (Fig. 1c), $b \to u$ exchange diagrams (Fig. 1d), annihilation diagrams (Fig. 1e), or vertical W loop diagrams (Fig. 1f), these contributions are expected to be negligible in most cases.

This paper reports preliminary results on the decays $B \to K\pi$, $B \to K^*\pi$, $B \to K\rho$, $B \to \pi\rho$, and $B \to X_s\phi$. We place upper limits on these modes in the range 10^{-4} to 10^{-5}.

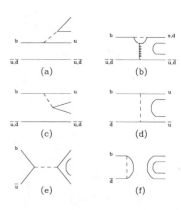

Fig. 1. Feynman diagrams: (a) external W emission, (b) loop or gluonic penguin, (c) internal W, (d) W exchange, (e) annihilation, and (f) vertical W loop.

We also include updated results on the decays $B^0 \to K^+\pi^-$, $B^0 \to \pi^+\pi^-$, and $B^0 \to K^+K^-$. A search is made for the decay $B \to \tau\bar{\nu}_\tau$ which proceeds through an annihilation diagram similar to that shown in Fig. 1e, and hence provides a measurement of $f_B|V_{ub}|$. This decay also probes physics beyond the standard model, such as annihilation to a charged Higgs. We set a limit of $BR(B \to \tau\bar{\nu}_\tau) < 2.2 \times 10^{-3}$ at the 90% confidence level.

2.   Data sample

The data set used in this analysis was collected with the CLEO-II detector[1] at the Cornell Electron Storage Ring (CESR). It consists of 2.0 fb^{-1} taken on the $\Upsilon(4S)$ (on-resonance) and 1.0 fb^{-1} taken at a center of mass energy ~ 55 MeV below the $\Upsilon(4S)$ (off-resonance). The on-resonance sample contains 2.2 million $B\overline{B}$ pairs. The off-resonance sample, taken below the $B\overline{B}$ threshold, is used for continuum background estimates.

3.   Charmless Hadronic Decays

We provide here a brief overview of the analysis issues and tools used to address these concerns. A more detailed description of this analysis appears in.[2]

3.1.   Background suppression

In general, $b \to c$ decays do not produce back-to-back particles with sufficient momentum to be a source of background to the rare decay modes under study; the dominant background comes from non-resonant (continuum) production of light quarks, $e^+e^- \to q\bar{q}$, $q = u, d, s, c$. Since B mesons from $\Upsilon(4S)$ decay are produced nearly at rest in the laboratory frame, the B decays are isotropic and the resulting event shape is rather spherical. Continuum background events naturally contain high momentum back-to-back particles in the two opposing light quark jets. We require $|\cos\theta_{thrust}| < 0.7$, where θ_{thrust} is the angle between the thrust axis of the candidate and that of the rest of the event. This cut removes $\sim 90\%$ of the continuum background with $\sim 65\%$ efficiency for signal events. Additional continuum suppression is achieved using a Fisher discriminant which reduces continuum backgrounds by a factor of 2 to 3, after requiring $|\cos\theta_{thrust}| < 0.7$, with efficiencies for signal events $\sim 80\%$.

In the process $e^+e^- \to \Upsilon(4S) \to B\overline{B}$, the B mesons are produced with the same energy as the colliding $e^\pm$ beams, E_{beam}. We define the variable $\Delta E = E_{candidate} - E_{beam}$. A signal region is defined by $|\Delta E| < 2.0 - 2.5\sigma$, where σ is the rms resolution determined from Monte Carlo studies. We also define the beam-constrained B mass as $M_B = \sqrt{E_{beam}^2 - \vec{p}_{candidate}^2}$, where $\vec{p}_{candidate}$ is the vector sum of the momenta of the candidate daughters. The resolution in the beam-constrained mass, $\sigma = 2.6 \pm 0.2$ MeV, is dominated by the spread in beam energy. The signal region is defined as $|M_B - 5280| < 6 - 8$ MeV, depending on the decay mode under consideration.

3.2.   Results

Results and a range of theoretical predictions for these decays appear in Table 1. The expected background in the signal region is obtained by scaling the number of events seen in a large sideband region in the ΔE vs. M_B plane.

Table 1. Event yields, estimated background in the signal region, upper limits on the branching ratios, previous upper limits, and theoretical predictions. Upper limits are at the 90% confidence level. For the decays $B \to X_s\phi$, particles in parentheses indicate a $K^*(892)$ decaying to those particles.

Mode	Event Yield	Est. Bkgr.	Upper Limit on BR (10^{-5})	Previous UL (10^{-5})	Theory (10^{-5})
$\pi^0\pi^0$	1	1.7 ± 0.4	1.0	–	0.03-0.10
$\pi^\pm\rho^\mp$	6	2.4 ± 0.6	9.5	52	1.9-8.8
$\pi^0\rho^0$	1	1.5 ± 0.5	2.9	40	0.07-0.23
$\pi^+\pi^0$	8	5.8 ± 0.6	2.3	24	0.6-2.1
$\pi^+\rho^0$	3	1.8 ± 0.3	4.1	15	0.0-1.4
$K^0\pi^0$	2		6.3	–	0.5-0.8
$K^{*+}\pi^-$	3	0.6 ± 0.2	23.8	44	0.1-1.9
$K^{*0}\pi^0$	0	0.9 ± 0.3	3.5	–	0.3-0.5
$K^+\rho^-$	2	1.7 ± 0.4	4.3	–	0.0-0.2
$K^0\phi$	1		10.7	49	0.1-1.3
$(K^+\pi^-)\phi$	1				
$(K^0\pi^0)\phi$	0				
$K^{*0}\phi$ (combined)	1		3.9	32	0.0-3.1
$\phi\phi$	0		4.8	–	–
$K^+\pi^0$	10	5.0 ± 0.5	3.2	–	0.3-1.3
$K^0\pi^+$	6	2.0 ± 0.5	6.8	9	1.1-1.2
$K^{*0}\pi^+$	2	0.6 ± 0.2	6.0	13	0.6-0.9
$K^+\rho^0$	1	1.1 ± 0.2	2.6	7	0.01-0.06
$K^+\phi$	0		1.4	8	0.1-1.5
$(K^+\pi^0)\phi$	0				
$(K^0\pi^+)\phi$	1				
$K^{*+}\phi$ (combined)	1		9.0	130	0.0-3.1

Our previously published results[3] for $B^0 \to K^+\pi^-$ and $B^0 \to \pi^+\pi^-$ showed a significant signal in the sum of the two modes. Since publication, we have added 50% more data. Updated results appear in Table 2. Our sensitivity is at the level of theoretical predictions for the modes $\pi^+\pi^-, K^+\pi^-, \pi^+\pi^0, \pi^\pm\rho^\mp, K^+\phi$, and $K^{*0}\phi$.

Table 2. Updated results for $B^0 \to K^+\pi^-$, $B^0 \to \pi^+\pi^-$, and $B^0 \to K^+K^-$. Upper limits are at the 90% confidence level.

Mode	Event Yield	$BR\,(10^{-5})$	Theoretical Predictions (10^{-5})[4]
$\pi^+\pi^-$	$8.5^{+4.9}_{-4.0}$	< 2.2	1.0-2.6
$K^+\pi^-$	$7.1^{+4.2}_{-3.4}$	< 1.9	1.0-2.0
K^+K^-	$0.0^{+1.6}_{-0.0}$	< 0.7	−
$\pi^+\pi^- + K^+\pi^-$	$15.7^{+5.3}_{-4.5}$	$1.8^{+0.6}_{-0.5} \pm 0.2$	

4. $B \to \tau\bar{\nu}_\tau$

This analysis is described in detail in.[5] We search for events in which the tau decays leptonically, and for which the other B-meson decay products are all hadronic. For these events the missing energy-momentum 4-vector for the event is equivalent, within detector resolution, to the 4-momentum of the three neutrinos. The single lepton is removed from the event, and then ΔE and M_B are calculated, as described above, using the remaining tracks and showers. This is equivalent to reconstructing the $B \to \tau\bar{\nu}_\tau$ decay. We subtract continuum background using scaled off-resonance data. A 2-dimensional fit is done in ΔE versus M_B for three components:(1)$B \to \tau\bar{\nu}_\tau$ signal, (2)$B\bar{B}$ background with primary leptons from semileptonic B decay, and (3)other $B\bar{B}$ backgrounds. The yield of $B \to \tau\bar{\nu}_\tau$ is found to be -14 ± 37 events, corresponding to an upper limit of $BR(B \to \tau\bar{\nu}_\tau) < 2.2 \times 10^{-3}$ at the 90% confidence level, two orders of magnitude above standard model predictions.

References

1. Y. Kubota *et al.* (CLEO), *Nucl. Instr. Methods A* **320** (1992) 66.
2. J. Gronberg *et al.* (CLEO), *ICHEP94 Ref. GSL0391* (1994).
3. M. Battle *et al.* (CLEO), *Phys. Rev. Lett.* **71** (1993) 3922.
4. A. Deandrea, N. Di Bartolomeo, R. Gatto, F. Feruglio, and G. Nardulli, *Phys. Lett. B* **320** (1994) 170; N. G. Deshpande and Xiao-Gang He, *OITS-538* (1994); A. Deandrea, N. Di Bartolomeo, R. Gatto, and G. Nardulli, *Phys. Lett. B* **318** (1993) 549; A.J. Davies, T. Hayashi, M. Matsuda, and M. Tanimoto, *AUE-02-93*; L. L. Chau, H-Y Cheng, W.K. Sze, H. Yao, and B. Tseng, *Phys. Rev. D* **43** (1991) 2176; N. G. Deshpande and J. Trampetic, *Phys. Rev. D* **41** (1990) 895; M. Bauer, B. Stech, and M. Wirbel, *Z. Phys. C* **43** (1987) 103.
5. J.P. Alexander *et al.* (CLEO), *ICHEP94 Ref. GSL0160* (1994).

CP Violation and Rare Decays

GERMAN E. VALENCIA, *Iowa State*

GEORGE G. GOLLIN, *Illinois (Urbana-Champaign)*

OBSERVATION OF CHARMLESS HADRONIC BEAUTY DECAYS

A. M. LITKE and G. TAYLOR

Institute for Particle Physics, University of California,
Santa Cruz, CA 95064
Representing the ALEPH Collaboration

ABSTRACT

In a sample of 1.55 million hadronic Z^0 decays, three candidates for charmless hadronic beauty decay have been observed in the ALEPH detector at LEP. The probability that these events come from background sources has been estimated from Monte Carlo studies to be 1.8×10^{-4}, a significance of over $3.5\,\sigma$. Upper limits, of order 10^{-4}, have been obtained for a variety of exclusive final states involving charmless decays of B_d^0, B_s^0, and Λ_b hadrons.

1. Introduction

Decays of beauty hadrons into hadronic final states which do not contain a charm quark can proceed either via $b \to u$ transitions, or through $b \to s$ or $b \to d$ hadronic "penguin" transitions. These transitions can give rise to decays such as $B_d^0 \to \pi^+\pi^-, K^+\pi^-$; $B_s^0 \to K^-\pi^+, K^-K^+$; and $\Lambda_b \to p\pi^-, pK^-$. Only limited experimental data is available concerning these decays, but their study is important. For example, the decay $B_d^0 \to \pi^+\pi^-$ is expected to play a prominent role in future CP violation studies.

Recent measurements[1] with the CLEO II detector have found the first evidence for charmless hadronic B_d^0 decays. The CLEO Collaboration measures a combined branching ratio $BR(B_d^0 \to \pi^+\pi^-, K^+\pi^-) = (2.4^{+0.8}_{-0.7} \pm 0.2) \times 10^{-5}$, but does not obtain statistically significant signals in the two individual channels. A search at LEP is of interest, as it may provide independent confirmation of the CLEO results, and give new information on B_s^0 and Λ_b decays.

2. Summary of the Analysis Method

Charmless hadronic beauty decays may be identified by looking for two oppositely charged tracks, originating from a displaced vertex, which have an invariant mass above the kinematic limit for any two tracks coming from beauty decays containing charm particles. Such events were selected using a combination of cuts based on kinematics and topological vertexing.

Kinematic Cuts

- Each track was required to have a momentum above 3 GeV/c

- Neither track was an identified lepton

- The momentum sum of the two tracks had to be at least 20 GeV/c

Topological Vertexing

- Each track was required to have at least one three-dimensional coordinate in the vertex detector

- The three-dimensional impact parameter of each track had to be inconsistent, by more than $3\,\sigma$, with the interaction point

- Each track had to cross the flight path of the beauty hadron candidate in front of the interaction point

- The two tracks had to form a consistent vertex with a χ^2 probability $> 1\%$

- Their vertex had to be at least $6\,\sigma$ in front of the interaction point

- The beauty hadron candidate had to point back from its vertex to the interaction point with a χ^2 probability $> 1\%$

Monte Carlo studies have shown that the maximum invariant mass for two tracks coming from a b meson decay involving charm, which pass the above kinematic cuts, and with each track assigned the pion mass, is around 4.8 GeV/c^2. The corresponding limit from b baryon decay is 5.0 GeV/c^2. A signal region was defined by requiring that the invariant mass of the two tracks, under the pion hypothesis, had to be at least $3\,\sigma$ above 5 GeV/c^2, and when one track was given the proton hypothesis the mass had to be less than $3\,\sigma$ above the Λ_b mass (the ± 50 MeV/c^2 uncertainty[2] on the Λ_b mass was added in quadrature). The efficiency of these cuts was measured using Monte Carlo to be $(27.5 \pm 0.7)\%$.

3. Results

The above selection criteria were applied to a sample of 1.55 million observed hadronic Z^0 decays (corresponding to 1.63 million hadronic Z^0 decays produced). Each event was required to have a precisely determined three-dimensional interaction point. The invariant mass distribution, under the pion hypothesis, is shown for both unlike-sign and like-sign pairs in Figure 1. In the unlike-sign pair sample there are three events which satisfy the selection criteria outlined above, with no events above the signal region. In the like-sign pairs there is one event with a mass of 5.73 GeV/c^2, under the pion hypothesis.

A background estimate was obtained using Monte Carlo simulation. Three background events, one like-sign and two unlike-sign, were observed in the mass region

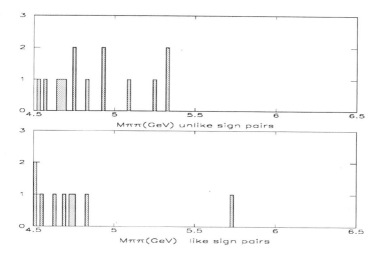

Figure 1: Invariant mass distribution, under the pion hypothesis, for both unlike-sign and like-sign candidates.

between 5.0 and 6.5 GeV/c^2 in a sample equivalent to 36 million hadronic Z^0 decays. The ability of the Monte Carlo to reproduce accurately the level of background observed in the data was tested by studying events in which two like-sign tracks pass the kinematic cuts, and each track has at least one three-dimensional vertex detector hit and a significant ($> 3\sigma$) positive impact parameter. In the mass region between 5.0 and 6.5 GeV/c^2 there are 33 ± 6 events observed in the data compared with a Monte Carlo prediction of 17 ± 3. This underestimate of the background in the Monte Carlo, by a factor of two, was confirmed with more statistics by including also tracks with significant negative impact parameters. To be conservative it was assumed that the probability to observe a background event in the data was two times higher than in the Monte Carlo. It was also assumed that the background is roughly flat in the above mass region, and that like-sign and unlike-sign background events are equally produced (both these assumptions are true in both data and Monte Carlo for the loose cuts described above).

With these assumptions, the corresponding binomial probability for observing three or more background events in the data in the signal region, when three background events have been observed in the Monte Carlo sample, is 1.8×10^{-4}. The corresponding binomial probability for observing one or more background events in the data is 0.22, indicating consistency between data and Monte Carlo background estimates. With somewhat less statistical power, the data itself was used to estimate the background probability. The binomial probability for observing three or more events in the signal region, when a total of four events is observed, is 0.05.

Decay Hypothesis	Event 13366 1385	Event 16396 312	Event 22026 6311	# Events with $P_{\chi^2} > 0.1$	90% Confidence Level BR Upper Limit
$B_d \to \pi\pi$	0.53	0.20	0.06	2	7.5×10^{-5}
$B_d \to K\pi$	0.03	0.79	0.21	2	7.5×10^{-5}
$B_d \to KK$	< 0.01	< 0.01	< 0.01	0	3.2×10^{-5}
$B_d \to p\bar{p}$	< 0.01	< 0.01	< 0.01	0	3.2×10^{-5}
$B_s \to \pi\pi$	0.90	< 0.01	0.11	2	2.5×10^{-4}
$B_s \to K\pi$	0.18	0.06	0.86	2	2.5×10^{-4}
$B_s \to KK$	< 0.01	0.03	< 0.01	0	1.1×10^{-4}
$B_s \to p\bar{p}$	< 0.01	< 0.01	< 0.01	0	1.1×10^{-4}
$\Lambda_b \to p\pi$	0.01	< 0.01	0.03	0	1.6×10^{-4}
$\Lambda_b \to pK$	< 0.01	0.02	< 0.01	0	1.6×10^{-4}

Table 1: χ^2 probabilities P_{χ^2} for the candidate events for the different decay hypotheses together with the total number of candidate events which are consistent with each hypothesis. Also shown are the 90% confidence level upper limits obtained for the corresponding branching ratios.

The event sample selected above was used to set branching ratio upper limits for exclusive decays of beauty hadrons into two charged hadrons. For each individual final state a χ^2 probability was obtained from the invariant mass and mass resolution of the system, and the dE/dx values and resolutions of the two tracks, if available. This probability was required to be greater than 0.1 for the corresponding hypothesis. The consistency of the three candidate events with each given exclusive final state is shown in Table 1. From the total number of events consistent with each final state, the 90 % confidence level upper limits for the individual branching ratios were obtained. These limits were calculated assuming that $\Gamma(Z^0 \to b\bar{b})/\Gamma(Z^0 \to q\bar{q}) = 0.22$, and that the fraction of b quarks fragmenting into B_d, B_s, and Λ_b are 0.40, 0.12 and 0.08, respectively.

4. Conclusions

In a sample of 1.55 million hadronic Z^0 decays three candidates for charmless hadronic beauty decay have been observed. The probability that these events come from background sources has been estimated from Monte Carlo to be 1.8×10^{-4}, a significance of over $3.5\,\sigma$. Branching ratio upper limits, of order 10^{-4}, have been obtained for a variety of exclusive final states involving charmless hadronic decays of beauty hadrons.

References

1. CLEO Collab., M. Battle et al., Phys. Rev. Lett. **71** (1993) 3922.
2. Particle Data Group, Phys. Rev. D **45** (1992).

RADIATIVE LEPTONIC DECAYS OF HEAVY MESONS *

GUSTAVO BURDMAN

Fermi National Accelerator Laboratory, Batavia, Illinois 60510, USA

TERRY GOLDMAN

Los Alamos National Laboratory, Los Alamos, New Mexico 87545, USA

and

DANIEL WYLER

Institut für Theoretische Physik, Universität Zürich, Zürich, Switzerland

ABSTRACT

We compute the photon spectrum and the rate for the decays $B(D) \to l\nu_l\gamma$ These photonic modes constitute a potentially large background for the purely leptonic decays which are used to extract the heavy meson decay constants. While the rate for $D \to l\nu\gamma$ is small, the radiative decay in the B meson case could be of comparable magnitude or even larger than $B \to \mu\nu$. This would affect the determination of f_B if the τ channel cannot be identified. We obtain theoretical estimates for the photonic rates and disscuss their possible experimental implications.

The experimental difficulty in the measurement of purely leptonic decays of heavy pseudoscalars is due mostly to the effect of helicity suppression: back-to-back leptons must make a spin 0 final state, but the anti-neutrino is right-handed and forces the charged lepton to this helicity which introduces a factor of the lepton mass in the amplitude. In the end the decay rate is suppressed by the factor $(m_l/m_H)^2$, where m_H is the heavy meson mass. For instance for the B meson

$$\Gamma(B \to l\bar{\nu}_l) = \frac{G_F^2}{8\pi}|V_{ub}|^2 f_B^2 \left(\frac{m_l}{m_B}\right)^2 m_B^3 \left(1 - \frac{m_l^2}{m_B^2}\right), \tag{1}$$

where the pseudoscalar meson decay constant is defined by

$$\langle 0|\bar{q}\gamma_\mu(1 - \gamma_5)b|B(P)\rangle = if_B P_\mu \tag{2}$$

and analogous expressions can be written for D mesons. Thus, when the charged lepton is an electron the purely leptonic decay is practically inaccessible. At the other extreme,

---

*Presented by G. Burdman at DPF'94, Univ. of New Mexico, Albuquerque, New Mexico, August2-6, 1994.

Fig. 1. Structure Dependent contribution to $B^- \to \mu^- \bar{\nu} \gamma$.

There are other decays that indirectly involve the heavy meson decay constants. For instance, the decay $B \to \pi l \nu_l$ is expected to be largely dominated by a B^* pole diagram at very low recoiling pion energies.[1-3] This implies the presence of the vector-meson decay constant f_{B^*} which can be related to f_B by Heavy Quark Spin Symmetry.[7] However this region of phase space is difficult to access due to kinematic suppression.

Here we investigate the decay modes $B^-(D^-) \to l^- \nu_l \gamma$.[5] We will see that certain contributions to this mode avoid helicity suppression. Thus, what in principle could be regarded as a mere radiative correction to the purely leptonic decays has the potential to be of comparable magnitude and in some cases even much larger.

Given that the μ modes are the most interesting from the point of view of the extraction of heavy meson decay constants we will concentrate on them. The treatment for $l = e$ is analogous and the numerical differences between these two cases will be stressed when relevant.

The emission of a real photon in leptonic decays of heavy mesons can proceed via two mechanisms: Internal Bremsstrahlung (IB) and Structure Dependent (SD) photon emission.[6] The IB contributions include photon radiation from the decaying charged particle as well as from the charged lepton in the final state. The important feature is that they are still suppressed by a factor of the lepton mass.[5]

On the other hand, the SD diagrams involve the contributions from heavy intermediate states coupling to the initial heavy pseudoscalar and the photon. In Fig. 1 we show the contributions from vector and axial-vector mesons. The helicity suppression is avoided because the meson directly coupling to the lepton pair has spin one. These types of diagrams were previously considered in the context of light pseudoscalar decays, in particular in $\pi \to l \nu \gamma$ and $K \to l \nu \gamma$.[6]

For instance, the contribution corresponding to a vector-meson in Fig. 1 requires the knowledge of the $B^* B \gamma$ coupling. This is defined in the process $B^* \to B \gamma$ with the amplitude given by

$$\mathcal{A}(B^* \to B\gamma) = e \mu_V \epsilon^{\alpha\beta\gamma\delta} \epsilon_\alpha^* v_\beta k_\gamma \lambda_\delta \tag{3}$$

where v is the velocity four-vector of the decaying particle, k is the photon four-momentum, λ is the polarization four-vector of the B^* and μ_V is a constant characterizing the strength of the transition. In this case, however, one expects this coupling to have an energy dependence that suppresses it as the photon energy increases.

There will also be contributions from heavy axial-vector meson states ($J^P = 1^+$). Heavy Quark Symmetry (HQS) tells us that there will be two of such states, belonging to the $j_l = 1/2$ and $j_l = 3/2$ doublets (j_l is the spin of the light degrees of freedom).

At this point we do not differentiate them and call B_1 the generic axial-vecton meson. Their photonic couplings to the pseudoscalars is of the form

$$\mathcal{A}_A^{(i)}(B_1 \to B\gamma) = e\mu_A^i(v \cdot k\lambda \cdot \epsilon - v \cdot \epsilon\lambda\cdot, k) \tag{4}$$

with $i = 1/2, 3/2$.

Defining $x = 2E_\gamma/m_B$ and $y = 2E_l/m_B$ as the rescaled photon and charged lepton energies in the B rest frame the double differential decay rate is given by

$$\frac{d^2\Gamma}{dxdy} = \frac{G_F^2|V_{ub}|^2}{32\pi^2}\alpha m_B^3 f_{B^*}^2 \cdot \mu_V^2 \times \left\{ \frac{1}{(x + \frac{2\Delta}{m_B})^2} + \left(\sum_i \frac{\gamma_i}{x + \frac{2\Delta_i}{m_B}}\right)^2 \right\}$$
$$\times \left[y^2(1-x) + y(3x - x^2 - 2) + \frac{1}{2}(3x^2 - 4x - x^3 + 2) \right] \tag{5}$$

where we defined

$$\gamma_i = \frac{\mu_A^{(i)} f_A^i}{\mu_V f_{B^*}} \tag{6}$$

as the relative axial-vector to vector meson coulping strength and the f_A^i's are the axial-vector meson decay constants. The fact that there are no relations among decay constants of states of different spin-parity doublets will be reflected in the persistence of the unknown γ_i's when we normalize to $\Gamma(B \to \mu\nu)$. For now let us assume that μ_V and $\mu_A^{(i)}$ do not depend on the photon energy so we can integrate and compare with the purely leptonic decay. To this end we define the ratio

$$R_B^\mu = \frac{\Gamma(B \to \mu\nu\gamma)}{\Gamma(B \to \mu\nu)} \tag{7}$$

To have an idea of the potential importance of the photonic mode we take a definite value for the mass differences $\Delta_i = 600$ MeV as suggested by the charmed meson system.[8] This gives approximately

$$R_B^\mu \approx 2\mu_V^2 \left(1 + \frac{(\gamma_{1/2} + \gamma_{3/2})^2}{2}\right) \text{GeV}^2 \tag{8}$$

where we made use of the HQS relation $f_{B^*} = m_B f_B$. Eqn. (8) shows that unless there are unnaturally small B-photon couplings, the photonic decay, which in principle could have been considered a small radiative correction, will dominate the leptonic decay or will be at least of comparable magnitude. Therefore it is important to have a good theoretical understanding of the couplings of all the relevant intermediate states in order to substract these events as a background for $B \to \mu\nu$. The fact that, to this order in $(m_\mu/m_B)^2$, the result of Eqn. (5) is independent of the lepton mass implies that

$$\Gamma(B \to e\nu\gamma) = \Gamma(B \to \mu\nu\gamma) \tag{9}$$

and that

$$\Gamma(B \to e\nu\gamma) \gg \Gamma(B \to e\nu) \tag{10}$$

allows for the separation of both effects by using the electronic modes as well. Thus integrating (5) over the photon energy and integrating the resulting μ spectrum around the end point over a region the size of the experimental μ energy resolution in $B \to \mu\nu$ will eliminate the background. Perhaps even more interesting, given the richness of the physics that enters in them, is the possibility of observing the photonic modes at branching ratios that will soon be accesible ($\approx 10^{-5} - 10^{-6}$).

The treatement of charmed mesons is entirely analogous to B mesons. With the obvious replacements in (7) we obtain

$$R_D^\mu \approx 4 \times 10^{-2} \mu_V^2 \left(1 + \frac{(\gamma_{1/2} + \gamma_{3/2})^2}{2}\right), \text{GeV}^2 \tag{11}$$

which is suppressed by the factor $(m_D/m_B)^4$ in rescaling (7). Therefore the effect is, as expected, less spectacular in the D mesons, although it potentially important.

Estimates of μ_V and the γ_i's can be obtained in variuos different approaches. The simplest prediction is that of the Non-relativistic Quark Model (NRQM) and is given by

$$\mu_V = \frac{Q_Q}{m_Q} + \frac{Q_q}{m_q}. \tag{12}$$

The first term if fixed by HQS whereas the second is the NRQM coupling of the photon to the light degrees of freedom. Here m_q has to be taken as a constituent quark mass (≈ 300 MeV) which gives

$$\mu_V \approx 2GeV^{-1}, \tag{13}$$

indicating that the ratio (7) could be sizeable. The axial-vector meson couplings are expected to be of the same order as μ_V. In fact the NRQM gives $\mu_A^{(1/2)} = \mu_V/\sqrt{3}$ and $\mu_A^{(3/2)} = \sqrt{2/3}\mu_V$, where the superscripts indicate an axial-vector meson belonging to the $(0^+, 1^+)$ and $(1^+, 2^+)$ doublets repectively. These contributions are also proportional to f_A^i, the axial-vector meson decay constants.

Corrections to this simple picture can be addressed in an effective theory incorporating HQS and Chiral Symmetry.[2,3] In this theory one-loop diagrams where the coupling to the photon occurs via the goldstone bosons in the loop, could give large corrections to (12).[5] Althought these corrections somehow reduce the value of μ_V, in principle could enhance $\mu_A^{(i)}$.

An important question in terms of the phenomenological impact of these decays is the possibility of a form-factor suppression affecting μ_V and $\mu_A^{(i)}$. This would soften the photon spectrum, that peaks at an energy of $\approx 1.2\text{GeV}$ when the couplings are approximated to be constant. When an energy suppresion is taken into account the average photon energy would drop typically to a few hundred MeV. This suppression would result in a smaller value of R_B^μ by a factor that strongly depends on the energy dependence chosen. Typical energy dependences would reduce R_B^μ in (7) by a factor of 2 to 4. The first estimate is obtained with a monopole type suppression whereas the second case corresponds to an exponential suppression. In both cases the energy scale was chosen to be 1GeV.

To summarize, for the charmed mesons D and D_s the effect can be a fraction of the purely muonic decay. Allowing for a typical form-factor suppression in μ_V and $\mu_A^{(i)}$ and assuming a large range for the axial-vector decay constants we have

$$R_D^\mu \approx (1 - 10) \times 10^{-2} \mu_V^2 , \text{GeV}^2 \tag{14}$$

which would translate into $Br(D_s \to \mu\nu\gamma) \approx 10^{-5} - 10^{-4}$ and $Br(D \to \mu\nu\gamma) \approx 10^{-6} - 10^{-5}$.

The situation is very different for the B meson decays. There we obtain

$$R_B^\mu \approx (0.5 - 6)\mu_V^2 \text{GeV}^2. \tag{15}$$

Therefore, it is reasonable to expect $R_B^\mu = 1 - 20$, which would translate into branching fractions that could be in the order of 10^{-6}.

There are in principle several other decays of heavy mesons that proceed via the SD mechanism. For instance, the decay $B_s \to \mu^+\mu^-\gamma$ will be enhanced over the helicity suppressed $B_s \to \mu^+\mu^-$ by an expression similar to (15). On the other hand, $B \to D^{(*)}\gamma l\nu$ and any other exclusive semileptonic decay with a photon added to the final state can be considered. Similar results may hold for non-leptonic decays. Although these decays are indeed suppressed by α relative to the corresponding non-photonic decays, some of them are of interest in their own right and might became observable in the near future.

References

1. N. Isgur and M. B. Wise, Phys. Rev. **D41**, 151 (1990).
2. M.B. Wise, Phys. Rev. **D45**, 2188 (1992).
3. G. Burdman and J.F. Donoghue, Phys. Lett. **B280**, 287 (1992).
4. T. M. Yan, H. Y. Chen, C. Y. Cheung, G. L. Lin, Y. C. Lin and H. L. Yu, Phys. Rev. **D46**, 1148 (1992).
5. G. Burdman, T. Goldman and D. Wyler, preprint FERMILAB-Pub-94/120-T, submitted to Phys. Rev. **D**.
6. T. Goldman and W. J. Wilson, Phys. Rev. **D15**, 709 (1977), and references therein.
7. N. Isgur and M.B. Wise, Phys. Lett **B237**, 527 (1990).
8. Particle Data Group, Phys. Rev. **D45**, (1990).

SPONTANEOUS CP VIOLATION AND CP ASYMMETRIES IN B DECAYS

C. N. LEUNG

Department of Physics and Astronomy, University of Delaware,
Newark, DE 19716

ABSTRACT

CP asymmetries in B decays are computed in the aspon model of spontaneous CP violation. These asymmetries are generally smaller than those predicted for the standard model by two orders of magnitude or more. Thus, the model may be ruled out if these asymmetries are observed in the upcoming B factories.

The study of CP violation involves two fundamental questions: what is the underlying mechanism of CP violation in the weak interaction (so far observed only in the neutral kaon system) and why is CP violation not observed in the strong interaction (the strong CP problem[1])? In the standard model with three fermion families, CP is explicitly broken in the weak interaction lagrangian by the Kobayashi-Maskawa mechanism.[2] In this case the strong CP problem may be solved by assuming a massless up quark or by invoking the Peccei-Quinn mechanism.[3] The first solution seems to conflict with current algebra results, while the second solution requires a fine tuning to hide the axion, the pseudo-Goldstone boson of the spontaneously broken Peccei-Quinn symmetry.

An alternative solution of the strong CP problem is to assume that CP is only spontaneously broken. Then the physical $\bar{\theta} = \theta_{QCD} + Arg(detM_u + detM_d)$ can be made vanishing at tree level provided that the up-quark mass matrix, M_u, and the down-quark mass matrix, M_d, have real determinants, and a small $\bar{\theta}$ (less than 2×10^{-10}) can be achieved by assuring that the loop corrections to $\bar{\theta}$ are sufficiently small. This can be naturally implemented in grand unified models via the Nelson-Barr mechanism.[4]

A variant of the Nelson-Barr models, the aspon model,[5] was introduced several years ago. The model contains a new $U(1)_X$ symmetry. Unlike the Peccei-Quinn symmetry, new fermions are introduced such that $U(1)_X$ is anomaly free and hence can be gauged. Scalar fields are introduced whose vacuum expectation values spontaneously break CP as well as the $U(1)_X$ symmetry, resulting in a massive vector boson, the aspon.

This talk reports on a recent study[6] of the prospects for distinguishing the aspon model from the standard model through CP asymmetry measurements in B decays which will be undertaken in future B factories. We shall consider the specific model described in Ref. 7.

The time-dependent CP asymmetry is defined as

$$a_f(t) = \frac{\Gamma(B^0(t) \to f) - \Gamma(\bar{B}^0(t) \to f)}{\Gamma(B^0(t) \to f) + \Gamma(\bar{B}^0(t) \to f)} \tag{1}$$

where the final state f is a CP eigenstate. We consider the specific examples of $f = \pi^+\pi^-$ and ψK_S from B_d decay and $f = \rho K_S$ from B_s decay. Let us define $\lambda(f)$ by

$$\lambda(\pi^+\pi^-) = \left(\frac{q}{p}\right)_{B_d} \left(\frac{\bar{A}}{A}\right)_{B_d \to \pi^+\pi^-} \tag{2}$$

$$\lambda(\psi K_S) = \left(\frac{q}{p}\right)_{B_d} \left(\frac{\bar{A}}{A}\right)_{B_d \to \psi K} \left(\frac{q}{p}\right)_K \tag{3}$$

$$\lambda(\rho K_S) = \left(\frac{q}{p}\right)_{B_s} \left(\frac{\bar{A}}{A}\right)_{B_s \to \rho K} \left(\frac{q}{p}\right)_K^* \tag{4}$$

where q, p are the mixing parameters in the B^0–$\bar{B}^0$ system ($B_{1,2}$ are the mass eigenstates): $|B_{1,2}\rangle = p|B^0\rangle \pm q|\bar{B}^0\rangle$, and similarly for the neutral kaon system, and A, $\bar{A}$ are the decay amplitudes: $A, \bar{A} = \langle f|H|B^0, \bar{B}^0\rangle$. If $|q/p| = 1$ and $|\bar{A}/A| = 1$, then $\lambda(f)$ is related to the CP asymmetry through the $B_1 - B_2$ mass difference ΔM by

$$a_f(t) = -\text{Im}\lambda(f)\sin(\Delta M t). \tag{5}$$

It can be shown for the standard model that[8]: $\text{Im}\lambda(\pi^+\pi^-) = \sin 2\alpha$, $\text{Im}\lambda(\psi K_S) = -\sin 2\beta$, and $\text{Im}\lambda(\rho K_S) = -\sin 2\gamma$, where α, β, and γ are the angles of the unitarity triangle defined by the relation

$$V_{ub}^* V_{ud} + V_{tb}^* V_{td} + V_{cb}^* V_{cd} = 0. \tag{6}$$

Here V is the CKM matrix. Such relations no longer hold in the aspon model.

As will be shown below, Eq. 5 is also valid in the aspon model. Consequently, to evaluate the CP asymmetries for the aspon model we need only compute the different factors in the $\lambda(f)$ given in Eqs. 2–4 above. Consider first the three mixing factors (q/p) for the B_d, B_s, and K neutral meson systems. They are given by

$$\left(\frac{q}{p}\right)_\xi = \left(\frac{M_{12}^*(\xi)}{M_{12}(\xi)}\right)^{\frac{1}{2}} \tag{7}$$

where $M_{12}(\xi)$ is the amplitudes for $\bar{\xi} \to \xi$. The aspon model introduces additional Feynman diagrams involving aspon exchange to those involving W exchange already present in the standard model. For instance, $M_{12}(\xi)$ have contributions from both one-loop W^+W^- exchange (first term) and tree-level aspon exchange (second term) amplitudes:

$$M_{12}(B_d) \propto \frac{G_F m_t^2}{32\pi^2} f(z_t) (V_{td}^* V_{tb})^2 \frac{g^2}{M_W^2} + (x_3^* x_1)^2 \frac{g_A^2}{M_A^2} \tag{8}$$

$$M_{12}(B_s) \quad \propto \quad \frac{G_F m_t^2}{32\pi^2} f(z_t) \left(V_{ts}^* V_{tb}\right)^2 \frac{g^2}{M_W^2} + \left(x_3^* x_2\right)^2 \frac{g_A^2}{M_A^2} \tag{9}$$

$$M_{12}(K) \quad \propto \quad \frac{G_F m_c^2}{32\pi^2} \left(V_{cd}^* V_{cs}\right)^2 \frac{g^2}{M_W^2} + \left(x_2^* x_1\right)^2 \frac{g_A^2}{M_A^2} \tag{10}$$

where m_t (m_c) is the mass of the top (charm) quark, $z_t = m_t^2/M_W^2$, g is the $SU(2)_L$ coupling constant, g_A is the new $U(1)_X$ coupling, M_A is the mass of the aspon, and the x_i are given in the notation of Ref. 7 as $x_i = F_i/M$. The function $f(z_t)$ in Eqs. 8–9 is defined as

$$f(z_t) = \frac{1}{4} \left[1 + \frac{9}{1 - z_t} - \frac{6}{(1 - z_t)^2} - \frac{6 z_t^2 \ln z_t}{(1 - z_t)^3} \right] \tag{11}$$

The decay amplitudes A also have both W-exchange (first term) and aspon-exchange (second term) contributions:

$$A_{B_d \to \pi^+ \pi^-} \quad \propto \quad 3 \left(V_{ud} V_{ub}^*\right) \frac{g^2}{M_W^2} + x_1^* x_3 |\tilde{x}_1|^2 \frac{g_A^2}{M_A^2} \zeta_{\pi\pi} \tag{12}$$

$$A_{B_d \to \psi K} \quad \propto \quad \left(V_{cs} V_{cb}^*\right) \frac{g^2}{M_W^2} + 3 x_2^* x_3 |\tilde{x}_2|^2 \frac{g_A^2}{M_A^2} \zeta_{\psi K} \tag{13}$$

$$A_{B_s \to \rho K} \quad \propto \quad \left(V_{ud} V_{ub}^*\right) \frac{g^2}{M_W^2} + x_1^* x_3 |\tilde{x}_1|^2 \frac{g_A^2}{M_A^2} \zeta_{\rho K} \tag{14}$$

and similarly for the amplitudes $\bar{A}$. The ζ factors, which are of order 1, account for the fact that the aspon-exchange and the W-exchange amplitudes involve different strong interaction dynamics. The factors of 3 in Eqs. 12–13 are color factors. In Eqs. 12-14, $\tilde{x}_i = C_{ij} x_j$, where C is a real matrix related to the CKM matrix of the aspon model (see Refs. 6 and 7 for details). In all cases, W-exchange dominates both the real and imaginary parts of the decay amplitudes. Hence, $|\bar{A}/A| = 1$ and, together with Eq. 7, we see that Eq. 5 is indeed valid for the aspon model.

Since CP is spontaneously broken, the CKM matrix elements that appear in the W-exchange contributions in Eqs. 8-10 and Eqs. 12- 14 are predominantly real. Consequently the W-exchange amplitudes have very small phases. In comparison, the aspon contributions have a much smaller magnitude but an unpredicted arbitrary phase. As a result, $|\mathrm{Im}\lambda(f)|$ and correspondingly the CP asymmetries in the aspon model are expected to be smaller than that in the standard model. Indeed we find

$$|\mathrm{Im}\lambda(\pi^+\pi^-)| \quad \lesssim \quad 1 \times 10^{-5} \tag{15}$$

$$|\mathrm{Im}\lambda(\psi K_S)| \quad \lesssim \quad 2 \times 10^{-3} \tag{16}$$

$$|\mathrm{Im}\lambda(\rho K_S)| \quad \lesssim \quad 2 \times 10^{-3} \tag{17}$$

The numerical values are obtained by using the central values of the V_{ij} listed in the Review of Particle Properties and an aspon mass of 300 GeV, and the magnitudes of the x_i have been restricted by the upper limit on $\bar{\theta}$ to be*: $|x_1| \lesssim 10^{-2}$, $|x_2| \lesssim 10^{-3}$, $|x_3| \lesssim 10^{-4}$. We note that the final states with a kaon typically have larger asymmetries.

These decays also have relatively small branching ratios (the upper limits are current experimental limits at 90% C.L.): BR($B_d \to \pi^+\pi^-$) < 2.9×10^{-5}, BR($B_d \to \psi K_S$) = $(7.5\pm2.1)\times10^{-4}$, and BR($B_s \to \rho K_S$) < 3.2×10^{-4}. In fact, the branching ratio for the decay $B_s \to \rho K_S$ is estimated[10] to be 5×10^{-7}. Thus, if the CP asymmetries in B decays discussed here are observed in the planned B factories, the aspon model may be ruled out. On the other hand, if the CP asymmetries are not observed at the level expected for the standard model, it may be a signal that spontaneous CP violation is the underlying mechanism of CP violation.

We mention in closing the relation

$$
\begin{aligned}
\frac{\lambda(\psi K_S)\lambda(\rho K_S)}{\lambda(\pi^+\pi^-)} &= \left(\frac{q}{p}\right)_{B_s}\left(\frac{\bar{A}}{A}\right)_{B_d \to \psi K} \\
&= \left(\frac{q}{p}\right)_{B_s}\left(\frac{\bar{A}}{A}\right)_{B_s \to D_s^+ D_s^-} \\
&= \lambda(D_s^+ D_s^-)
\end{aligned}
\tag{18}
$$

which can be derived from Eqs. 2-4 by virtue of the fact that the aspon-exchange terms in Eqs. 12-14 are negligible. This relation also holds in the standard model. In view of the large branching ratio (of order 10^{-2}) for the decay $B_s \to D_s^+ D_s^-$, it may be more effective to measure the CP asymmetry for this decay (instead of the decay $B_s \to \rho K_S$ which has a much smaller branching ratio) and then use Eq. 18 to determine the CP asymmetry for $B_s \to \rho K_S$ and the angle γ of the unitarity triangle.

References

1. For a review of the strong CP problem, see R. D. Peccei, in *CP Violation*, ed. C. Jarlskog (World Scientific, Singapore, 1989) p. 503.
2. M. Kobayashi and T. Maskawa, *Prog. Theor. Phys.* **49** (1973) 652.
3. R. D. Peccei and H. Quinn, *Phys. Rev. Lett.* **38** (1977) 1440.
4. A. Nelson, *Phys. Lett.* **136B** (1984) 387; S. M. Barr, *Phys. Rev.* **D37** (1984) 1805.
5. P. H. Frampton and T. W. Kephart, *Phys. Rev. Lett.* **66** (1991) 1666.
6. A. W. Ackley, P. H. Frampton, B. Kayser and C. N. Leung, *Outcome from Spontaneous CP Violation for B Decays*, to appear in *Phys. Rev. D*.
7. P. H. Frampton and D. Ng, *Phys. Rev.* **D43** (1991) 3034.
8. See, e.g., Y. Nir, in *Perspectives in the Standard Model*, ed. R. K. Ellis, C. T. Hill and J. D. Lykken (World Scientific, Singapore, 1990) p. 339, and B. Kayser, in *Proceedings of the Second International Symposium on Particles, Strings, and Cosmology*, ed. P. Nath and S. Reucroft (World Scientific, Singapore, 1992) p. 837.
9. P. H. Frampton and T. W. Kephart, *Phys. Rev.* **D47** (1993) 3655.
10. R. Aleksan, B. Kayser and D. London, preprint DAPNIA/SPP 94-06 (NSF-PT-94-2, UdeM-LPN-TH-94-189), March, 1994.

---

*These limits are obtained with the assumption that the masses of the new heavy quarks are of the same order as the scale of spontaneous CP breaking. This is a weak assumption as the gauge invariant masses of the vector-like quarks are not protected to be light by any symmetry. The situation can be remedied by implementing the chiral aspon model discussed in Ref. 9.

DIRECT CP VIOLATION IN $b \rightarrow dJ/\psi$ DECAYS

ISARD DUNIETZ

Fermilab, P.O. Box 500, Batavia, IL 60510

and

JOÃO M. SOARES

TRIUMF, 4004 Wesbrook Mall, Vancouver, BC Canada V6T 2A3

ABSTRACT

Direct CP violation in self-tagging B-meson decays of the type $b \rightarrow dc\bar{c}$, where the charm-anticharm pair forms a J/ψ, is investigated. The CP asymmetry requires the contribution to the amplitude from decays into other states, which rescatter into dJ/ψ *via* final state interactions. In this case, the intermediate states that contribute are those with quark content $du\bar{u}$ or $dc\bar{c}$. The former contribution is OZI suppressed and gives an asymmetry of *a few* $\times 10^{-3}$, in the standard model. The latter can occur at order α_s, but it requires both tree and penguin contributions to the amplitude; it gives an asymmetry of about 1%. This is within close reach of the experiments that are planned for hadron colliders in the near future.

Direct CP violation may occur in decays of the type $b \rightarrow qc\bar{c}$ ($q = s, d$), where the charm-anticharm pair forms a J/ψ.[1] We find that the Standard Model predicts a CP asymmetry

$$a_{CP} = \frac{\Gamma(b \rightarrow dJ/\psi) - \Gamma(\bar{b} \rightarrow \bar{d}J/\psi)}{\Gamma(b \rightarrow dJ/\psi) + \Gamma(\bar{b} \rightarrow \bar{d}J/\psi)} \tag{1}$$

at the level of 1%[2,3] (whereas the analogous asymmetry in $b \rightarrow sJ/\psi$ is suppressed by a factor of $\sin^2 \theta_C$). This is roughly the expected reach of an experiment with a sample of 10^{10} B-mesons.[2] Because self-tagging decay modes can be used, and $J/\psi \rightarrow l^+l^-$ provides a very clean signature, these decays are particularly suitable for hadronic accelerators, where such a large number of B-mesons can be produced.

The amplitude for the decay $b \rightarrow dJ/\psi$, in the absence of final state scattering, is

$$A_{d\psi}^{(0)} = V_{cb}V_{cd}^* T_{d\psi} + V_{tb}V_{td}^* P_{d\psi}, \tag{2}$$

where the dominant tree amplitude, $T_{d\psi}$, and the small penguin term, $P_{d\psi}$, can be calculated from the effective Hamiltonian in ref. 4 (we have used the factorization prescription and dropped all non-leading terms in the $1/N_c$ expansion[5]). Due to the effect of final state interactions, the decay amplitude will have additional contributions from processes of the type $b \to i \to dJ/\psi$. When the intermediate state i is on-mass-shell, this contributes to the decay amplitude with an absorptive term; it corresponds to the convolution of the amplitude $A_i^{(0)}$, for the weak decay $b \to i$, with the amplitude $A(i \to dJ/\Psi)$, for the final state scattering. The $b \to dJ/\psi$ decay amplitude is then

$$A_{d\psi} = A_{d\psi}^{(0)} + i\frac{1}{2}\sum_i A_i^{(0)}A(i \to dJ/\Psi). \tag{3}$$

If the weak decay amplitudes $A_{d\psi}^{(0)}$ and $A_i^{(0)}$ have a relative CP-odd phase, the interference between the dispersive and absorptive terms in eq. 3 gives rise to the CP violating quantity

$$\Delta_{d\psi} \equiv |A_{d\psi}|^2 - |\bar{A}_{d\psi}|^2 = 2\sum_i Im\{A_{dJ/\psi}^{(0)}A_i^{(0)\dagger}\}A(i \to dJ/\psi). \tag{4}$$

The CP asymmetry in eq. 1 is then

$$a_{CP} = \frac{|A_{d\psi}|^2 - |\bar{A}_{d\psi}|^2}{2|A_{d\psi}^{(0)}|^2} \tag{5}$$

(the denominator was approximated for the case where the dispersive amplitude in eq. 3 is dominant).

The intermediate states that are relevant to this case are $i = du\bar{u}$ and $dc\bar{c}$; they contribute in very different ways to the CP asymmetry. The weak decay amplitude into the intermediate state $du\bar{u}$ is

$$A_{du\bar{u}}^{(0)} = V_{ub}V_{ud}^*T_{du\bar{u}} \tag{6}$$

(the small penguin amplitudes in $A_{d\psi}^{(0)}$ or $A_{du\bar{u}}^{(0)}$ are irrelevant for the contribution of $du\bar{u}$ to the CP asymmetry). The corresponding term in eq. 4 is then

$$\Delta_{d\psi}^{du\bar{u}} = 2Im\{V_{cb}V_{cd}^*V_{ub}^*V_{ud}\}T_{d\psi}T_{du\bar{u}}A(u\bar{u} \to J/\psi), \tag{7}$$

where the final state scattering $u\bar{u} \to J/\psi$ is OZI suppressed: it can only occur at the order α_s^3, or via the electromagnetic interaction. We have calculated the electromagnetic 1-photon scattering diagram and found that the resulting CP asymmetry is

$$a_{CP} = \eta \times 0.7\%, \tag{8}$$

where the CKM parameter η ranges presently between 0.1 and 0.6.[6] A rough estimate shows that the order α_s^3 QCD scattering gives a contribution of very similar strength.

In the case of the intermediate state $dc\bar{c}$, the final state scattering does not suffer from the OZI suppression, and it is dominated by the QCD scattering at the order α_s: a $c\bar{c}$ pair produced in a color octet state scatters to J/ψ by exchanging a gluon with the d-quark. Notice that the absorptive amplitude due to the width of the J/ψ cannot contribute to the CP asymmetry as it would violate CPT[7]; other intermediate states with $c\bar{c}$ in a color singlet can only contribute at higher orders in α_s, and they will be neglected. The amplitude for the weak decay $b \to d(c\bar{c})_8$, where the charm-anticharm pair forms a color octet, is

$$A_8^{(0)} = V_{cb}V_{cd}^*(T_8^V + T_8^A) + V_{tb}V_{td}^*(P_8^V + P_8^A). \tag{9}$$

The different symbols designate tree (T_8) and penguin (P_8) terms, and terms that correspond to the $c\bar{c}$ pair in a vector (superscript V) and in an axial-vector (superscript A) state. They are obtained from the same effective Hamiltonian as before. The terms T_8^V and P_8^V do not contribute to the absorptive part of the $b \to dJ/\psi$ amplitude (at this order in α_s): according to Furry's theorem, if the $c\bar{c}$ pair is in a vector state, it cannot scatter to the J/ψ by exchanging a gluon with the d-quark. The contribution of the intermediate state $dc\bar{c}$ to eq. 4 is then

$$\Delta_{d\psi}^{dc\bar{c}} = 2Im\{V_{cb}V_{cd}^*V_{tb}^*V_{td}\}A(d(c\bar{c})_8 \to dJ/\psi)(P_8^A T_{d\psi} - T_8^A P_{d\psi}). \tag{10}$$

Here the suppression comes from the need to include the penguin amplitudes in the weak decays, so that a relative CP-odd phase appears between the dispersive and absorptive amplitudes.[7] From eq. 10, one obtains a CP asymmetry

$$a_{CP} = \eta \times 2.8\%. \tag{11}$$

We wish to thank Lincoln Wolfenstein, Howard Trottier and Per Ernstrom for enlightening discussions and helpful criticism. This work was partly supported by the Natural Science and Engineering Research Council of Canada.

References

1. I. Dunietz, Phys. Lett. **B316**, 561 (1993).
2. I. Dunietz and J. M. Soares, Phys. Rev. **D 49**, 5904 (1994).
3. J. M. Soares, TRIUMF report No. TRI-PP-94-20, 1994.
4. See, for example, A. J. Buras *et al.*, Nucl. Phys. **B370**, 69 (1992); **B375**, 501 (1992).
5. M. Wirbel, B. Stech and M. Bauer, Z. Phys. **C29**, 637 (1985).
6. See, for example, B. Winstein and L. Wolfenstein, Rev. Mod. Phys. **65**, 1113 (1993).
7. L. Wolfenstein, Phys. Rev. **D 43**, 151 (1991).

PATI-SALAM MODEL AND RARE MESON DECAYS

G. VALENCIA

Department of Physics, Iowa State University, Ames, IA 50011

and

S. WILLENBROCK*

Department of Physics, University of Illinois, 1110 W. Green St., Urbana, IL 61801

ABSTRACT

We study meson decays mediated by the heavy gauge bosons of the Pati-Salam model of quark-lepton unification. We consider the scenarios in which the τ lepton is associated with the third, second, and first generation of quarks. The most sensitive probes, depending on the scenario, are rare K, π, and B decays.

1. Introduction

One of the unexplained features of the standard model of the strong and electroweak interactions is why some fermions, the quarks, experience the strong interaction while others, the leptons, do not. Experience has taught us to look for symmetry even when it is not apparent, and this leads one to speculate that, at some deeper level, quarks and leptons are identical. Perhaps there exists a symmetry between quarks and leptons which is broken at high energy, in much the same way that the electroweak symmetry is broken at an energy scale of $(\sqrt{2}G_F)^{-1/2} \approx 250$ GeV.

If we further speculate that the quark-lepton symmetry is a local gauge symmetry, we are led to predict a new force of nature which mediates transitions between leptons and quarks. The simplest model which incorporates this idea is the Pati-Salam model,[1] based on the group $SU(4)_c$. The subgroup $SU(3)_c$ is the ordinary strong interaction, and lepton number is the fourth "color". At some high energy scale, the group $SU(4)_c$ is spontaneously broken to $SU(3)_c$, liberating the leptons from the influence of the strong interaction and breaking the symmetry between quarks and leptons.

In a recent paper we explore signals for quark-lepton unification à la Pati-Salam.[2] We show that rare K, π, and B decays are the most sensitive probes of the presence of quark-lepton transitions mediated by heavy Pati-Salam bosons. A new feature of our analysis is that we do not restrict ourselves to the assumption that the τ lepton is associated with the third generation of quarks, but also consider the possibility that it is associated with the second, or even first, generation. Another recent paper on Pati-Salam bosons also considers this possibility.[3] Our analyses overlap for $K_L \rightarrow \mu^{\pm}e^{\mp}$ and

---

*Presenter

$\Gamma(\pi^+ \to e^+\nu)/\Gamma(\pi^+ \to \mu^+\nu)$, and agree. We further show that $\Gamma(K^+ \to e^+\nu)/\Gamma(K^+ \to \mu^+\nu)$ and rare B decays are the most sensitive probes in the scenario in which the τ lepton is associated with the first generation of quarks.

Pati and Salam proposed a class of unified models which incorporate quark-lepton unification.[1] A common feature of these models is the group $SU(4)_c$, with the subgroup $SU(3)_c$ corresponding to the strong interaction, and with lepton number identified as the fourth "color". We consider the minimal model which embodies quark-lepton unification via the $SU(4)_c$ Pati-Salam group, based on the group $SU(4)_c \times SU(2)_L \times U(1)_{T_{3R}}$.

The particle content of the $SU(4)_c \times SU(2)_L \times U(1)_{T_{3R}}$ model is

$$\begin{pmatrix} u_R & u_G & u_B & \nu \\ d_R & d_G & d_B & e \end{pmatrix}_L \quad (4, 2, 0)$$

$$\begin{pmatrix} u_R & u_G & u_B & \nu \end{pmatrix}_R \quad (4, 1, +\tfrac{1}{2})$$

$$\begin{pmatrix} d_R & d_G & d_B & e \end{pmatrix}_R \quad (4, 1, -\tfrac{1}{2})$$

where the subscripts on the quarks denote color (red, green, blue), and the subscripts L, R denote chirality. The model is free of gauge and mixed gravitational anomalies. The $U(1)_{T_{3R}}$ quantum numbers of the $SU(2)_L$ singlet fields, $\pm\tfrac{1}{2}$, suggest that $U(1)_{T_{3R}}$ is a subgroup of an $SU(2)_R$ group; hence the notation. We will not make this additional assumption, since it does not affect our analysis.

One canonically associates the τ lepton with the third generation of quarks, both for reasons of mass (they are the heaviest known fermions of their respective classes), and for historical reasons (the τ lepton and the b quark were the last fundamental fermions discovered; evidence for the top quark has recently been presented). This is certainly a natural assumption. However, the flavor-symmetry-breaking mechanism, which is responsible for fermion mass generation, is a mystery. One should keep an open mind to the possibility that the τ lepton is actually associated with the second or first generation of quarks.

Generically, one would expect that there is a mixing matrix, analogous to the Cabibbo-Kobayashi-Maskawa (CKM) matrix, which describes the mixing of the lepton generations with the quark generations. We will make the assumption that this matrix is nearly diagonal, as is the CKM matrix, but consider the scenarios where the τ lepton is most closely associated with the third, second or first generations in the following sections.

Because the Pati-Salam interaction conserves $B-L$ and fermion number, it cannot mediate nucleon decay. Purely leptonic transitions, such as $\mu \to e\gamma$ and $\mu N \to eN$, and meson-antimeson mixing, are induced only at one loop, and vanish in the limit of zero intergenerational mixing. The natural place to search for the Pati-Salam interaction is therefore in meson decays.

2. Tau lepton associated with third generation of quarks

The long-lived kaon, due to its longevity, is a sensitive probe of suppressed interactions which produce unusual decays. Pati and Salam observed that the decay $K_L \to \mu^\pm e^\mp$, shown in Fig. 1, provides the best bound on the mass of the Pati-Salam bosons.[1] Here we update this bound, based on the recent upper bound BR $(K_L \to \mu^\pm e^\mp) < 3.9 \times 10^{-11}$ (90% C.L.) from Brookhaven E791.[4] Combined with previous experiments, this yields

$$BR(K_L \to \mu^\pm e^\mp) < 3.3 \times 10^{-11} \quad (90\% \text{ C.L.}) . \tag{1}$$

This implies a lower bound on the mass of the Pati-Salam boson of

$$M_c > 1400 \text{ TeV.} \tag{2}$$

It is remarkable that physics at such a high scale can be probed by this decay. Future experiments may probe branching ratios as small as 10^{-14}, increasing the lower bound on M_c by a factor of about 7.

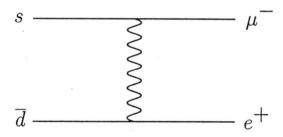

Fig. 1. Diagram for $\overline{K}^0 \to \mu^- e^+$, mediated by a heavy Pati-Salam boson.

3. Tau lepton associated with second generation of quarks

At first sight, associating the τ lepton with the second generation of quarks and, say, the muon with the third generation seems unnatural. However, the τ lepton is comparable in mass to the second-generation charm quark. Although the muon is a factor of about 40 less massive than the bottom quark, the bottom quark is at least a factor of 30 less massive than the top quark ($m_t > 131$ GeV), so large intragenerational mass ratios do occur.

Because the strange quark is associated with the τ lepton, the decay of K_L to leptons does not occur via the Pati-Salam interaction. Pati-Salam bosons also mediate transitions between up quarks and neutrinos, so if we replace the s quark and muon in Fig. 1 with an up quark and electron neutrino, we obtain the diagram for $\pi^+ \to e^+ \nu_e$. This process involves only first-generation quarks. Since the decay $\pi^+ \to e^+ \nu_e$ also proceeds via the weak interaction, the presence of a contribution from the Pati-Salam interaction manifests itself as a violation of lepton universality in $R_{e/\mu} = \Gamma(\pi^+ \to e^+\nu)/\Gamma(\pi^+ \to \mu^+\nu)$.[5] The theoretical prediction from the weak interaction is[6]

$$R_{e/\mu}^{theory} = (1.2352 \pm .0005) \times 10^{-4} \tag{3}$$

while the combined current experimental measurement is[7,8]

$$R_{e/\mu} = (1.2310 \pm .0037) \times 10^{-4} . \tag{4}$$

The theoretical uncertainty is much less than the experimental uncertainty.

The contribution of the Pati-Salam interaction to $\pi^+ \to e^+ \nu_e$ is obtained via the interference of the Pati-Salam and weak amplitudes. The absence of a deviation of the theoretical prediction from the experimental measurements yields a lower bound on the mass of the Pati-Salam boson of

$$M_c > 250 \text{ TeV} . \tag{5}$$

This bound is a factor of about five less stringent than the bound from $K_L \to \mu^\pm e^\mp$. Nevertheless, it is the strongest bound for the scenario considered here.

If we replace the s quark in Fig. 1 with a b quark, we obtain the diagram for $\overline{B}_d^0 \to \mu^- e^+$. The present upper bound on this decay from CLEO[9]

$$BR(\overline{B}_d^0 \to \mu^\pm e^\mp) < 5.9 \times 10^{-6} \tag{6}$$

places a lower bound on the Pati-Salam-boson mass of

$$M_c > 16 \text{ TeV} . \tag{7}$$

The upper bound on this decay can be significantly improved with B_d^0 mesons produced in hadron colliders. A lower bound on the branching ratio of 10^{-9} translates into $M_c > 140$ TeV.

4.   Conclusions

In a recent paper we have studied rare meson decays induced by the heavy gauge bosons of the Pati-Salam model of quark-lepton unification.[2] We have considered the scenarios in which the leptons are associated with the quark generations in all six permutations. The lower bounds obtained on the mass of the Pati-Salam bosons are given in Table 1. Bounds from $K_L \to \mu^\pm e^\mp$ and lepton universality in charged pions decays are well known, and we have updated them. We have shown that in the two scenarios in which the τ lepton is associated with the first generation of quarks, the best bounds come from $B^+ \to e^+\nu$ and lepton universality in charged kaon decays. All of these measurements have the potential for improvement.

At present, the bounds from $B_d^0, B_s^0 \to \mu^\pm e^\mp$ are not the strongest in any of the scenarios. However, the large number of these mesons which are produced in hadron colliders can potentially be used to probe branching ratios as small as 10^{-9}. The resulting bound on the Pati-Salam-boson mass would be the best for three of the scenarios. A high-resolution silicon vertex detector is essential for such a measurement.

Table 1. Lower bound on the mass of the Pati-Salam boson (TeV) from rare K, π, and B decays. The first column indicates how the leptons are associated with the first, second, and third generation of quarks. The best bound for each scenario is enclosed in a box. The bounds assuming $BR(B_d^0, B_s^0 \to \mu^\pm e^\mp) < 10^{-9}$ are shown in parentheses. A dash indicates the decay does not occur via the Pati-Salam interaction.

	$K_L \to \mu^\pm e^\mp$	$\frac{\pi^+ \to e^+\nu}{\pi^+ \to \mu^+\nu}$	$\frac{K^+ \to e^+\nu}{K^+ \to \mu^+\nu}$	$B_d^0 \to \mu^\pm e^\mp$	$B_s^0 \to \mu^\pm e^\mp$	$B^+ \to e^+\nu, \mu^+\nu$
$e\mu\tau$	1400	250	4.9	-	-	-
$\mu e\tau$	1400	76	130	-	-	-
$e\tau\mu$	-	250	-	16(140)	-	12
$\mu\tau e$	-	76	-	16(140)	-	13
$\tau\mu e$	-	-	4.9	-	(140)	13
$\tau e\mu$	-	-	130	-	(140)	12

References

1. J. Pati and A. Salam, Phys. Rev. D **10**, 275 (1974).
2. G. Valencia and S. Willenbrock, ILL-(TH)-94-17, to appear in Phys. Rev. D.
3. A. Kuznetsov and N. Mikheev, Phys. Lett. **B329**, 295 (1994).
4. K. Arisaka *et al.*, Phys. Rev. Lett. **70**, 1049 (1993).
5. O. Shanker, Nucl. Phys. **B204**, 375 (1982).
6. W. Marciano and A. Sirlin, Phys. Rev. Lett. **71**, 3629 (1993).
7. D. Britton *et al.*, Phys. Rev. Lett. **68**, 3000 (1992).
8. C. Czapek *et al.*, Phys. Rev. Lett. **70**, 17 (1993).
9. R. Ammar *et al.*, Phys. Rev. D **49**, 5701 (1994).

POSSIBLE EXPERIMENT TO SEARCH FOR THE NEUTRON ELECTRIC DIPOLE MOMENT USING NEUTRON INTERFEROMETRY

THOMAS DOMBECK

Astronomy and Astrophysics Center, University of Chicago,
5640 S. Ellis Avenue, Chicago, IL 60637, USA

ABSTRACT

The most sensitive neutron Electric Dipole Moment (EDM) experiments to date look for a change in the polarization state in a beam of neutrons induced by the EDM interaction with an applied electric field. The method suggested in this paper looks for a change in phase in the two legs of an interferometer caused by multiple Bragg scatters of a polarized room-temperature neutron from a perfect silicon crystal. Depending on the neutron polarization state, each scatter will cumulatively advance or retard the phase of the neutron due to an interaction of the EDM with the atomic electric fields. Calculations indicate that this method can result in sensitivities competitive with present limits on the EDM, but with different systematic errors from the beam experiments. Prospects are discussed for reducing the present limits on the neutron EDM using this interferometry method.

1. Introduction

The most recent beam experiment[1] to search for the neutron Electric Dipole Moment (EDM) achieved a 95% confidence level upper limit on the EDM of $<1.2 \times 10^{-25}$ e-cm. Another technique employed by Shull[2] used the EDM interaction with atomic electric fields in a single Bragg scattering but achieved an upper limit of $<8.4 \times 10^{-22}$ e-cm at a 99% confidence level which was considerably less sensitive than the beam experiments. In this paper, I describe a potentially sensitive method to measure the EDM based on the well-established principle of room-temperature neutron interferometry[3] combined with multiple Bragg scattering from a perfect crystal of Silicon to induce a neutron phase shift.

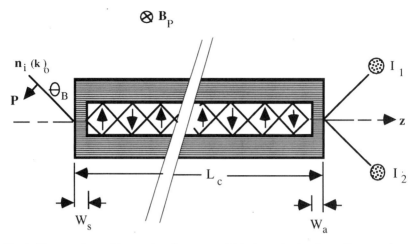

Fig. 1. The experimental layout (not drawn to scale) is shown in the above diagram. A perfect crystal of Si has a narrow slot cut along its length. A polarized neutron is split into two beams upon entry into the slot. Each beam will undergo multiple Bragg scatters and recombine at the end of the slot. The neutron spin precesses in the plane of the interferometer due to the applied **B** field such that one beam always points into and the other beam always away from the slot surface during the Bragg scatters. Opposite phase differences will accumulate between the two beams due to the interaction between the EDM and the atomic electric charges in the crystal. A change in counting rates at I_1 and I_2 that correlates with a change in the incident polarization is then evidence for an EDM.

2. Experimental Technique and Sensitivity

The experimental arrangement is shown in the diagram in Fig. 1 and the method of inducing the neutron phase shift is described in the caption. Assuming that the average potential energy $<\Delta V>$ is small compared to the neutron kinetic energy E_0, the neutron phase shift is given by:[3]

$$\Delta\beta = \Delta k \, S = -1/2 \, |k_0| \, S <\Delta V>/E_0 = (2\pi <\Delta V>/2hv) \, S \qquad (1)$$

where h is Planck's constant, k_0 is the incident neutron wave vector and v is its velocity. S is the path length in either interferometer leg over which the neutron senses this change in its potential energy.

In the experiment, the change in potential energy is induced by the interaction between the EDM, μ_e, and the atomic electric field sensed by the neutron as it undergoes Bragg scattering. An expression for this can be obtained by making use of the Born approximation to average over the incident and scattered waves:[4]

$$<\Delta V> = -<\mu_e \cdot [R \ Ze(1-f)/(4\pi\varepsilon_0 R^3)]> = \mu_e \ [Ze(1-f)q^2/(8\pi\varepsilon_0)] \tag{2}.$$

$Ze(1-f)$ is the screened atomic electric charge and $q=k_s-k_0$ is the difference between the scattered and incident wave vectors, having a magnitude $|q|=2k_0\sin\theta_B=4\pi\sin\theta_B/\lambda$. The term in the brackets in Eq. 2 is the average electric field, which can be in excess of 10^{10} volts/cm as the neutron wavefunction is localized near the atomic nucleus during the scattering.

Counting over N neutrons in the experiment, will result in a sensitivity to the neutron EDM given by:

$$N^{-1/2} = 4\pi \ (hv)^{-1} \ [Ze(1-f)q^2/(8\pi\varepsilon_0)] \ (2N_BL_p/\cos\theta_B) \ \Delta\mu_e \tag{3},$$

where the total interaction path length S was substituted using N_B, the number of multiple Bragg scatters, and L_p, the penetration depth into the crystal in each Bragg scatter.[4] For 1.44Å neutrons, this depth is 33μm. Assuming a 100-cm-long crystal with a 2.4mm slot yields about 1000 scatters. Inserting these values into Eq. 3 yields a one-standard deviation sensitivity near 10^{-26} e-cm for 4×10^8 neutrons counted. This slot width is wide enough to contain the beam width (1.5mm) at the University of Missouri reactor where previous interferometry experiments have been performed, and would require about 200 running days.

A source of systematic error could result if B_p (570 Gauss in the above example) is not perpendicular to the plane of scattering. This would allow a component of spin out of the plane inducing a phase shift due to Schwinger scattering.[4] For the EDM experiment discussed in this paper, an alignment tolerance on the 20μrad level may be required which is about five times better than achieved in the older EDM beam experiments.[4]

References

1. K. F. Smith et al. *Phys. Let.* **B234** (1990) 191.

2. C. Shull and R. Nathans. *Phys. Rev. Let.* **19** (1967) 384.

3. Descriptions of room-temperature neutron interferometry can be found in H. Rauch et al. *Phys. Let.* **A47** (1974) 369; Colella et al. *Phys. Rev. Let.* **34** (1975) 1472.

4. R. Golub and J. Pendlebury. *Contemp. Phys.* **13** (1972) 519.

PROBING NEW PHYSICS IN RARE CHARM PROCESSES

J.L. HEWETT

Stanford Linear Accelerator Center, Stanford University, Stanford, CA 94309, USA

ABSTRACT

The possibility of using the charm system to search for new physics is addressed. Phenomena such as $D^0 - \bar{D}^0$ mixing and rare decays of charmed mesons are first examined in the Standard Model to test our present understanding and to serve as benchmarks for signals from new sources. The effects of new physics from various classes of non-standard dynamical models on $D^0 - \bar{D}^0$ mixing are investigated.

We examine the prospect of using one-loop processes in the charm sector as a laboratory for probing new physics. Similar processes in the K system have played a strong and historical role[1] in constraining new physics, while corresponding investigations have just begun[2] in the b-quark sector. In contrast, charm has played a lesser role in the search for new physics. Due to the effectiveness of the GIM mechanism, short distance Standard Model (SM) contributions to rare charm processes are very small. Most reactions are thus dominated by long distance effects which are difficult to reliably calculate. However, a recent estimation[3] of such effects indicates that there is a window for the clean observation of new physics in some interactions. In fact, it is precisely because the SM flavor changing neutral current (FCNC) rates are so small, that charm provides an important opportunity to discover new effects, and offers a detailed test of the SM in the up-quark sector. Due to space-time limitations,[3] this talk will concentrate on $D^0 - \bar{D}^0$ mixing. First, the SM predictions are reviewed, and then the expectations in various extensions of the SM are discussed. We conclude with a brief summary of SM rates for rare D meson decays.

Currently, the best limits[4] on $D^0 - \bar{D}^0$ mixing are from fixed target experiments, with $x_D \equiv \Delta m_D/\Gamma < 0.083$ (where $\Delta m_D = m_2 - m_1$ is the mass difference), yielding $\Delta m_D < 1.3 \times 10^{-13}$ GeV. The bound on the ratio of wrong-sign to right-sign final states is $r_D \equiv \Gamma(D^0 \to \ell^- X)/\Gamma(D^0 \to \ell^+ X) < 3.7 \times 10^{-3}$, where

$$r_D \approx \frac{1}{2}\left[\left(\frac{\Delta m_D}{\Gamma}\right)^2 + \left(\frac{\Delta\Gamma}{2\Gamma}\right)^2\right], \tag{1}$$

in the limit $\Delta m_D/\Gamma, \Delta\Gamma/\Gamma \ll 1$. Several high volume charm experiments are planned for the future, with 10^8 charm mesons expected to be reconstructed. Several rare processes, including $D^0 - \bar{D}^0$ mixing, can then be probed another $1-2$ orders of magnitude below present sensitivities.

The short distance SM contributions to Δm_D proceed through a W box diagram with internal d, s, b-quarks. In this case the external momentum, which is of order m_c, is communicated to the light quarks in the loop and can not be neglected. The effective Hamiltonian is

$$\mathcal{H}_{eff}^{\Delta c=2} = \frac{G_F \alpha}{8\sqrt{2}\pi x_w} \left[|V_{cs}V_{us}^*|^2 \left(I_1^s \mathcal{O} - m_c^2 I_2^s \mathcal{O}' \right) + |V_{cb}V_{ub}^*|^2 \left(I_3^b \mathcal{O} - m_c^2 I_4^b \mathcal{O}' \right) \right], \quad (2)$$

where the I_j^q represent integrals[5] that are functions of m_q^2/M_W^2 and m_q^2/m_c^2, and $\mathcal{O} = [\bar{u}\gamma_\mu(1-\gamma_5)c]^2$ is the usual mixing operator while $\mathcal{O}' = [\bar{u}(1+\gamma_5)c]^2$ arises in the case of non-vanishing external momentum. The numerical value of the short distance contribution is $\Delta m_D \sim 5 \times 10^{-18}$ GeV (taking $f_D = 200\,\text{MeV}$). The long distance contributions have been computed via two different techniques: (i) the intermediate particle dispersive approach yields[3,6] $\Delta m_D \sim 10^{-4}\Gamma \simeq 10^{-16}$ GeV, and (ii) heavy quark effective theory which results[7] in $\Delta m_D \sim (1-2) \times 10^{-5}\Gamma \simeq 10^{-17}$ GeV. Clearly, the SM predictions lie far below the present experimental sensitivity!

One reason the SM expectations for $D^0 - \bar{D}^0$ mixing are so small is that there are no heavy particles participating in the box diagram to enhance the rate. Hence the first extension to the SM that we consider is the addition[8] of a heavy $Q = -1/3$ quark which may be present, e.g., as an iso-doublet fourth generation b'-quark, or as a singlet quark in E_6 grand unified theories. The current bound[4] on the mass of such an object is $m_{b'} > 85\,\text{GeV}$, assuming that it decays via charged current interactions. We can now neglect the external momentum and Δm_D is given by the usual expression,[9]

$$\Delta m_D = \frac{G_F^2 M_W^2 m_D}{6\pi^2} f_D^2 B_D |V_{cb'}V_{ub'}^*|^2 F(m_{b'}^2/M_W^2). \quad (3)$$

The value of Δm_D is displayed in this model in Fig. 1a as a function of the overall CKM mixing factor for various values of the heavy quark mass. We see that Δm_D approaches the experimental bound for large values of the mixing factor. A naive estimate in the four generation SM yields[4] the restrictions $|V_{cb'}| < 0.571$ and $|V_{ub'}| < 0.078$.

Another simple extension of the SM is to enlarge the Higgs sector by an additional doublet. First, we examine two-Higgs-doublet models which avoid tree-level FCNC by introducing a global symmetry. Two such models are Model I, where one doublet (ϕ_2) generates masses for all fermions and the second (ϕ_1) decouples from the fermion sector, and Model II, where ϕ_2 gives mass to the up-type quarks, while the down-type quarks and charged leptons receive their mass from ϕ_1. Each doublet receives a vacuum expectation value v_i, subject to the constraint that $v_1^2 + v_2^2 = v_{SM}^2$. The charged Higgs boson present in these models will participate in the box diagram for Δm_D. The $H^\pm$ interactions with the quark sector are governed by the Lagrangian

$$\mathcal{L} = \frac{g}{2\sqrt{2}M_W} H^\pm [V_{ij} m_{u_i} A_u \bar{u}_i(1-\gamma_5)d_j + V_{ij} m_{d_j} A_d \bar{u}_i(1+\gamma_5)d_j] + h.c., \quad (4)$$

with $A_u = \cot\beta$ in both models and $A_d = -\cot\beta(\tan\beta)$ in Model I(II), where $\tan\beta \equiv v_2/v_1$. The expression for Δm_D in these models can be found in Ref. (9). From the Lagrangian it is clear that Model I will only modify the SM result for Δm_D for very small values of $\tan\beta$, and this region is already excluded[2,10] from $b \to s\gamma$ and $B_d^0 - \overline{B}_d^0$ mixing. However, enhancements can occur in Model II for large values of $\tan\beta$, as demonstrated in Fig. 1b.

Next we consider the case of extended Higgs sectors without natural flavor conservation. In these models the above requirement of a global symmetry which restricts each fermion type to receive mass from only one doublet is replaced[11] by approximate flavor symmetries which act on the fermion sector. The Yukawa couplings can then possess a structure which reflects the observed fermion mass and mixing hierarchy. This allows the low-energy FCNC limits to be evaded as the flavor changing couplings to the light fermions are small. We employ the Cheng-Sher ansatz,[11] where the flavor changing couplings of the neutral Higgs are $\lambda_{h^0 f_i f_j} \approx (\sqrt{2}G_F)^{1/2}\sqrt{m_i m_j}\Delta_{ij}$, with the $m_{i(j)}$ being the relevant fermion masses and Δ_{ij} representing a combination of mixing angles. h^0 can now contribute to Δm_D through tree-level exchange and the result is displayed in Fig. 2a as a function of the mixing factor. $D^0 - \bar{D}^0$ mixing can also be mediated by h^0 and t-quark virtual exchange in a box diagram, however these contributions only compete with those from the tree-level process for large values of Δ_{ij}. In Fig. 2b we show the constraints placed on the parameters of this model from the present experimental bound on Δm_D for both the tree-level and box diagram contributions.

The last contribution to $D^0 - \bar{D}^0$ mixing that we will discuss here is that of scalar leptoquark bosons. Leptoquarks are color triplet particles which couple to a lepton-quark pair and are naturally present in many theories beyond the SM which relate leptons and quarks at a more fundamental level. We parameterize their *a priori* unknown couplings as $\lambda_{\ell q}^2/4\pi = F_{\ell q}\alpha$. They participate in Δm_D via virtual exchange inside a box diagram,[12] together with a charged lepton or neutrino. Assuming that there is no leptoquark-GIM mechanism, and taking both exchanged leptons to be the same type, we obtain the restriction

$$\frac{F_{\ell c}F_{\ell u}}{m_{LQ}^2} < \frac{196\pi^2\Delta m_D}{(4\pi\alpha f_D)^2 m_D}. \tag{5}$$

The resulting bounds in the leptoquark coupling-mass plane are presented in Fig. 3.

We close our discussion by displaying the expected branching fractions for various rare charm decay modes in the SM in Table 1. We present both the short distance predictions (neglecting QCD corrections, which may be important in some decay modes), an upper bound on the long distance estimates, as well as the current experimental limits.[4,13] For more details we refer the reader to Ref. (3). We urge our experimental colleagues to continue the search for rare charm processes!

References

1. See, for example, G. Beall, M. Bander, and A. Soni, Phys. Rev. Lett. **48**, 848 (1982); J. Ellis and D.V. Nanopoulos, Phys. Lett. **110B**, 44 (1982).
2. E. Thorndike, in *27th International Conference on High Energy Physics*, Glasgow, Scotland, August, 1994; J.L. Hewett in *21st SLAC Summer Institute on Particle Physics*, Stanford, CA, July 1993.

Decay Mode	Experimental Limit	$B_{S.D.}$	$B_{L.D.}$
$D^0 \to \mu^+\mu^-$	$< 1.1 \times 10^{-5}$	$(1-20) \times 10^{-19}$	$< 3 \times 10^{-15}$
$D^0 \to \mu^\pm e^\mp$	$< 1.0 \times 10^{-4}$	0	0
$D^0 \to \gamma\gamma$	—	10^{-16}	$< 3 \times 10^{-9}$
$D \to X_u + \gamma$		1.4×10^{-17}	
$D^0 \to \rho^0\gamma$	$< 1.4 \times 10^{-4}$		$< 2 \times 10^{-5}$
$D^0 \to \phi^0\gamma$	$< 2.0 \times 10^{-4}$		$< 10^{-4}$
$D^+ \to \rho^+\gamma$	—		$< 2 \times 10^{-4}$
$D \to X_u + \ell^+\ell^-$		4×10^{-9}	
$D^0 \to \pi^0\mu\mu$	$< 1.7 \times 10^{-4}$		
$D^0 \to \bar{K}^0 ee/\mu\mu$	$< 17.0/2.5 \times 10^{-4}$		$< 2 \times 10^{-15}$
$D^+ \to \pi^+ ee/\mu\mu$	$< 250/4.6 \times 10^{-5}$	few$\times 10^{-10}$	$< 10^{-8}$
$D^+ \to K^+ ee/\mu\mu$	$< 480/8.5 \times 10^{-5}$		$< 10^{-15}$
$D^0 \to X_u + \nu\bar{\nu}$		2.0×10^{-15}	
$D^0 \to \pi^0\nu\bar{\nu}$	—	4.9×10^{-16}	$< 6 \times 10^{-16}$
$D^0 \to \bar{K}^0\nu\bar{\nu}$	—		$< 10^{-12}$
$D^+ \to X_u + \nu\bar{\nu}$	—	4.5×10^{-15}	
$D^+ \to \pi^+\nu\bar{\nu}$	—	3.9×10^{-16}	$< 8 \times 10^{-16}$
$D^+ \to K^+\nu\bar{\nu}$	—		$< 10^{-14}$

Table 1. Standard Model predictions for the branching fractions due to short and long distance contributions for various rare D meson decays. Also shown are the current experimental limits.

3. For further details, see, G. Burdman, E. Golowich, J. Hewett, and S. Pakvasa, Phys. Rep. (in preparation).
4. L. Montanet *et al.*, Particle Data Group, Phys. Rev. **D50**, 1173 (1994).
5. A. Datta, Z. Phys. **C27**, 515 (1985).
6. G. Burdman in *CHARM2000 Workshop*, Fermilab, June 1994; J. Donoghue *et al.*, Phys. Rev. **D33**, 179 (1986).
7. H. Georgi, Phys. Lett. **B297**, 353 (1992); T. Ohl *et al.*, Nucl. Phys. **B403**, 605 (1993).
8. K.S. Babu *et al.*, Phys. Lett. **B205**, 540 (1988); T.G. Rizzo, Int. J. Mod. Phys. **A4**, 5401 (1989).
9. T. Inami and C.S. Lim, Prog. Theor. Phys. **65**, 297 (1981).
10. V. Barger, J.L. Hewett, and R.J.N. Phillips, Phys. Rev. **D41**, 3421 (1990).
11. S. Pakvasa and H. Sugawara, Phys. Lett. **73B**, 61 (1978); T.P. Cheng and M. Sher, Phys. Rev. **D35**, 3484 (1987); L. Hall and S. Weinberg, Phys. Rev. **D48**, 979 (1993).
12. S. Davidson, D. Bailey, and B.A. Campbell, Z. Phys. **C61**, 613 (1994).
13. M. Selen, CLEO Collaboration, talk presented at *APS Spring Meeting*, Washington D.C., April 1994; J. Cumalet, talk presented at *Workshop on the Tau-Charm Factory in the Era of B-Factories and CESR*, SLAC, Stanford, CA, August 1994.

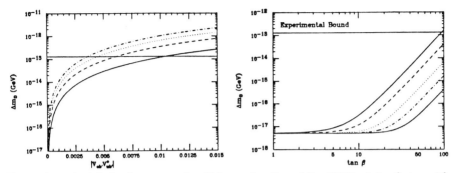

Fig. 1: Δm_D in (a) the four generation SM as a function of the CKM mixing factor with the solid, dashed, dotted, dash-dotted curve corresponding to $m_{b'} = 100, 200, 300, 400$ GeV, respectively. (b) in two-Higgs-doublet model II as a function of $\tan\beta$ with, from top to bottom, the solid, dashed, dotted, dash-dotted, solid curve representing $m_{H\pm} = 50, 100, 250, 500, 1000$ GeV. The solid horizontal line corresponds to the present experimental limit.

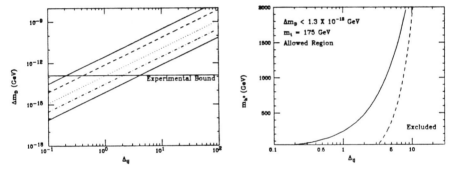

Fig. 2: (a) Δm_D in the flavor changing Higgs model described in the text as a function of the mixing factor with $m_{h^0} = 50, 100, 250, 500, 1000$ GeV corresponding to the solid, dashed, dotted, dash-dotted, solid curve from top to bottom. (b) Constraints in the mass-mixing factor plane from Δm_D from the tree-level process (solid curve) and the box diagram (dashed).

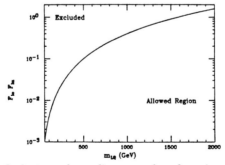

Fig. 3: Constraints in the leptoquark coupling-mass plane from Δm_D.

SEARCH FOR THE FLAVOR CHANGING NEUTRAL CURRENT DECAY $D^\circ \to \mu^+\mu^-$ IN 800 GeV/c PROTON-SILICON INTERACTIONS

G.H. MO[16]

For the E771 collaboration

T. Alexopoulos[1], L. Antoniazzi[2], M. Arenton[3], H.C. Ballagh[4], H. Bingham[4],
A. Blankman[5], M. Block[6], A. Boden[7], G. Bonomi[2], S.V. Borodin[5], J. Budagov[8],
Z.L. Cao[3], G. Cataldi[9], T.Y. Chen[10], K. Clark[11], D. Cline[7], S. Conetti[3], M. Cooper[12],
G. Corti[3], B. Cox[3], P. Creti[9], C. Dukes[3], C. Durandet[1], V. Elia[9], A.R. Erwin[1],
E. Evangelista[9], L. Fortney[13], V. Golovatyuk[3], E. Gorini[9], F. Grancagnolo[9],
K. Hagan-Ingram[3], M. Haire[14], P. Hanlet[3], M. He[15], G. Introzzi[2], M. Jenkins[11],
J. Jennings[1], D. Judd[14], W. Kononenko[5], W. Kowald[13], K. Lau[16], T. Lawry[3],
A. Ledovskoy[3], G. Liguori[2], J. Lys[4], P.O. Mazur[17], A. McManus[3], S. Misawa[4],
C.T. Murphy[17], K. Nelson[3], M. Panareo[9], V. Pogosian[3], S. Ramachandran[7],
M. Recagni[3], J. Rhoades[7], J. Segal[3], W. Selove[5], R.P. Smith[17], L. Spiegel[17],
J.G. Sun[3], S. Tokar[18], P. Torre[2], J. Trischuk[19], L. Turnbull[14], I. Tzamouranis[3],
D.E. Wagoner[14], C.R. Wang[15], C. Wei[13], W. Yang[17], N. Yao[10], N.J. Zhang[15],
S.N. Zhang[5] and B.T. Zou[13]

1. University of Wisconsin, Madison, Wisconsin, USA
2. Pavia INFN and University, Pavia, Italy
3. University of Virginia, Charlottesville, Virginia, USA
4. University of California at Berkeley, Berkeley, California, USA
5. University of Pennsylvania, Philadelphia, Pennsylvania, USA
6. Northwestern University, Evanston, Illinois, USA
7. University of California at Los Angeles, Los Angeles, California, USA
8. JINR, Dubna, Russia
9. Lecce INFN and University, Lecce, Italy
10. Nanjing University, Nanjing, People's Republic of China
11. University of South Alabama, Mobile, Alabama, USA
12. Vanier College, Montreal, Quebec, Canada
13. Duke University, Durham, North Carolina, USA
14. Prairie View A&M, Prairie View, Texas, USA
15. Shandong University, Jinan, Shandong, People's Republic of China
16. University of Houston, Houston, Texas, USA
17. Fermilab, Batavia, Illinois, USA
18. Comenius University, Bratislava, Slovakia
19. McGill University, Montreal, Quebec, Canada

ABSTRACT

We report on the preliminary result of a search for the flavor changing neutral current (FCNC) decay $D^0 \to \mu^+ \mu^-$ in 800 GeV/c proton-silicon interactions in Fermilab experiment E771. FCNC is highly suppressed in the standard model, but is a very sensitive probe for physics beyond the standard model. Based on 50% of the data collected by the E771 experiment in the 1991 fixed-target run, we searched for a D^0 signal in the dimuon mass spectrum from events accumulated with dimuon triggers. No evidence for $D^0 \to \mu^+ \mu^-$ was found. An upper limit on the branching ratio of 1.7×10^{-5} at 90% confidence level was obtained. We expect to achieve a limit of 5×10^{-6} using the entire E771 data and by including information from the silicon vertex detector. This limit will be a factor of two improvement over the existing limit of 1.1×10^{-5}.

The study of flavor changing neutral current (FCNC) processes has played an important role in the development of the standard model (SM).[1] The suppression of the dimuon decay rate for $K_L \to \mu^+ \mu^-$ has led to the discovery of the GIM mechanism to explain the suppression of strangeness changing neutral current.[2] The observed branching ratio of 7.3×10^{-9} agrees well with the prediction of the SM.[3] The mechanism has been generalized to the standard six-quark model. The dimuon decay of D^0, which is mediated by charm changing neutral current, is expected to be highly suppressed as well. The present limit of 1.1×10^{-5} is many orders of magnitude above the SM prediction.[3,4]

In the framework of the SM, $D^0 \to \mu^+ \mu^-$ proceeds via higher order electroweak interactions. Figure 1 shows some of the contributing Feynman diagrams for this decay. In the free-quark limit, the decay rate is given by[4]

$$\Gamma \sim \frac{G_F^2 M_D^5}{192\pi^3} \cdot \frac{f_D^2}{M_D^2} \cdot \frac{m_\mu^2}{M_D^2} \left(1 - \frac{4m_\mu^2}{M_D^2}\right)^{1/2} \left(\frac{\alpha}{4\pi \sin^2 \theta_W}\right)^2 \left| \sum_{i=2}^{3} V_{ci}^* V_{ui} C(x_i) \right|^2 . \quad (1)$$

$x_i = (m_i/M_W)^2$, where m_i ($i = s, b$) is the quark mass and M_W is the mass of $W^\pm$. The first factor is the decay rate of a free charm quark. The second factor results from the annihilation of the quarks. The last factor involves the CKM matrix elements for the coupling of the quarks in the loop. The function C is the result of integrating over the

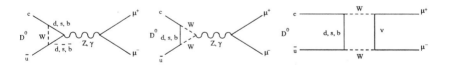

Fig. 1. Feynman diagrams for the FCNC decay $D^0 \to \mu^+ \mu^-$ in the standard model.

internal momenta in the loop, summed over all the contributing Feynman diagrams. For $D^0 \to \mu^+\mu^-$, $C \simeq x_i \sim 10^{-5}$ is the main source of suppression. The contribution of the b quark is negligible due to the small CKM matrix elements, $V_{cb}^* V_{ub} \simeq 1.6 \times 10^{-4}$. The decay is also suppressed by angular momentum conservation, as a result of the zero spin of D^0, giving rise to the factor m_μ^2/m_D^2 in the decay rate. Combining all these factors, the SM branching ratio is expected to be $\sim 10^{-19}$.

Due to its small branching ratio, this decay mode is a sensitive probe to new physics beyond the SM. For example, some models involving heavy fourth generation quarks enhance the branching ratio by several orders of magnitude. Supersymmetric models allow wino and squarks in the loop. The squarks are expected to be heavy, resulting in a branching fraction of the order of $\sim 10^{-9}$. The limit on the branching ratio will put constraint on these models.[5]

E771, a fixed target experiment at Fermilab,[6] is primarily designed to study B particle production and decay. In the 1991 run, the experiment accumulated (1.65 ± 0.17) × 10^{13} protons on target, and recorded about 127 million dimuon triggers, together with 62 million high p_t single muon triggers.

A schematic of the E771 detector is shown in Fig. 2. The detector features an open geometry large acceptance spectrometer, equipped with a silicon microstrip vertex detector (SMVD). The target consists of 12 layers of silicon foils, each 2 mm thick and 5 cm in diameter. The target foils are separated by 4 mm, to allow for decay of heavy quark particles. The SMVD consists of 18 silicon planes, 6 planes located upstream of the target to measure the beam protons, and 12 planes downstream of the target forming a high precision detector for primary and secondary vertices. The momentum analysis magnet is a large aperture window-frame dipole magnet with a vertical field of about 1.43 Tesla providing a transverse momentum kick of 0.821 GeV/c. The tracking system consists of 19 wire chambers (31 wire planes) located in front of the magnet and 9 wire chambers (24 wire, 6 stripe, and 6 pad planes) behind the magnet. The

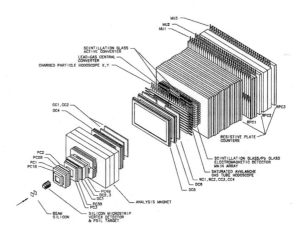

Fig. 2. Schematic of the E771 spectrometer.

electromagnetic calorimeter is a scintillating glass and lead glass array to detect electrons and photons. The muon detector consists of three planes of resistive plate counters (RPC's), each embedded behind a hadron absorption wall. The pad signals from RPC's provide two-dimensional positions of muon hits. A triple coincidence of three quasi-projective RPC pads defines a muon signal.

The online dimuon trigger requires two muon signals from the RPC's, in conjunction with a beam signal and an interaction signal. When running at an interaction rate close to 2 MHz, the dimuon trigger rate was about 200 Hz.

For this report, 64 million dimuon triggers (50% of the entire dimuon data) were analyzed by a muon tracking program. Events with very high multiplicity were rejected. The program reconstructs muon tracks first in the muon detector, which defines search regions for track segments in the wire chambers behind the magnet (rear tracks). The rear muon tracks are then matched to track segments in the front wire chambers. For the analysis described in this paper, no silicon track information has been used.

An asymmetric p_t cut was applied to the muon pairs: $p_t > 0.4$ GeV/c for one muon and $p_t > 1.0$ GeV/c for the other muon. These cuts are effective in rejecting π/K decay background, but retain a large fraction of D^0 dimuon decays. To reduce background further, the two muon tracks are required to intersect within some distance from the target: the distance of closest approach between two muon tracks has to be less than 4 mm, and the z-distance between the point of closest approach and the target center has to be less than 8 cm.

After applying these cuts, we were left with 135707 unlike-sign and 111977 like-sign dimuon events. The invariant mass spectra are shown in Fig. 3(A). The solid histogram is for unlike-sign μ pairs, and the dashed histogram is for the like-sign μ pairs. The overall shape of both histograms is formed by the p_t cuts which impose a strong suppression on the low mass dimuons. For the unlike-sign dimuons, sharp peaks at the J/ψ, ϕ, ρ^0/ω mass regions are clearly visible. Since the ρ^0 and ω are close in mass

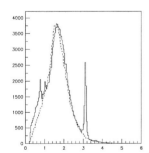 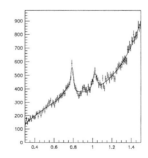 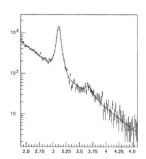

Fig. 3. (A) Dimuon invariant mass spectra below 6 GeV/c^2. The solid histogram is for unlike-sign dimuons, and the dashed line is for like-sign; (B) The unlike-sign mass spectrum from 0.3 to 1.5 GeV/c^2. The solid line is a least χ^2 fit to the data for a continuum plus ρ^0/ω and ϕ hypothesis; (C) The unlike-sign mass spectrum from 2.4 to 4.6 GeV/c^2 in logarithmic scale. The solid line is a least χ^2 fit for a continuum plus J/ψ and $\psi(2S)$ hypothesis.

and the ρ^0 is very broad, ρ^0 and ω are not seen separately. $\psi(2S)$ is also observed (see Fig. 3(C)). We fitted the unlike-sign spectrum to a model composed of these known resonances superimposed on a continuum as shown in Figs. 3(B) and 3(C). We modeled the continuum by a quartic polynomial in Fig. 3(B) and by an exponential function in Fig. 3(C). The ρ^0 is represented by a Breit-Wigner function, and the other resonances by Gaussians.

The sensitivity of the search for a D^0 signal depends on the trigger efficiency, geometric acceptance, and dimuon tracking efficiency. We have determined these quantities by a hit-level Monte Carlo (MC) simulation which overlays dimuon MC events on dimuon trigger events. The production cross section for the J/ψ, based on this analysis, is 324 ± 53 nb/nucleon assuming an A dependence of $\alpha = 0.92 \pm 0.008$.[7] This is in good agreement with the result, $330 \pm 5 \pm 35$ nb/nucleon, of a more detailed analysis based on the entire dimuon data.[8] We have also determined the cross sections for $\psi(2S)$, ϕ, ω, and ρ^0. The results are listed in Table 1. For the ω cross section, since the branching ratio has not yet been measured, e-μ universality (with a phase space correction) was assumed.

Table 1. Measurement results for the observed resonances. The fourth column lists the full width for the ρ^0 resonance and, otherwise, the detector mass resolutions for the other resonances.

Resonance	Events recovered	Mass (MeV/c^2)	Mass resolution or Full width (MeV/c^2)	Cross section (per nucleon)
ρ^0	2142 ± 180	753 ± 15	147 ± 12	101 ± 20 mb
ω	495 ± 53	788 ± 3	10.1 ± 2.5	15.7 ± 3.8 mb
ϕ	622 ± 50	1023 ± 3	22.9 ± 2.7	1.93 ± 0.40 mb
J/ψ	6258 ± 91	3108 ± 4	47.3 ± 0.6	324 ± 53 nb
$\psi(2S)$	109 ± 17	3684 ± 8	64.5 ± 4.2	37.8 ± 7.2 nb

The invariant mass spectrum for unlike-sign dimuons, from 1.45 to 2.50 GeV/c^2, is shown in Fig. 4(A). There is no visible enhancement at the D^0 mass. The spectrum is well described by a quartic polynomial with a χ^2 of 0.85 per degree of freedom (DOF) for 66 DOF. To search for a possible D^0 (and $\bar{D}^0$) signal, this spectrum was fitted to a quartic polynomial continuum plus a D^0 signal as shown in Fig. 4(B). The D^0 signal is represented by a Gaussian, with its mass fixed at 1864.5 MeV/c^2.[3] We allowed the D^0 mass to vary by ± 10 MeV/c^2, and found no significant change in the result. The mass resolution at D^0 was determined to be 30.7 ± 2.8 MeV/c^2 by interpolating the mass resolutions of the resonances in Table 1 as a linear function of the mass. This best fit is consistent with a downward fluctuation of the continuum by 48 events which is less than one standard deviation. Figure 4(C) shows a fit in which the number of D^0's assumed in the fit is increased until the probability of the fit drops to 10%. This gives the maximum number of D^0's in the data at 90 % confidence level (CL). The maximum number of D^0's is 95.4, with a χ^2 of 1.57/DOF for 11 DOF.

To arrive at an upper limit for the $D^0 \rightarrow \mu^+ \mu^-$ decay branching ratio, we normalize the D^0 yield to that of the J/ψ dimuon decay. In this way, some systematic uncertainties

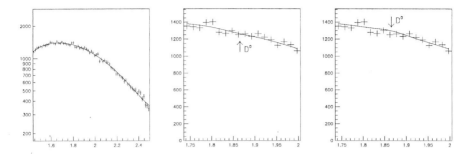

Fig. 4. (A) The unlike-sign dimuon mass spectrum from 1.45 to 2.50 GeV/c^2 with a least χ^2 fit for a quartic polynomial continuum; (B) The best fit for a continuum plus D^0 resonance hypothesis around the D^0 mass region; (C) The fit for a maximum number of D^0 events.

are canceled out.

$$BR(D^0 \rightarrow \mu^+\mu^-) = \frac{N_{D^0} + N_{\bar{D}^0}}{N_{J/\psi}} \cdot \frac{\sigma(pSi \rightarrow J/\psi) \cdot BR(J/\psi \rightarrow \mu^+\mu^-)}{\sigma(pSi \rightarrow D^0) + \sigma(pSi \rightarrow \bar{D}^0)} \cdot \frac{A_{J/\psi}}{A_{D^0}} . \quad (2)$$

In Eq. 2, the numbers of dimuons from D^0 and J/ψ decays, $N_{J/\psi}$ and $N_{D^0} + N_{\bar{D}^0}$, are obtained from the fit. We used the production cross section of J/ψ in Ref. 8. A J/ψ dimuon decay branching ratio of 0.0597 ± 0.0025 was used.[3] The production cross section for $D^0 + \bar{D}^0$, $22 \, ^{+9}_{-7} \pm 5.5$ μb/nucleon, was taken from the measurement of E743 which used a hydrogen bubble chamber in 800 GeV/c proton beam.[9] A nuclear dependence of $\alpha = 1$ was assumed for the D^0 production cross section. $A_{J/\psi}$ and A_{D^0} stand for the overall efficiencies of geometric acceptance and kinematic cuts for J/ψ and D^0, respectively. The ratio was determined by Monte Carlo simulation to be 1.58. From these numbers, an upper limit (at 90% CL) for the branching ratio of $D^0 \rightarrow \mu^+\mu^-$ $< 1.7 \times 10^{-5}$ was obtained.

This result is consistent with the non-observation of the $\Delta C = 1$ FCNC decay $D^0 \rightarrow \mu^+\mu^-$ at the present sensitivity. This limit is compatible with the existing limit of 1.1×10^{-5}.[3] It should be noted that this result is preliminary, and is based on 50% of the entire data. Further improvement can be made by using the entire data sample and the information from SMVD. We expect to improve the limit by about a factor of two in the near future.

In summary, we searched for the charm changing neutral current decay $D^0 \rightarrow \mu^+\mu^-$ in our dimuon data. No evidence for this decay was found. The 90% CL upper limit is 1.7×10^{-5}.

We thank the Fermilab staff including the Research and Computing Divisions. This work is supported by DOE, NSF, Natural Science and Engineering Research Council of Canada, Instituto Nazionale di Fisica Nucleare (INFN), and the Texas Advanced Research Program (ARP).

References

1. S.L. Glashow, *Nucl. Phys.* **22** (1961) 579; S. Weinberg, *Phys. Rev. Lett.* **19** (1967) 1264; A. Salam, in *Elementary Particle Theory: Relativistic Groups and Analyticity (Nobel Symposium No. 8)*, edited by N. Svartholm (Almqvist and Wiksell, Stockholm, 1968) p. 367.
2. S.L. Glashow, J. Illiopoulos, and L. Maiani, *Phys. Rev.* **D2** (1970) 1285.
3. Particle Data Group, in *Review of Particle Properties, Phys. Rev.* **D45** (1992).
4. T. Inami and C.S. Lim, *Prog. Theor. Phys.* **65** (1981) 297; M.J. Savage, *Phys. Lett.* **266B** (1991) 135; M. Gorn, *Phys. Rev.* **D20** (1979) 2380; J.M. Soares and A. Barroso, *Phys. Rev.* **D39** (1989) 1973; K.S. Babu et al., *Phys. Lett.* **205B** (1988) 540; L.L. Chau, *Phys. Rep.* **95** (1983) 1.
5. K.S. Babu et al., *Phys. Lett.* **205B** (1988) 540; T.G. Rizzo, in *Mixing of Ordinary and Exotic Quarks*, The Fourth Family of Quarks and Leptons, Second International Symposium (Academy of Sciences, New York, 1989) p. 411; T. Inami and C. S. Lim, *Nucl. Phys.* **B207** (1982) 533.
6. T. Alexopoulos et al., *Nucl. Phys.* (Proc. Suppl.) **B27** (1992) 257.
7. D.A. Alde et al., FERMILAB-Pub-90/156-E, and *Phys. Rev. Lett.* **64** (1990) 2479.
8. T. Alexopoulos et al., presented by G. Corti, *High Mass Dimuon States Produced in 800* GeV/c *pSi interactions*, in the same proceedings.
9. R. Ammar et al., *Phys. Rev. Lett.* **61** (1988) 2185.

$D^0 \bar{D}^0$ MIXING AND DCSD - SEARCH FOR $D^0 \to K^+\pi^-(\pi^0)$

TIEHUI (TED) LIU

High Energy Physics Laboratory, Harvard University,
42 Oxford Street, Cambridge, MA 02138, USA

(Representing the CLEO collaboration)

ABSTRACT

The reaction $D^0 \to K^+\pi^-(\pi^0)$ can occur either through $D^0 \bar{D}^0$ mixing or Doubly Cabibbo Suppressed Decay (DCSD). Using the CLEO II data sample, we have observed a signal for $D^0 \to K^+\pi^-$ and find $\mathcal{B}(D^0 \to K^+\pi^-)/\mathcal{B}(D^0 \to K^-\pi^+) = 0.0077 \pm 0.0025\,(\text{stat.}) \pm 0.0025\,(\text{sys.})$. No signal events have been found for $D^0 \to K^+\pi^-\pi^0$, leading to an upper limit on inclusive $D^0 \to K^+\pi^-\pi^0$ as $\mathcal{B}(D^0 \to K^+\pi^-\pi^0)/\mathcal{B}(D^0 \to K^-\pi^+\pi^0) < 0.0068$. Some new ideas, applicable to future mixing searches, are introduced.

1. Motivation

Following the discovery of the D^0 meson at SPEAR in 1976, experimenters began to search for $D^0 \bar{D}^0$ mixing in either hadronic decays $D^0 \to \bar{D}^0 \to K^+\pi^-(X)$,[1] or semileptonic decays $D^0 \to \bar{D}^0 \to X^+ l^- \nu$. The past decade has seen an exponential rise in sensitivity.[2] Much of the enthusiasm for $D^0 \bar{D}^0$ mixing search stems from the belief that it carries a large discovery potential for new physics, since the mixing rate $R_{\text{mixing}} \equiv \mathcal{B}(D^0 \to \bar{D}^0 \to \bar{f})/\mathcal{B}(D^0 \to f)$ is expected to be very small in the Standard Model. Theoretical calculations of $D^0 \bar{D}^0$ mixing in the Standard Model are plagued by large uncertaities. Recent stduies[3] have shown that long distance contributions to $D^0 \bar{D}^0$ mixing are smaller than previously estimated and the prevailing conclusion within the Standard Model seems to be that[4] $R_{\text{mixing}} < 10^{-7}$. One can characterize $D^0 \bar{D}^0$ mixing in terms of two dimensionless variables: $x = (m_2 - m_1)/(\gamma_1 + \gamma_2)$ and $y = (\gamma_1 - \gamma_2)/(\gamma_1 + \gamma_2)$, where m_i and γ_i ($i = 1, 2$) are the masses and decay rates of the two CP eigenstates (even and odd) respectively. Assuming $x, y \ll 1$, we have $R_{\text{mixing}} = (x^2 + y^2)/2$. Mixing can be caused either by $x \neq 0$ (meaning mixing is genuinely caused by the $D^0 - \bar{D}^0$ transition) or by $y \neq 0$ (meaning the fast decaying component quickly disapears, leaving the slow decaying component which is a mixture of D^0 or $\bar{D}^0$). A measurement of a very small mixing rate (below 10^{-7}) is not possible with present experimental sensitivity. However, the observation of a larger value for R_{mixing} caused by $x \neq 0$ would imply the existence of new physics beyond the Standard Model.

Mixing has a unique attribute, namely the decay time-dependence, which can be used to distinguish mixing from other processes that yield the same final state. For example, in hadronic decays there is another process, DCSD, which can also yield the same final state $D^0 \rightarrow K^+\pi^-(X)$. The time-dependence in the case of $D^0 \rightarrow K^+\pi^-$ is $I(D^0 \rightarrow K^+\pi^-)(t) \propto \left(R_{DCSD} + \sqrt{2R_{DCSD}R_{mixing}} \; t\cos\phi + \frac{1}{2}R_{mixing}t^2\right)e^{-t}$, where t is measured in units of the average D^0 lifetime $2/(\gamma_1 + \gamma_2)$, ϕ is an unknown phase, and $R_{DCSD} = |\rho|^2$ where $\rho = \text{Amp}(D^0 \rightarrow K^+\pi^-)/\text{Amp}(\bar{D}^0 \rightarrow K^+\pi^-)$, denoting the relative strength of DCSD. R_{DCSD} is expected to be of the order of $\tan^4\theta_C \sim 0.3\%$ (where θ_C is the Cabibbo angle), which is much higher than the expected value of R_{mixing}. It is, therefore, believed by many that this method is inherently limited by the presence of "annoying DCSD background"; that the signature of mixing appears only at longer decay times therefore it will suffer from DCSD fluctuation, destructive interference could wipe out the signature of mixing. That this observation is not necessarily true can be easily seen from the interference term. Except in the extreme case $\cos\phi = 0$, the signature of small mixing ($R_{mixing} \ll R_{DCSD}$) will be greatly enhanced by the very presence of DCSD through interference. In fact, the true value of the interfernce term is $\frac{\cos\phi}{t}\sqrt{8R_{DCSD}/R_{mixing}}$ times larger than that of the mixing term. For semileptonic method, one does not have this advantage. The enhancement reaches its maximum when $|\cos\phi| = 1$ (even better with $\cos\phi = -1$). In addition, the fact that the interference term peaks at $t = 1$ shows that the richest signature of potentially small mixing is at shorter decay times. One may find the details of the above discussion elsewhere.[2] Therefore, the signature of mixing is a deviation from a perfect exponential time distribution with the slope of the average D^0 decay rate $(\gamma_1 + \gamma_2)/2$. If mixing is indeed small, hadronic decays would have greater sensitivity to mixing than semileptonic decays, as long as $\cos\phi \neq 0$. Our ability to observe the signature of a potentially small mixing signal depends on the number of $D^0 \rightarrow K^+\pi^-(X)$ events we will have. Therefore, observing $D^0 \rightarrow K^+\pi^-(X)$ would be a crucial step on the way to observing a potential small mixing signal by using this technique. It is under this strong motivation that CLEO II has searched for $D^0 \rightarrow K^+\pi^-(\pi^0)$ and found a signal for $D^0 \rightarrow K^+\pi^-$. The description of this search follows.

2. Technique

We consider the decay chain $D^{*+} \rightarrow D^0\pi_s^+ \rightarrow K\pi(\pi^0)\pi_s^+$, where the π_s^+ has a soft momentum spectrum and is thus referred to as the slow pion. The charge of the slow pion is correlated with the charm quantum number of the D^0 meson and can be used to tag whether a D^0 or $\bar{D}^0$ meson was produced. Our technique is to search for wrong sign $D^{*+} \rightarrow D^0\pi_s^+ \rightarrow K^+\pi^-(\pi^0)\pi_s^+$ decays in which the slow pion has the same charge as the kaon. The right sign signal $D^{*+} \rightarrow D^0\pi_s^+ \rightarrow K^-\pi^+(\pi^0)\pi_s^+$ is used for normalization. Identical cuts are applied to both the right sign and wrong sign samples. Without a precision vertex detector, CLEO II can only in effect measure the rate $\mathcal{B}(D^0 \rightarrow K\pi(X))$ integrated over all times of a pure D^0 decaying to a final state $K\pi(X)$.

In the case of $D^0 \rightarrow K^+\pi^-$, $R = \mathcal{B}(D^0 \rightarrow K^+\pi^-)/\mathcal{B}(D^0 \rightarrow K^-\pi^+) = R_{DCSD} + \sqrt{2R_{DCSD}R_{mixing}}\cos\phi + R_{mixing}$, which is then given by $R = N(\text{wrong sign})/N(\text{right sign})$, where N refers to the number of events observed. The ratio is insensitive to uncertainties in tracking efficiency and particle indentification efficiency. Our analysis relies on the fact that a signal will manifest itself as a peak in both the distributions of D^0 mass and of the mass difference $\Delta M \equiv M(K\pi\pi_s) - M(K\pi) - M(\pi_s)$.

In the case of $D^0 \to K^+\pi^-\pi^0$ (or $K^+\pi^-(X)$), things are more complicated. The decay $D^0 \to K^+\pi^-\pi^0$ includes four submodes: $K^+\rho^-$, $K^{*+}\pi^-$, $K^{*0}\pi^0$ as well as non-resonant $K^+\pi^-\pi^0$, and these four submodes interfere with each other. The main complication faced by this analysis is that the resonant substructure is not necessarily the same for DCSD and for Cabibbo favored decay (CFD). For $D^0 \to K\pi\pi^0$, the true yield density $n(p)$ at a point p in the Dalitz plot can be written as: $n(p) \propto |f_1\, e^{i\phi_1} A_{3b} + f_2\, e^{i\phi_2} BW_{\rho^+}(p) + f_3\, e^{i\phi_3} BW_{K^{*-}}(p) + f_4\, e^{i\phi_4} BW_{\bar{K}^{*0}}(p)|^2$, where f_i are the relative amplitudes for each component and ϕ_i are the interference phases between each submode. A_{3b} is the S-wave three-body decay amplitude, which is flat across the Dalitz plot. The various $BW(p)$ terms are Breit-Wigner amplitudes for the $K^*\pi$ and $K\rho$ sub-reactions. Note that in general:[7] $f_i^{DCSD}/f_i^{CFD} \neq f_j^{DCSD}/f_j^{CFD}$ $(i \neq j)$ and $\phi_i^{DCSD} \neq \phi_i^{CFD}$. As both DCSD and mixing (CFD) contribute to the wrong sign decay, the yield density for the wrong sign events $n_w(p)$ will have a complicated form, as for each submode DCSD and mixing may interfere with each other. The inclusive rate is defined as $R = \mathcal{B}(D^0 \to K^+\pi^-\pi^0)/\mathcal{B}(D^0 \to K^-\pi^+\pi^0)$. Let $m_r(p)$ ($m_w(p)$) be the *measured* density at the point p in the Dalitz plot for the right sign (wrong sign) events, and let $\epsilon(p)$ be the absolute detection efficiency at the point p. The quantity R is then given by $R = \int n_w(p)\, da/\int n_r(p)\, da = \int \frac{m_w(p)}{\epsilon(p)}\, da/\int \frac{m_r(p)}{\epsilon(p)}\, da$, where $\int$ indicates integration over the whole Dalitz plot.

3. Event Selection

The data sample used in this study consists of 1.8 (2.5 for $K\pi\pi^0$) fb^{-1} of integrated luminosity near the $\Upsilon(4S)$ resonance, collected with the CLEO II detector. The CLEO II detector has been described in detail elsewhere.[8] All D^{*+} candidates are required to have $x_{D^{*+}} > 0.64$ (0.60 for $K\pi\pi^0$), where $x_{D^{*+}} = p/p_{max}$ and $p_{max} = \sqrt{E_{beam}^2 - M_{D^{*+}}^2}$. This cut significantly reduces combinatorial background and keeps about 30% (35% for $K\pi\pi^0$) of the continuum D^{*+}s. The particle identification cuts combine information from time-of-flight scintillation counters and dE/dx in the drift chamber. The most serious background is the doubly misidentified $D^0 \to K^-\pi^+(\pi^0)$, where the particle type assignments of K^- and π^+ are interchanged, and the D^0 is from $D^{*+} \to D^0\pi_s^+$ decay. This background peaks in the mass difference plot, and will show up in the D^0 mass plot as a broad enhancement around the D^0 peak. When the kaon and the pion momenta are nearly equal, the doubly misidentified kaon-pion combination may still give a good D^0 mass. To remove this background, we invert the kaon and pion assignments and recalculate the D^0 mass, denoted M_{flip}. If M_{flip} is within 4 σ (standard deviation) of the nominal D^0 mass, the combination is discarded. In the case of $D^0 \to K^+\pi^-\pi^0$, isolated photons, detected by the barrel CsI calorimeter, with a minimum energy of 30 MeV/c, are paired to form π^0 candidates. Combinations within 2σ of the nominal π^0 mass are selected as π^0 candidates. The background from random π^0 candidates peaks in the low momentum region and can be effectively removed by a cut $p(\pi^0) > 400$ MeV/c.

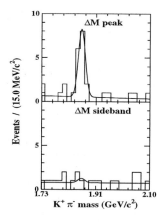
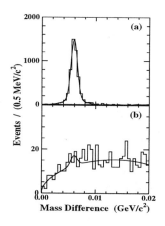

Fig. 1. The CLEO signal for $D^0 \to K^+\pi^-$. The D^0 mass for wrong sign events. (a) for events in the ΔM peak; (b) for events in the ΔM sideband.

Fig. 2. The CLEO result on $D^0 \to K^+\pi^-\pi^0$. The mass difference plots for (a) right sign and (b) wrong sign events after all the cuts are applied.

4. Observation of $D^0 \to K^+\pi^-$

One can find the details of this analysis elsewhere.[9] Figure 1 shows the wrong sign D^0 mass distributions (a) for events in the mass difference peak and (b) for the sideband with a tight particle identification cut. The signal is clearly present with very little background. We find R = 0.0077 ± 0.0025 (stat.) ± 0.0025 (sys.). This value corresponds to 2.92 ± 0.95 (stat.) ± 0.95 (sys.) $\tan^4 \theta_C$, where $\tan^4 \theta_C = 0.00264$ is used. As we do not measure the decay time distribution, this signal could be due to either $D^0\bar{D}^0$ mixing or DCSD, or a combination of the two. The theoretical prediction for R in DCSD is about $2\tan^4 \theta_C$,[5] which is quite consistent with the measured value. It is, therefore, believed by many that the signal is mostly due to DCSD, although it remains consistent with the current best experimental upper limit on mixing,[6] which is 0.0037-0.007 where the range reflects the posssible effects of interference between DCSD and mixing with an unknown phase.

5. Upper Limit on $D^0 \to K^+\pi^-\pi^0$

We have used a large Monte Carlo sample ($D^{*+} \to D^0\pi^+$ followed by $D^0 \to K^-\pi^+\pi^0$ non-resonant decay) and used the distribution of the accepted Monte Carlo events to study the detection efficiency $\epsilon(p)$ across the Dalitz plot. Since the events were generated according to three-body phase space (flat across the Dalitz plot), any deviation from flatness is a direct indication of efficiency variation. Figure 2 shows the mass difference plots for the right sign and wrong sign after all the cuts are applied. Fitting the distributions indicates a yield of 5149.3 ± 84.6 right sign and 14.2 ± 11.1 wrong sign events. After taking into account the detection efficiency $\epsilon(p)$,[10] we find

R $= 0.0034 \pm 0.0027$, leading to a 90% C.L. upper limit of R < 0.0068. One can find the details of this analysis elsewhere.[10] This upper limit includes the possible effects of the interference between $D^0 \bar{D}^0$ mixing and DCSD for each submode as well as the interference between submodes. This upper limit is for the inclusive rate. Theoretical predictions for R in $D^0 \to K^+\pi^-\pi^0$ have only been made for some exclusive modes. One estimate gives R for $D^0 \to K^+\rho^-$ and $D^0 \to K^{*+}\pi^-$ to be $0.5\tan^4\theta_C$ and $3\tan^4\theta_C$ respectively.[5] We are attempting to obtain separate limits on the individual submodes, but our lack of knowledge of the various possible submodes (their amplitude and interference phases) precludes a quantitative analysis of the exclusive rate for each submode.

6. Possible New Methods for Future Mixing Searches

As was discussed in the introduction, the observation of $D^0 \to K^+\pi^-$ is a crucial step on the way to observing a potentially small mixing signal by using this technique. With the CLEO II signal at the level R $= 0.0077 \pm 0.0025$ (stat.) ± 0.0025 (sys.), any experimenter can estimate the number of reconstructed $D^0 \to K^+\pi^-$ events their data sample will have in the future and what kind of sensitivity their experiment could have. Eighteen years after the search for $D^0 \to K^+\pi^-$ started, we have finally arrived at the point where we can take advantage of possible DCSD-mixing interference to make the mixing search easier.

Although it is true that more-than-two body D^0 hadronic decays, such as $D^0 \to K^+\pi^-\pi^0$, are very complicated due to the possible difference in the resonant substructure of the DCSD and CFD (mixing) decays, this unique attribute could, in principle, provide additional information which could allow one to distinguish DCSD and mixing in the future.[2] For instance, combined with the decay time information, one can perform a multi-dimensional fit to the data by using the information on ΔM, $M(D^0)$, proper decay time t, and the yield density on Dalitz plot $n_w(p,t)$. The extra information on the resonant substructure will, in principle, put a much better constraint on the amount of mixing. Of course, precise knowlegde of the resonant substructure for DCSD is needed here and so far we do not know anything about it. This means that understanding DCSD in D decays could be a very important step on the way to observing mixing using this technique. In principle, one can use the wrong sign sample at very low decay times (which is almost pure DCSD) to study the resonant substructure of the DCSD decays. It is also worth pointing out that it may be possible that a good understanding of DCSD could be reached by measuring the pattern of D^+ DCSD decays where the signature is not confused by a potential mixing component. In the near future, we should have a good understanding of DCSD decays and this method could become a feasible way to search for $D^0 \bar{D}^0$ mixing.

Moreover, we can also use the Singly Cabibbo Suppressed Decays (SCSD) to study mixing.[2] This is because the SCSD decays, such as $D^0 \to K^+K^-, \pi^+\pi^-$ (or $K^+K^-\pi^0$), occur only through the CP even (or odd) eigenstate, which means the decay time distribution is a perfect exponential time distribution with the slope of γ_1 (or γ_2), assuming CP conservation. Therefore, one can use these modes to measure γ_1 (or γ_2). Thus one can then measure $y = (\gamma_1 - \gamma_2)/(\gamma_1 + \gamma_2)$. Unlike in the case of $D^0 \to K^+\pi^-(X)$, the mixing signature is not a deviation from a perfect exponential time distribution, but rather a deviation of the slope from the average D^0 decay rate

$(\gamma_1 + \gamma_2)/2$. Observation of a non-zero y would demonstrate mixing caused by the decay rate difference $(R_{\text{mixing}} = (x^2 + y^2)/2)$. It is worth pointing out that there are many advantages with this method. For example, one can use Cabibbo favored decay modes, such as $D^0 \rightarrow K^- \pi^+$, to measure the average D^0 decay rate $(\gamma_1 + \gamma_2)/2$. This, along with many SCSD CP even (or odd) final states, would allow for valuable cross checks on systematic uncertainties. Furthermore, there is no need to tag the D^0 nor to know the primary vertex location, since we only need to determine the slope. Of course, the sensitivity of this method depends on how well we can determine the slope. In the future, if one could measure the slope difference (y) at 10^{-3} level using $D^0 \rightarrow K^+ K^-, \pi^+ \pi^-, K^- \pi^+$ etc., the sensitivity to mixing caused by the decay rate difference $(\sim y^2/2)$ would be already close to 10^{-6} level.

In the case of $D^0 \rightarrow K^+ \pi^- (X)$ and $D^0 \rightarrow X^+ l^-$, we are only measuring $R_{\text{mixing}} = (x^2 + y^2)/2$. Since many extensions of the Standard Model predict[11] large $x = (m_2 - m_1)/(\gamma_1 + \gamma_2)$, it is very important to measure x and y separately. Fortunately, SCSD can provide us the information on y, so together with the information on R_{mixing} obtained from other methods, we can in effect measure x. In this sense, it is best to think of the quest to observe mixing (new physics) as a program rather than a single effort.

References

1. We discuss D^0 decays explicitly in the text; its charge conjugate decays are also implied throughout the text unless otherwise stated.
2. T. Liu, "The $D^0 \bar{D}^0$ Mixing Search - Current Status and Future Prospects", in Proceedings of the Charm 2000 Workshop, Fermilab, June 7-9, 1994.
3. H. Georgi, *Phys. Lett.* B **297**, 353 (1992); T. Ohl, G. Ricciardi and E.H. Simmons, *Nucl. Phys.* B **403**, 603 (1993).
4. G. Burdman, "Charm mixing and CP violation in the Standard Model", S. Pakvasa, "Charm as Probe of New Physics", in Proceedings of the Charm 2000 Workshop, Fermilab, June 7-9, 1994.
5. I. Bigi and A.I. Sanda, *Phys.Lett.* **B171**,320(1986). R.C. Verma and A.N. Kamal, *Phys.Rev.* **D43**,829(1991). L.L. Chau and H.Y. Cheng, ITP-SB-93-49,UCD-93-31.
6. J.C. Anjos *et al.*,(E691 Collaboration), *Phys.Rev.Lett.* **60**, 1239 (1988). For details of this study, see T. Browder, Ph.D Thesis, UCSB-HEP-88-4,(1988).
7. It has been pointed out[5] that it is unlikely that just one universal suppression factor will affect the individual DCSD. For example, SU(3) breaking can introduce a significant enhancement for D^0 DCSD decays $D^0 \rightarrow K^+ \pi^-, K^{*+} \pi^-$, while SU(6) breaking can introduce a sizeable suppression relative to the naive expectation for D^0 DCSD decay $D^0 \rightarrow K^+ \rho^-$.
8. Y. Kubota, *et al.*,(CLEO Collaboration), Nucl. Inst. and Meth. A **320** 66(1992).
9. D. Cinabro, *et al.*,(CLEO Collaboration), *Phys.Rev.Lett.* **72**, 1406(1994).
10. G. Grawford, *et al.*,(CLEO Collaboration), "Search for $D^0 \rightarrow K^+ \pi^- \pi^{0}$", in Proceedings of the 27th International Conference on High Energy Physics, Glasgow, Scotland, 20-27 July, 1994.
11. A. Datta, *Phys. Lett.* **B154**, 287(1985). A.M. Hadeed and B. Holdom, *Phys. Lett.* **B159**, 379(1985). B. Mukhopadhyaya, A. Raychaudhuri and A. Ray, *Phys. Lett.* **B190**, 93(1987). K.S. Babu *et al. Phys. Lett.* **B205**, 540(1988). E.Ma *et al. Mod. Phys. Lett.* A**3**, 319(1988). L. Hall and S. Weinberg, *et al. Phys. Rev.* **D48**,979(1993).

PRELIMINARY RESULTS ON THE DECAYS $D^+ \to K^+\pi^+\pi^-$, $D^+ \to K^+K^+K^-$

MILIND PUROHIT*

and

JAMES WIENER†

Department of Physics, Princeton University, Princeton, NJ 08544 U.S.A.

ABSTRACT

Preliminary results on the Doubly Cabibbo Suppressed Decays (DCSD) $D^+ \to K^+\pi^+\pi^-$ and $D^+ \to K^+K^+K^-$ from roughly 40% of the E791 data set are summarized. We present an analysis of candidate events from E791 data in these decay modes. We show that there is a clear signal in the $D^+ \to K^+\pi^+\pi^-$ mode and discuss backgrounds and sources of error in the results. We disagree with a published signal level in the $D^+ \to K^+K^+K^-$ mode.

1. Introduction

Most models for the decay of D-mesons explain the longer D^+ lifetime (relative to the D^0, D_s^+ lifetimes) as being due to interference of the color-allowed and color-suppressed spectator diagrams. This interference only occurs for the D^+ meson, but not for the D^0 and D_s^+ spectator decays. Thus, Doubly Cabibbo Suppressed Decays (DCSD) of the D^+ are interesting both because they are rare and have never been confirmed and because they do not suffer interference in the quark level weak decay spectator diagram. Definite predictions have been made about their rates, based on models of D mesons and their decay mechanisms. Naively, the ratio of the DCSD decay rates to corresponding Cabibbo-favored rates is expected to be of the order of $\tan^4 \theta_C$. The lack of interference in the spectator diagrams further enhances this ratio.

Preliminary analyses of $\sim$40% of E791's data set have now been completed.[1] Since the DCSD signal is expected (and known) to be small compared to the Cabibbo-favored signal, the analysis requires much tighter cuts than those used for the Cabibbo-favored decay $D^+ \to K^-\pi^+\pi^+$. The tighter cuts ensure lower backgrounds. The analysis is further complicated by reflections due to particle misidentification from decays such as $D^+ \to K^-\pi^+\pi^+$ and $D_s^+, D^+ \to K^-K^+\pi^+$. Below, we first describe the cuts used in the analysis and then describe the treatment of the 3-body backgrounds.

---

*Representing the E791 Collaboration: Centro Brasileiro de Pesquisas Fisicas, Brazil, University of California, Santa Cruz, University of Cincinnati, CINVESTAV, Mexico, Fermilab, Illinois Institute of Technology, Kansas State University, University of Mississippi, Princeton University, Universidad Autonoma de Puebla, Mexico, Tel Aviv University, Israel, Tufts University, University of Wisconsin, Yale University.

Now at the University of South Carolina, Columbia, SC 29212 U.S.A.

†Also representing the E791 Collaboration

2.   Signal selection cuts

Since we are searching for the DCSD signal, we tuned our analysis cuts using the Cabibbo-favored signal. Usually, cuts are chosen to maximize the ratio $S/\sqrt{S+B}$ where S is the signal and B is the background. However, we maximized instead the quantity $5 \times \tan^4\theta_C \times S/\sqrt{5 \times \tan^4\theta_C \times S + B}$, (where S and B are the signal and background in the Cabibbo-favored mode) since we naively expected the DCSD signal to be a few times $\tan^4\theta_C$ of the Cabibbo-favored signal. After this optimization, exactly the same cuts are applied (with one exception discussed below) to the Cabibbo-favored and DCSD samples.

The main cuts can be summarized as follows. All decay tracks are required to travel through at least one magnet, to have a χ^2/DOF less than 5 and a contribution of greater than 25 to the χ^2 if they are forcibly included in the primary vertex fit. The decay vertex is required to be at least 20σ downstream of the production vertex, the p_T-balance around the D flight direction is required to be less than 250 MeV/c, the impact parameter of the D-meson is required to be less than 30 microns at the production vertex, no more than one additional track is allowed within 50 microns of the D decay vertex and the decay vertex is required to be at least 2σ outside any target. Further, the product of the ratios of each decay track's impact parameter to the secondary and primary vertices is required to be less than 0.001.

In the Cabibbo-favored decays, the kaon was identified by charge alone as the particle with charge opposite to that of the D-meson. In the DCSD decays, of the two particles with the same charge as the D-meson the kaon was taken to be the one with larger Čerenkov probability to be a kaon. In both cases however, tracks considered pions are required to have a pion Čerenkov probability greater than 0.55 and similarly the kaon is required to have a kaon probability greater than 0.30. In Figure 1 we show the invariant mass distribution for the Cabibbo-favored decays $D^+ \rightarrow K^-\pi^+\pi^+$.

3.   Treatment of backgrounds

A large number of possible sources of backgrounds were studied. Using Monte Carlo we showed that the background contributions from the decays $\overline{D}^0 \rightarrow K^+\pi^+\pi^-\pi^-$, $\overline{D}^0 \rightarrow K^+\pi^-$, $\overline{D}^0 \rightarrow K^+\pi^-\pi^0$, $D^+ \rightarrow \pi^+\pi^-\pi^+$, $D^+ \rightarrow K^-\pi^+\pi^+\pi^0$, $D_s^+ \rightarrow \phi e^+\nu_e$ and $\Lambda_c^+ \rightarrow pK^-\pi^+$ were all negligible. However, the 3-body decays $D^+ \rightarrow K^-\pi^+\pi^+$ and $D_s^+, D^+ \rightarrow K^-K^+\pi^+$ contribute significantly to the background. In this analysis we chose to *explicitly cut out* such decays by discarding events which have a candidate $KK\pi$ or Cabibbo-favored $K\pi\pi$ mass in the signal region. Figures 2 and 3 show the candidate DCSD events with the alternate 3-body mass hypotheses just described. It is clear that there is significant contamination. Events that fall in the shaded regions are removed. However, these cuts remove not only the true $K^+\pi^+\pi^-$ and $K^+K^-\pi^+$ decays from D-mesons but also background under those peaks. Hence parts of the "true" (or combinatoric) background under the candidate DCSD signal are also removed.

We have simulated the resulting background shape in three different ways: (1) by mixing D decay tracks from different data events, (2) from Monte Carlo and (3) by fitting the background away from peaks in two-dimensional plots with the DCSD mass on one axis and the reflected mass on the other. All three methods give identical

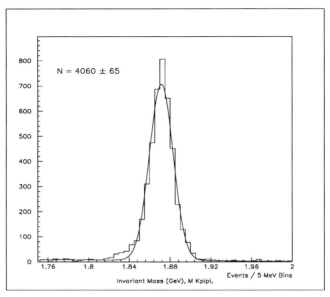

Fig. 1. Invariant mass distribution for the Cabibbo-favored decay $D^+ \to K^-\pi^+\pi^+$ from ~40% of the E791 data sample. This is used as normalization for the doubly Cabibbo-suppressed signal.

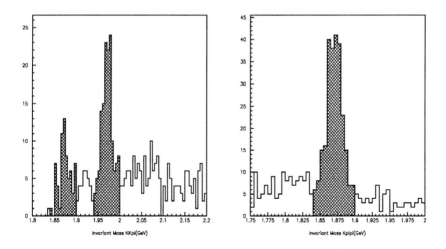

Fig. 2. The DCSD candidate events with a) a $K^-K^+\pi^+$ mass hypothesis and b) a $K^-\pi^+\pi^+$ mass hypothesis. Clear reflected signals can be seen in both cases. Events in the shaded regions are explicitly removed from the DCSD sample.

results within errors and we choose the fitting technique to obtain our central values. Figure 4 shows the DCSD invariant mass with the shaded region showing the expected background shape. It is clear that there is a 4σ signal.

4. Results and Systematic errors

From fits to data displayed in the figures above, we obtain the ratio[1]

$$\frac{\Gamma(D^+ \to K^+\pi^-\pi^+)}{\Gamma(D^+ \to K^-\pi^+\pi^+)} = (3.9 \pm 0.9 \pm 0.5) \times \tan^4\theta_c$$

This is already a much better limit/signal than the Particle Data booklet[2] limit of $20 \times \tan^4\theta_c$.

A preliminary examination of the resonant subcomponents has been done and we find that

$$\frac{\Gamma(D^+ \to K^{*0}\pi^+)}{\Gamma(D^+ \to K^-\pi^+\pi^+)} < 2.9 \times \tan^4\theta_c$$

If this is considered to be a signal, we obtain

$$\frac{\Gamma(D^+ \to K^{*0}\pi^+)}{\Gamma(D^+ \to K^-\pi^+\pi^+)} = (1.9 \pm 0.6) \times \tan^4\theta_c$$

(statistical error only). Figure 5 shows the DCSD signal after the K^* selection.

Similarly, an examination of decays to three charged kaons has revealed that

$$\frac{\Gamma(D^+ \to K^+K^-K^+)}{\Gamma(D^+ \to K^-\pi^+\pi^+)} < 1.7 \times \tan^4\theta_c$$

and the resonant decay can be compared to the $\phi\pi^+$ decay mode giving

$$\frac{\Gamma(D^+ \to \phi K^+)}{\Gamma(D^+ \to \phi\pi^+)} < 20.3 \times \tan^4\theta_c$$

These limits are lower than and inconsistent with the level at which WA82 has claimed a signal (in the $K^+K^-K^+$ mode, see[3]) and a little lower than but consistent with a signal observed by E691 (see[4]).

Systematic errors in the above quantities have also been studied. The shift in the central values due to the different background shapes indicates that the systematic error due to the background shape uncertainty is $\sim$10% in the $D^+ \to K^+\pi^+\pi^-$ study and is $\sim$11% in the $D^+ \to K^+K^+K^-$ study. The relative efficiency of the Čerenkov cuts in the DCSD and Cabibbo-favored $D^+ \to K\pi\pi$ analyses is $99.0 \pm 0.2\%$ and we attribute a systematic error of no more than 4% to this source. We have checked that the lifetime of the $D^+ \to K^+\pi^+\pi^-$ sample is consistent with the Cabibbo-favored sample and that the acceptance is uniform across the Dalitz plot.

In the case of the $D^+ \to K^+K^+K^-$ sample, there is some uncertainty in the width of the candidate signal, which contributes an uncertainty of around 3%. Finally, also for the $D^+ \to K^+K^+K^-$ study, the uncertainty in the relative acceptance to the $D^+ \to K^-\pi^+\pi^+$ decays contributes $\sim$5% systematic error and there is a $\sim$5% statistical error due to Monte Carlo statistics.

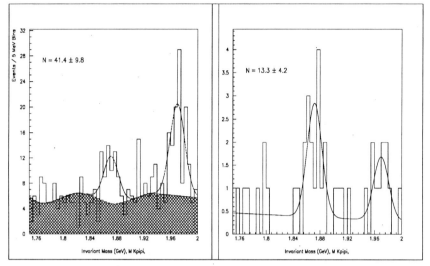

Fig. 3. Invariant mass distributions for a) the doubly Cabibbo-suppressed decay $D^+ \to K^+\pi^+\pi^-$ and b) for the doubly Cabibbo-suppressed decay $D^+ \to K^+\pi^+\pi^-$ with the $K\pi$ mass required to lie in the interval 840 - 945 MeV/c^2 and the cosine of the angle the Kaon makes with the D-direction in the K^* rest frame required to be greater than 0.5. Both figures are for data from $\sim$40% of the E791 sample.

References

1. J. S. Wiener "Doubly Cabibbo Suppressed Decays of the Charged D Meson", Princeton University Ph. D. thesis, unpublished (1994).
2. The Particle Data Group in the "Review of Particle Properties", Phys. Rev. **D50**, (1994) 1.
3. M. Adamovich et al., Phys. Lett. **B305** (1993) 177.
4. J. C. Anjos et al., Phys. Rev. Lett. **69** (1992) 2892.

RESULTS ON RARE AND FORBIDDEN CHARM DECAYS FROM FERMILAB E687

J.R.WILSON *

*Department of Physics and Astronomy
University of South Carolina
Columbia, SC 29208*

Representing the E687 Collaboration

ABSTRACT

Rare and forbidden decays are a sensitive test of the standard model and a likely place to observe new physics. Preliminary limits on direct CP violation, flavor changing neutral currents and lepton number violation from analysis of E687 data are presented.

1. Introduction

Rare processes are prime places to search for physics beyond the Standard Model. By choosing decay modes where the expected branching fraction is extremely small (or identically zero) any observation could be a signature of new physics. Although many of the rare decay processes discussed here have been studied extensively in the kaon sector it is possible that new physics is not uniform across quark generations and may be enhanced in the heavy quark sector.

The results presented here are from data collected by Fermilab experiment E687 during the 1990-1991 fixed target running period. E687 utilized a photon beam of average energy ≈ 200 GeV. The data yielded a sample of about 80000 fully reconstructed charm particle decays. A detailed description of the spectrometer and general data analysis techniques can be found elsewhere.[1]

2. CP Violation

Direct CP violation can be observed through asymmetries in the decay of particle and antiparticle which come about through interference between strong and weak components of the decay amplitude.[2,3] The CP violating term of the weak lagrangian is expected to be small so these asymmetries will be easiest to observe in decay modes with a small weak amplitude so that the interference effects induced by CP violation are enhanced. We have searched for asymmetries in singly Cabibbo suppressed decays of $D^0-\bar{D}^0$ and D^+-D^- mesons.

---

*For a complete list of co-authors, see H.W.K.Cheung, "*The Physics of Charm Lifetimes*", these proceedings.

Table 1. CP Violating Asymmetry and 90% Confidence Level Limits.

Decay Mode	Measured Asymmetry	90% C.L. limit
$D^0 \to K^+K^-$	0.024 ± 0.084	$-.11 < A_{CP} < .16$
$D^+ \to K^-K^+\pi^+$	-0.031 ± 0.068	$-.14 < A_{CP} < .081$
$D^+ \to \overline{K}^{*0}K^+$	-0.12 ± 0.13	$-.33 < A_{CP} < .094$
$D^+ \to \phi\pi^+$	0.066 ± 0.086	$-.075 < A_{CP} < .21$

CP asymmetries can be masked by production asymmetries so we normalized the singly suppressed decay modes to the Cabibbo favored modes. The singly suppressed modes and normalization modes were selected with similar particle ID and vertex cuts to minimize systematic effects. For the D^0 case we looked for asymmetries in the K^+K^- mode and normalized to $\Gamma(D^0 \to K^-\pi^+)$. The D^0 ($\bar{D}^0$) was tagged by the soft pion from the decay of a D^{*+} (D^{*-}). The asymmetry is given by the ratio $A_{CP} = \frac{\eta_+ - \eta_-}{\eta_+ - \eta_-}$, where $\eta_+ = \frac{N_{corr}(D^0 \to K^+K^-)}{N_{corr}(D^0 \to K^-\pi^+)}$ and η_- is the charge conjugate of η_+. N_{corr} is the yield corrected by acceptance and reconstruction efficiency. The CP violation limit we extracted for this decay mode appears in Table 1.

The analysis of the D^+ proceeds in a manner analogous to that of the D^0 except that the CP mode is $D^+ \to K^-K^+\pi^+$ and the normalization mode was $D^+ \to K^-\pi^+\pi^+$ (charge conjugate is similar). The resonant submodes have different hadronic properties and hence could have very different CP asymmetries so they were considered separately from the inclusive mode. The only difference in the analysis of the resonant submodes is that a sideband subtraction was done to account for the non-resonant D^+s which appear in the background under the peak of the resonance. The CP violation limits for the D^+ are summarized in Table 1.

3. Other Rare Decays

Like CP violation, flavor changing neutral currents (FCNC) are allowed by the standard model. They are second order electroweak effects and have been calculated to have branching ratios smaller then 10^{-8}.[4] Lepton number violating decay modes (LNV) however, are forbidden by the standard model.

The three FCNC decays $D^0 \to \mu^+\mu^-$, $D^+ \to \pi^+\mu^+\mu^-$, and $D^+ \to K^+\mu^+\mu^-$ were all studied using E687 data. The branching ratios were measured relative to $K^-\pi^+$ for the D^0 and to $K^-\pi^+\pi^+$ for the D^+. A primary background for these modes is reflection of the Cabibbo favored modes by misidentification of the pions as muons. In E687 this mis-ID rate was about 1% per track (negligible) and there were typically no events in the D mass region for the FCNC modes. These preliminary limits are summarized in Table 2 along with comparisons to the current PDG values[5] and other recent experimental results (also mostly preliminary).

The LNV decays investigated were $D^+ \to K^-\mu^+\mu^+$, and $D^+ \to \pi^-\mu^+\mu^+$ (+ c.c.). The preliminary limits from these searches are also listed in Table 2.

Table 2. 90% Confidence Level Upper Limits ($\times 10^5$) on FCNC and LNV Charm Decay Modes.

Type	Decay Mode	E687	E653	E771	E789	E791	PDG
FCNC	$D^0 \to \mu^+\mu^-$	2.7		1.2	3.1		1.1
	$D^+ \to \pi^+\mu^+\mu^-$	9.7	22			4.6	290
	$D^+ \to K^+\mu^+\mu^-$	8.5	33				920
LNV	$D^+ \to \pi^-\mu^+\mu^+$	17	20				680
	$D^+ \to K^-\mu^+\mu^+$	20	33				430

References

1. P.L.Frabetti et al., NIM **A320** (1992) 519.
2. F. Buccella et al., Phys. Lett. **B302** (1993) 319.
3. M. Golden and B. Grinstein, Phys. Lett. **B222** (1989) 501.
4. A. J. Schwartz, Modern Phys. Lett. A, Vol. 8, No. 11 (1993) 967.
5. L. Montanet et al., "Review of Particle Properties", Phys. Rev. D50, No. 3 (1994) 1173.

A SEARCH FOR $D^+ \to \pi^+ l^+ l^-$ BY FERMILAB E791

AI NGUYEN *

Department of Physics, Kansas State University, Manhattan, KS 66506, U.S.A.

ABSTRACT

We present preliminary results of a search for the charm-changing neutral current decays $D^+ \to \pi^+ \mu^+ \mu^-$ and $D^+ \to \pi^+ e^+ e^-$, based on the analysis of one half of the data from Fermilab experiment E791, accumulated during the 1991-92 fixed target run with a 500 GeV π^- beam incident on a segmented target. Assuming that the decay kinematics are the same for $\pi\mu\mu$ as for $K\pi\pi$, we set an upper limit on the branching ratio for this channel, $B(D^+ \to \pi^+ \mu^+ \mu^-) < 1.9 \times 10^{-5}$ at the 90% confidence level.

Fermilab E791 is a high statistics heavy quark experiment with excellent sensitivity for rare charm decays. The world's largest sample of charm decays was recorded in the 1991-1992 fixed-target run in a 500 GeV/c π^- beam. We have searched 54% of these data for the flavour-changing neutral current (FCNC) decays $D^+ \to \pi^+ \mu^+ \mu^-$ and $D^+ \to \pi^+ e^+ e^-$. In the Standard Model these decays are allowed in second order and the branching ratio for the inclusive process $c \to u l^+ l^-$ is estimated to be 1.8×10^{-8}.[1]

The E791 spectrometer is an upgraded version of the apparatus used in Fermilab experiments E516, E691, and E769.[2] Twenty-three planes of silicon microstrip detectors provided information for efficient track and vertex reconstruction. Muon identification was based on two planes of scintillation counters located behind shielding with a thickness equivalent to 2.5 meters of steel. Electron identification was based on the agreement between the energy deposited in the calorimeter and the momentum from the tracking chambers (E/p), and on the electromagnetic shower shape.

The strategy of the experiment was to collect data with an open trigger and then select interesting events offline. During the 1991-1992 run the data acquisition system accepted 10^4 events/sec during a typical beam spill and wrote a total of 2×10^{10} interactions onto 24,000 8 mm tapes.[3] Approximately 96% of our data set has now been reconstructed. The number of fully reconstructable charm decays is estimated to be 2.5×10^5.

Our search for $D^+ \to \pi^+ \mu^+ \mu^-$ requires pairs of oppositely-charged muons from a three-prong secondary vertex significantly detached from the primary vertex. The muon identification efficiency is 97% within our geometric acceptance, and the pion

---

*Representing the E791 Collaboration: Centro Brasileiro de Pesquisas Fisicas, Brazil, University of California, Santa Cruz, University of Cincinnati, CINVESTAV, Mexico, Fermilab, Illinois Institute of Technology, Kansas State University, University of Mississippi, Princeton University, Universidad Autonoma de Puebla, Mexico, Tel Aviv University, Israel, Tufts University, University of Wisconsin, Yale University.

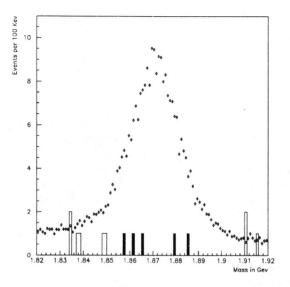

Fig. 1. Invariant mass distributions for our normalization channel $K\pi\pi$ (diamonds, scaled by $1/100$) and for our search channel $\pi\mu\mu$ (histogram, events in search window are shaded).

misidentification probability varies between 5% and 10%, depending on the particle's momentum.

We require secondary tracks and vertices to be of good quality. The decay vertex is required to be well separated from material in the target foils. Decay tracks are required to have a significant impact parameter relative to the primary vertex. To ensure that no tracks are missing from the secondary vertex and, in particular to eliminate background from charm semileptonic decays, the net momentum transverse to the candidate D^+ line of flight must be less than 250 MeV/c.

Candidate di-muon decays are those selected secondary vertices that contain two oppositely charged tracks identified as muons and whose $\pi\mu\mu$ invariant mass is in the D^+ mass region. As shown in figure 1, we find five such candidate vertices with $\pi\mu\mu$ mass between 1.85 and 1.89 GeV/c^2.

We calculate the expected number of background events as a product of two terms, $N_{bg} = N_0 \times R$, where the normalization $N_0 = 1390$ is the number of three-prong candidates selected prior to any muon identification, with $\pi\mu\mu$ mass between 1.85 and 1.89 GeV/c^2, and $R = (0.68\pm0.06)\%$ is the probability that a pair of oppositely-charged tracks is misidentified as a muon pair. We measure R using the background-subtracted rate of such events in our sample of $D^+ \rightarrow K^-\pi^+\pi^+$ decays with identical vertex cuts. The resultant background estimate is $N_{bg} = 9.5 \pm 0.9$ events.

To obtain an upper limit for the decay $D^+ \to \pi^+\mu^+\mu^-$ we normalize to the well established decay mode $D^+ \to K^-\pi^+\pi^+$. Using event selection criteria identical to those employed for the FCNC search but without muon identification, we find a total of $N(K^-\pi^+\pi^+) = 20327 \pm 150$ events.

Assuming that $\pi\mu\mu$ and $K\pi\pi$ are both non-resonant phase space decays we obtained a Monte-Carlo estimate of the relative efficiency $\eta = \epsilon(\pi\mu\mu)/\epsilon(K\pi\pi) = 96\%$, excluding our efficiency for muon identification. We compute the upper limit as

$$B(D^+ \to \pi^+\mu^+\mu^-) = \frac{U(\pi^+\mu^+\mu^-)}{N(K^-\pi^+\pi^+)}\frac{1}{\eta\epsilon^2}B(D^+ \to K^-\pi^+\pi^+) \tag{1}$$

where $U = 3.9$ is the 90%-confidence-level upper limit for an observation of 5 events when a background of 9.5 is expected. We determine the muon identification efficiency to be $\epsilon = (97 \pm 1)\%$ within our geometrical acceptance. The Particle Data Group value for the hadronic branching ratio is $B(D^+ \to K^-\pi^+\pi^+) = (9.1 \pm 0.6)\%$.[4] The resulting upper limit on the branching ratio is $B(D^+ \to \pi^+\mu^+\mu^-) = 1.9 \times 10^{-5}$ at the 90% confidence level. We investigated possible degradations of this limit due to the uncertainty in our background and found such effects to be negligible.

Thus our search for $D^+ \to \pi^+\mu^+\mu^-$, for the same assumptions made in the analysis leading to the current best limit,[5] results in an improvement by more than two orders of magnitude. The important common assumption is that the $\pi\mu\mu$ and $K\pi\pi$ decay modes have the same kinematic decay distributions. However, the FCNC process, if it exists, is new physics, with unknown properties. There is no a priori reason to expect identical decay kinematics.

Our search for $D^+ \to \pi^+e^+e^-$ employs vertex selection criteria identical to those used in our $\pi\mu\mu$ search. Our electron identification efficiency varies from 62% for momenta below 9 GeV/c to 45% for momenta above 20 GeV/c, with a momentum-independent pion misidentification probability of 0.8%.

Candidate di-electron decays are those selected secondary vertices that contain two oppositely-charged tracks identified as electrons and whose πee invariant mass is in the D^+ mass region. Since on average electrons lose 23% of their energy in material upstream of the calorimeters we expect a degradation of our mass resolution for semi-electronic decays and adjust the lower edge of our D mass window to compensate for this effect. We find no candidates with πee mass between 1.83 and 1.89 GeV/c^2. Preliminary analysis indicates that our sensitivity to the πee channel is comparable to our sensitivity to the $\pi\mu\mu$ channel. Our limit on $B(D^+ \to \pi^+e^+e^-)$ will be available later this year.

References

1. A. J. Schwartz, *Mod. Phys. Lett.* **A8** (1993) 967.
2. J. A. Appel, *Ann. Rev. Nucl. Part. Sci.* **42** (1992) 367.
3. S. Amato et al., *Nucl. Instr. and Meth.* **A324** (1993) 535.
4. Review of Particle Properties, *Phys. Rev.* **D50** (1994) 1203.
5. P. Haas et al., *Phys. Rev. Lett.* **60** (1988) 1614.

A SEARCH FOR $D^0\overline{D}^0$ MIXING AT THE FERMILAB E791 EXPERIMENT

GUY BLAYLOCK*

Santa Cruz Institute for Particle Physics,
University of California,
Santa Cruz, CA 95064, U.S.A.

ABSTRACT

We describe two searches for $D^0\overline{D}^0$ mixing which use the decay mode $D^* \to \pi D^0$ with $D^0 \to \pi(K3\pi)$ or $D^0 \to \pi(K\pi)$. Results are presented for 25% of the total E791 data sample for the $K3\pi$ mode and 33% for the $K\pi$ mode. When we allow arbitrary interference between mixing and doubly-Cabibbo-suppressed amplitudes, our mixing limits from measurement of the lifetime distributions are $r_{mix}(K3\pi) < 3.0\%$ and $r_{mix}(K\pi) < 0.9\%$ at 90% CL. If, as in other reported limits, we assume no interference, the limits drop to $r_{mix}(K3\pi) < 0.3\%$ and $r_{mix}(K\pi) < 0.5\%$.

1.   Introduction

The strength of D meson mixing is usually characterized by the parameter r_{mix}, which describes the rate of $D^0\overline{D}^0$ mixing relative to the rate of normal D^0 decays:

$$r_{mix} \equiv \frac{R(D^0 \to \overline{D}^0 \to \overline{X})}{R(D^0 \to X)}. \tag{1}$$

In the Standard Model, r_{mix} is expected to be several orders of magnitude below the current experimental limit[4] of 0.37×10^{-2}. However, a number of other models predict measurable rates for $D^0\overline{D}^0$ mixing. These include models with a fourth generation bottom quark, supersymmetry, multiple Higgs doublets, and many others.[1] Evidence of $D^0\overline{D}^0$ mixing near the current limit could therefore be interpreted as an indication of new physics. Conversely, if mixing remains undetected below the current experimental limit, the parameters of many new physics theories can be constrained.

In order to search for $D^0\overline{D}^0$ mixing, it is necessary to measure the charm quantum number of neutral D mesons at both their production and decay points. A discrepancy between the charm quantum number of the D at birth and at decay is a possible indication of mixing. At E791, we analyze the decay chains $D^* \to \pi D \to \pi(K\pi)$ and

---

*Representing the E791 Collaboration: Centro Brasileiro de Pesquisas Fisicas, University of California at Santa Cruz, University of Cincinnati, CINVESTAV, Fermilab, Illinois Institute of Technology, Kansas State University, University of Mississippi, Princeton University, Universidad Autonoma de Puebla, Tel Aviv University, Tufts University, University of Wisconsin, Yale University.

$D^* \to \pi D \to \pi(K3\pi)$. The charge of the pion from the D^* decay identifies the D at birth, while the charge of the daughter kaon identifies the D at decay.

By convention, events with the charge of the kaon opposite the charge of the pion from the D^* decay are called "right-sign" decays. Events with the kaon charge equal to the pion charge are labelled "wrong-sign" decays, and may include events in which the D has mixed. Unfortunately, the wrong-sign samples for the hadronic final states contain contributions from doubly-Cabibbo-suppressed decays (DCSD's) as well as mixing.[2] However, these two contributions can be distinguished statistically by looking at the decay time distribution. In the limit of small mixing (and no CP violation), the rate of wrong-sign decays has the time dependence:[3]

$$R(wrong\ sign) \propto e^{-\Gamma t}\left\{(\Delta M^2 + \frac{1}{4}\,\Delta\Gamma^2)\,t^2 + 4\,|\rho|^2 + 2\,Re(\rho)\Delta\Gamma\,t\right\} \qquad (2)$$

where ρ represents the amplitude for DCSD's relative to Cabibbo-favored decays. The term proportional to t^2 denotes mixing, while the constant term represents DCSD's and the term proportional to t corresponds to interference between the mixing and DCSD amplitudes.

2. The E791 Experiment

E791 is a high statistics, charm production experiment which ran at the Fermilab Taggged Photon Lab in 1991 and 1992. A 500 GeV/c π^- beam was incident on a segmented foil target. The features of the detector which are most important to the mixing measurement are the silicon microstrip detector, the drift chamber system and the Cherenkov detectors. The 23 planes of silicon microstrip detectors provided precise charged particle tracking near the target region and allowed reconstruction of the charm decay vertex. Two analysis magnets and 35 planes of drift chambers measured charged particle momenta. Two threshold Cherenkov counters allowed charged particle identification over a momentum range of 6 to 200 GeV/c. More than 20 billion events were recorded on 24,000 8mm data tapes in the course of five months. Event reconstruction will be completed by fall 1994 and is expected to yield about 200,000 fully reconstructed charm decays.

3. Data Samples

A neutral D meson is identified primarily by the secondary decay vertex, which is typically separated from the primary interaction by a few millimeters. The reconstructed D momentum is used to insure that the D came from the primary interaction, and also provides a determination of the proper lifetime. The identities of the decay products as kaons or pions are determined by the Cherenkov counters. The possibility of misidentifying two decay products, thereby turning a right-sign decay into a wrong-sign decay, is ruled out by mass reflection cuts.

Figure 1 shows the distribution of $m_{K3\pi}$ versus $\Delta m = m_{(K3\pi)\pi} - m_{K3\pi}$ for our selected sample of $D^* \to \pi D \to \pi(K3\pi)$ candidates. Approximately 25% of the full E791 data sample is represented. In the right-sign plot (left), a clear signal is seen at $m_{K3\pi} = m_{D^0} = 1.87$ GeV/c^2 and $\Delta m = m_{D^*} - m_{D^0} = 0.145$ GeV/c^2. No significant signal is seen in the wrong-sign plot (right). In both plots, the dominant background comes from real D^0 decays which combine with a random track in the event to form a false D^* candidate. This is evident in each case by the band of excess events at the D^0 mass.

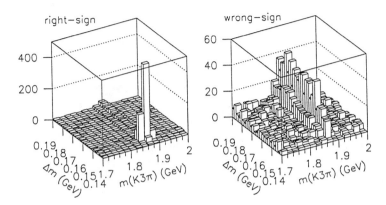

Fig. 1. A plot of $\Delta m \equiv m_{(K3\pi)\pi} - m_{K3\pi}$ versus $m_{K3\pi}$ for right-sign candidates (left) and for wrong-sign candidates (right). In both plots, the bands of events at $m_{K3\pi} \approx 1.87$ GeV/c^2 are due to real D^0 decays combining with random pions to make false D^* candidates.

    Figure 2 shows the decay time distribution of the candidates from the wrong-sign signal region defined by $1.845 < m_{K3\pi} < 1.885$ GeV/^{2} and $0.144 < \Delta m < 0.147$ GeV/c^2. The distribution has been corrected for efficiency as determined from the right-sign data. The dashed line shows the expected background, indicating that the events in this region are indeed consistent with pure background.

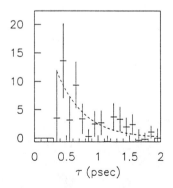

Fig. 2. A plot of the decay time distribution of wrong-sign events, corrected for efficiency. The dashed line shows the expected background.

4. Results

By fitting the distribution in Fig. 2 to the form of Eq. 2 (plus background), we can extract upper limits for the mixing parameter r_{mix}. In the case of $D^0 \to K3\pi$, we calculate $r_{mix}(K3\pi) < 3.0 \times 10^{-2}$ at 90% CL. A similar analysis of $D^0 \to K\pi$ candidates has been performed on 33% of the full E791 data sample. From a fit to the distributions in $m_{K\pi}$, Δm and lifetime, we calculate $r_{mix}(K\pi) < 0.9 \times 10^{-2}$.

In the $K\pi$ final state, unlike the $K3\pi$ final state, there is a small excess of events in the wrong-sign signal region (not shown) which are consistent with doubly-Cabibbo-suppressed decays. Our estimate of the rate is $r_{DCSD} = 1.9^{+0.6}_{-0.8} \times 10^{-2}$.

The previous best experimental limit on $D^0\overline{D^0}$ mixing comes from the E691 experiment,[4] which quotes $r_{mix} < 0.37 \times 10^{-2}$ from a study of both $D \to K\pi$ and $D \to K3\pi$ decays. This limit assumes no interference between DCSD and mixing amplitudes. If, for comparison, we calculate limits on mixing using the same assumption, we obtain $r_{mix}(K3\pi) < 0.3 \times 10^{-2}$ and $r_{mix}(K\pi) < 0.5 \times 10^{-2}$.

5. Prospects

E791 is currently setting the world's best limits on $D^0\overline{D^0}$ mixing. The full E791 data sample is 3 to 4 times larger than the samples shown in this paper. We will have completed analysis of the full data sample by fall 1994.

References

1. For a summary of predictions for $D^0\overline{D^0}$ mixing see the contribution by JoAnne Hewett in these proceedings.
2. It should be noted that there is no DCSD amplitude to wrong-sign semileptonic final states. This is one important advantage of using the semileptonic modes which partially offsets the disadvantage of a missing neutrino.
3. I.I. Bigi, *Proceedings of the XXIII International Conference on High Energy Physics*, Berkeley, ed. S.C. Loken, World Scientific, 1986, p. 857.
   In the formula shown here, a term which explicitly violates CP ($\mp 4Im(\rho)\Delta M\, t$), and a term which is third order in small quantities ($-2|\rho|^2\Delta\Gamma\, t$) have both been neglected.
4. J.C. Anjos *et al.*, *Phys. Rev. Lett.* **60**, 1239 (1988).

CP VIOLATION IN HYPERON DECAYS

XIAO-GANG HE*

Institute of Theoretical Science, University of Oregon, Eugene
OR 97403, USA

and

SANDIP PAKVASA

Department of Physics and Astronomy, University of Hawaii, Honolulu
HI 96822, USA

ABSTRACT

The present status for CP violation in hyperon decays is reviewed.

1. Introduction

Non-leptonic hyperon decays of Λ, Σ and Ξ[1-8] are interesting processes to test CP conservation outside the neutral Kaon sysytem. Measurements of CP violation in hyperon decays will provide us with useful information about the origin of CP violation. Several proposals have been made to look for CP violation in hyperon decays.[6,7] Recently the E871 proposal at Fermilab has been approved.[8] The expected sensitivity for CP violation test is about the same order of magnitude as the Standard Model (SM) prediction. This experiment will measure CP violation in $\Xi^- \to \Lambda \pi^-$ and $\Lambda \to p\pi^-$. The E871 experiment will start to take data as early as 1996. In this talk we will review the present status for CP violation in hyperon decays. We will concentrate on Ξ and Λ decays because there is not hope to measure CP violation in Σ decays in the near future.

Non-leptonic hyperon decays can proceed into both S-wave and P-wave final states with amplitudes S and P, respectively. One can write the amplitude as

$$Amp(B_i \to B_f \pi) = S + P\vec{\sigma} \cdot \vec{q}, \qquad (1)$$

where $\vec{q}$ is the momentum of the final baryon B_f. Experimental observables are: the decay width Γ, and the parameters in the decay angular distribution. In the rest frame of the initial hyperon, the angular distribution is given by

$$\frac{4\pi}{\Gamma}\frac{d\Gamma}{d\Omega} = 1 + \alpha \hat{s}_i \cdot \hat{q} + \hat{s}_f \cdot [(\alpha + \hat{s}_i \cdot \hat{q})\hat{q} + \beta \hat{s}_i \times \hat{q} + \gamma(\hat{q} \times (\hat{s}_i \times \hat{q}))], \qquad (2)$$

---

*Talk presented by Xiao-Gang He

where $s_{i,f}$ are the spins of the initial and final baryons, respectively. $\hat{v}$ indicates the direction of the corresponding vector. The parameters α, β and γ are defined as

$$\alpha = \frac{2Re(S^*P)}{|S|^2 + |P|^2}, \quad \beta = \frac{2Im(S^*P)}{|S|^2 + |P|^2}, \quad \gamma = \frac{|S|^2 - |P|^2}{|S|^2 + |P|^2}. \tag{3}$$

Only two of them are independent. We will discuss α and β. In the literature, β is sometimes parametrized as $\beta = \sqrt{1 - \alpha^2}\sin\phi$.

It is convenient to write the amplitudes as

$$S = \sum_i S_i e^{i(\phi_i^S + \delta_i^S)}, \quad P = \sum_i P_i e^{i(\phi_i^P + \delta_i^P)} \tag{4}$$

to explicitly separate the strong rescattering phases δ_i and the weak CP violating phases ϕ_i.

The decay amplitudes $\bar{S}$ and $\bar{P}$ for anti-hyperon can be parametrized in a similar way. Then

$$\bar{S} = -\sum_i S_i e^{i(-\phi_i^S + \delta_i^S)}, \quad \bar{P} = \sum_i P_i e^{i(-\phi_i^P + \delta_i^P)}. \tag{5}$$

We will denote the observables in anti-hyperon decays with a bar on the corresponding ones in hyperon decays.

2. CP Violating Observables

Several CP violating observables can be constructed using the observables discussed in the previous section. The interesting ones are[2]

$$\Delta = \frac{\Gamma - \bar{\Gamma}}{\Gamma + \bar{\Gamma}}, \quad A = \frac{\Gamma\alpha + \bar{\Gamma}\bar{\alpha}}{\Gamma\alpha - \bar{\Gamma}\bar{\alpha}} \approx \frac{\alpha + \bar{\alpha}}{\alpha - \bar{\alpha}} + \Delta,$$

$$B = \frac{\Gamma\beta + \bar{\Gamma}\bar{\beta}}{\Gamma\beta - \bar{\Gamma}\bar{\beta}} \approx \frac{\beta + \bar{\beta}}{\beta - \bar{\beta}} + \Delta. \tag{6}$$

All these CP violating observables can, in principle, be measured experimentally. It has been shown that the low energy reaction $p_i\bar{p}_i \to \Lambda\bar{\Lambda} \to p_f\pi^- \bar{p}_f\pi^+$ can be used to measure A for Λ.[9] The measurement

$$\tilde{A} = \frac{N_p^+ - N_p^- + N_{\bar{p}}^+ - N_{\bar{p}}^-}{N_{total}}, \tag{7}$$

is equal to $P_\Lambda\alpha_\Lambda A(\Lambda)$. Here $N_p^\pm$ indicates events with $(\hat{p}_i \times \hat{P}_\Lambda) \cdot \hat{p}_f > 0$ or < 0, and similarly for anti-particles. P_Λ is the polarization of the Λ produced in the $p\bar{p}$ collision.

The measurement of B requires the analysis of the polarization of the final baryon. Low energy $p\bar{p}$ collision can also measure B for Ξ decays.[9] In the process, $p_i\bar{p}_i \to \Xi\bar{\Xi} \to \Lambda\pi^-\bar{\Lambda}\pi^+ \to p_f\pi^-\pi^- \bar{p}_f\pi^+\pi^+$, one can measure

$$\tilde{B} = \frac{\tilde{N}_p^+ - \tilde{N}_p^- + \tilde{N}_{\bar{p}}^+ - \tilde{N}_{\bar{p}}^-}{N_{total}}, \tag{8}$$

where $\tilde{N}_p^{\pm}$ indicates events with $\hat{P}_\Xi \cdot (\hat{p}_f \times \hat{p}_\Lambda) > 0$ or < 0, and similarly for anti-particles. $\tilde{B}$ is given by $(\pi/8)P_\Xi\alpha_\Lambda\beta_\Xi(A(\Lambda) + B(\Xi))$, and hence a measurement of $\tilde{A}$ and $\tilde{B}$ yields $A(\Lambda)$ and $B(\Xi)$.

There have been some measurements for CP violation in hyperon decays by several groups,

$$
\begin{aligned}
A(\Lambda \to p\pi^-) &= -0.02 \pm 0.14 , &&\text{From } p\bar{p} \to \Lambda X \text{ and } p\bar{p} \to \bar{\Lambda}X. &&\text{Ref.[10]} \\
A(\Lambda \to p\pi^-) &= -0.07 \pm 0.09 , &&\text{From } p\bar{p} \to \Lambda\bar{\Lambda} \to p\pi^-\bar{p}\pi^+. &&\text{Ref.[11]} \\
A(\Lambda \to p\pi^-) &= 0.01 \pm 0.10 , &&\text{From } J/\psi \to \Lambda\bar{\Lambda} \to p\pi^-\bar{p}\pi^+. &&\text{Ref.[12]}
\end{aligned}
$$

$$(9)$$

The experiment at E871 will measure $\alpha_\Lambda\alpha_\Xi$ in the decay $\Xi^- \to \Lambda\pi^- \to p\pi^-\pi^-$, and similar measurement for anti-Ξ decays.[8] An asymmetry A_{asy} can be extracted

$$
A_{asy} = \frac{\alpha_\Lambda\alpha_\Xi - \bar{\alpha}_\Lambda\bar{\alpha}_\Xi}{\alpha_\Lambda\alpha_\Xi + \bar{\alpha}_\Lambda\bar{\alpha}_\Xi} \approx A(\Lambda) + A(\Xi) . \tag{10}
$$

The expected sensitivity for A_{asy} is 10^{-4} and may reach 10^{-5} which would test the SM predictions.

3. Theoretical Calculations

There are large uncertainties in theoretical calculations for the CP violating observables due to our poor understanding of the hadronic matrix elements. To reduce errors it is best to use experimental measurements for CP conserving quantities and to calculate CP violating parameters theoretically, that is, we calculate the weak phases $\phi^{s,p}$. The experimental data on CP conserving quantities are summarized below.

The isospin decomposition of Λ and Ξ are given by[13]

$$
\begin{aligned}
\Lambda \to p\pi^- , \quad S(\Lambda_-^0) &= -\sqrt{\frac{2}{3}}S_{11}e^{i(\delta_1+\phi_1^s)} + \sqrt{\frac{1}{3}}S_{33}e^{i(\delta_3+\phi_3^s)}) , \\
P(\Lambda_-^0) &= -\sqrt{\frac{2}{3}}P_{11}e^{i(\delta_{11}+\phi_1^p)} + \sqrt{\frac{1}{3}}P_{33}e^{i(\delta_{33}+\phi_3^p)}) , \\
\Lambda \to n\pi^0 , \quad S(\Lambda_0^0) &= \sqrt{\frac{1}{3}}S_{11}e^{i(\delta_1+\phi_1^s)} + \sqrt{\frac{2}{3}}S_{33}e^{i(\delta_3+\phi_3^s)}) , \\
P(\Lambda_0^0) &= \sqrt{\frac{1}{3}}P_{11}e^{(i\delta_{11}+\phi_1^p)} + \sqrt{\frac{2}{3}}P_{33}e^{i(\delta_{33}+\phi_3^p)}) , \\
\Xi^- \to \Lambda\pi^- , \quad S(\Xi_-^-) &= S_{12}e^{i(\delta_2+\phi_{12}^s)} + \frac{1}{2}S_{32}e^{i(\delta_2+\phi_{32}^s)}) , \\
P(\Xi_-^-) &= P_{12}e^{i(\delta_{21}+\phi_{12}^p)} + \frac{1}{2}P_{32}e^{i(\delta_{21}+\phi_{32}^p)}) , \\
\Xi^0 \to \Lambda\pi^0 , \quad S(\Xi_-^-) &= \sqrt{\frac{1}{2}}(S_{12}e^{i(\delta_2+\phi_{12}^s)} - S_{32}e^{i(\delta_2+\phi_{32}^s)}) , \\
P(\Xi_-^-) &= \sqrt{\frac{1}{2}}(P_{12}e^{i(\delta_{21}+\phi_{12}^p)} - P_{32}e^{i(\delta_{21}+\phi_{32}^p)}) .
\end{aligned}
$$

$$(11)$$

These decays are dominated by the $\Delta I = 1/2$ amplitudes. Experimental measurements give:[14] $S_{33}/S_{11} = 0.027\pm0.008$, $P_{33}/P_{11} = 0.03\pm0.037$; $S_{32}/S_{12} = -0.046\pm0.014$, and $P_{32}/P_{12} = -0.01 \pm 0.04$. From $N\pi$ scattering, the strong rescattering phase for Λ decays are determined to be[15]: $\delta_1 \approx 6.0^0$, $\delta_3 \approx -3.8$, $\delta_{11} \approx -1.1^0$ and $\delta_{31} = -0.7^0$ with errors of order 1^0. The strong rescattering phases for Ξ decays are not exprimentially determined. Theoretical predictions very a large range. Nath and Kumer[16] obtained: $\delta_{21} = -2.7^0$ and $\delta_2 = -18.7$. Martin[17] obtained $\delta_{21} = -1.2^0$. Recently, Lu, Savage and Wise,[18] using chiral pertubation theory, obtained $\delta_{21} = -1.7^0$ and $\delta_2 = 0$ to the lowest order. In this last estimate, contributions from $1/2^-$ and $3/2^-$ states are not included, which can give rise to a significant δ_2.

To a very good approximation, the CP violating observables can be simplified to yield,[2]

$$
\begin{aligned}
\Delta(\Lambda_-^0) &= -2\Delta(\Lambda_0^0) = \sqrt{2}\frac{S_{33}}{S_{11}}\sin(\delta_3 - \delta_1)\sin(\phi_3^s - \phi_1^s) , \\
A(\Lambda_-^0) &= A(\Lambda_0^0) = -\tan(\delta_{11} - \delta_1)\sin(\phi_1^p - \phi_1^s) , \\
B(\Lambda_-^0) &= B(\Lambda_0^0) = \cot(\delta_{11} - \delta_1)\sin(\phi_1^p - \phi_1^s) , \\
\Delta(\Xi_-^-) &= \Delta(\Xi_0^0) = 0 , \\
A(\Xi_-^-) &= A(\Xi_0^0) = -\tan(\delta_{21} - \delta_2)\sin(\phi_{12}^p - \phi_{12}^s) \\
B(\Xi_-^-) &= B(\Xi_0^0) = \cot(\delta_{21} - \delta_2)\sin(\phi_{12}^p - \phi_{12}^s) .
\end{aligned}
\tag{12}
$$

It is well known that for large top quark mass, there is considerable cancellation for the $I = 0$ and $I = 1$ contributions to ϵ'/ϵ, and ϵ'/ϵ can be quite small.[19] Such cancellation does not happen to the quantities A and B because they are dominated by $I = 1/2$ quantities. Hence hyperon decays probe a somewhat different operators, even in the SM, from ϵ'/ϵ.

3.1.   The Standard Model Predictions.

In the SM, the origin of CP violation is the non-trivial phase in the KM matrix.[20] The effective Hamiltonian responsible for non-leptonic hyperon decays is given by

$$
H_{eff} = \frac{G_F}{\sqrt{2}}V_{ud}^*V_{us} \sum_i C_i(\mu)Q_i ,
\tag{13}
$$

where the sum is over all the Q_i four-quark operators, and the $C_i(\mu)$ is the Wilson coefficients. C_i contains both the CP conserving and CP violating parts. To separate KM mixings and other dependences, C_i is usually parametrized as $C_i = z_i + \tau y_i$, where $\tau = -V_{td}^*V_{ts}/V_{ud}^*V_{us}$. CP violating part is proportional to $Im(\tau)$. To obtain the weak phases, we need to evaluate

$$
ImM = \frac{G_F}{\sqrt{2}}V_{ud}^*V_{us}Im(\tau) < \pi B_f|\sum_i y_i(\mu)Q_i(\mu)|B_i > .
\tag{14}
$$

The quantity $y_i(\mu)$ is calculated by taking $y(m_W)$ as an initial value and then using renormalization group equation to reach the scale μ.[19] The most difficult part of the calculation is to evaluate $< \pi B_f |Q_i| B_i >$. At present, there is no convincing method to calculate these matrix elements. There are many models which can give estimates. However, it is known that all these models can not satisfactorily explain the $I = 1/2$ dominance in hyperon decays. They can at most produce the experimental amplitudes up to a factor of 2. It is therefore expected that the estimate for CP violation in hyperon decays can easily off by a factor of 2. The following is the result obtained by using the vacuum saturation and factorization approximation,[4]

$$ImM = \frac{G_F}{\sqrt{2}} V_{ud}^* V_{us} Im(\tau)[(M_1^s + M_3^s)V + (M_1^p + M_3^p)P] , \tag{15}$$

with

$$M_1^s = \frac{y_1 - 2y_2}{3} - \frac{y_7}{2} + \xi(\frac{-2y_1 + y_2}{3} - y_3 - \frac{y_8}{2}) - Y(y_6 + \frac{y_8}{2} + \xi(y_5 + \frac{y_7}{2})) ,$$

$$M_1^p = \frac{y_1 - 2y_2}{3} - \frac{y_7}{2} + \xi(\frac{-2y_1 + y_2}{3} - y_3 - \frac{y_8}{2}) + Z(y_6 + \frac{y_8}{2} + \xi(y_5 + \frac{y_7}{2})) ,$$

$$M_3^s = -\frac{y_1 + y_2}{3}(1 + \xi) + \frac{y_7}{2} + \frac{y_8}{2} - Y(\xi y_7 + y_8) ,$$

$$M_3^p = -\frac{y_1 + y_2}{3}(1 + \xi) + \frac{y_7}{2} + \frac{y_8}{2} + Z(\xi y_7 + y_8) , \tag{16}$$

where $Y = 2m_\pi^2/(m_u + m_d)(m_s - m_u)$, $Z = 2m_\pi^2/(m_u + m_d)(m_s + m_u)$, and $\xi = 1/N$ with N the number of color.

For $\Lambda \to p\pi^-$,

$$V = i\sqrt{2}F_\pi(m_\Lambda - m_p)\sqrt{\frac{3}{2}}\bar{p}\Lambda , \quad P \approx -i\frac{2F_\pi F_K m_\pi^2}{m_K^2 - m_\pi^2} g_{\Lambda pK} \bar{p}\gamma_5\Lambda , \tag{17}$$

where m_i are the masses of the particle i, $F_\pi = 93$ MeV, $F_K \approx 1.3F_\pi$ and $g_{\Lambda pK} \approx -13.3$. Similarly one can obtain the decay apmlitudes for Ξ decays.

Using these estimates for the matrix elements and information for the CP violating parameter $Im(\tau)$ from ϵ and other constraints,[4] predictions for the CP violating observables Δ, A and B can be obtained. Δ is predicted to be less than 10^{-6}. It is very small. The parameter $A(\Lambda)$ is in the range $-(0.5 \sim 0.1) \times 10^{-4}$. $B(\Lambda)$ is about 60 times larger.

The same calculation has been done using MIT bag model[2] and other models for hadronic matrix elements.[3,4] The MIT bag model predicts the same orders of magnitude for A and B as the vacumm saturation predictions. Larger values are possible in other models.[4] It has recently been shown that the gluon dipole operator also has significant contributions to $A(\Xi)$.[5] In Table 1, We list the allowed ranges of Δ, A and B using MIT bag model for Λ and Ξ decays. The ranges include uncertainties from the KM matrix elements and uncertainties in top quark mass.[4] In particular, $A(\Xi)$ is in the range $-(0.1 \sim 1) \times 10^{-4}$ and hence the quantity A_{asy} to be measured by E871, $A(\Lambda) + A(\Xi)$, is expected to be in the range $-(0.2 \sim 1.5) \times 10^{-4}$.

3.2. The Multi-Higgs Model Predicitions.

I will consider multi-Higgs model with neutral flavour current conservation at the tree level and CP is violated spontaneously. This is the model proposed by Weinberg.[21] In this model the most important operator related to CP violation is

$$L_{CPV} = i\tilde{f}\bar{d}T^a\sigma_{\mu\nu}(1 - \gamma_5)sG_a^{\mu\nu} , \tag{18}$$

where $G_a^{\mu\nu}$ is the gluon field strength, T^a is the $SU(3)_C$ generator, and $\tilde{f}$ is a constant depending on several parameters. This operator can reproduce CP violation in the neutral Kaon sector provided that[22]

$$m_K\sqrt{2}|\epsilon|\Delta m_{K_L-K_S} \approx 10^{-7} < \pi^0|L_{CPV}|K^0 > . \tag{19}$$

This fixes the strength of CP violation in this model. The predictions for CP violation in hyperon decays have been carried out in Ref.[2] using bag model and pole model calculations. The results are listed in Table 1.

In models in which flavor changing neutral currents are responsible for CP non-conservation, all effect in hyperon decays as well as ϵ'/ϵ are essentially zero.

3.3. The Left-Right Symmetric Model Predictions.

The Left-Right symmetric models are based on the gauge group $SU(3)_C \times SU(2)_L \times SU(2)_R \times U(1)_{B-L}$. In this model there are additional CP violating phases from the right-handed KM matrix. Here we consider a simple model of this type, the "isoconjugate" Left-Right model.[23] In this model there is no mixing between W_L and W_R. There is no CP violation in the left-handed sector. All CP violations are coming from the right-handed KM matrix. And $|V_{Lij}| = |V_{Rij}|$ for the KM matrices. The full $\Delta S = 1$ Hamiltonian has the form

$$H_W = \frac{G_F}{\sqrt{2}}(O_{LL} + \eta e^{i\beta}O_{RR}) . \tag{20}$$

The operators O_{LL} and O_{RR} are identical operators, except that O_{LL} is a product of two left-handed currents whrease O_{RR} has two right-handed currents. Because this structure, one can easily see that parity-nonconserving processes have an identical phase factor $1 + i\eta\beta$, while all parity-conserving ones have phase $1 - i\eta\beta$. We have: $\phi_i^s = \eta\beta$ and $\phi_i^p = -\eta\beta$ for all decays. The strength is fixed by requiring this phase to explain CP violation in the neutral Kaon system.[24] From this consideration, $\eta\beta$ is determined to be about 4.4×10^{-5}. Because the simple phase structure, Δ is always zero in this model. The predictions for A and B are given in Table 1.

Table 1. Predictions of CP violation in hyperon decays.

Λ decay	KM model	Weinberg Model	Left-Right Model
$\Delta(\Lambda^0_-)$	$< 10^{-6}$	-0.8×10^{-5}	0
$A(\Lambda^0_-)$	$-(5 \sim 1) \times 10^{-5}$	-2.5×10^{-5}	-1.1×10^{-5}
$B(\Lambda^0_-)$	$(3 \sim 0.6) \times 10^{-4}$	1.6×10^{-3}	7.0×10^{-4}
Ξ decay			
$\Delta(\Xi^-_-)$	0	0	0
$A(\Xi^-_-)$	$-(10 \sim 1) \times 10^{-5}$	-3.2×10^{-4}	2.5×10^{-5}
$B(\Xi^-_-)$	$(10 \sim 1) \times 10^{-3}$	3.8×10^{-3}	-3.1×10^{-4}

From Table 1 we see that,in general, Δ is very small. It may be difficult to measure it experimentally. The prediction for the CP violating observable A is close to the region which will be probed by the E871 experiment, with experimental sensitivity 10^{-4} to 10^{-5} in A_{asy}, new and useful information about CP violation will be obtained.

We thank Drs. N. Deshpande J. Donoghue, H. Steger and G. Valencia for conllaboratotions on the related topic. This work was supported in part by the U.S. D.O.E. under grant DE-FG06-85ER40224 and DE-FG03-94ER40833.

References

1. T. Brown, S.F. Tuan and S. Pakvasa, Phys. Rev. Lett. **51** (1983) 1823; L.-L. Chau and H.Y. Cheng, Phys. Lett. **B131** (1983) 202; L. Wolfenstein and D. Chang, Preprint, CMU-HEP83-5; C. Kounnas, A.B. Lahanas, P. Pavlopoulos, Phys. Lett. **B127** (1983) 381.
2. J. Donoghue and S. Pakvasa. Phys. Rev. Lett. **55** (1985) 162; J. Donoghue, Xiao-Gang He and S. Pakvasa, Phys. Rev. **D34** (1986)833.
3. M.J. Iqbal and G. Miller, Phys. Rev. **D41**, (1990) 2817.
4. Xiao-Gang He, H. Steger and G. Valencia, Phys. Lett. **B272** (1991) 411.
5. N. Deshpande, Xiao-Gang He and S. Pakvasa, Phys. Lett. **326** (1994) 307.
6. N. Hamann et al., CERN/SPSPLC 92-19, SPSLC/M491.
7. S.Y. Hsueh and P. Rapidis, FNAL proposal.
8. G. Gidal, P. Ho, K.B. Luk and E.C. Dukes, Fermilab Proposal P-871.
9. J. Donoghue, B. Holstein and G. Valencia, Phys. Lett. **178** (1986) 319.
10. P. Chauvat et al., Phys. Lett. **B163** (1985) 273.
11. P. Barnes et al., Phys. Lett. **B199** (1987) 147.
12. M. Tixier et al., Phys. Lett. **B212** (1988) 523.
13. O. E. Overseth and S. Pakvasa, Phys. Rev. **184** (1969) 1663.
14. O. E. Overseth, Phys. lett. **B111** (1982) 286.
15. L. Roper et al., Phys. Rev. **138** (1965) 190.
16. R. Nath and B. Kumar, Nuov. Cim. **36** (1965) 669.
17. B. Martin, Phys. Rev. **138** (1965) 1136
18. M. Lu, M. Savage and M. Wise, Preprint, CALT-68-1940, CMU-HEP94-22.
19. G. Buchalla, A. Buras and M. Harlander, Nucl. Phys. **B337** (1990) 313; E.A. Paschos and Y.L. Wu, Mod. Phys. Lett. **A6** (1991) 93.
20. M. Kobayashi and T. Maskawa, Prog. Theor. Phys. **49** (1973) 652.
21. S. Weinberg, Phys. Rev. Lett. **37** (1976) 657.
22. J. Donoghue, B. Holstein, Phys. Rev. **D32** (1985) 1152.
23. R. N. Mohapatra and J.C. Pati, Phys. Rev. **D11** (1975) 566.
24. G. Beal, M. Bander and A. Soni, Phys. Rev. Lett. **47** (1981) 552.

PROBING VIOLATION OF QUANTUM MECHANICS IN THE K_0-$\bar{K}_0$ SYSTEM

P. HUET[*]

Stanford Linear Accelerator Center, Stanford University,
Stanford, CA 94309

ABSTRACT

I present a recent study made in collaboration with M.E. Peskin, on the time dependence of a kaon beam propagating according to the "$\alpha\beta\gamma$" generalization of quantum mechanics due to Ellis, Hagelin, Nanopoulos and Srednicki, in which CP- and CPT-violating signatures arise from the evolution of pure states to mixed states. The magnitude of two of its parameters β and γ are constrained on the basis of existing experimental data. New facilities such as ϕ factories are shown to be particularly adequate to study this generalization from quantum mechanics and to disentangle its parameters from other CPT violating perturbations of the kaon system.

In the early 70's, unitary evolution predicted by quantum mechanics had already been given an experimental scrutiny.[2,3] Theoretical motivations arose from developments in the quantum theory of gravity which led S. Hawking to propose a generalization of quantum mechanics which allows the evolution of pure states to mixed states.[4] This formulation was shown by D. Page to conflict with CPT conservation.[5] Subsequently, Ellis, Hagelin, Nanopoulos, and Srednicki (EHNS)[6] observed that systems which exhibit quantum coherence over a macroscopic distance are most appropriate to probe the violation of quantum mechanics of the type proposed by S. Hawking. One of the simplest system exhibiting this property is a beam of neutral kaons. EHNS set up a generalized evolution equation for the K_0-$\bar{K}_0$ system in the space of density matrices which contains three new CPT violating parameters α, β and γ. These parameters have dimension of mass and could be as large as $m_K^2/m_{\text{Pl}} \sim 10^{-19}$ GeV. This equation was subsequently used by Ellis, Mavromatos, and Nanopoulos[7] who exploited experimental data on K_L and K_S to delineate an allowed region in the space of the parameters α, β, and γ. This region is compatible with the expected order of magnitude above and with the possibility that violation of quantum mechanics accounts for all CP violation observed in the K_0-$\bar{K}_0$ system.

The present talk is a brief outline of a recent work,[1] which develops a general parameterization to incorporate CPT violation from both within and outside quantum mechanics and uses it to analyze past, present and future experiments on the K_0-$\bar{K}_0$ system. Studies of the time dependence of the kaon system of the early 1970's are

[*]This work was supported by the Department of Energy, contract DE-AC03-76SF00515.

combined with recent results from CPLEAR to constrain the EHNS parameters β and γ, limiting the contribution of violation of quantum mechanics to no more than 10% of the CP violation observed in the K_0-$\bar{K}_0$ system.

ϕ factories are new facilities[8] dedicated to the study of the properties of the kaon system; they are expected to give particularly incisive tests of CPT violation.[9] It is shown that these future facilities are especially suitable to test violation of quantum mechanics and to disentangle the EHNS parameters from other CPT violating perturbations.

1.   Violation of quantum mechanics: the formalism.

The composition of a beam of neutral kaons is uniquely characterized by the density matrix $\rho_K(\tau)$ as function of the proper time τ. The value of the density matrix $\rho_K(\tau)$ is obtained from its value at the source – given a certain production mechanism, an effective Hamiltonian $H = M - \frac{i}{2}\Gamma$, which incorporates the natural width of the system,[*] and the laws of generalized quantum mechanics

$$i\frac{d}{d\tau}\rho_K = H\,\rho_K - \rho_K\,H^\dagger + \delta\!\!\!/\!h\,\rho_K. \tag{1}$$

The first term of the RHS of this equation accounts for the quantum mechanical evolution, while the second term accounts for the loss of coherence in the evolution of the beam and is written so to preserve the linearity of the time-evolution. This term was written by EHNS so to require that it does not break conservation of probability and does not decrease the entropy of the system; that makes $\delta\!\!\!/\!h$ expressible in terms of six parameters. In order to lower this number to a more tractable one, EHNS further assumed that this term conserves strangeness, reducing $\delta\!\!\!/\!h$ to three unknown parameters α, β and γ. These parameters have the dimensions of mass and might be as large as $m_K^2/m_{\rm Pl} \sim 10^{-19}$ GeV.

The solution of Eq. (1) is generally expressible in terms of two real eigenvectors ρ_L, ρ_S and one complex eigenvector ρ_I and its complex conjugate $\rho_{\bar{I}}$, as

$$\rho_K(\tau) = A_L\rho_L^{(\diamond)}e^{-\Gamma_L\tau} + A_S\rho_S^{(\diamond)}e^{-\Gamma_S\tau} + A_I\rho_I^{(\diamond)}e^{-\bar{\Gamma}\tau}e^{-i\Delta m\tau} + A^\dagger\rho_{\bar{I}}^{(\diamond)}e^{-\bar{\Gamma}\tau}e^{+i\Delta m\tau}. \tag{2}$$

In the absence of the quantum mechanics violating term $\delta\!\!\!/\!h\,\rho_K$ in Eq. (1), the eigenmodes ρ_L, ρ_S and ρ_I are expressible in terms of the pure states $|K_L\rangle$ and $|K_S\rangle$ as[†] $\rho_L^{(\diamond)} = |K_L\rangle\langle K_L|$, $\rho_S^{(\diamond)} = |K_S\rangle\langle K_S|$ and $\rho_I^{(\diamond)} = |K_S\rangle\langle K_L|$, while $\bar{\Gamma} = (\Gamma_L + \Gamma_S)/2$, $\Delta\Gamma = \Gamma_S - \Gamma_L$ and $\Delta m = m_L - m_S$.

In presence of the quantum mechanics violating term $\delta\!\!\!/\!h\,\rho_K$, the eigenmodes are changed to, in first order in small quantities $(d = \Delta m + i\Delta\Gamma/2)$

$$\rho_L = \rho_L^{(\diamond)} + \frac{\gamma}{\Delta\Gamma}\rho_S^{(\diamond)} + \frac{\beta}{d}\rho_I^{(\diamond)} + \frac{\beta}{d^\star}\rho_{\bar{I}}^{(\diamond)}, \quad \rho_S = \rho_S^{(\diamond)} - \frac{\gamma}{\Delta\Gamma}\rho_L^{(\diamond)} - \frac{\beta}{d^\star}\rho_I^{(\diamond)} - \frac{\beta}{d}\rho_{\bar{I}}^{(\diamond)} \tag{3}$$

$$\rho_I = \rho_I^{(\diamond)} - \frac{\beta}{d^\star}\rho_S^{(\diamond)} + \frac{\beta}{d}\rho_L^{(\diamond)} - \frac{i\alpha}{2\Delta m}\rho_I^{(\diamond)} \qquad \rho_{\bar{I}} = \rho_I^\dagger. \tag{4}$$

---

[*]as well as CP and CPT perturbations compatible with quantum mechanics.

[†]We use a diamond superscript to label quantum mechanical quantities.

The corresponding eigenvalues are corrected by the shifts $\Gamma_{L,S} \to \Gamma_{L,S} + \gamma$, $\bar{\Gamma} \to \bar{\Gamma} + \alpha$ and $\Delta m \to \Delta m \cdot (1 - \frac{1}{2}(\beta/\Delta\Gamma)^2)$. The experimental relevance of these shifts is discussed in Ref. 1.

Any observable of the kaon beam can be computed by tracing ρ_K with an appropriate operator $\mathcal{P}$, we write $\langle \mathcal{P} \rangle = \mathrm{Tr}[\rho_K \mathcal{O}_\mathcal{P}]$. The major effect of violation of quantum mechanics is embodied in the eigenmodes ρ_L, ρ_S, ρ_I. These density matrices are no longer pure density matrices in contrast to their quantum mechanical counterparts. This loss of purity alters the decay properties of the beam. For example, the properties of the beam at large time, $\tau \gg 1/\Gamma_S$, are dominated by the properties of ρ_L. The second term on the RHS of the equation for ρ_L as given in Eq. (3) is proportional to $\rho_S^{(\Diamond)}$ and is even under CP conjugation. That results in an enhancement of the rate of decay into two pions at late time in the evolution of the beam, proportional to $\frac{\gamma}{\Delta\Gamma}$. A similar argument leads to expect an enhancement by an amount $\propto \beta/|d| \cos\phi_{SW}$ and $\beta/|d| \sin\phi_{SW}$ in the intermediate time region, $\tau \sim 1/\bar{\Gamma}$.

2.   Experimental constraints on β and γ.

We can exploit these facts to establish some experimental bounds on violation of quantum mechanics. Two observables are used in the analysis. The time dependent two-pion decay rate

$$\frac{\Gamma(K(\tau) \to \pi^+\pi^-)}{\Gamma(K(0) \to \pi^+\pi^-)} = \frac{\mathrm{Tr}\rho_K(\tau)\mathcal{O}_{+-}}{\mathrm{Tr}\rho_K(0)\mathcal{O}_{+-}} = e^{-\Gamma_S\tau} + 2|\eta_{+-}|^2 \cos(\Delta m\tau - \phi_{+-}) + R_L e^{-\Gamma_L\tau} \quad (5)$$

and the semileptonic decay rate at large time $\tau \gg 1/\Gamma_S$

$$\delta_L = \frac{\Gamma(K_L \to \pi^-\ell^+\nu) - \Gamma(K_L \to \pi^+\ell^-\nu)}{\Gamma(K_L \to \pi^-\ell^+\nu) + \Gamma(K_L \to \pi^+\ell^-\nu)} = \frac{\mathrm{Tr}\rho_K(\tau)(\mathcal{O}_{\ell^+} - \mathcal{O}_{\ell^-})}{\mathrm{Tr}\rho_K(\tau)(\mathcal{O}_{\ell^+} + \mathcal{O}_{\ell^-})}. \quad (6)$$

The relevant measurable quantities are R_L, δ_L and η_{+-}. R_L and δ_L reflect the large time properties of the beam; the complex number $\eta_{+-} = |\eta_{+-}| \exp(i\phi_{+-})$ is a property of the intermediate time region. In quantum mechanics, these quantities relate according to $R_L = |\eta_{+-}|^2$ and $\delta_L/2 = \mathrm{Re}\,\eta_{+-}$. After allowance has been made for violation of quantum mechanics, they relate according to $R_L \simeq |\eta_{+-}|^2 + \gamma/\Delta\Gamma + 4|\eta_{+-}|\beta/|d|$, $\delta_L/2 = \mathrm{Re}\,(\eta_{+-} - 2\beta/d)$. The geometry of these corrections is given in Fig. 1a . The current experimental situation is discussed in Ref. 1 and shown on Fig. 1b. The parameter β is proportional to the distance of the ellipse to the vertical band while the distance of the ellipse to the arc provides a measurement of γ. This comparison leads to the bounds $\beta = (0.12 \pm 0.44) \times 10^{-18}$ GeV and $\gamma = (-1.1 \pm 3.6) \times 10^{-21}$ GeV. In obtaining these bounds, we made allowance for CP and CPT violation in the Hamiltonian time evolution of the beam but we set to zero the CPT quantum mechanics perturbations of the decay amplitudes in the two-pion and semi-leptonic channels. In general,[1] the previous constraints on β and γ appear as constraints on combinations of CPT-violating parameters. Unless, unnatural cancellations occur among these parameters, they can be independently constrain, in which case, neither of them contributes more than 10% of the total CP violation observed in the K_0–$\bar{K}_0$ system.

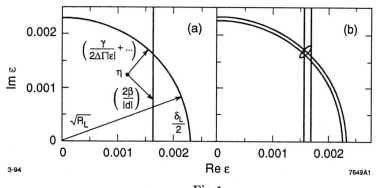

-Fig.1-

Theoretical predictions (a) and experimental data (b).

3.   Tests of quantum mechanics at a ϕ-factory.

At a ϕ factory, a spin-1 meson decays to an antisymmetric state of two kaons which propagates with opposite momentum. If the kaons are neutral, the resulting wavefunction, in the basis of CP eigenstates $|K_1\rangle$, $|K_2\rangle$, is $\phi \to (|K_1, p > \otimes|K_2, -p > -|K_2, p > \otimes|K_1, -p >)\sqrt{2}$. The two-kaon density matrix resulting from this decay is a 4×4 matrix P, which, in the context of generalized quantum mechanics, evolves according to Eq. (1). When expressed in terms of the eigenmodes ρ_L, ρ_S and ρ_I, it takes the form

$$
\begin{aligned}
P = \ &\frac{1}{2}\Big[\rho_S \otimes \rho_L + \rho_L \otimes \rho_S - \rho_I \otimes \rho_I - \rho_{\bar{I}} \otimes \rho_{\bar{I}}\Big] \\
&- 2\frac{\beta}{d}(\rho_S \otimes \rho_I + \rho_I \otimes \rho_S) - 2\frac{\beta}{d^*}(\rho_S \otimes \rho_{\bar{I}} + \rho_{\bar{I}} \otimes \rho_S) \\
&+ 2\frac{\beta}{d^*}(\rho_L \otimes \rho_I + \rho_I \otimes \rho_L) + 2\frac{\beta}{d}(\rho_L \otimes \rho_{\bar{I}} + \rho_{\bar{I}} \otimes \rho_L) \\
&+ i\frac{\alpha}{\Delta m}(\rho_I \otimes \rho_{\bar{I}} - \rho_{\bar{I}} \otimes \rho_I) + \frac{2\gamma}{\Delta\Gamma}(\rho_L \otimes \rho_L - \rho_S \otimes \rho_S).
\end{aligned} \tag{7}
$$

The time dependence of each term is obtained from the substitutions $\rho_i \otimes \rho_j \to \rho_i \otimes \rho_j exp(-\lambda_i\tau_1 - \lambda_j\tau_2)$ with $\lambda_L = \Gamma_L$, $\lambda_S = \Gamma_S$ and $\lambda_i = \bar{\Gamma} + i\Delta m$.

The first term in the brackets has the canonical form predicted by quantum mechanics after the replacement $\rho_K^{(\Diamond)} \to \rho_K$, while the remaining terms give systematic corrections to this result. These new terms signal the breakdown of the antisymmetry of the final state wave function, that is, the breakdown of angular momentum conservation. This is expected in the framework of density matrix evolution equations, as was explained in Ref. 10.

The above peculiar dependence on τ_1 and τ_2 is a unique signature of violation of quantum mechanics and provides an unambiguous method to isolate the EHNS parameters from the quantum mechanics CPT violating perturbations of the decay rates. As in the one kaon system, any observable is obtained by tracing the density matrix with a suitable hermitian operator. The basic observables computed from P are double differential decay rates, the probabilities that the kaon with momentum p decays into the final state f_1 at proper time τ_1 while the kaon with momentum $(-p)$ decays to the final state f_2 at proper time τ_2. We denote this quantity as $\mathcal{P}(f_1, \tau_1; f_2, \tau_2)$. A situation of particular importance is the decay into two identical final states $f_1 = f_2$. This quantity has no dependence on the CP and CPT parameters and depends on the two times in a manner completely fixed by quantum mechanics irrespective of the properties of the decay amplitudes. This characteristic is lost when violation of quantum mechanics is incorporated as in Eq. (7). One can, for instance, interpolate the double decay rates into identical final states $\mathcal{P}(f, \tau_1; f, \tau_2)$ on the line of equal time $\tau_1 = \tau_2$. This quantity vanishes identically according to the principles of quantum mechanics and thus is of order α, β and γ. As an illustration, the semileptonic double decay rate at equal time yields($\ell^\pm \equiv \pi^\mp \ell^\pm \nu$)

$$\mathcal{P}(\ell^\pm, \tau; \ell^\pm, \tau)/\mathcal{P}(\ell^\pm, \tau; \ell^\mp, \tau) = \frac{1}{2}[1 - e^{-2(\alpha-\gamma)\tau}(1 - \frac{\alpha}{\Delta m}\sin 2\Delta m\tau)] + \frac{1}{2}\frac{\gamma}{\Delta\Gamma}[e^{+\Delta\Gamma\tau} - e^{-\Delta\Gamma\tau}]$$
(8)

$$\pm 4\frac{\beta}{|d|}[\sin(\Delta m\tau - \phi_{SW})e^{-\Delta\Gamma\tau/2} + \sin(\Delta m\tau + \phi_{SW})e^{+\Delta\Gamma\tau/2}].$$
(9)

The three coefficients α, β, and γ are selected by terms which are monotonic in τ, oscillatory with frequency Δm, and oscillatory with frequency $2\Delta m$.

There seems to be no difficulty in constraining CPT violation from outside quantum mechanics in a ϕ factory independently of other CPT violating perturbation of quantum mechanics . The reverse is not true: any observable at a ϕ factory is expected to receive α, β and γ corrections. These corrections are, however, easily computed and can be systematically taken into account.[1]

References

1. P.Huet and M.E. Peskin, *preprint* SLAC-PUB-6454 (1994), (*submitted to Nucl.Phys.B*).
2. P.H. Eberhard, *CERN Report No.* CERN 72-1,(1972), (unpublished).
3. W.C. Carithers et al., *Phys. Rev.* **D14**, 290 (1976).
4. S.W. Hawking, *Phys. Rev.* **D 14**, 2460 (1975); *Commun. math. Phys.* **87**, 395 (1982).
5. D.N. Page, *Gen. Rel. Grav.* **14**, (1982)
6. J. Ellis, J.S. Hagelin, D.V. Nanopoulos and M. Srednicki, *Nucl. Phys.* **B241**,381 (1984).
7. J. Ellis, N. E. Mavromatos, and D. V. Nanopoulos, CERN-TH.6755/92 (1992).
8. *The DAΦNE PHYSICS HANDBOOK*, edited by L. Maiani, G. Pancheri and N. Paver, (INFN, Frascati).
9. For a review, see R.D. Peccei, preprint UCLA/93/TEP/19 (1993).
10. T. Banks, M. Peskin and L. Susskind, *Nucl. Phys.* **B 244** 125 (1984).

RECENT RESULTS ON CPT TESTS IN THE NEUTRAL KAON SYSTEM

ELLIOTT CHEU*

*Enrico Fermi Institute, University of Chicago, 5640 S. Ellis Ave.,
Chicago, IL 60637, USA*

ABSTRACT

Using the E773 detector at Fermilab, we have measured several of the CP violating parameters in neutral kaon decay. In the decay $K \to \pi\pi$, we find the phase of η_{+-} to be $\Phi_{+-} = 43.35° \pm 0.70° \pm 0.79°$ and the difference of the phases of η_{+-} and η_{00} to be $\Delta\Phi = 0.67° \pm 0.85° \pm 1.10°$ where the first errors are statistical and the second are systematic. The values of these parameters provide tests of CPT conservation. Improvements upon these measurements will be a major focus of the KTEV experiment.

1.   Introduction

Understanding how various interactions behave under the transformations of charge conjugation, C, parity, P and time reversal, T, has played a major role in particle physics. In particular, the discovery of parity violation[1-3] in the weak interaction was a major step in formulating the standard model. Since local field theories are symmetric under the combined operation CPT,[4] searching for violations of CPT is a good way of testing the validity of the standard model. One prediction of CPT is the equality of particle/anti-particle masses. In the kaon system $(m_{K°} - m_{\overline{K°}})/m_K \lesssim 10^{-20}$ which makes it a particularly nice system in which to test CPT.

Mixing occurs between the strangeness eigenstates, $K°$ and $\overline{K°}$, because they can both decay to the same state, $\pi\pi$. If the weak intereaction conserved CP, the weak eigenstates would be the same as the CP eigenstates, simple linear combinations of $K°$ and $\overline{K°}$. However, with the discovery of CP violation,[5] one has to modify this view and write the weak eigenstates as,

$$K_{S,L} = \frac{1}{\sqrt{2(1 + |\epsilon|^2)}} \left[(1 + \epsilon)K° \pm (1 - \epsilon)\overline{K°} \right], \tag{1}$$

where ϵ characterizes the level of CP violation. It is useful to write the ratios of the CP violating amplitude divided by the CP conserving amplitude,

$$\eta_{+-} \equiv \frac{A(K_L \to \pi^+\pi^-)}{A(K_S \to \pi^+\pi^-)} \tag{2}$$

---

*Representing the E773 Collaboration.

$$\eta_{oo} \equiv \frac{A(K_L \to \pi^o\pi^o)}{A(K_S \to \pi^o\pi^o)} \tag{3}$$

CPT invariance predicts[6] that the difference between the phases of these two ratios should be approximately zero.

$$\Delta\Phi \equiv \Phi_{oo} - \Phi_{+-} \simeq 0. \tag{4}$$

In addition, the value of the phase, Φ_{+-}, should obey,

$$\Phi_{+-} \simeq \tan^{-1}\left(\frac{2\Delta m}{\Gamma_S - \Gamma_L}\right) \tag{5}$$

The term on the right hand side of the equation is often called the "superweak" phase, Φ_{SW}.

2. The E773 Detector and Trigger

In this experiment we collected data simultaneously from two neutral kaon beams produced by 800 GeV/c protons which impinged upon a beryllium target. In each beam K_L particles struck a regenerator which produced a coherent mixture of K_L and K_S in the forward direction. The rate of $\pi\pi$ decays after a regenerator is described by the following equation:

$$R \sim |\rho/\eta|^2 e^{-\Gamma_S t} + e^{-\Gamma_L t} + |\rho/\eta| e^{-(\Gamma_S + \Gamma_L)t/2} \cos(\Delta m t + \Phi_\rho - \Phi_\eta), \tag{6}$$

where ρ is the amplitude for forward regeneration, Φ_ρ is the phase of this regeneration and Φ_η is the phase we want to measure. The magnitude of ρ has been found to have a power law dependence on the energy of the kaon. In addition, by assuming that the elastic kaon-nucleon scattering amplitudes are analytic, one can relate the phase of the regeneration amplitude to the power law. By fitting to Eq. 6 one can then determine the phases, Φ_η.

One of the regenerators, the upstream regenerator or UR, was approximately 1.2 meters in length and located 117 meters downstream of the target. The second regenerator, the DR, was located 128 meters from the target and was 0.4 meters in length. Each regenerator was made of plastic scintillator which allowed us to reject most events in which a kaon inelastically scattered in one of the regenerators. The regenerators alternated between the two beams once a minute between each beam spill.

The E773 detector has been described elsewhere.[7] The kaons decay region was defined by a vacuum tank which extended from the downstream end of each of the regenerators to the vacuum window at 159 meters. The E773 detector is very similar to the E731 detector[8] used to measure $Re(\epsilon'/\epsilon)$. The E773 experiment took data for a period of approximately two months and in that period collected over 400 million triggers.

Charged events were triggered by requiring at least two hits in the trigger hodoscope located at 180 meters. We required these hits to have an up-down, left-right pattern. An additional trigger hodoscope was located 141 m from the target and was removed halfway through the run to increase the decay volume. The data collected prior to the removal of this hodoscope is called Data Set 1 while that collected after its removal is called Data Set 2. In addition to the the hodoscopes, the trigger for charged events required that no charged particles hit one of the twelve photon veto counters arranged around the detector. Charged events were reconstructed using the four drift chamber magnetic spectrometer. The momentum resolution of this system is parametrized by $(\sigma_p/p)^2 = (0.45\%)^2 + (0.012p\%)^2$ with p measured in GeV/c. In the charged mode trigger patterns of hits in the spectrometer were required to be consistent with the symmetric decay of $K \rightarrow \pi^+\pi^-$. Muons from $K_L \rightarrow \mu\pi\nu$ ($K_{\mu 3}$) events were vetoed in the trigger by a muon hodoscope located approximately 190 meters from the target.

Electromagnetic showers were detected in the calorimeter located at 181 meters. The calorimeter consisted of 804 Lead-Glass blocks which were 18.7 radiation lengths deep. In the neutral mode trigger, a first-level trigger required at least 25 GeV to be deposited in the calorimeter. At the second level the trigger required four or six distinct clusters of energy with at least 1 GeV of energy in each cluster. Events in which charged particles hit one of the photon vetoes were vetoed in the neutral mode trigger. A hodoscope located behind a 21 radiation length thick lead wall was used to veto hadronic showers which originated in the lead glass calorimeter.

3. Data Analysis

In both the neutral and charged mode analysis we first needed to isolate a clean sample of coherent $K \rightarrow \pi\pi$ decays. For charged decays we selected events in which only two tracks were found in both the X and Y views of the spectrometer. The X and Y tracks were matched to each other by determining which tracks pointed to the same clusters in the electromagnetic spectrometer. To remove backgrounds from $K \rightarrow e\pi\nu$ (K_{e3}) events we required that the energy determined in the calorimeter divided by the track momentum (E/p) was less than 0.8 for each track. Although $K_{\mu 3}$ events were vetoed in the trigger, low momentum muons could have ranged out in the muon steel. Therefore, we required that each of the tracks had a momentum greater than 7.0 GeV/c. For each two-track event, track and vertex quality cuts were performed. The invariant mass of the two tracks was formed under the assumption $\Lambda \rightarrow p\pi$ and events within 6MeV/c^2 of the Λ mass were discarded. The $\pi^+\pi^-$ mass distribution is shown in Figure 1. Events within 14 MeV/c^2 of the known kaon mass were kept.

The largest remaining background to the coherent $\pi^+\pi^-$ sample comes from non-coherent K decays in which the kaon scatters in the regenerator. Many of these decays are removed by requiring that there is no activity in either of the two regenerators. Since these decays are not described by Eq. 6, it is necessary to remove them. We do so by looking at the square of the transverse momentum of the reconstructed kaon, p_T^2. Coherent decays, as shown in Fig.1, peak at very low p_T^2 while scattered kaons have a broad distribution. We required $p_T^2 < 250$ (MeV/c)^2 and determined the remaining background by extrapolating the distribution in Fig. 1 under the coherent peak. For the both beams the background was small, 0.3% for the UR and 0.8% for the DR.

For the $K \rightarrow \pi^\circ\pi^\circ$ analysis the four electromagnetic clusters were paired up to form π° pairs. For the four clusters there are three possible combinations. For each of the three combinations the decay vertex of each π° pair was calculated and a χ^2 was

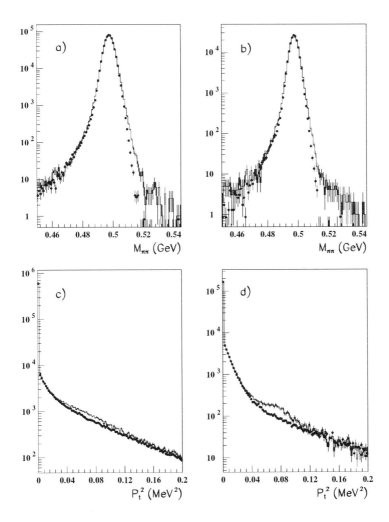

Fig. 1. The $\pi^+\pi^-$ mass distributions for the a) UR and b) DR beams from data set 2. Figures c) and d) are the p_T^2 distributions for the UR and DR beams, respectively, also from data set 2. In the p_T^2 distributions, the bump at 0.08 MeV^2 is due to K^* decays. In all four plots the histogram is the data and the points are the Monte Carlo simulation.

calculated under the assumption that the two pairs came from the same decay vertex. We chose the combination with the best χ^2 and only kept those events with $\chi^2 < 4$. To reduce the background in the $\pi^\circ\pi^\circ$ sample from $K_L \to \pi^\circ\pi^\circ\pi^\circ$ decays, we made cuts on energy deposited in the photon veto counters and the shape of the clusters in the electromagnetic calorimeter. Events in which the four photon invariant mass was between 474 MeV/c^2 and 522 MeV/c^2 were retained.

The backgrounds which remained in the above sample were due to $K \to \pi^\circ\pi^\circ\pi^\circ$ decays, kaon interactions in the material and scattered kaons. The level of background from $3\pi^\circ$ decays was determined from running a Monte Carlo in which two photons are lost or fused with other clusters in the calorimeter. The background level from $K_L \to \pi^\circ\pi^\circ\pi^\circ$ was approximately 0.6% for the UR beam and 1.4% for the DR beam. Interactions of the beam with material was understood by interpolating the non-$3\pi^\circ$ background in the invariant mass plot under the kaon mass peak. This background was approximately 0.1%. The final background is due to non-coherent decays of kaons. As in the charged mode analysis, this background is significantly reduced by requiring that the activity in the regenerators is low. Unlike the charged mode analysis, the transverse momentum cannot be calculated since the transverse position of the decay was not known. Instead we use the center-of-energy of the four photons to define a variable called "ring number." The ring number is the area of the smallest square centered on one of the beam holes which encloses the center-of-energy. Non-coherent events will have a large ring number since the center-of-energy will not lie in one of the beam holes. We determined the amount of background due to scattered kaons by generating Monte Carlo events where the kaon scattered. By normalizing these Monte Carlo events to the data we can extrapolate under the coherent peak to determine the level of background. The level of background from these studies was 3.0% for the upstream regenerator and 9.7% for the downstream regenerator.

The determination of the phases, Φ_{+-} and Φ_{oo}, depends upon knowing the number of coherent $K \to \pi\pi$ decays which occur downstream of the regenerators. To make this determination we have to understand the detector and analysis acceptance very well as a function of the kaon momentum and decay vertex position. This is done by relying upon a Monte Carlo simulation of the detector in which the main parameters that we tune are the positions of various apertures as well as the response of the drift chambers and lead glass calorimeter. To check that we understand our simulation, we use high statistics kaon decays, K_{e3} and $K \to \pi^\circ\pi^\circ\pi^\circ$. As shown in Figure 2, the reconstructed decay position in the K_{e3} mode agrees well with the Monte Carlo simulation, indicating that we do have a good understanding of the detector acceptance. The slope of the ratio of the data/MC comparison is used to determine the systematic error due to the detector acceptance.

The final data sample used in this analysis is shown in Table 1. For the neutral mode analysis, only Data Set 2 was used.

4. Φ_{+-} and $\Delta\Phi$ Fits

Using the kaon decay rate described by Eq. 1, we fit for the following quantities, Φ_{+-}, $\Delta\Phi$, Δm and τ_S. We performed three separate fits: 1) Fit for Φ_{+-}. 2) Fit for Δm and τ_S. 3) Fit for $\Delta\Phi$. The data was divided into kaon momentum and decay vertex bins of 10 GeV/c and 2 meters, respectively. For the UR beam the z region

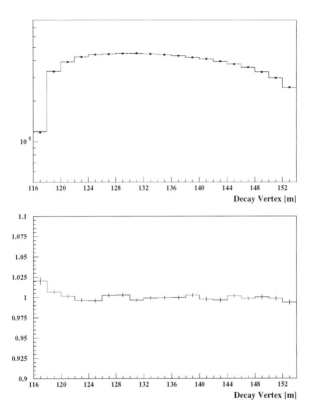

Fig. 2. The decay vertex distribution for K_{e3} decays. The top plot shows the data (histogram) and the Monte Carlo prediction (dots). The lower plot shows the ratio of the data divided by the Monte Carlo.

Table 1. E773 Data Set

Sample	Set One		Set Two	
	UR	DR	UR	DR
$\pi^+\pi^-$	825K	185K	745K	205K
$\pi^\circ\pi^\circ$	85K	35K	180K	75K
$\pi^+\pi^-\gamma$	4.5K	1.2K	3.8K	1.2K
$\pi^\circ\pi^\circ\pi^\circ$	384K	281K	676K	492K
$\pi^+e^-\nu$	7,000K	3,300K	11,150K	7,374K

Table 2. Systematic errors.

	Φ_{+-} fit	$\Delta m/\tau_S$ fit		$\Delta\Phi$ fit
	Degrees	$10^{10}\hbar/s$	$10^{-10}s$	Degrees
Regenerator positions, lengths	0.10	0.0008	0.0004	-
Charged mode acceptance	0.57	0.0017	0.0010	0.2
Neutral mode acceptance	-	-	-	0.2
Background subtraction	0.20	0.0005	0.0002	0.6
Deviations from a power law for the reg. amplitude	0.50	0.0011	0.0008	-
Lead glass energy scale	-	-	-	0.6
Lead glass energy resolution	-	-	-	0.5
lead glass nonlinear response	-	-	-	0.5
Total (added in quadrature)	0.79	0.0022	0.0014	1.1

extended from 118.5 to 127m and 129 to 154m, excluding the region surrounding the downstream regenerator. In the DR beam we used the region from 130 to 154 meters. The momentum region was between 30 and 160 GeV/c. For the Φ_{+-} and $\Delta\Phi$ fits, we fixed the values of Δm and τ_S to be $0.5286 \times 10^{10}\hbar s^{-1}$ and $0.8922 \times 10^{-10}s$, respectively. In the first fit we allowed the values of the two parameters which describe ρ to float as well as the kaon flux, two energy spectrum parameters and the value of Φ_{+-}. The χ^2 of this fit was 598 for 578 degrees of freedom. For the fit for Δm and τ_S we allowed all of the same parameters to float except that we fixed the value of Φ_{+-} to $\tan^{-1}(2\Delta m/\tau_S)$. In this fit the χ^2 was 598 for 577 degrees of freedom. In our fit to $\Delta\Phi$ additional parameters describing the energy shape and normalization were allowed to float as well as the two phases, Φ_{+-} and Φ_{oo}, and $|\epsilon'/\epsilon|$. In this fit only the data from set 2 was used. The χ^2 for this fit was 682 for 619 degrees of freedom.

The systematic errors in our determination of Φ_{+-}, $\Delta\Phi$, Δm and τ_S are listed in Table 2. For the Φ_{+-} fit the largest source of error comes from our understanding of the detector acceptance. We plan to reduce the error from this source as we improve our understanding of the K_{e3} data. In the case of the $\Delta\Phi$ fit, the largest source of error results from the cut on the minimum energy of an individual cluster. This sensitivity is included in the error due to the nonlinear response of the lead glass. At the moment the sensitivity to this cut is not yet understood.

The results of our fits are as follows:

$$
\begin{aligned}
\Phi_{+-} = {} & 43.35° \pm 0.70°(stat) \pm 0.79°(syst) \\
& - 0.62° \frac{\tau_S - 0.8922 \times 10^{-10}s}{0.0020 \times 10^{-10}s} + 0.38° \frac{\Delta m - 0.5286 \times 10^{10}\hbar/s}{0.0024 \times 10^{10}\hbar/s}.
\end{aligned} \tag{7}
$$

The fits to Δm and τ_S yield

$$
\Delta m = [0.5286 \pm 0.0029(stat) \pm 0.0022(syst)] \times 10^{10}\hbar s^{-1} \tag{8}
$$

$$
\tau_S = [0.8929 \pm 0.0014(stat) \pm 0.0014(syst)] \times 10^{-10}s. \tag{9}
$$

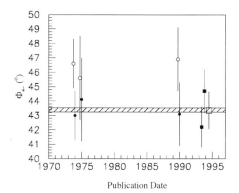

Fig. 3. Recent measurements of Φ_{+-}. Our result is shown as an open square. The open circles are the results used to determine $\Phi_{+-,PDG}$, the solid points are the same measurements corrected using our Δm and the solid squares are more recent results. The band is the $\pm 1 \sigma$ band of Φ_{sw}. The τ_S and Δm dependences are included in the errors. The references to the results are in time order.[9–13]

We quote the dependence of Φ_{+-} upon Δm and τ_S because there is a strong correlation between this phase and Δm and τ_S. Previous measurements of Φ_{+-} have hinted at CPT violation due to the deviation of Φ_{+-} from Φ_{SW} as shown in Figure 3. However, if we modify the values of Φ_{+-} using the value of Δm found in our second fit, we find that there is no discrepancy between Φ_{SW} and all measurements of Φ_{+-}. The value of Δm found in our fit, while lower than the PDG value, is consistent with the value found by E731. For the $\Delta\Phi$ fit, there is no Δm or τ_S dependence. We find

$$\Delta\Phi = 0.67° \pm 0.85°(stat) \pm 1.1°(syst). \qquad (10)$$

This result is consistent with zero, in agreement with predictions of CPT.

5. The KTEV Experiment

The KTEV experiment is a continuation of the kaon program at Fermilab. It has two main goals. The first is to measure $Re(\epsilon'/\epsilon)$ to a level of 1×10^{-4}. The second goal is to measure or improve the limits on the branching ratios of many rare kaon decays. In particular we hope to be able to reach the single event sensitivity of 1×10^{-11} for the decay $K_L \to \pi^°e^+e^-$. This mode is interesting since the decay may be dominated by a contribution from direct CP violation.

For E832 the KTEV detector will be similar to that of E731, utilizing the double beam technique with a vacuum beam and a regenerator beam. For E731 some of the largest sources of systematic error were the neutral mode energy scale and the neutral mode acceptance. In KTEV we will reduce these sources of systematic error by utilizing a calorimeter composed of approximately 3100 blocks of undoped CsI. As shown in

Table 3. Calorimeter Performance

	E731	KTEV
Energy Resolution	$\frac{\Delta E}{E} = 1.5\% \oplus \frac{5.0\%}{\sqrt{E}}$	$\frac{\Delta E}{E} = \frac{1.05\%}{E^{0.35}} \oplus \frac{0.7\%}{\sqrt{E}}$
Position Resolution	~ 2.8mm	~ 1.0 mm
Depth	18.7 r.l.	27.0 r.l.
Light Output	Fast, Cerenkov	25 ns & 1000ns Scintillator

Table 3 both the energy and position resolution of the KTEV calorimeter will be a great improvement over the E731 detector. A recent beam test confirmed that we can expect the energy resolution of the calorimeter to be much better than 1%. The readout electronics of the calorimeter are also much improved over the E731 electronics since the calorimeter information is digitized every 19 ns. This helps in understanding both the energy resolution and any accidental activity which occurs either before or after the real event.

In addition to the measurements of $Re(\epsilon'/\epsilon)$ we expect to be able to substantially improve the measurements of Φ_{+-} and $\Delta\Phi$. The improvements will come from KTEV's improved calorimetry as well as the higher statistics in the $K \to \pi\pi$ modes. Scaling from E731 we expect to be able to reach $\Delta\Phi \sim 0.2°$. We also plan to perform a high statistics measurement of the charge asymmetry in semileptonic kaon decays which provides a further test of CPT.

References

1. C.S. Wu et al., Phys. Rev. **105**, 1413 (1957).
2. R.L Garwin, L.M. Lederman and M. Weinrich, Phys. Rev. **105**, 1415 (1957).
3. J.I. Friedman and V.L. Telegdi, Phys. Rev. **105**, 1681 (1957).
4. R.F. Streater and A.S. Wightman, *PCT, Spin and Statistics, and All That*, Benjamin, New York, 1964.
5. J.H. Christenson, J.W. Cronin, V.L. Fitch and R. Turlay, Phys. Rev. Lett. **13**, 138 (1964).
6. V.V. Barmin et al., Nucl. Phys. **B247**, 293 (1984), **B254**, 747(E) (1985).
7. B. Schwingenheuer, Talk given at the 5th International Conference of the Intersection of Nuclear and Particle Physics, St. Petersberg, FL, 1994.
8. L.K. Gibbons, Ph.D. Thesis, University of Chicago (1993), unpublished.
9. S. Gjesdal et al., Phys. Lett. **52B**, 119 (1974).
10. W.C. Carithers et al., Phys. Rev. Lett. **34**, 1244 (1975).
11. R. Carosi et al., Phys. Lett. **B 237**, 303 (1990).
12. L.K. Gibbons et al., Phys. Rev. Lett. **70**, 1199 (1993).
13. P. Sanders for CPLEAR, talk given at the ASPEN Conference 1994.

RECENT RESULTS OF CP, T AND CPT TESTS WITH THE CPLEAR EXPERIMENT

The CPLEAR Collaboration *

Presented by KERSTIN JON-AND

Department of Physics Frescati, Royal Institute of Technology, Frescativägen 24, S-104 05 Stockholm, Sweden

ABSTRACT

The CPLEAR experiment at CERN has been designed to measure CP, T and CPT violation parameters in the neutral kaon system by observing time-dependent decay rate asymmetries between K^0's and $\overline{K}^0$'s. The method is described as well as the detector. First results of the analysis of high-statistics data samples from 1992 and 1993 are reported. The determination of the CP violation parameters for K_S and K_L, and the observation of T violation in the mixing matrix are discussed, as well as a precise measurement of Δm and a search for $\Delta S = \Delta Q$ rule violation. Present precisions are at the level of the world average values or better, and some parameters have not been measured before.

R. Adler[2], T. Alhalel[2], A. Angelopoulos[1], A. Apostolakis[1], E. Aslanides[11], G. Backenstoss[2], C.P. Bee[11], O. Behnke[17], J. Bennet[9], V. Bertin[11], F. Blanc[7,13], P. Bloch[4], Ch. Bula[13], P. Carlson[15], M. Carroll[9] J. Carvalho[5], E. Cawley[9], S. Charalambous[16], M. Chardalas[16], G. Chardin[14], M.B. Chertok[3], M. Danielsson[15], A. Cody[9], S. Dedoussis[16], M. Dejardin[4], J. Derre[14], M. Dodgson[9], J. Duclos[14], A. Ealet[11], B. Eckart[2], C. Eleftheriadis[16], I. Evangelou[8], L. Faravel[7,11], P. Fassnacht[11], J.L. Faure[14], C. Felder[2], R. Ferreira-Marques[5], W. Fetscher[17], M. Fidecaro[4], A. Filipčič[10], D. Francis[3], J. Fry[9], E. Gabathuler[9], R. Gamet[9], D. Garreta[14], T. Geralis[13], H.-J. Gerber[17], A. Go[3], P. Gumplinger[17], C. Guyot[14], A. Haselden[9], P.J. Hayman[9], F. Henry-Couannier[11], R.W. Hollander[6], E. Hubert[11], K. Jansson[15], H.U. Johner[7], K. Jon-And[15], P.R. Kettle[13], C. Kochowski[14], P. Kokkas[8], R. Kreuger[6], T. Lawry[3], R. Le Gac[11], F. Leimgruber[2], A. Liolios[16], E. Machado[5], P. Maley[9], I. Mandić[10], N. Manthos[8], G. Marel[14], M. Mikuž[10], J. Miller[3], F. Montanet[11], T. Nakada[13], A. Onofre[5], B. Pagels[17], P. Pavlopoulos[2], F. Pelucchi[11], J. Pinto da Cunha[5], A. Policarpo[5], G. Polivka[2], H. Postma[6], R. Rickenbach[2], B.L. Roberts[3], E. Rozaki[1], T. Ruf[4], L. Sacks[9], L. Sakeliou[1], P. Sanders[9], C. Santoni[2], K. Sarigiannis[1], M. Schäfer[17], L.A. Schaller[7], A. Schopper[4], P. Schune[14], A. Soares[14], L. Tauscher[2], C. Thibault[12], F. Touchard[4], C. Touramanis[9], F. Triantis[8], D.A. Tröster[2], E. Van Beveren[5], C.W.E. Van Eijk[6], S. Vlachos[2], P. Weber[17], O. Wigger[13], C. Witzig[17], M. Wolter[17], C. Yeche[14], D. Zavrtanik[10] and D. Zimmerman[3]

[1]University of Athens, [2]University of Basle, [3]Boston University, [4]CERN, [5]LIP and University of Coimbra, [6]Delft University of Technology, [7]University of Fribourg, [8]University of Ioannina, [9]University of Liverpool, [10]J. Stefan Inst. and Phys. Dep. University of Ljubljana, [11]CPPM, IN2P3-CNRS et Université d'Aix-Marseille II, [12]CSNSM, IN2P3-CNRS, [13]Paul-Scherrer-Institut(PSI), [14]DAPNIA/SPP, CE Saclay, [15]KTH Stockholm, [16]University of Thessaloniki, [17]ETH-ITP Zürich

1.   Introduction

The CPLEAR experiment has been designed to study CP, T and CPT violation in the neutral kaon system. We recall that the K_S and K_L states are expressed in terms of the CP eigenstates K_1 and K_2 by $|K_S> = |K_1> + \epsilon_S |K_2>$ and $|K_L> = |K_2> + \epsilon_L |K_1>$. The parameters ϵ_S and ϵ_L can be written as $\epsilon_S = \epsilon_T + \delta_{CPT}$, and $\epsilon_L = \epsilon_T - \delta_{CPT}$, where ϵ_T and δ_{CPT} express a possible violation of the T and CPT symmetries respectively in the $K^0 - \overline{K}^0$ oscillations.[1] If CPT holds the relations $\epsilon_S = \epsilon_L = \epsilon_T$ and $\delta_{CPT} = 0$ are valid. So far CP violation has only been observed for K_L. The parameters measured are $Re(\epsilon_L)$, and the ratios between the CP-forbidden amplitude for $K_L \to \pi^+\pi^-$, $\pi^0\pi^0$ and $\pi^+\pi^-\gamma$, and the corresponding CP-allowed amplitude for K_S (η_{+-}, η_{00}, $\eta_{+-\gamma}$ respectively).

The CPLEAR experiment is unique in using K^0 and $\overline{K}^0$, which are tagged at the production time, $t = 0$. It measures time-dependent asymmetries between the rates, $R_f(t) \equiv R(K^0(t = 0) \to f; t)$ and $\overline{R}_{\overline{f}} \equiv \overline{R}(\overline{K}^0(t = 0) \to \overline{f}; t)$, where f and $\overline{f}$ are CP-conjugate final states, with $f = \pi^+\pi^-$, $\pi^0\pi^0$, $\pi^+\pi^-\pi^0$, $\pi^{+(-)}e^{-(+)}\overline{\nu}(\nu)$. With the non-leptonic final states, these measurements lead to the determination of the parameters η_{+-}, η_{00} and η_{+-0}, where the latter denotes the ratio between the CP-forbidden amplitude for $K_S \to \pi^+\pi^-\pi^0$(CP=-1) and the corresponding CP-allowed amplitude for K_L. With the semileptonic final states, CPLEAR performs direct measurements of the parameters $Re(\epsilon_T)$, $Re(\delta_{CPT})$ and $Re(\epsilon_S)$; moreover, as will be shown later, it is sensitive to the parameter x, which describes the possible violation of the $\Delta S = \Delta Q$ rule.

2.   The CPLEAR Experiment

The neutral kaons are produced in $p\overline{p}$ annihilations at rest: $p\overline{p} \to K^0 K^- \pi^+$, $\overline{K}^0 K^+ \pi^-$. These two "golden" channels have an equal branching ratio of about 2×10^{-3}. The strangeness of the neutral kaon at production time is tagged by measuring the charge of the accompanying charged kaon. The antiprotons annihilate in a gaseous, 16 bar pressure, hydrogen target. The detector has a cylindrical geometry. It is contained inside a solenoidal magnet with a field of 0.44 T. The charged tracks are measured by two multi-wire proportional chambers, six drift chambers and two layers of streamer tubes. The particle identification detector (PID) consists of a Cherenkov counter sandwiched between two scintillator counters and is placed behind the streamer tubes. It is designed to distinguish between charged kaons and pions. It is also used to identify electrons produced in semileptonic decays of neutral kaons. The photons from the neutral pions in $\pi^+\pi^-\pi^0$ or $\pi^0\pi^0$ decays are detected in the 18 layer gas sampling electro-magnetic calorimeter.

Due to the small branching ratio of the "golden" channels the experiment requires a high annihilation rate (≈ 1 MHz) and a fast online event selection. A sophisticated multi-level trigger with hardwired processors has been designed to distinguish the "golden" events from the dominant multipionic background. The processors use information from all the elements of the detector. The trigger requires the presence of a charged kaon and imposes constraints on the number of charged tracks, the number of clusters in the calorimeter and the kinematics of the event. The decision is taken in

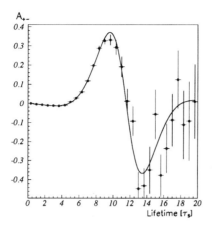

Figure 1: The A_{+-} asymmetry. The line is a fit to the data.

about 34 μs, reducing the rate by about a factor 1000. More details on the detector can be found elsewhere.[2]

In the offline analysis, the "golden" events are selected by refining the particle identification and demanding the missing mass at the primary vertex to be the K^0 mass.

Here preliminary results based on the analysis of high statistics data samples taken in 1992 and 1993 are reported.

3. Physics Analysis

3.1. The Final State $\pi^+\pi^-$

The main physics aim of measuring this decay channel is to determine, with higher precision than previously obtained, the phase ϕ_{+-} of the CP violating parameter η_{+-}. If CPT holds, the phase ϕ_{+-} should be close to the superweak phase $\phi_{SW} = \arctan(2\Delta m/\Delta\Gamma)$, where $\Delta m = m_L - m_S$ and $\Delta\Gamma = \Gamma_S - \Gamma_L$ denote the differences between the K_L and K_S masses and decay widths respectively. The decays of neutral kaons into two charged pions are selected by fits imposing kinematic and geometrical constraints to the events. The selection procedure leaves a ratio of background to signal of less than 2×10^{-4}, the background mainly being due to semileptonic decays at long lifetimes ($t > 10\tau_S$ where τ_S is the K_S lifetime).

The decay rate asymmetry is given by:

$$A_{+-}(t) = \frac{\bar{R}_{\pi^+\pi^-}(t) - R_{\pi^+\pi^-}(t)}{\bar{R}_{\pi^+\pi^-}(t) + R_{\pi^+\pi^-}(t)} \approx 2Re(\epsilon_L) - \frac{2|\eta_{+-}|e^{\Delta\Gamma t/2}cos(\Delta mt - \phi_{+-})}{1 + |\eta_{+-}|^2 e^{-\Delta\Gamma t}}. \quad (1)$$

Since K^+ and K^-, as well as π^- and π^+, have different interaction cross-sections with the detector material, the detection efficiency for $K^+\pi^-$, $\epsilon(K^+\pi^-)$, is different from that of $K^-\pi^+$, $\epsilon(K^-\pi^+)$. The experimental asymmetry has to be corrected for the difference in the strangeness tagging. The strangeness normalisation, $\alpha = \epsilon(K^+\pi^-)/\epsilon(K^-\pi^+)$, is computed using decays in the low lifetime region, where the effect of CP violation is negligible, and turns out to be close to one. The statistical accuracy of α is 0.001.

In Fig. 1 the experimental asymmetry is shown together with a fit of Eq. 1 to the data. The plot is based on a sample of 20×10^6 $\pi^+\pi^-$ decays with a decay time greater than 1 τ_S. The obtained parameter values from the fit are:

$$
\begin{aligned}
|\eta_{+-}| &= (2.163 \pm 0.045_{stat} \pm 0.030_{syst} \pm 0.060_{MC}) \times 10^{-3} \\
\phi_{+-} &= (44.7 \pm 0.9_{stat} \pm 0.4_{syst} \pm 0.7_{\Delta m} \pm 1.0_{MC})^{\circ}
\end{aligned}
\tag{2}
$$

The last error arises from the systematic uncertainty of α. This error is expected to become negligible when an increased number of simulated events will be available. The systematic errors left are mainly due to the uncertainty in the calculation of regeneration effects and the statistical error of α. The value of ϕ_{+-} is found to be consistent with the value of ϕ_{SW} ($43.7^\circ \pm 0.2^\circ$). The statistical precision of ϕ_{+-} is slightly better than the world average value[3]; including data taking until 1995 the total error is expected to decrease by a factor of two.

3.2. The Final State $\pi^\pm e^\mp \nu$

The semileptonic events are selected by imposing kinematic and geometrical constraints demanding the missing mass at the primary vertex to be the K^0 mass, and the total missing mass of the event to be the neutrino mass. One of the decay tracks is required to be identified as an electron using the Cherenkov and scintillator signals from the PID. The number of selected semileptonic decay events is about 60×10^4.

There are four semileptonic decay rates: $R^+(t) \equiv R(K^0(t=0) \to \pi^- e^+ \nu_e; t)$, $\overline{R}^-(t) \equiv R(\overline{K}^0(t=0) \to \pi^+ e^- \overline{\nu_e}; t)$, $R^-(t) \equiv R(K^0(t=0) \to \pi^+ e^- \overline{\nu_e}; t)$ and $\overline{R}^+(t) \equiv R(\overline{K}^0(t=0) \to \pi^- e^+ \nu_e; t)$. According to the $\Delta S = \Delta Q$ rule a K^0 is only allowed to decay into a positive lepton and a $\overline{K}^0$ into a negative one, and $R^-(t)$ and $\overline{R}^+(t)$ are due to $K^0 - \overline{K}^0$ oscillations.

By forming the difference of the allowed and the forbidden rates according to the $\Delta S = \Delta Q$ rule, one gets the asymmetry

$$
A_1(t) = \frac{\overline{R}^-(t) + R^+(t) - (\overline{R}^+(t) + R^-(t))}{\overline{R}^-(t) + R^+(t) + (\overline{R}^+(t) + R^-(t))} = \frac{2e^{-\overline{\Gamma}t}\cos(\Delta mt)}{(1 + 2Re(x))e^{-\Gamma_S t} + (1 - 2Re(x))e^{-\Gamma_L t}}
\tag{3}
$$

($\overline{\Gamma}$ is the average value of the K_S and K_L decay widths) which is sensitive to the oscillation frequency Δm. It is also sensitive to $Re(x)$ where x is the ratio of the amplitudes for the $\Delta S = \Delta Q$ forbidden and allowed decays, $x = <\pi^- l^+ \nu_l |H| \overline{K}^0 > / < \pi^- l^+ \nu_l |H| K^0 >$. The asymmetry A_2 formed from the difference between $\overline{K}^0$ and K^0 decay rates (allowed and forbidden) is sensitive to $Im(x)$ and $Re(\epsilon_S)$:

$$
A_2(t) = \frac{\overline{R}^+(t) + \overline{R}^-(t) - (R^+(t) + R^-(t))}{\overline{R}^+(t) + \overline{R}^-(t) + (R^+(t) + R^-(t))}.
\tag{4}
$$

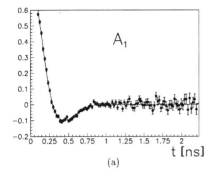
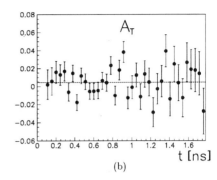

Figure 2: (a) The asymmetry A_1. (b) The asymmetry A_T. The results of the fits are superimposed to the data.

In Fig. 2(a) the asymmetry A_1 is shown together with the result from the fit of Eq. 3 to the data. The parameters obtained from fits to the asymmetries A_1 and A_2 are:

$$\Delta m = (0.5347 \pm 0.0032_{stat} \pm 0.0010_{syst} \pm 0.0030_{MC}) \times 10^{10} \hbar s^{-1}$$
$$Re(x) = (4.3 \pm 7.6_{stat} \pm 4.0_{syst} \pm 8.0_{MC}) \times 10^{-3}$$
$$Im(x) = (10.1 \pm 4.1_{stat} \pm 1.3_{syst} \pm 5.0_{MC}) \times 10^{-3}$$
$$Re(\epsilon_S) = (0.9 \pm 0.7_{stat} \pm 0.3_{syst} \pm 0.7_{MC}) \times 10^{-3}. \tag{5}$$

The most important sources of the systematic errors are the uncertainty in the amount and possible asymmetry of the remaining background and the statistical uncertainty of α. The precision of x is the best ever reached and $Re(\epsilon_S)$ has never been measured before. The statistical precision of Δm is close to that of the world average value.[3] The statistical precision will be further improved by a factor of two including data taking until 1995. A precise value of Δm is of importance for determining the phase ϕ_{+-}.

Analysis of the semileptonic decays enables for the first time a direct test of T and CPT violation. By forming the difference between the "forbidden" decay rates, $\overline{R}^+(t)$ and $R^-(t)$, which become allowed through $K^0-\overline{K}^0$ oscillation, one measures directly the difference of the probabilities $P(K^0 \to \overline{K}^0)$ and $P(\overline{K}^0 \to K^0)$. These probabilities being different is a signal of T violation. Assuming the $\Delta S = \Delta Q$ rule the following relation holds:

$$A_T(t) = \frac{\overline{R}^+(t) - R^-(t)}{\overline{R}^+(t) + R^-(t)} = 4Re(\epsilon_T). \tag{6}$$

The asymmetry

$$A_{CPT}(t) = \frac{R^+(t) - \overline{R}^-(t)}{R^+(t) + \overline{R}^-(t)} = 4Re(\delta_{CPT}) \quad (x = 0; t >> \tau_S) \tag{7}$$

gives a direct measurement of CPT violation since it measures the difference of the probabilities $P(K^0 \to K^0)$ and $P(\overline{K}^0 \to \overline{K}^0)$. In Fig. 2 (b) the A_T asymmetry is shown together with the result of a fit of Eq. 6 to the data. The fits to the asymmetries A_T and A_{CPT} yield the values:

$$A_T = (4.5 \pm 2.1_{stat} \pm 2.0_{syst} \pm 1.5_{MC}) \times 10^{-3}$$
$$Re(\delta_{CPT}) = (-0.1 \pm 0.5_{stat} \pm 0.5_{syst} \pm 0.4_{MC}) \times 10^{-3}. \tag{8}$$

If CPT holds one expects the value of A_T to be 6.54×10^{-3} which is compatible with our result. The value of $Re(\delta_{CPT})$ is compatible with zero. The most important contribution to the systematic errors is a possible difference in the detection efficiency of the decay products $\pi^+ e^-$ and $\pi^- e^+$.

3.3.   The Final State $\pi^+\pi^-\pi^0$

The aim of the analysis is to measure η_{+-0}, the amplitude ratio between the CP forbidden K_S decay and the CP allowed K_L decay, and the branching ratio of the CP allowed $K_S \to \pi^+\pi^-\pi^0$. To select the $\pi^+\pi^-\pi^0$ decays at least one photon is required in the calorimeter apart from kinematic and geometrical constraints. The final state $\pi^+\pi^-\pi^0$ is not a CP eigenstate. The CP eigenvalue depends on the relative angular momentum of the charged pion system and the neutral pion. The CP even K_S amplitude is antisymmetric with respect to the Dalitz plot variable X; its contribution to the asymmetry vanishes when integrating over the Dalitz plot ($X \gtrless 0 \Rightarrow E_{\pi^+} \gtrless E_{\pi^-}$ where $E_{\pi^\pm}$ is the energy of the charged pion in the K^0 rest frame). Thus the asymmetry

$$A_{+-0}(t) = \frac{\overline{R}(t) - R(t)}{\overline{R}(t) + R(t)} \approx 2Re(\epsilon_S) - 2e^{-\Delta\Gamma t/2}[Re(\eta_{+-0})cos(\Delta mt) - Im(\eta_{+-0})sin(\Delta mt)]$$
$$\tag{9}$$

is a function of the real and imaginary part of η_{+-0} only. In Fig. 3(a) the data is shown together with the best fit of Eq. 9 yielding the values:

$$Re(\eta_{+-0}) = (5 \pm 22_{stat} \pm 7_{syst}) \times 10^{-3}$$
$$Im(\eta_{+-0}) = (16 \pm 24_{stat} \pm 18_{syst}) \times 10^{-3}. \tag{10}$$

If $A_{+-0}(t)$ is analyzed separately for the two halves of the Dalitz plot with $E_{\pi^+} > E_{\pi^-}$ ($X > 0$) and $E_{\pi^+} < E_{\pi^-}$ ($X < 0$), respectively, the CP-allowed contribution to $K_S \to \pi^+\pi^-\pi^0$ can be extracted. In this case, the $cos(\Delta mt)$ term takes the form $[Re(\eta_{+-0}) \pm \lambda]cos(\Delta mt)$, where the parameter λ ($+\lambda$ for $X > 0$, $-\lambda$ for $X < 0$) is the ratio of the CP allowed K_S and the CP allowed K_L amplitude. In Fig. 3(b) the A_{+-0} asymmetry is shown separately for each half of the Dalitz plot. The lines represent a simultaneous fit of the expected asymmetries to the two sets of data yielding a value of $\lambda = 0.051 \pm 0.016_{stat} \pm 0.020_{syst}$. From the value of λ one can compute the branching ratio, $BR(K_S \to \pi^+\pi^-\pi^0) = (8.2^{+6.0+7.3}_{-4.4-4.9}) \times 10^{-7}$. This branching ratio has not been measured before.

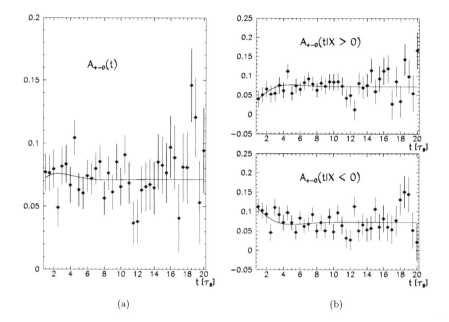

(a)   (b)

Figure 3: (a) Asymmetry $A_{+-0}(t)$. (b) Asymmetries $A_{+-0}(t|X \gtrless 0)$. The lines are fits to the data.

4. Conclusions and Outlook

CPLEAR has access to CP, T and CPT tests based on the comparison of tagged K^0 and $\overline{K}^0$ decay rates. Preliminary results from the data up to 1993 have been presented. These include measurements of parameters which have not been directly measured before, $Re(\epsilon_T)$, $Re(\delta_{CPT})$, $Re(\epsilon_S)$ and the branching ratio of the CP allowed $K_S \to \pi^+\pi^-\pi^0$, as well as measurements of ϕ_{+-}, x and η_{+-0} with better accuracy than the current world average values. Data taking until the end of 1995 will improve the statistical precision on the reported parameters by a factor of two; the systematic errors are expected to be reduced to a lower level than the statistical ones.

References

1. T.D.Lee and C.S.Wu, Annual Rev. of Nucl. Science **11** (1966) 511.
2. R.Adler et al., *Phys. Lett.* **B286** (1992) 180.
3. Review of Particle Properties, *Phys. Rev.* **D45** (1992).

DIRECT MEASUREMENT OF THE BRANCHING RATIO FOR THE DECAY OF THE ETA MESON INTO TWO PHOTONS

D. WHITE[1], R. ABEGG[2], R. ABELA[3], A. BOUDARD[4], W. BRISCOE[5],
M. CLAJUS[1], J.M. DURAND[6], A. EFENDIEV[6,7], B. FABBRO[4], P. FUCHS[6,2],
M. GARÇON[4], L. LYTKIN[7], B. MAYER[4], T. MORRISON[5], B. NEFKENS[1],
V. NIKULINE[4,6,8], J.F. PILLOT[4], J. PROKOP[5],
E. TOMASI-GUSTAFSSON[6,4], and W. VAN OERS[2]

[1]*Department of Physics, UCLA, Los Angeles, California 90024, USA*
[2]*TRIUMF, 4004 Wesbrook Mall, Vancouver, British Columbia, Canada V6T 2A3*
[3]*Paul Scherrer Institut, CH-5232 Villigen, Switzerland*
[4]*DAPNIA/SPhN, CE-Saclay, 91191 Gif-sur-Yvette, France*
[5]*Department of Physics, The George Washington University, Washington, D.C. 20052, USA*
[6]*Laboratoire National Saturne, 91191 Gif-sur-Yvette, France*
[7]*JINR, Laboratory of Nuclear Problems, 141980 Dubna, HPO-Box 79, Moscow district, Russia*
[8]*SPNPI, 188350 Gatchina, St. Petersburg district, Russia*

ABSTRACT

The SPES2 η collaboration has made the first direct measurement of $\mathrm{BR}(\eta \to \gamma\gamma) \equiv \Gamma(\eta \to \gamma\gamma)/\Gamma(\eta \to all)$. The systematic uncertainties are quite different than in the previous indirect measurements. The η mesons were produced in the reaction p d $\to$ ^{3}He η, 1.5 MeV above threshold. The recoil ^{3}He was identified and its vector momentum was measured by the SPES2 spectrometer. The η is tagged by its recoil ^{3}He. Two electromagnetic calorimeters, each consisting of 61 hexagonal BGO crystals located 65° left and right of the beam axis, measured the photon energies and directions with resolutions $\sigma_{E_\gamma}/E_\gamma \sim 4\%$ and $\sigma_\theta \sim 0.7°$ respectively. A lead collimator placed in front of the right calorimeter defined the detector acceptance. The left calorimeter was not collimated. Based on a clean sample of $(6.44 \pm 0.03) \times 10^4$ $\eta \to \gamma\gamma$ events, the *preliminary* result of our analysis is $\mathrm{BR}(\eta \to \gamma\gamma) = 0.400 \pm 0.002(stat) \pm 0.007(syst)$.

1. Motivation.

The branching ratio $\mathrm{BR}(\eta \to \gamma\gamma) \equiv \Gamma(\eta \to \gamma\gamma)/\Gamma(\eta \to all)$ is fundamental in several respects:

- The $\mathrm{BR}(\eta \to \gamma\gamma)$ is needed for the determination of $\Gamma(\eta \to all)$, the decay width of the η meson. No direct measurement of $\Gamma(\eta \to all)$ is feasible. Rather, the partial width of the inverse reaction $\gamma\gamma \to \eta$ is measured via $e^+e^- \to e^+e^-\gamma^*\gamma^* \to e^+e^-\eta$ or γN Primakoff production. Using the $\eta \to \gamma\gamma$ branching ratio presented in this report, we find

$$\Gamma(\eta \to all) = \frac{\Gamma(\eta \to \gamma\gamma)}{BR(\eta \to \gamma\gamma)} = \frac{0.46 \pm 0.04 \text{ keV}^1}{0.400 \pm 0.007} = 1.2 \pm 0.1 \text{ keV}. \quad (1)$$

- The BR($\eta \to \gamma\gamma$) is needed to calculate the branching ratios of certain electromagnetic η decays for which $\eta \to \gamma\gamma$ is the intermediate state: $\eta \to \mu^+\mu^-$, $\eta \to e^+e^-$, $\eta \to e^+e^-\gamma$, $\eta \to \mu^+\mu^-\gamma$, $\eta \to e^+e^-e^+e^-$, $\eta \to \mu^+\mu^-\mu^+\mu^-$, and $\eta \to e^+e^-\mu^+\mu^-$.

- The BR($\eta \to \gamma\gamma$) is needed in many η production cross section measurements. In these experiments, the cross-sections for η production are measured by identifying the η meson via its two-photon decay. The η production cross section results are then inversely proportional to BR($\eta \to \gamma\gamma$).

Furthermore, the direct measurement of BR($\eta \to \gamma\gamma$), when combined with results of other experiments, can be used to place upper limits on branching ratios of rare neutral decay modes of the η.

The world average prior to this experiment, BR($\eta \to \gamma\gamma$) = 0.389 ± 0.005,[1] depended on the measurement $\Gamma(\eta \to \text{neutrals})/\Gamma(\eta \to all)$ = 0.705 ± 0.008 by Basile et al.[2] In that experiment, the measurement of the number of η's produced relied on a subtraction of a large number of background events, which were estimated using a Monte-Carlo simulation. No discussion of the systematic uncertainties of this background subtraction was presented. It is possible that the systematic uncertainty may have been underestimated.

LNS Experiment 258 is the first *direct* measurement of BR($\eta \to \gamma\gamma$). The systematic uncertainties are quite different than in the previous indirect measurements.

2. Setup.

The experimental setup is shown in Fig. 1. The η mesons were produced in the reaction p d $\to$ ^{3}He η, 1.5 MeV above threshold. The recoil ^{3}He was identified and its vector momentum measured by the SPES2 spectrometer. The reaction p d $\to$ ^{3}He η was selected by cutting on the ^{3}He momentum and angle.

The two photons from an $\eta \to \gamma\gamma$ decay were detected by an electromagnetic calorimeter with two arms located at 65° left and right of the beam axis; 65° is the angle of the largest Jacobian, where the $\eta \to \gamma\gamma$ signal to background ratio is largest. The right photon detector arm consisted of a collimator, which defined the $\eta \to \gamma\gamma$ acceptance; the veto counter "V", which helped identify charged particles; and a calorimeter "R" which measured the energy and angle of one of the photons with resolutions $\sigma_{E_\gamma}/E_\gamma \sim 4\%$ and $\sigma_\theta \sim 0.7°$ respectively. The left photon detector arm consisted of the calorimeter "L" which was identical to R. In order to minimize the systematic uncertainty in the $\eta \to \gamma\gamma$ detector acceptance, the L calorimeter was not collimated, and was positioned to accept all of the photons from $\eta \to \gamma\gamma$ which were in coincidence with photons accepted on the right arm. Either calorimeter had 61 BGO counters, 20 cm (17.8 radiation lengths) thick.

To aid our understanding of the collimator edge effects, we interchanged the collimator with other collimators of different acceptance. The distance between the calorimeters and the target was varied to aid in estimating the systematic uncertainty in the $\eta \to \gamma\gamma$ acceptance associated with the EM shower energy leakage out of the sides of the calorimeter.

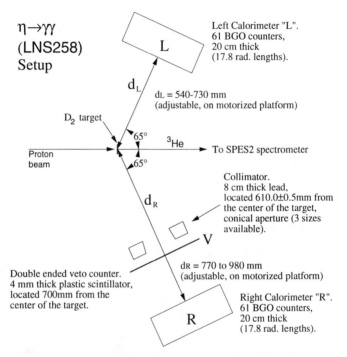

$\eta \to \gamma\gamma$
(LNS258)
Setup

Left Calorimeter "L".
61 BGO counters,
20 cm thick
(17.8 rad. lengths).

d_L

d_L = 540-730 mm
(adjustable, on motorized platform)

D_2 target

65°

^{3}He

Proton beam

To SPES2 spectrometer

65°

Collimator.
8 cm thick lead,
located 610.0±0.5mm from
the center of the target,
conical aperture (3 sizes
available).

d_R

V

Double ended veto counter.
4 mm thick plastic scintillator,
located 700mm from the
center of the target.

d_R = 770 to 980 mm
(adjustable, on motorized platform)

Right Calorimeter "R".
61 BGO counters,
20 cm thick
(17.8 rad. lengths).

R

Fig. 1. Top view of the experimental setup.

3.   Selection of p d → ^{3}He η.

The ^{3}He from p d → ^{3}He η near threshold always fall within a small band of ^{3}He momentum and angle. First, cuts are applied to the ^{3}He angle. The number of η mesons produced is then obtained using the spectrum of the ^{3}He momentum dispersion $\delta_{He} \equiv 100 \times (p_{He} - p_0)/p_0$, where p_0 is the central value of the ^{3}He momentum p_{He} dictated by p d → ^{3}He η kinematics near threshold. One observes a clean signal of p d → ^{3}He η events over a flat background (Fig. 2). The shape of the background (mostly p d → ^{3}He $\pi^+\pi^-$) under the peak was obtained from a measurement below threshold (the dashed histogram in Fig. 2). The background-to-signal ratio is 7.6%.

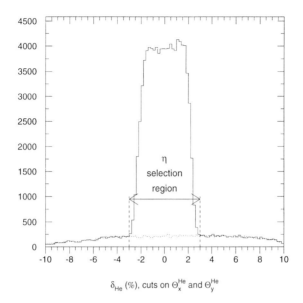

δ_{He} (%), cuts on Θ_x^{He} and Θ_y^{He}

Fig. 2. Spectrum of δ_{He}. Cuts on the ^{3}He emission angle have been applied. There is a clean signal of p d $\rightarrow$ ^{3}He η events above the flat background. The vertical dashed lines illustrate the cut used to select p d $\rightarrow$ ^{3}He η. Sub-threshold data, indicated by the dashed histogram, is subtracted. This histogram shows all data for one run condition, 11% of the p d $\rightarrow$ ^{3}He η events recorded on tape.

Since the p d $\rightarrow$ ^{3}He η data is acquired simultaneously with the $\eta \rightarrow \gamma\gamma$ data, the resulting branching ratio measurement is independent of systematic errors in proton beam intensity, target thickness, η tagging efficiency, knowledge of the production cross section σ(p d $\rightarrow$ ^{3}He η), and computer dead time.

4. Selection of $\eta \rightarrow \gamma\gamma$.

The first step in the selection of $\eta \rightarrow \gamma\gamma$ is applying the same cuts as in §3 to select p d $\rightarrow$ ^{3}He η. The $\eta \rightarrow \gamma\gamma$ candidates are then subject to timing cuts and to a test of energy conservation. We then define $\Delta\theta_{\text{LR}} \equiv \theta_{\text{LR}}^{\text{CALC}} - \theta_{\text{LR}}^{\text{MEAS}}$ where $\theta_{\text{LR}}^{\text{MEAS}}$ is the 2-photon opening angle as measured by the calorimeters and $\theta_{\text{LR}}^{\text{CALC}}$ is the expected $\eta \rightarrow \gamma\gamma$ opening angle calculated from the measured ^{3}He 3-momentum. See [3],[4], or [5] for details regarding the calculation of $\theta_{\text{LR}}^{\text{CALC}}$. Figure 3 shows a spectrum of $\Delta\theta_{\text{LR}}$ with cuts on the other variables. One observes a clean $\eta \rightarrow \gamma\gamma$ signal with almost no background.

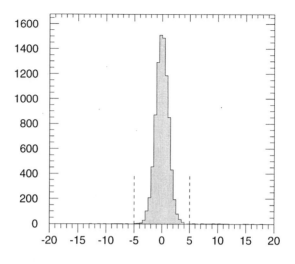

$\Delta\Theta_{LR}$ (degrees), all other cuts applied

Fig. 3. Spectrum of $\Delta\theta_{LR}$ showing a clean $\eta \to \gamma\gamma$ signal. There is almost no background. The cut used to select $\eta \to \gamma\gamma$ events is illustrated with vertical dashed lines.

5. Results.

Based on a sample of $(6.44 \pm 0.03) \times 10^4$ $\eta \to \gamma\gamma$ events, the *preliminary* result is

$$\mathrm{BR}(\eta \to \gamma\gamma) = 0.400 \pm 0.002(stat) \pm 0.007(syst). \qquad (2)$$

Table 1 lists the systematic uncertainties.

6. Upper limit on rare neutral η decays.

To date, $\eta \to \gamma\gamma$, $\eta \to \pi^0\pi^0\pi^0$, and $\eta \to \pi^0\gamma\gamma$ are the only *neutral* η decays which have been observed. By combining the results of this experiment with other experiments, one may place an upper limit on the branching ratios of all other neutral η decays, including the CP-violating decay $\eta \to \pi^0\pi^0$ and the C-violating decay $\eta \to \pi^0\pi^0\gamma$. No upper limits have previously been set for these forbidden decays.

Taking X to be all neutral η decays except $\eta \to \gamma\gamma$, $\eta \to \pi^0\pi^0\pi^0$, and $\eta \to \pi^0\gamma\gamma$, we may calculate

$$\mathrm{BR}(X) = \mathrm{BR}(\eta \to neutrals) - \mathrm{BR}(\eta \to \gamma\gamma) \times [1 + \frac{\Gamma(\eta \to \pi^0\pi^0\pi^0)}{\Gamma(\eta \to \gamma\gamma)} + \frac{\Gamma(\eta \to \pi^0\gamma\gamma)}{\Gamma(\eta \to \gamma\gamma)}] \quad (3)$$

where $\mathrm{BR}(\eta \to \gamma\gamma)$ is reported in this paper and the other ratios are reported in [2] and [6]. In §1 we questioned the low systematic uncertainty reported in [2] for the

Table 1. The systematic uncertainties.

Source of the systematic uncertainty	Preliminary Estimated Uncertainty
Statistical uncertainty in the Monte-Carlo	0.2%
Uncertainty in the beam properties	1%
Uncertainty in the position and machining of the target and collimator	0.6%
Simulation of collimator edge effects and simulation of physical processes	1%
η background subtraction uncertainty	0.6%
$\eta \to \gamma\gamma$ background subtraction uncertainty	0.2%
Electronics	0.2%
Calibration of experiment and simulation	0.5%
TOTAL	1.8%

$BR(\eta \to neutrals)$ experiment. Suppose we triple the uncertainty reported in that experiment. Then, substituting the experiment results for the quantities in Eq. 3,

$$BR(X) = (0.705 \pm 0.024) - (0.400 \pm 0.007)[1 + (0.822 \pm 0.009) + (1.8 \pm 0.4) \times 10^{-3}]$$

$$= -0.025 \pm 0.027 \quad Preliminary \qquad (4)$$

If we approximate Gaussian probability density functions for the uncertainties given above,

$$BR(X) < \sim 0.03 \quad (90\% \text{ confidence level}) \quad Preliminary \qquad (5)$$

This provides the first measured upper limits

$$BR(\eta \to \pi^0\pi^0) < \sim 0.03 \quad and \quad BR(\eta \to \pi^0\pi^0\gamma) < \sim 0.03 \quad Preliminary \qquad (6)$$

References

1. Particle Data Group, *Review of Particle Properties, Phys. Rev.* **D45** (1992).
2. M. Basile, D. Bollini, P. Dalpiaz, P.L. Frabetti, T. Massam, F. Navach, F.L. Navarria, M.A. Schneegans and A. Zichichi, *Il Nuovo Cimento* **3A** (1971) 796.
3. R.S. Kessler *et al, Phys. Rev. Lett.* **70** (1993) 892.
4. R. Abegg *et al,* accepted for publication in *Phys. Rev.* **D**.
5. R.S. Kessler. PhD thesis (University of California Los Angeles, 1992). Unpublished.
6. D. Alde *et al, Sov. J. Nucl. Phys.* **40** (1984) 918.

A MEASUREMENT OF THE BRANCHING RATIO
$K_L \to \mu^+\mu^-\gamma$

M. B. SPENCER*

*Physics Department, University of California at Los Angeles,
405 Hilgard Ave, Los Angeles, CA 90024, USA.*

ABSTRACT

We present a measurement of the K_L muon Dalitz decay $K_L \to \mu^+\mu^-\gamma$. We find a branching ratio $BR(K_L \to \mu^+\mu^-\gamma) = (3.55 \pm 0.25(stat) \pm 0.23(sys)) \times 10^{-7}$, and $\alpha_{K^*} = -0.15^{+0.14}_{-0.12}$.

1.   Physics

The decay $K_L \to l^+l^-\gamma$ occurs through a two photon intermediate state, where one of the photons is off-shell. The $K\gamma^*\gamma$ vertex strength is a function of the invariant mass of the off-shell photon, $F(q^2, 0; M_K)$. The integral over final state phase space gives a differential decay rate of

$$\frac{d\Gamma(K_L \to l^+l^-\gamma)}{d(q^2/m_K^2)} \propto \frac{|F|^2}{q^2/m_K^2} \left(1 - \frac{q^2}{m_K^2}\right)^{3/2} \left(1 + \frac{2m_l^2}{q^2}\right) \left(1 - \frac{4m_l^2}{q^2}\right)^{1/2}. \tag{1}$$

Bergstrom et al[1] have given a description of the physics at the vertex as the sum of two terms, one through $K_L \to (\pi^0, \eta, \eta') \to \gamma^*\gamma$, the other through $K_L \to K^*\gamma \to (\rho, \omega, \phi)\gamma \to \gamma^*\gamma$. The relative contribution of these two channels is parametrized through α_{K^*},

$$F(q^2, 0; M_K) = \frac{|A_{\gamma\gamma}|_{exp}}{1 - q^2/m_\rho^2} + \alpha_{K^*}A_K(q^2). \tag{2}$$

where A_K is a sum over pole terms, and $A_{\gamma\gamma}$ is the experimental amplitude for $K_L \to \gamma\gamma$ both of which can be found in ref 1. The value of α_{K^*} in various models is $|\alpha_{K^*}| = 0,$[2] 0.2-0.3,[1] 1.0.[3] Pure QED ($F = 1$) predicts $BR(K_L \to \mu^+\mu^-\gamma) = (2.30 \pm 0.10) \times 10^{-7}$, the error coming from the uncertainty in $BR(K_L \to \gamma\gamma)$.

2.   Data

The data set considered here is from the experiment E799, which ran during the 1991 Fermilab fixed target run. Protons of $800 GeV$ were incident on a Be target producing two nearly parallel neutral Kaon beams with momenta between $20 - 220 GeV$.

---

*Representing the Fermilab E799 Collaboration.

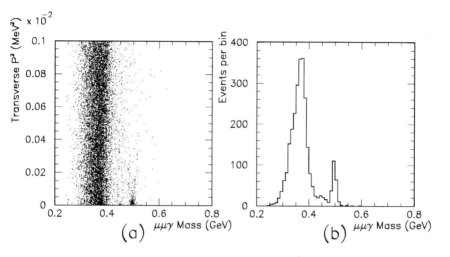

Fig. 1. (a) $P_t^2 - vs - \mu^+\mu^-\gamma$ mass distribution of events after cuts, (b) $\mu^+\mu^-\gamma$ mass distrubtion after cuts for events with $P_t^2 < 200MeV^2$.

The decay volume was 90-170m downstream of the target. A magnet with a transverse momentum kick (P_t) of $200MeV$ and four drift chambers with $100\mu m$ resolution form the spectrometer. Downstream of the spectrometer was a lead glass calorimeter with energy resolution of 4.4%. Beam-holes in the calorimeter allowed the beams to pass through to a dump. Events with photons falling outside the acceptance of the calorimeter were rejected with photon vetos surrounding the decay volume. Just upstream of the calorimeter were two hodoscope banks used for triggering on one or more charged tracks. A hadron veto counter was located downstream of the calorimeter. Muon identification was performed by two scintillator hodoscopes situated behind 3m of steel. One of these was an overlapping series of counters giving hermetic coverage, the other consisted of 16 non-overlapping vertical counters. The non-overlapping bank had trigger logic (at least two non-adjacent counters hit) to identify candidate 2μ events.

The 2μ trigger for this data set consisted of: no photon veto hits, two hits each in the trigger hodoscopes, hits in the drift chambers consistent with 2 tracks, di-muon trigger logic, at least $6GeV$ deposited in the calorimeter and finally events firing the hadron veto were rejected. The 2μ trigger collected about 60 million candidate events during the run.

3. Analysis

Events were selected with two good reconstructed tracks, one extra in-time cluster in the calorimeter and other calorimeter requirements such as no fused clusters, no clusters within 2.8cm of the beam-holes. Finer cuts were then made on kinematic variables reconstructed fom the tracks and extra cluster. The track momenta were required to be greater than $7GeV$ to ensure candidate muons had adequate momentum to reach the muon banks, the extra cluster energy was required to be greater than $8GeV$ to keep well away from possibles biases from the $6GeV$ calorimeter trigger.

At this stage a signal of ≈ 200 events emerges, clustered in the region of low transverse momentum squared (P_t^2), and about the Kaon mass (Fig. 1a).

4.   Backgrounds and Normalization

In Fig. 1b is the $\mu^+\mu^-\gamma$ invariant mass distribution for events with $P_t^2 < 200MeV^2$. We cut in the region $0.482 - 0.512GeV$, and after using a high P_t^2 sideband to subtract the 7.6 background events, we have 199.4 signal events. The background is uniquely identified by a kinematic parameter that measures the longitudinal momentum of the π^0 from tracking information alone.[4] This parameter cleanly separates $K_L \to \pi^+\pi^-\pi^0$ from $K_L \to \pi^\pm\mu^\mp\nu$ events. The $K_L \to \pi^+\pi^-\pi^0$ events fall very sharply with increasing $\mu^+\mu^-\gamma$ mass and the background remaining under the mass peak is solely $K_L \to \pi^\pm\mu^\mp\nu$ where the pion is misidentified as a muon and there is an accidental cluster in the calorimeter. Using the Monte Carlo simulation we calculate the acceptance for signal events to be 1.80%.

We used fully reconstructed $K_L \to \pi^+\pi^-\pi^0$ events for normalization. This data was collected in parallel with the 2μ data and was prescaled by 3600, but did not include the muon identification in the trigger. We used a similar set of cuts to identify $K_L \to \pi^+\pi^-\pi^0$ events, except requiring 2 extra clusters, and that they reconstruct to the π^0 mass. This left 19400 events, with an acceptance of 1.81%.

5.   Systematic Errors

The largest source of systematic error arose from muon identification. We determined the following systematic uncertainties: absolute muon identification efficiency (5.4%), background subtraction (1.9%), acceptance uncertainty (2.2%), normalization statistics (0.7%), uncertainty in the $\pi^+\pi^-\pi^0$ branching ratio (1.6%). Thus the total systematic error was $\pm 6.4\%$.

6.   Branching Ratio

The final result is:

$$BR(K_L \to \mu^+\mu^-\gamma) = (3.55 \pm 0.25(stat) \pm 0.23(sys)) \times 10^{-7}. \qquad (3)$$

The only previous evidence for this decay was a single candidate event, where 0.1 were expected.[5]

From the branching ratio one can extract a range for α_{K^*} as shown in Fig. 2a. We find $\alpha_{K^*} = -0.15^{+0.14}_{-0.12}$. It is seen to be consistent with the current world average[6] from $K_L \to e^+e^-\gamma$ di-electron invarant mass distributions.

Lastly, there may be additional information on α_{K^*} from the di-muon invariant mass spectrum shown in Fig. 2b, and that work is currently in progress.

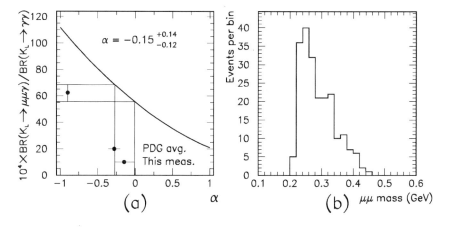

Fig. 2. (a) Determination of α_{K^*} from branching ratio and the particle data groups current world average. The curve is from the integral of eqn. 1. (b) The di-muon mass distribution for the 200 $K_L \to \mu^+ \mu^- \gamma$ events in this analysis.

7. The Future

The next generation of this experiment $(\mathrm{KTeV}/\mathrm{E799\text{-}II})$[7] will have 30 times more flux and better acceptance than phase-I and thus should be able to accumulate more than $10{,}000$ $K_L \to \mu^+ \mu^- \gamma$ events. Extracting α_{K^*} from the branching fraction is limited by knowledge of $BR(K_L \to \gamma\gamma)$ currently known to 4.7%. Future improvements on measuring α_{K^*} will require either a better determination of $BR(K_L \to \gamma\gamma)$ or $BR(K_L \to e^+ e^- \gamma)$.

References

1. L. Bergstrom, E. Masso, and P. Singer, *Phys. Lett.* **131B** (1983) 229.
2. L. M. Seghal, *Phys. Rev.* **D11** (1973) 3303.
3. J. J. Sakurai, *Phys. Rev.* **156** (1968) 1508.
4. D. Leurs *et al*, *Phys. Rev.* **133B** (1964) 1276.
5. A. S. Carroll *et al*, *Phys. Rev. Lett.* **44,8** (1980) 525.
6. M. Aguilar-Benitez *et al* , *Phys. Rev.* **D45** (1992).
7. K. Arisaka *et al*, *KTeV Design Report* (Fermi National Accelerator Laboratory, Batavia, Illinois) (1992).

A MEASUREMENT OF THE BRANCHING RATIO OF $K_L \rightarrow e^+e^-\gamma\gamma$

TSUYOSHI NAKAYA*

Reseach Division, Fermilab, PO Box 500
Batavia, Illinois 60510 USA

for the

FNAL E799 COLLABORATION

ABSTRACT

A new measurement of the $K_L \rightarrow e^+e^-\gamma\gamma$ branching ratio was carried out in Fermilab experiment E799. This process is an important background to searches for the decay $K_L \rightarrow \pi^0 e^+e^-$, which is expected to have a large direct CP violating component. We observed 58 $K_L \rightarrow e^+e^-\gamma\gamma$ events. The measured branching ratio is $BR(K_L \rightarrow e^+e^-\gamma\gamma, E_\gamma^* > 5MeV) = (6.5 \pm 1.2(stat.) \pm 0.6(sys.)) \times 10^{-7}$.

1. Introduction

The process $K_L \rightarrow e^+e^-\gamma\gamma$ is dominated by a K_L Dalitz decay, $K_L \rightarrow e^+e^-\gamma$, with an internal bremsstrahlung photon. This radiative K_L Dalitz decay provides an excellent testing ground for QED radiative corrections. These radiative corrections are particularly important for the precise measurement of the branching ratio of the parent $K_L \rightarrow e^+e^-\gamma$ decay, and for studies of the non-trivial $K_L\gamma^*\gamma$ vertex which contributes to the $K_L \rightarrow e^+e^-\gamma$ form factor. In addition, the radiative K_L Dalitz decay is expected to be the most serious background[1] in experiments searching for the CP violating decay $K_L \rightarrow \pi^0 e^+e^-$ beyond the current experimental sensitivity($\sim 10^{-9}$).[2] The expected $K_L \rightarrow e^+e^-\gamma\gamma$ branching ratio is calculated to be 5.8×10^{-7} with an infrared cutoff of $5MeV$ in the center of mass frame of the kaon.[1] The previous measurement of $BR(K_L \rightarrow e^+e^-\gamma\gamma) = (6.6 \pm 3.2) \times 10^{-7}$ is based on 17 ± 8 events.[3] It is important to measure this branching ratio more precisely to compare it with the expected branching ratio and to better establish the background level for $K_L \rightarrow \pi^0 e^+e^-$.

2. Experimental Apparatus and Event Selection

The goal of Fermilab experiment E799 phase-I was to search for the decay $K_L \rightarrow \pi^0 e^+e^-$ and other multi-body rare K_L decays. A detailed description of the E799 detector can be found elsewhere.[4]

Two types of triggers were used for this analysis to accept both $K_L \rightarrow e^+e^-\gamma\gamma$ decay and $K_L \rightarrow e^+e^-\gamma$ decay as a normalization sample. Both triggers required two

---

*Also belongs to Department of Physics, Osaka University, Toyonaka, Osaka, 560 Japan

hits in each hodoscope, drift chamber hits consistent with two tracks. In addition, to satisfy the $K_L \to e^+e^-\gamma\gamma$ trigger there had to be a minimum total energy in the calorimeter of $55 GeV$, and four clusters of energy in the calorimeter; each cluster having an energy threshold of $2.5 GeV$. Likewise the $K_L \to e^+e^-\gamma$ trigger demanded a minimum total energy of $6 GeV$ and three identified clusters. The latter trigger had to be prescaled by 14 because of the high rate.

Offline analysis of the data required two reconstructed tracks, each pointing to a cluster, which formed a good vertex in the detector volume. Electrons were identified by requiring the track momentum (p) to match the calorimeter cluster energy (E) to within 15% ($0.85 \le E/p \le 1.15$). Clusters not associated with tracks were considered as photon candidates. Events with exactly one(two) photon candidate(s) were used for the $K_L \to e^+e^-\gamma$ ($K_L \to e^+e^-\gamma\gamma$) sample. Kinematic quantities were then calculated assuming the photons in the event originated from the two-track vertex. Events with the square of the transverse momentum of $e^+e^-\gamma\gamma$ with respect to the K_L direction (P_t^2) less than $1000(MeV/c)^2$ were kept. The $\pi^+\pi^-\gamma\gamma$ invariant mass $(M_{\pi\pi\gamma\gamma})$ was calculated by assuming that charged particles were pions, and events with this mass between $450 MeV/c^2$ and $550 MeV/c^2$ were rejected as $K_L \to \pi^+\pi^-\pi^0$ events.

The remaining backgrounds at this stage were the K_{e3} with two extra clusters, $K_L \to 2\pi^0$ with π^0 Dalitz decay and the $K_L \to e^+e^-\gamma$ decay with an external bremsstrahlung photon. In order to further reject these backgrounds, the following two quantities were defined;

$$\min\Sigma\cos = \text{Minimum}(\cos\theta_{11} + \cos\theta_{21}, \cos\theta_{12} + \cos\theta_{22}), \text{and}$$
$$\theta_{min} = \text{Minimum}(\theta_{11}, \theta_{21}, \theta_{12}, \theta_{22}),$$

where θ_{ij} is the angle between the ith electron and the jth photon in the center of mass frame of the kaon. These distributions are shown in Figure 1 and Figure 2 for the data, the $K_L \to e^+e^-\gamma\gamma$ Monte Carlo and the background Monte Carlo. The $\min\Sigma\cos$ is required to be less than -0.6 in order to reject the K_{e3} background. The value of θ_{min} is required to be larger than 0.06 radians to suppress the $K_L \to e^+e^-\gamma$ background, and required to be smaller than 0.5 radians mainly to suppress the $K_L \to 2\pi^0$ background. More detailed descriptions of these cuts are found elsewhere.[5]

Figure 3 shows the invariant mass distribution of the $e^+e^-\gamma\gamma$ and $e^+e^-\gamma$ events. The $e^+e^-\gamma$ events were selected with the same cuts as the $e^+e^-\gamma\gamma$ events except for the $M_{\pi\pi\gamma\gamma}$ and the θ_{min} cuts. There is a clear peak in Figure 3 at the kaon mass with 69 and 275 events in the kaon mass window for the $e^+e^-\gamma\gamma$ and $e^+e^-\gamma$ events, respectively. We used the Monte Carlo to estimate the levels of backgrounds. Using the reconstructed kaon mass we normalized each background to the region below the K_L mass and extrapolated the number we expect in the K_L mass window. The number of background events is 2.1 ± 0.3 events from $K_L \to 2\pi^0$, 0.6 ± 0.2 events from K_{e3}, and 0.5 ± 0.2 events from $K_L \to 3\pi^0$. Correcting for these backgrounds reduces the number of $e^+e^-\gamma\gamma$ ($N_{ee\gamma\gamma}$) and $e^+e^-\gamma$ ($N_{ee\gamma}$) events to 65.8 ± 8.3 events and 272.7 ± 16.6 events, respectively.

3. Branching Ratio Results

In order to determine the $K_L \to e^+e^-\gamma\gamma$ branching ratio, it is necessary to determine the amount of $K_L \to e^+e^-\gamma$ background which remains in the $e^+e^-\gamma\gamma$ sample. In

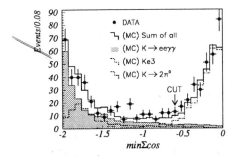
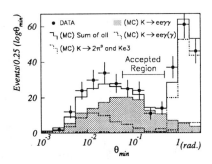

Fig. 1. The $\min\Sigma\cos$ distributions. The level of the background Monte Carlo was estimated by normalizing to the region of reconstructed kaon mass below the K_L mass.

Fig. 2. The minimum angle distribution (θ_{min}) between each electron, photon combination in the center of mass frame of the kaon.

addition, the normalization sample, $e^+e^-\gamma$ events, consists of the radiative K_L Dalitz decay with a missing photon, and the K_L Dalitz decay. We use the following method to disentangle the number of radiative and non-radiative K_L Dalitz decays in the $e^+e^-\gamma\gamma$ and $e^+e^-\gamma$ samples.

The ratio of the number of remaining $e^+e^-\gamma\gamma$ events to $e^+e^-\gamma$ events is related to $R = BR(K_L \to e^+e^-\gamma\gamma)/BR(K_L \to e^+e^-\gamma)$ as:

$$\frac{N_{ee\gamma\gamma}}{N_{ee\gamma}} = \frac{(1-R) \times A_{ee\gamma\to ee\gamma\gamma} + R \times A_{ee\gamma\gamma\to ee\gamma\gamma}}{(1-R) \times A_{ee\gamma\to ee\gamma} + R \times A_{ee\gamma\gamma\to ee\gamma}}, \tag{1}$$

where $BR(K_L \to e^+e^-\gamma)$ is the branching ratio of the $K_L \to e^+e^-\gamma$ including the radiative decay, $e^+e^-\gamma\gamma$, and $A_{x\to y}$ is the Monte Carlo probability that an event generated as x is accepted as y. Each acceptance was determined from a Monte Carlo simulation with energies of kaons between 35 and 220 GeV. The calculated values of $A_{ee\gamma\to ee\gamma}$, $A_{ee\gamma\gamma\to ee\gamma}$, $A_{ee\gamma\to ee\gamma\gamma}$ and $A_{ee\gamma\gamma\to ee\gamma\gamma}$ were 9.2×10^{-4}, 13.1×10^{-4}, 0.44×10^{-4} and 7.6×10^{-4}, respectively, where $A_{x\to ee\gamma}$ includes a trigger prescale factor of $1/14$.

By solving Eq. 1, R is found to be $0.29 \pm 0.06(stat.)$. By using the numerator of Eq. 1, the number of $K_L \to e^+e^-\gamma\gamma$ events is calculated to be $57.5 \pm 8.5(stat.)$, and the number of $K_L \to e^+e^-\gamma$ background is $8.3 \pm 2.0(stat.)$. An infrared cutoff of $5MeV(E_\gamma^* \geq 5MeV)$ is used to calculate the branching ratio of $K_L \to e^+e^-\gamma\gamma$, to allow direct comparison of this measurement to theoretical predictions[1] as well as the previous published measurement.[3] The Monte Carlo predicts the effect of this cutoff to be $BR(K_L \to e^+e^-\gamma\gamma, E_\gamma^* \geq 5MeV)/BR(K_L \to e^+e^-\gamma\gamma) = 0.247$. The main sources of systematic errors came from uncertainties in normalization, detector resolution and background estimation. The total systematic error was calculated to be 8.8%. By using the inclusive branching ratio, $BR(K_L \to e^+e^-\gamma) = (9.1 \pm 0.5) \times 10^{-6}$, the branching ratio of $K_L \to e^+e^-\gamma\gamma$ is measured to be $BR(K_L \to e^+e^-\gamma\gamma, E_\gamma^* \geq 5MeV) = (6.5 \pm 1.2(stat.) \pm 0.6(sys.)) \times 10^{-7}$. This measurement is more precise than previous measurements, and is consistent with both theoretical expectations and the previous measurement.

A less restrictive analysis was performed to check the above result, and to study the kinematic distributions of $K_L \rightarrow e^+e^-\gamma\gamma$ events with higher statistics. Removing the θ_{min} cut at 0.06 radians increases the number of $e^+e^-\gamma\gamma$ events to 198.8 ± 14.3. The branching ratio of $K_L \rightarrow e^+e^-\gamma\gamma$ obtained in this less restrictive analysis is $BR(K_L \rightarrow e^+e^-\gamma\gamma, E_\gamma^* \geq 5MeV) = (7.7 \pm 1.4(stat.) \pm 1.0(sys.)) \times 10^{-7}$ based on 151.6 ± 17.4 $K_L \rightarrow e^+e^-\gamma\gamma$ signal and 47.2 ± 9.9 $K_L \rightarrow e^+e^-\gamma$ background. This result is consistent with the more restrictive analysis. Figure 4 shows the M_{ee} and $M_{\gamma\gamma}$ distributions of the $e^+e^-\gamma\gamma$ events from the less restrictive analysis. The data distributions are consistent with Monte Carlo distributions. This is the first time these kinematic distributions were compared for the $K_L \rightarrow e^+e^-\gamma\gamma$ decay.

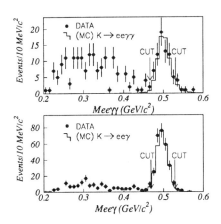

Fig. 3. The invariant mass distributions of $e^+e^-\gamma\gamma$ events and $e^+e^-\gamma$ events.

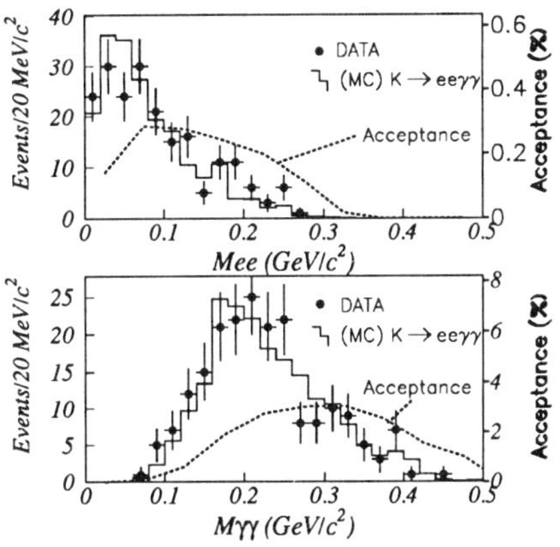

Fig. 4. The e^+e^- invariant mass and the $\gamma\gamma$ invariant mass distributions without the minimum angle cut ($\theta_{min} \geq 0.06$). The Monte Carlo includes the sum of the $K_L \rightarrow e^+e^-\gamma\gamma$ events and $K_L \rightarrow e^+e^-\gamma$ background in the proportions described in the text. Also shown is the $K_L \rightarrow e^+e^-\gamma\gamma$ acceptance as a function of M_{ee} and $M_{\gamma\gamma}$.

References

1. H.B. Greenlee, Phys. Rev. **D42**, (1990) 3724.
2. D.A. Harris, *et al.*, Phys. Rev. Lett. **71**, (1993) 3918.
3. W.M.Morse *et al.*, Phys. Rev. **D45**, (1992) 36.
4. K. S. McFarland *et al.*, Phys. Rev. Lett. **71**, (1993) 31.
5. T. Nakaya *et al.*, OULNS93-09, submitted to PRL.

A LIMIT ON THE BRANCHING RATIO FOR $K_L \to \pi^0 \nu \bar{\nu}$

MATTHEW WEAVER

UCLA Department of Physics
Los Angeles, CA 90024 USA

for the FNAL E799 Collaboration

ABSTRACT

The measurement of an upper limit on the branching ratio for $K_L \to \pi^0 \nu \bar{\nu}$ by the Fermilab E799 Collaboration is presented. The $K_L \to \pi^0 \nu \bar{\nu}$ decay rate is dominated by direct CP violating amplitudes. Consequently, the branching ratio is related to the CP violating parameters of the Standard Model in a calculable manner.

1. Introduction

The phenomenon of CP violation in the neutral kaon system has for many years been consistent with the hypothesis that the physical state K_L (K_S) is composed mostly of a CP odd (even) eigenstate with a small admixture of the CP even (odd) eigenstate, i.e. indirect CP violation. Hence, the experimental search for direct CP violating phenomena is a natural program for extending our knowledge of particle interactions. Listed in Table 1 are three particular modes of K_L decay in which direct CP violating effects are sought. The table shows a comparison of the ratio of direct and indirect CP violating contributions (ϵ'/ϵ) to the measurements. Clearly, the decay mode $K_L \to \pi^0 \nu \bar{\nu}$ would be the purest evidence for direct CP violation if a branching ratio is measured near the predicted level. Furthermore, determination of the CP violating parameter η of the CKM matrix is straightforward from the branching ratio measurement, since calculation of the $K_L \to \pi^0 \nu \bar{\nu}$ decay rate is free from theoretical uncertainties; i.e., the hadronic matrix element for the decay has been measured. The

Table 1. Direct CP Violation Searches

Mode	$K_{L,S} \to \pi\pi$	$K_L \to \pi^0 l^+ l^-$	$K_L \to \pi^0 \nu \bar{\nu}$
ϵ'/ϵ	$10^{-4} - 10^{-3}$	≈ 1	10^3
predicted branching ratio	$10^3 (K_L)$ (measured)	10^{-12}	10^{-11}

predicted branching ratio is[1]

$$B(K_L \to \pi^\circ \nu\bar{\nu}) = 4.8 \times 10^{-11} D(\frac{M_{top}}{M_W})^2 A^4 \eta^2, \tag{1}$$

where the function D varies from 2.7 to 3.7 for values of M_{top} from $150 GeV/c^2$ to $200 GeV/c^2$. Herein is a report upon a search for the $K_L \to \pi^\circ \nu\bar{\nu}$ decay process performed by the Fermilab E799 Collaboration.[2]

2. Experimental Signature

In the decay process $K_L \to \pi^\circ \nu\bar{\nu}$, only the secondary π° decay products are detectable. While there are several common sources of π° decay in a neutral beam, the π° from a $K_L \to \pi^\circ \nu\bar{\nu}$ decay typically possesses a large value for P_T, the component of momentum transverse to the K_L line of flight. It is particularly important to measure values of P_T which distinguish the signal mode from the $\Lambda \to n\pi^\circ$ decays, whose P_T is bounded below $104 MeV/c^2$. Hence, the region of P_T greater than $160 MeV/c^2$ is chosen as a signal region. If the experimental signature is focused upon detecting the two photon state of the secondary decay $\pi^\circ \to \gamma\gamma$, the K_L line of flight cannot be determined. The result is a P_T measurement uncertainty which is directly dependent upon the transverse size of the neutral beam. An effective measurement of P_T from the $\pi^\circ \to \gamma\gamma$ secondary decay requires a beamline optimized towards the $K_L \to \pi^\circ \nu\bar{\nu}$ search, which was not the case for the E799 experiment, whose primary goal was the measurement of the $K_L \to \pi^\circ e^+ e^-$ branching ratio.

The experimental signature was chosen to be the detection of an isolated $\pi^\circ \to e^+ e^- \gamma$ decay with sufficiently large P_T. The presence of charged particle tracking in the detector design allows measurement of the decay vertex and, consequently, a precise measurement of P_T. Furthermore, additional kinematic quantities may be measured to verify the presence of a π° decay. Although the choice to search exclusively for $\pi^\circ \to e^+ e^- \gamma$ secondary decays incurs a sensitivity loss by a factor of 80 due to the small branching ratio of $\pi^\circ \to e^+ e^- \gamma$ equal to 1.2%, the additional experimental constraints allow adequate background control for a more sensitive measurement of the $K_L \to \pi^\circ \nu\bar{\nu}$ branching ratio.

3. The E799 Detector and Beams

The E799 detector[3] is composed of four major elements. First, a 40 meter long evacuated tank housing several photon veto modules is used as the decay region from which K_L decay products proceed to the detector. Next follows a charged particle spectrometer consisting of a pair of drift chambers before and a pair after an analysing magnet of 200 MeV/c momentum kick. An electromagnetic calorimeter composed of 804 lead glass blocks is located downstream of the spectrometer. Finally, a muon identification system follows the calorimeter.

The two neutral beams were produced from the interaction of an 800 GeV/c proton beam incident upon a 50cm thick Be target located roughly 120 meters upstream of the decay volume. A set of collimators defined the two rectangular beams. The momenta of K_L's in the beams covered a wide range, with a peak momentum of 50 GeV/c for K_L's decaying in the decay volume. Only 2.7% of K_L's in the beam decayed within this region.

4.   Trigger and Data Collection

The data were collected during 8 weeks of the 1991-92 Fermilab fixed target run. The trigger was formed from the combination of the following conditions.

- Detection of two charged particles, consisting of two hits in the trigger hodoscope and hit counting in the drift chambers.

- Three calorimeter clusters from electron and photon showers and a minimum calorimeter energy sum of 3 GeV.

- A veto upon signals from the photon veto modules and hadron veto.

In addition, a prescale factor of $1/14$ was applied to the trigger to lower the otherwise irreducible rate so that the data collection for other analyses was not affected.

5.   Analysis

The analysis for the $K_L \rightarrow \pi^\circ \nu \bar{\nu}$ search consisted of particle identification and event reconstruction. Charged particle trajectories were retraced from the drift chamber information, and the momenta were calculated from the bend angle through the analysing magnetic field. Electrons were then identified by requiring that the track momentum measurement and the associated calorimeter cluster energy measurement agree within 15%. Photons were identified by a significant calorimeter energy deposit ($E > 1 GeV$) which could not be associated with any track incident upon the calorimeter. The reconstructed vertex was required to be within the beam and fiducial region of $120m < Z_{vtx} < 159m$ and have an estimated uncertainty $\sigma_z < 3m$. Signals of energy measurement in the photon veto modules were also enforced.

The remaining events with an identified electron, positron, and photon were subjected to several kinematic constraints in order to remove background from $K_L \rightarrow \pi^\circ \pi^\circ \pi^\circ$, $\Lambda \rightarrow n\pi^\circ$, $K_L \rightarrow \pi^+ \pi^- \pi^\circ$, and $K_L \rightarrow \pi^\pm e^\mp \nu(\gamma)$ decays. Dalitz decays of the π° ($\pi^\circ \rightarrow e^+ e^- \gamma$) were selected by requiring $M_{ee}/M_{ee\gamma} < 0.3$ and $cos\theta^*_{e^+\gamma} + cos\theta^*_{e^-\gamma} < -1.5$ in the $e^+ e^- \gamma$ rest frame. To remove the remaining background from K_{e3} decays, P_T was required to be greater than the maximum kinematically allowed value for a $K_L \rightarrow \pi^\pm e^\mp \nu \gamma$ decay, given by

$$max P_T = \frac{M_K^2 - M_{\pi e \gamma}^2}{2 M_K}. \tag{2}$$

Figure 1 shows the final signal requirements $160 < P_T(MeV/c) < 231$ and $125 < M_{ee\gamma}(MeV/c^2) < 145$ imposed upon the $K_L \rightarrow \pi^\circ \nu \bar{\nu}$ Monte Carlo and upon the data. No signal candidates were found in the data.

The $K_L \rightarrow \pi^\circ \nu \bar{\nu}, \pi^\circ \rightarrow e^+ e^- \gamma$ search was normalized by reconstructing the topologically similar mode $K_L \rightarrow e^+ e^- \gamma$. A clean sample of 233 $K_L \rightarrow e^+ e^- \gamma$ events were detected. Hence, an upper limit on the branching ratio for $K_L \rightarrow \pi^\circ \nu \bar{\nu}$ of 5.8×10^{-5} at the 90% confidence level was determined. This includes a small correction due to the systematic uncertainty, dominated by the uncertainty in the $K_L \rightarrow e^+ e^- \gamma$ branching ratio measurement.

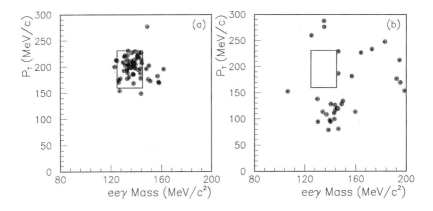

Fig. 1. Scatter plot of P_T versus $ee\gamma$ mass for (a) $K_L \to \pi^\circ \nu\bar{\nu}$ simulation and (b) data.

6. Conclusion

The FNAL E799 Collaboration has measured an upper limit on the branching ratio for $K_L \to \pi^\circ \nu\bar{\nu}$ of 5.8×10^{-5} at the 90% confidence level. Although the sensitivity of this search for the $K_L \to \pi^\circ \nu\bar{\nu}$ decay is well short of the predicted branching ratio level, future searches will do much better based upon the experience of this analysis. A new collaboration (KTeV) has designed and constructed an experiment which will be able to reach a sensitivity at the 10^{-8} level for the $K_L \to \pi^\circ \nu\bar{\nu}$ decay mode. In addition, a KEK experiment dedicated to search for the $K_L \to \pi^\circ \nu\bar{\nu}, \pi^\circ \to \gamma\gamma$ decay at an even greater sensitivity has been proposed. Because of the clear interpretation that a $K_L \to \pi^\circ \nu\bar{\nu}$ branching ratio measurement would receive, this decay mode is the target of a great deal of experimental design effort.

7. Collaboration List

E799 University of California - Los Angeles, University of Chicago, University of Colorado, Elmhurst College, Fermilab, University of Illinois, Osaka University, Rutgers University

References

1. L. Littenberg, *Phys. Rev.* **D39** (1989) 3322.
2. M. Weaver et al, *Phys. Rev. Lett.* **72** (1994) 3758.
3. K. S. McFarland et al, *Phys. Rev. Lett.* **71** (1993) 31. The E799 detector is illustrated in Figure 1 of this reference. A detailed description can also be found in M. Weaver, UCLA Ph. D. Thesis, 1994 (unpublished).

**D P F '94
REGISTRATIONS,
COMMUNICATIONS,
AND
ABSTRACTS CENTER**

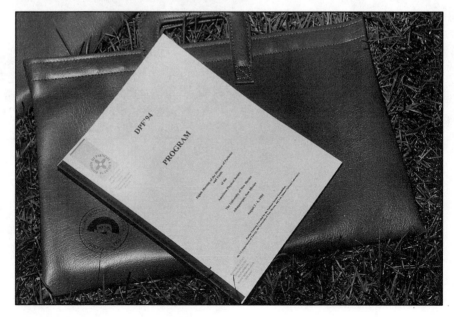

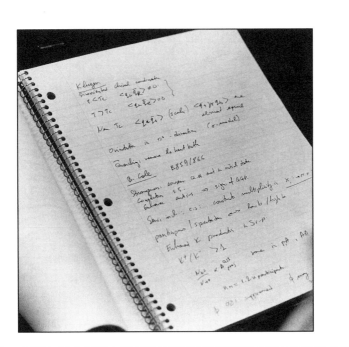

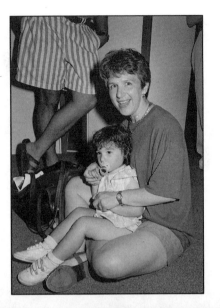

- Do RENORMALIZATION USING WKB TECHNIQUE

- <u>RENORMALIZED HARTREE EQUATIONS</u> :

$$\ddot{\phi} \pm m_R^2 \phi + \frac{\lambda_R}{2} c_L \phi^3 + \frac{\lambda_R}{2} \phi \langle \Sigma^2(t) \rangle_R = 0$$

$$\left[\frac{d^2}{dt^2} + k^2 \pm m_R^2 + \frac{\lambda_R}{2} \phi^2(t) + \frac{\lambda_R}{2} \langle \Sigma^2(t) \rangle_R \right] \cdot U_k^H(t) = 0$$

$$\langle \Sigma^2(t) \rangle_R = \frac{1}{\left(1 - \frac{\lambda_R}{16\pi^2} \ln \frac{\Lambda}{k}\right)} \cdot \left\{ \int \frac{d^3k}{(2\pi)^3} \frac{(|U_k^H|^2 - 1)}{2\omega_k(t_0)} + \frac{\lambda_R}{16\pi^2} \ln \frac{\Delta}{k} (\phi^2(t) - \phi_0^2) \right\}$$

+ : Unbroken Symmetry ; − : broken symmetry